Introduction To Operator Algebras

INTRODUCTION TO OPERATOR ALGEBRAS

LI BING-REN
Academia Sinica, Beijing

World Scientific
Singapore • New Jersey • London • Hong Kong

Published by

W□ □

5 Toh Tuck Link, Singapore 596224

USA□ □ □ □27 Warren Street, Suite 401-402, Hackensack, NJ 07601

□ □ □ □ □57 Shelton Street, Covent Garden, London WC2H 9HE

British Library Cataloguing-in-Publication Data
A catalogue record for this book is available from the British Library.

INTRODUCTION TO OPERATOR ALGEBRAS

ISBN-13 978-981-02-0941-4
ISBN-10 981-02-0941-X

Introduction

An operator algebra here, precisely speaking, is a $*$ algebra consisting of bounded linear operators on some Hilbert space, i.e., a $*$ subalgebra of $B(H)$, where $B(H)$ is the collection of all bounded linear operators on some Hilbert space H. Since it is an infinite dimensional object (generally, H is infinite dimensional), so for studying it we must ask that it is closed under some topology. Under usual linear topologies in $B(H)$, we find that the closures of operator algebras are just two classes: weak closure and uniform closure. Hence, we need mainly to study the weakly closed operator algebras (Von Neumann algebras or W^*-algebras) and the uniformly closed operator algebras (C^*-algebras).

When J. Von Neumann studied the spectral theory of operators and the quantum physics, he recognized that we need new mathematical tools for studying the infinite dimensional systems in the physical world. In 1929, he introduced the concept of weakly closed rings, renamed Von Neumann algebras by J. Dixmier latter. He pointed out two essential characteristics: 1) such rings must be self-adjoint; 2) such rings are weakly closed. So that each spectral projection of any self-adjoint operator in such ring belongs still to this ring. J. Von Neumann and his collaborator F. Murray laid down the foundation for this new field of Mathematics during the period of the 1930's and early in the 1940's. Their results are still important up to now. Then, many important results and powerful techniques were added to this theory, for example, the structure theory of factors, the general theory of weights and traces, Tomita-Takesaki theory, the Connes classification of type (III) factors, Jones index theory, and etc. The theory of Von Neumann algebras got great progress and many applications.

In 1943, I.M. Gelfand and M.A. Naimark laid down the foundation for the theory of C^*-algebras. They showed that a Banach $*$ algebra can be isometrically $*$ isomorphic to a uniformly closed operator algebra on some Hilbert space under few conditions. Then, I.E. Segal gave its perfected form, i.e., the famous GNS construction. The theory of C^*-algebras can be divided into two parts: the essential structure of the algebras, and the theory of representations. Of course, these two parts have very closed connections. As well as the theory of Von Neumann algebras, the developments of the theory of C^*-algebras are also very great.

Now the theory of operator algebras becomes one of the most important fields of mathematics.

This book consists of 17 Chapters and an appendix.

Chapter 1 is the fundamentals of Von Neumann algebras. It contains the definition of Von Neumann algebras; Von Neumann's double commutation theorem; the commutation theorem of tensor products of Von Neumann algebras; Kaplansky's density theorem; the comparison of projections; the normalness, polar decomposition and the orthogonal decomposition of linear functionals; the Radon-Nikodym theorem; the structure of normal $*$ homomorphisms; the comparison of cyclic projections; the spatial theory; σ-finite Von Neumann algebras, and etc. In particular, the analysis of topologies is given in detail. Moreover, by the Van Daele-Rieffel approach, we can put the commutation theorem of tensor products in this chapter, and it is convenient for the latter.

Chapter 2 is the fundamentals of C^*-algebras. It contains the definition of C^*-algebras; the Gelfand-Naimark-Segal construction; the characterization of the extreme points in unit ball; Kadison's transitivity theorem; the theory of representations; the enveloping Von Neumann algebras; the multiplier algebras; the axioms of complex and real C^*-algebras, and etc. From the GNS construction, every (abstract) C^*-algebra can be isometrically $*$ isomorphic to a uniformly closed operator algebra on some Hilbert space (concrete C^*-algebra). Hence, the definition of C^*-algebras is independent of the Hilbert spaces. Moreover, last section of this chapter is devoted to the famous Gelfand-Naimark conjecture: the axioms for C^*-algebras. This conjecture is very interesting and important, and in a long period of time many mathematicians studied it.

In Chapter 3, we discuss the tensor products of C^*-algebras. The tensor products are a useful method to construct new C^*-algebras from given C^*-algebras. On the algebraic tensor product of C^*-algebras, the first natural C^*-norm is the spatial C^*-norm. But M. Takesaki discovered that there exist more than one C^*-norm possibly. So we must study the properties of general C^*-norm. Here, the important result is that each C^*-norm on the algebraic tensor product of C^*-algebras is a cross–norm. We also discuss the completely positive maps on C^*–algebras, inductive limit of C^*–algebras, infinite tensor products of C^*-algebras. In last section of this chapter, we give an introduction to an important class of C^*-algebras: nuclear C^*-algebras.

W^*-algebras are the "abstract" Von Neumann algebras. In Chapter 4, using Tomiyama's projections of norm one, the Sakai's theorem (any W^*-algebra can be $*$ isomorphic to some Von Neumann algebra) is proved. We also discuss the normal part and the singular part of a linear functional, the characterizations of weakly compact subsets in the predual, and etc. In some sense, this chapter is the continuance of Chapter 1.

In Chapter 5, we study the abelian operator algebras. It contains the general forms of abelian W^*-algebras; the characterizations of the spectral spaces of abelian W^*-algebras; the measure description of an important class of representations for an abelian C^*-algebra, and etc. Moreover, the section 5.1 can be seen as preliminaries of this chapter.

Chapter 6 is the traditional program: the classification of Von Neumann algebras. First, we give the Murray-Von Neumann decomposition of a Von Neumann algebra. Then we discuss the properties and characterizations of the finite, the semi-finite, the pure infinite, the discrete (or type (I)), the continuous and type (II) Von Neumann algebras. Finally, we study the classification of the tensor products of Von Neumann algebras.

The theory of factors occupies the central place in the theory of Von Neumann algebras. In Chapter 7, first we describe the classification of factors with dimension theory. Then we point out the uniqueness of the hyperfinite type (II_1) factor up to $*$ isomorphism; and the existences of type (II) and (III) factors by the standard method of group measure spaces. Moreover, the examples of non-hyperfinite type (II_1) factors and non-nuclear C^*-algebra are given.

Chapter 8 is the Tomita-Takesaki theory. It is an important part of modern theory of operator algebras. Following Van Daele-Rieffel approach, we discuss the KMS condition, and the modular automorphism group of a W^*-algebra. It also contains the innerness of the modular automorphism group in semi-finite case; the Connes unitary cocycle theorem; and etc. Since we don't study the general theory of weights and generalized Hilbert algebras, so the results are restricted to σ-finite case.

In 1930's, Murray-von Neumann divided the factors into five classes. Later, we know that such classification is not complete. Even there exist uncountably many non-isomorphic type (II_1), (II_∞) ad (III) factors on a separable Hilbert space. 1973, A. Connes got a great progress on the classification of type (III) factors. Chapter 9 is devoted to this theory. It contains the Arveson spectrum; the Connes spectrum; the type (III_λ) ($0 \leq \lambda \leq 1$) classification of type (III) factors in σ-finite case; and the examples of type (III_λ) factors.

We discuss the Borel structure in Chapter 10. This chapter can be seen as preliminaries for some later parts.

E. Effros introduced a standard Borel structure into the collection of all Von Neumann algebras on a separable Hilbert space. Chapter 11 is devoted to this theory.

Chapter 12 is the important traditional program: Reduction theory. The aim of this theory is to simplify the study of general Von Neumann algebras.

By the view of Borel maps, we discuss the measurable fields of Hilbert spaces, operators and Von Neumann algebras. Then we discuss the decompositions of Hilbert spaces and Von Neumann algebras. Moreover, the Borel subsets of the Von Neumann algebras space and the state spaces on separable C^*-algebras are also studied in this chapter.

In Chapter 13, we discuss the spectrum of a C^*-algebra, CCR, GCR, NGCR algebras. The main result is that GCR$\Longleftrightarrow$ type I for a C^*-algebra (Kaplansky-Glimm-Sakai).

Chapter 14 is the decomposition theory for the states on C^*-algebras. By Choquet theory and the Sakai-Ruelle construction, any state on a C^*-algebra can be represented as an integral of pure states or factorial states (roughly speaking). This theory is also very useful for the mathematical physics.

In 1960, J. Glimm introduced a class of C^*-algebras: uniformly hyperfinite (UHF) algebras. Then in 1972, O. Bratteli introduced approximately finite (AF) algebras, which are the generalization of (UHF) algebras. Chapter 15 is devoted to this theory. It contains the equivalent definitions; isomorphism theorem; Bratteli diagrams; dimension groups (K_0-groups), stable isomorphism theorem, and etc. Now the K-theory is very important in the study of C^*-algebras (see [11]), and the dimension group is an excellent example.

Chapter 16 is the theory of crossed products. It contains W^*-crossed products, Takesaki duality theorem; C^*-crossed products, Takai duality theorem; and some examples. Crossed products have long been used to construct interesting C^*-and W^*-algebras. But it is difficult to obtain some good information about the internal structure of crossed products. Hence, we should pay more attention to this field, and study it furthermore.

Chapter 17 is an introduction to the Jones index theory. 1930's Von Neumann and Murray introduced the concept of coupling constant. In 1983, V. Jones gave a new form to this concept. Then he defined the index for subfactors, and proved a surprising theorem on the values of index for subfactors. And also he introduced a very important technique in the proof of this theorem: the towers of algebras. Now this theory becomes a focus of many fields in mathematics and physics.

The appendix is a survey on weak topology and weak $*$ topology in a Banach space and its conjugate space. Maybe, it is helpful to understand the topologies in a Von Neumann algebra.

This book is an introduction, written in a self-contained manner (for reading it, just need the general knowledge of functional analysis and measure theory, [178], [67], and some foundation of abstract harmonic analysis). We hope that it can offer the basic concepts, techniques, structures and important results of the theory of operator algebras. And we also expect that it can serve as a text for graduate students.

The author would like to express here his sincere gratitude to Professor K.

Shiraiwa, Professor J. Tomiyama and Professor Xia Daoxing for their recommendation and the moral support. He acknowledges gratefully the support of the NSF of China. And the author is also very grateful to his home Institute for its support and typing the manuscript.

Contents

Chapter 1
Fundamentals of Von Neumann Algebras

1.1. Banach spaces of operators on a Hilbert space

Let H be a Hilbert space over the complex field $\mathbb{C}$. We shall denote by $F(H), C(H)$ and $B(H)$ the sets of all linear operators of *finite rank*, all *compact* linear operators, and all *bounded* linear operators on H respectively. The identity operator on H is denoted by 1_H, or 1 simply if no confusion arises.

Proposition 1.1.1. If H is separable and infinite dimensional, then $C(H)$ is the unique non-zero proper closed two-sided ideal of $B(H)$.

Proof. Let I be a non-zero closed two-sided ideal of $B(H)$, and $0 \neq a \in I$. Then there are $\xi, \eta \in H$ such that $a\xi = \eta \neq 0$. For any $\xi', \eta' \in H$, there is $b \in B(H)$ such that $b\eta = \xi'$. Then

$$ba(\xi \otimes \eta') = (ba\xi) \otimes \eta' = \xi' \otimes \eta' \in I,$$

where $\xi' \otimes \eta'$ is the operator of one rank on H : $\xi' \otimes \eta'(\cdot) = \langle \cdot, \eta' \rangle \xi'$. Thus $F(H) \subset I$ and $C(H) \subset I$.

If $C(H) \neq I$, we have $t \in I \backslash C(H)$. Then $h = (t^*t)^{1/2} \in I \backslash C(H)$. Let $\{e_\lambda\}$ be the spectral family of h. Since h is not compact, it follows that there exists $\varepsilon > 0$ such that $\dim(1 - e_\varepsilon)H = \infty$. Now we can take an isometry v from H onto $(1 - e_\varepsilon)H$. Then $v^*hvH = H$ and v^*hv is invertible. Since $v^*hv \in I$, we have $I = B(H)$. Q.E.D.

Proposition 1.1.2. If H is infinite dimensional, then $C(H)$ is not the conjugate space of any Banach space.

Proof. By the Krein-Milman theorem, the closed unit ball of the conjugate space of any Banach space has an extreme point at least. So it suffices to show that the closed unit ball of $C(H)$ has no any extreme point, i.e., for any $a \in C(H), \|a\| \leq 1$, we only need to find a non-zero $b \in C(H)$ such that $\|a \pm b\| \leq 1$.

If the rank of a is finite, let H_1 be the linear span of aH and a^*H, and $H_2 = H_1^\perp$, then $\dim H_1 < \infty, H_1$ and H_2 are invariant for a and a^*, and $a = a^* = 0$ on H_2. Thus it is easily verified that b exists.

Now suppose that the rank of a is infinite. Let $a = wh$ be the polar decomposition of a, and write $h = \sum_n \lambda_n p_n$, where $\{p_n\}$ is a sequence of projections of one rank on H, and $p_n p_m = 0, \forall n \neq m, 0 < \lambda_n \leq 1, \forall n$, and $\lambda_n \to 0$. Pick N and $\varepsilon > 0$ such that $\lambda_N \in (0,1)$, and $|\lambda_N \pm \varepsilon| \leq 1$. Now let $b = \varepsilon w p_N$. Then $b \neq 0$ and $\|a \pm b\| \leq 1$. Q.E.D.

Remark. A *projection* p on a Hilbert space, we always mean that it is *self-adjoint*, i.e., $p^* = p = p^2$.

We shall denote by $S(H)$ the set of all operators of *Hilbert-Schmidt class* on H, i.e., $a \in S(H)$ if a is compact, and $\sum_n |\lambda_n|^2 < \infty$, where $\{\lambda_n\}$ is the set of all eigenvalues of $(a^*a)^{1/2}$ (counting the multiplicity).

Proposition 1.1.3. Let $a \in B(H)$. Then $a \in S(H)$ if and only if for some (then for any) normalized orthogonal basis $\{\xi_l\}$ of H we have $\sum_l \|a\xi_l\|^2 < \infty$.

Proof. let $\{\xi_l\}, \{\eta_r\}$ be two normalized orthogonal bases of H. Then

$$\begin{aligned}\sum_l \|a\xi_l\|^2 &= \sum_l \|(a^*a)^{1/2}\xi_l\|^2 = \sum_{l,r} |\langle (a^*a)^{1/2}\xi_l, \eta_r\rangle|^2 \\ &= \sum_{l,r} |\langle \xi_l, (a^*a)^{1/2}\eta_r\rangle|^2 = \sum_r \|(a^*a)^{1/2}\eta_r\|^2 = \sum_r \|a\eta_r\|^2.\end{aligned}$$

Hence it suffices to prove this proposition for some fixed basis.

Let $a \in S(H)$, and $\{\xi_l\}$ be a normalized orthogonal basis of H such that all eigenvectors corresponding to positive eigenvalues of $(a^*a)^{1/2}$ are contained in $\{\xi_l\}$. Then $\sum_l \|a\xi_l\|^2 = \sum_n \lambda_n^2 < \infty$.

Conversely, if $\sum_{l \in \Lambda} \|a\xi_l\|^2 < \infty$, where $\{\xi_l\}_{l \in \Lambda}$ is a normalized orthogonal basis of H, then for any $\varepsilon > 0$ we can find a finite subset F of Λ such that $\|p_F a p_F - a\| < \varepsilon$, where p_F is the projection from H onto the linear span of $\{\xi_l | l \in F\}$. Thus $a \in C(H)$. Now pick a basis $\{\xi_l\}$ containing all eigenvectors corresponding to positive eigenvalues of $(a^*a)^{1/2}$. Then we can see $a \in S(H)$. Q.E.D.

Let $a \in S(H). \|a\|_2 = (\sum_n \lambda_n^2)^{1/2} = \sum_l \|a\xi_l\|^2$ is called the *Hilbert-Schmidt* *norm* of a, where $\{\lambda_n\}$ is the set of all positive eigenvalues of $(a^*a)^{1/2}$ (counting the multiplicity), and $\{\xi_l\}$ is any normalized orthogonal basis of H.

Proposition 1.1.4. For any $a \in S(H), b \in B(H)$, we have

$$\|a\| \le \|a\|_2 = \|a^*\|_2, \quad \|ba\|_2 \le \|a\|_2\|b\|, \quad \|ab\|_2 \le \|a\|_2\|b\|.$$

In particular, $S(H)$ is a $*$ two-sided ideal of $B(H)$.

Proof. Since $\sum_l \|a^*\xi_l\|^2 = \sum_{l,l'} |\langle a^*\xi_l, \xi_{l'}\rangle|^2 = \sum_{l,l'} |\langle \xi_l, a\xi_{l'}\rangle|^2 = \sum_{l'} \|a\xi_{l'}\|^2$, it follows that $\|a\|_2 = \|a^*\|_2$. Let $a = wh$ be the polar decomposition of a, where $h = (a^*a)^{1/2}$. Then

$$\|a\| \le \|(a^*a)^{1/2}\| = \max_n \lambda_n \le \|a\|_2.$$

Moreover, since $\sum_l \|ba\xi_l\|^2 \le \|b\|^2 \sum_l \|a\xi_l\|^2$, we have $\|ba\|_2 \le \|a\|_2\|b\|$. Furthermore, $\|ab\|_2 = \|b^*a^*\|_2 \le \|a^*\|_2\|b^*\| = \|a\|_2\|b\|$. Q.E.D.

Proposition 1.1.5. Define $\langle a, b\rangle_2 = \sum_l \langle a\xi_l, b\xi_l\rangle$, where $\{\xi_l\}$ is a normalized orthogonal basis of H. Then $(S(H), \langle,\rangle_2)$ is a Hilbert space, and $F(H)$ is dense in it.

Proof. First, we must show that the inner product $\langle,\rangle_2$ in $S(H)$ is independent of the choice of basis $\{\xi_l\}$. Let $\{\eta_r\}$ be another normalized orthogonal basis of H. Since $\sum_{l,r} |\langle a\xi_l, \eta_r\rangle\langle \eta_r, b\xi_l\rangle| \le \|a\|_2\|b\|_2$, it follows that the series $\sum_{l,r} \langle a\xi_l, \eta_r\rangle\langle \eta_r, b\xi_l\rangle$ is convergent absolutely. Then

$$\begin{aligned}\sum_l \langle a\xi_l, b\xi_l\rangle &= \sum_{l,r} \langle a\xi_l, \eta_r\rangle\langle \eta_r, b\xi_l\rangle \\ &= \sum_{r,l} \langle b^*\eta_r, \xi_l\rangle\langle \xi_l, a^*\eta_r\rangle = \sum_r \langle b^*\eta_r, a^*\eta_r\rangle.\end{aligned}$$

Similarly, $\sum_r \langle a\eta_r, b\eta_r\rangle = \sum_r \langle b^*\eta_r, a^*\eta_r\rangle$. Hence the definition of $\langle,\rangle_2$ is independent of the choice of $\{\xi_l\}$.

Now let $\{a_n\}$ be a Cauchy sequence in $(S(H), \langle,\rangle_2)$. Since $\|\cdot\|_2 \ge \|\cdot\|$, it follows that there exists $a \in C(H)$ such that $\|a_n - a\| \to 0$. From $\sum_l \|(a_n - a_m)\xi_l\|^2 \to 0 (n, m \to \infty)$, it is easily verified that $a \in S(H)$ and $\|a_n - a\|_2 \to 0$. Thus $(S(H), \langle,\rangle_2)$ is a Hilbert space.

Finally, for any $a \in S(H)$ and $\varepsilon > 0$, if $\{\xi_l\}_{l\in\Lambda}$ is a normalized orthogonal basis of H, then there is a finite subset F of Λ, such that $\sum_{l\notin F}(\|a\xi_l\|^2+\|a^*\xi_l\|^2) < \varepsilon^2$. Let p_F be the projection from H onto the linear span of $\{\xi_l | l \in F\}$. Then

$$\|p_F a p_F - a\|_2^2 = \sum_{l\notin F}\|a\xi_l\|^2 + \sum_{l\in F}\|(1-p_F)a\xi_l\|^2.$$

But

$$\sum_{l\in F}\|(1-p_F)a\xi_l\|^2 = \sum_{l\in F}\sum_{l'\notin F}|\langle a\xi_l, \xi_{l'}\rangle|^2 \le \sum_{l'\notin F}\|a^*\xi_{l'}\|^2,$$

it follows that $\|p_F a p_F - a\|_2 < \varepsilon$. Q.E.D.

We shall denote by $T(H)$ the set of all operators of *trace class* on H, i.e., $a \in T(H)$ if a is compact and $\sum_n \lambda_n < \infty$, where $\{\lambda_n\}$ is the set of all positive eigenvalues of $(a^*a)^{1/2}$ (counting the multiplicity). For $a \in T(H), \|a\|_1 = \sum_n \lambda_n$ is called the *trace norm* of a.

Proposition 1.1.6. 1) Let $a \in B(H)$. Then $a \in T(H)$ if and only if $(a^*a)^{1/4} \in S(H)$. Moreover, if $\{\xi_l\}$ is a normalized orthogonal basis of H, and $a \in T(H)$, then

$$\|a\|_1 = \|(a^*a)^{1/4}\|_2^2 = \sum_l \langle (a^*a)^{1/2}\xi_l, \xi_l\rangle.$$

2) If $a, b \in S(H)$, then $a, b \in T(H)$.
3) Let $a \in C(H)$. Then $a \in T(H)$ if and only if

$$\sup\left\{\sum_n |\langle a\xi_n, \eta_n\rangle| \;\middle|\; \begin{array}{c}\{\xi_n\} \text{ and } \{\eta_n\} \text{ are arbitrary}\\ \text{normalized orthogonal sequences in } H\end{array}\right\} < \infty.$$

Moreover, if $a \in T(H)$, then $\|a\| \le \|a\|_2 \le \|a\|_1$, and $\|a\|_1$ is equal to above sup.
4) $T(H)$ is a $*$ two-sided ideal of $B(H)$, and $T(H) \subset S(H)$, and $T(H)$ is the linear span of $T(H)_+$.

Proof. If $\{\lambda_n\}$ is the set of all positive eigenvalues of $(a^*a)^{1/2}$, then $\{\lambda_n^{1/2}\}$ is the set of all positive eigenvalues of $(a^*a)^{1/4}$. That comes to the conclusion 1).

2) Let $a, b \in S(H)$, and $c = w(c^*c)^{1/2}$ be the polar decomposition of c, where $c = ab$. Since

$$\sum_l \langle (c^*c)^{1/2}\xi_l, \xi_l\rangle = \sum_l \langle b\xi_l, a^*w\xi_l\rangle = \langle b, a^*w\rangle_2 \le \|a\|_2\|b\|_2,$$

it follows that $c = ab \in T(H)$.

3) Let $a \in T(H)$, and $a = w(a^*a)^{1/2} = w(a^*a)^{1/4} \cdot (a^*a)^{1/4}$ be the polar decomposition of a. Then for any normalized orthogonal sequences $\{\xi_n\}, \{\eta_n\}$ of H,

$$\begin{aligned}\sum_n |\langle a\xi_n, \eta_n\rangle| \leq & \ (\sum_n \|(a^*a)^{1/4}\xi_n\|^2)^{1/2} \cdot (\sum_n \|(a^*a)^{1/4}w^*\eta_n\|^2)^{1/2} \\ \leq & \ \|(a^*a)^{1/4}\|_2 \cdot \|(a^*a)^{1/4}w^*\|_2 \leq \|a\|_1.\end{aligned}$$

Moreover, if ξ_n is the eigenvector corresponding to λ_n, and $\eta_n = w\xi_n, \forall n$, then $\sum_n |\langle a\xi_n, \eta_n\rangle| = \sum_n \lambda_n = \|a\|_1$. Of course, $(\sum_n \lambda_n)^2 \geq \sum_n \lambda_n^2$. Hence $\|a\|_1 \geq \|a\|_2 \geq \|a\|$.

4) From $a = w(a^*a)^{1/4} \cdot (a^*a)^{1/4}$ and 2), it is obvious. Q.E.D.

Proposition 1.1.7. $(T(H), \|\cdot\|_1)$ is a Banach space, and $F(H)$ is its dense subset.

Proof. By Proposition 1.1.6, $\|\cdot\|_1$ is a norm on $T(H)$ exactly. Now let $\{a_n\}$ be a Cauchy sequence in $(T(H), \|\cdot\|_1)$. Since $\|\cdot\|_1 \geq \|\cdot\|$, there exists $a \in C(H)$ such that $\|a_n - a\| \to 0$.

For any finite normalized orthogonal sequences $\{\xi_k\}, \{\eta_k\}$, since

$$\begin{aligned}\sum_k |\langle (a_n - a)\xi_k, \eta_k\rangle| = & \ \lim_m \sum_k |\langle (a_n - a_m)\xi_k, \eta_k\rangle| \\ \leq & \ \overline{\lim_m} \|a_n - a_m\|_1 \to 0 \quad (\text{as } n \to \infty),\end{aligned}$$

it follows that $a \in T(H)$ and $\|a_n - a\|_1 \to 0$, and $(T(H), \|\cdot\|_1)$ is a Banach space.

Let $a \in T(H)$, and $a = w(a^*a)^{1/2}$ be its polar decomposition. Write $(a^*a)^{1/2} = \sum_n \lambda_n p_n$, where $\{p_n\}$ is a sequence of projections of one rank on H, and $p_n p_m = 0, \forall n \neq m$, and $\{\lambda_n\}$ is the set of all positive eigenvalues of $(a^*a)^{1/2}$. Then

$$\|a - \sum_{n=1}^N \lambda_n w p_n\|_1 = \sum_{n>N} \lambda_n \to 0,$$

i.e., $F(H)$ is dense in $(T(H), \|\cdot\|_1)$. Q.E.D.

Remark. The completions of $F(H)$ according to the norms $\|\cdot\|_1 \geq \|\cdot\|_2 \geq \|\cdot\|$ are $T(H) \subset S(H) \subset C(H)$ respectively.

For each $a \in T(H), tr(a) = \sum_l \langle a\xi_l, \xi_l\rangle$ is called the *trace* of a, where $\{\xi_l\}$ is any normalized orthogonal basis of H. Since a is the product of two elements of $S(H)$, the trace of a is well-defined.

Proposition 1.1.8. For any $a \in T(H), b \in B(H)$, we have

$$tr(ab) = tr(ba), \quad |tr(ba)| \leq \|a\|_1 \|b\|.$$

Proof. From the proof of Proposition 1.1.5, we have $tr(cd) = tr(dc)(\forall c, d \in S(H))$ indeed. Since a is the product of two elements of $S(H)$, conclude that $tr(ab) = tr(ba)$.

Let $a = w(a^*a)^{1/2} = w(a^*a)^{1/4} \cdot (a^*a)^{1/4}$ be the polar decomposition of a. Then

$$\begin{aligned} |tr(ab)| = & \; |\langle (a^*a)^{1/4}, (bw(a^*a)^{1/4})^* \rangle_2| \\ \leq & \; \|(a^*a)^{1/4}\|_2^2 \|b\| = \|a\|_1 \|b\|. \end{aligned}$$

Q.E.D.

Theorem 1.1.9. 1) $C(H)^* = T(H)$, i.e., for any continuous linear functional f on $C(H)$ there exists unique $a \in T(H)$ such that

$$\|f\| = \|a\|_1, \quad \text{and} \quad f(c) = tr(ac), \forall c \in C(H);$$

conversely, for any $a \in T(H)$ $tr(a\cdot)$ will be a continuous linear functional with norm $\|a\|_1$ on $C(H)$.

2) $T(H)^* = B(H)$, i.e., for any continuous linear functional f on $(T(H), \|\cdot\|_1)$ there exists unique $b \in B(H)$ such that

$$\|f\| = \|b\|, \quad \text{and} \quad f(a) = tr(ab), \forall a \in T(H);$$

conversely, for any $b \in B(H)$ $tr(\cdot b)$ will be a continuous linear functional with norm $\|b\|$ on $(T(H), \|\cdot\|_1)$.

Proof. 1) Let $a \in T(H), \{\lambda_n\}$ be the set of all positive eigenvalues of $(a^*a)^{1/2}$, and ξ_n be the eigenvector corresponding to $\lambda_n, \|\xi_n\| = 1, \forall n$. For any positive integer N, we define an operator c of finite rank as follows:

$$cw\xi_i = \xi_i, \quad 1 \leq i \leq N; \quad c = 0 \text{ on } \{w\xi_1, \cdots, w\xi_N\}^\perp,$$

where $a = w(a^*a)^{1/2}$ is the polar decomposition of a. It is obvious that $\|c\| = 1$, and

$$|tr(ac)| = |\sum_i \langle (a^*a)^{1/2} cw\xi_i, \xi_i \rangle| = \sum_{i=1}^N \lambda_i \to \|a\|_1$$

(as $N \to \infty$). Thus $tr(a\cdot)$ determines a continuous linear functional with norm $\|a\|_1$ on $C(H)$.

Now let $f \in C(H)^*$. Since $|f(\xi \otimes \eta)| \leq \|f\| \|\xi\| \|\eta\|$, there exists unique $a \in B(H)$ such that

$$f(\xi \otimes \eta) = \langle a\xi, \eta \rangle, \quad \forall \xi, \eta \in H, \quad \text{and} \quad \|a\| \leq \|f\|.$$

Further from $\langle a\xi,\eta\rangle = tr(a(\xi\otimes\eta))$, we have

$$f(c) = tr(ac), \quad \forall c \in F(H).$$

So it suffices to prove $a \in T(H)$. Let $a = w(a^*a)^{1/2}$ be the polar decomposition of a, and $\{\xi_l\}_{l\in\Lambda}$ be a normalized orthogonal basis of w^*H. For any finite subset F of Λ, define $c_F = \sum\limits_{l\in F} \xi_l \otimes w\xi_l \in F(H)$. Clearly $\|c_F\| = 1$. Then

$$|\sum_{l\in F}\langle (a^*a)^{1/2}\xi_l, \xi_l\rangle| = |tr(ac_F)| = |f(c_F)| \leq \|f\|.$$

Since F is arbitrary, and $w^*H = (a^*a)^{1/2}H$, it follows that $a \in T(H)$.

2) Let $b \in B(H)$. By Proposition 1.1.8 and

$$tr(b(\xi\otimes\eta)) = \langle b\xi,\eta\rangle, \forall \xi,\eta \in H, \qquad \|b\| = \sup_{\|\xi\|=\|\eta\|=1} |\langle b\xi,\eta\rangle|,$$

$tr(b\cdot)$ is a continuous linear functional with norm $\|b\|$ on $(T(H), \|\cdot\|_1)$.

Now let $f \in T(H)^*$. Since

$$|f(\xi\otimes\eta)| \leq \|f\|\|\xi\otimes\eta\|_1 = \|f\|\|\xi\|\|\eta\|, \quad \forall \xi,\eta \in H,$$

there exists unique $b \in B(H)$, such that

$$f(\xi\otimes\eta) = \langle b\xi,\eta\rangle = tr(b(\xi\otimes\eta)), \quad \forall \xi,\eta \in H.$$

Further from the density of $F(H)$ in $T(H)$, we have $f(a) = tr(ab), \forall a \in T(H)$.

Q.E.D.

Remark. By natural embedding, $B(H)$ can be regarded as the second conjugate space $C(H)^{**}$ of $C(H)$.

If a Banach space X is the conjugate space of a Banach space Y, we shall say that Y is the *predual* of X, and write $Y = X_*$. Hence, $T(H)$ is the predual of $B(H)$, and $T(H) = B(H)_*$.

Remark. Let $\mathbb{R}$ and $\mathbb{C}$ denote the real number field and the complex number field, respectively.

In this book, a linear space always means that it is over $\mathbb{C}$. In special case, we shall write a real linear space.

If E is a subset of a linear space, we shall denote by $[E]$ the *linear span* of E.

The symbol "1" will be used extensively in this book. It can be the number 1, or the identity operator on a linear space, or the identity of an algebra. Please don't confuse them.

Notes. The results in this section were obtained by R.Schatten and J. Von Neumann in the study of the tensor product of Banach spaces. A further detailed account of the theory was presented by Schatten.

References. [19],[151],[152],

1.2. Locally convex topologies in $B(H)$

Let H be a Hilbert space. We introduce the following locally convex topologies in $B(H)$:

1) *weak (operator) topology.* A net $a_l \to 0$, if

$$\langle a_l \xi, \eta \rangle \to 0, \quad \forall \xi, \eta \in H;$$

2) *strong (operator) topology.* A net $a_l \to 0$, if

$$\|a_l \xi\| \to 0, \quad \forall \xi \in H;$$

3) *strong $*$ (operator) topology.* A net $a_l \to 0$, if

$$(\|a_l \xi\| + \|a_l^* \xi\|) \to 0, \quad \forall \xi \in H;$$

4) *σ-weak (operator) topology.* A net $a_l \to 0$, if

$$\sum_n \langle a_l \xi_n, \eta_n \rangle \to 0, \quad \forall \quad \sum_n (\|\xi_n\|^2 + \|\eta_n\|^2) < \infty;$$

5) *σ-strong (operator) topology.* A net $a_l \to 0$, if

$$\sum_n \|a_l \xi_n\|^2 \to 0, \quad \forall \quad \sum_n \|\xi_n\|^2 < \infty;$$

6) *σ-strong $*$ (operator) topology.* A net $a_l \to 0$, if

$$\sum_n (\|a_l \xi_n\|^2 + \|a_l^* \xi_n\|^2) \to 0, \quad \forall \quad \sum_n \|\xi_n\|^2 < \infty;$$

7) $\sigma(B(H), T(H))$. A net $a_l \to 0$, if

$$tr(a_l a) \to 0, \quad \forall a \in T(H),$$

i.e., $\sigma(B(H), T(H))$ is the w^*-topology in $B(H)$, seeing $B(H)$ as the conjugate space of $T(H)$;

8) $s(B(H), T(H))$. A net $a_l \to 0$, if net $a_l^* a_l \to 0$ with respect to $\sigma(B(H), T(H))$;

9) $s^*(B(H), T(H))$. A net $a_l \to 0$, if nets $a_l^* a_l$ and $a_l a_l^* \to 0$ with respect to $\sigma(B(H)), T(H)$;

10) *Mackey topology* $\tau(B(H),T(H))$. A net $a_l \to 0$, if

$$tr(a_l a) \to 0, \quad \text{uniformly for} \quad a \in E,$$

where E is any $\sigma(T(H),B(H))$-compact subset of the Banach space $(T(H), \|\cdot\|_1)$;

11) *uniform (operator) topology*. It is the topology generated by the operator norm $\|\cdot\|$.

In the following, a subset (or a net) of $B(H)$ is called bounded, if the function of operator norm is bounded on it.

Proposition 1.2.1. 1) The $*$-operation of operators is continuous with respect to above topologies 1), 3), 4), 6), 7), 9), 10), 11).

2) For any fixed $b \in B(H)$, the maps $\cdot \to b\cdot$ and $\cdot \to \cdot b$ are continuous in $B(H)$ with respect to all above topologies.

3) If nets $\{a_l\}$ and $\{b_r\}$ of $B(H)$ are bounded, and $a_l \to a$ (strongly), $b_r \to b$ (strongly), then we have $a_l b_r \to ab$ (strongly).

Proof. If net $\{a_l\} \subset T(H)$ and $a_l \to 0(\sigma(T(H),B(H)))$, i.e., $tr(a_l b) \to 0, \forall b \in B(H)$, then we also have $a_l^* \to 0(\sigma(T(H),B(H)))$ since $tr(a_l^* b) = \overline{tr(b^* a_l)}, \forall b \in B(H)$. Also, for any $b \in B(H), a_l b \to 0$ and $ba_l \to 0$ with respect to $\sigma(T(H),B(H))$. Thus, if E is a $\sigma(T(H),B(H))$-compact subset of $T(H)$, then E^*, bE and Eb are also $\sigma(T(H),B(H))$-compact subsets of $T(H), \forall b \in B(H)$. Now the conclusions with respect to $\tau(B(H),T(H))$ are clear.

The rest conclusions are left to the reader. Q.E.D.

Proposition 1.2.2. Let f be a linear functional on $B(H)$. Then the following conditions are equivalent:

1) f is $\sigma(B(H),T(H))$-continuous;

2) There exists unique $a \in T(H)$ such that

$$f(b) = tr(ab), \quad \forall b \in B(H);$$

3) There are $\{\xi_n\}, \{\eta_n\} \subset H$ with $\sum_n (\|\xi_n\|^2 + \|\eta_n\|^2) < \infty$ such that $f(b) = \sum_n \langle b\xi_n, \eta_n\rangle, \forall b \in B(H)$.

Moreover, $f \geq 0$ (i.e. for any $b \in B(H)$ and $b \geq 0, f(b) \geq 0$) if and only if $a \geq 0$. In this case, we can take $\xi_n = \eta_n, \forall n$.

Proof. Since $T(H)^* = B(H)$, the equivalence of 1) and 2) is obvious.

Let $f(\cdot) = tr(a\cdot)$ for some $a \in T(H)$, and $f \geq 0$. For any $\xi \in H$, let p be the projection from H onto $[\xi]$. Then

$$\|\xi\|^{-2}\langle a\xi, \xi\rangle = tr(ap) = f(p) \geq 0.$$

This implies $a \geq 0$. Conversely, if $f(\cdot) = tr(a\cdot)$ and $a \geq 0$, it is clear that $f \geq 0$ since $tr(a\cdot) = tr(a^{1/2} \cdot a^{1/2})$.

2)$\Longrightarrow$3). Write $a = a_1 a_2$ such that $a_1 \in S(H)$ and $a_2^* = a_2 \in S(H)$. Let $\{\xi_l\}_{l\in\Lambda}$ be a normalized orthogonal basis of H. Then there is a countable subset J of Λ such that $a_1\xi_l = a_2\xi_l = 0, \forall l \in \Lambda\backslash J$. Further,

$$f(b) = tr(a_2 b a_1) = \sum_{l\in J}\langle ba_1\xi_l, a_2\xi_l\rangle, \quad \forall b \in B(H),$$

and $\sum_{l\in J}(\|a_1\xi_l\|^2 + \|a_2\xi_l\|^2) < \infty$. Moreover, if $f \geq 0$, we can take $a_1 = a_2 = a^{1/2}$. Then $a_1\xi_l = a_2\xi_l, \forall l \in J$.

3)$\Longrightarrow$ 2). Take a normalized orthogonal sequence $\{\varsigma_n\}$ of H. Define

$$a_1\varsigma_n = \xi_n, \quad a_2\varsigma_n = \eta_n, \forall n; \text{ and } a_1 = a_2 = 0 \text{ on } \{\varsigma_n\}^\perp.$$

Obviously, a_1 and $a_2 \in S(H)$. Then $a = a_1 a_2^* \in T(H)$ and

$$f(b) = \sum_n \langle ba_1\varsigma_n, a_2\varsigma_n\rangle = tr(a_2^* b a_1) = tr(ab), \quad \forall b \in B(H).$$

Q.E.D.

Theorem 1.2.3. The relations between topologies 1)–11) are as follows:

$$\begin{array}{ccccccc} & & \text{top.3)} & \supset & \text{top.2)} & \supset & \text{top.1)} \\ & & \cap & & \cap & & \cap \\ \text{top.11)} \supset \text{top.10)} & \supset & \text{top.9)} & \supset & \text{top.8)} & \supset & \text{top.7)} \\ & & \wr & & \wr & & \wr \\ & & \text{top.6)} & \supset & \text{top.5)} & \supset & \text{top.4)}, \end{array}$$

where "$\supset$" means that the left side is finer than the right side.

Proof. Obviously, we have the following relations: top.3) $\supset$ top.2) $\supset$ top.1), top.6) $\supset$ top.5) $\supset$ top.4), top.4) $\supset$ top.1), top.5) $\supset$ top.2), top.6) $\supset$ top.3), top.11) $\supset$ top.10) $\supset$ top.7) (by the Mackey theorem).

By Proposition 1.2.2, the equivalence of top.7) and top.4) is clear.

Now we prove the equivalence of top.8) and top.5). Let a net $a_l \to 0(s(B(H), T(H)))$. Since top.7) $\sim$ top.4), we have that $a_l^* a_l \to 0$ (σ-weakly) and $a_l \to 0$ (σ-strongly). Conversely, let a net $a_l \to 0$ (σ-strongly). For any $0 \leq a \in T(H)$, by Proposition 1.2.2 there exists a sequence $\{\xi_n\} \subset H$ with $\sum_n \|\xi_n\|^2 < \infty$ such that

$$tr(ab) = \sum_n \langle b\xi_n, \xi_n\rangle, \quad \forall b \in B(H).$$

Hence $tr(a_l^* a_l a) = \sum_n \|a_l \xi_n\|^2 \to 0, \forall a \in T(H)_+$, and $a_l \to 0(s(B(H),T(H)))$.

Since $*$ operation is continuous with respect to top.9) and top.6), it follows from top.8) $\sim$ top.5) that top.9) $\sim$ top.6). Further we have top.9) $\supset$ top.8) $\supset$ top.7) from top.6) $\supset$ top.5) $\supset$ top.4).

Now we prove that top.10) $\supset$ top.8) $\supset$ top.7). By the Mackey theorem and top.5) $\sim$ top.8) $\supset$ top.7), it suffices to show that any top.5)-continuous linear functional f on $B(H)$ must be $\sigma(B(H),T(H))$-continuous. For such f, there is a top.5)- neighborhood $U = U(0; \{\xi_n^{(1)}\}, \cdots, \{\xi_n^{(k)}\}; 1)$ of zero such that

$$|f(b)| \le 1, \quad \forall b \in U.$$

There $\{\xi_n^{(i)}\} \subset H, \sum_n \|\xi_n^{(i)}\|^2 < \infty, 1 \le i \le k$, and $b \in U$ means that $\sum_n \|b\xi_n^{(i)}\|^2 \le 1, 1 \le i \le k$. Let $\{\xi_n\} = \bigcup_{i=1}^k \{\xi_n^{(i)}\}$. Then $\sum_n \|\xi_n\|^2 < \infty$ and

$$|f(b)| \le (\sum_n \|b\xi_n\|^2)^{1/2}, \quad \forall b \in B(H).$$

Define $\widetilde{H} = \bigoplus_{i=1}^\infty H_i$, where $H_i = H, \forall i$. Then $\tilde{\xi} = (\xi_n) \in \widetilde{H}$, and $|f(b)| \le \|\tilde{b}\tilde{\xi}\|, \forall b \in B(H)$, where $\tilde{b} \in B(\widetilde{H})$ is defined by $\tilde{b}\tilde{\eta} = (b\eta_n), \forall \tilde{\eta} = (\eta_n) \in \widetilde{H}$. In particular, $f(b) = 0$ if $\tilde{b}\tilde{\xi}=0$. Then we can define a linear functional $\tilde{f}$ on the linear subspace $\{\tilde{b}\tilde{\xi} | b \in B(H)\}$ of $\widetilde{H}$ as follows:

$$\tilde{f}(\tilde{b}\tilde{\xi}) = f(b), \quad \forall b \in B(H).$$

Since $|\tilde{f}(\tilde{b}\tilde{\xi})| = |f(b)| \le \|\tilde{b}\tilde{\xi}\|, \forall b \in B(H)$, it follows that there is $\tilde{\eta} = (\tilde{\eta}_n) \in \widetilde{H}$ such that

$$f(b) = \tilde{f}(\tilde{b}\tilde{\xi}) = \langle \tilde{b}\tilde{\xi}, \tilde{\eta} \rangle = \sum_n \langle b\xi_n, \eta_n \rangle, \quad \forall b \in B(H).$$

Now by Proposition 1.2.2, f is $\sigma(B(H),T(H))$-continuous.

Finally, since $*$ is $\tau(B(H),T(H))$-continuous, it follows that top.10) $\supset$ top.9). Q.E.D.

Theorem 1.2.4. In any bounded ball of $B(H)$, we have weak(operator) topology$\sim \sigma(B(H),T(H))$, strong (operator) topology $\sim s(B(H),T(H))$, and strong $*$ (operator) topology $\sim s^*(B(H),T(H))$.

Proof. Let a net $a_l \to 0$ (weakly), and $\|a_l\| \le K, \forall l$. Then for any $\{\xi_n\}, \{\eta_n\} \subset H$ with $\sum_n (\|\xi_n\|^2 + \|\eta_n\|^2) < \infty$, we have

$$|\sum_n \langle a_l\xi_n, \eta_n \rangle| \le \sum_{n=1}^N |\langle a_l\xi_n, \eta_n \rangle| + \frac{K}{2} \sum_{n>N} (\|\xi_n\|^2 + \|\eta_n\|^2).$$

It follows that $a_l \to 0$ (σ-weakly). Similarly, we can prove the rest parts.
Q.E.D.

We shall prove that $\tau(B(H),T(H)) \sim s^*(B(H),T(H))$ in any bounded ball of $B(H)$ (see Section 1.11). However, we have the following.

Proposition 1.2.5. If H is infinite dimensional, then in whole $B(H), s^*(B(H), T(H))$ is not equivalent to $\tau(B(H),T(H))$.

Proof. Let $\{p_n\}$ be an infinite sequence of non-zero projections on H and $p_n p_m = 0, \forall n \neq m$. Put $K = \{\sqrt{n}p_n | n = 1,2,\cdots\}$. We say that $K \cap U(0; a_1,\cdots,a_k;\varepsilon) \neq \emptyset$ for any $\varepsilon > 0, 0 \leq a_i \in T(H), 1 \leq i \leq k$, where $U(0;a_1,\cdots,a_k,\varepsilon) = \{b \in B(H) | tr((b^*b + bb^*)a_i) < \varepsilon, 1 \leq i \leq k\}$. In fact, if there are $a_1,\cdots,a_k \in T(H)_+$ and $\varepsilon > 0$, such that $K \cap U(0;a_1,\cdots,a_k;\varepsilon) = \emptyset$, then

$$2ntr(p_n a) \geq \varepsilon, \quad \forall n,$$

where $a = \sum_{i=1}^{k} a_i$. Let $p = \sum_n p_n$. Then

$$tr(pa) = \lim_N \sum_{n=1}^{N} tr(p_n a) \geq \lim_N \sum_{n=1}^{N} \frac{\varepsilon}{2n} = \infty,$$

a contradiction.

Since $\{U(0;a_1,\cdots,a_k;\varepsilon) \mid \varepsilon > 0, k = 1,2,\cdots, a_i \in T(H)_+\}$ is a $s^*(B(H), T(H))$-neighborhood basis of zero, it follows that 0 belongs to the $s^*(B(H), T(H))$-closure of K.

Now it suffices to show that 0 is not contained in the $\tau(B(H),T(H))$-closure of K.

For each n, take $\xi_n \in p_n H, \|\xi_n\| = 1$. Then, $c_n = \frac{2}{\sqrt{n}}\xi_n \otimes \xi_n \in T(H)$, and $\|c_n\|_1 = \frac{2}{\sqrt{n}} \to 0$. Thus $L = \{0, c_n | n = 1,2,\cdots\}$ is a compact subset of $(T(H), \|\cdot\|_1)$. Further L is $\sigma(T(H),B(H))$-compact. It follows that

$$L^0 = \{b \in B(H) \mid |tr(bc_n)| \leq 1, \forall n\}$$

is a $\tau(B(H),T(H))$-neighborhood of 0. Obviously, $\sqrt{n}p_n \notin L^0$ since $tr(\sqrt{n}p_n c_n) = 2, \forall n$. This implies that $K \cap L^0 = \emptyset$ and 0 is not contained in the $\tau(B(H),T(H))$-closure of K.
Q.E.D.

By Theorems 1.2.3, 1.2.4 and the standard results of duality theory (see Appendix), we have the following.

Proposition 1.2.6. Let f be a linear functional on $B(H)$. Then the following statements are equivalent: 1) f is $\sigma(B(H),T(H))$-continuous; 2)

f is $s(B(H),T(H))$-continuous; 3) f is $s^*(B(H),T(H))$-continuous; 4) f is $\tau(B(H),T(H))$-continuous; 5) f is weakly continuous in any bounded ball of $B(H)$; 6) f is strongly continuous in any bounded ball of $B(H)$; 7) f is strongly $*$ continuous in any bounded ball of $B(H)$.

On the weak (or strong) continuity of linear functionals, we have the following.

Proposition 1.2.7. Let f be a linear functional on $B(H)$. Then the following conditions are equivalent:

1) f is weakly continuous;
2) f is strongly continuous;
3) There exists unique $v \in F(H)$ such that

$$f(b) = tr(bv), \quad \forall b \in B(H);$$

4) There are $\xi_1, \cdots, \xi_n$, and $\eta_1, \cdots, \eta_n \in H$ such that

$$f(b) = \sum_{i=1}^{n} \langle b\xi_i, \eta_i \rangle, \quad \forall b \in B(H).$$

Moreover, $f \geq 0$ if and only if $v \geq 0$. In this case, we can also take $\xi_i = \eta_i, 1 \leq i \leq n$.

Proof. It is clears that 3) and 4) are equivalent, and 3) or 4) implies 1), 1) implies 2). Thus it suffices to show that 2) forces 4).

Let f be strongly continuous. Then there is a strong neighborhood $U = U(0; \eta_1, \cdots, \eta_m; 1) = \{b \in B(H) | \|b\eta_i\| \leq 1, 1 \leq i \leq m\}$ of 0 such that $|f(b)| \leq 1, \forall b \in U$. Let $\{\xi_1, \cdots, \xi_n\}$ be a normalized orthogonal basis of the linear span of $\{\eta_1, \cdots, \eta_m\}$. Then for enough small $\varepsilon > 0$, the strong neighborhood $V = U(0; \xi_1, \cdots, \xi_n, \varepsilon)$ will be contained in U. In particular, if $b \in B(H)$ and $b\xi_i = 0, 1 \leq i \leq n$, then we have $f(b) = 0$.

Of course, f is also $s(B(H),T(H))$-continuous. By Proposition 1.2.6, we have $a \in T(H)$ such that $f(b) = tr(ab), \forall b \in B(H)$. Let $\{\xi_l\}$ be a normalized orthogonal basis of H containing $\{\xi_1, \cdots, \xi_n\}$. Take $c \in B(H)$ such that $c\xi_i = 0, 1 \leq i \leq n$, and $c\xi_l = a^*\xi_l, \forall \xi_l \neq \xi_1, \cdots, \xi_n$. Then $0 = f(c) = tr(ac) = \sum_l \langle ac\xi_l, \xi_l \rangle = \sum_{\xi_l \neq \xi_1, \cdots, \xi_n} \|a^*\xi_l\|^2$. Hence, $a^*\xi_l = 0, \forall \xi_l \neq \xi_1, \cdots, \xi_n$. Let $a^*\xi_i = \eta_i, 1 \leq i \leq n$. Then $f(b) = tr(ab) = \sum_l \langle ab\xi_l, \xi_l \rangle = \sum_{i=1}^{n} \langle b\xi_i, \eta_i \rangle$.

The last statement on $f \geq 0$ is contained in the proof of Proposition 1.2.2 indeed. Q.E.D.

On the closedness of a convex subset of $B(H)$, by Propositions 1.2.6, 1.2.7, and the separation theorem, the Krein-Šmulian theorem (see Appendix), we have the following.

Proposition 1.2.8. Let K be a convex subset of $B(H)$. Then the following statements are equivalent: 1) K is $\sigma(B(H), T(H))$-closed; 2) K is $s(B(H), T(H))$-closed; 3) K is $s^*(B(H), T(H))$-closed; 4) K is $\tau(B(H), T(H))$-closed; 5) $K \cap \lambda S$ is weakly closed, $\forall \lambda > 0$; 6) $K \cap \lambda S$ is strongly closed, $\forall \lambda > 0$; 7) $K \cap \lambda S$ is strongly $*$ closed, $\forall \lambda > 0$. There S is the closed unit ball of $(B(H), \|\cdot\|)$. Moreover, if K is bounded, we can replace $K \cap \lambda S$ by K in 5), 6) and 7).

Proposition 1.2.9. Let K be a convex subset of $B(H)$. Then the weak closure of K is equal to the strong closure of K, or K is weakly closed if and only if K is strongly closed.

As the end of this section, we mention the following proposition which we shall use often.

Proposition 1.2.10. Let $\{a_l\}$ be a bounded increasing net of self-adjoint elements of $B(H)$, i.e., $a_l^* = a_l, \|a_l\| \leq M$ (some constant), $\forall l$, and $a_{l'} \geq a_l, \forall l' \geq l$. Then we have $a_l \to a = \sup_l a_l$ (strongly)

Proof. Since $\{\langle a_l\xi, \xi\rangle\}$ is a bounded increasing net of real numbers for any $\xi \in H$, it follows that $\lim\langle a_l\xi, \eta\rangle$ exists for any $\xi, \eta \in H$. Then by $|\langle a_l\xi, \eta\rangle| \leq M\|\xi\|\|\eta\|$, we have $a \in B(H)$ such that $\lim\langle a_l\xi, \eta\rangle = \langle a\xi, \eta\rangle, \forall \xi, \eta \in H$. Of course, $\langle a\xi, \xi\rangle = \sup\limits_l \langle a_l\xi, \xi\rangle, \forall \xi \in H$. Thus we have $a = \sup\limits_l a_l$ and $a_l \to a$ (weakly). Moreover, for any $\xi \in H$ we also have

$$\begin{aligned} \|(a - a_l)\xi\|^2 \leq & \ \|(a - a_l)^{1/2}\|^2 \cdot \|(a - a_l)^{1/2}\xi\|^2 \\ \leq & \ 2M\langle (a - a_l)\xi, \xi\rangle \to 0. \end{aligned}$$

Therefore $a_l \to a$ (strongly). Q.E.D.

Notes. In this section, we used many standard results of duality theory in Banach spaces. These can be found in Appendix.

The weak topology, the strong topology, the σ-weak topology and the σ-strong topology were introduced by J. Von Neumann.

Moreover, Propposition 1.2.5 is due to F.J.Yeadon.

References. [19], [31], [89], [146], [151], [152], [200].

1.3. Von Neumann's double commutation theorem

Definition 1.3.1. Let H be a Hilbert space. A $*$ subalgebra M of $B(H)$ is called a *Von Neumann algebra* (or *VN algebra* simply), if

$$M = M'',$$

where $M' = \{b \in B(H) \mid ba = ab, \forall a \in M\}$ is the *commutant* of M, and $M'' = (M')'$ is the double commutant of M.

If E is a subset of $B(H)$, and M is the smallest VN algebra containing E, then M is called the VN algebra generated by E.

Proposition 1.3.2. Let E be a subset of $B(H)$. Then $(E \sqcup E^*)'$ is a VN algebra, and the VN algebra generated by E is $(E \cup E^*)''$. In particular, the commutant of any VN algebra is a VN algebra.

Proof. Obviously, we have $(E \sqcup E^*) \subset (E \sqcup E^*)''$ and $(E \sqcup E^*)' \subset (E \sqcup E^*)'''$. Now if $a \in (E \sqcup E^*)'''$, then $ab = ba, \forall b \in (E \sqcup E^*)$, and $a \in (E \sqcup E^*)'$. Thus $(E \sqcup E^*)' = (E \sqcup E^*)'''$ is a VN algebra. Furthermore, $(E \sqcup E^*)'' = ((E \sqcup E^*)')'$ is also a VN algebra.

Now suppose that N is a VN algebra, and $N \supset E$. Then $N \supset (E \sqcup E^*)$. Since $N' \subset (E \sqcup E^*)'$, we have $N'' = N \supset (E \sqcup E^*)''$. Therefore, $(E \sqcup E^*)''$ is the VN algebra generated by E. Q.E.D.

Proposition 1.3.3. 1) Let M be a VN algebra on H. Then M is weakly closed. In particular, M is the conjugate space of quotient Banach space $T(H)/M_\perp$, where

$$M_\perp = \{a \in T(H) \mid tr(ab) = 0, \forall b \in M\}.$$

2) Let $\{M_l\}$ be a set of VN algebras on H. Then $M = \bigcap_l M_l$ is also a VN algebra, and M' is generated by $\sqcup_l M_l'$.

Proof. 1) is clear. Now we prove 2). Obviously the following conditions are equivalent: (1) $a \in M$; (2) $a \in M_l, \forall l$; (3) $a \in (M_l')', \forall l$; (4) $a \in (\sqcup_l M_l')'$. Thus $M = (\sqcup_l M_l')'$ is a VN algebra, and M' is equal to $(\sqcup_l M_l')''$, the VN algebra generated by $\sqcup_l M'_l$.

Q.E.D.

Proposition 1.3.4. Let M be a VN algebra on H.

1) Let $a = vh$ be the *polar decomposition* of a, where $a \in M$. Then $v, h \in M$. In particular, the projection vv^* from H onto $\overline{aH}$ belongs to M.

2) Let a be a *normal element* of M, i.e., $a^*a = aa^*$, and $\{e(\cdot)\}$ be the spectral measure of a. Then $e(\Delta) \in M$ for any Borel subset Δ of $\mathbb{C}$.

3) M is the uniform closure of the linear span of Proj (M), where Proj (M) is the set of all projections of M. And also M is the linear span of $U(M)$, where $U(M)$ is the set of all *unitary elements* of M(i.e.$u \in U(M)$ if $u^*u = uu^* = 1$).

Proof. 1) Obviously, we have $h = (a^*a)^{1/2} \in M$. Let $b' \in M'$. If $\xi \in (a^*H)^\perp$, then $a\xi = v\xi = 0, ab'\xi = b'a\xi = 0$, and $b'\xi \in (a^*H)^\perp$. Thus $vb'\xi = 0 = b'v\xi$. For any $\eta \in H$, we have

$$b'v(a^*a)^{1/2}\eta = b'a\eta = ab'\eta = v(a^*a)^{1/2}b'\eta = vb'(a^*a)^{1/2}\eta.$$

Since $(a^*a)^{1/2}H$ is dense in $\overline{a^*H}$, it follows that $b'v\varsigma = vb'\varsigma, \forall \varsigma \in \overline{a^*H}$. Therefore $b'v = vb', \forall b' \in M'$, i.e., $v \in M'' = M$.

2) It is clear since $e(\Delta)b' = b'e(\Delta), \forall b' \in M'$.

3) Since any spectral projection of any self-adjoint element of M belongs to M, it follows that M is the uniform closure of the linear span of Proj (M). Now if $h^* = h \in M$ and $\|h\| \le 1$, then $(1-h^2)^{1/2} \in M$ and $(h \pm i(1-h^2)^{1/2}) \in U(M)$. Therefore M is the linear span of $U(M)$. Q.E.D.

Proposition 1.3.5. Let M be a VN algebra on H. Then Proj (M) is a *complete lattice* with respect to the inclusion relation. Moreover, if $\{p_l\}_{l\in\Lambda} \subset$ Proj (M), then

$$\begin{aligned}\sup_{l\in\Lambda} p_l &= \text{(strongly) - }\lim_F \sup_{l\in F} p_l \\ &= \text{the projection from } H \text{ onto } \overline{[\cup_{l\in\Lambda} p_l H]},\end{aligned}$$

and

$$\begin{aligned}\inf_{l\in\Lambda} p_l &= \text{(strongly)- }\lim_F \inf_{l\in F} p_l \\ &= \text{the projection from } H \text{ onto } \bigcap_{l\in\Lambda} p_l H,\end{aligned}$$

where F is any finite subset of Λ, and is directed by the inclusion relation.

Proof. Let $p.q \in$ Proj (M). By Proposition 1.3.4, the projection from H onto $\overline{[(1-q)pH]}$ belongs to M. Now noticing

$$\overline{[pH + qH]} = qH \oplus \overline{[(1-q)pH]},$$

so the projection from H onto $\overline{[pH + qH]}$ also belongs to M, i.e., $\sup\{p,q\} \in M$. Moreover, since

$$\overline{[pH + qH]} = (pH \cap qH) \oplus \overline{[(1-p)qH]} \oplus \overline{[(1-q)pH]},$$

it follows that $\inf\{p,q\} \in M$.

Further, for any finite subset F of Λ, we have

$$\sup_{l\in F} p_l \in M, \quad \text{and} \quad \inf_{l\in F} p_l \in M.$$

By Proposition 1.2.10, we have

$$\sup_{l\in\Lambda} p_l = \sup_F \sup_{l\in F} p_l = \text{(strongly)-}\lim_F \sup_{l\in F} p_l \in M$$

since M is strongly closed.

Considering the family of projections $\{(1-\inf_{l\in F} p_l) \mid F$ is any finite subset of $\Lambda\}$, we can get the rest conclusion. Q.E.D.

Proposition 1.3.6. Let M be a VN algebra on $H, p \in \text{Proj }(M)$. Then $M_p = pMp$ and $M'_p = M'p$ are two VN algebras on pH, and also $(M_p)' = M'_p$.

Proof. Obviously, $(M'_p)' \supset M_p$. For $a \in (M'_p)' \subset B(pH)$, define

$$\overline{a} = \begin{cases} a, & \text{on } pH, \\ 0, & \text{on } (1-p)H. \end{cases}$$

Then we have $\overline{a} \in B(H)$. For any $b' \in M'$, since pH and $(1-p)H$ are two invariant subspaces of b', it follows from $b'p \in M'_p$ that $\overline{a}b' = b'\overline{a}$. Thus $\overline{a} \in M$, and $a = p\overline{a}p \in M_p$. Further, $M_p = (M'_p)'$ is a VN algebra on pH.

Now it suffices to show that $M'_p = (M_p)'$, or to prove that $a' \in M'_p$ for any $a' \in (M_p)'$ since $(M_p)' \supset M'_p$. By Proposition 1.3.4, we may assume that a' is a unitary operator on pH. Let q be a projection on H such that $qH = \overline{[MpH]}$. It is easily verified that $q \in M \cap M'$. Define

$$\begin{cases} v'\sum_i a_i\xi_i = \sum_i a_ia'\xi_i, & \forall \xi_i \in pH, a_i \in M; \\ v'(1-q)\xi = 0, & \forall \xi \in H. \end{cases}$$

Since $a' \in (M_p)'$ and a' is unitary on pH, it follows that

$$\begin{aligned} \|\sum_i a_ia'\xi_i\|^2 &= \sum_{i,j}\langle a_ipa'\xi_i, a_jpa'\xi_j\rangle \\ &= \sum_{i,j}\langle pa_j^*a_ipa'\xi_i, a'\xi_j\rangle \\ &= \sum_{i,j}\langle pa_j^*a_ip\xi_i, a'^*a'\xi_j\rangle = \|\sum_i a_i\xi_i\|^2. \end{aligned}$$

Hence v' can be extended to a partial isometry on H, which is still denoted by v', and $v'^*v' = q$. For any $a \in M$, since

$$v'a\sum_i a_i\xi_i = \sum_i aa_ia'\xi_i = av'\sum_i a_i\xi_i, \forall \xi_i \in pH, a_i \in M,$$

and

$$v'a(1-q)\xi = v'(1-q)a\xi = 0 = av'(1-q)\xi, \quad \forall \xi \in H,$$

we have $v' \in M'$. Obviously, $a' = v'p \in M'_p$. This completes the proof.

Q.E.D.

Definition 1.3.7. Let M be a VN algebra on H. $Z = Z(M) = M \cap M'$ is called the *center* of M. If $Z = \mathbb{C}1_H$ (or write $Z = \mathbb{C}$ simply), then M is called a *factor*.

Proposition 1.3.8. Let M be a VN algebra on H, Z be its center, and $p \in$Proj (M). Then $M_p \cap M'_p = Zp$. In particular, if M is a factor, then M_p and M'_p are also factors (on pH). Moreover, if q is a central projection of M_p, then there is a central projection z of M such that $q = zp$.

Proof. Obviously, $Zp \subset M_p \cap M'_p$. Conversely, if $a \in M_p \cap M'_p$, we have $b' \in M'$ such that $a = b'p$. Let r be the projection from H onto $\overline{[MpH]}$. Then $r \in Z$ and $r \geq p$. Hence $a = b'p = (b'r)p$. If b' is replaced by $b'r$, then we may assume that $b' = b'r$. For any $a' \in M,'$ noticing $b'a'p = b'p \cdot a'p = aa'p = a'ap = a'b'p$ and by the definition of r, we have $(a'b' - b'a')r = 0$. But $b'r = b'$ and $r \in Z$, so $a'b' = b'a', \forall a' \in M'$. Therefore $b' \in Z$ and $a = b'p \in Zp$, i.e., $M_p \cap M'_p = Zp$.

Now let q be a central projection of M_p. From preceding paragraph, there is $z \in Z$ such that $q = zp$. Of course, $q = \frac{1}{2}(z + z^*)p$, so we may assume $z^* = z$. Since $q^2 = q$, this implies $(z^2 - z)p = 0$, and $(z^2 - z)r = 0$. Again by the discussion of preceding paragraph, we may assume $z = zr$. Then $z^2 = z$ since $(z^2 - z)r = 0$. Therefore, z is a central projection of M, and also $q = zp$.

Q.E.D.

Theorem 1.3.9. Let H be a Hilbert space, and M be a $*$ subalgebra of $B(H), \overline{M}^w$ be the weak closure of M. Then

$$\overline{M}^w = \{a \in M'' \mid ap_0 = p_0 a = a\} = M'' p_0,$$

where p_0 is the projection from H onto $\overline{[MH]}$, and $p_0 \in M' \cap M'', (1 - p_0)H = \{\xi \in H \mid a\xi = 0, \forall a \in M\}$. In particular, if M is also nondegenerate (i.e. $p_0 = 1$), then we have $\overline{M}^w = M''$.

Proof. By the definition of p_0, it is clear that $p_0 a = a, \forall a \in M$. Thus $p_0 a = a, \forall a \in \overline{M}^w$. Obviously, $\overline{M}^w \subset M'', p_0 \in M' \cap M''$, and $(1 - p_0)H$ is the null subspace of M.

Let $a \in M'', p_0 a = ap_0 = a$, and $U(a; \xi_1, \cdots, \xi_n, \varepsilon) = \{b \in B(H) \mid \|(a - b)\xi_i\| < \varepsilon, 1 \leq i \leq n\}$ be a strong neighborhood of a. Put

$$\widetilde{H} = H \oplus \cdots \oplus H \quad (n \text{ times})$$

and let $\tilde{p}' = (p'_{ij})_{1\le i,j\le n}$ be the projection from $\widetilde{H}$ onto $\overline{[(b\xi_1,\cdots,b\xi_n)|b\in M]}$, where $p'_{ij} \in B(H), \forall i,j$. For any $b \in M$, define

$$\tilde{b}\tilde{\eta} = (b\eta_1,\cdots,b\eta_n), \quad \forall\tilde{\eta} = (\eta_1,\cdots,\eta_n) \in \widetilde{H}.$$

Then $\tilde{b} \in B(\widetilde{H})$, and $\tilde{p}'\widetilde{H} = \overline{[\tilde{b}\tilde{\xi}|b\in M]}$, where $\tilde{\xi} = (\xi_1,\cdots,\xi_n)$. Clearly, $\tilde{b}\tilde{p}' = \tilde{p}'\tilde{b}, \forall b \in M$. Thus $p'_{ij} \in M', \forall i,j$. Since $\tilde{p}'\tilde{b}\tilde{\xi} = \tilde{b}\tilde{\xi}, \forall b \in M$, it follows that

$$b\xi_i = \sum_{k=1}^{n} p'_{ik} b\xi_k = b\sum_{k=1}^{n} p'_{ik}\xi_k, \quad \forall b \in M, 1\le i\le n.$$

Hence $(\xi_i - \sum_{k=1}^{n} p'_{ik}\xi_k) \in (1-p_0)H, 1\le i\le n$. But $a(1-p_0) = 0$, so

$$a\xi_i - \sum_{k=1}^{n} p'_{ik} a\xi_k = 0, \quad 1\le i\le n,$$

i.e., $(a\xi_1,\cdots,a\xi_n) \in \tilde{p}'\widetilde{H}$. Now by the definition of $\tilde{p}'$, for the above $\varepsilon > 0$ we can find $b \in M$ such that

$$\|b\xi_i - a\xi_i\| < \varepsilon, \quad 1\le i\le n.$$

This implies that $M \cap U(a;\xi_1,\cdots,\xi_n;\varepsilon) \ne \emptyset, \forall \xi_1,\cdots,\xi_n \in H$, and $\varepsilon > 0$. Hence a belongs to the strong closure of M. By Proposition 1.2.9, we have $a \in \overline{M}^w$.

Q.E.D.

Now we come to the *Von Neumann's double commutation theorem.*

Theorem 1.3.10. Let H be a Hilbert space, and M be a $*$ subalgebra of $B(H)$. Then M is a VN algebra on H if and only if M is weakly closed and $1_H \in M$.

As the end of this section, we consider the topological problem about a VN algebra. Let M be a VN algebra on H. We have introduced many locally convex topologies in $B(H)$ (see Section 1.2). Naturally, these topologies can be restricted into M. Moreover, from Proposition 1.3.3 M is the conjugate space of the Banach space $M_* = T(H)/M_\perp$. Using M_*, we can also introduce the following topologies:

1) $\sigma(M,M_*)$. The weak $*$ topology of M for M_*;
2) $s(M,M_*)$. A net $a_l \to 0$, if net $a_l^*a_l \to 0$ with respect to $\sigma(M,M_*)$;
3) $s^*(M,M_*)$. A net $a_l \to 0$, if nets $a_l^*a_l$ and $a_la_l^* \to 0$ with respect to $\sigma(M,M_*)$;
4) $\tau(M,M_*)$. The Mackey topology of M for M_*.

Clearly, we have the following relations:

$$\sigma(M, M_*) \sim (\sigma(B(H), T(H))|M) \sim (\sigma\text{-weak top.}|M);$$

$$s(M, M_*) \sim (s(B(H), T(H))|M) \sim (\sigma\text{-strong top.}|M);$$

$$s^*(M, M_*) \sim (s^*(B(H), T(H))|M) \sim (\sigma\text{-strong} * \text{top.}|M);$$

$$\tau(M, M_*) \supset (\tau(B(H), T(H))|M).$$

Moreover, the relations $\sigma(M, M_*) \sim$ (weak top. $|M$), $s(M, M_*) \sim$ (strong top. $|M$) and $s^*(M, M_*) \sim$ (strong $*$ top. $|M$) are possible. See Sections 1.10, 1.13, and 2.11.

Finally, the results of Propositions 1.2.6, 1.2.8 and 1.2.9 can be moved to M, but we shall not discuss these in detail.

Notes. Theorem 1.3.10 was obtained by J. Von Neumann. This is the first important result on the theory of Von Neumann algebras. It says that the algebraic definition is equivalent to the topological definition for Von Neumann algebras.

It was J.Dixmier who recognized the σ-weak topology as the weak $*$ topology in the duality between $B(H)$ and $T(H)$, and thus proved that a Von Neumann algebra is the conjugate space of a Banach space.

About the topologies in a VN algebra M on H, we have pointed out that $\tau(M, M_*) \supset (\tau(B(H), T(H))|M)$. A natural question is that these two topologies are equivalent? It seems still open.

References.[1], [19], [114].

1.4. Tensor products of Von Neumann algebras

First consider *tensor products of Hilbert spaces.* Let H_1, H_2 be two complex Hilbert spaces. Let

$$H_1 \odot H_2 = \left\{u = \sum_{j=1}^{n} \xi_j^{(1)} \otimes \xi_j^{(2)} \;\middle|\; \begin{array}{c} n = 1, 2, \cdots, \\ \xi_j^{(i)} \in H_i, i = 1, 2, 1 \le j \le n \end{array}\right\}.$$

If we can introduce a definition of zero element (an equivalent relation), then $H_1 \odot H_2$ will be a linear space. We say that $u = \sum_{j=1}^{n} \xi_j^{(1)} \otimes \xi_j^{(2)}$ is the zero element, denote it by $u = 0$, if for any $\eta_1 \in H_1, \eta_2 \in H_2$,

$$\langle u, \eta_1 \otimes \eta_2 \rangle = \sum_{j=1}^{n} \langle \xi_j^{(1)}, \eta_1 \rangle \langle \xi_j^{(2)}, \eta_2 \rangle = 0.$$

Equivalently, $u = 0$ if in H_1,

$$\sum_{j=1}^{n} \lambda_{jk}\xi_j^{(1)} = 0, \quad 1 \le k \le m,$$

where $\xi_j^{(2)} = \sum_{k=1}^{m} \lambda_{jk}\varsigma_k, 1 \le j \le n$, and $\{\varsigma_1, \cdots, \varsigma_m\}$ is a basis of the linear span of $\{\xi_1^{(2)}, \cdots, \xi_n^{(2)}\}$.

For any $u = \sum_{j=1}^{n} \xi_j^{(1)} \otimes \xi_j^{(2)}$ and $v = \sum_{k=1}^{m} \eta_k^{(1)} \otimes \eta_k^{(2)}$, we define

$$\langle u, v\rangle = \sum_{j,k} \langle \xi_j^{(1)}, \eta_k^{(1)}\rangle\langle \xi_j^{(2)}, \eta_k^{(2)}\rangle.$$

Since for any $\lambda_1, \cdots, \lambda_n \in \mathbb{C}$,

$$\sum_{i,j} \langle \xi_i^{(1)}, \xi_j^{(1)}\rangle \lambda_i \bar{\lambda}_j = \|\sum_i \lambda_i \xi_i^{(1)}\|^2 \ge 0,$$

it follows that the matrix

$$(\langle \xi_i^{(1)}, \xi_j^{(1)}\rangle)_{1\le i,j\le n}$$

is non-negative. Then we have a $n \times n$-unitary matrix (u_{ij}) such that

$$u_{ij}{}^* \cdot (\langle \xi_i^{(1)}, \xi_j^{(1)}\rangle) \cdot (u_{ij}) = \begin{pmatrix} \mu_1 & & 0 \\ & \ddots & \\ 0 & & \mu_n \end{pmatrix}$$

where $\mu_i \ge 0, 1 \le i \le n$. Further we can write

$$\langle \xi_i^{(1)}, \xi_j^{(1)}\rangle = \sum_{k=1}^{n} u_{ik}\mu_k \overline{u_{jk}} = \sum_{k=1}^{n} \alpha_{ik}\bar{\alpha}_{jk},$$

where $\alpha_{ik} = u_{ik}\mu_k^{1/2}, \forall 1 \le i, j, k \le n$. Similarly, we can write

$$\langle \xi_i^{(2)}, \xi_j^{(2)}\rangle = \sum_{k=1}^{n} \beta_{ik}\bar{\beta}_{jk}, \quad 1 \le i, j \le n.$$

Then

$$\begin{aligned} \langle u, u\rangle &= \sum_{i,j} \langle \xi_i^{(1)}, \xi_j^{(1)}\rangle\langle \xi_i^{(2)}, \xi_j^{(2)}\rangle \\ &= \sum_{k,l} (\sum_i \alpha_{ik}\beta_{il}) \cdot \overline{(\sum_j \alpha_{jk}\beta_{jl})} \ge 0. \end{aligned}$$

Now by the Schwartz inequality:

$$|\langle u, \eta_1 \otimes \eta_2\rangle| \le \langle u, u\rangle^{1/2} \cdot \|\eta_1\| \cdot \|\eta_2\|, \quad \forall \eta_1 \in H_1, \eta_2 \in H_2,$$

we can see that $u = \sum_{j=1}^{n} \xi_j^{(1)} \otimes \xi_j^{(2)} = 0$ if $\langle u, u \rangle = 0$. This means that $\langle , \rangle$ is a inner product on the linear space $H_1 \odot H_2$.

Definition 1.4.1. The completion of $H_1 \odot H_2$ with respect to the inner product $\langle , \rangle$ is denoted by $H_1 \otimes H_2$, and called the *tensor product* of H_1 and H_2.

Proposition 1.4.2. Let $a_i \in B(H_i), i = 1, 2$. Then there exists unique $a_1 \otimes a_2 \in B(H_1 \otimes H_2)$ such that

$$(a_1 \otimes a_2)(\xi_1 \otimes \xi_2) = a_1\xi_1 \otimes a_2\xi_2, \quad \forall \xi_i \in H_i, i = 1, 2,$$

and $\|a_1 \otimes a_2\| = \|a_1\| \cdot \|a_2\|$.

Proof. For any $u = \sum_j \xi_j^{(1)} \otimes \xi_j^{(2)} \in H_1 \odot H_2$, where $\{\xi_j^{(2)}\}$ is a normalized orthogonal family in H_2, we have

$$\|(a_1 \otimes 1_2)u\|^2 = \sum_j \|a_1\xi_j^{(1)}\|^2 \le \|a_1\|^2 \sum_j \|\xi_j^{(1)}\|^2 \le \|a_1\|^2\|u\|^2.$$

Thus $a_1 \otimes 1_2$ can be extended uniquely to a bounded operator on $H_1 \otimes H_2$. Similarly we have $1_1 \otimes a_2 \in B(H_1 \otimes H_2)$. Now we can define $a_1 \otimes a_2 = (a_1 \otimes 1_2)(1_1 \otimes a_2) \in B(H_1 \otimes H_2)$. Obviously, $\|a_1 \otimes a_2\| \le \|a_1\| \cdot \|a_2\|$. However, since

$$\begin{aligned} \|a_1 \otimes a_2\| &\ge \sup\{\|a_1\xi_1\| \cdot \|a_2\xi_2\| \mid \xi_i \in H_i, \|\xi_i\| = 1, i = 1, 2\} \\ &= \|a_1\| \cdot \|a_2\|, \end{aligned}$$

it follows that $\|a_1 \otimes a_2\| = \|a_1\| \cdot \|a_2\|$. Q.E.D.

Using the method of Hilbert direct sum, we can also describe $H_1 \otimes H_2$. Let $\{e_l | l \in \Lambda\}$ be a normalized orthogonal basis of H_2, where the *cardinal number* $\sharp\Lambda$ of Λ is $\dim H_2$. For each $l \in \Lambda$, define

$$H^{(l)} = \{\xi \otimes e_l \mid \xi \in H_1\}.$$

Then $H^{(l)}$ is a closed linear subspace of $H_1 \otimes H_2$, and isomorphic to $H_1, \forall l \in \Lambda$. Moreover,

$$H_1 \otimes H_2 = \sum_{l \in \Lambda} \oplus H^{(l)}.$$

Suppose that $u_l\xi = \xi \otimes e_l, \forall \xi \in H_1, l \in \Lambda$. Then u_l is an isometric map from H_1 into $H_1 \otimes H_2$; and its image is $H^{(l)}, \forall l \in \Lambda$. Clearly, u_l^* is a linear map from $H_1 \otimes H_2$ onto H_1 such that $u_l^* H^{(l')} = \{0\}, \forall l' \ne l$, and u_l^* maps isometrically $H^{(l)}$ onto $H_1, \forall l \in \Lambda$. Moreover, $u_l^* u_l$ is the identity map on H_1; and $u_l u_l^* = p_l$ is the projection from $H_1 \otimes H_2$ onto $H^{(l)}$; and $\sum_{l \in \Lambda} p_l = 1$.

Let $a \in B(H_1 \otimes H_2)$, and $a_{ll'} = u_l^* a u_{l'} (\in B(H_1)), \forall l, l' \in \Lambda$. Then the operator a will be determined by the operator matrix $(a_{ll'})_{l,l' \in \Lambda}$. In fact, for any $\xi \in H_1 \otimes H_2$, we have

$$a\xi = \sum_l (u_l^* a \xi) \otimes e_l = \sum_l (\sum_{l'} a_{ll'} \xi_{l'}) \otimes e_l,$$

where $\xi_{l'} = u_{l'} \xi \in H_1, \forall l'$. Thus, identifying $H^{(l)}$ with $H_1, \forall l \in \Lambda$, we can write $a = (a_{ll'})_{l,l' \in \Lambda}$.

Lemma 1.4.3. Let $a = (a_{ll'})$ and $b = (b_{ll'}) \in B(H_1 \otimes H_2)$. Then $ab = (\sum_{l''} a_{ll''} b_{l''l'})_{l,l' \in \Lambda}$, where for any $l, l' \in \Lambda$ the series $\sum_{l''} a_{ll''} b_{l''l'}$ is convergent with respect to the strong (operator) topology in $B(H_1)$.

Proof. Since $\sum_{l \in \Lambda} p_l = \sum_{l \in \Lambda} u_l u_l^* = 1$ with respect to the strong (operator) topology in $B(H_1 \otimes H_2)$, it follows that with respect to strong (operator) topology

$$u_l^* a b u_{l'} = u_l^* a (\sum_{l'' \in \Lambda} u_{l''} u_{l''}^*) b u_{l'} = \sum_{l'' \in \Lambda} a_{ll''} b_{l''l'},$$

$\forall l, l' \in \Lambda$. Q.E.D.

Lemma 1.4.4. Let $a_i \in B(H_i), i = 1, 2$ and $(\lambda_{ll'})_{l,l' \in \Lambda}$ be the matrix representation of a_2 with respect to the basis $\{e_l\}_{l \in \Lambda}$. Then $a_1 \otimes a_2 = (\lambda_{ll'} a_1)_{l,l' \in \Lambda}$. In particular, $a_1 \otimes 1_2 = (\delta_{ll'} a_1)_{l,l' \in \Lambda}$.

Proof. For any $\xi \in H_1$, we have

$$\begin{aligned} u_l^*(a_1 \otimes a_2) u_{l'} \xi &= u_l^*(a_1 \otimes a_2)(\xi \otimes e_{l'}) \\ &= u_l^*(a_1 \xi \otimes a_2 e_{l'}) \\ &= u_l^*(a_1 \xi \otimes \sum_{l''} \lambda_{l''l'} e_{l''}) \\ &= \sum_{l''} u_l^*(a_1 \xi \otimes \lambda_{l''l'} e_{l''}) = \lambda_{ll'} a_1 \xi, \end{aligned}$$

$\forall, l, l' \in \Lambda$. Hence $u_l^*(a_1 \otimes a_2) u_{l'} = \lambda_{ll'} a_1, \forall l, l' \in \Lambda$. Q.E.D.

Lemma 1.4.5. In $B(H_1 \otimes H_2)$, we have

$$\{u_l u_{l'}^* | l, l' \in \Lambda\}' = \{a_1 \otimes 1_2 | a_1 \in B(H_1)\}.$$

Proof. By Lemma 1.4.4, we have

$$\begin{aligned} u_s^*(a_1 \otimes 1_2)(u_l u_{l'}^*)u_t = & \; u_s^*(a_1 \otimes 1_2)u_l \cdot u_{l'}^* u_t = \delta_{sl}\delta_{tl'}a_1 \\ = & \; u_s^*(u_l u_{l'}^*)(a_1 \otimes 1_2)u_t, \qquad \forall s,t,l,l' \in \Lambda. \end{aligned}$$

Thus $a_1 \otimes 1_2 \in \{u_l u_{l'}^* | l,l' \in \Lambda\}', \forall a_1 \in B(H_1)$.

Conversely, let $a \in \{u_l u_{l'}^* |\, l,l' \in \Lambda\}'$. If $l \neq l'$, then

$$\begin{cases} u_l^* a u_{l'} = u_l^* a(u_{l'} u_{l'}^*)u_{l'} = (u_l^* u_{l'}) \cdot u_{l'}^* a u_{l'} = 0, \\ u_l^* a u_l = u_{l'}^*(u_{l'} u_l^*)a u_l = u_{l'}^* a u_{l'} \cdot u_l^* u_l = u_{l'}^* a u_{l'}. \end{cases}$$

Thus $a_1 = u_l^* a u_l \in B(H_1)$ is well-defined (i.e. independent of the choice of $l \in \Lambda$). Now by Lemma 1.4.4, we have

$$a = (\delta_{ll'}a_1) = a_1 \otimes 1_2.$$

Q.E.D.

Now we study the *tensor product of VN algebras*.

Definition 1.4.6. Let M_i be a VN algebra on a Hilbert space $H_i, i = 1,2$. The VN algebra generated by the set $\{a_1 \otimes a_2 | a_i \in M_i, i = 1,2\}$ (on $H_1 \otimes H_2$) is called the *tensor product* of M_1 and M_2, and denoted by $M_1 \overline{\otimes} M_2$, i.e.,

$$M_1 \overline{\otimes} M_2 = \{a_1 \otimes a_2 \mid a_i \in M_i, i = 1,2\}''$$

For example, by Lemma 1.4.5 we have

$$B(H_1) \overline{\otimes} \mathbb{C} 1_2 = \{a_1 \otimes 1_2 \mid a_1 \in B(H_1)\}.$$

Proposition 1.4.7. Let M_1 be a VN algebra on H_1. Then

$$M_1 \overline{\otimes} B(H_2) = \{a \in B(H_1 \otimes H_2) \mid a = (a_{l,l'})_{l,l' \in \Lambda} \text{ and } a_{ll'} \in M, \forall l,l' \in \Lambda\}.$$

Proof. By Lemma 1.4.3, the right side is a VN algebra on $H_1 \otimes H_2$. And also by Lemma 1.4.4 we have

$$M_1 \overline{\otimes} B(H_2) \subset \text{the right side}.$$

Now let $a = (a_{ll'})_{l,l' \in \Lambda} \in B(H_1 \otimes H_2)$, and $a_{l,l'} \in M_1, \forall l,l' \in \Lambda$. For any finite subsets E,F of Λ, define

$$a_{ll'}^{(E,F)} = \begin{cases} a_{ll'}, & \text{if } l \in E, l' \in F, \\ 0, & \text{otherwise,} \end{cases}$$

and $a_{E,F} = (a_{ll'}^{(E,F)})_{l,l'\in\Lambda}$. By Lemma 1.4.4, we have

$$a_{E,F} = \sum_{s\in E, t\in F} a_{st} \otimes (\delta_{sl}\delta_{tl'})_{l,l'\in\Lambda} \in M_1 \overline{\otimes} B(H_2).$$

Moreover, it is easy to see $a_{E,F} \to a$ (weakly). Therefore $a \in M_1 \overline{\otimes} B(H_2)$.

Q.E.D.

We have defined $M_1 \overline{\otimes} M_2$. Naturally we guess that

$$(M_1 \overline{\otimes} M_2)' = M_1' \overline{\otimes} M_2'.$$

For this purpose, we need to do some analysis in detail.

Let H be a complex Hilbert space, $\langle , \rangle$ be its inner product. If we regard H as a real linear space, and define $\langle , \rangle_r = Re\langle , \rangle$, then $(H, \langle , \rangle_r)$ will be a real Hilbert space. In the rest part of this section, "$\perp$" is understood with respect to $\langle , \rangle_r$.

Lemma 1.4.8. Let M, N be two $*$ subalgebras containing $1 (= 1_H)$ of $B(H)$, and $M \subset N'$. Moreover, suppose that M admits a *cyclic vector* ξ (i.e. $\overline{M\xi} = H$). Then the following statements are equivalent:

1) $M' = N''$;

2) $\overline{(M_h\xi + iN_h\xi)}$ is dense in H;

3) $(M_h\xi)^\perp = \overline{iN_h\xi}$,

where M_h, N_h are the subsets of all self-adjoint elements of M, N respectively, and bar "—" is the norm closure.

Proof. Let $t' \in (M')_h, a \in M_h$. Then $t'a = at'$ is still self-adjoint. Thus $\mathrm{Im}\langle a\xi, t'\xi\rangle = 0$, and $\langle a\xi, it'\xi\rangle_r = 0$. This implies $i(M')_h\xi \subset (M_h\xi)^\perp$. Moreover, by $M \subset N'$ and $N \subset N'' \subset M'$ we have $iN_h\xi \subset (M_h\xi)^\perp$.

Suppose that 3) holds. Then

$$\overline{(M_h\xi + iN_h\xi)} \supset M_h\xi + \overline{iN_h\xi} = M_h\xi + (M_h\xi)^\perp.$$

Hence 3) implies 2).

Suppose that 2) holds. Since $iN_h\xi \subset (M_h\xi)^\perp$, it follows from 2) that $iN_h\xi$ is dense in $(M_h\xi)^\perp$. Hence 2) implies 3).

Suppose that 2) or 3) holds. We have pointed out that $(M')_h\xi \subset i(M_h\xi)^\perp$. So by 3), $(M')_h\xi \subset \overline{N_h\xi}$. Then for any $t' \in (M')_h$, there is a sequence $\{b_n\} \subset N_h$ such that $\|t'\xi - b_n\xi\| \to 0$. Let $s' \in N', a, c \in M$. From $M \subset N'$, it follows that

$$\begin{aligned}\langle s't'a\xi, c\xi\rangle &= \lim_n \langle s'ab_n\xi, c\xi\rangle = \lim_n \langle s'a\xi, cb_n\xi\rangle \\ &= \langle s'a\xi, ct'\xi\rangle = \langle t's'a\xi, c\xi\rangle.\end{aligned}$$

Since ξ is cyclic for M and $a, c \in M$ are arbitrary, we have $s't' = t's', \forall s' \in N'$. Thus $t' \in N''$ and $M' \subset N''$. But $M \subset N'$, so $M' = N''$. This means that 2) or 3) implies 1).

Now suppose that 1) holds, i.e., $M' = N''$. We must prove that $\eta = 0$ if $\eta \in (M_h\xi + iN_k\xi)^\perp$.

In the complex Hilbert space $H \oplus H$, define

$$M_2 = \left\{ \begin{pmatrix} a & 0 \\ 0 & a \end{pmatrix} \mid a \in M \right\}.$$

It is easy to see that

$$M_2' = \left\{ \begin{pmatrix} b_1 & b_2 \\ b_3 & b_4 \end{pmatrix} \mid b_j \in M', 1 \le j \le 4 \right\}.$$

Fix $\eta \in (M_h\xi + iN_h\xi)^\perp$. Let P be the projection from $H \oplus H$ onto $\overline{M_2\binom{\xi}{\eta}}$. Then $P \in M_2'$, and we can write $P = \begin{pmatrix} p & r \\ r^* & q \end{pmatrix}$, where $p = p^*, q = q^*$ and $r \in M'$. Since $P\binom{\xi}{\eta} = \binom{\xi}{\eta}$, it follows that

$$p\xi + r\eta = \xi. \tag{1}$$

By $\eta \perp M_h\xi$, i.e., $\mathrm{Re}\langle \eta, a\xi \rangle = 0, \forall a \in M_h$, we have

$$\langle \eta, a\xi \rangle = -\langle a\xi, \eta \rangle = -\langle \xi, a\eta \rangle, \forall a \in M_h.$$

Further, $\langle \eta, a\xi \rangle = -\langle \xi, a\eta \rangle$, $\forall a \in M$, i.e.,

$$\langle \begin{pmatrix} \eta \\ \xi \end{pmatrix}, \begin{pmatrix} a & 0 \\ 0 & a \end{pmatrix} \begin{pmatrix} \xi \\ \eta \end{pmatrix} \rangle = 0, \forall a \in M.$$

Thus $P\binom{\eta}{\xi} = 0$, and

$$p\eta = r\xi = 0. \tag{2}$$

By $\eta \perp iN_h\xi$, i.e., $\mathrm{Re}\langle \eta, ib\xi \rangle = Im\langle \eta, b\xi \rangle = 0, \forall b \in N_h$, we have $\langle \eta, b\xi \rangle = \langle b\xi, \eta \rangle = \langle \xi, b\eta \rangle, \forall b \in N_h$. Further, we have

$$\langle \eta, b\xi \rangle = \langle \xi, b\eta \rangle, \forall b \in N'' \tag{3}$$

From (2) (3), (1) and $M' = N''$ it follows that

$$\langle \eta, p\eta \rangle = -\langle \eta, r\xi \rangle = -\langle \xi, r\eta \rangle = -\langle \xi, (1-p)\xi \rangle. \tag{4}$$

Since P and $(1 - P)$ are projections, this implies that $p = p^2 + rr^*$ and $(1-p) = (1-p)^2 + rr^*$. Then by (4), we get

$$0 \le \|p\eta\|^2 + \|r^*\eta\|^2 = -(\|(1-p)\xi\|^2 + \|r^*\xi\|^2) \le 0.$$

This implies that $p\eta = (1-p)\xi = 0$. Since $(1-p) \in M'$ and ξ is a cyclic vector for M, it follows that $(1-p) = 0, p = 1, \eta = 0$. Q.E.D.

Lemma 1.4.9. Let H_j be a complex Hilbert space and K_j be a real closed linear subspace of H_j such that $(K_j + iK_j)$ is dense in $H_j, j = 1, 2$. Then

$$K_1 \otimes K_2 + i(K_1^\perp \otimes K_2^\perp) = H_1 \otimes H_2,$$

where $K_1 \otimes K_2, K_1^\perp \otimes K_2^\perp$ are the real closed linear spans of $\{\xi_1 \otimes \xi_2 \mid \xi_1 \in K_1, \xi_2 \in K_2\}, \{\eta_1 \otimes \eta_2 \mid \eta_1 \in K_1^\perp, \eta_2 \in K_2^\perp\}$ respectively, and $K_j^\perp$ is the orthogonal complement of K_j in $(H_j, \langle , \rangle_r), j = 1, 2$.

Proof. It is easily verified that $(K_1 \otimes K_2) \perp i(K_1^\perp \otimes K_2^\perp)$. Thus it suffices to show $\xi = 0$ if $\xi \in H_1 \otimes H_2$ and $\xi \perp (K_1 \otimes K_2 + iK_1^\perp \otimes K_2^\perp)$.

For the above ξ, define a map t from H_1 into H_2 such that

$$\langle t\xi_1, \xi_2 \rangle_r = \langle \xi, \xi_1 \otimes \xi_2 \rangle_r, \quad \forall \xi_1 \in H_1, \xi_2 \in H_2.$$

Clearly, t is bounded and real linear. Since

$$\begin{aligned} \langle t(i\xi_1), \xi_2 \rangle_r = & \langle \xi, i\xi_1 \otimes \xi_2 \rangle_r = \langle \xi, \xi_1 \otimes i\xi_2 \rangle_r \\ = & \langle t\xi_1, i\xi_2 \rangle_r = \langle -it\xi_1, \xi_2 \rangle_r, \end{aligned}$$

it follows that

$$t(i\xi_1) = -it\xi_1, \quad t^*(i\xi_2) = -it^*\xi_2, \quad \forall \xi_1 \in H_1, \xi_2 \in H_2. \tag{1}$$

For any $\xi_j \in K_j, j = 1, 2$, from $\xi \perp K_1 \otimes K_2$ we have $\langle t\xi_1, \xi_2 \rangle_r = \langle \xi, \xi_1 \otimes \xi_2 \rangle_r = 0$. Thus

$$tK_1 \subset K_2^\perp, \quad t^*K_2 \subset K_1^\perp. \tag{2}$$

By $\xi \perp iK_1^\perp \otimes K_2^\perp$, we have $\langle t(i\eta_1), \eta_2 \rangle_r = \langle \xi, i\eta_1 \otimes \eta_2 \rangle_r = 0, \forall \eta_j \in K_j^\perp, j = 1, 2$. Hence

$$t(iK_1^\perp) \subset K_2, \quad t^*K_2^\perp \subset iK_1. \tag{3}$$

Now by (1), (2), (3), we have

$$t^*tK_1^\perp \subset iK_1^\perp, \text{ and } (t^*t)^2 K_1^\perp \subset K_1^\perp. \tag{4}$$

Since t^*t is a non-negative operator on $(H_1, \langle , \rangle_r)$ and t^*t can be approximated uniformly by the polynomials of $(t^*t)^2$, it follows that $t^*tK_1^\perp \subset K_1^\perp$. Then from (4), we have $t^*tK_1^\perp \subset K_1^\perp \cap iK_1^\perp$. It is easy to see $K_1^\perp \cap iK_1^\perp = \{0\}$ since $(K_1 + iK_1)$ is dense in H_1. Thus

$$tK_1^\perp = \{0\}. \tag{5}$$

Then $\langle t^*K_2, K_1^\perp \rangle_r = \{0\}$, i.e., $t^*K_2 \subset K_1$. Again by (2), we have

$$t^*K_2 = \{0\}. \tag{6}$$

From (1) and (6), it follows that $\langle itK_1, K_2\rangle_r = \langle tK_1, K_2\rangle_r = \{0\}$. Thus $tK_1 \subset K_2^{\perp} \cap iK_2^{\perp}$. Similarly, we have $K_2^{\perp} \cap iK_2^{\perp} = \{0\}$ since $(K_2 + iK_2)$ is dense in H_2. Hence $tK_1 = \{0\}$. Now by (5), $t = 0$. Therefore, $\langle \xi, \xi_1 \otimes \xi_2\rangle_r = \langle t\xi_1, \xi_2\rangle_r = 0, \forall \xi_1 \in H_1, \xi_2 \in H_2$, and $\xi = 0$. Q.E.D.

Lemma 1.4.10. Let M_j be a VN algebra on H_j, and $\xi_j (\in H_j)$ be a cyclic vector for $M_j, j = 1, 2$. Then we have $(M_1 \overline{\otimes} M_2)' = M_1' \overline{\otimes} M_2'$.

Proof. Let $M = M_1 \overline{\otimes} M_2$, and $N = M_1' \overline{\otimes} M_2'$. Clearly, $M \subset N'$. Since $\xi = \xi_1 \otimes \xi_2$ is a cyclic vector for M, by Lemma 1.4.8 it suffices to show that

$$\overline{M_h \xi} + i\overline{N_h \xi} = H_1 \otimes H_2.$$

Let $K_j = \overline{(M_j)_h \xi_j}, j = 1, 2$. Clearly, $K_1 \otimes K_2 \subset \overline{M_h \xi}$. Using Lemma 1.4.8 to the case of $\{H_j, M_j, N_j = M_j'\}$, then we have $((M_j)_h \xi_j)^{\perp} = i\overline{(M_j')_h \xi_j}, j = 1, 2$. Thus

$$K_1^{\perp} \otimes K_2^{\perp} = \overline{(M_1')_h \xi_1} \otimes \overline{(M_2')_h \xi_2} \subset \overline{N_h \xi}.$$

Now it suffices to prove that

$$K_1 \otimes K_2 + i(K_1^{\perp} \otimes K_2^{\perp}) = H_1 \otimes H_2,$$

or by Lemma 1.4.9, only to prove that $(K_j + iK_j)$ is dense in $H_j, j = 1, 2$. Noticing that $(K_j + iK_j) = \overline{(M_j)_h \xi_j} + i\overline{(M_j)_h \xi_j} \supset M_j \xi_j$, it follows that $(K_j + iK_j)$ is dense in H_j since ξ_j is a cyclic vector for $M_j, j = 1, 2$. Q.E.D.

Theorem 1.4.11. Let H_j be a complex Hilbert space, and M_j be a VN algebra on $H_j, j = 1, 2$. Then we have $(M_1 \overline{\otimes} M_2)' = M_1' \overline{\otimes} M_2'$ on $H_1 \otimes H_2$.

Proof. For fixed $\xi_j \in H_j$. Let p_j' be the projection from H_j onto $\overline{M_j \xi_j}$, and $K_j = p_j' H_j = \overline{M_j \xi_j}, N_j = M_j p_j', j = 1, 2$. Then $p_j' \in M'_j, N_j$ is a VN algebra on K_j, and also N_j asmits a cyclic vector $\xi_j, j = 1, 2$. By Lemma 1.4.10, we have $(N_1 \overline{\otimes} N_2)' = N_1' \overline{\otimes} N_2'$ on $K_1 \otimes K_2$.

Let $p' = p_1' \otimes p_2'$. Then it is the projection from $H_1 \otimes H_2$ onto $K_1 \otimes K_2$. Moreover, $p' \in M'_1 \overline{\otimes} M'_2$ and

$$(M_1 \overline{\otimes} M_2) p' = N_1 \overline{\otimes} N_2, \quad p'(M_1' \overline{\otimes} M_2') p' = N_1' \overline{\otimes} N_2'.$$

Let $a \in (M_1 \overline{\otimes} M_2)', b \in (M_1' \overline{\otimes} M_2')'$. Since $p' \in M_1' \overline{\otimes} M_2'$, it follows that

$$\begin{aligned} p'bp' = bp' \in (M_1' \overline{\otimes} M_2')' p' &= (p'(M_1' \overline{\otimes} M_2') p')' \\ &= (N_1' \overline{\otimes} N_2')' = N_1 \overline{\otimes} N_2. \end{aligned}$$

Also since $p' \in M_1' \overline{\otimes} M_2' \subset (M_1 \overline{\otimes} M_2)'$, we have

$$p'ap' \in p'(M_1 \overline{\otimes} M_2)' p' = ((M_1 \otimes M_2) p')' = (N_1 \overline{\otimes} N_2)'.$$

Thus $p'ap'$ and $p'bp'$ commute. Let $\xi = \xi_1 \otimes \xi_2$. Then

$$\begin{aligned}\langle ab\xi, \xi\rangle &= \langle p'ap' \cdot p'bp'\xi, \xi\rangle \\ &= \langle p'bp' \cdot p'ap'\xi, \xi\rangle = \langle ba\xi, \xi\rangle.\end{aligned}$$

Therefore, we have

$$\langle ab\xi_1 \otimes \xi_2, \xi_1 \otimes \xi_2\rangle = \langle ba\xi_1 \otimes \xi_2, \xi_1 \otimes \xi_2\rangle, \quad \forall \xi_1 \in H_1, \xi_2 \in H_2.$$

Further, $ab = ba, \forall a \in (M_1 \overline{\otimes} M_2)', b \in (M_1' \overline{\otimes} M_2')'$. This implies that $(M_1 \overline{\otimes} M_2)' \subset (M_1' \overline{\otimes} M_2')'' = M'_1 \overline{\otimes} M_2'$, and $(M_1 \overline{\otimes} M_2)' = M_1' \overline{\otimes} M_2'$. Q.E.D.

Proposition 1.4.12. Let M_j be a VN algebra on H_j, and Z_j be its center, $j = 1, 2$. Then $Z = Z_1 \overline{\otimes} Z_2$, where Z is the center of $M_1 \overline{\otimes} M_2$. In particular, the tensor product of two factors (see Definition 1.3.7) is still a factor.

Proof. Clearly, $Z_1 \overline{\otimes} Z_2 \subset Z$. However, $Z \subset M_1 \overline{\otimes} M_2 \subset M_1 \overline{\otimes} B(H_2)$, and $Z \subset (M_1 \overline{\otimes} M_2)' = M_1' \overline{\otimes} M_2' \subset M_1' \overline{\otimes} B(H_2)$. Now by Proposition 1.4.7, we have $Z \subset Z_1 \overline{\otimes} B(H_2)$. Thus $Z' \supset (Z_1 \overline{\otimes} B(H_2))' = Z_1' \overline{\otimes} \mathbb{C} 1_2$. Similarly, $Z' \supset \mathbb{C} 1_1 \overline{\otimes} Z_2'$. Therefore, $Z' \supset Z_1' \overline{\otimes} Z_2', Z = Z'' \subset (Z_1' \overline{\otimes} Z_2')' = Z_1 \overline{\otimes} Z_2$, and $Z = Z_1 \overline{\otimes} Z_2$.
Q.E.D.

Proposition 1.4.13. Let M_j, N_j be two VN algebras on $H_j, j = 1, 2$. Then

$$((M_1 \overline{\otimes} M_2) \cup (N_1 \overline{\otimes} N_2))'' = (M_1 \cup N_1)'' \overline{\otimes} (M_2 \cup N_2)'',$$

$$(M_1 \overline{\otimes} M_2) \cap (N_1 \overline{\otimes} N_2) = (M_1 \cap N_1) \overline{\otimes} (M_2 \cap N_2).$$

Proof. The first equality is clear by the definition of tensor product of VN algebras. Similarly,

$$((M_1' \overline{\otimes} M_2') \cup (N_1' \overline{\otimes} N_2'))'' = (M_1' \cup N_1')'' \overline{\otimes} (M_2' \cup N_2')''.$$

Picking commutants of above equality and by Theorem 1.4.11, we get the second equality. Q.E.D.

Notes. The tensor product commutation theorem, Theorem 1,4.11, was first proved for semifinite Von Neumann algebras by Y. Misonou. But the general case remained unsolved for a long time. Using the concepts of generalized Hilbert algebras and modular Hilbert algebras, M. Tomita solved the problem at first. After Tomita's solution, there were several simple proofs depending on the theory of unbounded linear operators. The present proof here is very elementary, and follows the approach of M. Rieffel and A. Van Daele.

References. [109], [111], [134], [149], [150], [181].

1.5. Comparison of projections and central cover

Definition 1.5.1. Let M be a VN algebra on a Hilbert space H, and p and q be two projections of M. p and q are said to be equivalent (relative to M), and we write $p \sim q$, if there exists an element $v \in M$ such that $v^*v = p$, and $vv^* = q$. We write $p \precsim q$, if there exists a projection q_1 of M such that $q_1 \leq q$ and $p \sim q_1$.

It is easily verified that " $\sim$" is an equivalent relation. And also we shall see that " $\precsim$" is an order relation (Proposition 1.5.3).

Let M, N be two VN algebras on Hilbert spaces H, K respectively. M and N are said to be *spatial $*$ isomorphic*, if there is a unitary operator u from H onto K such that $uMu^* = N$.

Proposition 1.5.2. Let M be a VN algebra on a Hilbert space H.

1) Suppose that p, q are two projections of M, and $p \sim q$. Then the VN algebras M_p, M'_p on pH are spatially $*$ isomorphic to the VN algebras M_q, M'_q on qH respectively.

2) Let $\{p_l\}, \{q_l\}$ be two orthogonal families of projections of M, and $p_l \sim q_l, \forall l$. Then $p = \sum_l p_l \sim q = \sum_l q_l$.

3) Let $a \in M$, p be the projection from H onto $\overline{aH}$, and q be the projection from H onto $\overline{a^*H}$. Then $p \sim q$.

4) Let p, q be two projections of M. Then

$$(p - \inf\{p, 1-q\}) \sim (q - \inf\{q, 1-p\}),$$

$$(\sup\{p, q\} - q) \sim (p - \inf\{p, q\}).$$

Proof. 1) Let $p = v^*v$ and $q = vv^*$. Then v is a unitary operator from pH onto qH. Moreover,

$$vpapv^* = qaq, \quad va'pv^* = a'q, \quad \forall a \in M, a' \in M'.$$

Therefore, the VN algebras M_p, M'_p on pH are spatially $*$ isomorphic to the VN algebras M_q, M'_q on qH respectively.

2) Let $p_l = v_l^*v_l, q_l = v_lv_l^*, \forall l$. Define $v = \sum_l v_l$. Then $v \in M$ and $p = v^*v, q = vv^*$. Thus $p \sim q$.

3) It is clear by Proposition 1.3.4.

4) Since

$$\overline{qpH} = \overline{(q - \inf\{q, 1-p\})H},$$

$$\overline{pqH} = \overline{(p - \inf\{p, 1-q\})H},$$

and $(pq)^* = qp$, it follows by 3) that $(q - \inf\{q, 1-p\}) \sim (p - \inf\{p, 1-q\})$. Furthermore, if write $q_1 = 1 - q$, then

$$\begin{aligned} \sup\{p, q\} - q = & \sup\{1 - q_1, p\} - (1 - q_1) = q_1 - \inf\{q_1, 1 - p\} \\ \sim & \; p - \inf\{p, 1 - q_1\} = p - \inf\{p, q\}. \end{aligned}$$

Q.E.D.

Proposition 1.5.3. Suppose that p, q are two projections of a VN algebra M, and $p \preceq q, q \preceq p$. Then $p \sim q$.

Proof. Let $p \sim q_1 \leq q$ and $q \sim p_1 \leq p$. From $q_1 \leq q \sim p_1$, we have a projection $p_2 \in M$ such that $p_2 \leq p_1$ and $p_2 \sim q_1$. Thus $p \geq p_1 \geq p_2, p \sim p_2$. From $p_1 \leq p \sim p_2$, we also have a projection $p_3 \in M$ such that $p_3 \leq p_2, p_1 \sim p_3$, i.e., $p \geq p_1 \geq p_2 \geq p_3, p \sim p_2, p_1 \sim p_3$. Further from $p_2 \leq p_1 \sim p_3$, we have p_4 such that $p_4 \leq p_3, p_4 \sim p_2$. Generally, we get

$$p = p_0 \geq p_1 \geq p_2 \geq \cdots,$$

and

$$p_0 \sim p_2 \sim \cdots \sim p_{2n} \sim \cdots, \quad p_1 \sim p_3 \sim \cdots \sim p_{2n+1} \sim \cdots.$$

Let $p_n = u_n^* u_n, p_{n+2} = u_n u_n^*, n = 0, 1, \cdots$. By the above discussion, it follows that

$$p_{n+1} = (u_n p_{n+1})^*(u_n p_{n+1}), \quad p_{n+2} = (u_n p_{n+1})(u_n p_{n+1})^*, \quad \forall n.$$

Thus we have $p_n - p_{n+1} = (u_n(p_n - p_{n+1}))^*(u_n(p_n - p_{n+1}))$ and

$$(u_n(p_n - p_{n+1}))(u_n(p_n - p_{n+1}))^* = u_n(p_n - p_{n+1})u_n^* = p_{n+2} - p_{n+3},$$

i.e.,

$$(p_n - p_{n+2}) \sim (p_{n+2} - p_{n+3}), \quad \forall n.$$

Noticing that

$$\begin{aligned} p = & \; ((p_0 - p_1) \oplus (p_1 - p_2)) \oplus ((p_2 - p_3) \oplus (p_3 - p_4)) \oplus \cdots \oplus \inf\{p_n | n\}, \\ p_1 = & \; ((p_1 - p_2) \oplus (p_2 - p_3)) \oplus ((p_3 - p_4) \oplus (p_4 - p_5)) \oplus \cdots \oplus \inf\{p_n | n\} \end{aligned}$$

and by Proposition 1.5.2, it follows that $p \sim p_1$. But $p_1 \sim q$, thus we have $p \sim q$.

Q.E.D.

Theorem 1.5.4. Let M be a VN algebra on a Hilbert space H, p and q be two projections of M. Then there is a central projection z of M such that

$$qz \precsim pz, \text{ and } p(1-z) \precsim q(1-z).$$

In particular, if M is a factor, then we have either $q \precsim p$ or $p \precsim q$.

Proof. Let $c(p), c(q)$ be the projections from H onto $\overline{[MpH]}, \overline{[MqH]}$ respectively. Clearly, $c(p), c(q) \in M \cap M'$ and $c(p) \geq p, c(q) \geq q$.

If $c(p)c(q) = 0$, then $z = c(p)$ satisfies our condition. Thus we may assume that $c(p)c(q) \neq 0$. Since M is the linear span of its unitary elements, it follows that there are two unitary elements u, v of M, such that $PQ \neq 0$, where P, Q are the projections from H onto $\overline{upH}, \overline{vqH}$ respectively. Clearly, we have $p \sim P$ and $q \sim Q$. Let g', h' be the projections from H onto $\overline{PQH}, \overline{QPH}$ respectively. Then $0 \neq g' \leq P, 0 \neq h' \leq Q$, and $g' \sim h'$. Since $p \sim P, q \sim Q$, it follows that there are two projections g, h of M such that $0 \neq g \leq p, 0 \neq h \leq q$, and $g \sim h$.

By the Zorn lemma, there are two maximal orthogonal families $\{g_l\}, \{h_l\}$ of projections of M such that $0 \neq g_l \leq p, 0 \neq h_l \leq q, g_l \sim h_l, \forall l$. Let $g = \sum_l g_l, h = \sum_l h_l$. Then $0 \neq g \leq p, 0 \neq h \leq q$, and $g \sim h$. Since the families $\{p_l\}$ and $\{q_l\}$ are maximal, it follows that $c(p_1)c(q_1) = 0$, where $p_1 = p - g, q_1 = q - h$, and $c(p_1), c(q_1)$ are the projections from H onto $\overline{[Mp_1H]}, \overline{[Mq_1H]}$ respectively.

Finally, let $z = c(p_1)$. Then

$$\begin{aligned} qz &= q_1c(q_1)z + hz = hz \sim gz \leq pz, \\ p(1-z) &= p_1c(p_1)(1-z) + g(1-z) = g(1-z) \sim h(1-z) \leq q(1-z). \end{aligned}$$

Q.E.D.

Proposition 1.5.5. Let M be a VN algebra, p and q be two projections of M.

1) There exists a central projection z such that

$$pz \precsim qz, \quad (1-p)(1-z) \precsim (1-q)(1-z).$$

2) There exist three central projections z_1, z_2, z_3, with $z_iz_j = 0, \forall i \neq j$, and $\sum_{i=1}^{3} z_i = 1$ such that: (1) $pz \sim qz$ for any central projection z of M and $z \leq z_1$; (2) $pz \precsim qz$ and $pz \not\sim qz$ for any non-zero central projection z of M and $z \leq z_2$; (3) $qz \precsim pz$ and $qz \not\sim pz$ for any non-zero central projection z of M and $z \leq z_3$.

Proof. 1) By Theorem 1.5.4, there is a central projection z of M such that

$$z\inf\{p,1-q\}\preceq z\inf\{1-p,q\}, \tag{1}$$

$$(1-z)\inf\{1-p,q\}\preceq(1-z)\inf\{p,1-q\}. \tag{2}$$

By Proposition 1.5.2, we also have

$$(zp-z\inf\{p,1-q\})\sim(zq-z\inf\{q,1-p\}), \tag{3}$$

$$\begin{aligned}&((1-z)(1-p)-(1-z)\inf\{1-p,q\})\\ \sim\ &((1-z)(1-q)-(1-z)\inf\{1-q,p\}).\end{aligned} \tag{4}$$

Consider (1)+(3), (2)+(4). That comes to

$$zp\preceq zq,(1-z)(1-p)\preceq(1-z)(1-q).$$

2) By the Zorn lemma, there is a maximal orthogonal family $\{z_l\}$ of central projections of M such that $pz_l\sim qz_l,\forall l$. Let $z_1=\sum_l z_l$. Clearly, $pz_1\sim qz_1$. Suppose that z is a central projection of M such that $pz\sim qz$. Obviously, $pz(1-z_1)\sim qz(1-z_1)$. Since $z(1-z_1)z_l=0,\forall l$, and the family $\{z_l\}$ is maximal, it follows that $z\leq z_1$. Thus z_1 is the maximal central projection of M with the condition $pz\sim qz$.

Now in the VN algebra $M(1-z_1)$, using Theorem 1.5.4 for the projections $p(1-z_1)$ and $q(1-z_1)$, then we shall find z_2 and z_3. Q.E.D.

Theorem 1.5.6 Let $\{p_l\}_{l\in\Lambda}$ be a family of projections of a VN algebra M with conditions:

$$p_lp_{l'}=0,\quad p_l\sim p_{l'},\quad \forall l\neq l',\text{ and }\sum_{l\in\Lambda}p_l=1.$$

Then M is spatially $*$ isomorphic to $M_p\overline{\otimes}B(K)$, where $p\sim p_l,\forall l$, and K is a Hilbert space with $\dim K=\#\Lambda$.

Proof. Let H be the action space of $M,L=pH,p=v_l^*v_l,p_l=v_lv_l^*,\forall l$. In terms of $\{v_l\}$, we can define a unitary operator from $H=\sum_{l\in\Lambda}\oplus H_l$ onto $L\otimes K$, where $H_l=p_lH,\forall l$, and K is a Hilbert space with $\dim K=\#\Lambda$. Thus we let $H=L\otimes K$.

For any $a\in M,a'\in M',l,l'\in\Lambda$, we have

$$a_{ll'}=v_l^*av_{l'}\in M_p,$$

$$a'_{ll'}=v_l^*a'v_{l'}=v_l^*v_{l'}a'=\delta_{ll'}a'p\in M'_p.$$

By Lemma 1.4.4 and Proposition 1.4.7, it follows that

$$M\subset M_p\overline{\otimes}B(K),\quad M'\subset M'_p\overline{\otimes}\mathbb{C}1_K.$$

Now by Theorem 1.4.11, we have $M = M_p \overline{\otimes} B(K)$. Q.E.D.

Definition 1.5.7. Let M be a VN algebra on a Hilbert space H, and p be a projection of M. Denote by $c(p)$ the projection from H onto $\overline{[MpH]}$. Clearly, $c(p) \in M \cap M'$. And $c(p)$ is called the *central cover* (or *central support*) of p (relative to M).

Proposition 1.5.8. Let M be a VN algebra on a Hilbert space H.

1) If p is a projection of M, then $c(p)$ is the minimal central projection containing p, and

$$c(p) = \sup\{q \mid q \text{ is a projection of } M, \text{ and } q \sim p\}.$$

2) Let p, q be two projections of M, and $p \preceq q$. Then $c(p) \leq c(q)$. In particular, if $p \sim q$, then $c(p) = c(q)$.

3) Let $\{p_l\}$ be a family of projections of M, and $p = \sup\limits_l p_l$. Then $c(p) = \sup\limits_l c(p_l)$.

4) Let p be a projection of M, and z be a central projection of M. Then $zc(p) = c(pz)$.

5) Let p, q be two projections of M, and $p \geq q$. Then the central cover of q in the VN algebra M_p is $pc(q)$, where $c(q)$ is the central cover of q in M.

Proof. 2) Let $p = v^*v$ and $vv^* \leq q$. Then

$$c(p)H = \overline{[Mv^*vH]} \subset \overline{[MvH]} \subset \overline{[MqH]} = c(q)H,$$

i.e., $c(p) \leq c(q)$.

1) Let z be a central projection of M, and $z \geq p$. Since $zap\xi = ap\xi, \forall a \in M, \xi \in H$, it follows that $z \geq c(p)$.

Now by 2), $c(p) \geq \sup\{q|q \sim p\}$. Thus, it suffices to show $\sup\{q|q \sim p\}$ is a central projection of M. Obviously, $u \cdot \sup\{q|q \sim p\} = \sup\{q|q \sim p\} \cdot u$ for any unitary element u of M. By Proposition 1.3.4, $\sup\{q|q \sim p\}$ is central.

3) Clearly, $c(p) \geq \sup\limits_l c(p_l) \geq p$. However, $\sup\limits_l c(p_l)$ is still a central projection of M. Thus $c(p) = \sup\limits_l c(p_l)$.

4) It is clear from $zc(p)H = \overline{[zMpH]} = \overline{[MpzH]} = c(pz)H$.

5) It is obvious from $\overline{[M_p qpH]} = \overline{[pMpqpH]} = \overline{[pMqH]} = p\overline{[MqH]} = pc(q)H$. Q.E.D.

Proposition 1.5.9. Let p, q be two projections of a VN algebra M. Then the following statements are equivalent: 1) $c(p)c(q) \neq 0$; 2) $pMq \neq \{0\}$; 3) there exist two projections p_1, q_1 of M such that $0 \neq p_1 \leq p, 0 \neq q_1 \leq q$, and $p_1 \sim q_1$.

Proof. If $pMq = \{0\}$, then $[\overline{MpH}]\perp[\overline{MqH}]$, i.e., $c(p)\cdot c(q) = 0$. Conversely, if $c(p)c(q) = 0$, then

$$paq = pc(p)aqc(p) = paqc(p)c(q) = 0, \quad \forall a \in M,$$

i.e. $pMq = \{0\}$. Thus statements 1) and 2) are equivalent.

1) $\Longrightarrow$ 3). It is contained in the proof of Theorem 1.5.4 indeed.

3) $\Longrightarrow$ 1). Since $z = c(p_1) = c(q_1) \neq 0$, and $z \leq c(p), z \leq c(q)$, it follows that $c(p)c(q) \neq 0$. Q.E.D.

Proposition 1.5.10. Let p be a projection of a VN algebra M. Then:

1) $a'p \to a'c(p)(\forall a' \in M')$ is a $*$ isomorphism from $M'p$ onto $M'c(p)$;

2) $a' \to a'p(\forall a' \in M')$ is a $*$ isomorphism from M' onto $M'p$ if and only if $c(p) = 1$.

Proof. It is easily verified that $a'p = 0$ if and only if $a'c(p) = 0(a' \in M')$. Thus 1) and the sufficiency of 2) are obvious. Now let $a' \to a'p$ be a $*$ isomorphism from M' onto $M'p$. Since $(1 - c(p)) \in M'$ and $(1 - c(p))p = 0$, it follows that $1 - c(p) = 0$, $c(p) = 1$. Q.E.D.

Notes. The material presented in this section and Sections 6.1, 7.1 is today called the Murray-Von Neumann *dimension theory*. When F.J. Murray and J. Von Neumann laid the foundation for the theory, they first developed it for factors. Despite its importance, it has remained unchanged ever since Murray-Von Neumann's time.

References.[28], [82], [111].

1.6 Kaplansky's density theorem

Theorem 1.6.1. Let N, M be two $*$ subalgebras of $B(H)$, and $N \subset M$ here H is a Hilbert space. And also suppose that N is weakly dense in M. Then $(N)_1$ is $\tau(B(H), T(H))$-dense in $(M)_1$, where $(N)_1, (M)_1$ are the closed unit balls of N, M respectively.

Proof. By weak (operator) continuity of $*$ operation, N_h is weakly dense in M_h, where N_h, M_h are the sets of all self-adjoint elements of N, M respectively. Futher, from Proposition 1.2.9, N_h is also strongly dense in M_h.

We may assume that N and M are uniformly closed. For any $a^* = a \in (M)_1$, let $a' = a(1 + (1 - a^2)^{1/2})^{-1}$. Then $a' \in M_h$ and $a = 2a'(1 + a'^2)^{-1}$. Pick a net $\{b'_l\} \subset N_h$ such that $b'_l \to a'$ strongly. Let $b_l = 2b'_l(1+b'^2_l)^{-1}$. Then $b_l^* = b_l \in (N)_1, \forall l$. We prove that

$$b_l \to a \quad \text{(strongly)}.$$

In fact,

$$\begin{aligned} \frac{1}{2}(b_l - a) = & \ (1+b'^2_l)^{-1}[b'_l(1+a'^2) - (1+b'^2_l)a'](1+a'^2)^{-1} \\ = & \ (1+b'^2_l)^{-1}(b'_l - a')(1+a'^2)^{-1} \\ & +(1+b'^2_l)^{-1}b'_l(a' - b'_l)a'(1+a'^2)^{-1} \\ = & \ (1+b'^2_l)^{-1}(b'_l - a')(1+a'^2)^{-1} + b_l(a' - b'_l)a. \end{aligned}$$

Since $b'_l \to a'$ (strongly), $\|b_l\| \le 1, \|(1+b'^2_l)^{-1}\| \le 1, \forall l$, it follows from Proposition 1.2.1 that $b_l \to a$ (strongly). This implies that $(N_h)_1$ is strongly dense in $(M_h)_1$.

In $H \oplus H$, define

$$M^{(2)} = \left\{ \begin{pmatrix} a_1 & a_2 \\ a_3 & a_4 \end{pmatrix} \mid a_i \in M, 1 \le i \le 4 \right\},$$

$$N^{(2)} = \left\{ \begin{pmatrix} b_1 & b_2 \\ b_3 & b_4 \end{pmatrix} \mid b_i \in N, 1 \le i \le 4 \right\}.$$

Similarly, $(N_h^{(2)})_1$ is strongly dense in $(M_h^{(2)})_1$. Thus for any $a \in (M)_1$, we have a net $\begin{pmatrix} b_1^{(l)} & b_2^{(l)} \\ b_3^{(l)} & b_4^{(l)} \end{pmatrix}$ in $(N_h^{(2)})_1$ such that $\begin{pmatrix} b_1^{(l)} & b_2^{(l)} \\ b_3^{(l)} & b_4^{(l)} \end{pmatrix} \to \begin{pmatrix} 0 & a \\ a^* & 0 \end{pmatrix}$ (strongly). In particular, $b_2^{(l)} \to a$ (strongly) and $\|b_2^{(l)}\| \le 1, \forall l$. This implies that $(N)_1$ is strongly dense in $(M)_1$. Now by Proposition 1.2.8, $(N)_1$ is $\tau(B(H),T(H))$-dense in $(M)_1$. Q.E.D.

Corollary 1.6.2. Let M be a $*$ subalgebra of $B(H)$. Then the weak closure of M is equal to the $\tau(B(H),T(H))$-closure of M.

Corollary 1.6.3. Let M be a $*$ subalgebra of $B(H)$, and $1 \in M$. Then the following statements are equivalent: 1) M is a VN algebra; 2) M is $\sigma(B(H),T(H))$-closed; 3) $(M)_1$ is weakly closed.

Proof. By Corollary 1.6.2 and Theorem 1.3.10, the statements 1) and 2) are equivalent. It is clear that 2) implies 3); and $(M)_1$ is $\sigma(B(H),T(H))$-closed if $(M)_1$ is weakly closed. Thus by the Krein-Šmulian theorem, it follows that 3) implies 2). Q.E.D.

Proposition 1.6.4 Let M, N be two uniformly closed $*$ subalgebras of $B(H), N \subset M$ and N is weakly dense in M. Then $(N_h)_1$ is strongly dense in $(M_h)_1$, and $(N_+)_1$ is strongly dense in $(M_+)_1$, where N_+, M_+ are the sets of all positive elements of N, M respectively.

Proof. In the proof of Theorem 1.6.1, we have pointed out that $(N_h)_1$ is strongly dense in $(M_h)_1$.

Now let $a \in (M_+)_1$. Then there is a net $\{b_l\} \subset (N_h)_1$ such that $b_l \to a$ (strongly). Let

$$f(t) = \begin{cases} t, & 0 \le t \le 1, \\ 0, & -1 \le t \le 0. \end{cases}$$

Then $f(a) = a, f(b_l) \in (N_+)_1, \forall l$. Pick a sequence $\{p_n(t)\}$ of polynomials on $[-1, 1]$ such that

$$p_n(0) = 0, \quad \forall n; \quad \text{and} \quad \max_{-1 \le t \le 1} |f(t) - p_n(t)| \to 0.$$

Then for any $\xi \in H$, we have

$$\begin{aligned} &\|f(b_l)\xi - a\xi\| = \|(f(b_l) - f(a))\xi\| \\ \le \ &\|f(b_l) - p_n(b_l)\| \cdot \|\xi\| + \|f(a) - p_n(a)\| \cdot \|\xi\| + \|(p_n(b_l) - p_n(a))\xi\|. \end{aligned}$$

That implies $f(b_l) \to a$ (strongly). Q.E.D.

Proposition 1.6.5. Let M_l be a VN algebra on a Hilbert space $H_l, l \in \Lambda$. Then $M = \sum_{l \in \Lambda} \oplus M_l = \{(a_l)_{l \in \Lambda} \mid a_l \in M_l, \forall l \in \Lambda, \text{ and } \sup_{l \in \Lambda} \|a_l\| < \infty\}$ is a VN algebra on the *Hilbert direct sum* $H = \sum_{l \in \Lambda} \oplus H_l$, *and* M *is called the direct sum of the VN algebras family* $\{M_l\}_{l \in \Lambda}$. Moreover, $M' = \sum_{l \in \Lambda} \oplus M_l'$, and the projection z_l from H onto H_l is a central projection of M such that $M_l = M z_l, M_l' = M' z_l, \forall l \in \Lambda$.

Proof. It is obvious form Corollary 1.6.3. Q.E.D.

Notes. Theorem 1.6.1 is called the *Kaplansky's density theorem*, and indeed due to him. It is a powerful tool of the theory of Von Neumann algebras.

From Corollary 1.6.2, let M be a $*$ subalgebra of $B(H)$, then the closures of M with respect to all topologies introduced in Section 1.2 are only two: the uniform closure of M and the weak closure of M. That is why in the theory of operator algebras we study C^*-algebras and W^*-algebras mainly.

References. [83].

1.7 Ideals in Von Neumann algebras

Let M be a VN algebra on a Hilbert space H. From Section 1.3, we have

$$(\sigma(B(H),T(H))|M) \sim \sigma(M,M_*) \sim (\sigma\text{-weak top. }|M),$$

$$(s(B(H),T(H))|M) \sim s(M,M_*) \sim (\sigma\text{-strong top. }|M),$$

$$(s^*(B(H),T(H))|M) \sim s^*(M,M_*) \sim (\sigma\text{-strong }*\text{ top. }|M).$$

In the following, we shall denote simply these topologies by σ-top, s-top. or σ-strong top., s^*-top. or σ-strong $*$ top. respectively.

Proposition 1.7.1. Let M be a VN algebra on a Hilbert space H, I be a σ-closed left (or right) ideal of M. Then there exists unique projection p of M such that $I = Mp$ (or pM), and also I is weakly closed. Furthermore, if I is a σ-closed two-sided ideal of M, then there exists unique central projection z of M such that $I = Mz$, in particular, $I^* = I$.

Proof. Let I be a σ-closed left ideal of M. Then $E = I \cap I^*$ is a σ-closed $*$ subalgebra of M. By Corollary 1.6.2, E is also weakly closed. From Theorem 1.3.9, E has an identity p. Clearly $Mp \subset I$. Let $a \in I$ and $a = wh$ be the polar decomposition of a. It is easy to see $h = (a^*a)^{1/2} \in E$. Thus $hp = h, a = whp = ap \in Mp$. Therefore we have $Mp = I$.

If there is another projection q of M such that $I = Mp = Mq$, then we have $p = pq, p = p^* = qp, p = qpq \leq q$. Similarly, $q \leq p$. Thus $p = q$.

Now let I be two-sided. From preceding paragraph, there are two projections p and q of M such that

$$I = Mp = qM.$$

Then we have $a, b \in M$ such that $q = ap, p = qb$. Thus $q = qp = p$. Put $z = p = q$. Then $I = Mz = zM$. Hence for any $c \in M$, there exist $d, e \in M$ such that $cz = zd, zc = ez$. Then $cz = zcz = zc, \forall c \in M$, i.e., z belongs to the center of M. Q.E.D.

Remark. It M is a factor,then each non-zero two-sided ideal of M must be σ-dense in M.

Proposition 1.7.2. Let I be a two-sided ideal of a VN algebra M. Then $I^* = I$, and I is the linear span of its positive elements, i.e., $I = [I_+]$. Moreover, if $0 \leq a \in \overline{I}^w$, then there is a family $\{a_l\}_{l \in \Lambda} \subset I_+$ such that

$a = \lim_F \sum_{l \in F} a_l = \sum_{l \in \Lambda} a_l$, where $\overline{I}^w$ is the weak closure of I; F is any finite subset of Λ, and directed by the inclusion relation.

Proof. Let $b \in I$ and $b = wh$ be the polar decomposition of b. Then $h = w^*b \in I, b^* = hw^* \in I$. Thus $I^* = I$.

Let $h^* = h \in I$. Then there exist two projections $p, q \in M$ such that $ph \geq 0, qh \leq 0$, and $p + q = 1$. Hence $I = [I_+]$.

For $0 \leq a \in \overline{I}^w$, by the Zorn lemma we have a maximal family $\{a_l\}_{l \in \Lambda} \subset I_+$ such that $\sum_{l \in F} a_l \leq a$ for any finite subset F of Λ. By Proposition 1.2.10,

$$\sup_F \sum_{l \in F} a_l = \text{(strongly)-}\lim_F \sum_{l \in F} a_l = \sum_{l \in \Lambda} a_l \in \overline{I}^w.$$

Denote this element by a_1, and let $b = a - a_1$, so it suffices to prove that $b = 0$.

By Proposition 1.7.1, $\overline{I}^w = Mz$, where z is a central projection of M. From Theorem 1.6.1, we have a net $\{b_t\} \subset I$ such that

$$\|b_t\| \leq 1, \forall t, \quad \text{and } b_t \to z \text{ (strongly)}.$$

Then $b^{1/2} b_t b^{1/2} \to b$ (strongly).

If $b \neq 0$, then there is an index t such that $b^{1/2} b_t b^{1/2} \neq 0$. Thus $b^{1/2} c b^{1/2} \neq 0$, where $c = b_t b b_t^*$. We may assume that $0 \leq c \leq 1$. Then $0 \neq b^{1/2} c b^{1/2} \leq b = a - a_1 = a - \sum_{l \in \Lambda} a_l$. It is impossible since the family $\{a_l\}_{l \in \Lambda}$ is maximal. Therefore $b = 0, a = \sum_{l \in \Lambda} a_l$. Q.E.D.

Proposition 1.7.3. Let M be a VN algebra on a Hilbert space H, Z be its center, $\{t_{ij} \mid 1 \leq i, j \leq 1\} \subset M$, and $\{t'_{ij} \mid 1 \leq i, j \leq n\} \subset M'$. Then the following statements are equivalent:

1) $\sum_{k=1}^n t_{ik} t'_{kj} = 0, \forall 1 \leq i, j \leq n$;

2) There exists $\{z_{ij} \mid 1 \leq i, j \leq n\} \subset Z$ such that

$$\sum_{k=1}^n t_{ik} z_{kj} = 0, \quad \sum_{k=1}^n z_{ik} t'_{kj} = t'_{ij}, \quad \forall 1 \leq i, j \leq n.$$

Proof. It is clear that 2) implies 1).

Now let 1) hold. Suppose that K is a n-dimensional Hilbert space. Then

$$t = (t_{ij})_{1 \leq i,j \leq n} \in M \overline{\otimes} B(K),$$

$$t' = (t'_{ij})_{1 \leq i,j \leq n} \in M' \overline{\otimes} B(K).$$

Let

$$I' = \{x' \in M' \overline{\otimes} B(K) \mid t x' = 0\}.$$

Obviously, I' is a σ-closed right ideal of $M'\overline{\otimes}B(K)$. By Proposition 1.7.1, there is a projection $z' = (z_{ij})_{1\le i,j\le n}$ of $M'\overline{\otimes}B(K)$, where $z_{ij} \in M', 1 \le i,j \le n$, such that $I' = z'(M'\overline{\otimes}B(K))$.

For any $a' \in M', x' \in I'$, since

$$t(a' \otimes 1)x' = (a' \otimes 1)tx' = 0,$$

it follows that $(a' \otimes 1)I' \subset I'$, where $1 = 1_K$ is the identity operator on K. In particular, $(a' \otimes 1)z' \in I'$. Thus $z'(a' \otimes 1)z' = (a' \otimes 1)z'; (a' \otimes 1)z' = z'(a' \otimes 1); z_{ij}a' = a'z_{ij}, \forall a' \in M', 1 \le i,j \le n$; and $z_{ij} \in Z, 1 \le i,j \le n$.

Since $z' \in I'$, it follows that $tz' = 0$, i.e.

$$\sum_{k=1}^{n} t_{ik}z_{kj} = 0, \quad \forall 1 \le i,j \le n.$$

From condition 1), we have $tt' = 0$. Thus $t' \in I', z't' = t'$, i.e.,

$$\sum_{k=1}^{n} z_{ik}t'_{kj} = t'_{ij}, \forall 1 \le i,j \le n.$$

Q.E.D.

References.[28], [156].

1.8 Normal positive linear functionals

Definition 1.8.1. Let M be a VN algebra on a Hilbert space H, φ be a linear functional on M, φ is said to be *positive*, and denoted by $\varphi \ge 0$, if $\varphi(a) \ge 0$ for any $a \in M_+$ (the set of all positive elements of M).

For two linear functionals φ, ψ on M, the relation $\psi \le \varphi$ means that $(\varphi - \psi) \ge 0$. Moreover, a positive linear functional φ on M is said to be *faithful*, if $\varphi(a) = 0$ for some $a \in M_+$, then we have $a = 0$.

Clearly, if $\varphi \ge 0$, then we have $\varphi(a^*) = \overline{\varphi(a)}, \forall a \in M$; and the Schwartz inequality:

$$|\varphi(b^*a)|^2 \le \varphi(a^*a) \cdot \varphi(b^*b), \quad \forall a,b \in M;$$

and $\|\varphi\| = \varphi(1)$.

Definition 1.8.2. A positive linear functional φ on a VN algebra M is said to be *normal*, if for any bounded increasing net $\{a_l\} \subset M_+$, we have

$$\sup_l \varphi(a_l) = \varphi(\sup_l a_l).$$

A linear functional φ on M is called a *normal state* (on M), if it is positive, normal and $\varphi(1) = 1$.

Definition 1.8.3. A positive linear functional φ on a VN algebra M is said to be *completely additive*, if for any orthogonal family $\{p_l\}$ of projections of M, we have

$$\varphi(\sum_l p_l) = \sum_l \varphi(p_l).$$

Lemma 1.8.4. Let φ, ψ be two completely additive positive linear functionals on a VN algebra M, and p be a non-zero projection of M such that $\varphi(p) \leq \psi(p)$. Then there exists a non-zero projection q of $M, q \leq p$, such that

$$\varphi(a) \leq \psi(a), \quad \forall a \in M_+ \quad \text{and} \quad a \leq q.$$

Proof. Let

$$\mathcal{L} = \left\{ (q_l) \,\middle|\, \begin{array}{c} (q_l) \text{ is an orthogonal family of non-zero projections of } M, \\ \text{and } \varphi(q_l) > \psi(q_l), q_l \leq p, \forall l \end{array} \right\}.$$

Then $\mathcal{L}$ is a partially ordered set with respect to the inclusion relation. If $\mathcal{L}$ is non-empty, by the Zorn lemma $\mathcal{L}$ has a maximal element $\{q_l\}_{l\in\Lambda}$ at least. Let $q_0 = \sum_{l\in\Lambda} q_l$. Clearly $0 \neq q_0 \leq p$. Since φ and ψ are completely additive, it follows that $\varphi(q_0) > \psi(q_0)$. However, $\varphi(p) \leq \psi(p)$, thus $0 \neq q = p - q_0 \leq p$ and $q \neq p$. Now let r be any projection of M, and $r \leq q$. Since $\{q_l\}_{l\in\Lambda}$ is maximal, we must have $\varphi(r) \leq \psi(r)$. Further, if $a \in M_+$ and $a \leq q$, then it is easy to see $a \in qMq$. Now we can write $a = \int_0^1 \lambda de_\lambda$, where $e_\lambda \in qMq, \forall \lambda$. Suppose that

$$a_n = \sum_{k=0}^{n-1} \frac{k}{n}(e_{(k+1)/n} - e_{k/n}).$$

Then $0 \leq a - a_n \leq \frac{1}{n}q$ and $\varphi(a_n) \leq \psi(a_n)$ since $(e_{(k+1)/n} - e_{k/n}) \leq q$ for each $k, \forall n$. From $\varphi(a_n) \to \varphi(a), \psi(a_n) \to \psi(a)$, we have $\varphi(a) \leq \psi(a), \forall a \in M_+$ and $a \leq q$. Therefore q satisfies our conditions.

If $\mathcal{L}$ is empty, then for any projection r of M with $r \leq p$ we have $\varphi(r) \leq \psi(r)$. Similar to the preceding paragraph we have

$$\varphi(a) \leq \psi(a), \quad \forall a \in M_+ \text{ and } a \leq p.$$

So the projection p is just what we want to find. Q.E.D.

Proposition 1.8.5. Let φ be a positive linear functional on a VN algebra M. Then φ is σ-continuous if and only if φ is completely additive.

Proof. The necessity is obvious. Now let φ be completely additive. First we can find a maximal orthogonal family $\{q_l\}_{l\in\Lambda}$ of projections of M such that $\varphi(\cdot q_l)$ is σ-continuous, $\forall l \in \Lambda$. Let $q = \sum_{l\in\Lambda} q_l$. We claim that $\varphi(\cdot q)$ is σ-continuous also. In fact, for any $a \in M$ with $\|a\| \le 1$ and any finite subset F of Λ, by the Schwartz inequality we have

$$|\varphi(aq) - \varphi(a\sum_{l\in F} q_l)|^2 \le \varphi(\sum_{l\notin F} q_l) \cdot \varphi(a\sum_{l\notin F} q_l a^*).$$

Since $0 \le a\sum_{l\notin F} q_l a^* \le aa^* \le 1$, and φ is completely additive, it follows that

$$\varphi(aq) = \lim_F \varphi(a\sum_{l\in F} q_l) = \sum_{l\in\Lambda} \varphi(aq_l),$$

uniformly for $a \in M$ and $\|a\| \le 1$. Therefore, $\varphi(\cdot q)$ is σ-continuous.

Now it suffices to show that $q = 1$. If $q \ne 1$, then $p = 1 - q \ne 0$. Pick $\xi \in H$ (the action space of M) such that $\varphi(p) \le \psi(p)$, where $\psi(\cdot) = \langle \cdot\xi, \xi\rangle$. By Lemma 1.8.4, there is a non-zero projection q_0 of $M, q_0 \le p$, such that

$$\varphi(a) \le \psi(a), \qquad \forall a \in M_+ \text{ and } a \le q_0.$$

If $a \in M$ and $\|a\| \le 1$, then $0 \le q_0 a^* a q_0 \le q_0$ and

$$|\varphi(aq_0)|^2 \le \varphi(1)\varphi(q_0 a^* a q_0) \le \varphi(1)\psi(q_0 a^* a q_0) = \varphi(1)\|aq_0\xi\|^2.$$

This implies that $\varphi(\cdot q_0)$ is strongly continuous on any bounded ball of M. By Proposition 1.2.6, $\varphi(\cdot q_0)$ is σ-continuous on M. It is impossible since the family $\{q_l\}_{l\in\Lambda}$ is maximal. Therefore we have $q = 1$. Q.E.D.

Theorem 1.8.6. Let φ be a positive linear functional on a VN algebra M. Then the following statements are equivalent:

1) φ is σ-continuous;
2) φ is normal;
3) φ is completely additive.

Proof. By Proposition 1.8.5, 1) and 3) are equivalent. Moreover, it is obvious that 1) implies 2) and 2) implies 3). Q.E.D.

Corollary 1.8.7. Let φ, ψ be two positive linear functionals on a VN algebra M, and $\varphi \ge \psi$. If $\varphi \in M_*$, then $\psi \in M_*$.

Proof. Let $\{a_l\}$ be a bounded increasing net in M_+, and $a = \sup_l a_l$. Since $0 \le \psi(a - a_l) \le \varphi(a - a_l) \to 0$, it follows that $\psi(a) = \lim_l \psi(a_l) = \sup_l \psi(a_l)$, i.e., ψ is normal. Q.E.D.

Proposition 1.8.8. Let φ be a normal positive linear functional on a VN algebra M. Define

$$p_\varphi = \sup\{p \mid p \text{ is a projection of } M, \text{ such that } \varphi(p) = 0\}.$$

Then $Mp_\varphi = \{a \in M | \varphi(a^*a) = 0\}$, and $\varphi(p_\varphi a) = \varphi(ap_\varphi) = 0, \forall a \in M$.

Proof. Let $L = \{a \in M | \varphi(a^*a) = 0\}$. Clearly, L is a s-closed left ideal of M. By Propositions 1.2.8 and 1.7.1, there is a projection p_0 of M such that $L = Mp_0$. If p is a projection of M such that $\varphi(p) = 0$, then $p \in L$ and $p \leq p_0$. Thus $p_0 = p_\varphi$. Now by the Schwartz inequality, we have $\varphi(p_\varphi a) = \varphi(ap_\varphi) = 0, \forall a \in M$. Q.E.D.

Definition 1.8.9. Let φ, p_φ be as in Proposition 1.8.8. Then $(1 - p_\varphi)$ is called the *support* of φ, and denote it by $s(\varphi) = 1 - p_\varphi$.

Clearly, by Proposition 1.8.8, we have

$$\varphi(s(\varphi)a) = \varphi(as(\varphi)) = \varphi(s(\varphi)as(\varphi)) = \varphi(a), \quad \forall a \in M.$$

Proposition 1.8.10. Let φ be a normal positive linear functional on a VN algebra $M, a \in M_+$ such that $\varphi(a) = 0$. Then $s(\varphi)as(\varphi) = 0$. Moreover, φ is faithful if and only if $s(\varphi) = 1$.

Proof. If $s(\varphi)as(\varphi) \neq 0$, by its spectral decomposition we can find a non-zero projection p of M and $\lambda > 0$ such that $p \leq s(\varphi)$ and $\lambda p \leq s(\varphi)as(\varphi)$. Then $\varphi(p) = 0$, and $0 \neq p \leq p_\varphi = 1 - s(\varphi)$. It is impossible since $p \leq s(\varphi)$. Therefore $s(\varphi)as(\varphi) = 0$. The rest conclusion is obvious. Q.E.D.

Proposition 1.8.11. Let M be a VN algebra on a Hilbert space H, and $\xi \in H, \varphi(\cdot) = \langle \cdot\xi, \xi\rangle$. Then $s(\varphi)H = \overline{M'\xi}$.

Proof. Let p be a projection of M. Then $\varphi(p) = 0 \Longleftrightarrow p\xi = 0 \Longleftrightarrow pM'\xi = 0$. That comes to the conclusion. Q.E.D.

Now let M be a VN algebra on a Hilbert space H, and φ be a positive linear functional on M. Let

$$L_\varphi = \{a \in M \mid \varphi(a^*a) = 0\}.$$

L_φ is called the *left kernel* of φ. By the Schwartz inequality, we can see that L_φ is a left ideal of M. Let $a \to a_\varphi = a + L_\varphi(\forall a \in M)$ be the quotient map from M onto M/L_φ. Then we can define

$$\langle a_\varphi, b_\varphi\rangle = \varphi(b^*a), \quad \forall a, b \in M.$$

Clearly, $\langle , \rangle$ will be an inner product on M/L_φ. Denote its completion by H_φ. For any $a \in M$, define

$$\pi_\varphi(a)b_\varphi = (ab)_\varphi, \quad \forall b \in M.$$

It is clear that $\pi_\varphi(a)$ is a linear map on M/L_φ, and

$$\|\pi_\varphi(a)b_\varphi\|^2 = \varphi(b^*a^*ab) \le \|a\|^2\varphi(b^*b) = \|a\|^2 \cdot \|b_\varphi\|^2, \quad \forall b \in M.$$

Then $\pi_\varphi(a)$ can be uniquely extended to a bounded linear operator on H_φ. Denote this extension still by $\pi_\varphi(a)$. Then we have a $*$ (algebraic) homomorphism

$$\pi_\varphi : M \to B(H_\varphi)$$

i.e.,

$$\begin{cases} \pi_\varphi(\lambda a + \mu b) = \lambda\pi_\varphi(a) + \mu\pi_\varphi(b) \\ \pi_\varphi(ab) = \pi_\varphi(a)\pi_\varphi(b), \pi_\varphi(a^*) = \pi_\varphi(a)^*, \end{cases}$$

$\forall a, b \in M, \lambda, \mu \in \mathbb{C}$. Moreover, we also have pointed out that $\|\pi_\varphi(a)\| \le \|a\|$, and $\varphi(a) = \langle \pi_\varphi(a)1_\varphi, 1_\varphi \rangle, \forall a \in M$, where 1_φ is the canonical image of the identity 1 of M in M/L_φ.

Definition 1.8.12. Let M be a VN algebra, and K be a Hilbert space. If π is a $*$ (algebraic) homomorphism from M into $B(K)$, then $\{\pi, K\}$ is called a $*$ *representation* of M.

From the preceding paragraph, for any positive linear functional φ on M we can get a $*$ representation $\{\pi_\varphi, H_\varphi\}$ of M. And this representation admits a *cyclic vector* 1_φ (i.e. $\overline{\pi_\varphi(M)1_\varphi} = H_\varphi$) such that $\varphi(a) = \langle \pi_\varphi(a)1_\varphi, 1_\varphi \rangle, \forall a \in M$. This is called the *GNS construction*. We shall discuss it again in Chapter 2 in detail.

Proposition 1.8.13. Let M be a VN algebra on a Hilbert space H, φ be a normal positive linear functional on M, and $\{\pi_\varphi, H_\varphi\}$ be the $*$ representation of M generated by φ. Then $\pi_\varphi(M)$ is a VN algebra on H_φ, and π_φ is σ-σ continuous. Moreover, if φ is faithful, then π_φ is isometric, and 1_φ is also a cyclic vector for $\pi_\varphi(M)'$.

Proof. Let $I = \{a \in M \mid \pi_\varphi(a) = 0\}$. Clearly, I is a two-sided ideal of M. We claim that the unit ball of I is strongly closed. In fact, if a net $\{a_l\} \subset I$ with $\|a_l\| \le 1, \forall l$, and $a_l \to a$ (strongly), then $a_l^* a_l \xrightarrow{\sigma} a^*a$. Further,

$$\begin{aligned} \|\pi_\varphi(a)b_\varphi\|^2 &= \varphi(b^*a^*ab) = \lim_l \varphi(b^*a_l^*a_lb) \\ &= \lim_l \|\pi_\varphi(a_l)b_\varphi\|^2 = 0, \qquad \forall b \in M. \end{aligned}$$

Thus $a \in I$.

Now by Propositions 1.2.8 and 1.7.1, there is a central projection z of M such that $I = M(1-z)$. Therefore, π_φ is a $*$ isomorphism from Mz onto $\pi_\varphi(M)$.

If $a \in Mz$ such that $\pi_\varphi(a) \geq 0$, we claim that $a \geq 0$. In fact, since π_φ is injective on Mz, it follows from $\pi_\varphi(a)^* = \pi_\varphi(a)$ that $a^* = a$. So we can write $a = a_+ - a_-$, where $0 \leq a_\pm \in Mz$ and $a_+ \cdot a_- = 0$. Suppose that $a_- \neq 0$. Then $B = \pi_\varphi(a_-)$ is non-zero and positive on H_φ, and there is $\xi \in H_\varphi$ such that $\eta = B^{3/2}\xi \neq 0$. Now we get a contradiction:

$$0 \leq \langle \pi_\varphi(a)B\xi, B\xi\rangle = -\|\eta\|^2 < 0.$$

Thus a_- must be zero, and $a = a_+ \geq 0$.

Let $h^* = h \in Mz$. Since $\pi_\varphi(z) = 1$ and

$$-\|\pi_\varphi(h)\| \cdot 1 \leq \pi_\varphi(h) \leq \|\pi_\varphi(h)\| \cdot 1,$$

it follows from the discussion of preceding paragraph that $-\|\pi_\varphi(h)\| \cdot 1 \leq h \leq \|\pi_\varphi(h)\| \cdot 1$, i.e. $\|h\| \leq \|\pi_\varphi(h)\|$. Thus we have $\|h\| = \|\pi_\varphi(h)\|, \forall h^* = h \in Mz$. Furthermore, from $\|\pi_\varphi(a)\|^2 = \|\pi_\varphi(a^*a)\|$ and $\|a^*a\| = \|a\|^2$, we have $\|a\| = \|\pi_\varphi(a)\|, \forall a \in Mz$. So π_φ is an isometric $*$ isomorphism from Mz onto $\pi_\varphi(M)$.

In order to prove that $\pi_\varphi(M)$ is a VN algebra on H_φ, by Corollary 1.6.3 it suffices to show that the unit ball of $\pi_\varphi(M)$ is weakly closed. Let $\{A_l\}$ be a net of $\pi_\varphi(M)$ with $\|A_l\| \leq 1, \forall l$, and $A_l \to A$ (weakly). From preceding paragraph, there exists a net $\{a_l\} \subset Mz$ such that $\pi_\varphi(a_l) = A_l, \|a_l\| = \|A_l\|, \forall l$. Since the unit ball of Mz is weakly compact, we may assume that $a_l \to a$ (weakly) and $a \in Mz$. Then

$$\begin{aligned}\langle Ab_\varphi, c_\varphi\rangle &= \lim_l \langle \pi_\varphi(a_l)b_\varphi, c_\varphi\rangle = \lim_l \varphi(c^*a_lb)\\ &= \varphi(c^*ab) = \langle \pi_\varphi(a)b_\varphi, c_\varphi\rangle,\end{aligned}$$

$\forall b, c \in M$. Therefore $A = \pi_\varphi(a) \in \pi_\varphi(M)$, and the unit ball of $\pi_\varphi(M)$ is weakly closed.

About the σ-σ continuity of π_φ, by Proposition 1.2.6 it suffices to show that π_φ is weakly continuous on the unit ball of M. This is easy.

Finally, let φ be faithful. Then it is easy to see that π_φ is faithful (injective), and also π_φ is isometric. Moreover, let P be the projection from H_φ onto $\overline{\pi_\varphi(M)'1_\varphi}$. Then $P \in \pi_\varphi(M)'' = \pi_\varphi(M)$, and there is unique projection p of M such that $\pi_\varphi(p) = P$. Since $\varphi(1-p) = \|\pi_\varphi(1-p)1_\varphi\|^2 = \|(1-P)1_\varphi\|^2 = 0$, it follows that $P = 1$, i.e. 1_φ is a cyclic vector for $\pi_\varphi(M)'$. Q.E.D.

References.[22], [28], [150].

1.9 Polar decomposition and orthogonal decomposition

Let M be a VN algebra. For any $\varphi \in M_*$ and $a \in M$, define

$$(R_a\varphi)(b) = \varphi(ba), \quad (L_a\varphi)(b) = \varphi(ab), \quad \forall b \in M.$$

Obviously, $R_a\varphi$ and $L_a\varphi \in M_*$.

Lemma 1.9.1. Let $\varphi \in M_*$, and p be projection of M such that $\|R_p\varphi\| = \|\varphi\|$. Then $\varphi = R_p\varphi$.

Proof. We may assume that $\|\varphi\| = 1$. If $R_{(1-p)}\varphi \neq 0$, then there exists $a \in M$ with $\|a\| \leq 1$ such that $(R_{(1-p)}\varphi)(a) = \delta > 0$. Since $M = (M_*)^*$, we can find $b \in M$ with $\|b\| = 1$ such that $(R_p\varphi)(b) = \|R_p\varphi\| = \|\varphi\| = 1$. Notice that

$$\|bp + \delta a(1-p)\|^2 = \|bpb^* + \delta^2 a(1-p)a^*\| \leq 1 + \delta^2.$$

Thus $\|bp + \delta a(1-p)\| < 1 + \delta^2$. However,

$$\varphi(bp + \delta a(1-p)) = (R_p\varphi)(b) + \delta(R_{(1-p)}\varphi)(a) = 1 + \delta^2.$$

This is a contradiction since $\|\varphi\| = 1$. Therefore $R_{(1-p)}\varphi = 0, \varphi = R_p\varphi$.

Q.E.D.

Lemma 1.9.2. Let $f \in M^*$ and $a \in M_+$ with $\|a\| \leq 1$ be such that $f(a) = \|f\|$. Then $f \geq 0$.

Proof. Pick a real number θ such that $e^{i\theta} f(1-a) \geq 0$. Since $\|a + e^{i\theta}(1-a)\| \leq 1$, it follows that

$$\|f\| \leq f(a) + e^{i\theta} f(1-a) \leq \|f\|.$$

Thus $f(1-a) = 0$, i.e. $f(1) = f(a) = \|f\|$.

Now let $b \in M_+$. We need to prove $f(b) \geq 0$. We may assume that $\|b\| \leq 1$ and $\|f\| = 1$. If $f(b) = \lambda + i\mu$, where $\lambda, \mu \in \mathbb{R}$, then

$$1 \geq \|1 - b\| \geq |f(1-b)| = ((1-\lambda)^2 + \mu^2)^{1/2}.$$

That implies $\lambda \geq 0$. Moreover, since for any $r \in \mathbb{R}$

$$\|b\|^2 + r^2 = \|b + ir\|^2 \geq |f(b + ir)|^2 \geq \mu^2 + 2r\mu + r^2,$$

it follows that $\mu = 0$. Therefore $f(b) = \lambda \geq 0$. Q.E.D.

Theorem 1.9.3. Let M be a VN algebra, and $\varphi \in M_*$. Then we have unique $\omega \in M_*$ with $\omega \geq 0$ and unique partial isometry v of M such that

$$\varphi = R_v\omega, \qquad v^*v = s(\omega).$$

Moreover, $\omega = R_{v^*}\varphi, \|\varphi\| = \|\omega\|$.

Proof. Since $M = (M_*)^*$, we can find $a \in M, \|a\| \leq 1$ such that $\varphi(a) = \|\varphi\|$. Let $a^* = uh$ be the polar decomposition of a^* and $\omega = R_{u^*}\varphi$. From $\|\omega\| \leq \|\varphi\| = \varphi(a) = \omega(h) \leq \|\omega\|$ and by Lemma 1.9.2, it follows that $\|\varphi\| = \|\omega\|$ and $\omega \geq 0$. Let $p = uu^*$. By $a = ap$,

$$\|R_p\varphi\| \leq \|\varphi\| = \varphi(ap) = (R_p\varphi)(a) \leq \|R_p\varphi\|.$$

Then by Lemma 1.9.1, we have $\varphi = R_p\varphi = R_u\omega$. Since $R_{u^*u}\omega = R_{u^*}\varphi = \omega$, it follows that $u^*u \geq s(\omega)$. Let $v = us(\omega)$. By Definition 1.8.9, we have

$$\varphi = R_v\omega, \quad v^*v = s(\omega), \quad \omega = R_{v^*}\varphi.$$

Now suppose that there is another $0 \leq \omega' \in M_*$ and $v' \in M$ with $\varphi = R_{v'}\omega'$ and $v'^*v' = s(\omega')$. It is easy to see that $\omega' = R_{v'^*}\varphi$ and $\|\omega'\| = \|\varphi\|$. We may assume $\|\varphi\| = 1$. Then $1 = \omega'(1) = \varphi(v'^*) = \omega(v'^*v), 1 = \omega(1) = \varphi(v^*) = \omega'(v^*v')$. By the Schwartz inequality

$$1 = \omega(v'^*v)^2 \leq \omega(v^*v'v'^*v) \leq 1,$$

thus we have $\omega((v'^*v - 1)^*(v'^*v - 1)) = 0$. Then again by the Schwartz inequality, $\omega(b(v'^*v - 1)) = 0, \forall b \in M$, i.e.,

$$\omega(b) = \varphi(bv'^*) = \omega'(b), \quad \forall b \in M, \text{ or } \omega = \omega'.$$

Let $q = v'^*v$. Then $s(\omega)qs(\omega) = s(\omega')qs(\omega) = q$ since $\omega' = \omega$. From $\omega((q-1)^*(q-1)) = 0$ and Proposition 1.8.10, we have

$$s(\omega)(1-q^*)(1-q)s(\omega) = 0.$$

Thus $s(\omega) = qs(\omega) = q$. Now by $v'^*v' = s(\omega') = s(\omega) = q$, it follows that

$$v'^*(v' - v) = 0. \tag{1}$$

Again by $v^*v = s(\omega) = q^* = v^*v'$, we get

$$v^*(v' - v) = 0. \tag{2}$$

Considering (1) – (2), we obtain $v' = v$. Q.E.D.

Definition 1.9.4. The unique expression $\varphi = R_v\omega$ (see Theorem 1.9.3) is called the *polar decomposition* of $\varphi(\in M_*)$, and ω is called the ***absolute value*** of φ and denoted by $|\varphi|$.

If $M = B(H), \varphi(\cdot) = tr(\cdot t) \in M_* = T(H)$, where $t \in T(H)$, let $t = vh$ be the polar decomposition of t, then the polar decomposition of φ is $R_v\omega$, and $\omega(\cdot) = tr(\cdot h)$.

Proposition 1.9.5. For any $\varphi \in M_*$, define $\varphi^*(a) = \overline{\varphi(a^*)}, \forall a \in M$. Then the polar decomposition of φ^* is $\varphi^* = R_{v^*}\psi$ if the polar decomposition of φ is $\varphi = R_v\omega$. Here $\psi = L_{v^*}\varphi$.

Proof. Clearly, $\psi \geq 0, \varphi^* = R_{v^*}\psi$,and $\|\psi\| = \|\omega\| = \|\varphi\| = \|\varphi^*\|$. By Theorem 1.9.3, it suffices to show $s(\psi) = vv^*$. For any projection p of M, the following conditions are equivalent: (1) $\psi(p) = 0$; (2) $\omega(v^*pv) = 0$; (3)$s(\omega)v^*pvs(\omega) = 0$; (4) $v^*pv = 0$; (5) $pv = 0$; (6) $pvv^* = 0$. That implies $s(\psi) = vv^*$. Q.E.D.

Definition 1.9.6. $s(|\varphi|)$ and $s(|\varphi^*|)$ are called the *left* and *right support* of $\varphi(\in M_*)$, denoted by $s_l(\varphi)$ and $s_r(\varphi)$ respectively.

Clearly, when φ is hermitian, i.e., $\varphi^* = \varphi$, we have $s_l(\varphi) = s_r(\varphi)$. If $\varphi \geq 0$, then $s_l(\varphi) = s_r(\varphi) = s(\varphi)$.

Proposition 1.9.7. Let $\varphi \in M_*$, and $\varphi = R_v\omega$ be the polar decomposition of φ. Then

$$s_l(\varphi) = v^*v = \inf\{p \mid p \text{ is a projection of } M, \text{ and } L_p\varphi = \varphi\},$$

$$s_r(\varphi) = vv^* = \inf\{p \mid p \text{ is a projection of } M, \text{ and } R_p\varphi = \varphi\}.$$

Proof. Let p be a projection of M. The following conditions are equivalent: (1) $L_p\varphi = \varphi$; (2) $L_p\omega = \omega$; (3) $p \geq s(\omega) = s_l(\varphi)$. So we can get the expression of $s_l(\varphi)$ immediately. The proof of the expression of $s_r(\varphi)$ is similar.

Q.E.D.

Now we study *hermitian functionals.* Let φ be a linear functional on a VN algebra M. Then φ is said to be hermitian, if $\varphi^* = \varphi$, where $\varphi^*(a) = \overline{\varphi(a^*)}, \forall a \in M$. Obviously, φ is hermitian if and only if $\varphi(h) \in \mathbb{R}, \forall h^* = h \in M$.

Theorem 1.9.8. Let M be a VN algebra, $\varphi = \varphi^* \in M_*$. Then there exist unique $\varphi_+, \varphi_- \in M_*, \varphi_\pm \geq 0$ such that

$$\varphi = \varphi_+ - \varphi_-, \quad \|\varphi\| = \|\varphi_+\| + \|\varphi_-\|.$$

Moreover, let $q_+ = s(\varphi_+), q_- = s(\varphi_-)$. Then

$$|\varphi| = \varphi_+ + \varphi_-, \quad q_+ \cdot q_- = 0, \quad s(|\varphi|) = q_+ + q_-$$

and $\varphi = R_{(q_+ - q_-)}|\varphi|$ is the polar decomposition of φ.

Proof. Let $\varphi = R_v|\varphi|, \varphi^* = R_{v^*}|\varphi^*|$ be the polar decompositions of φ, φ^* respectively. Since $\varphi^* = \varphi$ and by the uniqueness of the polar decomposition, we have $v^* = v$. Then we can write

$$v = q_+ - q_-,$$

where q_+, q_- are two projections of M, and $q_+ \cdot q_- = 0$. Since $s(|\varphi|) = v^*v = q_+ + q_-$, it follows that

$$|\varphi| = L_{q_+}R_{q_+}|\varphi| + L_{q_-}R_{q_-}|\varphi| + L_{q_+}R_{q_-}|\varphi| + L_{q_-}R_{q_+}|\varphi|.$$

In addition, by Proposition 1.9.5, $|\varphi^*| = L_vR_v|\varphi| = |\varphi|$. Thus $L_{q_+}R_{q_-}|\varphi| = L_{q_+}R_{q_-}L_vR_v|\varphi| = -L_{q_+}R_{q_-}|\varphi|$, i.e., $L_{q_+}R_{q_-}|\varphi| = 0$. Similarly, $L_{q_-}R_{q_+}|\varphi| = 0$. So we have

$$|\varphi| = L_{q_+}R_{q_+}|\varphi| + L_{q_-}R_{q_-}|\varphi|,$$

and

$$\varphi = R_v|\varphi| = L_{q_+}R_{q_+}|\varphi| - L_{q_-}R_{q_-}|\varphi|.$$

Write $\varphi_+ = L_{q_+}R_{q_+}|\varphi|$, $\varphi_- = L_{q_-}R_{q_-}|\varphi|$. Then $\varphi_\pm \in M_*, \varphi_\pm \geq 0$ and $\varphi = \varphi_+ - \varphi_-, |\varphi| = \varphi_+ + \varphi_-$. By Theorem 1.9.3, $\|\varphi\| = \|\ |\varphi|\ \| = |\varphi|(1) = \|\varphi_+\| + \|\varphi_-\|$. Moreover, it is obvious that $\varphi_+(1 - q_+) = 0$. And also, if p is a projection of M such that $\varphi_+(p) = 0$, then $|\varphi|(q_+pq_+) = 0, s(|\varphi|)q_+pq_+s(|\varphi|) = 0, q_+pq_+ = 0, pq_+ = 0$, i.e., $p \leq 1 - q_+$. Thus $q_+ = s(\varphi_+)$. Similarly, $q_- = s(\varphi_-)$.

Now suppose that $\varphi_1, \varphi_2 \in M_*, \varphi_1, \varphi_2 \geq 0$, such that

$$\varphi = \varphi_1 - \varphi_2, \quad \|\varphi\| = \|\varphi_1\| + \|\varphi_2\|.$$

Since $\|\varphi_+\| = \varphi(q_+) \leq \varphi_1(q_+) \leq \|\varphi_1\|$ and $\|\varphi_-\| = -\varphi(q_-) \leq \varphi_2(q_-) \leq \|\varphi_2\|$, it follows that

$$\|\varphi_+\| = \varphi_1(q_+) = \|\varphi_1\| = \varphi_1(1), \quad \|\varphi_-\| = \varphi_2(q_-) = \|\varphi_2\| = \varphi_2(1).$$

Further, $s(\varphi_1) \leq q_+, s(\varphi_2) \leq q_-$, and $s(\varphi_+) \cdot s(\varphi_-) = 0$. So we get $\varphi = R_{(s(\varphi_1)-s(\varphi_2))}(\varphi_1 + \varphi_2)$. By the uniqueness of the polar decomposition, it follows that

$$s(\varphi_1) - s(\varphi_2) = q_+ - q_-, \quad \varphi_1 + \varphi_2 = \varphi_+ + \varphi_-.$$

But $s(\varphi_1) \cdot s(\varphi_2) = q_+ \cdot q_- = 0$, it must be that $s(\varphi_1) = q_+, s(\varphi_2) = q_-$. Furthermore, $\varphi_1 = L_{s(\varphi_1)}(\varphi_1 + \varphi_2) = L_{q_+}(\varphi_+ + \varphi_-) = \varphi_+, \varphi_2 = L_{s(\varphi_2)}(\varphi_1 + \varphi_2) = L_{q_-}(\varphi_+ + \varphi_-) = \varphi_-$. Q.E.D.

Corollary 1.9.9. Any σ-continuous linear functionals on a VN algebra must be a linear sum of normal positive linear functionals.

Definition 1.9.10. The unique expression $\varphi = \varphi_+ - \varphi_-$ of a hermitian functional φ (see Theorem 1.9.8) is called the *orthogonal* (or *Jordan*) *decomposition* of φ.

The result in 1.9.8 generalizes the ordinary Jordan decomposition of a signed measure.

References. [64], [145], [150], [180].

1.10 Radon-Nikodyn theorems

Lemma 1.10.1. Let N be a VN algebra on a Hilbert space K; and N admits a cyclic vector ξ (i.e. $\overline{N\xi} = K$). Define

$$\varphi(a) = \langle a\xi, \xi\rangle, \quad \forall a \in N.$$

If ψ is a positive linear function on N with $\psi \le \varphi$, then there exists unique $t' \in N'$ with $0 \le t' \le 1$ such that

$$\psi(a) = \langle at'\xi, \xi\rangle, \quad \forall a \in N.$$

Proof. On the dense linear subspace $N\xi$ of K, define

$$[a\xi, b\xi] = \psi(b^*a), \quad \forall a, b \in N.$$

Since $\psi \le \varphi$, it follows that $\|[a\xi, b\xi]\| \le \|a\xi\| \cdot \|b\xi\|, \forall a, b \in N$. Thus there exists unique $t' \in B(K)$ such that

$$\psi(b^*a) = [a\xi, b\xi] = \langle t'a\xi, b\xi\rangle, \forall a, b \in N.$$

From $o \le \psi \le \varphi$, we have $0 \le t' \le 1$. Further, since

$$\langle t'ab\xi, c\xi\rangle = \psi(c^*ab) = \psi((a^*c)^*b) = \langle at'b\xi, c\xi\rangle$$

$\forall a, b, c \in N$ and $\overline{N\xi} = K$, it follows that $t' \in N'$. That comes to the conclusion.

Q.E.D.

Lemma 1.10.2. Let φ_0, φ_1 be two positive linear functionals on a VN algebra N, and $\varphi_1 = R_a\varphi_0$ for some $a \in N$. Then $\varphi_1 \le \|a\|\varphi_0$.

Proof. Let $\varphi_{n+1} - R_{a^{2^n}}\varphi_0, n = 0, 1, \cdots$. For any $b \in N_+$, by the Schwartz inequality we have

$$0 \le \varphi_1(b) = \varphi_0(b^{1/2} \cdot b^{1/2}a) \le \varphi_0(b)^{1/2} \cdot \varphi_0(a^*ba)^{1/2}.$$

Since $\varphi_2(b) = \varphi_1(ba) = \overline{\varphi_1(a^*b)} = \overline{\varphi_0(a^*ba)} \geq 0, \forall b \in N_+$ it follows that $\varphi_2 \geq 0$ and $0 \leq \varphi_1(b) \leq \varphi_0(b)^{1/2} \cdot \varphi_2(b)^{1/2}, \forall b \in N_+$. Similarly, we can prove that $\varphi_3 \geq 0$ and $0 \leq \varphi_2(b) \leq \varphi_0(b)^{1/2}\varphi_3(b)^{1/2}, \forall b \in N_+$. Thus $0 \leq \varphi_1(b) \leq \varphi_0(b)^{\frac{1}{2}+\frac{1}{4}} \cdot \varphi_3(b)^{1/4}, \forall b \in N_+$. Generally, we have

$$\begin{aligned} 0 \leq \varphi_1(b) \leq & \ \varphi_0(b)^{\frac{1}{2}+\cdots+\frac{1}{2^n}} \cdot \varphi_0(ba^{2^n})^{\frac{1}{2^n}} \\ \leq & \ \varphi_0(b)^{\frac{1}{2}+\cdots+\frac{1}{2^n}} \cdot \|a\|(\|\varphi_0\| \cdot \|b\|)^{\frac{1}{2^n}}, \quad \forall b \in N_+, \forall n. \end{aligned}$$

Now let $n \to \infty$, then we get $\varphi_1 \leq \|a\|\varphi_0$. Q.E.D.

Theorem 1.10.3. Let M be a VN algebra, $\varphi, \psi \in M_*$ and $\varphi \geq \psi \geq 0$. Then there exists $t_0 \in M$ with $0 \leq t_0 \leq 1$ such that

$$\psi(a) = \varphi(t_0 a t_0), \quad \forall a \in M.$$

Proof. First, let $s(\varphi) = 1$. If $\{\pi_\varphi, H_\varphi\}$ is the $*$ representation of M generated by φ, by Proposition 1.8.13 then $N = \pi_\varphi(M)$ is a VN algebra on H_φ, and N is $*$ isomorphic to M. Using Lemma 1.10.1, we have $t' \in N'$ with $0 \leq t' \leq 1$ such that

$$\psi(a) = \langle \pi_\varphi(a) t' 1_\varphi, t' 1_\varphi \rangle, \quad \forall a \in M.$$

Define two functionals on N' : $\varphi'(a') = \langle a' 1_\varphi, 1_\varphi \rangle, \forall a' \in N'$, and $\psi' = R_{t'}\varphi'$. Let $\psi' = R_{v'}\omega'$ be the polar decomposition of ψ'. Then $\omega' = R_{v'^* t'}\varphi'$. By Lemma 1.10.2, we have

$$\omega' \leq \|v'^* t'\| \varphi' \leq \varphi'.$$

Further, by Lemma 1.10.1 there is $t_0 \in M$ with $0 \leq t_0 \leq 1$ such that

$$\omega'(a') = \langle \pi_\varphi(t_0) a' 1_\varphi, 1_\varphi \rangle, \forall a' \in N'.$$

Since $\omega' = R_{v'^* t'}\varphi'$, it follows that

$$\langle a' \pi_\varphi(t_0) 1_\varphi, 1_\varphi \rangle = \langle a' v'^* t' 1_\varphi, 1_\varphi \rangle, \quad \forall a' \in N'.$$

By Proposition 1.8.13, 1_φ is a cyclic vector for N'. Thus

$$\pi_\varphi(t_0) 1_\varphi = v'^* t' 1_\varphi. \tag{1}$$

From (1),

$$\begin{aligned} \langle a' t' 1_\varphi, 1_\varphi \rangle = & \ \psi'(a') = (R_{v'}\omega')(a') = (R_{v' v'^* t'}\varphi')(a') \\ = & \ \langle a' v' v'^* t' 1_\varphi, 1_\varphi \rangle = \langle a' v' \pi_\varphi(t_0) 1_\varphi, 1_\varphi \rangle, \end{aligned}$$

$\forall a' \in N'$. Hence

$$t' 1_\varphi = v' \pi_\varphi(t_0) 1_\varphi. \tag{2}$$

From (1), (2), for any $a \in M$

$$\begin{aligned}\psi(a) = & \langle \pi_\varphi(a)t'1_\varphi, t'1_\varphi\rangle = \langle \pi_\varphi(a)v'\pi_\varphi(t_0)1_\varphi, t'1_\varphi\rangle \\ = & \langle \pi_\varphi(a)\pi_\varphi(t_0)1_\varphi, \pi_\varphi(t_0)1_\varphi\rangle = \varphi(t_0 a t_0).\end{aligned}$$

That comes to the conclusion for $s(\varphi) = 1$.

Now consider general φ. Let $p = s(\varphi), M_p = pMp$. On the VN algebra M_p, φ is faithful, and also $\varphi \geq \psi \geq 0$. From the preceding paragraph, there exists $t_0 \in M$ with $0 \leq t_0 \leq p$ such that

$$\psi(a) = \varphi(t_0 a t_0), \quad \forall a \in M_p.$$

Clearly, $p = s(\varphi) \geq s(\psi)$. Then for any $a \in M$,

$$\psi(a) = \psi(pap) = \varphi(t_0 pap t_0) = \varphi(t_0 a t_0).$$

Q.E.D.

Theorem 1.10.4. Let M be a VN algebra, $\varphi, \psi \in M_*$ and $\varphi \geq \psi \geq 0$. If λ is a complex number with $\mathrm{Re}\lambda \geq \frac{1}{2}$, then there exists $h \in M$ with $0 \leq h \leq 1$ such that

$$\psi(a) = \lambda\varphi(ha) + \overline{\lambda}\varphi(ah), \quad \forall a \in M.$$

Moreover, if φ is faithful, then h is unique.

Proof. Since $\{h \in M \mid 0 \leq h \leq 1\}$ is a $\sigma(M, M_*)$-compact convex subset of M, it follows that

$$\mathcal{L} = \{\lambda\varphi(h\cdot) + \overline{\lambda}\varphi(\cdot h) \mid h \in M, 0 \leq h \leq 1\}$$

is a $\sigma(M_*, M)$-compact convex subset of M_*. If $\psi \notin \mathcal{L}$, then there exists a $a = a^* \in M$ and a $\mu \in \mathbb{R}$ such that

$$\psi(a) > \mu \geq f(a), \quad \forall f \in \mathcal{L}.$$

Write $a = a_+ - a_-$, where $a_\pm \in M_+$ and $a_+ \cdot a_- = 0$, and pick a projection p of M such that $pa_+ = a_+, pa_- = 0$. Then

$$\psi(a_+) > \mu \geq \lambda\varphi(pa) + \overline{\lambda}\varphi(ap) = 2\mathrm{Re}\lambda \cdot \varphi(a_+) \geq \varphi(a_+).$$

It is impossible since $\varphi \geq \psi$. Thus $\psi \in \mathcal{L}$, i.e., there exists $h \in M$ with $0 \leq h \leq 1$ such that

$$\psi(a) = \lambda\varphi(ha) + \overline{\lambda}\varphi(ah), \quad \forall a \in M.$$

Moreover let φ be faithful, and suppose that h and k satisfy our condition simultaneously. Since

$$\begin{aligned}(\lambda + \overline{\lambda})(h-k)^2 = & [\lambda h(h-k) + \overline{\lambda}(h-k)h] \\ & -[\lambda k(h-k) + \overline{\lambda}(h-k)k],\end{aligned}$$

it follows that $\varphi((h-k)^2) = 0$ and $h = k$. Q.E.D.

Proposition 1.10.5. Let M be a VN algebra on a Hilbert space H, φ be a σ-(or weakly, or strongly) continuous linear functional on M. Then φ can be extended to a σ-(or weakly, or strongly) continuous linear functional ψ on $B(H)$, and $\|\psi\| = \|\varphi\|$. Moreover, if $\varphi \geq 0$, it can be choosed that $\psi \geq 0$.

Proof. First, we assume $0 \leq \varphi \in M_*$. Since $M_* = T(H)/M_\perp$, we have $t^* = t \in T(H)$ such that

$$\varphi(a) = tr(ta), \quad \forall a \in M.$$

Write $t = t_+ - t_-$, where $t_\pm \in T(H)_+$ and $t_+ \cdot t_- = 0$. Then

$$tr(t_+ a) \geq \varphi(a) \geq 0, \quad \forall a \in M_+.$$

By Theorem 1.10.3, there is $t_0 \in M$ with $0 \leq t_0 \leq 1$ such that

$$\varphi(a) = tr(t_+ t_0 a t_0) = tr(t_0 t_+ t_0 a), \quad \forall a \in M.$$

Let $\psi(\cdot) = \mathrm{tr}(t_0 t_+ t_0 \cdot)$. Then ψ is a σ-continuous linear functional on $B(H)$. Clearly, ψ is an extension of φ, and $\psi \geq 0$, and $\|\varphi\| = \varphi(1) = \psi(1) = \|\psi\|$.

For general $\varphi \in M_*$, let $\varphi = R_v \omega$ be the polar decomposition of φ. From the preceding paragraph, there is a σ-continuous positive linear functional g on $B(H)$ such that $g|M = \omega, \|g\| = \|\omega\| = \|\varphi\|$. Let $\psi = R_v g$. Then ψ is σ-continuous on $B(H)$, and is an extension φ. Clearly, $\|\varphi\| \leq \|\psi\| \leq \|v\|\|g\| \leq \|\varphi\|$. Thus $\|\psi\| = \|\varphi\|$.

When φ is weakly (or strongly) continuous, according to the above procedure it suffices to show there exists $t \in F(H)$ such that $\varphi(a) = tr(ta), \forall a \in M$.

If φ is weakly continuous, then there exists a weak neighborhood $U = U(0; \xi_1, \cdots, \xi_n; \eta_1, \cdots, \eta_n; 1) = \{a \in M \mid |\langle a\xi_i, \eta_i\rangle| \leq 1, 1 \leq i \leq n\}$ of zero such that $|\varphi(a)| \leq 1, \forall a \in U$. Define a semi-norm $p(\cdot)$ on $B(H)$:

$$p(b) = \sum_{i=1}^{n} |\langle b\xi_i, \eta_i\rangle|, \quad \forall b \in B(H).$$

Then $|\varphi(a)| \leq p(a), \forall a \in M$. By the Hahn-Banach theorem, φ can be extended to a linear functional ψ on $B(H)$ such that $|\psi(b)| \leq p(b), \forall b \in B(H)$. Clearly, ψ is weakly continuous on $B(H)$. By Proposition 1.2.7, there is $t \in F(H)$ such that $\psi(b) = \mathrm{tr}(tb), \forall b \in B(H)$. In particular, $\varphi(a) = \mathrm{tr}(ta), \forall a \in M$.

If φ is strongly continuous, replacing above $U = U(0, \xi_1, \cdots, \xi_n; \eta_1, \cdots, \eta_n; 1)$ and $p(\cdot) = \sum_{i=1}^{n} |\langle \cdot \xi_i, \eta_i\rangle|$ by $U = U(0; \xi_1, \cdots, \xi_n; 1) = \{a \in M \| \|a\xi_i\| \leq 1, 1 \leq i \leq n\}$ and $p(\cdot) = \sum_{i=1}^{n} \| \cdot \xi_i\|^2$, we can get the same conclusion by the same argument. Q.E.D.

Remark. By Propositions 1.2.2, 1.2.7 and 1.10.5, each σ-continuous functional φ on M has the following form:

$$\varphi(a) = \sum_{i=1}^{\infty} \langle a\xi_i, \eta_i \rangle, \forall a \in M,$$

where $\sum_{i=1}^{\infty}(\|\xi_i\|^2 + \|\eta_i\|^2) < \infty$, and if $\varphi \geq 0$, then we can choose $\xi_i = \eta_i, \forall i$; each weakly or strongly continous functional φ on M has the following form:

$$\varphi(a) = \sum_{i=1}^{n} \langle a\xi_i, \eta_i \rangle, \quad \forall a \in M,$$

and if $\varphi \geq 0$, then we can choose $\xi_i = \eta_i, 1 \leq i \leq n$.

As an application of Proposition 1.10.5, we have the following.

Proposition 1.10.6. Let M be a VN algebra on a Hilbert space H. Then the following statements are equivalent:

1) $\sigma(M, M_*) \sim$ (weak topology $|M$);
2) $s(M, M_*) \sim$ (strong topology $|M$);
3) $s^*(M, M_*) \sim$ (strong $*$ topology $|M$);
4) Each σ-continuous linear functional on M has the form $\sum_{i=1}^{n} \langle \cdot \xi_i, \eta_i \rangle$;
5) Each σ-continuous positive linear functional on M has the form $\sum_{i=1}^{n} \langle \cdot \xi_i, \xi_i \rangle$.

Proof. Clearly, 1), 2) and 3) are equivalent; 4) and 5) are equivalent.

From 4), each σ-continuous functional on M is weakly continuous. Thus (weak topology $|M) \sim \sigma(M, M_*)$.

By Proposition 1.10.5, it is clear that 1) implies 4). Q.E.D.

Remark. The conditions in Proposition 1.10.6 are possible. For example, see Sections 1.13 and 2.11.

References.[32], [127], [147], [159].

1.11. The equivalence of the topologies s^* and τ in a bounded ball

Let M be a VN algebra on a Hilbert space H, $(M)_1 = \{x \in M \mid \|x\| \le 1\}$ be its unit ball. In this section, we shall prove that $s^*(M, M_*) \sim \tau(MM_*)$ in any bounded ball of M.

Lemma 1.11.1. Let $\{a_n^* = a_n\}$ be in $(M)_1$ and $a_n \to 0$ (strongly). Then for any $\delta > 0$ there is a sequence $\{p_n\}$ of projections of M such that

$$p_n \to 1(\text{strongly}), \quad \text{and} \quad \|a_n p_n\| \le \delta, \forall n.$$

Proof. Let $a_n = \int_1^1 \lambda de_\lambda^{(n)}$ be the spectral decomposition of a_n, and define

$$p_n = \int_{-\delta}^{\delta} de_\lambda^{(n)}, \quad q_n = 1 - p_n, \quad \forall n.$$

Then

$$\delta^{-2} a_n^2 \ge \left(\int_{-1}^{-\delta} + \int_{\delta}^{1}\right) \frac{\lambda^2}{\delta^2} de_\lambda^{(n)} \ge q_n, \forall n.$$

Since $a_n^2 \to 0$ (weakly), it follows that $q_n \to 0$ (strongly), $p_n \to 1$ (strongly). Moreover, $\|a_n p_n\| = \|\int_{-\delta}^{\delta} \lambda de_\lambda^{(n)}\| \le \delta, \forall n$. So the sequence $\{p_n\}$ is just what we want to find. Q.E.D.

Lemma 1.11.2. Let φ be a faithful normal positive linear functional on M, and define

$$d(a, b) = \varphi((a-b)^*(a-b))^{1/2}, \quad \forall a, b \in (M)_1.$$

Then $((M)_1, d)$ is a complete metric space, and the topology generated by d is equivalent to $s(M, M_*)$ in $(M)_1$.

Proof. By Propositions 1.2.2 and 1.10.5, there is a sequence $\{\xi_n\} \subset H$ such that

$$\sum_n \|\xi_n\|^2 < \infty, \quad \varphi(a) = \sum_n \langle a\xi_n, \xi_n \rangle, \forall a \in M.$$

Thus $d(a, b) = (\sum_n \|(a-b)\xi_n\|^2)^{1/2}$ is a metric on $(M)_1$.

We claim that $\overline{[M'\xi_n|n]}$ is dense in H. In fact, let p be projection from H onto $\overline{[M'\xi_n|n]}$. Then $p \in M$, and $p\xi_n = \xi_n, \forall n$. Since $\varphi(1-p) = \sum_n \langle (1-p)\xi_n, \xi_n \rangle = 0$ and φ is faithful, it follows that $p = 1$, i.e., $[M'\xi_n|n]$ is dense in H.

From this claim, our lemma is easy. Q.E.D.

Lemma 1.11.3. Let φ_k, φ_0 be in M_*, and $\varphi_k \to \varphi_0(\sigma(M_*, M))$. In addition, suppose that $\{a_n\}$ is a sequence of $(M)_1$ and $a_n \to 0(s^*(M, M_*))$. Then $\lim_n \varphi_k(a_n) = 0$ uniformly for k.

Proof. Obviously, $\{\|\varphi_k\| \mid k\}$ is bounded. So we may assume that $\|\varphi_k\| \leq 1/2, \forall k$. By Theorem 1.9.8, for any k we can write

$$\varphi_k = (\varphi_k^{(1)} - \varphi_k^{(2)}) + i(\varphi_k^{(3)} - \varphi_k^{(4)}),$$

where $0 \leq \varphi_k^{(j)} \in M_*, 1 \leq j \leq 4$, and

$$\frac{1}{2} \geq \|\varphi_k^{(1)}\| + \|\varphi_k^{(2)}\| = \|\varphi_k^{(1)} - \varphi_k^{(2)}\|, \quad \varphi_k^{(1)} - \varphi_k^{(2)} = \frac{1}{2}(\varphi_k + \varphi_k^*),$$
$$\frac{1}{2} \geq \|\varphi^{(3)}\| + \|\varphi_k^{(4)}\| = \|\varphi_k^{(3)} - \varphi_k^{(4)}\|, \quad \varphi_k^{(3)} - \varphi_k^{(4)} = \frac{1}{2i}(\varphi_k - \varphi_k^*).$$

Clearly, $\|[\varphi_k]\| \leq 1$ and $0 \leq \varphi \in M_*$, where $[\varphi_k] = \sum_{j=1}^{4} \varphi_k^{(j)}, \forall k$, and $\varphi = \sum_{k=1}^{\infty} \frac{1}{2^k}[\varphi_k]$. Let $p = s(\varphi)$. Then $\varphi_k^{(j)}(1-p) = 0$, and $\varphi_k^{(j)}(a) = \varphi_k^{(j)}(pap), \forall k, j$ and $a \in M$. Further,

$$\varphi_k(a) = \varphi_k(pap), \forall k, \text{ and } a \in M.$$

Clearly, $pa_np \to 0(s^*(M, M_*))$. Considering the problem to pMp, we may assume that $p = 1$, or φ is faithful.

Let d be the metric on $(M)_1$ as in Lemma 1.11.2. For any fixed $\varepsilon > 0$ and m, define

$$H_m = \{a \in (M)_1 \mid |\varphi_k(a) - \varphi_0(a)| \leq \varepsilon, \quad \forall k \geq m\}.$$

Clearly, H_m is a closed subset of $((M)_1, d)$. Since $\varphi_k \to \varphi_0(\sigma(M_*, M))$, it follows that

$$(M)_1 = \bigcup_{m=1}^{\infty} H_m.$$

Now by the Baire category theorem and Lemma 1.11.2, there is a $a_0 \in (M)_1$ and a $\mu > 0$ and a m_0 such that

$$\{b \in (M)_1 \mid d(a_0, b) \leq \mu\} \subset H_{m_0}.$$

Since $\{\frac{1}{2}(a_n + a_n^*)\}_n$ and $\{\frac{1}{2i}(a_n - a_n^*)\}_n \subset (M)_1$ and they converge to 0 with respect to $s(M, M_*)$, we may assume that $a_n = a_n^*, \forall n$. Now pick $\delta = \frac{1}{3}\varepsilon$. By Lemma 1.11.1, there is a sequence $\{p_n\}$ of projections of M such that

$$p_n \to 1(s(M, M_*)), \quad \|a_n p_n\| \leq \delta, \quad \forall n.$$

Let $\psi_k = \varphi_k - \varphi_0, \forall k$. Then

$$\begin{aligned}|\psi_k(a_n)| \leq\ & |\psi_k(p_n a_n p_n)| + |\psi_k((1-p_n)a_n p_n)| \\ & + |\psi_k(p_n a_n (1-p_n))| + |\psi_k((1-p_n)a_n(1-p_n))| \qquad (1) \\ \leq\ & 3\delta + |\psi_k((1-p_n)a_n(1-p_n))|, \quad \forall k, n.\end{aligned}$$

Let $b_n = p_n a_0 p_n + (1-p_n)a_n(1-p_n), \forall n$. Then $b_n \in (M)_1, \forall n$, and $b_n \to a_0(s(M, M_*))$. Thus there is n_1 such that $b_n \in H_{m_0}, \forall n \geq n_1$. By the definition of H_{m_0}, we have

$$|\psi_k(b_n)| \leq \varepsilon, \quad \forall k \geq m_0, \quad n \geq n_1. \qquad (2)$$

Since $p_n a_0 p_n \to a_0(s(M, M_*))$, it follows that there is n_2 such that $p_n a_0 p_n \in H_{m_0}, \forall n \geq n_2$. Thus

$$|\psi_k(p_n a_0 p_n)| \leq \varepsilon, \quad \forall k \geq m_0, n \geq n_2. \qquad (3)$$

Now if $n \geq n_1, n_2$, and $k \geq m_0$, by (1), (2), (3),

$$\begin{aligned}|\varphi_k(a_n) - \varphi_0(a_n)| =\ & |\psi_k(a_n)| \\ \leq\ & 3\delta + |\psi_k((1-p_n)a_n(1-p_n))| \\ \leq\ & 3\delta + |\psi_k(b_n)| + |\psi_k(p_n a_0 p_n)| \leq 3\varepsilon.\end{aligned}$$

This means that $\lim_n \varphi_k(a_n) = 0$ uniformly for k. Q.E.D.

Lemma 1.11.4. Let A be a $\sigma(M_*, M)$-compact subset of M_*. For any $\varepsilon > 0$ there exists a $\delta > 0$ and a finite subset F of A such that if $a \in (M)_1$ with $[\varphi](a^*a + aa^*) < \delta, \forall \varphi \in F$, then $|\varphi(a)| < \varepsilon, \forall \varphi \in A$. There $[\varphi] = \varphi_1 + \varphi_2 + \varphi_3 + \varphi_4$ and $(\varphi_1 - \varphi_2), (\varphi_3 - \varphi_4)$ are the orthogonal decompositions of $\frac{1}{2}(\varphi + \varphi^*), \frac{1}{2i}(\varphi - \varphi^*)$ respectively.

Proof. Suppose that our assertion is false for some $\varepsilon > 0$. Then for $\frac{1}{2}$ and any fixed $\varphi_0 \in A$, we have $a_1 \in (M)_1$ and $\varphi_1 \in A$ such that

$$[\varphi_0](a_1^* a_1 + a_1 a_1^*) < \frac{1}{2}, \quad |\varphi_1(a_1)| \geq \varepsilon.$$

For $\frac{1}{2^2}$ and $\{\varphi_0, \varphi_1\} \subset A$, we also have $a_2 \in (M)_1$ and $\varphi_2 \in A$ such that

$$[\varphi_i](a_2^* a_2 + a_2 a_2^*) < \frac{1}{2^2}, \quad i = 0, 1, \quad |\varphi_2(a_2)| \geq \varepsilon.$$

$\cdots\cdots$. Generally, we have $\{\varphi_n | n = 0, 1, \cdots\} \subset A, \{a_n | n = 1, 2, \cdots\} \subset (M)_1$ such that

$$\begin{cases} [\varphi_i](a_j^* a_j + a_j a_j^*) \leq \frac{1}{2^j}, & 0 \leq i \leq j-1, \\ |\varphi_j(a_j)| \geq \varepsilon, & j = 1, 2, \cdots \end{cases}$$

Since A is $\sigma(M_*,M)$-compact, by the Eberlein-Šmulian theorem there is a subsequence $\{n_k\} \subset \{0,1,\cdots\}$ and $\psi \in A$ such that $\varphi_{n_k} \to \psi(\sigma(M_*,M))$. Clearly, A is bounded. Thus $0 \leq \varphi = \sum_{k=1}^{\infty} 2^{-k}[\varphi_{n_k}] \in M_*$. Notice that

$$\begin{aligned}|\varphi(a_j^*a_j + a_ja_j^*)| \leq & \sum_{k=1}^{m} 2^{-k}|[\varphi_{n_k}](a_j^*a_j + a_ja_j^*)| \\ & + \sum_{k>m} 2^{-k+1}\|[\varphi_{n_k}]\| \\ < & \ 2^{-j} + 2^{-m+2}\sup_n \|\varphi_n\|,\end{aligned}$$

where m is an integer such that $n_1,\cdots,n_m \leq j-1$, and $n_{m+1} > j-1$. Since $m \to \infty$ as $j \to \infty$, it follows that

$$\varphi(a_j^*a_j + a_ja_j^*) \to 0, \quad \text{as} \quad j \to \infty.$$

Let $p = s(\varphi)$. By the Schwartz inequality, it is easy to see that

$$\varphi((pa_jp)^*(pa_jp)) \to 0, \quad \varphi((pa_jp)(pa_jp)^*) \to 0.$$

Since φ is faithful on pMp, by Lemma 1.11.2 we have

$$pa_jp \to 0(s^*(M,M_*)).$$

By Lemma 1.11.3 and $p \geq s([\varphi_{n_k}])$ $(\forall k)$, it follows that

$$\lim_j \varphi_{n_k}(pa_jp) = \lim_j \varphi_{n_k}(a_j) = 0$$

uniformly for k. Then we get a contradiction since $|\varphi_{n_k}(a_{n_k})| \geq \varepsilon$, $\forall k$.

Q.E.D.

Lemma 1.11.5. Let A be a $\sigma(M_*,M)$-compact subset of M_*. Then we have $\psi \in M_*, \psi \geq 0$ with the following property: for any $\varepsilon > 0$, there is $\delta > 0$ such that if $a \in (M)_1$ and $\psi(a^*a + aa^*) < \delta$, then $|\varphi(a)| < \varepsilon, \forall \varphi \in A$.

Proof. For $\frac{1}{n}$, by Lemma 1.11.4 there is a finite subset F_n of A and $\delta_n > 0$ such that if $a \in (M)_1$ and $|\varphi|(a^*a + aa^*) < \delta_n, \forall \varphi \in F_n$, then $|\varphi(a)| < \frac{1}{n}, \forall \varphi \in A$. Define

$$\psi = \sum_{n=1}^{\infty} 2^{-(n+m_n)} \sum_{\varphi \in F_n} |\varphi|.$$

where $m_n = {}^\sharp F_n$,and the definition of $|\varphi|$ is as in Lemma 1.11.4. For any $\varepsilon > 0$, pick $\frac{1}{n_0} < \varepsilon$, and let $\delta = \delta_{n_0}/2^{n_0+m_{n_0}}$. If $a \in (M)_1$ and $\psi(a^*a + aa^*) < \delta$, then

$$|\varphi|(a^*a + aa^*) < \delta_{n_0}, \quad \forall \varphi \in F_{n_0}.$$

By the definitions of F_{n_0} and δ_{n_0}, we have $|\varphi(a)| < \frac{1}{n_0} < \varepsilon, \forall \varphi \in A$.

Q.E.D.

Theorem 1.11.6 . Let M be a VN algebra. Then in any bounded ball of M, we have $\tau(M, M_*) \sim s^*(M, M_*)$.

Proof. Let $\{a_l\}$ be a net of $(M)_1$ with $a_l \to 0(s^*(M, M_*))$, and A be a $\sigma(M_*, M)$-compact subset of M_*. It suffices to show that $\varphi(a_l) \to 0$ uniformly for $\varphi \in A$.

For that A and any $\varepsilon > 0$, pick $0 \le \psi \in M_*$ and $\delta > 0$ as in Lemma 1.11.5. Since $a_l \to 0(s^*(M, M_*))$, it follows that there is an index l_0 such that

$$\psi(a_l^* a_l + a_l a_l^*) < \delta, \quad \forall l \ge l_0.$$

Then by Lemma 1.11.5, $|\varphi(a_l)| < \varepsilon, \forall l \ge l_0, \varphi \in A$. This means that $\varphi(a_l) \to 0$ uniformly for $\varphi \in A$.

Q.E.D.

Proposition 1.11.7. Let M be a VN algebra, and $\dim M = \infty$. Then in whole $M, \tau(M, M_*) \not\sim s^*(M, M_*)$. Moreover, in $(M)_1$, uniform topology $\not\sim \tau(M, M_*)$. In consequence, if a VN algebra M is reflexive as a Banach space, then $\dim M < \infty$.

Proof. Let M be infinite dimensional. First we claim that M contains an infinite orthogonal sequence $\{p_n\}$ of non-zero projections. In fact, if $\dim Z = \infty$, where Z is the center of M, then we can easily find such $\{p_n\} \subset Z$. If $\dim Z < \infty$, then we may assume that M is a factor. By the theory of factors (see Section 7.1), such $\{p_n\}$ exists.

Now using above $\{p_n\}$, similar to the proof of Proposition 1.2.5 we can see that $\tau(M, M_*)$ is not equivalent to $s^*(M, M_*)$ in whole M.

Again using above $\{p_n\}$, let $p = \sum_{n=1}^{\infty} p_n$. By Proposition 1.2.10 and Theorem 1.11.6, $\sum_{n=1}^{m} p_n \to p$ with respect to $\tau(M, M_*)$. On the other hand, we have

$$\|p - \sum_{n=1}^{m} p_n\| = 1, \quad \forall m.$$

Thus, in $(M)_1$ the uniform topology is not equivalent to $\tau(M, M_*)$.

If M is reflexive as a Banach space, then $\tau(M, M_*) \sim$ uniform topology. Therefore, M must be finite dimensional.

Q.E.D.

Remark. If A is a C^*-algebra, and A is reflexive as a Banach space, then $\dim A < \infty$. Indeed, A^{**} is a VN-algebra (see Section 2.11), and is reflexive as a Banach space.

Notes. It was S. Sakai who initiated the study of the Mackey topology of a Von Neumann algebra. He showed that the Mackey topology on any bounded ball of a finite Von Neumann algebra M agrees with $s^*(M, M_*)$. Theorem 1.11.6 was due to C. Akemann who gave an affirmative answer for Sakai conjecture. Further in Section 4.5, we shall study the characterizations of $\sigma(M_*, M)$-compact subsets of M_*.

References.[2], [146].

1.12. Normal $*$ homomorphisms

Definition 1.12.1. A $*$ (algebraic) homomorphism Φ from a VN algebra M into a VN algebra N is said to be *normal*, if for any bounded increasing net $\{a_l\} \subset M_+$ we have

$$\sup_l \Phi(a_l) = \Phi(\sup_l a_l).$$

Let Φ be a $*$ homomorphism from M to N. First, Φ will preserve the order, i.e., $\Phi(M_+) \subset N_+$. This is obvious because any element of M_+ has the form a^*a (for some $a \in M$). Secondly, $\|\Phi\| \leq 1$. In fact, $\Phi(1_M) = p$ is a projection of N. For any $h = h^* \in M$, since $-\|h\| \cdot 1_M \leq h \leq \|h\| \cdot 1_M$, it follows that $-\|h\|p \leq \Phi(h) \leq \|h\|p$. Thus $\|\Phi(h)\| \leq \|h\|$. Further, $\|\Phi(a)\|^2 = \|\Phi(a^*a)\| \leq \|a\|^2, \forall a \in M$. Hence $\|\Phi\| \leq 1$. From these facts, if $\{a_l\}$ is a bounded increasing net on M_+, then $\{\Phi(a_l)\}$ is also a bounded increasing net of N_+. Now by Proposition 1.2.10, $\sup_l a_l$ and $\sup_l \Phi(a_l)$ exist.

Proposition 1.12.2. Let Φ be a $*$ homomorphism from a VN algebra M to a VN algebra N. Then the following statements are equivalent:

1) Φ is σ-σ continuous;

2) Φ is normal;

3) Φ is *completely additive*, i.e., $\Phi(\sum_l p_l) = \sum_l \Phi(p_l)$ for any orthogonal family $\{p_l\}$ of projections of M.

Moreover, $\Phi(M)$ is a σ-closed $*$ subalgebra of N if Φ is normal.

Proof. By Proposition 1.2.10, we can easily see that 1) implies 2). Moreover, it is clear that 2) implies 3). Now let Φ be completely additive. Then for any $0 \leq \varphi \in N_*, \varphi \circ \Phi$ is a completely additive positive functional on M, and by Theorem 1.8.6, $\varphi \circ \Phi \in M_*$. Further, by Corollary 1.9.9, $\varphi \circ \Phi \in M_*, \forall \varphi \in N_*$. Therefore Φ is σ-σ continuous.

Now if Φ is normal and $N \subset B(K)$, let $I = \{a \in M \mid \Phi(a) = 0\}$, then I is a σ-closed two-sided ideal of M. Thus there is a central projection z of M such that $I = M(1-z)$, and Φ is a $*$ isomorphism from Mz into $B(K)$. Similar to the proof of Proposition 1.8.13, the unit ball of $\Phi(M)$ is weakly closed. Therefore $\Phi(M)$ is σ-closed. Q.E.D.

Proposition 1.12.3. Let Φ be a $*$ isomorphism from a VN algebra M onto a VN algebra N. Then Φ is normal and isometric.

Proof. Let $\{a_l\}$ be a bounded increasing net of M_+, and $a = \sup_l a_l$. Then $b = \sup_l \Phi(a_l) \leq \Phi(a)$. Moreover, since Φ^{-1} is a $*$ isomorphism from N onto M, it follows that

$$\sup_l \Phi^{-1}(\Phi(a_l)) = a \leq \Phi^{-1}(b).$$

Thus $b = \Phi(a)$, i.e., Φ is normal. Another conclusion is clear. Q.E.D.

Theorem 1.12.4. Let M, N be two VN algebras on Hilbert spaces H, K respectively, Φ be a normal $*$ homomorphism from M onto N. Then

$$\Phi = \Phi_3 \circ \Phi_2 \circ \Phi_1,$$

where Φ_1 is an *ampliation* of M, i.e., there exists a Hilbert space L such that $\Phi_1(a) = a \otimes 1_L, \forall a \in M$; Φ_2 is an *induction* of $M \otimes \otimes \mathbb{C} 1_L$, i.e., there is a projection p' of $(M \overline{\otimes} \mathbb{C} 1_L)'$ such that $\Phi_2(a \otimes 1_L) = (a \otimes 1_L)p', \forall a \in M$; Φ_3 is a spatial $*$ isomorphism from $(M \overline{\otimes} \mathbb{C} 1_L)p'$ (a VN algebra on $p'(H \otimes L)$) onto N.

Proof. First we assume that N admits a cyclic vector η. Define

$$\varphi(a) = \langle \Phi(a)\eta, \eta \rangle, \quad \forall a \in M.$$

Obviously, φ is a normal positive functional on M. By Proposition 1.10.5, there is a sequence $\{\xi_n\} \subset H$ with $\sum_n \|\xi_n\|^2 < \infty$ such that $\varphi(a) = \sum_n \langle a\xi_n, \xi_n \rangle$, $\forall a \in M$. Let $L = l^2, \xi = (\xi_n) \subset H \otimes L, \Phi_1(a) = a \otimes 1_L, \forall a \in M$. Then

$$\varphi(a) = \langle \Phi_1(a)\xi, \xi \rangle, \quad \forall a \in M.$$

Suppose that p' is the projection from $H \otimes L$ onto $\overline{\Phi_1(M)\xi}$. Then

$$p' \in \Phi_1(M)' = (M \overline{\otimes} \mathbb{C} 1_L)'.$$

Let $\Phi_2(a \otimes 1_L) = (a \otimes 1_L)p', \forall a \in M$. Clearly,

$$\varphi(a) = \langle (\Phi_2 \circ \Phi_1)(a)\xi, \xi \rangle, \quad \forall a \in M.$$

Define a linear map u from $\Phi(M)\eta$ to $p'(H\otimes L)$ as follows:

$$u\Phi(a)\eta=(\Phi_2\circ\Phi_1)(a)\xi=p'(a\xi_n)=(a\xi_n),$$

$\forall a\in M$. Since $\langle\Phi(a)\eta,\eta\rangle=\varphi(a)=\langle(\Phi_2\circ\Phi_1)(a)\xi,\xi\rangle,\forall a\in M$, it follows that u is isometric. Moreover, since $\Phi(M)\eta=N\eta$ and $(\Phi_2\circ\Phi_1)(M)\xi=\Phi_1(M)\xi$ are dense in K and $p'(H\otimes L)$ respectively, u can be extended to a unitary operator from K onto $p'(H\otimes L)$. Clearly,

$$u\Phi(a)u^{-1}=\Phi_2\circ\Phi_1(a),\quad \forall a\in M.$$

Then we can define a spatial $*$ isomorphism Φ_3 by the operator u, and $\Phi=\Phi_3\circ\Phi_2\circ\Phi_1$.

For general case, write

$$K=\sum_l\oplus K_l,\quad K_l=\overline{N\eta_l},\forall l.$$

Let q_l' be the projection from K onto K_l; then $q_l'\in N',\forall l$. For each $l,\Phi_l=q_l'\Phi$ is a normal $*$ homomorphism from M onto $N_l=Nq_l'$. By the above argument $\Phi_l=\Phi_3^{(l)}\circ\Phi_2^{(l)}\circ\Phi_1^{(l)},\forall l$. Define $\Phi_i=\sum_l\oplus\Phi_i^{(l)},i=1,2,3$. Then $\Phi=\Phi_3\circ\Phi_2\circ\Phi_1$, and Φ_1,Φ_2,Φ_3 satisfy our conditions. Q.E.D.

Proposition 1.12.5. Let Φ be a $*$ isomorphism from a VN algebra M onto a VN algebra N. Then there exists a VN algebra V and two projections $p',q'\in V'$ with $c(p')=c(q')=1$ such that M,N are spatially $*$ isomorphic to Vq',Vp' respectively, and Φ corresponds to the $*$ isomorphism: $vq'\to vp'(\forall v\in V)$.

Proof. Keep the notations in Theorem 1.12.4, and let $V=M\overline{\otimes}\mathbb{C}1_L$. Since Φ,Φ_1,Φ_3 are $*$ isomorphisms, it follows that Φ_2 is also a $*$ isomorphism. By Proposition 1.5.10, the central cover of p' in V' is 1. Further from Theorem 1.12.4, N is spatial $*$ isomorphic to Vp'.

Let $q'=1_H\otimes q$, where q is a rank one projection on L. Then $q'\in V'$. Since

$$\overline{V'q'(H\otimes L)}\supset H\otimes B(L)qL,$$

it follows that the central cover of q' in V' is 1. Clearly, M is spatially $*$ isomorphic to Vq', and Φ corresponds to the $*$ isomorphism: $vq'\to vp'(\forall v\in V)$. Q.E.D.

Theorem 1.12.6. Let M_i,N_i be the VN algebras on the Hilbert spaces H_i,K_i respectively, Φ_i be a normal $*$ homomorhism from M_i onto $N_i,i=1,2$. Then there exists unique normal $*$ homomorphism Φ from $M_1\overline{\otimes}M_2$ onto $N_1\overline{\otimes}N_2$ such that

$$\Phi(a_1\otimes a_2)=\Phi_1(a_1)\otimes\Phi_2(a_2),\quad \forall a_1\in M_1,a_2\in M_2.$$

Moreover, if Φ_1 and Φ_2 are $*$ isomorphisms (or spatial $*$ isomorphisms),then Φ is also a $*$ isomiphism (or spatial $*$ isomorphism).

Proof. By Theorem 1.12.4, we can write

$$\Phi_i(a_i) = u_i(a_i \otimes 1_i)p_i' u_i^{-1}, \quad \forall a_i \in M_i,$$

where 1_i is the identity operator on a Hilbert space $L_i, p_i' \in (M_i \overline{\otimes} \mathbb{C} 1_i)', u_i$ is a unitary operator from $p_i'(H_i \otimes L_i)$ onto $K_i, i = 1,2$. Let $L = L_1 \otimes L_2$. Then

$$p' = p_1' \otimes p_2' \in (M_1 \overline{\otimes} \mathbb{C} 1_1)' \overline{\otimes} (M_2 \overline{\otimes} \mathbb{C} 1_2)' = (M_1 \overline{\otimes} M_2 \overline{\otimes} \mathbb{C} 1_L)'$$

and $u = u_1 \otimes u_2$ is a unitary operator from $p_1'(H_1 \otimes L_1) \otimes p_2'(H_2 \otimes L_2) = p'(H_1 \otimes H_2 \otimes L)$ onto $K_1 \otimes K_2$. Now define

$$\Phi(a) = u(a \otimes 1_L)p' u^{-1}, \quad \forall a \in M_1 \overline{\otimes} M_2.$$

Clearly. Φ is a normal $*$ homomorphism from $M_1 \overline{\otimes} M_2$ onto $u(M_1 \overline{\otimes} M_2 \overline{\otimes} \mathbb{C} 1_L)p'$ u^{-1} such that

$$\Phi(a_1 \otimes a_2) = \Phi_1(a_1) \otimes \Phi_2(a_2), \quad \forall a_1 \in M_1, \quad a_2 \in M_2.$$

Since Φ is normal and $M_1 \overline{\otimes} M_2$ is generated by $\{a_1 \otimes a_2 \mid a_1 \in M_1, a_2 \in M_2\}$, it follows that $u(M_1 \overline{\otimes} M_2 \otimes \mathbb{C} 1_L)p' u^{-1}$ will be generated by $\{\Phi_1(a_1) \otimes \Phi_2(a_2) | a_1 \in M_1, a_2 \in M_2\}$. Therefore,

$$N_1 \overline{\otimes} N_2 = u(M_1 \overline{\otimes} M_2 \otimes \mathbb{C} 1_L)p' u^{-1},$$

i.e., Φ is the normal $*$ homomorphism from $M_1 \overline{\otimes} M_2$ onto $N_1 \overline{\otimes} N_2$ which we want to find. From the normality, obviously Φ is unique.

Now suppose that Φ_i is a $*$ isomorphism. Then $c(p_i') = 1$ by the proof of Proposition 1.12.5, $i = 1,2$. Further $c(p') = 1$. By Proposition 1.5.10, Φ is also a $*$ isomorphism. Q.E.D.

References. [28], [109], [113], [185].

1.13 Comparison of cyclic projections and spatial $*$ isomorphic theorem

Definition 1.13.1. Let M be a VN algebra on a Hilbert space H. For each $\xi \in H$, we denote by p_ξ (resp. p_ξ') the projection from H onto $\overline{M'\xi}$ (resp. $\overline{M\xi}$), and call it the *cyclic projection* of M (resp. M') determined by ξ.

Theorem 1.13.2. Let M be a VN algebra on a Hilbert space $H, \xi, \eta \in H$. Then the relations $p_\eta \precsim p_\xi$ (in M) and $p'_\eta \precsim p'_\xi$ (in M') are equivalent.

Proof. Let $p'_\eta \precsim p'_\xi$ (in M'), i.e. there is a $u' \in M'$ such that

$$u'^* u' = p'_\eta, \qquad u'u'^* \le p'_\xi.$$

Clearly, $u'u'^* = p'_{u'\eta}, p_\eta = p_{u'\eta}$. Replacing η by $u'\eta$, we may assume that $\eta \in \overline{M\xi}$.

Let $\varphi(a) = \langle a\eta, \xi\rangle, \forall a \in M$, and $\varphi = R_v\omega$ be the polar decomposition of φ. Then $\omega = R_{v^*}\varphi, \varphi = R_{vv^*}\varphi$, and $(\eta - vv^*\eta) \in (M\xi)^\perp$. On the other hand, $\eta \in \overline{M\xi}, (\eta - vv^*\eta) \in \overline{M\xi}$. Thus

$$\eta = vv^*\eta. \tag{1}$$

Since $p_\xi\xi = \xi$ and $\omega = R_{v^*}\varphi$, it follows that $\omega = L_{p_\xi}\omega$. In addition $\omega \ge 0$, so $\omega = R_{p_\xi}\omega$, i.e.,

$$\langle v^*\eta, a\xi\rangle = \langle p_\xi v^*\eta, a\xi\rangle, \quad \forall a \in M.$$

Thus $(v^*\eta - p_\xi v^*\eta) \in (M\xi)^\perp$. But $\eta \in \overline{M\xi}, (v^*\eta - p_\xi v^*\eta) \in \overline{M\xi}$, so we have

$$v^*\eta \in p_\xi H = \overline{M'\xi}. \tag{2}$$

By (2), $v^*\overline{M'\eta} = \overline{M'v^*\eta} \subset \overline{M'\xi}$; by (1), $vv^*\overline{M'\eta} = \overline{M'\eta}$. It follows that v^*p_η is a partial isometry from $\overline{M'\eta}$ into $\overline{M'\xi}$. Therefore, $p_\eta \precsim p_\xi$ (in M).

Similarly, we can prove the converse. Q.E.D.

Definition 1.13.3. Let M be a VN algebra on a Hilbert space H. A vector ξ of H is said to be *cyclic* for M, if $\overline{M\xi} = H$ (see Lemma 1.4.9).

A vector η of H is said to be *separating* for M, if $a \in M$ is such that $a\eta = 0$, then $a = 0$.

It is easy to see that a vector η of H is separating for M if and only if η is cyclic for M'.

Proposition 1.13.4. Let M be a VN algebra on a Hilbert space H, ξ be a cyclic vector for M, and η be a separating vector for M. Then there is a vector ς of H such that ς is cyclic and separating for M.

Proof. Since $p'_\xi = 1 \ge p'_\eta$, it follows from Theorem 1.13.2 that $p_\eta \precsim p_\xi$ (in M). But $p_\eta = 1$, so by Proposition 1.5.3 we have $p_\xi \sim p_\eta$, i.e., there exists $v \in M$ such that $v^*v = p_\xi, vv^* = p_\eta = 1$. Let $\varsigma = v\xi$. Then $\overline{M'\varsigma} = v\overline{M'\xi} = H$, and $\overline{M\varsigma} \supset \overline{Mv^*v\xi} = \overline{M\xi} = H$, i.e., $\overline{M\varsigma} = H$. Therefore, ς is cyclic and separating for M. Q.E.D.

Theorem 1.13.5. Let M_i be a VN algebra on a Hilbert space $H_i, i = 1, 2$. Suppose that M_1 and M_2 admit a cyclic and separating vector. Then every $*$ isomorphism Φ from M_1 onto M_2 is spatial.

Proof. By Proposition 1.12.3, Φ is normal. Thus

$$M = \{a \oplus \Phi(a) \mid a \in M_1\}$$

is a VN algebra on $H_1 \oplus H_2$. Let p_i' be the projection from $(H_1 \oplus H_2)$ onto $H_i, i = 1, 2$. Clearly,

$$p_1', p_2' \in M', \quad M_{p_1'} = M_1, \quad M_{p_2'} = \Phi(M_1) = M_2.$$

Now by Proposition 1.5.2, it suffices to show that $p_1' \sim p_2'$ in M'.

Let $\xi_i (\in H_i)$ be the cyclic-separating vector for $M_i, i = 1, 2$. Then $p_i'(H_1 \oplus H_2) = H_i = \overline{M_i \xi_i} = \overline{M \xi_i}, i = 1, 2$. From Theorem 1.13.2, it suffices to show that $p_1 \sim p_2$ in M, where p_i is the projection from $(H_1 \oplus H_2)$ onto $\overline{M' \xi_i}, i = 1, 2$.

Now we prove a more stronger result: $p_1 = p_2 = 1$. In fact, since

$$\{a_1' \oplus a_2' \mid a_i' \in M_i', i = 1, 2\} \subset M', \quad \overline{M_i' \xi_i} = H_i, i = 1, 2,$$

it follows that $p_i \geq p_i', (1 - p_i) p_i' = 0, i = 1, 2$. From $(1 - p_i) \in M, i = 1, 2$, and the definition of M, there are $a, b \in M_1$ such that

$$a \oplus \Phi(a) = 1 - p_1, \quad b \oplus \Phi(b) = 1 - p_2.$$

Therefore, $a = \Phi(b) = 0$. Further, $a = b = 0$ and $p_1 = p_2 = 1$. Q.E.D.

Proposition 1.13.6. Let M be a VN algebra on a HIlbert space $H, \xi (\in H)$ be a separating vector for M. Then for any normal positive linear functional φ on M, there is $\eta \in \overline{M \xi}$ such that $\varphi(a) = \langle a\eta, \eta \rangle, \forall a \in M$. In consequence, we have $\sigma(M, M_*) \sim$ (weak topology $|M$), $s(M, M_*) \sim$ (strong topology $|M$), $s^*(M, M_*) \sim$ (strong $*$ topology $|M$).

Proof. Let $\{\pi_\varphi, H_\varphi\}$ be the $*$ representation of M generated by $\varphi, \widetilde{H} = H \oplus H_\varphi$, and $\widetilde{M} = \{\tilde{a} = a \oplus \pi_\varphi(a) \mid a \in M\}$. By Proposition 1.8.13, $\widetilde{M}$ is a VN algebra on $\widetilde{H}$. Put

$$\tilde{\varphi}(\tilde{a}) = \langle \tilde{a} \tilde{1}_\varphi, \tilde{1}_\varphi \rangle = \langle \pi_\varphi(a) 1_\varphi, 1_\varphi \rangle = \varphi(a), \quad \forall a \in M,$$

where $\tilde{a} = a \oplus \pi_\varphi(a), \tilde{1}_\varphi = (0, 1_\varphi)$. Let $\tilde{\xi} = (\xi, 0) \in \widetilde{H}$. Obviously, $\tilde{\xi}$ is a separating vector for $\widetilde{M}$, so $\overline{\widetilde{M}' \tilde{\xi}} = \widetilde{H} \supset \overline{\widetilde{M}' \tilde{1}_\varphi}$. By Theorem 1.13.2, $p'_{\tilde{1}_\varphi} \preceq p'_{\tilde{\xi}}$ in $\widetilde{M}'$. Since $p'_{\tilde{1}_\varphi} \widetilde{H} = \overline{\widetilde{M} \tilde{1}_\varphi} = \overline{\pi_\varphi(M) 1_\varphi} = H_\varphi$, it follows that $p'_{\tilde{1}_\varphi} = p'_\varphi$ is the projection from $\widetilde{H}$ onto H_φ. Then there exists $v' \in \widetilde{M}'$ such that $v'^*v' = p'_\varphi$ and $v'v'^* \le p'_{\tilde{\xi}}$. Noticing that $p'_{\tilde{\xi}}\widetilde{H} = \overline{\widetilde{M}\tilde{\xi}} = (\overline{M\xi}, 0)$, we can write

$$v'\tilde{1}_\varphi = (\eta, 0) = \tilde{\eta}$$

for some $\eta \in \overline{M\xi}$. Further,

$$\begin{aligned}\varphi(a) = &\ \langle \tilde{a}\tilde{1}_\varphi, \tilde{1}_\varphi\rangle = \langle \tilde{a}\tilde{1}_\varphi, v'^*v'\tilde{1}_\varphi\rangle \\ = &\ \langle \tilde{a}v'\tilde{1}_\varphi, v'\tilde{1}_\varphi\rangle = \langle \tilde{a}\tilde{\eta}, \tilde{\eta}\rangle = \langle a\eta, \eta\rangle, \quad \forall a \in M.\end{aligned}$$

Q.E.D.

Corollary 1.13.7. Let M be a VN algebra on a Hilbert space H. If M admits a separating vector, then for any $\varphi \in M_*$ there are $\xi, \eta \in H$ such that $\varphi(a) = \langle a\xi, \eta\rangle, \forall a \in M$.

Proof. It is obvious by the polar decomposition of φ. Q.E.D.

Notes. The main results (1.13.2, 1.13.5 and 1.13.6) in this section are the so-called *spatial theory* of Von Neumann algebras. It is due to F. Murray and J. Von Neumann. B.J. Vowden once gave another proof. The present proof here is taken from R. Herman and M. Takesaki.

References.[69], [111], [177], [192].

1.14. σ-Finite Von Neumann algebras

Definition 1.14.1. A VN algebra M is said to be *σ-finite* (or *countably decomposable*), if every family of non-zero pairwise orthogonal projections of M must be countable.

Proposition 1.14.2. Let M be a VN algebra on a Hilbert space H. Then the following statements are equivalent:

1) M is σ-finite;

2) M admits a *separating sequence of vectors* $\{\xi_n\}(\subset H)$, i.e. if $a \in M$ with $a\xi_n = 0, \forall n$, then $a = 0$;

3) M' admits a *cyclic sequence of vectors* $\{\eta_n\}(\subset H)$, i.e., $[a'\eta_n | a' \in M', n]$ is dense in H;

4) There is a faithful normal positive linear functional on M.

Proof. Since a sequence $\{\xi_n\}$ is separating for M if and only if $\{\xi_n\}$ is cyclic for M', it follows that 2) and 3) are equivalent.

Now let $\{\xi_n\}$ be separating for M, and $\{p_l\}_{l\in\Lambda}$ be an orthogonal family of projections of M. Since $\sum_{l\in\Lambda}\|p_l\xi_n\|^2 \leq \|\xi_n\|^2, \forall n$, it follows that there is a countable subset J of Λ such that $p_l\xi_n = 0, \forall n$ and $l \notin J$. From $\{\xi_n\}$ is separating for M, then $p_l = 0, \forall l \notin J$. Thus M is σ-finite.

Suppose that M is σ-finite. Write

$$H = \sum_l \oplus H_l, \quad H_l = \overline{M'\eta_l} = p_l H, \quad \forall l.$$

Clearly, $\{l | p_l \neq 0\}$ is countable. Thus $H = \sum_n \oplus \overline{M'\eta_n}$, and the sequence $\{\eta_n\}$ is cyclic for M'. Therefore 1), 2) and 3) are equivalent.

Suppose that there is a faithful normal positive linear functional φ on M. Let $\{p_l\}_{l\in\Lambda}$ be any orthogonal family of projections of M. Since $\sum_{l\in\Lambda}\varphi(p_l) = \varphi(\sum_{l\in\Lambda} p_l) < \infty$, it follows that there is a countable subset J of Λ such that $\varphi(p_l) = 0, \forall l \notin J$. Then $p_l = 0$, $\forall l \notin J$, i.e., M is σ-finite.

Finally, if M is σ-finite, then M admits a separating sequence $\{\xi_n\}$ of vectors. We may assume that $\sum_n \|\xi_n\|^2 < \infty$, and let

$$\varphi(a) = \sum_n \langle a\xi_n, \xi_n\rangle, \quad \forall a \in M.$$

Clearly, φ will be a faithful normal positive linear functional on M.

Q.E.D.

Proposition 1.14.3. Let M be a VN algebra on a Hilbert space H. If M_* is separable, then M is σ-finite. In particular, if H is separable, then M is σ-finite.

Proof. Let $\{\varphi_n\}$ be a countable dense subset of $\{\psi \in M_* \mid \psi \geq 0, \|\psi\| \leq 1\}$, and $\varphi = \sum_n 2^{-n}\varphi_n$. Clearly, $0 \leq \varphi \in M_*$. If $a \in M_+$ is such that $\varphi(a) = 0$, then $\varphi_n(a) = 0, \forall n$. Further, $\psi(a) = 0, \forall 0 \leq \psi \in M_*$. By Corollary 1.9.9, $a = 0$. Therefore φ is faithful, and M is σ-finite.

Now let H be separable. We prove that M_* is separable. In fact, for a countable dense subset $\{\xi_n\}$ of H, Let $\omega_{nm}(a) = \langle a\xi_n, \xi_m\rangle, \forall a \in M, n, m$. If $\{\omega_{nm} \mid n, m\}$ is not dense in M_*, then there exists $0 \neq a \in M$ such that $\omega_{nm}(a) = 0, \forall n, m$, i.e., $\langle a\xi_n, \xi_m\rangle = 0, \forall n, m$. Since $\{\xi_n\}$ is dense in H, it follows that $a = 0$. This is a contradiction. So $\{\omega_{nm} \mid n, m\}$ is dense in M_*.

Q.E.D.

An algebra A is called *abelian* or *commutative*, if $ab = ba, \forall a, b \in A$.

Proposition 1.14.4. Let M be a σ-finite abelian VN algebra on a Hilbert space H. Then M admits a separating vector.

Proof. From the proof of 1.14.2, we can write

$$H = \sum_n \oplus H_n, \quad H_n = p_n H = \overline{M'\eta_n}, \forall n.$$

Then $\{\eta_n\}$ is a separating sequence of vectors for M. We may assume that $\sum_n \|\eta_n\|^2 < \infty$. Put $\eta = \sum_n \eta_n$. Since M is abelian, it is easily verified that η is a separating vector for M. Q.E.D.

Proposition 1.14.5. Let M be a σ-finite VN algebra, and $(M)_1$ be its unit ball. Then $((M)_1, s(M, M_*))$ and $((M)_1, \tau(M, M_*))$ are metrizable and can become two complete metric spaces.

Proof. By the proof of Proposition 1.14.2, $\varphi(\cdot) = \sum_n \langle \cdot \xi_n, \xi_n \rangle$ is a faithful normal positive linear functional on M, where $\{\xi_n\}$ is a separating sequence of vectors for M with $\sum_n \|\xi_n\|^2 < \infty$. From Lemma 1.11.2, $((M_1), d_s)$ is a complete metric space where $d_s(a, b) = \varphi((a - b)^*(a - b))^{1/2} = (\sum_n \|(a - b)\xi_n\|^2)^{1/2}, \forall a, b \in (M)_1$, and the topology generated by d_s is equivalent to $s(M, M_*)$ in $(M)_1$. Moreover, put

$$d_\tau(a, b) = \{\sum_n \|(a - b)\xi_n\|^2 + \sum_n \|(a - b)^*\xi_n\|^2\}^{1/2}, \quad \forall a, b \in (M)_1.$$

Clearly, $((M)_1, d_\tau)$ is a complete metric space, and the topology generated by d_τ is equivalent to $s^*(M, M_*) \sim \tau(M, M_*)$ in $(M)_1$. Q.E.D.

Definition 1.14.6. Let$M$ be a VN algebra on a HIlbert space H, p be a projection of M. p is said to be *σ-finite*, if $M_p = pMp$ is a σ-finite VN algebra on pH.

Proposition 1.14.7. Let M be a VN algebra on a Hilbert space H.

1) If φ is a normal positive linear functional on M, then $s(\varphi)$ is a σ-finite projection.

2) Let p, q be two projections of M, and $p \sim q$. If p is σ-finite, then q is also σ-finite.

3) If $\{p_n\}$ is a sequence of σ-finite projections of M, then $p = \sup_n p_n$ is also a σ-finite projection.

Proof. 1) It is obvious since φ is faithful on $s(\varphi)Ms(\varphi)$.

2) Since M_p and M_q are $*$ isomorphic and M_p is σ-finite, it follows that M_q is σ-finite, or q is σ-finite.

3) For each n, if $\{\xi_k^{(n)}\}_k(\subset p_nH)$ is a cyclic sequence of vectors for $M'p_n$, then $\{\xi_k^{(n)} \mid n,k\}$ will be a cyclic sequence of vectors for $M'p$. Thus p is σ-finite.

Q.E.D.

Proposition 1.14.8. Let p,q be σ-finite projections of VN algebra M,N respectively. Then $p\otimes q$ is a σ-finite projection of $M\overline{\otimes}N$.

Proof. Since $p\otimes q(M\overline{\otimes}N)p\overline{\otimes}q = pMp\overline{\otimes}qNq$, it suffices to show that the tensor product of two σ-finite VN algebra is σ-finite.

Let M_i be a σ-finite VN algebra on a Hilbert space $H_i, i=1,2$. If $\{\xi_n^{(i)}\}(\subset H_i)$ is a cyclic sequence of vectors for $M_i', i=1,2$, then $\{\xi_n^{(1)}\otimes\xi_m^{(2)} \mid n,m\}$ will be a cyclic sequence of vectors for $M_1'\overline{\otimes}M_2' = (M_1\overline{\otimes}M_2)'$. Therefore $M_1\overline{\otimes}M_2$ is σ-finite.

Q.E.D.

Theorem 1.14.9. Let M be a VN algebra on a Hilbert space H. Then there is a decomposition $M = \sum_l \oplus M_l$. Here M_l is σ-finite or is spatially $*$ isomorphic to $N_l\overline{\otimes}B(K_l)$, where N_l is some σ-finite VN algebra and K_l is some Hilbert space, $\forall l$.

Proof. Let φ be a nor-zero normal positive linear functional on M, and $s(\varphi)$ be its support. Then $s(\varphi)$ is a σ-finite projection of M. Suppose that $\{p_l\}_{l\in\Lambda}$ is a maximal orthogonal family of projections of M such that $p_l \sim s(\varphi), \forall l$. Let $q = 1-\sum_{l\in\Lambda} p_l$. By Theorem 1.5.4, there is a central projection z of M such that

$$qz \preceq s(\varphi)z, \quad s(\varphi)(1-z) \preceq q(1-z).$$

Since the family $\{p_l\}_{l\in\Lambda}$ is maximal, it follows that $s(\varphi)z \neq 0$.

Assume that $v_l \in Mz$ is such that $v_l^*v_l = s(\varphi)z, v_lv_l^* = p_lz, \forall l\in\Lambda$. Pick $v_0\in Mz$ such that $v_0v_0^* = qz, v_0^*v_0 \leq s(\varphi)z$.

If $\Lambda = \{1,2,\cdots,\}$ is countable, define

$$\psi(a) = \sum_{n=0}^{\infty} 2^{-n}\varphi(v_n^*av_n), \quad \forall a\in Mz.$$

Then $0\leq\psi\in M_*$. If $a\in Mz$ is such that $\psi(a^*a)=0$, then

$$\varphi(v_n^*a^*av_n) = 0, \quad \forall n\geq 0.$$

By the definition of v_n, $v_n^* a^* a v_n \in s(\varphi)Ms(\varphi), \forall n \geq 0$. Thus $av_n = 0, \forall n \geq 0$. Further, $ap_l z = aqz = 0, \forall l \in \Lambda$, and $a = az = a(qz + \sum_{l \in \Lambda} p_l z) = 0$. Hence ψ is faithful on Mz, and Mz is σ-finite.

If Λ is not countable, we can write

$$\Lambda = \bigsqcup_{\beta \in \mathbb{I}} \Lambda_\beta, \quad \Lambda_\beta \cap \Lambda_{\beta'} = \emptyset, \quad \forall \beta \neq \beta' \in \mathbb{I},$$

where each Λ_β is countably infinite. Fix an index $\beta_0 \in \mathbb{I}$, and let

$$\gamma_{\beta_0} = q + \sum_{l \in \Lambda_{\beta_0}} p_l, \quad \gamma_\beta = \sum_{l \in \Lambda_\beta} p_l, \quad \forall \beta \neq \beta_0.$$

Then $\{\gamma_\beta z | \beta \in \mathbb{I}\}$ is an orthogonal family of projection with $\gamma_\beta z \sim \gamma_{\beta'} z, \forall \beta, \beta' \in \mathbb{I}$, and $\sum_{\beta \in \mathbb{I}} \gamma_\beta z = z$. By Theorem 1.5.6, Mz will be spatially $*$ isomorphic to $N \overline{\otimes} B(K)$, where K is a Hilbert space with $\dim K = {}^\# \mathbb{I}, N = \gamma_{\beta_0} M z \gamma_{\beta_0}$. Since Λ_{β_0} is countable, it follows from preceding paragraph that N is σ-finite.

Through above procedure, we find a non-zero central projection z of M such that Mz is σ-finite or is spatially $*$ isomorphic to $N \overline{\otimes} B(K)$, where N is σ-finite and K is some Hilbert space. Finally, by the Zorn lemma, the proof will be completed. Q.E.D.

Proposition 1.14.10. Let M be a VN algebra. Then there is an orthogonal family $\{p_l\}$ of σ-finite projections of M such that $\sum_l p_l = 1$.

Proof. Let $\{p_l\}$ be a maximal orthogonal family of σ-finite projections of M, and $p = \sum_l p_l$. If $(1-p) \neq 0$, then by Theorem 1.14.9 $(1-p)M(1-p)$ contains a non-zero σ-finite projection q. Clearly, q is still a σ-finite projection of M and $qp_l = 0, \forall l$. This is a contradiction. Therefore $\sum_l p_l = 1$. Q.E.D.

References.[18], [28], [82], [150].

Chapter 2
Fundamentals of C^*-Algebras

2.1. Definition and basic properties of C^*-algebras

Definition 2.1.1. Let A be a *Banach algebra* over $\mathbb{C}$. If A admits a map: $x \to x^* \in A$ with the following properties:

$$(\lambda x + \mu y)^* = \overline{\lambda} x^* + \overline{\mu} y^*, \quad (xy)^* = y^* x^*, \quad (x^*)^* = x,$$

$\forall x, y \in A, \lambda, \mu \in \mathbb{C}$, then A is called a *Banach $*$ algebra*, and the map: $x \to x^*$ is called the *$*$-operation* of A. If the $*$-operation of A satisfies the following additional condition:

$$\|x^* x\| = \|x\|^2,$$

then A is called a *C^*-algebra*.

If A is a C^*-algebra, it is easy to see that $\|x^*\| = \|x\|, \forall x \in A$. Thus the $*$-operation of A is continuous. In general, a C^*-algebra need not have an identity. If however, the C^*-algebra A has an identity (denoted by 1_A, or 1 simply if no confusion arises), then $1_A^* = 1_A, \|1_A\| = 1$. Let A be a C^*-algebra, E be a subset of A. B is called the C^*-subalgebra of A generated by E, if B is the smallest C^*-subalgebra of A containing E.

Let Ω be a locally compact Hausdorff space, $C_0^\infty(\Omega)$ be the set of all continuous functions on Ω vanishing at infinity. With the usual structure, $C_0^\infty(\Omega)$ is a C^*-algebra. The algebra $C_0^\infty(\Omega)$ has an identity if and only if Ω is compact. The C^*-algebra $C_0^\infty(\Omega)$ is *abelian (commutative)*.

Let H be a Hilbert space, A be a uniformly closed $*$ subalgebra of $B(H)$. Then A is a (*concrete*) C^*-algebra.

Later we shall show that: any (abstract) C^*-algebra is isometrically $*$ isomorphic to a concrete C^*-algebra.

Proposition 2.1.2. Let A be a C^*-algebra, and A has no identity. On $(A\dot{+}\mathbb{C})$, define

$$\|x+\lambda\| = \sup\{\|xy+\lambda y\| \mid y \in A, \|y\| \le 1\}$$

$\forall x \in A, \lambda \in \mathbb{C}$. Then $(A\dot{+}\mathbb{C})$ is a C^*-algebra with identity 1, and A with original norm is a C^*-subalgebra of $(A\dot{+}\mathbb{C})$.

Proof. It suffices to show that for any $x \in A, \lambda \in \mathbb{C}$,

$$\|(x+\lambda)^*(x+\lambda)\| = \|x+\lambda\|^2.$$

Let $0 < \mu < 1$. Then there is $y \in A, \|y\| \le 1$, such that

$$\begin{aligned} \mu^2\|x+\lambda\|^2 \le & \ \|xy+\lambda y\|^2 = \|y^*(x+\lambda)^*(x+\lambda)y\| \\ \le & \ \|(x+\lambda)^*(x+\lambda)\|. \end{aligned}$$

Again let $\mu \to 1-$, then we get

$$\|x+\lambda\|^2 \le \|(x+\lambda)(x+\lambda)\| \le \|(x+\lambda)^*\| \cdot \|x+\lambda\|. \tag{1}$$

Thus $\|x+\lambda\| \le \|(x+\lambda)^*\|$. Further $\|x+\lambda\| = \|(x+\lambda)^*\|$. From (1), $\|(x+\lambda)^*(x+\lambda)\| = \|x+\lambda\|^2$. Q.E.D.

Remark. If A has an identity e, on $(A\dot{+}\mathbb{C})$, define

$$\|x+\lambda\| = \max\{\|x+\lambda e\|, |\lambda|\}$$

$\forall x \in A, \lambda \in \mathbb{C}$. Then $(A\dot{+}\mathbb{C})$ is a C^*-algebra with a new identity 1, and A is its C^*-subalgebra.

Let A be a Banach algebra, $a \in A$. Denote the *spectrum* of a by $\sigma(a)$, and the *spectral radius* of a by $\nu(a)$.

Proposition 2.1.3. Let A be a C^*-algebra, h be a *self-adjoint* element of A (i.e. $h^* = h$). Then $\sigma(h) \subset \mathbb{R}$, and $\|h\| = \nu(h)$.

Proof. We may assume that A has an identity. Since $*$-operation is continuous, it follows that

$$(e^{ith})^* = e^{-ith}, \quad \forall t \in \mathbb{R}.$$

For any $\lambda \in \sigma(h)$, by

$$|e^{it\lambda}|^2 \le \|e^{ith}\|^2 = \|(e^{ith})^* \cdot e^{ith}\| = 1, \quad \forall t \in \mathbb{R}$$

Then $\overline{\lambda} = \lambda$. Thus $\sigma(h) \subset \mathbb{R}$. Moreover, by

$$\|h\| = \|h^*h\|^{1/2} = \|h^2\|^{1/2} = \cdots = \|h^{2^n}\|^{1/2^n} \to \nu(h),$$

so $\|h\| = \nu(h)$. Q.E.D.

Theorem 2.1.4. Let A be an abelian C^*-algebra, Ω be its spectral space. Then A is isometrically $*$ isomorphic to $C_0^\infty(\Omega)$. Moreover, A has an identity if and only if Ω is compact.

Proof. Replacing A by $(A\dot{+}\mathbb{C})$ if it is necessary, we may assume that A has an identity. Then Ω is compact. Let $x \to x(\cdot)$ be the *Gelfand transformation* from A into $C(\Omega)$. By Proposition 2.1.3, $x^*(t) = \overline{x(t)}, \forall x \in A, t \in \Omega$, and

$$\|x\|^2 = \|x^*x\| = \nu(x^*x) = \max_{t\in\Omega} |x^*x(t)| = \max_{t\in\Omega} |x(t)|^2, \forall x \in A.$$

Thus the Gelfand transformation is an isometrical $*$ isomorphism from A to $C(\Omega)$. Moreover, it is clear that the subalgebra $\{x(\cdot) \mid x \in A\}$ separates the points of Ω (i.e., for any $s \neq t \in \Omega$, there is $x \in A$ such that $x(s) \neq x(t)$). So $\{x(\cdot) \mid x \in A\} = C(\Omega)$ by the Stone-Weierstrass theorem. Q.E.D.

Lemma 2.1.5. Let A be a C^*-algebra with an identity, $h = h^* \in A$ and $0 \notin \sigma(h)$. Then there is a sequence $\{p_n(\cdot)\}$ of polynomials with a zero constant term such that $\|p_n(h) - h^{-1}\| \to 0$.

Proof. Let B be the C^*-subalgebra of A generated by $\{(h-\lambda), (h-\lambda)^{-1} \mid \lambda \notin \sigma(h)\}$. Then B is abelian and $1 \in B$. By Theorem 2.1.4, $B \cong C(\Omega)$, where Ω is the spectral space of B. Clearly, $\sigma_B(h) = \sigma(h) = \{h(t) \mid t \in \Omega\}$. From Proposition 2.1.3, $h(t) = \overline{h(t)}, \forall t \in \Omega$. Moreover, since $0 \notin \sigma(h)$, it follows that $\min_{t\in\Omega} |h(t)| = \varepsilon > 0$. Define a continuous function f on $[-\|h\|, \|h\|]$ such that

$$f(0) = 0, f(\lambda) = \lambda^{-1}, \quad \forall \lambda \in [-\|h\|, \|h\|] \backslash (-\varepsilon, \varepsilon),$$

and pick a sequence $\{p_n(\cdot)\}$ of polynomials with a zero constant term such that

$$\max\{|p_n(\lambda) - f(\lambda)| \mid -\|h\| \leq \lambda \leq \|h\|\} \to 0.$$

Then

$$\begin{aligned}\|p_n(h) - h^{-1}\| &= \max_{t\in\Omega} |p_n(h(t)) - h(t)^{-1}| \\ &\leq \max_{-\|h\|\leq\lambda\leq\|h\|} |p_n(\lambda) - f(\lambda)| \to 0.\end{aligned}$$

Q.E.D.

Proposition 2.1.6. Let A be a C^*-algebra with an identity 1, B be a C^*-subalgebra of A and $1 \in B$. Then for any $b \in B, \sigma_B(b) = \sigma_A(b)$.

Proof. By Lemma 2.1.5, $\sigma_B(b^*b) = \sigma_A(b^*b), \sigma_B(bb^*) = \sigma_A(bb^*)$. Thus if b is invertible in A, then b^*b and bb^* are invertible in B. So b has a left and a right inverses in B. Therefore, b is invertible in B. Q.E.D.

Proposition 2.1.7. Let A be a C^*-algebra with an identity 1, u be a *unitary* element of A (i.e., $u^*u = uu^* = 1$). Then $\sigma(u) \subset \{\lambda \in \mathbb{C} | \, |\lambda| = 1\}$. Moreover, A is the linear span of all unitary elements of A.

Proof. From Theorem 2.1.4 and Proposition 1.3.4, it is easy. Q.E.D.

Proposition 2.1.8. Let A be a C^*-algebra with an identity 1, a be a *normal* element of A (i.e., $a^*a = aa^*$), B be the C^*-subalgebra generated by $\{1, a\}$. Then $B \cong C(\sigma(a))$, and the function corresponding to a is $a(\lambda) = \lambda, \forall \lambda \in \sigma(a)$. In particular, $\|a\| = \nu(a)$.

Proof. By Theorem 2.1.4, $B \cong C(\Omega)$. Clearly, the map $t \to a(t)$ is injective and continuous from Ω onto $\sigma(a)$. Since Ω and $\sigma(a)$ are compact, it follows that $\Omega \cong \sigma(a)$. Q.E.D.

Proposition 2.1.9. Let A be an abelian C^*-algebra, $\|\cdot\|_1$ be another norm on A such that

$$\|ab\|_1 \le \|a\|_1 \cdot \|b\|_1, \quad \forall a, b \in A.$$

Then $\|x\| \le \|x\|_1, \forall x \in A$.

Proof. We may assume that A has an identity. Then $A \cong C(\Omega)$. Let

$$\Omega_1 = \{t \in \Omega \mid \text{if } \{x_n\} \subset A \text{ and } \|x_n\|_1 \to 0, \text{ then } x_n(t) \to 0\}.$$

We claim that Ω_1 is dense in Ω. In fact, suppose that there is a non-empty open subset U of Ω such that $\overline{\Omega}_1 \cap \overline{U} = \emptyset$. Then we have $a \in A$ with $a(t) = 1, \forall t \in \Omega_1$ and $a(s) = 0, \forall s \in U$. Let A_1 the completion of $(A, \|\cdot\|_1)$. If a is not invertible in A_1, then there is a non-zero multiplicative linear functional ρ on A_1 such that $\rho(a) = 0$. Let $\rho|A = t$. Then $t \in \Omega_1$. Thus $a(t) = 1$ and $\rho(a) = a(t) = 0$. This is a contradiction. So a is invertible in A_1. On the other hand, there is $b \in A$ such that $\text{supp}b(\cdot) \subset U$. Then $ab = 0$. It is impossible since a is invertible in A_1. Therefore Ω_1 is dense in Ω.

Finally, let X be the spectral space of A_1. Then for any $x \in A$,

$$\begin{aligned} \|x\|_1 &\ge \max\{|\rho(x)| \mid \rho \in X\} \\ &= \max_{t \in \Omega_1} |x(t)| = \max_{t \in \Omega} |x(t)| = \|x\|. \end{aligned}$$

Q.E.D.

Proposition 2.1.10. Let A be a C^*-algebra, $\|\cdot\|_1$ be another norm on A such that

$$\|ab\|_1 \le \|a\|_1 \cdot \|b\|_1, \quad \|a^*a\|_1 = \|a\|_1^2, \quad \forall a, b \in A.$$

Then $\|a\| = \|a\|_1, \forall a \in A$.

Proof. By Proposition 2.1.2, we may assume that A has an identity. Let $h = h^* \in A$, and B be the C^*-subalgebra generated by $\{1, h\}$, B_1 be the completion of $(B, \|\cdot\|_1)$. Suppose that Ω_1 and Ω are the spectral spaces of B_1 and B respectively. By the proof of Proposition 2.1.9, $\{(\rho|B) \mid \rho \in \Omega_1\}$ is dense in Ω. Thus

$$\|h\|_1 = \max_{\rho \in \Omega_1} |\rho(h)| = \max_{t \in \Omega} |h(t)| = \|h\|.$$

Further, for any $a \in A, \|a\|^2 = \|a^* a\| = \|a^* a\|_1 = \|a\|_1^2$, i.e., $\|a\| = \|a\|_1$.
Q.E.D.

Notes. Theorem 2.1.4 is due to I.M. Gelfand. The aim of the theory of C^*-algebras is to understand a uniformly closed self-adjoint algebra of operators on a Hilbert space (concrete C^*-algebra). A Banach $*$ algebra satisfying the axioms in Definition 2.1.1 is still sometimes called a B^*-algebra. The name C^*-algebra was coined by I.E. Segal. Presumably the C is meant to indicate that a C^*-algebra is a non-commucative analogue of $C(\Omega)$, whereas the $*$ recalls the importance of the $*$-operation.

References. [51], [81].

2.2. Positive cones of C^*-algebras

Definition 2.2.1. Let A be a C^*-algebra. An element a of A is said to be *positive*, denoted by $a \geq 0$, if $a^* = a$, and $\sigma(a) \subset \mathbb{R}_+ = [0, \infty)$. We shall denote by A_+ the subset of all positive elements of A.

Proposition 2.2.2. Let A be a C^*-algebra, $h^* = h \in A$. Then there are unique $h_+, h_- \in A_+$ such that

$$h = h_+ - h_-, \quad h_+ \cdot h_- = 0.$$

Proof. We may assume that A has an identity 1. Let B be the abelian C^*-subalgebra generated by $\{1, h\}$. By Theorem 2.1.4, there are $h_+, h_- \in B_+$ such that $h = h_+ - h_-$ and $h_+ \cdot h_- = 0$. From Proposition 2.1.6, $h_+, h_- \in A_+$.

Now if there exist other $h'_+, h'_- \in A_+$ such that $h = h'_+ - h'_-$ and $h'_+ \cdot h'_- = 0$, then the set $\{h, h'_+, h'_-\}$ is commutative. Let C be the abelian C^*-subalgebra generated by $\{1, h, h'_+, h'_-\}$. Clearly $C \supset B$. Using Theorem 2.1.4 for C, we can see that $h'_+ = h_+, h'_- = h_-$.
Q.E.D.

Proposition 2.2.3. Let A be a C^*-algebra with an identity 1, and $h = h^* \in A$ with $\|h\| \le 1$. Then $h \ge 0$ if and only if $\|1 - h\| \le 1$.

Proof. Using Theorem 2.1.4 for the abelian C^*-subalgebra generated by $\{1, h\}$, it is easy to get the conclusion. Q.E.D.

Proposition 2.2.4. Let A be a C^*-algebra. Then A_+ is a cone, i.e., if $a, b \in A_+$, we have $(a+b) \in A_+$. Moreover, $A_+ \cap (-A_+) = \{0\}$.

Proof. We may assume that A has an identity 1. By Proposition 2.2.3,

$$\|1 - \frac{a}{\|a\|}\| \le 1, \quad \|1 - \frac{b}{\|b\|}\| \le 1.$$

Then

$$\|1 - \frac{a+b}{\|a\| + \|b\|}\| \le \frac{\|a\| \cdot \|1 - \frac{a}{\|a\|}\| + \|b\| \cdot \|1 - \frac{b}{\|b\|}\|}{\|a\| + \|b\|} \le 1.$$

Again by Proposition 2.2.3, $(a+b) \in A_+$.

Moreover, if $h \in A_+ \cap (-A_+)$, then $\sigma(h) = \{0\}$. From Proposition 2.1.3, $h = 0$, i, e., $A_+ \cap (-A_+) = \{0\}$. Q.E.D.

Remark. We introduce a partial order "$\ge$" in $A_H = \{h \in A \mid h^* = h\}$, i.e., $a \ge b$ if $(a - b) \ge 0$.

Proposition 2.2.5. Let A be a C^*-algebra, and $a \in A_+$. Then there exists unique $a^{1/2} \in A_+$ such that $a^{1/2} \cdot a = a \cdot a^{1/2}$ and $(a^{1/2})^2 = a$. Moreover, this $a^{1/2}$ can be approximated arbitrarily by the polynomials of a with zero constant terms.

Proof. It is similar to Proposition 2.2.2. Q.E.D.

Lemma 2.2.6. Let B be an algebra with an identity, and $a, b \in B$. Then $\sigma(ab) \sqcup \{0\} = \sigma(ba) \sqcup \{0\}$.

Proof. Let $0 \ne \lambda \notin \sigma(ab)$, and $u = (ab - \lambda)^{-1}$. Since

$$(bua - 1) \cdot (ba - \lambda) = (ba - \lambda) \cdot (bua - 1) = \lambda,$$

it follows that $(ba - \lambda)$ is invertible. Q.E.D.

Proposition 2.2.7. Let A be a C^*-algebra, and $a \in A$. Then $a \in A_+$ if and only if there exists $b \in A$ such that $a = b^*b$.

Proof. The necessity is clear by Proposition 2.2.5. Now let $a = b^*b$. Clearly $a^* = a$. By Proposition 2.2.2 and 2.2.5, we can write $a = u^2 - v^2$, where $u, v \in A_+$ and $uv = 0$. Then

$$(bv)^*(bv) = vav = -v^4 \leq 0. \tag{1}$$

Let $bv = h + ik$, where $h^* = h, k^* = k$. Then by Proposition 2.2.4,

$$(bv)(bv)^* = -(bv)^*(bv) + 2(h^2 + k^2) = v^4 + 2(h^2 + k^2) \geq 0. \tag{2}$$

From (1), (2) and Lemma 2.2.6, $(bv)^*(bv) = -v^4 \in A_+ \cap (-A_+) = \{0\}$. Further by Proposition 2.2.5, $v = 0$, and $a = u^2 \in A_+$. Q.E.D.

Proposition 2.2.8. Let A be a C^*-algebra on a Hilbert space H (i.e. A is a uniformly closed $*$ subalgebra of $B(H)$), and $a \in A$. Then $a \in A_+$ if and only if a is a positive operator on H.

Proof. The necessity is clear from Proposition 2.2.7. Now let $a (\in A)$ be a positive operator on H. At least we have $a^* = a$. By Proposition 2.2.2, $a = a_+ - a_-$, where $a_+, a_- \in A_+$ and $a_+ \cdot a_- = 0$. For any $\xi \in H$,

$$0 \leq \langle aa_-\xi, a_-\xi \rangle = -\langle a_-^3 \xi, \xi \rangle \leq 0.$$

Therefore, $a_-^3 = 0, a_- = 0$, and $a = a_+ \in A_+$. Q.E.D.

Proposition 2.2.9. Let A be a C^*-algebra.

1) If $a, b \in A_+$ and $a \leq b$, then $\|a\| \leq \|b\|$, and $c^*ac \leq c^*bc, \forall c \in A$.

2) A_+ is a closed subset of A.

3) If A has an identity, $a, b \in A_+, a \leq b$,and a, b are invertible, then $b^{-1} \leq a^{-1}$.

Proof. 1) we may assume that A has an identity 1. Then $0 \leq a \leq b \leq \|b\| \cdot 1$. Using Theorem 2.1.4 for the abelian C^*-subalgebra generated by $\{1, a\}$, we can get that $\|a\| \leq \|b\|$. From Proposition 2.2.7, it is clear that $c^*ac \leq c^*bc, \forall c \in A$.

2) Let $\{a_n\} \subset A_+$, and $a_n \to a$. We need to prove $a \in A_+$. Clearly, $a^* = a$. Write $a = a_+ - a_-$, where $a_+, a_- \in A_+$ and $a_+ \cdot a_- = 0$. Let $b_n = a_- a_n a_- \in A_+$. Then $b_n \to b = a_- a a_- = -a_-^3 \leq 0$. Further $0 \leq -b \leq b_n - b \leq \|b_n - b\| \to 0$. Thus $b = 0, a_- = 0$, and $a = a_+ \in A_+$.

3) Since $(a^{-1})^{1/2}(b - a)(a^{-1})^{1/2} \geq 0$, it follows that $(a^{-1})^{1/2} b (a^{-1})^{1/2} \geq 1$. By the function representation, $a^{1/2} b^{-1} a^{1/2} \leq 1 = a^{1/2} a^{-1} a^{1/2}$. Therefore, $b^{-1} \leq a^{-1}$. Q.E.D.

Proposition 2.2.10. Let A be a C^*-algebra, $a, b \in A_+$ and $a \leq b$. Then for any $\lambda \in [0,1]$, we have $a^\lambda \leq b^\lambda$, where a^λ, b^λ are defined as in Proposition 2.1.8.

Proof. We may assume that A has an identity 1. First, suppose that a, b are invertible. Let $E = \{\lambda \in [0,1] \mid a^\lambda \leq b^\lambda\}$. Since A_+ is closed, it follows that E is closed. Clearly, $0, 1 \in E$. So it suffices to show that for any $\lambda, \mu \in E$, we have $\frac{\lambda+\mu}{2} \in E$. Fix $\lambda, \mu \in E$. Since

$$b^{-\frac{\lambda}{2}} a^\lambda b^{-\frac{\lambda}{2}} \leq 1, \quad b^{-\frac{\mu}{2}} a^\mu b^{-\frac{\mu}{2}} \leq 1,$$

it follows that $\|b^{-\frac{\lambda}{2}} a^\lambda b^{-\frac{\lambda}{2}}\| \leq 1$ and $\|b^{-\frac{\mu}{2}} a^\mu b^{-\frac{\mu}{2}}\| \leq 1$. Further, $\|a^{\lambda/2} b^{-\lambda/2}\| \leq 1$ and $\|a^{\mu/2} b^{-\mu/2}\| \leq 1$. By Lemma 2.2.6,

$$\begin{aligned} 1 \geq\ & \|b^{-\mu/2} a^{\frac{\lambda+\mu}{2}} b^{-\lambda/2}\| \geq \nu(b^{-\mu/2} a^{\frac{\lambda+\mu}{2}} b^{-\lambda/2}) \\ =\ & \nu(b^{-\frac{\lambda+\mu}{4}} a^{\frac{\lambda+\mu}{2}} b^{-\frac{\lambda+\mu}{4}}) = \|b^{-\frac{\lambda+\mu}{4}} a^{\frac{\lambda+\mu}{2}} b^{-\frac{\lambda+\mu}{4}}\|, \end{aligned}$$

i.e.

$$0 \leq b^{-\frac{\lambda+\mu}{4}} a^{\frac{\lambda+\mu}{2}} b^{-\frac{\lambda+\mu}{4}} \leq 1, \quad \text{or} \quad a^{\frac{\lambda+\mu}{2}} \leq b^{\frac{\lambda+\mu}{2}}.$$

Therefore $\frac{\lambda+\mu}{2} \in E$.

For general case, from the preceding paragraph we have $(a+\varepsilon)^\lambda \leq (b+\varepsilon)^\lambda, \forall \varepsilon > 0$. Let $\varepsilon \to 0+$, then $a^\lambda \leq b^\lambda$. Q.E.D.

Proposition 2.2.11. Let A be a C^*-algebra, $S = \{a \in A \mid \|a\| \leq 1\}$ be its (closed) unit ball. Then $\text{Int}(S) \cap A_+ = \{a \in A_+ \mid \|a\| < 1\}$ is directed with respect to the partial order "$\geq$" in A_+, i.e. for any $x, y \in \text{Int(S)} \cap A_+$ there is $z \in \text{Int(S)} \cap A_+$ such that $z \geq x$ and $z \geq y$.

Proof. Consider the problem in $(A \dot{+} \mathbb{C})$. Let

$$a = x(1-x)^{-1}, b = y(1-y)^{-1},$$

$$z = (a+b)(\frac{1}{2} + a + b)^{-1} = 1 - \frac{1}{2}(\frac{1}{2} + a + b)^{-1}.$$

Then $a, b, z \in A_+$, and $z \in \text{Int(S)} \cap A_+$.

The inequality $z \geq x$ is equivalent to the following inequality

$$1 - \frac{1}{2}(\frac{1}{2} + a + b)^{-1} \geq x \quad \text{or} \quad (1-x) \geq \frac{1}{2}(\frac{1}{2} + a + b)^{-1}.$$

By Proposition 2.2.9, the latter is equivalent to the

$$(1-x)^{-1} \leq (1 + 2a + 2b).$$

It is clear that $(1+2a) \geq (1-x)^{-1}$ by the definition of a. Thus $(1+2a+2b) \geq (1-x)^{-1}$. So $z \geq x$. Similarly, $z \geq y$. Q.E.D.

Notes. Proposition 2.2.4 is due to M. Fukamiya, J.L. Kelley and R.L. Vaught. Proposition 2,2,7 is due to I. Kaplansky. These results are the key lemmas for the Gelfand-Naimark conjecture (see Section 2.14). Proposition 2.2.10 is known as the Löwner-Heinz theorem, and the proof here is due to G.K. Pedersen.

References. [50], [81], [90], [126], [177].

2.3. States and the Gelfand-Naimark-Segal construction

Definition 2.3.1. Let A be a C^*-algebra, ρ be a linear functional on A. ρ is said to be *positive*, denoted by $\rho \geq 0$, if $\rho(a) \geq 0, \forall a \in A_+$. ρ is called a *state* if $\rho \geq 0$ and $\|\rho\| = 1$. The set of states of A is denoted by $S(A)$.

Clearly, if $\rho \geq 0$, then we have $\rho(a^*) = \overline{\rho(a)}, \forall a \in A$; and the Schwartz inequality:

$$|\rho(b^*a)|^2 \leq \rho(b^*b) \cdot \rho(a^*a), \quad \forall a, b \in A.$$

Proposition 2.3.2. Let A be a C^*-algebra, and ρ be a positive linear functional on A. Then ρ is continuous. Moreover, if A has an identity 1, then $\|\rho\| = \rho(1)$.

Proof. First, we say that ρ is bounded on $S \cap A_+ = \{a \in A_+ \mid \|a\| \leq 1\}$. In fact, suppose that there is a sequence $\{x_n\} \subset S \cap A_+$ such that

$$\rho(x_n) \geq n^2, \quad \forall n.$$

By Proposition 2.2.9, A_+ is closed. Thus $x = \sum_n \frac{1}{n^2} x_n \in A_+$. Then for any positive integer N,

$$N \leq \sum_{n=1}^{N} \frac{1}{n^2} \rho(x_n) \leq \rho(\sum_{n=1}^{N} \frac{1}{n^2} x_n) \leq \rho(x).$$

It is impossible. Therefore, there is a constant K such that

$$\rho(a) \leq K, \quad \forall a \in S \cap A_+.$$

Further, by Proposition 2.2.2, $\|\rho\| \leq 4K$, i.e., ρ is continuous.

Now if A has an identity 1, then by the Schwartz inequality we have

$$|\rho(a)| \leq \rho(1)^{1/2} \rho(a^*a)^{1/2} \leq \|a^*a\|^{1/2} \rho(1) = \|a\| \rho(1),$$

$\forall a \in A$. Thus $\|\rho\| = \rho(1)$. Q.E.D.

Proposition 2.3.3. Let A be a C^* -algebra, and $\rho \in A^*$.

1) If there is $a \in A_+$ with $\|a\| \leq 1$ such that $\rho(a) = \|\rho\|$, then $\rho \geq 0$.

2) Suppose that $\|\rho\| \leq 1$, and there is $0 \neq a \in A_+$ such that $\rho(a) = \|a\|$. Then $\rho \in S(A)$.

Proof, 1) Since ρ can be extended to a continuous linear functional on $(A\dot{+}\mathbb{C})$ with the same norm, we may assume that A has an identity. Now by the proof of Lemma 1.9.2, $\rho \geq 0$.

2) By the hypothesis, we have $\|\rho\| = 1$, and $\rho(\frac{a}{\|a\|}) = \|\rho\|$. Again from 1), it follows that $\rho \in S(A)$. Q.E.D.

Proposition 2.3.4. Let A be a C^*-algebra, ρ be a positive linear functional on A, and $a \in A_+$ with $\|a\| < 1$. Then

$$\begin{aligned}\|\rho\| &= \sup\{\rho(b) \mid b \in A_+, \|b\| \leq 1\} \\ &= \sup\{\rho(b) \mid b \in A_+, \|b\| \leq 1 \text{ and } b \geq a\}.\end{aligned}$$

Proof. Since $\rho(x^*) = \overline{\rho(x)}, \forall x \in A$, it follows that there is a sequence $\{h_n\}(\subset A)$ with $h_n^* = h_n, \|h_n\| \leq 1, \forall n$, such that $\rho(h_n) \to \|\rho\|$. By Proposition 2.2.2, $h_n = h_n^+ - h_n^-$, where $h_n^\pm \in A_+$ and $\|h_n^\pm\| \leq \|h_n\| \leq 1, \forall n$. Replacing $\{h_n^\pm\}$ by a subsequence if it is necessary, we may assume that $\lim_n \rho(h_n^+)$ and $\lim_n \rho(h_n^-)$ exist. Since $0 \leq \lim_n \rho(h_n^+) \leq \|\rho\|$, it follows that

$$\lim_n \rho(h_n^+) = \|\rho\|, \quad \lim_n \rho(h_n^-) = 0.$$

Therefore $\|\rho\| = \sup\{\rho(b) \mid b \in A_+, \|b\| \leq 1\}$.

Now for any $\varepsilon > 0$, there is $c \in A_+$ with $\|c\| < 1$ such that $\rho(c) \geq \|\rho\| - \varepsilon$. From Proposition 2.2.11, we have $b \in A_+$ with $\|b\| < 1$ such that $b \geq c, b \geq a$. It is sure that $\rho(b) \geq \rho(c) \geq \|\rho\| - \varepsilon$. Therefore $\|\rho\| = \sup\{\rho(b) \mid b \in A_+, \|b\| \leq 1, b \geq a\}$. Q.E.D.

Proposition 2.3.5. Let A be a C^*-algebra, and ρ be a positive linear functional on A. Define

$$\tilde{\rho}(a + \lambda) = \rho(a) + \lambda\mu_0, \quad \forall a \in A, \lambda \in \mathbb{C}$$

where μ_0 is a fixed positive number with $\mu_0 \geq \|\rho\|$. Then $\tilde{\rho}$ is a positive linear functional on $(A\dot{+}\mathbb{C})$.

Proof. It suffices to show that $\rho(a) + \lambda\mu_0 \geq 0$ for any positive element $(a + \lambda)$ of $(A\dot{+}\mathbb{C})$.

Since $(a+\lambda) \geq 0$ in $(A\dot{+}\mathbb{C})$, it follow that $a^* = a, \overline{\lambda} = \lambda$. Let B be the abelian C^*-subalgebra of $(A\dot{+}\mathbb{C})$ generated by $\{1, a\}$. Then $B \cong C(\Omega)$, and $\lambda + a(t) \geq 0, \forall t \in \Omega$. Since a is not invertible in $(A\dot{+}\mathbb{C})$, and $\sigma_B(a) = \sigma_{(A\dot{+}\mathbb{C})}(a)$, so there is $t_0 \in \Omega$ such that $a(t_0) = 0$. Thus $\lambda = \lambda + a(t_0) \geq 0$.

If $a \geq 0$ or $\|\rho\| = 0$, then $\rho(a) + \lambda\mu_0 \geq 0$ immediately. Now let $\rho \neq 0$ and $a = a_+ - a_-$ with $a_- \neq 0$, where $a_+, a_- \in A_+$ and $a_+ \cdot a_- = 0$. Clearly $\inf\{a(t) \mid t \in \Omega\} = -\|a_-\|$. Further,

$$\begin{aligned} 0 \leq\ & \lambda + \inf\{a(t) \mid t \in \Omega\} = \lambda - \|a_-\| \\ \leq\ & \lambda - \|\rho\|^{-1}\rho(a_-) \leq \|\rho\|^{-1}\{\lambda\mu_0 + \rho(a_+) - \rho(a_-)\} \end{aligned}$$

i.e. $\rho(a) + \lambda\mu_0 \geq 0$. Q.E.D.

Corollary 2.3.6. Let ρ be a state on a C^*-algebra A. Define $\tilde{\rho}(a+\lambda) = \rho(a) + \lambda, \forall a \in A, \lambda \in \mathbb{C}$. Then $\tilde{\rho}$ is a state on $(A\dot{+}\mathbb{C})$, and $(\tilde{\rho}|A) = \rho$.

Proposition 2.3.7. Let A be a C^*-algebra. Then the state space $S(A)$ of A is a convex subset of A^*.

Proof. For any $\varphi_1, \varphi_2 \in S(A)$ and $\lambda \in (0,1)$, we need to show that $\lambda\varphi_1 + (1-\lambda)\varphi_2 \in S(A)$.

For any $\varepsilon > 0$, pick $a \in A_+$ with $\|a\| < 1$ such that $\varphi_1(a) \geq 1 - \varepsilon$. Now by Proposition 2.3.4,

$$\begin{aligned} 1 \geq\ & \|\lambda\varphi_1 + (1-\lambda)\varphi_2\| \\ =\ & \sup\{\lambda\varphi_1(b) + (1-\lambda)\varphi_2(b) \mid b \in A_+, \|b\| \leq 1, b \geq a\} \\ \geq\ & \lambda\varphi_1(a) + (1-\lambda)\sup\{\varphi_2(b) \mid b \in A_+, \|b\|| \leq 1, b \geq a\} \\ \geq\ & \lambda(1-\varepsilon) + (1-\lambda). \end{aligned}$$

Since ε is arbitrary, it follows that $\lambda\varphi_1 + (1-\lambda)\varphi_2 \in S(A)$. Q.E.D.

Definition 2.3.8. Let A be a C^*-algebra, $S(A)$ be its state space. Each extreme point of $S(A)$ is called a *pure state* on A. The set of *pure states* is denoted by $P(A)$.

Proposition 2.3.9. Let A be a C^*-algebra.

1) For $\rho \in P(A)$, let $\tilde{\rho}(a+\lambda) = \rho(a) + \lambda, \forall a \in A, \lambda \in \mathbb{C}$. Then $\tilde{\rho}$ is a pure state on $(A\dot{+}\mathbb{C})$.

2) If $\tilde{\rho}$ is a pure state on $(A\dot{+}\mathbb{C}), \rho = (\tilde{\rho}|A)$, then either $\rho = 0$ or $\rho \in P(A)$.

Proof. 1) Suppose that there are two states $\tilde{\rho}_1, \tilde{\rho}_2$ on $(A\dot{+}\mathbb{C})$ and $\lambda \in (0,1)$ such that $\tilde{\rho} = \lambda\tilde{\rho}_1 + (1-\lambda)\tilde{\rho}_2$. Let $\rho_i = (\tilde{\rho}_i|A), i = 1, 2$. Since $\|\rho_1\| \leq 1, \|\rho_2\| \leq 1$

and

$$1 = \|\rho\| \leq \lambda\|\rho_1\| + (1-\lambda)\|\rho_2\| \leq 1,$$

it follows that $\rho_i \in S(A), i = 1,2$. By $\rho \in P(A)$ and $\rho = \lambda\rho_1 + (1-\lambda)\rho_2$ on A, we have $\rho = \rho_1 = \rho_2$. Moreover, $\tilde{\rho}(1) = \tilde{\rho}_1(1) = \tilde{\rho}_2(1) = 1$. Thus $\tilde{\rho} = \tilde{\rho}_1 = \tilde{\rho}_2$, i.e., $\tilde{\rho}$ is a pure state on $(A\dot{+}\mathbb{C})$.

2) Suppose that $\rho \neq 0$. Let $\|\rho\| = \mu_0 < 1$. By Proposition 2.3.5, $\tilde{\sigma}$ is a state on $(A\dot{+}\mathbb{C})$, where $\tilde{\sigma}(a+\lambda) = \mu_0^{-1}\rho(a) + \lambda, \forall a \in A, \lambda \in \mathbb{C}$. Let $\tilde{\sigma}_0(a+\lambda) = \lambda(\forall a \in A, \lambda \in \mathbb{C})$. Obviously, $\tilde{\sigma}_0$ is a (pure) state on $(A\dot{+}\mathbb{C})$. Moreover, we have the equality:

$$\tilde{\rho} = \mu_0\tilde{\sigma} + (1-\mu_0)\tilde{\sigma}_0.$$

It is a contradiction since $\tilde{\rho}$ is pure. Therefore $\|\rho\| = 1$, i.e. $\rho \in S(A)$.

Now let $\rho_1, \rho_2 \in S(A)$ and $\lambda \in (0,1)$ be such that $\rho = \lambda\rho_1 + (1-\lambda)\rho_2$. ρ_1 and ρ_2 can be naturally extended to states $\tilde{\rho}_1$ and $\tilde{\rho}_2$ on $(A\dot{+}\mathbb{C})$. Clearly, $\tilde{\rho} = \lambda\tilde{\rho}_1 + (1-\lambda)\tilde{\rho}_2$. Since $\tilde{\rho}$ is pure, it follows that $\tilde{\rho} = \tilde{\rho}_1 = \tilde{\rho}_2$ and $\rho = \rho_1 = \rho_2$. Therefore $\rho \in P(A)$. Q.E.D.

Proposition 2.3.10. Let A be an abelian C^*-algebra. Then $\rho \in P(A)$ if and only if ρ is a non-zero multiplicative linear functional on A.

Proof. By Proposition 2.3.9, we may assume that A has an identity 1.

Let $\rho \in P(A)$. Since $A = [A_+]$, it suffices to show that $\rho(ab) = \rho(a)\rho(b)$ for any $a, b \in A_+$. If $\rho(a) = 0$, then by the Schwartz inequality,

$$|\rho(ab)|^2 = |\rho(a^{1/2} \cdot a^{1/2}b)|^2 \leq \rho(a)\rho(bab) = 0.$$

Thus $\rho(ab) = 0 = \rho(a)\rho(b)$. Now we may assume that

$$0 \leq a \leq 1, \quad 0 < \rho(a) < 1.$$

Let $\rho_1(\cdot) = \rho(a)^{-1}\rho(a\cdot), \rho_2(\cdot) = \rho(1-a)^{-1}\rho((1-a)\cdot)$. Clearly, $\rho_1, \rho_2 \in S(A)$, and $\rho = \rho(a)\rho_1 + (1-\rho(a))\rho_2$. Since ρ is pure, it follows that $\rho = \rho_1 = \rho_2$. Therefore $\rho(ab) = \rho(a)\rho(b)$.

Conversely, let ρ be a non-zero multiplicative linear functional on A. If $A \cong C(\Omega)$, then there is $t_0 \in \Omega$ such that

$$\rho(a) = a(t_0) = \int_\Omega a(t)d\delta_{t_0}(t), \quad \forall a \in A.$$

By the Riesz representation theorem, each state on A corresponds to a probability measure on Ω. Therefore $\rho \in P(A)$ since ρ corresponds to a point measure. Q.E.D.

Remark. There is another direct proof (without using the Riesz representative theorem) as follows. Suppose that ρ is a non-zero multiplicative linear functional on A. Clearly $\rho \in S(A)$. Let $\rho_1, \rho_2 \in S(A)$ and $\lambda \in (0,1)$ such that

$$\rho = \lambda\rho_1 + (1-\lambda)\rho_2.$$

For any $h = h^* \in A$,

$$\begin{aligned} \lambda\rho_1(h^2) + (1-\lambda)\rho_2(h^2) &= \rho(h^2) = \rho(h)^2 \\ &= [\lambda\rho_1(h) + (1-\lambda)\rho_2(h)]^2. \end{aligned}$$

By the Schwartz inequality, $\rho_i(h)^2 \le \rho_i(h^2), i = 1,2$. Then

$$\begin{aligned} 0 = &\ -[\lambda\rho_1(h) + (1-\lambda)\rho_2(h)]^2 + \lambda\rho_1(h^2) + (1-\lambda)\rho_2(h^2) \\ \ge &\ -[\lambda^2\rho_1(h)^2 + (1-\lambda)^2\rho_2(h)^2 + 2\lambda(1-\lambda)\rho_1(h)\rho_2(h)] \\ &\ +\lambda\rho_1(h)^2 + (1-\lambda)\rho_2(h)^2 \\ = &\ \lambda(1-\lambda)[\rho_1(h) - \rho_2(h)]^2. \end{aligned}$$

Thus $\rho_1(h) = \rho_2(h), \forall h = h^* \in A$. Further $\rho = \rho_1 = \rho_2$ and $\rho \in P(A)$.

Proposition 2.3.11. Let A be a C^*-algebra with an identity 1, E be a $*$ linear subspace (i.e., if $a \in E$, then $a^* \in E$) of A and $1 \in E$. Let

$$\mathcal{F} = \left\{ f \,\middle|\, \begin{array}{c} f \text{ is a state on } E, \text{ i.e., } f \text{ is a linear functional on } E, \\ f(a^*) = \overline{f(a)}, \forall a \in E, f(b) \ge 0, \forall b \in E \cap A_+, \text{ and } f(1) = 1 \end{array} \right\}.$$

Then we have that:
1) Each element of $\mathcal{F}$ can be extended to a state on A;
2) Each extreme point of $\mathcal{F}$ can be extended to a pure state on A.

Proof. 1) Let $f \in \mathcal{F}$. For any $h = h^* \notin E$, since $-\|h\|\cdot 1 \le h \le \|h\|\cdot 1, 1 \in E$, and A_+ is a cone, we can define $f(h)$ such that

$$\begin{aligned} &\sup\{f(b) \mid b = b^* \in E, b \le h\} \\ \le &\ f(h) \le \inf\{f(c) \mid c = c^* \in E, c \ge h\}. \end{aligned}$$

Then f is a state on $E \dot{+} [h]$. In fact, for $(a + \lambda h) \in A_+$, where $a \in E, \lambda \in \mathbb{C}$, we need to prove $f(a + \lambda h) \ge 0$. Clearly $a^* = a, \bar{\lambda} = \lambda$. When $\lambda = 0$, it is clear. If $\lambda > 0$, then $h \ge -\lambda^{-1}a$. By the definition of $f(h), f(h) \ge -\lambda^{-1}f(a)$, i.e., $f(a + \lambda h) \ge 0$. If $\lambda < 0$, then $h \le -\lambda^{-1}a$. By the definition of $f(h), f(h) \le -\lambda^{-1}f(a)$, i.e., $f(a + \lambda h) \ge 0$.

By the above procedure and the Zorn lemma, f can be extended to a state on A.

2) Let f be an extreme point of $\mathcal{F}$. Put

$$\mathcal{L} = \{\rho \in S(A) \mid (\rho|E) = f\}.$$

By 1), $\mathcal{L} \neq \emptyset$. It is easy to see that $\mathcal{L}$ is a weak $*$ compact convex subset of A^*. From the Krein-Milmann theorem, $\mathcal{L}$ has an extreme point ρ at least. Now it suffices to show $\rho \in P(A)$. Let $\rho_1, \rho_2 \in S(A)$ and $\lambda \in (0,1)$ be such that $\rho = \lambda\rho_1 + (1-\lambda)\rho_2$. Clearly, $f_i = (\rho_i|E) \in \mathcal{F}, i = 1,2$. Since $\lambda f_1 + (1-\lambda)f_2 = (\rho|E) = f$ and f is an extreme point of $\mathcal{F}$, it follows that $f = f_1 = f_2$. Thus $\rho_1, \rho_2 \in \mathcal{L}$. Since ρ is an extreme point of $\mathcal{L}, \rho = \rho_1 = \rho_2$. Therefore $\rho \in P(A)$.

Q.E.D.

Corollary 2.3.12. Let A be a C^*-algebra, and B be a C^*-subalgebra of A. Then each state (or pure state) on B can be extended to a state (or pure state) on A.

Proof. Each state (or pure state) on B can be extended to a state (or pure state) on $(B\dot{+}\mathbb{C})$ (see 2.3.6 and 2.3.9). Further it can be extended to a state (or pure state) on $(A\dot{+}\mathbb{C})$ (see 2.3.11). By Proposition 2.3.9, its restriction is still a state (or pure state) on A. Q.E.D.

Remark. Suppose that A has an identity 1, and $1 \in B, \varphi \in S(B)$. By Proposition 2.3.3, each extension of φ preserving the norm is a state on A.

Proposition 2.3.13. Let A be a C^*-algebra, and $h = h^* \in A$. If $0 \neq \lambda \in \sigma(h)$, then there is a pure state ρ on A such that $\rho(h) = \lambda$.

Proof. By Proposition 2.3.9, we may assume that A has an identity 1. Let $B \cong C(\Omega)$ be the abelian C^*-subalgebra generated by $\{1, h\}$. Then there is $t \in \Omega$ such that $h(t) = \lambda$. Define $f(b) = b(t), \forall b \in B$. By Proposition 2.3.10, f is pure state on B. By Corollary 2.3.12, f can be extended to a pure state ρ on A. Clearly, $\rho(h) = f(h) = h(t) = \lambda$. Q.E.D.

Remark. If A itself has an identity, then the condition "$\lambda \neq 0$" is not necessary.

Corollary 2.3.14. Let A be a C^*-algebra, and $h = h^* \in A$. Then there is a pure state ρ on A such that $|\rho(h)| = \|h\|$. Consequently, $\|h\| = \sup\{|\rho(h)| \mid \rho \in P(A)\}$.

Corollary 2.3.15. Let A be a C^*-algebra, and $a \in A$ such that $\rho(a) \geq 0, \forall \rho \in P(A)$. Then $a \in A_+$.

Proof. It suffices to show $a^* = a$. If $\frac{a-a^*}{2i} \neq 0$, then there is $\rho \in P(A)$ such that $\rho(\frac{a-a^*}{2i}) \neq 0$. Thus $\rho(a) \notin \mathbb{R}$, a contradiction. Therefore $a^* = a$. Q.E.D.

The GNS construction was once discussed in Section 1.8. The same procedure can be carried for the C^*-algebras. Due to its importance, we shall study it again in detail.

Definition 2.3.16. Let A be a C^*-algebra. $\{\pi, H\}$ is called a $*$ *representation of* A, if π is a $*$ homomorphism from A into $B(H)$, where H is a Hilbert space, i.e.,

$$\pi(\lambda a + \mu b) = \lambda\pi(a) + \mu\pi(b),$$

$$\pi(ab) = \pi(a)\pi(b),$$

$$\pi(a^*) = \pi(a)^*,$$

$\forall a, b \in A, \lambda, \mu \in \mathbb{C}$.

If there is a vector $\xi \in H$ such that $\overline{\pi(A)\xi} = H$, then ξ is called a *cyclic vector* for $\{\pi, H\}$, and $\{\pi, H\}$ is said to be *cyclic*.

The $*$ representation $\{\pi, H\}$ of A is said to be *faithful*, if $\pi(a) = 0$ implies $a = 0$.

Two $*$ representations $\{\pi_1, H_1\}$ and $\{\pi_2, H_2\}$ of A are *unitarily equivalent*, denoted by $\{\pi_1, H_1\} \cong \{\pi_2, H_2\}$, if there is a unitary operator u from H_1 onto H_2 such that

$$u\pi_1(a)u^{-1} = \pi_2(a), \forall a \in A.$$

Proposition 2.3.17. Let $\{\pi, H\}$ be a $*$ representation of a C^*-algebra A. Then $\|\pi\| \leq 1$, and π preserves order, i.e., $\pi(A_+) \subset B(H)_+$. Moreover, if π is faithful, then π is isometric, and $\pi^{-1}(\pi(A)_+) = A_+$.

Proof. Consider $(A\dot{+}\mathbb{C})$, and put $\pi(1) = 1_H$. Thus we may assume that A has an identity 1, and $\pi(1) = 1_H$. Then $\sigma(\pi(a)) \subset \sigma(a), \forall a \in A$. Further,

$$\begin{aligned} \|\pi(h)\| &= \sup\{|\lambda| \mid \lambda \in \sigma(\pi(h))\} \\ &\leq \sup\{|\lambda| \mid \lambda \in \sigma(h)\} = \|h\|, \quad \forall h = h^* \in A. \end{aligned}$$

Hence $\|\pi(a)\| = \|\pi(a^*a)\|^{1/2} \leq \|a^*a\|^{1/2} = \|a\|, \forall a \in A$, i.e., $\|\pi\| \leq 1$. Clearly, $\pi(A_+) \subset B(H)_+$.

Now let π be faithful. If there is $e \in A$ such that $\pi(e) = 1_H$, then e is just the identity of A. If $1_H \notin \pi(A)$, considering $(A\dot{+}\mathbb{C})$ and putting $\pi(1) = 1_H$, then π is still faithful on $(A\dot{+}\mathbb{C})$. In other words, we may assume that A has an identity 1, and $\pi(1) = 1_H$. Now our conclusion can be obtained from the proof of Proposition 1.8.13. Q.E.D.

Let A be a C^*-algebra, and $\varphi \in S(A)$. Put

$$L_\varphi = \{a \in A \mid \varphi(a^*a) = 0\}.$$

L_φ is called the *left kernel* of φ. By the Schwartz inequality, L_φ is closed left ideal of A. Let

$$a \to a_\varphi = a + L_\varphi \qquad (\forall a \in A)$$

be the quotient map from A onto A/L_φ. On A/L_φ, define

$$\langle a_\varphi, b_\varphi \rangle = \varphi(b^*a), \forall a, b \in A.$$

Then $\langle , \rangle$ is well-defined, and is an inner product on A/L_φ. Denote by H_φ the completion of $(A/\mathrm{L}_\varphi, \langle , \rangle)$. For any $a \in A$, define a linear map $\pi_\varphi(a)$: $A/L_\varphi \to A/L_\varphi$ as follows:

$$\pi_\varphi(a)b_\varphi = (ab)_\varphi, \quad \forall b \in A.$$

Since $b^*a^*ab \le \|a\|^2 b^*b$, it follow that

$$\|\pi_\varphi(a)b_\varphi\|^2 = \varphi(b^*a^*ab) \le \|a\|^2 \cdot \|b_\varphi\|^2$$

$\forall\ b \in A$. Therefore $\pi_\varphi(a)$ can be uniquely extended to a bounded linear operator on H_φ, still denoted by $\pi_\varphi(a)$. It is easy to show that $\{\pi_\varphi, H_\varphi\}$ is a $*$ representation of A.

Proposition 2.3.18. Let A be a C^*-algebra, and $\varphi \in S(A)$.

1) If $\{\pi_\varphi, H_\varphi\}$ is the $*$ representation of A generated by φ as above, then $\{\pi_\varphi, H_\varphi\}$ admits a cyclic vector ξ_φ, and ξ_φ can be chosen such that

$$\pi_\varphi(a)\xi_\varphi = a_\varphi, \quad \varphi(a) = \langle \pi_\varphi(a)\xi_\varphi, \xi_\varphi \rangle, \quad \forall a \in A.$$

2) Let $\tilde{\varphi}$ be the natural extension of φ on $(A\dot{+}\mathbb{C})$ (i.e., $\tilde{\varphi}(a+\lambda) = \varphi(a) + \lambda, \forall a \in A, \lambda \in \mathbb{C}$), and $\{\pi_{\tilde{\varphi}}, H_{\tilde{\varphi}}\}$ be the $*$ representation of $(A\dot{+}\mathbb{C})$ generated by $\tilde{\varphi}$. Then there is a unitary operator u from H_φ on $H_{\tilde{\varphi}}$ such that $v\pi_\varphi(a)u^{-1} = \pi_{\tilde{\varphi}}(a), \forall a \in A$.

Proof. Define $ua_\varphi = a_{\tilde{\varphi}}, \forall a \in A$. Then u can be extended to an isometry from H_φ into $H_{\tilde{\varphi}}$.

By Proposition 2.3.4, there is a sequence $\{a_n\} \subset A_+$ with $\|a_n\| \le 1, \forall n$, such that $\varphi(a_n) \to 1$. From the Schwartz inequality, $\varphi(a_n) = \tilde{\varphi}(a_n) \le \varphi(a_n^2)^{1/2} \le 1$, it follows that $\varphi(a_n^2) \to 1$. Further $\tilde{\varphi}((1-a_n)^2) \to 0$, i.e., $u(a_n)_\varphi \to 1_{\tilde{\varphi}}$ in $H_{\tilde{\varphi}}$. Therefore, u is a unitary operator from H_φ onto $H_{\tilde{\varphi}}$. Moreover, since $u\pi_\varphi(a)b_\varphi = u(ab)_\varphi = (ab)_{\tilde{\varphi}} = \pi_{\tilde{\varphi}}(a)b_{\tilde{\varphi}} = \pi_{\tilde{\varphi}}(a)ub_\varphi, \forall a, b \in A$, we have $u\pi_\varphi(a)u^{-1} = \pi_{\tilde{\varphi}}(a), \forall a \in A$. Finally, pick $\xi_\varphi = u^{-1}1_{\tilde{\varphi}}$. That comes to the conclusion. Q.E.D.

Proposition 2.3.19. Let A be a C^*-algebra, and Δ be a subset of $S(A)$ such that $\sup\{\varphi(a) \mid \varphi \in \Delta\} = \|a\|, \forall a \in A_+$. Then

$$\{\pi_\Delta = \sum_{\varphi\in\Delta} \oplus \pi_\varphi, H_\Delta = \sum_{\varphi\in\Delta} \oplus H_\varphi\}$$

is a faithful $*$ representation of A.

Proof. For any $a \in A$, by Propositions 2.3.17 and 2.3.18

$$\begin{aligned} \|a\|^2 \geq & \ \|\pi_\Delta(a)\|^2 = \sup\{\|\pi_\varphi(a^*a)\| \mid \varphi \in \Delta\} \\ \geq & \ \sup\{\langle \pi_\varphi(a^*a)\xi_\varphi, \xi_\varphi\rangle \mid \varphi \in \Delta\} \\ = & \ \sup\{\varphi(a^*a) \mid \varphi \in \Delta\} = \|a\|^2. \end{aligned}$$

Therefore, $\|a\| = \|\pi_\Delta(a)\|, \forall a \in A$. Q.E.D.

Remark. By Corollary 2.3.14, $P(A)$ or any $\sigma(A^*, A)$-dense subset of $S(A)$ can be chosen as Δ.

Theorem 2.3.20. Any C^*-algebra can be isometrically $*$ isomorphic to a concrete C^*-algebra on some Hilbert space.

Proposition 2.3.21. Let A be a C^*-algebra, $\{\pi, H\}$ be a $*$ representation of A.

1) If π admits a cyclic vector ξ, let $\rho(a) = \langle \pi(a)\xi, \xi\rangle, \forall a \in A$, then $\{\pi_\rho, H_\rho\} \cong \{\pi, H\}$.

2) There exists $\Delta \subset S(A)$ such that $\{\pi, H\}$ is unitarily equivalent to the direct sum of some zero representation and $\{\pi_\varphi, H_\varphi\}(\varphi \in \Delta)$.

Proof. 1) Let $u\pi(a)\xi = a_\rho, \forall a \in A$. Then u can be extended to a unitary operator from H onto H_ρ; and it is easy to see that $u\pi(a)u^{-1} = \pi_\rho(a), \forall a \in A$.

2) By the Zorn lemma, we can write

$$H = \sum_{l\in\Lambda} \oplus H_l \oplus H_0,$$

where $H_l = \overline{\pi(A)\xi_l}$, and $\|\xi_l\| = 1, \forall l \in \Lambda$, and $H_0 = \{\xi \in H \mid \pi(a)\xi = 0, \forall a \in A\}$. For any $l \in \Lambda$, let $\varphi_l(a) = \langle \pi(a)\xi_l, \xi_l\rangle, \forall a \in A$. Then $\Delta = \{\varphi_l \mid l \in \Lambda\}$ satisfies our condition. Q.E.D.

The following proposition is a version of the Radon-Nikodym theorem.

Proposition 2.3.22. Let φ, ψ be two positive linear functionals on a C^*-algebra A, and $\varphi \leq \psi$ (i.e. $\varphi(a) \leq \psi(a), \forall a \in A_+$). Then there exists unique $t' \in \pi_\psi(A)', 0 \leq t' \leq 1$, such that $\varphi(a) = \langle \pi_\psi(a)t'\xi_\psi, \xi_\psi\rangle, \forall a \in A$, where $\{\pi_\psi, H_\psi, \xi_\psi\}$ is the cyclic $*$ representation of A generated by ψ (as in Proposition 2.3.18).

Proof. On the dense subspace A/L_ψ of H_ψ, define

$$[a_\psi, b_\psi] = [\pi_\psi(a)\xi_\psi, \pi_\psi(b)\xi_\psi] = \varphi(b^*a), \quad \forall a, b \in A.$$

Since $\varphi \leq \psi$, it follows that $\|[a_\psi, b_\psi]\| \leq \|a_\psi\| \cdot \|b_\psi\|, \forall a, b \in A$. Thus there is unique $t' \in B(H_\psi)$ such that

$$\varphi(b^*a) = \langle t'\pi_\psi(a)\xi_\psi, \pi_\psi(b)\xi_\psi\rangle, \quad \forall a, b \in A.$$

Now by the proof of Lemma 1.10.1, we can get the conclusion. Q.E.D.

Now we study the *orthogonal (Jordan) decomposition* of a *hermiatian* functional. Let A be a C^*-algebra, and $X = \{\rho \in A^* \mid \rho \geq 0 \text{ and } \|\rho\| \leq 1\}$. Clearly, X is a compact Hausdorff space with respect to $\sigma(A^*, A)$. Denote by $C_r(X)$ the set of all real continuous functions on X. For $a \in A_H$ (the set of all *self-adjoint* elements of A), define $a(\rho) = \rho(a), \forall \rho \in X$; then $a(\cdot) \in C_r(X)$. By 2.3.14 and 2.3.15, the map: $a \to a(\cdot)$ is isometric (i.e. $\|a\| = \sup_{\rho \in X} |a(\rho)|$) and preserves order (i.e. $a(\cdot) \geq 0$ if $a \in A_+$) from A_H into $C_r(X)$, and also its inverse preserves the order (i.e. $a \in A_+$ if $a(\cdot) \geq 0$).

Suppose that f is a *hermiatian* continuous linear functional on A, i.e., $f \in A^*$ and $f^* = f$, where f^* is defined by $f^*(a) = \overline{f(a^*)}, \forall a \in A$. Then $\|f\| = \|f|A_H\|$. Put $F(a(\cdot)) = f(a), \forall a \in A_H$. Clearly, F can be extended to a continuous linear functional on $C_r(X)$, still denoted by F, with the same norm $\|f\|$. By the Riesz representation theorem, we can write

$$F = F_+ - F_-, \quad \|f\| = \|F\| = \|F_+\| + \|F_-\|,$$

where F_+ and F_- are positive on $C_r(X)$. Restricting F_+, F_- to $\{a(\cdot) \mid a \in A_H\}$, we get positive functionals f_+, f_- on A_H. Let $f_\pm(a + ib) = f_\pm(a) + if_\pm(b), \forall a, b \in A_H$. Then f_+ and f_- are positive on A, and $f = f_+ - f_-$. Moreover, since $\|f\| = \|F\| = \|F_+\| + \|F_-\|, \|F_\pm\| \geq \|f_\pm\|$, and $\|f\| = \|f_+ - f_-\| \leq \|f_+\| + \|f_-\|$, it follows that $\|f\| = \|f_+\| + \|f_-\|$.

The above decomposition is called the *orthogonal decomposition* (or *Jordan decomposition*) of the herniation functional f. When A is commutative, it is the ordinary Jordan decomposition of a signed measure exactly.

Now we prove that the above decomposition is unique.

By Proposition 2.3.18, for each $\rho \in X$, there is a cyclic $*$ representation $\{\pi_\rho, H_\rho, \xi_\rho\}$ of A such that

$$\rho(a) = \langle \pi_\rho(a)\xi_\rho, \xi_\rho\rangle, \quad \forall a \in A.$$

Let

$$\pi = \sum_{\rho \in X} \oplus \pi_\rho, \quad H = \sum_{\rho \in X} \oplus H_\rho.$$

Then $\{\pi, H\}$ is a faithful $*$ representation of A. Let $M = \pi(A)''$. Then M is a VN algebra on H. We may assume $\|f\| \leq 1$; so $f_\pm \in X$. Write $\xi_\pm = \xi_{f_\pm}$; then $f_\pm(a) = \langle \pi(a)\xi_\pm, \xi_\pm \rangle, \forall a \in A$. Identifying A with $\pi(A)$, f and $f_\pm$ can be naturally extended as follows:

$$f_\pm(b) = \langle b\xi_\pm, \xi_\pm \rangle, \quad f(b) = f_+(b) - f_-(b), \quad \forall b \in M.$$

Denote by $\|f\|_M, \|f_\pm\|_M$ the norms of $f, f_\pm$ as the functionals on M respectively. By the Kaplansky density theorem, we have $\|f_\pm\|_M = \|f_\pm\|$. Since $\|f\| \leq \|f\|_M \leq \|f_+\|_M + \|f_-\|_M = \|f_+\| + \|f_-\| = \|f\|$, it follows that $\|f\|_M = \|f_+\|_M + \|f_-\|_M$. Now by Theorem 1.9.8, we obtain the following theorem.

Theorem 2.3.23. Let A be a C^*-algebra, f be a hermition continuous linear functional on A, i.e., $f \in A^*$ and $f(a^*) = \overline{f(a)}, \forall a \in A$. Then there exist unique positive linear functionals f_+ and f_- on A such that

$$f = f_+ - f_-, \quad \text{and} \quad \|f\| = \|f_+\| + \|f_-\|.$$

Corollary 2.3.24. Let A be a C^*-algebra. Then A^* is the linear span of $S(A)$.

Notes. The GNS construction was studied first by I.M. Gelfand and M.A. Naimark. Then I.E. Segal gave its perfected form.

References. [52], [155].

2.4. Approximate identities and quotient C^*-algebras

Proposition 2.4.1. Let A be a C^*-algebra, and L be a left ideal of A. Then there is a net $\{d_l\} \subset L$ with $d_l \in A_+, \|d_l\| \leq 1, \forall l$, and $d_l \leq d_{l'}, \forall l \leq l'$, such that

$$\|xd_l - x\| \to 0, \forall x \in \overline{L}.$$

Proof. Let Λ be the set of all finite subsets of L. And Λ is directed by the inclusion relation. For any $l = \{x_1, \cdots, x_n\} \in \Lambda$, put

$$h_l = \sum_{i=1}^n x_i^* x_i, \quad d_l = nh_l(1 + nh_l)^{-1}.$$

Clearly, $h_l, d_l \in L \cap A_+$, and $\|d_l\| \leq 1$.

Let $l' = \{x_1, \cdots, x_n, \cdots, x_m\} \geq l = \{x_1, \cdots, x_n\}$, where $m \geq n$, and $x_i \in L, 1 \leq i \leq m$. Then $(\frac{1}{n} + h_l) \leq (\frac{1}{n} + h_{l'})$. By Proposition 2.2.9, we have $(\frac{1}{n} + h_{l'})^{-1} \leq (\frac{1}{n} + h_l)^{-1}$. Since $\frac{1}{n}(\frac{1}{n} + h_{l'})^{-1} \geq \frac{1}{m}(\frac{1}{m} + h_{l'})^{-1}$, it follows that $\frac{1}{n}(\frac{1}{n} + h_l)^{-1} \geq \frac{1}{m}(\frac{1}{m} + h_{l'})^{-1}$. Further,

$$d_l = 1 - \frac{1}{n}(\frac{1}{n} + h_l)^{-1} \leq 1 - \frac{1}{m}(\frac{1}{m} + h_{l'})^{-1} = d_{l'}.$$

Now suppose $l = \{x_1, \cdots, x_n\} \in \Lambda$. Clearly,

$$\|(1 - d_l)h_l(1 - d_l)\| = \|h_l(1 + nh_l)^{-2}\| \leq \frac{1}{4n}.$$

Moreover, $(1 - d_l)h_l(1 - d_l) = \sum_{i=1}^{n}(x_i(1 - d_l))^*(x_i(1 - d_l))$. Thus $\|x_i - x_i d_l\| \leq (2\sqrt{n})^{-1}, 1 \leq i \leq n$.

For any $x \in L$ and $\varepsilon > 0$, pick $l_\varepsilon \in \Lambda$ such that $x \in l_\varepsilon$ and $\#l_\varepsilon > (4\varepsilon^2)^{-1}$. From preceding paragraph, we have $\|x - xd_l\| < \varepsilon, \forall l \geq l_\varepsilon$. Therefore, $\|x - xd_l\| \to 0, \forall x \in \overline{L}$. Q.E.D.

Definition 2.4.2. Let A be a C^*-algebra. A net $\{d_l\}(\subset A)$ is called an *approximate identity* for A, if $0 \leq d_l \leq 1, d_l \leq d_{l'}, \forall l \leq l'$, and $\|ad_l - a\| \to 0, \|d_l a - a\| \to 0, \forall a \in A$.

By Proposition 2.4.1, we have the following.

Theorem 2.4.3. Every C^*-algebra admits an approximate identity.

Proposition 2.4.4. Let A be a C^*-algebra, and $\{d_l\}$ be an approximate identity for A.

1) For any $\varphi \in S(A)$, we have

$$\lim_l \varphi(d_l) = \lim_l \varphi(d_l^2) = 1.$$

2) If A has no identity, then the C^*-norm on $(A \dot{+} \mathbb{C})$ (see Proposition 2.1.2) is also represented by

$$\|x + \lambda\| = \lim_l \|xd_l + \lambda d_l\| = \lim_l \|d_l x + \lambda d_l\|,$$

$\forall x \in A, \lambda \in \mathbb{C}$.

Proof. 1) For any $a \in A$ with $\|a\| \leq 1$, by the Schwartz inequality we have

$$1 \geq \varphi(d_l) \geq \varphi(d_l^2) \geq \varphi(d_l^2)\varphi(a^*a) \geq |\varphi(d_l a)|^2 \to |\varphi(a)|^2.$$

Since $1 = \|\varphi\| = \sup\{|\varphi(a)| \mid a \in A, \|a\|| \leq 1\}$, it follows that $\lim_l \varphi(d_l) = \lim_l \varphi(d_l^2) = 1$.

2) Fix $x \in A, \lambda \in \mathbb{C}$. For any $\varepsilon > 0$, we can pick $y \in A$ with $\|y\|| \leq 1$ such that

$$\|x + \lambda\| \geq \|xy + \lambda y\| > \|x + \lambda\| - \varepsilon.$$

Since $d_l y \to y$, it follows that

$$\|x + \lambda\| \geq \|(x + \lambda)d_l\| \geq \|(x + \lambda)d_l y\| \geq \|x + \lambda\| - \varepsilon$$

for l enough late. Therefore, $\|x+\lambda\| = \lim_l \|xd_l + \lambda d_l\|$ and $\|x+\lambda\| = \|x^* + \overline{\lambda}\| = \lim_l \|x^* d_l + \overline{\lambda} d_l\| = \lim_l \|d_l x + \lambda d_l\|, \forall x \in A, \lambda \in \mathbb{C}$. Q.E.D.

Definition 2.4.5. Let $\{\pi, H\}$ be a $*$ representation of a C^*-algebra A. The closed linear span of $\{\pi(a)\xi \mid a \in A, \xi \in H\}$ is called the *essential subspace* of $\{\pi, H\}$. The $*$ representation $\{\pi, H\}$ of A is said to be *nondegenerate*, if its essential subspace is H.

Clearly, the orthogonal complement of the essential subspace is the null subspace, i.e.

$$\{\pi(a)\xi \mid a \in A, \xi \in H\}^{\perp} = \{\eta \in H \mid \pi(a)\eta = 0, \forall a \in A\}.$$

Therefore, the null subspace of a nondegenerate $*$ representation is trivial. In this case, the weak closure of $\pi(A)$ is a VN algebra on H (see Theorem 1.3.9).

Proposition 2.4.6. Let A be a C^*-algebra, and $\{d_l\}$ be an approximate identity for A, $\{\pi, H\}$ be a $*$ representation of A. Then $\pi(d_l) \to p$ (strongly), where p is the projection from H onto the essential subspace $\overline{[\pi(A)H]}$ of $\{\pi, H\}$. In particular, if $\{\pi, H\}$ is nondegenerate, then $\pi(d_l) \to 1$ (strongly).

Proof. By Proposition 1.2.10, we have $\pi(d_l) \to p = \sup_l \pi(d_l)$ (strongly). Let $K = \overline{[\pi(A)H]}$. Then for any $\eta \in K^{\perp}, \pi(d_l)\eta = 0, \forall l$, and $p\eta = 0$. On the other hand, for any $a \in A, \xi \in H$, since $\|\pi(d_l a)\xi - \pi(a)\xi\| \leq \|d_l a - a\| \cdot \|\xi\| \to 0$, it follows that $p\pi(a)\xi = \pi(a)\xi$. Therefore, $pH = K$.

Q.E.D.

Remark. If $\{\pi, H\}$ is a nondegenerate $*$ representation of A, by 2.4.4 and 2.4.6, then for any $\xi \in H$ with $\|\xi\| = 1, \langle \cdot \xi, \xi \rangle \in S(A)$.

Proposition 2.4.7. Let I be a closed two-sided ideal of a C^*-algebra A. Then $I^* = \{a^* \mid a \in I\} = I$.

Proof. By Proposition 2.4.1, there is a net $\{d_l\} \subset I$ such that $ad_l \to a, \forall a \in I$. Then for any $a \in I$,

$$\|d_l a^* - a^*\| = \|(ad_l - a)^*\| = \|ad_l - a\| \to 0.$$

Since $d_l a^* \in I, \forall l$ and I is closed, it follows that $a^* \in I, \forall a \in I$. Q.E.D.

Now let A be a C^*-algebra, and I be a closed two-sided ideal of A. By Proposition 2.4.7, A/I is a Banach $*$ algebra with respect to the quotient norm. Let $\{d_l\}$ be an approximate identity for I, and $a \to \tilde{a} = a + I$ be the cononical map from A onto A/I. We claim that

$$\|\tilde{a}\| = \lim_l \|ad_l - a\|, \quad \forall a \in A.$$

In fact, fix $a \in A$. For any $b \in I$, since $bd_l \to b$, it follows that

$$\begin{aligned} \overline{\lim_l}\|ad_l - a\| &= \overline{\lim_l}\|ad_l - a + bd_l - b\| \\ &= \overline{\lim_l}\|(a+b)(1-d_l)\| \leq \|a+b\|. \end{aligned}$$

Thus $\overline{\lim_l}\|ad_l - a\| \leq \inf\{\|a+b\| \mid b \in I\} = \|\tilde{a}\|$. On the other hand, since $ad_l \in I$, we have

$$\overline{\lim_l}\|ad_l - a\| \geq \underline{\lim_l}\|ad_l - a\| \geq \|\tilde{a}\|.$$

Therefore, $\|\tilde{a}\| = \lim_l \|ad_l - a\|, \quad \forall a \in A$.

Now for any $a \in A, b \in I$, since $bd_l \to b$, we have

$$\begin{aligned} \|\tilde{a}\|^2 &= \lim_l \|ad_l - a\|^2 = \lim_l \|(a - ad_l)^*(a - ad_l)\| \\ &= \lim_l \|(1-d_l)a^*a(1-d_l)\| \\ &= \lim_l \|(1-d_l)(a^*a+b)(1-d_l)\| \leq \|a^*a+b\|. \end{aligned}$$

Hence

$$\|\tilde{a}\|^2 \leq \inf\{\|a^*a+b\| \mid b \in I\} = \|\widetilde{a^*a}\| \leq \|\tilde{a}^*\| \cdot \|\tilde{a}\|, \quad \forall a \in A.$$

Furthermore, $\|\tilde{a}^*\tilde{a}\| = \|\tilde{a}\|^2, \forall a \in A$. Therefore we have the following.

Theorem 2.4.8. Let A be a C^*-algebra, and I be a closed two-sided ideal of A. Then A/I is a C^*-algebra in a natural way.

Proposition 2.4.9. Let Φ be a $*$ homomorphism from a C^*-algebra A into another C^*-algebra B. Then $\Phi(A)$ is a C^*-subalgebra of B. In particular, if

$\{\pi, H\}$ is a $*$ representation of a C^*-algebra A, then $\pi(A)$ is a C^*-algebra on H.

Proof. By Theorem 2.3.20, it suffices to consider the case of $\{\pi, H\}$ and A. Let $I = \{a \in A \mid \pi(a) = 0\}$. Then I is a closed two-sided ideal of A. Define

$$\tilde{\pi}(\tilde{a}) = \pi(a), \quad \forall \tilde{a} \in A/I \text{ and } a \in \tilde{a}.$$

Clearly, $\{\tilde{\pi}, H\}$ is a faithful $*$ representation of the quotient C^*-algebra A/I. By Proposition 2.3.17, $\pi(A) = \tilde{\pi}(A/I)$ is a C^*-algebra on H. Q.E.D.

Proposition 2.4.10. Let A be a C^*-algebra, I be a closed two-sided ideal of A and B be a C^*-subalgebra of A. Then $(B + I) = \{(b + c) \mid b \in B, c \in I\}$ is a C^*-subalgebra of A, and the C^*-algebras $(B + I)/I$ and $B/(B \cap I)$ are canonically isomorphic.

Proof. Let $a \rightarrow \tilde{a} = a + I$ be the canonical map from A onto A/I. Clearly, it is also a $*$ homomorphism. By Proposition 2.4.9, $\tilde{B} = \{\tilde{b} \mid b \in B\}$ is a C^*-subalgebra of A/I.

It suffices to show that $(B + I)$ is closed. Let $\{x_n\} \subset (B + I)$ and $x_n \rightarrow x$. Then $\tilde{x}_n \rightarrow \tilde{x}$. Since $\tilde{B} = \widetilde{B + I}$ is a C^*-subalgebra of A/I, it follows that $\tilde{x} \in \tilde{B}$, i.e., $x \in (B + I)$.

Now it is easy to see that

$$b + (B \cap I) \longrightarrow b + I \quad (\forall b \in B)$$

is a $*$ isomorphicm from $B/(B \cap I)$ onto $(B + I)/I$. Q.E.D.

Proposition 2.4.11. Let A be a C^*-algebra, and I be a closed two-sided ideal of A. If ρ is a state (or pure state) on A and $\rho(I) = \{0\}$, let $\tilde{\rho}(\tilde{a}) = \rho(a)(\forall \tilde{a} \in A/I, a \in \tilde{a})$, then $\tilde{\rho}$ is a state (or pure state) on A/I. Conversely, if $\tilde{\rho}$ is a state (or pure state) on A/I, then there is unique state (or pure state) ρ on A such that $\rho(I) = \{0\}$ and $\rho(a) = \tilde{\rho}(\tilde{a}), \forall a \in A$.

Proof. Let ρ be a state on A with $\rho(I) = \{0\}$. Then we can define $\tilde{\rho}(\tilde{a}) = \rho(a), \forall \tilde{a} \in A/I, a \in \tilde{a}$. Clearly, $\tilde{\rho}$ is a positive linear functional on A/I, and $\|\tilde{\rho}\| \leq 1$. On the other hand,

$$\begin{aligned} 1 = & \sup\{|\rho(a)| \mid a \in A, \|a\| \leq 1\} \\ = & \sup\{|\tilde{\rho}(\tilde{a})| \mid a \in A, \|a\| \leq 1\} \leq \|\tilde{\rho}\|. \end{aligned}$$

Thus $\|\tilde{\rho}\| = 1$ and $\tilde{\rho}$ is a state on A/I.

Conversely, let $\tilde{\rho}$ be a state on A/I. Define $\rho(a) = \tilde{\rho}(\tilde{a}), \forall a \in A$. Clearly, ρ is a positive linear functional on A, and $\rho(I) = \{0\}$. From the preceding paragraph, we can see that $\|\rho\| = \|\tilde{\rho}\|$, so $\|\rho\| = 1$ and ρ is a state on A.

Now let ρ be a pure state on A with $\rho(I) = \{0\}$. By the preceding paragraph, $\tilde{\rho}$ is a state on A/I. Suppose that there are states $\tilde{\rho}_1, \tilde{\rho}_2$ on A/I and $\lambda \in (0,1)$ such that $\tilde{\rho} = \lambda\tilde{\rho}_1 + (1-\lambda)\tilde{\rho}_2$. Define $\rho_i(a) = \tilde{\rho}_i(\tilde{a}), \forall a \in A, i = 1,2$. Then ρ_1 and ρ_2 are two states on A with $\rho_1(I) = \rho_2(I) = \{0\}$. Clearly, $\rho = \lambda\rho_1 + (1-\lambda)\rho_2$. Since ρ is pure, it follows that $\rho = \rho_1 = \rho_2$. Further $\tilde{\rho} = \tilde{\rho}_1 = \tilde{\rho}_2$. Therefore $\tilde{\rho}$ is pure on A/I.

Finally, let $\tilde{\rho}$ be a pure state on A/I. Then there is unique state ρ on A such that $\rho(I) = \{0\}$ and $\rho(a) = \tilde{\rho}(\tilde{a}), \forall a \in A$. Suppose that there two states ρ_1, ρ_2 on A and $\lambda \in (0,1)$ such that $\rho = \lambda\rho_1 + (1-\lambda)\rho_2$. For any $a \in I \cap A_+$, from $\rho(a) = 0$ we have $\rho_1(a) = \rho_2(a) = 0$. Further, $\rho_1(I) = \rho_2(I) = \{0\}$. Thus $\tilde{\rho} = \lambda\tilde{\rho} + (1-\lambda)\tilde{\rho}_2$, where $\tilde{\rho}_i$ is defined by $\tilde{\rho}_i(\tilde{a}) = \rho_i(a), \forall \tilde{a} \in A/I, a \in \tilde{a}, i = 1,2$. Since $\tilde{\rho}$ is pure, it follows that $\tilde{\rho} = \tilde{\rho}_1 = \tilde{\rho}_2$. Further $\rho = \rho_1 = \rho_2$. Therefore ρ is pure on A. Q.E.D.

Notes. I.E. Segal showed the existence of an approximate identity in a C^*-algebra.

References. [25], [52], [81], [155], [156].

2.5. Extreme points of the unit ball and the existence of an identity

Theorem 2.5.1. Let A be a C^*-algebra, and $S = \{a \in A \mid \|a\| \le 1\}$ be its unit ball, $x \in S$. Then x is an extreme point of S if and only if $(1 - x^*x)A(1 - xx^*) = \{0\}$.

Moreover, if x is an extreme point of S, then x is a partial isometry, i.e. x^*x and xx^* are projections.

Proof. Let x be an extreme point of S. First we prove that x^*x is a projection. In fact, let B be the abelian C^*-subalgebra generated by x^*x. Then $B \cong C_0^\infty(\Omega)$. If there is $t_0 \in \Omega$ such that $x^*x(t_0) \in (0,1)$. By the continuity, we can find an open neighborhood $U(\subset \Omega)$ of t_0 and $\varepsilon \in (0,1)$ such that

$$0 < x^*x(t) < 1 - \varepsilon, \quad \forall t \in U.$$

Pick $d \in B$ such that

$$0 \le d(t) \le 1, \forall t \in \Omega, d(t_0) = 1, d(\Omega \backslash U) = \{0\},$$

and $\eta \in (0,1)$ such that $2\eta + \eta^2 \le \varepsilon$. Then

$$0 \le (1 \pm \eta d(t))^2 x^*x(t) = \begin{cases} x^*x(t)(\le 1), & \forall t \notin U, \\ \le (1+\eta)^2(1-\varepsilon)(< 1), & \forall t \in U. \end{cases}$$

Since d and x^*x commute, it follows that

$$\begin{aligned}\|x \pm \eta xd\|^2 &= \|(x(1 \pm \eta d))^* \cdot (x(1 \pm \eta d))\| \\ &= \|(1 \pm \eta d)^2 \cdot x^*x\| \leq 1.\end{aligned}$$

Now from $x = \frac{1}{2}(x + \eta xd) + \frac{1}{2}(x - \eta xd)$, we get $xd = 0$. Further $x^*x \cdot d = 0$. It is impossible since $(x^*x \cdot d)(t_0) = x^*x(t_0) > 0$. Therefore $x^*x(t) = 0$ or 1, $\forall t \in \Omega$, i.e. x^*x is a projection. Similarly, xx^* is a projection since x^* is still an extreme point of S.

Let $p = x^*x, q = xx^*$. If $y \in (1-p)A(1-q)$ with $\|y\| \leq 1$, then $py = 0$. Further, $0 = y^*py = (xy)^* \cdot (xy)$, and $xy = 0$. By Theorem 2.3.20,

$$\begin{aligned}\|x \pm y^*\|^2 = \ &\|x^* \pm y\|^2 = \|(x^* \pm y)^* \cdot (x^* \pm y)\| \\ = \ &\|xx^* + y^*y\| = \|qxx^*q + (1-q)y^*y(1-q)\| \\ = \ &\max\{\|xx^*\|, \|y^*y\|\} \leq 1.\end{aligned}$$

Since x is an extreme point of S and $x = \frac{1}{2}(x + y^*) + \frac{1}{2}(x - y^*)$, it follows that $y^* = 0$. Therefore, $(1-p)A(1-q) = \{0\}$.

Conversely, suppose that $(1 - x^*x)A(1 - xx^*) = \{0\}$. Then

$$0 = x^*(1 - xx^*)x(1 - x^*x) = x^*x \cdot (1 - x^*x)^2.$$

Thus $\sigma(x^*x) \subset \{0, 1\}$, i.e., x^*x is a projection. Similarly, xx^* is also a projection. Let $p = x^*x, q = xx^*$. Since $(xp - x)^*(xp - x) = px^*xp - px^*x - x^*xp + x^*x = 0$, it follows that

$$xp = x, \quad px^* = x^*. \tag{1}$$

Suppose that there are $a, b \in S$ and $\lambda \in (0,1)$ such that $x = \lambda a + (1-\lambda)b$. Then $p = x^*xp = \lambda x^*ap + (1-\lambda)x^*bp$. By (1), $p \cdot x^*ap = x^*ap \cdot p$. Thus the set $\{p, x^*ap, x^*bp\}$ is commutative, and it can generate an abelian C^*-subalgebra with an identity p. By the Gelfand transformation, we can see that

$$p = x^*ap = x^*bp. \tag{2}$$

By (1), (2) and $q = xx^*$, we have

$$x = qap = qbp. \tag{3}$$

From (2), (3), $pa^*qap = pa^*x = (x^*ap)^* = p$, hence

$$\begin{aligned}1 \geq \|pa^*ap\| &= \|pa^*qap + pa^*(1-q)ap\| \\ &= \|p + pa^*(1-q)ap\|.\end{aligned}$$

But $pa^*(1-q)ap$ is a positive element of the C^*-subalgebra pAp, and pAp has an identity p, so $pa^*(1-q)ap = 0, (1-q)ap = 0, ap = qap$. By (3), we obtain

$$x = ap. \tag{4}$$

Since $x = \lambda a + (1-\lambda)b$, it follows that $y = \lambda c + (1-\lambda)d$, where $y = x^*, c = a^*, d = b^*$. Replacing $\{x, a, b, p, q\}$ by $\{y, c, d, q, p\}$ in above procedure (1)–(4), we obtain $y = cq$ since $y^*y = q$ and $yy^* = p$. Thus

$$x = qa. \tag{5}$$

Since $(1-q)a(1-p) \in (1-q)A(1-p) = \{0\}$, it follows that $a = ap + qa - qap$. Now by (3), (4), (5), $x = a$. Further, $x = a = b$. Therefore, x is an extreme point of S. Q.E.D.

Corollary 2.5.2. If a C^*-algebra has an identity, then the identity is an extreme point of its unit ball.

Theorem 2.5.3. Let A be a C^*-algebra, and $S = \{a \in A \mid \|a\| \leq 1\}$ be its unit ball. Then A has an identity if and only if S has an extreme point at least.

Proof. The necessity is clear from Corollary 2.5.2. Now suppose that S has an extreme point x. Let $p = x^*x, q = xx^*$, and $\{d_l\}$ be an approximate identity for A. By Theorem 2.5.1, $(1-q)d_l(1-p) = 0, \forall l$. Thus

$$d_l \to p + q - qp.$$

Clearly, $e = p + q - qp$ is an identity of A. Q.E.D.

Proposition 2.5.4. Let A be a C^*-algebra, and $S(A)$ be its state space. Then A has an identity if and only if $S(A)$ is compact with respect to the w^*-topology $\sigma(A^*, A)$.

Proof. The necessity is clear. Now suppose that A has no identity. We prove that $(S(A), \sigma(A^*, A))$ is not compact. It suffice to show that $0 \in \overline{S(A)}^\sigma$, where $\overline{S(A)}^\sigma$ is the $\sigma(A^*, A)$-closure of $S(A)$ in A^*. Let $U = U(0; a_1, \cdots, a_n; \varepsilon) = \{f \in A^* \mid |f(a_i)| < \varepsilon, 1 \leq i \leq n\}$ be any $\sigma(A^*, A)$-neighborhood of 0, we need to prove that $U \cap S(A) \neq \emptyset$. Since $A = [A_+]$, we may assume that $a_i \in A_+, 1 \leq i \leq n$. Let $a = a_1 + \cdots + a_n$. It suffices to show $U(0; a; \varepsilon) \cap S(A) \neq \emptyset$. By Theorem 2.3.20, we may assume that $A \subset B(H)$ (some Hilbert space H), and A is nondegenerate on H. Since A has no identity, it follows that a is not invertible in $B(H)$. Thus there is $\xi \in H$ with $\|\xi\| = 1$, such that $\langle a\xi, \xi \rangle < \varepsilon$. Let $\rho(\cdot) = \langle \cdot \xi, \xi \rangle$. By Proposition 2.4.6, $\rho \in S(A) \cap U(0; a; \varepsilon)$. Therefore, $S(A) \cap U(0; a; \varepsilon) \neq \emptyset$. Q.E.D.

Let E be a linear space, K be a convex subset of E, and F be a subset of E. We shall denote by ExK and CoF the sets of all extreme points of K and convex hull of F respectively.

Now let A be a C^*-algebra, $S(A)$ and $P(A)$ be its state space and pure state space respectively. For any $E \subset A^*$, the $\sigma(A^*, A)$-closure of E in A^* is denoted by $\overline{E}^\sigma$. Clearly, $\{\rho \in A^* \mid \rho \geq 0, \|\rho\| \leq 1\}$ is a $\sigma(A^*, A)$-compact convex subset of A^*. And also it is easily verified that $Ex\{\rho \in A^* \mid \rho \geq 0, \|\rho\| \leq 1\} = \{0, P(A)\}$. By the Kreim-Milmann theorem, we have

$$\{\rho \in A^* \mid \rho \geq 0, \|\rho\| \leq 1\} = \overline{Co\{0, P(A)\}}^\sigma.$$

If A has no identity, then from the proof of Proposition 2.5.4, $0 \in \overline{S(A)}^\sigma$. Thus $\overline{Co\{0, P(A)\}}^\sigma \subset \overline{S(A)}^\sigma$. Clearly, $\overline{S(A)}^\sigma \subset \{\rho \in A^* \mid \rho \geq 0, \|\rho\| \leq 1\}$. Therefore, we get

$$\overline{S(A)}^\sigma = \overline{Co\{0, P(A)\}}^\sigma = \{\rho \in A^* \mid \rho \geq 0, \|\rho\| \leq 1\}.$$

Furthermore, for any $\varphi \in S(A)$ there is a net $\{\sum_i \lambda_i^{(l)} \varphi_i^{(l)}\}_l \subset Co\{0, P(A)\}$, where $\lambda_i^{(l)} \geq 0, \varphi_i^{(l)} \in P(A), \forall i, l$ and $\sum_i \lambda_i^{(l)} \leq 1$, such that

$$\sum_i \lambda_i^{(l)} \varphi_i^{(l)} \to \varphi \quad (\sigma(A^*, A)).$$

Since $\|\varphi\| = 1$, considering a subnet, we may assume that $\sum_i \lambda_i^{(l)} \to 1$. Then $(\sum_i \lambda_i^{(l)})^{-1} \sum_i \lambda_i^{(l)} \varphi_i^{(l)} \to \varphi \ (\sigma(A^*, A))$. Thus $S(A) \subset \overline{CoP(A)}^\sigma$, and we have the following.

Proposition 2.5.5. Let A be a C^*-algebra. Then

$$Ex\{\rho \in A^* \mid \rho \geq 0, \|\rho\| \leq 1\} = \{0, P(A)\}$$

and

$$\{\rho \in A^* \mid \rho \geq 0, \|\rho\| \leq 1\} = \overline{Co\{0, P(A)\}}^\sigma.$$

Moreover, if A has no identity, then we have

$$\{\rho \in A^* \mid \rho \geq 0, \|\rho\| \leq 1\} = \overline{CoP(A)}^\sigma = \overline{S(A)}^\sigma.$$

Notes. Theorem 2.5.1 is due to R.V. Kadison. Proposition 2.5.4 was obtained by I.E. Segal.

References. [77], [156].

2.6. Transitivity theorem and irreducible * representations

Definition 2.6.1. Let A be a C^*-algebra, and $\{\pi, H\}$ be a * representation of A. $\{\pi, H\}$ is said to be *algebraically irreducible*, if E is a linear subspace of H such that $\pi(a)\xi \in E, \forall a \in A, \xi \in E$ (i.e., E is invariant for $\pi(A)$), then either $E = \{0\}$ or $E = H$.$\{\pi, H\}$ is said to be *topologically irreducible*, if E is a closed linear subspace of H such that $\pi(a)\xi \in E, \forall a \in A, \xi \in E$, then either $E = \{0\}$ or $E = H$.

Clearly, an algebraically irreducible * representation must be topologically irreducible. In this section, we shall show that the converse is also true for a C^*-algebra.

Proposition 2.6.2. Let A be a C^*-algebra, and $\{\pi, H\}$ be a * representation of A. Then $\{\pi, H\}$ is topologically irreducible if and only if $\pi(A)$ is dense in $B(H)$ with respect to the weak (operator) topology.

Proof. Let $\{\pi, H\}$ be topologically irreducible. Then $\pi(A)'$ does not contain any projection which is not equal to 0 or 1. Hence, $\pi(A)' = \mathbb{C}, \pi(A)'' = B(H)$. Of course, π is nondegenerate. By Theorem 1.3.9, $\pi(A)$ is weakly dense in $B(H)$. Coversely, if $\pi(A)$ is weakly dense in $B(H)$, then it is clear that $\pi(A)' = \mathbb{C}$. Therefore $\{\pi, H\}$ is topologically irreducible. Q.E.D.

Lemma 2.6.3. Let H be a Hilbert space, $\xi_i, \eta_i \in H, 1 \le i \le n$, and $\langle \xi_i, \xi_j \rangle = \delta_{ij}, 1 \le i, j \le n$. Then there exists $b \in B(H)$ with $b^* = b$ such that

$$b\xi_i = \eta_i, \quad 1 \le i \le n, \text{ and } \|b\|^2 \le \sum_{i=1}^{n} \|\eta_i\|^2.$$

Moreover, if there is $h^* = h \in B(H)$ such that $h\xi_i = \eta_i, 1 \le i \le n$, then we can take $b \in B(H)$ with $b^* = b$ such that

$$b\xi_i = \eta_i, \quad 1 \le i \le n, \text{ and } \|b\|^2 \le 2\sum_{i=1}^{n} \|\eta_i\|^2.$$

Proof. Let K be the linear span of $\{\xi_1, \cdots, \xi_n, \eta_1, \cdots, \eta_n\}$, and $\{\xi_1, \cdots, \xi_m\}$ be a normalized orthogonal basis of $K (m \ge n)$. Now take $b \in B(H)$ such that $bK^\perp = \{0\}, bK \subset K$, and $(b|K)$ has a matrix representation with respect to the basis $\{\xi_1, \cdots, \xi_m\}$ as follows:

$$\begin{pmatrix} \alpha_{11} & \cdots & \cdots & \alpha_{1n} & \\ \cdots & \cdots & \cdots & \cdots & \Delta \\ \alpha_{n1} & \cdots & \cdots & \alpha_{nn} & \\ \alpha_{n+1,1} & \cdots & \cdots & \alpha_{n+1,n} & \\ \cdots & \cdots & \cdots & \cdots & 0 \\ \alpha_{m1} & \cdots & \cdots & \alpha_{mn} & \end{pmatrix},$$

where $\eta_i = \sum_{j=1}^{m} \alpha_{ji}\xi_j, 1 \le i \le n$. Then we have $b\xi_i = \eta_i, 1 \le i \le n$.

For first case, pick $\Delta = (0)$. Then

$$\begin{aligned} \|b\| = & \|b|K\| = \sup\{\|b\sum_{i=1}^{m} \lambda_i\xi_i\| \mid \sum_{i=1}^{m}|\lambda_i|^2 \le 1\} \\ = & \sup\{\|\sum_{i=1}^{n} \lambda_i\eta_i\| \mid \sum_{i=1}^{m}|\lambda_i|^2 \le 1\} \le (\sum_{i=1}^{n}\|\eta_i\|^2)^{1/2}. \end{aligned}$$

For second case, pick

$$(\Delta) = \begin{pmatrix} \overline{\alpha_{n+1,1}} & \cdots & \cdots & \overline{\alpha_{m1}} \\ \cdots & \cdots & \cdots & \cdots \\ \overline{\alpha_{n+1,n}} & \cdots & \cdots & \overline{\alpha_{mn}} \end{pmatrix}.$$

Then we have $b^* = b$ since $\alpha_{ji} = \langle \eta_i, \xi_j\rangle = \langle h\xi_i, \xi_j\rangle = \langle \xi_i, h\xi_j\rangle = \langle \xi_i, \eta_j\rangle = \overline{\alpha_{ij}}, \forall 1 \le i,j \le n$. Moreover,

$$\begin{aligned} \|b\|^2 = & \|b^*b\| = \max\{\lambda \mid \lambda \in \sigma(b^*b)\} \\ \le & tr(b^*b) = \sum_{i=1}^{m} \|b\xi_i\|^2 \\ = & \sum_{i=1}^{n}\|\eta_i\|^2 + \sum_{i=n+1}^{m}\sum_{j=1}^{n}|\alpha_{ij}|^2 \le 2\sum_{i=1}^{n}\|\eta_i\|^2. \end{aligned}$$

Q.E.D.

Lemma 2.6.4. Let $H_1, \cdots, H_n$ be Hilbert spaces, $H = \sum_{j=1}^{n} \oplus H_j$ be the Hilbert direct sum of $H_1, \cdots, H_n$, and $M = \sum_{j=1}^{n} \oplus B(H_j)$ (a VN algebra on H). Suppose that A is a C^*-algebra on H with $A \subset M$, and A is weakly dense in M. Assume that $t_j \in B(H_j), e_j$ is a projection of finite rank on H_j, and p_j is the projection from H_j onto $[e_jH_j, t_je_jH_j], 1 \le j \le n$. Then for any $\varepsilon > 0$, there is $b \in A$ such that

$$be_j = t_je_j, 1 \le j \le n, \|b\| \le \varepsilon + \max_{1\le j\le n} \|p_jt_jp_j\|$$

Moreover, if $t_j^* = t_j, 1 \le j \le n$, then we can choose $b \in A$ such that

$$b^* = b, \quad be_j = t_j e_j, 1 \le j \le n, \quad \|b\| \le \max_{1 \le j \le n} \|p_j t_j p_j\|.$$

Proof. Let $t = \sum_{j=1}^n \oplus t_j, p = \sum_{j=1}^n \oplus p_j$. Then $t, p \in M$. Pick a normalized orthogonal basis $\{\xi_1, \cdots, \xi_m\}$ of $pH = \sum_{j=1}^n \oplus p_j H_j$, such that ξ_i belongs to some $p_j H_j, \forall i$.

For $\varepsilon_1 > 0$, by Theorem 1.6.1 there is $b_0 \in A$ such that

$$\|b_0 \xi_i - ptp\xi_i\| < \varepsilon_1, \quad 1 \le i \le m, \quad \|b_0\| \le \|ptp\|.$$

By the definition of $M, b_0\xi_i, ptp\xi_i$ and ξ_i will belong to the same H_j (some j depending on i), $\forall i$. Then from Lemma 2.6.3 we can find $a_1 \in M$, such that

$$a_1 \xi_i = ptp\xi_i - b_0 \xi_i, \quad 1 \le i \le m,$$

$$\|a_1\|^2 \le 2 \sum_{i=1}^m \|ptp\xi_i - b_0\xi_i\|^2 < 2m\varepsilon_1^2.$$

Similarly, for $\varepsilon_2 > 0$ we can find $b_1 \in A$ such that

$$\|b_1 \xi_i - a_1 \xi_i\| < \varepsilon_2, \quad 1 \le i \le m, \quad \|b_1\| \le \|a_1\| < \sqrt{2m}\varepsilon_1.$$

$\cdots\cdots$. Generally, we have $\{a_0 = ptp, a_1, \cdots\} \subset M$ and $\{b_0, b_1, \cdots\} \subset A$ such that

$$\|b_k \xi_i - a_k \xi_i\| < \varepsilon_{k+1}, \quad 1 \le i \le m, k = 0, 1, \cdots,$$

$$a_{k+1} \xi_i = a_k \xi_i - b_k \xi_i, \quad 1 \le i \le m, k = 0, 1, \cdots,$$

$$\|b_k\| \le \|a_k\| < \sqrt{2m}\varepsilon_k, \quad k = 1, 2, \cdots, \|b_0\| \le \|a_0\|.$$

Moreover, if $t_j^* = t_j, 1 \le j \le n$, then by Lemma 2.6.3 and Proposition 1.6.4 the above a_k and b_k can be chosen self-adjoint, $k = 0, 1, \cdots$.

Now take $\varepsilon_k = (2m)^{-\frac{1}{2}} 2^{-k} \varepsilon$. Then $\|a_k\| < 2^{-k}\varepsilon, k = 1, 2, \cdots$. Let $b = \sum_{k=0}^\infty b_k$. Obviously, $b \in A$ and

$$\|b\| < \varepsilon + \|ptp\| = \varepsilon + \max_{1 \le j \le n} \|p_j t_j p_j\|,$$

and

$$b\xi_i = \lim_N \sum_{k=0}^N b_k \xi_i = \lim_N (a_0 \xi_i - a_{N+1} \xi_i) = ptp\xi_i,$$

$1 \le i \le m$. In particular, $be_j = ptpe_j = t_j e_j, 1 \le j \le n$.

Finally, if $t_j^* = t_j, 1 \le j \le n$, we can choose a normalized orthogonal basis $\{\xi_1, \cdots, \xi_m\}$ of pH satisfying

$$ptp\xi_i = \lambda_i \xi_i, \quad 1 \le i \le m,$$

where $\lambda_1, \cdots, \lambda_m$ are real. From $b_k^* = b_k, \forall k$, we have $b^* = b$ and $b\xi_i = ptp\xi_i = \lambda_i\xi_i, 1 \le i \le m$. Define a real continuous function f as follows:

$$f(\lambda) = \begin{cases} \lambda, & \text{if } |\lambda| \le \|ptp\|, \\ -\|ptp\|, & \text{if } \lambda \le -\|ptp\|, \\ \|ptp\|, & \text{if } \lambda \ge \|ptp\|. \end{cases}$$

Since $|\lambda_i| \le \|ptp\|$, it follows that $f(b)\xi_i = \lambda_i\xi_i, 1 \le i \le m$. Then $f(b)^* = f(b) \in A, \|f(b)\| \le \|ptp\| = \max_{1\le j\le n} \|p_j t_j p_j\|$, and $f(b)e_j = ptpe_j = t_j e_j, \quad 1 \le j \le n$. Q.E.D.

Theorem 2.6.5. Let A be a C^*-algebra, $\{\pi_j, H_j\}$ be a topologically irreducible $*$ representation of $A, t_j \in B(H_j)$, and e_j be a projection of finite rank on $H_j, 1 \le j \le n$. In addition suppose that $\{\pi_i, H_i\}$ is not unitarily equivalent to $\{\pi_j, H_j\}, \forall 1 \le i \ne j \le n$.

1) For any $\varepsilon > 0$, there is $a \in A$ such that

$$\pi_j(a)e_j = t_j e_j, \quad 1 \le j \le n, \quad \|a\| \le \varepsilon + \max_{1\le j\le n} \|p_j t_j p_j\|,$$

where p_j is the projection from H_j onto $[e_j H_j, t_j e_j H_j], 1 \le j \le n$. Moreover, if $t_j^* = t_j, 1 \le j \le n$, then a can be chosen self-adjoint.

2) If t_j is a unitary operator on $H_j, 1 \le j \le n$, then there is a unitary element $u = e^{ih}$ of $(A \dot{+} \mathbb{C})$, where $h^* = h \in A$, such that

$$\pi_j(u)e_j = t_j e_j, 1 \le j \le n.$$

Moreover, if A has an identity, then we can choose $u \in A$.

Proof. Let $H = \sum_{j=1}^n \oplus H_j, \pi = \sum_{j=1}^n \oplus \pi_j$. By Proposition 2.4.9, $\pi(A)$ is a C^*-algebra on H, and $\pi(A) \subset M = \sum_{j=1}^n \oplus B(H_j)$.

Denote the projection from H onto H_j by p_j'; clearly, $p_j' \in \pi(A)', 1 \le j \le n$. Let z_j be the central cover of p_j' in $\pi(A)', 1 \le j \le n$. We claim that $z_i z_j = \delta_{ij} z_i, \forall i, j$. In fact, if $z_j z_k \ne 0$ for some $j \ne k$, by Proposition 1.5.9 there are non-zero projections p_j'', p_k'' of $\pi(A)'$ such that $p_k'' \le p_k', p_j'' \le p_j'$ and $p_j'' \sim p_k''$ in $\pi(A)'$. But the $*$ representations π_j and π_k are irreducible, so $p_j'' = p_j', p_k'' = p_k'$. Thus $p_j' \sim p_k'$ in $\pi(A)'$. This is a contradiction since π_j is not unitarily equivalent to π_k. Therefore $z_i z_j = \delta_{ij} z_i, \forall 1 \le i, j \le n$. Further, since

$\sum_{j=1}^{n} p'_j = 1$ and $z_j \geq p'_j, \forall j$, it follows that $p'_j = z_j, 1 \leq j \leq n$. Now we have

$$\{\sum_{j=1}^{n} \oplus \pi_j(a_j) \mid a_j \in A, 1 \leq j \leq n\} \subset \pi(A)'' \subset M.$$

By Proposition 2.6.2, $\pi_j(A)$ is weakly dense in $B(H_j), 1 \leq j \leq n$. Hence $\pi(A)$ is weakly dense in M. By Lemma 2.6.4, we get the conclusion 1).

Now let t_j be a unitary operator on H_j, then $\dim e_j H_j = \dim t_j e_j H_j$, and there is a unitary operator u_j on $p_j H_j$ such that $u_j e_j = t_j e_j, 1 \leq j \leq n$. Pick a normalized basis $\{\xi_k^{(j)}\}$ of $p_j H_j$ such that $u_j \xi_k^{(j)} = exp(i\lambda_k^{(j)})\xi_k^{(j)}$, where $\lambda_k^{(j)}$ is real, $\forall 1 \leq k \leq \dim p_j H_j, 1 \leq j \leq n$. Define $h_j^* = h_j \in B(H_j)$ such that

$$h_j \xi_k^{(j)} = \lambda_k^{(j)} \xi_k^{(j)}, 1 \leq k \leq \dim p_j H_j, \text{ and } h_j(1 - p_j) = 0,$$

$\forall 1 \leq j \leq n$. By 1), we can find $h^* = h \in A$ such that

$$\pi(h) p_j = h_j p_j, \quad 1 \leq j \leq n.$$

Since $p_j \geq e_j$, it follows that

$$\pi_j(e^{ih}) e_j = \pi_j(e^{ih}) p_j e_j = e^{ih_j} p_j e_j = u_j e_j = t_j e_j,$$

$\forall 1 \leq j \leq n$. Q.E.D.

Theorem 2.6.6. Any topological irreducible $*$ representation of a C^*-algebra is also algebraically irreducible.

Proof. Let $\{\pi, H\}$ be a topologically irreducible $*$ representation of a C^*-algebra A. Suppose that K is a non-zero proper linear subspace of H such that $\pi(a)K \subset K, \forall a \in A$. Pick $0 \neq \xi \in K$ and $\eta \in H \backslash K$. By Theorem 2.6.5, there is $a \in A$ such that $\pi(a)\xi = \eta$. That contradicts $\pi(a)\xi \in K$. Therefore, such K does not exist. In other words, $\{\pi, H\}$ is algebraically irreducible. Q.E.D.

Remark. From now on, an *irreducible $*$ representation* of a C^*-algebra means that it is algebraically irreducible or topologically irreducible.

Notes. The *transitivity theorem*, Theorem 2.6.5, is due to R.V. Kadison.

References. [79].

2.7. Pure states and regular maximal left ideals

Theorem 2.7.1. Let A be a C^*-algebra, and ρ be a state on A. Then ρ is pure if and only if the $*$ representation $\{\pi_\rho, H_\rho\}$ generated by ρ is irreducible. Moreover, if ρ is pure, then $H_\rho = A/L_\rho$, where $L_\rho = \{a \in A \mid \rho(a^*a) = 0\}$ is the left kernel of ρ.

Proof. Let π_ρ be irreducible. Suppose that there are states ρ_1 and ρ_2 on A and $\lambda \in (0,1)$ such that

$$\rho = \lambda\rho_1 + (1-\lambda)\rho_2.$$

On A/L_ρ, define

$$[a_\rho, b_\rho] = \rho_1(b^*a), \quad \forall a, b \in A,$$

where $a \to a_\rho$ is the canonical map from A onto A/L_ρ. Since

$$\|[a_\rho, b_\rho]\|^2 \leq \rho_1(b^*b)\rho_1(a^*a) \leq \lambda^{-2}\|a_\rho\| \cdot \|b_\rho\|,$$

it follows that there is $h = h^* \in B(H_\rho)$ such that

$$\rho_1(b^*a) = \langle ha_\rho, b_\rho \rangle, \quad \forall a, b \in A.$$

It is easy to see $h \in \pi_\rho(A)'$. But π_ρ is irreducible, so that $h = \mu$, where μ is some real number, i.e., $\rho_1(b^*a) = \mu\rho(b^*a), \forall a, b \in A$. From $\rho, \rho_1 \in S(A)$, it follows that $\mu = 1$, and $\rho = \rho_1 = \rho_2$. Therefore, ρ is pure.

Conversely, let ρ be a pure state on A, $\xi_\rho (\in H_\rho)$ be a cyclic vector for $\pi_\rho(A)$ as in Proposition 2.3.18. Suppose that there is a projection p' of $\pi_\rho(A)'$ with $p' \neq 0, 1$. We claim that $p'\xi_\rho \neq 0$. In fact, if $p'\xi_\rho = 0$, then

$$\{\pi_\rho(a)\xi_\rho \mid a \in A\} = \{\pi_\rho(a)(1-p')\xi_\rho \mid a \in A\} \subset (1-p')H_\rho.$$

It is impossible since ξ_ρ is a cyclic vector for $\pi_\rho(A)$. Thus $p'\xi_\rho \neq 0$. Similarly, $(1-p')\xi_\rho \neq 0$. Put $\lambda = \|p'\xi_\rho\|^2$. Then $\lambda \in (0,1)$. Define

$$\rho_1(a) = \lambda^{-1}\langle \pi_\rho(a)p'\xi_\rho, p'\xi_\rho \rangle,$$

$$\rho_2(a) = (1-\lambda)^{-1}\langle \pi_\rho(a)(1-p')\xi_\rho, (1-p')\xi_\rho \rangle,$$

$\forall a \in A$. By Proposition 2.4.6, ρ_1 and ρ_2 are two states on A. Clearly, $\rho = \lambda\rho_1 + (1-\lambda)\rho_2$. Since ρ is pure, it follows that $\rho = \rho_1 = \rho_2$. Then

$$\|\lambda^{-\frac{1}{2}}p'\pi_\rho(a)\xi_\rho\|^2 = \rho_1(a^*a) = \rho(a^*a) = \|\pi_\rho(a)\xi_\rho\|^2, \quad \forall a \in A.$$

Hence $\lambda^{-\frac{1}{2}}p'$ is an isometry on H_ρ. But p' is a projection and $p' \neq 0, 1$, so this is a contradiction. Therefore $\pi_\rho(A)' = \mathbb{C}$, i.e. π_ρ is irreducible.

Finally, since A/L_ρ is dense in H_ρ and is invariant for $\pi_\rho(A)$, it follows from Theorem 2.6.6 that $H_\rho = A/L_\rho$. Q.E.D.

Definition 2.7.2. A left ideat L of a C^*-algebra A is said to be *regular*, if there is $x_0 \in A$ such that $(ax_0 - a) \in L, \forall a \in A$. In this case, x_0 is called a *modular unit* for L.

Of course, if A itself has an identity, then any left ideal of A is regular. Now suppose that A has no identity. Let ρ be a state on $A, \tilde{\rho}$ be the natural extension of ρ on $(A\dot{+}\mathbb{C})$, and $L, \tilde{L}$ be the left kernels of $\rho, \tilde{\rho}$ respectively. Clearly, $L \subset \tilde{L}$, and $\tilde{L}$ may be a direct sum of L and a linear subspace of one dimension. If $L \neq \tilde{L}$, then L is a regular left ideal of A; and if $L = \tilde{L}$, then L is not a regular left ideal of A.

Theorem 2.7.3. Let ρ be a pure state on a C^*-algebra A, and L_ρ be its left kernel. Then L_ρ is a regular maximal left ideal of A, and

$$N(\rho) = \{a \in A \mid \rho(a) = 0\} = L_\rho + L_\rho^*.$$

Proof. Let L be a left ideal of A containing L_ρ, and $\{\pi_\rho, H_\rho\}$ be the irreducible $*$ representation of A generated by ρ. Since L/L_ρ is an invariant subspace for $\pi_\rho(A)$ and $L \neq A$, it follows that $L = L_\rho$, i.e. L_ρ is a maximal left ideal of A.

Let $\xi_\rho(\in H_\rho)$ be a cyclic vector for $\pi_\rho(A)$ as in Proposition 2.3.18. If $b = b^* \in N(\rho)$, then $\langle b_\rho, \xi_\rho\rangle = \rho(b) = 0$. Thus we can find a self-adjoint element of $B(H_\rho)$, which maps ξ_ρ to 0 and maps b_ρ to b_ρ. By Theorem 2.6.5, there is $h = h^* \in A$ such that

$$\pi_\rho(h)\xi_\rho = 0, \quad \pi_\rho(h)b_\rho = b_\rho.$$

Let $c = b - hb$. Since $\rho(c^*c) = \|\pi_\rho(h)b_\rho - b_\rho\|^2 = 0$, it follows that $c \in L_\rho$. Further, $b = c + hb, b = b^* = bh + c^*$, and $c^* \in L_\rho^*$. Noticing

$$\rho((bh)^* \cdot (bh)) \leq \|b\|^2\rho(h^2) = \|b\|^2 \cdot \|\pi_\rho(h)\xi_\rho\|^2 = 0,$$

we have $bh \in L_\rho$ and $b = bh + c^* \in L_\rho + L_\rho^*$. From $N(\rho) = N(\rho)^*$, it follows that $N(\rho) \subset L_\rho + L_\rho^*$. On the other hand, it is clear that $L_\rho + L_\rho^* \subset N(\rho)$ by the Schwartz inequality. Therefore, $N(\rho) = L_\rho + L_\rho^*$.

By Theorem 2.7.1, $H_\rho = A/L_\rho$. Thus there is $a \in A$ such that $a_\rho = \xi_\rho$. Further, $\pi_\rho(b)a_\rho = \pi_\rho(b)\xi_\rho = b_\rho$, and

$$\rho((ba - b)^* \cdot (ba - b)) = \|\pi_\rho(b)a_\rho - b_\rho\|^2 = 0,$$

i.e., $(ba - b) \in L_\rho, \forall b \in A$. Therefore, L_ρ is regular. Q.E.D.

Lemma 2.7.4. Let A be a C^*-algebra with an identity, L be a closed left ideal of A, and $a \in A_+$. If for any $\varepsilon > 0$ there is $a_\varepsilon \in L \cap A_+$ such that $a \leq a_\varepsilon + \varepsilon$, then $a \in L$.

Proof. By Proposition 2.2.5, $a_\varepsilon^{1/2} \in L$. Since $0 \le a \le a_\varepsilon + \varepsilon$, it follows that

$$\begin{aligned} 0 \le\ & (a_\varepsilon^{1/2} + \varepsilon^{1/2})^{-1} a (a_\varepsilon^{1/2} + \varepsilon^{1/2})^{-1} \\ \le\ & (a_\varepsilon^{1/2} + \varepsilon^{1/2})^{-1} (a_\varepsilon + \varepsilon)(a_\varepsilon^{1/2} + \varepsilon^{1/2})^{-1} \\ =\ & (a_\varepsilon + \varepsilon)(a_\varepsilon + 2\varepsilon^{1/2} a_\varepsilon^{1/2} + \varepsilon)^{-1} \le 1. \end{aligned}$$

Thus we have

$$\begin{aligned} & \|a^{1/2}(a_\varepsilon^{1/2} + \varepsilon^{1/2})^{-1} a_\varepsilon^{1/2} - a^{1/2}\|^2 \\ =\ & \|\varepsilon^{1/2} a^{1/2} (a_\varepsilon^{1/2} + \varepsilon^{1/2})^{-1}\|^2 \\ =\ & \varepsilon \|(a_\varepsilon^{1/2} + \varepsilon^{1/2})^{-1} a (a_\varepsilon^{1/2} + \varepsilon^{1/2})^{-1}\| \le \varepsilon \end{aligned}$$

and

$$a^{1/2}(a_\varepsilon^{1/2} + \varepsilon^{1/2})^{-1} a_\varepsilon^{1/2} \to a^{1/2} \quad \text{as} \quad \varepsilon \to 0+ .$$

But $a^{1/2}(a_\varepsilon^{1/2} + \varepsilon^{1/2})^{-1} a_\varepsilon^{1/2} \in L$, and L is closed, therefore $a \in L$. Q.E.D.

Lemma 2.7.5. Let L_1 and L_2 be two closed left ideals of a C^*-algebra A, and $L_1 \subset L_2$. If for any state ρ on A with $\rho(L_1) = \{0\}$, we have also $\rho(L_2) = \{0\}$. Then $L_1 = L_2$.

Proof. We may assume that A has an identity. Let $a \in L_2 \cap A_+$. For any $\varepsilon > 0$, let $\Omega_\varepsilon = \{\rho \in S(A) \mid \rho(a) \ge \varepsilon\}$. Then Ω_ε is a $\sigma(A^*, A)$-compact subset of A^*. For any $\rho \in \Omega_\varepsilon$, since $\rho(L_2) \ne \{0\}$, it follows from the assumption that $\rho(L_1) \ne \{0\}$. Thus there is $a^{(\rho)} \in L_1$ such that $|\rho(a^{(\rho)})| > 1$. By the continuity, we can find a $\sigma(A^*, A)$-neighborhood V_ρ of ρ such that $|f(a^{(\rho)})| > 1, \forall f \in V_\rho$. Since $\bigcup_{\rho \in \Omega_\varepsilon} V_\rho \supset \Omega_\varepsilon$, it follows from the compactness of Ω_ε that there are $\rho_1, \cdots, \rho_n \in \Omega_\varepsilon$ such that $\Omega_\varepsilon \subset \bigcup_{i=1}^n V_i$, where $V_i = V_{\rho_i}, 1 \le i \le n$. Let $a_i = a^{(\rho_i)}$, then $1 < |f(a_i)| \le f(a_i^* a_i), \forall f \in V_i \cap \Omega_\varepsilon, 1 \le i \le n$. In particular, $\rho(\sum_{i=1}^n a_i^* a_i) > 1, \forall \rho \in \Omega_\varepsilon$. Put $a_\varepsilon = \|a\| \sum_{i=1}^n a_i^* a_i$. Clearly, $a_\varepsilon \in L_1 \cap A_+$ and

$$\rho(a_\varepsilon) \ge \|a\| \ge \rho(a) \ge \varepsilon, \quad \forall \rho \in \Omega_\varepsilon.$$

Thus $\rho(a_\varepsilon + \varepsilon - a) \ge 0, \forall \rho \in S(A)$. By Corollary 2.3.15, $a_\varepsilon + \varepsilon \ge a$. Since $\varepsilon > 0$ is arbitrary and $a_\varepsilon \in L_1 \cap A_+$ it follows from Lemma 2.7.4 that $a \in L_1$. Thus we have

$$L_2 \cap A_+ \subset L_1 \cap A_+.$$

By Proposition 2.4.1, there is a net $\{d_l\} \subset L_2 \cap A_+$ such that $ad_l \to a, \forall a \in L_2$. But $d_\varepsilon \in L_2 \cap A_+ \subset L_1 \cap A_+, \forall l$, so it must be $L_2 \subset L_1$. Further $L_1 = L_2$.

Q.E.D.

Q.E.D.

Theorem 2.7.6. Let L be a closed left ideal of a C^*-algebra A. Then L is the intersection of all regular maximal left ideals containing L.

Proof. Let $\Omega = \{\rho \in A^* \mid \rho \geq 0, \|\rho\| \leq 1, \text{ and } \rho(L) = \{0\}\}$. Clearly,

$$\cap\{L_\rho \mid \rho \in \Omega\} \supset L,$$

where L_ρ is the left kernel of $\rho, \forall \rho \in \Omega$. By Lemma 2.7.5, we have $L = \cap\{L_\rho \mid \rho \in \Omega\}$. Since Ω is a $\sigma(A^*, A)$-compact convex subset of A^*, it follows from the Krein-Milmann theorem that $Co(Ex\Omega)$ is $\sigma(A^*, A)$-dense in Ω, where $Ex\Omega$ is the set of all extreme points of $\Omega, Co(\cdots)$ is the convex hull of $(\cdots)$. If $a \in A$ such that $\rho(a^*a) = 0, \forall \rho \in Ex\Omega$, then $\rho(a^*a) = 0, \forall \rho \in Co(Ex\Omega)$, and further $\rho(a^*a) = 0, \forall \rho \in \Omega$. Therefore, we have

$$\cap\{L_\rho \mid \rho \in Ex\Omega\} = \cap\{L_\rho \mid \rho \in \Omega\} = L. \tag{1}$$

Noticing that $\Omega \neq \{0\}$ (otherwise, by Lemma 2.7.5, we have $L = A$, a contradiction), from Theorem 2.7.3 it suffices to show that ρ is a pure state on A for each $\rho \in Ex\Omega$ and $\rho \neq 0$.

Now let $\varphi \in Ex\Omega$ and $\varphi \neq 0$. Clearly, φ is a state on A. Suppose that there are two states φ_1, φ_2 on A and $\lambda \in (0,1)$ such that $\varphi = \lambda\varphi_1 + (1-\lambda)\varphi_2$. For any $a \in L$, from (1) it must be $a \in L_\varphi$, i.e., $\varphi(a^*a) = 0$. Further, $\varphi_i(a^*a) = 0, i = 1,2$, and by the Schwartz inequality, $\varphi_i(a) = 0, i = 1,2$. Hence $\varphi_i(L) = \{0\}$ and $\varphi_i \in \Omega, i = 1,2$. Since $\varphi \in Ex\Omega$, it follows that $\varphi = \varphi_1 = \varphi_2$. Therefore, φ is a pure state on A. Q.E.D.

Theorem 2.7.7. Let L be a maximal left ideal of a C^*-algebra A. Then L is regular if and only if L is closed.

Proof. The sufficiency is clear from Theorem 2.7.6. Now let L be a regular maximal left ideal of A. So there is $x_0 \in A$ such that $(bx_0 - b) \in L, \forall b \in A$. Let $\tilde{L} = L \dot{+} \mathbb{C}(1 - x_0)$.

We claim that $\tilde{L}$ is a maximal left ideal of $(A \dot{+} \mathbb{C})$. In fact, suppose that J is a left ideal of $(A \dot{+} \mathbb{C})$ and $J \supset \tilde{L}$. If $y = a + \lambda \in J$, where $a \in A, \lambda \in \mathbb{C}$, then $\lambda x_0 + a = y - \lambda(1 - x_0) \in J \cap A$ since $(1 - x_0) \in \tilde{L} \subset J$. Since L is a maximal left ideal of A, it follows that $J \cap A = L$. Thus $y \in L \dot{+} \mathbb{C}(1 - x_0) = \tilde{L}$, and $J = \tilde{L}$.

Now $\tilde{L}$ is closed since $(A \dot{+} \mathbb{C})$ has an identity. Therefore, $L = \tilde{L} \cap A$ is closed. Q.E.D.

Theorem 2.7.8. Let L be a regular maximal left ideal of a C^*-algebra A. Then there is unique state ρ on A such that $N(\rho) = \{a \in A \mid \rho(a) = 0\} \supset L$. Moreover, this ρ is pure, and its left kernel is L, and $N(\rho) = L + L^*$.

Proof. By Theorem 2.7.7 and the proof of Theorem 2.7.6, there is a pure state ρ on A such that $L_\rho \supset L$ and $\rho(L) = \{0\}$. Since L is maximal, it follows that $L = L_\rho$. Further, by Theorem 2.7.3, $N(\rho) = L + L^*$.

Now suppose that φ is another state on A with $\varphi(L) = \{0\}$. Then $\varphi(N(\rho)) = \{0\}$. If $x_0(\in A)$ is a modular unit for L and $\{d_l\}$ is an approximate identity for A, then we have

$$\rho(x_0) = \lim_l \rho(d_l x_0) = \lim_l \rho(d_l)$$

and

$$\varphi(x_0) = \lim_l \varphi(d_l x_0) = \lim_l \varphi(d_l).$$

By Proposition 2.4.4, we get $\rho(x_0) = \varphi(x_0) = 1$. Moreover, it is clear that $A = N(\rho)\dot{+}\mathbb{C}x_0$. Therefore, $\varphi = \rho$. Q.E.D.

Corollary 2.7.9. For any C^*-algebra A, there is an one to one correspondence between the set of all pure states on A and the set of all regular maximal left ideals of A.

Theorem 2.7.10. Let A be a C^*-algebra, ρ be a state on A, and L be the left kernel of ρ. If L is regular, then the following statements are equivalent:

1) ρ is a pure state on A;
2) L is a maximal left ideal of A;
3) $N(\rho) = \{a \in A \mid \rho(a) = 0\} = L + L^*$.

Proof. By Theorems 2.7.3 and 2.7.8, it is clear that 1) and 2) are equivalent and 1) implies 3).

Now let $N(\rho) = L + L^*$ and $x_0(\in A)$ be a modular unit for L. Suppose that there are two states ρ_1, ρ_2 on A and $\lambda \in (0,1)$ such that $\rho = \lambda\rho_1 + (1-\lambda)\rho_2$. If $x \in L$, then we have $x^*x \in N(\rho), \rho_i(x^*x) = 0$, and $\rho_i(x) = 0, i = 1,2$. Further, $\rho_i(N(\rho)) = \{0\}, i = 1,2$. By the proof of Theorem 2.7.8, we can get $\rho(x_0) = \rho_1(x_0) = \rho_2(x_0) = 1$. Moreover, it is clear that $A = N(\rho)\dot{+}\mathbb{C}x_0$. Therefore, $\rho = \rho_1 = \rho_2$. So ρ is a pure state on A. Q.E.D.

References. [79], [81], [155], [156].

2.8. Ideals and quotient C^*-algebras

Definition 2.8.1. Let A be a C^*-algebra, and $P(A)$ be the set of pure states on A (see Definition 2.3.8). Denote by $\hat{A}$ the set of all unitarily equivalent classes of irreducible $*$ representations of A, and let $\text{Prim}(A)$ be the set of all primitive ideals of A. J is called a *primitive ideal* of A, if there is an irreducible $*$ representation π of A such that $\ker\pi = J$.

Clearly, a primitive ideal is a closed two-sided $*$ ideal. Since two unitarily equivalent representations have the same kernel, so we have a natural map from $\hat{A}$ onto $\text{Prim}(A)$. Moreover, for each $\rho \in P(A)$, the $*$ representation generated by ρ is irreducible from Theorem 2.7.1. Conversely, if $\{\pi, H\}$ is an irreducible $*$ representation of A, then by Proposition 2.4.6, $\rho(\cdot) = \langle\pi(\cdot)\xi, \xi\rangle$ is a state on A for any $\xi \in H$ with $\|\xi\| = 1$. And also from Proposition 2.3.21, the $*$ representation $\{\pi_\rho, H_\rho\}$ generated by ρ is unitarily equivalent to $\{\pi, H\}$, and is irreducible. Furthermore, ρ is pure by Theorem 2.7.1. Therefore, we have a map from $P(A)$ onto $\hat{A}$ by the GNS construction.

Proposition 2.8.2. Let J be a closed two-sided ideal of a C^*-algebra A. Then

$$\begin{aligned} J = & \cap\{I \in \text{Prim}(A) \mid I \supset J\} \\ = & \cap\{\ker\pi_\rho \mid \rho \in P(A), \text{ and } (\rho|J) = 0\}. \end{aligned}$$

Proof. Since J is also a closed left ideal of A, it follows from the proof of Theorem 2.7.6 that

$$\begin{aligned} J = & \cap\{L_\rho \mid \rho \in P(A), \text{ and } (\rho|J) = 0\} \\ \supset & \cap\{\ker \pi_\rho \mid \rho \in P(A), \text{ and } (\rho|J) = 0\}. \end{aligned}$$

Now let $a \in J$. For any $\rho \in P(A)$ and $(\rho|J) = 0$, since J is two-sided, it follows that

$$\|\pi_\rho(a)b_\rho\|^2 = \rho(b^*a^*ab) = 0, \forall b \in A.$$

Hence, $\pi_\rho(a) = 0$, and $a \in \ker\pi_\rho$. Therefore, we obtain

$$\begin{aligned} J = & \cap\{\ker\pi_\rho \mid \rho \in P(A), \text{ and } (\rho|J) = 0\} \\ = & \cap\{I \in \text{Prim}(A) \mid I \supset J\}. \end{aligned}$$

Q.E.D.

Corollary 2.8.3. Let Ω be a compact Hausdorff space, and J be a closed ideal of $C(\Omega)$. Then there is a closed subset Ω_0 of Ω such that

$$J = \{f \in C(\Omega) \mid f(t) = 0, \forall t \in \Omega_0\}.$$

Proof. It is clear from Propositions 2.8.2 and 2.3.10. Q.E.D.

Definition 2.8.4. Let A be a C^*-algebra, and J be a closed two-sided ideal of A. Define

$$P_J(A) = \{\rho \in P(A) \mid (\rho|J) = 0\},$$

$$P^J(A) = \{\rho \in P(A) \mid (\rho|J) \neq 0\} = P(A)\backslash P_J(A),$$

$$\hat{A}_J = \{\pi \in \hat{A} \mid \ker \pi \supset J\}, \quad \hat{A}^J = \hat{A}\backslash\hat{A}_J,$$

$$\text{Prim}_J(A) = \{I \in \text{Prim }(A) \mid I \supset J\},$$

$$\text{Prim}^J(A) = \text{Prim }(A)\backslash\text{Prim}_J(A).$$

Theorem 2.8.5. Let J be a closed two-sided ideal of a C^*-algebra A.

1) For any $\pi \in \hat{A}_J$, let $\tilde{\pi}(\tilde{a}) = \pi(a), \forall \tilde{a} \in A/J, a \in \tilde{a}$, where $a \to \tilde{a} = a + J$ is the canonical map from A onto A/J. Then $\pi \to \tilde{\pi}$ is a bijection from $\hat{A}$ onto $(A/J)^\wedge$.

2) $\pi \to (\pi|J)$ is a bijection from $\hat{A}^J$ onto $\hat{J}$.

Proof. 1) It is obvious.

2) Let $\{\pi, H\}$ be an irreducible $*$ representation of A, and $\pi|J \neq 0$. Since J is a two-sided ideal, it follows that the linear span K of $\{\pi(a)\xi \mid a \in J, \xi \in H\}$ is a non-zero invariant subspace for $\pi(A)$. Now since π is irreducible, it must be that $K = H$, and $\{\pi|J, H\}$ is a nondegenerate $*$ representation of J. If $\{d_l\}(\subset J)$ is an approximate identity for J, by Proposition 2.4.6, we have $\pi(d_l) \to 1$ (strongly). Then $\pi(ad_l) \to \pi(a)$ (strongly), $\forall a \in A$, i.e., $\pi(J)$ is strongly dense in $\pi(A)$. So $\{\pi|J, H\}$ is an irreducible $*$ representation of J by Proposition 2.6.2.

Conversely, let $\{\pi, H\}$ be an irreducible $*$ representation of J. By Theorem 2.6.6, $H = [\pi(b)\xi \mid b \in J, \xi \in H]$. For any $b_1, \cdots, b_n \in J$ and $\xi_1, \cdots, \xi_n \in H$,

$$\sum_{1 \leq i,j \leq n} \langle \pi(b_j^* \cdot b_i)\xi_i, \xi_j \rangle$$

is a positive linear functional on $(A\dot{+}\mathbb{C})$. Thus

$$\|\sum_{i=1}^n \pi(ab_i)\xi_i\| \leq \|a\| \cdot \|\sum_{i=1}^n \pi(b_i)\xi_i\|,$$

$\forall a \in A, b_1, \cdots, b_n \in J, \xi_1, \cdots, \xi_n \in H$. Further, $\{\pi, H\}$ can be uniquely extended to an irreducible $*$ representation $\{\pi', H\}$ of A such that

$$\pi'(a)\pi(b)\xi = \pi(ab)\xi, \forall a \in A, b \in J, \xi \in H.$$

Moreover, if $\{\pi_1, H_1\} \cong \{\pi_2, H_2\}$, then it is easy to see that $\{\pi_1', H_1\} \cong \{\pi_2', H_2\}$.

Therefore, $\pi \to (\pi|J)$ is a bijection from $\hat{A}^J$ onto $\hat{J}$. Q.E.D.

Definition 2.8.6. A C^*-algebra A is said to be *prime*, if $xAy = \{0\}$ implies either $x = 0$ or $y = 0$. A closed two-sided ideal J of a C^*-algebra A is said to be *prime*, if the quotient C^*-algebra A/J is prime.

Lemma 2.8.7. Let A be a C^*-algebra.

1) Suppose that J_1 and J_2 are closed two-sided ideals of A. Then $[J_1 J_2] = J_1 \cap J_2$.

2) A is prime if and only if for any non-zero closed two-sided ideals J_1, J_2 of A we have $J_1 \cap J_2 \neq \{0\}$.

3) Suppose that J is a closed two-sided ideal of A. Then the following statements are equivalent:

(a) J is prime;

(b) If $x, y \in A$ such that $xAy \subset J$, then either $x \in J$ or $y \in J$;

(c) If J_1, J_2 are closed two-sided ideals of A and $J_1 J_2 \subset J$, then either $J_1 \subset J$ or $J_2 \subset J$;

(d) If J_1, J_2 are closed two-sided ideals of A, and $J_i \supset J$, and $J_i \neq J$, $i = 1, 2$, then $J_1 \cap J_2 \neq J$.

Proof. 1) Clearly, $[J_1 J_2] \subset J_1 \cap J_2$. Now let $a \in J_1 \cap J_2$. Since $J_1 \cap J_2$ is a C^*-subalgebra of A, we can write $a = b_1^2 - b_2^2 + ib_3^2 - ib_4^2$, where $b_j \in (J_1 \cap J_2)_+, 1 \leq j \leq 4$. Obviously, $b_j^2 \in J_1 J_2, 1 \leq j \leq 4$. Thus $a \in [J_1 J_2]$, and $[J_1 J_2] = J_1 \cap J_2$.

2) Let A be prime, and J_1, J_2 be non-zero closed two-sided ideals of A. Pick $0 \neq x \in J_1$ and $0 \neq y \in J_2$, and let $J_x(= \overline{[AxA]}), J_y(= \overline{[AyA]})$ be the closed two-sided ideals generated by x, y respectively. Clearly, $\{0\} \neq J_x \subset J_1, \{0\} \neq J_y \subset J_2$, and $xAy \subset J_x J_y \subset J_1 J_2 \subset J_1 \cap J_2$. Since A is prime and x, y are non-zero, it follows that $xAy \neq \{0\}$, and $J_1 \cap J_2 \neq \{0\}$.

Conversely, if there are x and $y \in A$ with $x \neq 0, y \neq 0$ such that $xAy = \{0\}$, then $J_x J_y \subset \overline{[AxAyA]} = \{0\}$, where $J_x = \overline{[AxA]}$ and $J_y = \overline{[AyA]}$ are non-zero closed two-sided ideals of A. By 1), we have $J_x \cap J_y = [J_x J_y] = \{0\}$. Therefore, if A is not prime, then there are non-zero closed two-sided ideals J_1 and J_2 such that $J_1 \cap J_2 = \{0\}$.

3) By Definition 2.8.6 and 2), it is clear that (a), (b) and (d) are equivalent.

(d)$\Longrightarrow$(c). Let J_1, J_2 be closed two-sided ideals of A and $J_1 J_2 \subset J$. If $J_i \not\subset J, i = 1, 2$, then $I_i = J + J_i \neq J, i = 1, 2$. Now by (d), we have

$$J \neq I_1 \cap I_2, \quad J \subset I_1 \cap I_2.$$

However, by 1) and $J_1 J_2 \subset J$ we have $I_1 \cap I_2 = [I_1 I_2] \subset J$, a contradiction. Thus it must be that either $J_1 \subset J$ or $J_2 \subset J$.

Now let (c) hold, and $x, y \in A$ be such that $xAy \subset J$. Then $J_x J_y \subset J$, where $J_x = \overline{[AxA]}, J_y = \overline{[AyA]}$. By (c), either $J_x \subset J$ or $J_y \subset J$. Thus either $x \in J$ or $y \in J$, and J is prime. Q.E.D.

Proposition 2.8.8. Let A be a C^*-algebra, and $J \in \text{Prim}(A)$. Then J is prime.

Proof. Let $\{\pi, H\}$ be an irreducible $*$ representation of A such that $\ker\pi = J$. Suppose that $x, y \in A$ are such that $xAy = \{0\}$. Then $\pi(x)\pi(A)\pi(y) = \{0\}$. By Proposition 2.6.2 we have $\pi(x)B(H)\pi(y) = \{0\}$. Thus either $\pi(x)$ or $\pi(y)$ is zero, i.e., either x or y is in $\ker\pi = J$. Therefore, J is prime. Q.E.D.

Theorem 2.8.9. Let J be a closed two-sided ideal of a C^*-algebra A.

1) $I \to I/J$ is a bijection from $\text{Prim}_J(A)$ onto $\text{Prim}(A/J)$.

2) $I \to I \cap J$ is a bijection from $\text{Prim}^J(A)$ onto $\text{Prim}(J)$.

Proof. 1) Let $I \in \text{Prim}_J(A)$, and $\{\pi, H\}$ be an irreducible $*$ representation of A such that $\ker\pi = I$. For any $\tilde{a} \in A/J$, since $J \subset I$, we can define $\tilde{\pi}(\tilde{a}) = \pi(a)$, here $a \in \tilde{a} = a + J$. Then $\{\tilde{\pi}, H\}$ is an irreducible $*$ representation of A/J, and $\ker\tilde{\pi} = I/J$. Thus $I/J \in \text{Prim}(A/J)$.

Moreover, let $\{\tilde{\pi}, H\}$ be an irreducible $*$ representation of A/J. Define $\pi(a) = \tilde{\pi}(\tilde{a}), \forall a \in A$, here $a \to \tilde{a}$ is the canonical map from A onto A/J. Then $\{\pi, H\}$ is an irreducible $*$ representation of A and $\ker\pi = I \supset J$, $\ker\tilde{\pi} = I/J$. Thus, $I \to I/J$ is a map from $\text{Prim}_J(A)$ onto $\text{Prim}(A/J)$.

Now let $I_1/J = I_2/J$, where $I_1, I_2 \in \text{Prim}_J(A)$. For any $a \in I_1$, then we have $b \in I_2$ such that $(a - b) \in J$. Since $J \subset I_2$, it follows that $a \in I_2$. Thus $I_1 \subset I_2$. Similarly, $I_2 \subset I_1$. Thus $I_1 = I_2$. Therefore, $I \to I/J$ is a bijection from $\text{Prim}_J(A)$ onto $\text{Prim}(A/J)$.

2) Let $I \in \text{Prim}^J(A)$, and $\{\pi, H\}$ be an irreducible $*$ representation of A such that $\ker\pi = I$. Since $\ker\pi = I \not\supset J$, it follows by Theorem 2.8.5 that $\{(\pi|J), H\}$ is an irreducible $*$ representation of J, and $\ker(\pi|J) = I \cap J \in \text{Prim}(J)$. Now if $\{\sigma, H\}$ is an irreducible $*$ representation of J, then by the proof of Theorem 2.8.5 $\{\sigma, H\}$ can be uniquely extended to an irreducible $*$ representation of A. So $I \to I \cap J$ is a map from $\text{Prim}^J(A)$ onto $\text{Prim}(J)$.

Now let $I_1 \cap J = I_2 \cap J$, where $I_1, I_2 \in \text{Prim}^J(A)$. Then $I_2 \supset I_1 \cap J \supset I_1 J$. But $I_2 \not\supset J$, by Proposition 2.8.8 we have $I_2 \supset I_1$. Similarly, $I_1 \supset I_2$. Thus $I_1 = I_2$. Therefore, $I \to I \cap J$ is a bijection from $\text{Prim}^J(A)$ onto $\text{Prim}(J)$.

Q.E.D.

Theorem 2.8.10. Let J be a closed two-sided ideal of a C^*-algebra A.

1) $\rho \to \tilde{\rho}$ is a bijection from $P_J(A)$ onto $P(A/J)$, where $\tilde{\rho}(\tilde{a}) = \rho(a), \forall \tilde{a} \in A/J, a \in \tilde{a}$.

2) $\rho \to (\rho|J)$ is a bijection from $P^J(A)$ onto $P(J)$.

Proof. 1) It is Proposition 2.4.11 exactly.

2) Let $\rho \in P^J(A)$, and $\{\pi, H, \xi\}$ be the irreducible cyclic $*$ representation generated by ρ. If $J \subset \ker\pi$, then $\rho(a) = \langle\pi(a)\xi, \xi\rangle = 0, \forall a \in J$. This is a contradiction since $\rho(J) \neq \{0\}$. Thus $J \not\subset \ker\pi$. Then by Theorem 2.8.5, $\{\pi|J, H\}$ is an irreducible $*$ representation of J. Further by Proposition 2.3.21, $\{\pi|J, H\}$ is unitarily equivalent to the $*$ representation generated by the state $\langle(\pi|J)(\cdot)\xi, \xi\rangle = (\rho|J)(\cdot)$. Thus $(\rho|J) \in P(J)$.

Moreover, if $\{d_l\}(\subset J)$ is an approximate identity for J, by Proposition 2.4.6 we have $\pi(d_l) \to 1$ (strongly). Then $\rho(a) = \langle\pi(a)\xi, \xi\rangle = \lim_l \langle\pi(ad_l)\xi, \xi\rangle$, $\forall a \in A$. This means that the behaviour of ρ on A is determined by $(\rho|J)$. Hence, if $\rho_1, \rho_2 \in P^J(A)$ and $(\rho_1|J) = (\rho_2|J)$, then $\rho_1 = \rho_2$.

Finally, let $\sigma \in P(J)$, and $\{\pi_\sigma, H, \xi\}$ be the irreducible cyclic $*$ representation of J generated by σ. By Theorem 2.8.5, $\{\pi_\sigma, H\}$ can be uniquely extended to an irreducible $*$ representation $\{\pi, H\}$ of A. Then $\rho(\cdot) = \langle\pi(\cdot)\xi, \xi\rangle(\forall \cdot \in A)$ is an extension of σ, and the $*$ representation of A generated by ρ is unitarily equivalent to $\{\pi, H\}$ (see Proposition 2.3.11). Thus $\rho \in P^J(A)$ and $(\rho|J) = \sigma$.

Therefore, $\rho \to (\rho|J)$ is a bijection from $P^J(A)$ onto $P(J)$. Q.E.D.

From Theorem 2.8.5, 2.8.9 and 2.8.10, we have the following diagram:

$$\begin{array}{ccccc} P(A/J) & \to & P(A) & \longleftarrow & P(J) \\ \downarrow & & \downarrow & & \downarrow \\ (A/J)^\wedge & \to & \hat{A} & \longleftarrow & \hat{J} \\ \downarrow & & & & \downarrow \\ \mathrm{Prim}(A/J) & \to & \mathrm{Prim}(A) & \longleftarrow & \mathrm{Prim}(J) \end{array}$$

It is easily checked that this diagram is commutative.

Notes. About the converse of Proposition 2.8.8, J. Dixmier gave the following result. Let A be a separable or Type I (GCR) C^*-algebra, and J be a prime closed two-sided ideal of A, then J is also primitive. But the question for the general case is still open.

References. [23], [28], [33], [156].

2.9. Hereditary C^*-subalgebras

Lemma 2.9.1. Let A be a C^*-algebra, $a, x, y \in A$ and $a \geq 0$. Suppose that there are number $\lambda, \mu > 0$ with $(\lambda + \mu) > 1$ such that

$$x^*x \leq a^\lambda, \quad yy^* \leq a^\mu.$$

Let $u_n = x(\frac{1}{n} + a)^{-\frac{1}{2}}y, \forall n$. Then there is $u \in A$, such that $\|u_n - u\| \to 0$ and $\|u\| \leq \|a^{\frac{\lambda+\mu-1}{2}}\|$.

Proof. Let $d_{nm} = (\frac{1}{n} + a)^{-\frac{1}{2}} - (\frac{1}{m} + a)^{-\frac{1}{2}}, \forall n, m$. Then

$$\begin{aligned} \|u_n - u_m\|^2 = & \|xd_{nm}y\|^2 = \|y^* d_{nm} x^* x d_{nm} y\| \\ \leq & \|y^* d_{nm} a^\lambda d_{nm} y\| = \|a^{\frac{\lambda}{2}} d_{nm} y\|^2 \\ = & \|a^{\frac{\lambda}{2}} d_{nm} y y^* d_{nm} a^{\frac{\lambda}{2}}\| \\ \leq & \|a^{\frac{\lambda}{2}} d_{nm} a^\mu d_{nm} a^{\frac{\lambda}{2}}\| = \|d_{nm} a^{\frac{\lambda+\mu}{2}}\|^2. \end{aligned}$$

We may assume that A has an identity 1. Let B be the abelian C^*-subalgebra generated by $\{a, 1\}$. Then $B \cong C(\Omega)$. For each $t \in \Omega, ((\frac{1}{n} + a)^{-\frac{1}{2}} a^{\frac{\lambda+\mu}{2}})(t) \nearrow a^{\frac{\lambda+\mu-1}{2}}(t)$. By the Dini theorem, this convergence is uniform for $t \in \Omega$. Thus $\|d_{nm} a^{\frac{\lambda+\mu}{2}}\| \to 0$ and $\|u_n - u_m\| \to 0$, and there is $u \in A$ such that $\|u_n - u\| \to 0$. Similarly, we can prove that $\|u_n\| \leq \|(\frac{1}{n} + a)^{-\frac{1}{2}} a^{\frac{\lambda+\mu}{2}}\|, \forall n$. Therefore, $\|u\| \leq \|a^{\frac{\lambda+\mu-1}{2}}\|$. Q.E.D.

Proposition 2.9.2. Let A be a C^*-algebra, $x, a \in A$ and $a \geq 0, x^*x \leq a$. Then for any $\lambda \in (0, \frac{1}{2})$, there is $u \in A$ such that

$$x = ua^\lambda, \text{ and } \|u\| \leq \|a^{\frac{1}{2}-\lambda}\|.$$

Proof. Let $u_n = x(\frac{1}{n} + a)^{-\frac{1}{2}} a^{\frac{1}{2}-\lambda}, \forall n$. By Lemma 2.9.1, there is $u \in A$ such that

$$\|u_n - u\| \to 0, \text{ and } \|u\| \leq \|a^{\frac{1}{2}-\lambda}\|.$$

Moreover, since $x^*x \leq a$, it follows that

$$\|x - u_n a^\lambda\| \leq \|a^{\frac{1}{2}}[1 - (\frac{1}{n} + a)^{-\frac{1}{2}} a^{\frac{1}{2}}]\| \to 0.$$

Therefore, $x = ua^\lambda$. Q.E.D.

Corollary 2.9.3. Let A be a C^*-algebra, $x \in A$, and $\alpha \in (0,1)$. Then there is $u \in A$ such that

$$x = u(x^*x)^{\alpha/2}, \text{ and } \|u\| \leq \|(x^*x)^{\frac{1-\alpha}{2}}\|.$$

Proof. It is clear by Proposition 2.9.2 and picking $a = x^*x, \lambda = \alpha/2$. Q.E.D.

Remark. A *factorization* with $\alpha = 1$ (*polar decomposition*) is not possible in a general C^*-algebra.

Definition 2.9.4. Let A be a C^*-algebra. A cone $M(\subset A_+)$ is said to be *hereditary*, if $a \in A_+$ with $a \leq b$ for some $b \in M$ implies $a \in M$.

For a hereditary cone M, we define

$$L(M) = \{x \in A \mid x^*x \in M\}.$$

It is easy to see that $L(M)$ is a left ideal of A.

A C^*-algebra B of A is said to be *hereditary*, if B_+ is hereditary.

Theorem 2.9.5. Let A be a C^*-algebra.

1) $B \to B_+$ is a bijection from the set of all hereditary C^*-subalgebras of A onto the set of all closed hereditary cones of A_+. Its inverse is $M \to L(M) \cap L(M)^*$.

2) $M \to L(M)$ is a bijection from the set of all closed hereditary cones of A_+ onto the set of all closed left ideals of A, and $M = L(M)_+$. Its inverse is $L \to L_+$.

3) $L \to L \cap L^*$ is a bijection from the set of all closed left ideals of A onto the set of all hereditary C^*-subalgebras of A, and $L_+ = (L \cap L^*)_+$. Its inverse is $B \to L(B_+)$.

Proof. Let B be a hereditary C^*-subalgebra of A. By the definition, B_+ is a closed hereditary cone of A_+. Since $B = [B_+]$, so the map $B \to B_+$ is injective.

Let M be a closed hereditary cone of A_+. Clearly, $L(M)$ is a closed left ideal of A. Assume that $x \in A$ is such that $x^*x \in L(M)$. By Corollary 2.9.3, we can write $x = u(x^*x)^{1/4}$ for some $u \in A$. Thus $x \in L(M)$, and $L(M)_+ = \{x^*x \mid x \in L(M)\}$. By the definition of $L(M)$, we have $M = L(M)_+$. Thus the map $M \to L(M)$ is injective.

Let L be a closed left ideal of A. Clearly, $L \cap L^*$ is a C^*-subalgebra of A. Since $L_+ \subset L \cap L^* \subset L$, it follows that $L_+ = (L \cap L^*)_+$. So in order to prove that $L \cap L^*$ is hereditary, it suffices to show that L_+ is hereditary. Suppose that $a \in A_+$ and $a \leq b$ for some $b \in L_+$. By Proposition 2.9.2, we have a factorization $a^{\frac{1}{2}} = vb^{1/3}$, where $v \in A$. Thus $a^{\frac{1}{2}} \in L_+$, and $a \in L_+$, i.e., L_+ is hereditary.

Suppose that L is a closed left ideal of A. From the preceding paragraph, $M = L_+$ is a closed hereditary cone of A_+. Let $x \in A$ be such that $x^*x \in M$. By Corollary 2.9.3, we can write $x = u(x^*x)^{1/3}$. Thus $x \in L$. Further, by the definition we have $L = L(M)$. So 2) is proved.

Suppose that M is a closed hereditary cone of A_+. From preceding paragraphs, $L(M)$ is a closed left ideal of A; $L(M) \cap L(M)^*$ is a hereditary C^*-subalgebra of A; and $M = L(M)_+ = (L(M) \cap L(M)^*)_+$. So 1) is proved.

Finally, let B be a hereditary C^*-subalgebra of A. Thus B_+ is a closed hereditary cone of A_+; $L(B_+)$ is a closed left ideal of A; $L(B_+) \cap L(B_+)^*$ is a hereditary C^*-subalgebra of A. Since $(L(B_+) \cap L(B_+)^*)_+ = L(B_+)_+ = B_+$, it follows that $B = L(B_+) \cap L(B_+)^*$. So the map $L \to L \cap L^*$ is surjective. Moreover, if L is a closed left ideal of A such that $B = L \cap L^*$, then $B_+ = (L \cap L^*)_+ = L_+$, and $L = L(B_+)$. So the map $L \to L \cap L^*$ is also injective; and its inverse is $B \to L(B_+)$. Therefore, 3) is proved. Q.E.D.

Lemma 2.9.6. Let Φ be a $*$ homomorphism from a C^*-algebra A onto a C^*-algebra $B, a \in A_+, b \in B$ with $b^*b \leq \Phi(a)$. Then there is $x \in A$ such that $b = \Phi(x)$ and $x^*x \leq a$.

Proof. Pick $y \in A$ such that $\Phi(y) = b$. Write

$$y^*y - a = h - k,$$

where $h, k \in A_+$ and $hk = 0$. Since $\Phi(y)^*\Phi(y) = b^*b \leq \Phi(a)$, it follows that $0 \leq \Phi(h) \leq \Phi(k)$. But $\Phi(h)^{\frac{1}{2}}\Phi(k)\Phi(h)^{1/2} = 0$, so we have $\Phi(h) = 0$.

Clearly, $y^*y \leq a + h, a^{\frac{1}{2}}a^{\frac{1}{2}} = a$. Let

$$x_n = y(\frac{1}{n} + a + h)^{-\frac{1}{2}}a^{\frac{1}{2}}, \forall n.$$

Then by Lemma 2.9.1, there is $x \in A$ such that $\|x_n - x\| \to 0$. From $\Phi(h) = 0$, we have

$$\Phi(x) = \lim_n \Phi(x_n) = b \lim_n (\frac{1}{n} + \Phi(a))^{-\frac{1}{2}}\Phi(a)^{\frac{1}{2}} = b.$$

Moreover, since $y^*y \leq a + h$, it follows that for any n

$$x_n^*x_n \leq a^{\frac{1}{2}}(\frac{1}{n} + a + h)^{-\frac{1}{2}}(a + h)(\frac{1}{n} + a + h)^{-\frac{1}{2}}a^{\frac{1}{2}} \leq a.$$

Therefore, $x^*x \leq a$, and x is what we want to find. Q.E.D.

Proposition 2.9.7. Let A, B be two C^*-algebras, C be a hereditary C^*-subalgebra of A, and Φ be a $*$ homomorphism from A onto B. Then $\Phi(C)$ is also a hereditary C^*-subalgebra of B.

Proof. The result is clear by Lemma 2.9.6. Q.E.D.

Proposition 2.9.8. Let A be a C^*-algebra, B be a hereditary C^*-subalgebra of A, and φ be a state on B. Then there exists unique state ψ on A such that $(\psi|B) = \varphi$.

Proof. Let $\{d_l\}(\subset B_+)$ be an approximate identity for B. For any $a \in A_+$, since $0 \leq d_lad_l \leq \|a\|^2 d_l^2 \in B_+$ and B is hereditary, it follows that $d_lad_l \in B, \forall l$. Further, we have $d_lAd_l \subset B, \forall l$.

Now let ψ be a state on A such that $(\psi|B) = \varphi$. Then $\|\psi\| = \|\varphi\| = 1 = \lim_l \varphi(d_l) = \lim_l \varphi(d_l^2)$ (see Proposition 2.4.4). Further, by the Schwartz inequality we have $\psi(a(1-d_l)) \to 0, \psi((1-d_l)a) \to 0, \forall a \in A$. Thus $\psi(a) = \lim_l \psi(d_l a d_l), \forall a \in A$. From the preceding paragraph $d_l A d_l \subset B$, so we have

$$\psi(a) = \lim_l \varphi(d_l a d_l), \quad \forall a \in A.$$

Therefore, the extension ψ is unique. Q.E.D.

Notes. Theorem 2.9.5 is due to E.G. Effros.

References. [33], [127].

2.10. Comparison, disjunction and quasi-equivalence of * representations

Definition 2.10.1. Let A be a C^*-algebra, and $\{\pi, H\}$ be a * representation of A. If K is a closed linear subspace of H, and K is invariant for π (i.e., $\pi(a)\xi \in K, \forall a \in A, \xi \in K$), then $\{\pi, K\}$ is also a * representation of A, and $\{\pi, K\}$ is called a * *subrepresentation* of $\{\pi, H\}$.

Suppose that $\{\pi_1, H_1\}$ and $\{\pi_2, H_2\}$ are two * representation of A. The symbol "$\pi_1 \preceq \pi_2$" means that $\{\pi_1, H_1\}$ is unitarily equivalent to a * subrepresentation of $\{\pi_2, H_2\}$.

Proposition 2.10.2. Let $\{\pi_1, H_1\}$ and $\{\pi_2, H_2\}$ be two * representations of a C^*-algebra A.

1) Let $\pi = \pi_1 \oplus \pi_2, H = H_1 \oplus H_2$, and let p_i' be the projection from H onto H_i (clearly, $p_i' \in \pi(A)'$), i=1, 2. Then $\pi_1 \preceq \pi_2$ if and only if $p_1' \preceq p_2'$ in $\pi(A)'$.

2) If $\pi_1 \preceq \pi_2$ and $\pi_2 \preceq \pi_1$, then we have $\{\pi_1, H_1\} \cong \{\pi_2, H_2\}$, i.e., $\{\pi_1, H_1\}$ and $\{\pi_2, H_2\}$ are unitarily equivalent.

Proof. 1) It is clear. Moreover, from 1) and Proposition 1.5.3, we can get 2) immediately. Q.E.D.

Definition 2.10.3. Let $\{\pi_1, H_1\}$ and $\{\pi_2, H_2\}$ be two * representations of a C^*-algebra A. π_1 and π_2 are said to be *disjoint*, denoted by $\pi_1 \perp \pi_2$, if any non-zero * subrepresentation of π_1 is not unitarily equivalent to any non-zero * subrepresentation of π_2.

Proposition 2.10.4. Let $\{\pi_1, H_1\}$ and $\{\pi_2, H_2\}$ be two * representations of a C^*-algebra A. Let $\pi = \pi_1 \oplus \pi_2, H = H_1 \oplus H_2$; and let p_i' be the projection from H on H_i (clearly, $p'_i \in \pi(A)'$), $i = 1,2$. Then the following statements are equivalent:

1) $\pi_1 \perp \pi_2$;
2) $c(p'_1) \cdot c(p'_2) = 0$, where $c(p'_i)$ is the central cover of p'_i in $\pi(A)', i = 1,2$;
3) p'_i is a central projection of $\pi(A)', i = 1,2$.

Proof. Since $p'_1 \oplus p'_2 = 1$, it follows that the statements 2) and 3) are equivalent.

Clearly, $\pi_1 \perp \pi_2$ if and only if there are no projections q'_1 and q'_2 of $\pi(A)'$ such that $0 \neq q'_i \leq p'_i, i = 1,2$, and $q'_1 \sim q'_2$ in $\pi(A)'$. Then by Proposition 1.5.9, the statements 1) and 2) are equivalent. Q.E.D.

Definition 2.10.5. A nondegenerate $*$ representation $\{\pi, H\}$ of a C^*-algebra A is said to be *factorial*, if the VN algebra on H generated by $\pi(A)$ (i.e. $\pi(A)''$) is a factor.

Proposition 2.10.6. Let $\{\pi_1, H_1\}$ and $\{\pi_2, H_2\}$ be two factorial $*$ representations of a C^*-algebra A. Then one of the relations $\pi_1 \perp \pi_2, \pi_1 \precsim \pi_2, \pi_2 \precsim \pi_1$ holds.

Proof. Let $\pi = \pi_1 \oplus \pi_2, H = H_1 \oplus H_2, M = \pi(A)''$ and p'_i be the projection from H onto $H_i, i = 1,2$. Then $p'_1, p'_2 \in M'$. By the assumption, Mp'_i is a factor on $H_i, i = 1,2$. If $c(p'_i)$ is the central cover of p'_i in M', by Proposition 1.5.10 Mp'_i and $Mc(p'_i)$ are $*$ isomorphic, $i = 1,2$. Thus $Mc(p_i)$ is also a factor on $Hc(p'_i)$; and $c(p'_i)$ is a minimal central projection of M (i.e., if z is a central projection of M and $z \leq c(p'_i)$, then either $z = 0$ or $z = c(p'_i)), i = 1,2)$. Therefore, we have either $c(p'_1) \cdot c(p'_2) = 0$ or $c(p'_1) = c(p'_2)$.

When $c(p'_1) \cdot c(p'_2) = 0$, by Proposition 2.10.4 $\pi_1 \perp \pi_2$ holds.

Now let $c(p'_1) = c(p'_2) = z$. Since Mz is a factor, by Proposition 1.3.8 $M'z$ is also a factor. From Theorem 1.5.4, we have either $p'_1 \precsim p'_2$ or $p'_2 \precsim p'_1$ in $M'z$ (also in M'). Further by Proposition 2.10.2, either $\pi_1 \precsim \pi_2$ or $\pi_2 \precsim \pi_1$ holds. Q.E.D.

Proposition 2.10.7. Let $\{\pi_1, H_1\}$ and $\{\pi_2, H_2\}$ be two irreducible $*$ representations of a C^*-algebra A. Then $\pi_1 \perp \pi_2$ if and only if π_1 and π_2 are not unitarily equivalent.

Proof. Since any non-zero $*$ subrepresentation of π_i must be π_i itself, $i = 1,2$, the conclusion is obvious from Definition 2.10.3. Q.E.D.

Proposition 2.10.8. If $\pi \perp \pi_l, \forall l$, then $\pi \perp \sum_l \oplus \pi_l$.

Proof. Let H, H_l be the action spaces of π, π_l respectively, $\forall l$. Let $\sigma = \pi \oplus \sum_l \oplus \pi_l, K = H \oplus \sum_l \oplus H_l$; and let p', p'_l be the projections from K onto H, H_l respectively, $\forall l$. Then $p', p'_l \in \sigma(A)'$, $\forall l$. By Proposition 2.10.4 we have $c(p') \cdot c(p'_l) = 0, \forall l$. From Proposition 1.5.8,

$$c(p') \perp \sup_l c(p'_l) = c(\sup_l p'_l) = c(\sum_l p'_l).$$

Again by Proposition 2.10.4, $\pi \perp \sum_l \oplus \pi_l$. Q.E.D.

Definition 2.10.9. Let $\{\pi_1, H_1\}$ and $\{\pi_2, H_2\}$ be two nondegenerate $*$ representations of a C^*-algebra A, and $M_i = \pi_i(A)'', i = 1, 2$. π_1 and π_2 are said to be *quasi-equivalent*, denoted by $\pi_1 \approx \pi_2$, if there is a $*$ isomorphism Φ from M_1 onto M_2, such that $\Phi(\pi_1(a)) = \pi_2(a), \forall a \in A$.

Proposition 2.10.10. Let $\{\pi_1, H_1\}$ and $\{\pi_2, H_2\}$ be two nondegenerate $*$ representations of a C^*-algebra A. Then the following statements are equivalent:

1) $\pi_1 \approx \pi_2$;

2) No non-zero $*$ subrepresentation of π_i is disjoint from $\pi_j, 1 \leq i \neq j \leq 2$;

3) Let $\pi = \pi_1 \oplus \pi_2, H = H_1 \oplus H_2$, and p'_i be the projection from H onto $H_i (\in \pi(A)'), i = 1, 2$. Then $c(p'_1) = c(p'_2)$;

4) There exists an ampliation π of π_1 (i.e., there is a Hilbert space K such that $\pi(a) = \pi_1(a) \otimes 1_K, \forall a \in A$) and a projection p' of $\pi(A)'$ with central cover $1 (= 1_{H_1 \otimes K})$ such that $\pi p' \cong \pi_2$;

5) There exist ampliations of π_1 and π_2 which are equivalent.

Proof. 1)$\Longrightarrow$ 4). It is clear Theorem 1.12.4 and Proposition 1.12.5.

4) $\Longrightarrow$ 1). Define Φ_3, Φ_2, Φ_1 as in Theorem 1.12.4; and let $\Phi = \Phi_3 \circ \Phi_2 \circ \Phi_1$. Then it is immediate that $\pi_1 \approx \pi_2$.

4) $\Longrightarrow$ 5). From the condition 4), π_2 is unitarily equivalent to a $*$ subrepresentation of $\pi_1 \otimes 1_K$, where K is some Hilbert space. Since 1) and 4) are equivalent, it follows that π_1 is also unitarily equivalent to a $*$ subrepresentation of $\pi_2 \otimes 1_L$, where L is some Hilbert space. Let R be an infinite dimensional Hilbert space, and dim$R \geq$dim K, dim$L$. Then

$$\pi_2 \otimes 1_R \preceq \pi_1 \otimes 1_K \otimes 1_R \cong \pi_1 \otimes 1_R \preceq \pi_2 \otimes 1_L \otimes 1_R \cong \pi_2 \otimes 1_R.$$

By Proposition 2.10.2, we have $\pi_1 \otimes 1_R \cong \pi_2 \otimes 1_R$. That comes to 5).

5) $\Longrightarrow$ 2). Suppose that $\pi_1 \otimes 1_R \cong \pi_2 \otimes 1_R$, where R is a Hilbert space. If σ_i is a non-zero $*$ subrepresentation of π_i, then we can regard σ_i as a non-zero $*$ subrepresentation of $\pi_i \otimes 1_R$. By Definition 2.10.3, σ_i is not disjoint from $\pi_j \otimes 1_R$. Further, by Proposition 2.10.8 σ_i is not disjoint from π_j.

2) $\Longrightarrow$ 3). If $c(p_1') \neq c(p_2')$, we may assume that $c(p_2') \not\leq c(p_1')$. Then $z = c(p_2') - c(p_1') \cdot c(p_2')$ is a non-zero central projection of $\pi(A)'$, and $z \leq c(p_2')$ and $z \perp c(p_1')$. By Proposition 1.5.8, $zp_2' \neq 0$. Clearly, $c(zp_2') \perp c(p_1')$. By Proposition 2.10.4, $\{\pi_2, zH_2\} \perp \{\pi_1, H_1\}$. That contradicts the condition 2). Thus $c(p_1') = c(p_2')$.

3) $\Longrightarrow$ 1). Let $z = c(p_1') = c(p_2')$, and $M = \pi(A)''$. By Proposition 1.5.10, Mp_i' and Mz are $*$ isomorphic, $i = 1, 2$. Thus we have a $*$ isomophism Φ from $M_1 = Mp_1'$ onto $M_2 = Mp_2'$ such that $\Phi(bp_1') = bp_2', \forall b \in M$. In particular, for any $a \in A$, since $\pi_i(a) = \pi(a)p_i', i = 1, 2$, it follows that $\Phi(\pi_1(a)) = \pi(a)p_2' = \pi_2(a)$. Therefore, $\pi_1 \approx \pi_2$. Q.E.D.

Proposition 2.10.11. Let $\{\pi_1, H_1\}$ and $\{\pi_2, H_2\}$ be two nondegenerate $*$ representations of a C^*-algebra A.

1) If $\pi_1 \cong \pi_2$, then $\pi_1 \approx \pi_2$.

2) If π_1 and π_2 are irreducible, and $\pi_1 \approx \pi_2$, then $\pi_1 \cong \pi_2$.

Proof. 1) It is obvious. Now we prove 2). By Proposition 2.10.10, π_1 is not disjoint from π_2. Since π_1 and π_2 are irreducible, it follows from Definition 2.10.3 that π_1 is unitarily equivalent to π_2. Q.E.D.

Proposition 2.10.12. Let $\{\pi_1, H_1\}$ and $\{\pi_2, H_2\}$ be two factorial $*$ representations of a C^*-algebra A. Then we have either $\pi_1 \perp \pi_2$ or $\pi_1 \approx \pi_2$.

Proof. By Proposition 2.10.6, we may assume that $\pi_1 \precsim \pi_2$. Then there is a projection $p' \in \pi_2(A)'$ such that $\pi_1 \cong \pi_2 p'$. But $\pi_2(A)'$ is a factor, so the central cover of p_2' in $\pi_2(A)'$ is 1. By Proposition 1.5.10, $\pi_2 \approx \pi_2 p'$. Therefore $\pi_1 \approx \pi_2$. Q.E.D.

References. [28], [104], [105].

2.11 The enveloping Von Neumann algebra

Definition 2.11.1. Let A be a C^*-algebra, and $S(A)$ be its state space. For each $\varphi \in S(A)$, we have a cyclic $*$ representation $\{\pi_\varphi, H_\varphi, \xi_\varphi\}$ of A (see Proposition 2.3.18). Then the faithful $*$ representation

$$\pi_u = \sum_{\varphi \in S(A)} \oplus \pi_\varphi, \quad H_u = \sum_{\varphi \in S(A)} \oplus H_\varphi$$

is called the *universal $*$ representation* of A. And $\pi_u(A)''$ is called the *enveloping VN algebra* of the C^*-algebra A, denoted by $\overline{A} = \pi_u(A)''$.

Suppose that ψ is a normal state on $\overline{A}$. Since A and $\pi_u(A)$ are $*$ isomorphic, it follows that there is a state φ on A such that $\varphi(a) = \psi(\pi_u(a)), \forall a \in A$. By the GNS construction, we have $\xi_\varphi \in H_\varphi \subset H_u$ such that

$$\psi(\pi_u(a)) = \varphi(a) = \langle \pi_\varphi(a)\xi_\varphi, \xi_\varphi\rangle = \langle \pi_u(a)\xi_\varphi, \xi_\varphi\rangle, \forall a \in A.$$

Further, $\psi(b) = \langle b\xi_\varphi, \xi_\varphi\rangle, \forall b \in \overline{A} = \pi_u(A)''$. Therefore, every normal state on the VN algebra $M = \pi_u(A)''$ is a vector state. From Proposition 1.10.6, we have $\sigma(M, M_*) \sim$ (weak operator top. $|M$), $s(M, M_*) \sim$ (strong operator top. $|M$) and $s^*(M, M_*) \sim$ (strong $*$ operator top. $|M$) .

Now we study the relation between the enveloping VN algebra $\overline{A}$ and the second conjugate space A^{**}. By Proposition 1.3.3, $\overline{A}$ is the conjugate space of the Banach space $\overline{A}_* = T(H_u)/\overline{A}_\perp$, where

$$\overline{A}_\perp = \{t \in T(H_u) \mid tr(tb) = 0, \forall b \in \overline{A}\}.$$

Through the following way, the Banach space $\overline{A}_*$ and the conjugate space A^* of A are isometrically isomorphic. For any $f \in \overline{A}_*$, let

$$F(a) = \pi_u(a)(f), \quad \forall a \in A.$$

Then $F \in A^*$ and $\|F\| = \|f\|$. Conversely, any element of A^* must be of above form. In fact, if $f \in \overline{A}_*$, by density theorem 1.6.1 $\|f\| = \sup\{|\pi_u(a)(f)| \mid a \in A, \|a\| \leq 1\} = \|F\|$. Now let $\varphi \in S(A)$. Then $\varphi(a) = \langle \pi_u(a)\xi_\varphi, \xi_\varphi\rangle, \forall a \in A$. Let p_φ be the one rank projection of H_u onto $[\xi_\varphi]$, and f be the canonical image of p_φ in $\overline{A}_* = T(H_u)/\overline{A}_\perp$. Then

$$\pi_u(a)(f) = tr(\pi_u(a)p_\varphi) = \langle \pi_u(a)\xi_\varphi, \xi_\varphi\rangle = \varphi(a), \forall a \in A.$$

Since A^* is the linear span of $S(A)$, it follows that for each $F \in A^*$ there is unique $f \in \overline{A}_*$ such that

$$F(a) = \pi_u(a)(f), \quad \forall a \in A.$$

Denote the above isomorphism from $\overline{A}_*$ onto A^* by π_*, i.e.,

$$\pi_*(f)(a) = \pi_u(a)(f), \quad \forall f \in \overline{A}_*, a \in A.$$

Then $(\pi_*)^*$ is an isometrical isomorphism from A^{**} onto $\overline{A}$, and it is $\sigma(A^{**}, A^*)$-$\sigma(\overline{A}, \overline{A}_*)$ continuous. Moreover, since for any $a \in A \subset A^{**}$

$$(\pi_*)^*(a)(f) = \pi_*(f)(a) = \pi_u(a)(f), \forall f \in \overline{A}_*,$$

it follows that $(\pi_*)^*|A = \pi_u$, i.e., $(\pi_*)^*$ is an extension of the $*$ isomorphism π_u from A onto $\pi_u(A)$. So we can write $(\pi_*)^* = \pi_u$ simply. Now we have the following theorem.

Theorem 2.11.2. Let A be a C^*-algebra. Then the second conjugate space A^{**} of A is isometrically isomorphic to the enveloping VN algebra $\overline{A}$ of A. So we can introduce a multiplication and a $*$ operation on A^{**} such that A^{**} becomes a C^*-algebra, and A becomes a C^*-subalgebra of A^{**}. Moreover, if A has an identity, then this identity is also an identity of A^{**}.

In the above discussion, the multiplication and the $*$ operation on A^{**} are defined through $\overline{A}$. But we have another way. It depends on A and A^* directly.

Theorem 2.11.3. Let A be a C^*-algebra, and define a $*$ operation and a multiplication (*Arens multiplication*) on A^{**} as follows:

$$X^*(F) = \overline{X(F^*)}, \quad F^*(a) = \overline{F(a^*)},$$
$$XY(F) = X([Y,F]), [Y,F](a) = Y(L_aF), (L_aF)(b) = F(ab),$$

$\forall a, b \in A, F \in A^*, X, Y \in A^{**}$. Then this $*$ operation and multiplication on A^{**} are the same as in Theorem 2.11.2.

Proof. Keep the above notations: $\overline{A}, \overline{A}_*, \overline{A}_\perp, \pi_* : \overline{A}_* \to A^*$, and $(\pi_*)^* = \pi_u : A^{**} \to \overline{A}$.

For any $X \in A^{**}$, pick a net $\{x_l\} \subset A$ such that $x_l \to X(\sigma(A^{**}, A^*))$. Since $(X - x_l)^*(F) = (X - x_l)(F^*), \forall F \in A^*$, it follows that $x_l^* \to X^*(\sigma(A^{**}, A^*))$. But $\pi_u = (\pi_*)^*$ is $\sigma(A^{**}, A^*)$-$\sigma(\overline{A}, \overline{A}_*)$ continuous, thus we have

$$\pi_u(X^*) = \pi_u(X)^*.$$

For any $t \in T(H_u)$ and $a \in A$, denote the canonical images of t and $t\pi_u(a)(\in T(H_u))$ in $\overline{A}_* = T(H_u)/\overline{A}_\perp$ by f and L_af respectively. Since for any $b \in A$

$$\begin{aligned}(L_a\pi_*(f))(b) &= \pi_*(f)(ab) = \pi_u(ab)(f)\\ &= tr(t\pi_u(ab)) = \pi_u(b)(L_af),\end{aligned}$$

it follows that $\pi_*(L_af) = L_a\pi_*(f)$.

For $Y \in A^{**}$, let g be the canonical image of $\pi_u(Y)t(\in T(H_u))$ in $\overline{A}_* = T(H_u)/\overline{A}_\perp$. From

$$\begin{aligned}[Y, \pi_*(f)](a) &= Y(L_a\pi_*(f)) = Y(\pi_*(L_af))\\ &= \pi_u(Y)(L_af) = tr(\pi_u(Y)t\pi_u(a))\\ &= \pi_u(a)(g) = \pi_*(g)(a), \quad \forall a \in A,\end{aligned}$$

we have $[Y, \pi_*(f)] = \pi_*(g)$.

Now for any $X, Y \in A^{**}$ by

$$\begin{aligned}\pi_u(XY)(f) = & \ (XY)(\pi_*(f)) = X([Y, \pi_*(f)]) \\ = & \ X(\pi_*(g)) = \pi_u(X)(g) = tr(\pi_u(X)\pi_u(Y)t) \\ = & \ (\pi_u(X)\pi_u(Y))(f), \quad \forall f \in \overline{A}_*,\end{aligned}$$

we have $\pi_u(XY) = \pi_u(X)\pi_u(Y)$. Q.E.D.

Proposition 2.11.4. Let A be a C^*-algebra, and B be a C^*-subalgebra of A. Then the C^*-algebra B^{**} is $*$ isomorphic to the $\sigma(A^{**}, A^*)$-closure $\overline{B}^{\sigma}$ of B in A^{**}.

Proof. For any $X \in B^{**}$, let $\Phi(X)(F) = X(F|B), \forall F \in A^*$. Then Φ is an isometric linear isomorphism from B^{**} onto $\overline{B}^{\sigma}$. By Theorem 2.11.3, Φ also keeps the $*$ operation and multiplication. Q.E.D.

Notes. The second conjugate space of a C^*-algebra is very important since it is a W^*-algebra (see Chapter 4). The Theorem 2.11.2 ($A^{**} \cong \overline{A}$) is due to S. Sherman and Z. Takeda.

Moreover, let B be a Banach algebra. We can introduce two kinds of Arens multiplication on B^{**}. The first Arens multiplication is as in Theorem 2.11.3, i.e.,

$$(XY)(F) = X([Y, F]), \quad [Y, F](a) = Y(L_a F), \quad (L_a F)(b) = F(ab),$$

$\forall a, b \in B, F \in B^*, X, Y \in B^{**}$.

The second Arens multiplication is as follows:

$$(X \cdot Y)(F) = Y([X, F]'), \quad [X, F]'(a) = X(R_a F), \quad (R_a F)(b) = F(ba),$$

$\forall a, b \in B, F \in B^*, X, Y \in B^{**}$.

A natural question is when we have $XY = X \cdot Y, \forall X, Y \in B^{**}$.

Definition (P. Civin and B. Yood). A Banach algebra B is said to be *regular*, if $XY = X \cdot Y, \forall X, Y \in B^{**}$. Let

$$Z(B^{**}) = \{X \in B^{**} \mid XY = X \cdot Y, \forall Y \in B^{**}\}.$$

$Z(B^{**})$ is called the *topological center* of B^{**}. Clearly, B is regular $\Longleftrightarrow$ $Z(B^{**}) = B^{**}$; $X \in Z(B^{**}) \Longleftrightarrow$ the map $\cdot \to X \cdot$ is continuous in $(B^{**}, \sigma(B^{**}, B^*))$; and $B \subset Z(B^{**}) \subset B^{**}$.

By Sakai theorem (see Section 4.2), any C^*-algebra A is regular, i.e., $Z(A^{**}) = A^*$. Therefore, for any C^*-algebra A, two kinds of Arens multiplication are the same on A^{**}.

References. [16], [64], [161], [168].

2.12. The multiplier algebra

Definition 2.12.1. Let A be a C^*-algebra, and see the second conjugate space A^{**} of A as the enveloping VN algebra. Let

$$M(A) = \{a \in A^{**} \mid aA \cup Aa \subset A\}.$$

Then $M(A)$ is called the *multiplier algebra* of A.

Clearly, $M(A)$ is a C^*-subalgebra of $A^{**}; A \subset M(A)$ and $A = M(A) \Longleftrightarrow A$ has an identity; A is a closed two-sided ideal of $M(A)$. The C^*-algebra $Q(A) = M(A)/A$ is called the *out multiplier algebra* of A.

Definition 2.12.2. Let A be a nondegenerate C^*-algebra on a Hilbert space H. Let

$$L_H(A) = \{x \in B(H) \mid xA \subset A\},$$

$$R_H(A) = \{x \in B(H) \mid Ax \subset A\},$$

and

$$M_H(A) = L_H(A) \cap R_H(A).$$

$L_H(A)$ is called the set of *left multipliers* of A on H; $R_H(A)$ is called the set of *right multipliers* of A on H; $M_H(A)$ is called the *multiplier algebra* of A on H.

Suppose that $\{d_l\}$ is an approximate identity for A. Since A is nondegenerate on H, it follows that $d_l \to 1(= 1_H)$ (strongly). Thus

$$L_H(A), R_H(A), M_H(A) \subset \overline{A}^s = A''.$$

Clearly, $A \subset M_H(A)$, and $A = M_H(A) \Longleftrightarrow 1 \in A$.

Definition 2.12.3. Let A be a C^*-algebra. A linear map $\rho : A \to A$ is called a *left* (or *right*) *centralizer*, if

$$\rho(xy) = \rho(x)y \quad (\text{or} = x\rho(y)), \quad \forall x, y \in A.$$

Proposition 2.12.4. Let ρ be a left (or right) centralizer of a C^*-algebra A. Then ρ is continuous (bounded).

Proof. Suppose that there is a sequence $\{x_n\} \subset A$ such that

$$\|x_n\| < \frac{1}{n}, \quad \text{and } \|\rho(x_n)\| > n, \quad \forall n.$$

Let

$$a = \begin{cases} \sum_n x_n x_n^*, & \text{if } \rho \text{ is left,} \\ \sum_n x_n^* x_n, & \text{if } \rho \text{ is right.} \end{cases}$$

Since $(x_n^*)^* x_n^* \leq a$ (ρ left) or $x_n^* x_n \leq a$ (ρ right), by Proposition 2.9.2 we can write

$$x_n = a^{1/3} u_n \quad (\rho \text{ left})$$

or

$$x_n = u_n a^{1/3} \quad (\rho \text{ right}),$$

where $\|u_n\| \leq \|a^{1/6}\|, \forall n$. Thus when ρ is left, we have

$$n < \|\rho(x_n)\| \leq \|\rho(a^{1/3})\| \cdot \|u_n\| \leq \|a^{1/6}\| \cdot \|\rho(a^{1/3})\|, \quad \forall n;$$

when ρ is right, we have

$$n < \|\rho(x_n)\| \leq \|u_n\| \cdot \|\rho(a^{1/3})\| \leq \|a^{1/6}\| \cdot \|\rho(a^{1/3})\|, \forall n.$$

This is a contradiction. Therefore, ρ is bounded. Q.E.D.

Definition 2.12.5. (ρ_1, ρ_2) is called a *double centralizer* of a C^*-algebra A, if ρ_1 and ρ_2 are two maps from A into A such that $x\rho_1(y) = \rho_2(x)y, \forall x, y \in A$.

Proposition 2.12.6. Let (ρ_1, ρ_2) be a double centralizer of a C^*-algebra A. Then ρ_1 and ρ_2 are linear; ρ_1 is a left centralizer; ρ_2 is a right centralizer; and $\|\rho_1\| = \|\rho_2\|$.

Proof. Let $\{d_l\}$ be an approximate identity for A. Since for any $x, y \in A$ and $\lambda, \mu \in \mathbb{C}$

$$\begin{aligned} d_l\rho_1(\lambda x + \mu y) &= \rho_2(d_l)(\lambda x + \mu y) \\ &= \lambda\rho_2(d_l)x + \mu\rho_2(d_l)y \\ &= d_l(\lambda\rho_1(x) + \mu\rho_1(y)), \quad \forall l, \end{aligned}$$

it follows that $\rho_1(\lambda x + \mu y) = \lambda\rho_1(x) + \mu\rho_1(y)$. Thus ρ_1 is linear. Similarly, ρ_2 is also linear. Further, from

$$d_l\rho_1(xy) = (\rho_2(d_l)x)y = d_l\rho_1(x)y, \quad \forall l,$$

we have $\rho_1(xy) = \rho_1(x)y, \forall x, y \in A$. Thus ρ_1 is a left centralizer of A. Similarly, ρ_2 is a right centralizer of A.

Moreover, from $d_l\rho_1(x) = \rho_2(d_l)x, \rho_2(x)d_l = x\rho_1(d_l), \forall l$ and $x \in A$, we have $\|\rho_1\| \leq \|\rho_2\|$ and $\|\rho_2\| \leq \|\rho_1\|$ respectively. Therefore, $\|\rho_1\| = \|\rho_2\|$. Q.E.D.

Proposition 2.12.7. Let A be a nondegenerate C^*-algebra on a Hilbert space H, and $\{d_l\}$ be an approximate identity for A. Then the map

$$\rho \rightarrow x = (\text{strongly-}) \lim_l \rho(d_l)$$

is an isometric bijection from left (or right) centralizer set of A onto $L_H(A)$ (or $R_H(A)$).

Moreover, the map

$$(\rho_1, \rho_2) \rightarrow x = (\text{strongly-}) \lim_l \rho_1(d_l) = (\text{strongly-}) \lim_l \rho_2(d_l)$$

is an isometric bijection from the set of double centralizers of A onto $M_H(A)$, here $\|(\rho_1, \rho_2)\|$ is defined by $\|\rho_1\| = \|\rho_2\|$.

Proof. Let ρ be a left centralizer of A. By Proposition 2.12.4, $\{\rho(d_l)\}$ is a bounded net of $B(H)$. Since any bounded closed ball of $B(H)$ is weakly compact, it follows that $\{\rho(d_l)\}$ has a weak cluster point x at least. From $\rho(d_l y) = \rho(d_l)y$, it is easy to see that $\rho(y) = xy, \forall y \in A$. If x' is another weak cluster point of $\{\rho(d_l)\}$, then we have

$$(x - x')y = 0, \quad \forall y \in A.$$

Thus $x = x'$, and (weakly)-$\lim_l \rho(d_l)$ exists. Moreover, from $\rho(y) = xy \in A, \forall y \in A$, it follows that the map $\rho \rightarrow x$ is injective and $x \in L_H(A)$.

For any $x \in L_H(A)$, define

$$\rho(y) = xy, \quad \forall y \in A.$$

Clearly, ρ is a left centralizer of A, and

$$x = (\text{strongly-}) \lim_l x d_l = (\text{strongly-}) \lim_l \rho(d_l)$$

since A is nondegenerate on H. Moreover, from

$$\begin{aligned} \|\rho\| &= \sup\{\|\rho(y)\| \mid y \in A, \|y\| \leq 1\} \\ &= \sup\{\|xy\| \mid y \in A, \|y\| \leq 1\} \leq \|x\| \end{aligned}$$

and $x =$(strongly)-$\rho(d_l)$, we have $\|\rho\| = \|x\|$. Therefore, $\rho \rightarrow x$ is an isometric bijection from the set of left centralizers of A onto $L_H(A)$.

For the right case, the proof is similar.

Now let (ρ_1, ρ_2) be a double centralizer of A. From preceding paragraph, $x_i =$(strongly)-$\lim_l \rho_i(d_l)$ exists, $i = 1, 2$. We say that $x_1 = x_2$. In fact, for any $y, z \in A$, we have

$$y x_1 z = y \rho_1(z) = \rho_2(y) z = y x_2 z.$$

Thus $x_1 = x_2$. Put $x = x_1 = x_2$, then $x \in L_H(A) \cap R_H(A) = M_H(A)$, and $\|x\| = \|\rho_1\| = \|\rho_2\|$. Moreover, if $x \in M_H(A)$, let $\rho_1(y) = xy$ and $\rho_2(y) = yx, \forall y \in A$, then (ρ_1, ρ_2) is a double centralizer of A. Therefore, $(\rho_1, \rho_2) \to x$ is an isometric bijection from the set of double centralizers of A onto $M_H(A)$.

Q.E.D.

Definition 2.12.8. For any two double centralizers (ρ_1, ρ_2) and $(\rho_1', \rho_2'), \lambda, \mu \in \mathbb{C}$, let

$$\lambda(\rho_1, \rho_2) + \mu(\rho_1', \rho_2') = (\lambda\rho_1 + \mu\rho_1', \lambda\rho_2 + \mu\rho_2'),$$

$$(\rho_1, \rho_2) \cdot (\rho_1', \rho_2') = (\rho_1 \circ \rho_1', \rho_2' \circ \rho_2)$$

and $(\rho_1, \rho_2)^* = (\rho_2^*, \rho_1^*)$, where $\rho_i^*(a) = \rho_i(a^*)^*, \forall a \in A, i = 1, 2$. Then the set of double centralizers of A is a $*$ algebra. Denote it by $D(A)$.

Proposition 2.12.9. Let A be a C^*-algebra, and $\{\pi, H\}$ be a nondegenerate faithful $*$ representation of A. Then $M(A), M_H(\pi(A))$ and $D(A)$ are $*$ isomorphic. Moreover, $\{\pi, H\}$ can be uniquely extended to a faithful $*$ representation of $M(A)$; denote this extension still by $\{\pi, H\}$, then $\pi(M(A)) = M_H(\pi(A))$.

Proof. Let $B = \pi(A)$. Then B is a nondegenerate C^*-algebra on H. By Proposition 2.12.7, $x \to (L_x, R_x)$ is a bijection from $M_H(B)$ onto $D(B)$, where $L_x(y) = xy, R_x(y) = yx, \forall y \in B$. Since $(L_{x^*}, R_{x^*}) = (R_x^*.L_x^*) = (L_x, R_x)^*$ and $(L_{xy}, R_{xy}) = (L_x L_y, R_y R_x) = (L_x, R_x) \cdot (L_y, R_y), \forall x, y \in M_H(B)$, it follows that $M_H(B)$ is $*$ isomorphic to $D(B)$ as $*$ algebras. Clearly, $D(A)$ and $D(B)$ are $*$ isomorphic. Further by Theorem 2.11.2 $M(A), M_H(\pi(A))$ and $D(A)$ are $*$ isomorphic.

From the preceding paragraph, we can describe the $*$ isomorphism from $M(A)$ onto $M_H(\pi(A))$ as follows:

$$\begin{aligned} x \in M(A) \to \quad & (L_x, R_x) \in D(A) \\ \to \quad & (\pi \circ L_x \circ \pi^{-1}, \pi \circ R_x \circ \pi^{-1}) \in D(\pi(A)) \\ \to \quad & (\text{strongly-}) \lim_l \pi \circ L_x \circ \pi^{-1}(\pi(d_l)) \\ = \quad & (\text{strongly-}) \lim_l \pi(x d_l) \in M_H(\pi(A)), \end{aligned}$$

where $\{d_l\}$ is an approximate identity for A. Clearly, (strongly)-$\lim_l \pi(xd_l) = \pi(x), \forall x \in A$. Thus the $*$ representation $\{\pi, H\}$ of A can be extended to a faithful $*$ representation of $M(A)$. If this extension is denoted by $\{\pi, H\}$ still, then $\pi(x) =$(strongly)-$\lim_l \pi(xd_l), \forall x \in M(A)$, and $\pi(M(A)) = M_H(\pi(A))$. Moreover, since A is a two-sided ideal of $M(A)$ and $\{\pi, H\}$ is a nondegenerate $*$ representation of A, the extension must be unique. Q.E.D.

Example 1. Let H be a Hilbert space, and $A = C(H)$. By Proposition 2.12.9, $M(A)$ is $*$ isomorphic to $B(H)$.

Example 2. Let X be a locally compact Hausdorff space, and $A = C_0^\infty(X)$. Define

$$H = l^2(X), \quad \pi(f)\xi(x) = f(x)\xi(x),$$

$\forall f \in A, \xi \in H, x \in X$. Then $\{\pi, H\}$ is a nondegenerate faithful$*$ representation of A. If $T \in M_H(\pi(A))$, then it is easily verified that there is a bounded function g on X such that $(T\xi)(x) = g(x)\xi(x), \forall \xi \in H, x \in X$. Since $fg \in C_0^\infty(X), \forall f \in C_0^\infty(X)$, it follows that g is a bounded continuous function on X. Denote the set of all bounded continuous functions on X by $C^b(X)$. Then $C^b(X)$ is $*$ isomorphic to $C(\beta X)$, where βX is the Stone-Čech compactification of X. Now by Proposition 2.12.9, $M(A)$ is $*$ isomorphic to $C(\beta X)$.

Proposition 2.12.10. Let A be a closed two-sided ideal of a C^*-algebra B. Then there is a unique $*$ homomorphism $\sigma : B \to M(A)$ such that $\sigma(a) = a, \forall a \in A$.

Moreover, σ is injective if and only if A is *essential* in B, i.e., for any non-zero closed two-sided ideal J of B, $A \cap J \neq \{0\}$.

In particular, A is an essential ideal of $M(A)$; and $M(A)$ is the maximal C^*-algebra containing A as an essential ideal, i.e., if B is a C^*-algebra containing A as an essential ideal, then B is $*$ isomorphic to a C^*-subalgebra of $M(A)$.

Proof. By Proposition 2.11.4, A^{**} is a σ-closed two-sided ideal of B^{**}. Thus there is a unique central projection z of B^{**} such that $A^{**} = B^{**}z$. Clearly, $b \to bz$ is a $*$ homomorphism from B into $M(A)$, and this $*$ homomorphism satisfies our condition.

Now if a $*$ homomorphism σ from B into $M(A)$ satisfies our condition, then for any $b \in B, a \in A$, we have

$$ba = \sigma(ba) = \sigma(b)\sigma(a) = \sigma(b)a.$$

Pick a net which converges to z σ-weakly, then we get $\sigma(b) = bz, \forall b \in B$. Therefore, σ is unique.

Suppose that σ is injective. If there is a non-zero closed two-sided ideal J of B such that $J \cap A = \{0\}$, then $JA = \{0\}$, and $\sigma(b)a = 0, \forall b \in J, a \in A$. By $M(A) \subset A^{**}$ and Theorem 2.11.2, we have $\sigma(b) = 0, \forall b \in J$. It is impossible since σ is injective and J is non-zero. Thus A is essential in B. Conversely, suppose that A is essential in B. If $\ker\sigma$ is non-zero, then there is a non-zero element $a \in A \cap \ker\sigma$. But $a = \sigma(a) = 0$, we get a contradiction. Therefore, $\ker\sigma = \{0\}$, i.e., σ is injective. Q.E.D.

Definition 2.12.11. Let A be a C^*-algebra. For any $a \in A$, define a semi-norm

$$\|\cdot\|_a = \|\cdot a\| + \|a \cdot\|,$$

on $M(A)$. Then the topology generated by the seminorm family $\{\|\cdot\|_a \mid a \in A\}$ is called the *strict topology* in $M(A)$, denoted by $s = s(M(A), A)$.

Clearly, $s(M(A), A)$ is a locally convex Hausdorff linear topology in $M(A)$.

Proposition 2.12.12. Let A be a C^*-algebra.

1) A is dense in $(M(A), s)$.

2) $(M(A), s)$ is complete. Consequently, $M(A)$ is the completion of $(A, s|A)$.

Proof. 1) Let $\{d_l\}$ be an approximate identity for A. For any $x \in M(A)$, since $xd_l \in A$ and

$$\begin{aligned} \|xd_l - x\|_a &= \|(xd_l - x)a\| + \|a(xd_l - x)\| \\ &\le \|x\| \cdot \|d_l a - a\| + \|(ax)d_l - ax\| \to 0 \end{aligned}$$

$\forall a \in A$, it follows that $xd_l \xrightarrow{s} x$. Thus A is dense in $(M(A), s)$.

2) Let $\{x_l\}$ be a s-Cauchy net of $M(A)$. Then for any $a \in A, \{x_l a\}$ and $\{ax_l\}$ are two Cauchy nets of A with respect to the norm. Hence we have two linear maps ρ_1 and ρ_2 in A such that

$$x_l a \xrightarrow{\|\cdot\|} \rho_1(a), \qquad ax_l \xrightarrow{\|\cdot\|} \rho_2(a),$$

$\forall a \in A$. Clearly,

$$a\rho_1(b) = \lim_l ax_l b = \rho_2(a)b,$$

$\forall a, b \in A$. Thus (ρ_1, ρ_2) is a double centralizer of A. By Proposition 2.12.9, there is a unique $x \in M(A)$ such that $(\rho_1, \rho_2) = (L_x, R_x)$. Further, since

$$\begin{aligned} \|x_l - x\|_a &= \|(x_l - x)a\| + \|a(x_l - x)\| \\ &= \|x_l a - \rho_1(a)\| + \|ax_l - \rho_2(a)\| \to 0, \end{aligned}$$

$\forall a \in A$, it follows that $x_l \to x(s(M(A), A))$. Therefore, $(M(A), s)$ is complete. Furthermore, $M(A)$ is the completion of $(A, s|A)$. Q.E.D.

Notes. Propositions 2.12.9 and 2.12.12 are due to R. Busby. About further developments, see references.

References. [4], [5], [14].

2.13. Finite dimensional C^*-algebras

Lemma 2.13.1. Let M be a finite dimensional factor on a Hilbert space H. Then M is spatial $*$ isomorphic to $B(H_n)\overline{\otimes}\mathbb{C}1_K$, where H_n is a n-dimensional Hilbert space, n^2 =dim M, and K is some Hilbert space.

Proof. A non-zero projection p of M is said to be minimal, if a projection $q \in M$ with $q \leq p$ implies either $q = 0$ or $q = p$. Since M is finite dimensional, it follows that there is an orthogonal family $\{p_1, \cdots, p_n\}$ of minimal projections of M such that $\sum_{i=1}^n p_i = 1$.

For each $i \in \{1, \cdots, n\}, M_i = p_i M p_i$ is a factor on $p_i H$, and any projection of M_i is either zero or $p_i (= 1_{H_i})$ since p_i is minimal. By Proposition 1.3.4, $M_i = \mathbb{C} p_i$.

For any $i, j \in \{1, \cdots, n\}$, since M is a factor, it follows from Theorem 1.5.4 that we have either $p_i \preceq p_j$ or $p_j \preceq p_i$. But p_i and p_j are minimal, thus p_i is equivalent to p_j (relative to M).

Let $p = p_1$ and $K = p_1 H$. Since $M_p = \mathbb{C}1_K$, it follows by Theorem 1.5.6 that M is spatial $*$ isomorphic to $B(H_n)\overline{\otimes}\mathbb{C}1_K$. Q.E.D.

Corollary 2.13.2. Let M be a finite dimensional factor. Then M is $*$ isomorphic to the algebra of $n \times n$-matrices, where n^2=dim$M$.

Theorem 2.13.3. If A is a finite dimensional C^*-algebra, then A is $*$ isomorphic to a direct sum $\sum_{i=1}^m \oplus M_{n_i}$, where M_{n_i} is the algebra of $n_i \times n_i$-matrices, $1 \leq i \leq m$. Moreover, the sequence $\{n_1, \cdots, n_m\}$ is a complete invariant for the algebraic structure of A, i.e., if B is another finite dimensional C^*-algebra with the associated sequence $\{n'_1 \cdots, n'_{m'}\}$, then A and B are $*$ isomorphic if and only if $\{n_1, \cdots, n_m\} = \{n'_1, \cdots, n'_{m'}\}$.

Proof. Let $\{\pi, H\}$ be a nondegenerate faithful $*$ representation of A. Then $\pi(A)$ is nondegenerate on H and dim$\pi(A) < \infty$. By Theorem 1.3.10, $\pi(A) = M$ is a VN algebra on H. Since the center Z of M is finite dimensional, it follows that there is an orthogonal family $\{z_1, \cdots, z_m\}$ of minimal central projections of Z such that $\sum_{i=1}^m z_i = 1$. For each $i \in \{1, \cdots, m\}, Mz_i$ is a finite dimensional factor on $z_i H$. By Corollary 2.13.2, Mz_i is $*$ isomorphic to $M_{n_i}, 1 \leq i \leq m$. Therefore, A is $*$ isomorphic to $\sum_{i=1}^m \oplus M_{n_i}$.

The proof of the rest conclusion is easy, and is left to the reader. Q.E.D.

Proposition 2.13.4. Let G be a finite group, and $R(G) = \{\sum_{s\in G} c_s s \mid c_s \in \mathbb{C}, \forall s \in G\}$ be its group algebra. Then $R(G)$ is isomorphic to a direct sum $\sum_{i=1}^m \oplus M_{n_i}$, and $\sharp G = n_1^2 + \cdots n_m^2$.

Proof. Let $\cdot \to \lambda_\cdot$ be the left regular representation of G on $l^2(G)$, i.e.,

$$(\lambda_s f)(t) = f(s^{-1}t),$$

$\forall s, t \in G, f \in l^2(G)$. Since $\lambda_s^* = \lambda_{s^{-1}}, \lambda_s\lambda_t = \lambda_{st}, \forall s, t \in G$, it follows that $[\lambda_s \mid s \in G]$ is a finite dimensional C^*-algebra on $l^2(G)$. Clearly, $R(G)$ is isomorphic to $[\lambda_s \mid s \in G]$. Therefore, by Theorem 2.13.3 we can get the conclusion. Q.E.D.

Remark. This result is well-known in the text-book of algebras.

2.14. The axioms for C^*-algebras

The unit ball of a C^*-algebra

Proposition 2.14.1. Let A be a C^*-algebra with an identity 1, $S = \{a \in A \mid \|a\| \leq 1\}$ be its closed unit ball, and $A_u = \{v \in A \mid v^*v = vv^* = 1\}$ be the set of unitary elements of A. Then $S = \overline{CoA_u}$, i.e., S is the closed convex hull of A_u.

Proof. Let $a \in A$ and $\|a\| < 1$. Then

$$f(a, \lambda) = (1 - aa^*)^{-\frac{1}{2}}(1 + \lambda a)$$

is invertible in A for any $\lambda \in \mathbb{C}$ and $|\lambda| = 1$. Since

$$\begin{aligned} a^*(1 - aa^*)^{-1} &= a^* \sum_{n=0}^{\infty}(aa^*)^n \\ &= \sum_{n=0}^{\infty}(a^*a)^n a^* = (1 - a^*a)^{-1}a^*, \end{aligned}$$

it follows that for any $\lambda \in \mathbb{C}$ and $|\lambda| = 1$,

$$\begin{aligned} f(a,\lambda)^* f(a,\lambda) + 1 &= (1 + \overline{\lambda}a^*)(1 - aa^*)^{-1}(1 + \lambda a) + 1 \\ &= (1 - aa^*)^{-1} + \overline{\lambda}a^*(1 - aa^*)^{-1} \\ &\quad + \lambda(1 - aa^*)^{-1}a + [a^*(1 - aa^*)^{-1}a + 1] \\ &= (1 - aa^*)^{-1} + \overline{\lambda}(1 - a^*a)^{-1}a^* \\ &\quad + \lambda(1 - aa^*)^{-1}a + (1 - a^*a)^{-1}. \end{aligned}$$

Thus we can see that

$$f(a,\lambda)^*f(a,\lambda) = f(a^*,\overline{\lambda})^*f(a^*,\overline{\lambda}), \quad \forall|\lambda| = 1.$$

Further,

$$u_\lambda = f(a,\lambda)f(a^*,\overline{\lambda})^{-1} \in A_u, \quad \forall|\lambda| = 1.$$

Now let

$$u(\lambda) = (1 - aa^*)^{-1/2}(\lambda + a)(1 + \lambda a^*)^{-1}(1 - a^*a)^{1/2}, \forall|\lambda| \le 1.$$

Clearly, $u(\lambda)$ is analytic in $|\lambda| < 1$, and $u(\lambda) = \lambda u_{\overline{\lambda}} \in A_u, \forall|\lambda| = 1$. Moreover, $u(0) = a$. Thus

$$\begin{aligned} a &= \frac{1}{2\pi i}\int_{|\lambda|=1} \frac{u(\lambda)}{\lambda} d\lambda = \frac{1}{2\pi}\int_0^{2\pi} u(e^{i\theta})d\theta \\ &= \lim_n \frac{1}{n}\sum_{k=1}^{n} u(e^{2\pi i k/n}). \end{aligned}$$

Since $u(\lambda) \in A_u, \forall|\lambda| = 1$, $a \in \overline{CoA_u}$. Further, $S = \overline{CoA_u}$. Q.E.D.

Theorem 2.14.2. Let A be a C^*-algebra with an identity, and $S = \{a \in A \mid \|a\| \le 1\}$ be its closed unit ball. Then $Co\{e^{ih} \mid h^* = h \in A\}$ is dense in S.

Proof. Let $\{\pi_u, H_u\}$ be the universal $*$ representation of A (see Definition 2.11.1), and $\overline{A} = \pi_u(A)''$ be the enveloping VN algebra of A.

If u is a unitary element of $\overline{A}$, then we have the following spectral decomposition:

$$\begin{aligned} u &= \int_0^{2\pi} e^{i\theta}dp(\theta) = \lim_n \sum_{k=1}^{n} e^{2\pi i k/n}[p(\frac{2\pi k}{n}) - p(\frac{2\pi(k-1)}{n})] \\ &= \lim_n \exp(i\sum_{k=1}^{n} \frac{2\pi k}{n} p_k^{(n)}), \end{aligned}$$

where $\{p_k^{(n)} = p(\frac{2\pi k}{n}) - p(\frac{2\pi(k-1)}{n}) \mid 1 \le k \le n\}$ is an orthogonal family of projections of $\overline{A}$, and $\sum_{k=1}^{n} p_k^{(n)} = 1$. By Proposition 2.14.1, $Co\{e^{ih} \mid h^* = h \in \overline{A}\}$ is dense in the closed unit ball of $\overline{A}$ with respect to the uniform topology.

From density Theorem 1.6.1, $Co\{\pi_u(e^{ih}) \mid h^* = h \in A\}$ is strongly dense in the closed unit ball of $\overline{A}$. Consequently, $Co\{\pi_u(e^{ih}) \mid h^* = h \in A\}$ is strongly dense in $\pi_u(S)$.

Now suppose that $Co\{e^{ih} \mid h^* = h \in A\}$ is not dense in S. Then we can find $a \in S$ and $f \in A^*$ such that

$$\sup\{\text{Re}f(e^{ih}) \mid h^* = h \in A\} < \text{Re}f(a).$$

However, from preceding paragraph there is a net $\{a_l\} \subset Co\{e^{ih} \mid h^* = h \in A\})$ such that $\pi_u(a_l) \to \pi_u(a)$ (strongly). Then for any state ρ on A we have

$$\rho(a_l) = \langle \pi_u(a_l)\xi_\rho, \xi_\rho\rangle \to \langle \pi_u(a)\xi_\rho, \xi_\rho\rangle = \rho(a).$$

Thus $a_l \to a(\sigma(A, A^*))$. In particular, $f(a_l) \to f(a)$. This is a contradiction. Therefore, $Co\{e^{ih} \mid h^* = h \in A\}$ is dense in S. Q.E.D.

Theorem 2.14.3. Let A be a C^*-algebra with an identity. Then for any $a \in A$, we have

$$\begin{aligned} \|a\| &= \inf\{\sum_j |\lambda_j| \mid a = \sum_j \lambda_j e^{ih_j}, \text{ where } h_j^* = h_j \in A, \forall j\} \\ &= \inf\{\sum_g |\lambda_j| \mid a = \sum_j \lambda_j u_j, \text{ where } u_j \in A_u, \forall j\}. \end{aligned}$$

Proof. It suffices to show the first equality. Let

$$\|a\|_1 = \inf\{\sum_j |\lambda_j| \mid a = \sum_j \lambda_j e^{ih_j}, \text{ where } h_j^* = h_j \in A, \forall j\}.$$

Clearly, $\|a\|_1 \geq \|a\|$.

For any $a^* = a \in A$ and $\|a\| \leq 1, v_\pm = a \pm i(1-a^2)^{1/2} \in A_u$. Since $\sigma(a) \subset [-1,1]$, it follows that $\sigma(v_\pm) \neq \{\lambda \| \lambda| = 1\}$. Thus there are $h_\pm^* = h_\pm \in A$ such that $v_\pm = e^{ih_\pm}$ respectively, and

$$a = \frac{1}{2}v_+ + \frac{1}{2}v_- \in Co\{e^{ih} \mid h^* = h \in A\}.$$

Now suppose that $x \in A$ and $\|x\| \leq 1/2$. From preceding paragraph, it follows that $(x + x^*)$ and $i(x - x^*) \in Co\{e^{ih} \mid h^* = h \in A\}$. But we can write $x - x^* = e^{-\frac{\pi}{2}i} \cdot i(x - x^*)$, so $(x - x^*) \in Co\{e^{ih} \mid h^* = h \in A\}$. Further $x = \frac{1}{2}(x + x^*) + \frac{1}{2}(x - x^*) \in Co\{e^{ih} \mid h^* = h \in A\}$, i.e.,

$$\frac{1}{2}S \subset Co\{e^{ih} \mid h^* = h \in A\},$$

where $S = \{a \in A \mid \|a\| \leq 1\}$ is the closed unit ball of A. Hence we obtain

$$\|a\| \leq \|a\|_1 \leq 2\|a\|, \quad \forall a \in A.$$

For any $0 \neq a \in A$, by Theorem 2.14.2 there is a sequence $\{a_n\} \subset Co\{e^{ih} \mid h^* = h \in A\}$ such that

$$\|a_n - \|a\|^{-1}a\| \to 0.$$

Since $\|\cdot\| \sim \|\cdot\|_1$, it follows that $\|a_n - \|a\|^{-1}a\|_1 \to 0$. Clearly, $\|a_n\|_1 \leq 1$. Thus $\|a\|_1 \leq \|a\|$. Further, we have $\|a\|_1 = \|a\|, \forall a \in A$. Q.E.D.

Proposition 2.14.4. Let A be a C^*-algebra with an identity, $S = \{a \in A \mid \|a\| \leq 1\}$ be its closed unit ball, and Int(S)= $\{a \in A \mid \|a\| < 1\}$ be its open unit ball. Then we have

$$\text{Int(S)} \subset Co\{e^{ih} \mid h^* = h \in A\} \subset S.$$

Proof. Suppose that $a \in$Int(S). By Theorem 2.14.3, we can write $a = \sum_j \lambda_j e^{ih_j}$, where $h_j^* = h_j \in A, \lambda_j > 0, \forall j$, and $\sum_j \lambda_j < 1$. Then

$$\begin{aligned} a = & \sum_j \lambda_j e^{ih_j} + \frac{1 - \sum_j \lambda_j}{2} e^{i \cdot 0} + \frac{1 - \sum_j \lambda_j}{2} e^{i \cdot \pi} \\ \in & \ Co\{e^{ih} \mid h^* = h \in A\}. \end{aligned}$$

Q.E.D.

Theorem 2.14.5. Let A be a C^*-algebra with an identity, B be a normed space, and Φ be a bounded linear map from A into B. Then

$$\|\Phi\| = \sup\{\|\Phi(e^{ih})\| \mid h^* = h \in A\}.$$

Proof. By Proposition 2.14.4, it follows that

$$\begin{aligned} \|\Phi\| = & \sup_{\|a\|<1} \|\Phi(a)\| \\ = & \sup\{\|\sum_j \lambda_j \Phi(e^{ih_j})\| \mid h_j^* = h_j \in A, \lambda_j > 0, \forall j, \text{ and } \sum_j \lambda_j = 1\} \\ \leq & \sup\{\|\Phi(e^{ih})\| \mid h^* = h \in A\} \leq \|\Phi\|. \end{aligned}$$

Q.E.D.

Strictly positive elements

Definition 2.14.6. Let A be a C^*-algebra, and $S(A)$ be its state space. An element $a \in A_+$ is said to be *strictly positive*, if $\varphi(a) > 0, \forall \varphi \in S(A)$.

If A has an identity, by Proposition 2.3.13 then $a (\in A_+)$ is strictly positive if and only if a is invertible in A.

Lemma 2.14.7. Let A be a C^*-algebra, $a \in A_+$ be strictly positive, and $\{\pi, H\}$ be a nondegenerate $*$ representation of A. Then $\pi(a)H$ is dense in H.

Proof. Suppose that there is $\xi \in H$ with $\|\xi\| = 1$ such that $\langle \pi(a)\eta, \xi \rangle = 0, \forall \eta \in H$. Let $\rho(\cdot) = \langle \pi(\cdot)\xi, \xi \rangle$. Then $\rho \in S(A)$. But $\rho(a) = 0$, this is a contradiction since a is strictly positive. Therefore, $\pi(a)H$ is dense in$H$.

Q.E.D.

Theorem 2.14.8. Let A be a C^*-algebra. Then there is a strictly positive element in A at least if and only if A admits an approximate identity $\{d_n\}_{n=1}^\infty$ such that $d_n d_m = d_m d_n, \forall n, m$.

Proof. Suppose that $\{d_n\}_{n=1}^\infty$ is an approximate identity for A, and $d_n d_m = d_m d_n, \forall n, m$. Let $a = \sum_{n=1}^\infty 2^{-n} d_n \in A$. For any state ρ on A, since $\rho(d_n) \to 1$ (Proposition 2.4.4), it follows that $\rho(a) > 0$. Thus a is strictly positive.

Conversely, suppose that $a \in A_+$ is strictly positive. We may assume that $\|a\| = 1$. Put $d_n = a^{\frac{1}{n}}, n = 1, 2, \cdots$. Clearly, $d_n d_m = d_m d_n, d_m \geq d_n \geq 0, \|d_n\| = 1, \forall m \geq n$. Now it suffices to show that

$$\|x d_n - x\| \to 0, \quad \forall x \in A_+.$$

Fix $x \in A_+$, and let $z_n = x - x^{\frac{1}{2}} d_n x^{\frac{1}{2}}$. Clearly, $z_n \geq z_m \geq 0, \forall m \geq n$. Let

$$\Omega = \{\rho \in A^* \mid \rho \geq 0, \|\rho\| \leq 1\}.$$

It is a $\sigma(A^*, A)$-compact subset of A^*. Let $z_n(\rho) = \rho(z_n), \forall n, \rho \in \Omega$. Then $z_n(\cdot) \in C(\Omega), \forall n$, and $z_1(\cdot) \geq \cdots \geq z_n(\cdot) \geq \cdots$. We claim that

$$\lim_n z_n(\rho) = 0, \quad \forall \rho \in \Omega.$$

In fact, for any $\rho \in \Omega \backslash \{0\}$, let $\{\pi_\rho, H_\rho, \xi_\rho\}$ be the cyclic $*$ representation of A generated by ρ. Then

$$\begin{aligned} z_n(\rho) &= \langle \pi_\rho(z_n) \xi_\rho, \xi_\rho \rangle \\ &= \langle \pi_\rho(x) \xi_\rho, \xi_\rho \rangle - \langle \pi_\rho(x^{\frac{1}{2}} d_n x^{\frac{1}{2}}) \xi_\rho, \xi_\rho \rangle. \end{aligned}$$

By Lemma 2.14.7, $\pi_\rho(a) H_\rho$ is dense in H_ρ. Moreover, $\pi_\rho(d_n)\pi_\rho(a)\eta = \pi_\rho(a^{1+\frac{1}{n}})\eta \to \pi_\rho(a)\eta, \forall \eta \in H_\rho$. Thus $\pi_\rho(d_n) \to 1$ (strongly), and $z_n(\rho) \to 0, \forall \rho \in \Omega$. Now by the Dini theorem, we get

$$\max\{|z_n(\rho)| \mid \rho \in \Omega\} \to 0$$

Further, $\|z_n\| \to 0$ by Corollary 2.3.14, i.e. $x^{\frac{1}{2}} d_n x^{\frac{1}{2}} \to x$. Therefore,

$$\begin{aligned} \|x d_n - x\|^2 &= \|(1 - d_n)x\|^2 \leq 4\|x\| \cdot \|(1 - d_n)^{1/2} x^{1/2}\|^2 \\ &= 4\|x\| \cdot \|x^{1/2}(1 - d_n) x^{1/2}\| \to 0. \end{aligned}$$

Q.E.D.

Theorem 2.14.9. Let A be a separable C^*-algebra. Then A has a strictly positive element at least.

Proof. Let $\{x_n\}$ be a countable dense subset of $A_+ \cap S$ and $a = \sum_n 2^{-n} x_n$, where $S = \{b \in A \mid \|b\| \le 1\}$ is the closed unit ball of A. For any state ρ on A, since $\rho(x_n) > 0$ for some n, it follows that $\rho(a) > 0$. Thus a is strictly positive. Q.E.D.

Proposition 2.14.10. If A has a strictly positive element, then the set of strictly positive elements is dense in A_+.

Proof. Let a be strictly positive. For any $b \in A_+, (b + \frac{1}{n}a)$ is also strictly positive, and $(b + \frac{1}{n}a) \to b$. Therefore, the set of strictly positive elements is dense in A_+. Q.E.D.

Banach * algebras

Definition 2.14.11. A is called a *Banach * algebra* if A is a complex Banach algebra and admits a map: $x \to x^*(\in A)$ with the following properties:

$$(\lambda x + \mu y)^* = \overline{\lambda} x^* + \overline{\mu} y^*, (xy)^* = y^* x^*, (x^*)^* = x,$$

$\forall x, y \in A, \lambda, \mu \in \mathbb{C}$.

The * operation on A or A itself is said to be *hermitian*, if for any $x^* = x \in A$, its spectrum $\sigma(x) \subset \mathbb{R}$.

$x \in A$ is said to be *positive*, denoted by $x \ge 0$, if $x^* = x$ and $\sigma(x) \subset \mathbb{R}_+ = [0, \infty)$. Moreover, $a \ge b$ if $(a - b) \ge 0$.

Lemma 2.14.2. Let A be a Banach * algebra with an identity, and B be a *maximal abelian * subalgebra* of A. Then B is closed, and $\sigma_B(b) = \sigma_A(b), \forall b \in B$.

Proof. It is easily verified that B is closed. Now suppose that $b \in B, \lambda \in \mathbb{C}$ and $(b - \lambda)^{-1}$ exists in A. Since $\{(b - \lambda)^{-1}, (b^* - \overline{\lambda})^{-1}, B\}$ is commutative and B is maximal abelian, it follows that $(b - \lambda)^{-1} \in B$. Therefore, for any $b \in B$ we have $\sigma_B(b) = \sigma_A(b)$. Q.E.D.

Lemma 2.14.13. Let A be an abelian *semi-simple* Banach * algebra with an identity. Then the * operation is continuous automatically.

Proof. Suppose that Ω is the spectral space of A. For any $\rho \in \Omega$, define $\overline{\rho}(a) = \overline{\rho(a^*)}, \forall a \in A$. It is easy to see that $\overline{\rho} \in \Omega$.

Now let $\{x_n\} \subset A$ and $x, y \in A$ be such that

$$\|x_n - x\| \to 0, \quad \text{and } \|x_n^* - y\| \to 0.$$

Then for any $\rho \in \Omega$,

$$\begin{aligned}|\rho(x-y^*)| \leq & \ |\rho(x_n - x)| + |\rho(x_n - y^*)| \\ = & \ |\rho(x_n - x)| + |\overline{\rho}(x_n^* - y)| \\ \leq & \ \|x_n - x\| + \|x_n^* - y\| \to 0,\end{aligned}$$

i.e., $\rho(x-y^*) = 0$. Since A is semi-simple, it follows that $x = y^*$. Thus the $*$ operation is a closed linear operator on the real Banach space A. Further, the $*$ operation is continuous. Q.E.D.

Theorem 2.14.14. Let A be a Banach $*$ algebra with an identity, $a \in A, a \geq 0$, and a be invertible in A. Then there is $u \in A$ such that: 1) $u \geq 0$ and u is invertible in A; 2) $u^2 = a$; 3) if B is any maximal abelian $*$ subalgebra of A, and $a \in B$, then $u \in B$ too.

Proof. We may assume that $\|a\| < 1$. Thus $\nu(1-a) < 1$, and there is $\varepsilon \in (0,1)$ and a positive integer n_0 such that $\|(1-a)^n\|^{1/n} \leq 1-\varepsilon, \forall n \geq n_0$. Since the complex function

$$(1+z)^{1/2} = \sum_{n=0}^{\infty} \lambda_n z^n$$

is analytic in $|z| < 1$, it follows that the sequence

$$\{a_k = \sum_{n=0}^{k} \lambda_n (a-1)^n \mid k = 0, 1, \cdots\}$$

is convergent. Suppose that

$$a_k \to u + iv,$$

where $u^* = u, v^* = v$. Then we have

$$(u+iv)^2 = a. \tag{1}$$

Since $a^* = a$, it follows that

$$uv = -vu. \tag{2}$$

Now let B be a maximal abelian $*$ subalgebra of A and $a \in B$. Clearly, $a_k \in B, \forall k$. From Lemma 2.14.12, we have $(u+iv) \in B$. So it is obvious that

$$u, v \in B, \quad \text{and } uv = vu. \tag{3}$$

Summing up (1), (2), (3), we obtain

$$a = u^2 - v^2, \quad u, v \in B, \text{ and } uv = 0. \tag{4}$$

Let R be the radical of B. Clearly, $R^* = R$. Hence B/R is an abelian semi-simple Banach $*$ algebra. By Lemma 2.14.13, the $*$ operation is continuous on B/R. Suppose that $b \to \tilde{b} = b + R$ is the canonical map from B onto B/R. Then

$$(\widetilde{a_k - u})^* = (\widetilde{a_k - u}) \to (\widetilde{iv}) = (\widetilde{iv})^*.$$

Thus $\tilde{v} = \tilde{0}$, and $v \in R$.

If $0 \in \sigma(u)$, then by Lemma 2.14.12 there is $\rho \in \Omega(B)$ such that $\rho(u) = 0$, where $\Omega(B)$ is the spectral space of B. Since $v \in R$, so $\rho(v) = 0$. Then $\rho(a) = \rho(u^2 - v^2) = 0$, and $0 \in \sigma_B(a)$. This is a contradiction since a is invertible in A and $\sigma_B(a) = \sigma_A(a)$. Therefore, u is invertible in A, and $u^{-1} \in B$. So from (4) we can see that $v = u^{-1}uv = 0$, and $a = u^2$.

Finally, for any $\rho \in \Omega(B)$, by Lemma 2.14.12 we have that $\lambda = \rho(a) \in (0,1)$. Then

$$\rho(a_k) = \sum_{n=1}^{k} \lambda_n(\lambda - 1)^n \to (1 + (\lambda - 1))^{1/2} = \lambda^{1/2} \geq 0,$$

and $\rho(u) = \lim_k \rho(a_k) \geq 0$. Again by Lemma 2.14.12, $\sigma(u) \subset \mathbb{R}_+$, i.e. $u \geq 0$.

Q.E.D.

Theorem 2.14.15. Let A be a hermitian Banach $*$ algebra. Then $A_+ = \{a \in A \mid a \geq 0\}$ is a cone, i.e., if $a, b \in A_+$, then $(a + b) \in A_+$.

Proof. We may assume that A has an identity.

First step. To show the following inequality:

$$\nu(x) \leq \nu(x^*x)^{1/2}, \quad \forall x \in A.$$

In fact, fix $x \in A$ and $\varepsilon > 0$, and let $y = (\nu(x^*x) + \varepsilon)^{-1/2}x$. Then $\nu(y^*y) < 1$. Since the $*$ operation is hermitian, it follows that $(1 - y^*y) \geq 0$, and $(1 - y^*y)$ is invertible in A. By Theorem 2.14.14, we have an invertible element w of A such that $w \geq 0$ and $w^2 = 1 - y^*y$. Notice the equality:

$$(1 + y^*)(1 - y) = w[1 + w^{-1}(y^* - y)w^{-1}]w.$$

Since $\sigma(iw^{-1}(y^* - y)w^{-1}) \subset \mathbb{R}$, it follows that the right side of above quality is invertible. Further, $(1 - y)$ has a left inverse.

Suppose that $\nu(y) > 1$. Pick $\lambda \in \sigma(y)$ such that $|\lambda| = \nu(y)$. Since $\nu(y^*y) < 1$, it follows that $(1 - |\lambda|^{-2}y^*y)$ is positive and invertible. Similar to the preceding paragraph, $(1 - \lambda^{-1}y)$ has a left inverse. Let z be the left inverse of $(y - \lambda)$. Since λ is a boundary point of $\sigma(y)$, we can pick a sequence $\{\lambda_n\}$ of regular points of y such that $\lambda_n \to \lambda$. Then $\|(y - \lambda_n)^{-1}\| \to \infty$, and

$$\begin{aligned} 1 = & \|z(y - \lambda)(y - \lambda_n)^{-1}\| \cdot \|(y - \lambda_n)^{-1}\|^{-1} \\ = & \|z + (\lambda_n - \lambda)z(y - \lambda_n)^{-1}\| \cdot \|(y - \lambda_n)^{-1}\|^{-1} \\ \leq & \|z\| \cdot \|(y - \lambda_n)^{-1}\|^{-1} + |\lambda_n - \lambda| \cdot \|z\| \to 0. \end{aligned}$$

This is impossible. Thus $\nu(y) \leq 1$, and $\nu(x) \leq (\nu(x^*x) + \varepsilon)^{1/2}$. Since ε is arbitrary, we have $\nu(x) \leq \nu(x^*x)^{1/2}, \forall x \in A$.

Second step. To prove that

$$\nu(hk) \leq \nu(h)\nu(k), \quad \forall h^* = h, k^* = k \in A.$$

In fact, from the first step we have

$$\begin{aligned} \nu(hk)^2 \leq \ & \nu(kh^2k) = \lim_n \|(kh^2k)^n\|^{\frac{1}{n}} \\ = \ & \lim_n \|k(h^2k^2)^{n-1}h^2k\|^{\frac{1}{n}} \leq \nu(h^2k^2). \end{aligned}$$

Generally, we have

$$\nu(hk) \leq \nu(h^{2^n}k^{2^n})^{1/2^n} \leq \|h^{2^n}\|^{1/2n} \cdot \|k^{2^n}\|^{1/2^n}.$$

Let $n \to \infty$, we obtain $\nu(hk) \leq \nu(h)\nu(k)$ immediately.

Finally, we return to prove the theorem. Let $a, b \in A_+$. Notice that

$$\begin{aligned} 1 + a + b = \ & (1+a)(1+b) - ab \\ = \ & (1+a)(1-uv)(1+b), \end{aligned}$$

where $u = (1+a)^{-1}a, v = (1+b)^{-1}b$. Clearly, $\nu(u) < 1, \nu(v) < 1$. By the second step, $\nu(uv) < 1$. Thus $(1 + a + b)$ is invertible, i.e., $(-1) \notin \sigma(a+b)$. For $\lambda > 0$, similarly we have $(-1) \notin \sigma(\frac{a}{\lambda} + \frac{b}{\lambda})$, i.e., $-\lambda \notin \sigma(a+b)$. Moreover, $\sigma(a+b) \subset \mathbb{R}$ since A is hermitian. Therefore, $(a+b) \geq 0$. Q.E.D.

Theorem 2.14.16. Let A be a Banach $*$ algebra. Then A is hermitian if and only if $a^*a \geq 0, \forall a \in A$.

Proof. Suppose that $a^*a \geq 0, \forall a \in A$. If there is $h^* = h \in A$ such that $\sigma(h) \not\subset \mathbb{R}$, then $\sigma(h^2) \not\subset \mathbb{R}_+$. This is a contradiction since $h^2 = h^*h \geq 0$. Thus A is hermitian.

Conversely, suppose that A is hermitian. We may assume that A has an identity. Suppose that there is $x \in A$ such that

$$\delta = \inf\{\lambda \mid \lambda \in \sigma(x^*x)\} < 0.$$

Replacing x by μx (some $\mu > 0$), we may assume that $\delta \in (-1, \frac{-1}{3})$. Put

$$y = 2x(1 + x^*x)^{-1}.$$

Then $1 - y^*y = (1 - x^*x)^2(1 + x^*x)^{-2} \geq 0$, and $\sigma(y^*y) \subset (-\infty, 1]$. Write $y = h + ik$, where $h^* = h, k^* = k$. By Theorem 2.14.15,

$$1 + yy^* = 2(h^2 + k^2) + (1 - y^*y) \geq 0,$$

and $\sigma(yy^*) \subset [-1,\infty)$. Since $\sigma(y^*y)\backslash\{0\} = \sigma(yy^*)\backslash\{0\}$ (see Lemma 2.2.6), it follows that

$$\sigma(y^*y) \subset [-1,1].$$

From $\delta \in \sigma(x^*x)$ and $y^*y = 4x^*x(1+x^*x)^{-2}$, we have

$$4\delta/(1+\delta)^2 \in \sigma(y^*y).$$

Thus $|\frac{4\delta}{(1+\delta)^2}| \le 1$, *i.e.*, $6|\delta| \le 1+\delta^2$. Further $|\delta| < 1/3$ since $1+\delta^2 < 2$. Then we obtain a contradiction since $\delta \in (-1,-1/3)$. Therefore, we have $x^*x \ge 0, \forall x \in A$. Q.E.D.

C^*-equivalent algebras

Definition 2.14.17. Let A be a Banach $*$ algebra with an identity. A linear functional ρ on A is called a *state*, if

$$\rho(1) = 1, \quad \text{and } \rho(a) \ge 0, \quad \forall a \in A_+.$$

If A is also hermitian, then for any $h^* = h \in A$, $\rho(h) \in I\!R$ since $\|h\|+h \ge 0$. Further $\rho(a^*) = \overline{\rho(a)}, \forall a \in A$. Moreover, by Theorem 2.14.16 we have also the Schwartz inequality:

$$|\rho(b^*a)|^2 \le \rho(a^*a)\rho(b^*b), \quad \forall a,b \in A.$$

Lemma 2.14.18. Let A be a hermitian Banach $*$ algebra with an identity, and $h^* = h \in A$. Then for each $\lambda \in [\lambda_1,\lambda_2]$, where $\lambda_1 = \min\{\mu \mid \mu \in \sigma(h)\}, \lambda_2 = \max\{\mu \mid \mu \in \sigma(h)\}$, there is a state ρ on A such that $\rho(h) = \lambda$.

Proof. On the linear subspace $[1,h]$ of A, define

$$\rho(\alpha+\beta h) = \alpha + \beta\lambda, \quad \forall \alpha,\beta \in \mathbb{C}.$$

Suppose that $\alpha+\beta h \ge 0$ for some $\alpha,\beta \in \mathbb{C}$. In particular, $\alpha,\beta \in I\!R$. Then the real number $(\alpha+\beta\lambda)$ is between $(\alpha+\beta\lambda_1)$ and $(\alpha+\beta\lambda_2)$. Since $(\alpha+\beta\lambda_j) \in \sigma(\alpha+\beta h)$, it follows that $(\alpha+\beta\lambda_j) \ge 0, j = 1,2$. Thus $(\alpha+\beta\lambda) \ge 0$. This means that ρ is a state on $[1,h]$.

Now by Theorem 2.14.15 and the fact that the $*$ operation is hermitian, and by a similar proof of Proposition 2.3.11, ρ can be extended to a state on A.

Q.E.D.

Lemma 2.14.19. Let Δ be a compact subset of $\mathbb{C}$, and $0 \in \Delta$. Then for each $\lambda \in \mathbb{C}$, we have

$$\max\{|\lambda+\mu| \mid \mu \in \Delta\} \ge \frac{1}{3}(\max\{|\mu| \mid \mu \in \Delta\} + |\lambda|).$$

Proof. Since $|\lambda+\mu| \geq |\mu| - |\lambda|$, it follows that

$$\max\{|\lambda+\mu| \mid \mu \in \Delta\} \geq \max\{|\mu| \mid \mu \in \Delta\} - |\lambda|.$$

In addition, by $0 \in \Delta$ we have $2\max\{|\lambda+\mu| \mid \mu \in \Delta\} \geq 2|\lambda|$. Therefore, $\max\{|\lambda+\mu| \mid \mu \in \Delta\} \geq \frac{1}{3}(\max\{|\mu| \mid \mu \in \Delta\} + |\lambda|)$. Q.E.D.

Lemma 2.14.20. Let A be a Banach $*$ algebra. Suppose that there is a positive constant K such that

$$K\|h\| \leq \nu(h), \quad \forall h^* = h \in A.$$

Then the $*$ operation on A is continuous.

Proof. Let $H = \{a \in A \mid a^* = a\}$. It suffices to show that H is closed.

For any $h \in \overline{H}$, there is a sequence $\{h_n\} \subset H$ such that $\|h_n - h\| \to 0$. Then for each $\varepsilon > 0$, we have that $\nu(h) + \varepsilon \geq \nu(h_n) \geq K\|h_n\|$ if n is sufficiently large. Hence we get

$$\nu(h) \geq K\|h\|, \forall h \in \overline{H}. \tag{1}$$

Now let $\{h_n\} \subset H, h_n \to k$, and $k^* = -k$. Since $(h_m + h_n)^2 \in H$ and $(h_m+h_n)^2 \to (k+h_n)^2$ as $m \to \infty$, it follows from (1) that

$$\begin{aligned} K\|(k+h_n)^2\| \leq \; & \nu((k+h_n)^2) = \nu(k+h_n)^2 \\ = \; & \nu((k+h_n)^*)^2 = \nu(h_n - k)^2 \\ \leq \; & \|h_n - k\|^2 \to 0, \quad \text{as } n \to \infty. \end{aligned}$$

Thus

$$\begin{aligned} \|k^2 + h_n^2\| = \; & \frac{1}{2}\|(k+h_n)^2 + (k-h_n)^2\| \\ \leq \; & \frac{1}{2}(\|(k+h_n)^2\| + \|k-h_n\|^2) \to 0. \end{aligned}$$

Further, $2\|k^2\| \leq \|k^2 + h_n^2\| + \|k^2 - h_n^2\| \to 0$, so $\|k^2\| = 0$. Again by (1) we have

$$0 = \|k^2\| \geq \nu(k^2) = \nu(k)^2 \geq K^2\|k\|^2,$$

and $k = 0$. Therefore, H is closed. Q.E.D.

Definition 2.14.21. A Banach $*$ algebra $(A, \|\cdot\|)$ is said to be C^*-equivalent , if there is a new norm $\|\cdot\|_1$ on A such that $\|\cdot\|_1 \sim \|\cdot\|$ and $(A, \|\cdot\|_1)$ is a C^*-algebra.

Theorem 2.14.22. Let A be a hermitian Banach $*$ algebra. If there exists a positive constant K such that

$$K\|h\| \leq \nu(h), \quad \forall h^* = h \in A,$$

then A is C^*-equivalent.

Proof. If A has no identity, then we consider the Banach $*$ algebra $A\dot{+}\mathbb{C}$. Clearly, $(A\dot{+}\mathbb{C})$ is still hermitian. Suppose that $(h+\lambda)$ is a self-adjoin element of $(A\dot{+}\mathbb{C})$. Obviously, $h^* = h(\in A)$, and $\overline{\lambda} = \lambda(\in \mathbb{R})$. Since A has no identity, it follows that $0 \in \sigma(h)$. By Lemma 2.14.19 and $K \leq 1$, we have

$$\begin{aligned} \nu(h+\lambda) = &\max\{|\mu+\lambda| \mid \mu \in \sigma(h)\} \\ \geq &\frac{1}{3}(\nu(h)+|\lambda|) \geq \frac{K}{3}(\|h\|+|\lambda|) \\ = &\frac{K}{3}(\|h+\lambda\|). \end{aligned}$$

Thus, we may assume that A has an identity.

By Lemma 2.14.20, there is a positive constant M such that

$$\|a^*\| \leq M^2\|a\|, \qquad \forall a \in A.$$

Let ρ be a state on A, and $L_\rho = \{a \in A \mid \rho(a^*a) = 0\}$. By the Schwartz inequality, L_ρ is a left ideal of A. Suppose that $a \to a_\rho = a + L_\rho$ is the canonical map from A onto A/L_ρ, and define an inner product on A/L_ρ:

$$\langle a_\rho, b_\rho\rangle = \rho(b^*a), \qquad \forall a,b \in A.$$

Denote the completion of $(A/L_\rho, \langle,\rangle)$ by H_ρ. For any $a \in A$, define a linear map $\pi_\rho(a)$ on A/L_ρ:

$$\pi_\rho(a)b_\rho = (ab)_\rho, \qquad \forall b \in A.$$

For any $\varepsilon > 0$, by Theorem 2.14.14, there is $u^* = u \in A$ such that

$$\|a^*a\| + \varepsilon - a^*a = u^2.$$

Then by Theorem 2.14.16, $b^*(\|a^*a\| + \varepsilon - a^*a)b = (ub)^*(ub) \geq 0, \forall b \in A$. Hence, $\|a^*a\|\rho(b^*b) + \varepsilon\rho(b^*b) \geq \rho(b^*a^*ab)$. Let $\varepsilon \to 0+$, then we get

$$\begin{aligned} \|\pi_\rho(a)b_\rho\|^2 = &\rho(b^*a^*ab) \\ \leq &\|a^*a\| \cdot \|b_\rho\|^2 \leq M^2\|a\|^2 \cdot \|b_\rho\|^2, \qquad \forall b \in A. \end{aligned}$$

So $\pi_\rho(a)$ can be uniquely extended to a bounded linear operator on H_ρ, still denoted by $\pi_\rho(a)$. Clearly,

$$\|\pi_\rho(a)\| \leq M\|a\|, \quad \rho(a) = \langle \pi_\rho(a)1_\rho, 1_\rho\rangle, \qquad \forall a \in A,$$

and $\{\pi_\rho, H_\rho\}$ is a $*$ representation of A.

Let $S(A)$ be the state space of A. Construct the universal $*$ representation of A:

$$\pi = \sum_{\rho\in S(A)} \oplus\pi_\rho, \qquad H = \sum_{\rho\in S(A)} \oplus H_\rho;$$

and let $\|a\|_1 = \|\pi(a)\|, \forall a \in A$. Suppose that there is $a \in A$ such that $\pi(a) = 0$. Write $a = a_1 + ia_2$, where $a_1^* = a_1, a_2^* = a_2$. Then $\pi(a_1) = \pi(a_2) = 0$. In particular, $\rho(a_1) = \rho(a_2) = 0, \forall \rho \in S(A)$. By Lemma 2.14.18, $\nu(a_1) = \nu(a_2) = 0$. But $\nu(a_j) \geq K\|a_j\|, j = 1,2$, so $a_1 = a_2 = 0$. Therefore, $\|\cdot\|_1$ is a norm on A. Clearly $\|a^*a\|_1 = \|a\|_1^2, \forall a \in A$.

Now it suffices to show $\|\cdot\| \sim \|\cdot\|_1$ on A. Obviously, $\|\cdot\|_1 \leq M\|\cdot\|$. Moreover, let $\{a_n\} \subset A$ and $\|a_n\|_1 \to 0$. We may assume that $a_n^* = a_n$ since $\|a_n^*\|_1 = \|a_n\|_1, \forall n$. By Lemma 2.14.18,

$$\begin{aligned}\|a_n\|_1 = & \sup\{\|\pi_\rho(a_n)\| \mid \rho \in S(A)\} \\ \geq & \sup\{|\rho(a_n)| \mid \rho \in S(A)\} = \nu(a_n) \geq K\|a_n\|, \quad \forall n.\end{aligned}$$

Further, $\|a_n\| \to 0$ too. Therefore, $\|\cdot\| \sim \|\cdot\|_1$ on A. Q.E.D.

Theorem 2.14.23. Let A be a Banach $*$ algebra. If there is a positive constant K such that

$$K\|a^*a\| \geq \|a^*\| \cdot \|a\|$$

for any normal element a of A (i.e. $a^*a = aa^*$), then A is C^*-equivalent.

Proof. For any $h^* = h \in A$, by the assumption $K\|h^2\| \geq \|h\|^2$. Generally, we have

$$K^{2^n-1}\|h^{2^n}\| \geq \|h\|^{2^n}, \quad \forall n.$$

Thus $K\nu(h) \geq \|h\|, \forall h^* = h \in A$. Now by Theorem 2.14.22, it suffices to prove that the $*$ operation on A is hermitian.

Let $h^* = h \in A$. By Lemma 2.14.20, the $*$ operation is continuous on A. Thus $f(th) = e^{ith} - 1 = \sum_{n=1}^\infty \frac{(ith)^n}{n!}$ is a normal element of A, and $f(th)^* = f(-th), \forall t \in \mathbb{R}$. Then from preceding paragraph and the assumption, we have

$$\begin{aligned}K\nu(2 - e^{ith} - e^{-ith}) = & K\nu(f(th)^*f(th)) \\ \geq & \|f(th)^*f(th)\| \\ \geq & K^{-1}\|f(th)^*\| \cdot \|f(th)\| \\ \geq & K^{-1}\nu(f(th))^2, \qquad \forall t \in \mathbb{R}.\end{aligned}$$

Let $\beta = \max\{|\text{Im}\lambda| \mid \lambda \in \sigma(h)\}$. Since $\sigma(h) = \overline{\sigma(h)}$, it follows that there is $\alpha \in \mathbb{R}$ such that $(\alpha \pm i\beta) \in \sigma(h)$. Then for $t > 0$,

$$\begin{aligned}2(1 + e^{\beta t}) \geq & \nu(2 - e^{ith} - e^{-ith}) \geq K^{-2}\nu(f(th))^2 \\ \geq & K^{-2}|1 - e^{it(\alpha - i\beta)}|^2 \\ = & K^{-2}(1 + e^{2\beta t} - 2e^{\beta t}\cos \alpha t).\end{aligned}$$

This is impossible if $\beta > 0$, so it must be that $\beta = 0$, i.e., $\sigma(h) \subset \mathbb{R}$, and the $*$ operation on A is hermitian. Q.E.D.

The axioms for C^*-algebras

Theorem 2.14.24. Let A be a Banach $*$ algebra with an identity. If there is a positive constant K such that $\|e^{ih}\| \leq K, \forall h^* = h \in A$, then A is C^*-equivalent. Moreover, if $K = 1$, then A itself is a C^*-algebra.

Proof. First step. To show that A is hermitian. Suppose that $h^* = h \in A$, and $(\alpha + i\beta) \in \sigma(h)$, where $\alpha, \beta \in \mathbb{R}$. By $\sigma(h) = \overline{\sigma(h)}$, we may assume that $\beta \leq 0$. Then for any $t > 0$,

$$K \geq \|e^{ith}\| > |e^{it(\alpha+i\beta)}| = e^{-\beta t}.$$

Thus $\beta = 0$, and $\sigma(h) \subset \mathbb{R}$.

Second step. To prove that

$$\inf\{\|h^2\| \mid h^* = h \in A, \|h\| = 1\} = \varepsilon > 0.$$

In fact, let $h^* = h$ and $\|h\| = 1$. Put $\|h^2\| = \eta$. Clearly $0 \leq \eta \leq 1$, and

$$\|h^{2n}\| \leq \|h^2\|^n = \eta^n, \quad \|h^{2n+1}\| \leq \|h^{2n}\| \leq \eta^n, \quad \forall n.$$

Let $\delta = \eta^{1/3}$. For any $n \geq 1$,

$$\|h^{2n}\| \leq \delta^{3n} \leq \delta^{2n}, \qquad \|h^{2n+1}\| \leq \delta^{3n} \leq \delta^{2n+1}.$$

Thus $\|h^n\| \leq \delta^n, \forall n \geq 2$. Suppose that $t > 0$. Then

$$\begin{aligned} K \geq \ & \|e^{ith}\| \geq \|th\| - 1 - \sum_{n=2}^{\infty} t^n \|h^n\|/n! \\ \geq \ & t - 1 - \sum_{n=2}^{\infty} \frac{1}{n!} t^n \delta^n \geq t - e^{t\delta}, \end{aligned}$$

i.e., $K + e^{t\delta} \geq t$. Pick $t = K + 2$, then $e^{\delta(K+2)} \geq 2$. Therefore,

$$\varepsilon \geq ((K+2)^{-1} \ln 2)^3 > 0.$$

Third step. We claim that

$$\nu(h) \geq \varepsilon \|h\|, \qquad \forall h^* = h \in A,$$

where ε is as in second step. In fact, from second step we have $\|h^2\| \geq \varepsilon \|h\|^2$ for any $h^* = h \in A$. Generally, $\|h^{2^n}\| \geq \varepsilon^{2^n - 1} \|h\|^{2^n}$. Therefore, $\nu(h) \geq \varepsilon \|h\|, \forall h^* = h \in A$.

Now by first, third steps and Theorem 2.14.22, A is C^*-equivalent.

If $K = 1$, consider identity map $I : (A, \|\cdot\|_1) \to (A, \|\cdot\|)$, where $\|\cdot\|_1 \sim \|\cdot\|$ on A, and $(A, \|\cdot\|_1)$ is a C^*-algebra. By Theorem 2.14.5, we have that $\|I\| \leq 1$, i.e., $\|a\| \leq \|a\|_1, \forall a \in A$. Suppose that there is $a_0 \in A$ such that $\|a_0\|_1 > \|a_0\|$. By Proposition 2.1.8,

$$\nu(a_0^* a_0) \leq \|a_0^*\| \cdot \|\|a_0\| < \|a_0^*\|_1 \cdot \|a_0\|_1 = \|a_0^* a_0\| = \nu(a_0^* a_0).$$

This is a contradiction. Therefore, $\|a\| = \|a\|_1, \forall a \in A$. Q.E.D.

Lemma 2.14.25. Let A be a Banach $*$ algebra with an identity, and $\|a^* a\| = \|a^*\| \cdot \|a\|$ for any normal element $a \in A$. Then A is a C^*-algebra.

Proof. Let $h^* = h \in A$, and $\sigma_n(h) = \sum_{k=0}^{n} \frac{1}{k!}(ih)^k, \forall n$. Clearly, $\sigma_n(h)$ is normal, and $\sigma_n(h)^* = \sigma_n(-h)$. Then

$$\|\sigma_n(h) \cdot \sigma_n(-h)\| = \|\sigma_n(h)\| \cdot \|\sigma_n(-h)\|, \quad \forall n.$$

Let $n \to \infty$, we get $\|e^{ih}\| \cdot \|e^{-ih}\| = 1$. By Theorem 2.14.23 and the assumption, A is C^*-equivalent. In particular, the $*$ operation is hermitian. Thus $\sigma(h) \subset \mathbb{R}$. Further, we have $\|e^{ih}\| \geq 1, \|e^{-ih}\| \geq 1$. Therefore, $\|e^{ih}\| = 1, \forall h^* = h \in A$. Now by Theorem 2.14.24, A is a C^*-algebra. Q.E.D.

Theorem 2.14.26. Let A be a Banach $*$ algebra. If $\|a^* a\| = \|a^*\| \cdot \|a\|$ for any normal element $a \in A$, then A is a C^*-algebra.

Proof. By Lemma 2.14.25, we may assume that A has no identity.

By Theorem 2.14.23, A is C^*-equivalent. Suppose that $\|\cdot\|'$ is a norm on A such that $\|\cdot\| \sim \|\cdot\|'$ and $(A, \|\cdot\|')$ is a C^*-algebra. By the assumption, we have

$$\|h\| = \nu(h) = \|h\|', \quad \forall h^* = h \in A.$$

In particular, $\|d_l\| = \|d_l\|', \forall l$, where $\{d_l\}$ is an approximate identity for $(A, \|\cdot\|')$.

We say that for any $a \in A$,

$$\|a\| = \sup\{\|ab\| \mid b \in A, \|b\| \leq 1\}.$$

Indeed, since $\|\cdot\|'$ and $\|\cdot\|$ are equivalent, it follows that $\|ad_l - a\| \to 0$. Then

$$\|a\| \geq \sup\{\|ab\| \mid b \in A, \|b\| \leq 1\} \geq \|ad_l\| \to \|a\|.$$

Thus $\|a\| = \sup\{\|ab\| \mid b \in A, \|b\| \leq 1\}$.

On $A \dot{+} \mathbb{C}$, define

$$\|a + \lambda\| = \sup\{\|ab + \lambda b\| \mid b \in A, \|b\| \leq 1\},$$

$\forall a \in A, \lambda \in \mathbb{C}$. Suppose that for some $a \in A$ and $\lambda \in \mathbb{C}$ we have $ab + \lambda b = 0, \forall b \in A$. Since A has no identity and $(A, \|\cdot\|')$ is a C^*-algebra, it follows from Proposition 2.1.2 that $a = 0$ and $\lambda = 0$. Thus $(A \dot{+} \mathbb{C}, \|\cdot\|)$ is a Banach $*$ algebra with an identity 1, and $(A, \|\cdot\|)$ is a Banach $*$ subalgebra of $(A \dot{+} \mathbb{C}, \|\cdot\|)$.

We need to prove that $(A \dot{+} \mathbb{C}, \|\cdot\|)$ is a C^*-algebra. By Theorem 2.14.24, it suffices to show that

$$\|e^{ih}\| \leq 1, \qquad \forall h^* = h \in A.$$

Fix $h^* = h \in A$. By the definition of the norm on $(A \dot{+} \mathbb{C})$ there is a sequence $\{b_n\} \subset A$ with $\|b_n\| \leq 1, \forall n$, such that $\|e^{ih}\| = \lim_n \|e^{ih} b_n\|$.

Let B be the closed $*$ subalgebra of A generated by $\{h, b_n.b_n^* \mid n\}$. Then $(B, \|\cdot\|')$ is a separable C^*-subalgebra of $(A, \|\cdot\|')$.

If B has an identity p, by Lemma 2.14.25 then $(B, \|\cdot\|)$ is a C^*-algebra. Thus

$$\begin{aligned} \|e^{ih}\| &= \lim_n \|e^{ih} b_n\| = \lim_n \|(p + \sum_{j=1}^{\infty} \frac{(ih)^j}{j!}) b_n\| \\ &\leq \|p + \sum_{j=1}^{\infty} \frac{(ih)^j}{j!}\| \leq 1. \end{aligned}$$

Now suppose that B has no identity. On $B \dot{+} \mathbb{C}$, define

$$\|b + \lambda\|_1 = \sup\{\|bc + \lambda c\| \mid c \in B, \|c\| \leq 1\},$$

$\forall b \in B, \lambda \in \mathbb{C}$. Similarly, $(B \dot{+} \mathbb{C}, \|\cdot\|_1)$ is a Banach $*$ algebra with an identity 1, and $\|b\| = \|b\|_1, \forall b \in B$.

By Theorem 2.14.9, $(B, \|\cdot\|')$ has a strictly positive element a. And by the proof of Theorem 2.14.8, $\{d_n = (a/\|a\|')^{\frac{1}{n}}\}_n$ is an approximate identity for $(B, \|\cdot\|')$. Since

$$\|b + \lambda\|_1 \geq \|(b + \lambda) d_n\| \geq \|(b + \lambda) d_n c\| \to \|(b + \lambda) c\|$$

$\forall c \in B$ and $\|c\| \leq 1$, it follows that

$$\|b + \lambda\|_1 = \lim_n \|(b + \lambda) d_n\|,$$

$\forall b \in B, \lambda \in \mathbb{C}$. In particular,

$$\begin{aligned} \|e^{-ia}\|_1 \cdot \|e^{ia}\|_1 &= \lim_n \|e^{-ia} d_n\| \cdot \|e^{ia} d_n\| \\ &= \lim_n \|(e^{ia} d_n)^* (e^{ia} d_n)\| \\ &= \lim_n \|d_n^2\| = \lim_n \|d_n^2\|' = 1. \end{aligned}$$

Moreover, since $\sigma(a) \subset \mathbb{R}$ and $\|e^{\pm ia}\|_1 \geq \nu(e^{\pm ia}) = 1$, it follows that $\|e^{\pm ia}\|_1 = 1$. By Theorem 2.14.9 and Proposition 2.14.10, the set of strictly positive elements is dense in $(B, \|\cdot\|')_+$. Noticing that

$$\|e^{ib_1} - e^{ib_2}\|_1 \le \sum_{k=1}^{\infty} \frac{1}{k!} \|b_1^k - b_2^k\|, \quad \forall b_1, b_2 \in B$$

and $\|\cdot\| \sim \|\cdot\|'$, we have

$$\|e^{\pm ib}\|_1 = 1, \qquad \forall b \in (B, \|\cdot\|')_+.$$

Now we come back to consider h. Write $h = h_+ - h_-$ where $h_\pm \in (B, \|\cdot\|')_+$, and $h_+ \cdot h_- = 0$. Since

$$1 = \|e^{ih_+}\|_1 = \|e^{ih} \cdot e^{ih_-}\|_1 \le \|e^{ih}\|_1 \le \|e^{ih_+}\|_1 \cdot \|e^{-ih_-}\|_1 = 1,$$

it follows that $\|e^{ih}\|_1 = 1$. Further,

$$\begin{aligned} \|e^{ih}\| = & \lim_n \|e^{ih} b_n\| \le \sup\{\|e^{ih} b\| \mid b \in B, \|b\| \le 1\} \\ = & \|e^{ih}\|_1 = 1. \end{aligned}$$

This completes the proof. Q.E.D.

The Submultiplication of a linear C^*-norm

Definition 2.14.27. Let A be a $*$ algebra over the complex field $\mathbb{C}$. A norm $\|\cdot\|$ on A is called a C^*-*norm*, if it satisfies the following conditions:

i) (*submultiplication*) $\|xy\| \le \|x\| \cdot \|y\|, \forall x, y \in A$;

ii) (C^*-*condition*) $\|x^* x\| = \|x\|^2, \forall x \in A$.

Clearly, the completion of $(A, \|\cdot\|)$ is a C^*-algebra if $\|\cdot\|$ is a C^*-norm on A.

Now let $\|\cdot\|$ be a norm on A. If it only satisfies the above condition ii), i.e., $\|x^* x\| = \|x\|^2, \forall x \in A$, then $\|\cdot\|$ is called a *linear C^*-norm* on A.

Theorem 2.14.28. Let A be a (complex) $*$ algebra, and $\|\cdot\|$ be a linear C^*-norm on A, such that $(A, \|\cdot\|)$ is a Banach space. Then $(A, \|\cdot\|)$ is a C^*-algebra, i.e., $\|\cdot\|$ satisfies the submultiplication ($\|xy\| \le \|x\| \cdot \|y\|, \forall x, y \in A$) automatically.

Proof 1) We claim that

$$\|xy\| \le 4\|x^*\|\|y\|, \qquad \forall x, y \in A.$$

In fact, for $x, y \in A$ and numbers $\lambda, \mu \ge 0$, noticing that

$$\begin{aligned} & (\lambda y^* + \mu x)(\lambda y + \mu x^*) - (\lambda y^* - \mu x)(\lambda y - \mu x^*) \\ = & \; 2\lambda\mu(y^* x^* + xy), \end{aligned}$$

and

$$\begin{aligned}&(\lambda y^* + i\mu x)(\lambda y - i\mu x^*) - (\lambda y^* - i\mu x)(\lambda y + i\mu x^*)\\ =\ & i2\lambda\mu(-y^*x^* + xy),\end{aligned}$$

and by the C^*-condition, we have

$$\begin{aligned}4\lambda\mu\|xy\| \leq\ & \|\lambda y + \mu x^*\|^2 + \|\lambda y - \mu x^*\|^2\\ & + \|\lambda y - i\mu x^*\|^2 + \|\lambda y + i\mu x^*\|^2\\ \leq\ & 4(\lambda\|y\| + \mu\|x^*\|)^2.\end{aligned}$$

Let $\lambda = \|y\|^{-1}, \mu = \|x^*\|^{-1}$, then we get

$$\|xy\| \leq 4\|x^*\|\|y\|, \qquad \forall x, y \in A.$$

2) We prove that

$$\|x^*\| = \|x\|, \quad \text{and} \quad \|xy\| \leq 4\|x\|\|y\|, \qquad \forall x, y \in A.$$

In fact, if $h^* = h \in A$, then $\|h^2\| = \|h\|^2$ by the C^*-condition. Generally, we have

$$\|h^{2^n}\| = \|h\|^{2^n}, \qquad \forall n.$$

Thus for any $a \in A$, by 1)

$$\begin{aligned}\|a^*a\|^{2^n} =\ & \|(a^*a)^{2^n}\| = \|a^*(aa^*)^{2^n-1}a\|\\ \leq\ & 4^2\|a\|^2 \cdot \|(aa^*)^{2^n-1}\|, \forall n.\end{aligned}$$

Noticing $\|h^{2^k}\| = \|h\|^{2^k} (\forall k = 1, 2, \cdots,$ and $h^* = h \in A)$ and the claim 1), we have that

$$\begin{aligned}\|(aa^*)^{2^n-1}\| =\ & \|(aa^*)^{1+2+\cdots+2^{n-1}}\|\\ \leq\ & 4^{n-1}\|aa^*\| \cdot \|(aa^*)^2\| \cdots \|(aa^*)^{2^{n-1}}\|\\ =\ & 4^{n-1}\|aa^*\|^{2^n-1}, \forall n.\end{aligned}$$

Thus

$$\|a^*a\|^{2^n} \leq 4^{n+1}\|a\|^2 \cdot \|aa^*\|^{2^n-1}, \qquad \forall n.$$

Further, $\|a^*a\| \leq \|aa^*\|, \forall a \in A$; and $\|a^*a\| = \|aa^*\|, \forall a \in A$. By the C^*-condition, $\|x\|^2 = \|x^*x\| = \|xx^*\| = \|x^*\|^2$, i.e., $\|x^*\| = \|x\|, \forall x \in A$. Again by 1), $\|xy\| \leq 4\|x\|\|y\|, \forall x, y \in A$.

Now define a new norm $\|\cdot\|' = 4\|\cdot\|$ on A. Then by 1), 2), $(A, \|\cdot\|')$ is a Banach $*$ algebra, and $\|x^*\|' = \|x\|', \forall x \in A$. Moreover, by the C^*-condition

$$\|a^*a\|' = 4\|a^*a\| = 4\|a\|^2 = \frac{1}{4}\|a^*\|' \cdot \|a\|', \qquad \forall a \in A.$$

From Theorem 2.14.23, $(A, \|\cdot\|')$ is C^*-equivalent, i.e.,

$$\|\cdot\| \sim \|\cdot\|' = |\cdot|$$

on A, where $|a| = \nu(a^*a)^{1/2} (\forall a \in A)$ is the C^*-norm on A. Furthermore, by the C^*-condition of $\|\cdot\|$

$$\begin{aligned} |a|^2 = \nu(a^*a) &= \lim_n |(a^*a)^{2^n}|^{2^{-n}} \\ &= \lim_n \|(a^*a)^{2^n}\|^{2^{-n}} = \|a^*a\| = \|a\|^2, \end{aligned}$$

$\forall a \in A$. Therefore, $(A, \|\cdot\|)$ is a C^*-algebra. Q.E.D.

Notes. In 1943, I.M. Gelfand and M. Naimark proved the following theorem: If A is a Banach $*$ algebra with an identity, and satisfies: 1) $\|a^*a\| = \|a^*\| \cdot \|a\|, \forall a \in A$; 2) $\|a^*\| = \|a\|, \forall a \in A$; 3) $(1 + a^*a)$ is invertible, $\forall a \in A$, then A is isometrically $*$ isomorphic to a uniformly closed $*$ operator algebra on some Hilbert space. This is a fundamental theorem for the theory of C^*-algebras. Also, they conjectured that the conditions 2) and 3) are not necessary. I. Kaplansky pointed out that the condition 3) can be canceled easily. Then we have the usual definition of C^*-algebras (2.1.1) and the Theorem 2.3.20. Proposition 2.14.1 is due to J. Glimm and R.V.Kadison. Using this proposition, they answered affirmatively the Gelfand-Naimark conjecture for unital case. Theorem 2.14.14 is due to J.W.M. Ford, Theorem 2.14.15 is due to V.Pták; and Theorem 2.14.16 is due to S.Shirali and J.W.M. Ford. These results also have their own interest in the theory of Banach algebras. Theorem 2.14.22 and 2.14.23 are due to R.Arens. Theorem 2.14.24 is due to B.W. Glickfeld. And Theorem 2.14.26 is due to G.A. Elliott.

The question on the submultiplication of a linear C^*-norm was presented by R.S. Doran. Theorem 2.14.28 is due to H. Araki and G.A. Elliott.

References. [6], [7], [30], [41], [48], [50], [52], [53], [56], [66], [87], [90], [131], [162], [191], [201].

2.15 Real C^*–algebras

Definition 2.15.1. Let A be a real Banach $*$ algebra (i.e., a Banach $*$ algebra over the real field $\mathbb{R}$) .A is called a *real C^*–algebra,* if $A_c = A \dot{+} iA$ can be normed to become a (complex) C^*–algebra such that the original norm on A remains unchanged.

Proposition 2.15.2. A real Banach $*$ algebra A is a real C^*-algebra if and only if A can be isometrically $*$ isomorphic to a unifomly closed $*$ algebra of operators on a real Hilbert space.

Proof. Let π be an isometrical $*$ isomorphism from A into $B(H)$, where H is some real Hilbert space. Consider the complex Hilbert space $H_c = H\dot{+}iH$, and define

$$\|a + ib\| = \|\pi(a) + i\pi(b)\|_{H_c}, \quad \forall a, b \in A$$

It is easily verified that $A_c = A\dot{+}iA$ is a (complex) C^*-algebra by this norm. Moreover, it is obvious that $\|a\| = \|\pi(a)\| = \|\pi(a)\|_{H_c}, \forall a \in A$. Therefore, A is a real C^*-algebra.

Conversely, let A be a real C^*-algebra. Then $A_c = A\dot{+}iA$ is a (complex) C^*-algebra. We may assume that $A_c \subset B(H_c)$, where H_c is a complex Hilbert space. Consider $H_c = H$ as a real linear space, and define

$$\langle \xi, \eta \rangle_r = \text{Re } \langle \xi, \eta \rangle, \quad \forall \xi, \eta \in H.$$

Then H is a real Hilbert space, and A is a uniformly closed $*$ algebra of operators on H. Q.E.D.

Proposition 2.15.3. Let A be an abelian real C^*-algebra. Then A is isometrically $*$ isomorphic to

$$C_0^\infty(\Omega, -) = \{f \in C_0^\infty(\Omega) | f(\bar{t}) = \overline{f(t)}, \forall t \in \Omega\},$$

where Ω is a locally compact Hausdorff space, and bar "—" is a homeomorphism of Ω such that $\bar{\bar{t}} = t, \forall t \in \Omega$.

Proof. Assume that $A_c = A\dot{+}iA \cong C_0^\infty(\Omega)$, and for each $t \in \Omega$ define

$$\bar{t}(a + ib) = \overline{a(t)} + i\overline{b(t)}, \quad \forall a, b \in A,$$

where $x \to x(\cdot)$ is the Gelfand transformation from A onto $C_0^\infty(\Omega)$. Clearly, $t \to \bar{t}$ is a homeomorphism of $\Omega; \bar{\bar{t}} = t, \forall t \in \Omega$; and $\{a(\cdot) | a \in A\} \subset \{f \in C_0^\infty(\Omega) | f(\bar{t}) = \overline{f(t)}, \forall t \in \Omega\}$. Now let $f \in C_0^\infty(\Omega)$ with $f(\bar{t}) = \overline{f(t)}, \forall t \in \Omega$. Suppose that $f(\cdot) = a(\cdot) + ib(\cdot)$, where $a, b \in A$. Then

$$\begin{aligned} \overline{a(t)} - i\overline{b(t)} &= \overline{(a + ib)(t)} = \overline{f(t)} \\ &= f(\bar{t}) = a(\bar{t}) + ib(\bar{t}) = \overline{a(t)} + i\overline{b(t)}, \forall t \in \Omega. \end{aligned}$$

Hence , $b(t) = 0, \forall t \in \Omega, b = 0$, and $f(\cdot) = a(\cdot)$. Q.E.D.

Now let A be a real C^*-algebra, and define

$$J(a + ib) = a - ib, \ \forall a, b \in A.$$

Then J is a conjugate linear isometric $*$ isomorphism of the (complex) C^*-algebra $A_c = A\dot{+}iA$, i.e.,

$$\begin{aligned} J(\lambda x + \mu y) &= \overline{\lambda} Jx + \overline{\mu} Jy, Jx^* = (Jx)^*, Jxy = Jx \cdot Jy, \\ J^2 &= id, \ \|Jx\| = \|x\|, \ \forall x, y \in A_c, \lambda, \mu \in \mathbb{C}. \end{aligned}$$

In fact, it suffices to show that J is isometric. We may assum that $A \subset B(H)$, where H is a real Hilbert space. Since for any $a, b \in A$, $\xi, \eta \in H$,

$$\begin{aligned} \|(a+ib)(\xi+i\eta)\| &= \|a\xi - b\eta\|^2 + \|b\xi + a\eta\|^2 \\ &= \|(a-ib)(\xi - i\eta)\|^2, \end{aligned}$$

and $\|\xi + i\eta\|^2 = \|\xi - i\eta\|^2 = \|\xi\|^2 + \|\eta\|^2$, we have $\|Jx\| = \|x\|, \forall x \in A_c$.

Conversely, let J be a conjugate linear isometric $*$ isomorphism of a (complex) C^*-algebra A_c, and $A = \{x \in A_c | Jx = x\}$. Then it is easy to see that A is a real C^*-algebra, and $A_c = A\dot{+}iA$. Therefore, we have the following

Proposition 2.15.4. There is a bijection between the collection of all real C^*-algebras and the collection $\{(A_c, J) | A_c$ is a complex C^*-algebra, J is a conjugate linear isometric $*$ isomorphism of $A_c\}$.

Proposition 2.15.5. Let A be a real C^*-algebra, and define

$$f(a+ib) = f(a) + if(b), \ \forall a, b \in A, f \in A^*.$$

Then A^* can be embedded isometrically into A_c^* such that $A_c^* = A^*\dot{+}iA^*$, where $A_c = A\dot{+}iA$.

Proof. We may assume that $A \subset B(H)$, where H is a real Hilbert space. Since $f(\in A^*)$ can be extended to a linear functional on $B(H)$ with the same norm, we may also assume that $A = B(H)$ and $\|f\| = 1$. Then there exists a net $\{t_l\} \subset B(H)_* = T(H)$ such that

$$\|t_l\|_1 \leq 1, \forall l; \text{ and } f_l(b) = tr(t_l b) \to f(b), \forall b \in B(H).$$

Clearly, we have $f_l \longrightarrow f$ in topology $\sigma(B(H_c)^*, B(H_c))$, where $H_c = H\dot{+}iH$ and t_l as an element of $T(H_c)$ maintains its trace norm, so f_l as an element of $B(H_c)^*$ maintains also its norm, i.e., ≤ 1. Hence, f as an element of $B(H_c)^*$ has the norm one. Therefore, A^* can be embedded isometrically into A_c^*.

Q.E.D.

Proposition 2.15.6. Let A be a real C^*-algebra. $\rho \in A^*$ is called a *state* , if $\rho(a^*a) \geq 0, \rho(a^*) = \rho(a), \forall a \in A$, and $\|\rho\| = 1$. Denote the state space of A by $S(A)$. Then $S(A) = \text{Re } S(A_c) = \{\text{Re } \rho_c | \rho_c$ is a state on $A_c\}$, where

$A_c = A\dot{+}iA$ is a (complex) C^*-algebra. Moreover, for each $\rho \in S(A)$, by the GNS construction there is a cyclic $*$ representation $\{\pi_\rho, H_\rho, \xi_\rho\}$ of A such that

$$\rho(a) = \langle \pi_\rho(a)\xi_\rho, \xi_\rho \rangle, \quad \forall a \in A,$$

where H_ρ is a real Hilbert space; and

$$\{\pi = \oplus_{\rho \in S(A)} \pi_\rho, \quad H = \oplus_{\rho \in S(A)} H_\rho\}$$

is a faithful $*$ representation of A.

The proof is similar to Section 2.3.

Definition 2.15.7. Let A be a real Banach $*$ algebra. For each $x \in A$, let $\sigma(x)$ be the *spectrum* of x in the (complex) algebra $A_c = A\dot{+}iA$, and $\nu(x) = \max\{|\lambda| | \lambda \in \sigma(x)\}$ be the *spectral radius* of x.

An element a of A is said to be *positive* , denoted by $a \geq 0$, if $a^* = a$ and $\sigma(a) \subset [0, \infty)$.

A is said to be *hermitian,* if for each $h^* = h \in A$ we have $\sigma(h) \subset \mathbb{R}$; A is said to be *skew-hermitian,* if for each $k^* = -k \in A$ we have $\sigma(k) \subset i\mathbb{R}$; A is said to be *symmetric*, if for each $a \in A$ we have $a^*a \geq 0$.

Theorem 2.15.8. Let A be a real Banach $*$ algebra. Then A is symmetric if and only if A is hermitian and skew-hermitian.

Moreover, if A is symmetric, then $A_+ = \{a \in A | a \geq 0\}$ is a cone, i.e., if $a, b \in A_+$, then we have $(a+b) \in A_+$.

The proos is similar to Theorem 2.14.5 and 2.14.6.

Lemma 2.15.9. Let A be a real symmetric Banach $*$ algebra with an identity 1. A linear functional ρ on A is called a *state,* if

$$\rho(1) = 1, \rho(x^*x) \geq 0, \rho(x^*) = \rho(x), \forall x \in A.$$

Then for each $h^* = h \in A$ there is a state ρ on A such that $\rho(h) = \nu(h)$.

Proof. Let $A_H = \{h \in A | h^* = h\}$ and $A_K = \{k \in A | k^* = -k\}$. Then $A = A_H \dot{+} A_K$.

Now for $h \in A_H$, define a linear functional ρ on $[1, h] = \{\alpha + \beta h | \alpha, \beta \in \mathbb{R}\}$:

$$\rho(\alpha + \beta h) = \alpha + \beta\nu(h), \quad \forall \alpha, \beta \in \mathbb{R}.$$

Clearly, $\rho(a) \geq 0, \forall\, 0 \leq a \in [1, h]$, and $\rho(h) = \nu(h)$. Let

$$\mathcal{L} = \left\{ (E, \rho_E) \;\middle|\; \begin{array}{c} E \text{ is a linear subspace of } A_H, \text{and} 1, h \in E; \\ \rho_E \text{ is linear on } E, \text{and} \rho(a) \geq 0, \forall 0 \leq a \in E; \\ \text{and } \rho_E(\alpha + \beta h) = \alpha + \beta\nu(h), \forall \alpha, \beta \in \mathbb{R} \end{array} \right\},$$

and $(E,\rho_E) \leq (F,\rho_F)$, if $E \subset F$ and $\rho_F|E = \rho_E$. By the Zorn lemma, $\mathcal{L}$ contains a maximal element (E,ρ_E) . From Theorem 2.15.8 and Proposition 2.3.11 , it is easy to see that $E = A_H$. Further, define $\rho(A_K) = \{0\}$. Then this ρ will satisfy our conditions. Q.E.D.

Theorem 2.15.10. Let A be a real hermitian Banach $*$ algebra, and $\|x^*x\| = \|x\|^2, \forall x \in A$. Then A is a real C^*-algebra.

Proof. By Proposition 2.1.2, we may assume that A has an identity 1. For any $k^* = -k$, since $*$ is isormetric, it follows that

$$(e^{tk})^* = e^{-tk},\ 1 = \|(e^{tk})^* \cdot e^{tk}\| = \|e^{tk}\|^2,\ \forall t \in \mathbb{R}.$$

Hence, $\sigma(k) \subset i\mathbb{R}$, i.e., A is also skew–hermitian. By Theorem 2.15.8, A is symmetric.

For any $a \in A$ and $\varepsilon > 0$, by Theorem 2.14.14 (it also holds for real case) there is $b \in A$ such that

$$b^*b = \|a\|^2 + \varepsilon - a^*a.$$

Hence, for any state ρ on A, by the GNS construction there is cyclic $*$ representation $\{\pi_\rho, H_\rho, \xi_\rho\}$ of A such that

$$\rho(a) = \langle \pi_\rho(a)\xi_\rho, \xi_\rho \rangle, \quad \forall a \in A.$$

Now by Lemma 2.15.9, A admits a faithful $*$ representation $\{\pi, H\}$, where H is a real Hilbert space. Define

$$\|x\|_1 = \|\pi(x)\|, \quad \forall x \in A.$$

Then $\|\cdot\|_1$ is a norm on A, and $\|x^*x\|_1 = \|x\|_1^2, \forall x \in A$. For any $h^* = h \in A$, clearly we have $\sigma(\pi(h)) \subset \sigma(h)$. Hence, $\|h\|_1 = \nu(\pi(h)) \leq \nu(h) = \|h\|, \forall h \in A$. On the other hand, for any state ρ on A we have

$$\|\pi(h)\| \geq |\langle \pi_\rho(h)\xi_\rho, \xi_\rho \rangle| = |\rho(h)|.$$

By Lemma 2.15.9, $\|h\|_1 = \|\pi(h)\| \geq \nu(h) = \|h\|, \forall h^* = h \in A$.

So $\|h\|_1 = \|h\|, \forall h^* = h \in A$. Further,

$$\|x\|_1 = \|x^*x\|_1^{1/2} = \|x^*x\|^{1/2} = \|x\|, \forall x \in A.$$

Therefore, A is a real C^*–algebra. Q.E.D.

Remark. In above theorem, the hermitian condition is necessary. Indeed, consider $\mathbb{C}$ with norm $\|\lambda\| = |\lambda|$ and $*$ operation $\lambda^* = \lambda (\forall \lambda \in \mathbb{C})$, then the real Banach $*$ algebra $\mathbb{C}$ is not hermitian. So $\mathbb{C}$ is not a real C^*–algebra.

Definitioln 2.15.11. A real Banach $*$ algebra $(A, \|\cdot\|)$ is said to be *real C^*-equivalent*, if there is a new norm $\|\cdot\|_1$ on A such that $\|\cdot\| \sim \|\cdot\|_1$, and $(A, \|\cdot\|_1)$ is a real C^*-algebra.

Theorem 2.15.12. Let A be a real Banach $*$ algebra. Suppose that A satisfies one of the following conditions:

1) A is symmetric, and there is a positive constatnt K such that $K\nu(x) \geq \|x\|$ for each $x^* = x$ or $x^* = -x$ of A

2) A is hermitian, and there is a positive constant K such that $K\|x^*x\| \geq \|x^*\| \cdot \|x\|$ for each normal x of A (i.e., $x^*x = xx^*$);

3) there is a positive constant K such that $K\|x^*x + y^*y\| \geq \|x^*\| \cdot \|x\|$ for any normal $x, y \in A$ and $xy = yx$.

Then A is real C^*-equivalent.

Proof. 1) By Lemma 2.14.19, we may assume that A has an identity 1. By Lemma 2.14.20, $A_H = \{h \in A | h^* = h\}$ is closed in A. Now let $\{k_n\} \subset A_K = \{k \in A | k^* = -k\}$ be such that $k_n \to h \in A_H$. For any $k \in A_K$, let $z_n = kk_n + k_n k, \forall n$. Then $z_n \in A_H$ and $z_n \to z = kh + hk$. Since A_H is closed, it follows that $z \in A_H \cap A_K = \{0\}$. Hence,

$$kh + hk = 0, \forall k \in A_K.$$

In particular, $k_n h + hk_n = 0, \forall n$, and $h^2 = 0$. But $0 = \nu(h^2) = \nu(h)^2 \geq K^{-2}\|h\|^2$, so $h = 0$, i.e., A_K is also closed. Further, the $*$ operation is continuous on A.

For each state ρ on A, by the GNS construction there is a cyclic $*$ representation $\{\pi_\rho, H_\rho, \xi_\rho\}$ of A such that

$$\rho(x) = \langle \pi_\rho(x)\xi_\rho, \xi_\rho \rangle, \forall x \in A.$$

where H_ρ is a real Hilbert space. Further, let

$$\{\pi = \oplus_{\rho \in S(A)} \pi_\rho, \quad H = \oplus_{\rho \in S(A)} H_\rho\},$$

where $S(A)$ is the state space of A. Since $*$ is continuous, it follows that

$$\begin{aligned} \|\pi(x) = \|\pi(x^*x)\|^{1/2} &= \nu(\pi(x^*x))^{1/2} \\ &\leq \nu(x^*x)^{1/2} \leq \|x^*x\|^{1/2} \leq K'\|x\|, \end{aligned}$$

$\forall x \in A$, where K' is some positive constant. Define $\|x\|_1 = \|\pi(x)\|, \forall x \in A$. We say that $\|\cdot\|_1$ is a norm on A. In fact, let $\|x\|_1 = 0$ and $x = h + k$, where $h^* = h, k^* = -k$. Then for each state ρ on A we have

$$|\rho(x)| = |\rho(h)| = |\langle \pi_\rho(x)\xi_\rho, \xi_\rho \rangle| \leq \|\pi(x)\| = 0.$$

By Lemma 2.15.9, we can see that $h = 0$. Further, from $0 = \pi(x^*x) = -\pi(-k)^2$ we have $k = 0$, and $x = 0$.

Now it suffices to show that $\|\cdot\|$ is continuous with respect to $\|\cdot\|_1$. Let $\|x_n\|_1 \to 0$. Since $\|x_n^*\|_1 = \|x_n\|_1, \forall n$, it follows that $\|h_n\|_1 \to 0$ and $\|k_n\|_1 \to 0$, where $h_n^* = h_n, k_n^* = -k_n, x_n = h_n + k_n, \forall n$. By $\|x\|_1 \geq \nu(x) \geq K^{-1}\|x\|, \forall x^* = \pm x \in A$, we can see that $\|h_n\| \to 0, \|k_n\| \to 0$, and $\|x_n\| \to 0$. Therefore, $\|\cdot\|$ is continuous with respect to $\|\cdot\|_1$, and A is real C^*-equivalent.

2) By 1), it suffices to show that A is skew-hermitian. Similar to the 1), $*$ is continuous. For $k^* = -k \in A$, consider $f(tk) = e^{tk} - 1, \forall t \in \mathbb{R}$. Then $f(tk)$ is normal and $f(tk)^* = f(-tk), \forall t \in \mathbb{R}$. Now by the proof of Theorem 2.14.23, we can see that $\sigma(k) \subset i\mathbb{R}$.

3) For $h^* = h \in A$, let $x = \cos h - 1, y = \sin h$. Then x, y are normal and $xy = yx$. Hence

$$\|\cos h - 1\|^2 \leq K\|(\cos h - 1)^2 + \sin^2 h\| = 2K\|\cos h - 1\|,$$

i.e.$\|\cos h - 1\| \leq 2K, \forall h^* = h \in A$. Consequently,

$$\|\cos(\lambda t) - 1\| \leq 2K, \forall t \in \mathbb{R}, \lambda \in \sigma(h), \text{and } h^* = h.$$

Therefore, $\sigma(h) \subset \mathbb{R}, \forall h^* = h \in A$, i.e., A is hermitian. Now by 2), A is real C^*-equivalent. Q.E.D.

Now consider the unit ball of a real C^*-algebra.

Proposition 2.15.13. Let A be a real C^*-algebra, S be its closed unit ball, i.e., $S = \{x \in A | \|x\| \leq 1\}, S_c$ be the closed unit ball of the (complex) C^*-algebra $A_c = A + iA$, and $x_0 \in S$. Then the following conditions are equivalent:

1) $x_0 \in ExS$;

2) $x_0 \in ExS_c$;

3) $(1 - x_0^*x_0)A(1 - x_0x_0^*) = \{0\}$.

In the case of $x_0 \in ExS, x_o$ is a partial isometry of A, i.e., $x_0^*x_0$ and $x_0x_0^*$ are projections of A.

Moreover, S has an extreme point at least if and only if A has an identity.

Proof. Let $x_0 \in ExS$. From Theorem 2.5.1 (and notice Proposition 2.15.3), we can prove that x_0 is a partial isometry of A, and $(1 - x_0^*x_0)A(1 - x_0x_0^*) = \{0\}$. Hence, $(1 - x_0^*x_0)A_c(1 - x_0x_0^*) = \{0\}$. Now by Theorem 2.5.1, we have $x_0 \in ExS_c$. The other conclusions are easy. Q.E.D.

Corollary 2.15.14. Let $x_0 \in ExS$, and $0 \neq b \in A$. Then we have $\|x_0 + ib\| > 1$ in A_c.

Proof. If $\|x_0+ib\| \le 1$, then by Proposition 2.15.2 we can see that $\|x_0-ib\| = \|x_0+ib\| \le 1$. This contradicts the fact $x_0 \in ExS_c$. Therefore, $\|x_0+ib\| > 1$.
Q.E.D.

Lemma 2.15.15. Let H be a complex Hilbert space, and $A(\subset B(H))$ be a real uniformly closed $*$ algebra. If $A \cap iA = \{0\}$, then $A_c = A\dot{+}iA$ is a(complex) C^*-algebra on H, and

$$\|a+ib\| \ge \max(\|a\|,\|b\|), \forall a, b \in A.$$

Moreover, if M is a real weakly closed $*$ operator algebra on H, and $M \cap iM = \{0\}$, then $M_c = M\dot{+}iM$ is also weakly closed.

Proof. Since $(H, \mathrm{Re}\langle,\rangle)$ is a real Hilbert space, it follows from Proposition 2.15.2 that A is a real C^*-algebra. Then there is a norm $\|\cdot\|_1$ on A_c such that $(A_c, \|\cdot\|_1)$ is a (complex) C^*-algebra, and $\|x\|_1 = \|x\|, \forall x \in A$. Clearly, the operator norm is also a C^*-norm on A_c. By Proposition 2.1.10, $\|\cdot\|_1 = \|\cdot\|$ on A_c , i.e., A_c is a C^*-algebra on H.

Now suppose that $\{\pi, K\}$ is a faithful $*$ representation of A, where K is a real Hiblert space. Then $\{\pi, K_c\}$ is a faithful $*$ representation of A_c, where $K_c = K\dot{+}iK$. Hence, for every $a, b \in A$,

$$\begin{aligned}\|a+ib\| &= \|\pi(a)+i\pi(b)\| \\ &\ge \sup\{\|\pi(a)\xi + i\pi(b)\xi\| \,|\, \xi \in K, \|\xi\| \le 1\} \\ &\ge \max\left\{\sup_{\|\xi\|\le 1}\|\pi(a)\xi\|, \sup_{\|\xi\|\le 1}\|\pi(b)\xi\|\right\} = \max(\|a\|,\|b\|).\end{aligned}$$

Finally, it suffices to show that the closed unit ball of M_c is weakly closed. Suppose that $(a_l+ib_l) \to x \in B(H)$ weakly, where $a_l, b_l \in M, \forall l$, and $\|a_l + ib_l\| \le 1, \forall l$. Since $\max(\|a_l\|,\|b_l\|) \le 1, \forall l$, and the closed unit ball of M is weakly compact, we may assume that $a_l \to a$ and $b_l \to b$ weakly, where $a, b \in M$. Therefore, $x = a+ib \in M_c$.
Q.E.D.

Lemma 2.15.16. Let H be a complex Hilbert space, M be a real weakly closed $*$ operator algebra on H, and $1 = 1_H \in M$. Then Co $\{u | u \in M$ is unitary $\}$ is weakly dense in $S = \{x \in M | \|x\| \le 1\}$.

Proof. Since S is weakly compact, so by the Krein–Milmann theorem it suffices to show that

$$x_0 \in \overline{Co\{u \in M | u \text{ is unitary }\}}^w$$

for any $x_o \in ExS$, where "$\overline{\quad}^w$" means the weak closure.

Now let $x_o \in ExS$. By Proposition 2.15.13, x_o is a partial isometry of M, and $(1 - x_o^* x_o)M(1 - x_o x_o^*) = \{0\}$. For any projection p of M, denote by $c(p)$

the minimal central projection of M containing p, i.e., $c(p)$ is the projection from H onto $\overline{[MpH]}$. By Proposition 1.5.9, we have

$$c(1 - x_0^* x_0) \cdot c(1 - x_0 x_0^*) = 0.$$

Let $z = 1 - c(1 - x_0^* x_0)$. Then $x_0^* x_0 \geq z, x_0 x_0^* \geq 1 - z$. Replacing $\{x_0, M, H\}$ by $\{x_0 z, Mz, zH\}$ and $\{x_0^*(1-z), M(1-z), (1-z)H\}$ respectively, we may assume that $x_0^* x_0 = 1$, and $x_0 x_0^* = p < 1$. Now it suffices to show that for any $\xi_1, \cdots, \xi_m, \eta_1, \cdots, \eta_m \in H$ there is a unitary element u of M such that

$$|\langle (x_0 - u)\xi_i, \eta_i \rangle| < 1, \quad 1 \leq i \leq m.$$

We can write that

$$H = H_0 \oplus \sum_{k=0}^{\infty} \oplus x_0^k (1-p) H$$

and let q_k be the projection from H onto $x_0^k(1-p)H, k = 0, 1, \cdots$, and q be the projection from H onto H_0. Pick a positive integer n such that

$$\| \sum_{k>n} q_k \xi_i \| < \frac{1}{2}(1 + \max_{1 \leq i \leq m} \|\eta_i\|)^{-1}, \quad 1 \leq i \leq m,$$

and let $u = x_0$ on $H_0 \oplus (1-p)H \oplus \cdots \oplus x_0^n(1-p)H, u = x_0^{*n+1}$ on $x_0^{n+1}(1-p)H, u = 1$ on $\sum_{k>n+1} \oplus x_0^k(1-p)H$. Then u is a unitary element of M and

$$\begin{aligned} |\langle (x_0 - u)\xi_i, \eta_i \rangle| &= |\langle \sum_{k>n} (x_0 - u) q_k \xi_i, \eta_i \rangle| \\ &\leq \|x_0 - u\| \cdot \| \sum_{k>n} q_k \xi_i \| \cdot \|\eta_i\| < 1, \end{aligned}$$

$1 \leq i \leq m.$ Q.E.D.

Proposition 2.15.17. Let A be a real C^*–algebra with an identity, and x_0 be a normal element of A with $\|x_0\| \leq 1$. Then x_o belongs to the closure of

$$Co\{\cos b \cdot e^a | a, b \in A, a^* = -a, b^* = b, a \sim b \sim x_o\}$$

where "$x \sim y$" means that $xy = yx$.

Proof. We may assume that A is abelian. Then by Theorem 2.14.2, x_0 belongs to the closure of

$$Co\{e^k | k \in A_c, k^* = -k\}$$

i.e., there are $\lambda_j^{(n)} > 0, \sum_j \lambda_j^{(n)} = 1$ and $k_j^{(n)*} = -k_j^{(n)} \in A_c$ such that

$$\| \sum_j \lambda_j^{(n)} \exp(k_j^{(n)}) - x_o \| \longrightarrow 0.$$

Write $k_j^{(n)} = a_j^{(n)} + ib_j^{(n)}$, where $a_j^{(n)*} = -a_j^{(n)}, b_j^{(n)*} = b_j^{(n)} \in A, \forall n, j$. Since A is abelian, and $A_c = A\dot{+}iA$, it follows that

$$\|\sum_j \lambda_j^{(n)} \cos b_j^{(n)} \cdot \exp(a_j^{(n)}) - x_0\| \longrightarrow 0.$$

Q.E.D.

Corollary 2.15.18. Let H be a complex Hilbert space, M be a real weakly closed $*$ operator algebra on H, $1 = 1_H \in M$, and $S = \{x \in M | \|x\| \leq 1\}$ be the closed unit ball of M. Then the subset

$$Co\{\cos b \cdot e^a | a, b \in M, a^* = -a, b^* = b\}$$

is weakly dense in S.

Proof. This comes directly from Lemma 2.15.16 and Proposition 2.15.17.

Q.E.D.

Theorem 2.15.19. Let A be a real C^*-algebra with an identity, and $S = \{x \in A | \|x\| \leq 1\}$ be its closed unit ball. Then

$$Co\{\cos b \cdot e^a | a, b \in A, a^* = -a, b^* = b\}$$

is dense in S.

Proof. Let $\{\pi, H\}$ be the universal $*$ representation of the complex C^*-algebra $A_c = A\dot{+}iA$ (see Definition 2.11.1), and $M = \overline{\pi(A)}^w$.

We say that $M \cap iM = \{0\}$. In fact, if $x \in M \cap iM$, then there exist two nets $\{a_\alpha\}, \{b_\beta\}$ of A such that

$$\pi(a_\alpha) \xrightarrow{w} x, \quad \pi(ib_\beta) \xrightarrow{w} x.$$

Hence $\pi(a_\alpha - ib_\beta) \longrightarrow 0$ weakly. In particular, for any $c, d \in A$ and any state ρ on A_c we have

$$\langle \pi(a_\alpha - ib_\beta)\pi(c)\xi_\rho, \pi(d^*)\xi_\rho\rangle = \rho(d(a_\alpha - ib_\beta)c) \to 0.$$

Since $A_c^* = A^*\dot{+}iA^*$ (Proposition 2.15.5.) and A_c^* is the linear span of its state space, we have

$$f(d(a_\alpha - ib_\beta)c) \to 0, \quad \forall f \in A^*,$$

i.e., $f(da_\alpha c) \to 0, f(db_\beta c) \to 0, \forall f \in A^*$. Furthermore, for each state ρ on A_c we have

$$\langle \pi(a_\alpha)\pi(c)\xi_\rho, \pi(d^*)\xi_\rho\rangle = \rho(da_\alpha c) \to 0,$$

$$\langle \pi(b_\beta)\pi(c)\xi_\rho, \pi(d^*)\xi_\rho\rangle = \rho(db_\beta c) \to 0,$$

i.e., $\langle x\pi(c)\xi_\rho, \pi(d^*)\xi_\rho\rangle = 0, \forall c, d \in A$ and state ρ on A_c. By the construction of $\{\pi, H\}$, we have $x = 0$.

Now we point out that $\pi(S)$ is τ–dense in the closed unit ball S_M of M. Indeed, for any $x \in S_M$ by Theorem 1.6.1 there is a net $\{a_l + ib_l\}$ of A_c such that

$$\pi(a_l) + i\pi(b_l) \longrightarrow x \text{ weakly,}$$

and $\|a_l + ib_l\| \leq 1, \forall l$, where $a_l, b_l \in A, \forall l$. By Lemma 2.15.15, $\|a_l\| \leq 1, \|b_l\| \leq 1, \forall l$. Since S_M is weakly compact, we may assume that

$$\pi(a_l) \xrightarrow{w} y \in M, \quad \pi(b_l) \xrightarrow{w} z \in M.$$

Hence , $x = y + iz$, i.e., $x - y = iz \in M \cap iM = \{0\}$. Therefore, $x = y = w$-$\lim \pi(a_l)$ belongs to the weak closure of $\pi(S)$. On the other hand, $\pi(S)$ is a bounded convex set, and its weak closure is equal to its τ–closure. So $\pi(S)$ is τ–dense in S_M.

Moreover, we have that $Co\{\pi(\cos b \cdot e^a) | a, b \in A, a^* = -a, b^* = b\}$ is weakly dense in S_M. In fact, by Corollary 2.15.18 it suffices to show that its weak closure contains $\cos y \cdot e^x, \forall x, y \in M$ with $y^* = y, x^* = -x$. From the preceding paragraph, there are two nets $\{a_\alpha\}, \{b_\beta\}$ of A such that

$$\pi(a_\alpha) \xrightarrow{\tau} x, \quad \pi(b_\beta) \xrightarrow{\tau} y,$$

and $\|a_\alpha\| \leq \|x\|, \forall \alpha, \|b_\beta\| \leq \|y\|, \forall \beta$. Since $*$ operation is continuous in τ–topology, we may assume that $a_\alpha^* = -a_\alpha, b_\beta^* = b_\beta, \forall \alpha, \beta$. Clearly, for any polynomials P and Q,

$$\pi(P(a_\alpha)) \xrightarrow{s} P(x), \quad \pi(Q(b_\beta)) \xrightarrow{s} Q(y).$$

Therefore we have

$$\pi(e^{a_\alpha}) \xrightarrow{s} e^x, \quad \pi(\cos b_\beta) \xrightarrow{s} \cos y,$$

and $\pi(\cos b_\beta \cdot e^{a_\alpha}) \to \cos y \cdot e^x$ (weakly).

Now if there exists $x_0 \in S \backslash \overline{Co\{\cos b \cdot e^a | a, b \in A, a^* = -a, b^* = b\}}$, then by the Asoli theorem we can find $f \in A_c^*$ such that

$$\text{Re } f(x_0) > \sup\{\text{Re } f(\cos b \cdot e^a) | a, b \in A, a^* = -a, b^* = b\}.$$

On the other hand, from the preceding paragraph there is a net $\{x_l\} \subset Co\{\cos b \cdot e^a | a, b \in A, a^* = -a, b^* = b\}$ such that

$$\pi(x_l) \longrightarrow \pi(x_0)(\text{ weakly}) .$$

In particular, for any state ρ on A_c we have

$$\rho(x_l - x_0) = \langle \pi(x_l - x_0)\xi_\rho, \xi_\rho\rangle \to 0.$$

Hence, $f(x_l) \to f(x_0)$. This is a contradiction. Therefore, $Co\{\cos b \cdot e^a | a, b \in A, a^* = -a, b^* = b\}$ is dense in S. Q.E.D.

Lemma 2.15.20. Let A be a real C^*-algebra with an identity 1 , and S be its closed unit ball. Then

$$\frac{1}{2}S \subset Co\{\cos b \cdot e^a | a, b \in A, a^* = -a, b^* = b\}.$$

Proof. Let $x \in S$ and write $x = h + k$, where $h, k \in A$ and $h^* = h, k^* = -k$.
Suppose that B is the real abelian C^*-subalgebra of A generated by $\{1, h\}$. Then , $B \cong C(\Omega, -)$, and $h \to h(t) = \overline{h(t)} = h(\bar{t}), -1 \leq h(t) \leq 1, \forall t \in \Omega$. Clearly, $\arccos h(t) \in C(\Omega, -)$, where $\arccos \lambda \in [0, \pi]$ if $\lambda \in [-1, 1]$. Hence , there is $b \in B$ with $b^* = b$ such that $b(t) = \arccos h(t), \forall t \in \Omega$, i.e., $h = \cos b$.
Suppose that C is the real abelian C^*-subalgebra of A generated by $\{1, k\}$. Then $C \cong C(\Omega', -), k \to k(t) = -\overline{k(t)} = -k(\bar{t}), |k(t)| \leq 1, \forall t \in \Omega'$. Clearly, $-i\arcsin(ik(t)) \in C(\Omega', -)$, where $\arcsin \lambda \in [-\pi/2, \pi/2]$ if $\lambda \in [-1, 1]$. Hence , there is $a \in C$ with $a^* = -a$ such that $a(t) = -i\arcsin(ik(t))$, i.e., $k = \frac{1}{2}(e^a - e^{-a})$. Therefore,

$$x = \cos b + \frac{1}{2}(e^a - e^{-a}) = 2\left(\frac{\cos b}{2} + \frac{e^a}{4} + \frac{\cos\pi \cdot e^{-a}}{4}\right).$$

That comes to our conclusion. Q.E.D.

Proposition 2.15.21. Let A be a real C^*-algebra with an identity. Then for any $x \in A$ we have

$$\|x\| = \inf\left\{\sum_j |\lambda_j| \,\middle|\, \begin{array}{l} x = \sum_j \lambda_j \cos b_j \cdot e^{a_j}, \text{where } a_j, b_j \in A \\ \text{and } a_j^* = -a_j, b_j^* = b_j, \forall j \end{array}\right\}.$$

Proof. By Lemma 2.15.20, we can define

$$\|x\|_1 = \inf\{\sum_j |\lambda_j| | \cdots\}, \forall x \in A.$$

Clearly, $\|\cdot\|_1$ is a norm on A and $\|\cdot\| \leq \|\cdot\|_1$.
Now by Lemma 2.15.20 we have $\|x/2\|x\|\|_1 \leq 1$, so that

$$\|x\| \leq \|x\|_1 \leq 2\|x\|, \forall x \in A.$$

If $\|x\| = 1$, then by Theorem 2.15.19 there exists a sequence $\{x_n\} \subset Co\{\cos b \cdot e^a | a, b \in A, a^* = -a, b^* = b\}$ such that $\|x_n - x\| \to 0$. Hence, $\|x_n - x\|_1 \to 0$. Since $\|x_n\|_1 \leq 1, \forall n$, it follows that $\|x\|_1 \leq 1$. Therefore $\|x\|_1 \leq \|x\|$, and $\|x\|_1 = \|x\|, \forall x \in A$. Q.E.D.

Proposition 2.15.22. Let A be a real C^*-algebra with an identity 1 , and S be its closed unit ball. Then

$$\{h \in A | h^* = h, \|h\| \leq 1\} \cup \{k \in A | k^* = -k, \|k\| \leq 1\}$$
$$\subset \; Co\{\cos b \cdot e^a | a, b \in A, a^* = -a, b^* = b\}$$

and

$$\mathrm{Int}(S) \subset Co\{\cos b \cdot e^a | a, b \in A, a^* = -a, b^* = b\} \subset S,$$

where $\mathrm{Int}(S) = \{x \in A | \|x\| < 1\}$.

Proof. The former conclusion is contained in the proof of Lemma 2.15.20 indeed.

Now let $x \in \mathrm{Int}(S)$. By Proposition 2.15.21 we can write

$$x = \sum_k \lambda_k \cos b_k \cdot e^{a_k},$$

where $\lambda_k > 0, a_k^* = -a_k, b_k^* = b_k, \forall k$, and $\sum_k \lambda_k < 1$. Then

$$x = \sum_k \lambda_k \cos b_k \cdot e^{a_k} + \frac{1 - \sum_k \lambda_k}{2} \cos 0 + \frac{1 - \sum_k \lambda_k}{2} \cos \pi$$

belongs to $Co\{\cos b \cdot e^a | a, b \in A, a^* = -a, b^* = b\}$. Q.E.D.

Proposition 2.15.23. Let A be a real C^*-algebra with an identity, and Φ be a bounded real linear mapping from A to a real normed linear space B. Then

$$\|\Phi\| = \sup\{\|\Phi(\cos b \cdot e^a)\| | a, b \in A, a^* = -a, b^* = b\}.$$

Proof. From Proposition 2.15.22 we have

$$\text{the right side } \leq \|\Phi\| = \sup_{x \in \mathrm{Int}(S)} \|\Phi(x)\| \leq \text{the right side.}$$

Q.E.D.

Now we discuss the axioms of real C^*-algebras.

Theorem 2.15.24. Let A be a real Banach $*$ algebra with an idenity. If there exists a constant $K (\geq 1)$ such that

$$\|e^a\| \leq K, \quad \|\cos b\| \leq K,$$

$\forall a, b \in A$ and $a^* = -a, b^* = b$, then A is real C^*-equivalent.

Moreover, if the constant $K = 1$, then A itself is a real C^*–algebra.

Proof. For any $a, b \in A$ with $a^* = -a, b^* = b$, we have

$$|e^{\lambda t}| \le \|e^{ta}\| \le K, \quad |\cos(\mu t)| \le \|\cos(tb)\| \le K,$$

$\forall t \in I\!R, \lambda \in \sigma(a), \mu \in \sigma(b)$. Hence, $\sigma(a) \subset iI\!R, \sigma(b) \subset I\!R$, i.e., A is hermitian and skew–hermitian.

Similar to the proof of Theorem 2.14.24, we can show that

$$\inf\{\|x^2\| | x \in A, x^* = x \text{ or } (-x), \text{ and } \|x\| = 1\} > 0.$$

Hence, there exists a constant $C > 0$ such that

$$C\nu(x) \ge \|x\|, \forall x \in A \text{ with } x^* = \pm x.$$

Then by Theorem 2.15.12, A is real C^*–equivalent.

Now let $K = 1$. From the preceding paragraph, there is a new norm $\|\cdot\|_1$ on A such that $\|\cdot\|_1 \sim \|\cdot\|$ and $(A, \|\cdot\|_1)$ is a real C^*–algebra. Consider the identity map $I : (A, \|\cdot\|_1) \to (A, \|\cdot\|)$. By Proposition 2.15.23, we have $\|I\| \le 1$, i.e., $\|x\| \le \|x\|_1, \forall x \in A$. If there is $x_0 \in A$ such that $\|x_0\| < \|x_0\|_1$, then

$$\begin{aligned} \nu(x_0^* x_0) &= \|x_0^* x_0\|_1 = \|x_0^*\|_1 \cdot \|x_0\|_1 \\ &> \|x_0^*\| \cdot \|x_0\| \ge \|x_0^* x_0\| \ge \nu(x_0^* x_0). \end{aligned}$$

This is a contradiction. Therefore, $\|x\| = \|x\|_1, \forall x \in A$, and A is a real C^*–algebra itself. Q.E.D.

Theorem 2.15.25. Let A be a real hermitian Banach $*$ algebra with an identity 1. If for any normal element x of A we have $\|x^*x\| = \|x^*\| \cdot \|x\|$, then A is a real C^*–algebra.

Proof. From our assumption, we have $\|x\| = \nu(x), \forall x^* = \pm x \in A$. In particular, $\|\cos b\| = \nu(\cos b) \le 1, \forall b^* = b \in A$. By the proof of Theorem 2.15.12.1) , the $*$ operation is continuous on A. Hence, for every $a^* = -a \in A$ we have

$$1 = \nu(1) = \|1\| = \|(e^a)^* \cdot e^a\| = \|e^a\| \cdot \|e^{-a}\|.$$

By the proof of Theorem 2.15.12.2), A is skew–hermitian. Thus, $\|e^{\pm a}\| \ge |e^{\pm\lambda}| = 1, \forall \lambda \in \sigma(a)$, and $\|e^a\| = 1$. Then from Theorem 2.15.24 A is a real C^*–algebra. Q.E.D.

Theorem 2.15.26. Let A be a real hermitian Banach $*$ algebra. If for any $x \in A$ we have

$$\|x^*x\| = \|x^*\| \cdot \|x\|,$$

then A is a real C^*–algebra.

Proof. By Theorem 2.15.12, A is real C^* –equivalent, i.e., there is a new norm $\|\cdot\|_1$ on A such that $\|\cdot\|_1 \sim \|\cdot\|$, and $(A, \|\cdot\|_1)$ is a real C^*–algebra. Clarly,

$$\|x\|_1 = \|x\| = \nu(x), \quad \forall x^* = \pm x \in A.$$

Let $\{e_l\}$ be an approximate identity for A, and define

$$\|x + \lambda\| = \sup\{\|xy + \lambda y\| | y \in A, \|y\| \le 1\}$$

$\forall x \in A, \lambda \in I\!R$ (here we assume that A has no identity, otherwise, the conclusion is obvious by Theorem 2.15.25). Then $(A \dot{+} I\!R)$ is a real hermitian Banach $*$ algebra with an identity 1, and the original norm $\|\cdot\|$ on A remains unchanged (since $\|e_l\| = \|e_l\|_1 \le 1, \forall l$.) Similar to the proof of Proposition 2.4.4, we have

$$\|x + \lambda\| = \lim_l \|e_l(\lambda + x)\| = \lim_l \|(\lambda + x)e_l\|,$$

$\forall x \in A, \lambda \in I\!R$. Hence

$$\begin{aligned} \|(x+\lambda)^*\| \cdot \|x + \lambda\| &= \lim_l \|\bar{\lambda} e_l + e_l x^*\| \cdot \|x e_l + \lambda e_l\| \\ &= \lim_l \|e_l (x+\lambda)^*(x+\lambda) e_l\| \\ &\le \lim_l \|(x+\lambda)^*(x+\lambda)e_l\| = \|(x+\lambda)^*(x+\lambda)\| \\ &\le \|(x+\lambda)^*\| \cdot \|x+\lambda\|, \end{aligned}$$

and $\|(x+\lambda)^* \cdot (x+\lambda)\| = \|(x+\lambda)^*\| \cdot \|x+\lambda\|, \forall x \in A, \lambda \in I\!R$. Now by Theorem 2.15.25, $(A \dot{+} I\!R))$ and A are real C^*–algebras. Q.E.D.

Notes. Theorem 2.15.10 is due to L. Ingelstam and T.W.Palmer. Lemma 2.15.16 is due to B.Russo and H.A. Dye. Theorems 2.15.19 and 2.15.26 are due to B.R.Li, and Theorem 2.15.26 gives an affirmative answer for the Gelfand–Naimark conjecture in real case. However, it is still an opern question: does Theorem 2.15.25 hold in the absence of the identity?

References. [72], [100], [101], [124], [141].

Chapter 3
Tensor Products of C^*-Algebras

3.1. Tensor products of Banach spaces and cross-norms

Let $X_1, \cdots, X_n$ be (complex) Banach spaces, and

$$\otimes_{i=1}^n X_i = \{\sum_j \otimes_{i=1}^n x_j^{(i)} \mid x_j^{(i)} \in X_i, \forall i, j\}.$$

$u = \sum_j \otimes_{i=1}^n x_j^{(i)}$ is called zero, if for any $f_i \in X_i^*, 1 \le i \le n$,

$$(\otimes_{i=1}^n f_i)(u) = \sum_j \prod_{i=1}^n f_i(x_j^{(i)}) = 0.$$

Then $\otimes_{i=1}^n X_i$ is a linear space, and is called the *algebraic tensor product* of $X_1, \cdots, X_n$. This is a generalization of the algebraic tensor product of Hilbert spaces (see Section 1.4).

If $\alpha(\cdot)$ is a norm on $\otimes_{i=1}^n X_i$, then the completion of $(\otimes_{i=1}^n X_i, \alpha(\cdot))$, denoted by α -$\otimes_{i=1}^n X_i$, is called the *tensor product* of $X_1, \cdots, X_n$ with respect to $\alpha(\cdot)$.

Definition 3.1.1. A norm $\alpha(\cdot)$ on the algebraic tensor product of Banach spaces $X_1, \cdots, X_n$ is called a *cross-norm*, if $\alpha(\otimes_{i=1}^n x_i) = \|x_1\| \cdots \|x_n\|, \forall x_i \in X_i, 1 \le i \le n$.

Proposition 3.1.2. Let $X_1, \cdots, X_n$ be Banach spaces, and $\otimes_{i=1}^n X_i$ be their algebraic tensor product.

1) $\lambda(u) = \sup\{|\otimes_{i=1}^n f_i(u)| \mid f_i \in X_i^*, \|f_i\| \le 1, 1 \le i \le n\}(\forall u \in \otimes_{i=1}^n X_i)$ is a cross-norm on $\otimes_{i=1}^n X_i$.

2) $\gamma(u) = \inf\{\sum_j \prod_{i=1}^n \|x_j^{(i)}\| \mid u = \sum_j \otimes_{i=1}^n x_j^{(i)}\}(\forall u \in \otimes_{i=1}^n X_i)$ is the largest cross-norm on $\otimes_{i=1}^n X_i$.

Proof. It is easy, and we leave it to the reader. Q.E.D.

Let $X_1, \cdots, X_n$ be Banach spaces; and consider the algebraic tensor product $\otimes_{i=1}^n X_i^*$ of conjugate spaces $X_1^*, \cdots, X_n^*$. Since X_i is $\sigma(X_i^{**}, X_i^*)$-dense in $X_i^{**}, 1 \le i \le n$, it follows that $u^*(\in \otimes_{i=1}^n X_i^*)$ is zero if and only if $u^*(\otimes_{i=1}^n x_i) = 0, \forall x_i \in X_i, 1 \le i \le n$. Now if $\alpha(\cdot)$ is a norm on $\otimes_{i=1}^n X_i$ such that $\alpha^*(u^*) = \sup\{|u^*(u)| \mid u \in \otimes_{i=1}^n X_i, \alpha(u) \le 1\} < \infty, \forall u^* \in \otimes_{i=1}^n X_i^*$, then $\alpha^*(\cdot)$ becomes a norm on $\otimes_{i=1}^n X_i^*$; and $\alpha^*(\cdot)$ is called the *dual norm* of $\alpha(\cdot)$.

Proposition 3.1.3. Let $X_1, \cdots, X_n$ be Banach spaces.

1) The dual norm of $\gamma(\cdot)$ on $\otimes_{i=1}^n X_i$ is just the norm $\lambda(\cdot)$ on $\otimes_{i=1}^n X_i^*$ (about the definitions of $\gamma(\cdot), \lambda(\cdot)$, see Proposition 3.1.2) .

2) Let $\alpha(\cdot)$ be a cross-norm on $\otimes_{i=1}^n X_i$. Then $\alpha^*(\cdot)$ is a cross-norm on $\otimes_{i=1}^n X_i^*$ if and only if $\lambda(\cdot) \le \alpha(\cdot) \le \gamma(\cdot)$. In this case, we also have $\gamma(\cdot) \ge \lambda^*(\cdot) \ge \alpha^*(\cdot) \ge \lambda(\cdot)$ on $\otimes_{i=1}^n X_i^*$.

Proof. 1) For any $u^* \in \otimes_{i=1}^n X_i^*$, since the unit ball of X_i is $\sigma(X_i^{**}, X_i^*)$-dense in the unit ball of $X_i^{**}, 1 \le i \le n$, it follows that

$$\begin{aligned}
\gamma^*(u^*) &= \sup\{|u^*(u)| \mid u \in \otimes_{i=1}^n X_i, \gamma(u) \le 1\} \\
&\ge \sup\{|u^*(\otimes_{i=1}^n x_i)| \mid x_i \in X_i, \|x_i\| \le 1, 1 \le i \le n\} \\
&= \sup\{|\otimes_{i=1}^n x_i(u^*)| \mid x_i \in X_i^{**}, \|x_i\| \le 1, 1 \le i \le n\} \\
&= \lambda(u^*).
\end{aligned}$$

However, for any $u \in \otimes_{i=1}^n X_i$ with $\gamma(u) \le 1$ and $\varepsilon > 0$, we can write $u = \sum_j \otimes_{i=1}^n x_j^{(i)}$ such that $\sum_j \prod_{i=1}^n \|x_j^{(i)}\| - \gamma(u) \le \varepsilon$. Then

$$\begin{aligned}
|u^*(u)| \le & \sum_j |u^*(\otimes_{i=1}^n x_j^{(i)})| \le \sum_j \lambda(u^*) \prod_{i=1}^n \|x_j^{(i)}\| \\
\le & \lambda(u^*)(\gamma(u) + \varepsilon) \le \lambda(u^*)(1 + \varepsilon).
\end{aligned}$$

Thus $\gamma^*(u^*) \le (1+\varepsilon)\lambda(u^*)$. Since ε is arbitrary, it follows that $\gamma^*(u^*) \le \lambda(u^*)$. Therefore, $\gamma^*(u^*) = \lambda(u^*), \forall u^* \in \otimes_{i=1}^n X_j^*$.

2) For any $u \in \otimes_{i=1}^n X_i$, by the definition of $\lambda(\cdot)$ it is easy to see that

$$\lambda(u) \le \alpha(u) \sup\{\alpha^*(\otimes_{i=1}^n f_i) \mid f_i \in X_i^*, \|f_i\| \le 1, 1 \le i \le n\}.$$

Hence if $\alpha^*(\cdot)$ is a cross-norm on $\otimes_{i=1}^n X_i^*$, then we have $\lambda(\cdot) \le \alpha(\cdot)$. Obviously, $\alpha(\cdot) \le \gamma(\cdot)$ since $\gamma(\cdot)$ is the largest cross-norm on $\otimes_{i=1}^n X_i$.

Conversely, suppose that $\lambda(\cdot) \leq \alpha(\cdot) \leq \gamma(\cdot)$ on $\otimes_{i=1}^n X_i$. For any $u^* \in \otimes_{i=1}^n X_i^*$, by 1) we have

$$\begin{aligned}
\lambda(u^*) = &\ \sup\{|u^*(u)| \mid u \in \otimes_{i=1}^n X_i, \gamma(u) \leq 1\} \\
\leq &\ \sup\{|u^*(u)| \mid \alpha(u) \leq 1\} = \alpha^*(u^*) \\
\leq &\ \sup\{|u^*(u)| \mid \lambda(u) \leq 1\} = \lambda^*(u^*) \\
\leq &\ \sup\{\sum_j \mid \otimes_{i=1}^n f_j^{(i)}(u)| \mid \lambda(u) \leq 1\} \leq \sum_j \prod_{i=1}^n \|f_j^{(i)}\|,
\end{aligned}$$

where $u^* = \sum_j \otimes_{j=1}^n f_j^{(i)}$ and $f_j^{(i)} \in X_j^*, \forall i, j$, i.e., on $\otimes_{i=1}^n X_i^*$,

$$\gamma(\cdot) \geq \lambda^*(\cdot) \geq \alpha^*(\cdot) \geq \lambda(\cdot).$$

Since $\gamma(\cdot)$ and $\lambda(\cdot)$ are cross-norms, it follows that $\alpha^*(\cdot)$ is also a cross-norm.

Q.E.D.

Notes. The tensor products of Banach spaces was first studied by R. Schatten and J. Von Neumann. The general theory of tensor products was further developed by A. Grothendieck, which led him to the discovery of *nuclear spaces.*

References.[63], [151].

3.2. Tensor products of C^*-algebras and the spatial C^*-norm

Let $A_1, \cdots, A_n$ be C^*-algebras. Naturally, define

$$(\otimes_{i=1}^n a_i) \cdot (\otimes_{i=1}^n b_i) = \otimes_{i=1}^n a_i b_i, \quad (\otimes_{i=1}^n a_i)^* = \otimes_{i=1}^n a_i^*$$

$\forall a_i, b_i \in A_i, 1 \leq i \leq n$. Then $\otimes_{i=1}^n A_i$ is a $*$ algebra.

Definition 3.2.1. Let $A_1, \cdots, A_n$ be C^*-algebras. A norm $\alpha(\cdot)$ on $\otimes_{i=1}^n A_i$ is called a *C^*-norm*, if

$$\alpha(uv) \leq \alpha(u)\alpha(v), \quad \alpha(u^*u) = \alpha(u)^2,$$

$\forall u, v \in \otimes_{i=1}^n A_i$. The completion of $\otimes_{i=1}^n A_i$ with respect to $\alpha(\cdot)$, denoted by α-$\otimes_{i=1}^n A_i$, is called the *tensor product* of $A_1, \cdots, A_n$ with respect to $\alpha(\cdot)$. Clearly, α-$\otimes_{i=1}^n A_i$ is also a C^*-algebra.

Proposition 3.2.2. Let $\alpha(\cdot)$ be a C^*-norm on $\otimes_{i=1}^n A_i$. Then $\alpha(\cdot) \le \gamma(\cdot)$, where $\gamma(\cdot)$ is as in Proposition 3.1.2.

Proof. First, notice the following fact: Let A be a C^*-algebra, and $a \in A_+$. Then $\|a\| \le 1$ if and only if $a^2 \le a$.

Now let $b_i \in A_i, 0 \le b_i \le 1, 1 \le i \le n$. Since

$$\otimes_{i=1}^n b_i^2 \le b_1 \otimes \otimes_{i=2}^n b_i^2 \le \cdots \le \otimes_{i=1}^n b_i$$

in α-$\otimes_{i=1}^n A_i$, it follows from the above fact that $\alpha(\otimes_{i=1}^n b_i) \le 1$.

Further, for any $a_i \in A_i, 1 \le i \le n$, we have

$$\begin{aligned}\alpha(\otimes_{i=1}^n a_i)^2 &= \alpha(\otimes_{i=1}^n a_i^* a_i)\\ &\le \|a_1^* a_1\| \cdots \|a_n^* a_n\| = (\|a_1\| \cdots \|a_n\|)^2.\end{aligned}$$

Therefore, $\alpha(\cdot) \le \gamma(\cdot)$ on $\otimes_{i=1}^n A_i$. Q.E.D.

Proposition 3.2.3. Let $\alpha(\cdot)$ be a C^*-norm on $\otimes_{i=1}^n A_i$, and $\{d_{l_i}^{(i)}\}$ be an approximate identity for $A_i, 1 \le i \le n$. Then $\{d_l = \otimes_{i=1}^n d_{l_i}^{(i)} \mid l = (l_1, \cdots, l_n)\}$ is an approximate identity for α-$\otimes_{i=1}^n A_i$.

Proof. From Proposition 3.2.2, $\alpha(d_l) \le \gamma(d_l) = \prod_{i=1}^n \|d_{l_i}^{(i)}\| \le 1, \forall l = (l_1, \cdots, l_n)$. So it suffices to show that

$$\alpha(d_l \otimes_{i=1}^n a_i - \otimes_{i=1}^n a_i) \to 0, \text{ and } \alpha(\otimes_{i=1}^n a_i d_l - \otimes_{i=1}^n a_i) \to 0,$$

$\forall a_i \in A_i, 1 \le i \le n$. Notice the following equalities:

$$\otimes_{i=1}^n b_i a_i - \otimes_{i=1}^n a_i = \sum_{i=1}^n \otimes_{j=1}^{i-1} a_j \otimes (b_i a_i - a_i) \otimes \otimes_{j=i+1}^n b_j a_j$$

and

$$\otimes_{i=1}^n a_i b_i - \otimes_{i=1}^n a_i = \sum_{i=1}^n \otimes_{j=1}^{i-1} a_j \otimes (a_i b_i - a_i) \otimes \otimes_{j=i+1}^n a_j b_j.$$

Now from $\alpha(\cdot) \le \gamma(\cdot)$ and the property of $\{d_{l_i}^{(i)}\}_{(1 \le i \le n)}$, we can get the conclusion immediately. Q.E.D.

Definition 3.2.4. Let $A_1, \cdots, A_n$ be C^*-algebras, and $\otimes_{i=1}^n A_i$ be their algebraic tensor product. $u \in \otimes_{i=1}^n A_i$ is said to be *positive*, denoted by $u \ge 0$, if $u = \sum_{j=1}^m u_j^* u_j$, where $u_1, \cdots, u_m \in \otimes_{i=1}^n A_i$. Moreover, let $(\otimes_{i=1}^n A_i)_+ = \{u \in \otimes_{i=1}^n A_i \mid u \ge 0\}$.

A linear functional φ on $\oplus_{i=1}^n A_i$ is said to be *positive*, denoted by $\varphi \geq 0$, if $\varphi(u) \geq 0, \forall u \in (\otimes_{i=1}^n A_i)_+$.

Clearly, $(\otimes_{i=1}^n A_i)_+$ is a cone; $\otimes_{i=1}^n A_i$ is the linear span of $(\otimes_{i=1}^n A_i)_+$, and for any C^*-norm $\alpha(\cdot)$ on $\otimes_{i=1}^n A_i$,

$$(\otimes_{i=1}^n A_i)_+ \subset (\alpha\text{-} \otimes_{i=1}^n A_i)_+.$$

Moreover, if φ is a positive linear functional on $\otimes_{i=1}^n A_i$, then we have $\varphi(u^*) = \overline{(\varphi(u))}$ and the Schwartz inequality:

$$|\varphi(v^*u)|^2 \leq \varphi(u^*u)\varphi(v^*v),$$

$\forall u, v \in \otimes_{i=1}^n A_i$.

Proposition 3.2.5. Let φ_i be a positive linear functional on a C^*-algebra $A_i, 1 \leq i \leq n$. Then $\otimes_{i=1}^n \varphi_i$ is positive on $\otimes_{i=1}^n A_i$, where $\otimes_{i=1}^n \varphi_i$ is defined by $(\otimes_{i=1}^n \varphi_i)(\otimes_{j=1}^n a_j) = \prod_{i=1}^n \varphi_i(a_i), \forall a_i \in A_i, 1 \leq i \leq n.$

Proof. It is similar to the Section 1.4. Q.E.D.

Theorem 3.2.6. Let A_i be a C^*-algebra, and $S_i = S(A_i)$ be the state space of $A_i, 1 \leq i \leq n$. For any $u \in \otimes_{i=1}^n A_i$, define

$$\alpha_0(u)^2 = \sup\left\{ \frac{\otimes_{i=1}^n \varphi_i(v^*u^*uv)}{\otimes_{i=1}^n \varphi_i(v^*v)} \middle| \begin{array}{c} \varphi_i \in S_i, 1 \leq i \leq n, \\ v \in \otimes_{i=1}^n A_i \text{ and } \otimes_{i=1}^n \varphi_i(v^*v) > 0 \end{array} \right\}.$$

Then $\alpha_0(\cdot)$ is a C^*-norm on $\otimes_{i=1}^n A_i$, and $\lambda(\cdot) \leq \alpha_0(\cdot) \leq \gamma(\cdot)$. In particular, $\alpha_0(\cdot)$ is a cross-norm on $\otimes_{i=1}^n A_i$.

Proof. Let $\varphi_i \in S_i$, and $\{\pi_{\varphi_i}, H_{\varphi_i}, \xi_{\varphi_i}\}$ be the cyclic $*$ representation of A_i generated by $\varphi_i, 1 \leq i \leq n$. Naturally, we can define a $*$ representation $\{\otimes_{i=1}^n \pi_{\varphi_i}, \otimes_{i=1}^n H_{\varphi_i}\}$ of $\otimes_{i=1}^n A_i$. Since A_i/L_{φ_i} is dense in H_{φ_i}, where L_{φ_i} is the left kernel of $\varphi_i, 1 \leq i \leq n$, it follows that

$$\| \otimes_{i=1}^n \pi_{\varphi_i}(u)\|^2 = \sup\left\{ \frac{\otimes_{i=1}^n \varphi_i(v^*u^*uv)}{\otimes_{i=1}^n \varphi_i(v^*v)} \middle| v \in \otimes_{i=1}^n A_i, \text{ and } \otimes_{i=1}^n \varphi_i(v^*v) > 0 \right\},$$

$\forall u \in \otimes_{i=1}^n A_i$. Thus we have

$$\alpha_0(u) = \sup\{\| \otimes_{i=1}^n \pi_{\varphi_i}(u)\| \mid \varphi_i \in S_i, 1 \leq i \leq n\},$$

$\forall u \in \otimes_{i=1}^n A_i$. If $u \in \otimes_{i=1}^n A_i$ is such that $\alpha_0(u) = 0$, then

$$0 = \langle \otimes_{i=1}^n \pi_{\varphi_i}(u) \otimes_{j=1}^n \xi_{\varphi_j}, \otimes_{k=1}^n \xi_{\varphi_k} \rangle = \otimes_{i=1}^n \varphi_i(u),$$

$\forall \varphi_i \in S_i, 1 \leq i \leq n$. Hence $u = 0$ since A_i^* is the linear span of $S_i, 1 \leq i \leq n$. Therefore, $\alpha_0(\cdot)$ is a C^*-norm on $\otimes_{i=1}^n A_i$. By Proposition 3.2.2, $\alpha_0(\cdot) \leq \gamma(\cdot)$ on $\otimes_{i=1}^n A_i$.

Notice the following facts. Let A be a C^*-algebra, and $f \in A^*$ with $\|f\| = 1$. By Theorem 2.11.2 and 1.9.3, there is the polar decomposition $f = R_u\varphi$, where φ is a state on A, and u is a partial isometry of A^{**}. Since the unit ball of A is $\sigma(A^{**}, A^*)$-dense in the unit ball of A^{**}, it follows that there is a net $\{a_l\} \subset A$ with $\|a_l\| \leq 1, \forall l$, such that $a_l \to u(\sigma(A^{**}, A^*))$. Consequently, $R_{a_l}\varphi \to R_u(\varphi) = f(\sigma(A^*, A^{**}))$.

Now for any $u \in \otimes_{i=1}^n A_i$, by the above facts and the Schwartz inequality

$$\begin{aligned}
\lambda(u)^2 &= \sup\{|\otimes_{i=1}^n f_i(u)|^2 \mid f_i \in A_i^*, \|f_i\| = 1, 1 \leq i \leq n\} \\
&= \sup\left\{|\otimes_{i=1}^n R_{a_i}\varphi_i(u)|^2 \;\middle|\; \begin{array}{ll} \varphi_i \in S_i, & a_i \in A_i, \text{and} \\ \|a_i\| \leq 1, & 1 \leq i \leq n \end{array}\right\} \\
&\leq \sup\left\{\otimes_{i=1}^n \varphi_i(uu^*) \cdot \otimes_{j=1}^n \varphi_j(\otimes_{k=1}^n a_k^* a_k) \;\middle|\; \begin{array}{l} \varphi_i \in S_i, a_i \in A_i, \text{and} \\ \|a_i\| \leq 1, 1 \leq i \leq n \end{array}\right\} \\
&= \sup\{\langle(\otimes_{i=1}^n \pi_{\varphi_i}(uu^*) \otimes_{j=1}^n \xi_{\varphi_j}, \otimes_{k=1}^n \xi_{\varphi_k}\rangle \mid \varphi_i \in S_i, 1 \leq i \leq n\} \\
&\leq \alpha_0(u)^2.
\end{aligned}$$

Q.E.D.

Corollary 3.2.7. Let $f_i \in A_i^*, 1 \leq i \leq n$. Then the linear functional $\otimes_{i=1}^n f_i$ on $\otimes_{i=1}^n A_i$ can be uniquely extended to a linear functional on α_0-$\otimes_{i=1}^n A_i$ with the norm $\prod_{i=1}^n \|f_i\|$.

Proof. For any $u \in \otimes_{i=1}^n A_i$, by the definition of $\lambda(\cdot)$ and $\lambda(\cdot) \leq \alpha_0(\cdot)$ on $\otimes_{i=1}^n A_i$ we have

$$|\otimes_{i=1}^n f_i(u)| \leq \prod_{i=1}^n \|f_i\|\lambda(u) \leq \prod_{i=1}^n \|f_i\|\alpha_0(u).$$

That comes to the conclusion. Q.E.D.

Proposition 3.2.8. The above norm $\alpha_0(\cdot)$ on $\otimes_{i=1}^n A_i$ is also represented by

$$\begin{aligned}
\alpha_0(u) &= \sup\{\|\otimes_{i=1}^n \pi_i(u)\| \mid \pi_i \text{ is a } * \text{ representation of } A_i, 1 \leq i \leq n\} \\
&= \sup\{\|\otimes_{i=1}^n \pi_{\rho_i}(u)\| \mid \rho_i \in \Delta_i, 1 \leq i \leq n\},
\end{aligned}$$

$\forall u \in \otimes_{i=1}^n A_i$, where Δ_i is any set of positive linear functionals on A_i such that S_i is contained in the $\sigma(A_i^*, A_i)$-closure of $\{\sum_j \lambda_j \rho_j^{(i)} \mid \lambda_j \geq 0, \rho_j^{(i)} \in \Delta_i, \forall j\}$, and π_{ρ_i} is the $*$ representation of A_i generated by $\rho_i, \forall \rho_i \in \Delta_i, 1 \leq i \leq n$.

Proof. Let

$$\alpha(u) = \sup\{\|\otimes_{i=1}^n \pi_{\rho_i}(u)\| \mid \rho_i \in \Delta_i, 1 \leq i \leq n\},$$

$\forall u \in \otimes_{i=1}^n A_i$.

Since any $*$ representation of a C^*-algebra is a direct sum of a zero $*$ representation and a family of cyclic $*$ representations, and any cyclic $*$ representation is unitarily equivalent to a $*$ representation generated by a state, it follows from Theorem 3.2.6 that

$$\begin{aligned}\alpha_0(u) = & \sup\{\| \otimes_{i=1}^n \pi_{\varphi_i}(u)\| \mid \varphi_i \in S_i, 1 \le i \le n\} \\ = & \sup\{\| \otimes_{i=1}^n \pi_i(u)\| \mid \pi_i \text{ is a } * \text{ representation of } A_i, 1 \le i \le n\} \\ \ge & \alpha(u),\end{aligned}$$

$\forall u \in \otimes_{i=1}^n A_i$.

$\alpha(\cdot)$ is also a C^*-norm on $\otimes_{i=1}^n A_i$. In fact, if $\alpha(u) = 0$ for some $u \in \otimes_{i=1}^n A_i$, then $\otimes_{i=1}^n \rho_i(u) = 0, \forall \rho_i \in \Delta_i, 1 \le i \le n$. By the definitions of $\Delta_1, \cdots, \Delta_n$, we have $u = 0$.

For any $\rho_i \in \Delta_i, 1 \le i \le n$, since

$$| \otimes_{i=1}^n \rho_i(u)| \le \| \otimes_{i=1}^n \pi_{\rho_i}(u)\| \le \alpha(u), \quad \forall u \in \otimes_{i=1}^n A_i,$$

it follows that $\otimes_{i=1}^n \rho_i$ can be extended to a positive linear functional on α-$\otimes_{i=1}^n A_i$. Further, for any $u, v \in \otimes_{i=1}^n A_i$, $v^*(\alpha(u^*u) - u^*u)v$ is a positive element of α-$\otimes_{i=1}^n A_i$. Thus we have

$$\otimes_{i=1}^n \rho_i(v^*(\alpha(u^*u) - u^*u)v) \ge 0,$$

$\forall \rho_i \in \Delta_i, 1 \le i \le n, u, v \in \otimes_{i=1}^n A_i$. By the definitions of $\Delta_1, \cdots, \Delta_n$, it is clear that

$$\otimes_{i=1}^n \varphi_i(v^*(\alpha(u^*u) - u^*u)v) \ge 0,$$

$\forall \varphi_i \in S_i, 1 \le i \le n$, and $u, v \in \otimes_{i=1}^n A_i$. Now from Theorem 3.2.6, $\alpha(u) \ge \alpha_0(u), \forall u \in \otimes_{i=1}^n A_i$. Therefore, $\alpha_0(\cdot) = \alpha(\cdot)$ on $\otimes_{i=1}^n A_i$. Q.E.D.

Lemma 3.2.9. Let A be a C^*-algebra on a Hilbert space H, and $\Delta = \{\omega_\xi(\cdot) = \langle \cdot \xi, \xi\rangle \mid \xi \in H, \|\xi\| = 1\}$. Then the $\sigma(A^*, A)$-closure of $Co\Delta$ contains the state space $S(A)$ of A.

Proof. Let $\varphi \in S(A)$. By Corollary 2.3.12, φ can be extended to a state on $B(H)$, still denoted by φ. Since the unit ball of $T(H)$ is w^*-dense in the unit ball of $B(H)^*$, it follows that there is a net $\{t_l\} \subset T(H)$ with $\|t_l\|_1 \le 1, \forall l$, such that

$$tr(t_l b) \to \varphi(b), \quad \forall b \in B(H).$$

By $\varphi^* = \varphi$, we may assume that $t_l^* = t_l, \forall l$. Write $t_l = t_l^+ - t_l^-$, where $t_l^\pm \in T(H)_+$ and $t_l^+ \cdot t_l^- = 0, \forall l$. Clearly, $\|t_l^\pm\|_1 \le \|t_l\|_1 \le 1, \forall l$. Since the unit ball of $B(H)^*$ is w^*-compact, so we may assume that the nets $\{t_l^+\}$ and $\{t_l^-\}$ are w^*-convergent in $B(H)^*$. Let $|t_l| = t_l^+ + t_l^-, \forall l$. Then there is $\psi \in B(H)^*$ such that

$tr(|t_l|b) \to \psi(b), \forall b \in B(H)$. Obviously, ψ is positive on $B(H)_+$. Moreover, since $\varphi(1) = 1 = \lim_l tr(t_l) \le \lim_l tr(|t_l|) = \psi(1)$ and $\||t_l|\|_1 = \|t_l\|_1 \le 1, \forall l$, it follows that ψ is a state on $B(H)$ and $\lim_l tr(t_l^-) = 0$. Further, for any $a \in B(H)_+, 0 \le tr(t_l^- a) \le \|a\| tr(t_l^-) \to 0$. Thus, $tr(t_l^+ b) \to \varphi(b), \forall b \in B(H)$, and we may assume that $t_l \ge 0, \forall l$. Now write

$$tr(t_l b) = \sum_n \lambda_n^{(l)} \langle b\xi_n^{(l)}, \xi_n^{(l)} \rangle, \quad \forall b \in B(H),$$

where $\|\xi_n^{(l)}\| = 1, \lambda_n^{(l)} > 0, \forall n, l$, and let

$$\omega_{N,l}(\cdot) = \sum_{n=1}^{N} (\sum_{k=1}^{N} \lambda_k^{(l)})^{-1} \lambda_n^{(l)} \langle \cdot \xi_n^{(l)}, \xi_n^{(l)} \rangle, \quad \forall N, l.$$

Clearly, $\omega_{N,l} \in Co\Delta$, and $\omega_{N,l}(a) \to \varphi(a), \forall a \in A$. Q.E.D.

Theorem 3.2.10. Let $\{\pi_i, H_i\}$ be a faithful $*$ representation of a C^*-algebra $A_i, 1 \le i \le n$. Then

$$\alpha_0(u) = \| \otimes_{i=1}^n \pi_i(u) \|, \quad \forall u \in \otimes_{i=1}^n A_i,$$

where $\alpha_0(\cdot)$ is defined by Theorem 3.2.5.

Proof. By Proposition 3.2.8,

$$\begin{aligned}
&\alpha_0(u)^2 \ge \| \otimes_{i=1}^n \pi_i(u) \|^2 \\
\ge\ & \sup \left\{ \frac{\| \otimes_{i=1}^n \pi_i(uv) \otimes_{j=1}^n \xi_j \|^2}{\| \otimes_{i=1}^n \pi_i(v) \otimes_{j=1}^n \xi_j \|^2} \,\middle|\, \begin{array}{l} \xi_i \in H_i, \|\xi_i\| = 1, 1 \le i \le n, \\ v \in \otimes_{i=1}^n A_i \text{ and } \otimes_{i=1}^n \pi_i(v) \otimes_{j=1}^n \xi_j \ne 0 \end{array} \right\} \\
=\ & \sup \left\{ \frac{\otimes_{i=1}^n \rho_i(v^* u^* u v)}{\otimes_{i=1}^n \rho_i(v^* v)} \,\middle|\, \begin{array}{l} \rho_i \in \Delta_i, \quad 1 \le i \le n, \\ v \in \otimes_{i=1}^n A_i \text{ and } \otimes_{i=1}^n \rho_i(v^* v) > 0 \end{array} \right\} \\
=\ & \sup\{ \| \otimes_{i=1}^n \pi_{\rho_i}(u) \|^2 \mid \rho_i \in \Delta_i, 1 \le i \le n \},
\end{aligned}$$

$\forall u \in \otimes_{i=1}^n A_i$, where $\Delta_i = \{ \langle \pi_i(\cdot)\xi_i, \xi_i \rangle \mid \xi_i \in H_i, \|\xi_i\| = 1 \}$, and π_{ρ_i} is the $*$ representation of A_i generated by $\rho_i, \forall \rho_i \in \Delta_i, 1 \le i \le n$. Further, from Proposition 3.2.8 and Lemma 3.2.9 it follows that

$$\sup\{ \| \otimes_{i=1}^n \pi_{\rho_i}(u) \| \mid \rho_i \in \Delta_i, 1 \le i \le n \} = \alpha_0(u),$$

$\forall u \in \otimes_{i=1}^n A_i$. That comes to the conclusion. Q.E.D.

From this theorem, the geometric sense of the C^*-norm $\alpha_0(\cdot)$ on $\otimes_{i=1}^n A_i$ is given. Therefore, $\alpha_0(\cdot)$ is called the *spatial C^*-norm* on $\otimes_{i=1}^n A_i$. Later, we shall see that $\alpha_0(\cdot)$ is the minimal C^*-norm on $\otimes_{i=1}^n A_i$ indeed. So sometimes we also denote α_0 - $\otimes_{i=1}^n A_i$ by min -$\otimes_{i=1}^n A_i$.

Proposition 3.2.11. $\otimes_{i=1}^n A_i^*$ is w^*-dense in $(\alpha_0\text{-}\otimes_{i=1}^n A_i)^*$.

Proof. From Corollary 3.2.7, $\otimes_{i=1}^n A_i^* \subset (\alpha_0\text{-}\otimes_{i=1}^n A_i)^*$. We may assume that A_i is a C^*-algebra on a Hilbert space $H_i, 1 \le i \le n$. By Theorem 3.2.10, α_0-$\otimes_{i=1}^n A_i$ is the uniform closure of $\otimes_{i=1}^n A_i$ in $B(\otimes_{i=1}^n H_i)$. Further, from Lemma 3.2.9, the w^*-closure of $Co\{\otimes_{i=1}^n \langle \cdot \xi_i, \xi_i \rangle \mid \xi_i \in H_i, 1 \le i \le n\}$ contains the state space of α_0-$\otimes_{i=1}^n A_i$. Therefore, $\otimes_{i=1}^n A_i^*$ is w^*-dense in $(\alpha_0\text{-}\otimes_{i=1}^n A_i)^*$. Q.E.D.

Notes. The tensor products of C^*-algebras was studied first by T. Turumaru. α_0-$\otimes_{i=1}^n A_i$ is also called the *injective tensor product* of $A_1, \cdots, A_n$. Proposition 3.2.2 seems very simple, but we got it quite later. B.J. Vowden gave first proof. The present proof here is taken from C. Lance. Theorem 3.2.10 is due to A. Wolfsohn.

References. [95], [171], [185], [193], [198].

3.3. The maximal C^*-norm

Let $A_1, \cdots, A_n$ be C^*-algebras. From Theorem 3.2.5, there is a C^*-norm $\alpha_0(\cdot)$ on $\otimes_{i=1}^n A_i$. By Proposition 3.2.2, we have $\alpha(\cdot) \le \gamma(\cdot)$ for any C^*-norm $\alpha(\cdot)$ on $\otimes_{i=1}^n A_i$. Thus we can define the *maximal C^*-norm* $\alpha_1(\cdot)$ on $\otimes_{i=1}^n A_i$:

$$\alpha_1(u) = \sup\{\alpha(u) \mid \alpha \text{ is a } C^*\text{-norm on } \otimes_{i=1}^n A_i\},$$

$\forall u \in \otimes_{i=1}^n A_i$. And sometimes we also denote α_1-$\otimes_{i=1}^n A_i$ by max-$\otimes_{i=1}^n A_i$. Clearly, $\alpha_1(\cdot) \le \gamma(\cdot)$ on $\otimes_{i=1}^n A_i$.

As in Definition 2.4.5, a $*$ representation $\{\pi, H\}$ of $\otimes_{i=1}^n A_i$ is said to be *nondegenerate*, if $[\pi(u)\xi \mid \xi \in H, u \in \otimes_{i=1}^n A_i]$ is dense in H.

Lemma 3.3.1. Let $\{\pi, H\}$ be a nondegenerate $*$ representation of $\otimes_{i=1}^n A_i$. Then there is unique nondegenerate $*$ representation $\{\pi_i, H\}$ of $A_i, 1 \le i \le n$, such that

$$\begin{cases} \pi_i(a_i)\pi_j(a_j) = \pi_j(a_j)\pi_i(a_i) \\ \pi(\otimes_{i=1}^n a_i) = \pi_1(a_1)\cdots\pi_n(a_n) \end{cases}$$

$\forall a_i \in A_i, 1 \le i \ne j \le n$. Consequently, $\|\pi(u)\| \le \gamma(u), \forall u \in \otimes_{i=1}^n A_i$. Moreover, if π is faithful, then π_i is also faithful, $1 \le i \le n$.

Proof. Let $A_i^{(1)} = A_i \dot{+} \mathbb{C}1_i, 1 \leq i \leq n$. For any $i \in \{1, \cdots, n\}$ and any $\xi = \sum_k \pi(u_k)\xi_k$, where $u_k \in \otimes_{i=1}^n A_i$, $\xi_k \in H, \forall k$, define

$$\varphi_i(b_i) = \sum_{k,l} \langle \pi(\otimes_{j \neq i} 1_j \otimes b_i \cdot u_k)\xi_k, \pi(u_l)\xi_l \rangle$$

$\forall b_i \in A_i^{(1)}$. Clearly, φ_i is a positive linear functional on $A_i^{(1)}, \forall i$. Then

$$\begin{aligned} \|\sum_k \pi(\otimes_{j \neq i} 1_j \otimes b_i \cdot u_k)\xi_k\| &= \varphi_i(b_i^* b_i)^{1/2} \\ &\leq \|b_i\| \varphi_i(1_i)^{1/2} = \|b_i\| \cdot \|\xi\|. \end{aligned}$$

Since π is nondegenerate, it follows that there is unique $\pi_i(b_i) \in B(H)$ such that

$$\pi_i(b_i)\xi = \sum_k \pi(\otimes_{j \neq i} 1_j \otimes b_i \cdot u_k)\xi_k,$$

$\forall b_i \in A_i^{(1)}$. It is easy to see that $\{\pi_i, H\}$ is a $*$ representation of $A_i^{(1)}, 1 \leq i \leq n$, and

$$\pi_1(b_1) \cdots \pi_n(b_n)\xi = \sum_k \pi(\otimes_{i=1}^n b_i \cdot u_k)\xi_k$$

$\forall b_i \in A_i^{(1)}, 1 \leq i \leq n$ and $\xi = \sum_k \pi(u_k)\xi_k$. Therefore

$$\pi(\otimes_{i=1}^n a_i) = \pi_1(a_1) \cdots \pi_n(a_n),$$

$$\pi_i(a_i)\pi_j(a_j) = \pi_j(a_j)\pi_i(a_i),$$

$\forall a_i \in A_i, 1 \leq i \neq j \leq n$.

Suppose that $\xi \in H$ is such that $\pi_i(a_i)\xi = 0, \forall a_i \in A_i$ for some i. Then $\pi(\otimes_{j=1}^n a_j)\xi = \prod_{j \neq i} \pi_j(a_j)\pi_i(a_i)\xi = 0, \forall a_j \in A_j, 1 \leq j \leq n$. But π is nondegenerate, so $\xi = 0$. Thus π_i is nondegenerate, $1 \leq i \leq n$.

If $\{\pi'_i\}_{i=1}^n$ satisfies the same conditions as well as $\{\pi_i\}_{i=1}^n$, then we have

$$\pi'_i(a_i)\xi = \sum_k \pi(\otimes_{j \neq i} 1_j \otimes a_i \cdot u_k)\xi_k = \pi_i(a_i)\xi,$$

$\forall a_i \in A_i$ and $\xi = \sum_k \pi(u_k)\xi_k$. Since π is nondegenerate, it follows that $\pi'_i = \pi_i, 1 \leq i \leq n$, i.e., $\{\pi_i\}_{i=1}^n$ is unique.

Finally, let π be faithful. By $\pi(\otimes_{i=1}^n a_i) = \pi_1(a_1) \cdots \pi_n(a_n) (\forall a_i \in A_i, 1 \leq i \leq n)$, clearly π_i is also faithful, $1 \leq i \leq n$. Q.E.D.

In Lemma 3.3.1, if we don't assume that π is nondegenerate, then we have still a decomposition $\{\pi_i\}_{i=1}^n$ of π, but $\{\pi_i\}_{i=1}^n$ is not unique possibly.

Proposition 3.3.2. For any $u \in \otimes_{i=1}^n A_i$, we have

$$\alpha_1(u) = \sup\{\|\pi(u)\| \mid \pi \text{ is a } * \text{ representation of } \otimes_{i=1}^n A_i\}.$$

Proof. Let π_0 be a faithful $*$ representation of α_1-$\otimes_{i=1}^n A_i$. Then $\alpha_1(u) = \|\pi_0(u)\|$. Hence

$$\alpha_1(u) \leq \sup\{\|\pi(u)\| \mid \pi \text{ is a } * \text{ representation of } \otimes_{i=1}^n A_i\}.$$

By Lemma 3.3.1, $\|\pi(\cdot)\| \leq \gamma(\cdot)$ for any $*$ representation π of $\otimes_{i=1}^n A_i$, so

$$\sup\{\|\pi(\cdot)\| \mid \pi \text{ is a } * \text{ representation of } \otimes_{i=1}^n A_i\}$$

is a C^*-norm on $\otimes_{i=1}^n A_i$. Further the conclusion follows from the definition of $\alpha_1(\cdot)$, Q.E.D.

Proposition 3.3.3. For any $u \in \otimes_{i=1}^n A_i$, we have

$$\alpha_1(u) = \sup\{\alpha(u) \mid \alpha(\cdot) \text{ is a } C^*\text{-seminorm on } \otimes_{i=1}^n A_i\}.$$

Here $\alpha(\cdot)$ is called a C^*-*seminorm* on $\otimes_{i=1}^n A_i$, if $\alpha(\cdot)$ is a seminorm on $\otimes_{i=1}^n A_i$, and $\alpha(uv) \leq \alpha(u)\alpha(v), \alpha(u^*u) = \alpha(u)^2, \forall u, v \in \otimes_{i=1}^n A_i$.

Proof. Clearly, the right side $\geq$ the left side. Now let $\alpha(\cdot)$ be a C^*-seminorm on $\otimes_{i=1}^n A_i$. Then

$$I = \{v \in \otimes_{i=1}^n A_i \mid \alpha(v) = 0\}$$

is a two-sided $*$ ideal of $\otimes_{i=1}^n A_i$. Suppose that

$$v \to \tilde{v} = v + I$$

is the canonical map from $\otimes_{i=1}^n A_i$ onto $\otimes_{i=1}^n A_i/I$, and let

$$\tilde{\alpha}(\tilde{v}) = \alpha(v), \quad \forall \tilde{v} \in \otimes_{i=1}^n A_i/I, \quad v \in \tilde{v}.$$

Then $\tilde{\alpha}(\cdot)$ is a C^*-norm on $\otimes_{i=1}^n A_i/I$. Pick a faithful $*$ representation $\tilde{\pi}$ of $\tilde{\alpha}$-$(\otimes_{i=1}^n A_i/I)$, and define

$$\pi(v) = \tilde{\pi}(\tilde{v}), \quad \forall v \in \otimes_{i=1}^n A_i.$$

Then π is a $*$ representation of $\otimes_{i=1}^n A_i$. Now from Proposition 3.3.2,

$$\alpha_1(u) \geq \|\pi(u)\| = \|\tilde{\pi}(\tilde{u})\| = \tilde{\alpha}(\tilde{u}) = \alpha(u),$$

$\forall u \in \otimes_{i=1}^n A_i$. That comes to the conclusion. Q.E.D.

Proposition 3.3.4. There is a bijection from the set $\{\alpha(\cdot) \mid \alpha(\cdot) \text{ is a } C^*$ - norm on $\otimes_{i=1}^n A_i\}$ onto the set $\{I \mid I$ is a closed two-sided ideal of α_1-$\otimes_{i=1}^n A_i$, and $I \cap \otimes_{i=1}^n A_i = 0\}$ as follows:

Let $\alpha(\cdot)$ be a C^*-norm on $\otimes_{i=1}^n A_i$. Then there exists unique closed two-sided ideal I_α of α_1-$\otimes_{i=1}^n A_i$ such that the map $u \to \tilde{u} = u + I_\alpha (\forall u \in \otimes_{i=1}^n A_i)$

can be extended to a $*$ isomorphism from $\alpha\text{-}\otimes_{i=1}^n A_i$ onto $\alpha_1\text{-}\otimes_{i=1}^n A_i/I_\alpha$. In addition, this I_α satisfies the condition: $I_\alpha \cap \otimes_{i=1}^n A_i = \{0\}$. Conversely, for each closed two-sided ideal I of $\alpha_1\text{-}\otimes_{i=1}^n A_i$ with $I \cap \otimes_{i=1}^n A_i = \{0\}$, there is unique C^*-norm $\alpha(\cdot)$ on $\otimes_{i=1}^n A_i$ such that $I = I_\alpha$.

Consequently, for any C^*-norm $\alpha(\cdot)$ on $\otimes_{i=1}^n A_i$,

$$\alpha(u) = \inf\{\alpha_1(u+v) \mid v \in I_\alpha\}, \quad \forall u \in \otimes_{i=1}^n A_i.$$

Proof. Let $\alpha(\cdot)$ be a C^*-norm on $\otimes_{i=1}^n A_i$. Clearly, $\alpha(\cdot) \le \alpha_1(\cdot)$. Thus the identity map id on $\otimes_{i=1}^n A_i$ can be uniquely extended to a $*$ homomorphism from $\alpha_1\text{-}\otimes_{i=1}^n A_i$ to $\alpha\text{-}\otimes_{i=1}^n A_i$. But $id(\alpha_1\text{-}\otimes_{i=1}^n A_i)$ is a C^*-subalgebra of $\alpha\text{-}\otimes_{i=1}^n A_i$ containing $\otimes_{i=1}^n A_i$, so $id\ (\alpha_1\text{-}\otimes_{i=1}^n A_i) = \alpha\text{-}\otimes_{i=1}^n A_i$. Denote the kernel of this $*$ homomorphism by I_α. Clearly, I_α is a closed two-sided ideal of $\alpha_1\text{-}\otimes_{i=1}^n A_i$. Moreover, since $id(u) = u, \forall u \in \otimes_{i=1}^n A_i$, it follows that $I_\alpha \cap \otimes_{i=1}^n A_i = \{0\}$. Then we get a natural $*$ isomorphism from $\alpha\text{-}\otimes_{i=1}^n A_i$ onto $\alpha_1\text{-}\otimes_{i=1}^n A_i/I_\alpha$ such that $u \to \tilde{u} = u + I_\alpha, \forall u \in \otimes_{i=1}^n A_i$.

Now let I be a closed two-sided ideal of $\alpha_1\text{-}\otimes_{i=1}^n A_i$, and Φ be a $*$ isomorphism from $\alpha\text{-}\otimes_{i=1}^n A_i$ onto $\alpha_1\text{-}\otimes_{i=1}^n A_i/I$ such that $\Phi(u) = \tilde{u} = u + I, \forall u \in \otimes_{i=1}^n A_i$. Define

$$\Psi(a) = \Phi^{-1}(\tilde{a}), \quad \forall a \in \alpha_1\text{-} \otimes_{i=1}^n A_i,$$

where $\tilde{a} = a + I$ is the canonical image of $a (\in \alpha_1\text{-}\otimes_{i=1}^n A_i)$ in $\alpha_1\text{-}\otimes_{i=1}^n A_i/I$. Then Ψ is a $*$ homomorphism from $\alpha_1\text{-}\otimes_{i=1}^n A_i$ onto $\alpha\text{-}\otimes_{i=1}^n A_i$, and its kernel is I, and $\Psi(u) = u, \forall u \in \otimes_{i=1}^n A_i$. Thus Ψ is the extension of the identity map on $\otimes_{i=1}^n A_i$, and $I = \ker\Psi = I_\alpha$.

Conversely, let I be a closed two-sided ideal of $\alpha_1\text{-}\otimes_{i=1}^n A_i$, and $I \cap \otimes_{i=1}^n A_i = \{0\}$. Then $\otimes_{i=1}^n A_i$ can be embedded into $\alpha_1\text{-}\otimes_{i=1}^n A_i/I$, and the norm on $\alpha_1\text{-}\otimes_{i=1}^n A_i/I$ determines a C^*-norm $\alpha(\cdot)$ on $\otimes_{i=1}^n A_i$. Then $\alpha\text{-}\otimes_{i=1}^n A_i$ and $\alpha_1\text{-}\otimes_{i=1}^n A_i/I$ are $*$ isomorphic, and this $*$ isomorphism maps u to $\tilde{u} = u + I, \forall u \in \otimes_{i=1}^n A_i$. Then from preceding paragraph, we obtain that $I = I_\alpha$. Q.E.D.

Remark. In Section 3.9, we shall discuss the maximal C^*-norm $\alpha_1(\cdot)$ furthermore.

Notes. Lemma 3.3.1 is due to M. Takesaki. $\alpha_1\text{-}\otimes_{i=1}^n A_i$ is also called the *projective tensor product* of $A_1, \cdots, A_n$; it was considered first by A. Guichardet. The maximal C^*-norm $\alpha_1(\cdot)$ is important and is more natural than $\alpha_0(\cdot)$. But very little is known about $\alpha_1(\cdot)$. The theory of *nuclear C^*-algebras* will be important in this aspect (see Section 3.9).

References. [65], [95], [171].

3.4. States on algebraic tensor product

Let $A_1, \cdots, A_n$ be C^*-algebras, and $\otimes_{i=1}^n A_i$ be their algebraic tensor product.

Proposition 3.4.1. If φ is a positive linear functional on $\otimes_{i=1}^n A_i$, then there is a positive constant K such that

$$|\varphi(\otimes_{i=1}^n a_i)| \leq K\|a_1\| \cdots \|a_n\|, \quad \forall a_i \in A_i, 1 \leq i \leq n.$$

Proof. Fix $a_i \in (A_i)_+, 2 \leq i \leq n$. Then $\varphi(\cdot \otimes \otimes_{i=2}^n a_i)$ is a positive linear functional on A_1, so it is continuous (Proposition 2.3.2). Similarly, $\varphi(\otimes_{i=1}^n a_i)$ is continuous for each variable. Now the result follows from the principle of uniform boundedness. Q.E.D.

Proposition 3.4.2. Let $A_i^{(1)} = A_i \dot{+} \mathbb{C} 1_i, 1 \leq i \leq n$, and φ be a positive linear functional on $\otimes_{i=1}^n A_i$. Then φ can be extended to a positive linear functional $\tilde{\varphi}$ on $\otimes_{i=1}^n A_i^{(1)}$ such that for any subset $\mathbb{I}$ of $\{1, \cdots, n\}$ and $a_i \in A_i, i \notin \mathbb{I}$,

$$\tilde{\varphi}(\otimes_{i\in\mathbb{I}} 1_i \otimes_{i\notin\mathbb{I}} a_i) = \lim_{l_i, i\in\mathbb{I}} \varphi(\otimes_{i\in\mathbb{I}} d_{l_i}^{(i)} \otimes \otimes_{i\notin\mathbb{I}} a_i),$$

where $\{d_{l_i}^{(i)}\}$ is an approximate identify for $A_i, 1 \leq i \leq n$.

Proof. For any $a_i \in A_i, 2 \leq i \leq n$, by Proposition 2.4.4 the following two limits exist and are equal:

$$\lim_{l_1} \varphi(d_{l_1}^{(1)} \otimes \otimes_{i=2}^n a_i) = \lim_{l_1} \varphi(d_{l_1}^{(1)2} \otimes \otimes_{i=2}^n a_i).$$

Now φ can be extended to a linear functional $\tilde{\varphi}$ on $A_1^{(1)} \otimes \otimes_{i=2}^n A_i$ such that

$$\tilde{\varphi}(1_1 \otimes \otimes_{i=2}^n a_i) = \lim_{l_1} \varphi(d_{l_1}^{(1)} \otimes \otimes_{i=2}^n a_i),$$

$\forall a_i \in A_i, 2 \leq i \leq n$. For any $x = \sum_j (\lambda_j 1_1 + a_j^{(1)}) \otimes \otimes_{i=2}^n a_j^{(i)} \in A_1^{(1)} \otimes \otimes_{i=2}^n A_i$, where $a_j^{(i)} \in A_i, \lambda_j \in \mathbb{C}, \forall i, j$,

$$\begin{aligned} \tilde{\varphi}(x^* x) = & \ \varphi(\sum_{j,k}(\overline{\lambda}_j a_k^{(1)} + \lambda_k a_j^{(1)*} + a_j^{(1)*} a_k^{(1)}) \otimes \otimes_{i=2}^n a_j^{(i)*} a_k^{(i)}) \\ & + \lim_{l_1} \varphi(\sum_{j,k} d_{l_1}^{(1)2} \otimes \otimes_{i=2}^n \overline{\lambda}_j \lambda_k a_j^{(i)*} a_k^{(i)}) \\ = & \ \lim_{l_1} \varphi(y_{l_1}^* y_{l_1}) \geq 0, \end{aligned}$$

where $y_{l_1} = \sum_j (\lambda_j d_{l_1}^{(1)} + a_j^{(1)}) \otimes \otimes_{i=2}^n a_j^{(i)}$. In this way, we can obtain a positive linear functional $\tilde{\varphi}$ on $\otimes_{i=1}^n A_i^{(1)}$ satisfying the conditions. Q.E.D.

Proposition 3.4.3. Let φ be a positive linear functional on $\otimes_{i=1}^n A_i$, and $\{d_{l_i}^{(i)}\}$ be an approximate identity for $A_i, 1 \le i \le n$. Then

$$\begin{aligned}\lim_{l_1,\cdots,l_n} \varphi(\otimes_{i=1}^n d_{l_i}^{(i)}) &= \sup\{\varphi(\otimes_{i=1}^n a_i) \mid a_i \in (A_i)_+, \|a_i\| \le 1, 1 \le i \le n\} \\ &= \sup\{|\varphi(\otimes_{i=1}^n a_i)| \,\big|\, a_i \in A_i, \|a_i\| \le 1, 1 \le i \le n\}.\end{aligned}$$

Proof. Let $\tilde{\varphi}$ be an extension of φ as in Proposition 3.4.2.. Then

$$\lim_{l_1,\cdots,l_n} \varphi(\otimes_{i=1}^n d_{l_i}^{(i)}) = \tilde{\varphi}(\otimes_{i=1}^n 1_i).$$

If $a_i \in (A_i)_+, \|a_i\| \le 1, 1 \le i \le n$, then we have

$$\otimes_{i=1}^n 1_i \ge a_1 \otimes_{i=2}^n 1_i \ge \cdots \ge \otimes_{i=1}^n a_i,$$

in $\otimes_{i=1}^n A_i^{(1)}$. Thus $\tilde{\varphi}(\otimes_{i=1}^n 1_i) \ge \varphi(\otimes_{i=1}^n a_i)$. Since $d_{l_i}^{(i)} \in (A_i)_+, \|d_{l_i}^{(i)}\| \le 1, \forall l_i, 1 \le i \le n$, it follows that

$$\lim_{l_1,\cdots,l_n} \varphi\left(\otimes_{i=1}^n d_{l_i}^{(i)}\right) = \sup\left\{\varphi\left(\otimes_{i=1}^n a_i\right) \mid a_i \in (A_i)_+, \|a_i\| \le 1, 1 \le i \le n\right\}.$$

Moreover, for any $a_i \in A_i, \|a_i\| \le 1, 1 \le i \le n$, it follows that

$$\begin{aligned}|\varphi(\otimes_{i=1}^n a_i)|^2 &= |\tilde{\varphi}(\otimes_{i=1}^n a_i)|^2 \\ &\le \tilde{\varphi}(\otimes_{i=1}^n 1_i)\,\varphi(\otimes_{i=1}^n a_i^* a_i) \\ &\le \sup\left\{\varphi\left(\otimes_{i=1}^n b_i\right)^2 \Big|\, b_i \in (A_i)_+, \|b_i\| \le 1, 1 \le i \le n\right\}\end{aligned}$$

by the Schwartz inequality. That comes to the conclusion. Q.E.D.

Definition 3.4.4. A positive linear functional φ on $\otimes_{i=1}^n A_i$ is called a *state*, if

$$\sup\left\{\varphi\left(\otimes_{i=1}^n a_i\right) | a_i \in (A_i)_+, \|a_i\| \le 1, 1 \le i \le n\right\} = 1.$$

Denote by $S(\otimes_{i=1}^n A_i)$ the set of states on $\otimes_{i=1}^n A_i$.

Proposition 3.4.5. Let φ be a state on $\otimes_{i=1}^n A_i$. Then there exists unique state $\tilde{\varphi}$ on $\otimes_{i=1}^n A_i^{(1)}$ which is an extension of φ, where $A_i^{(1)} = A_i \dot{+} \mathbb{C} 1_i, 1 \le i \le n$.

Proof. The existence of $\tilde{\varphi}$ follows from Proposition 3.4.2. Now let ψ be a state on $\otimes_{i=1}^n A_i^{(1)}$, and also be an extension of φ. Then

$$\psi(1_1 \otimes \otimes_{i=2}^n a_i) \geq \psi(d_l \otimes \otimes_{i=2}^n a_i)) = \varphi(d_l \otimes \otimes_{i=2}^n a_i), \quad \forall a_i \in (A_i)_+, \quad 2 \leq i \leq n,$$

where $\{d_l\}$ is an approximate identity for A_1. By Proposition 3.4.2,

$$\psi(1_1 \otimes \otimes_{i=2}^n a_i) \geq \tilde{\varphi}(1_1 \otimes \otimes_{i=2}^n a_i), \quad \forall a_i \in (A_i)_+, \quad 2 \leq i \leq n.$$

Suppose that there is a $a_i \in (A_i)_+$ with $\|a_i\| < 1, 2 \leq i \leq n$, such that

$$\psi(1_1 \otimes \otimes_{i=2}^n a_i) - \tilde{\varphi}(1_1 \otimes \otimes_{i=2}^n a_i) = \delta > 0.$$

Then for $b_i \in (A_i)_+, \|b_i\| < 1, b_i \geq a_i, 2 \leq i \leq n$,

$$\begin{aligned}
& \psi(1_1 \otimes \otimes_{i=2}^n b_i) - \tilde{\varphi}(1_1 \otimes \otimes_{i=2}^n b_i) \\
= \ & [\psi(1_1 \otimes (b_2 - a_2) \otimes \otimes_{i=3}^n b_i) - \tilde{\varphi}(1_1 \otimes (b_2 - a_2) \otimes \otimes_{i=3}^n b_i)] \\
& + [\psi(1_1 \otimes a_2 \otimes (b_3 - a_3) \otimes \otimes_{i=4}^n b_i) - \tilde{\varphi}(1_1 \otimes a_2 \otimes (b_3 - a_3) \otimes \otimes_{i=4}^n b_i)] \\
& + \cdots + [\psi(1_1 \otimes \otimes_{i=2}^n a_i) - \tilde{\varphi}(1_1 \otimes \otimes_{i=2}^n a_i)] \geq \delta > 0,
\end{aligned}$$

i.e.,

$$1 \geq \psi(1_1 \otimes \otimes_{i=2}^n b_i) \geq \tilde{\varphi}(1_1 \otimes \otimes_{i=2}^n b_i) + \delta.$$

Further,

$$1 \geq \sup\{\tilde{\varphi}(1_1 \otimes \otimes_{i=2}^n b_i) | \ b_i \in (A_i)_+, b_i \geq a_i, \|b_i\| < 1, 2 \leq i \leq n\} + \delta.$$

But by Proposition 2.11 and Definition 3.4.4,

$$\sup\{\tilde{\varphi}(1_1 \otimes \otimes_{i=2}^n b_i) | a_i \leq b_i \in A_i, \|b_i\| < 1, 2 \leq i \leq n\} = 1.$$

We get a contradiction. Thus ψ and $\tilde{\varphi}$ are equal on $A_1^{(1)} \otimes \otimes_{i=2}^n A_i$. By this procedure, we can see that $\psi = \tilde{\varphi}$. Q.E.D.

Proposition 3.4.6. Let φ be a state on $\otimes_{i=1}^n A_i$. Then φ can be uniquely extended to a state on α_1-$\otimes_{i=1}^n A_i$. Consequently,

$$\alpha_1(u)^2 = \sup\{\varphi(u^* u) | \ \varphi \in S(\otimes_{i=1}^n A_i)\}, \forall u \in \otimes_{i=1}^n A_i.$$

Proof. By Proposition 3.4.5, φ can be uniquely extended to a state $\tilde{\varphi}$ on $\otimes_{i=1}^n A_i^{(1)}$, where $A_i^{(1)} = A_i \dot{+} \mathbb{C} 1_i, 1 \leq i \leq n$. Let

$$L = \{x \in \otimes_{i=1}^n A_i^{(1)} | \ \tilde{\varphi}(x^* x) = 0\}.$$

It is a left ideal of $\otimes_{i=1}^n A_i^{(1)}$ by the Schwartz inequality. Suppose that $x \to \tilde{x} = x + L$ is the canonical map from $\otimes_{i=1}^n A_i^{(1)}$ onto $\otimes_{i=1}^n A_i^{(1)}/L$. Define an inner product on the quotient space:

$$\langle \tilde{x}, \tilde{y} \rangle = \tilde{\varphi}(y^* x), \quad \forall x, y \in \otimes_{i=1}^n A_i^{(1)}.$$

Denote the completion of $(\otimes_{i=1}^n A_i^{(1)}/L, \langle , \rangle)$ by H. For any $x \in \otimes_{i=1}^n A_i^{(1)}$, let

$$\pi(x)\tilde{y} = \widetilde{xy}, \quad \forall y \in \otimes_{i=1}^n A_i^{(1)}.$$

Since $\otimes_{i=1}^n x_i^* x_i \leq \prod_{i=1}^n \|x_i\|^2 \otimes_{i=1}^n 1_i$ in $\otimes_{i=1}^n A_i^{(1)}$, it follows that

$$\|\pi(\otimes_{i=1}^n x_i)\tilde{y}\|^2 = \tilde{\varphi}(y^* \cdot \otimes_{i=1}^n x_i^* x_i \cdot y) \leq \prod_{i=1}^n \|x_i\|^2 \cdot \|\tilde{y}\|^2,$$

$\forall y \in \otimes_{i=1}^n A_i^{(1)}$. Thus $\pi(\otimes_{i=1}^n x_i)$ can be uniquely extended to a bounded linear operator on $H, \forall x_i \in A_i^{(1)}, 1 \leq i \leq n$. Consequently, we obtain a $*$ representation $\{\pi, H\}$ of $\otimes_{i=1}^n A_i^{(1)}$ such that

$$\tilde{\varphi}(x) = \langle \pi(x)\tilde{1}, \tilde{1} \rangle, \quad \forall x \in \otimes_{i=1}^n A_i^{(1)},$$

where $\tilde{1} = \otimes_{i=1}^n 1_i + L = \widetilde{\otimes_{i=1}^n 1_i}$. Furthermore, by Proposition 3.3.2

$$|\varphi(u)| = |\tilde{\varphi}(u)| \leq \|\pi(u)\| \leq \alpha_1(u), \quad \forall u \in \otimes_{i=1}^n A_i^{(1)}.$$

Therefore, φ can be uniquely extended to a state on α_1-$\otimes_{i=1}^n A_i^{(1)}$.

From Proposition 3.2.3, $(\psi| \otimes_{i=1}^n A_i)$ is a state on $\otimes_{i=1}^n A_i$ for any state ψ on α_1-$\otimes_{i=1}^n A_i$. Then by the preceding paragraph, we have

$$\alpha_1(u)^2 = \sup\{\varphi(u^* u) | \ \varphi \in S(\otimes_{i=1}^n A_i)\}, \quad \forall u \in \otimes_{i=1}^n A_i.$$

Q.E.D.

From Proposition 3.4.6 and 3.2.3, $\psi \to (\psi| \otimes_{i=1}^n A_i)$ is a bijection from $S(\alpha_1$-$\otimes_{i=1}^n A_i)$ onto $S((\otimes_{i=1}^n A_i)$.

Proposition 3.4.7. Let φ be a state on $\otimes_{i=1}^n A_i$. Then by the GNS construction, we can get a cyclic $*$ representation $\{\pi, H, \xi\}$ of $\otimes_{i=1}^n A_i$ such that

$$\varphi(u) = \langle \pi(u)\xi, \xi \rangle, \quad \forall u \in \otimes_{i=1}^n A_i.$$

Proof. By Proposition 3.4.6, φ can be uniquely extended to a state on α_1-$\otimes_{i=1}^n A_i$, which is still denoted by φ. Let

$$L = \{u \in \otimes_{i=1}^n A_i | \ \varphi(u^* u) = 0\}$$

and

$$L_\varphi = \{a \in \alpha_1\text{-} \otimes_{i=1}^n A_i |\ \varphi(a^* a) = 0\}.$$

Clearly, $L \subset L_\varphi$. Suppose that $u \to \tilde{u} = u + L$ and $a \to a_\varphi = a + L_\varphi$ are the canonical maps from $\otimes_{i=1}^n A_i$ onto $\otimes_{i=1}^n A_i / L$ and from $\alpha_1\text{-}\otimes_{i=1}^n A_i$ onto $\alpha_1\text{-}\otimes_{i=1}^n A_i^{(1)} / L_\varphi$ respectively. Then we have the following diagram:

$$\begin{array}{ccccc} & & \tilde{u} \in A_i/L & \longrightarrow & H \\ & \nearrow & & & \\ u \in \otimes A_i & & & & \downarrow U \\ & \searrow & & & \\ & & u_\varphi \in \alpha_1\text{-} \otimes A_i / L_\varphi & \longrightarrow & H_\varphi \end{array}$$

where $\{\pi, H\}$ is the $*$ representation of $\otimes_{i=1}^n A_i$ generated by φ (the GNS construction, $H = (\otimes_{i=1}^n A_i/L, \langle,\rangle)^-$, see the proof of Proposition 3.4.6), and $\{\pi_\varphi, H_\varphi, \xi_\varphi\}$ is the cyclic $*$ representation of $\alpha_1\text{-}\otimes_{i=1}^n A_i$ generated by φ. It is easily verified that U can be uniquely extended to a unitary operator from H onto H_φ, and

$$\pi(u) = U^{-1}\pi_\varphi(u)U, \quad \forall u \in \otimes_{i=1}^n A_i.$$

Now let $\xi = U^{-1}\xi_\varphi$. Then $\{\pi, H, \xi\}$ satisfies our conditions. Q.E.D.

Proposition 3.4.8. Let $\alpha(\cdot)$ be a C^*-norm on $\otimes_{i=1}^n A_i$, and $\Gamma = \{\varphi \in S(\otimes_{i=1}^n A_i) \mid \varphi$ is continuous with respect to $\alpha(\cdot)\}$. Then for any $u \in \otimes_{i=1}^n A_i$,

$$\alpha(u) = \sup\{\varphi(u^* u)^{1/2} | \varphi \in \Gamma\} = \sup\{\|\pi_\varphi(u)\| | \varphi \in \Gamma\},$$

where $\{\pi_\varphi, H_\varphi, \xi_\varphi\}$ is the cyclic $*$ representation of $\otimes_{i=1}^n A_i$ generated by $\varphi, \forall \varphi \in \Gamma$.

Proof. By Proposition 3.2.3, $\Gamma = \{(\psi | \otimes_{i=1}^n A_i^{(1)}) | \psi \in S(\alpha\text{-} \otimes_{i=1}^n A_i)\}$. So we can get the first equality. Replacing $\alpha_1(\cdot)$ by $\alpha(\cdot)$ in the proof of Proposition 3.4.7, we get $\|\pi_\varphi(u)\| \leq \alpha(u), \forall \varphi \in \Gamma$. However, $\varphi(u^* u) = \langle \pi_\varphi(u^* u)\}\xi_\varphi, \xi_\varphi\rangle = \|\pi_\varphi(u)\xi_\varphi\|^2 \leq \|\pi_\varphi(u)\|^2, \forall \varphi \in \Gamma$. Therefore, $\alpha(u) = \sup\{\|\pi_\varphi(u)\|\ \varphi \in \Gamma\}$.
Q.E.D.

References. [37], [95], [96].

3.5. The inequality $\lambda(\cdot) \leq \alpha_0(\cdot) \leq \alpha(\cdot) \leq \gamma(\cdot)$

Lemma 3.5.1. Let $A_1, \cdots, A_n$ be C^*-algebras, and A_n be with no identity. If $x \in \otimes_{i=1}^{n-1} A_i \otimes A_n^{(1)}$, where $A_n^{(1)} = A_n \dot{+} \mathbb{C} 1_n$, such that $xv = 0, \forall v \in \otimes_{i=1}^n A_i$, then $x = 0$.

Proof. We may assume that $A_i \subset B(H_i)$ and A_i is nondegenerate on $H_i, 1 \leq i \leq n$. Since A_n has no identity, we may also assume that 1_n is the identity operator on H_n. Then

$$\langle xv \otimes_{i=1}^n \xi_i, \otimes_{i=1}^n \eta_i \rangle = 0, \quad \forall v \in \otimes_{i=1}^n A_i, \quad \xi_i, \eta_i \in H_i, \quad 1 \leq i \leq n.$$

Thus x is a zero operator on $\otimes_{i=1}^n H_i$. By Theorem 3.2.10, $\|x\| = 0 = \alpha_0(x)$. Therefore, $x = 0$. Q.E.D.

Proposition 3.5.2. Let $A_1, \cdots, A_n$ be C^*-algebras, and A_n be with no identity. Suppose that $\alpha(\cdot)$ is a C^*-norm on $\otimes_{i=1}^n A_i$. Define

$$\tilde{\alpha}(x) = \sup\{\alpha(xu) | u \in \otimes_{i=1}^n A_i, \alpha(u) \leq 1\}, \quad \forall x \in \otimes_{i=1}^{n-1} A_i \otimes A_n^{(1)},$$

where $A_n^{(1)} = A_n \dot{+} \mathbb{C} 1_n$. Then $\tilde{\alpha}(\cdot)$ is a C^*-norm on $\otimes_{i=1}^{n-1} A_i \otimes A_n^{(1)}$, and is an extension of $\alpha(\cdot)$. Moreover, if $\{d_{l_i}^{(i)}\}$ is an approximate identity for $A_i, 1 \leq i \leq n$, then

$$\tilde{\alpha}(x) = \lim_{l_1, \cdots, l_n} \alpha(x \cdot \otimes_{i=1}^n d_{l_i}^{(i)}), \quad \forall x \in \otimes_{i=1}^{n-1} A_i \otimes A_n^{(1)}.$$

Proof. By Proposition 3.2.2, $\alpha(\otimes_{i=1}^n a_i)$ is continuous for each variable a_i in $A_i, 1 \leq i \leq n$. Then

$$\alpha(xv) = \lim_{l_1, \cdots, l_n} \alpha(x \cdot \otimes_{i=1}^n d_{l_i}^{(i)} \cdot v),$$

$\forall x \in \otimes_{i=1}^{n-1} A_i \otimes A_n^{(1)}, v \in \otimes_{i=1}^n A_i$. For any $v \in \otimes_{i=1}^n A_i$ with $\alpha(v) \leq 1$ and any $\varepsilon > 0$, if $(l_1, \cdots, l_n)$ is sufficiently later, then we have

$$\begin{aligned} \alpha(xv) \leq & \ \alpha(x \cdot \otimes_{i=1}^n d_{l_i}^{(i)} \cdot v) + \varepsilon \leq \alpha(x \cdot \otimes_{i=1}^n d_{l_i}^{(i)}) + \varepsilon \\ \leq & \ \sup\{\alpha(xu) | \ u \in \otimes_{i=1}^n A_i, \alpha(u) \leq 1\} + \varepsilon. \end{aligned}$$

Therefore,

$$\begin{aligned} \tilde{\alpha}(x) = & \ \sup\{\alpha(xu) | \ u \in \otimes_{i=1}^n A_i, \alpha(u) \leq 1\} \\ = & \ \lim_{l_1, \cdots, l_n} \alpha(x \cdot \otimes_{i=1}^n d_{l_i}^{(i)}), \quad \forall x \in \otimes_{i=1}^{n-1} A_i \otimes A_n^{(1)}. \end{aligned}$$

Consequently, $\tilde{\alpha}(\cdot)$ is an extension of $\alpha(\cdot)$.

Similarly, from $\tilde{\alpha}(x^*)$ $=\sup\{\alpha(ux) | u \in \otimes_{i=1}^n A_i, \alpha(u) \leq 1\}$ and $\alpha(ux) \leq \alpha(ux \cdot \otimes_{i=1}^n d_{l_i}^{(i)}) + \varepsilon \leq \alpha(x \cdot \otimes_{i=1}^n d_{l_i}^{(i)}) + \varepsilon$ we can see that

$$\tilde{\alpha}(x^*) = \tilde{\alpha}(x), \quad \forall x \in \otimes_{i=1}^{n-1} A_i \otimes A_n^{(1)}.$$

By Lemma 3.5.1, $\tilde{\alpha}(\cdot)$ is a norm on $\otimes_{i=1}^{n-1} A_i \otimes A_n^{(1)}$. Moreover, since for any $u \in \otimes_{i=1}^{n} A_i$ with $\alpha(u) \leq 1$

$$\alpha(xyu) = \lim_{l_1,\cdots,l_n} \alpha(x \cdot \otimes_{i=1}^{n} d_{l_i}^{(i)} \cdot yu) \leq \tilde{\alpha}(x)\tilde{\alpha}(y),$$

it follows that $\tilde{\alpha}(xy) \leq \tilde{\alpha}(x)\tilde{\alpha}(y), \forall x, y \in \otimes_{i=1}^{n-1} A_i \otimes A_n^{(1)}$. Noticing that

$$\begin{aligned} \tilde{\alpha}(x)^2 = \ & \tilde{\alpha}(x^*)\tilde{\alpha}(x) \geq \tilde{\alpha}(x^*x) = \sup\{\alpha(x^*xu) | u \in \otimes_{i=1}^{n} A_i, \alpha(u) \leq 1\} \\ \geq \ & \sup\{\alpha(u^*x^*xu) | u \in \otimes_{i=1}^{n} A_i, \alpha(u) \leq 1\} = \tilde{\alpha}(x)^2, \end{aligned}$$

we get $\tilde{\alpha}(x^*x) = \tilde{\alpha}(x)^2, \forall x \in \otimes_{i=1}^{n-1} A_i \otimes A_n^{(1)}$. So $\tilde{\alpha}(\cdot)$ is a C^*-norm on $\otimes_{i=1}^{n-1} A_i \otimes A_n^{(1)}$. Q.E.D.

Proposition 3.5.3. Let $A_i \cong C(\Omega_i)$, where Ω_i is a compact Hausdorff space, $1 \leq i \leq n$. Then there is only one C^*-norm $\alpha_0(\cdot)$ on $\otimes_{i=1}^{n} A_i$, and $\alpha_0(\cdot) = \lambda(\cdot)$, and $\lambda\text{-}\otimes_{i=1}^{n} A_i \cong C(\Omega_1 \times \cdots \times \Omega_n)$.

Proof. Let $\alpha(\cdot)$ be any C^*-norm on $\otimes_{i=1}^{n} A_i$. Then $\alpha\text{-}\otimes_{i=1}^{n} A_i$ is an abelian C^*-algebra with an identity. Suppose that Ω is the spectral space of $\alpha\text{-}\otimes_{i=1}^{n} A_i$. If $\varphi \in \Omega$, then $\varphi(\otimes_{j\neq i} 1_j \otimes \cdot)$ is a non-zero multiplicative linear functional on A_i(where 1_j is the identity of A_j), denoted by $\varphi_i(\in \Omega_i), 1 \leq i \leq n$. Further, $\varphi(u) = \otimes_{i=1}^{n} \varphi_i(u)$, $\forall u \in \otimes_{i=1}^{n} A_i$. So we can view Ω as a closed subset of $\Omega_1 \times \cdots \times \Omega_n$. If $\Omega \neq \Omega_1 \times \cdots \Omega_n$, then there is a non-empty open subset U_i of $\Omega_i, 1 \leq i \leq n$, such that $(U_1 \times \cdots \times U_n) \cap \Omega = \emptyset$. Pick $0 \neq a_i \in A_i$ such that

$$\text{supp} a_i(\cdot) \subset U_i, \qquad 1 \leq i \leq n.$$

Then $\varphi(\otimes_{i=1}^{n} a_i) = 0, \forall \varphi \in \Omega$. This is a contradiction since $\otimes_{i=1}^{n} a_i \neq 0$ as an element of $\alpha\text{-}\otimes_{i=1}^{n} A_i$. Therefore, $\Omega = \Omega_1 \times \cdots \times \Omega_n$, i.e.,

$$\alpha\text{-} \otimes_{i=1}^{n} A_i \cong C(\Omega_1 \times \cdots \times \Omega_n),$$

and so $\alpha(\cdot) = \alpha_0(\cdot)$. Moreover, for any $u \in \otimes_{i=1}^{n} A_i$,

$$\begin{aligned} \alpha_0(u) = \ & \sup\{| \otimes_{i=1}^{n} \varphi_i(u)| \mid \varphi_i \in \Omega_i, 1 \leq i \leq n\} \\ = \ & \sup\{| \otimes_{i=1}^{n} f_i(u)| \mid f_i \in A_i^*, \|f_i\| \leq 1, 1 \leq i \leq n\} \\ = \ & \lambda(u). \end{aligned}$$

By Theorem 3.2.6, $\alpha_0(\cdot) = \lambda(\cdot)$ on $\otimes_{i=1}^{n} A_i$. Q.E.D.

Proposition 3.5.4. Let A_i be a C^*-algebra with an identity $1_i, 1 \leq i \leq n$, and $\alpha(\cdot)$ be a C^*-norm on $\otimes_{i=1}^{n} A_i$. Suppose that $\beta(\cdot)$ is the restriction of $\alpha(\cdot)$ on $1_1 \otimes \otimes_{i=2}^{n} A_i$. If φ is a state on $\alpha\text{-}\otimes_{i=1}^{n} A_i$ such that $\chi(\cdot) = \varphi(\cdot \otimes \otimes_{i=2}^{n} 1_i)$

is a pure state on A_1, then there is unique state ψ on $\beta\text{-}\otimes_{i=2}^n A_i$ such that $\varphi = \chi \otimes \psi$ on $\otimes_{i=1}^n A_i$.

Proof. Define $\psi(v) = \varphi(1_1 \otimes v), \forall v \in \otimes_{i=2}^n A_i$. Clearly ψ is a state on $\otimes_{i=2}^n A_i$, and $|\psi(v)| \leq \beta(v), \forall v \in \otimes_{i=1}^n A_i$. Thus ψ can be uiniquely extended to a state on $\beta\text{-}\otimes_{i=2}^n A_i$. Now it suffices to prove that

$$\varphi(\otimes_{i=1}^n a_i) = \chi(a_1)\psi(\otimes_{i=2}^n a_i), \quad \forall 0 \neq a_i \in (A_i)_+, \quad 1 \leq i \leq n.$$

Fix $0 \neq a_i \in (A_i)_+, 1 \leq i \leq n$.

If $\psi(\otimes_{i=2}^n a_i) = 0$, then by the Schwartz inequality we have

$$\begin{aligned} 0 \leq\ & \varphi(\otimes_{i=1}^n a_i) = \varphi(a_1 \otimes \otimes_{i=2}^n a_i^{1/2} \cdot 1_1 \otimes \otimes_{i=2}^n a_i^{1/2}) \\ \leq\ & \varphi(a_1^2 \otimes \otimes_{i=2}^n a_i)^{1/2} \cdot \psi(\otimes_{i=2}^n a_i) = 0. \end{aligned}$$

Thus $\varphi(\otimes_{i=1}^n a_i) = \chi(a_1)\psi(\otimes_{i=2}^n a_i) = 0$.

If $\psi(\otimes_{i=2}^n a_i) = \|a_2\| \cdots \|a_n\|$, let $v = \prod_{i=2}^n \|a_i\| \otimes_{i=2}^n 1_i - \otimes_{i=2}^n a_i$, then $v \in (\otimes_{i=2}^n A_i)_+$, and $\varphi(a_1 \otimes v) = 0$ since $0 \leq \varphi(a_1 \otimes v) \leq \|a_1\|\varphi(1_1 \otimes v) = \|a_1\|\psi(v) = 0$. Thus

$$\varphi(\otimes_{i=1}^n a_i) = \varphi(a_1 \otimes \otimes_{i=2}^n 1_i) \cdot \prod_{i=2}^n \|a_i\| = \chi(a_1)\psi(\otimes_{i=2}^n a_i).$$

Now suppose that $0 < \lambda = \psi(\otimes_{i=2}^n a_i) < \|a_2\| \cdots \|a_n\|$. Let $\mu = \|a_2\| \cdots \| a_n\| - \lambda, \rho_1(\cdot) = \lambda^{-1}\varphi(\cdot \otimes \otimes_{i=2}^n a_i)$, and $\rho_2(\cdot) = \mu^{-1}[\prod_{i=2}^n \|a_i\|\varphi(\cdot \otimes \otimes_{i=2}^n 1_i) - \varphi(\cdot \otimes \otimes_{i=2}^n a_i)]$. Then ρ_1 and ρ_2 are two states on A_1, and

$$\chi(\cdot) = (\prod_{i=2}^n \|a_i\|)^{-1}\lambda\rho_1(\cdot) + (\prod_{i=2}^n \|a_i\|)^{-1}\mu\rho_2(\cdot)$$

on A_1. Since $\chi(\cdot)$ is a pure state on A_1, it follows that $\chi = \rho_1 = \rho_2$. Therefore, we obtain

$$\varphi(\otimes_{i=1}^n a_i) = \lambda\rho_1(a_1) = \lambda\chi(a_1) = \chi(a_1)\psi(\otimes_{i=2}^n a_i).$$

Q.E.D.

Corollary 3.5.5. Let A_i be a C^*-algebra with an identity $1_i, 1 \leq i \leq n$, and φ be a state on $\otimes_{i=1}^n A_i$. If $\chi_i(\cdot) = \varphi(\cdot \otimes \otimes_{j \neq i} 1_j)$ is a pure state on $A_i, 1 \leq i \leq k$ (some $k \leq n$), then $\varphi = \chi_1 \otimes \cdots \otimes \chi_k \otimes \psi$, where ψ is a state on $\otimes_{k+1}^n A_i$. Moreover, if φ is continuous with respect to a C^*-norm $\alpha(\cdot)$ on $\otimes_{i=1}^n A_i$, then ψ is continuous with respect to the C^*-norm $\beta(\cdot)$ on $\otimes_{k+1}^n A_i$, where $\beta(v) = \alpha(\otimes_{i=1}^k 1_i \otimes v), \forall v \in \otimes_{k+1}^n A_i$.

Proposition 3.5.6. Let A_i be a C^*-algebra with an identity $1_i, 1 \leq i \leq n$,and $A_1 \cdots, A_k$ be abelian (some $k \leq n$). Suppose that $\alpha(\cdot)$ is a C^*-norm on $\otimes_{i=1}^n A_i$ and φ is a pure state on α-$\otimes_{i=1}^n A_i$. Then

$$\varphi = \chi_1 \otimes \cdots \otimes \chi_k \otimes \psi,$$

where $\chi_i(\cdot) = \varphi(\cdot \otimes \otimes_{j\neq i} 1_j)$ is a pure state on $A_i, 1 \leq i \leq k$, and $\psi(v) = \varphi(\otimes_{i=1}^k 1_i \otimes v)(\forall v \in \otimes_{k+1}^n A_i)$ can be uniquely extended to a pure state on β-$\otimes_{k+1}^n A_i$, and $\beta(v) = \alpha(\otimes_{i=1}^k 1_i \otimes v), \forall v \in \otimes_{k+1}^n A_i$.

Proof. Let $\{\pi, H, \xi\}$ be the irreducible cyclic $*$ representation of $A = \alpha$-$\otimes_{i=1}^n A_i$ generated by φ. Since $\pi(A)' = \mathbb{C} 1_H$ and A_i is abelian $(1 \leq i \leq k)$, it follows that there is a pure state (non-zero multiplicative linear functional) χ_i on A_i such that

$$\pi(\cdot \otimes \otimes_{j\neq i}^n 1_j) = \chi_i(\cdot) 1_H, \quad \forall \cdot \in A_i,$$

$1 \leq i \leq k$. By Corollary 3.5.5 and $\chi_i(\cdot) = \langle \pi(\cdot \otimes \otimes_{j\neq i} 1_j)\xi, \xi\rangle = \varphi(\cdot \otimes \otimes_{j\neq i} 1_j), 1 \leq i \leq k$, we obtain $\varphi = \chi_1 \otimes \cdots \otimes \chi_k \otimes \psi$, where ψ is a state on $\otimes_{k+1}^n A_i$, and is continuous with respect to $\beta(\cdot)$.

Now it suffices to prove that ψ is a pure state on β-$\otimes_{k+1}^n A_i$. Let ψ_1, ψ_2 be two states on β-$\otimes_{k+1}^n A_i$ and $\lambda \in (0,1)$ such that

$$\psi = \lambda\psi_1 + (1-\lambda)\psi_2.$$

Then for any $u \in \otimes_{i=1}^n A_i$ and $j = 1, 2$,

$$|\chi_1 \otimes \cdots \otimes \chi_k \otimes \psi_j(u)|^2 \leq \chi \otimes \cdots \otimes \chi_k \otimes \psi_j(u^* u)$$

$$\leq \left(\frac{1}{\lambda} + \frac{1}{1-\lambda}\right)\varphi(u^* u) \leq \left(\frac{1}{\lambda} + \frac{1}{1-\lambda}\right)\alpha(u)^2.$$

Hence $\chi_1 \otimes \cdots \otimes \chi_k \otimes \psi_j$ can be uniquely extended to a state φ_j on α-$\otimes_{i=1}^n A_i, j = 1, 2$. Clearly, $\varphi = \lambda\varphi_1 + (1-\lambda)\varphi_2$. Since φ is a pure state on α-$\otimes_{i=1}^n A_i$, it follows that $\varphi = \varphi_1 = \varphi_2$. Furthermore, $\psi = \psi_1 = \psi_2$, i.e., ψ is a pure state on β-$\otimes_{k+1}^n A_i$. Q.E.D.

Lemma 3.5.7. Let A be a C^*-algebra with an identity, and $S(A)$ be the state space of A. Suppose that E is a $\sigma(A^*, A)$-compact convex subset of $S(A)$ such that for any $h^* = h \in A$ there is $\varphi \in E$ with $\varphi(h) = \max\{\lambda \mid \lambda \in \sigma(h)\}$. Then $E = S(A)$.

Proof. Suppose that $\rho \in S(A) \backslash E$. By separation theorem, there is $h^* = h \in A$ such that

$$\rho(h) > \sup\{\varphi(h) \mid \varphi \in E\}.$$

Clearly, $\rho(h) \leq \max\{\lambda | \lambda \in \sigma(h)\}$. However, by the assumption $\sup\{\varphi(h) \mid \varphi \in E\} = \max\{\lambda \mid \lambda \in \sigma(h)\}$. This is a contradiction. Therefore, $E = S(A)$.

Q.E.D.

Lemma 3.5.8. Let $A_1, \cdots .A_n$ be C^*-algebras, and $\alpha(\cdot)$ be a C^*-norm on $\otimes_{i=1}^n A_i$. Then $\alpha(\cdot) \geq \alpha_0(\cdot)$ on $\otimes_{i=1}^n A_i$ if and only if for each $\varphi_i \in S_i$ (the state space of A_i), $1 \leq i \leq n, \otimes_{i=1}^n \varphi_i$ is continuous with respect to $\alpha(\cdot)$.

Proof. The necessity is clear from Corollary 3.2.7. Conversely, if $\otimes_{i=1}^n \varphi_i$ is continuous with respect to $\alpha(\cdot), \forall \varphi_i \in S_i, 1 \leq i \leq n$, then

$$\otimes_{i=1}^n \varphi_i(v^*(\alpha(u^*u) - u^*u)v) \geq 0, \quad \forall u, v \in \otimes_{i=1}^n A_i,$$

$\forall \varphi_i \in S_i, 1 \leq i \leq n$. Now by Theorem 3.2.6, we obtain that $\alpha(\cdot) \geq \alpha_0(\cdot)$.

Q.E.D.

Proposition 3.5.9. Let $A_i \cong C(\Omega_i)$, where Ω_i is a compact Hausdorff space, $1 \leq i \leq n-1$,and A_n be a C^*-algebra with an identity 1_n. Then there is only one C^*-norm $\alpha_0(\cdot)$ on $\otimes_{i=1}^n A_i$, and $\alpha_0(\cdot) = \lambda(\cdot)$, and

$$\lambda\text{-} \otimes_{i=1}^n A_i \cong C(\Omega_1 \times \cdots \times \Omega_{n-1}, A_n).$$

Proof. Let $\alpha(\cdot)$ be a C^*-norm on $\otimes_{i=1}^n A_i$. Fix $\chi_i \in \Omega_i$ (i.e., χ_i is a pure state on A_i), $1 \leq i \leq n-1$, and put

$$E = \left\{ \chi_n \middle| \begin{array}{c} \chi_n \text{is a state on } A_n \text{ such that} \\ \otimes_{i=1}^n \chi_i \text{ is continuous with respect to } \alpha(\cdot) \end{array} \right\}.$$

Clearly, E is a $\sigma(A_n^*, A_n)$-compact convex subset of S_n, where S_n is the state space of A_n. For any $h^* = h \in A_n$, let B be the abelian C^*-subalgebra of A_n generated by $\{1_n, h\}$. Pick a state ψ_B on B such that $\psi_B(h) = \max\{\lambda \mid \lambda \in \sigma(h)\}$. By Proposition 3.5.3, there is only one C^*-norm on $\otimes_{i=1}^{n-1} A_i \otimes B$. Further by Corollary 3.2.7, $\otimes_{i=1}^{n-1} \chi_i \otimes \psi_B$ is continuous on $\otimes_{i=1}^{n-1} A_i \otimes B$ with respect to $\alpha(\cdot)$. Thus $\otimes_{i=1}^{n-1} \chi_i \otimes \psi_B$ can be extended to a state φ on $\alpha\text{-}\otimes_{i=1}^n A_i$. Clearly, $\varphi(\cdot \otimes \otimes_{j \neq i} 1_j) = \chi_i(\cdot), 1 \leq i \leq n-1$. By Corollary 3.5.5, $\varphi = \otimes_{i=1}^n \chi_i$, where χ_n is a state on A_n, and is an extension of ψ_B. In particular,, $\chi_n \in E$ and $\chi_n(h) = \psi_B(h) = \max\{\lambda \mid \lambda \in \sigma(h)\}$. From Lemma 3.5.7, $E = S(A_n) = S_n$.

From the preceding paragraph, $\otimes_{i=1}^n \chi_i$ is continuous with respect to $\alpha(\cdot), \forall \chi_i \in \Omega_i, 1 \leq i \leq n-1$, and $\chi_n \in S_n$. Further, $\otimes_{i=1}^n \varphi_i$ is continuous with respect to $\alpha(\cdot), \forall \varphi_i \in S_i$ (the state space of A_i), $1 \leq i \leq n$. By Lemma 3.5.8, we get $\alpha(\cdot) \geq \alpha_0(\cdot)$ on $\otimes_{i=1}^n A_i$.

However, if φ is a pure state on $\alpha\text{-}\otimes_{i=1}^n A_i$, then by Proposition 3.5.6 we have $\varphi = \otimes_{i=1}^n \chi_i$, where χ_i is a pure state on $A_i, 1 \leq i \leq n$. Further for any

$u \in \otimes_{i=1}^n A_i$,

$$\begin{aligned} \alpha(u)^2 = & \ \alpha(u^*u) \\ = & \ \sup\{\varphi(u^*u) \mid \varphi \text{ is a pure state on } \alpha\text{-} \otimes_{i=1}^n A_i\} \\ = & \ \sup\{\otimes_{i=1}^n \chi_i(u^*u) \mid \chi_i \text{ is a pure state on } A_i, 1 \le i \le n\} \\ \le & \ \lambda(u^*u) \le \alpha_0(u^*u) = \alpha_0(u)^2. \end{aligned}$$

Therefore, there is only one C^*-norm $\alpha_0(\cdot)$ on $\otimes_{i=1}^n A_i$.

For each $u \in \otimes_{i=1}^n A_i$, we can uniquely write that $u = u(t_1, \cdots, t_{n-1})$, where $u(t_1, \cdots, t_{n-1})$ is a continuous map from $\Omega_1 \times \cdots \times \Omega_{n-1}$ into A_n. Clearly,

$$\|u\| = \max\{\|u(t_1, \cdots, t_{n-1})\| \mid t_i \in \Omega_i, 1 \le i \le n-1\}$$

is a C^*-norm on $\otimes_{i=1}^n A_i$. Therefore, $\alpha_0(u) = \|u\|, \forall u \in \otimes_{i=1}^n A_i$, i.e.,

$$\alpha_0\text{-} \otimes_{i=1}^n A_i \cong C(\Omega_1 \times \cdots \times \Omega_{n-1}, A_n).$$

Finally, since for any $u \in \otimes_{i=1}^n A_i$

$$\begin{aligned} \alpha_0(u) = & \ \sup\left\{ |f_n(u(t_1, \cdots, t_n))| \;\middle|\; \begin{matrix} f_n \in A_n^*, \|f_n\| \le 1 \text{ and} \\ t_i \in \Omega_i, 1 \le i \le n-1 \end{matrix} \right\} \\ \le & \ \sup\left\{ |\otimes_{i=1}^n f_i(u)| \;\middle|\; \begin{matrix} f_i \in A_i^*, \|f_i\| \le 1, \\ 1 \le i \le n \end{matrix} \right\} = \lambda(u), \end{aligned}$$

it follows that $\lambda(\cdot) = \alpha_0(\cdot)$ on $\otimes_{i=1}^n A_i$. Q.E.D.

Theorem 3.5.10. Let A_i be a C^*-algebra, $1 \le i \le n$, and $\alpha(\cdot)$ be a C^*-norm on $\otimes_{i=1}^n A_i$. Then on $\otimes_{i=1}^n A_i$,

$$\lambda(\cdot) \le \alpha_0(\cdot) \le \alpha(\cdot) \le \gamma(\cdot).$$

Consequently, $\alpha(\cdot)$ is a cross-norm on $\otimes_{i=1}^n A_i$.

Proof. It suffices to prove that $\alpha_0(\cdot) \le \alpha(\cdot)$. By Proposition 3.5.2, $\alpha(\cdot)$ can be extended to a C^*-norm on $\otimes_{i=1}^n A'_i$, where $A'_i = A_i$ if A_i has an identity; and $A'_i = A_i \dot{+} \mathbb{C} 1_i$ if A_i has no idenity. Again by Theorem 3.2.10, $\alpha_0(\cdot)$ can be extended to the $\alpha_0(\cdot)$ on $\otimes_{i=1}^n A'_i$. Thus we may assume that A_i has an identity $1_i, 1 \le i \le n$.

If $(n-1)$ C^*-algebras of $\{A_1, \cdots, A_n\}$ are abelian, then $\alpha(\cdot) = \alpha_0(\cdot)$ on $\otimes_{i=1}^n A_i$ by Proposition 3.5.9. Now we assume that $\alpha(\cdot) \ge \alpha_0(\cdot)$ on $\otimes_{i=1}^n A_i$ if $k(\le n-1)$ C^*-algebras of $\{A_1, \cdots, A_n\}$ are abelian. And we want to prove that the assersion holds for $(k-1)$.

Let $A_1, \cdots, A_{k-1}$ be abelian. Fix a pure state χ_i on $A_i, 1 \le i \le n-1$, and put $E = \{\chi_n \mid \chi_n \in S_n$, and $\otimes_{i=1}^n \chi_i$ is continuous with respect to $\alpha(\cdot)\}$, where

$S_n = S(A_n)$ is the state space of A_n. Then E is a $\sigma(A_n^*, A_n)$-compact convex subset of S_n. For any $h^* = h \in A_n$, let B be the abelian C^*-subalgebra of A_n generated by $\{1_n, h\}$. Pick a state ψ_B on B such that $\psi_B(h) = \max\{\lambda \mid \lambda \in \sigma(h)\}$. Since k C^*-algebras of $\{A_1, \cdots, A_{n-1}, B\}$ are abelian, it follows from the induction that $\alpha(\cdot) \geq \alpha_0(\cdot)$ on $\otimes_{i=1}^{n-1} A_i \otimes B$. Further by Lemma 3.5.8, $\otimes_{i=1}^{n-1} \chi_i \otimes \psi_B$ is continuous on $\otimes_{i=1}^{n-1} A_i \otimes B$ with respect to $\alpha(\cdot)$. Thus $\otimes_{i=1}^{n-1} \chi_i \otimes \psi_B$ can be extended to a state φ on α-$\otimes_{i=1}^n A_i$. From Corollary 3.5.5, $\varphi = \otimes_{i=1}^n \chi_i$, where χ_n is a state on A_n, and is an extension of ψ_B. In particular, $\chi_n \in E$ and $\chi_n(h) = \psi_B(h) = \max\{\lambda \mid \lambda \in \sigma(h)\}$. By Lemma 3.5.7, $E = S(A_n) = S_n$. Therefore $\otimes_{i=1}^n \chi_i$ is continuous with respect to $\alpha(\cdot)$, where χ_i is any pure state on $A_i, 1 \leq i \leq n$. Furthermore, $\otimes_{i=1}^n \varphi_i$ is continuous with respect to $\alpha(\cdot), \forall \varphi_i \in S_i$ (the state space of A_i), $1 \leq i \leq n$. Finally, by Lemma 3.5.8 we obtain $\alpha(\cdot) \geq \alpha_0(\cdot)$ on $\otimes_{i=1}^n A_i$. Q.E.D.

Lemma 3.5.11. Let Φ be a $*$ homomorphism from a C^*-algebra A onto a C^*-algebra B. Then Φ^* is an isometric map from B^* to A^*.

Proof. Let $I = \{a \in A \mid \Phi(a) = 0\}$. Then I is a closed two-sided ideal of A, and A/I is $*$ isomorphic to B. Thus for any $b \in B, \|b\| = \inf\{\|a\| \mid a \in A, \Phi(a) = b\}$. In consequence,

$$\{b \in B \| \|b\| < 1\} \subset \Phi(\{a \in A \mid \|a\| \leq 1\}).$$

Now for any $g \in B^*$,

$$\begin{aligned} \|\Phi^*(g)\| &= \sup\{|g(\Phi(a))| \mid a \in A, \|a\| \leq 1\} \\ &\geq \sup\{|g(b)| \mid b \in B, \|b\| < 1\} = \|g\|. \end{aligned}$$

However, $\|\Phi^*\| = \|\Phi\| \leq 1$. Therefore, Φ^* is isometric. Q.E.D.

Proposition 3.5.12. Let $\alpha(\cdot)$ be a C^*-norm on $\otimes_{i=1}^n A_i$. Then $\alpha^*(\cdot)$ is a cross-norm on $\otimes_{i=1}^n A_i^*$, and is independent of the choice of $\alpha(\cdot)$.

Proof. By Theorem 3.5.10 and Proposition 3.1.3, $\alpha^*(\cdot)$ is a cross-norm on $\otimes_{i=1}^n A_i^*$. From $\alpha_0^*(\cdot) \geq \alpha^*(\cdot) \geq \alpha_1^*(\cdot)$, it suffices to show that $\alpha_1^*(\cdot) = \alpha_0^*(\cdot)$ on $\otimes_{i=1}^n A_i^*$. Clearly, there is a $*$ homomorphism Φ from α_1-$\otimes_{i=1}^n A_i$ onto α_0-$\otimes_{i=1}^n A_i$ such that $\Phi(u) = u, \forall u \in \otimes_{i=1}^n A_i$. For any $\omega \in \otimes_{i=1}^n A_i^*$, it is easy to see that $\alpha_0^*(\omega)$ is the norm of ω as an element of $(\alpha_0$-$\otimes_{i=1}^n A_i)^*$. Now by Lemma 3.5.11,

$$\begin{aligned} \alpha_1^*(\omega) &= \sup\{|\omega(u)| \mid u \in \otimes_{i=1}^n A_i, \alpha_1(u) \leq 1\} \\ &= \sup\{|\omega(\Phi(u))| \mid u \in \otimes_{i=1}^n A_i, \alpha_1(u) \leq 1\} \\ &= \sup\{|\Phi^*(\omega)(u)| \mid u \in \otimes_{i=1}^n A_i, \alpha_1(u) \leq 1\} \\ &= \|\Phi^*(\omega)\| = \alpha_0^*(\omega), \quad \forall \omega \in \otimes_{i=1}^n A_i^*. \end{aligned}$$

Q.E.D.

Notes. Propositions 3.5.6, 3.5.9 and Theorem 3.5.10 are due to M. Takesaki.

References. [96], [150], [171], [193].

3.6. Completely positive maps

Let n be a positive integer, H_n be a n-dimensional Hilbert space, and $M_n = B(H_n)$ be the algebra of $n \times n$ matrices.

Lemma 3.6.1. Let A be a C^*-algebra, and n be a positive integer. Then there is only one C^*-norm $\alpha_0(\cdot)$ on $M_n \otimes A$, and α_0 -$(M_n \otimes A) = M_n \otimes A$. Moreover, if A is a C^*-algebra on a Hilbert space H, then $M_n \otimes A$ is $*$ isomorphic to the C^*-algebra $M_n(A)$ on $H \oplus \cdots \oplus H(n$ times), where

$$M_n(A) = \{(a_{ij})_{1 \leq i,j \leq n} | a_{ij} \in A, \forall i, j\},$$

and

$$M_n(A)^* = M_n(A^*) = \{(f_{ij})_{1 \leq i,j \leq n} | f_{ij} \in A^*, \forall i, j\},$$

where $\langle (f_{ij}), (a_{ij}) \rangle = \sum_{ij} f_{ij}(a_{ij})$.

Proof. Let $\{e_{ij} | 1 \leq i, j \leq n\}$ be a matrix unit of M_n, i.e.,

$$e_{ij}^* = e_{ij}, \quad e_{ij}e_{kl} = \delta_{jk}e_{il}, \quad \forall i.j, k, l.$$

Then each $u \in M_n \otimes A$ can be uniquely expressed by $u = \sum_{ij} a_{ij} \otimes e_{ij}$, and we get a $*$ isomorphism Φ from $M_n \otimes A$ onto $M_n(A) : \Phi(u) = (a_{ij})$.

Clearly, $M_n(A)$ is a C^*-algebra on $H \oplus \cdots \oplus H(n$ times). Define $\|u\| = \|\Phi(u)\| = \|(a_{ij})\|, \forall u \in M_n \otimes A$. Then $M_n \otimes A$ is a C^*-algebra. By Proposition 2.1.10, there is only one C^*-norm $\alpha_0(\cdot) = \|\cdot\|$ on $M_n \otimes A$, and α_0 - $(M_n \otimes A) = M_n \otimes A$.

The rest conclusion is obvious. Q.E.D.

From now on, we shall identify $M_n \otimes A$ with $M_n(A)$.

Proposition 3.6.2. Let n be a positive integer, A be a C^*-algebra, and $a = (a_{ij}) \in M_n(A)$. Then the following statements are equivalent:

1) a is a positive element of $M_n(A)$;
2) a is a sum of matrices of the form $(a_i^* a_j)$ with $a_1, \cdots a_n \in A$;
3) $\sum_{i,j} x_i^* a_{ij} x_j \geq 0$ in A, $\forall x_1, \cdots, x_n \in A$.

Proof. 1) $\Rightarrow$ 2) . Let $(a_{ij}) = (b_{ij})^*(b_{ij})$. Then

$$a_{ij} = \sum_k b_{ki}^* b_{kj}, \forall i, j.$$

Put $c_k = (b_{ki}^* b_{kj}), 1 \leq k \leq n$. Then $a = c_1 + \cdots + c_n$.

2) $\Rightarrow$ 3). It is obvious.

3) $\Rightarrow$ 1). For any cyclic $*$ representation $\{\pi, K, \xi\}$ of A , define a $*$ representation $\{\tilde{\pi}, K \oplus \cdots \oplus K(n \text{ times }) \}$ of $M_n(A)$:

$$\tilde{\pi}((b_{ij})) = (\pi(b_{ij})), \forall (b_{ij}) \in M_n(A).$$

For any $\xi_1, \cdots \xi_n \in K$, Pick $x_m^{(i)} \in A$ such that

$$\pi(x_m^{(i)})\xi \longrightarrow \xi_i, 1 \leq i \leq n.$$

Then by the condition 3),

$$\langle \tilde{\pi}(a)(\xi_i), (\xi_i) \rangle = \lim_m \langle \pi(\sum_{ij} x_m^{(i)*} a_{ij} x_m^{(j)})\xi, \xi \rangle \geq 0.$$

Thus $\tilde{\pi}(a)$ is a positive operator on $K \oplus \cdots \oplus K(n \text{ times })$.

Now let $\{\pi_l\}$ be a family of cyclic $*$ representations of A such that $\pi = \sum_l \oplus \pi_l$ is faithful for A. Then $\tilde{\pi} = \sum_l \oplus \tilde{\pi}_l$ is also faithful for $M_n(A)$. From the preceding paragraph, $\tilde{\pi}(a) \geq 0$. Therefore, a is a positive element of $M_n(A)$.

Q.E.D.

Definition 3.6.3. Let Φ be a linear map from a C^*-algebra A to a C^*-algebra B, and n be a positive integer. Naturally, define a linear map Φ_n from $M_n(A)$ to $M_n(B)$:

$$\Phi_n((a_{ij})) = (\Phi(a_{ij})), \quad \forall (a_{ij}) \in M_n(A).$$

Φ is said to be *n-positive*, if $\Phi_n(M_n(A)_+) \subset M_n(B)_+$.$\Phi$ is said to be *completely positve*, if Φ is n-positive for any positive integer n.

Proposition 3.6.4. 1) If Φ is a $*$ homomorphism from A to B, then Φ is completely positive.

2) The composition of completely positive maps is completely positive.

3) Let $\{\pi, H\}$ be a $*$ representation of A, and v be a bounded linear map from a Hilbert space K to H. Then $\Phi(\cdot) = v^*\pi(\cdot)v$ is a completely positive map from A to $B(K)$.

Proof. 1) It is clear since Φ_n is also a $*$ homomorphism from $M_n(A)$ to $M_n(B)$ for any positive integer n. 2) is obvious.

3) For any $n, a_1, \cdots a_n \in A$ and $b_1, \cdots, b_n \in B(K)$,

$$\sum_{i,j} b_i^* \Phi(a_i^* a_j) b_j = (\sum_i \pi(a_i) v b_i)^* \cdot (\sum_i \pi(a_i) v b_i) \geq 0.$$

Now by Proposition 3.6.2, we can see that Φ is completely positive. Q.E.D.

Lemma 3.6.5. Let A, B be C^*-algebra, and Φ be a positive (i.e. 1-positive) linear map from A to B. Then Φ is continuous.

Proof. It suffices to show that Φ is a closed operator. Suppose that $a_n \longrightarrow 0$ in A, and $\Phi(a_n) \longrightarrow b$ in B. For any positive linear functional f on B, $f \circ \Phi$ is a positive linear functional on A. Thus $f \circ \Phi$ is continuous (Proposition 2.3.2) , and $f \circ \Phi(a_n) \longrightarrow 0$. Therefore $f(b) = 0, \forall f \in B^*$, and $b = 0$, i.e., Φ is closed.

Q.E.D.

Proposition 3.6.6. Let A, B be C^*-algebras, and Φ be a positive linear map from A to B. If either A or B is abelian , then Φ is completely positive.

Proof. Let $B \cong C_0^\infty(\Omega)$, where Ω is a locally compact Hausdorff space. For any $n, a_1, \cdots, a_n \in A, b_1, \cdots b_n \in B$ and $t \in \Omega$, notice that

$$\begin{aligned}(\textstyle\sum_{i,j} b_i^* \Phi(a_i^* a_j) b_j)(t) &= \textstyle\sum_{i,j} \overline{b_i(t)} \Phi(a_i^* a_j)(t) b_j(t) \\ &= \Phi((\textstyle\sum_i b_i(t) a_i)^* \cdot (\textstyle\sum_i b_i(t) a_i))(t) \geq 0.\end{aligned}$$

Therefore, Φ is completely positive.

Now suppose that $A \cong C_0^\infty(\Omega)$, and $B \subset B(H)$. For any $n, a_1, \cdots, a_n \in A$, and $\xi_1, \cdots, \xi_n \in H$, we need to prove that

$$\sum_{i,j} \langle \Phi(a_i^* a_j) \xi_j, \xi_i \rangle \geq 0.$$

By Lemma 3.6.5, for any $i, j \in \{1, \cdots, n\}$ there is a finite Radon measure μ_{ij} such that

$$\langle \Phi(a) \xi_j, \xi_i \rangle = \int_\Omega a(t) d\mu_{ij}(t), \forall a \in A.$$

Let $\mu = \sum_{i,j} |\mu_{ij}|$. Then there is $f_{ij} \in L^1(\Omega, \mu)$ such that $\mu_{ij} = f_{ij} \cdot \mu, \forall i, j$. Fix $\lambda_1, \cdots, \lambda_n \in \mathbb{C}$. Since Φ is positive, it follows that

$$\begin{aligned}&\textstyle\int_\Omega |a(t)|^2 d(\sum_{i,j} \overline{\lambda_i} \lambda_j \mu_{ij}(t)) \\ = \;& \langle \Phi(a^* a)(\textstyle\sum_i \lambda_i \xi_i), (\sum_i \lambda_i \xi_i) \rangle \geq 0,\end{aligned}$$

$\forall a \in A$. Thus $\sum_{i,j} \overline{\lambda_i} \lambda_j \mu_{ij}$ is a positive measure on Ω. Further, $\sum_{i,j} \overline{\lambda_i} \lambda_j f_{ij}(t) \geq 0$, a.e.$\mu$. So we can find a Borel subset Ω_0 of Ω such that

$$\mu(\Omega_0) = 0, \quad \text{and} \quad \sum_{i,j} \overline{\lambda_i} \lambda_j f_{ij}(t) \geq 0,$$

$\forall t \notin \Omega_0$ and any complex rational numbers $\lambda_1, \cdots, \lambda_n$. But any complex number can be approximated arbitrarily by complex rational numbers, hence we have

$$\sum_{i,j} \overline{\lambda}_i \lambda_j f_{ij}(t) \geq 0, \forall t \notin \Omega_0, \lambda_1, \cdots, \lambda_n \in \mathbb{C}.$$

Therefore,

$$\begin{aligned} & \textstyle\sum_{i,j} \langle \Phi(a_i^* a_j)\xi_j, \xi_i \rangle \\ = & \textstyle\sum_{i,j} \int_\Omega (a_i^* a_j)(t) f_{ij}(t) d\mu(t) \\ = & \int_\Omega (\textstyle\sum_{i,j} \overline{a_i(t)} a_j(t) f_{ij}(t)) d\mu(t) \geq 0. \end{aligned}$$

Q.E.D.

Theorem 3.6.7. Let A be a C^*-algebra, K be a Hilbert space, and Φ be a completely positive linear map from A to $B(K)$. Then there exists a $*$ representation $\{\pi, H\}$ of A, a normal $*$ homomorphism Ψ from the VN algebra $B = \Phi(A)'$ to $B(H)$,and a bounded linear operator v from K to H, such that

$$\Phi(a) = v^* \pi(a) v, \forall a \in A, \Phi(b)v = vb, \forall b \in B,$$

and $\Phi(B) \subset \pi(A)', H = \overline{[\pi(A)vK]}, \|v\| = \|\Phi\|^{1/2}$.

Moreover, if A has an identity 1, and $\Phi(1) = 1_K$, then v can be isometric.

Proof. Let $A \otimes K$ be the algebraic tensor product of the Banach spaces A and K. Define

$$\langle \sum_i a_i \otimes \xi_i, \sum_j b_j \otimes \eta_j \rangle = \sum_{i,j} \langle \Phi(b_j^* a_i)\xi_i, \eta_j \rangle,$$

$\forall a_i, b_j \in A, \xi_i, \eta_j \in K$. Since Φ is completely positive, if follows that $\langle , \rangle$ is a non-negative inner product on $A \otimes K$. Let $N = \{x \in A \otimes K \mid \langle x, x \rangle = 0\}$, and let $x \longrightarrow \tilde{x} = x + N$ be the canonical map from $A \otimes K$ onto $(A \otimes K)/N$. Then we get an inner product on $(A \otimes K)/N$,

$$\langle \tilde{x}, \tilde{y} \rangle = \langle x, y \rangle, \forall \tilde{x}, \tilde{y} \in A \otimes K/N, x \in \tilde{x}, y \in \tilde{y}.$$

Denote the completion of $(A \otimes K/N, \langle , \rangle)$ by H, and let

$$\begin{aligned} \pi(a) \textstyle\sum_i \widetilde{a_i \otimes \xi_i} &= \textstyle\sum_i \widetilde{aa_i \otimes \xi_i}, \\ \Psi(b) \textstyle\sum_i \widetilde{a_i \otimes \xi_i} &= \textstyle\sum_i \widetilde{a_i \otimes b\xi_i}, \end{aligned}$$

$\forall a \in A, b \in B = \Phi(A)' (\subset B(K))$, and $a_i \in A, \xi_i \in K, \forall i$. Since Φ is completely positive, it follows that

$$\left\|\pi(a)\sum_i \widetilde{a_i \otimes \xi_i}\right\|^2 = \sum_{i,j=1}^n \langle \Phi(a_j^* a^* a a_i)\xi_i, \xi_j\rangle$$

$$= \langle \Phi_n\left(\begin{pmatrix} a_1 & \cdots & a_n \\ 0 & \cdots & 0 \\ \cdots & \cdots & \cdots \\ 0 & \cdots & 0 \end{pmatrix}^* \begin{pmatrix} a & & 0 \\ & \vdots & \\ 0 & & a \end{pmatrix}^* \begin{pmatrix} a & & 0 \\ & \vdots & \\ 0 & & a \end{pmatrix} \begin{pmatrix} a_1 & \cdots & a_n \\ 0 & \cdots & 0 \\ \cdots & \cdots & \cdots \\ 0 & \cdots & 0 \end{pmatrix}\right)$$

$$\begin{pmatrix} \xi_1 \\ \ddots \\ \xi_n \end{pmatrix}, \begin{pmatrix} \xi_1 \\ \ddots \\ \xi_n \end{pmatrix}\rangle$$

$$\leq \left\|\begin{pmatrix} a & & 0 \\ & \vdots & \\ 0 & & a \end{pmatrix}\right\|^2 \langle \Phi_n\left(\begin{pmatrix} a_1 & \cdots & a_n \\ 0 & \cdots & 0 \\ \cdots & \cdots & \cdots \\ 0 & \cdots & 0 \end{pmatrix}^* \begin{pmatrix} a_1 & \cdots & a_n \\ 0 & \cdots & 0 \\ \cdots & \cdots & \cdots \\ 0 & \cdots & 0 \end{pmatrix}\right) \begin{pmatrix} \xi_1 \\ \ddots \\ \xi_n \end{pmatrix}, \begin{pmatrix} \xi_1 \\ \ddots \\ \xi_n \end{pmatrix}\rangle$$

$$= \|a\|^2 \cdot \left\|\sum_i \widetilde{a_i \otimes \xi_i}\right\|^2.$$

Thus $\pi(a)$ can be uniquely extended to a bounded linear operator on H, still denoted by $\pi(a)$. Clearly, $\{\pi, H\}$ is a $*$ representation of A.

From $B = \Phi(A)'$, we have

$$\left\|\Psi(b)\sum_{i=1}^n \widetilde{a_i \otimes \xi_i}\right\|^2 = \sum_{i,j=1}^n \langle b^* b \Phi(a_i^* a_j)\xi_j, \xi_i\rangle.$$

But $(\Phi(a_i^* a_j))_{1\leq i,j\leq n}$ is a positive element of $M_n(B')$, so we can write that

$$(\Phi(a_i^* a_j)) = (b'_{ij})^* \cdot (b'_{ij}),$$

where $b'_{ij} \in B', 1 \leq i,j \leq n$. Then

$$\left\|\Psi(b)\sum_{i=1}^n \widetilde{a_i \otimes \xi_i}\right\|^2$$

$$= \langle \begin{pmatrix} b^* b & & \\ & \vdots & \\ & & b^* b \end{pmatrix} (b'_{ij}) \begin{pmatrix} \xi_1 \\ \ddots \\ \xi_n \end{pmatrix}, (b'_{ij}) \begin{pmatrix} \xi_1 \\ \ddots \\ \xi_n \end{pmatrix}\rangle$$

$$\leq \|b\|^2 \langle (b'_{ij}) \begin{pmatrix} \xi_1 \\ \ddots \\ \xi_n \end{pmatrix}, (b'_{ij}) \begin{pmatrix} \xi_1 \\ \ddots \\ \xi_n \end{pmatrix}\rangle = \|b\|^2 \cdot \left\|\sum_{i=1}^n \widetilde{a_i \otimes \xi_i}\right\|^2.$$

Hence $\Psi(b)$ can be also extended to a bouned linear operator on H, still denoted by $\Psi(b)$. Clearly, Ψ is a $*$ homomorphism from B to $B(H)$, and $\Psi(B) \subset \pi(A)'$.

If $\{b_l\}$ is a bounded increasing net of B_+, then $\{\Psi(b_l)\}$ is a bounded increasing net of $B(H)_+$. For any $a_i \in A, \xi_i \in K$,

$$\langle \Psi(b_l)\sum_i \widetilde{a_i \otimes \xi_i}, \sum_j \widetilde{a_j \otimes \xi_j}\rangle = \sum_{i,j}\langle \Phi(a_j^* a_i)b_l\xi_i, \xi_j\rangle.$$

Picking the limit for l, we can see that

$$\sup_l \Psi(b_l) = \Psi(\sup_l b_l),$$

i.e., Ψ is normal.

Now suppose that $\{d_l\}$ is an approximate identity for A . By Lemma 3.6.5, $\{\Phi(d_l)\}$ is a bounded increasing net of $B(K)_+$. Thus $\sup_l \Phi(d_l) =$ (strongly) - $\lim_l \Phi(d_l)$. Let $v_l : K \longrightarrow H$ be as follows

$$v_l\xi = \widetilde{d_l \otimes \xi}, \forall \xi \in K.$$

It is clear that $\|v_l\xi\|^2 = \langle \Phi(d_l^2)\xi, \xi\rangle \leq \|\Phi\| \cdot \|\xi\|^2$, and $\|v_l\| \leq \|\Phi\|^{1/2}, \forall l$. If $l' \geq l$, then $(d_{l'} - d_l)^2 \leq d_{l'} - d_l$ and

$$\|(v_{l'} - v_l)\xi\|^2 \leq \langle (\Phi(d_{l'}) - \Phi(d_l))\xi, \xi\rangle \xrightarrow{(l',l)} 0, \forall \xi \in K.$$

Thus there is $v : K \longrightarrow H$ with $\|v\| \leq \|\Phi\|^{1/2}$ such that $v_l \longrightarrow v$ (strongly) . Since

$$\langle v_l^* \widetilde{a \otimes \xi}, \eta\rangle = \langle \Phi(d_l a)\xi, \eta\rangle \longrightarrow \langle \Phi(a)\xi, \eta\rangle$$

$\forall \eta \in K, \forall l$, it follows that $v^*\widetilde{a \otimes \xi} = \Phi(a)\xi, \forall a \in A, \xi \in K$. Thus $v^*\pi(a)v_l\xi = v^*\widetilde{ad_l \otimes \xi} = \Phi(ad_l)\xi, \forall l, \forall \xi \in K$ and $a \in A$. Further,

$$\Phi(a) = v^*\pi(a)v, \forall a \in A.$$

Consequently, $\|\Phi\| \leq \|v\|^2$. Hence $\|v\| = \|\Phi\|^{1/2}$. For any $a, b \in A, \xi, \eta \in K$,

$$\begin{aligned}\langle \pi(a)v\xi, \widetilde{b \otimes \eta}\rangle &= \lim_l \langle \widetilde{ad_l \otimes \xi}, \widetilde{b \otimes \eta}\rangle \\ &= \langle \Phi(b^*a)\xi, \eta\rangle = \langle \widetilde{a \otimes \xi}, \widetilde{b \otimes \xi}\rangle.\end{aligned}$$

Hence $\pi(a)v\xi = \widetilde{a \otimes \xi}, \forall a \in A, \xi \in K$. Consequently, $\overline{[\pi(A)vK]} = H$, and $\{\pi, H\}$ is nondegenerate for A. By Proposition 2.4.6 $\pi(d_l) \longrightarrow 1_H$ (strongly). Notice that

$$\pi(d_l)\Psi(b)v\xi = \Psi(b)\pi(d_l)v\xi = \widetilde{d_l \otimes b\xi} = \pi(d_l)vb\xi,$$

$\forall b \in B, \xi \in K, \forall l$. Therefore , $\Psi(b)v = vb, \forall b \in B$.

Finally, if A has an identity 1, and $\Phi(1) = 1_K$, we can pick $d_l = 1, \forall l$. Then $v\xi = \widetilde{1 \otimes \xi}, \forall \xi \in K$. Thus v is isometric. Q.E.D.

Proposition 3.6.8. Let A, B be C^*-algebras, and Φ be a completely positive linear map from A to B. Then

$$\Phi(a)^*\Phi(a) \le \|\Phi\| \, \Phi(a^*a), \forall a \in A.$$

Proof. We may assume that $B \subset B(K)$ for some Hilbert space K. Then by Theorem 3.6.7,

$$\begin{aligned} \Phi(a)^*\Phi(a) &= v^*\pi(a^*)vv^*\pi(a)v \\ &\le \|v\|^2 v^*\pi(a^*a)v = \|\Phi\| \, \Phi(a^*a), \quad \forall a \in A. \end{aligned}$$

Q.E.D.

Lemma 3.6.9. Let A be a C^*-algebra, B be a C^*-subalgebra of A, and $\{\pi, H\}$ be a $*$ representation of B . Then there is a Hilbert space H_1 and a $*$ representation $\{\pi_1, H_1\}$ of A such that $H_1 \supset H$ and

$$\pi_1(b)\xi = \pi(b)\xi, \forall b \in B, \xi \in H.$$

Proof. We may assume that $\{\pi, H\}$ is generated by a state φ on B. Then φ can be extended to a state ψ on A. Let $\{\pi_1, H_1\}$ be the $*$ representation of A generated by ψ. Then $\{\pi_1, H_1\}$ is what we want to find. Q.E.D.

Proposition 3.6.10. Let A be a C^*-algebra, B be a C^* –subalgebra of A , and Φ be a completely positive linear map from B to $B(K)$ (K some Hilbert space). Then Φ can be extended to a completely positive linear map from A to $B(K)$.

Proof. By Theorem 3.6.7, there is a $*$ representation $\{\pi, H\}$ of B and a $v : K \longrightarrow H$ such that $\Phi(b) = v^*\pi(b)v, \forall b \in B$. Let $\{\pi_1, H_1\}$ be a $*$ representation of A satisfying Lemma 3.6.9, and P be the projection from H_1 onto H. Define

$$\Psi(a) = v^*P\pi_1(a)Pv, \forall a \in A.$$

By Proposition 3.6.4, Ψ is a completely positive linear map from A to $B(K)$. Clearly , Ψ is also an extension of Φ. Q.E.D.

Proposition 3.6.11. Let Φ_i be a completely positive linear map from A_i to $B_i, 1 \le i \le n$. Then $\otimes_{i=1}^n \Phi_i$ can be extended to a completely positive linear map from α_0 - $\otimes_{i=1}^n A_i$ to α_0 - $\otimes_{i=1}^n B_i$.

Proof. Let $B_i \subset B(K_i), 1 \le i \le n$. By Theorem 3.6.7, there is a $*$ representation $\{\pi_i, H_i\}$ of A_i and $v_i : K_i \longrightarrow H_i$ such that

$$\Phi_i(a_i) = v_i^*\pi_i(a_i)v_i, \forall a_i \in A_i, 1 \le i \le n.$$

By Proposition 3.2.8, $\otimes_{i=1}^n \pi_i$ can be extended to a $*$ representation of α_0 - $\otimes_{i=1}^n A_i$. Suppose that

$$\Phi(a) = (\otimes_{i=1}^n v_i)^* \cdot \otimes_{i=1}^n \pi_i(a) \cdot \otimes_{i=1}^n v_i, \forall a \in \alpha_0\text{-} \otimes_{i=1}^n A_i.$$

Then Φ is a completley positive linear map from α_0 - $\otimes_{i=1}^n A_i$ to $B(\otimes_{i=1}^n K_i)$, and is an extension of $\otimes_{i=1}^n \Phi_i$. Moreover, since $\Phi(\otimes_{i=1}^n A_i) \subset \otimes_{i=1}^n B_i$, and α_0 - $\otimes_{i=1}^n B_i$ is the uniform closure of $\otimes_{i=1}^n B_i$ in $B(\otimes_{i=1}^n K_i)$, it follows that $\Phi(\alpha_0$ $-\otimes_{i=1}^n A_i) \subset \alpha_0\text{-}\otimes_{i=1}^n B_i$. Q.E.D.

Lemma 3.6.12. Let $a_s = (a_{ij}^{(s)}) \in M_n(A)_+, 1 \le s \le m, a_{ij}^{(s)} a_{kl}^{(s')} = a_{kl}^{(s')} a_{ij}^{(s)}, \forall i$, j, k, l and $s \ne s'$. Then

$$a = (a_{ij}^{(1)} \cdots a_{ij}^{(m)}) \in M_n(A)_+.$$

Proof. It suffices to prove this lemma for $m = 2$. Let $x = (x_{ij}), y = (y_{ij}) \in M_n(A)_+$, and $x_{ij} y_{kl} = y_{kl} x_{ij}, \forall i, j, k, l$, we need to prove $(x_{ij} y_{ij}) \in M_n(A)_+$.

Suppose that B and C are the C^*-subalgebras of A generated by $\{x_{ij} \mid i, j\}$ and $\{y_{i,j} \mid i, j\}$ respectively. Since $x_{ij}^* = x_{ji}, y_{kl}^* = y_{lk} \forall i, j, k, l$, it follows that $bc = cb, \forall b \in B, c \in C$. Clearly , $x \in M_n(B)_+, y \in M_n(C)_+$. Then we can write

$$x_{ij} = \sum_k b_{ki}^* b_{kj}, y_{ij} = \sum_k c_{ki}^* c_{kj}, \forall i, j,$$

where $b_{ki} \in B, c_{kj} \in C, \forall i, j, k$. Therefore,

$$(x_{ij} y_{ij}) = \sum_{k,l} ((b_{ki} c_{li})^* \cdot (b_{kj} c_{lj}))$$

is a positive element of $M_n(A)$ by Proposition 3.6.2. Q.E.D.

Proposition 3.6.13. Let Φ_i be a completely positive linear map from A_i to B , and $\Phi_i(a_i)\Phi_j(a_j) = \Phi_j(a_j)\Phi_i(a_i), \forall a_i \in A_i, a_j \in A_j, 1 \le i \ne j \le n$. Denfine $\Phi(\otimes_{i=1}^n a_i) = \Pi_{i=1}^n \Phi(a_i), \forall a_i \in A_i, 1 \le i \le n$. Then Φ can be extended to a completely positive linear map from α_1 - $\otimes_{i=1}^n A_i$ to B.

Proof. Let B_i be the C^*-subalgebra of B generated by $\Phi(A_i), 1 \le i \le n$. Then $b_i b_j = b_j b_i, \forall b_i \in B_i, b_j \in B_j, 1 \le i \ne j \le n$.

First we say that Φ is positive from $\otimes_{i=1}^n A_i$ to B. In fact, let $u = \sum_{j=1}^m \otimes_{i=1}^n a_j^{(i)}$, where $a_j^{(i)} \in A_i, \forall i, j$. Then

$$\Phi(u^* u) = \sum_{j,k=1}^m \Phi_1(a_j^{(1)*} a_k^{(1)}) \cdots \Phi_n(a_j^{(n)*} a_k^{(n)}).$$

Define $b_{jk} = \Phi_1(a_j^{(1)*} a_k^{(1)}) \cdots \Phi_n(a_j^{(n)*} a_k^{(n)}), \forall j, k$. By Lemma 3.6.12, $(b_{jk}) \in M_m(B)_+$. Further by Proposition 3.6.2, $\sum_{j,k=1}^m d_l b_{jk} d_l \in B_+, \forall l$, where $\{d_l\}$ is an approximate identity for B. Thereofore, $\Phi(u^* u) = \sum_{j,k} b_{jk} \in B_+$.

Now for any positive linear functional ρ on B, $\rho \circ \Phi$ is positive on $\otimes_{i=1}^n A_i$. By Proposition 3.4.6, $\rho \circ \Phi$ can be uniquely extended to a positive linear functional on α_1 -$\otimes_{i=1}^n A_i$. Generally, $f \circ \Phi$ can be uniquely extended to a bounded linear functional on α_1 -$\otimes_{i=1}^n A_i$ for any $f \in B^*$. Then we can define a linear map $\Phi' : B^* \longrightarrow (\alpha_1 \text{-}\otimes_{i=1}^n A_i)^*$ as follows

$$\Phi'(f) = f \circ \Phi, \forall f \in B^*.$$

We claim that Φ' is closed . In fact, let $f_k (\in B^*) \longrightarrow 0$, and $\Phi'(f_k) = f_k \circ \Phi \longrightarrow F(\in (\alpha_1 \text{-}\otimes_{i=1}^n A_i)^*)$. Since for any $a_i \in A_i, 1 \leq i \leq n$,

$$F(\otimes_{i=1}^n a_i) = \lim_k f_k(\Pi_{i=1}^n \Phi_i(a_i)) = 0,$$

it follows that $F = 0$, i.e., Φ' is closed. So Φ' must be continuous .

For any $u \in \otimes_{i=1}^n A_i$, notice that

$$\begin{aligned} \|\Phi(u)\| &= \sup\{|f \circ \Phi(u)| \mid f \in B^*, \|f\| \leq 1\} \\ &= \sup\{|\Phi'(f)(u)| \mid f \in B^*, \|f\| \leq 1\} \\ &\leq \|\Phi'\| \, \alpha_1(u). \end{aligned}$$

Hence, Φ can be uniquely extended to a bounded linear map from α_1 -$\otimes_{i=1}^n A_i$ to B, still denoted by Φ.

Finally , we prove that Φ is completely positive from α_1-$\otimes_{i=1}^n A_i$ to B . By the continuity of Φ and Proposition 3.6.2, it suffices to show that for any $m, u_1, \cdots, u_m \in \otimes_{i=1}^n A_i$ and $b_1, \cdots, b_m \in B$,

$$\sum_{i,j=1}^m b_i^* \Phi(u_i^* u_j) b_j \in B_+.$$

Suppose that $u_i = \sum_{k=1}^p \otimes_{s=1}^n a_{ik}^{(s)}, 1 \leq i \leq m$, where $a_{ik}^{(s)} \in A_s, \forall i, k, s$. Since Φ_i is completely positive , it follows that

$$(\Phi_s(a_{ik}^{(s)*} a_{jl}^{(s)}))_{1 \leq i,j \leq m, 1 \leq k,l \leq p} \in M_{mp}(B)_+,$$

$1 \leq s \leq n$. By Lemma 3.6.12.

$$\left(\Pi_{s=1}^n \Phi_s(a_{ik}^{(s)*} a_{jl}^{(s)})\right)_{1 \leq i,j \leq m, 1 \leq k,l \leq p} \in M_{pm}(B)_+.$$

Let $b_{ik} = b_i, \forall 1 \leq i \leq m, 1 \leq k \leq p$. Then

$$\sum_{i,j=1}^m b_i^* \Phi(u_i^* u_j) b_j = \sum_{i,j=1}^m \sum_{k,l=1}^p b_{ik}^* \prod_{s=1}^n \Phi_s(a_{ik}^{(s)*} a_{jl}^{(s)}) b_{jl}.$$

By Proposition 3.6.2, it is a positive element of B. Q.E.D.

Notes. Theorem 3.6.7 is due to W. Stinespring. The recognition of the importance of completely positive maps in the tensor products was due to E.Effros and C.Lance. Proposition 3.6.10 and its generalization are due to W.Arveson.

References. [8], [37], [95], [163].

3.7. The inductive limit of C^*–algebras

Let Π be a directed index set , and A_α be a C^*–algebra for each $\alpha \in \Pi$. And also there is a $*$ isomorphism $\Phi_{\beta\alpha}$ from A_α to A_β if $\alpha \leq \beta (\alpha, \beta \in \Pi)$ such that

$$\Phi_{\gamma\beta}\Phi_{\beta\alpha} = \Phi_{\gamma\alpha}, \forall \alpha, \beta, \gamma \in \Pi \text{ and } \alpha \leq \beta \leq \gamma.$$

Let

$$\mathcal{J} = \times_{\alpha \in \Pi} A_\alpha = \{(a_\alpha)_{\alpha \in \Pi} \mid a_\alpha \in A_\alpha, \forall \alpha \in \Pi\}.$$

By the addition, multiplication and $*$ operation on each component, $\mathcal{J}$ becomes naturally a $*$ algebra. Let

$$\mathcal{L} = \{(a_l) \in \mathcal{J} \mid \text{ there is an index } \alpha \text{ such that } a_\beta = \Phi_{\beta\alpha}(a_\alpha), \forall \beta \geq \alpha.\}$$

Clearly, $\mathcal{L}$ is a $*$ subalgebra of $\mathcal{J}$. For $a = (a_\alpha) \in \mathcal{L}$, we can define that $\|a\| = \lim_\alpha \|a_\alpha\|$. Obviously , $\|\cdot\|$ is a C^*–seminorm on $\mathcal{L}$. Further , let

$$\begin{aligned} \vartheta &= \{a \in \mathcal{L} \mid \|a\| = 0\} \\ &= \{(a_l) \in \mathcal{J} \mid \text{there is an index } \alpha \text{ such that } a_\beta = 0, \forall \beta \geq \alpha.\} \end{aligned}$$

Clearly, ϑ is a $*$ two–sided ideal of $\mathcal{L}$. Denote the canonical map from $\mathcal{L}$ onto $\mathcal{L}/\vartheta$ by $a \longrightarrow \tilde{a} = a + \vartheta (\forall a \in \mathcal{L})$. Then $\|\tilde{a}\| = \|a\| \, (\forall \tilde{a} \in \mathcal{L}/\vartheta, a \in \tilde{a})$ is a C^*–norm on $\mathcal{L}/\vartheta$. Further , we get a C^*–algebra $A = (\mathcal{L}/\vartheta, \|\cdot\|)^-$.

Definition 3.7.1. Denote the above C^*–algebra A by

$$\varinjlim_\alpha \{A_\alpha, \Phi_{\beta\alpha} \mid (\alpha, \beta) \in \Pi \times \Pi, \text{and } \alpha \leq \beta\}.$$

It is called the *inductive limit* of $\{A_\alpha \mid \alpha \in \Pi\}$ defined by the family of $*$ isomorphisms $\{\Phi_{\beta,\alpha} \mid (\alpha, \beta) \in \Pi \times \Pi, \text{and } \alpha \leq \beta\}$.

Now for any $\alpha \in \Pi$, let

$$\mathcal{L}_\alpha = \{(a_l) \in \mathcal{J} \mid a_\beta = \Phi_{\beta\alpha}(a_\alpha), \forall \beta \geq \alpha\}.$$

Clearly, $\mathcal{L}_\alpha$ is a $*$ subalgebra of $\mathcal{L}$, and

$$\widetilde{A}_\alpha = \mathcal{L}_\alpha/\vartheta = \{\tilde{a} \in \mathcal{L}/\vartheta \mid \text{ there is } (a_l) \in \tilde{a} \text{ such that } a_l = \Phi_{l\alpha}(a_\alpha) \text{ if } l \geq \alpha; a_l = 0 \text{ if } l \not\geq \alpha.\}$$

For any $a_\alpha \in A_\alpha$, define

$$a_l = \begin{cases} \Phi_{l\alpha}(a_\alpha), & \text{if } l \geq \alpha, \\ 0, & \text{if } l \not\geq \alpha. \end{cases}$$

Then $\Phi_\alpha(a_\alpha) = \widetilde{(a_l)_{l\in\mathbb{I}}}(\forall a_\alpha \in A_\alpha)$ determines a $*$ isomorphism from A_α onto $\widetilde{A}_\alpha$. Consequently, $\widetilde{A}_\alpha$ becomes a C^*–subalgebra of A.

We claim that

$$\Phi_\alpha = \Phi_\beta \Phi_{\beta\alpha}, \quad \forall \alpha \leq \beta.$$

In fact , for any $a_\alpha \in A_\alpha$ and $\beta \geq \alpha$ we have

$$\Phi_\alpha(a_\alpha) = \left(a_l = \begin{cases} \Phi_{l\alpha}(a_\alpha), & \text{if } l \geq \alpha, \\ 0, & \text{if } l \not\geq \alpha \end{cases} \right)_{l\in\mathbb{I}} + \vartheta$$

and

$$\Phi_\beta \Phi_{\beta\alpha}(a_\alpha) = \left(b_l = \begin{cases} \Phi_{l\beta}\Phi_{\beta\alpha}(a_\alpha) = \Phi_{l\alpha}(a_\alpha), & \text{if } l \geq \beta \\ 0, & \text{if } l \not\geq \beta \end{cases} \right)_{l\in\mathbb{I}} + \vartheta.$$

Therefore, $\Phi_\alpha = \Phi_\beta \Phi_{\beta\alpha}, \forall \alpha \leq \beta$. Consequently,

$$\widetilde{A}_\alpha = \Phi_\alpha(A_\alpha) = \Phi_\beta\Phi_{\beta\alpha}(A_\alpha) \subset \Phi_\beta(A_\beta) = \widetilde{A}_\beta, \forall \alpha \leq \beta.$$

Moreover, since $\mathcal{L} = \cup_{\alpha\in\mathbb{I}} \mathcal{L}_\alpha$, it follows that $\mathcal{L}/\vartheta = \cup_{\alpha\in\mathbb{I}} \widetilde{A}_\alpha$. Thus , we have the following.

Theorem 3.7.2. Let $A = \varinjlim_\alpha \{A_\alpha, \Phi_{\beta\alpha} \mid (\alpha, \beta) \in \mathbb{I} \times \mathbb{I}, \text{and } \alpha \leq \beta\}$. Then there is a family $\{\widetilde{A}_\alpha \mid \alpha \in \mathbb{I}\}$ of C^*–subalgebras of A, and a $*$ isomorphism Φ_α from A_α onto $\widetilde{A}_\alpha$ for each $\alpha \in \mathbb{I}$, such that : 1) $\widetilde{A}_\alpha \subset \widetilde{A}_\beta, \forall \alpha \leq \beta$; 2) $\Phi_\alpha = \Phi_\beta \Phi_{\beta\alpha}, \forall \alpha \leq \beta$; 3) $\cup_{\alpha\in\mathbb{I}} \widetilde{A}_\alpha$ is dense in A.

Conversely, we also have the following.

Theorem 3.7.3. Let $A = \varinjlim_\alpha \{A_\alpha, \Phi_{\beta,\alpha} \mid \alpha, \beta \in \mathbb{I}, \text{ and } \alpha \leq \beta\}, B$ be a C^* algebra, $\{B_\alpha \mid \alpha \in \mathbb{I}\}$ be a family of C^* –subalgebras of B, and Ψ_α be a $*$ isomorphism from A_α onto $B_\alpha, \forall \alpha \in \mathbb{I}$. Assume that : 1) $B_\alpha \subset B_\beta, \forall \alpha \leq \beta$; 2) $\Psi_\alpha = \Psi_\beta \Phi_{\beta\alpha}, \forall \alpha \leq \beta$; 3) $\cup_{\alpha\in\mathbb{I}} B_\alpha$ is dense in B. Then there is a $*$ isomorphism Ψ from A onto B such that $\Psi(\widetilde{A}_\alpha) = B_\alpha, \Psi\Phi_\alpha = \Psi_\alpha, \forall \alpha \in \mathbb{I}$, where $\{\widetilde{A}_\alpha, \Phi_\alpha \mid \alpha \in \mathbb{I}\}$ is defined by Theorem 3.7.2.

Proof. For any $\alpha \in \mathbb{I}, \Psi_\alpha \Phi_\alpha^{-1}$ is a $*$ isomorphism from $\tilde{A}_\alpha$ onto B_α. Now we claim that

$$\Psi_\beta \Phi_\beta^{-1} \mid \tilde{A}_\alpha = \Psi_\alpha \Phi_\alpha^{-1}, \forall \alpha \leq \beta.$$

In fact , $\Psi_\beta \Phi_\beta^{-1}(a) = \Psi_\beta \Phi_\beta^{-1} \Phi_\alpha \Phi_\alpha^{-1}(a) = \Psi_\beta \Phi_\beta^{-1} \cdot \Phi_\beta \Phi_{\beta\alpha} \Phi_\alpha^{-1}(a) = (\Psi_\beta \Phi_{\beta\alpha}) \Phi_\alpha^{-1}(a) = \Psi_\alpha \Phi_\alpha^{-1}(a), \forall a \in \tilde{A}_\alpha$. Then we can define a $*$ isomorphism Ψ from $\cup_{\alpha \in \mathbb{I}} \tilde{A}_\alpha$ onto $\cup_{\alpha \in \mathbb{I}} B_\alpha$ such that

$$\Psi \mid \tilde{A}_\alpha = \Psi_\alpha \Phi_\alpha^{-1}, \forall \alpha \in \mathbb{I}.$$

Clearly, Ψ is isometric. Thus Ψ can be uniquely extended to a $*$ isomorphism from A onto B, still denoted by Ψ. Then Ψ is what we want to find.

Q.E.D.

Corollary 3.7.4. Let A be a C^*–algebra, and $\{A_\alpha \mid \alpha \in \mathbb{I}\}$ be a family of C^*–subalgebras of A with $A_\alpha \subset A_\beta, \forall \alpha \leq \beta$, and $\overline{\cup_\alpha A_\alpha} = A$. Denote the embedding map from A_α into A_β by $\Phi_{\beta\alpha}, \forall \alpha \leq \beta$. Then A is $*$ isomorphic to $\varinjlim_\alpha \{A_\alpha, \Phi_{\beta\alpha} \mid \alpha, \beta \in \mathbb{I}, \alpha \leq \beta\}$.

Proof. Using Theorem 3.7.3 to the case: $B = A, B_\alpha = A_\alpha$, and $\Psi_\alpha = 1_\alpha$ (the identity map on A_α) , $\forall \alpha \in \mathbb{I}$, we can get the conclusion immediately.

Q.E.D.

Theorem 3.7.5. Let $A = \varinjlim_\alpha \{A_\alpha, \Phi_{\beta\alpha} \mid \alpha, \beta \in \mathbb{I}, \alpha \leq \beta\}$, and $B = \varinjlim_\alpha \{B_\alpha, \Psi_{\beta\alpha} \mid \alpha, \beta \in \mathbb{I}, \alpha \leq \beta\}$. Suppose that there exists a $*$ isomorphism $\wedge_\alpha$ from A_α onto $B_\alpha (\forall \alpha \in \mathbb{I})$ such that $\wedge_\beta \Phi_{\beta\alpha} = \Psi_{\beta\alpha} \wedge_\alpha$ $(\forall \alpha \leq \beta)$. Then A is $*$ isomorphic to B.

Proof. By Theorem 3.7.2, $A = \overline{\cup_\alpha \tilde{A}_\alpha}, B = \overline{\cup_\alpha \tilde{B}_\alpha}$, and there is a $*$ isomorphism Φ_α from A_α onto $\tilde{A}_\alpha$, and a $*$ isomorphism Ψ_α from B_α onto $\tilde{B}_\alpha, \forall \alpha \in \mathbb{I}$, such that

$$\Phi_\alpha = \Phi_\beta \Phi_{\beta\alpha}, \Psi_\alpha = \Psi_\beta \Psi_{\beta\alpha}, \forall \alpha \leq \beta.$$

Then $\Psi_\alpha \wedge_\alpha$ is a $*$ isomorphism from A_α onto $\tilde{B}_\alpha, \forall \alpha \in \mathbb{I}$, and

$$(\Psi_\beta \wedge_\beta) \Phi_{\beta\alpha} = (\Psi_\beta \Psi_{\beta\alpha}) \wedge_\alpha = \Psi_\alpha \wedge_\alpha, \forall \alpha \leq \beta.$$

Now by Theorem 3.7.3, A is $*$ isomorphic to B. Q.E.D.

Let C, D be two algebras with the identities $1_C, 1_D$ respectively. A linear map Φ from C to D is said to be *unital* , if $\Phi(1_C) = 1_D$.

Proposition 3.7.6. Let $A = \varinjlim_n\{A_n, \Phi_{mn} \mid m,n = 1,2,\cdots,m \geq n\}$, and $B = \varinjlim_n\{B_n, \Psi_{mn} \mid m,n = 1,2,\cdots,m \geq n\}$.

Suppose that A_n and B_n are $*$ isomorphic to M_{p_n}, where M_{p_n} is the algebra of $p_n \times p_n$ matrices, $\forall n$, and Φ_{mn} and Ψ_{mn} are unital , $\forall m \geq n$. Then A and B are $*$ isomorphic.

Proof. By Theorem 3.7.2, we can write $A = \overline{\cup_n \widetilde{A}_n}$ and $B = \overline{\cup_n \widetilde{B}_n}$, where $\widetilde{A}_n$ and $\widetilde{B}_n$ are $*$ isomorphic to $M_{p_n}, \forall n$, and

$$1_A \in \widetilde{A}_1 \subset \cdots \subset \widetilde{A}_n \subset \cdots, \quad 1_B \in \widetilde{B}_1 \subset \cdots \subset \widetilde{B}_n \subset \cdots.$$

Now it suffices to construct a $*$ isomorphism $\wedge_n$ from $\widetilde{A}_n$ onto $\widetilde{B}_n$ for each n such that

$$\wedge_{n+1} \mid \widetilde{A}_n = \wedge_n, \forall n.$$

Suppose that $\wedge_1, \cdots, \wedge_n$ satisfy the conditions. We need to construct $\wedge_{n+1}$ such that $\wedge_{n+1} \mid \widetilde{A}_n = \wedge_n$.

Let $\{e_{ij} \mid 1 \leq i,j \leq p_n\}$ be the matrix unit of $\widetilde{A}_n$, i.e.,

$$e_{ij}^* = e_{ji}, e_{ij}e_{kl} = \delta_{jk}e_{ie}, \forall i,j,k,l.$$

We identify $\widetilde{A}_{n+1}$ with $M_{p_{n+1}} = B(H)$, where H is a Hilbert space of p_{n+1}-dimension. Then $\{e_{ii} \mid 1 \leq i \leq p_n\}$ is a family of pairwise orthogonal projections on H, and $\sum_i e_{ii} = 1_H$, and $e_{ii} \sim e_{jj}$ (relative to the VN algebra $B(H)$) , $\forall i,j$. Put $H_i = e_{ii}H, \forall i$, then dim $H_i = m$ is independent of i . Clearly, $m = p_n^{-1}p_{n+1}, i.e., p_n \mid p_{n+1}$. Now we can pick $\{v_{1i,1j} \mid 1 \leq i,j \leq m\} \subset \widetilde{A}_{n+1}$ such that

$$v_{1i,1j}^* = v_{1j,1i}, v_{1i,1j}v_{1k,1l} = \delta_{jk}v_{1i,1l},$$

$\forall 1 \leq i,j,k,l \leq m$, and $\sum_{i=1}^m v_{1i,1i} = e_{11}$. Further , let

$$v_{ij,kl} = e_{i1}v_{1j,1l}e_{1k},$$

$\forall 1 \leq i,k \leq p_n, 1 \leq j,l \leq m$. It is easy to check that $\{v_{ij,kl} \mid 1 \leq i,k \leq p_n, 1 \leq j,l \leq m\}$ is a matrix unit of $\widetilde{A}_{n+1}$.

Put $f_{ij} = \wedge_n(e_{ij}), \forall i,j$. Clearly , $\{f_{ij} \mid 1 \leq i,j \leq p_n\}$ is a matrix unit of $\widetilde{B}_n$. Similarly , we can find $\{u_{1i,1j} \mid 1 \leq i,j \leq m\} \subset \widetilde{B}_{n+1}$ such that

$$u_{1i,1j}^* = u_{1j,1i}, u_{1i,1j}u_{1k,1l} = \delta_{jk}, u_{1i,1l},$$

$\forall 1 \leq i,j,k,l \leq m$, and $\sum_{i=1}^m u_{1i,1i} = f_{11}$. Furhter , let

$$u_{ij,kl} = f_{i1}u_{1j,1l}f_{1k}, 1 \leq i,k \leq p_n, 1 \leq j,l \leq m.$$

Then it is a matrix unit of $\widetilde{B}_{n+1}$.

Finally, define

$$\wedge_{n+1} v_{ij,kl} = u_{ij,kl}, 1 \le i,k \le p_n, 1 \le j,l \le m.$$

Then $\wedge_{n+1}$ is a $*$ isomorphism from $\tilde{A}_{n+1}$ onto $\tilde{B}_{n+1}$, and $\wedge_{n+1} \mid \tilde{A}_n = \wedge_n$.

Q.E.D.

Now let $A = \varinjlim_{\alpha}\{A_\alpha, \Phi_{\beta\alpha} \mid \alpha, \beta \in \mathit{II}, \alpha \le \beta\}$. And suppose that φ_α is a state on $A_\alpha, \forall \alpha \in \mathit{II}$, such that

$$\varphi_\beta(\Phi_{\beta\alpha}(a_\alpha)) = \varphi_\alpha(a_\alpha), \forall a_\alpha \in A_\alpha, \alpha \le \beta.$$

Further , define a state $\tilde{\varphi}_\alpha$ on $\tilde{A}_\alpha$ as follows:

$$\tilde{\varphi}_\alpha(a_\alpha) = \varphi_\alpha(\Phi_\alpha^{-1}(a_\alpha)), \forall a_\alpha \in \tilde{A}_\alpha,$$

$\forall \alpha \in \mathit{II}$, where $\{\tilde{A}_\alpha, \Phi_\alpha \mid \alpha \in \mathit{II}\}$ is defined in Theorem 3.7.2. Then we have

$$\begin{aligned} \tilde{\varphi}_\beta(a_\alpha) &= \varphi_\beta(\Phi_\beta^{-1}(a_\alpha)) = \varphi_\beta(\Phi_\beta^{-1}\Phi_\alpha\Phi_\alpha^{-1}(a_\alpha)) \\ &= \varphi_\beta(\Phi_\beta^{-1}\Phi_\beta\Phi_{\beta\alpha}\Phi_\alpha^{-1}(a_\alpha)) = \varphi_\alpha(\Phi_\alpha^{-1}(a_\alpha)) = \tilde{\varphi}_\alpha(a_\alpha), \end{aligned}$$

$\forall a_\alpha \in \tilde{A}_\alpha, \alpha \le \beta$. Thus , we obtain a linear functional φ on $\cup_\alpha \tilde{A}_\alpha$ defined by $\{\tilde{\varphi}_\alpha \mid \alpha \in \mathit{II}\} : \varphi \mid \tilde{A}_\alpha = \tilde{\varphi}_\alpha, \forall \alpha \in \mathit{II}$. Clearly, φ can be uniquely extended to a state on A, still denoted by φ.

Definition 3.7.7. The above state φ on $A = \varinjlim_{\alpha}\{A_\alpha, \Phi_{\beta\alpha} \mid \alpha, \beta \in \mathit{II}, \alpha \le \beta\}$ is called the *inductive limit* of $\{\varphi_\alpha, \Phi_{\beta\alpha} \mid \alpha, \beta \in \mathit{II}, \alpha \le \beta\}$, and denoted by

$$\varphi = \varinjlim_{\alpha}\{\varphi_\alpha, \Phi_{\beta\alpha} \mid \alpha, \beta \in \mathit{II}, \alpha \le \beta\}.$$

Notes. The inductive limit of C^*-algebras was studied first by Z.Takeda.

References. [150], [167].

3.8. Infinite tensor products of C^*–algebras

Let $\wedge$ be an index set , and A_l be a C^*–algebra with an identity $1_l, \forall l \in \wedge$. Let

$$\otimes_{l\in\wedge} A_l = \{u_F \otimes \otimes_{l \notin F} 1_l \mid F \text{ is any finite subset of } \wedge, u_F \in \otimes_{l\in F} A_l\}.$$

We say that $u = u_F \otimes \otimes_{l \notin F} 1_l = 0$, if $\otimes_{l \in \wedge} \varphi_l(u) = \otimes_{l \in F} \varphi_l(u_F) = 0, \forall \varphi_l \in S_l$ (the state space of A_l), $l \in \wedge$. Clearly, $\otimes_{l \in \wedge} A_l$ becomes a $*$ algebra in a natural way , and is called the *algebraic tensor product* of $\{A_l \mid l \in \wedge\}$.

A norm $\|\cdot\|$ on $\otimes_{l \in \wedge} A_l$ is called a *cross-norm*, if for any finite subset F of $\wedge$ and $a_l \in A_l, \forall l \in F$, we have

$$\|\otimes_{l \in F} a_l \otimes \otimes_{l \notin F} 1_l\| = \Pi_{l \in F} \|a_l\| .$$

For example,

$$\begin{aligned} \lambda(u) = \quad & \sup\{|\otimes_{l \in F} f_l \otimes \otimes_{l \notin F} \varphi_l(u)| \mid F \text{ is any finite subset of } \wedge; \\ & f_l \in A_l^*, \|f_l\| \leq 1, \forall l \in F; \varphi_l \in S_l, \forall l \notin F\} \end{aligned}$$

and

$$\gamma(u) = \inf\{\sum_j \Pi_{l \in F} \left\|a_j^{(l)}\right\| \mid u = \sum_j \otimes_{l \in F} a_j^{(l)} \otimes \otimes_{l \notin F} 1_l\}$$

are two cross-norms on $\otimes_{l \in \wedge} A_l$, and $\gamma(\cdot)$ is the largest cross-norm.

A norm $\alpha(\cdot)$ on $\otimes_{l \in \wedge} A_l$ is called a *C^*-norm* , if

$$\alpha(uv) \leq \alpha(u)\alpha(v), \alpha(u^* u) = \alpha(u)^2,$$

$\forall u, v \in \otimes_{l \in \wedge} A_l$. Then the completion of $(\otimes_{l \in \wedge} A_l, \alpha(\cdot))$, denoted by α - $\otimes_{l \in \wedge} A_l$, is called the *tensor product* of $\{A_l \mid l \in \wedge\}$ with respect to $\alpha(\cdot)$.

Similar to the discussion in Sections 3.1 –3.5, we can see that: there is a spatial C^*-norm $\alpha_0(\cdot)$ on $\otimes_{l \in \wedge} A_l$; and any C^* -norm $\alpha(\cdot)$ must satisfy the inequality $\lambda(\cdot) \leq \alpha_0(\cdot) \leq \alpha(\cdot) \leq \gamma(\cdot)$ on $\otimes_{l \in \wedge} A_l$; in particular, any C^*-norm $\alpha(\cdot)$ is a cross-norm ; and etc.

Now let $\mathbb{F} = \{F \mid F \text{is any finite subset of } \wedge\}$. $\mathbb{F}$ is a directed index set with respect to the inclusion relation. Suppose that $\alpha(\cdot)$ is a C^*-norm on $\otimes_{l \in \wedge} A_l$. For any $F \in \mathbb{F}$, let

$$\alpha_F(u_F) = \alpha(u_F \otimes \otimes_{l \notin F} 1_l), \forall u_F \in \otimes_{l \in F} A_l.$$

Then $\alpha_F(\cdot)$ is a C^*-norm on $\otimes_{l \in F} A_l, \forall F \in \mathbb{F}$. Put $B_F = \alpha_F$ - $\otimes_{l \in F} A_l, \forall F \in \mathbb{F}$, and for any $F', F \in \mathbb{F}$ with $F' \supset F$, define

$$\Phi_{F'F}(u_F) = u_F \otimes \otimes_{l \in F' \setminus F} 1_l, \forall u_F \in \otimes_{l \in F} A_l.$$

Then $\Phi_{F'F}$ can be uniquely extended to a unital $*$ isomorphism from B_F into $B_{F'}$, and

$$\Phi_{F''F} = \Phi_{F''F'} \Phi_{F'F}, \forall F'' \supset F' \supset F.$$

By Theorem 3.7.3, it is easy to prove the following.

Proposition 3.8.1. α - $\otimes_{l \in \wedge} A_l$ is $*$ isomorphic to

$$\varinjlim_F \{B_F, \Phi_{F'F} \mid F', F \in \mathbb{F}, \text{and } \ F \subset F'\}.$$

As an example , we consider the following.

Definition 3.8.2. Let A be a C^*-algebra with an identity. A is said to be (UHF) (*uniformly hyperfinite*) , if there exists an increasing sequence (i.e. $A_n \subset A_m, \forall n \leq m$) of C^*-subalgebras A_n containing the identity such that the uniform closure of $\cup_n A_n$ is A and A_n is $*$ isomorphic to $B(H_n)($ $\dim H_n < \infty), \forall n$.

If $\dim H_n = p_n, \forall n, A$ is also called a (UHF) C^*-algebra of type $\{p_n \mid n = 1, 2, \cdots\}$.

Let Φ_{mn} be the embedding map from A_n into $A_m, \forall n \leq m$. Then A is $*$ isomorphic to

$$\varinjlim_n \{A_n, \Phi_{mn} \mid m, n = 1, 2, \cdots, n \leq m\}$$

by Corollary 3.7.4. Moreover, from Proposition 3.7.6 the (UHF) C^*-algebras of same type are $*$ isomorphic.

Proposition 3.8.3. A (UHF) C^*-algebra of type $\{p_n\}$ is existential if and only if $p_n \mid p_{n+1}, \forall n$. And it is $*$ isomorphic to the infinite tensor product α_0 - $\otimes_{n=1}^{\infty} M_{m_n}$, where M_{m_n} is the algebra of $m_n \times m_n$ matrices, and $m_1 = p_1, m_n = p_{n-1}^{-1} p_n, \forall n \geq 2$.

Proof. The necessity is contained in the proof of Proposition 3.7.6 indeed. Conversely, let $p_n \mid p_{n+1}, \forall n$. Since α_0 - $\otimes_{i=1}^{n} M_{m_i} = M_{p_n}, \forall n$, it follows that $\otimes_{n=1}^{\infty} M_{m_n}$ is a $(UHF) C^*$-algebra of type $\{p_n\}$.

Moreover, all $(UHF) C^*$-algebras of same type are $*$ isomorphic. Therefore, the $(UHF) C^*$-algebra of type $\{p_n\}$ is $*$ isomorphic to α_0 -$\otimes_{n=1}^{\infty} M_{m_n}$.

Q.E.D.

Notice that $\otimes_{i=1}^{n} M_{m_i}$ is finite dimensional, $\forall n$. Thus from Proposition 2.1.10 there is only one C^*-norm $\alpha_0(\cdot)$ on $\otimes_{n=1}^{\infty} M_{m_n}$ indeed.

In chapter 15, we shall discuss the (UHF) algebras furthermore.

Definition 3.8.4. Let Λ be an index set , and H_l be a Hilbert space , $\forall l \in \Lambda$. Fix $\xi_l \in H_l$ with $\|\xi_l\| = 1, \forall l \in \Lambda$. The *tensor product* of $\{H_l \mid l \in \Lambda\}$ with respect to the vector $\xi = (\xi_l)_{l \in \Lambda}$, denoted by $\otimes_{l \in \Lambda}^{\xi} H_l$, is the completion of $(\odot_{l \in \Lambda}^{\xi} H_l, \langle , \rangle)$, where

$$\odot_{l \in \Lambda}^{\xi} H_l = \cup \{\odot_{l \in F} H_l \otimes \otimes_{l \notin F} \xi_l \mid F \subset \Lambda, {}^{\#}F < \infty\}$$

and

$$\langle \otimes_{l \in F} \eta_l \otimes \otimes_{l \notin F} \xi_l, \otimes_{l \in F} \varsigma_l \otimes \otimes_{l \notin F} \xi_l \rangle = \Pi_{l \in F} \langle \eta_l, \varsigma_l \rangle,$$

$\forall \eta_l, \varsigma_l \in H_l, l \in F$, and $F \subset \Lambda, {}^{\#}F < \infty$.

Lemma 3.8.5 Let M, N be VN algebras on Hilbert spaces H, K respectively, and $f \in M_*, g \in N_*$. Then $f \otimes g$ can be uniquely extended to a σ-continuous linear functional on $M \overline{\otimes} N$ with norm $\|f\| \cdot \|g\|$, where $(f \otimes g)(x \otimes y) = f(x)g(y), \forall x \in M, y \in N$.

Proof. Pick $s \in T(H)$, and $t \in T(K)$ such that

$$f(x) = tr(sx), g(y) = tr(ty), \forall x \in M, y \in N.$$

It is obvious that $s \otimes t \in T(H \otimes K), \|s \otimes t\|_1 = \|s\|_1 \cdot \|t\|_1$, and $tr((s \otimes t)(x \otimes y)) = f(x)g(y), \forall x \in M, y \in N$. Clearly, $\varphi(\cdot) = tr((s \otimes t)\cdot) \in (M \overline{\otimes} N)_*$, and φ is the unique extension of $f \otimes g$ on $M \otimes N$ with $\|\varphi\| \geq \|f\| \cdot \|g\|$. However, $\|\varphi\| \leq \|s \otimes t\|_1 = \|s\|_1 \cdot \|t\|_1$, and $\|s\|_1, \|t\|_1$ can be approximated arbitrarily to $\|f\|, \|g\|$ respectively. Therefore, $\|\varphi\| = \|f\| \cdot \|g\|$. Q.E.D.

Proposition 3.8.6. Let $H = \otimes_{l \in \wedge}^{\xi} H_l, M_l$ be a VN algebra on $H_l (\forall l \in \wedge)$, and M be the VN algebra on H generated by $\{M_l \otimes \otimes_{l' \neq l} 1_{l'} \mid l \in \wedge\}$. Then we have the following:

1) M' is generated by $\{M_l' \otimes \otimes_{l' \neq l} 1_{l'} \mid l \in \wedge\}$;
2) $M = B(H)$ if and only if $M_l = B(H_l), \forall l \in \wedge$;
3) M is a factor if and only if M_l is a factor, $\forall l \in \wedge$.

Proof. 1) Let F be a finite subset of $\wedge$. Clearly,

$$H = (\otimes_{l \in F} H_l) \otimes (\otimes_{l \notin F}^{\xi} H_l), \quad M = M_F \overline{\otimes} M_{\wedge \setminus F},$$

where M_F is the tensor product of $\{M_l \mid l \in F\}$, i.e., $M_F = \overline{\otimes}_{l \in F} M_l$, and $M_{\wedge \setminus F}$ is generated by $\{M_l \otimes \otimes_{l' \notin F, l' \neq l} 1_{l'} \mid l \notin F\}$. By Theorem 1.4.12,

$$M' = M_F' \overline{\otimes} M_{\wedge \setminus F}'.$$

Let $g_F(\cdot) = \langle \cdot \otimes_{l \notin F} \xi_l, \otimes_{l \notin F} \xi_l \rangle \in (M_{\wedge \setminus F}')_*$. By Lemma 3.8.5, there is a linear map Φ_F from M' to M_F' such that

$$\Phi_F(x)(f_F) = (f_F \otimes g_F)(x), \forall x \in M', f_F \in (M_F')_*.$$

For any $\eta = \eta_F \otimes \otimes_{l \notin F} \xi_l, \varsigma = \varsigma_F \otimes \otimes_{l \notin F} \xi_l \in \odot_{l \in \wedge}^{\xi} H_l$, where $\eta_F, \varsigma_F \in \odot_{l \in F} H_l$, define

$$t \cdot = \langle \cdot, \eta \rangle \varsigma, \forall \cdot \in H,$$

$$t_F \cdot = \langle \cdot, \eta_F \rangle \varsigma_F, \forall \cdot \in \otimes_{l \in F} H_l$$

and

$$t_{\wedge \setminus F} \cdot = \langle \cdot, \otimes_{l \notin F} \xi_l \rangle \otimes_{l \notin F} \xi_l, \forall \cdot \in \otimes_{l \notin F}^{\xi} H_l.$$

Clearly, $t, t_F, t_{\wedge \setminus F}$ are one rank operators on $H, \otimes_{l \in F} H_l, \otimes_{l \notin F}^{\xi} H_l$ respectively, and $t = t_F \otimes t_{\wedge \setminus F}$, and $g_F(\cdot) = tr(t_{\wedge \setminus F} \cdot), \forall \cdot \in M_{\wedge \setminus F}'$. Let

$$f_F(\cdot) = tr(t_F \cdot) \in (M_F')_*, \text{and} \quad p_F = \otimes_{l \notin F} 1_l.$$

Then for any $x \in M'$, we have

$$t_r(t \cdot \Phi_F(x) \otimes p_F) = tr(t_F \cdot \Phi_F(x)) \cdot tr(t_{\wedge \setminus F} \cdot p_F)$$
$$= \Phi_F(x)(f_F) = (f_F \otimes g_F)(x) = tr(tx).$$

From the preceding paragraph, we can see that

$$\lim_F tr(t \cdot \Phi_F(x) \otimes p_F) = tr(tx),$$

$\forall x \in M'$, and any one rank operator t on H with form $t \cdot = \langle \cdot, \eta \rangle \varsigma$ for some $\eta, \varsigma \in \odot^{\xi}_{l \in \wedge} H_l$. Since $\{\Phi_F(x) \otimes p_F \mid F\}$ is bounded, it follows that

$$\Phi_F(x) \otimes p_F \longrightarrow x(\sigma(B(H), T(H)), \forall x \in M'.$$

By Theorem 1.4.12, $\Phi_F(x) \subset M'_F = \overline{\otimes}_{l \in F} M'_l, \forall x \in M'$. Thus

$$M' \subset \{M'_l \otimes \otimes_{l' \neq l} 1_{l'} \mid l \in \wedge\}''.$$

However, it is obvious that $M'_l \otimes \otimes_{l' \neq l} 1_{l'} \subset M', \forall l \in \wedge$. Therefore , M' is generated by $\{M'_l \otimes \otimes_{l' \neq l} 1_{l'} \mid l \in \wedge\}$.

2) $M = B(H)$ if and only if $M' = \mathbb{C} 1_H$. By 1), This is equivalent to $M'_l = \mathbb{C} 1_l, \forall l \in \wedge$, i.e., $M_l = B(H_l), \forall l \in \wedge$.

3) M is a factor on H if and only if $\{M, M'\}'' = B(H)$. By 1) , this is equaivalent to the fact

$$\{(M_l \cup M'_l) \otimes \otimes_{l' \neq l} 1_{l'} \mid l \in \wedge\}'' = B(H).$$

By 2) , the fact is equivalent to $(M_l \cup M'_l)'' = B(H_l), \forall l \in \wedge$, i.e., M_l is a factor on $H_l, \forall l \in \wedge$. Q.E.D.

A state φ on a C^*-algebra is said to be *factorial* , if the $*$ representation π_φ is factorial (Definition 2.10.5) , where π_φ is generated by φ.

Proposition 3.8.7. Let $\wedge$ be an index set , A_l be a C^*-algebra with an identity 1_l , and φ_l be a state on $A_l, \forall l \in \wedge$. If $\alpha(\cdot)$ is a C^*-norm on $\otimes_{l \in \wedge} A_l$, and $A = \alpha\text{-}\otimes_{l \in \wedge} A_l$, then $\otimes_{l \in \wedge} \varphi_l$ can be uniquely extended to a state φ on A. Moreover, φ is pure or factorial if and only if φ_l is pure or factorical , $\forall l \in \wedge$.

Proof. Let $\{\pi_l, H_l, \xi_l\}$ be the cyclic $*$ representation of A_l generated by $\varphi_l, \forall l \in \wedge$, and

$$H = \otimes^{\xi}_{l \in \wedge} H_l, \quad \pi = \otimes_{l \in \wedge} \pi_l,$$

where $\xi = (\xi_l)_{l \in \wedge}$. By Proposition 3.2.8 and $\alpha(\cdot) \geq \alpha_0(\cdot), \{\pi, H, \xi\}$ is a cyclic $*$ representation of A.

By Corollary 3.2.7 and $\alpha(\cdot) \geq \alpha_0(\cdot), \otimes_{l\in\Lambda}\varphi_l$ can be uniquely extended to a state φ on A. Let $\{\pi_\varphi, H_\varphi, \xi_\varphi\}$ be the cyclic $*$ representation of A generated by φ. Since

$$\langle \pi(u)\xi, \xi\rangle = \otimes_{l\in\Lambda}\varphi_l(u) = \varphi(u) = \langle \pi_\varphi(u)\xi_\varphi, \xi_\varphi\rangle,$$

$\forall u \in \otimes_{l\in\Lambda}A_l, \{\pi, H\}$ is unitarily equivalent to $\{\pi_\varphi, H_\varphi\}$. That comes to the conclusion from Proposition 3.8.6 immediately. Q.E.D.

Remark. By Proposition 3.8.1 , $A = \alpha$ - $\otimes_{l\in\Lambda}A_l$ is $*$ isomorphic to

$$B = \varinjlim_F\{B_F, \Phi_{F'F} \mid F', F \in \mathbb{F} \quad \text{and} \quad F \subset F'\}.$$

Clearly, $\otimes_{l\in F}\varphi_l$ can be uniquely extended to a state φ_F on $B_F = \alpha_F$-$\otimes_{l\in F}\varphi_l$, $\forall\ F \in \mathbb{F}$. By Definition 3.7.7, the state $\varphi = \otimes_{l\in\Lambda}\varphi_l$ on A is corresponding to the inductive limit state

$$\varinjlim_F\{\varphi_F, \Phi_{F'F} \mid F, F' \in \mathbb{F}, \text{and} \quad F \subset F'\}$$

on B

Notice that the algebra M_n of $n \times n$ matrices is simple, i.e., M_n contains no non-zero closed two-sided ideals (see Proposition 1.1.1) . So each non-zero $*$ representation of M_n is faithful. In consequence, each state on M_n is factorial.

Therefore, we have the following.

Proposition 3.8.8. Each product state on a (UHF) C^*-algebra α_0 - $\otimes_{n=1}^{\infty}M_{m_n}$ is factorial.

Notes. (UHF) algebras first appeared in the thesis of J.Glimm. The corresponding notion for VN algebra goes back to F.J. Murray and J.Von Neumann, see Chapter 7. J.Dixmier considered inductive limits of matrix algebras without the demand that the embeddings preserve units. Later, O. Bratteli gave a study of C^*-algebras which are inductive limits of arbitrary finite-dimensional C^*-algebras with arbitrary embeddings, see Chapter 15.

References. [26], [54], [116], [150], [167].

3.9 Nuclear C^*-algebras

Definition 3.9.1. A C^*-algebra A is said to be *nuclear*, if for every C^*-algebra B, there is only one C^*-norm on $A \otimes B$ (equivalently, the maximal C^*-norm $\alpha_1(\cdot)$ on $A \otimes B$ coincides with the spatial C^*-norm $\alpha_0(\cdot)$).

Proposition 3.9.2. Let A be a finite dimensional C^*–algebra. Then A is nuclear, and for any C^*–algebra $B, A \otimes B$ is complete with respect to the unique C^*–norm on $A \otimes B$.

Proof. If $A = M_n$, the algebra of $n \times n$ matrices, then the conclusion is the Lemma 3.6.1 exactly, and $M_n \otimes B = M_n(B)$ for any C^*–algebra.

Now if $A = \oplus_{k=1}^m M_{n_k}$, then for any C^*–algebra B, $A \otimes B = \oplus_{k=1}^m M_{n_k}(B)$ is a C^*–algebra with respect to a natural C^*–norm. Further by Proposition 2.1.10 there is only one C^*–norm on $A \otimes B$. Q.E.D.

Lemma 3.9.3. Let A be a C^*–algebra with no identity, $A^{(1)} = A \dot{+} \mathbb{C}1$, and $\alpha(\cdot)$ be a C^*–norm on $A^{(1)} \otimes B$, where B is a C^*–algebra. Then

$$\alpha\text{-}(A^{(1)} \otimes B) = \alpha\text{-}(A \otimes B) \dot{+} 1 \otimes B,$$

and

$$\alpha(x) = \tilde{\alpha}(x) = \sup\{\alpha(xu) | u \in A \otimes B, \alpha(u) \leq 1\},$$

$\forall x \in A^{(1)} \otimes B$. In consequence, any C^*–norm on $A \otimes B$ can be uniquely extended to a C^*–norm on $A^{(1)} \otimes B$.

Proof. Clearly, α - $(A \otimes B)$ is a closed two–sided ideal of α - $(A^{(1)} \otimes B)$, and $1 \otimes B$ is a C^*–subalgebra of α-$(A^{(1)} \otimes B)$. Hence, by Proposition 2.4.10 we have

$$\alpha\text{-}(A^{(1)} \otimes B) = \alpha\text{-}(A \otimes B) + 1 \otimes B.$$

Now let $\{u_n\}$ be a sequence of $A \otimes B$, and $b \in B$ such that $\alpha(u_n - 1 \otimes b) \longrightarrow 0$. Suppose that f is a pure state on $A^{(1)}$ with $f|A = 0$. By Corollary 3.2.6 and Theorem 3.5.10, $f \otimes g \in (\alpha\text{-}(A^{(1)} \otimes B))^*, \forall g \in B^*$. Then we have

$$g(b) = f \otimes g(1 \otimes b - u_n) \longrightarrow 0, \quad \forall g \in B^*,$$

and $b = 0$. Therefore,

$$\alpha\text{-}(A^{(1)} \otimes B) = \alpha\text{-}(A \otimes B) \dot{+} 1 \otimes B.$$

For any $x \in \alpha\text{-}(A \otimes B)$ and $b \in B$, pick a sequence $\{u_n\}$ of $(A \otimes B)$ such that $\alpha(u_n - x) \longrightarrow 0$. Then

$$\alpha((u_n - u_m) \cdot (1 \otimes b)) \leq \alpha(u_n - u_m)\tilde{\alpha}(1 \otimes b) \longrightarrow 0,$$

i.e., $\{u_n \cdot (1 \otimes b) | n\}$ is a Cauchy sequence of α-$(A \otimes B)$. Clearly, the limit of $\{u_n \cdot (1 \otimes b) | n\}$ is independent of the choice of $\{u_n\}$ with $\alpha(u_n - x) \longrightarrow 0$. Thus we can define $x \cdot (1 \otimes b) = \lim_n u_n \cdot (1 \otimes b)$, and $C = \alpha$ -$(A \otimes B) \dot{+} 1 \otimes B$ becomes a $*$ algebra naturally. Now C is a C^*–algebra with respect to $\alpha(\cdot)$ and $\tilde{\alpha}(\cdot)$. By Proposition 2.1.10, it must be $\alpha(\cdot) = \tilde{\alpha}(\cdot)$ on $A^{(1)} \otimes B$.

Q.E.D.

Lemma 3.9.4. Let A be a C^*-algebra with no identity, $A^{(1)} = A \dot{+} \mathbb{C}1$, and B be any C^*-algebra.

(i) If $\alpha(\cdot)$ is the maximal C^*-norm on $A^{(1)} \otimes B$, then $\alpha(\cdot)|(A \otimes B)$ is the maximal C^*-norm on $A \otimes B$.

(ii) If $\beta(\cdot)$ is the maximal C^*-norm on $A \otimes B$, then $\tilde{\beta}(\cdot)$ is the maximal C^*-norm on $A^{(1)} \otimes B$, where $\tilde{\beta}(\cdot)$ is the unique extendsion of $\beta(\cdot)$ on $A^{(1)} \otimes B$ (see Lemma 3.9.3).

(iii) If $\alpha(\cdot)$ is the minimal (spatial) C^*-norm on $A^{(1)} \otimes B$, then $\alpha(\cdot)|(A \otimes B)$ is the minimal C^*-norm on $A \otimes B$.

(iv) If $\beta(\cdot)$ is the minimal C^*-norm on $A \otimes B$, then $\tilde{\beta}(\cdot)$ is the minimal C^*-norm on $A^{(1)} \otimes B$.

Proof. (i) Let $\beta(\cdot)$ be any C^*-norm on $A \otimes B$. Then by $\tilde{\beta}(\cdot) \leq \alpha(\cdot)$ on $A^{(1)} \otimes B$, we have $\beta(\cdot) \leq \alpha(\cdot)|(A \otimes B)$. Hence, $\alpha(\cdot)|(A \otimes B)$ is the maximal C^* -norm on $A \otimes B$.

(ii) Let $\alpha(\cdot)$ be the maximal C^*-norm on $A^{(1)} \otimes B$. Then by (i) $\alpha(\cdot)|(A \otimes B) = \beta(\cdot)$. Further, from the uniqueness of extension we must have $\alpha(\cdot) = \tilde{\beta}(\cdot)$ on $A^{(1)} \otimes B$.

(iii) It is obvious by Theorem 3.2.9.

(iv) Let $\alpha(\cdot)$ be the minimal C^*-norm on $A^{(1)} \otimes B$. Then by (iii) $\alpha(\cdot)|(A \otimes B) = \beta(\cdot)$. Again by the uniqueness of extension, we obtain that $\alpha(\cdot) = \tilde{\beta}(\cdot)$ on $A^{(1)} \otimes B$. Q.E.D.

Proposition 3.9.5. Let A be a C^*-algebra with no identity. Then A is nuclear if and only if $A^{(1)} = A \dot{+} \mathbb{C}1$ is nuclear.

Proof. Let $A^{(1)}$ be nuclear, and B be any C^*-algebra. If $\alpha(\cdot)$ is the unique C^*-norm on $A^{(1)} \otimes B$, then by Lemma 3.9.4 (i) and (iii) $\alpha(\cdot)|(A \otimes B)$ is the maximal and the minimal C^*-norm on $A \otimes B$. Therefore, there is only one C^*-norm on $A \otimes B$, and A is nuclear.

Conversely, let A be nuclear, B be any C^*-algebra, and $\beta(\cdot)$ be the unique C^*-norm on $A \otimes B$. Then by Lemma 3.9.4 (ii) and (iv) $\tilde{\beta}(\cdot)$ is the maximal and the minimal C^*-norm on $A^{(1)} \otimes B$. Therefore, there is only one C^*-norm on $A^{(1)} \otimes B$, and $A^{(1)}$ is nuclear. Q.E.D.

Proposition 3.9.6. Let A be an abelian C^*-algebra. Then A is nuclear.

Proof. By Proposition 3.9.5, we may assume that A has an identity. If B is a C^*-algebra with an identity, then by Proposition 3.5.9 there is only one C^*-norm on $A \otimes B$. Now let B be a C^*-algebra with no identity, and $B^{(1)} = B \dot{+} \mathbb{C}1$.

Then there is only one C^*-norm $\alpha(\cdot)$ on $A\otimes B^{(1)}$. By Lemma 3.9.4 (i) and (iii) $\alpha(\cdot)|(A\otimes B)$ is the maximal and the minimal C^*-norm on $A\otimes B$. Therefore, there is only one C^*-norm on $A\otimes B$, and A is nuclear. Q.E.D.

Now consider the maximal C^*-norm.

Let $A_1,\cdots,A_n$ be C^*-algebras, and $A_i^{(1)}=A_i$ if A_i has an identity 1_i; $A_i^{(1)}=A_i\dotplus\mathbb{C}1_i$ if A_i has no identity, $1\le i\le n$. If $\alpha(\cdot)$ is a C^*-norm on $\otimes_{i=1}^n A_i$, then by Proposition 3.5.2,

$$\begin{aligned}\tilde{\alpha}(x) &= \sup\{\alpha(xu)|u\in\otimes_{i=1}^n A_i,\alpha(u)\le 1\}\\ &= \lim_{l_1,\cdots,l_n}\alpha\left(x\cdot\otimes_{i=1}^n d_{l_i}^{(i)}\right),\quad \forall x\in\otimes_{i=1}^n A_i^{(1)},\end{aligned}$$

is a C^*-norm on $\otimes_{i=1}^n A_i^{(1)}$ and is an extension of $\alpha(\cdot)$, where $\{d_{l_i}^{(i)}|l_i\}$ is an approximate identity for $A_i, 1\le i\le n$. We shall say that $\tilde{\alpha}(\cdot)$ is the canonical extension of $\alpha(\cdot)$.

If $\alpha(\cdot)$ is a C^*-norm on $\otimes_{i=1}^n A_i$, and $\mathbb{I}$ is a nonempty subset of $\{1,\cdots,n\}$, for any $v\in\otimes_{i\in\mathbb{I}}A_i$ define

$$\alpha_{\mathbb{I}}(v)=\tilde{\alpha}(v\otimes\otimes_{i\notin\mathbb{I}}1_i)=\lim_{l_i,i\notin\mathbb{I}}\alpha(v\otimes\otimes_{i\notin\mathbb{I}}d_{l_i}^{(i)}),$$

then $\alpha_{\mathbb{I}}(\cdot)$ is a C^*-norm on $\otimes_{i\in\mathbb{I}}A_i$. We shall say that $\alpha_{\mathbb{I}}(\cdot)$ is the C^*-norm induced by $\alpha(\cdot)$ on $\otimes_{i\in\mathbb{I}}A_i$.

Lemma 3.9.7. Let $\{\mathbb{I}_t|1\le t\le k\}$ be a partition of $\{1,\cdots,n\}$, i.e., $\emptyset\ne\mathbb{I}_t\subset\{1,\cdots,n\},\mathbb{I}_t\cap\mathbb{I}_{t'}=\emptyset,\forall 1\le t\ne t'\le k$, and $\sqcup_{t=1}^k\mathbb{I}_t=\{1,\cdots,n\},\beta_t(\cdot)$ be a C^*-norm on $\otimes_{i\in\mathbb{I}_t}A_i, 1\le t\le k$, and $\alpha(\cdot)$ be a C^*-norm on $\otimes_{t=1}^k B_t$, where $B_t=\beta_t$ - $\otimes_{i\in\mathbb{I}_t}A_i, 1\le t\le k$. Then $\alpha(\cdot)|\otimes_{i=1}^n A_i$ is a C^*-norm on $\otimes_{i=1}^n A_i$.

Proof. It suffices to show that: if $u\in\otimes_{i=1}^n A_i$, then $\alpha(u)=0$ if and only if $u=0$.

First, let $\alpha(u)=0$. By Corollary 3.2.6 and Theorem 3.5.10 we have $\otimes_{t=1}^k F_t\in(\alpha\text{-}\otimes_{t=1}^k B_t)^*,\forall F_t\in B_t^*, 1\le t\le k$. Hence, $\otimes_{t=1}^k F_t(u)=0,\forall F_t\in B_t^*,1\le t\le k$. Since $\otimes_{i\in\mathbb{I}_t}f_i\in B_t^*,\forall f_i\in A_i^*,i\in\mathbb{I}_t,1\le t\le k$, it follows that

$$\otimes_{i=1}^n f_i(u)=0,\quad \forall f_i\in A_i^*,1\le i\le n,$$

and $u=0$.

Now let $u=0$. We need to prove that u is also the zero–element of $\otimes_{t=1}^k B_t$. It is equivalent to show that : if $\mathbb{I}$ is a non–empty subset of $\{1,\cdots,n\}$, and β is a C^*-norm on $\otimes_{i\in\mathbb{I}}A_i$, then u is also the zero –element of $\otimes_{i\notin\mathbb{I}}A_i\otimes B$, where $B=\beta$ - $\otimes_{i\in\mathbb{I}}A_i$.

We use the induction for m, where $m=n-{}^{\#}\mathbb{I}$. If $m=0$, i.e., $\mathbb{I}=\{1,\cdots,n\}$, then the conclusion is obvious. Now let the conclusion hold for

$(m-1)$, where $m \geq 1$. For m and $\mathbb{I} \subset \{1, \cdots, n-1\}$, we can write $u = \sum_j u_j \otimes x_j$, where $\{x_j|j\}$ is linearly independent in A_n, and $u_j \in \otimes_{i=1}^{n-1} A_i, \forall j$. Since $u = 0$, it follows that $u_j = 0$ in $\otimes_{i=1}^{n-1} A_i, \forall j$. By induction assumption, u_j is also the zero-element of $\otimes\{A_i|i \notin \mathbb{I}, 1 \leq i \leq n-1\} \otimes B, \forall j$, and hence,

$$(\otimes\{f_i|i \notin \mathbb{I}, 1 \leq i \leq n-1\} \otimes F)(u_j) = 0, \quad \forall f_i \in A_i^*, F \in B^*, \forall j.$$

Further, for any $f_i \in A_i^*, i \notin \mathbb{I}, 1 \leq i \leq n$, and $F \in B^*$,

$$\begin{aligned} &(\otimes\{f_i|i \notin \mathbb{I}, 1 \leq i \leq n\} \otimes F)(u) \\ = &\sum_j (\otimes\{f_i|i \notin \mathbb{I}, 1 \leq i \leq n-1\} \otimes F)(u_j) \cdot f_n(x_j) = 0. \end{aligned}$$

Therefore, u is also the zero -element of $\otimes_{i \notin \mathbb{I}} A_i \otimes B$. Q.E.D.

Proposition 3.9.8. Let $\alpha(\cdot)$ be the maximal C^*-norm on $\otimes_{i=1}^n A_i$, $\mathbb{I}$ be a non-empty subset of $\{1, \cdots, n\}$, and $\alpha_{\mathbb{I}}(\cdot)$ be the C^*-norm on $\otimes_{i \in \mathbb{I}} A_i$ induced by $\alpha(\cdot)$. Then $\alpha_{\mathbb{I}}(\cdot)$ is the maximal C^*-norm on $\otimes_{i \in \mathbb{I}} A_i$.

Proof. Let $\beta(\cdot)$ be a C^*-norm on (max-$\otimes_{i \in \mathbb{I}} A_i$) $\otimes$ (max -$\otimes_{i \notin \mathbb{I}} A_i$). By Lemma 3.9.7, $\beta(\cdot)| \otimes_{i=1}^n A_i$ is a C^*-norm on $\otimes_{i=1}^n A_i$. Let $\beta_{\mathbb{I}}(\cdot)$ be the C^*-norm on $\otimes_{i \in \mathbb{I}} A_i$ induced by $(\beta(\cdot)| \otimes_{i=1}^n A_i)$.
Then

$$\begin{aligned} \beta_{\mathbb{I}}(v) &= \lim \beta(v \otimes \otimes_{i \notin \mathbb{I}} d_{l_i}^{(i)}) \\ &\leq \lim \alpha(v \otimes \otimes_{i \notin \mathbb{I}} d_{l_i}^{(i)}) = \alpha_{\mathbb{I}}(v), \end{aligned}$$

$\forall v \in \otimes_{i \in \mathbb{I}} A_i$. However,let $\alpha'_{\mathbb{I}}(\cdot)$ be the maximal C^*-norm on $\otimes_{i \in \mathbb{I}} A_i$. Since $\beta(\cdot)$ is a C^*-norm on (max-$\otimes_{i \in \mathbb{I}} A_i$) $\otimes$ (max -$\otimes_{i \notin \mathbb{I}} A_i$), it follows that

$$\alpha'_{\mathbb{I}}(v) = \lim \beta(v \otimes d_l), \quad \forall v \in \otimes_{i \in \mathbb{I}} A_i,$$

where $\{d_l\}$ is an approximate identity for max - $\otimes_{i \notin \mathbb{I}} A_i$. By Proposition 3.2.2, we can pick $\{d_l\} = \{\otimes_{i \notin \mathbb{I}} d_{l_i}^{(i)}\}$. Therefore,

$$\alpha_{\mathbb{I}}(v) \leq \alpha'_{\mathbb{I}}(v) = \beta_{\mathbb{I}}(v) \leq \alpha_{\mathbb{I}}(v), \quad \forall v \in \otimes_{i \in \mathbb{I}} A_i,$$

i.e., $\alpha_{\mathbb{I}}(\cdot)$ is the maximal C^*-norm on $\otimes_{i \in \mathbb{I}} A_i$. Q.E.D.

Theorem 3.9.9. Let $\{\mathbb{I}_t|1 \leq t \leq k\}$ be a partition of $\{1, \cdots, n\}$; $\beta(\cdot)$ be the maximal C^*-norm on $\otimes_{t=1}^k B_t$, where $B_t =$ max- $\otimes_{i \in \mathbb{I}_t} A_i, 1 \leq t \leq k$; and $\alpha(\cdot)$ be the maximal C^*- norm on $\otimes_{i=1}^n A_i$. Then $\beta(\cdot)| \otimes_{i=1}^n A_i = \alpha(\cdot)$, and $\beta(\cdot)$ is the unique extension of $\alpha(\cdot)$ on$\otimes_{t=1}^k B_t$. Formally, we can write that

$$\text{max} - \otimes_{i=1}^n A_i = \text{max} - \otimes_{t=1}^k (\text{max} - \otimes_{i \in \mathbb{I}_t} A_i).$$

Proof. By Proposition 3.2.2, the extension of $\alpha(\cdot)$ on$\otimes_{t=1}^k B_t$ is unique.

Assume that $\alpha(\cdot)$ can be extended to a C^*–norm $\overline{\alpha}(\cdot)$ on $\otimes_{t=1}^k B_t$. Clearly, $\overline{\alpha}(\cdot) \leq \beta(\cdot)$. Then $\beta(\cdot)|\otimes_{i=1}^n A_i \geq \alpha(\cdot)$. But $\alpha(\cdot)$ is maximal on $\otimes_{i=1}^n A_i$, so we have $\beta(\cdot)|\otimes_{i=1}^n A_i = \alpha(\cdot)$. Further, by the uniqueness of extension we get $\beta(\cdot) = \overline{\alpha}(\cdot)$ on $\otimes_{t=1}^k B_t$.

Now it suffices to show that there is an extension of $\alpha(\cdot)$ on $\otimes_{t=1}^k B_t$.

Let f_t be a state on $B_t, 1 \leq t \leq k$, and $u \in \otimes_{i=1}^n A_i$. Write $u = \sum_j \otimes_{t=1}^k u_j^{(t)}$, where $u_j^{(t)} \in \otimes_{i \in I_t} A_i, \forall j$. Similar to the definition of the tensor product of Hilbert spaces (see Section 1.4), we have

$$\otimes_{t=1}^k f_t(u^* u) = \sum_{i,j} \prod_{t=1}^k f_t(u_i^{(t)*} u_j^{(t)}) \geq 0.$$

Hence, $\otimes_{t=1}^k f_t$ is positive on $\otimes_{i=1}^n A_i$. By Propositios 3.4.3 and 3.4.6, $\otimes_{t=1}^k f_t \in (\alpha\text{-}\otimes_{i=1}^n A_i)^*$. Further, $\otimes_{t=1}^k B_t^* \subset (\alpha\text{-}\otimes_{i=1}^n A_i)^*$.

For any $u = \sum_i \otimes_{t=1}^k x_j^{(t)} \in \otimes_{t=1}^k B_t$, where $x_j^{(t)} \in B_t\ \forall j$ and t, then there exist $x_{jm}^{(t)} \in \otimes_{i \in I_t} A_i$ such that

$$\alpha_t(x_j^{(t)} - x_{jm}^{(t)}) \longrightarrow 0, \quad \forall j \text{ and } t,$$

where $\alpha_t(\cdot)$ is the maximal C^*–norm on $\otimes_{i \in I_t} A_i$, i.e., $B_t = \alpha_t\text{-}\otimes_{i \in I_t} A_i, 1 \leq t \leq k$. By Proposition 3.9.8, $\alpha_t(\cdot)$ is exactly the C^*–norm induced by $\alpha(\cdot)$ on $\otimes_{i \in I_t} A_i, \forall t$. Noticing that

$$\begin{aligned} \alpha(\otimes_{i=1}^n a_i) &= \alpha(\prod_{t=1}^k (\otimes_{i \in I_t} a_i \otimes \otimes_{i \notin I_t} 1_i)) \\ &\leq \prod_{t=1}^k \alpha_t(\otimes_{i \in I_t} a_i), \quad \forall a_i \in A_i, 1 \leq i \leq n, \end{aligned}$$

we have $\alpha(u_m - u_{m'}) \longrightarrow 0$, where $u_m = \sum_j \otimes_{t=1}^k x_{jm}^{(t)} \in \otimes_{i=1}^n A_i, \forall m$. So that we can define

$$\overline{\alpha}(u) = \lim_m \alpha(u_m)$$

which is obviously independent of the choice of equivalent $\{x_{jm}^{(t)}|m\}(\forall j, t)$.

Now it suffices to prove that $\overline{\alpha}(u) = 0$ if and only if $u = 0$.

Let $\overline{\alpha}(u) = 0$ i.e., $\alpha(u_m) \longrightarrow 0$. For each $f_t \in B_t^*, 1 \leq t \leq k$, since $\otimes_{t=1}^k f_t \in$

$(\alpha\text{-}\otimes_{i=1}^n A_i)^*$, it follows that $\otimes_{t=1}^k f_t(u_m) \longrightarrow 0$, and

$$\begin{aligned} 0 = & \lim_m \otimes_t f_t(\sum_j \otimes_t x_{jm}^{(t)}) = \sum_j \prod_{t=1}^k \lim_m f_t(x_{jm}^{(t)}) \\ = & \sum_j \prod_{t=1}^k f_t(x_j^{(t)}) = \otimes_{t=1}^k f_t(u), \end{aligned}$$

i.e., $u = 0$ in $\otimes_{t=1}^k B_t$.

Conversely, let u be the zero-element of $\otimes_{t=1}^k B_t$. If u has the form $\otimes_{t=1}^k x^{(t)}$, where $x^{(t)} \in B_t, 1 \le t \le k$, then there must exist t_0 such that $x^{(t_0)} = 0$. Now let $x_m^{(t)} \in \otimes_{i \in \mathbb{I}_t} A_i$ such that $\alpha_t(x^{(t)} - x_m^{(t)}) \longrightarrow 0, 1 \le t \le k$. Then $\alpha_{t_0}(x_m^{(t_0)}) \longrightarrow 0$. Therefore, we have

$$\overline{\alpha}(u) = \lim_m \alpha(\otimes_{t=1}^k x_m^{(t)}) \le \prod_{t=1}^k \lim_m \alpha_t(x_m^{(t)}) = 0.$$

We make the induction assumption that if u is the zero-element of $\otimes_{t=1}^k B_t$ which has the form $\sum_j \otimes_{t=1}^s x_j^{(t)} \otimes \otimes_{t=s+1}^k x^{(t)}$, where $x_j^{(t)} \in B_t, \forall j, 1 \le t \le s$, and $x^{(t)} \in B_t, s+1 \le t \le k$, then $\overline{\alpha}(u) = 0$. Now let

$$u = \sum_j \otimes_{t=1}^{s+1} x_j^{(t)} \otimes \otimes_{t=s+2}^k x^{(t)} = 0.$$

Thus we can find linearly independent elements $\{e_r\}$ in B_{s+1} such that

$$x_j^{(s+1)} = \sum_r \lambda_{jr} e_r, \quad \forall j$$

and then

$$u = \sum_r (\sum_j \lambda_{jr} \otimes_{t=1}^s x_j^{(t)} \otimes e_r \otimes \otimes_{t=s+2}^k x^{(t)}) = 0.$$

We may assume that $x^{(t)} \ne 0, s+2 \le t \le k$, so that

$$u_r = \sum_j \lambda_{jr} \otimes_{t=1}^s x_j^{(t)} \otimes e_r \otimes \otimes_{t=s+2}^k x^{(t)} = 0, \forall r.$$

By the induction assumption, we have $\overline{\alpha}(u_r) = 0, \forall r$. Therefore, $\overline{\alpha}(u) \le \sum_r \overline{\alpha}(u_r) = 0$. Q.E.D.

Remark. By Theorem 3.2.9, it is obvious that

$$\min\text{-}\otimes_{i=1}^n A_i = \min\text{-} \otimes_{t=1}^k (\min\text{-}\otimes_{i \in \mathbb{I}_t} A_i)$$

for any partition $\{\mathbb{I}_t | 1 \le t \le k\}$ of $\{1, \cdots, n\}$.

Now we study the tensor product of nuclear C^*–algebras.

Proposition 3.9.10. Let $A_1, \cdots, A_n$ be nuclear C^*–algebras, and B be any C^*–algebra. Then there is only one C^*–norm on $\otimes_{i=1}^n A_i \otimes B$. In particular, min - $\otimes_{i=1}^n A_i$ = max - $\otimes_{i=1}^n A_i$ is a nuclear C^*–algebra.

Proof. When $n = 1$, the conclusion is apparent. Now assuming that there is only one C^*–norm on $\otimes_{i=2}^n A_i \otimes B$, then by Theorem 3.9.9 we have

$$\begin{aligned} \max\text{-} (\otimes_{i=1}^n A_i \otimes B) &= \max\text{-} (A_1 \otimes \max\text{-} (\otimes_{i=2}^n A_i \otimes B)) \\ &= \min\text{-} (A_1 \otimes \min\text{-}(\otimes_{i=2}^n A_i \otimes B)) \\ &= \min\text{-}(\otimes_{i=1}^n A_i \otimes B), \end{aligned}$$

and hence there is only one C^*–norm on $\otimes_{i=1}^n A_i \otimes B$. Moreover,

$$\begin{aligned} \max\text{-} ((\max\text{-} \otimes_{i=1}^n A_i) \otimes B) &= \max\text{-}(\otimes_{i=1}^n A_i \otimes B) \\ &= \min\text{-}(\otimes_{i=1}^n A_i \otimes B) \\ &= \min\text{-}((\min\text{-} \otimes_{i=1}^n A_i) \otimes B), \end{aligned}$$

and therefore max - $\otimes_{i=1}^n A_i$ = min - $\otimes_{i=1}^n A_i$ is also a nuclear C^*–algebra.

Q.E.D.

Theorem 3.9.11. Let $A_1, \cdots, A_n$ be C^*–algebras. Then min-$\otimes_{i=1}^n A_i$ is nuclear if and only if each A_i is nuclear.

Proof. By Propositio 3.9.10, it suffices to prove the necessity.

Let min - $\otimes_{i=1}^n A_i$ be nuclear. Since min - $\otimes_{i=1}^n A_i$ = min -($A_1 \otimes$ min - $\otimes_{i=2}^n A_i$), we may assume that $n = 2$.

Suppose that $\alpha_0(\cdot), \alpha_1(\cdot)$ are the minimal C^*–norm, the maxmimal C^*–norm on $A_1 \otimes B$ respectively, where B is any C^*–algebra . Fix any $u \in A_1 \otimes B$. We need to prove that

$$\alpha_0(u) = \alpha_1(u).$$

If $\beta(\cdot)$is the minimal C^*–norm on $A_1 \otimes A_2 \otimes B$, then we have

$$\alpha_0(u) = \lim \beta(u \otimes d_l),$$

where $\{d_l\}$ is an approximate identity for A_2. Since min -($A_1 \otimes A_2$) is nuclear and

$$\begin{aligned} \min\text{-}(A_1 \otimes A_2 \otimes B) = &\ \min\text{-}((\min\text{-}(A_1 \otimes A_2)) \otimes B) \\ = &\ \max\text{-} ((\min\text{-} (A_1 \otimes A_2)) \otimes B), \end{aligned}$$

it follows from Proposition 3.4.6 that

$$\begin{aligned}\alpha_0(u) &= \lim \beta(u \otimes d_l) \\ &= \lim\sup\{g(u^*u \otimes d_l^* d_l)^{1/2} | g \in S(\min - (A_1 \otimes A_2) \otimes B)\},\end{aligned}$$

where $S(\otimes_{i=1}^m C_i)$ is the state space of $\otimes_{i=1}^m C_i$ (see Definition 3.4.4) for any C^*-algebras $C_1, \cdots, C_m$. For any $f \in S(A_1 \otimes B)$ and $h \in S(A_2)$, it is easy to see that $f \otimes h$ can be uniquely extended to a state on min - $(A_1 \otimes A_2) \otimes B$, still denoted by $f \otimes h$. Hence, by Proposition 3.4.6 we have

$$\begin{aligned}\alpha_0(u) &\geq \lim_l \sup \left\{ (f \otimes h)(u^*u \otimes d_l^2)^{1/2} \middle| \begin{array}{l} f \in S(A_1 \otimes B) \text{ and} \\ h \in S(B) \end{array} \right\} \\ &= \sup\{f(u^*u)^{1/2} | f \in S(A_1 \otimes B)\} = \alpha_1(u).\end{aligned}$$

Therefore, $\alpha_0(u) = \alpha_1(u), \forall u \in A_1 \otimes B$, and A_1 is nuclear. Similarly, A_2 is nuclear. Q.E.D.

Theorem 3.9.12. Let $\wedge$ be any index set; A_l be a C^*-algebra with an identity, $\forall l \in \Lambda$; and B be any C^*-algebra with an identity.

If A_l is nuclear, $\forall l \in \Lambda$, then there is only one C^*-norm on $\otimes_{l\in\Lambda} A_l \otimes B$. In particular, min - $\otimes_{l\in\Lambda} A_l =$ max - $\otimes_{l\in\Lambda} A_l$ is a nuclear C^*-algebra.

Moreover, min - $\otimes_{l\in\Lambda} A_l$ is nuclear if and only if A_l is nuclear, $\forall l \in \Lambda$.

Proof. It is immediate from Theorem 3.9.11 and the definition of $\otimes_{l\in\Lambda} A_l$. Q.E.D.

Theorem 3.9.13. The inductive limit of nuclear C^*-algebras is nuclear.

Proof. Let $A = \lim_{\overrightarrow{\alpha}}\{A_\alpha, \Phi_{\beta\alpha} | \alpha, \beta \in \mathbb{I}, \alpha \leq \beta\}$, and A_α be nuclear, $\forall \alpha \in \mathbb{I}$. By Theorem 3.7.2, there is a family $\{\tilde{A}_\alpha | \alpha \in \mathbb{I}\}$ of C^*-subalgebras of A such that : 1) $\tilde{A}_\alpha$ is $*$ isomorphic to A_α(so $\tilde{A}_\alpha$ is also nuclear) , $\forall \alpha \in \mathbb{I}$; 2) $\tilde{A}_\alpha \subset \tilde{A}_\beta, \forall \alpha \leq \beta$; 3) $\sqcup_{\alpha\in\mathbb{I}} \tilde{A}_\alpha$ is dense in A.

If B is any C^*-algebra, and $\alpha(\cdot), \beta(\cdot)$ are two C^*-norms on $A \otimes B$, then it must be that $\alpha(\cdot) = \beta(\cdot)$ on $(\sqcup_{\alpha\in\mathbb{I}} \tilde{A}_\alpha) \otimes B$. Further, $\alpha(\cdot) = \beta(\cdot)$ on $A \otimes B$. Therefore, there is only one C^*-norm on $A \otimes B$, and A is nuclear. Q.E.D.

Remark. By Theorem 3.9.13, we can see that any (UHF) algebra and any (AF) algebra (see Chapter 15) are nuclear. In particular, $C(H)$ is nuclear. Moreover, we can prove that any type I (GCR) C^*-algebra (see Chapter 13) is nuclear. In Section 7.4, there are the examples of non-nuclear C^*-algebras.

Notes. The tensor product of VN algebras is defined naturally. M. Takesaki discovered first that there are two possible C^*-norms on the algebraic tensor product of C^*-algebras. Thus, the theory of C^*-tensor products is complicated. M. Takesaki introduced a class of C^*-algebras with the property (T). Later, C. Lance gave such class of C^*-algebras another name: nuclear C^*-algebras, since this class of C^*-algebras has some connections with the theory of nuclear spaces. The theory of nuclear C^*-algebras is very rich and deep. C. Lance once gave a nice survey.

Theorem 3.9.9 is due to B.-R. Li .

References. [37], [95], [97], [102], [171].

Chapter 4

W^*-Algebras

4.1 Projections of norm one

Definition 4.1.1. Let A be a C^*-algebra with an identity 1, and B be a C^*-subgebra of A, and $1 \in B$. P is called a *projection of norm one* from A onto B, if P is linear, $PA = B, Pb = b, \forall b \in B$, and $\|Pa\| \le \|a\|, \forall a \in A$.

Lemma 4.1.2. Let Φ be a positive linear map from a C^*-algebra A to a C^*-algebra B. If A has an identity 1, then $\|\Phi\| = \|\Phi(1)\|$.

Proof. By Theorem 2.14.5, it suffices to show that

$$\|\Phi(e^{ih})\| \le \|\Phi(1)\|, \forall h^* = h \in A.$$

Thus we may assume that A is commutative. By Proposition 3.6.6, Φ is completely positive. Let $B \subset B(K)$. By Theorem 3.6.7, there is a nondegenerate $*$ representation $\{\pi, H\}$ of A and an operator $v : K \to H$ such that

$$\Phi(a) = v^*\pi(a)v, \forall a \in A, \quad \text{and } \|v\| = \|\Phi\|^{1/2}.$$

Then $\|\Phi(1)\| = \|v^*v\| = \|v\|^2 = \|\Phi\|$. Q.E.D.

Proposition 4.1.3. Let A be a C^*-algebra with an identity 1, P be a positive linear map on $A, P1 = 1$ and

$$P(Pb_1 \cdot a \cdot Pb_2) = Pb_1 \cdot Pa \cdot Pb_2, \quad \forall a, b_1, b_2 \in A.$$

Then $PA = B$ is a C^*-subalgebra of $A, 1 \in B$ and P is a projection of norm one from A onto B.

Proof. By Lemma 4.1.2, $\|P\| = 1$. Clearly, $P^2 a = Pa$ ($\forall a \in A$) by the assumptions. Now it suffices to show that B is a C^*-subalgebra of A. Since P is positive, it follows that $B^* = B$. Moreover, we have

$$Pa \cdot Pb = Pa \cdot P1 \cdot Pb = P(Pa \cdot 1 \cdot Pb), \quad \forall a, b \in A.$$

Thus, B is a $*$ subalgebra of A. Finally, if $a_n, a \in A$, and $Pa_n \to a$, then $P^2 a_n = Pa_n \to Pa = a$. Therefore, B is closed. Q.E.D.

Lemma 4.1.4. Let P be a projection of norm one from A onto B. Then P can be extended to a projection of norm one from A^{**} onto B^{**}, and this extension is σ-σ continuous.

Proof. By Proposition 2.11.4, B^{**} is the $\sigma(A^{**}, A^*)$-closure of B in A^{**}. Hence B^{**} is C^*-subalgebra of A^{**}.

Clearly, P^{**} is a linear map of norm one from A^{**} onto B^{**}, and is σ-σ continuous. Moreover, P^{**} is a projection since $P^{**}|A = P$. Q.E.D.

Theorem 4.1.5. Let A be a C^*-algebra with an identity 1, and B be a C^*-subalgebra of $A, 1 \in B$, and P be a projection of norm one from A onto B. Then:

1) P is completely positive, in particular, $PA_+ \subset B_+$ and $Pa^* = (Pa)^*, \forall a \in A$;

2) $P(Pa \cdot b) = Pa \cdot Pb = P(a \cdot Pb), \forall a, b \in A$;

3) $(Px)^* \cdot (Px) \leq P(x^* x), \forall x \in A$.

Proof. First we claim that P is positive. In fact, let $B \subset B(K)$ and $1 = 1_K$. For any $\xi \in K$, put

$$\omega_\xi(a) = \langle P(a)\xi, \xi\rangle, \quad \forall a \in A.$$

Since $P1 = 1, \|P\| = 1$, it follows that $\omega_\xi(1) = \|\xi\|^2 = \|\omega_\xi\|$. By Proposition 2.3.3, $\omega_\xi(\cdot)$ is a positive linear functional on A. Thus P is positive.

Further, we have $Pa^* = (Pa)^*, \forall a \in A$.

By Lemma 4.1.4, we may assume that B is the closed linear span of projections of B. So for the conclusion 2), it suffices to show that

$$P(pa) = p \cdot Pa,$$

where p is a projection of B, and $a \in A$.

If $y \in A_+$ and $\|y\| \leq 1$, then $p \geq pyp$. Further, $p = Pp \geq P(pyp)$. Thus $pP(pyp)p = P(pyp)$. Generally, we have

$$P(pxp) = pP(pxp)p, \quad \forall x \in A. \tag{1}$$

Replacing p by $(1-p)$, we get

$$\begin{aligned} & P((1-p)x(1-p)) \\ = \ & (1-p)P((1-p)x(1-p))(1-p), \quad \forall x \in A. \end{aligned} \tag{1'}$$

Let $a \in A$ and $\|a\| \leq 1$. Then

$$\begin{aligned} & \|pa(1-p) \pm np\| \\ = \ & \|(pa(1-p) \pm np) \cdot (pa(1-p) \pm np)^*\|^{1/2} \\ = \ & \|pa(1-p)a^*p + n^2p\|^{1/2} \leq (1+n^2)^{1/2}. \end{aligned}$$

Put $a' = P(pa(1-p))$ and $b = \frac{1}{2}(pa'p + pa'^*p)$. If $b \neq 0$, then there is $0 \neq \overline{\lambda} = \lambda \in \sigma(b)$. Since

$$\begin{aligned} \|a' \pm np\| \geq \ & \|pa'p \pm np\| \\ \geq \ & \|b \pm np\| \geq \lambda \pm n, \end{aligned}$$

it follows that $(1+n^2)^{1/2} \geq \|pa(1-p) \pm np\| \geq \|P(pa(1-p) \pm np)\| = \|a' \pm np\| \geq \lambda \pm n$. But this is impossible when $|n|$ is sufficiently large. Therefore, $\frac{1}{2}(pa'p + pa'^*p) = 0$. Replacing n by in, we can prove that $\frac{1}{2}(pa'p - pa'^*p) = 0$ similarly. So we obtain that

$$pa'p = 0. \tag{2}$$

Since $a'^* = P((1-p)a^*p)$, it follows from a similar discussion that $(1-p)a'^*(1-p) = 0$. Further we get that

$$(1-p)a'(1-p) = 0. \tag{3}$$

Suppose that $(1-p)a'p \neq 0$. By (2), (3), we have

$$\begin{aligned} \|a' + n(1-p)a'p\| = \ & \|pa'(1-p) + (n+1)(1-p)a'p\| \\ = \ & \max\{\|pa'(1-p)\|, (n+1)\|(1-p)a'p\|\} \\ = \ & (n+1)\|(1-p)a'p\| \end{aligned}$$

if n is sufficiently large. However, since $(1-p)a'p \in B$, it follows that

$$\begin{aligned} \|a' + n(1-p)a'p\| = \ & \|P(pa(1-p) + n(1-p)a'p)\| \\ \leq \ & \|pa(1-p) + n(1-p)a'p\| = n\|(1-p)a'p\| \end{aligned}$$

if n is sufficiently large. That is a contradiction. Therefore, we have

$$(1-p)a'p = 0. \tag{4}$$

By (2), (3), (4), $a' = pa'(1-p)$, i.e.,

$$P(pa(1-p)) = pP(pa(1-p))(1-p). \tag{5}$$

Replacing p by $(1-p)$, similarly we have

$$P((1-p)ap) = (1-p)P((1-p)ap)p. \tag{6}$$

By $Pa = P(pap) + P(pa(1-p)) + P((1-p)ap) + P((1-p)a(1-p))$ and (1), (1'), (5), (6), we can see that

$$p \cdot Pa \cdot (1-p) = P(pa(1-p))$$

and

$$p \cdot Pa \cdot p = P(pap).$$

Therefore $p \cdot Pa = P(pa)$. That comes to the conclusion 2).

For any $n, b_1, \cdots, b_n \in B, a_1, \cdots, a_n \in A$, by the conclusion 2) and $P(A_+) \subset B_+$,

$$\begin{aligned}\sum_{i,j} b_i^* P(a_i^* a_j) b_j &= \sum_{i,j} P(b_i^* a_i^* a_j b_j) \\ &= P((\sum_i a_i b_i)^* \cdot (\sum_i a_i b_i)) \geq 0.\end{aligned}$$

Therefore, P is completely positive.

Finally, for any $x \in A$,

$$\begin{aligned}& P(x^*x) - (Px)^*P(x) \\ = \; & P(x^*x) - P(Px^* \cdot x) - P(x^* \cdot Px) + P(Px^* \cdot 1 \cdot Px) \\ = \; & P((x - Px)^* \cdot (x - Px)) \geq 0,\end{aligned}$$

i.e. $(Px)^* \cdot (Px) \leq P(x^*x)$. Q.E.D.

Proposition 4.1.6. Let M, N be VN algebras on Hilbert spaces H, K respectively. Then there is a σ-σ continuous projection Φ of norm one from $M\overline{\otimes}N$ onto $1_H \otimes N \cong N$.

Proof. Fix a normal state φ on M. By Lemma 3.8.5, we can define

$$\Phi(x)(f) = x(\varphi \otimes f), \quad \forall x \in M\overline{\otimes}N, f \in N_*.$$

Then Φ satisfies the conditions. Q.E.D.

Notes. In general, a linear map from a C^*-algebra A onto its C^*-subalgebra B satisfying the conditions 1), 2), 3) of Theorem 4.1.5, is called a *conditional expectation*, and was studied first by H. Umegaki. A conditional expectation is clearly a projection of norm one from A onto B. Conversely, J.Tomiyama proved that a projection of norm one from A onto B is automatically a conditional expectation.

References. [183], [184], [187].

4.2. W^*-algebras and their $*$ representations

Definition 4.2.1. A C^*-algebra M is called a *W^*-algebra*, if there is a Banach space M_* such that $(M_*)^* = M$.

For a W^*-algebra M, M_* is called the *predual* of M (see Section 1.1.)

From Proposition 1.3.3, every VN algebra is a W^*-algebra. By Theorem 2.11.2, if A is a C^*-algebra, then A^{**} is a W^*-algebra.

Lemma 4.2.2. Let M be a W^*-algebra. Then M has an identity.

Proof. Since the closed unit ball S of M is $\sigma(M, M_*)$-compact and convex, it follows from the Krein-Milmann theorem that S admits an extreme point at least. Now from Theorem 2.5.3, M has an identity. Q.E.D.

Now let M be a W^*-algebra, 1 be its identity. By Theorem 2.11.2, M^{**} is also a W^*-algebra; 1 is also an identity of M^{**}; and M is a C^*-subalgebra of M^{**}. Let M_* be the predual of M. See M_* as a closed linear subspace of M^*, and let $P : M^{**} \to M$ as follows

$$P(X) = X|M_*, \quad \forall X \in M^{**}.$$

Clearly, P is a projection of norm one from M^{**} onto M, and is $\sigma(M^{**}, M^*)$ - $\sigma(M, M_*)$ continuous. Let

$$I = \{X \in M^{**} \mid PX = o\}.$$

Clearly, I is just the orthogonal complement $M_*^{\perp}$ of M_* as a closed linear subspace of M^*. Thus, I is $\sigma(M^{**}, M^*)$-closed. By Theorem 4.1.5,

$$P(aXb) = a \cdot PX \cdot b, \quad \forall X \in M^{**}, a, b \in M.$$

By Theorem 2.11.2, $M^{**} \cong \overline{M}$ (the enveloping VN algebra of the C^*-algebra M). Thus, the multiplication on M^{**} is $\sigma(M^{**}, M^*)$-continuous for each variable. Moreover, M is a $\sigma(M^{**}, M^*)$-dense subset of M^{**}. Therefore, I is a $\sigma(M^{**}, M^*)$-closed $*$ two-sided ideal of M^{**}. By Proposition 1.7.1, there is a unique central projection z of M^{**} such that

$$M_*^{\perp} = I = M^{**}(1 - z).$$

Since $P = P^2$ and I is an ideal, it follows that

$$(PX - X) \in I, \quad \text{and} \quad (PX - X)Y \in I$$

$\forall X, Y \in M^{**}$. By Theorem 4.1.5, we can see that

$$P(XY) = PX \cdot PY, \quad \forall X, Y \in M^{**}.$$

Thus, P is a $*$ isomorphism from $M^{**}z$ onto M.

Let $Q(: M \to M^{**}z)$ be the inverse of $(P|M^{**}z)$. Since $P(xz) = x, \forall x \in M$, it follows that

$$Q(x) = xz, \quad \forall x \in M.$$

For any $X \in M^{**}$, we can write $X = PX + (X - PX)$, where $PX \in M, (X - PX) \in I = M_*^{\perp} = M^{**}(1 - z)$. If $x \in M \cap I$, then $x = Px = 0$. Therefore, we get

$$M^{**} = M \dot{+} M_*^{\perp}.$$

For any $F \in M^*, R_z F$ and $R_{(1-z)} F$ (see Section 1.9) $\in M^*$ since the multiplication on M^{**} is $\sigma(M^{**}, M^*)$-continuous for each variable. Hence, $M^* = R_z M^* \dot{+} R_{(1-z)} M^*$. Now we claim that

$$M_* = R_z M^*.$$

In fact, since M_* is a closed linear subspace of M^*, it follows that $M_* = (M_*^{\perp})_{\perp} = (M^{**}(1 - z))_{\perp}$. Thus, $M_* \supset R_z M^*$. Conversely, if $f \in M_*$, then by the definition of $z, f(X(1 - z)) = 0, f(X) = f(Xz) = (R_z f)(X), \forall X \in M^{**}$. Thus $f = R_z f \in R_z M^*$, and $M_* = R_z M^*$.

We say that the $*$ isomorphism $Q(: M \to M^{**}z)$ is also $\sigma(M, M_*)$ - $\sigma(M^{**}, M^*)$ continuous. In fact, let $\{x_l\}$ be a net of M, and $x_l \to 0(\sigma(M, M_*))$. Then for each $F \in M^*$,

$$F(Q(x_l)) = F(x_l z) = f(x_l) \to 0$$

since $f = R_z F \in M_*$.

From the above discussion, we obtain the following.

Proposition 4.2.3. Let M be a W^*-algebra, and M_* be its predual. Embedding M, M_* canonically into M^{**}, M^* respectively, then we can find a central projection z of M^{**} and a projection P of norm one from M^{**} onto M such that:

1) P is also a $*$ homomorphism from M^{**} onto M, and is $\sigma(M^{**}, M^*)$ - $\sigma(M, M_*)$ continuous;

2) P is a $*$ isomorphism from $M^{**}z$ onto M. If $Q(: M \to M^{**}z)$ is the inverse of $(P|M^{**}z)$,then $Q(x) = xz, \forall x \in M$, and Q is also $\sigma(M, M_*)$ - $\sigma(M^{**}, M^*)$ continuous;

3) $M_*^{\perp} = M^{**}(1 - z), M_* = R_z M^*$, and

$$M^{**} = M \dot{+} M_*^{\perp}, \quad M^* = M_* \dot{+} R_{(1-z)} M^*.$$

Definition 4.2.4. Let M be a W^*-algebra, and M_* be its predual. $\{\pi, H\}$ is called a *W^*-representation* of M, if π is a $*$ homomorphism from M to $B(H)$, and π is $\sigma(M, M_*)$-$\sigma(B(H), T(H))$ continuous.

If $\{\pi, H\}$ is a W^*-representation of M, then $\pi(M)$ is a weakly closed $*$ subalgebra of $B(H)$ by the proof of Proposition 1.8.13. Moreover, if π is nondegenerate, then $\pi(1_M) = 1_H$, and $\pi(M)$ is a VN algebra on H. If π is faithful, by Proposition 1.2.6 and $\sigma(M, M_*)$-compactness of the closed unit ball of M, then the $*$ isomorphism π^{-1} from $\pi(M)$ onto M is also $\sigma(B(H), T(H))$-$\sigma(M, M_*)$ continuous.

Theorem 4.2.5. Let M be a W^*-algebra. Then M admits a faithful nondegenerate W^*-representation. In consequence, M is $*$ isomorphic to a VN algebra on some Hilbert space, and this $*$ isomorphism is σ-σ continuous.

Proof. Let $\{\pi, H\}$ be the universal $*$ representation of M as a C^*-algebra. By the discussion of Section 2.11, $\{\pi, H\}$ can be extended to a faithful nondegenerate W^*-representation of the W^*-algebra M^{**}, which is denoted by $\{\pi, H\}$ still. By Proposition 4.2.3, there is a σ-σ continuous $*$ isomorphism Q from M onto $M^{**}z$, where z is a central projection of M^{**}. Then $\{\pi\circ Q, \pi(z)H\}$ is a faithful nondegenerate W^*-representation of M. Q.E.D.

By Theorem 4.2.5, we can regard a W^*-algebra as a VN algebra. In particular, we have Proposition 4.2.6. Let M be a W^*-algebra, and M_* be its predual. Then the $*$ operation on M is $\sigma(M, M_*)$-continuous; the multiplication is $\sigma(M, M_*)$-continuous for each variable; M_* is the linear span of normal positive linear functional on M, in consequence, M_* is unique; for any normal positive linear functional φ on M, by the GNS construction there is a cyclic W^*-representation $\{\pi_\varphi, H_\varphi, 1_\varphi\}$ of M.

Proposition 4.2.6. Denote the normal state space on M by $S_n(M)$. Then the *normal universal $*$ representation*

$$\left\{\pi = \sum_{\varphi\in S_n(M)} \oplus\pi_\varphi, \quad H = \sum_{\varphi\in S_n(M)} \oplus H_\varphi\right\}$$

is a faithful nondegenerate W^*-representation of M.

Now we discuss some properties of W^*-representations.

Theorem 4.2.7. Let A be a C^*-algebra, and $\{\pi, H\}$ be a $*$ representation of A. Then there exists a unique W^*-representation $\{\tilde{\pi}, H\}$ of A^{**} such that $\tilde{\pi}$ is an extension of π, and $\tilde{\pi}(A^{**})$ is the weak closure of $\pi(A)$. Consequently, there is a bijection between the set of $*$ representations of A and the set of W^*-representations of A^{**}.

Proof. Notice that $\pi : A \to B(H)$ and $\pi^* : B(H)^* \to A^*$. Regard $T(H)$ as a closed linear subspace of $B(H)^*$ (since $T(H)^* = B(H)$), and let $\pi_* = \pi^*|T(H)$. We claim that $\tilde{\pi} = (\pi_*)^*(A^{**} \to T(H)^* = B(H)\ )$ satisfies the conditions.

In fact, since $\pi_* : T(H) \to A^*$ and $\tilde{\pi} = (\pi_*)^*$, it follows that $\tilde{\pi}$ is $\sigma(A^{**}, A^*)$-$\sigma(B(H), T(H))$ continuous. Notice that

$$\tilde{\pi}(a)(t) = a(\pi_*(t)) = a(\pi^*(t)) = \pi(a)(t),$$

$\forall t \in T(H), a \in A$. Thus $\tilde{\pi}$ is an extension of π. Further, $\tilde{\pi}$ is a W^*-representation of A^{**} since A is $\sigma(A^{**}, A^*)$-dense in A^{**} and $\tilde{\pi}$ is σ - σ continuous. Clearly, $\tilde{\pi}$ is the unique σ - σ continuous extension of π, and $\tilde{\pi}(A^{**})$ is the weak closure of $\pi(A)$. Q.E.D.

Proposition 4.2.8. Let $\{\pi_1, H_1\}$ and $\{\pi_2, H_2\}$ be two nondegenerate W^*-representations of a W^*-algebra M, and $\ker\pi_i = \{a \in M | \pi_i(a) = 0\}, i = 1, 2$. If $\ker\pi_1 \subset \ker\pi_2$, then $\{\pi_2, H_2\}$ is unitarily equivalent to an induction of some amplication of $\{\pi_1, H_1\}$, i.e. there is a Hilbert space K and a projection p' of $(\pi_1(M)\overline{\otimes}\mathbb{C}1_K)'$ such that $\{\pi_2, H_2\} \cong \{\pi, p'(H_1 \otimes K)\}$, where $\pi(a) = (\pi_1(a) \otimes 1_K)p', \forall a \in M$.

Proof. Let $M_i = \pi_i(M), i = 1, 2$. Then M_i is a VN algebra on $H_i, i = 1, 2$. Since $\ker\pi_1 \subset \ker \pi_2$, there is a normal $*$ homomorphism Φ from M_1 onto M_2 such that $\Phi \circ \pi_1 = \pi_2$. Now by Theorem 1.12.4, we can get the conclusion. Q.E.D.

Proposition 4.2.9. Let $\{\pi_1, H_1\}$ and $\{\pi_2, H_2\}$ be two W^*-representations of a W^*-algebra M, and $\ker \pi_i = \{a \in M \mid \pi_i(a) = 0\}, i = 1, 2$. If $\pi_i(M)$ admits a cyclic-separating vector in $H_i, i = 1, 2$, and $\ker \pi_1 = \ker \pi_2$, then $\{\pi_1, H_1\} \cong \{\pi_2, H_2\}$.

Proof. Let $M_i = \pi_i(M)$, then M_i is a VN algebra on $H_i, i = 1, 2$. Since $\ker\pi_1 = \ker\pi_2$, thus there is a $*$ isomorphism Φ from M_1 onto M_2, such that $\Phi \circ \pi_1 = \pi_2$. Now by Theorem 1.13.5, the conclusion can be obtained. Q.E.D.

Notes. We have a definition of abstract C^*-algebras (see Chapter 2). A natural question is how to define abstract VN algebras. This question received considerable attention during 1950's. Theorem 4.2.5 is due to S. Sakai, and it gives an answer for the above question. However, the proof of Theorem 4.2.5 presented here is due to J. Tomiyama based on his result, Theorem 4.1.5. The uniqueness of the predual of a W^*-algebra, due to J. Dixmier, answered completely the question concerning to what extent the algebraic structure of a W^*-algebra determines its topological structure.

References. [19], [143], [183], [184].

4.3. Tensor products of W^*-algebras

Let M, N be two W^*-algebras. We want to define the tensor product $M\bar{\otimes}N$ of M and N such that $M\bar{\otimes}N$ is still a W^*-algebra. If we regard M, N as VN algebras, using the tensor product of VN algebras and by Theorem 1.12.6, we can define $M\overline{\otimes}N$. But in this section, we shall define $M\bar{\otimes}N$ from M and N themselves.

Let M, N be two W^*-algebras, and M_*, N_* be their preduals respectively. As C^*-algebras, there is a spatial C^*-norm $\alpha_0(\cdot)$ on the algebraic tensor product $M \otimes N$. Then we get a C^*-algebra α_0 - $(M \otimes N)$. Let $\alpha_0^*(\cdot)$ be the dual norm of $\alpha(\cdot)$ on $M^* \otimes N^*$. Then

$$(\alpha_0\text{-}(M \otimes N))^* \supset \alpha_0^*\text{-}(M^* \otimes N^*) \supset \alpha_0^*\text{-}(M_* \otimes N_*),$$

where $\alpha_0^*\text{-}(M^* \otimes N^*)$ is the completion of $(M^* \otimes N^*, \alpha_0^*(\cdot))$; and $\alpha_0^*\text{-}(M_* \otimes N_*)$ is the completion of $(M_* \otimes N_*, \alpha_0^*(\cdot))$, and is equal to the closure of $M_* \otimes N_*$ in $\alpha_0^*\text{-}(M^* \otimes N^*)$. Let

$$I = (\alpha_0^*\text{-}(M_* \otimes N_*))^\perp (\subset (\alpha_0\text{-}(M \otimes N))^{**}),$$

i.e., I is the orthogonal complement of $\alpha_0^*\text{-}(M_* \otimes N_*)$ which is regarded as a closed linear subspace of $(\alpha_0\text{-}(M \otimes N))^*$. Suppose that $Y \in I, X \in (\alpha_0\text{-}(M \otimes N))^{**}$. Pick a net $\{x_l\} \subset M \otimes N$ such that $x_l \to X$ with respect to the w^*-topology in $(\alpha_0\text{-}(M \otimes N))^{**}$. For any $f \in M_* \otimes N_*$, since $L_{x_l}f$ and $R_{x_l}f \in M_* \otimes N_*$, it follows that

$$f(x_lY) = (L_{x_l}f)(Y) = 0, \quad f(Yx_l) = (R_{x_l}f)(Y) = 0,$$

$\forall l$. Taking the limits, we get XY and $YX \in I$. So I is a σ-closed two-sided ideal of the W^*-algebra $(\alpha_0\text{-}(M \otimes N))^{**}$. Therefore, $(\alpha_0\text{-}(M \otimes N))^{**}/I$ is a W^*-algebra, and its predual is $\alpha_0^*\text{-}(M_* \otimes N_*)$.

Definition 4.3.1. The W^*-algebra $(\alpha_0\text{-}(M \otimes N))^{**}/I$ is called the *tensor product* of W^*-algebras M and N, which is denoted by $M\bar{\otimes}N$.

From preceding paragraph, $M\bar{\otimes}N = (\alpha_0^*\text{-}(M_* \otimes N_*)^*$, and $(M\bar{\otimes}N)_* = \alpha_0^*\text{-}(M_* \otimes N_*)$.

Lemma 4.3.2. $M_* \otimes N_*$ is w^*-dense in $(\alpha_0\text{-}(M \otimes N))^*$.

Proof. By Proposition 3.2.10, $M^* \otimes N^*$ is w^*-dense in $(\alpha_0\text{-}(M \otimes N))^*$. Notice that the unit balls of M_*, N_* are w^*-dense in the unit balls of M^*, N^* respectively, and $\alpha_0^*(\cdot)$ is a cross-norm on $M^* \otimes N^*$. Then it is easy to see that $M_* \otimes N_*$ is dense in $M^* \otimes N^*$ with respect to the w^*-topology in $(\alpha_0\text{-}(M \otimes N))^*$. Therefore, $M_* \otimes N_*$ is w^*-dense in $(\alpha_0\text{-}(M \otimes N))^*$. Q.E.D.

Proposition 4.3.3. $\alpha_0\text{-}(M \otimes N) \cap I = \{0\}$, in consequence, $\alpha_0\text{-}(M \otimes N)$ can be embedded in $M \bar{\otimes} N$. Moreover, $\alpha_0\text{-}(M \otimes N)$ is w^*-dense in $M \overline{\otimes} N$.

Proof. Let $x \in \alpha_0\text{-}(M \otimes N) \cap I$. Then

$$f \otimes g(x) = 0, \quad \forall f \in M_*, \quad g \in N_*.$$

By Lemma 4.3.2, we have $x = 0$.

Now if $\widetilde{X} \in M \bar{\otimes} N = (\alpha_0\text{-}(M \otimes N))^{**}/I$, and $X \in \widetilde{X}$, then there is a net $\{x_l\} \subset \alpha_0\text{-}(M \otimes N)$ such that $x_l \to X$ with respect to the w^*-topology in $\alpha_0\text{-}(M \otimes N))^{**}$. Further, for any $F \in (M \bar{\otimes} N)_* = \alpha_0^*\text{-}(M_* \otimes N_*) \subset (\alpha_0\text{-}(M \otimes N))^*$,

$$|(\tilde{x}_l - \widetilde{X})(F)| = |(x_l - X)(F)| \to 0.$$

Therefore, $\alpha_0\text{-}(M \otimes N)$ is w^*-dense in $M \bar{\otimes} N$. Q.E.D.

Theorem 4.3.4. Let $\{\pi_i, H_i\}$ be a nondegenerate W^*-representation of a W^*-algebra $M_i, i = 1, 2$. Then there exists a unique W^*-representation $\{\pi, H\}$ of $M_1 \bar{\otimes} M_2$, where $H = H_1 \otimes H_2$ such that

$$\pi(a_1 \otimes a_2) = \pi_1(a_1) \otimes \pi_2(a_2), \quad \forall a_i \in M_i, \quad i = 1, 2,$$

and $\pi(M \bar{\otimes} N) = \pi(M) \bar{\otimes} \pi(N)$ (the tensor product of VN algebras $\pi(M)$ and $\pi(N)$). Moreover, if π_i is faithful, $i = 1, 2$, then π is also faithful.

Proof. By Proposition 3.2.7, there is a unique $*$ representation $\{\pi_0, H\}$ of $\alpha_0\text{-}(M_1 \otimes M_2)$ such that

$$\pi_0(a_1 \otimes a_2) = \pi_1(a_1) \otimes \pi_2(a_2), \forall a_1 \in M_1, a_2 \in M_2.$$

By Theorem 4.2.7, $\{\pi_0, H\}$ can be uniquely extended to a W^*-representation $\{\tilde{\pi}_0, H\}$ of $(\alpha_0\text{-}(M_1 \otimes M_2))^{**}$. For any $\xi_i, \eta_i \in H_i$, let $f_i(\cdot) = \langle \pi_i(\cdot)\xi_i, \eta_i\rangle \in (M_i)_*, i = 1, 2$. Then $f_1 \otimes f_2 \in (M_1)_* \otimes (M_2)_* \subset (\alpha_0\text{-}(M_1 \otimes M_2))^*$. By the definition of I (see 4.3.1), we have

$$f_1 \otimes f_2(I) = \{0\}.$$

Since $H_1 \odot H_2$ is dense in $H_1 \otimes H_2$, it follows that $\tilde{\pi}_0(I) = \{0\}$. Thus $\{\tilde{\pi}_0, H\}$ induces a W^*-representation $\{\pi, H\}$ of $M_1 \bar{\otimes} M_2 = (\alpha_0\text{-}(M_1 \otimes M_2))^{**}/I$. Clearly,

$\{\pi, H\}$ satisfies the conditions. Moreover, the uniqueness of such $\{\pi, H\}$ is also obvious.

Now suppose that π_i is faithful, $i = 1,2$. For any $f_i \in (M_i)_*$, since M_i is $*$ isomorphic to $\pi_i(M_i)$, there are two sequences $\{\xi_n^{(i)}\}$ and $\{\eta_n^{(i)}\}(\subset H_i)$ with $\sum_n(\|\xi_n^{(i)}\|^2 + \|\eta_n^{(i)}\|^2) < \infty$ such that

$$f_i(\cdot) = \sum_n \langle \pi_i(\cdot)\xi_n^{(i)}, \eta_n^{(i)}\rangle, \quad \forall \cdot \in M_i, \quad i = 1,2.$$

Thus for any $x \in M_1 \bar{\otimes} M_2$, we have

$$(f_1 \otimes f_2)(x) = \sum_{j,k} \langle \pi(x)\xi_j^{(1)} \otimes \xi_k^{(2)}, \quad \eta_j^{(1)} \otimes \eta_k^{(2)}\rangle.$$

If $\pi(x) = 0$, then $(f_1 \otimes f_2)(x) = 0, \forall f_i \in (M_i)_*, i = 1,2$. But $(M_1)_* \otimes (M_2)_*$ is dense in α_0^*-$((M_1)_* \otimes (M_2)_*) = (M_1 \bar{\otimes} M_2)_*$, so $x = 0$, and π is also faithful.

Q.E.D.

Corollary 4.3.5. Let M_i be a VN algebra on a Hilbert space $H_i, i = 1,2$. Then the W^*-tensor product of W^*-algebras M_1 and M_2 is $*$ isomorphic to the VN tensor product of VN algebras M_1 and M_2.

Proposition 4.3.6. Let φ_i be a normal positive linear functional on a W^*-algebra $M_i, i = 1,2$. Then there is a unique normal positive linear functional φ on $M_1 \bar{\otimes} M_2$ such that

$$\varphi(a_1 \otimes a_2) = \varphi_1(a_1)\varphi_2(a_2), \quad \forall a_i \in M_i, \quad i = 1,2,$$

and $s(\varphi) = s(\varphi_1) \otimes s(\varphi_2)$.

Proof. Let $\{\pi_i, H_i, \xi_i\}$ be the cyclic W^*-representation of M_i generated by $\varphi_i, i = 1,2$. By Theorem 4.3.4, $\pi_1 \otimes \pi_2$ can be extended to a W^*-representation $\{\pi, H\}$ of $M_1 \bar{\otimes} M_2$, where $H = H_1 \otimes H_2$. Now let

$$\varphi(x) = \langle \pi(x)\xi_1 \otimes \xi_2, \xi_1 \otimes \xi_2\rangle, \quad \forall x \in M_1 \bar{\otimes} M_2.$$

Then φ is what we want to find. Moreover, by Proposition 1.8.11 and Theorem 1.4.12, we can see that $s(\varphi) = s(\varphi_1) \otimes s(\varphi_2)$. Q.E.D.

Proposition 4.3.7. Let Φ_i be a completely positive linear map from a W^*-algebra M_i to a W^*-algebra N_i, and also Φ_i be σ - σ continuous, $i = 1,2$. Then there exists a σ- σ continuous completely positive linear map Φ from $M_1 \bar{\otimes} M_2$ to $N_1 \bar{\otimes} N_2$ such that

$$\Phi(a_1 \otimes a_2) = \Phi_1(a_1) \otimes \Phi_2(a_2), \quad \forall a_1 \in M_1, \quad a_2 \in M_2.$$

Proof. By Proposition 3.6.11, there is a completely positive linear map Φ_0 from α_0-$(M_1 \otimes M_2)$ to α_0-$(N_1 \otimes N_2)$ such that

$$\Phi_0(a_1 \otimes a_2) = \Phi_1(a_1) \otimes \Phi_2(a_2), \quad \forall a_1 \in M_1, \quad a_2 \in M_2.$$

For any $f_i \in (N_i)_*, i = 1,2$, since Φ_i is σ - σ continuous, $i = 1,2$, it follows that

$$\Phi_0^*(f_1 \otimes f_2) = \Phi_1^*(f_1) \otimes \Phi_2^*(f_2) \in (M_1)_* \otimes (M_2)_*.$$

Further, $\Phi_0^*(\alpha_0^*\text{-}((N_1)_* \otimes (N_2)_*)) \subset \alpha_0^*\text{-}((M_1)_* \otimes (M_2)_*)$. Let

$$\Phi = (\Phi_0^* | \alpha_0^*\text{-}((N_1)_* \otimes (N_2)_*))^*.$$

Then Φ is a σ-σ continuous linear map from $M_1 \bar{\otimes} M_2$ to $N_1 \bar{\otimes} N_2$ (see Definition 4.3.1), and

$$\Phi(a_1 \otimes a_2) = \Phi_1(a_1) \otimes \Phi_2(a_2), \quad \forall a_i \in M_i, \quad i = 1,2.$$

Finally, we prove that Φ is completely positive. Assume that $N_1 \bar{\otimes} N_2 \subset B(H)$. Then we need to prove

$$\sum_{i,j} \langle \Phi(x_i^* x_j) \xi_j, \xi_i \rangle \geq 0$$

for any n, $x_1, \cdots, x_n \in M_1 \bar{\otimes} M_2$ and $\xi_1, \cdots, \xi_n \in H$. This is immediate from $\Phi | M_1 \otimes M_2 = \Phi_0$ and Theorem 1.6.1. Q.E.D.

References. [109], [150].

4.4. Completely additive functionals and singular functionals

Definition 4.4.1. Let M be a W^*-algebra, and M_* be its predual. By Proposition 4.2.3, there is a central projection z of M^{**} such that

$$M^* = M_* + R_{(1-z)} M^*, \quad M_* = R_z M^*.$$

Any element of M_* (a $\sigma(M, M_*)$-continuous functional on M) is called a *normal functional* on M, and any element of $R_{(1-z)} M^*$ is called a *singular functional* on M.

For any $F \in M^*$, we have the unique decomposition

$$F = F_n + F_s, \quad F_n = R_z F \in M_*, \quad F_s = R_{(1-z)} F,$$

F_n, F_s are the normal, singular functionals on M respectively. It is easy to see that $\|F\| = \|F_n\| + \|F_s\|$.

Theorem 4.4.2. Let F be a positive linear functional on a W^*-algebra M. Then F is singular if and only if for any non-zero projection p of M, there is a non-zero projection q of M with $q \leq p$ such that $F(q) = 0$.

Proof. By Proposition 2.3.2, $F \in M^*$. Write $F = F_n + F_s$ as Definition 4.4.1.

Sufficiency. If $F_n \neq 0$, then $s(F_n) = p$ is a non-zero projection of M. By the assumption, there is a non-zero projection q of M with $q \leq p$ such that $F(q) = 0$. By Definition 1.8.9, $F_n(q) > 0$. Clearly, $F_s(q) \geq 0$. Then we get a contradiction. Therefore, $F_n = 0$, and $F = F_s$ is singular.

Necessity. Let $F_n = 0, F = F_s$, and p be a non-zero projection of M. We may assume that $F(p) > 0$. Pick a normal positive linear functional f on M such that $f(p) > F(p)$. Suppose that

$$\mathcal{L} = \{q | q \text{ is a projection of } M, q \leq p, \text{ and } f(q) \leq F(q)\}.$$

With the inclusion relation of projections, $\mathcal{L}$ is a non-empty partially ordered set. Let $\{q_l\}$ be a totally ordered subset of $\mathcal{L}$, and $q = \sup_l q_l$. Since f is normal, it follows that

$$F(q) \geq \sup_l F(q_l) \geq \sup_l f(q_l) = f(q).$$

Thus $q \in \mathcal{L}$. By the Zorn lemma, $\mathcal{L}$ has a maximal element p_0. But $p \notin \mathcal{L}$, so $q_0 = p - p_0 \neq 0$. For any non-zero projection q of M and $q \leq q_0$, we have

$$F(q) < f(q)$$

since p_0 is maximal. Further, $F(q_0 x q_0) \leq f(q_0 x q_0), \forall x \in M_+$. By Proposition 1.6.4, $F(q_0 X q_0) \leq f(q_0 X q_0)), \forall X \in M_+^{**}$. In particular, $F(q_0(1-z)) \leq f(q_0(1-z))$. Since $f \in M_* = R_z M^*$, it follows that $f(q_0(1-z)) = 0$ and $F(q_0(1-z)) = 0$. Moreover, F is singular, i.e., $F = R_{(1-z)} F$. Therefore, $F(q_0) = F(q_0(1-z)) = 0$, and q_0 satisfies the condition. Q.E.D.

Corollary 4.4.3. Let F be a singular positive linear functional on a W^*-algebra M, and p be a projection of M. Then there is an orthogonal family $\{p_l\}$ of projections of M such that $\sum_l p_l = p$, and $F(p_l) = 0, \forall l$.

Definition 4.4.4. Let M be a W^*-algebra, and $f \in M^*$. f is said to be *completely additive*, if for any orthogonal family $\{p_l\}$ of projections of M, we have $f(p) = \sum_l f(p_l)$, where $p = \sum_l p_l$.

The following theorem is a generalization of Proposition 1.8.5.

Theorem 4.4.5. Let M be a W^*-algebra, and $f \in M^*$. Then f is normal if and only if f is completely additive.

Proof. The necessity is obvious. Now let f be completely additive and $f = f_n + f_s$. We need to prove that $f_s = 0$. By Theorem 2.3.23. Write $f = f^{(1)} - f^{(2)} + if^{(3)} - if^{(4)}$, where $f^{(j)} \geq 0, 1 \leq j \leq 4$. Then $f_s = f_s^{(1)} - f_s^{(2)} + if_s^{(3)} - if_s^{(4)}$, where $f_s^{(j)}$ is singular and positive, $\forall j$. Define $g_s = \sum_{j=1}^4 f_s^{(j)}$. Then g_s is also singular and positive on M. Let p be a projection of M. By Corollary 4.4.3, there is an orthogonal family $\{p_l\}$ of projections of M such that $p = \sum_l p_l$ and $g_s(p_l) = 0, \forall l$. Then $f_s(p_l) = 0, \forall l$. Since f is completely additive and $f_n \in M_*$, it follows that

$$\begin{aligned} f_s(p) = & \ f(p) - f_n(p) = \sum_l [f(p_l) - f_n(p_l)] \\ = & \ \sum_l f_s(p_l) = 0. \end{aligned}$$

Therefore, $f_s = 0$ since p is arbitrary. Q.E.D.

Now let Λ be a set, and $\nu(\cdot)$ be a bounded additive complex valued function defined on all subsets of Λ, i.e., $\sup_{J\subset\Lambda} |\nu(J)| < \infty$, and

$$\nu(\Lambda_1 \cup \Lambda_2) = \nu(\Lambda_1) + \nu(\Lambda_2),$$

$\forall \Lambda_1, \Lambda_2 \subset \Lambda$ and $\Lambda_1 \cap \Lambda_2 = \emptyset$. Denote the set of all such ν by $BV(\Lambda)$. Clearly, $BV(\Lambda)$ is a linear space.

1) Let $\nu \in BV(\Lambda)$. Define

$$v(\nu)(J) = \sup\{\sum_i |\nu(J_i)| \,|\, J_i \subset J, J_i \cap J_j = \emptyset, \forall i \neq j\},$$

$\forall J \subset \Lambda$. Then $v(\nu) \in BV(\Lambda)$.

In fact, let $J_1, \cdots, J_n \subset J$ and $J_i \cap J_j = \emptyset, \forall i \neq j$. Write $\{1, \cdots, n\} = I_1 \cup I_2 = I_3 \cup I_4$, where

$$I_1 = \{i \mid \text{Re } \nu(J_i) \geq 0\}, \quad I_2 = \{i \mid \text{Re } \nu(J_i) < 0\},$$
$$I_3 = \{i \mid \text{Im } \nu(J_i) \geq 0\}, \quad I_4 = \{i \mid \text{Im } \nu(J_i) < 0\}.$$

Then

$$\begin{aligned} \sum_{i=1}^n |\nu(J_i)| \leq & \ \sum_{i\in I_1} \text{Re}\nu(J_i) - \sum_{i\in I_2} \text{Re}\nu(J_i) \\ & + \sum_{i\in I_3} \text{Im}\nu(J_i) - \sum_{i\in I_4} \text{Im}\nu(J_i) \\ = & \ \text{Re}\nu(\bigcup_{i\in I_1} J_i) - \text{Re}\nu(\bigcup_{i\in I_2} J_i) \\ & + \text{Im}\nu(\bigcup_{i\in I_3} J_i) - \text{Im}\nu(\bigcup_{i\in I_4} J_i) \\ \leq & \ 4 \sup_{J'\subset\Lambda} |\nu(J')| < \infty. \end{aligned}$$

Clearly, $v(\nu)$ is additive. Thus $v(\nu) \in BV(\Lambda)$.

2) For any $\nu \in BV(\Lambda)$, define $\|\nu\| = v(\nu)(\Lambda)$. Then $(BV(\Lambda), \|\cdot\|)$ is a Banach space.

The proof is easy. We leave it to the reader.

3) Denote the set of all bounded complex functions on Λ by $l^\infty(\Lambda)$, and define $\|f\| = \sup_{l\in\Lambda} |f(l)|, \forall f \in l^\infty(\Lambda)$. Then $l^\infty(\Lambda)^* = BV(\Lambda)$.

First, let $f \in l^\infty(\Lambda)$ and f be simple, i.e., there is a partition $\Lambda = \cup_{i=1}^n \Lambda_i$ (where $\Lambda_I \cap \Lambda_j = \emptyset, \forall i \neq j$), and complex numbers $\lambda_1, \cdots, \lambda_n$ such that

$$f(l) = \lambda_i, \quad \forall l \in \Lambda_i, \quad 1 \leq i \leq n.$$

For any $\nu \in BV(\Lambda)$, define

$$\nu(f) = \int_\Lambda f(l) d\nu(l) = \sum_{i=1}^n \lambda_i \nu(\Lambda_i).$$

Clearly, $|\nu(f)| \leq \|\nu\| \cdot \|f\|$.

For any $f \in l^\infty(\Lambda)$, pick a sequence $\{f_n\}(\subset l^\infty(\Lambda))$ of simple functions such that $\|f_n - f\| \to 0$. Since $|\nu(f_n - f_m)| \leq \|\nu\| \cdot \|f_n - f_m\| \to 0$, then we can define

$$\nu(f) = \lim_n \nu(f_n), \qquad \forall \nu \in BV(\Lambda).$$

Clearly, this definition is independent of the choice of $\{f_n\}$, and $|\nu(f)| \leq \|\nu\| \cdot \|f\|, \forall f \in l^\infty(\Lambda), \nu \in BV(\Lambda)$. Also, it is easy to see that

$$\|\nu\| = \sup\{|\nu(f)| \mid f \in l^\infty(\Lambda), \|f\| \leq 1\}.$$

Thus, $BV(\Lambda)$ can be isometrically embedded in $l^\infty(\Lambda)^*$.

On the other hand, let $F \in l^\infty(\Lambda)^*$, and define $\nu(J) = F(\chi_J)$, where χ_J is the characteristic function of $J, \forall J \subset \Lambda$. Then $\nu \in BV(\Lambda)$,and $\nu(f) = F(f), \forall f \in l^\infty(\Lambda)$. So $l^\infty(\Lambda)^* = BV(\Lambda)$.

4) Let $\nu \in BV(\Lambda)$. Clearly, we have

$$\sum_{l\in\Lambda} |\nu(\{l\})| \leq \|\nu\|.$$

5) Let $\{\nu_n\} \subset BV(\Lambda)$,and $\sup_n \|\nu_n\| < \infty$, and

$$\lim_n \nu_n(J) = 0, \qquad \forall J \subset \Lambda.$$

Then $\lim_n \sum_{l\in\Lambda} |\nu_n(\{l\})| = 0$.

In fact, suppose that there is a $\varepsilon > 0$ such that

$$\sum_{l\in\Lambda} |\nu_n(\{l\})| \geq \varepsilon, \qquad \forall n \tag{1}$$

(replacing $\{n\}$ by a subsequence in necessary case). For $n_1 = 1$, there is a finite subset F_1 of Λ such that

$$\sum_{l\in F_1} |\nu_{n_1}(\{l\})| > \sum_{l\in\Lambda} |\nu_{n_1}(\{l\})| - \frac{\varepsilon}{10}.$$

Since $\nu_n(l) \to 0, \forall l \in \Lambda$, there exists n_2 such that

$$\sum_{l\in F_1} |\nu_{n_2}(\{l\})| < \frac{\varepsilon}{20}.$$

Thus there is a finite subset $F_2 \subset \Lambda\backslash F_1$ such that

$$\sum_{l\in F_2} |\nu_{n_2}(\{l\})| > \sum_{l\in\Lambda} |\nu_{n_2}(\{l\})| - \frac{\varepsilon}{10}.$$

$\cdots$. Generally, we can find a sequence $\{F_k\}$ of finite subsets of Λ and a subsequence $\{n_k\}$ such that

$$\sum_{l\in F_k} |\nu_{n_k}(\{l\})| > \sum_{l\in\Lambda} |\nu_{n_k}(\{l\})| - \frac{\varepsilon}{10}, \quad \forall k$$

and $F_k \cap F_j = \emptyset, \forall k \neq j$.

Fix m such that $m > \frac{10}{\varepsilon} \sup_n \|\nu_n\|$.

Let $E_1 = F_1, \mu_1 = \nu_{n_1}$. If

$$v(\mu_1)(\bigcup_{j=1}^{\infty} F_{mj+p}) \geq \frac{\varepsilon}{10}, \qquad 1 \leq p \leq m,$$

then

$$\begin{aligned} \|\mu_1\| \geq\ & v(\mu_1)(\sqcup_{p=1}^{m} \sqcup_{j=1}^{\infty} F_{mj+p}) \\ =\ & \sum_{p=1}^{m} v(\mu_1)(\sqcup_{j=1}^{\infty} F_{mj+p}) \geq m \cdot \frac{\varepsilon}{10} \\ >\ & \sup_n \|\nu_n\| \geq \|\mu_1\|. \end{aligned}$$

This is a contradiction. Thus, there is an integer p_1 with $1 \leq p_1 \leq m$ such that

$$v(\mu_1)(\sqcup_{j=1}^{\infty} F_{mj+p_1}) < \frac{\varepsilon}{10}.$$

Let $E_2 = F_{m+p_1}, \mu_2 = \nu_{n_{m+p_1}}$, and $F'_j = F_{mj+p_1}, \forall j$. Similarly, there is an integer p_2 with $1 \leq p_2 \leq m$ such that

$$\begin{aligned} & v(\mu_2)(\sqcup_{j=1}^{\infty} F'_{mj+p_2}) \\ =\ & v(\mu_2)(\sqcup_{j=1}^{\infty} F_{m(mj+p_2)+p_1}) < \frac{\varepsilon}{10}. \end{aligned}$$

$\cdots$. Generally, we have $\{p_s \mid s = 1, 2, \cdots\}$ such that

$$1 \leq p_s \leq m,$$

$$\sum_{l\in E_s}|\mu_s(\{l\})| > \sum_{l\in\Lambda}|\mu_s(\{l\})| - \frac{\varepsilon}{10},$$

$$v(\mu_s)(\sqcup_{j>s}E_j) < \frac{\varepsilon}{10} \tag{2}$$

(noticing (1)), $\forall s$, where $\mu_s = \nu_{n_{b_s}}, E_s = F_{b_s}$, and $b_1 = 1, b_2 = mb_1 + p_1, \cdots, b_s = mb_{s-1} + p_s, \cdots$.

Now define a $f \in l^\infty(\Lambda)$ as follows:

$$f(l) = \begin{cases} 0, & \text{if } l \notin \sqcup_{j=1}^\infty E_j, \\ \overline{arg\mu_j(\{l\})}, & \text{if } l \in E_j, \forall j. \end{cases}$$

By $^{\#}E_j < \infty(\forall j)$ and (2), for any s we have

$$\begin{aligned} &|\mu_s(f) - \sum_{l\in E_s}|\mu_s(\{l\})|| \\ \leq\ & |\sum_{j<s}\int_{E_j} f d\mu_s| + |\int_{E_s} f d\mu_s - \sum_{l\in E_s}|\mu_s(\{l\})|| \\ &+|\int_{\sqcup_{j>s}E_j} f d\mu_s| \\ \leq\ & \sum_{j<s}\sum_{l\in E_j}|\mu_s(\{l\})| + v(\mu_s)(\sqcup_{j>s}E_j) < \frac{\varepsilon}{5}. \end{aligned}$$

Thus from (1), (2),

$$\begin{aligned} |\mu_s(f)| \geq\ & \sum_{l\in E_s}|\mu_s(\{l\})| - \frac{\varepsilon}{5} \\ >\ & \sum_{l\in\Lambda}|\mu_s(\{l\})| - \frac{3}{10}\varepsilon \geq \frac{7}{10}\varepsilon, \end{aligned}$$

$\forall s$. However, from $\sup_n \|\nu_n\| < \infty$ and $\nu_n(J) \to 0, \forall J \subset \Lambda$, it must be that $\nu_n(f) \to 0$. Then $\mu_s(f) \to 0$, a contradiction. Therefore,

$$\lim_n \sum_{l\in\Lambda}|\nu_n(\{l\})| = 0.$$

Proposition 4.4.6. Let M be a W^*-algebra, $\{f, f_k \mid k = 1, 2, \cdots\} \subset M^*$ and $f_k \to f(\sigma(M^*, M))$. Then the normal part $\{f_k^n\}$ and the singular part $\{f_k^s\}$ of $\{f_k\}$ converge to the normal part f^n and the singular part f^s of f respectively in $\sigma(M^*, M)$-topology.

Proof. We may assume that $f = 0$. So it suffices to show that $f_k^n \to 0(\sigma(M^*, M))$. Since $\sup_k \|f_k^n\| < \infty$, it is enough to prove that

$$f_k^n(p) \to 0$$

for any projection p of M.

For any $h \in M^*$, by Theorem 2.3.23 we can uniquely write $h = h_1 - h_2 + ih_3 - ih_4$, where $h_j \geq 0, 1 \leq j \leq 4$. Define $[h] = h_1 + h_2 + h_3 + h_4$.

Now let $g = \sum_{k=1}^{\infty} \frac{1}{2^k}[f_k^s]$. Then g is a singular positive functional on M. Fix a projection p of M. By Corollary 4.4.3, there is an orthogonal family $\{p_l\}_{l\in\Lambda}$ of projections of M such that

$$p = \sum_{l\in\Lambda} p_l, \qquad g(p_l) = 0, \qquad \forall l \in \Lambda.$$

Thus $f_k^s(p_l) = 0, \forall k, l$.

Define $\nu_k \in BV(\Lambda)$ as follows

$$\nu_k(J) = f_k(\sum_{l\in J} p_l), \qquad \forall J \subset \Lambda.$$

Since $\sup_k \|f_k\| < \infty$, it follows that

$$\sup\{|\nu_k(J)| \mid k, J \subset \Lambda\} < \infty.$$

Clearly, $\lim_k \nu_k(J) = 0, \forall J \subset \Lambda$. Then by the discussion 5) about $BV(\Lambda)$ and $f_k^s(p_l) = 0, \forall k, l$, we have

$$\begin{aligned}\lim_k \sum_{l\in\Lambda} |f_k^n(p_l)| &= \lim_k \sum_{l\in\Lambda} |f_k(p_l)| \\ &= \lim_k \sum_{l\in\Lambda} |\nu_k(\{l\})| = 0.\end{aligned}$$

Further, from $f_k^n \in M_*(\forall k)$,

$$\lim_n f_k^n(p) = \lim_k \sum_{l\in\Lambda} f_k^n(p_l) = 0.$$

Q.E.D.

Theorem 4.4.7. Let M be a W^*-algebra,and M_* be its predual. Then M_* is weakly sequentially complete.

Proof. Let $\{f_k\}$ be a weakly Cauchy sequence in M_*. Since $\sup_k \|f_k\| < \infty$, it follows that there is a $f \in M^*$ such that $f_k \to f(\sigma(M^*, M))$. By Proposition 4.4.6

$$f_k^n \to f^n(\sigma(M^*, M)), \quad \text{and} \quad f_k^s \to f^s(\sigma(M^*, M)).$$

But $f_k^n = f_k, f_k^s = 0, \forall k$, therefore $f^s = 0$,and $f = f^n \in M_*$. Q.E.D.

Notes. The decomposition of M^* into the normal part and the singular part was given by M. Takesaki. Theorem 4.4.2 is due to M.Takesaki. Theorem

4.4.7 is due to S.Sakai and C.A. Akemann. Moreover, J.F. Aarnes proved that $(M, \tau(M, M_*))$ is a complete locally convex space for a W^*-algebra M.

References. [1], [2], [3], [170].

4.5. The characterizations of weakly compact subsets in predual

Let M ba a W^*-algebra. Lemma 1.11.5 gives a charaterization of a weakly compact subset of its predual M_*. Further, we have the following.

Theorem 4.5.1. Let M be a W^*-algebra, M_* be its predual, and $A \subset M_*$. Then the following statements are equivalent:

1) The $\sigma(M_*, M)$-closure of A is $\sigma(M_*, M)$-compact;

2) There exists a normal positive linear functional ψ on M with the property that for any $\varepsilon > 0$ there exists $\delta = \delta(\varepsilon)(> 0)$ such that $|\varphi(a)| < \varepsilon$ for each $\varphi \in A$ if $a \in M$ with $\|a\| \leq 1$ and $\psi(a^*a + aa^*) < \delta$;

3) A is bounded, and for any sequence $\{a_n\} \subset M$ with $a_n \to 0(s^*(M, M_*))$, $\lim\limits_n \varphi(a_n) = 0$ uniformly for $\varphi \in A$;

4) A is bounded, and for any decreasing sequence $\{p_n\}$ of projections of M with $\inf_n p_n = 0, \lim\limits_n \varphi(p_n) = 0$ uniformly for $\varphi \in A$;

5) A is bounded, and for any orthogonal sequence $\{p_n\}$ of projections of $M, \lim\limits_n \varphi(p_n) = 0$ uniformly for $\varphi \in A$;

6) For any maximal abelian W^*-subalgebra N of M, the $\sigma(N_*, N)$-closure of $\{(\varphi|N) \mid \varphi \in A\}$ is $\sigma(N_*, N)$-compact;

7) A is bounded, and for any increasing net $\{p_l\}$ of projections of M, $\lim\limits_l \varphi(p_l) = \varphi(p)$ uniformly for $\varphi \in A$, where $p = \sup_l p_l$;

8) A is bounded, and for any increasing net $\{p_l\}$ of projections of M with $\sup_l p_l = 1, \|L_{(1-p_l)}R_{(1-p_l)}\varphi\| \to 0$ uniformly for $\varphi \in A$.

Proof. 1) $\Longrightarrow$ 2). It is Lemma 1.11.5 exactly.

2) $\Longrightarrow$ 3). Pick ψ as in the Condition 2) and $\varepsilon = 1$. For any $a \in M$, suppose that m is sufficiently large such that

$$\psi(b^*b + bb^*) < \delta = \delta(1),$$

where $b = a/m$, and $\|b\| \leq 1$. Then by 2), we have

$$\sup\{|\varphi(a)| \mid \varphi \in A\} \leq m.$$

By the principle of uniform boundedness, A is bounded.

Now let $\{a_n\}$ be a sequence of M with $a_n \to 0(s^*(M, M_*))$. Clearly, $\sup_n \|a_n\| \leq K$ (some constant). For any $\varepsilon > 0$, since $(a_n^* a_n + a_n a_n^*) \to 0(\sigma(M, M_*))$, there is n_0 such that

$$K^{-2}\psi(a_n^* a_n + a_n a_n^*) < \delta = \delta(\varepsilon), \quad \forall n \geq n_0.$$

By 2), $|\varphi(a_n)| < \varepsilon K, \forall \varphi \in A$ and $n \geq n_0$. Therefore, $\lim_n \varphi(a_n) = 0$ uniformly for $\varphi \in A$.

3) $\Longrightarrow$ 4). It is obvious.

4) $\Longrightarrow$ 5). Let $\{p_n\}$ be an orthogonal sequence of projections of M. Then $\{q_n = \sum_{k \geq n} p_k\}_n$ is a decreasing sequence of projections of M with $\inf_n q_n = 0$. By 4), $\lim_n \varphi(q_n) = 0$ uniformly for $\varphi \in A$. Then

$$\varphi(p_n) = \varphi(q_n) - \varphi(q_{n+1}) \to 0$$

uniformly for $\varphi \in A$.

5) $\Longrightarrow$ 1). Regard A as a subset of M^*, and let $\overline{A}$ be the $\sigma(M^*, M)$-closure of A in M^*. Since A is bounded, it follows that $\overline{A}$ is $\sigma(M^*, M)$-compact. Now it suffices to show that $\overline{A} \subset M_*$. By Theorem 4.4.5, we need to prove that f is completely additive for any $f \in \overline{A}$. Also this is equivalent to prove that

$$\sum_{l \in F} \varphi(p_l) \to \varphi(p) \quad \text{uniformly for} \quad \varphi \in A,$$

where $\{p_l \mid l \in \Lambda\}$ is any given orthogonal family of projections of M, $\{F \mid F \subset \Lambda, {}^{\#}F < \infty\}$ is directed by the inclusion relation, and $p = \sum_{l \in \Lambda} p_l$. In fact, let $f \in \overline{A}$ and $\{\varphi_\alpha\}$ be a net of A with $\varphi_\alpha \to f(\sigma(M^*, M))$. For any $\varepsilon > 0$, then there is a finite subset F_0 such that

$$|\sum_{l \in F} \varphi_\alpha(p_l) - \varphi_\alpha(p)| < \varepsilon, \quad \forall F \supset F_0 \quad \text{and } \alpha.$$

Thus $|\sum_{l \in F} f(p_l) - f(p)| < \varepsilon, \forall F \supset F_0$. Further, f is completely additive.

Now suppose that there is a $\varepsilon > 0$ such that for any finite subset F of Λ we can find $\varphi_F \in A$ with

$$|\sum_{l \notin F} \varphi_F(p_l)| > \varepsilon.$$

Then we can pick a disjoint sequence $\{F_n\}$ of finite subsets of Λ and $\{\varphi_n\} \subset A$ such that

$$|\sum_{l \in F_n} \varphi_n(p_l)| \geq \varepsilon, \qquad \forall n.$$

Since $\{q_n = \sum_{l \in F_n} p_l\}$ is an orthogonal sequence of projections of M, it follows from 5) that

$$\lim_n \varphi(q_n) = 0 \quad \text{uniformly for} \quad \varphi \in A.$$

This is a contradiction. Therefore,

$$\sum_{l\in F}\varphi(p_l)\to\varphi(p)\quad\text{uniformly for }\varphi\in A.$$

6) $\Longrightarrow$ 5). Let $\{p_n\}$ be an orthogonal sequence of projections of M, and N be a maximal abelian W^*-subalgebra containing $\{p_n\}$. By 6), the $\sigma(N_*,N)$-closure of

$$A_N=\{(\varphi|N)\mid\varphi\in A\}$$

is $\sigma(N_*,N)$-compact. Since 1) implies 5), it follows that $\varphi(p_n)=(\varphi|N)(p_n)\to 0$ uniformly for $\varphi\in A$.

Similarly, we can prove that $\sup\{|\varphi(h)|\mid\varphi\in A\}<\infty$ for any $h^*=h\in M$. Thus, A is bounded.

1) $\Longrightarrow$ 6). Let $\overline{A}$ be the $\sigma(M_*,M)$-closure of A. Then $\overline{A}$ is a $\sigma(M_*,M)$-compact subset of M_*. Suppose that N is a maximal abelian W^*-subalgebra of M. Clearly, $\overline{A}_N=\{(\varphi|N)\mid\varphi\in\overline{A}\}$ is a $\sigma(N_*,N)$-compact subset of N_*. But $\{(\varphi|N)\mid\varphi\in A\}\subset\overline{A}$, so 6) holds.

2) $\Longrightarrow$ 7). Let $\{p_l\}$ be an increasing net of projections of M,and $p=\sup_l p_l$. Pick ψ as in the condition 2). For any $\varepsilon>0$, since

$$\psi((p-p_l)^*(p-p_l)+(p-p_l)(p-p_l)^*)=2\psi(p-p_l)\to 0,$$

there is l_0 such that $\psi(p-p_l)<\delta=\delta(\varepsilon),\forall l\geq l_0$. By 2), we have $|\varphi(p-p_l)|<\varepsilon,\forall\varphi\in A,l\geq l_\varepsilon$. That is $\lim_l\varphi(p_l)=\varphi(p)$ uniformly for $\varphi\in A$. Moreover, by the same proof of 2) $\Longrightarrow$ 3), A is bounded.

7) $\Longrightarrow$ 4). Let $\{p_n\}$ be a decreasing sequence of projections of M, and $\inf_n p_n=0$. Then $\{(1-p_n)\}$ is an increasing sequence of projections of M and $\sup_n(1-p_n)=1$. By 7), $\lim_n\varphi(1-p_n)=\varphi(1)$ uniformly for $\varphi\in A$, i.e., $\varphi(p_n\to 0)$ uniformly for $\varphi\in A$.

8) $\Longrightarrow$ 7). It suffices to notice that: if $\{p_l\}$ is an increasing net of projections of M, then $\{q_l=p_l+(1-p)\}$ is an increasing net of projections of M, and $\sup_l q_l=1$, where $p=\sup_l p_l$.

2) $\Longrightarrow$ 8) . Let $\{p_l\}$ be an increasing net of projections of M, and $\sup_l p_l=1$. Pick ψ as in the condition 2). For any $\varepsilon>0$, there is $l(\varepsilon)$ such that $\psi(1-p_l)<\frac{1}{2}\delta(\varepsilon),\forall l\geq l(\varepsilon)$. Then for any $a\in M$ with $\|a\|\leq 1$ and $l\geq l(\varepsilon)$, we have

$$\psi((1-p_l)a^*(1-p_l)a(1-p_l)+(1-p_l)a(1-p_l)a^*(1-p_l))\leq 2\psi(1-p_l)<\delta(\varepsilon)$$

By 2), $|\varphi(1-p_l)a(1-p_l)|<\varepsilon,\forall\varphi\in A$ and $l\geq l(\varepsilon),\|a\|\leq 1$. Therefore, $\|L_{(1-p_l)}R_{(1-p_l)}\varphi\|\to 0$ uniformly for $\varphi\in A$. Moreover, A is bounded by the same proof of 2) $\Longrightarrow$ 3). Q.E.D.

Proposition 4.5.2. Let M be a W^*-algebra, M_* be its predual, and $A\subset(M_*)_+$. Suppose that the $\sigma(M_*,M)$-closure of A is $\sigma(M_*,M)$-compact. Then the $\sigma(M_*, M)$-closure of $E = \{R_a\varphi \mid a \in M, \|a\| \leq 1, \varphi \in A\}$ is also $\sigma(M_*, M)$-compact.

Proof. Clearly, E is bounded. Let $\{p_n\}$ be a decreasing sequence of projections of M and $\inf_n p_n = 0$. By Theorem 4.5.1, $\varphi(p_n) \to 0$ uniformly for $\varphi \in A$. From the Schwartz inequality,

$$|R_a\varphi(p_n)| \leq \varphi(a^*a)^{1/2}\varphi(p_n)^{1/2} \leq \|\varphi\|^{1/2}\varphi(p_n)^{1/2}$$

$\forall a \in M$ and $\|a\| \leq 1$. Thus $\rho(p_n) \to 0$ uniformly for $\rho \in E$. Now again by Theorem 4.5.1, the $\sigma(M_*, M)$-closure of E is $\sigma(M_*, M)$-compact. Q.E.D.

Notes. Theorem 4.5.1 is a combination of results due to several mathematicians: A. Grothendieck, S.Sakai, M.Takesaki, H. Umegaki and finally, C.A. Akemann.

References. [2], [62], [146], [169], [188].

Chapter 5

Abelian Operator Algebras

5.1. Measure theory on locally compact Hausdorff spaces

Let Ω be a localy compact Hausdorff space, and $\mathcal{B}$ be the collection of all *Borel subsets* of Ω (i.e. the σ-Bool ring generated by compact subsets of Ω). Define

$$\mathcal{B}_{loc} = \{E \subset \Omega \mid E \cap K \in \mathcal{B}, \forall K \text{ compact} \subset \Omega\}.$$

$\mathcal{B}_{loc}$ is a σ-Bool algebra. Each subset in $\mathcal{B}_{loc}$ is called a *locally Borel subset.* Clearly, $E \in \mathcal{B}_{loc}$ if and only if $E \cap F \in \mathcal{B}, \forall F \in \mathcal{B}$.

A complex function f on Ω is said to be *measurable*, if it is $\mathcal{B}$-measurable. f is said to be *locally measurable*, if it is $\mathcal{B}_{loc}$-measurable. Clearly, A measurable function is locally measurable. And a locally measurable function f is measurable if and only if $\{t \in \Omega \mid f(t) \neq 0\} \in \mathcal{B}$.

Let ν be a regular Borel measure on Ω. $F(\subset \Omega)$ is called *ν-zero*, if $F \in \mathcal{B}$ and $\nu(F) = 0$; $E(\subset \Omega)$ is called *locally ν-zero*, if $E \in \mathcal{B}_{loc}$ and $\nu(E \cap K) = 0, \forall K$ compact $\subset \Omega$. A Proposition about $P(t)$ on Ω holds *almost everywhere* with respect to ν (a.e.ν), if $\{t \in \Omega \mid P(t) \text{does not hold}\}$ is a subset of some ν-zero set; $P(t)$ on Ω holds *locally almost everywhere* with respect to ν (l.a.e.ν), if $\{t \in \Omega \mid P(t) \text{ does not hold}\}$ is a subset of some locally ν-zero set.

Let ν be a regular Borel measure on Ω. Then

$$V_0 = \cup\{V \subset \Omega \mid V \text{ is open and locally } \nu\text{-zero}\}$$

is the maximal locally ν-zero open subset. Let

$$\text{supp}\nu = (\Omega \backslash V_0).$$

It is called the *support* of ν, and clearly it has the following property. Let $U(\subset \Omega)$ be a Borel open subset. Then $\nu(U) = 0$ if and only if $U \cap \text{supp}\nu = \emptyset$.

Lemma 5.1.1. Let ν be a non-zero regular Borel measure on Ω. Then there is a non-empty compact subset $K(\subset \Omega)$ such that $\nu(K \cap U) > 0$ for any open subset U of Ω with $U \cap K \neq \emptyset$.

Proof. Since supp ν is a non-empty closed subset of Ω, we can find an open subset V such that $\overline{V}$ compact and $K = \overline{V \cap \text{supp}\nu} \neq \emptyset$. Then K is what we want to find. In fact, suppose that there is an open subset U with $U \cap K \neq \emptyset$ such that $\nu(K \cap U) = 0$. Then $\nu(U \cap V \cap \text{supp}\nu) = 0$ and $\nu(U \cap V) = \nu(E)$, where $E = (U \cap V)\backslash\text{supp}\nu$. But E is open and $E \cap \text{supp}\nu = \emptyset$, so we have

$$\nu(U \cap V) = \nu(E) = 0.$$

From the definition of supp $\nu, U \cap V \cap \text{supp } \nu = \emptyset$. On the other hand, pick $t \in U \cap K$. Since U is an open neighborhood of t and $t \in K = \overline{V \cap \text{supp}\nu}$, it follows that

$$U \cap V \cap \text{supp}\nu \neq \emptyset.$$

We get a contradiction. Therefore, K is what we want to find. Q.E.D.

Proposition 5.1.2. Let ν be a non-zero regular Borel measure on Ω. Then there is a disjoint family $\{K_l\}_{l\in\Lambda}$ of non-empty compact subsets of Ω such that $N = \Omega\backslash \sqcup_{l\in\Lambda} K_l$ is a locally ν-zero subset, and the family $\{K_l\}_{l\in\Lambda}$ has the *locally countable property*, i.e., for any compact subset K of Ω the index set $\{l \in \Lambda \mid K_l \cap K \neq \emptyset\}$ is countable.

Proof. By Lemma 5.1.1 and the Zorn lemma, there is a maximal disjoint family $\{K_l\}_{l\in\Lambda}$ of non-empty compact subsets of Ω such that $\nu(K_l \cap U) > 0$ for any open subset U of Ω with $U \cap K_l \neq \emptyset, \forall l$.

Suppose that V is an open subset of Ω and $\overline{V}$ is compact. Then

$$\sum_{l\in\Lambda} \nu(K_l \cap V) \leq \nu(V) < \infty.$$

Thus $\{l \in \Lambda \mid \nu(K_l \cap V) > 0\}$ is countable. However if some $l \in \Lambda$ is such that $\nu(K_l \cap V) = 0$, then $K_l \cap V = \emptyset$ by the property of K_l. Thus $\{l \in \Lambda \mid K_l \cap V \neq \emptyset\}$ is countable. From this discussion, it is easily verified that the family $\{K_l\}_{l\in\Lambda}$ has the locally countable property. In consequence, $\sqcup_{l\in\Lambda} K_l \in \mathcal{B}_{loc}$ and $N = \Omega\backslash \sqcup_{l\in\Lambda} K_l \in \mathcal{B}_{loc}$.

Now we prove that N is locally ν-zero. Suppose that there is a compact subset $H \subset N$ such that $\nu(H) > 0$. Applying Lemma 5.1.1 to H and $(\nu|H)$, we can find a non-empty compact subset $K \subset H$ such that $\nu(U_H \cap K) > 0$ for each open subset U_H of H with $U_H \cap K \neq \emptyset$. Thus for any open subset U of Ω with $U \cap K \neq \emptyset$ we have also $\nu(U \cap K) = \nu((U \cap H) \cap K) > 0$. Clearly, $K \cap K_l = \emptyset, \forall l \in \Lambda$. This is a contradiction since the family $\{K_l\}_{l\in\Lambda}$ is maximal. Therefore, N is locally ν-zero. Q.E.D.

Let f be a locally measurable function on Ω, and ν be a regular Borel measure on Ω. f is said to be *locally essentially bounded* with respect to ν, if there is a constant C such that

$$|f(t)| \leq C, \qquad l.a.e.\nu.$$

The minimum of such C is called locally essentially supremum of f, denoted by $\|f\|_\infty$. Let

$$L^\infty(\Omega,\nu) = \left\{ f \,\middle|\, \begin{array}{c} f \text{ is locally measurable on } \Omega, \\ \text{and is locally essentially bounded} \end{array} \right\}.$$

Clearly, $(L^\infty(\Omega,\nu), \|\cdot\|_\infty)$ is an abelian C^*-algebra. Indeed, it is a W^*-algebra.

Theorem 5.1.3. $L^1(\Omega,\nu)^* = L^\infty(\Omega,\nu)$.

Proof. Suppose that $f \in L^\infty(\Omega,\nu)$. Define

$$F(g) = \int_\Omega f(t)g(t)d\nu(t), \qquad \forall g \in L^1(\Omega,\nu).$$

Clearly, $F \in L^1(\Omega,\nu)^*$ and $\|F\| = \|f\|_\infty$.

Now let $F \in L^1(\Omega,\nu)^*$. For any compact subset K of Ω, since $\nu(K) < \infty$, there is unique $f_K \in L^\infty(K, \nu|K)$ such that

$$|f_K(t)| \leq \|F\|, \quad \forall t \in K, \quad \text{and} \quad F(g) = \int_K f_K(t)g(t)d\nu(t),$$

$\forall g \in L^1(K,\nu|K)$ (see [178] Theorem 7.4-A). Then we can write $f_K = F|\ L^1(K, \nu|K)$.

By Proposition 5.1.2, $\Omega = N \sqcup \sqcup_{l\in\Lambda} K_l$. Then for any $l \in \Lambda$,

$$F|L^1(K_l, \nu|K_l) = f_l \in L^\infty(K_l, \nu|K_l).$$

Let

$$f(t) = \sum_{l\in\Lambda} \chi_{K_l}(t) f_l(t).$$

Then $|f(t)| \leq \|F\|, \forall t \in \Omega$, and $f \in L^\infty(\Omega,\nu)$. For any $g \in L^1(\Omega,\nu)$, since $\text{supp} g = \{t \in \Omega \mid g(t) \neq 0\} \in \mathcal{B}$, it follows that

$$J = \{l \in \Lambda \mid K_l \cap \text{supp}\, g \neq \emptyset\}$$

is countable. Let $g_l = \chi_{K_l} g$. Then $g = \sum_{l\in J} g_l$. Now by the continuity of F and the bounded convergence theorem, we have $F(g) = \int_\Omega f(t)g(t)d\nu(t)$. Therefore, $L^1(\Omega,\nu)^* = L^\infty(\Omega,\nu)$. Q.E.D.

Let ν be a regular Borel measure on Ω. A function f on Ω is said to be non-negative locally ν-integrable, if f is non-negative locally measurable, and for any compact subset K of $\Omega, \chi_K f \in L^1(\Omega,\nu)$. In this case, define

$$\mu(E) = \int_E f d\nu = \int f\chi_E d\nu, \quad \forall E \in \mathcal{B}.$$

Then μ, denoted by $\mu = f \cdot \nu$, is also a regular Borel measure on Ω, and is *absolutely continuous* with respect to ν, denoted by $\mu \prec \nu$, i.e., if $E \in \mathcal{B}$ with $\nu(E) = 0$, then $\mu(E) = 0$. Moreover, if g is a measurable function on Ω, then $g \in L^1(\Omega,\mu)$ if and only if $fg \in L^1(\Omega,\nu)$. And we have

$$\int_\Omega g d\mu = \int_\Omega f g d\nu.$$

Theorem 5.1.4. Let μ, ν be two regular Borel measures on Ω. Then the following statements are equivalent:

1) There is a non-negative locally ν-integrable function f such that $\mu = f \cdot \nu$;
2) If N is a locally ν-zero subset, then it is also locally μ-zero;
3) If K is a compact subset and $\nu(K) = 0$, then $\mu(K) = 0$, i.e. $\mu \prec \nu$.

Proof. The equivalence of 2) and 3) is obvious. And also it is clear that 1) implies 2). Now let 2) hold. By Proposition 5.1.2, $\Omega = N \sqcup \sqcup_{l\in\Lambda} K_l$. Since N is locally ν-zero, it follows from 2) that N is also locally μ-zero. For each $l \in \Lambda$, from $\nu(K_l) < \infty, \mu(K_l) < \infty$ and the Radon-Nikodym theorem there is $0 \le f_l \in L^1(K_l, \nu|K_l)$ such that

$$\mu(E) = \int_E f_l d\nu, \quad \forall E \in B \text{ and } E \subset K_l.$$

Let $f = \sum_{l\in\Lambda} \chi_{K_l} f_l$. Then f is non-negative locally measurable since the family $\{K_l\}_{l\in\Lambda}$ is locally countable. For any $E \in \mathcal{B}$, since $f|N = 0, \mu(E \cap N) = 0$ and $J = \{l \in \Lambda \mid K_l \cap E \neq \emptyset\}$ is countable, we have

$$\begin{aligned}\mu(E) &= \sum_{l\in J} \mu(K_l \cap E) \\ &= \sum_{l\in J} \int_{K_l\cap E} f d\nu = \int_E f d\nu,\end{aligned}$$

i.e., $\mu = f \cdot \nu$. Q.E.D.

μ and ν are said to be *equivalent*, denoted by $\mu \sim \nu$, if $\mu \prec \nu$ and $\nu \prec \mu$. In this case, clearly a.e.μ = a.e.ν, l.a.e.μ = l.a.e.ν, and there is a non-negative locally ν-integrable function f and a non-negative locally μ-integrable function g such that

$$\mu = f \cdot \nu, \quad \text{and} \quad \nu = g \cdot \mu.$$

And also, $f(t)g(t) = 1$, $l.a.e.\mu$ or $l.a.e.\nu$.

Let μ and ν be two regular Borel measures on Ω. μ and ν are said to be *singular* each other, denoted by $\mu \perp \nu$, if there is $A \in \mathcal{B}_{loc}$ such that A is locally μ-zero and $(\Omega \backslash A)$ is locally ν-zero.

Theorem 5.1.5. Let μ, ν be two regular Borel measure on Ω. Then we can uniquely write that $\mu = f \cdot \nu + \mu_1$, where f is non-negative locally ν-integrable, and $\mu_1 \perp \nu$.

Proof. By Theorem 5.1.4, there is a non-negative locally $(\mu + \nu)$-integrable function g such that $\mu = g \cdot (\mu + \nu)$ and $0 \le g(t) \le 1, \forall t \in \Omega$. Let

$$B = \{t \in \Omega \mid g(t) = 1\},$$

$$A = \{t \in \Omega \mid 0 \le g(t) < 1\},$$

$$\mu_1 = \mu | B, \qquad \mu_0 = \mu | A.$$

Clearly, A is locally μ_1-zero. If K is a compact subset and $K \subset B$, then

$$\mu(K) = \int_K g d(\mu + \nu) = \mu(K) + \nu(K)$$

and $\nu(K) = 0$. Thus B is locally ν-zero,and $\mu_1 \perp \nu$.

Now suppose that K is a ν-zero compact subset. Then

$$\begin{aligned} \mu_0(K) = &\ \mu(K \cap A) \\ = &\ \int_{K \cap A} g d\mu + \int_{K \cap A} g d\nu = \int_{K \cap A} g d\mu \end{aligned}$$

i.e., $\int_{K \cap A} (1 - g) d\mu$=0. By the definition of $A, \mu_0(K) = \mu(K \cap A) = 0$. Thus $\mu_0 \prec \nu$, and there is a non-negative locally ν-integrable function f such that $\mu_0 = f \cdot \nu$. So $\mu = f \cdot \nu + \mu_1$.

Finally, we prove the uniqueness. Let $\mu = f_i \cdot \nu + \mu_i$, where f_i is non-negative locally ν-integrable,and $\mu_i \perp \nu, i = 1, 2$. Then there is $A_i \in \mathcal{B}_{loc}$ such that A_i is locally ν-zero and $(\Omega \backslash A_i)$ is locally μ_i-zero, $i = 1, 2$. Clearly, $(A_1 \sqcup A_2)$ is locally ν-zero, and

$$\Omega \backslash (A_1 \sqcup A_2) = (\Omega \backslash A_1) \cap (\Omega \backslash A_2)$$

is locally μ_1-and μ_2-zero. Let K be a compact subset and $K \subset (A_1 \sqcup A_2)$. Then $\nu(K) = 0$, and $\mu_1(K) = \mu_2(K)$. Thus $\mu_1 | (A_1 \sqcup A_2) = \mu_2 | (A_1 \sqcup A_2)$,and $\mu_1 = \mu_2$. Further, $f_1 = f_2$ $l.a.e.\nu$. Q.E.D.

Notes. Proposition 5.1.2 is taken from N. Bourbaki. Theorem 5.1.3 is indeed a characterization of a *localizable measure space* (I.E. Segal).

Moreover, on the measurability (of subsets, functions and etc.) we follows the treatment of P.R. Halmos, i.e., the measurability is independent of the measures. So in this book, the expresion of some results are slightly different with some standard books.

References. [12], [67], [157], [178].

5.2. Stonean spaces

Definition 5.2.1. A Hausdorff space is said to be *extremely disconnected* if the closure of every open subset is also open. A compact extremely disconnected space is called a *Stonean space.*

Proposition 5.2.2. Let Ω be a Stonean space. Then the linear span of projections of $C(\Omega)$ is dense in $C(\Omega)$.

Proof. Let $f \in C(\Omega), f \geq 0$ and $\varepsilon > 0$. Consider the following partition:

$$0 = \lambda_0 < \lambda_1 < \cdots < \lambda_n = \|f\| + 1$$

such that $(\lambda_{i+1} - \lambda_i) < \varepsilon, 0 \leq i \leq n-1$. Clearly,

$$E_1 = \{t \in \Omega \mid f(t) < \lambda_1\}$$

is an open subset of Ω. Then $G_1 = \overline{E}_1$ is an open and closed subset of Ω. By induction, define

$$E_i = \{t \in \Omega \mid f(t) < \lambda_i, t \notin \bigsqcup_{j=1}^{i-1} G_j\}$$

and $G_i = \overline{E}_i, 2 \leq i \leq n$. Then E_i is open,and G_i is open and closed, $1 \leq i \leq n$.

We say that $\Omega = \bigsqcup_{i=1}^{n} G_i$. In fact, suppose that there is $t \in \Omega \backslash \bigsqcup_{i=1}^{n} G_i$. In particular, $t \notin G_n$, and $t \notin \bigsqcup_{i=1}^{n-1} G_i$. On the other hand, since $f(t) \leq \|f\| < \lambda_n$, it follows from the definition of E_n that $t \in E_n \subset G_n$. This is a contradiction. Thus $\Omega = \bigsqcup_{i=1}^{n} G_i$.

We have also $G_i \cap G_j = \emptyset, \forall i \neq j$. Indeed, we may assume $i > j$. Then $\bigsqcup_{k=1}^{i-1} G_k$ is an open subset containing G_j. Clearly, $(\Omega \backslash \bigsqcup_{k=1}^{i-1} G_k)$ is a closed subset containing E_i. Thus $G_i \subset (\Omega \backslash \bigsqcup_{k=1}^{i-1} G_k)$, and $G_i \cap G_j = \emptyset$.

Now let χ_i be the characteristic function of G_i. Then χ_i is a projection of $C(\Omega), 1 \leq i \leq n$. Notice $\lambda_{i-1} \leq f(t) \leq \lambda_i, \forall t \in G_i, 1 \leq i \leq n$. Therefore, $\|f - \sum_{i=1}^{n} \lambda_i \chi_i\| < \varepsilon$. Q.E.D.

Theorem 5.2.3. Let Ω be a compact Hausdorff space, and $C_r(\Omega)$ be the set of all real continuous functions on Ω. Then the following statements are equivalent:

1) Ω is a Stonean space.

2) For any bounded increasing net of non-negative function in $C_r(\Omega)$, there exists its least upper bound in $C_r(\Omega)$.

3) For any bounded subset of $C_r(\Omega)$, there exists its least upper bound in $C_r(\Omega)$.

4) Let g be a bounded real valued lower semicontinuous function on Ω. Then there is $f \in C_r(\Omega)$ such that $E = \{t \in \Omega \mid f(t) \neq g(t)\}$ is a first category Borel subset of Ω.

Moreover, the function f in 4) can be chosen as $f(t) = \overline{\lim_{t' \to t}} g(t'), \forall t \in \Omega$.

Proof. 4) $\Longrightarrow$ 3). Let A be a bounded subset of $C_r(\Omega)$. Then

$$g(t) = \sup\{f'(t) \mid f' \in A\}$$

is a bounded lower semicontinuous function on Ω. By 4), there is $f \in C_r(\Omega)$ and a first category subset E of Ω such that $f(t) = g(t), \forall t \notin E$. Clearly, $(g - f)$ is also lower semicontinuous. Then $G = \{t \in \Omega \mid g(t) > f(t)\}$ is open and $G \subset E$. Since Ω is a Baire space and E is first category, it follows that $G = \emptyset$, i.e., $f(t) \geq g(t), \forall t \in \Omega$. Suppose that $h \in C_r(\Omega)$ and $h \geq f', \forall f' \in A$. Then $h(t) \geq g(t), \forall t \in \Omega$, and $h(t) \geq f(t), \forall t \notin E$. In addition, $(\Omega \backslash E)$ is dense in Ω. Thus $h(t) \geq f(t), \forall t \in \Omega$, and f is the least upper bound of A.

3) $\Longrightarrow$ 2). It is obvious.

2) $\Longrightarrow$ 1). Let U be an open subset of Ω. Let

$$A = \{f' \in C_r(\Omega) \mid 0 \leq f' \leq 1, \text{supp } f' \subset U\}.$$

With respect to the partial order in $C_r(\Omega)$, A is a bounded increasing net of non-negative functions in $C_r(\Omega)$. From 2), A has its least upper bound f in $C_r(\Omega)$. Suppose that

$$g(t) = \sup\{f'(t) \mid f' \in A\}, \quad \forall t \in \Omega.$$

Clearly, $g(t) \le f(t), \forall t \in \Omega$. For any $t \in \Omega$, there is $f' \in A$ such that $f'(t) = 1$. Thus $g|U = 1$. On the other hand, since $f' \le 1, \forall f' \in A$, it follows that $f \le 1$. So $f|\overline{U} = 1$.

Suppose that there is $t_0 \notin \overline{U}$ such that $f(t_0) > 0$. Pick $h \in C_r(\Omega)$ such that $h \ge 0, h(t_0) = 0$ and $h(t) = 1, \forall t \in \overline{U}$. Then

$$f' \le \inf\{h, f\} \ne f, \quad \forall f' \in A.$$

This contradicts the fact that f is the least upper bound of A. Therefore, $f(t) = 0, \forall t \notin \overline{U}$. Further, $\overline{U}$ is open.

1) $\Longrightarrow$ 4). Let g be a bounded real valued lower semicontinuous function on Ω. We may assume that $0 \le g(t) \le 1, \forall t \in \Omega$. For any real number $\lambda, F(\lambda) = \{t \in \Omega \mid g(t) \le \lambda\}$ is a closed subset of Ω. Put $G(\lambda) = \text{Int}(F(\lambda))$. Noticing that

$$G(\lambda) = \Omega \backslash \overline{(\Omega \backslash F(\lambda))},$$

$G(\lambda)$ is an open and closed subset of Ω from 1). Thus the characteristic function χ_λ of $G(\lambda)$ belongs to $C_r(\Omega)$. Let

$$f_n = \sum_{k=1}^{2^n} \frac{k}{2^n}(\chi_{\frac{k}{2^n}} - \chi_{\frac{k-1}{2^n}}) = 1 - \sum_{k=1}^{2^n-1} \chi_{\frac{k}{2^n}}.$$

Fox fixed n and $t \in \Omega$, we assume that

$$k = \min\{i \mid 0 \le i \le 2^n, \chi_{\frac{i}{2^n}}(t) = 1\}.$$

Then $\chi_{\frac{m}{2^n}}(t) = 1, \forall m \ge k$, and $f_n(t) = \frac{k}{2^n}$. Moreover, clearly

$$\chi_{\frac{m}{2^{n+1}}}(t) = 1, \quad \forall m \ge 2k; \quad \text{and} \quad \chi_{\frac{m}{2^{n+1}}}(t) = 0, \forall m \le 2(k-1).$$

Thus

$$\frac{k}{2^n} - \frac{1}{2^{n+1}} \le f_{n+1}(t) \le \frac{k}{2^n}.$$

Further, we get

$$\|f_{n+1} - f_n\| \le \frac{1}{2^{n+1}}, \qquad \forall n.$$

Therefore, there is $f \in C_r(\Omega)$ such that $\|f_n - f\| \to 0$. Let

$$E = \bigsqcup_{n=1}^{\infty} \bigsqcup_{k=1}^{2^n} (F(\frac{k}{2^n}) - G(\frac{k}{2^n})).$$

Then E is a first category Borel subset of Ω. We claim that

$$f(t) = g(t), \qquad \forall t \notin E.$$

For fixed n, put $N = 2^n, F_k = F(\frac{k}{N}), G_k = G(\frac{k}{N}) = \text{Int}(F_k))$, and $E_k = \Omega \backslash F_k$. Then

$$F_1 \subset F_2 \subset \cdots \subset F_N = \Omega, \quad G_1 \subset G_2 \subset \cdots \subset G_N = \Omega$$

and

$$E_1 \supset E_2 \supset \cdots \supset E_N = \emptyset.$$

Thus $G_i \cap E_j = G_i \backslash F_j = \emptyset, \forall i \leq j$. By the formula

$$A \cap (B \cup C) = (A \cap B) \cup (A \cap C)$$

we have

$$\begin{aligned} (G_1 \bigsqcup E_1) \cap (G_2 \bigsqcup E_2) &= G_1 \cup E_2 \cup (E_1 \cap G_2), \\ [G_1 \bigsqcup E_2 \bigsqcup (E_1 \cap G_2)] \cap (G_3 \bigsqcup E_3) &= G_1 \bigsqcup E_3 \bigsqcup (G_2 \cap E_1) \bigsqcup (G_3 \cap E_2), \\ &\cdots\cdots, \end{aligned}$$

and

$$\bigcap_{k=1}^{N} (G_k \cup E_k) = G_1 \bigsqcup \bigsqcup_{k=1}^{N-1} (G_{k+1} \backslash F_k).$$

If $t \in G_1$, then $0 \leq g(t) \leq 1/N, \chi_{k/N}(t) = 1, \forall k \geq 1$ and

$$f_n(t) = \frac{1}{N} - \frac{1}{N} \chi_0(t).$$

If $t \in G_{k+1} \backslash F_k (1 \leq k \leq N-1)$, then

$$\frac{k}{N} < g(t) \leq \frac{k+1}{N};$$

$$\chi_{\frac{m}{N}}(t) = 1, \quad \forall m \geq k+1;$$

$$\chi_{\frac{m}{N}}(t) = 0, \quad \forall m \leq k;$$

and $f_n(t) = \frac{k+1}{N}$. In other words, we obtain

$$|g(t) - f_n(t)| \leq \frac{1}{2^n}, \quad \forall t \in \bigcap_{k=1}^{2^n} (G_k \bigsqcup E_k), \forall n.$$

Since $f_n \to f$ in $C_r(\Omega)$, it follows that $g(t) = f(t), \forall t \notin E$.

Now it suffices to show that

$$f(t) = \overline{\lim_{t' \to t}} g(t'), \quad \forall t \in \Omega.$$

From preceding discussion of $\|f_n - f_{n+1}\| \leq 1/2^{n+1} (\forall n), f_n(t) \searrow f(t), \forall t \in \Omega$. If $t' \in E$, then there is some n and some $k (1 \leq k \leq 2^n)$ such that $t' \in F(\frac{k}{2^n}) \backslash G(\frac{k}{2^n})$. Since

$$\chi_{\frac{2^p k}{2^{n+p}}}(t') = \chi_{\frac{k}{2^n}}(t') = 0 \quad \text{for any} \quad p,$$

it follows that

$$f_{n+p}(t') \geq 1 - \frac{1}{2^{n+p}} \sum_{j > 2^p k}^{2^{n+p}-1} 1 > \frac{k}{2^n} \geq g(t').$$

Thus $f(t) \geq g(t), \forall t \in \Omega$. Further

$$\overline{\lim_{t' \to t}} g(t') \leq \overline{\lim_{t' \to t}} f(t') = f(t), \quad \forall t \in \Omega.$$

On the other hand, for any $t \in \Omega$ and $\varepsilon > 0$ there is a neighborhood U of t such that $f(t'') > f(t) - \varepsilon, \forall t'' \in U$. Since E is first category, $U \backslash E \neq \emptyset$. Pick $t' \in U \backslash E$. Then $g(t') = f(t') > f(t) - \varepsilon$. Further

$$\overline{\lim_{t' \to t}} g(t') \geq f(t) - \varepsilon,$$

and $\overline{\lim_{t' \to t}} g(t') \geq f(t)$ since ε is arbitrary. Therefore

$$f(t) = \overline{\lim_{t' \to t}} g(t'), \quad \forall t \in \Omega.$$

Q.E.D.

Definition 5.2.4. Let Ω be a Stonean space, and μ be a regular Borel measure on Ω (i.e. a positive linear functional on $C(\Omega)$). μ is said to be *normal*, if $\mu(f) = \sup_l \mu(f_l)$ for any bounded increasing net $\{f_l\}$ of non-negative functions of $C_r(\Omega)$, where f is the least upper bound of $\{f_l\}$ in $C_r(\Omega)$.

Proposition 5.2.5. Let Ω be a Stonean space and μ be a normal regular Borel measure on Ω. Then $\mu(F) = \mu(E) = 0$ for any rare closed subset F and first category Borel subset E.

Proof. Let F be a rare closed subset of Ω. Then $(\Omega \backslash F)$ is open and dense in Ω, and

$$\Omega \backslash F = \bigsqcup \{\text{supp} f \mid f \in C(\Omega), 0 \leq f \leq 1, \ \text{supp} f \subset (\Omega \backslash F)\},$$

where $\text{supp} f = \overline{\{t \in \Omega \mid f(t) \neq 0\}}, \forall f \in C(\Omega)$. Since Ω is a Stonean space, it follows that

$$\Omega \backslash F = \bigsqcup \{G \subset \Omega \backslash F \mid G \text{ is open and closed}\}.$$

By the inclusion relation with respect to $G, \{\chi_G \mid G \subset \Omega \backslash F$, and G is open and closed $\}$ is a bounded increasing net in $C_r(G)$. Clearly, the least upper bound of $\{\chi_G\}$ in $C_r(G)$ is 1. Thus we have

$$\mu(\Omega) = \sup\{\mu(G) \mid G \text{ is as above}\}$$

since μ is normal. Further, $\mu(F) = 0$.

Now suppose that E is a first category Borel subset of Ω. We can write $E = \bigsqcup_n F_n$, where each F_n is rare. Then $\overline{F}_n$ is closed and rare, $\forall n$. Therefore $\mu(E) = 0$ from the preceding paragraph. Q.E.D.

Proposition 5.2.6. Let Ω be a Stonean space, and μ be a normal regular Borel measure on Ω. Then supp μ is an open and closed subset of Ω.

Proof. Let $F = \text{supp}\mu$. Then F is a closed subset, and $F\backslash \text{Int}(F)$ is a rare closed subset. By Proposition 5.2.5, $\mu(F) = \mu(\text{Int}(F))$. Let E be the closure of $\text{Int}(F)$. Then E is open and closed, and $\text{Int}(F) \subset E \subset F$. Thus $\mu(E) = \mu(F)$. By the definition of $\text{supp}\nu$, we have $E = F = \text{supp}\nu$. Q.E.D.

Proposition 5.2.7. Let Ω be a Stonean space, and h be a bounded measurable function on Ω. Then there is $f \in C(\Omega)$ such that

$$f(t) = h(t),, a.e.\mu$$

for any normal regular Borel measure μ on Ω.

Proof. We may assume that h is real valued. Then $g(t) = \overline{\lim_{t'\to t}} h(t')$ is a bounded real valued lower semicontinuous function on Ω. By Theorem 5.2.3, there is a $f \in C_r(\Omega)$ and a first category Borel subset E of Ω such that

$$f(t) = g(t), \qquad \forall t \notin E.$$

For any normal regular Borel measure μ on Ω, by the Lusin theorem there is a disjoint sequence $\{K_n\}$ of compact subsets of Ω such that h is continuous on $K_n, \forall n$, and

$$\mu(\Omega\backslash \bigsqcup_n K_n) = 0.$$

Then

$$h(t) = g(t), \qquad \forall t \in \bigsqcup_n \text{Int}(K_n).$$

Since $(K_n\backslash \text{ Int } (K_n))$ is rare and closed, it follows from Proposition 5.2.5 that $\mu(K\backslash\text{Int}(K_n)) = 0, \forall n$. Thus $\mu(E \bigsqcup(\Omega\backslash \bigsqcup_n \text{Int}(K_n)) = 0$ and

$$f(t) = h(t), \qquad \forall t \in (\bigsqcup_n \text{Int}(K_n)) \cap (\Omega\backslash E),$$

i.e., $f(t) = h(t)$, a.e.μ. Q.E.D.

Definition 5.2.8. Ω is called a *hyperstonean* space, if it is a stonean space,and for any $0 \leq f \in C(\Omega)$ and $f \neq 0$ there is a normal regular Borel measure μ on Ω such that $\mu(f) > 0$.

Proposition 5.2.9. Let Ω be a hyperstonean space. Then there is a family $\{\mu_l\}$ of normal regular Borel measure on Ω such that $\text{supp}\mu_l \cap \text{supp}\mu_{l'} = \emptyset, \forall l \neq l'$, and $\bigsqcup_l \text{supp}\mu_l$ is dense in Ω.

Proof. Let $\{\mu_l\}$ be a maximal family of normal regular Borel measures on Ω such that

$$\text{supp}\mu_l \cap \text{supp}\mu_{l'} = \emptyset, \qquad \forall l \neq l'.$$

Put $\Gamma = \bigsqcup_l \text{supp}\mu_l$. By Proposition 5.2.6, Γ is on open subset of Ω. Then $\overline{\Gamma}$ is open and closed. If $E = \Omega \backslash \overline{\Gamma} \neq \emptyset$, then $0 \leq \chi_E \in C(\Omega)$ and $\chi_E \neq 0$. From Definition 5.2.8, there is a normal regular Borel measure μ' on Ω such that $\mu'(E) > 0$. Let

$$\mu(\Delta) = \mu'(\Delta \backslash \overline{\Gamma}), \qquad \forall \text{ Borel subset } \quad \Delta.$$

Clearly, μ is a normal regular Borel measure on Ω, and

$$\emptyset \neq \text{supp}\mu \subset E = \Omega \backslash \overline{\Gamma}.$$

This is a contradiction since the family $\{\mu_l\}$ is maximal. Therefore, $\overline{\Gamma} = \Omega$.

Q.E.D.

Notes. The concept of Stonean spaces was introduced by M. Stone. The presentation here follows a treatise due to J. Dixmier.

References. [20], [164], [177].

5.3. Abelian W^*-algebras

Theorem 5.3.1. Let Z be a σ-finite abelian W^*-algebra, and Ω be its spectral space. Then Ω is a hyperstonean space, and there is a normal regular Borel measure ν on Ω such that

$$\text{supp}\nu = \Omega, \quad \text{and} \quad Z \cong C(\Omega) = L^\infty(\Omega, \nu).$$

Proof. Suppose that $Z \subset B(H)$, here H is some Hilbert space. By Proposition 1.14.5, Z admits a separating vector $\xi_0 (\in H)$. Let $f \to m_f$ be the $*$ isomorphism from $C(\Omega)$ onto Z. From Theorem 5.2.3 and Proposition 1.2.10, Ω is a Stonean space. Clearly, there is a regular Borel measure ν on Ω such that

$$\langle m_f \xi_0, \xi_0 \rangle = \int_\Omega f(t) d\nu(t), \qquad \forall f \in C(\Omega).$$

By Proposition 1.2.10, ν is normal.

Suppose that there is a non-empty open Borel subset U of Ω such that $\nu(U) = 0$. Pick $f \in C(\Omega), f \geq 0, f \neq 0$,and supp$f \subset U$. Then $\langle m_f \xi_0, \xi_0 \rangle = 0$.

Since ξ_0 is separating for Z, it follows that $f = 0$, a contradiction. Thus $\text{supp}\nu = \Omega$. In consequence, Ω is a hyperstonean space, and $C(\Omega)$ can be embedded into $L^\infty(\Omega, \nu)$.

Let $\{f_l\}$ be a net of $C(\Omega), \|f_l\| \leq 1, \forall l$ and $f_l \to f(\in L^\infty(\Omega, \nu))$ with respect to w^* -topology in $L^\infty(\Omega, \nu)$. Put $m_l = m_{f_l}(\in Z), \forall l$. Then $\|m_l\| \leq 1$. Replacing $\{m_l\}$ by its subset if necessary, we may assume that $m_l \to m_g$ weakly, where $g \in C(\Omega)$. Then for any $h \in C(\Omega)$,

$$|\int (f_l - g) h d\nu| = |\langle (m_l - m_g) m_h \xi_0, \xi_0 \rangle| \to 0.$$

Since $C(\Omega)$ is dense in $L^1(\Omega, \nu)$, it follows that $f_l \to g$ with respect to w^*-topology in $L^\infty(\Omega, \nu)$. Hence $f(t) = g(t)$, a.e.ν. From above discussion, $C(\Omega)$ is w^*-closed in $L^\infty(\Omega, \nu)$. Clearly, $C(\Omega)$ is w^*-dense in $L^\infty(\Omega, \nu)$. Therefore, $C(\Omega) = L^\infty(\Omega, \nu)$. Q.E.D.

Proposition 5.3.2. Let Ω be a compact Hausdorff space, and ν be a regular Borel measure on Ω. Then $L^\infty(\Omega, \nu)$ is a σ-finite abelian W^*-algebra.

Proof. By Theorem 5.1.3, $L^\infty(\Omega, \nu)$ is an abelian W^*-algebra. Let

$$\omega(f) = \int_\Omega f(t) d\nu(t), \qquad \forall f \in L^\infty(\Omega, \nu).$$

Since $1 \in L^1(\Omega, \nu)$, it follows that $\omega(\cdot)$ is a faithful σ-continuous positive functional on $L^\infty(\Omega, \nu)$. From Proposition 1.14.2, $L^\infty(\Omega, \nu)$ is σ-finite. Q.E.D.

Theorem 5.3.3. Let Ω be a hyperstonean space. Then $C(\Omega)$ is an abelian W^*-algebra. Moreover, if there is a normal regular Borel measure ν on Ω with $\text{supp}\nu = \Omega$, then $C(\Omega) = L^\infty(\Omega, \nu)$ is σ-finite.

Proof. First suppose that there is a normal regular Borel measure ν on Ω such that $\text{supp}\nu = \Omega$. Then $C(\Omega)$ can be embedded into $L^\infty(\Omega, \nu)$. Moreover, for any $h \in L^\infty(\Omega, \nu)$, by Proposition 5.2.7 there is $f \in C(\Omega)$ such that $f(t) = h(t)$, a.e.ν. Thus $C(\Omega) = L^\infty(\Omega, \nu)$. Further, $C(\Omega)$ is a σ-finite abelian W^*-algebra from Proposition 5.3.2.

Generally, by Proposition 5.2.9 there is a family $\{\nu_l\}$ of normal regular Borel measures on Ω such that $\text{supp}\nu_l \cap \text{supp}\, \nu_{l'} = \emptyset, \forall l \neq l'$, and $\Gamma = \bigsqcup_l \text{supp}\, \nu_l$ is dense in Ω. By Proposition 5.2.6, $\text{supp}\, \nu_l$ is open and closed, $\forall l$. Then Γ is a locally compact Hausdorff space. Let $\nu = \sum_l \oplus \nu_l$. Then ν is a regular Borel measure on Γ and $\text{supp}\, \nu = \Gamma$. Consequently, $f \to f|\Gamma$ is an injective map from $C(\Omega)$ to $L^\infty(\Gamma, \nu)$. Moreover, for any $h \in L^\infty(\Gamma, \nu)$, let $h(t) = 0, \forall t \in \Omega \backslash \Gamma$. Then by Proposition 5.2.7 there is $f \in C(\Omega)$ such that $f(t) = h(t)$, a.e.ν.

Thus $f(t) = h(t)$, l.a.e.ν on Γ. Further $C(\Omega)$ is $*$ isomorphic to $L^\infty(\Gamma, \nu)$, and is a W^*-algebra. Q.E.D.

Theorem 5.3.4. Let Z be an abelian W^*-algebra, and Ω be its spectral space. Then Ω is a hyperstonean space, and there is a locally compact Hausdorff space Γ and a regular Borel measure ν on Γ with supp$\nu = \Gamma$ such that Z is $*$ isomorphic to $L^\infty(\Gamma, \nu)$.

Proof. Let $Z \subset B(H)$, and $f \to m_f$ be the $*$ isomorphicm from $C(\Omega)$ onto Z. Then for any $\xi \in H$, there is a regular Borel measure ν_ξ such that

$$\langle m_f \xi, \xi \rangle = \int_\Omega f(t) d\nu_\xi(t), \quad \forall f \in C(\Omega).$$

From Theorem 5.2.3 and Proposition 1.2.10, Ω is a Stonean space, and ν_ξ is normal, $\forall \xi \in H$. If f is a non-zero positive element of $C(\Omega)$, then there is $\xi \in H$ such that $\langle m_f \xi, \xi \rangle > 0$, i.e., $\nu_\xi(f) > 0$. Therefore, Ω is hyperstonean. The rest conclusion is contained in the proof of Theorem 5.3.3 indeed. Q.E.D.

Definition 5.3.5. Let M be a W^*-algebra. $E(\subset M)$ is called a *generated subset* for M, if M is the smallest W^*-subalgebra containing E. Moreover, if M admits a countable generated subset, then M is called *countably generated.*

A generated subset for a C^*-algebra is understood similarly.

Lemma 5.3.6. Let Ω be a compact Hausdorff space. If the C^*-algebra $C(\Omega)$ is generated by a sequence $\{p_n\}$ of projections, then $C(\Omega)$ can be generated by an invertible positive element.

Proof. Let

$$h = \sum_{n=1}^{\infty} \frac{1}{3^n}(2p_n + \frac{1}{2}).$$

Then h is an invertible positive element of $C(\Omega)$. For any $t_1, t_2 \in \Omega$ and $t_1 \neq t_2$, there is a minimal positive integer k such that

$$p_k(t_1) \neq p_k(t_2)$$

since $\{p_n\}$ is a generated subset for $C(\Omega)$. Thus

$$\begin{aligned} |h(t_1) - h(t_2)| &= 2|\sum_{n=k}^{\infty} \frac{1}{3^n}(p_n(t_1) - p_n(t_2))| \\ &\geq \frac{2}{3^k} - 2\sum_{n=k+1}^{\infty} \frac{1}{3^n} = \frac{1}{3^k} > 0. \end{aligned}$$

Now by the Stone–Weierstrss theorem and Lemma 2.1.5, $C(\Omega)$ is generated by $\{h\}$. Q.E.D.

Theorem 5.3.7. Let Z be a countably generated abelian W^*-algebra. Then Z can be generated by an invertibel positive element. In particular, every abelian VN algebra on a separable Hilbert space is generated by a single operator.

Proof. Let $\{a_n\}$ be a generated subset for Z. Replacing a_n by $\frac{1}{2}(a_n+a_m^*)$, we may assume that $a_n^* = a_n, \forall n$. From the spectral decomposition of $\{a_n\}$, Z can be generated by a sequence $\{p_n\}$ of projections. Let A be the C^*-subalgebra of C^*-algebra Z generated by $\{p_n\}$. By Lemma 5.3.6, A is generated by an invertible positive element a. Clearly, A is also a generated subset for Z. Thus, Z is generated by a.

Moreover, each VN algebra on a separable Hilbert space is countably generated. That comes to the rest conclusion. Q.E.D.

Theorem 5.3.8. Suppose that Z is an abelian VN algebra on a separable Hilbert space H, and Z contains no minimal projection (a projection p of Z is said to be minimal, if $p \neq 0$ and any projection q of Z with $q \leq p$ implies either $q = 0$ or $q = p$). Then Z is $*$ isomorphic to $L^\infty([0,1])$, where measure on $[0, 1]$ is Lebesgue measure.

Proof. Let Ω be the spectral space of Z. By Theorem 5.3.1, Ω is a hyperstonean space, and there is a normal regular Borel measure ν on Ω with $\text{supp}\nu = \Omega$ such that $Z \cong C(\Omega) = L^\infty(\Omega,\nu)$. From Theorem 5.3.7, Z is generated by a positive element a. We may assume that $0 \leq a \leq 1$. Put $I = [0,1]$, and let $z \to z(\cdot)$ be the Gelfand transformation from Z to $C(\Omega)$. Then $a(\cdot)$ is a continuous map from Ω to I. Define a Borel measure μ on I and a $*$ homomorphism Φ from $L^\infty(I,\mu)$ to $L^\infty(\Omega,\nu)$ as follows:

$$\mu(E) = \nu(a^{-1}(E)), \quad \forall \text{ Borel subset } \quad E \subset I,$$

$$\Phi(f)(t) = f(a(t)), \quad \forall t \in \Omega, f \in L^\infty(I,\mu).$$

Clearly, $\Phi(p) = p(a)$ for any polynomial $p(\cdot)$ on I. Thus $\Phi(L^\infty(I,\mu))$ is w^*-dense in $L^\infty(\Omega,\nu)$ since Z is generated by $\{a\}$.

We claim that $\Phi(L^\infty(I,\mu))$ is dense in $L^1(\Omega,\nu)$. In fact, suppose that there is some $g \in L^\infty(\Omega,\nu)$ such that

$$\int_\Omega g(t)\Phi(f)(t)d\nu(t) = 0, \quad \forall f \in L^\infty(I,\mu).$$

Since $\Phi(L^\infty(I,\mu))$ is w^*-dense in $L^\infty(\Omega,\nu)$, there is a net $\{f_l\} \subset L^\infty(I,\mu)$ such that

$$\Phi(f_l) \to \bar{g} \quad (\sigma(L^\infty(\Omega,\nu), L^1(\Omega,\nu))).$$

Clearly, $g \in L^1(\Omega,\nu)$ too. Thus

$$0 = \int_\Omega g(t)\Phi(f_l)(t)d\nu(t) \to \int_\Omega |g(t)|^2 d\nu(t)$$

and $g = 0$. Therefore, $\Phi(L^\infty(I,\mu))$ is dense in $L^1(\Omega,\nu)$.

Now we say that Φ is σ-σ continuous. It suffices to show that $\Phi(f_l) \to 0$ (w^*-topology) for any net $\{f_l\} \subset L^\infty(I,\mu)$ and $\|f_l\| \le 1, \forall l$, and $f_l \to 0$ (w^*-topology). For any $g \in L^1(\Omega,\nu)$ and $\varepsilon > 0$, from preceding paragraph we can pick $f \in L^\infty(I,\mu)$ such that

$$\int_\Omega |g(t) - \Phi(f)(t)| d\nu(t) < \varepsilon.$$

Then

$$\begin{aligned} &|\int_\Omega g(t)\Phi(f_l)(t)d\nu(t)| \\ \le\ &|\int_I f_l(\lambda)f(\lambda)d\mu(\lambda)| + \int_\Omega |g(t) - \Phi(f)(t)|d\nu(t) < 2\varepsilon \end{aligned}$$

if l is sufficiently later. Thus $\Phi(f_l) \to 0$ (w^*-topology), and Φ is σ-σcontinuous.

Thus, we get $\Phi(L^\infty(I,\mu)) = L^\infty(\Omega,\nu)$.

Suppose that $f \in L^\infty(I,\mu)$ such that $\Phi(f) = 0$. Then

$$\Phi(fg) = 0, \qquad \forall g \in C(I),$$

and $\int_I f(\lambda)g(\lambda)d\mu(\lambda) = \int_\Omega \Phi(fg)(t)d\nu(t) = 0, \forall g \in C(I)$, and $f = 0$. Therefore, Φ is a $*$ isomorphism from $L^\infty(I,\mu)$ onto $L^\infty(\Omega,\nu)$.

The measure μ on I is not *atomic*, i.e., $\mu(\{\lambda\}) = 0, \forall\lambda \in I$. In fact, suppose that there is $\lambda \in I$ such that $\mu(\{\lambda\}) > 0$. Put $E = a^{-1}(\{\lambda\})$. Then $\nu(E) > 0$. So χ_E is a non-zero projection of $L^\infty(\Omega,\nu)$. Since $\Phi(\chi_{\{\lambda\}}) = \chi_E$ and $\chi_{\{\lambda\}}$ is a minimal projection of $L^\infty(I,\mu)$, it follows that χ_E is a minimal projection of $L^\infty(\Omega,\nu)(\cong Z)$. This contradicts the assumption.

Let $f(\lambda) = \mu([0,\lambda]), \forall\lambda \in I$. Then f is a continuous increasing function with $f(0) = 0$ and $f(1) = 1$ (we may assume that $\nu(\Omega) = 1$). Further, let

$$g(\lambda) = \min\{\lambda' \in I \mid f(\lambda') = \lambda\}, \quad \forall\lambda \in I.$$

Then g is a left continuous strictly increasing function on I, and has countable jump points at most. Suppose that $\{\lambda_1 < \lambda_2 < \cdots < \lambda_n < \cdots\}$ is the set of jump points of g. then there is a sequence $\{\lambda'_n\}$ with $\lambda_1 < \lambda'_1 < \lambda_2 < \lambda'_2 < \cdots < \lambda_n < \lambda'_n < \cdots$ such that for each n

$$f(\lambda) = f(\lambda_n), \quad \forall\lambda \in [\lambda_n, \lambda'_n]; \quad f(\lambda) > f(\lambda_n), \quad \forall\lambda > \lambda'_n.$$

Then $g \circ f(\lambda) = \lambda, \forall\lambda \in I \backslash \bigsqcup_n [\lambda_n, \lambda'_n]$. On the other hand

$$\mu([\lambda_n, \lambda'_n]) = \mu((\lambda_n, \lambda'_n]) = f(\lambda'_n) - f(\lambda_n) = 0,$$

$\forall n$. Thus $g \circ f(\lambda) = \lambda$, a.e.$\mu$.

Let m be the Lebesgue measure on I. For any $0 \le \lambda_1 \le \lambda_2 \le 1$, we have

$$m((f(\lambda_1), f(\lambda_2)]) = f(\lambda_2) - f(\lambda_1) = \mu((\lambda_1, \lambda_2]).$$

Thus $m = \mu \circ f^{-1}$. Further, $\mu = m \circ g^{-1}$.

Now define a $*$ homomorphism Ψ from $L^\infty(I) = L^\infty(I, m)$ to $L^\infty(I, \mu)$ as follows:

$$\Psi(h) = h \circ f, \quad \forall h \in L^\infty(I).$$

If $k \in L^\infty(I, \mu)$, then $k \circ g \in L^\infty(I)$, and

$$\Psi(k \circ g)(\lambda) = k(\lambda), a.e.\mu.$$

So $\Psi(L^\infty(I)) = L^\infty(I, \mu)$. Moreover, suppose that $h \in L^\infty(I)$ such that $\Psi(h) = 0$. Since $\int \Psi(h\bar{h})(\lambda) d\mu(\lambda) = \int |h(\lambda)|^2 dm(\lambda)$, it follows that $h = 0$. Thus Ψ is a $*$ isomorphism from $L^\infty(I)$ onto $L^\infty(I, \mu)$. Further, $\Phi \circ \Psi$ is a $*$ isomorphism from $L^\infty([0, 1])$ onto $L^\infty(\Omega, \nu)$. Q.E.D.

Corollary 5.3.9. Let H be a separable Hilbert space and $\mathcal{A}_a$ be the collection of all abelian VN algebras on H. For any $Z \in \mathcal{A}_a$, define $[Z] = \{Y \mid Y \in \mathcal{A}_a$, and Y is $*$ isomorphic to $Z\}$. Then $\{[Z] \mid Z \in \mathcal{A}_a\}$ is countable.

Definition 5.3.10. An abelian VN algebra Z on a Hilbert space H is said to be *maximal abelian*, if there is no abelian VN algebra on H which contains Z properly.

Clearly, Z is maximal abelian if and only if $Z = Z'$.

Definition 5.3.11. Let Ω be a locally compact Hausdorff space, and ν be a regular Borel measure on Ω. For any $f \in L^\infty(\Omega, \nu)$, define

$$\widehat{m}_f g = fg, \qquad \forall g \in L^2(\Omega, \nu).$$

Clearly, $\widehat{m}_f$ is a bounded linear operator on $L^2(\Omega, \nu)$. $\{\widehat{m}_f \mid f \in L^\infty(\Omega, \nu)\}$ is called the *multiplication algebra* on $L^2(\Omega, \nu)$.

Lemma 5.3.12. If Ω is a compact Hausdorff space, and ν is a regular Borel measure on Ω, then the multiplication algebra Z is a maximal abelian VN algebra on $L^2(\Omega, \nu)$.

Proof. Let $a' \in Z'$. Then for any $f \in L^\infty(\Omega, \nu) (\subset L^2(\Omega, \nu))$

$$a'f = a'\widehat{m}_f 1 = \widehat{m}_f a'1 = f \cdot a'1.$$

Put $a'1 = g (\in L^2(\Omega, \nu))$. Then $a'f = gf, \forall f \in L^\infty(\Omega, \nu)$.

We say that $|g(t) \leq \|a'\|$, a.e.ν. In fact, suppose that there is $\varepsilon > 0$ and a compact subset K of Ω such that

$$\nu(K) > 0, \text{ and } |g(t)| \geq \|a'\| + \varepsilon, \quad \forall t \in K.$$

Then

$$\begin{aligned}\nu(K)(\|a'\| + \varepsilon)^2 \leq & \int |g(t)\chi_K(t)|^2 d\nu(t) \\ = & \|a'\chi_K\|^2 \leq \|a'\|^2 \nu(K).\end{aligned}$$

This is a contradiction. Thus $|g(t)| \leq \|a'\|$, a.e.ν, and $g \in L^\infty(\Omega, \nu)$.

Now from $g \in L^\infty(\Omega, \nu)$ and $a'f = gf, \forall f \in L^\infty(\Omega, \nu)$, we have $a' = \widehat{m}_g$ since $L^\infty(\Omega, \nu)$ is dense in $L^2(\Omega, \nu)$. Therefore, $Z' = Z$. Q.E.D.

Theorem 5.3.13. Let Ω be a locally compact Hausdorff space, and ν be a regular Borel measure on Ω. Then the multiplication algebra Z is a maximal abelian VN algebra on $L^2(\Omega, \nu)$.

Proof. If $\widehat{m}_f = 0$ for some $f \in L^\infty(\Omega, \nu)$, then $\widehat{m}_f \chi_K = f\chi_K = 0$ (a.e.ν) for each compact subset K of Ω. Further, $f = 0, l.$a.e.ν. Thus $f \rightarrow \widehat{m}_f$ is a $*$ isomorphism from $L^\infty(\Omega, \nu)$ onto Z. Also, this $*$ isomorphism is σ-σ continuous. Consequently, Z is a VN algebra on $L^2(\Omega, \nu)$.

By Proposition 5.1.2, $\Omega = N \bigsqcup \bigsqcup_l K_l$, where N is locally ν-zero,and $\{K_l\}$ is a disjoint family of compact subsets of Ω with the locally countable property. Then

$$L^2(\Omega, \nu) = \sum_l \oplus L^2(K_l, \nu_l)$$

where $\nu_l = \nu | K_l, \forall l$. For any $a' \in Z'$, since $a'h_l = a'\chi_{K_l}h_l = \chi_{K_l}a'h_l \in L^2(K_l, \nu_l), \forall h_l \in L^2(K_l, \nu_l)$, it follows that $L^2(K_l, \nu_l)$ is invariant for $a', \forall l$. By Lemma 5.3.12, for each l there is $g_l \in L^\infty(K_l, \nu_l)$ such that

$$a'|L_2(K_l, \nu_l) = \widehat{m}_{g_l}.$$

Let

$$g = \sum_l \chi_{K_l} g_l.$$

Then $g \in L^\infty(\Omega, \nu)$, and $a' = \widehat{m}_g$. Therefore, $Z' = Z$. Q.E.D.

Proposition 5.3.14. Let Z be an abelian VN algebra on a Hilbert space $H, \xi_0 (\in H)$ be a cyclic vector for Z,and Ω be the spectral space of Z. Then there is a regular Borel measure ν on Ω, and a unitary operator u from H onto $L^2(\Omega, \nu)$ such that

$$\text{supp}\nu = \Omega, \quad Z \cong C(\Omega) = L^\infty(\Omega, \nu),$$

$$um_f u^{-1} = \widehat{m}_f, \quad \forall f \in L^\infty(\Omega, \nu),$$

where $f \to m_f$ is the Gelfand transformation from $C(\Omega)$ onto Z.

Proof. Since $Z \subset Z'$, it follows that ξ_0 is also separating for Z. Let ν be the regular Borel measure on Ω such that

$$\langle m_f \xi_0, \xi_0 \rangle = \int_\Omega f(t) d\nu(t), \quad \forall f \in C(\Omega).$$

Then by the proof of Theorem 5.3.1 we have

$$\text{supp}\nu = \Omega, \quad C(\Omega) = L^\infty(\Omega, \nu).$$

Now define $u m_f \xi_0 = f, \forall f \in C(\Omega)$. Then u can be extended to a unitary operator from H onto $L^2(\Omega, \nu)$ since ξ_0 is cyclic for Z. Further $u m_f u^{-1} = \widehat{m}_f, \forall f \in C(\Omega) = L^\infty(\Omega, \nu)$. Q.E.D.

Proposition 5.3.15. Let Z be an abelian VN algebra on a Hilbert space. Then Z is maximal abelian and σ-finite if and only if Z admits a cyclic vector.

Proof. The sufficiency is obvious from Proposition 5.3.14, Theorem 5.3.13 and Proposition 1.14.2.

Now suppose that Z is maximal abelian and σ-finite. By Proposition 1.14.5, Z admits a separating vector ξ_0. Further, ξ_0 is also cyclic for Z since $Z' = Z$. Q.E.D.

Corollary 5.3.16. Let Z be an abelian VN algebra on a separable Hilbert space. Then Z is maximal abelian if and only if Z admits a cyclic vector.

Theorem 5.3.17. Let Z be a maximal abelian VN algebra on a Hilbert space H. Then there is a locally compact Hausdorff space Ω and a regular Borel measure ν on Ω with $\text{supp}\nu = \Omega$ such that Z is unitarily equivalent to the multiplication algebra on $L^2(\Omega, \nu)$.

Proof. We can write

$$H = \sum_l \oplus H_l, \qquad H_l = \overline{Z\xi_l}, \quad \forall l.$$

Let p_l be the projection from H onto $H_l, \forall l$. Then $p_l \in Z' = Z, \forall l$. Suppose that Ω' is the spectral space of Z,and $f \to m_f$ is the $*$ isomorphism from $C(\Omega')$ onto Z. Then for each l there is an open and closed subset Ω_l of Ω' such that $p_l = m_{\chi_l}$, where χ_l is the characteristic function of Ω_l. Since $p_l p_{l'} = 0$, it follows that $\Omega_l \cap \Omega_{l'} = \emptyset, \forall l \neq l'$.

For each l, $Z_l = Z p_l$ admits a cyclic vector ξ_l, and Ω_l is its spectral space. By Proposition 5.3.14, there is a regular Borel measure ν_l on Ω_l with $\text{supp}\nu_l = \Omega_l$, and a unitary operator u_l from H_l onto $L^2(\Omega_l, \nu_l)$ such that

$$C(\Omega_l) = L^\infty(\Omega_l, \nu_l), \quad u_l m_f^{(l)} u_l^{-1} = \widehat{m}_f^{(l)}, \quad \forall f \in L^\infty(\Omega_l, \nu_l),$$

where $f \to m_f^{(l)}$ is the $*$ isomorphism from $C(\Omega_l)$ onto Z_l, and $\widehat{m}_f^{(l)}$ is the multiplication operator of f on $L^2(\Omega_l, \nu_l)$.

Put $\Omega = \bigsqcup_l \Omega_l$. Then Ω is an open dense subset of Ω'. So Ω is a locally compact Hausdorff space. Let $\nu = \sum_l \oplus \nu_l$. Then ν is a regular Borel measure on Ω, and $\text{supp}\nu = \Omega$. Further, let $u = \sum_l \oplus u_l$. Then u is a unitary operator from $H = \sum_l \oplus H_l$ onto $L^2(\Omega, \nu) = \sum_l \oplus L^2(\Omega_l, \nu_l)$. Denote the multiplication algebra on $L^2(\Omega, \nu)$ by $\widehat{Z} = \{\widehat{m}_g \mid g \in L^\infty(\Omega, \nu)\}$. For any $f \in C(\Omega')$, it is easy to see that $u m_f u^{-1} = \widehat{m}_g$, where $g = f|\Omega \in L^\infty(\Omega, \nu)$. Thus $uZu^{-1} \subset \widehat{Z}$. Since Z is maximal commutative, it follows that $uZu^{-1} = \widehat{Z}$. Q.E.D.

Definition 5.3.18. Let M be a W^*-algebra, and p be a projection of M. p is said to be *abelian (commutative)*, if pMp is abelian (commutative).

Proposition 5.3.19. Let M be a W^*-algebra, p and q be two projections of M, and p be abelian.

1) If $p \sim q$, then q is also abelian.
2) $pMp = Zp$, where Z is the center of M.
3) If $q \leq p$, then $q = c(q)p$, where $c(q)$ is the central cover of q in M.

Proof. we may assume that M is a VN algebra.

1) It is immediate from Proposition 1.5.2.

2) Since M_p is commutative, it follows that $M_p \subset M_p'$. Now by Proposition 1.3.8, $M_p = M_p \cap M_p' = Zp$.

3) By Proposition 1.5.8, the central cover of q in M_p is $c(q)p$. But M_p is commutative, so $q = c(q)p$. Q.E.D.

Notes. Theorem 5.3.8 is due to P. Halmos and J. Von Neumann.

Theorem 5.3.7 is due to J. Von Neumann. There is a general conjecture: if M is a VN algebra on a separable Hilbert space, then M is generated by a single operator? This conjecture is still open now, but we have rich results on it, see T. Saitô's Lectures.

References. [20], [68], [157], [142].

5.4. * Representations of abelian C^*-algebras

In this section, let A be an abelian C^*-algebra with an identity. Then $A \cong C(\Omega)$, where Ω is the spectral space of A, a compact Hausdorff space.

Theorem 5.4.1. Let $\{\pi, H, \xi\}$ be a cyclic * representation of A. Then there is unique (in the sense of equivalence) regular Borel measure μ on Ω such that

$$\{\pi, H\} \cong \{\Phi_\mu, L^2(\Omega, \mu)\},$$

where $(\Phi_\mu(a)f)(t) = a(t)f(t), \forall t \in \Omega, f \in L^2(\Omega, \mu)$, $a \in A$, and $a \to a(\cdot)$ is the Gelfand transformation from A onto $C(\Omega)$.

Proof. Let μ be the regular Borel measure on Ω such that

$$\langle \pi(a)\xi, \xi\rangle = \int_\Omega a(t)d\mu(t), \qquad \forall a \in A.$$

Further, define $u\pi(a)\xi = a(\cdot), \forall a \in A$. Then u can be extended to a unitary operator from H onto $L^2(\Omega, \mu)$, still denoted by u. Clearly, $u\pi(a)u^{-1} = \Phi_\mu(a), \forall a \in A$.

Now suppose that ν is a regular Borel measure on Ω, and v is a unitary operator from $L^2(\Omega, \mu)$ onto $L^2(\Omega, \nu)$ such that

$$v\Phi_\mu(a)v^{-1} = \Phi_\nu(a), \quad \forall a \in A.$$

Put $v1 = \alpha(\in L^2(\Omega, \nu))$. Then $va = v\Phi_\mu(a)1 = \Phi_\nu(a)\alpha$, and

$$\int |a(t)|^2 d\mu(t) = \int |a(t)\alpha(t)|^2 d\nu(t), \quad \forall a \in A.$$

Thus $\mu = |\alpha|^2 \cdot \nu$, and $\mu \prec \nu$. Similarly, $\nu \prec \mu$. Therefore, $\mu \sim \nu$. Q.E.D.

Each * representation of A is a direct sum of a zero representation and a family of cyclic * representations. In this section, we study a * representation $\{\pi, H\}$ of A such that π is a countable direct sum of cyclic * representations. By Proposition 1.14.2, π is as above if and only if $\pi(A)'$ is σ-finite.

Definition 5.4.2. Let $\{\pi, H\}$ be a * representation of A. For any $\xi \in H$, there is unique regular Borel measure μ_ξ on Ω such that

$$\langle \pi(a)\xi, \xi\rangle = \int_\Omega a(t)d\mu_\xi(t), \qquad \forall a \in A.$$

We introduce a partial order "$>$" on H as follows. $\xi > \eta$ means that $\mu_\xi \succ \mu_\eta$. Moreover, $\xi \in H$ is said to be *maximal*, if $\xi > \eta, \forall \eta \in H$.

Lemma 5.4.3. If $\eta \in H_\xi = \overline{\pi(A)\xi}$, then $\eta < \xi$.

Proof. By Theorem 5.4.1, there is a unitary operator u from H_ξ onto $L^2(\Omega, \mu_\xi)$ such that

$$u(\pi(a)|H_\xi)u^{-1} = \Phi_{\mu_\xi}(a), \quad \forall a \in A.$$

Let $f = u\eta(\in L^2(\Omega, \mu_\xi))$. Then

$$\langle \pi(a)\eta, \eta \rangle = \int a(t)|f(t)|^2 d\mu_\xi(t), \qquad \forall a \in A.$$

Thus, $\mu_\eta = |f|^2 \cdot \mu_\xi$ and $\mu_\eta \prec \mu_\xi, \eta < \xi$. Q.E.D.

Lemma 5.4.4. If $H = \sum_k \oplus H_k$, where $H_k = \overline{\pi(A)\xi_k}$ and $\|\xi_k\| \leq 1, \forall k$, then

$$\xi = \sum_k 2^{-k/2}\xi_k$$

is maximal.

Proof. Clearly, $\mu_\xi = \sum_k 2^{-k}\mu_{\xi_k}$. Fix $\eta \in H$,and write $\eta = \sum_k \eta_k$, where $\eta_k \in H_k, \forall k$. Then $\mu_\eta = \sum_k \mu_{\eta_k}$, and for each $k, \mu_{\eta_k} \prec \mu_{\xi_k}$ by Lemma 5.4.3. If E is a Borel subset of Ω such that $\mu_\xi(E) = 0$, then $\mu_{\xi_k}(E) = 0, \forall k$. Thus $\mu_{\eta_k}(E) = 0, \forall k$, and $\mu_\eta(E) = 0$. Therefore, $\mu_\eta \prec \mu_\xi$, i.e., $\eta < \xi$. Q.E.D.

Lemma 5.4.5. If $\pi(A)'$ is σ-finite, then the set of maximal vectors is dense in H.

Proof. By Lemma 5.4.4, there is a maximal vector $\xi(\in H)$ at least. Define $H_\xi = \overline{\pi(A)\xi}$, and fix $\eta \in H$. We can write $\eta = \eta_1 + \eta_2$, where $\eta_1 \in H_\xi, \eta_2 \in H_\xi^\perp$. Let $H_1 = \overline{\pi(A)\eta_1} \subset H_\xi$, and write

$$\xi = \xi_1 + \xi_2, \quad \text{where} \quad \xi_1 \in H_1, \xi_2 \in H_1^\perp.$$

Suppose that $\mu_i = \mu_{\xi_i}, i = 1, 2$. Then $\mu_\xi = \mu_1 + \mu_2$. For any $\varepsilon > 0$, clearly $\mu_{\eta_1 + \varepsilon\xi_2} = \mu_{\eta_1} + \varepsilon^2\mu_2$. By Lemma 5.4.3, $\mu_1 \prec \mu_{\eta_1}$. Thus

$$\mu_{\eta_1+\varepsilon\xi_2} \succ \mu_1 + \varepsilon^2\mu_2 \sim \mu_1 + \mu_2 = \mu_\xi$$

and $(\eta_1 + \varepsilon\xi_2)$ is also maximal since ξ is maximal. Notice that $\xi_1 \in H_1 \subset H_\xi$. So $\xi_2 = \xi - \xi_1 \in H_\xi$ and $\eta_1 + \varepsilon\xi_2 \in H_\xi$. Then

$$\mu_{\eta_1+\varepsilon\xi_2+\eta_2} = \mu_{\eta_1+\varepsilon\xi_2} + \mu_{\eta_2}$$

and $\eta + \varepsilon\xi_2 = \eta_1 + \varepsilon\xi_2 + \eta_2$ is maximal. Clearly,

$$\|(\eta + \varepsilon\xi_2) - \eta\| \leq \varepsilon\|\xi\|.$$

Therefore, the set of maximal vectors is dense in H. Q.E.D.

Lemma 5.4.6. If $\pi(A)'$ is σ-finite, then we have a decomposition

$$H = \sum_{k=1}^{\infty} \oplus H_k, \quad H_k = \overline{\pi(A)\xi_k},$$

and $\xi_1 > \xi_2 > \cdots > \xi_k > \cdots$.

Proof. Let $\{\varsigma_n\}$ be a cyclic sequence of vectors for $\pi(A)$, and

$$\begin{aligned} &\{\eta_k \mid k = 1, 2, \cdots\} \\ = \ &\{\varsigma_1, \varsigma_1, \varsigma_2, \varsigma_1, \varsigma_2, \varsigma_3, \varsigma_1, \varsigma_2, \varsigma_3, \varsigma_4, \cdots\}. \end{aligned}$$

By Lemma 5.4.5, pick a maximal vector $\xi_1 (\in H)$ such that

$$\|\xi_1 - \eta_1\| < 1.$$

Denote the projection from H onto $H_1 = \overline{\pi(A)\xi_1}$ by p_1. Similarly, there is a maximal vector ξ_2 in $H_1^{\perp} = (1 - p_1)H$ such that

$$\|\xi_2 - (1 - p_1)\eta_2\| < 1/2.$$

Again let p_2 be the projection from H onto $H_2 = \overline{\pi(A)\xi_2}$. $\cdots$. Generally, suppose that we have $\xi_1, \cdots, \xi_{k-1}$, and p_i is the projection from H onto $H_i = \overline{\pi(a)\xi_i}, 1 \leq i \leq k-1$. Then we can pick a maximal vector ξ_k in $(\sum_{i=1}^{k-1} \oplus H_i)^{\perp} = (1 - \sum_{i=1}^{k-1} p_i)H$ such that

$$\|\xi_k - (1 - \sum_{i=1}^{k-1} p_i)\eta_k\| < 1/k.$$

Further, let p_k be the projection from H onto $H_k = \overline{\pi(A)\xi_k}$. Clearly, the sequence $\{\xi_k\}$ satisfies:

$$\xi_1 > \xi_2 > \cdots > \xi_k > \cdots, \quad H_i \perp H_j, \quad \forall i \neq j.$$

Now it suffices to show that $H = \sum_k \oplus H_k$.

For fixed k, by the definition of $\{\eta_m\}$ there is a subsequence $\{k_n\}$ of $\{1,2,\cdots\}$ such that $\eta_{k_n} = \varsigma_k, \forall n$. Then

$$\begin{aligned} &\|(\xi_{k_n} + \sum_{j=1}^{k_n-1} p_j\varsigma_k) - \varsigma_k\| \\ = \ &\|\xi_{k_n} - (1 - \sum_{j=1}^{k_n-1} p_j)\eta_{k_n}\| < \frac{1}{k_n} \to 0. \end{aligned}$$

Thus $\varsigma_k \in \sum_{i=1}^{\infty} \oplus H_i, \forall k$. Since $[\pi(a)\varsigma_k \mid a \in A, k]$ is dense in H, it follows that $H = \sum_k \oplus H_k$. Q.E.D.

Lemma 5.4.7. Let μ, ν be two regular Borel measures on Ω, and v be a bounded linear operator from $L^2(\Omega,\mu)$ to $L^2(\Omega,\nu)$ such that

$$v\Phi_\mu(a) = \Phi_\nu(a)v, \quad \forall a \in A.$$

Then $vf = \alpha f, \forall f \in L^2(\Omega,\mu)$, where $\alpha = v1 \in L^2(\Omega,\nu)$.

Proof. Since $va = v\Phi_\mu(a)1 = \Phi_\nu(a)\alpha = \alpha a$, it follows that

$$\int |\alpha(t)a(t)|^2 d\nu(t) \le \|v\|^2 \int |a(t)|^2 d\mu(t), \quad \forall a \in C(\Omega).$$

Then $\|v\|^2\mu \ge |\alpha|^2 \cdot \nu$, and $\alpha f \in L^2(\Omega,\nu)$ for any $f \in L^2(\Omega,\mu)$. Further, from $va = \alpha a (\forall a \in C(\Omega))$ and the density of $C(\Omega)$ in $L^2(\Omega,\mu)$ we get $vf = \alpha f, \forall f \in L^2(\Omega,\mu)$. Q.E.D.

Lemma 5.4.8. Let $\{\mu_k\}, \{\nu_k\}$ be two sequences of regular Borel measures on Ω, and

$$H = \sum_k \oplus L^2(\Omega,\mu_k), \quad K = \sum_k \oplus L^2(\Omega,\nu_k).$$

Suppose that there is an isometry u from H to K such that

$$u\Phi_H(a) = \Phi_K(a)u, \quad \forall a \in A,$$

where $\Phi_H(a)(f_1,\cdots,f_k,\cdots) = (af_1,\cdots,af_k,\cdots)$ for any $a \in A$ and $(f_1,\cdots,f_k,\cdots) \in H$ (i.e., $f_k \in L^2(\Omega,\mu_k), \forall k$), and $\Phi_K(a)$ is defined similarly. Moreover, if we assume that

$$\mu_1 \succ \mu_2 \succ \cdots \succ \mu_k \succ \cdots, \quad \nu_2 \succ \nu_3 \succ \cdots \succ \nu_j \succ \cdots,$$

then $\nu_j \succ \mu_j, \forall j \ge 2$.

Proof. Let p_k be the projection from H onto $H_k = L^2(\Omega, \mu_k)$,and q_j be the projection from K onto $K_j = L^2(\Omega, \nu_j)$, and $u_{jk} = q_i u p_k, \forall j, k$. It is easy to see that

$$u_{jk}\Phi_{\mu_k}(a) = \Phi_{\nu_j}(a)u_{jk}, \quad \forall j, k, a \in A.$$

Put $\alpha_{jk} = u_{jk}1 (\in L^2(\Omega, \nu_j) = K_j)$. Then by Lemma 5.4.7,

$$\begin{aligned} & u(0, \cdots, f_k, 0 \cdots) \\ = \ & (\alpha_{1k} f_k, \cdots, \alpha_{jk} f_k, \cdots), \qquad \forall f_k \in H_k. \end{aligned}$$

Since u is isometric, it follows that

$$\int |f_k(t)|^2 d\mu_k(t) = \sum_j \int |\alpha_{jk}(t) f_k(t)|^2 d\nu_j(t) \tag{1}$$

$\forall f_k \in H_k$.

Now let E be a Borel subset of Ω such that $\nu_2(E) = 0$. Clearly, $\nu_j(E) = 0, \forall j \geq 2$. Then for any $a \in A$,

$$u : (k\text{-}th) \begin{pmatrix} 0 \\ \vdots \\ a\chi_E \\ 0 \\ \vdots \end{pmatrix} \longrightarrow \begin{pmatrix} a\alpha_{1k}\chi_E \\ \vdots \\ \vdots \\ a\alpha_{jk}\chi_E \\ \vdots \end{pmatrix} = \begin{pmatrix} a\alpha_{1k}\chi_E \\ 0 \\ \vdots \\ \vdots \\ \vdots \end{pmatrix} \tag{2}$$

Further, by (2)

$$\begin{aligned} & \int \alpha_{11}(t)\overline{\alpha_{12}(t)} a(t) \chi_E(t) d\nu_1(t) \\ = \ & \langle \begin{pmatrix} a\alpha_{11}\chi_E \\ 0 \\ \vdots \end{pmatrix}, \begin{pmatrix} a\alpha_{12}\chi_E \\ 0 \\ \vdots \end{pmatrix} \rangle = \langle u \begin{pmatrix} a\chi_E \\ 0 \\ \vdots \\ \vdots \end{pmatrix}, u \begin{pmatrix} 0 \\ a\chi_E \\ 0 \\ \vdots \end{pmatrix} \rangle \\ = \ & \langle \begin{pmatrix} a\chi_E \\ 0 \\ \vdots \\ \vdots \end{pmatrix}, \begin{pmatrix} 0 \\ a\chi_E \\ 0 \\ \vdots \end{pmatrix} \rangle = 0, \qquad \forall a \in A. \end{aligned}$$

Thus $\alpha_{11}(t)\overline{\alpha_{12}(t)} = 0$, a.e.$\nu_1, t \in E$. Put

$$E_1 = \{t \in E \mid \alpha_{11}(t) \neq 0\}, \quad E_2 = E \backslash E_1.$$

Then $\alpha_{12}(t) = 0$, a.e.$\nu_1, t \in E_1$. By (2),

$$u(\chi_{E_2}, 0, \cdots) = (\alpha_{11}\chi_{E_2}, 0, \cdots) = 0$$

since $\alpha_{11}(t) = 0$ on E_2. Thus $\mu_1(E_2) = 0$, and $\mu_2(E_2) = 0$ since $\mu_2 \prec \mu_1$. Again by (2), we have

$$0 = u(0, \chi_{E_2}, 0, \cdots) = (\alpha_{12}\chi_{E_2}, 0 \cdots),$$

so $\alpha_{12}(t) = 0$, a.e.$\nu_1, t \in E_2$, and $\alpha_{12}(t) = 0$, a.e.$\nu_1, t \in E$. Now from $u(0, \chi_E, 0, \cdots) = (\alpha_{12}\chi_E, 0, \cdots) = 0, \mu_2(E) = 0$. Therefore, we get

$$\nu_2 \succ \mu_2 \succ \mu_3 \succ \cdots. \tag{3}$$

By (1) for any $f_k \in H_k, k = 1, 2, \cdots$, we obtain

$$\int |\alpha_{1k} f_k(t)|^2 d\nu_1(t) \le \int |f_k(t)|^2 d\mu_k(t). \tag{4}$$

Suppose that E is a Borel subset of Ω such that $\nu_2(E) = 0$. Then by (3), $\mu_k(E) = 0, \forall k \ge 2$. From (4), we have

$$\int_E |\alpha_{1k}(t)|^2 d\nu_1(t) = 0, \quad \forall k \ge 2.$$

Thus $|\alpha_{1k}|^2 \cdot \nu_1 \prec \nu_2, \forall k \ge 2$. By Theorem 5.1.4, there is a non-negative measurable function β_k on Ω such that

$$|\alpha_{1k}|^2 \cdot \nu_1 = \beta_k \cdot \nu_2, \quad \forall k \ge 2. \tag{5}$$

Define $v : H \ominus H_1 \to K \ominus K_1$ as follows:

$$v(0, \cdots, f_k, 0 \cdots) = (0, (\beta_k + |\alpha_{2k}|^2)^{1/2} f_k, \alpha_{3k} f_k, \cdots)$$

$\forall f_k \in H_k, k \ge 2$. By (5) and (1), v is isometric. Clearly, we have

$$v\Phi_{H \ominus H_1}(a) = \Phi_{K \ominus K_1}(a) v, \quad \forall a \in A,$$

and $\mu_2 \succ \mu_3 \succ \cdots, \nu_3 \succ \nu_4 \succ \cdots$. From the discussion of preceding paragraphs, we obtain $\nu_3 \succ \mu_3$ and a relation between ν_2 and ν_3 which is similar to (5).

Cotinuing this process, we get $\nu_j \succ \mu_j (\forall j \ge 2)$ generally. Q.E.D.

Lemma 5.4.9. Let u be an isometry from $H = \sum_k \oplus L^2(\Omega, \mu_k)$ to $K = \sum_k \oplus L^2(\Omega, \nu_k)$,and

$$u\Phi_H(a) = \Phi_K(a) u, \quad \forall a \in A.$$

If $\mu_1 \succ \mu_2 \succ \cdots$, and $\nu_j \succ \nu_{j+1} \succ \cdots$, where j is an integer with $j \ge 2$, then $\nu_k \succ \mu_k, \forall k \ge j$.

Proof. When $j = 2$, it is exactly the Lemma 5.4.8. Now we assume that the Lemma holds for $(j-1), (j > 2)$.

$\sum_{k\geq j-1} \oplus L^2(\Omega,\nu_k)$ is invariant for the $*$ representation Φ_K. By Lemma 5.4.6 and Theorem 5.4.1, there is a sequence $\{\gamma_k \mid k \geq j-1\}$ of regular Borel measures on Ω with $\gamma_{j-1} \succ \gamma_j \succ \cdots$ such that the $*$ representations $\{K', \Phi_{K'}\}$ and $\{L, \Phi_L\}$ are unitarily equivalent, where

$$K' = \sum_{k\geq j-1} \oplus L^2(\Omega,\nu_k), \quad L = \sum_{k\geq j-1} \oplus L^2(\Omega,\gamma_k).$$

From Lemma 5.4.8, $\nu_k \succ \gamma_k, \forall k \geq j$. Now for H and

$$\sum_{k=1}^{j-2} \oplus L^2(\Omega,\nu_k) \oplus L$$

we have $\gamma_k \succ \mu_k (\forall k \geq j-1)$ by induction. Therefore, $\nu_k \succ \mu_k, \forall k \geq j$.

Q.E.D.

Lemma 5.4.10. Let u be a unitary operator from $H = \sum_k \oplus L^2(\Omega,\mu_k)$ onto $K = \sum_k \oplus L^2(\Omega,\nu_k)$,and

$$u\Phi_H(a)u^{-1} = \Phi_K(a), \quad \forall a \in A.$$

If $\mu_1 \succ \mu_2 \succ \cdots$ and $\nu_1 \succ \nu_2 \succ \cdots$, then $\nu_k \sim \mu_k, \forall k \geq 1$.

Proof. Let $\xi = u(1,0,\cdots)$. Then for any $a \in A$,

$$\begin{aligned}\int a(t)d\mu_1(t) &= \langle \Phi_H(a)(1,0,\cdots),(1,0,\cdots)\rangle \\ &= \langle \Phi_K(a)\xi,\xi\rangle = \int a(t)d\nu_\xi(t),\end{aligned}$$

where ν_ξ is the measure determined by ξ and Φ_K. Thus $\mu_1 = \nu_\xi$.

Clearly, $\langle \Phi_K(a)\eta_k,\eta_k\rangle = \int a(t)d\nu_k(t), \forall a \in A$, where $\eta_k = (0,\cdots,1,0\cdots)$ ($\in K$). Put $\eta = \sum_k (\|\eta_k\| 2^{\frac{k}{2}})^{-1}\eta_k$. Then η is maximal in K by Lemma 5.4.4. So $\nu_\eta \succ \nu_\xi = \mu_1$. Since $\nu_\eta = \sum_k (\|\eta_k\| 2^{\frac{k}{2}})^{-1}\nu_k$ and $\nu_1 \succ \nu_k (\forall k \geq 2)$, it follows that $\nu_\eta \sim \nu_1$. Thus $\nu_1 \succ \mu_1$. By Lemma 5.4.9, $\nu_k \succ \mu_k, \forall k \geq 2$. So we get $\nu_k \succ \mu_k, \forall k \geq 1$.

Similarly, $\mu_k \succ \nu_k (\forall k \geq 1)$ since u is unitary. Therefore, $\mu_k \sim \nu_k, \forall k \geq 1$.

Q.E.D.

Theorem 5.4.11. Let A be an abelian C^*-algebra with an identity, Ω be its spectral space, and $\{\pi, H\}$ be a $*$ representation of A such that $\pi(A)'$ is

σ-finite. Then there is a sequence $\{\mu_k\}$ of regular Borel measures on Ω with $\mu_1 \succ \mu_2 \succ \cdots$ such that

$$\{\pi, H\} \cong \{\Phi, \sum_k \oplus L^2(\Omega, \mu_k)\},$$

where $\Phi(a)(f_1, \cdots, f_k, \cdots) = (af_1, \cdots, af_k, \cdots)$ and $(af_k)(t) = a(t)f_k(t), \forall k, \forall a \in A$ and $(f_1, \cdots, f_k, \cdots) \in H$. And the sequence $\{\mu_k\}$ is unique in the sense of equivalence.

Moreover, for each $k \geq 1$ the measure μ_k is equivalent to

$$\min \left\{ \mu_\eta \;\middle|\; \begin{array}{c} \eta \text{ is a maximal vector in } (\sum_{j=1}^{k-1} \oplus \overline{\pi(a)\xi_j})^\perp \\ \forall \xi_1, \cdots, \xi_{k-1} \in H \text{ such that } \overline{\pi(A)\xi_i} \perp \overline{\pi(A)\xi_j}, \forall i \neq j. \end{array} \right\}$$

Where " min" is taken according to the absolute continuity of measures.

Proof. The result follows immediately from Lemma 5.4.10, 5.4.6., 5.4.9 and Theorem 5.4.1.

Q.E.D.

Remark. The determination of $\{\mu_k\}$ is very similar to the *Courant principle.* If a is a completely continuous non-negative operator on a Hilbert space H, and $\{\lambda_1 \geq \lambda_2 \geq \cdots\}$ is the sequence of eigenvalues of a, then for any k,

$$\lambda_k = \min_{\xi_1, \cdots, \xi_{k-1} \in H} \max_{0 \neq \eta \in [\xi_1, \cdots, \xi_{k-1}]^\perp} \frac{\langle a\eta, \eta \rangle}{\langle \eta, \eta \rangle}.$$

Proposition 5.4.12. With the assumptions and notations of Theorem 5.4.11, the $*$ representation $\{\pi, H\}$ is faithful if and only if $\text{Supp}\mu_1 = \Omega$.

Proof. Suppose that $\text{supp}\mu_1 = \Omega$. If $a \in A$ is such that $\Phi(a) = 0$, then $af = 0, \forall f \in L^2(\Omega, \mu_1)$. Picking $f = \bar{a}$, we can see that $a(t) = 0$, a.e.μ_1. Put $U = \{t \in \Omega \mid a(t) \neq 0\}$. Then U is open and $\mu_1(U) = 0$. But $\text{supp}\mu_1 = \Omega$, so $U = \emptyset$, i.e., $a = 0$. Thus π is faithful.

Conversely, if there is a non-empty open subset U such that $\mu_1(U) = 0$. Then $\mu_k(U) = 0, \forall k \geq 1$. Pick $a \in A$ such that $\text{supp}a(\cdot) \subset U$. Then $\Phi(a) = 0$. Thus π is not faithful.

Q.E.D.

Definition 5.4.13. A function $n(\cdot)$ on Ω is called a *multiplicity function*, if $n(\cdot)$ is measurable, and $n(t) \in \{1, 2, \cdots, \infty\}, \forall t \in \Omega$.

For any given regular Borel measure μ and multiplicity function $n(\cdot)$ on Ω, define a $*$ representation $\{\Phi_{\mu,n}, H_{\mu,n}\}$ of $A (\cong C(\Omega))$ as follows:

$$H_{\mu,n} = \sum_k \oplus H_k, \quad H_k = L^2(\Omega, \mu_k),$$
$$\mu_k = \chi_{E_k} \cdot \mu, \quad E_k = \{t \in \Omega \mid n(t) \geq k\}, \forall k \geq 1,$$

and

$$\Phi_{\mu,n}(a)(f_1,\cdots,f_k,\cdots) = (af_1,\cdots,af_k,\cdots),$$

$\forall (f_1,\cdots,f_k.\cdots) \in H$ and $a \in A$.

Lemma 5.4.14. Let μ be a regular Borel measure on $\Omega, 0 \le \rho \in L^1(\Omega,\mu)$, $\nu = \rho \cdot \mu$, and $E = \{t \in \Omega \mid \rho(t) > 0\}$. Then $\nu \sim \chi_E \cdot \mu$.

Proof. Let F be a Borel subset of Ω such that $\nu(F) = 0$. Since $\nu(F) = \int_F \rho(t)d\mu(t)$ and $\rho \ge 0$, it follows that $\rho(t) = 0, a.e.\mu, t \in F$. Thus there is a Borel subset $F_1 \subset F$ such that $\mu(F_1) = 0$ and $\rho(t) = 0, \forall t \in F\backslash F_1$. Then

$$(\chi_E \cdot \mu)(F) = \mu(E \cap F) = \mu(E \cap (F\backslash F_1)).$$

But $\rho(t) > 0, \forall t \in E$, and $\rho(t) = 0, \forall t \in F\backslash F_1$, so $E \cap (F\backslash F_1) = \emptyset$, and $(\chi_E \cdot \mu)(F) = \mu(E \cap (F\backslash F_1)) = 0$. Thus $\chi_E \cdot \mu \prec \nu$.

Conversely, let F be a Borel subset such that $(\chi_E \cdot \mu)(F) = \mu(E \cap F) = 0$. Clearly, $\rho(t) = 0, \forall t \in F\backslash E$. Then from $\rho \in L^1(\Omega,\mu)$,

$$\nu(F) = \int_F \rho(t)d\mu(t) = \int_{F\cap E} \rho(t)d\mu(t) = 0.$$

Hence, $\nu \prec \chi_E \cdot \mu$. Therefore $\nu \sim \chi_E \cdot \mu$. Q.E.D.

Theorem 5.4.15. Let A be an abelian C^*-algebra with an identity, Ω be its spectral space, and $\{\pi, H\}$ be a nondegenerate $*$ representation of A such that $\pi(A)'$ is σ-finite. Then there is unique (in the sense of equivalence) regular Borel measure μ on Ω, and unique (in the sense of a.e.μ) multiplicity function $n(\cdot)$ on Ω such that

$$\{\pi, H\} \cong \{\Phi_{\mu,n}, H_{\mu,n}\}.$$

Moreover, π is faithful if and only if $\text{supp}\mu = \Omega$.

Proof. Pick the sequence $\{\mu_k\}$ as in Theorem 5.4.11. By Theorem 5.1.4, there is $0 \le \rho_k \in L^1(\Omega,\mu)$ such that $\mu_k = \rho_k \cdot \mu, \forall k \ge 1$, where $\mu = \mu_1$ and $\rho_1 = 1$. Let

$$E_k = \{t \in \Omega \mid \rho_k(t) > 0\}, \quad \forall k.$$

Since $\mu_k \succ \mu_{k+1}$, we may assume that

$$\Omega = E_1 \supset E_2 \supset \cdots \supset E_k \supset \cdots.$$

Now let

$$n(t) = \begin{cases} k, & \text{if } t \in E_k\backslash E_{k+1}, \\ \infty, & \text{if } t \in \cap_k E_k. \end{cases}$$

Clearly, $n(\cdot)$ is a multiplicity function Ω, and $E_k = \{t \in \Omega \mid n(t) \geq k\}, \forall k$. By Lemma 5.4.14, $\mu_k \sim \chi_{E_k} \cdot \mu, \forall k$. Thus, $\{\pi, H\} \cong \{\Phi_{\mu,n}, H_{\mu,n}\}$.

From the uniqueness of $\{\mu_k\}$, it is easily verified that μ and $n(\cdot)$ are unique.

Finally, by Proposition 5.4.12, π is faithful if and only if $\Omega = \text{supp}\mu_1 = \text{supp}\mu$. Q.E.D.

Notes. Using the theory of type (I) VN algebras, we can also obtain the main results in this section. The presentation here follows a treatment due to A.A. Kirillov.

References. [10], [28], [91].

Chapter 6

The Classification of Von Neumann Algebras

6.1. The classification of Von Neumann algebras

Definite 6.1.1. Let M be a VN algebra. A projection p of M is said to be *finite*, if any projection q of M with $q \leq p$ and $q \sim p$ implies $q = p$. p is said to be *infinite*, if it is not finite, i.e., there exists a projection q of M such that $q \leq p, q \sim p$ and $q \neq p$. p is said to be *purely infinite*, if p contains no non-zero finite projection, i.e., if q is a projection of M with $q \leq p$ and $q \neq 0$, then q is infinite. Moreover, M is said to be *finite, infinite, purely infinite,* if its identity is a finite, infinite, purely infinite projection respectively.

Proposition 6.1.2. In a VN algebra M, there is a maximal finite central projection z_1.

Proof. Let $z_1 = \sup\{z \mid z \text{ is a finite central projection of } M\}$. It suffices to show that z_1 is finite. Suppose that p is a projection of M with $p \leq z_1$ and $p \sim z_1$. If z is any finite central projection of M, then $z = zz_1 \sim zp \leq z$. Thus $zp = z$, i.e. $p \geq z$. Further, $p = z_1$ and z_1 is finite. Q.E.D.

Proposition 6.1.3. Let p, q be two projections of a VN algebra M, $q \leq p$ and p be finite. Then q is also finite.

Proof. Let v be a partial isometry of M such that $v^*v = q$ and $vv^* = q_1 \leq q$. Define $u = v + (p - q)$. Then $u^*u = p, uu^* = (p-q) + q_1 \leq p$. Since p is finite, it follows that $(p-q) + q_1 = p$, i.e., $q_1 = q$. Therefore, q is also finite.

Q.E.D.

Proposition 6.1.4. In a VN algebra M, there is a maximal purely infinite central projection z_3.

Proof. Let $z_3 = \sup\{z|z$ is a purely infinite central projection of $M\}$. It suffices to show that z_3 is purely infinite. Suppose that p is a finite projection of M with $p \leq z_3$. If z is a purely infinite central projection of M, then pz is finite by Proposition 6.1.3. Since $pz \leq z$ and z is purely infinite, it follows that $pz = 0$. Further, $p = pz_3 = 0$. Therefore, z_3 is purely infinite. Q.E.D.

Definition 6.1.5. A VN algebra M is said to be *semifinite,* if $z_3 = 0$, where z_3 is defined by Proposition 6.1.4, i.e. any central projection of M is not purely infinite. M is said to be *properly infinite,* if $z_1 = 0$, where z_1 is defined by Proposition 6.1.2, i.e, any non-zero central projection of M is infinite. Moreover, a projection p of M is said to be *semi-finite,* or *properly infinite* , if the VN algebra M_p is semifinite, or properly infinite.

Theorem 6.1.6. Let M be a VN algebra. Then there is a unique decomposition:

$$M = M_1 \oplus M_2 \oplus M_3,$$

where $M_1 = Mz_1$ is finite, $M_3 = Mz_3$ is purely infinite, $M_2 = Mz_2$ is semi-finite and properly infinite, and $z_1 + z_2 + z_3 = 1$.

Proof. From Propositions 6.1.2 and 6.1.4, such decomposition exists. Now suppose that $M = Mp_1 \oplus Mp_2 \oplus Mp_3$ is another such decomposition. Clearly, $p_1 \leq z_1, p_3 \leq z_3$, and the central projection $(z_1 - p_1)p_i$ is finite , $i = 2, 3$. Then we have $(z_1 - p_1)p_i = 0 (i = 2, 3)$ since Mp_2 and Mp_3 are properly infinite. So $z_1 = p_1$. Moreover, if the central projection $(z_3 - p_3)p_i$ is not zero, then it is purely infinite, $i = 1$ or 2 . But Mp_1 and Mp_2 contain no purely infinite central projection, so $(z_3 - p_3)p_i$ must be zero, $i = 1, 2$, and $z_3 = p_3$. Therefore, $z_i = p_i, i = 1, 2, 3$. Q.E.D.

Definition 6.1.7. A VN algebra M is said to be *discrete,* if for any non-zero central projection z , there is a non-zero abelian projection q (see Definition 5.3.18) such that $q \leq z$. M is said to be *continuous,* if M contains no non-zero abelian projection. Moreover, a discrete VN algebra is also said to be *type (I)* ; a purely infinite VN algebra is also said to be *type (III)* ; a semi-finite and continuous VN algebra is said to be *type (II)* ; A finite type (II) VN algebra is also said to be *type* (II_1), and a properly infinite type (II) VN algebra is also said to be *type* (II_∞) .

Clearly, each abelian projection is finite. Thus a type (I) VN algebra is semi-finite.

Theorem 6.1.8. Let M be a VN algebra. Then there is a unique decomposition :

$$M = M_1 \oplus M_2 \oplus M_3,$$

where $M_i = Mz_i, i = 1,2,3$ are type (I), (II), (III) VN algebras respectively, and $z_1 + z_2 + z_3 = 1$.

Proof. By Proposition 6.1.4, there is a maximal purely infinite central projection z_3 in M. Then $M_3 = Mz_3$ is type (III).

Let $z_1 = \sup\{z | z$ is a central projection of M such that Mz is type (I)$\}$. We claim that Mz_1 is also type (I). In fact , suppose that p is a non-zero central projection of M with $p \leq z_1$. Then there is a central projection z of M such that $pz \neq 0$ and Mz is type (I) . So pz is a non-zero central projection of type (I) VN algebra Mz . By Definition 6.1.7, there is a non-zero abelian projection q of Mz such that $q \leq pz$. Clearly, q is also a non-zero abelian projection of Mz_1 . Therefore, Mz_1 is type (I) .

Since each abelian projection is finite, it follows that $z_1 z_3 = 0$.

Now let $z_2 = 1 - z_1 - z_3$. Clearly , $M_2 = Mz_2$ is semi-finite. If p is a non-zero abelian projection of M_2, then $c(p) \leq z_2$. We say that $Mc(p)$ is type (I) . In fact , suppose that z is a non-zero central projection of $Mc(p)$. By Proposition 1.5.8, $zp \neq 0$. Since $(pMp)z = zp(Mc(p))zp$, z contains a non-zero abelian projection zp . Thus , $Mc(p)$ is type (I). By the definition, we have $p \leq z_1$. This contradicts the fact that $c(p) \leq z_2$. Therefore , M_2 contains no non-zero abelian projection, i.e., M_2 is type (II).

Moreover, since z_1 and z_3 are maximal, this decomposition is unique.

Q.E.D.

Theorem 6.1.9. Let M be a VN algebra. Then there is a unique decomposition:

$$M = M_{11} \oplus M_{12} \oplus M_{21} \oplus M_{22} \oplus M_3,$$

where M_{11} is finite type (I) , M_{12} is properly infinite type (I) (and semi-finite also), M_{21} is type (II_1), M_{22} is type (II_∞) (and semi-finite also) , M_3 is type (III) (purely infinite).

Consequently, there are only five classes of factors.

References. [21], [28], [82], [111].

6.2. An ergodic type theorem for Von Neumann algebras

Let H be a Hilbert space, $h^* = h \in B(H)$, and p be a projection on H with $hp = ph$. Let

$$M_p(h) = \sup\{\langle h\xi, \xi\rangle | \xi \in pH, \|\xi\| = 1\},$$
$$m_p(h) = \inf\{\langle h\xi, \xi\rangle | \xi \in pH, \|\xi\| = 1\},$$
$$\omega_p(h) = M_p(h) - m_p(h).$$

Clearly, $M_p(h), m_p(h)$ are the maximal, minimal spectral points of $(h|pH)$ respectively. If $p = 1$, we denote $M_1(h), m_1(h), \omega_1(h)$ by $M(h), m(h), \omega(h)$ respectively. If $\mathcal{F}$ is a family of projections on H with $ph = hp, \forall p \in \mathcal{F}$, then we define

$$\omega_{\mathcal{F}}(h) = \sup\{\omega_p(h) | p \in \mathcal{F}\}.$$

Lemma 6.2.1. Let M be a VN algebra on H, $Z = M \cap M'$, and $h^* = h \in M$. Then there is a projection $z \in Z$ and a self-adjoint unitary operator $u \in M$ such that

$$\max\{\omega_z(\frac{1}{2}(h + uhu^{-1})), \omega_{1-z}(\frac{1}{2}(h + uhu^{-1}))\} \leq \frac{3}{4}\omega(h).$$

Proof. Let $n(h) = \frac{1}{2}(M(h) + m(h))$, be and let $h = \int \lambda de_\lambda$ be the spectral decomposition of h. Clearly, $e = e_{n(h)}$ and $f = 1 - e$ are two projections of M, and $M_e(h) \leq n(h) \leq m_f(h)$. By Theorem 1.5.4, there is a central projection z of M such that

$$ez \preceq fz, \quad fz' \preceq ez',$$

where $z' = 1 - z$. Thus there are partial isometries v, w of M such that

$$v^*v = ez, \quad vv^* = f_1 \leq fz,$$
$$w^*w = fz', \quad ww^* = e_1 \leq ez'.$$

Let $u = v + v^* + w + w^* + (1 - ez - f_1 - fz' - e_1)$. Since

$$\begin{aligned} H = \ & (ezH \oplus f_1H) \oplus (fz - f_1)H \\ & \oplus (fz'H \oplus e_1H) \oplus (ez' - e_1)H, \end{aligned}$$

it follows that u is a self-adjoint unitary element of M.

Now we prove that the above u and z satisfy the condition.

Since

$$\begin{aligned} hz &\geq m(h)ez + n(h)fz \\ &= m(h)ez + n(h)f_1 + n(h)(fz - f_1) \end{aligned}$$

and

$$(uhu^{-1})z \geq m(h)f_1 + n(h)ez + n(h)(fz - f_1),$$

it follows that

$$\begin{aligned}\tfrac{1}{2}(h+uhu^{-1})z &\geq \tfrac{1}{2}(m(h)+n(h))(f_1+ez)+n(h)(fz-f_1)\\ &\geq \tfrac{1}{2}(m(h)+n(h))z.\end{aligned}$$

Noticing that

$$\begin{aligned}M(h)-\tfrac{3}{4}\omega(h) &= \tfrac{1}{4}M(h)+\tfrac{3}{4}m(h)\\ &= \tfrac{1}{2}(m(h)+n(h)),\end{aligned}$$

we have

$$\begin{aligned}M(h)z &\geq \tfrac{1}{2}(h+uhu^{-1})z\\ &\geq (M(h)-\tfrac{3}{4}\omega(h))z,\end{aligned}$$

i.e.,

$$\omega_z(\frac{1}{2}(h+uhu^{-1}))\leq \frac{3}{4}\omega(h).$$

Similarly, from

$$\begin{aligned}hz' &\leq n(h)ez'+M(h)fz'\\ &= n(h)e_1+M(h)fz'+n(h)(ez'-e_1)\end{aligned}$$

and

$$(uhu^{-1})z'\leq n(h)fz'+M(h)e_1+n(h)(ez'-e_1),$$

we have

$$\begin{aligned}m(h)z' &\leq \tfrac{1}{2}(h+uhu^{-1})z'\\ &\leq \tfrac{1}{2}(n(h)+M(h))(fz'+e_1)+n(h)(ez'-e_1)\\ &\leq \tfrac{1}{2}(n(h)+M(h))z'=(m(h)+\tfrac{3}{4}\omega(h))z'.\end{aligned}$$

Thus

$$\omega_{1-z}(\frac{1}{2}(h+uhu^{-1}))\leq \frac{3}{4}\omega(h).$$

Q.E.D.

Lemma 6.2.2. Let M be a VN algebra, $Z=M\cap M', h^*=h\in M$, and $\mathcal{F}$ be a finite orthogonal family of projections of Z with $\sum_{z\in\mathcal{F}} z=1$. Then there is a finite orthogonal family $\mathcal{F}'$ of projections of Z with $\sum_{z'\in\mathcal{F}'} z'=1$ and a self-adjoint unitary element u of M such that

$$\omega_{\mathcal{F}'}(\frac{1}{2}(h+uhu^{-1}))\leq \frac{3}{4}\omega_{\mathcal{F}}(h).$$

Proof. Let $\mathcal{F} = \{z_1, \cdots, z_n\}$. For each $i \in \{1, \cdots, n\}$, by Lemma 6.2.1 there is a central projection c_{i1} of $M_i = Mz_i$ and a self-adjoint unitary element u_i of M_i such that

$$\omega_{c_{ij}}(\frac{1}{2}(h_i + u_i h_i u_i^{-1})) \leq \frac{3}{4}\omega_{z_i}(h), j = 1, 2,$$

where $h_i = hz_i, c_{i2} = z_i - c_{i1}$. Define $u = \sum_{i=1}^{n} u_i$. Then u is a self-adjoint unitary element of M and

$$\begin{aligned} \omega_{c_{ij}}(\tfrac{1}{2}(h + uhu^{-1})) &\leq \tfrac{3}{4}\omega_{z_i}(h) \\ &\leq \tfrac{3}{4}\omega_{\mathcal{F}}(h), \end{aligned}$$

$\forall 1 \leq i \leq n, j = 1, 2$. Now let $\mathcal{F}' = \{c_{ij} | 1 \leq i \leq n, j = 1, 2\}$. Then

$$\omega_{\mathcal{F}'}(\frac{1}{2}(h + uhu^{-1})) \leq \frac{3}{4}\omega_{\mathcal{F}}(h).$$

Q.E.D.

Definition 6.2.3. Let M be a VN algebra, and $G = U(M)$ be the set of all unitary elements of M. Denote by $\mathcal{Q}$ the set of all functions f on G satisfying: $f \geq 0, {}^{\#}\{u \in G | f(u) \neq 0\} < \infty$ and $\sum_{u \in G} f(u) = 1$.

For $f \in \mathcal{Q}$ and $a \in M$, let $f \cdot a = \sum_{u \in G} f(u) uau^{-1}$.

For $f, g \in \mathcal{Q}$, define $(f * g)(\cdot) = \sum_{u \in G} f(u) g(u^{-1} \cdot)$. Clearly , $f * g \in \mathcal{Q}$, and $(f * g) \cdot a = f \cdot (g \cdot a), \forall a \in M$.

Lemma 6.2.4. Let $h^* = h \in M$ and $\varepsilon > 0$. Then there is some $f \in \mathcal{Q}$ and some $z \in Z = M \cap M'$ such that

$$\|f \cdot h - z\| < \varepsilon.$$

Proof. By Lemma 6.2.1, there is a central projection p of M and $f_1 \in \mathcal{Q}$ such that

$$\omega_{\mathcal{F}_1}(f_1 \cdot h) < \frac{3}{4}\omega(h),$$

where $\mathcal{F}_1 = \{p, 1 - p\}$. Now we assume that for some positive integer j there is a finite orthogonal family $\mathcal{F}_j$ of projections of Z with $\sum_{p \in \mathcal{F}_j} p = 1$ and $f_j \in \mathcal{Q}$ such that

$$\omega_{\mathcal{F}_j}(f_j \cdot h) \leq (\frac{3}{4})^j \omega(h).$$

Using Lemma 6.2.2 for $f_j \cdot h$ and $\mathcal{F}_j$, then there is a finite orthogonal family $\mathcal{F}_{j+1}$ of projections of Z with $\sum_{p\in\mathcal{F}_{j+1}} p = 1$ and $g \in \mathcal{Q}$ such that

$$\begin{aligned}\omega_{\mathcal{F}_{j+1}}(g\cdot(f_j\cdot h)) \leq & \ \frac{3}{4}\omega_{\mathcal{F}_j}(f_j\cdot h)\\ \leq & \ (\tfrac{3}{4})^{j+1}\omega(h).\end{aligned}$$

Therefore, for any positive integer k there is a finite orthogonal family $\mathcal{F}_k$ of projections of Z with $\sum_{p\in\mathcal{F}_k} p = 1$ and $f_k \in \mathcal{Q}$ such that

$$\omega_{\mathcal{F}_k}(f_k\cdot h) \leq (\frac{3}{4})^k\omega(h).$$

Pick k such that $(\frac{3}{4})^k\omega(h) < \varepsilon$. For any $c \in \mathcal{F}_k$, let $\lambda_c = \|(f_k\cdot h)|cH\|$($H$ is the action space of M). Then $\|(f_k\cdot h)c - \lambda_c c\| \leq \omega_c(f_k\cdot h)$. Now let $f = f_k$ and $z = \sum_{c\in\mathcal{F}_k}\lambda_c c$. Then we have

$$\begin{aligned}\|f\cdot h - z\| &= \max\{\|(f\cdot h)c - \lambda_c c\| \mid c \in \mathcal{F}_k\}\\ &\leq \omega_{\mathcal{F}_k}(f_k\cdot h) \leq (\tfrac{3}{4})^k\omega(h) < \varepsilon.\end{aligned}$$

Q.E.D.

Lemma 6.2.5. Let $\{a_1,\cdots,a_n\} \subset M$ and $\varepsilon > 0$. Then there is a $f \in \mathcal{Q}$ and $\{z_1,\cdots,z_n\} \subset Z = M \cap M'$ such that $\|f\cdot a_k - z_k\| < \varepsilon, 1 \leq k \leq n$.

Proof. We may assume that $a_k^* = a_k, \forall k$. When $n = 1$, this is just Lemma 6.2.4. Now we assume that the conclusion holds for n.

For $a_1,\cdots,a_{n+1} \in M$ and $\varepsilon > 0$, first pick $z_1,\cdots,z_n \in Z$ and $f \in \mathcal{Q}$ such that

$$\|f\cdot a_k - z_k\| < \varepsilon, \quad 1 \leq k \leq n.$$

Again by Lemma 6.2.4, there is $g \in \mathcal{Q}$ and $z_{n+1} \in Z$ such that $\|g\cdot(f\cdot a_{n+1}) - z_{n+1}\| < \varepsilon$. Since $z_k \in Z$, it follows that

$$\begin{aligned}\|g\cdot(f\cdot a_k) - z_k\| &= \|g\cdot(f\cdot a_k - z_k)\|\\ &\leq \|f\cdot a_k - z_k\| < \varepsilon,\end{aligned}$$

$\forall 1 \leq k \leq n$. Therefore,

$$\|(g * f)\cdot a_k - z_k\| < \varepsilon, 1 \leq k \leq n+1.$$

Q.E.D.

Lemma 6.2.6. Let $\{a_k\} \subset M$. Then there is $\{z_k\} \subset Z = M \cap M'$ and $\{f_n\} \subset \mathcal{Q}$ such that

$$\|f_n \cdot a_k - z_k\| \xrightarrow{(n)} 0, \quad \forall k.$$

Proof. By Lemma 6.2.5, for a_1 we can pick $g_1 \in \mathcal{Q}$ and $z_{11} \in Z$ such that

$$\|f_1 \cdot a_1 - z_{11}\| < \frac{1}{2}.$$

For $g_1 \cdot a_1$ and $g_1 \cdot a_2$, there is $g_2 \in \mathcal{Q}$ and $z_{12}, z_{22} \in Z$ such that

$$\|(g_2 * g_1) \cdot a_k - z_{k2}\| < \frac{1}{2^2}, \quad k = 1, 2.$$

$\cdots$. Generally, we have $g_1, \cdots, g_n \in \mathcal{Q}$ and $z_{1n}, \cdots, z_{nn} \in Z$ such that

$$\|(g_n * \cdots * g_1) \cdot a_k - z_{kn}\| < \frac{1}{2^n}, \quad 1 \le k \le n.$$

Now let $f_n = g_n * \cdots * g_1$. Then for $1 \le k \le n$

$$\begin{aligned} \|f_{n+1} \cdot a_k - z_{kn}\| &= \|g_{n+1} \cdot (f_n \cdot a_k - z_{kn})\| \\ &\le \|f_n \cdot a_k - z_{kn}\| < \tfrac{1}{2^n}. \end{aligned}$$

Thus $\|f_{n+1} \cdot a_k - f_n \cdot a_k\| < \frac{1}{2^{n-1}}, 1 \le k \le n$, and $\{f_n \cdot a_k\}_n$ is a cauchy sequence for each fixed k. Further , $\{z_{kn}\}_n$ is also a cauchy sequence for each k. Suppose that $z_{kn} \xrightarrow{(n)} z_k (\in Z)$, then $\|f_n \cdot a_k - z_k\| \longrightarrow 0, \forall k$. Q.E.D.

Theorem 6.2.7. Let M be a VN algebra, $Z = M \cap M'$, and $a \in M$. Let

$$K(a) = \overline{\{f \cdot a \mid f \in \mathcal{Q}\}} \cap Z,$$

where the closure is taken with respect to uniform topology . Then $K(a) \neq \emptyset$.

Proof. By Lemma 6.2.6, there is $z \in Z$ and $\{f_n\} \subset \mathcal{Q}$ such that $\|f_n \cdot a - z\| \longrightarrow 0$. Therefore, $z \in K(a)$ and $K(a) \neq \emptyset$. Q.E.D.

Proposition 6.2.8. With the notations of Theorem 6.2.7, we have

1) $K(a_1 + a_2) \subset \overline{K(a_1) + K(a_2)}, \forall a_1, a_2 \in M$,
2) $K(za) \subset \overline{zK(a)}, \forall z \in Z, a \in M$.

Proof. 1) Let $z \in K(a_1 + a_2)$. Then for any $\varepsilon > 0$ there is $f \in \mathcal{Q}$ such that $\|f \cdot (a_1 + a_2) - z\| < \varepsilon$. By Lemma 6.2.6, there is $g \in \mathcal{Q}$ and $a_1 \in K(f \cdot a_i) \subset K(a_i)$ such that

$$\|g \cdot (f \cdot a_i) - z_i\| < \varepsilon, \quad i = 1, 2.$$

Since $\|g\cdot(f\cdot(a_1+a_2))-z\| \leq \|f\cdot(a_1+a_2)-z\| < \varepsilon$, it follows that $\|z-(z_1+z_2)\| < 3\varepsilon$. Therefore, $K(a_1+a_2) \subset \overline{K(a_1)+K(a_2)}$.

2) Let $c \in K(za)$. Then for any $\varepsilon > 0$ there is $f \in \mathcal{Q}$ such that

$$\|f\cdot(za)-c\| < \varepsilon.$$

By Theorem 6.2.7, there is $g \in \mathcal{Q}$ and $c_1 \in K(f\cdot a) \subset K(a)$ such that $\|g\cdot(f\cdot a)-c_1\| < \varepsilon$. Then

$$\begin{aligned}\|zc_1-c\| &\leq \|z((g*f)\cdot a)-c\| + \|z((g*f)\cdot a)-zc_1\| \\ &\leq \|g\cdot(f\cdot za-c)\| + \|z\|\cdot\|g\cdot(f\cdot a)-c_1\| \\ &< \varepsilon(1+\|z\|).\end{aligned}$$

Therefore, $K(za) \subset \overline{zK(a)}$. Q.E.D.

Notes . Theorem 6.2.7 is due to J. Dixmier. Through this approach he proved the existence of the central valued trace (see Section 6.3).

References. [18], [28].

6.3. Finite Von Neumann algebras

Proposition 6.3.1. Let M be a VN algebra.

1) M is finite if and only if $v \in M$ and $v^*v = 1$ imply $vv^* = 1$.

2) Suppose that M is finite, p and p' are projections of M and M' respectively. Then M_p and $M_{p'}$ are finite.

3) Suppose that $M = \sum_l \oplus M_l$. Then M is finite if and only if M_l is finite for each l.

Proof. 1) It is obvious since vv^* is also a projection.

2) By Proposition 6.1.3, p is a finte projection. Thus M_p is finite. Now let $c(p')$ be the central cover of p' in M'. Then $M_{p'}$ and $Mc(p')$ are $*$ isomorphic. Clearly, $Mc(p')$ is finite, so is $M_{p'}$.

3) The necessity is clear by 2). Conversely, let M_l be finite, and $M_l = Mz_l, \forall l$. If p is a projection of M with $p \sim 1$, then $pz_l \sim z_l, \forall l$. Since z_l is finite, it follows that $pz_l = z_l, \forall l$. Thus $p = 1$ and M is finite. Q.E.D.

Proposition 6.3.2. Let M be a finite VN algebra, and $\{p_1, p_2, q_1, q_2\}$ be projections of M satisfying

$$p_i \sim q_i, i = 1,2, \quad \text{and} \quad p_1 \leq p_2, \quad q_1 \leq q_2.$$

Then $(p_2 - p_1) \sim (q_2 - q_1)$.

Proof. By Theorem 1.5.4, there is a central projection z of M such that

$$(p_2 - p_1)z \precsim (q_2 - q_1)z,$$
$$(q_2 - q_1)(1 - z) \precsim (p_2 - p_1)(1 - z).$$

If $(p_2 - p_1)z \sim q \leq (q_2 - q_1)z$ and $q \neq (q_2 - q_1)z$, then

$$p_2 z = (p_1 z + (p_2 - p_1)z) \sim (q_1 z + q) \leq q_2 z, \text{ and } (q_1 z + q) \neq q_2 z$$

But $p_2 z \sim q_2 z$, so $q_2 z \sim (q_1 z + q) \leq q_2 z$ and $(q_1 z + q) \neq q_2 z$. This is impossible since $q_2 z$ is finite. Thus $(p_2 - p_1)z \sim (q_2 - q_1)z$. Similarly, $(p_2 - p_1)(1 - z) \sim (q_2 - q_1)(1 - z)$. Therefore, $(p_2 - p_1) \sim (q_2 - q_1)$. Q.E.D.

Definition 6.3.3. A positvie linear functional φ on a VN algebra M is called a *trace* , if

$$\varphi(a^* a) = \varphi(aa^*), \quad \forall a \in M.$$

If φ is a trace, then for any $a \in M_+$ and unitary element u of M, we have

$$\varphi(a) = \varphi((ua^{\frac{1}{2}})^* \cdot (ua^{\frac{1}{2}})) = \varphi(uau^*).$$

Therefore, $\varphi(ab) = \varphi(ba), \forall a, b \in M$.

Lemma 6.3.4. Let φ be a positive linear functional on M, and K be a positive constant such that

$$\varphi(p) \leq K\varphi(q)$$

for any projections p, q of M with $p \sim q$. Then

$$\varphi(a^* a) \leq K\varphi(aa^*), \quad \forall a \in M.$$

Proof. Let $a \in M$ and $\|a\| \leq 1$, and

$$a^* a = \int_0^1 \lambda de_\lambda = \lim_n \sum_{i=1}^n \frac{i}{n} p_i^{(n)}$$

be the spectral decomposition of $a^* a$, where $p_i^{(n)} = e_{\frac{i}{n}} - e_{\frac{i-1}{n}}, 1 \leq i \leq n$. If $a = uh$ is the polar decomposition of a, then $p_i^{(n)} \leq u^* u, \forall i, n$. Since

$$aa^* = ua^* au^* = \lim_n \sum_{i=1}^n \frac{i}{n} up_i^{(n)} u^*,$$

and $(up_i^{(n)})^*(up_i^{(n)}) = p_i^{(n)}, (up_i^{(n)})(up_i^{(n)})^* = up_i^{(n)}u^*$, it follows that

$$\begin{aligned}\varphi(a^*a) &= \lim_n \sum_{i=1}^n \frac{i}{n}\varphi(p_i^{(n)})\\ &\leq K \lim_n \sum_{i=1}^n \frac{i}{n}\varphi(up_i^{(n)}u^*) = K\varphi(aa^*).\end{aligned}$$

Q.E.D.

Corollary 6.3.5. Let φ be a positive linear functional on M. Then φ is a trace if and only if $\varphi(p) = \varphi(q)$ for any projections p, q of M with $p \sim q$.

Lemma 6.3.6. Let M be a finite VN algebra, p be a non-zero projection of M, and n be a positive integer. Then there exists a non-zero projection p_0 of M and a faithful normal state φ_0 on $M_0 = M_{p_0}$ such that

$$p_0 \leq p, \quad \varphi_0(a^*a) \leq (1 + \frac{1}{n})\varphi_0(aa^*), \forall a \in M_0.$$

Proof. Pick a normal state ψ on M_p, and let $\varphi(x) = \psi(pxp), \forall x \in M$. Then φ is a normal state on M and $s(\varphi) \leq p$. Replacing p by $s(\varphi)$, we may assume that $s(\varphi) = p$, i.e. there is a faithful normal state φ on M_p.

If for any projections q_1, q_2 of M_p with $q_1 \sim q_2$ we have $\varphi(q_1) = \varphi(q_2)$, then by Corollary 6.3.5, $\varphi_0 = \varphi$ and $p_0 = p$ satisfy the condition. Otherwise, by the Zorn lemma there are two maximal orthogonal families $\{e_l\}, \{f_l\}$ of projections of M_p such that

$$e_l \sim f_l, \quad \varphi(e_l) > \varphi(f_l), \forall l.$$

Put $e_1 = \sum_l e_l, f_1 = \sum_l f_l$. Then $e_1 \sim f_1$ and $\varphi(e_1) > \varphi(f_1)$. Consequently, $f_1 \leq p$ and $f_1 \neq p$. Since $e_1 \sim f_1$ and M is finite, it follows by Proposition 6.3.2 that $(p - e_1) \sim (p - f_1)$. Thus, we get $e_1 \leq p$ and $e_1 \neq p$. By the maximum of $\{e_l\}$ and $\{f_l\}$, we have $\varphi(e) \leq \varphi(f)$ for any projections e and f with $e \sim f$, and $e \leq p - e_1, f \leq p - f_1$.

Let

$$\mu_0 = \inf\left\{\mu \,\middle|\, \begin{array}{l}\mu > 0, \text{ and } \varphi(e) \leq \mu\varphi(f) \text{ for any projections}\\ e \text{ and } f \text{ with } e \sim f, \text{ and } e \leq p - e_1, f \leq p - f_1\end{array}\right\}$$

Clearly, $\mu_0 \leq 1$. We say that $0 < \varphi(p - e_1) \leq \mu_0$. In fact, if $\varphi(p - e_1) > \mu_0$, then there is $\mu \in [\mu_0, \varphi(p - e_1))$ such that $\varphi(e) \leq \mu\varphi(f)$ for any projections e and f with $e \sim f$, and $e \leq p - e_1, f \leq p - f_1$. In particular, $\varphi(p - e_1) \leq \mu\varphi(p - f_1) < \varphi(p - e_1)\varphi(p - f_1)$ and $\varphi(p - f_1) > 1$. But $\varphi(p - f_1) \leq 1$, a contradiction. Thus , $0 < \varphi(p - e_1) \leq \mu_0$.

Now pick $\varepsilon > 0$ such that $0 < (\mu_0 - \varepsilon)^{-1}\mu_0 \leq 1 + \frac{1}{n}$. By the definition of μ_0, there are projections e_2 and f_2 with $e_2 \sim f_2$, and $e_2 \leq p - e_1, f_2 \leq p - f_1$

such that $\varphi(e_2) > (\mu_0 - \varepsilon)\varphi(f_2)$. Clearly e_2 and f_2 are not zero. We claim that there are non-zero projections e_3 and f_3 with $e_3 \sim f_3$, and $e_3 \leq e_2, f_3 \leq f_2$, such that $\varphi(e) \geq (\mu_0 - \varepsilon)\varphi(f)$ for any projections e and f with $e \sim f$, and $e \leq e_3, f \leq f_3$. In fact , if such e_3 and f_3 don't exist, then e_2 and f_2 are not such e_3 and f_3 . Thus there are projections e and f with $e \sim f$, and $e \leq e_2, f \leq f_2$, such that $\varphi(e) < (\mu_0 - \varepsilon)\varphi(f)$. Further, $(e_2 - e)$ and $(f_2 - f)$ are not such e_3 and f_3 , we have also $\cdots\cdots$. By the Zorn lemma, we can write $e_2 = \sum_l \oplus e_l$ and $f_2 = \sum_l \oplus f_l$ such that

$$e_l \sim f_l, \quad \text{and} \quad \varphi(e_l) < (\mu_0 - \varepsilon)\varphi(f_l), \quad \forall l.$$

Since φ is normal , it follows that $\varphi(e_2) < (\mu_0 - \varepsilon)\varphi(f_2)$. This contradicts that $\varphi(e_2) > (\mu_0 - \varepsilon)\varphi(f_2)$. Thus e_3 and f_3 exist.

Let $v \in M_p$ be such that $v^*v = e, vv^* = f_3$, and define

$$\psi(x) = \varphi(v^*xv), \forall x \in f_3 M f_3.$$

Clearly , $\psi(f_3) = \varphi(e_3)$, and $\varphi(e_3) > 0$ since φ is faithful on M_p . If r and q are projections of $f_3 M f_3$ with $r \sim q$, by $(v^*q)^*(v^*q) = q$ we have $r \sim q \sim v^*qv$ in M_p and $v^*qv \leq e_3$. By the property of (e_3, f_3) and the definition of μ_0, we get $(\mu_0 - \varepsilon)\varphi(r) \leq \varphi(v^*qv) \leq \mu_0\varphi(r)$. In particular, $(\mu_0 - \varepsilon)\varphi(r) \leq \varphi(v^*rv) \leq \mu_0\varphi(r)$. Then

$$\begin{aligned}\psi(q) = \varphi(v^*qv) &\leq \mu_0\varphi(r)\\ &\leq \tfrac{\mu_0}{\mu_0-\varepsilon}\varphi(v^*rv) \leq (1 + \tfrac{1}{n})\psi(r).\end{aligned}$$

Finally, let $p_0 = f_3(\leq p)$ and

$$\varphi_0(x) = \psi(f_3)^{-1}\psi(x), \quad \forall x \in M_0 = M_{p_0}.$$

Clearly, φ_0 is a normal state on M_0. If $x \in M_0$ is such that $\varphi_0(x^*x) = 0$, then $vx = 0$ since φ is faithful on M_p . Further, $x = xf_3 = xvv^* = 0$, i.e., φ_0 is faithful on M_0. From the preceding paragraph, we have

$$\varphi_0(q) \leq (1 + \frac{1}{n})\varphi_0(r)$$

for any projections r and q of M_0 with $r \sim q$. By Lemma 6.3.4., we obtain $\varphi_0(a^*a) \leq (1 + \frac{1}{n})\varphi_0(aa^*), \forall a \in M_0$. Q.E.D.

Lemma 6.3.7. Let M be a finite VN algebra. Then for any positive integer n there is a normal state ψ_n on M such that

$$\psi_n(x^*x) \leq (1 + \frac{1}{n})\psi_n(xx^*), \forall x \in M.$$

Proof. By Lemma 6.3.7, for fixed n there is a non-zero projection p_0 of M and a faithful nonmal state φ_0 on M_{p_0} such that

$$\varphi_0(a^*a) \leq (1+\frac{1}{n})\varphi_0(aa^*), \forall a \in M_{p_0}.$$

Let $\{p_1, \cdots, p_m\}$ be a maximal orthogonal family of projections of M such that $p_i \sim p_0, 1 \leq i \leq m$ (notice that m is finite since M is finite). By Theorem 1.5.4, there is a central projection z of M such that

$$(1-\sum_i p_i)z \preceq p_0 z, \quad p_0(1-z) \preceq (1-\sum_i p_i)(1-z).$$

Since $\{p_i\}$ is maximal , it follows that $p_0 z \neq 0$.

Let $v_i^* v_i = p_0 z, v_i v_i^* = p_i z, 1 \leq i \leq m$, and

$$v_{m+1}v_{m+1}^* = (1-\sum_i p_i)z, \quad v_{m+1}^* v_{m+1} \leq p_0 z,$$

and define

$$\varphi_n(x) = \sum_{i=1}^{m+1} \varphi_0(v_i^* x v_i), \quad \forall x \in M.$$

Then for any $x \in M$,

$$\begin{aligned}\varphi_n(x^*x) &= \sum_{i=1}^{m+1} \varphi_0(v_i^* x^* x v_i) = \sum_{i,j=1}^{m+1} \varphi_0(v_i^* x^* v_j v_j^* x v_i) \\ &\leq (1+\tfrac{1}{n})\sum_{i,j} \varphi_0(v_j^* x v_i v_i^* x^* v_j) \\ &= (1+\tfrac{1}{n})\sum_j \varphi_0(v_j^* x x^* v_j) = (1+\frac{1}{n})\varphi_n(xx^*).\end{aligned}$$

Moreover, $\varphi_n(1) \geq m\varphi_0(p_0 z)$, and $\varphi_0(p_0 z) > 0$ since $p_0 z \neq 0$ and φ_0 is faithful on M_{p_0}. Therefore , $\psi_n(\cdot) = \varphi_n(1)^{-1}\varphi_n(\cdot)$ is what we want to find. Q.E.D.

Theorem 6.3.8. Let M be a finite VN algebra. Then for any $a \in M, {}^\# K(a) = 1$, where $K(a)$ is defined as in Theorem 6.2.7.

Proof. By Proposition 6.2.8 and Theorem 6.2.7, we may assume that $a \geq 0$ and $\|a\| \leq 1/2$.

Suppose that there are $c_1, c_2 \in K(a)$ and $c_1 \neq c_2$. Clearly, $c_1, c_2 \geq 0$ and $\|c_1 - c_2\| \leq 1$. Let $c_1 - c_2 = \int_{-1}^{1} \mu dz_\mu$ be the spectral decomposition of $(c_1 - c_2)$, where z_μ is a central projection of M, $\forall \mu$. Since $c_1 \neq c_2$, there is $\lambda > 0$ such that either $z_{-\lambda} \neq 0$ or $(1 - z_\lambda) \neq 0$. Let $z = z_{-\lambda}$ for the case of $z_{-\lambda} \neq 0$, or $z = 1 - z_\lambda$ otherwise. Then we have

$$c_2 z \geq c_1 z + \lambda z \quad \text{or} \quad c_1 z \geq c_2 z + \lambda z.$$

By the symmetry we may assume that $c_1 z \geq c_2 z + \lambda z$. Since $c_1 z \neq c_2 z$ and $c_1 z, c_2 z \in K(az)$, and replacing M by Mz , we may also assume that $z = 1$.

Pick $\{\psi_n\}$ as in Theorem 6.3.7. Then for any unitary element u of M,

$$\psi_n(u^* au) = \psi((a^{\frac{1}{2}}u)^*(a^{\frac{1}{2}}u)) \leq (1 + \frac{1}{n})\psi_n(a),$$

$$\psi_n(a) = \psi_n((a^{\frac{1}{2}}u)(a^{\frac{1}{2}}u)^*) \leq (1 + \frac{1}{n})\psi_n(u^* au).$$

Thus for any $f, g \in \mathcal{Q}$ (see Definition 6.2.3),

$$\psi_n(f \cdot a) \leq (1 + \frac{1}{n})\psi_n(a) \leq (1 + \frac{1}{n})^2 \psi_n(g \cdot a).$$

Let $\{f_k\}, \{g_k\} \subset \mathcal{Q}$ be such that $f_k \cdot a \longrightarrow c_1$, $g_k \cdot a \longrightarrow c_2$. Then

$$\psi_n(c_1) \leq (1 + \frac{1}{n})^2 \psi_n(c_2).$$

Further , from $c_1 \geq c_2 + \lambda$ it follows that

$$\psi_n(c_2) + \lambda \leq \psi_n(c_1) \leq (1 + \frac{1}{n})^2 \psi_n(c_2).$$

When n is sufficiently large , we get a contradiction since $\lambda > 0$.

Therefore, $K(a)$ contains only one element, $\forall a \in M$. Q.E.D.

Remark. In the end of section 6.4, we shall prove that : if $^\# K(a) = 1, \forall a \in M$, then M is finite.

Now we start to characterize finite VN algebras by normal tracial states.

Lemma 6.3.9. Let M be a finite VN algebra. Then there is a normal tracial state on M at least.

Proof. By Theorem 6.3.8, we can define a map $T(\cdot)$ from M to $Z = M \cap M'$ such that

$$K(a) = \{T(a)\}, \quad \forall a \in M.$$

From Proposition 6.2.8 and the definition of $K(\cdot), T$ is linear, and

$$T(z) = z, \quad \forall z \in Z, \quad T(M_+) \subset Z_+.$$

Since $K(u^* xu) = K(x)$ for any $x \in M$ and any unitary element u of M , it follows that $T(u^* xu) = T(x)$. Further , $T(xy) = T(yx), \forall x, y \in M$.

Let ψ be the same as the ψ_1 in Lemma 6.3.7, and define

$$\varphi(a) = \psi(T(a)), \quad \forall a \in M.$$

From the preceding paragraph, φ is a tracial state on M.

Now it suffices to show that φ is normal. Let $\{b_l\}$ be a bounded increasing net of M_+, and $b = \sup_l b_l$. Put $a_l = b - b_l, \forall l$. Then $a_l \longrightarrow 0(\sigma(M, M_*))$. We need to prove that $\varphi(a_l) \longrightarrow 0$. For any $\varepsilon > 0$, there is l_0 such that

$$0 \leq \psi(a_l) < \varepsilon, \quad \forall l \geq l_0$$

since ψ is normal. Pick $f_l \in \mathcal{Q}$ such that $\|f_l \cdot a_l - T(a_l)\| < \varepsilon, \forall l$. By Lemma 6.3.7, for $l \geq l_0$ we have

$$\begin{aligned} 0 \leq \varphi(a_l) &= \psi(T(a_l)) \leq \psi(f_l \cdot a_l) + \varepsilon \\ &= \sum_u f_l(u)\psi(u^* a_l u) + \varepsilon \\ &\leq 2\sum_u f_l(u)\psi(a_l) + \varepsilon < 3\varepsilon. \end{aligned}$$

Therefore, $\varphi(a_l) \longrightarrow 0$. Q.E.D.

Theorem 6.3.10 A VN algebra M is finite if and only if there is a faithful family of normal tracial states on M , i.e., for any non-zero $a \in M_+$ there is a normal tracial state φ on M such that $\varphi(a) > 0$.

Proof. Suppose that M is finite. Then there is a normal tracial state φ on M by Lemma 6.3.9. Let $z = s(\varphi)$. Then z is a non-zero central projection of M, and φ is faithful on Mz . Again we continue this process for finite VN algebra $M(1-z), \cdots$, and so on . By the Zorn lemma, there is a family $\{\varphi_l\}$ of normal tracial states on M such that

$$s(\varphi_l) \cdot s(\varphi_{l'}) = 0, \forall l \neq l', \quad \text{and} \quad \sum_l s(\varphi_l) = 1.$$

It is easily verified that the family $\{\varphi_l\}$ is faithful.

Conversely, let $\mathcal{F}$ be a faithful family of normal tracial states on M. If $w \in M$ is such that $w^*w = 1$, then $\varphi(1-p) = \varphi(w^*w) - \varphi(ww^*) = 0, \forall \varphi \in \mathcal{F}$, where $p = ww^*$. Since $\mathcal{F}$ is faithful, it follows that $p = 1$. Therefore, M is finite.

Q.E.D.

Another characterization of finite VN algebras is as follows.

Lemma 6.3.11. Let p be a projection of a VN algebra M, and let $v \in M$ be such that

$$v^*v = p, \quad vv^* \leq p \text{ and } vv^* \neq p.$$

Put

$$q_n = v^n v^{*n}, \quad n = 1, 2, \cdots, q_0 = p,$$

$$e_n = q_n - q_{n+1}, \quad n = 0, 1, 2, \cdots.$$

Then $\{e_n\}$ is an orthogonal sequence of non-zero projections of M with $e_n \sim e_m, \forall n, m$, and $e_n \longrightarrow 0$ (strongly).

Proof. Since $q_1 \leq p$, it follows that $pv = v$ and $v^{*n}v^n = p, \forall n \geq 1$. Thus q_n is a projection for each n. From $q_n q_{n+1} = q_{n+1}$ we have

$$p = q_0 \geq q_1 \geq q_2 \geq \cdots.$$

Further , $e_n e_m = 0, \forall n \neq m$, and $e_n \longrightarrow 0$ (strongly).

Let $u_n = vq_n, n = 0, 1, 2, \cdots$. Then

$$(u_n - u_{n+1})^*(u_n - u_{n+1}) = e_n,$$

$$(u_n - u_{n+1})(u_n - u_{n+1})^* = e_{n+1},$$

i.e., $e_n \sim e_{n+1}, \forall n \geq 0$. Moreover , by $e_0 = p - vv^* \neq 0$ we get $e_n \neq 0, \forall n \geq 0$.

Q.E.D.

Theorem 6.3.12. A VN algebra M is finite if and only if the $*$ operation is strongly continuous in any bounded ball of M.

Proof. Let M be finite. By Theorem 6.3.10, there is a faithful family $\mathcal{F}$ of normal tracial states on M. Suppose that $\{x_l\}$ is a net of M with $\|x_l\| \leq 1, \forall l$, and $x_l \longrightarrow 0$ (strongly). Then for any $a \in M_+$ and $\varphi \in \mathcal{F}$,

$$|L_a\varphi(x_l x_l^*)| = |\varphi(x_l^* a x_l)| \leq \|a\| \varphi(x_l^* x_l) \longrightarrow 0.$$

If $[L_a\varphi | a \in M, \varphi \in \mathcal{F}]$ is dense in M_* , then $x_l x_l^* \longrightarrow 0$ (weakly) . Further, $x_l^* \longrightarrow 0$ (strongly) . So the $*$ operation is strongly continuous in any bounded ball of M. Now we need to prove that $[L_a\varphi | a \in M, \varphi \in \mathcal{F}]$ is dense in M_* . Let $b \in M$ be such that $L_a\varphi(b) = 0, \forall a \in M, \varphi \in \mathcal{F}$. Then $\varphi(bb^*) = 0, \forall \varphi \in \mathcal{F}$. Since $\mathcal{F}$ is faithful, it follows that $b = 0$. So the above assertion holds.

Conversely, suppose that the $*$ operation is strongly continuous in any bounded ball of M. If M is not finite, then there is $v \in M$ such taht $v^*v = 1, vv^* \neq 1$. By Lemma 6.3.11, there is a sequence $\{e_n\}$ of non-zero projections of M with $e_n \cdot e_m = 0, e_n \sim e_m, \forall n \neq m$, and $e_n \longrightarrow 0$ (strongly). Let $w_n \in M$ be such that $w_n^* w_n = e_n, w_n w_n^* = e_1, \forall n$. Clearly, $w_n \longrightarrow 0$ (strongly), $\|w_n\| \leq 1, \forall n$. By the assumption, $w_n^* \longrightarrow 0$ (strongly). Thus , $e_1 = w_n w_n^* \longrightarrow 0$ (weakly) , a contradiction. Therefore, M is finite.

Q.E.D.

In the proof of Lemma 6.3.9, we introduce a map $T(\cdot)$ from a finite VN algebra to its center. Now we discuss the properties of that map in detail.

Definition 6.3.13. Let M be a finite VN algebra. The map T from M to $Z = M \cap M'$ is defined by $\{T(a)\} = K(a)(\forall a \in M)$ and is called the ***central valued trace*** on M, where $K(\cdot)$ is defined as in Theorem 6.2.7.

Proposition 6.3.14. Let M be a finite VN algebra, and $T : M \longrightarrow Z = M \cap M'$ be the central valued trace. Then:

1) T is a projection of norm one from M onto Z, and is σ-σ continuous. Consequently, $T(a) \geq 0, \forall a \in M_+; T(za) = zT(a), \forall a \in M, z \in Z; T(a)^*T(a) \leq T(a^*a), \forall a \in M$;

2) $T(ab) = T(ba), \quad \forall a, b \in M$;

3) $T(a^*a) = 0$ if and only if $a = 0$;

4) $\{\varphi(T(\cdot)) | \varphi$ is a normal state on $M\}$ is a faithful family of normal tracial states on M;

5) $p \precsim q$ if and only if $T(p) \leq T(q)$, where p and q are two projections of M.

Proof. 1) From the proof of Lemma 6.3.9, it suffices to show that T is σ -σ continuous. Since T is positive, this is equivalent to prove that $\varphi(T(\cdot))$ is normal for any normal state φ on M, i.e., to prove

$$\varphi(T(a)) = \sup{}_l \varphi(T(a_l))$$

for any bounded increasing net $\{a_l\}$ of M_+, where $a = \sup_l a_l$. But $\{T(a_l)\}$ is also a bounded increasing net of Z_+, by the normality of φ we need only to show that

$$T(a) = \sup_l T(a_l).$$

Clearly, $T(a) \geq \sup_l T(a_l)$. If they are not equal, then there is a non-zero central projection z and a positive number λ such that

$$zT(a) \geq z\sup_l T(a_l) + \lambda z.$$

Let $\mathcal{F}$ be a faithful family of normal tracial states on M. For any $\varepsilon > 0$ and l, pick $f_l \in \mathcal{Q}$ such that $\|f_l \cdot (a - a_l)z - T((a - a_l)z)\| < \varepsilon$. Then for any $\psi \in \mathcal{F}$,

$$\begin{aligned}|\psi(T((a - a_l)z))| &\leq |\psi(f_l \cdot (a - a_l)z)| + \varepsilon \\ &\leq \sum_u f_l(u)|\psi(u(a - a_l)zu^*)| + \varepsilon \\ &= |\psi((a - a_l)z)| + \varepsilon.\end{aligned}$$

Since ψ is normal, it follows that

$$\psi(T(a - a_l)z) \longrightarrow 0, \quad \forall \psi \in \mathcal{F},$$

i.e., $\psi(T(a)z) = \lim_l \psi(T(a_l)z) = \psi(z\sup_l T(a_l)), \forall \psi \in \mathcal{F}$. However, $\psi(T(a)z) \geq \psi(z\sup_l T(a_l)) + \lambda\psi(z), \forall \psi \in \mathcal{F}$. Thus, $\psi(z) = 0, \forall \psi \in \mathcal{F}$, and $z = 0$. This is a contradiction. Therefore, $T(a) = \sup_l T(a_l)$.

2) It is contained in the proof of Lemma 6.3.9.

3) Let $I = \{a \in M | T(a^*a) = 0\}$. Clearly , I is a $s(M, M_*)$-closed two-sided $*$ ideal of M. By Proposition 1.7.1, there is a central projection z of M such that $I = Mz$. In particular, $z \in I$, i.e.,

$$z = T(z) = T(z^*z) = 0.$$

Therefore , $I = 0$, This is just our conclusion.

4) If $a \in M_+$ is such that $\varphi(T(a)) = 0$ for any normal state on M , then $T(a) = 0$. By 3) , we have $a = 0$. Therefore, $\{\varphi \circ T | \varphi$ is a normal state on $M\}$ is faithful.

5) Suppose that $p \sim q_1 \leq q$. By 2) , we have $T(p) = T(q_1) \leq T(q)$. Conversely, let $T(p) \leq T(q)$. By Theorem 1.5.4, there is a central projection z such that

$$pz \precsim qz, \quad q(1-z) \precsim p(1-z).$$

Thus , $T(q)(1-z) \leq T(p)(1-z) \leq T(q)(1-z)$, i.e., $T(q)(1-z) = T(p)(1-z)$. Let $q(1-z) \sim p_1 \leq p(1-z)$. Then $T(p_1) = T(q)(1-z) = T(p)(1-z)$, and $T(p(1-z)-p_1) = 0$. By 3), $p_1 = (1-z)p$, i.e., $q(1-z) \sim p(1-z)$. Therefore, $p \precsim q$. Q.E.D.

Now consider some properties of $\sigma-$ finite and finite VN algebras.

Proposition 6.3.15. Let M be a VN algebra. Then the following statements are equivalent:

1) M is σ–finite and finite;
2) M is finite, and $Z = M \cap M'$ is σ–finite;
3) There is a faithful normal tracial state on M.

Proof. 3) $\Rightarrow$ 1). It is immediate from Theorem 6.3.10 and Proposition 1.14.2.

1) $\Rightarrow$ 2). It is obvious.

2) $\Rightarrow$ 3) . By Proposition 1.14.2, there is a faithful normal state ψ on Z. Let

$$\varphi(a) = \psi(T(a)), \quad \forall a \in M.$$

By Propsition 6.3.14, φ is a faithful normal tracial state on M. Q.E.D.

Proposition 6.3.16. Let M be a finite VN algebra. Then M is a direct sum of $\{M_l\}$, where M_l is σ–finite and finite, $\forall l$. Consequently, each finite factor is σ–finite.

Proof. From the proof of Theorem 6.3.10, there a family $\{\varphi_l\}$ of normal tracial states on M such that $s(\varphi_l) \cdot s(\varphi_{l'}) = 0, \forall l \neq l'$ and $\sum_l s(\varphi_l) = 1$. Then

$$M = \sum_l \oplus M_l, \quad M_l = Ms(\varphi_l), \quad \forall l.$$

Since φ_l is a faithful normal tracial state on M_l, it follows from Proposition 6.3.15 that M_l is σ-finite and finite, $\forall l$. Q.E.D.

Notes. Using a fixed point theorem , F. J. Yeadon gave another proof of the existence of a trace. Theorem 6.3.12 is due to S.Sakai.

References. [18], [78], [144], [150], [199].

6.4. Properly infinite Von Neumann algebras

Proposition 6.4.1. Let $M = \sum_l \oplus M_l$. Then M is properly infinite if and only if M_l is properly infinite, $\forall l$.

Proof. The necessity is obvious. Now let $M_l = Mz_l$ be properly infinite, $\forall l$, and z be some finite central projection of M. Then zz_l is a finite central projection of $M_l, \forall l$. Thus $zz_l = 0$ since M_l is properly infinite, $\forall l$, and $z = 0$. So M is properly infinite. Q.E.D.

Proposition 6.4.2. A VN algebra M is properly infinite if and only if there is no normal tracial state on M.

Proof. Let φ be a normal tracial state on M. Then $s(\varphi)$ is a non-zero central projection, and $Ms(\varphi)$ is finite by Proposition 6.3.15. Thus M is not properly infinite.

Conversely, if M is not properly infinite, then there is a non-zero central projection z of M such that Mz is finite. There is a normal tracial state ψ on Mz at least. Let $\varphi(\cdot) = \psi(\cdot z)$. Then φ is a normal tracial state on M. Q.E.D.

Proposition 6.4.3. If M is a properly infinite VN algebra, then the $*$ operation is not strongly continuous in unit ball of M.

Proof. It is immediate from Theorem 6.3.12. Q.E.D.

Theorem 6.4.4. Let M be a VN algebra. Then the following statements are equivalent:

1) M is properly infinite;

2) There is an orthogonal infinite sequence $\{p_n\}$ of projections of M such that

$$\sum_n p_n = 1, \quad p_n \sim 1, \forall n;$$

3) There is a projection p of M such that

$$p \sim (1-p) \sim 1.$$

Proof. 2) $\Rightarrow$ 3) . Pick $p = \sum_{n=0}^{\infty} p_{2n+1}$. Then we have

$$p \sim (1-p) \sim 1.$$

3) $\Rightarrow$ 1) . Let z be any non-zero central projection of M. Then $pz \sim (1-p)z \sim z$. Thus z is not finite, and M is properly infinite.

1) $\Rightarrow$ 2). Since the identity 1 is infinite, it follows that there is $v \in M$ such that $v^*v = 1$ and $vv^* \leq 1$, and $vv^* \neq 1$. Let

$$q_n = v^n v^{*n}, e_n = q_n - q_{n+1}, n = 0, 1, 2, \cdots.$$

By Lemma 6.3.11, $e_n \neq 0, e_n e_m = 0, e_n \sim e_m, \forall n \neq m$. By the Zorn lemma, there is a maximal orthogonal family $\{e_l | e \in \wedge\}$ of non-zero projections of M such that $\{e_l | e \in \wedge\} \supset \{e_n\}$, and $e_l \sim e_{l'}, \forall l, l' \in \wedge$. Define $p = 1 - \sum_{e \in \wedge} e_l$. By Theorem 1.5.4, there is a central projection z of M such that

$$pz \preceq e_0 z, \quad e_0(1-z) \preceq p(1-z).$$

Clearly , $z \neq 0$ by the maximum of the family $\{e_l | e \in \wedge\}$. Since $^{\#}\wedge$ is infinite , then we can write

$$\wedge = \bigcup_{i=1}^{\infty} \wedge_i,$$

where $\wedge_i \cap \wedge_j = \emptyset, \forall i \neq j$, and $^{\#}\wedge_i = {}^{\#}\wedge, \forall i$. Let

$$r_1 = pz + \sum_{l \in \wedge_1} e_l z, \quad r_j = \sum_{l \in \wedge_j} e_l z, \ \forall j \geq 2.$$

Clearly, $r_i r_j = 0, \forall i \neq j, r_1 \sim r_2 \sim \cdots \sim z$, and $z = \sum_{j=1}^{\infty} r_j$. Since $M(1-z)$ is still properly infinite, we can make the same process, $\cdots$. Then by the Zorn lemma, there is an orhtogonal family $\{z_l\}$ of non-zero central projections of M with $\sum_l z_l = 1$ such that for any l there exists an orthogonal infinite sequence $\{r_{ln} | n = 1, 2, \cdots\}$ of projections of M satisfying

$$r_{l1} \sim r_{l2} \sim \cdots \sim z_l, \quad \text{and} \quad \sum_n r_{ln} = z_l.$$

Now let $p_n = \sum_l r_{ln}, n = 1, 2, \cdots$. Then

$$p_n p_m = 0, \quad \forall n \neq m, \ \sum_n p_n = 1, \quad \text{and} \quad p_n \sim 1, \forall n.$$

Q.E.D.

Using Theorem 6.4.4, we can get a property of finite projections.

Proposition 6.4.5. Let p, q be two finite projections of a VN algebra M. Then $\sup\{p,q\}$ is also finite.

Proof. We may assume that $\sup\{p,q\} = 1$. By Proposition 1.5.2,

$$(1-p) \sim (q - \inf\{p,q\}) \leq q.$$

Thus $(1-p)$ is also finite. Suppose that z is a non-zero central projection of M such that Mz is properly infinite. Noticing that pz and qz are finite and

$$\sup\{pz, qz\} = z,$$

we may also assume that M is properly infinite

By Theorem 6.4.4, we can write $1 = r + (1-r)$, where $r \sim (1-r) \sim 1$. From Proposition 1.5.5, there is a central projection z such that

$$rz \precsim pz, \quad (1-r)(1-z) \precsim (1-p)(1-z).$$

Since pz and $(1-p)(1-z)$ are finite, it follows that $z(\sim rz)$ and $(1-z)(\sim (1-r)(1-z))$ are finite. This contradicts that M is properly infinite.

Therefore, $\sup\{p,q\}$ is finite. Q.E.D.

In the end of this section, we prove the conclusion of the Remark after Theorem 6.3.8.

Proposition 6.4.6. Let M be a VN algebra. If ${}^{\#}K(a) = 1, \forall a \in M$, then M is finite.

Proof. Suppose that z is a non-zero central projection of M such that Mz is properly infinite. By Theorem 6.4.4, there is a projection p of M such that $p \leq z$ and $p \sim (z-p) \sim z$. Let $u, v \in Mz$ be such that

$$u^*u = v^*v = z, \quad uu^* = p, \quad vv^* = z - p.$$

Define $\Phi : M \longrightarrow Z = M \cap M'$ such that $K(a) = \{\Phi(a)\}, \forall a \in M$. By the definition of $K(\cdot)$, it is clear that $\Phi(ab) = \Phi(ba), \forall a, b \in M$. Then

$$z = \Phi(z) = \Phi(p) = \Phi(z-p),$$

and $2z = 2\Phi(z) = \Phi(p) + \Phi(z-p) = \Phi(z) = z$. This contradicts that $z \neq 0$. Therefore, M is finite. Q.E.D.

References. [18], [82].

6.5. Semi–finite Von Neumann algebras

Definition 6.5.1. Let M be a VN algebra. A *trace* on M_+ is a function φ on M_+, taking non–negative, possibly infinite, real values, possessing the following properties:

1) $\varphi(a+b) = \varphi(a) + \varphi(b), \quad \forall a, b \in M_+;$
2) $\varphi(\lambda a) = \lambda\varphi(a), \forall a \in M_+, \lambda \geq 0$ (with the convention that $0 \cdot +\infty = 0$);
3) $\varphi(x^*x) = \varphi(xx^*), \forall x \in M.$

A trace φ on M_+ is said to be *faithful* , if $a \in M_+$ is such that $\varphi(a) = 0$, then $a = 0$.

A trace φ on M_+ is said to be *semi–finite,* if for any $0 \neq a \in M_+$, there is $0 \neq b \in M_+$ and $b \leq a$ such that $\varphi(b) < +\infty$.

A trace φ on M_+ is said to be *normal* , if for any bounded increasing net $\{a_l\}$ of M_+, we have

$$\varphi(\sup_l a_l) = \sup_l \varphi(a_l).$$

Proposition 6.5.2. Let φ be a trace on M_+, and

$$\mathcal{N} = \{x \in M | \varphi(x^*x) < \infty\},$$

$$\mathcal{M} = \mathcal{N}^2 = [xy | x, y \in \mathcal{N}]$$

(Later, $\mathcal{M}$ is called the *definition ideal* of φ) . Then $\mathcal{M}, \mathcal{N}$ are $*$ two–sided ideals of M; and

$$\mathcal{M} = [\mathcal{M}_+] = \{xy | x, y \in \mathcal{N}\},$$

$$\mathcal{M}_+ = \{a \in M_+ | \varphi(a) < \infty\},$$

where $\mathcal{M}_+ = \mathcal{M} \cap M_+$; and φ can be uniquely extended to a linear functional on $\mathcal{M}$, still denoted by φ; and also

$$\varphi(ab) = \varphi(ba), \quad a \in \mathcal{M}, b \in M, \quad \text{or} \quad a, b \in \mathcal{N}.$$

Moreover, the tracial condition $\varphi(x^*x) = \varphi(xx^*)(\forall x \in M)$ is equivalent to

$$\varphi(a) = \varphi(u^*au),$$

$\forall a \in M_+$ and unitary element $u \in M$.

Proof. Clearly, $\mathcal{N}$ is a $*$ two–sided ideal of M, and so is $\mathcal{M}$. If $a \in M_+$ with $\varphi(a) < \infty$, then $a^{\frac{1}{2}} \in \mathcal{N}$ and $a \in \mathcal{M}_+$. Conversely, let $a = \sum_j x_j^* y_j \in \mathcal{M}_+$, where $x_j, y_j \in \mathcal{N}, \forall j$. By the polarization,

$$\begin{aligned} 4x_j^* y_j = & (x_j + y_j)^*(x_j + y_j) - (x_j - y_j)^*(x_j - y_j) \\ & -i(x_j + iy_j)^*(x_j + iy_j) + i(x_j - iy_j)^*(x_j - iy_j). \end{aligned}$$

So we have that

$$a = \frac{1}{2}(a + a^*) \le \frac{1}{4}\sum_j (x_j + y_j)^*(x_j + y_j),$$

and $\varphi(a) < \infty$. Thus $\mathcal{M}_+ = \{a \in M_+ | \varphi(a) < \infty\}$. Further, from the polarization we can see that $\mathcal{M} = [\mathcal{M}_+]$. Now let $x \in \mathcal{M}$, and $x = uh$ be the polar decomposition of x. Clearly , $h = u^* x \in \mathcal{M}_+, h^{\frac{1}{2}} \in \mathcal{N}$, and $x = uh^{\frac{1}{2}} \cdot h^{\frac{1}{2}}$. Hence , $\mathcal{M} = \{xy | x, y \in \mathcal{N}\}$.

From $\mathcal{M} = [\mathcal{M}_+]$, it is easily verified that φ can be uniquely extended to a linear functional on $\mathcal{M}$, still denoted by φ .

If $a \in M_+$ and $u (\in M)$ is unitary, then

$$\varphi(uau^*) = \varphi((ua^{1/2})^* \cdot (ua^{\frac{1}{2}})) = \varphi(a).$$

In particular, $\varphi(a) = \varphi(uau^*), \forall a \in \mathcal{M}_+$ and unitary element $u \in M$. Further, $\varphi(a) = \varphi(uau^*), \forall a \in \mathcal{M}$ and unitary element $u \in M$. Since $\mathcal{M}$ is a two-sided ideal, it follows that $\varphi(ua) = \varphi(au), \forall a \in \mathcal{M}$ and unitary element $u \in M$. Hence, we have $\varphi(ab) = \varphi(ba), \forall a \in \mathcal{M}, b \in M$. Moreover, if $a, b \in \mathcal{N}$, by the polarization of ab and $\varphi(xx^*) = \varphi(x^*x) (\forall x \in M)$, we can obtain that $\varphi(ab) = \varphi(ba)$.

Finally, let $\varphi : M_+ \longrightarrow [0, +\infty]$ satisfy the conditions 1) , 2) of Definition 6.5.1 and

$$\varphi(a) = \varphi(u^* au),$$

$\forall a \in M_+$ and $u \in M$ is unitary . Similarly , we can also define $\mathcal{N}$ and $\mathcal{M}$, which possess the above properties. If $x \in M$ is such that $x^*x \in \mathcal{M}_+$, then $xx^* = w(x^*x)w^* \in \mathcal{M}_+$ also , where $x = wh$ is the polar decomposition of x. Thus, $\varphi(x^*x) < \infty$ if and only if $\varphi(xx^*) < \infty$. If $\varphi(x^*x)$ and $\varphi(xx^*)$ are finite, then we have

$$\varphi(xx^*) = \varphi(w(x^*x)w^*) = \varphi(w^*w(x^*x)) = \varphi(x^*x).$$

Therefore, φ is a trace on M_+. Q.E.D.

Proposition 6.5.3. Let φ be a normal trace on M_+, and $\mathcal{M}$ be the definition ideal of φ. Then $\varphi(a\cdot) \in M_*$ for any $a \in \mathcal{M}$.

Proof. We may assume that $a \in \mathcal{M}_+$. By Proposition 6.5.2, $\varphi(a\cdot) = \varphi(a^{\frac{1}{2}} \cdot a^{\frac{1}{2}})$. Since φ is normal , it follows that $\varphi(a^{\frac{1}{2}} \cdot a^{\frac{1}{2}})$ is normal. Therefore, $\varphi(a\cdot) \in M_*$. Q.E.D.

Propostiion 6.5.4. Let φ be a trace on M_+, and $\mathcal{M}$ be the definition ideal of φ. Then φ is semifinite if and only if $\mathcal{M}$ is $\sigma(M, M_*)$ -dense in M.

Moreover, if φ is a semi-finite trace on M_+, and p is a projection of M, then

$$p = \sup\{q | q \text{ is a projection of } M, q \geq p, \text{and } \varphi(q) < \infty\}.$$

Proof. Let φ be semi-finite, and $\overline{\mathcal{M}}$ be the $\sigma(M, M_*)$-closure of $\mathcal{M}$. Since $\overline{\mathcal{M}}$ is a σ -closed two-sided ideal of M, there is a central projection z of M such that $\overline{\mathcal{M}} = Mz$. If $1 - z \neq 0$, then there is $0 \neq a \in M_+$ such that $a \leq (1 - z)$, and $\varphi(a) < \infty$. So $a \in \mathcal{M}_+ \subset Mz$, a contradiction. Therefore, $z = 1$, i.e., $\mathcal{M}$ is $\sigma(M, M_*)$-dense in M.

Conversely, suppose that $\mathcal{M}$ is $\sigma(M, M_*)$-dense in M. By Proposition 1.7.2, for any $0 \neq a \in M_+$ there is an increasing net $\{a_l\} \subset \mathcal{M}_+$ such that $a = \sup_l a_l$. If l is sufficiently late, then $0 \neq a_l \leq a$ and $\varphi(a_l) < \infty$. Therefore, φ is semi-finite.

Now let φ be semi-finite, and p be a projection of M. Put

$$e = \sup\{q | q \text{ is a projection of } M, q \leq p, \varphi(q) < \infty\}.$$

If $e \leq p$ and $e \neq p$, then there is a non-zero projection q_1 with $q_1 \leq p - e$ such that $\varphi(q_1) < \infty$. This contradicts the definition of e. Therefore, $p = e$.

Q.E.D.

Proposition 6.5.5. Let φ be a normal trace on M_+. Then

$$z = \sup\{p | p \text{ is a projection of } M, \text{ and } \varphi(p) = 0\}$$

is a central projection of M, and

$$Mz = \{x \in M | \varphi(x^* x) = 0\},$$

and φ is faithful on $M_+(1 - z)$.

Proof. It is clear that $u^* z u = z$ for any unitary element u of M. Thus , z is a central projection of M.

By Proposition 1.5.2, $(\sup\{p, q\} - p) \sim (q - \inf\{p, q\})$. Thus $\varphi(\sup\{p.q\}) + \varphi(\inf\{p, q\}) = \varphi(p) + \varphi(q)$. Further, we have $\varphi(\sup\{p, q\}) = 0$ if $\varphi(p) = \varphi(q) = 0$. Hence

$$\{p | p \text{ is a projection of } M, \text{ and } \varphi(p) = 0\}$$

is an increasing net of projections with respect to the inclusion relation. Since φ is normal, it follows that $\varphi(z) = 0$. Further $\varphi | M_+ z = 0$.

Now if $0 \neq a \in M_+(1-z)$, then there is a non-zero projection p of $M(1-z)$ and a positive number λ such that $a \geq \lambda p$. Further, $\varphi(a) \geq \lambda\varphi(p) > 0$. Therefore, φ is faithful on $M_+(1-z)$.

Finally, from the preceding paragraphs, it is easy to see that $Mz = \{x \in M | \varphi(x^*x) = 0.\}$ Q.E.D.

Definition 6.5.6. Let M be a VN algebra, and φ be a normal trace on M_+. The central projection $(1-z)$ in Proposition 6.5.5 is called the *support* of φ, denoted by $s(\varphi)$.

Now we discuss the characterizations and properties of semi-finite VN algebras.

Proposition 6.5.7. A VN algbebra M is semi-finite if and only if there is a faithful family of semi-finite normal traces on M_+.

Proof. Let M be semi-finite. Then there is a non-zero finite projection p of M at least. Suppose that $\{p_l\}_{l\in\wedge}$ is a maximal orthogonal family of projections of M such that $p_l \sim p, \forall l \in \wedge$. By Theorem 1.5.4, there is a central projection z of M such that

$$p_0 z \preceq pz, \quad p(1-z) \preceq p_0(1-z),$$

where $p_0 = 1 - \sum_{l\in\wedge} p_l$. Since the family $\{p_l\}$ is maximal, it follows that $pz \neq 0$.

Let $v_l, v_0 \in M$ with

$$v_l^* v_l = pz, \quad v_l v_l^* = p_l z, \forall l \in \wedge; \quad v_0 v_0^* = p_0 z, \quad v_0^* v_0 \leq pz.$$

For a normal tracial state φ on M_{pz}, define

$$\psi(a) = \sum_{l\in\wedge\cup\{0\}} \varphi(v_l^* a v_l), \quad \forall a \in (Mz)_+.$$

Since $\sum_{l\in\wedge\cup\{0\}} v_l v_l^* = z$ and φ is normal , it follows that

$$\begin{aligned}\psi(a^*a) &= \sum_{l,r\in\wedge\cup\{0\}} \varphi(v_l^* a^* v_r v_r^* a v_l) \\ &= \sum_{l,r\in\wedge\cup\{0\}} \varphi(v^*{}_r a v_l v_l^* a v_r) = \psi(aa^*)\end{aligned}$$

$\forall a \in (Mz)_+$. Thus, ψ is a normal trace on $(Mz)_+$. Suppose that $\mathcal{M}$ is the definition ideal of ψ. It is clear that $p_l z \in \mathcal{M}_+, \forall l \in \wedge \cup \{0\}$. Moreover , from $\sum_{l\in\wedge} p_l z + p_0 z = z$, $\mathcal{M}$ is σ-dense in Mz, and from Proposition 6.5.4, ψ is semi-finite. Now , if $0 \neq a \in Mz$, then there is an index $l \in \wedge \cup \{0\}$ such that $av_l \neq 0$. Noticing that M_{pz} is finite, by Theorem 6.3.10 there is a normal tracial state φ on M_{pz} such that $\varphi(v_l^* a^* a v_l) \neq 0$. Then $\psi(a^* a) \neq 0$, where ψ is definied by φ as above. Therefore, there is a faithful family of semi-finite normal traces on $(Mz)_+$. Further, since $M(1-z)$ is still semi-finite, we can continue this process. Agaim by the Zorn lemma, there is a faithful family of semi-finite normal traces on M_+.

Conversely, let $\mathcal{F}$ be a faithful family of semi-finite normal traces on M_+. If z is a non-zero purely infinite central projection of M, then there is $\psi \in \mathcal{F}$ such that $\psi(z) > 0$. By Proposition 6.5.4, we can find a projection p of M with $p \leq z$ and $0 < \psi(p) < \infty$. Thus there exists a normal tracial state on M_p. From Propositon 6.4.2, M_p is not properly infinite. This is a contradiction since z is purely infinite and $0 \neq p \leq z$. Therefore, M must be semi-finite.

Q.E.D.

Theorem 6.5.8. A VN algebra M is semi-finite if and only if there is a faithful semi-finite normal trace on M_+.

Proof. The sufficiency is clear from Proposition 6.5.7. Now suppose that M is semi-finite. By the Zorn lemma, there is a family $\{\varphi_l\}$ of semi-finite normal traces on M_+ such that $s(\varphi_l) \cdot s(\varphi_{l'}) = 0, \forall l \neq l'$, and $\sum_l s(\varphi_l) = 1$. Then $\varphi = \sum_l \varphi_l$ is faithful semi-finite normal trace on M_+. Q.E.D.

Remark. If φ is a faithful semi-finite normal trace on M_+, then we can get a faithful W^*-representation of M by the GNS construction. In fact, let $\mathcal{N}, \mathcal{M}$ be as in Proposition 6.5.2. Define an inner product on $\mathcal{N}$:

$$\langle x, y \rangle = \varphi(y^* x) = \varphi(x y^*), \forall x, y \in \mathcal{N}.$$

Denote the completion of $(\mathcal{N}, \langle , \rangle)^-$ by H_φ . For any $a \in M$, define

$$\pi_\varphi(a) x_\varphi = (ax)_\varphi, \quad \forall x \in \mathcal{N},$$

where $x \longrightarrow x_\varphi (\forall x \in \mathcal{N})$ is the embedding of $\mathcal{N}$ in H_φ. It is easy to see that $\pi_\varphi(a)$ can be uniquely extended to a bounded linear operator on H_φ , still denoted by $\pi_\varphi(a)$. Then we obtain a $*$ representation $\{\pi_\varphi, H_\varphi\}$ of M.

If $a \in M$ is such that $\pi_\varphi(a) = 0$, then $ax = 0, \forall x \in \mathcal{N}$. Since $\mathcal{M}(\subset \mathcal{N})$ is σ-dense in M, it follows that $a = 0$. Thus π_φ is faithful. Moreover, if a net $\{a_l\} \subset M, \|a_l\| \leq 1, \forall l$, and $a_l \xrightarrow{\sigma} 0$, then for any $x, y \in \mathcal{N}$,

$$\langle \pi_\varphi(a_l) x_\varphi, y_\varphi \rangle = \varphi(y^* a_l x) = \varphi(x y^* a_l) \longrightarrow 0$$

by Proposition 6.5.3. Therefore, π_φ is also a W^*-representation of M.

Proposition 6.5.9. Let M be a VN algebra.

1) If M is semi-finite, p and p' are the projections of M and M' respectively, then M_p and $M_{p'}$ are also semi-finite.

2) If $M = \sum_l \oplus M_l$, then M is semi-finite if and only if M_l is semi-finite for each l.

Proof. 1) Let φ be a faithful semi-finite normal trace on M_+. Clearly ,φ is still a faithful semi-finite normal trace on $(M_p)_+$. Thus , M_p is semi-finite from Theorem 5.3.8. Moreover, $M_{p'}$ is a $*$ isomorphic to $Mc(p')$, where $c(p')$ is the central cover of p' in M'. Since $Mc(p')$ is semi-finite, it follows that $M_{p'}$ is also semi-finite.

2) It is obvious. Q.E.D.

Theorem 6.5.10. Let M be a VN algebra. Then the following statements are equivalent:

1) M is semi-finite;

2) There is an orthogonal family $\{p_l\}$ of finite projections of M such that $\sum_l p_l = 1$;

3) There is an increasing net $\{q_l\}$ of finite projections of M such that (strong)-lim $q_l = 1$;

4) There is a finite projection p of M such that $c(p) = 1$, where $c(p)$ is the central cover of p in M.

Proof. 1) $\Rightarrow$ 2) . By the Zorn lemma, there is a maximal orthogonal family $\{p_l\}$ of finite projections of M. If $p = 1 - \sum_l p_l \neq 0$, then there is a non-zero finite projection q of M_p since M_p is still semi-finite. Clearly , $qp_l = 0, \forall l$. This is a contradiction to the maximum of $\{p_l\}$. Therefore, $\sum_l p_l = 1$.

2) $\Rightarrow$ 3) . It is immediate from Proposition 6.4.5.

3) $\Rightarrow$ 1). Let z be any non-zero central projection of M. Then there is an index l such that $zq_l \neq 0$. Since zq_l is finite, it follows that z is not purely infinite. Therefore, M is semi-finite.

1) $\Rightarrow$ 4). By the Zorn lemma, there is a maximal family $\{p_l\}$ of finite projections of M such that $c(p_l) \cdot c(p_{l'}) = 0, \forall l \neq l'$. Put $p = \sum_l p_l$. We claim that p is also finite. In fact, if r is a projection of M with $r \leq p$ and $r \sim p$, then we have that

$$rc(p_l) \sim pc(p_l) = p_l, \quad rc(p_l) \leq pc(p_l) = p_l, \forall l.$$

Since p_l is finite , it follows that $rc(p_l) = p_l, \forall l$. Further,

$$p = \sum_l p_l = r \sum_l c(p_l) = rp \sum_l c(p_l) = rp = r.$$

Thus p is finite. Now it suffices to show $c(p) = 1$. If $1 - c(p) \neq 0$, then there is a non-zero finite projection $q \leq (1 - c(p))$ since $M(1 - c(p))$ is semi-finite. Clearly $c(q) \cdot c(p_l) = 0, \forall l$. This is a contradiction since the family $\{p_l\}$ is maximal.

4) $\Rightarrow$1). Let z be a non-zero central projection of M. By Proposition 1.5.8, $zp \neq 0$, where p is a finite projection with $c(p) = 1$. Thus z contains a non-zero finite projection zp, and z is not purely infinite. Therefore, M is semi-finite.

Q.E.D.

Lemma 6.5.11. Let N be a VN algebra on Hilbert space K, and ξ be a cyclic and separating vector for N with $\|\xi\| = 1$. If $\varphi(\cdot) = \langle \cdot, \xi, \xi \rangle$ is a tracial state on N, then there exists a conjugate linear isometry j with $j^2 = 1$ such that $a \longrightarrow jaj (\forall a \in N)$ is a conjugate linear $*$ algebraic isomorphism from N onto N'.

Proof. Define $ja\xi = a^*\xi, \forall a \in N$. Then j can be uniquely extended to a conjugate linear isometry on K, still denoted by j, since φ is a tracial state and ξ is cyclic for N. Clearly, $j^2 = 1$. For any $a, b, c \in N$, we have

$$jajbc\xi = bca^*\xi = bjajc\xi.$$

Thus $jNj \subset N'$. Further, noticing that

$$\langle (jaj)^* b\xi, c\xi \rangle = \varphi(ac^*b) = \varphi(c^*ba) = \langle ja^*jb\xi, c\xi \rangle,$$

$\forall a, b, c \in N$, we get $(jaj)^* = ja^*j, \forall a \in N$. It is clear that $jaj = 0$ implies $a = 0$.

Now it suffices to prove $jNj = N'$. Let $a' \in N'$ with $0 \leq a' \leq 1$. Define

$$\psi(a) = \langle aa'\xi, \xi \rangle, \quad \forall a \in N.$$

Then $0 \leq \psi \leq \varphi$. By Theorem 1.10.3, there is $t_0 \in N$ with $0 \leq t_0 \leq 1$ such that $\psi(a) = \varphi(t_0 a t_0), \forall a \in N$. Thus, $a'\xi = t_0^2\xi = jt_0^2 j\xi$. Since ξ is also a separating vector for N', it follows that $a' = jt_0^2 j$. Therefore, $jNj = N'$.

Q.E.D.

Lemma 6.5.12. Let M be a VN algebra on a Hilbert space $H, \xi \in H$, and p, p' be the projections from H onto $\overline{M'\xi}, \overline{M\xi}$ repectively. Then p is finite if and only if p' is finite.

Proof. Clearly, $p \in M, p' \in M'$ (see Definition 1.13.1) . Suppose that p is a finine projection of M. Consider the VN algebra $L = pp'Mp'p$ on $pp'H$. Then $\xi (\in pp'H)$ is cyclic and separating for L. By Proposition 6.3.1, L is finite. Clearly, L is also σ-finite. Hence, there is faithful normal tracial state

φ on L from Proposition 6.3.15. Let $\{\pi_\varphi, H_\varphi, \xi_\varphi\}$ be the faithful cyclic W^*-representation of L generated by φ. Put $N = \pi_\varphi(L)$. Then N is $*$ isomorphic to L, and is finite. Further, by Lemma 6.5.11, N' is also finite. Since N, L admit cyclic and separating vectors ξ_φ, ξ respectively, it follows from Theorem 1.13.5 that N is spatially $*$ isomorphic to L. Therefore, L' is also finite. Now if $x' \in M'$ is such that $pp'x'p'p = 0$, then

$$0 = yp'x'p'p\xi = yp'x'p'\xi = p'x'p'y\xi, \quad \forall y \in M.$$

But $\overline{M\xi} = p'H$, so $p'x'p' = 0$. Thus , $p'x'p' \longrightarrow pp'x'p'p(\forall x' \in M')$ is a $*$ isomorphism from $p'M'p'$ onto $pp'M'p'p$. Further, $p'M'p'$ is finite since $L' = pp'M'p'p$, i.e., p' is a finite projection of M'.

Similarly, p is finite if p' is finite. Q.E.D.

Propositon 6.5.13. Let M be a semi-finite VN algebra on a Hilbert space H. Then M' is also semi-finite.

Proof. Suppose that there is a non-zero central projection z of $M \cap M'$ such that $M'z$ is purely infinite. Clearly , Mz is still semi-finite. Therefore, we may assume that M is semi-finite , and also M' is purely infinite. By Theorem 6.5.10, there is a finite projection p of M with $c(p) = 1$. Then M' is $*$ isomorphic to M'_p. So we may further assume that M is finite, and also M' purely infinite.

Pick a non-zero vector $\xi \in H$, and let p, p' be cyclic projections of M, M' determined by ξ respectively. Clearly, p is finite. Then by Lemma 6.5.12, p' is also non-zero finite. This contradicts that M' is purely infinite.

Therefore, M' is semi-finite. Q.E.D.

Proposition 6.5.14. A VN algebra M is semi-finite if and only if M is $*$ isomorphic to some VN algebra N so that N' is finite.

Proof. The sufficiency is obvious from Proposition 6.5.13. Now let M be semi-finite. Then M' is also semi-finite. By Theorem 6.5.10, there is a finite projection p' of M' with $c(p') = 1$. Pick $N = M_{p'}$. Then M is $*$ isomorphic to N, and $N' = M'_{p'}$ is finite. Q.E.D.

Proposition 6.5.15. Let M be a semi-finite VN algebra, and p be a projection of M. Then p is finite if and only if there is a faithful family $\mathcal{F}$ of semi-finite normal traces on M_+ such that $\varphi(p) < \infty, \forall \varphi \in \mathcal{F}$.

Proof. Sufficiency. Let q be a projection of M with $q \leq p$ and $q \sim p$. Then

$$\varphi(q) = \varphi(p) < \infty, \quad \forall \varphi \in \mathcal{F},$$

i.e., $\varphi(p-q)=0, \forall\varphi\in\mathcal{F}$. But $\mathcal{F}$ is faithful, so $p=q$. Thus p is finite.

Necessity. Suppose that p is finite. By the proof of Proposition 6.5.7, there is a non-zero central projection z of M and a faithful family $\mathcal{F}_z$ of semi-finite normal traces on $(Mz)_+$ such that $\varphi(pz)<\infty, \forall\varphi\in\mathcal{F}_z$. By the Zorn lemma, we can find an orthogonal family $\{z_l\}$ of central projections of M and a faithful family $\mathcal{F}_l$ of semi-finite normal traces on $(Mz_l)_+$ for each l such that

$$\sum_l z_l = 1, \quad \varphi_l(pz_l)<\infty, \forall\varphi_l\in\mathcal{F}_l \text{ and } l.$$

Since for every l, each $\varphi_l(\in\mathcal{F}_l)$ can be naturally extended to a semi-finite normal trace on M_+, so $\mathcal{F}=\cup_l\mathcal{F}_l$ satisfies the conditions. Q.E.D.

Proposition 6.5.16. Let p be a finite projection of a VN algebra M. Then the $*$ operation is strongly continuous in any bounded ball of Mp.

Proof. Replacing M by $Mc(p)$, we may assume that M is semi-finite. By Propositon 6.5.15, there is a faithful family $\mathcal{F}$ of semi-finite normal traces on M_+ such that $\varphi(p)<\infty, \forall\varphi\in\mathcal{F}$.

Now let a net $x_l \longrightarrow 0$ (strongly), and $\|x_l\|\le 1, x_lp=x_l, \forall l$. For any $\varphi\in\mathcal{F}$ and $0\le a\in\mathcal{M}_\varphi$, where $\mathcal{M}_\varphi$ is the definition ideal of φ, by Propositions 6.5.2 and 6.5.3, we have

$$\begin{aligned}\|L_a\varphi(x_lx_l^*)\| &= |\varphi(ax_lx_l^*)| = |\varphi(x_l^*ax_l)| \\ &\le \|a\|\varphi(x_l^*x_l) = \|a\|(L_p\varphi)(x_l^*x_l) \longrightarrow 0.\end{aligned}$$

Hence, it suffices to show that the set $[L_a\varphi|\varphi\in\mathcal{F}, a\in(\mathcal{M}_\varphi)_+]$ is dense in M_*. Let $x\in M$ be such that

$$\varphi(ax)=0, \quad \forall\varphi\in\mathcal{F}, a\in\mathcal{M}_\varphi.$$

Then $\varphi(x^*ax)=0$ since $x^*a\in\mathcal{M}_\varphi$ for any $a\in\mathcal{M}_\varphi$, and $\varphi\in\mathcal{F}$. Since $\varphi(\in\mathcal{F})$ is semi-finite and normal. it follows from Propositions 6.5.4 and 1.7.2 that $\varphi(x^*x)=0, \forall\varphi\in\mathcal{F}$. But $\mathcal{F}$ is faithful, so $x=0$. That comes to the conclusion. Q.E.D.

As preliminaries of next section, we study the semi-finite and properly infinite VN algebras on a separable Hilbert space.

Lemma 6.5.17. Let H be a separable Hilbert space, and M be a semi-finite and properly infinite VN algebra on H. Then there is an orthogonal infinite sequence $\{p_n\}$ of finite projections of M with $p_n\sim p_m, \forall n,m$, such that $\sum_n p_n = 1$.

Proof. Let q be any non-zero finite projection of M, and $\{q_l\}_{l\in\wedge}$ be a maximal orthogonal family of projections of M with $q_l \sim q, \forall l \in \wedge$. Put

$$p = c(q) - \sum_{l\in\wedge} q_l.$$

By Theorem 1.5.4, there is a central projection z such that

$$pz \preceq qz, \quad q(1-z) \preceq p(1-z).$$

Clearly, $qz \neq 0$ since the family $\{q_l\}$ is maximal . Further, $z_1 = c(q)z \neq 0$, and

$$z_1 = \sum_{l\in\wedge} q_l z_1 + pz_1, \quad pz_1 \preceq qz_1.$$

If $^{\#}\wedge$ is finite, then by Proposition 6.4.5 z_1 is a non-zero finite central projection. It is impossible since M is properly infinite. Thus $^{\#}\wedge$ is infinite. Indeed $^{\#}\wedge$ is countably infinite since H is separable. Then $z_1 \sim \sum_{l\in\wedge} q_l z_1$, i.e., there is $v \in M$ such that

$$vv^* = z_1, \quad v^*v = \sum_{l\in\wedge} q_l z_1.$$

For each $l \in \wedge$, let $p_l = vq_l z_1 v^*$. Then $p_l p_{l'} = 0, \forall l \neq l'$, and

$$p_l \sim (vq_l z_1)^* \cdot (vq_l z_1) = q_l z_1, \forall l \in \wedge.$$

Clearly, $\{p_l\}_{l\in\wedge}$ is an orthogonal infinite sequence of finite projections of M, $p_l \sim p_{l'}, \forall l, l'$, and $\sum_{l\in\wedge} p_l = z_1 \neq 0$. Further, by the Zorn lemma we can get the conclusion. Q.E.D.

Proposition 6.5.18. Let M, N be two VN algebras on a separable Hilbert space H, and M', N' be semi-finite and properly infinite. If Φ is a $*$ isomorphism from M onto N, then Φ is also spatial.

Proof. By Proposition 1.12.5, we may assume that there is a VN algebra L on H, and projections p', q' of L' with $c(p') = c(q') = 1$ such that $M = L_{p'}, N = L_{q'}$, and $\Phi(ap') = aq', \forall a \in L$.

By Lemma 6.5.17, there are two orthogonal infinite sequences $\{p_i'\}, \{q_i'\}$ of finite projections of L' such that

$$p' = \sum_i p_i', \quad q' = \sum_i q_i',$$
$$p_i' \sim p_j', \quad q_i' \sim q_j', \quad \forall i, j.$$

Write $\{1, 2, \cdots\} = \cup_n \wedge_n$ such that $\Lambda_n \cap \Lambda_m = \emptyset, \forall n \neq m$, and each $\wedge_n$ is countably infinite.

Fix n. Then there is a central projection z of M such that

$$zq'_n \preceq z \sum_{i\in\wedge_n} p'_i, \quad (1-z)\sum_{i\in\wedge_n} p'_i \preceq (1-z)q'_n.$$

Fix $s \in \wedge_n$, and write $\wedge'_n = \wedge_n \backslash \{s\}$. Clearly , $\sum_{i\in\wedge_n} p'_i \sim \sum_{i\in\wedge'_n} p'_i$. Then

$$\begin{aligned}(1-z)\sum_{i\in\wedge_n} p'_i \sim \ & (1-z)\sum_{i\in\wedge'_n} p'_i \\ & \leq (1-z)\sum_{i\in\wedge_n} p'_i \preceq (1-z)q'_n.\end{aligned}$$

But q'_n is finite, so $(1-z)\sum_{i\in\wedge'_n} p'_i = (1-z)\sum_{i\in\wedge_n} p'_i$, i.e.

$$(1-z)p'_s = 0.$$

Moreover , $c(p'_s) = c(p') = 1$. Thus $z = 1$ and

$$q'_n \preceq \sum_{i\in\wedge_n} p'_i.$$

Since n is arbitrary, it follows that $q' \preceq p'$. Similarly, $p' \preceq q'$. Therefore, $p' \sim q'$, and Φ is spatial. Q.E.D.

References. [28], [144], [150].

6.6. Purely infinite Von Neumann algebras

Proposition 6.6.1. Let M be a VN algebra.

1) If $M = \sum_l \oplus M_l$, then M is purely infinite if and only if M_l is purely infinite, $\forall l$.

2) If M is purely infinite, then so is M'.

3) Let M be purely infinite, and p, p' be projections of M, M' respectively. Then M_p and $M_{p'}$ are also purely infinite.

Proof. 1) The necessity is obvious. Now let $M_l = Mz_l$ be purely infinite, $\forall l$. If p is a finite projection of M, then $pz_l = 0, \forall l$, and $p = 0$. Therefore, M is also purely infinite.

2) It is immediate from Proposition 6.5.13.

3) By 2) , it suffices to show that $M_{p'}$ is purely infinite. But $M_{p'}$ is $*$ isomorphic to $Mc(p')$, and $Mc(p')$ is purely infinite obviously, so $M_{p'}$ is purely infinite. Q.E.D.

Proposition 6.6.2. A VN algebra M is purely infinite if and only if there is no non-zero semi-finite normal trace on M_+.

Proof. If there is a non-zero semi-finite normal trace φ on M_+, then its support $s(\varphi)$ is a non-zero central projection of M. By Theorem 6.5.8, $Ms(\varphi)$ is semi-finite, i.e. M is not purely infinite. Conversely, if there is a non-zero central projection z of M such that Mz is semi-finite, then we have a non-zero semi-finite normal trace on $(Mz)_+$. Therefore, there is a non-zero semi-finite normal trace on M_+. Q.E.D.

Proposition 6.6.3. A VN algebra M is purely infinte if and only if the $*$ operation is not strongly continuous in any bounded ball of Mp for any non-zero projection p of M.

Proof. Let M be purely infinite, and p be a non-zero projection of M. Since p is infinite, there is $v \in M$ such that $v^*v = p, vv^* \le p$, and $vv^* \ne p$. By Lemma 6.3.11, we can find an orthogonal sequence $\{e_n\}$ of non-zero projections of M such that $e_n \sim e_m, \forall n, m, e_n \le p, \forall n$, and $e_n \longrightarrow 0$ (strongly) . Suppose that $w_n^* w_n = e_n, w_n w_n^* = e_1, \forall n$. Clearly , $w_n \longrightarrow 0$ (strongly) , $\|w_n\| \le 1, w_n p = w_n, \forall n$. However , $\{w_n^*\}$ does not converge to 0 strongly, so the $*$ operation is not strongly continuous in any bounded ball of Mp.

Conversely, if M is not purely infinite , then M contains a non-zero finite projection p. By Proposition 6.5.16, the $*$ operation is strongly continuous in any bounded ball of Mp. Q.E.D.

Proposition 6.6.4, Let M be a σ-finite and purely infinite VN algebra, and p, q be two projections of M with $c(p) = c(q)$. Then $p \sim q$.

Proof. From the Zorn lemma, we can take a maximal orthogonal family $\{z_l\}_{l\in\Lambda}$of central projections of M such that $qz_l \precsim pz_l, \forall l \in \Lambda$. Let

$$z = \sum_{l\in\Lambda} z_l, \quad p' = p(1-z), \quad q' = q(1-z).$$

By Theorem 1.5.4 and the maximum of the family $\{z_l\}_{l\in\Lambda}$,we can see that $p' \precsim q'$ (relative to $M(1-z)$).

Suppose that $p' \ne 0$. Pick a maximal orthogonal family $\{q'_s\}_{s\in\mathbb{I}}$ such that $q'_s \sim p', q'_s \le q', \forall s \in \mathbb{I}$. For Theorem 1.5.4, there is a central projection z' of M with $z' \le (1-z)$ such that

$$(q' - \sum_{s\in\mathbb{I}} q'_s)z' \precsim p'z',$$
$$p'(1-z') \precsim (q' - \sum_{s\in\mathbb{I}} q'_s)(1-z').$$

Clearly , $p'z' \neq 0$ since the family $\{q'_s\}_{s\in\Pi}$ is maximal . Since M_p is purely infinite, from Theorem 6.4.4 there is an orthogonal infinite sequnce $\{e_n\}$ of projections of M_p such that

$$\sum_{n=1}^{\infty} e_n = p, \quad e_n \sim p, \forall n.$$

Then

$$e_n z' \sim pz' = p'z' \sim q'_s z', \quad \forall n, s.$$

Since M is σ-finite, it follows that the index set Π is countable , and

$$\sum_{n=2}^{\infty} e_n z' \sim \sum_{s\in\Pi} q'_s z'.$$

In addition,

$$e_1 z' \sim p'z' \succeq (q' - \sum_{s\in\Pi} q'_s) z',$$

So we have $pz' \succeq q'z' = qz'$. Moreover, $0 \neq z' \leq 1 - z$. This is a contradiction since the family $\{z_l\}_{l\in\Lambda}$ is maximal. Therefore p' must be zero , i.e., $p \leq z$.

Now from $z \geq c(p) = c(q) \geq q$, we have

$$p = pz = \sum_{l\in\Lambda} pz_l \succeq \sum_{l\in\Lambda} qz_l = q.$$

Similarly, we can prove that $p \preceq q$. Therefore, $p \sim q$. Q.E.D.

Proposition 6.6.5. Let M be a σ-finite and purely infinite VN algebra, and a be a non-zero element of M. Then $K(a) \neq \{0\}$, where $K(a)$ is defined in Theorem 6.2.7.

Proof. By Proposition 6.2.8. and $K(a^*) = K(a)^*$, we may assume that $a^* = a$. Also we may assume that $\|a\| \leq 1$ and $a_+ \neq 0$ (otherwise, replace a by $-a$) . Then we can find a non-zero projection p of M and a positive integer n such that

$$q \geq \frac{1}{n} p - (1 - p).$$

If p contains a non-zero central projection z, then $a \geq \frac{n+1}{n} z - 1$. Further , for any $b \in K(a)$, we have $b \geq \frac{n+1}{n} z - 1$. Thus , $K(a) \neq \{0\}$.

Now suppose that p contains no non-zero central projection. Replacing M by $Mc(p)$, we may assume that $c(p) = 1$. Since $p \geq 1 - c(1-p)$, it follows that $c(1-p) = 1$. By Proposition 6.6.4, we have

$$p \sim (1 - p) \sim 1.$$

From Theorem 6.4.4, there are pairwise orthogonal projections $\{e_1, \cdots, e_{n+1}\}$ such that

$$p = \sum_{i=1}^{n+1} e_i, \quad \text{and} \quad e_i \sim p, \forall i.$$

Pick $v_i \in M$ such that

$$v_i^* v_i = e_i, \quad v_i v_i^* = e_{i+1}, \quad 0 \leq i \leq n,$$

where $e_0 = 1 - p$. Let $v_{n+1} = v_0^* \cdots v_n^*$. Then $u = v_1 + v_1 + \cdots + v_{n+1}$ is unitary, and

$$u e_i u^{-1} = e_{i+1}, \quad 0 \leq i \leq n, \quad u e_{n+1} u^{-1} = e_0.$$

Let

$$b = (n+2)^{-1} \sum_{j=0}^{n+1} u^j a u^{-j}.$$

Since

$$\sum_{j=0}^{n+1} u^j e_i u^{-j} = 1, \quad 0 \leq i \leq n+1,$$

it follows that

$$(n+2)b \geq \sum_{j=0}^{n+1} u^j \left(\frac{1}{n}(e_1 + \cdots + e_{n+1}) - e_0\right) u^{-j} = \frac{1}{n},$$

and $c \geq \frac{1}{n(n+2)}, \forall c \in K(b)$. Clearly , $K(b) \subset K(a)$. Therefore, $K(a) \neq \{0\}$.

Q.E.D.

Proposition 6.6.6. Let M be a purely infinite VN algebra on a separable Hilbert space H. Then M admits a cyclic and separating vector.

Proof. Let ξ be a non-zero vector of H, and p be the cyclic projection of M determined by ξ, and $\{p_l\}$ be a maximal orthogonal family of projections of M such that $p_l \sim p, \forall l$. From Theorem 1.5.4, there is a central projection z of M such that

$$(1 - \sum_l p_l) z \preceq pz,$$
$$p(1-z) \preceq (1 - \sum_l p_l)(1-z).$$

Clearly, $z \neq 0, zc(p) = c(pz) \geq \sum_l p_l z$. On the other hand.

$$c(p)z = c(pz) \geq (1 - \sum_l p_l) z.$$

Thus $c(p)z \geq z$, and $c(pz) = z$. By Proposition 6.4.4, we have $pz \sim z$. Further , there is $\eta \in H$ such that z is the cyclic projection of M determined by η.

By the Zorn lemma and the separability of H, there is an orthogonal sequence $\{z_n\}$ of central projections of M such that $\sum_n z_n = 1$ and $z_n H = \overline{M'\eta_n}$, where $\eta_n \in H, \forall n$. We may assume that $\|\eta_n\| \leq 2^{-n}, \forall n$. Then $\eta = \sum_n \eta_n$ is a cyclic vector for M'.

Similarly, there is a cyclic vector for M since M' is also purely infinite. Now from Proposition 1.13.4, M admits a cyclic and separating vector. Q.E.D.

The following proposition is a generalization of Proposition 6.5.18.

Proposition 6.6.7. Let M, N be two VN algebras on a separable Hilbert space H , and M', N' be properly infinite. Then each $*$ isomorphism Φ from M onto N is spatial.

Proof. Let z be the maximal central projection of M such that Mz is purely infinite. Then $N\Phi(z)$ is also purely infinite, and $M'(1-z), N'(1-\Phi(z))$ are semifinite and properly infinite.

From Proposition 6.6.6, both Mz and $N\Phi(z)$ admit a cyclic and separating vector. Thus $\Phi : Mz \longrightarrow N\Phi(z)$ is spatial by Theorem 1.13.5.

Moreover, $\Phi : M(1-z) \longrightarrow N(1-\Phi(z))$ is also spatial from Proposition 6.5.18. Therefore, the $*$ isomorphism Φ from M onto N is spatial. Q.E.D.

References. [18], [28], [144].

6.7. Discrete (type (I)) Von Neumann algebras

Theorem 6.7.1. Let M be a VN algebra on a Hilbert space H. Then the following statements are equivalent:

1) M is discrete (type (I));
2) M' is discrete (tyep (I));
3) M is $*$ isomorphic to some VN algebra N such that N' is abelian;
4) there is an abelian projection p of M such that $c(p) = 1$;
5) any non-zero projection of M contains a non-zero abelian projection of M.

Proof. 1) $\Rightarrow$ 4). Pick a maximal family $\{p_l\}$ of non-zero abelian projections of M such that $c(p_l) \cdot c(p_{l'}) = 0, \forall l \neq l'$, and put $p = \sum_l p_l$. If $c(p) \neq 1$, then $(1 - c(p))$ contains a non-zero abelian projection since M is discrete. This contradicts the maximum of the family $\{p_l\}$. Thus $c(p) = 1$. Moreover, by Proposition 1.5.9, $p_l M p_{l'} = \{0\}, \forall l \neq l'$. Therefore, p is also abelian.

2) $\Rightarrow$ 3). Suppose that M' is discrete. From the preceding paragraph, M' admits an abelian projection p' with $c(p') = 1$. Now let $N = M_{p'}$. Then M is $*$ isomorphic to N, and $N' = M'_{p'}$ is abelian.

3) $\Rightarrow$ 5). Suppose that Φ is a $*$ isomorphism from M onto N, where N is a VN algebra on a Hilbert space K, and N' is abelian. Let p be any non-zero projection of N, ξ be a non-zero vector of pK, and q be the cyclic projection of N determined by ξ. Since the VN algebra N'_q on $qK = \overline{N'\xi}$ is abelian, and admits a cyclic vector ξ, it follows from Proposition 5.3.15 that $N'_q = (N'_q)' = N_q$, and q is abelian. Clearly, $0 \neq q \leq p$. Therefore, any non-zero projection of M contains a non-zero abelian projecton of M.

5) $\Rightarrow$ 1) . It is obvious by Definition 6.1.7.

4) $\Rightarrow$ 2). Suppose that p is an abelian projection of M with $c(p) = 1$. Let $L = M'_p$. Then M' is $*$ isomorphic to L, and $L' = M_p$ is abelian. Now by 3) $\Rightarrow$ 5) $\Rightarrow$ 1), M' is discrete. Q.E.D.

Remark. For any VN algebra M, the VN algebra N generated by $M \cup M'$ is discrete. In fact, $N' = M \cap M'$ is abelian.

Proposition 6.7.2. Let M be a VN algebra.

1) If $M = \sum_l \oplus M_l$, then M is discrete if and only if M_l is discrete, $\forall l$.

2) Let M be discrete, and p, p' be projections of M, M' respectively. Then $M_p, M_{p'}$ are also discrete.

Proof. 1) It is abvious from Definition 6.1.7.

2) Since $M_{p'}$ is $*$ isomorphic to Mz, where $z = c(p')$, it follows that $M_{p'}$ is discrete. Moreover, M_p is also discrete by $(M_p)' = M'_p$. Q.E.D.

Proposition 6.7.3. Let M be a discrete (type (I)) factor. Then M is $*$ isomorphic to $B(K)$, where K is some Hilbert space.

Proof. From Theorem 6.7.1, M can be $*$ isomorphic to a VN algebra N on some Hilbert space K such that N' is abelian. Clearly, N is a factor, and N' is an abelian facor. Therefore, $N' = \mathbb{C}$, and $N = B(K)$. Q.E.D.

Remark. From this proposition, we can get the results on finite dimensional C^* -algebras (see Section 2.13), too.

Lemma 6.7.4. Let p, q be two projections of a VN algebra M, and p be abelian, $p \leq c(q)$. Then $p \precsim q$.

Proof. From Theorem 1.5.4, there is a central projection z such that

$$qz \preceq pz, \quad p(1-z) \preceq q(1-z).$$

Let $qz \sim p_1 \leq pz$. Then the central cover of p_1 in M_{pz} is $c(p_1)pz$ by Proposition 1.5.8. Since M_{pz} is abelian, it follows that

$$p_1 = c(p_1)pz = c(qz)pz = c(q)pz = pz.$$

Thus $qz \sim pz$, and $p \preceq q$. Q.E.D.

Lemma 6.7.5. Let M be a VN algebra on a Hilbert space H, and $\{p_l | l \in \wedge\}, \{q_r | r \in \varPi\}$ be two orthogonal families of abelian projections of M such that $p_l \sim p_{l'}, \forall l, l' \in \wedge, q_r \sim q_{r'}, \forall r, r' \in \varPi$, and $\sum_{l \in \wedge} p_l = \sum_{r \in \varPi} q_r = 1$. Then $^{\#}\wedge = {}^{\#}\varPi$.

Proof. Clearly, $c(p_l) = c(q_r) = 1, \forall l, r$. So by Lemma 6.7.4, we have $p_l \sim q_r, \forall l \in \wedge, r \in \varPi$.

If $^{\#}\wedge < \infty$, then M is finite by Proposition 6.4.5, and $^{\#}\varPi$ must also be finite. Thus $^{\#}\wedge$ and $^{\#}\varPi$ are finite or infinite simultaneously.

Consider the case that $^{\#}\wedge$ and $^{\#}\varPi$ are finite. We may assume that $^{\#}\wedge \leq {}^{\#}\varPi$. Then

$$1 = \sum_{l \in \wedge} p_l \sim \sum_{r \in \varPi'} q_r \leq \sum_{l \in \varPi} q_r = 1,$$

where $\varPi' \subset \varPi$ and $^{\#}\varPi' = {}^{\#}\wedge$. Since M is finite, it must be $\varPi' = \varPi$, and $^{\#}\wedge = {}^{\#}\varPi$.

Now let both $\wedge$ and $\varPi$ be infinite index sets. Fix $p \in \{p_l\}$. The abelian VN algebra M_p is finite obviously. So by Propositon 6.3.16 and 1.3.8, there is a non-zero central projection z of M such that M_{pz} is σ -finite. Considering $Mz, \{p_l z\}, \{q_r z\}$, we may assume that $z = 1$. For any $l \in \wedge, M_{p_l}$ is a σ-finite VN algebra on $p_l H$. By Proposition 1.14.2, there is a countable subset $\mathcal{M}_l$ of $p_l H$ such that $[M'\mathcal{M}_l]$ is dense in $p_l H$. Let $\varPi_l = \{r \in \varPi | q_r \mathcal{M}_l \neq \{0\}\}$. Since $\{q_r | r \in \varPi\}$ is pairwise orthogonal, $\varPi_l$ must be a countable subset of $\varPi, \forall l \in \wedge$. Moreover, if there exists $r \in \varPi \backslash \cup_{l \in \wedge} \varPi_l$, then $q_r \mathcal{M}_l = \{0\}, \forall l \in \wedge$. Further, $q_r M' \mathcal{M}_l = \{0\}$, i.e. $q_r p_l = 0, \forall l \in \wedge$. This is a contradiction since $\sum_l p_l = 1$ and $q_r \neq 0$. Therefore, $\varPi = \cup_{l \in \wedge} \varPi_l$, and $^{\#}\varPi \leq {}^{\#}\wedge$. Similarly, $^{\#}\wedge \leq {}^{\#}\varPi$. So $^{\#}\wedge = {}^{\#}\varPi$. Q.E.D.

Definition 6.7.6. A VN algebra M is called type (I_n) or n–*homogeneous* , where n is a finite or infinite cardinal number, if there is an orthogonal family $\{p_l | l \in \wedge\}$ of abelian projections of M such that $p_l \sim p_{l'}, \forall l, l' \in \wedge$ and $\sum_{l \in \wedge} p_l = 1$, and $^{\#}\wedge = n$.

By Lemma 6.7.5, the definition of type (I_n) is independent of the choice of the family $\{p_l | l \in \wedge\}$.

Proposition 6.7.7. Let M be a VN algebra.

1) If M is type (I_n) , then M is type (I).

2) M is type (I_n) if and only if M is spatially $*$ isomorphic to $N\overline{\otimes}B(H_n)$, where N is an abelian VN algebra, and H_n is a n–dimensional Hilbert space. Consequently, A factor of type (I_n) must be $*$ isomorphic to $B(H_n)$.

Proof. 1) Let $\{p_l\}$ be as in Definition 6.7.6. Then $c(p_l) = 1, \forall l$. By Theorem 6.7.1, M is discrete.

2) The necessity is immediate from Definition 6.7.6 and Theorem 1.5.6.

Conversely, suppose that $\{e_l|l \in \wedge\}$ is a normalized orthogonal basis of H_n , where $^{\#}\wedge = n$. Let p_l be the projection from H_n onto $[e_l]$. Then $\{p_l|l \in \wedge\}$ is an orthogonal family of abelian projections of $B(H_n)$ with $p_l \sim p_{l'}, \forall l, l'$ and $\sum\limits_l p_l = 1$. Therefore, $N\overline{\otimes}B(H_n)$ is type (I_n) since N is abelian. Q.E.D.

Lemma 6.7.8. Let M be a VN algebra, and $\{z_l\}$ be an orthogonal family of n–homogeneous central projections of M where n is a cardinal number. Then $z = \sum\limits_l z_l$ is also n–homogeneous.

Proof. Let $\{p_\alpha^{(l)}|\alpha \in \mathit{II}\}$ be an orthogonal family of abelian projections of Mz_l with $p_\alpha^{(l)} \sim p_{\alpha'}^{(l)}, \forall \alpha, \alpha'$, and $\sum\limits_\alpha p_\alpha^{(l)} = z_l, \forall l$, where $^{\#}\mathit{II} = n$. Put $p_\alpha = \sum\limits_l p_\alpha^{(l)}$. Then $\{p_\alpha|\alpha \in \mathit{II}\}$ is an orthogonal family of abelian projections of M such that $p_\alpha \sim p_{\alpha'}, \forall \alpha, \alpha'$, and $\sum\limits_\alpha p_\alpha = z$. Therefore, Mz is type (I_n).

Q.E.D.

Lemma 6.7.9. Let M be a VN algebra, and z_i be a n_i–homogeneous central projection of $M, i = 1, 2$. If $n_1 \neq n_2$, then $z_1z_2 = 0$.

Proof. Let $\{p_l^{(i)}|l \in \wedge_i\}$ be an orthogonal family of abelian projections of Mz_i such that $p_l^{(i)} \sim p_{l'}^{(i)}, \forall l, l'$, and $\sum\limits_{l\in\wedge_i} p_l^{(i)} = z_i$, where $^{\#}\wedge_i = n_i, i = 1, 2$. Then $\{p_l^{(1)}z_2|l \in \wedge_1\}$ and $\{p_l^{(2)}z_1|l \in \wedge_2\}$ are also two orthogonal family of pairwise equivalent abelian projections of Mz_1z_2 with $\sum\limits_{l\in\wedge_1} p_l^{(1)}z_2 = \sum\limits_{l\in\wedge_2} p_l^{(2)}z_1 = z_1z_2$.

If $z_1z_2 \neq 0$, then $^{\#}\wedge_1 = {}^{\#}\wedge_2$ by Lemma 6.7.5. This contradicts $n_1 \neq n_2$. Therefore, $z_1z_2 = 0$. Q.E.D.

Theorem 6.7.10. Let M be a type (I) VN algebra. Then there is unique decomposition $M = \sum\limits_{n\in E} \oplus M_n$, where E is some set of different candinal numbers, and M_n is type $(I_n), \forall n \in E$.

Proof. Pick a non-zero abelian projection p of M, and let $\{p_l\}$ be a maximal orthogonal family of projections of M such that $p_l \sim p, \forall l$. By Theorem 1.5.4, there is a central projection z such that

$$(1-q)z \precsim pz, \quad p(1-z) \precsim (1-q)(1-z),$$

where $q = \sum_l p_l$. By the maximum of $\{p_l\}, z$ is not zero. If $(1-q)z = 0$, then $z = \sum_l p_l z$ is homogeneous. Now let $(1-q)z \neq 0$. Then $z_1 = zc(1-q)$ is a non-zero central projection. Suppose that

$$(1-q)z \sim q_1 \leq pz.$$

Since pz is abelian , it follows from Proposition 1.5.8 that $q_1 = c(q_1)pz = c((1-q)z)pz = pz_1$. Clearly, $(1-q)z_1 = (1-q)z$. Thus

$$(1-q)z_1 \sim q_1 = pz_1 \sim p_l z_1, \forall l,$$

and $z_1 = \sum_l p_l z_1 + (1-q)z_1$ is homogeneous.

From above discussion, we get a non-zero homogeneous central projection of M. Further, by the Zorn lemma there is an orthogonal family $\{z_r\}$ of homogeneous central projections of M such that $\sum_r z_r = 1$. Finally, by Lemma 6.7.8 and 6.7.9, the conclusion can be obtained . Q.E.D.

References. [82], [86], [88].

6.8. Continuous Von Neumann algebras and type (II) Von Neumann algebras

Proposition 6.8.1. A VN algebra M is continuous if and only if there is no non-zero central projection z such that Mz is discrete. In consequence, a purely infinite VN algebra must be continuous.

Proof. If M is not continuous, then M contains a non-zero abelian projection p. Let $z = c(p)$. By Theorem 6.7.1, Mz is discrete. Conversely, suppose that z is a non-zero central projection such that Mz is discrete. Then Mz contains a non-zero abelian projection p. Clearly, p is also an abelian projection of M. Thus, M is not continuous. Q.E.D.

Proposition 6.8.2. Let M be a VN algebra.

1) If $M = \sum_l \oplus M_l$, then M is continuous (or type (II)) if and only if M_l is continuous (or type (II)) , $\forall l$.

2) Suppose that M is continuous (or type (II)), then so is M'.

3) Let M be continuous (or type (II)) and p, p' be projections of M, M' respectively. Then M_p and $M_{p'}$ are continuous (or type (II)).

Proof. 1) It is obvious by Definition 6.1.7.

2) Let M be continuous. If there is a non-zero central projection z of M' such that $M'z$ is discrete. Then by Theorem 6.7.1, $(M'z)' = Mz$ is also discrete, a contradiction. Therefore, M' is also continuous. Moreover, by Proposition 6.5.13, M' is type (II) if M is type (II).

3) From $(M_p)' = M'_p$ and the conclusion 2), it suffices to show that $M_{p'}$ is continuous. Since $M_{p'}$ is $*$ isomorphic to $Mc(p')$, the conclusion is obvious.

Q.E.D.

Theorem 6.8.3. A VN algebra M is continuous if and only if every projection p of M can be writen as $p = p_1 + p_2$, where p_1, p_2 are projections of M with $p_1 p_2 = 0$, and $p_1 \sim p_2$.

Proof. Sufficiency. If p is an abelian projection of M, then from the assumption we can write $p = p_1 + p_2$, where p_1, p_2 are projections of M with $p_1 p_2 = 0, p_1 \sim p_2$. Clearly , $c(p_1) = c(p)$. Since $p_1 \leq p$ and p is abelian, it follows from Proposition 1.5.8 that $p_1 = c(p_1)p = p$. Therefore, $p = 0$, and M is continuous.

Now let M be continuous, and p be any non-zero projection of M. Then M_p is not abelian, and there is a non-zero projection q of M_p such that $q \notin M_p \cap M'_p$. By Theorem 1.5.4, we can find a central projection z of M such that

$$qz \preceq (p-q)z, \quad (p-q)(1-z) \preceq q(1-z).$$

If $qz = (p-q)(1-z) = 0$, then $q = p(1-z) \in M_p \cap M'_p$, a contradiction. Thus we have either $qz \neq 0$ or $(p-q)(1-z) \neq 0$. If $qz \neq 0$, let $qz = r_1 \sim r_2 \leq (p-q)z$, then $r_1 r_2 = 0$, and $(r_1 + r_2) \leq p$. If $(p-q)(1-z) \neq 0$, let $(p-q)(1-z) = r_1 \sim r_2 \leq q(1-z)$, then $r_1 r_2 = 0$, and $r_1 + r_2 \leq p$. Therefore, p contains two non-zero projections r_1, r_2 such that $r_1 r_2 = 0$ and $r_1 \sim r_2$. Continue this process for $(p - (r_1 + r_2)) \cdot \cdots$. Then by the Zorn lemma, we can get a decomposition $p = p_1 + p_2$ with $p_1 p_2 = 0$ and $p_1 \sim p_2$.

Q.E.D.

Theorem 6.8.4. A VN algebra M is type (II) if and only if there is a decreasing sequence $\{p_n\}$ of projections of M such that p_1 is a finite projection with $c(p_1) = 1$, and $(p_n - p_{n+1}) \sim p_{n+1}, \forall n$.

Proof. Let M be type (II) . From Theorem 6.5.10, there is a finite projection

p_1 of M with $c(p_1) = 1$. By Theorem 6.8.3., we can write

$$p_1 = p_2 + q_2, \quad \text{where} \quad p_2 q_2 = 0, \quad \text{and} \quad p_2 \sim q_2,$$

$$\cdots\cdots$$

$$p_n = p_{n+1} + q_{n+1}, \quad \text{where} \quad p_{n+1} q_{n+1} = 0, \quad \text{and} \quad p_{n+1} \sim q_{n+1},$$

$$\cdots\cdots.$$

Then $\{p_n\}$ satisfies the conditions.

Conversely, suppose that $\{p_n\}$ is a decreasing sequence of projections of M, where p_1 is a finite projection with $c(p_1) = 1$, and $(p_n - p_{n+1}) \sim p_{n+1}, \forall n$. From Theorem 6.5.10, M is semi-finite. If M_{p_1} is continuous, then $(M_{p_1})' = M'_{p_1}$ is also continuous. Further, M' is continuous (since M' is $*$ isomorphic to M'_{p_1}) and M is type (II) . So it suffices to show that M_{p_1} is continuous, and we may assume that $p_1 = 1$, i.e., M is finite. By Propositions 6.3.16 and 6.8.2, we may also assume that M is σ-finite. Then there exists a faithful normal tracial state φ on M. If p is an abelian projection of M, then by Theorem 1.5.4 there is a sequence $\{z_n\}$ of central projections of M such that

$$p_n z_n \sim q_n \leq p z_n, \quad p(1 - z_n) \preceq p_n(1 - z_n), \forall n.$$

Since p is abelian, it follows that $q_n = c(q_n) p z_n = c(p_n) p z_n$. Noticing $p_n \sim (p_{n-1} - p_n)$, we have $c(p_n) \geq p_{n-1}$. Thus , $c(p_n) = 1$, and $q_n = p z_n$ i.e., $p_n z_n \sim p z_n$, and $p \preceq p_n, \forall n$. On the other hand, from

$$p_n = p_{n+1} + (p_n - p_{n+1}), \quad p_{n+1} \sim (p_n - p_{n+1}),$$

it follows that $\varphi(p_n) = 2\varphi(p_{n+1})$. Noticing that $\varphi(p_1) = \varphi(1) = 1$, we get that $\varphi(p_n) = 2^{-n+1}, \forall n$. In addition, $\varphi(p) \leq \varphi(p_n) = 2^{-n+1}, \forall n$, thus $\varphi(p) = 0$. Further $p = 0$ since φ is faithful. Therefore, M contains no non-zero abelian projection, and M is continuous.

Q.E.D.

References. [18], [82].

6.9. The types of tensor products of Von Neumann algebras

Let M_i be a VN algebra on a Hilbert space $H_i, i = 1, 2$. Their tensor product $M_1 \overline{\otimes} M_2$ is a VN algebra on $H_1 \otimes H_2$. In this section, we consider the relations between the types of M_1, M_2 and the type of $M_1 \overline{\otimes} M_2$.

Proposition 6.9.1. The VN algebra $M_1\overline{\otimes}M_2$ is finite if and only if both VN algebras M_1 and M_2 are finite.

Proof. Let $M_1\overline{\otimes}M_2$ be finite. Since M_1 is $*$ isomorphism to $M_1\otimes 1_2$, it must be that M_1 is finite. Similarly, M_2 is finite too.

Now suppose that both M_1 and M_2 are finite. By Proposition 6.3.16, we may also assume that both M_1 and M_2 are σ-finite. Then there are faithful normal tracial states φ_1, φ_2 on M_1, M_2 respectively. We can write

$$\varphi_i(\cdot) = \sum_n \langle \cdot \xi_n^{(i)}, \xi_n^{(i)}\rangle, \forall \cdot \in M_i,$$

where $\{\xi_n^{(i)}\} \subset H_i$ and $\sum_n \|\xi_n^{(i)}\|^2 < \infty, i = 1,2$. Consider $\varphi_1 \otimes \varphi_2(\cdot) = \sum_{n,m} \langle \cdot \xi_n^{(1)} \otimes \xi_m^{(2)}, \xi_n^{(1)} \otimes \xi_m^{(2)}\rangle$. Clearly, $\varphi_1\otimes\varphi_2$ is a normal tracial state on $M_1\overline{\otimes}M_2$. Since φ_i is faithful on M_i, $\{\xi_n^{(i)}\}$ is a cyclic sequence of vectors for $M_i', i = 1,2$. Thus $\{\xi_n^{(1)}\otimes\xi_m^{(2)}\}_{n,m}$ is cyclic for $(M_1\overline{\otimes}M_2)'$, and $\varphi_1\otimes\varphi_2$ is also faithful on $M_1\overline{\otimes}M_2$. Therefore, $M_1\overline{\otimes}M_2$ is finite. Q.E.D.

Proposition 6.9.2. $M_1\overline{\otimes}M_2$ is properly infinite if and only if either M_1 or M_2 is properly infinite.

Proof. The necessity is immediate from Proposition 6.9.1. Now assume that M_1 is properly infinite. If $M_1\overline{\otimes}M_2$ is not properly infinite, then there is a normal tracial state φ on $M_1\overline{\otimes}M_2$. Clearly, $(\varphi|M_1\otimes 1_2)$ is also a normal tracial state on M_1. It is impossible since M_1 is properly infinite. Therefore, it must be that $M_1\overline{\otimes}M_2$ is properly infinite. Q.E.D.

Proposition 6.9.3. If both M_1 and M_2 are semi-finite, then $M_1\overline{\otimes}M_2$ is also semi-finite.

Proof. By Proposition 6.5.14, we may assume that both M_1' and M_2' are finite. Then $(M_1\overline{\otimes}M_2)' = M_1'\overline{\otimes}M_2'$ is finite , and $M_1\overline{\otimes}M_2$ is semi-finite.

Q.E.D.

Lemma 6.9.4. Let N be a VN algebra, φ be a semi-finite normal trace on N_+, and $s(\varphi)$ be the support of φ. If $b \in Ns(\varphi)$ with $\varphi(b^*b) < \infty$, then the map $a \longrightarrow ba^*(\forall a \in N)$ is strongly continuous in any bounded ball of N.

Proof. Suppose that $\{a_l\}$ is a net of N with $\|a_l\| \leq 1$ and $a_l \longrightarrow 0$ (strongly) We need to prove that $a_l b^* b a_l^* \longrightarrow 0$ (weakly) .

Since $a_l b^* b a_l^* \in Ns(\varphi), \forall l$, we may assume that $s(\varphi) = 1$, i.e., φ is faithful. Then there is a faithful W^*-representation $\{\pi_\varphi, H_\varphi\}$ of N generated by φ (

see Section 6.5) . It suffices to show

$$\langle \pi_\varphi(a_l b^* b a_l^*) x_\varphi, x_\varphi \rangle = \varphi(x^* a_l b^* b a_l^* x) \longrightarrow 0$$

$\forall x \in \mathcal{N}$ by $\|a_l\| \leq 1, \forall l$ (see Propositon 6.5.2) . In fact, from Proposition 6.5.2., 6.5.3 and $b \in \mathcal{N}$, it follows that

$$\begin{aligned} \varphi(x^* a_l b^* b a_l^* x) &= \varphi(b a_l^* x x^* a_l b^*) \\ &\leq \|x\|^2 \varphi(b a_l^* a_l b^*) \\ &= \|x\|^2 \varphi(b^* b a_l^* a_l) \longrightarrow 0, \end{aligned}$$

$\forall x \in \mathcal{N}$. Q.E.D.

Proposition 6.9.5. $M_1 \overline{\otimes} M_2$ is purely infinite if and only if either M_1 or M_2 is purely infinite.

Proof. If both M_1 and M_2 are not purely infinite, then by Proposition 6.9.3, $M_1 \overline{\otimes} M_2$ is not purely infinite.

Now let M_1 be purely infinite. If $M_1 \overline{\otimes} M_2$ is not purely infinite, then there is a non-zero semifinite normal trace φ on $(M_1 \overline{\otimes} M_2)_+$. Pick $0 \neq b \in (M_1 \overline{\otimes} M_2)_+ s(\varphi)$ and $\varphi(b^2) < \infty$. Write

$$H_1 \overline{\otimes} H_2 = \sum_{l \in \wedge} \oplus H_l, \quad b = (b_{ll'})_{l,l' \in \wedge},$$

where H_1 and H_2 are the action spaces of M_1 and M_2 respectively, and $H_l = H_1, \forall l, {}^\# \wedge = \dim H_2, b_{ll'} \in B(H_1), \forall l, l'$. Since $b \geq 0$ and $b \neq 0$, there is an index l_0 such that $b_1 = b_{l_0 l_0} \geq 0$ and $b_1 \neq 0$. Now consider the following chain

$$M_1 \xrightarrow{\alpha} M_1 \overline{\otimes} M_2 \xrightarrow{\beta} M_1 \overline{\otimes} M_2 \xrightarrow{\gamma} M_1,$$

where $\alpha(a_1) = a_1 \otimes 1_2, \forall a_1 \in M_1, \beta(a) = ba^*$ and $\gamma(a) = a_{l_0 l_0}, \forall a \in M_1 \overline{\otimes} M_2$. By Lemma 6.9.4, the map

$$(\gamma \circ \beta \circ \alpha) : a_1 \longrightarrow b_1 a_1^*, \quad (\forall a_1 \in M_1)$$

is strongly continuous in any bounded ball of M_1 . Moreover, we can find a non-zero projection p_1 of M_1 and a positive number λ such that $b_1 \geq \lambda p_1$. Then the map $a_1 \longrightarrow p_1 a_1^* (\forall a_1 \in M_1)$ is also strongly continuous in any bounded ball of M_1 . Consequently, the map $a_1 \longrightarrow a_1^*$ is strongly continuous in any bounded ball of $M_1 p_1$. However, since $p_1 \neq 0$ and M_1 is purely infinite, we get a contradiction from Proposition 6.6.3. Therefore, $M_1 \overline{\otimes} M_2$ must be purely infinite. Q.E.D.

Corollary 6.9.6. If $M_1 \overline{\otimes} M_2$ is semi-finite, then M_1 and M_2 are semi-finite.

Proposition 6.9.7. Let M_i be type $(I_{n_i}), i = 1,2$. Then $M_1\overline{\otimes}M_2$ is type $(I_{n_1n_2})$. Consequently, if both M_1 and M_2 are discrete, then $M_1\overline{\otimes}M_2$ is also discrete.

Proof. By Proposition 6.7.7, we may assume that $M_i = N_i\overline{\otimes}B(K_i)$, where N_i is abelian, and dim $K_i = n_i, i = 1,2$. Then

$$M_1\overline{\otimes}M_2 = (N_1\overline{\otimes}N_2)\overline{\otimes}B(K_1 \otimes K_2).$$

Therefore, $M_1\overline{\otimes}M_2$ is type $(I_{n_1n_2})$. Q.E.D.

Proposition 6.9.8. Let M be a type (I_n) VN algebra. Then M is finite if and only if $n < \infty$.

Proof. By Proposition 6.7.7, we may assume $M = N\overline{\otimes}B(K)$, where N is abelian, and dim $K = n$. Clearly, N is finite. Thus by Proposition 6.9.1, M is finite if and only if $B(K)$ is finite, i.e., $n < \infty$. Q.E.D.

Proposition 6.9.9. Let M_2 be semi-finite, and M_1 be type (II). Then $M_1\overline{\otimes}M_2$ is type (II).

Proof. By Theorem 6.8.4, there is a decreasing sequence $\{p_n\}$ of finite projections of M_1 with $c(p_1) = 1$, and $p_{n+1} \sim (p_n - p_{n+1}), \forall n$. From Theorem 6.5.10, there is a finite projection q of M_2 with $c(q) = 1$. Now let $e_n = p_n \otimes q, \forall n$. Then $\{e_n\}$ is a decreasing sequence of finite projections of $M_1\overline{\otimes}M_2$ by Proposition 6.9.1. From Definition 1.5.7, it is easy to see that the central cover of e_1 in $M_1\overline{\otimes}M_2$ is 1 . Clearly , $e_{n+1} \sim (e_n - e_{n+1}), \forall n$. Therefore, $M_1\overline{\otimes}M_2$ is type (II) by Theorem 6.8.4. Q.E.D.

Proposition 6.9.10. $M_1\overline{\otimes}M_2$ is continuous if and only if either M_1 or M_2 is continuous.

Proof. Since any purely infinite VN algebra is continuous, we may assume that both M_1 and M_2 are semifinite. Thus the sufficiency is immediate from Proposition 6.9.9.

Now let $M_1\overline{\otimes}M_2$ be continuous, and M_1, M_2 be semi-finite. If M_1 and M_2 are not type (II), then by Proposition 6.9.7 $M_1\overline{\otimes}M_2$ is not continuous, a contradiction. Therefore, either M_1 or M_2 is continuous. Q.E.D.

Corollary 6.9.11. 1) If $M_1\overline{\otimes}M_2$ is discrete, then M_1 and M_2 are discrete; 2) If $M_1\overline{\otimes}M_2$ is type (II), then both M_1 and M_2 are semi-finite, and either M_1 or M_2 is continuous.

Summing up above, we have the following.

Theorem 6.9.12. 1) $M_1\overline{\otimes}M_2$ is finite, or semifinite, or discrete if and only if both M_1 and M_2 are fnite, or semi-finite, or discrete.

2) $M_1\overline{\otimes}M_2$ is properly infinite, or purely infinite, or continuous if and only if either M_1 or M_2 is properly infinite, or purely infinite, or continuous.

3) $M_1\overline{\otimes}M_2$ is type (II) if and only if both M_1 and M_2 are semi-finite, and either M_1 or M_2 is continuous.

Notes. The tensor product of semi-finite VN algebras was proved to be semi-finite by Y. Misonou. The case involving algebras of type III was settled by S.Sakai. Thus we have now the full result of Theorem 6.9.12.

References. [28], [109], [144].

Chapter 7
The Theory of Factors

7.1. Dimension functions

From the classification in Chapter 6, there are only five classes of factors:

1) Type (I_n) factors, i.e., discrete finite factors. It must be $*$ isomorphic to $B(H_n)$, where dim $H_n = n(< \infty)$;

2)Type(I_∞) factors, i.e. discrete infinite factors. It must be $*$ isomorphic to $B(H)$, where dim$H = \infty$;

3) Type (II_1) factors. i.e., continuous finite factors;

4)Type (II_∞) factors, i.e., continuous infinite factors;

5)Type (III) factors, i.e., purely infinite factors.

Definition 7.1.1. Let M be a factor. A trace φ on M_+ is called satisfying the condition (R), i.e., if M contains a non-zero finite projection, then there is a non-zero finite projection p_0 such that $\varphi(p_0) < \infty$.

Proposition 7.1.2. Let M be a factor, and φ be a faithful normal trace on M_+ satisfying (R).

1) Let p be a projection of M. Then p is finite or infinite if and only if $\varphi(p) < \infty$ or $\varphi(p) = +\infty$.

2) Let p, q be finite projections of M. Then $p \precsim q$ if and only if $\varphi(p) \leq \varphi(q)$.

3) If M contains a non-zero finite projection, then φ is semi-finite.

4) φ is uniquely determined up to multiplication by a positive constant.

Proof. 1) If p is infinite, then it must be properly infinite. By Theorem 6.4.4, we can write $p = p_1 + p_2$, where $p_1 p_2 = 0$ and $p_1 \sim p_2 \sim p$. Then $\varphi(p) = 2\varphi(p)$. Since φ is faithful, it follows that $\varphi(p) = +\infty$.

If p is finite, and $p \neq 0$, then by Definition 7.1.1, there is a non-zero finite projection p_0 of M such that $\varphi(p_0) < \infty$. We have either $p \precsim p_0$ or $p_0 \precsim p$

since M is a factor. Clearly, $\varphi(p) < \infty$ if $p \precsim p_0$. If $p_0 \precsim p$, then there is an orthogonal family $\{p_l | l \in \Lambda\}$ of projections of M such that

$$p_l \sim p_0, \quad p_l \leq p, \quad \forall l \in \Lambda, \quad \text{and} \quad (p - \sum_{l \in \Lambda} p_l) \precsim p_0.$$

Since p is finite, $^{\#}\Lambda$ must be finite. So it is easy to see that $\varphi(p) < \infty$.

Therefore, p is finite or infinite if and only if $\varphi(p) < \infty$ or $\varphi(p) = +\infty$.

2) If $p \precsim q$, then $\varphi(p) \leq \varphi(q)$ obviously. Conversely, let $\varphi(p) \leq \varphi(q)$. If $q \sim p_1 \leq p$ and $p_1 \neq p$, then $\varphi(q) = \varphi(p_1) < \varphi(p)$ since p is finite and φ is faithful. This is a contradiction. Thus $p \precsim q$.

3) For any $0 \neq a \in M_+$, there is a non-zero projection p of M and a positive number λ such that $a \geq \lambda p$. Since M is semi-finite, there is a non-zero finite projection q of M with $q \leq p$. Then, $a \geq \lambda q$, and $0 < \varphi(\lambda q) < \infty$ by 1). Therefore φ is semi-finite.

4) If M is purely infinite, then any non-zero projection of M is infinite. Therefore, $\varphi(a) = +\infty, \forall a \in M_+ \backslash \{0\}$, i.e., φ is uniquely determined.

Now suppose that M is semi-finite. By 3) and Theorem 6.5.8, such φ is existential. Let φ_1, φ_2 be two faithful semi-finite normal traces on M_+. We need to prove that $\varphi_1 = \lambda\varphi_2$ for some positive constant λ.

First, let M be finite, and put $\varphi = \varphi_1 + \varphi_2$. By 1) and Proposition 6.5.2, $\varphi, \varphi_1, \varphi_2$ can be extended to faithful normal traces on M. From Theorem 1.10.3, there is $t \in M$ with $0 \leq t \leq 1$ such that

$$\varphi_1(a) = \varphi(ta), \quad \forall a \in M.$$

Then

$$\varphi(tab) = \varphi_1(ab) = \varphi_1(ba) = \varphi(tba) = \varphi(atb),$$

i.e., $\varphi((ta - at)b) = 0, \forall a, b \in M$. Since φ is faithful, it follows that $t \in M \cap M' = \mathbb{C}$. Further, we get $\varphi_1 = \lambda\varphi_2$ for some positive constant λ.

Secondly, suppose that M is semi-finite and properly infinite. From Theorem 6.5.10, there is an increasing net $\{q_l\}$ of finite projections of M with $\sup_l q_l = 1$. From preceding paragraph, for each index l there is a positive constant λ_l such that

$$\varphi_1(a) = \lambda_l \varphi_2(a), \quad \forall a \in (M_{q_l})_+.$$

Since $\{q_l\}$ is increasing, it follows that λ_l is independent of the index l. Put $\lambda = \lambda_l, \forall l$. Then

$$\varphi_1(q_l a q_l) = \lambda\varphi_2(q_l a q_l), \quad \forall a \in M_+, \quad \forall l.$$

Moreover, by Proposition 6.5.2, $\varphi_i(q_l a q_l) = \varphi_i(a^{\frac{1}{2}} q_l a^{\frac{1}{2}}), \forall a \in M_+, i = 1, 2$. Further, from the normality of φ_1 and φ_2, we have $\varphi_1(a) = \lambda\varphi_2(a), \forall a \in M_+$.

Q.E.D.

Remark. For type (I) factor $B(H)$, the unique (up to multiplication by a positive constant) faithful semi-finite normal trace on $B(H)_+$ is as follows

$$\mathrm{tr}(\cdot) = \sum_l \langle \cdot\, \xi_l, \xi_l \rangle, \quad \forall \cdot \in B(H)_+,$$

where $\{\xi_l\}$ is a normalized orthogonal basis of H. For a finite factor, there is unique faithful normal tracial state on it. For any semi-finite factor, there is unique (up to multiplication by a positive constant) faithful semi-finite normal trace on its positive part. For purely infinite factors, the case is trivial.

Proposition 7.1.3. Let M be a factor, P =Proj(M) be the set of all projections of M, φ be as in Proposition 7.1.2, and $\mathcal{D} = \{\varphi(p) | p \in P\}$. Then multiplying φ by a proper positive constant, we can get the following:

1) $\mathcal{D} = \{0, 1, \cdots, n\}$, when M is type (I_n) (n finite or infinite). In particular, if $M = B(H_n)$, where $\dim H_n = n$, then

$$\varphi(p) = \dim pH_n, \quad \forall p \in P;$$

2) $\mathcal{D} = [0, 1]$, when M is type (II_1);
3) $\mathcal{D} = [0, +\infty]$, when M is type (II_∞);
4) $\mathcal{D} = \{0, +\infty\}$, when M is type (III).

Proof. 1) and 4) are obvious.

2) Let φ be the unique faithful normal tracial state on a type(II_1) factor M. By Theorem 6.8.3,

$$\{2^{-n}k | 1 \le k \le 2^n, n = 0, 1, \cdots\} \subset \mathcal{D}.$$

For any $\lambda \in [0, 1]$, pick $p_n \in P$ such that $\varphi(p_n) = \lambda_n \nearrow \lambda$. From Proposition 7.1.2, $p_n \precsim p_{n+1}, \forall n$.

Let $q_1 = p_1$, and $q_1 \sim q \le p_2$. By Proposition 6.3.2, $(1 - q_1) \sim (1 - q)$. Thus, $(p_2 - q) \precsim (1 - q_1)$. Let $(p_2 - q) \sim r \le (1 - q_1)$. Then $p_2 \sim q_1 + r$. Put $q_2 = q_1 + r$. Then $q_2 \ge q_1$, $q_i \sim p_i, i = 1, 2$. Generally, we can get $\{q_n\}$ with $q_n \le q_{n+1}, q_n \sim p_n, \forall n$. Let $q = \sup_n q_n$. Then $\varphi(q) = \sup_n \varphi(p_n) = \lambda$. Therefore, $\mathcal{D} = [0, 1\}$.

3) Suppose that $\{p_l\}$ is an increasing net of finite projections of M with $\sup_l p_l = 1$. Clearly, $\varphi(p_l) \nearrow \varphi(1) = +\infty$. By 2), $[0, \varphi(p_l)] \subset \mathcal{D}, \forall l$. Therefore, $\mathcal{D} = [0, +\infty]$. Q.E.D.

Lemma 7.1.4. Let M be a finite factor, and P =Proj(M) be the set of all projections of M. Suppose that $D : P \to [0, +\infty)$ satisfies:

1) if $p_1, p_2 \in P$ and $p_1 p_2 = 0$, then $D(p_1 + p_2) = D(p_1) + D(p_2)$;
2) for any unitary element $u \in M$ and $p \in P, D(upu^*) = D(p)$;

3) $D(1) > 0$.
Then $D = \varphi|P$, where φ is as in Proposition 7.1.2.

Proof. If $M = B(H_n)$, where $\dim H_n = n < \infty$, then by 2), there is some value λ such that $D(p) = \lambda$ for each minimal projection p of M. By 1), $D(1) = n\lambda$. By 3), $\lambda > 0$. We may assume that $\lambda = 1$. Since each projection of M is an orthogonal sum of several minimal projections of M, it follows that $D(P) = \{0, 1, \cdots, n\}$. By Proposition 7.1.3, $D = \varphi|P$.

Now let M be a type (II_1) factor. We may assume $D(1) = 1$. Suppose that φ is the unique faithful normal tracial state on M. We need to prove $D = \varphi|P$.

Notice the following fact: if $q \precsim p$, then $D(q) \leq D(p)$. Indeed, let $q \sim p_1 \leq p$. From Proposition 6.3.2, $(1-q) \sim (1-p_1)$. Thus, there is a unitary element $u \in M$ such that $p_1 = uqu^*$. Further, $D(q) = D(p_1) \leq D(p)$.

Fix $p \in P$. Since $D(1) = \varphi(1) = 1$, from Theorem 6.8.3 there is a subset $\{p_{n,k} | n = 0, 1, \cdots, 0 \leq k \leq 2^n\} \subset P$ such that $D(p_{n,k}) = \varphi(p_{n,k}) = 2^{-n}k, \forall n, k$. Thus, we can find $\{p_m\} \subset P$ such that $D(p_m) = \varphi(p_m) \nearrow \varphi(p)$. By Proposition 7.1.2, $p_m \precsim p, \forall m$. From preceding paragraph, $D(p) \geq D(p_m) = \varphi(p_m) \to \varphi(p)$. Hence, $D(p) \geq \varphi(p)$. Similarly, $D(1-p) \geq \varphi(1-p)$. Therefore, $D(p) = \varphi(p)$, and $D = \varphi|P$. Q.E.D.

Definition 7.1.5. Let M be factor, and P be the set of all projections of M. A function $D : P \to [0, +\infty]$ is called a *dimension function*, if: 1) $D(p) = 0 \Longleftrightarrow p = 0$; 2) for any unitary element $u \in M$ and $p \in P, D(upu^*) = D(p)$; 3) if $p, q \in P$ and $pq = 0$, then $D(p+q) = D(p) + D(q)$; 4) if M contains a non-zero finite projection, then there is a non-zero projection p_0 of M such that $D(p_0) < \infty$.

Theorem 7.1.6. Let M be a factor, P be the set of all projections of M, and $D(\cdot)$ be a dimension function on P. Then $D = \varphi|P$, where φ is as in Proposition 7.1.2.

Proof. First, we claim that: if $p (\in P)$ is infinite, then $D(p) = +\infty$. In fact, p is properly infinite. By Theorem 6.4.4, we can write $p = \sum_n p_n$, where

$$p_n p_m = \delta_{n,m} p_n, \quad p_n \sim p, \quad \forall n, m.$$

Clearly, we have a unitary element u_{nm} of M such that $u_{nm} p_n u_{nm}^* = p_m, \forall n, m$. Thus $D(p_n) = D(p_m), \forall n, m$. Of course, $D(p_n) > 0, \forall n$. Therefore, $D(p) = +\infty$.

From the preceding paragraph, we get $D = \varphi|P$ if M is purely infinite.

Now let M be semi-finite, and p_0 be as in Definition 7.1.5. Clearly, p_0 is finite (otherwise, $D(p_0) = +\infty$, a contradiction). Pick φ as in Proposition

7.1.2 such that $\varphi(p_0) = D(p_0)$. Now we prove that $D = \varphi|P$. It suffices to show that $D(p) = \varphi(p)$ for any finite projection p of M. Let $p(\in P)$ be finite, and $q = \sup\{p, p_0\}$. By Proposition 6.4.5, q is also finite. From Lemma 7.1.4, we have

$$D|(P \cap M_q) = \varphi|(P \cap M_q),$$

and $D(p) = \varphi(p)$ consequently. Q.E.D.

Corollary 7.1.7. The dimension function is uniquely determined up to multiplication by a positive constant.

References. [28], [111].

7.2. Hyperfinite type (II_1) factors

Let M be a type (II_1) factor. Then there is unique faithful normal tracial state φ on M. Define

$$\|x\|_2 = \varphi(x^*x)^{\frac{1}{2}}, \quad \forall x \in M.$$

Then $\|\cdot\|_2$ is a norm on M, and

$$\|x\|_2 = \|x^*\|_2 \le \|x\|, \quad \|xy\|_2 \le \min\{\|x\| \cdot \|y\|_2, \|x\|_2 \cdot \|y\|\}, \quad \forall x, y \in M.$$

From Lemma 1.11.2, the topology generated by $\|\cdot\|_2$ is equivalent to the strong (operator) topology in the unit ball $(M)_1$ of M.

Lemma 7.2.1. Let p be a projection of M, and $a^* = a \in (M)_1$. Then there is a spectral projection q of a such that $\|q - p\|_2 \le 9\|a - p\|_2^{\frac{1}{2}}$. Moreover, if $a \ge 0$, then

$$\|a^{\frac{1}{2}} - p\|_2 \le 13\|a - p\|_2^{\frac{1}{4}}.$$

Proof. Let $\varepsilon \in (0, 1/2), a = \int_{-1}^{1} \lambda de_\lambda$, and

$$q = 1 - e_{1-\varepsilon}, \quad q_1 = e_\varepsilon - e_{-\varepsilon}, \quad q_2 = 1 - q - q_1.$$

When $\lambda \notin [-\varepsilon, \varepsilon] \cup (1 - \varepsilon, 1]$, we have $|\lambda^2 - \lambda| \ge \varepsilon - \varepsilon^2 \ge \varepsilon/2$. Then

$$\begin{aligned} \frac{1}{2}\varepsilon\|q_2\|_2 \le\ & \|(a^2 - a)q_2\|_2 \le \|a^2 - a\|_2 \\ \le\ & \|(a - p)a\|_2 + \|p(a - p)\|_2 + \|p - a\|_2 \\ \le\ & 3\|p - a\|_2. \end{aligned}$$

i.e., $\|q_2\|_2 \le \frac{6}{\varepsilon}\|p-a\|_2$. On the other hand, $\|aq_1\| \le \varepsilon, \|aq-q\| \le \varepsilon$, so

$$\|a-q\|_2 \le \|aq-q\|_2 + \|aq_1\|_2 + \|aq_2\|_2 \le 2\varepsilon + \frac{6}{\varepsilon}\|p-a\|_2.$$

If $\|a-p\|_2^{\frac{1}{2}} < \frac{1}{2}$, put $\varepsilon = \|a-p\|_2^{\frac{1}{2}}$, then

$$\|p-q\|_2 \le \|a-p\|_2 + \|a-q\|_2 \le 9\|a-p\|_2^{\frac{1}{2}}.$$

If $\|a-p\|_2^{1/2} \ge 1/2$, then we have immediately

$$\begin{aligned}\|q-p\|_2 \le\ & \|q\|_2 + \|a-p\|_2 + \|a\|_2 \\ \le\ & 2 + (\|a\|_2 + \|p\|_2)^{1/2}\|a-p\|_2^{1/2} \\ \le\ & \|a-p\|_2^{1/2}(4 + (\|a\|_2 + \|p\|_2)^{1/2}) \le 9\|a-p\|_2^{1/2}.\end{aligned}$$

Now let $a \ge 0$, and keep above notations. Then

$$\|a^{\frac{1}{2}}q - q\| \le \varepsilon, \quad \|a^{\frac{1}{2}}q_1\| \le \varepsilon^{\frac{1}{2}}.$$

From $\|q_2\|_2 \le \frac{6}{\varepsilon}\|a-p\|_2$ (see preceding paragraph), we have that

$$\begin{aligned}\|a^{\frac{1}{2}} - q\|_2 \le\ & \|a^{\frac{1}{2}}q - q\|_2 + \|a^{\frac{1}{2}}q_1\|_2 + \|a^{\frac{1}{2}}q_2\|_2 \\ \le\ & \varepsilon + \varepsilon^{1/2} + \|q_2\|_2 \\ \le\ & \varepsilon + \varepsilon^{1/2} + \frac{6}{\varepsilon}\|a-pp\|_2.\end{aligned}$$

If $\|a-p\|_2^{1/2} < 1/2$, put $\varepsilon = \|a-p\|_2^{1/2}$, then

$$\|a^{1/2} - q\|_2 \le 7\|a-p\|_2^{1/2} + \|a-p\|_2^{1/4} \le 6\|a-p\|_2^{1/4}.$$

Since $\|q-p\|_2 \le 9\|a-p\|_2^{1/2}$ (see preceding paragraph), it follows that

$$\begin{aligned}\|a^{1/2} - p\| \le\ & \|a^{1/2} - q\|_2 + \|p-q\|_2 \\ \le\ & 6\|a-p\|_2^{\frac{1}{4}} + 9\|a-p\|_2^{\frac{1}{2}} \\ \le\ & (6 + \frac{9}{\sqrt{2}})\|a-p\|_2^{\frac{1}{4}} \le 13\|a-p\|_2^{\frac{1}{4}}.\end{aligned}$$

If $\|a-p\|_2^{\frac{1}{2}} \ge 1/2$, then we have immediately

$$\|a^{\frac{1}{2}} - p\|_2 \le \|a^{\frac{1}{2}}\|_2 + \|p\|_2 \le 2 \le 13\|a-p\|_2^{\frac{1}{4}}.$$

Q.E.D.

Lemma 7.2.2. Let p, q be two projections of M. Then there is a partial isometry w of M such that

$$w^*w \leq p, \quad ww^* \leq q, \quad \text{and} \quad \|w - p\|_2 \leq 14\|p - q\|_2^{\frac{1}{4}}.$$

Proof. Let $qp = wb$ be the polar decomposition of qp. Then $0 \leq b \leq 1, w^*w \leq p, ww^* \leq q$. Since

$$\|b^2 - p\|_2 = \|p(q - p)p\|_2 \leq \|q - p\|_2,$$

it follows from Lemma 7.2.1 that

$$\|b - p\|_2 \leq 13\|b^2 - p\|_2^{\frac{1}{4}} \leq 13\|q - p\|_2^{\frac{1}{4}}.$$

Noticing that $wp = w$, we have

$$\begin{aligned} \|w - p\|_2 \leq & \|w - qp\|_2 + \|qp - p\|_2 \\ = & \|w(p - b)\|_2 + \|(q - p)p\|_2 \\ \leq & \|p - b\|_2 + \|q - p\|_2 \\ \leq & 13\|q - p\|_2^{\frac{1}{4}} + \|q - p\|_2. \end{aligned}$$

If $\|q - p\|_2 \leq 1$, then $\|w - p\|_2 \leq 14\|q - p\|_2^{\frac{1}{4}}$. If $\|q - p\|_2 > 1$, then we have immediately

$$\|w - p\|_2 \leq 2 \leq 14\|q - p\|_2^{\frac{1}{4}}.$$

Q.E.D.

Lemma 7.2.3. Let u be a unitary element of M, w be a partial isometry of M, and $uw^*w = w$. Then $\|u - w\|_2^2 \leq 2\|w - 1\|_2$.

Proof. Since $(u - w)(u - w)^* = 1 - ww^*$, it follows that

$$\begin{aligned} \|u - w\|_2^2 = & \varphi(1 - ww^*) \\ \leq & |\varphi(1 - w)| + |\varphi(w(1 - w^*))| \\ \leq & \|1 - w\|_2 + \|1 - w^*\|_2 = 2\|w - 1\|_2. \end{aligned}$$

Q.E.D.

Lemma 7.2.4. Let p, q be projections of M, and $p \sim q$. Then there is a unitary element u of M such that

$$q = upu^*, \quad \|u-1\|_2 \le 36\|p-q\|_2^{\frac{1}{8}}.$$

Proof. By Lemma 7.2.2, there is a partial isometry w of M such that

$$w^*w \le p, \quad ww^* \le q, \quad \text{and} \quad \|w-p\|_2 \le 14\|q-p\|_2^{\frac{1}{4}}.$$

Since M is finite and $p \sim q$, by Proposition 6.3.2 there is $v \in M$ such that $v^*v = p - w^*w$, $vv^* = q - ww^*$.

Again from Lemma 7.2.2, there is a partial isometry w_1 of M such that

$$w_1^*w_1 \le 1-p, \quad w_1w_1^* \le 1-q,$$

and

$$\|w_1 - (1-p)\|_2 \le 14\|p-q\|_2^{\frac{1}{4}}.$$

By Proposition 6.3.2, $(1-p) \sim (1-q)$. Thus we can pick $v_1 \in M$ such that

$$v_1^*v_1 = 1-p-w_1^*w_1, \quad v_1v_1^* = 1-q-w_1w_1^*.$$

Now let $u = w + v + w_1 + v_1$. Then u is a unitary element of M, and $q = upu^*$. Notice that

$$\begin{aligned}\|w+w_1-1\|_2 &\le \|w-p\|_2 + \|w_1-(1-p)\|_2 \\ &\le 28\|p-q\|_2^{\frac{1}{4}}\end{aligned}$$

and $u(w+w_1)^*(w+w_1) = w+w_1$. By Lemma 7.2.3, it is clear that

$$\|w+w_1-u\|_2 \le \sqrt{2}\|w+w_1-1\|_2^{\frac{1}{2}} \le 8\|p-q\|_2^{\frac{1}{8}}.$$

Thus,

$$\begin{aligned}\|u-1\|_2 &\le \|w+w_1-u\|_2 + \|w+w_1-1\|_2 \\ &\le 8\|p-q\|_2^{\frac{1}{8}} + 28\|p-q\|_2^{\frac{1}{4}}\end{aligned}$$

If $\|p-q\|_2 \le 1$, then $\|u-1\|_2 \le 36\|p-q\|_2^{\frac{1}{8}}$. If $\|p-q\|_2 > 1$, then we have immediately

$$\|u-1\|_2 \le 2 < 36\|p-q\|_2^{\frac{1}{8}}.$$

Q.E.D.

In the following Lemmas 7.2.5–7.2.8, M possess the following property:

(*) For any elements $a_1, \cdots, a_m$ of M and $\varepsilon > 0$, there exists a finite dimensional $*$ subalgebra B of M and elements $b_1, \cdots, b_m \in B$ such that

$$\|a_i - b_i\|_2 \le \varepsilon, \quad 1 \le i \le m.$$

Moreover, N is called a *subfactor* of M, if N is a factor, $N \subset M$, and N contains the identity of M.

Lemma 7.2.5. For any $a_1, \cdots, a_m \in M$ and $\varepsilon > 0$, there is a type (I_{2^n}) subfactor N of M (n sufficiently large) and $b_1, \cdots, b_m \in N$ such that

$$\|a_i - b_i\|_2 \le \varepsilon, \quad 1 \le i \le m.$$

Proof. First for $\varepsilon/2$, there is a finite dimensional $*$ subalgebra A of M, and $c_1, \cdots, c_m \in A$ such that

$$\|a_i - c_i\|_2 \le \varepsilon/2, \quad 1 \le i \le m.$$

We may assume $1 \in A$, where 1 is the identity of M. By Section 2.13, there is an orthogonal finite set $\{z_i\}$ of central projections of A with $\sum_i z_i = 1$ such that $A_i = Az_i$ is a finite dimensional factor, $\forall i$. Suppose that $\{p_j^{(i)}\}_j$ is an orthogonal set of minimal projections of A_i with $\sum_j p_j^{(i)} = z_i, \forall i$. Clearly, $p_j^{(i)} \sim p_k^{(i)}$ (relative to A_i), $\forall j, k$. Thus, we have $\{w_j^{(i)}\} \subset A_i$ such that

$$w_1^{(i)} = p_1^{(i)}, \quad w_j^{(i)*} w_j^{(i)} = p_1^{(i)}, \quad w_j^{(i)} w_j^{(i)*} = p_j^{(i)}, \quad \forall j.$$

Then $\{w_j^{(i)} w_k^{(i)*}\}_{j,k}$ is a matrix unit of $A_i, \forall i$. Further, $\{e_{jk}^{(i)} = w_j^{(i)} w_k^{(i)*} | i, j, k\}$ is a basis of A. Now it suffices to show that: for enough small $\delta > 0$(δ depends on ε and $c_1, \cdots, c_m$), there exists a type (I_{2^n}) subfactor N of M(n sufficiently large) and $\{v_j^{(i)}\} \subset N$ such that

$$\|w_j^{(i)} - v_j^{(i)}\|_2 \le \delta, \quad \forall i, j.$$

Pick sufficiently large n, such that $2^{-n} < \delta^2$ and $2^{-n} < \varphi(p_1^{(i)}), \forall i$. From Proposition 7.1.3, for each i we can find an orthogonal set $\{q_k^{(i)}\}$ of projections of M such that

$$q_k^{(i)} \le p_1^{(i)}, \quad \varphi(q_k^{(i)}) = 2^{-n}, \quad \forall k,$$

and

$$\varphi(p_1^{(i)} - \sum_k q_k^{(i)}) < 2^{-n}.$$

Now let N be a type (I_{2^n}) subfactor of M such that $\{w_j^{(i)} q_k^{(i)} | i, j, k\} \subset N$, and define $v_j^{(i)} = w_j^{(i)} \sum_k q_k^{(i)}, \forall i, j$. Then

$$(w_j^{(i)} - v_j^{(i)})^*(w_j^{(i)} - v_j^{(i)}) = p_1^{(i)} - \sum_k q_k^{(i)}$$

and

$$\|w_j^{(i)} - v_j^{(i)}\|_2^2 = \varphi(p_1^{(i)} - \sum_k q_k^{(i)}) < 2^{-n} < \delta^2, \quad \forall i, j.$$

Q.E.D.

Lemma 7.2.6. For any $a_1, \cdots, a_m \in M$, any projection $p \in M$ and $\varepsilon > 0$, if $\varphi(p) = 2^{-n}$, then there exists a type (I_{2^r}) subfactor N of M (where $r \geq n$), $b_1, \cdots, b_m \in N$, and a projection $q \in N$ such that

$$\|a_i - b_i\|_2 \leq \varepsilon, \quad 1 \leq i \leq m, \quad \|p - q\|_2 \leq \varepsilon, \quad \varphi(q) = 2^{-n}.$$

Proof. By Lemma 7.2.5, we can find a type $((\mathrm{I}_{2^r})$ subfactor N of M and $b_1, \cdots, b_{m+1} \in N$ such that

$$r \geq n, \quad \|a_i - b_i\|_2 \leq \delta, \quad 1 \leq i \leq m, \quad \|p - b_{m+1}\|_2 \leq \delta,$$

where $\delta (> 0)$ will be determined later, and $b_{m+1}^* = b_{m+1}$. Let

$$b = 2b_{m+1}(1 + b_{m+1}^2)^{-1}.$$

Clearly, $b \in N, \|b\| \leq 1$. Since $p = 2p(1+p)^{-1}$, it follows that

$$\frac{1}{2}(b - p) = (1 + b_{m+1}^2)^{-1}(b_{m+1} - p)(1+p)^{-1} + \frac{b}{4}(p - b_{m+1})p.$$

Thus, $\|b - p\|_2 \leq \frac{5}{2}\delta$. By Lemma 7.2.1, there is a spectral projection q_1 of b such that

$$\|q_1 - p\|_2 \leq 9\|b - p\|_2^{\frac{1}{2}} \leq 15\delta^{\frac{1}{2}}.$$

Then,

$$|\varphi(q_1) - 2^{-n}| = |\varphi(p - q_1)| \leq \|p - q_1\|_2 \leq 15\delta^{\frac{1}{2}}.$$

Now pick a projection q of N such that $\varphi(q) = 2^{-n}$ and either $q \geq q_1$ or $q \leq q_1$. Then $\|q - q_1\|_2^2 = (\varphi(q_1) - 2^{-n}| \leq 15\delta^{\frac{1}{2}}$, and

$$\|q - p\|_2 \leq \|q - q_1\|_2 + \|q_1 - p\|_2 \leq 15\delta^{\frac{1}{2}} + \sqrt{15}\delta^{\frac{1}{4}}.$$

If take $\delta > 0$ is such that $\delta \leq \varepsilon$ and $15\delta^{\frac{1}{2}} + \sqrt{15}\delta^{\frac{1}{4}} \leq \varepsilon$, then we can get the conclusion.

Q.E.D.

Lemma 7.2.7. Let $a_1, \cdots a_m \in M$ and p be a projection of M with $\varphi(p) = 2^{-n}$, and $pa_i = a_i p = a_i, 1 \leq i \leq m$. Then for any $\varepsilon > 0$ there is a type $(I_2 r)$ subfactor N of M with $r \geq n$, and $b_1, \cdots, b_m \in N$ such that

$$p \in N, pb_i = b_i p = b_i, \quad \|a_i - b_i\|_2 \leq \varepsilon, \quad 1 \leq i \leq m.$$

Moreover, if $p \in L$, where L is a type $(I_2 n)$ subfactor of M, then we can choose the above $N \supset L$.

Proof. From Lemma 7.2.6, there is a type $(I_2 r)$ subfactor A of M, where $r \geq n$, and $c_1, \cdots, c_m, q \in A$, where q is a projection, such that

$$\|a_i - c_i\|_2 \leq \delta, \quad 1 \leq i \leq m, \quad \|p - q\|_2 \leq \delta, \quad \varphi(q) = 2^{-n},$$

where $\delta(> 0)$ will be determined later. Then $p \sim q$. By Lemma 7.2.4, we have a unitary element u of M such that

$$p = u^* q u, \quad \|u - 1\|_2 \leq 36\|p - q\|_2^{1/8} \leq 36\delta^{1/8}.$$

Let $N = u^* A u, b_i = p u^* c_i u p, 1 \leq i \leq m$. Then N is also a type $(I_2 r)$ subfactor of $M, b_1, \cdots, b_m, p \in N, p b_i = b_i p = b_i, 1 \leq i \leq m$ and

$$\begin{aligned} \|a_i - b_i\|_2 \leq & \ \|u^* c_i u - a_i\|_2 \leq \|c_i - u a_i u^*\|_2 \\ \leq & \ \|c_i - a_i\|_2 + \|a_i u - u a_i\|_2 \\ \leq & \ \|a_i - c_i\|_2 + 2\|a_i\| \cdot \|u - 1\|_2 \\ \leq & \ \delta + 72\|a_i\|\delta^{1/8}. \end{aligned}$$

It is enough to pick $\delta(> 0)$ such that

$$\delta + 72\delta^{1/8} \max_{1 \leq i \leq m} \|a_i\| \leq \varepsilon.$$

Now let $p \in L$, where L is a type $(I_2 n)$ subfactor of M. Suppose that $\{p_1 = p, p_2, \cdots, p_{2^n}\}$ is an orthogonal set of minimal projections of L. By Theorem 1.5.6, M is spatially $*$ isomorphic to $M_p \overline{\otimes} B(K)$, where $\dim K = 2^n$. This spatial $*$ isomorphism also maps L to $L_p \overline{\otimes} B(K) = \mathbb{C} 1_{pH} \overline{\otimes} B(K)$, where H is the action space of M. From the preceding paragraph, there is a type $(I_2 r)(r \geq n)$ subfactor A of M with $p \in A$, and $b_1, \cdots, b_m \in A$ such that $\|a_i - b_i\|_2 \leq \varepsilon, p b_i = b_i p = b, 1 \leq i \leq m$. Clearly, $p, b_1, \cdots, b_m \in A_p$. Since $\varphi(p) = 2^{-n}$, A_p should be $*$ isomorphic to a matrix algebra of order 2^{r-n}. Let $N = \Phi^{-1}(A_p \overline{\otimes} B(K))$, where Φ is the above spatial $*$ isomorphism from M onto $M_p \overline{\otimes} B(K)$. Clearly, $L \subset N$, and $p, b_1, \cdots, b_m \in N$, and N is type $(I_2 r)$.
Q.E.D.

Lemma 7.2.8. Let L be a type $(I_2 n)$ subfactor of $M, a_1, \cdots, a_m \in M$, and $\varepsilon > 0$. Then there is a type $(I_2 r)$ subfactor N of M, and $b_1, \cdots, b_m \in N$ such that

$$r \geq n, \quad L \subset N, \ \|a_i - b_i\|_2 \leq \varepsilon, \quad 1 \leq i \leq m$$

Proof. Suppose that $\{p_i \mid 1 \le i \le 2^n\}$ is an orthogonal set of minimal projections of L, and $\{w_j\} \subset L$ such that

$$w_1 = p_1, \quad w_j^* w_j = p_1, \quad w_j w_j^* = p_j, \quad \forall j.$$

Let $p = p_1, a_{ijk} = w_i^* a_k w_j$. Then $pa_{ijk} = a_{ijk}p = a_{ijk}, \forall 1 \le i,j \le 2^n, 1 \le k \le m$. From Lemma 7.2.7, there is a type (I_{2^r}) subfactor N of M with $r \ge n$, and $b_{ijk} \in N$ such that

$$L \subset N, \; pb_{ijk} = b_{ijk}p = b_{ijk}, \quad \|a_{ijk} - b_{ijk}\|_2 \le \delta,$$

$\forall i,j,k$, where $\delta > 0$ and $2^{2n}\delta \le \varepsilon$. Put

$$b_k = \sum_{1 \le i,j \le 2^n} w_i b_{ijk} w_j^*, \quad 1 \le k \le m.$$

Clearly, $b_1, \cdots, b_m \in N$. Notice that

$$\begin{aligned} a_k &= \sum_{i,j} p_i a_k p_j = \sum_{i,j} w_i w_i^* a_k w_j w_j^* \\ &= \sum_{i,j} w_i a_{ijk} w_j^*, \quad 1 \le k \le m. \end{aligned}$$

Thus, $\|a_k - b_k\|_2 \le \sum_{1 \le i,j \le 2^n} \|a_{ijk} - b_{ijk}\| \le 2^{2n}\delta \le \varepsilon, 1 \le k \le m.$ Q.E.D.

Proposition 7.2.9. Let M be a countably generated type (II_1) factor. If for any $a_1, \cdots, a_m \in M$ and $\varepsilon > 0$, there is a finite dimensional $*$ subalgebra B of M and $b_1, \cdots, b_m \in B$ such that $\|a_i - b_i\|_2 \le \varepsilon, 1 \le i \le m$, then we have an increasing sequence $\{M_n\}$ of subfactors of M such that: M_n is type $I_{2^n}, \forall n$, and $\sqcup_n M_n$ is $\sigma(M, M_*)$-dense in M.

Proof. Let $\{a_n\}$ be a generated subset of M. By Lemma 7.2.8, we can construct

$$M_{r_1} \subset \cdots \subset M_{r_k} \subset \cdots \subset M,$$

where for each k, M_{r_k} is a type $(I_{2^{r_k}})$ subfactor of M, and also there exists $b_1^{(k)}, \cdots, b_k^{(k)} \in M_{r_k}$ such that

$$\|b_i^{(k)} - a_i\|_2 \le \frac{1}{k}, \quad 1 \le i \le k.$$

Clearly, $\sqcup_k M_{r_k}$ is $\sigma(M, M_*)$-dense in M. Further, making a refinement of $\{M_{r_k}\}$, we can get the conclusion. Q.E.D.

Definition 7.2.10. A VN algebra M is said to be *hyperfinite*, if there is a sequence $\{p_n\}$ of positive integers and $1 \in M_{p_1} \subset \cdots \subset M_{p_n} \subset \cdots \subset M$, where M_{p_n} is a type (I_{p_n}) subfactor of $M, \forall n$, such that $\sqcup_n M_{p_n}$ is $\sigma(M, M_*)$-dense in M.

From Proposition 3.8.3, it must be $p_n | p_{n+1}, \forall n$.

Definition 7.2.11. A VN algebra M is said to be *approximately finite-dimensi- onal*, if there is an increasing sequence $\{A_n\}$ of finite dimensional $*$ subalgebras of M such that $\sqcup_n A_n$ is $\sigma(M, M_*)$-dense in M.

Theorem 7.2.12. Let M be a type (II_1) factor. Then the following statements are equivalent:

1) M is hyperfinite;

2) M is approximately finite-dimensional;

3) M is countably generated,and for any $a_1, \cdots, a_m \in M$ and $\varepsilon > 0$, there exists a finite dimensional $*$ subalgebra B of M and $b_1, \cdots, b_m \in B$ such that $\|a_i - b_i\|_2 \leq \varepsilon, 1 \leq i \leq m$;

4) M is countably generated, and for any $a_1, \cdots, a_m \in M$ and $\varepsilon > 0$, there exists a subfactor N of M and $b_1, \cdots, b_m \in N$ such that $\|a_i - b_i\|_2 \leq \varepsilon, 1 \leq i \leq m$.

Proof. It is clear that 1) implies 2), 2) implies 3), and 4) implies 3). From Lemma 7.2.5, 3) implies 4) obviously. Moreover, 3) implies 1) immediately from Proposition 7.2.9. Q.E.D.

Lemma 7.2.13. Let A be a (UHF) C^*-algebra. Then there exists unique tracial state φ on A, i.e., φ is a state on A and $\varphi(ab) = \varphi(ba), \forall a, b \in A$.

Proof. By Proposition 3.8.3. $A = \alpha_0$ - $\otimes_{n=1}^{\infty} M_{m_n}$. For each n, there is unique tracial state φ_n on M_{m_n}. Therefore, $\otimes_n \varphi_n$ is the unique tracial state on A.

Q.E.D.

Theorem 7.2.14. All hyperfinite type (II_1) factors are $*$ isomorphic.

Proof. Let M_i be a hyperfinite type (II_1) factor, φ_i be the unique faithful normal tracial state on M_i,and $\{\pi_i, H_i, \xi_i\}$ be the faithful cyclic W^*-representation of M generated by $\varphi_i, i = 1, 2$. Then $\pi_i(M_i)$ is also a hyperfinite type (II_1) factor on $H_i, i = 1, 2$.

Let A be a (UHF) C^*-algebra of type $\{2^n\}$. From Proposition 7.2.9 and Theorem 7.2.12, there is a $*$ isomorphism Φ_i from A into $\pi_i(M_i)$ such that $\Phi_i(A)$ is $\sigma(M, M_*)$-dense in $\pi_i(M_i), i = 1, 2$. Thus, $\langle \Phi_i(\cdot)\xi_i, \xi_i \rangle$ is a tracial state on $A, i = 1, 2$. By Lemma 7.2.13,

$$\langle \Phi_1(a)\xi_1, \xi_1 \rangle = \langle \Phi_2(a)\xi_2, \xi_2 \rangle, \quad \forall a \in A.$$

Let $u\Phi_1(a)\xi_1 = \Phi_2(a)\xi_2, \forall a \in A$. Then u can be uniquely extended to a unitary operator from H_1 onto H_2, still denoted by u. Clearly, $u\Phi_1(a)u^* =$

$\Phi_2(a), \forall a \in A$. Therefore,

$$u\pi_1(M_1)u^* = \pi_2(M_2),$$

and M_1 is $*$ isomorphic to M_2. Q.E.D.

Proposition 7.2.15. Let M be a finite VN algebra on a Hilbert space H. If M is also hyperfinite, then M is a factor.

Proof. Let z be a central projection of M, and $z \neq 0, 1$. Then there exist $\xi, \eta \in H$ with $\|\xi\| = \|\eta\| = 1$ such that

$$z\xi = \xi, \quad z\eta = 0.$$

Since M is hyperfinite, there is a $(UHF)C^*$-algebra $A \subset M$ with $1 \in A$, and A is $\sigma(M, M_*)$-dense in M. From Proposition 6.3.14, we have the central valued trace $T : M \to Z = M \cap M'$. Then $\langle T(\cdot)\xi, \xi\rangle$ and $\langle T(\cdot)\eta, \eta\rangle$ are two tracial states on A. By Lemma 7.2.13, $\langle T(a)\xi, \xi\rangle = \langle T(a)\eta, \eta\rangle, \forall a \in A$. Further, this equality holds on whole M. In particular,

$$\begin{aligned} 1 = \langle z\xi, \xi\rangle = & \ \langle T(z)\xi, \xi\rangle = \langle T(z)\eta, \eta\rangle \\ = & \ \langle z\eta, \eta\rangle = 0, \end{aligned}$$

a contradiction. Therefore, M is a factor. Q.E.D.

Proposition 7.2.16. Let M be a hyperfinite type (II_1) factor,and $\{p_n\}$ be any sequence of positive integers with $p_n | p_{n+1}, \forall n$, and $p_n \to \infty$. Then there exists an increasing sequence $\{M_{p_n}\}$ of subfactors of M, where M_{p_n} is type (I_{p_n}), $\forall n$, such that $\cup_n M_{p_n}$ is $\sigma(M, M_*)$-dense in M.

Proof. From Proposition 7.1.3, we can pick

$$1 \in N_1 \subset \cdots \subset N_n \subset \cdots \subset M,$$

where N_n is a type (I_{p_n}) subfactor of M, $\forall n$. Let N be the weak closure of $\cup_n N_n$. Clearly, $N \subset M$, and N is also finite. By Proposition 7.2.15 and $p_n \to \infty, N$ is also a hyperfinite type (II_1) factor. From Theorem 7.2.14, we have a $*$ isomorphism Φ from N onto M. Now let $M_{p_n} = \Phi(N_n), \forall n$. Then $\{M_{p_n}\}$ is what we want to find. Q.E.D.

Notes. Contrary to the W^*-case, there are uncountably many non-isomorphic (UHF) C^*-algebras (see Chapter 15).

References. [49], [110], [113], [196].

7.3. Construction of factors of type (II) and type (III)

Definition 7.3.1. (M,G,α) is called a *dynamical system*, if M is a VN algebra, G is a discrete group, and α is a (group) homorphism from G into $\mathrm{Aut}(M)$, where $\mathrm{Aut}(M)$ is the group of all $*$ automorphisms of M.

In Chapter 16, we shall study general W^*- and C^*-dynamical systems. For the aim of this section, Definition 7.3.1 is enough.

Now let (M,G,α) be a dynamical system, and H be the action space of M. Consider Hilbert space $H\otimes l^2(G)$, and define

$$(\pi(a)\xi)(g)=\alpha_{g^{-1}}(a)\xi(g),\quad (\lambda(h)\xi)(g)=\xi(h^{-1}g),$$

$\forall a\in M,\quad g,h\in G$, and $\xi(\cdot)\in H\otimes l^2(G)$.

Proposition 7.3.2. $\{\pi,H\otimes l^2(G)\}$ is a faithful W^*-representation of M, $\{\lambda,H\otimes l^2(G)\}$ is a unitary representation of G, and

$$\lambda(g)\pi(a)\lambda(g)^*=\pi(\alpha_g(a)),\quad \forall a\in M,\quad g\in G.$$

Proof. Clearly, π is faithful. Let a net $\{a_l\}\subset M, \|a_l\|\leq 1$ and $a_l\rightarrow 0$ (weakly). Since

$$\begin{aligned}|\langle\pi(a_l)\xi,\xi\rangle| &= |\sum_{g\in G}\langle\alpha_{g^{-1}}(a_l)\xi(g),\xi(g)\rangle|\\ &\leq \sum_{g\in F}|\langle\alpha_{g^{-1}}(a_l)\xi(g),\xi(g)\rangle|+\sum_{g\notin F}\|\xi(g)\|^2,\end{aligned}$$

$\forall\xi\in H\otimes l^2(G)$, where F is any finite subset of G, it follows that $\pi(a_l)\rightarrow 0$(weakly). Thus π is also a W^*-representation of M. Moreover, we can check the equality:

$$\lambda(g)\pi(a)\lambda(h)^*=\pi(\alpha_g(a)),\quad \forall a\in M,\quad g\in G$$

directly. Q.E.D.

Definition 7.3.3. The VN algebra on $H\otimes l^2(G)$ generated by $\{\pi(a),\lambda(g)|a\in M,g\in G\}$ is called the *crossed product* of dynamical system (M,G,α), denoted by $M\times_\alpha G$, i.e.,

$$M\times_\alpha G=\{\pi(a),\lambda(g)|a\in M,g\in G\}''.$$

Now let $\widetilde{H}=H\otimes l^2(G)$, and write

$$\widetilde{H}=\sum_{g\in G}\oplus H_g,\quad H_g=H,\quad \forall g\in G.$$

Let p_g be the projection from $\widetilde{H}$ onto $H_g, \forall g \in G$. Then any $x \in B(\widetilde{H})$ has a matrix representation

$$x = (x_{g,h})_{g,h \in G},$$

where $x_{g,h} = p_g x p_h^* \in B(H)), \forall g, h \in G$. For any $a \in M$ and $k \in G$, it is easy to see that

$$p_g \pi(a) p_h^* = \delta_{g,h} \alpha_{g^{-1}}(a), \quad p_g \lambda(k) p_h^* = \delta_{h,k^{-1}g},$$

$$p_g \pi(a) \lambda(k) p_h^* = \delta_{h,k^{-1}g} \alpha_{g^{-1}}(a), \quad \forall g, h \in G.$$

In the following, we assume that

$$\alpha_g(a) = u_g a u_g^*, \quad \forall a \in M, \quad g \in G,$$

where $g \to u_g (\forall g \in G)$ is a unitary representation of G on H, and $u_g M u_g^* = M, \forall g \in G$.

Lemma 7.3.4. For any $x \in M \times_\alpha G$, there is unique function $b : G \to M$ such that

$$p_g x p_h^* = u_g^* b_{gh^{-1}} u_g, \quad \forall g, h \in G.$$

If let $\Phi(x) = b_e$, where e is the unit of G, then Φ is a σ - σ continuous positive linear map from $M \times_\alpha G$ to M.

Proof. For $a \in M, k \in G$, since $p_g \pi(a) \lambda(k) p_h^* = \delta_{h,k^{-1}g} u_g^* a u_g$, we have

$$p_g \pi(a) \lambda(k) p_h^* = u_g^* b_{gh^{-1}} u_g, \quad \forall g, h \in G,$$

where

$$b_g = \begin{cases} a, & \text{if } g = k, \\ 0, & \text{if } g \neq k. \end{cases}$$

Generally, for $\sum_i \pi(a_i) \lambda(k_i)$, where $a_i \in M, k_i \in G$, and $k_i \neq k_j, \forall i \neq j$, let

$$b_g = \begin{cases} a_i, & \text{if } g = k_i \text{ for some } i, \\ 0, & \text{if } g \neq k_i, \forall i. \end{cases}$$

Then

$$p_g \sum_i \pi(a_i) \lambda(k_i) p_h^* = u_g^* b_{gh^{-1}} u_g, \quad \forall g, h \in G.$$

Since $\{\sum_i \pi(a_i) \lambda(k_i) \mid a_i \in M, k_i \in G\}$ is σ-dense in $M \times_\alpha G$ by Proposition 7.3.2, thus for any $x \in M \times_\alpha G$, there is $b : G \to M$ such that $p_g x p_h^* = u_g^* b_{gh^{-1}} u_g, \forall g, h \in G$.

Notice that

$$u_g p_g \sum_i \pi(a_i) \lambda(k_i) p_h^* u_g^* = \sum_i \delta_{k_i, gh^{-1}} a_i$$

$\forall a_i \in M, k_i \in G$. Then by the σ-density of $\{\sum_i \pi(a_i)\lambda(k_i) \mid a_i \in M, k_i \in G\}$ in $M \times_\alpha G$, we can see that

$$u_{g_1} p_{g_1} x p_{h_1}^* u_{g_1}^* = u_{g_2} p_{g_2} x p_{h_2}^* u_{g_2}^*$$

$\forall x \in M \times_\alpha G$ and $g_1 h_1^{-1} = g_2 h_2^{-1}$. Therefore, for each $x \in M \times_\alpha G$ the function $b : G \to M$ is unique.

Since $\Phi(x) = p_e x p_e^* (\forall x \in M \times_\alpha G)$, it follows that Φ is σ-σ continuous. Moreover, if $x = (u_g^* b_{gh^{-1}} u_g) \in M \times_\alpha G$, then $\Phi(xx^*) = \sum_{g \in G} b_g b_g^*$. Therefore, Φ is positive. Q.E.D.

Lemma 7.3.5. Let φ be a faithful semi-finite normal trace on M_+. If φ is G-invariant , i.e. $\varphi(\alpha_g(a)) = \varphi(a), \forall a \in M_+, g \in G$, then $\psi = \varphi \circ \Phi$ is a faithful semi-finite normal trace on $(M \times_\alpha G)_+$, and $\varphi = \psi \circ \pi$. Moreover, ψ is finite if and only if φ is finite.

Proof. Let $x = (u_g^* b_{gh^{-1}} u_g) \in M \times_\alpha G$. Then

$$\Phi(xx^*) = \sum_{g \in G} b_g b_g^*, \quad \Phi(x^* x) = \sum_{g \in G} u_g^* b_g^* b_g u_g.$$

Thus, $\psi = \varphi \circ \Phi$ is a trace on $(M \times_\alpha G)_+$. Clearly. ψ is normal since Φ is σ-σ continuous. If $\psi(xx^*) = 0$, since φ is faithful and Φ is positive, then $\Phi(xx^*) = 0$, i.e., $b_g = 0, \forall g \in G$, and $x = 0$. Hence, ψ is faithful. The equality $\varphi = \psi \circ \pi$ is obvious.

Since φ is semi-finite, it follows from Proposition 6.5.4 that there is an increasing net $\{a_l\}$ of M_+ such that $\sup_l a_l = 1$ and $\varphi(a_l) < \infty, \forall l$. Then $\{\pi(a_l)\}$ is also an increasing net of $(M \times_\alpha G)_+, \sup_l \pi(a_l) = 1$, and $\psi(\pi(a_l)) = \varphi(a_l) < \infty, \forall l$. For any $0 \neq x \in (M \times_\alpha G)_+$, there is an index l_0 such that $x^{\frac{1}{2}} \pi(a) x^{\frac{1}{2}} \neq 0$, where $a = a_{l_0}$. Then by Proposition 6.5.2,

$$\psi(x^{\frac{1}{2}} \pi(a) x^{\frac{1}{2}}) = \psi(\pi(a)^{\frac{1}{2}} x \pi(a)^{\frac{1}{2}}) \leq \|x\| \psi(\pi(a)) < \infty.$$

Hence, ψ is semi-finite also.

Finally, from $\psi = \varphi \circ \Phi$ and $\varphi = \psi \circ \pi, \psi$ is finite $\Longleftrightarrow \varphi$ is finite. Q.E.D.

Lemma 7.3.6. Suppose that M is abelian, and $\pi(M)$ is maximal commutative in $M \times_\alpha G$. Then $M \times_\alpha G$ is semi-finite if and only if there exists a G-invariant faithful semi-finite normal trace on M_+.

Proof. The sufficiency is immediate from Lemma 7.3.5 and Theorem 6.5.8. Now let $M \times_\alpha G$ be semi-finite. Then there is a faithful semi-finite normal trace ψ on $(M \times_\alpha G)_+$. Let $\varphi = \psi \circ \pi$. It is easy that φ is a faithful normal trace on M_+. By Proposition 7.3.2 and ψ is a trace, we have

$$\begin{aligned}\varphi(\alpha_g(a)) &= \psi(\pi(\alpha_g(a))) = \psi(\lambda(g)\pi(a)\lambda(g)^*) \\ &= \psi(\pi(a)) = \varphi(a),\end{aligned}$$

$\forall a \in M_+, g \in G$, i.e. φ is G-invarient.

We point out the following fact: if $x \in M \times_\alpha G$, then $\pi(\Phi(x)) \in \overline{K_x}^w$, where $K_x = Co\{\pi(u)^* x \pi(u) \mid u$ is a unitary element of $M\}$, and $\overline{K_x}^w$ is the weak closure of K_x. In fact, suppose that u is a unitary element of M, and $x = (u_g^* b_{gh^{-1}} u_g)$ is as in Lemma 7.3.4. Then

$$p_g \pi(u)^* x \pi(u) p_g^* = u_g^* \Phi(x) u_g, \quad \forall g \in G.$$

Further,

$$p_g y p_g^* = u_g^* \Phi(x) u_g, \quad \forall g \in G, \quad y \in \overline{K_x}^w.$$

Since $\overline{K_x}^w$ is a compact convex subset of $(B(\widetilde{H})$, weak top.) and M is abelian, it follows from the Kakutani-Markov fixed point theorem (see [31]) that there is $x_0 \in \overline{K_x^w}$ such that $x_0 = \pi(u)^* x_0 \pi(u)$ for any unitary element u of M. But $\pi(M)$ is maximal commutative in $M \times_\alpha G$, so $x_0 \in \pi(M)$, i.e. $x_0 = \pi(a)$ for some $a \in M$. Since $p_g x_0 p_g^* = u_g^* \Phi(x) u_g = \alpha_{g^{-1}}(\Phi(x)), \forall g \in G$, then $a = \Phi(x)$, i.e. $\pi(\Phi(x)) = x_0 \in \overline{K_x}^w$.

Now we prove that φ is semi-finite. Let $0 \neq a \in M_+$. Then $\pi(a)$ is also a non-zero positive element of $M \times_\alpha G$. Thus there is $0 \neq x \in (M \times_\alpha G)_+$ with $x \leq a$ such that $\psi(x) < \infty$. Since M is abelian, it follows that

$$0 \leq y \leq \pi(a), \quad \forall y \in \overline{K_x}^w.$$

In particular, $0 \leq \pi(\Phi(x)) \leq \pi(a)$. Hence $0 \leq \Phi(x) \leq a$. Φ is faithful on $(M \times_\alpha G)_+$ (see the proof of Lemma 7.3.4). So $\Phi(x) \neq o$. Now it suffices to show that $\varphi(\Phi(x)) < \infty$.

Put $\pi(\Phi(x)) = x_0$. Then there is a net $\{x_l\} \subset K_x$ such that $x_l \to x_0$ (weakly). By Proposition 6.5.4 and since ψ is semi-finite, there exists an increasing net $\{y_t\} \subset (M \times_\alpha G)_+$ with $\sup_t y_t = 1$, and $\psi(y_t) < \infty, \forall t$. By Proposition 6.5.2,

$$\psi(y_t x_l) = \psi(x_l^{\frac{1}{2}} y_t x_l^{\frac{1}{2}}) \leq \psi(x_l) = \psi(x).$$

From Proposition 6.5.3, $\psi(y_t x_0) = \lim_l \psi(y_t x_l) \leq \psi(x)$. Further, since ψ is normal, it follows from Proposition 6.5.2 that

$$\begin{aligned}\varphi(\Phi(x)) &= \psi(x_0) = \lim_t \psi(x_0^{1/2} y_t x_0^{1/2}) \\ &= \lim_t \psi(y_t x_0) < \infty.\end{aligned}$$

Q.E.D.

Lemma 7.3.7. If M is maximal commutative on H, and $M \cap Mu_g = \{0\}, \forall g \neq e$, then $\pi(M)$ is maximal commutative in $M \times_\alpha G$.

Proof. Let $x = (u_g^* b_{gh^{-1}} u_g) \in (M \times_\alpha G) \cap \pi(M)'$. From $x\pi(a) = \pi(a)x$, we get $b_g u_g a = a b_g u_g, \forall a \in M, g \in G$. Since M is maximal commutative on H, it follows that $b_g u_g \in M \cap Mu_g, \forall g \in G$. By the assumption, $b_g = 0, \forall g \neq e$. Therefore, $x = \pi(\Phi(x)) \in \pi(M)$. Q.E.D.

Lemma 7.3.8. Suppose that $\pi(M)$ is maximal commutative, and

$$\{a \in M \mid \alpha_g(a) = a, \forall g \in G\} = \mathbb{C}1_H.$$

Then $M \times_\alpha G$ is a factor.

Proof. Let x be a central element of $M \times_\alpha G$. In particular, $x\pi(b) = \pi(b)x, \forall b \in M$. Thus, $x = \pi(a)$ for some $a \in M$. By Lemma 7.3.4, it is easy to see

$$u(k) = (\delta_{k,gh^{-1}} u_k) \in (M \times_\alpha G)', \quad \forall k \in G.$$

Consequently, $u(k)\pi(a) = \pi(a)u(k)$. Hence,

$$u_k a u_k^* = a, \quad \forall k \in G.$$

By the assumption, $a = \lambda 1_H$ for some $\lambda \in \mathbb{C}$. Therefore, $M \times_\alpha G$ is a factor. Q.E.D.

Definition 7.3.9. (G, Ω, μ) is called a *group measure space*, if Ω is a locally compact Hausdorff space satisfying the second countability axiom, μ is a regular Borel measure on Ω, G is countable discrete group of homeomorphisms on Ω, and let μ be *quasi-invariant* under G, i.e., $\mu_g \prec \mu, \forall g \in G$, where $d\mu_g(\cdot) = d\mu(g^{-1}\cdot)$, and the action corresponding to g on t is $gt, \forall g \in G, t \in \Omega$.

Clearly, $\mu_g \sim \mu, \forall g \in G$. Then there exists a measurable function $r_g(\cdot)$ on Ω such that

$$0 < r_g(t) < \infty, a.e.\mu, \quad \text{and} \quad d\mu_g(\cdot) = r_g(\cdot)d\mu(\cdot),$$

$\forall g \in G$. Since G is countable, we may assume that

$$r_{gh}(\cdot) = r_h(g^{-1}\cdot)r_g(\cdot), \ a.e.\mu, \quad \forall g, h \in G.$$

Definition 7.3.10. Let (G, Ω, μ) be a group measure space.

1) (G, Ω, μ) is said to be *free*, if for each $g \in G$ with $g \neq e$, we have

$$\mu(\{t \in \Omega \mid gt = t\}) = 0;$$

2) (G,Ω,μ) is said to be *ergodic*, if a Borel subset E of Ω satisfies:

$$\mu((E\sqcup gE)\backslash(E\cap gE))=0,\quad \forall g\in G,$$

then we have either $\mu(E)=0$ or $\mu(\Omega\backslash E)=0$;

3) (G,Ω,μ) is said to be *measurable*, if there exists a σ-finite measure ν on all Borel subsets of Ω such that $\nu\sim\mu$ and ν is G-invariant, i.e., $d\nu(g\cdot)=d\nu(\cdot),\forall g\in G$;

4) (G,Ω,μ) is said to be *non-measurable*, if it is not measurable.

Now let (G,Ω,μ) be a group measure space. Let

$$H=L^2(\Omega,\mu),\quad M=\{m_f\mid f\in L^\infty(\Omega,\mu)\},$$

where $m_f\cdot=f\cdot,\forall f\in L^\infty(\Omega,\mu),\forall\cdot\in L^2(\Omega,\mu)$. From Theorem 5.3.13, the multiplication algebra M is a maximal commutative VN algebra on $H=L^2(\Omega,\mu)$. For each $g\in G$, define

$$(u_gf)(\cdot)=r_g(\cdot)^{1/2}f(g^{-1}\cdot),\quad \forall f\in L^2(\Omega,\mu)=H.$$

Clearly, $g\to u_g$ is a unitary representation of G on H, and

$$u_g^*m_fu_g=m_{f_g},\quad \forall f\in L^\infty(\Omega,\mu),\quad g\in G,$$

where $f_g(\cdot)=f(g\cdot)$. Let $\alpha_g(m_f)=u_gm_fu_g^*,\forall f\in L^\infty(\Omega,\mu),g\in G$. Then (M,G,α) is a dynamical system, and there is a VN algebra $M\times_\alpha G$ on $\widetilde{H}=H\otimes l^2(G)$.

Lemma 7.3.11. If (G,Ω,μ) is free and ergodic, then $\pi(M)$ is maximal commutative in $M\times_\alpha G,\{a\in M\mid\alpha_g(a)=a,\forall g\in G\}=\mathbb{C}1_H$, and $M\times_\alpha G$ is a factor.

Proof. First, we claim that

$$M\cap Mu_g=\{0\},\quad \forall g\in G\ \text{ and }\ g\neq e.$$

In fact, let $g\in G$ with $g\neq e$, and $F_g=\{t\in\Omega|gt=t\}$. Clearly, F_g is a μ-zero closed subset of Ω. Replacing Ω by $(\Omega\backslash F_g)$, we may assume that $F_g=\emptyset$. Now let $m_{f_1}=m_{f_2}u_g\in M\cap Mu_g$, where $f_1,f_2\in L^\infty(\Omega,\mu)$, and define $E=\{t\in\Omega|f_1(t)\neq0\}$. For each $t\in\Omega$, since $gt\neq t$ and g is a homeomorphism of Ω, there is an open neighborhood V_t of t such that $V_t\cap gV_t=\emptyset$. Clearly, $\{V_t\mid t\in\Omega\}$ is an open cover of Ω. But Ω satisfies the second countability axiom, so Ω admits a countable open cover $\{V_n\}$ such that $V_n\cap gV_n=\emptyset,\forall n$. If $\mu(E)>0$, then we have $\mu(V\cap E)>0$ for some n, where $V=V_n$. Further, pick a Borel subset F of Ω with $F\subset V\cap E$ and $0<\mu(F)<\infty$. From $m_{f_1}\chi_F=m_{f_2}u_g\chi_F$, we get

$$f_1(t)\chi_F(t)=f_2(t)r_g(t)^{1/2}\chi_F(g^{-1}t),\ a.e.\mu.$$

From $F \cap gF = \emptyset$ and $F \subset E$, the above equality does not hold at each $t \in F$. This is a contradiction since $\mu(F) > 0$. Therefore, $\mu(E) = 0$, i.e., $f_1 = 0$, and $M \cap Mu_g = \{0\}$.

Now from Lemma 7.3.7, $\pi(M)$ is maximal commutation in $M \times_\alpha G$.

Let $f \in L^\infty(\Omega, \mu)$ be such that $u_g^* m_f u_g = m_f, \forall g \in G$. Then $f(gt) = f(t)$, $a.e.\mu, \forall g \in G$. We may assume that f is real. If f is not a constant function, then there are real number r_1, r_2 with $r_1 < r_2$ such that $\mu(E) > 0$ and $\mu(\Omega \backslash E) > 0$, where $E = \{t \in \Omega \mid r_1 \leq f(t) < r_2\}$. On the other hand, since $f(t) = f(gt), a.e.\mu, \forall g \in G$, and G is countable, it follows that

$$\mu((E \sqcup gE) \backslash (E \cap gE)) = 0, \quad \forall g \in G.$$

Then we get either $\mu(E) = 0$ or $\mu(\Omega \backslash E) = 0$ since (G, Ω, μ) is ergodic, a contradiction. Therefore, f is a constant function, i.e.,

$$\{a \in M \mid \alpha_g(a) = a, \forall g \in G\} = \mathbb{C} 1_H.$$

Finarrly, by Lemma 7.3.8, $M \times_\alpha G$ is a factor. Q.E.D.

Lemma 7.3.12. Let (G, Ω, μ) be a free and ergodic group measure space, and ν be a G-invariant σ-finite measure on all Borel subsets of Ω with $\nu \sim \mu$ and $\nu(\{t\}) = 0, \forall t \in G$.

1) If $\nu(\Omega) < \infty$, then $M \times_\alpha G$ is a type (II_1) factor.
2) If $\nu(\Omega) = +\infty$, then $M \times_\alpha G$ is a type (II_∞) factor.

Proof. Define

$$\varphi(m_f) = \int_\Omega f(t) d\nu(t), \quad \forall f \in L^\infty(\Omega, \mu)_+.$$

Then φ is faithful on M_+ since $\nu \sim \mu$. Let $\{m_{f_l}\}$ be a bounded increasing net of M_+, and $m_f = \sup_l m_{f_l}$. By Theorem 5.3.13, $f_l \to f$ with respect to w^*-topology in $L^\infty(\Omega, \mu)$ or $L^\infty(\Omega, \nu)$. Since ν is σ-finite, we can write $\Omega = \sqcup_n E_n$, where $\{E_n\}$ is an increasing sequence of Borel subsets of Ω, and $\nu(E_n) < \infty, \forall n$. Thus $\chi_{E_n} \in L^1(\Omega, \nu)$ and

$$\int f_l \chi_{E_n} d\nu \to \int f \chi_{E_n} d\nu, \quad \forall n.$$

Further,

$$\sup_l \int f_l d\nu = \int \sup_l f_l d\nu,$$

i.e., φ is normal. The semi-finiteness of φ is obvious from the σ-finiteness of ν. Moreover, since ν is G-invariant, it follows that

$$\varphi(u_g^* m_f u_g) = \int f(gt) d\nu(t) = \int f(t) d\nu(t) = \varphi(m_f),$$

$\forall m_f \in M_+, g \in G$, i.e., φ is also G-invariant. Now by Lemma 7.3.6 and 7.3.11, $M \times_\alpha G$ is a semi-finite factor. If $\nu(\Omega) < \infty$, then φ is finite, and $M \times_\alpha G$ is also a finite factor from Lemma 7.3.5. If $\nu(\Omega) = +\infty$, then φ is not finite, and $M \times_\alpha G$ is an infinite factor from Lemma 7.3.5 and Proposition 7.1.2.

Now it suffices to show that $M \times_\alpha G$ is ciontinuous. Let p be any non-zero projection of M with $\varphi(p) < \infty$. Then by Lemma 7.3.5, $\psi = \varphi \circ \Phi$ is a faithful semi-finite normal trace on $(M \times_\alpha G)_+$ and $\varphi = \psi \circ \pi$. Thus $\psi(\pi(p)) < \infty$, and $\pi(p)$ is a non-zero finite projection of $M \times_\alpha G$ by Proposition 7.1.2. If $M \times_\alpha G$ is not continuous, we may assume that $M \times_\alpha G = B(K)$, where K is some Hilbert space. Then $\dim \pi(p) K < \infty$, and M contains a nonzero minimal projection $(\leq p)$. This contradicts the assumption: $\nu(\{t\}) = 0, \forall t \in \Omega$. Therefore, $M \times_\alpha G$ is continuous. Q.E.D.

Lemma 7.3.13. Let (G, Ω, μ) be free and ergodic. If (G, Ω, μ) is non-measurable, then $M \times_\alpha G$ is a type (III) factor.

Proof. If $M \times_\alpha G$ is semi-finite, then by Lemmas 7.3.11 and 7.3.6, there is a G-invariant faithful semi-finite normal trace φ on M_+. For any Borel subset E of Ω, define

$$\nu(E) = \varphi(m_{\chi_E}).$$

Then ν is a measure on all Borel subsets of Ω. Since φ is faithful, it follows that $\nu \sim \mu$. From the G-invariance of φ, ν is also G-invariant. By the Zorn lemma and the semi-finiteness of φ, there is an orthogonal family $\{p_l\}_{l \in \Lambda}$ of projections of M such that $\sum_{l \in \Lambda} p_l = 1$ and $\varphi(p_l) < \infty, \forall l$. Since $H = L^2(\Omega, \mu)$ is separable, Λ is countable. Suppose that $p_l = m_{\chi_{E_l}}$, where E_l is a Borel subset of $\Omega, \forall l$. Then $\nu(E_l) = \varphi(p_l) < \infty, \forall l$, and

$$\nu(\Omega \backslash \sqcup_{l \in \Lambda} E_l) = \varphi(1 - \sum_{l \in \Lambda} p_l) = 0.$$

Thus, ν is σ-finite. From Definition 7.3.10, (G, Ω, μ) is measurable, a contradiction. Therefore, $M \times_\alpha G$ is not semi-finite, and is a type (III) factor.

Q.E.D.

Lemma 7.3.14. Let (G, Ω, μ) be a group measure space, and

$$G_0 = \{g \in G | r_g(t) = 1, a.e.\mu\}$$

Then G_0 is a subgroup of G. If (G_0, Ω, μ) is ergodic, and $G_0 \neq G$, then (G, Ω, μ) is non-measurable.

Proof. Clearly, G_0 is a subgroup of G. Now let (G_0, Ω, μ) is ergodic, and $G_0 \neq G$. Suppose that ν is a σ-finite measure on all Borel subsets of Ω with

$\nu \sim \mu$, and ν is G-invariant. For any $g \in G_0$, since $\mu_g = \mu \sim \nu = \nu_g$, it follows that

$$\begin{aligned} \frac{d\mu}{d\nu}(t)d\nu(t) &= d\mu(t) = d\mu(g^{-1}t) \\ &= \frac{d\mu}{d\nu}(g^{-1}t)d\nu(g^{-1}t) = \frac{d\mu}{d\nu}(g^{-1}t)d\nu(t). \end{aligned}$$

Thus, $\frac{d\mu}{d\nu}(t) = \frac{d\mu}{d\nu}(g^{-1}t), \forall g \in G_0$. Now (G_0, Ω, μ) is ergodic, so $\frac{d\mu}{d\nu}(t) =$ constant $(a.e.\mu)$ by a similar discussion of Lemma 7.3.11. Further, μ is also G-invariant, i.e., $G = G_0$, a contradiction. Therefore, (G, Ω, μ) is non-measurable. Q.E.D.

From above discussions, we have the following.

Theorem 7.3.15. Let (G, Ω, μ) be a free and ergodic group measure space, and

$$H = L^2(\Omega, \mu), \quad M = \{m_f \mid f \in L^\infty(\Omega, \mu)\}$$

$$(u_g f)(t) = r_g(t)^{1/2} f(g^{-1}t), \quad \forall f \in H, g \in G,$$

$$\alpha_g(m_f) = u_g m_f u_g^*, \quad \forall f \in L^\infty(\Omega, \mu), g \in G,$$

where $r_g(\cdot) = (d\mu_g/d\mu)(\cdot)$ and $d\mu_g(\cdot) = d\mu(g^{-1}\cdot), \forall g \in G$.

1) If there is a σ-finite G-invariant measure on all Borel subsets of Ω with $\nu \sim \mu$ and $\nu(\{t\}) = 0, \forall t \in \Omega$, then $M \times_\alpha G$ is a type (II_1) factor when $0 < \nu(\Omega) < \infty$, and $M \times_\alpha G$ is a type (II_∞) factor when $\nu(\Omega) = +\infty$.

2) Let $G_0 = \{t \in G \mid r_g(t) = 1, a.e.\mu\}$ (a subgroup of G). If (G_0, Ω, μ) is ergodic and $G_0 \neq G$, then $M \times_\alpha G$ is a type (III) factor.

Example 1. Let Ω be one dimensional circle group (compact group), i.e., $\Omega = \{z \in \mathbb{C} \mid |z| = 1\}, \mu$ be the Haar measure on Ω with $\mu(\Omega) = 1, G$ be a countable infinite subgroup on Ω, and the action α of G to Ω be the multiplication of numbers.

Clearly, (G, Ω, μ) is free, μ is G-invariant, and $\mu(\{z\}) = 0, \forall z \in \Omega$.

Suppose that E is a Borel subset of Ω such that

$$\mu((E \sqcup gE)\backslash(E \cap gE)) = 0, \quad \forall g \in G.$$

Write

$$\chi_E(z) = \sum_n \lambda^n z^n,$$

where $\{z^n \mid n \in \mathbb{Z}\}$ is a normalized orthogonal basis of $L^2(\Omega, \mu)$. Then

$$\sum_n \lambda_n z^n = \chi_E(z) = \chi_E(gz) = \sum_n \lambda_n g^n z^n, \quad a.e.\mu,$$

$\forall g \in G$. Thus, $\lambda_n = 0, \forall n \neq 0$, i.e., either $\mu(E) = 0$ or $\mu(\Omega \backslash E) = 0$,and (G, Ω, μ) is ergodic.

Now by Theorem 7.3.15, $M \times_\alpha G$ is a type (II_1) factor.

Example 2. Let $\Omega = \mathbb{R}$ (a locally compact abelian group), μ be the Haar measure on Ω, G be a countable infinite dense subgroup of Ω (for example, $G = \{r \in \mathbb{R} \mid r \text{ is rational}\}$), and the action α of G on Ω be the addition of numbers.

Clearly, (G, Ω, μ) is free, μ is G-invariant, and $\mu(\Omega) = \infty, \mu(\{\eta\}) = 0, \forall \eta \in \Omega$.

Suppose that E is a Borel subset of Ω such that

$$\mu((E \sqcup (E + \eta)) \backslash (E \sqcap (E + \eta))) = 0, \quad \forall \eta \in G,$$

i.e., $u_\eta^* m_{\chi_E} u_\eta = m_{\chi_E}, \quad \forall \eta \in G$, where $\eta \to u_\eta$ is the regular representation of Ω on $L^2(\Omega, \mu)$. Since G is dense in Ω, it follows that $m_{\chi_E} u_\eta = u_\eta m_{\chi_E}, \forall \eta \in \Omega$. Thus, we have either $\mu(E) = 0$ or $\mu(\Omega \backslash E) = 0$, i.e., (G, Ω, μ) is ergodic.

Now by Theorem 7.3.15, $M \times_\alpha G$ is a type (II_∞) factor.

Example 3. Let (Ω, μ) be as in Example 2, $G = \{(\rho, \sigma) \mid \rho > 0, \rho, \sigma \text{ rational}\}$, and

$$\alpha(\rho, \sigma)\eta = \rho\eta + \sigma, \quad \forall (\rho, \sigma) \in G, \eta \in \Omega.$$

Clearly, (G, Ω, μ) is free, and μ is quasi-invariant under G.

Let $G_0 = \{(1, \sigma) \mid \sigma \text{ rational}\}$. By Example 2, (G_0, Ω, μ) is ergodic. Clearly, $G_0 \neq G$.

Now by Theorem 7.3.15, $M \times_\alpha G$ is a type (III) factor.

Theorem 7.3.16. On a separable Hilbert space, there exist five classes of factors: type (I_n), (I_∞) (II_1) (II_∞) (III) factors.

Type (II_∞) factors can be indeed constructed through type (II_1) factors.

Proposition 7.3.17. A factor M is type (II_∞) if and only if $M = N \overline{\otimes} B(H_\infty)$, where N is a type (II_1) factor, and H_∞ is a infinite dimensional Hilbert space.

Proof. The sufficiency is obvious from Theorem 6.9.12. Now suppose that M is a type (II_∞) factor. Pick a non-zero finite projection p of M, and let $\{p_l\}_{l \in \Lambda}$ be a maximal orthogonal family of projections of M such that $p_l \sim p, \forall l$. Then $q = 1 - \sum_{l \in \Lambda} p_l \precsim p$ by Proposition 6.4.5, $^\#\Lambda = \infty$. Thus,

$$1 = \sum_{l \in \Lambda} p_l + q \sim \sum_{l \in \Lambda} p_l.$$

Further, there exists an orthogonal family $\{q_l\}_{l \in \Lambda}$ of projections of M such that

$$\sum_{l \in \Lambda} q_l = 1, \quad q_l \sim p, \forall l.$$

Now by Theorem 1.5.6, $M = M_p \overline{\otimes} B(H_\infty)$, where M_p is a type (II_1) factor, and $\dim H_\infty = \sharp\Lambda = \infty$. Q.E.D.

We have another method to construct type (II_1) factors.

Let G be a discrete group, and $g \to \lambda_g, \rho_g$ be the left, right regular representations of G on $l^2(G)$ respectively, i.e.,

$$(\lambda_g f)(\cdot) = f(g^{-1}\cdot), \quad (\rho_g(f))(\cdot) = f(\cdot g),$$

$\forall f \in l^2(G), g \in G$. Let $R(G) = \{\lambda_g \mid g \in G\}''$.

Lemma 7.3.18. $R(G)$ is a σ-finite and finite VN algebra on $l^2(G)$.

Proof. For each $g \in G$, let $\varepsilon_g(k) = \delta_{g,k}$. Clearly ε_g is a unit vector of $l^2(G)$. Let e be the unit of G, and define

$$\varphi(a) = \langle a\varepsilon_e, \varepsilon_e\rangle, \forall a \in R(G).$$

Then φ is a normal state on $R(G)$. If $a \in R(G)$ satisfies $a\varepsilon_e = 0$, then

$$0 = \rho_{g^{-1}} a\varepsilon_e = a\rho_{g^{-1}}\varepsilon_e = a\varepsilon_g, \quad \forall g \in G.$$

But $[\varepsilon_g \mid g \in G]$ is dense in $l^2(G)$, so $a = 0$, i.e., φ is faithful. Moreover, since $\varphi(\lambda_g\lambda_h) = \varphi(\lambda_h\lambda_g), \forall g, h \in G$, it follows that $\varphi(ab) = \varphi(ba), \forall a, b \in R(G)$. Thus, φ is also a trace. Now by Proposition 6.3.15, $R(G)$ is σ-finite and finite. Q.E.D.

Definition 7.3.19. An infinite countable discrete group G is said to be of *infinite conjugacy class*, if for any $e \neq g \in G$, the conjugacy class $\{hgh^{-1} \mid h \in G\}$ of g is infinite. We often abbreviate such a group as an *ICC-group.*

For example, the group of all finite permutations of $\mathbb{N} = \{1, 2, \cdots\}$ is an ICC-group, and the free group of two or more generators is also an ICC-group,and etc.

Proposition 7.3.20. If G is an ICC-group, then $R(G)$ is a type (II_1) factor on $l^2(G)$.

Proof. Let $a \in R(G) \cap R(G)'$. Then for any $g \in G$,

$$a\varepsilon_e = \lambda_g a \lambda_{g^{-1}}\varepsilon_e = \lambda_g a \rho_g \varepsilon_e = \rho_g \lambda_g a \varepsilon_e,$$

i.e., $(a\varepsilon_e)(\cdot) = (a\varepsilon_e)(g^{-1}\cdot g), \forall g \in G$. Since $(a\varepsilon_e) \in l^2(G)$ and G is ICC, it follows that $(a\varepsilon_e)(h) = 0, \forall h \neq e$, i.e., $a\varepsilon_e = \lambda\varepsilon_e$ for some $\lambda \in \mathbb{C}$. By the proof of Lemma 7.3.18, $a = \lambda$. Thus, $R(G)$ is a factor. Moreover, $R(G)$ is infinite dimensional. Now by Lemma 7.3.18, $R(G)$ is type (II_1). Q.E.D.

Proposition 7.3.21. Let G be an ICC-group,and $\{G_n\}$ be an increasing sequence of finite subgroups of G with $G = \sqcup_n G_n$. Then $R(G)$ is a hyperfinite type (II_1) factor on $l^2(G)$.

Proof. Clearly, for each $n, [\lambda_g \mid g \in G_n]$ is a finite dimensional $*$ subalgebra of $R(G)$, and $\sqcup_n[\lambda_g \mid g \in G_n]$ is $\sigma(M, M_*)$-dense in M. Now by Theorem 7.2.12 and Proposition 7.3.20, we get the conclusion. Q.E.D.

Remark. Let G be the group of all finite permutations of $\mathbb{N} = \{1, 2, \cdots,\}$, and G_n be the finite subgroup of all permutations of $\{1, \cdots, n\}, \forall n$. Then $G = \sqcup_n G_n$.

Notes. the construction of factors in the section is standard. It is called the group measure space construction (of Murry-Von Neumann).

References. [113], [119], [132].

7.4 The existences of non–hyperfinite type (II_1) factors and non–nuclear C^*–algebras

Consider a discrete group G. Let $\varepsilon_g(h) = \delta_{g,h}, \forall g, h \in G$. Then $\{\varepsilon_g | g \in G\}$ is an orthogonal normalized basis of $l^2(G)$. Suppose that $g \longrightarrow \lambda_g, \rho_g$ are the left, right regular representations of G on $l^2(G)$ respectively, and $R(G) = \{\lambda_g | g \in G\}''$. By Lemma 7.3.18, $R(G)$ is a σ–finite and finite VN algebra on $l^2(G); \varphi(\cdot) = \langle \cdot\varepsilon_e, \varepsilon_e\rangle$ is a faithful normal tracial state on $R(G)$, where e is the unit of G; and ε_e is a cyclic–separating vector for $R(G)$.

Proposition 7.4.1. Define $jx\varepsilon_e = x^*\varepsilon_e, \forall x \in R(G)$. Then j can be uniquely extended to a conjugate linear isometry on $l^2(G)$, still denoted by j, and

$$j^2 = I; \quad \langle j\xi, j\eta\rangle = \langle \eta, \xi\rangle, \forall \xi, \eta \in l^2(G);$$

$$jx'\varepsilon_e = x'^*\varepsilon_e, \forall x' \in R(G)'; \ j\lambda_g j = \rho_g, \forall g \in G.$$

Moreover, $jR(G)j = R(G)' = \{\rho_g | g \in G\}''$.

Proof. Since $\varphi(\cdot)$ is tracial and ε_e is cyclic for $R(G)$, j can be uniquely extended to a conjugate linear isometry on $l^2(G)$. Clearly, $j^2 = I$, and $j\lambda_g j = \rho_g, \forall g \in G$. By $\langle jx\varepsilon_e, jy\varepsilon_e\rangle = \langle y\varepsilon_e, x\varepsilon_e\rangle, \forall x, y \in R(G)$, we have $\langle j\xi, j\eta\rangle =$

$\langle\eta,\xi\rangle, \forall\xi,\eta\in l^2(G)$. For any $x'\in R(G)', x\in R(G)$,

$$\begin{aligned}\langle jx'\varepsilon_e, x\varepsilon_e\rangle &= \langle jx'\varepsilon_e, jx^*\varepsilon_e\rangle \\ &= \langle x^*\varepsilon_e, x'\varepsilon_e\rangle = \langle x'^*\varepsilon_e, x\varepsilon_e\rangle.\end{aligned}$$

Hence, $jx'\varepsilon_e = x'^*\varepsilon_e, \forall x'\in R(G)'$. Further, by

$$\begin{aligned}jx'jy'z'\varepsilon_e = &\ jx'z'^*y'^*\varepsilon_e \\ = &\ y'z'x'^*\varepsilon_e = y'jx'jz'\varepsilon_e,\end{aligned}$$

$\forall x', y', z'\in R(G)$, we have $jR(G)'j\subset R(G)$. On the other hand , $jR(G)j = \{\rho_g|g\in G\}''\subset R(G)'$. Therefore, we obtain that $jR(G)j = R(G)' = \{\rho_g|g\in G\}''$. Q.E.D.

For any $\xi,\eta\in l^2(G)$, let

$$(\xi*\eta)(g) = \sum_{h\in G}\xi(h)\eta(h^{-1}g), \quad \xi^*(g) = \overline{\xi(g^{-1})}, \ \forall g\in G.$$

Clearly,

$$|(\xi*\eta)(g)|\leq\|\xi\|\cdot\|\eta\|, \quad \|\xi^*\| = \|\xi\|.$$

Proposition 7.4.2. Let

$$B = \left\{b\in l^2(G)\;\middle|\;\begin{array}{c}\text{there is a positive constant } K = K(b)\\ \text{such that } (b*c)\in l^2(G)\text{and}\|b*c\|\leq K\|c\|, \forall c\in l^2(G).\end{array}\right\}$$

Then:

(i)

$$\begin{aligned}B &= \{b\in l^2(G)|\ b*c\in l^2(G), \forall c\in l^2(G)\} \\ &= \left\{b\in l^2(G)\;\middle|\;\begin{array}{c}\text{there is } \lambda(b)\in B(l^2(G))\text{such that}\\ \lambda(b)\varepsilon_g = b*\varepsilon_g = \rho_g^* b, \forall g\in G\end{array}\right\};\end{aligned}$$

(ii) B is a $*$ algebra, and $b\longrightarrow\lambda(b)$ is a faithful $*$ representation of B on $l^2(G)$, where $\lambda(b)c = b*c, \forall b\in B, c\in l^2(G)$;

(iii) $R(G) = \lambda(B) = \{\lambda(b)|b\in B\}$. Consequently, if we define $\|b\| = \|\lambda(b)\|, \forall b\in B$, then B is a σ-finite and finite W^*-algebra, and $\varphi(b) = \langle\lambda(b)\varepsilon_e,\varepsilon_e\rangle(\forall b\in B)$ is a faithful normal tracial state on B;

(iv) $j\lambda(b)j = \rho(b)$, where $\rho(b)c = c*b^*, \forall b\in B, c\in l^2(G)$. And $R(G)' = \rho(B) = \{\rho(b)|b\in B\}$.

Proof. (i) Let $b\in l^2(G)$, and $b*c\in l^2(G), \forall c\in l^2(G)$. Then we can define a linear operator $\lambda(b)$ on $l^2(G)$:

$$\lambda(b)c = b*c, \quad \forall c\in l^2(G).$$

We claim that $\lambda(b)$ is continuous. It suffices to show that $\lambda(b)$ is a closed operator. Suppose that $\{\xi_n\}$ is sequence of $l^2(G)$ such that

$$\xi_n \longrightarrow 0, \text{and} \quad \lambda(b)\xi_n = b * \xi_n \longrightarrow \eta$$

in $l^2(G)$, where $\eta \in l^2(G)$. We need to prove $\eta = 0$. For any $g \in G$,

$$\begin{aligned}\langle \eta, \varepsilon_g \rangle &= \lim_n \langle b * \xi_n, \varepsilon_g \rangle \\ &= \lim_n \sum_{h \in G} b(h)\xi_n(h^{-1}g) = \lim_n \langle \xi_n, c \rangle = 0,\end{aligned}$$

where $c(\cdot) = b(g \cdot^{-1}) \in l^2(G)$. Hence, $\eta = 0$, and $\lambda(b)$ is continuous. Further, we have

$$B = \{b \in l^2(G) | b * c \in l^2(G), \forall c \in l^2(G)\}.$$

Now let $b \in l^2(G)$, and suppose that there is $\lambda(b) \in B(l^2(G))$ such that $\lambda(b)\varepsilon_g = b * \varepsilon_g, \forall g \in G$. Clearly, such $\lambda(b)$ is unique, and $\lambda(b)c = b * c, \forall c \in [\varepsilon_g | g \in G]$. For any $c \in l^2(G)$ and any finite subset F of G, let

$$c_F = \sum_{g \in G} \langle c, \varepsilon_g \rangle \varepsilon_g.$$

Then $c_F \longrightarrow c$ in $l^2(G)$, and $\lambda(b)c = \lim \lambda(b)c_F = \lim(b * c_F)$. In particular,

$$(\lambda(b)c)(g) = \langle \lambda(b)c, \varepsilon_g \rangle = \lim \langle b * c_F, \varepsilon_g \rangle = (b * c)(g)$$

since $|(b * c_F)(g) - (b * c)(g)| \leq \|b\| \cdot \|c_F - c\| \to 0, \forall g \in G$. Hence, $b * c = \lambda(b)c \in l^2(G), \forall c \in l^2(G)$, and $b \in B$.

(iii) For any $b \in B$ and $x \in R(G)$, we have

$$x\lambda(b)\varepsilon_s = x\rho_s^* b = \rho_s^* x b, \quad \forall s \in G.$$

By (i) , we get $xb \in B$ and $\lambda(xb) = x\lambda(b), \forall x \in R(G), b \in B$. In particular,

$$x = x\lambda(\varepsilon_e) = \lambda(x\varepsilon_e) \in \lambda(B), \ \forall x \in R(G).$$

On the other hand, for any $b \in B$,

$$\lambda(b)\rho_g\varepsilon_h = \lambda(b)\varepsilon_{hg^{-1}} = \rho_{hg^{-1}}^* b = \rho_g\lambda(b)\varepsilon_h,$$

$\forall g, h \in G$. Hence, $\lambda(b) \in \{\rho_g | g \in G\}' = R(G), \forall b \in B$.

(ii) For any $a, b \in B$,

$$\lambda(a)\lambda(b)\varepsilon_g = \lambda(a)\rho_g^* b = \rho_g^* \lambda(a)b, \ \forall g \in G.$$

Then by (i) we have $\lambda(a)b = a * b \in B$, and $\lambda(a * b) = \lambda(a)\lambda(b)$. Moreover,

$$\langle \lambda(b)^* \varepsilon_g, \varepsilon_h \rangle = \langle \varepsilon_g, b * \varepsilon_h \rangle = \langle b^* * \varepsilon_g, \varepsilon_h \rangle,$$

$\forall g,h \in G, b \in B$. Hence , $\lambda(b)^*\varepsilon_g = b^* * \varepsilon_g$, and by (i) we get $b^* \in B$ and $\lambda(b)^* = \lambda(b^*), \forall b \in B$. Therefore, B is a $*$ algebra, and $b \longrightarrow \lambda(b)$ is a $*$ representation of B on $l^2(G)$. Moreover, if $\lambda(B) = 0$ for some $b \in B$, then $0 = \lambda(b)\varepsilon_e = b * \varepsilon_e = b$.

(iv) It is obvious. Q.E.D.

Definition 7.4.3. Let M be a finite factor, and $\varphi(\cdot)$ be the (unique) faithful normal tracial state on M. We say that M has the *property* (Γ) , if for any $x_1, \cdots, x_m \in M$, and $\varepsilon > 0$, there is a unitary element u of M such that

$$\varphi(u) = 0, \quad \text{and} \quad \|u^* x_i u - x_i\|_2 \le \varepsilon, \ 1 \le i \le m,$$

where $\|x\|_2^2 = \varphi(x^*x), \forall x \in M$.

Proposition 7.4.4. If M is a hyperfinite type (II_1) factor, then M has the property (Γ).

Proof. By Proposition 7.2.10, there is an increasing sequence $\{M_n | n \ge 0\}$ of subfactors of M such that $\sqcup_n M_n$ is weakly dense in M, where M_n is type (I_{2^n}), $\forall n$. Now for any $x_1, \cdots, x_m \in M$, we can find n and $y_1, \cdots, y_m \in M_n$ such that $\|x_i - y_i\|_2 \le \varepsilon/2, 1 \le i \le m$. M_{n+1} is $*$ isomorphic to the tensor product of M_n and a type (I_2) subfactor. Hence, there is a unitary element u of M_{n+1} such that $\varphi(u) = 0$ and $uy_i = y_i u, 1 \le i \le m$. Further,

$$\|u^{-1}x_i u - x_i\|_2 \le \|u^{-1}(x_i - y_i)u\|_2 + \|x_i - y_i\|_2 \le \varepsilon,$$

$1 \le i \le m$. Therefore, M has the property (Γ). Q.E.D.

Proposition 7.4.5. Let G be an ICC group. If there is a non-empty subset F of G and elements g_1, g_2, g_3 of G with following properties:

(i) $F \sqcup g_1 F g_1^{-1} \sqcup \{e\} = G$,

(ii) the subsets $F, g_2 F g_2^{-1}$ and $g_3 F g_3^{-1}$ are disjoint,

then $R(G)$ has no the property (Γ). Therefore, by Proposition 7.3.20 and 7.4.4 $R(G)$ is a non-hyperfinite type (II_1) factor.

Proof. Suppose that $R(G)$ has the property (Γ). Then for $\varepsilon > 0$, by Proposition 7.4.2 there is $b \in B$ such that: $\lambda(b) = u$ is a unitary element of $R(G); \varphi(u) = \langle u\varepsilon_e, \varepsilon_e \rangle = 0$; and $\|x_i - u^* x_i u\|_2 \le \varepsilon$, where $x_i = \lambda(\varepsilon_{g_i}) = \lambda_{g_i}, 1 \le i \le 3$. Noticing that

$$(x_i^* - u^* x_i^* u)(x_i - u^* x_i u) = (u^* - x_i^* u^* x_i)(u - x_i^* u x_i),$$

we have

$$\begin{aligned} \|x_i - u^* x_i u\|_2 &= \|u - x_i^* u x_i\|_2 = \|\lambda(b - \varepsilon_{g_i}^* * b * \varepsilon_{g_i})\|_2 \\ &= \|b - \varepsilon_{g_i}^* * b * \varepsilon_{g_i}\| \le \varepsilon, \quad 1 \le i \le 3. \end{aligned}$$

Since $0 = \varphi(u) = b(e)$ and $\|b\| = \|u\|_2 = \varphi(u^*u)^{1/2} = 1$, it follows from the properties of F and g_1, g_2, g_3 that

$$\begin{aligned} 1 &\leq \sum_{g\in F} |b(g)|^2 + \sum_{g\in g_1 F g_1^{-1}} |b(g)|^2 \\ &= \sum_{g\in F} |b(g)|^2 + \sum_{g\in F} |(\varepsilon_{g_1}^* * b * \varepsilon_{g_1})(g)|^2 \end{aligned}$$

and

$$\begin{aligned} 1 &\geq \Big(\sum_{g\in F} + \sum_{g\in g_2Fg_2^{-1}} + \sum_{g\in g_3Fg_3^{-1}}\Big)|b(g)|^2 \\ &= \sum_{g\in F} |b(g)|^2 + \sum_{g\in F} |(\varepsilon_{g_2}^* * b * \varepsilon_{g_2})(g)|^2 + \sum_{g\in F} |(\varepsilon_{g_3}^* * b * \varepsilon_{g_3})(g)|^2. \end{aligned}$$

Let $b_i = \varepsilon_{g_1}^* * b * \varepsilon_{g_i}, 1 \leq i \leq 3$. Then by $\|b - b_i\| \leq \varepsilon, 1 \leq i \leq 3$, we have

$$\begin{aligned} & \sum_{g\in F} |b_1(g)|^2 = \sum_{g\in F} |(b_1(g) - b(g)) + b(g)|^2 \\ =\ & \sum_{g\in F} |b_1(g) - b(g)|^2 + \sum_{g\in F} |b(g)|^2 + 2Re \sum_{g\in F} \overline{(b_1(g) - b(g))} b(g) \\ \leq\ & \varepsilon^2 + \sum_{g\in F} |b(g)|^2 + 2\Big(\sum_{g\in F} |b_1(g) - b(g)|^2 \cdot \sum_{g\in F} |b(g)|^2\Big)^{1/2} \\ \leq\ & \varepsilon^2 + 2\varepsilon + \sum_{g\in F} |b(g)|^2, \end{aligned}$$

and

$$\sum_{g\in F} |b_i(g)|^2 \geq \sum_{g\in F} |b(g)|^2 - \varepsilon^2 - 2\varepsilon, \quad i = 2, 3.$$

Further,

$$3\sum_{g\in F} |b(g)|^2 - 4\varepsilon - 2\varepsilon^2 \leq 1 \leq 2\sum_{g\in F} |b(g)|^2 + 2\varepsilon + \varepsilon^2,$$

i.e.

$$\frac{1 - 2\varepsilon - \varepsilon^2}{2} \leq \sum_{g\in F} |b(g)|^2 \leq \frac{1 + 4\varepsilon + 2\varepsilon^2}{3}.$$

This is a contradiction if $\varepsilon(> 0)$ is small enough. Therefore, $R(G)$ has no the property (Γ). Q.E.D.

Theorem 7.4.6. There exists a non-hyperfinite type (II_1) factor on a separable Hilbert space.

Proof. Let $G = F_2$ be the free group generated by two elements g_1 and g_2, F be the subset of all reduced words of G ending with $g_1^n, n = \pm 1, \pm 2, \cdots$. If b is a reduced word of G ending with g_2^m and $m \neq 0$, then $c = g_1^{-1} b g_1 \in F$. Hence, $F \sqcup g_1 F g_1^{-1} \sqcup \{e\} = G$. Further, let $g_3 = g_2^{-1}$. Then $F, g_2 F g_2^{-1}$ and $g_3 F g_3^{-1}$ are disjoint obviously. Now by Proposition 7.4.5, the conclusion is obvious.

Q.E.D.

Now we construct examples of non-nuclear C^*-algebras. Let R be a type (II_1) factor on a Hilbert space H such that R' is also type (II_1). Suppose that A, B are C^*-subalgebras of R, R' respectively, and C is the C^*-algebra on H generated by $A \sqcup B$. Define a map $\Phi : A \otimes B \longrightarrow C$ as follows:

$$\Phi(\sum_i a_i \otimes b_i) = \sum_i a_i b_i, \quad \forall a_i \in A, b_i \in B.$$

The map Φ is well-defined. In fact, if $\sum_i a_i \otimes b_i = 0$, let $\{b'_j\}$ be a basis of $[b_i | i]$, and $b_i = \sum_j \lambda_{ij} b'_j, \forall i$, then we have $\sum_i \lambda_{ij} a_i = 0, \forall j$. Hence,

$$\sum_i a_i b_i = \sum_i a_i \sum_j \lambda_{ij} b'_j = \sum_j b_j (\sum_i \lambda_{ij} a_i) = 0.$$

Further, we say that Φ is injective. In fact, if $\sum_i a_i b_i = 0$, let $a_{ik} = a_k, \forall k; b_{kj} = b_k, \forall j$, then by Proposition 1.7.3 there are numbers $\{\lambda_{ij}\}$ such that

$$\begin{cases} \sum_k a_{ik}\lambda_{kj} = \sum_k \lambda_{kj} a_k = 0, \forall j, \\ \sum_k \lambda_{ik} b_{kj} = \sum_k \lambda_{ik} b_k = b_{ij} = b_i, \forall i. \end{cases}$$

Hence,

$$\begin{aligned} \sum_i a_i \otimes b_i &= \sum_i a_i \otimes (\sum_k \lambda_{ik} b_k) \\ &= \sum_k (\sum_i \lambda_{ik} a_i) \otimes b_k = 0. \end{aligned}$$

Define

$$\alpha(\sum_i a_i \otimes b_i) = \| \sum_i a_i b_i \|, \forall a_i \in A, b_i \in B.$$

Then $\alpha(\cdot)$ is a C^*-norm on $A \otimes B$, and α - $(A \otimes B)$ is $*$ isomorphic to C.

$R \overline{\otimes} R'$ is still a type (II_1) factor on $H \otimes H$. So there is a faithful normal tracial state on $R \overline{\otimes} R'$. $\alpha_{\overline{0}}$-$(A \otimes B)$ is a C^*-subalgebra of $R \overline{\otimes} R'$. Hence, there is a faithful tracial state on α_0- $(A \otimes B)$. If we can prove that there is no faithful tracial state on α-$(A \otimes B)$, then $\alpha(\cdot) \neq \alpha_0(\cdot)$ on $A \otimes B$. Therefore, A and B are non-nuclear C^*-algebras.

If there is a non-zero projection p of C and an infinite sequence $\{u_j\}$ of unitary elements of C such that $\{u_j p u_j^* | j\}$ is pairwise orthogonal , then we claim that there is no faithful tracial state on C (and α-$(A \otimes B)$). In fact, let τ be a faithful tracial state on C. Then by

$$\tau(1) \geq \sum_{j=1}^{n} \tau(u_j p u_j^*) = n\tau(p), \forall n$$

we get $\tau(p) = 0$. This is a contradiction since $p \neq 0$ and τ is faithful. Therefore, there is no faithful tracial state on C (and α - $(A \otimes B)$).

Consider $G = F_2$, the free group generated by two elements g_1 and g_2. Then $R = R(G)$ and $R' = R(G)'$ are tyep (II_1) factors on $H = l^2(G)$. Let A, B be the C^*–subalgebras of R, R' generated by $\{\lambda_g | g \in G\}, \{\rho_g | g \in G\}$ respectively. Since R is a factor, the C^*–algebra C generated by $(A \sqcup B)$ is irreducible on H. Let p be the projection from $H = l^2(G)$ onto $[\varepsilon_e]$.

1) Let $a = \lambda_{g_1}\rho_{g_1} + \lambda^*_{g_1}\rho^*_{g_1} + \lambda_{g_2}\rho_{g_2} + \lambda^*_{g_2}\rho^*_{g_2}$. Clearly, $a^* = a \in C$, and $\|a\| \leq 4$. Noticing that

$$\begin{aligned}(a\varepsilon_e)(g) &= \varepsilon_e(g_1^{-1}gg_1) + \varepsilon_e(g_1gg_1^{-1}) + \varepsilon_e(g_2^{-1}gg_2) + \varepsilon_e(g_2gg_2^{-1}) \\ &= 4\varepsilon_e(g),\end{aligned}$$

we have $ap = 4p, pa = (ap)^* = (4p)^* = 4p$, and $ap = pa = 4p$. Since $p \neq 0$, it follows that $\|a\| = 4$.

2) Let $\xi \in H = l^2(G), \xi(e) = 0$, and $\|\xi\| = 1$. Then for $j = 1$ or 2 we have

$$\sum_{g \in G} |\xi(g) - \xi(g_j g g_j^{-1})|^2 > \varepsilon^2,$$

where $\varepsilon = 1/25$.

In fact, for any subset E of G, denote the norm on $l^2(E)$ by $\|\cdot\|_E$, and let $\mu(E) = \|\xi\|_E^2 = \sum_{g\in E} |\xi(g)|^2, t_j^2 = \sum_{g \in G} |\xi(g) - \xi(g_j g g_j^{-1})|^2, \xi_j = \lambda^*_{g_j}\rho^*_{g_j}\xi, j = 1, 2.$
Then for any subset E of G we have

$$t_j^2 \geq \|\xi - \xi_j\|_E^2, \quad j = 1, 2. \tag{1}$$

By (1) and $\|\xi\| = 1$, for any subset E of G we get

$$\begin{aligned}|\|\xi_j\|_E^2 - \|\xi\|_E^2| &= |\mu(g_j E g_j^{-1}) - \mu(E)| \\ &= |\|\xi_j\|_E - \|\xi\|_E| \cdot (\|\xi_j\|_E + \|\xi\|_E) \leq 2t_j,\end{aligned} \tag{2}$$

$j = 1, 2$. Replacing E by $g_j^{-1}Eg_j$, we have

$$\max\{|\mu(E) - \mu(g_j E g_j^{-1})|, |\mu(E) - \mu(g_j^{-1}Eg_j)|\} \leq 2t_j, \tag{3}$$

$\forall j = 1, 2$, and any subset E of G. Now let E be the subset of all reduced words of $G = F_2$ beginning with g_1. Then the subsets $E, g_2Eg_2^{-1}, g_2^{-1}Eg_2$ are disjoint. By $\mu(G) = \|\xi\|^2 = 1$,

$$\min\{\mu(E), \mu(g_2Eg_2^{-1}), \mu(g_2^{-1}Eg_2)\} \leq 1/3. \tag{4}$$

Since $\xi(e) = 0$ and $E \sqcup g_1^{-1}Eg_1 = G\backslash\{e\}$, it follows that $\mu(E) + \mu(g_1^{-1}Eg) \geq \mu(E \sqcup g_1^{-1}Eg_1) = \mu(G) = 1$. Hence,

$$\max\{\mu(E), \mu(g_1^{-1}Eg_1)\} \geq 1/2. \tag{5}$$

Now by (4) and (3) , $\mu(E) \leq \frac{1}{3} + 2t_2$; and by (5) and (3) , $\mu(E) \geq \frac{1}{2} - 2t_1$. Hence , $\frac{1}{2} - 2t_1 \leq \frac{1}{3} + 2t_2$, i.e., $(t_1 + t_2) \geq \frac{1}{12}$. Therefore, for $j = 1$ or 2 we have

$$t_j = \sum_{g \in G} |\xi(g) - \xi(g_j g g_j^{-1})|^2 \geq \frac{1}{25} = \varepsilon.$$

3) $a(1-p) \leq (4 - \varepsilon^2)(1-p)$, where $\varepsilon = 1/25$, and a is the same as in 1).
In fact, by $ap = pa$ it suffices to show that

$$\langle a\xi, \xi \rangle \leq (4 - \varepsilon^2),$$

$\forall \xi \in (1-p)H$ and $\|\xi\| = 1$. Fix $\xi \in (1-p)H$ and $\|\xi\| = 1$. Clearly, $\xi(e) = \langle \xi, \varepsilon_e \rangle = \langle p\xi, \varepsilon_e \rangle = 0$. Since

$$\langle a\xi, \xi \rangle = 2\text{Re}[\langle \lambda_{g_1}^* \rho_{g_1}^* \xi, \xi \rangle + \langle \lambda_{g_2}^* \rho_{g_2}^* \xi, \xi \rangle]$$

and

$$\begin{aligned} & \sum_{g \in G} |\xi(g) - \xi(g_j g g_j^{-1})|^2 = \|\xi - \lambda_{g_j}^* \rho_{g_j}^* \xi\|^2 \\ = \quad & \|\xi\|^2 + \|\lambda_{g_j}^* \rho_{g_j}^* \xi\|^2 - 2\text{Re}\langle \lambda_{g_j}^* \rho_{g_j}^* \xi, \xi \rangle \\ = \quad & 2[1 - Re\langle \lambda_{g_j}^* \rho_{g_j}^* \xi, \xi \rangle], \quad j = 1, 2, \end{aligned}$$

it follows from the conclusion 2) that

$$\langle a\xi, \xi \rangle = 4 - \sum_{j=1}^{2} \sum_{g \in G} |\xi(g) - \xi(g_j g g_j^{-1})|^2 \leq 4 - \varepsilon^2.$$

Now we prove that $p \in C$.

Let $b = (4 + a)/8$. Then by $ap = pa = 4p, a^* = a \leq 4$ and the conclusion 3) we have

$$b \in C, bp = pb = p, 0 \leq b(1-p) \leq \delta(1-p),$$

where $\delta = (8 - \varepsilon^2)/8$, and $\varepsilon = 1/25$. Hence, for any positive integer n,

$$b^n p = p, \text{ and } 0 \leq b^n(1-p) \leq \delta^n(1-p).$$

Further, $\|b^n - p\| = \|b^n - b^n p\| = \|b^n(1-p)\| \leq \delta^n \longrightarrow 0$. Therefore, $p = \lim_n b^n \in C$.

Moreover, $\{\lambda_g p \lambda_g^* | g \in G = F_2\}$ is an infinite orthogonal sequence of projections of C obviously. Therefore, there is no faithful tracial state on C, and A, B are non-nuclear C^*-algebras.

Theorem 7.4.7. There exist separable non-nuclear C^*-algebras.

Notes. The property (Γ) was introdced by F.J.Murray and J.Von Neumann. Theorem 7.4.6 is also due to them. The examples of non-nuclear C^*-algebras presented here are due to M.Takesaki.

References. [28], [80], [113], [171], [194].

Chapter 8
Tomita-Takesaki Theory

8.1 The KMS condition

Definition 8.1.1. Let $(H, \langle\ ,\ \rangle)$ be a complex Hilbert space. Define $\langle\ ,\ \rangle_r = \mathrm{Re}\langle\ ,\ \rangle$. Then $H_r = (H, \langle\ ,\ \rangle_r)$ is a real Hilbert space (see H as a real linear space). Suppose that K is a closed real linear subspace of H, K is said to be *nondegenerate*, if $K \cap iK = \{0\}$, and $(K \dot{+} iK)$ is dense in H.

Lemma 8.1.2. Let K be a nondegenerate closed real linear subspace of H, p, q be the projections from H_r onto K, iK respectively (self-adjoint on H_r), $a = p + q$,and $p - q = jb$ be the polar decomposition of $(p - q)$ on H_r. Then

1) $pi = iq, ip = qi$;

2) a is a positive linear operator on $H, 0 \leq a \leq 2$, and $\{0, 2\}$ are not eigenvalues of A;

3) b is a positive linear operator on $H, b = a^{\frac{1}{2}}(2 - a)^{\frac{1}{2}}$, and 0 is not an eignevalue of b. Moreover b commutes with p, q, a and j;

4) j is a self-adjoint unitary operator on H_r and j is a conjugate linear operator on H, i.e., $ji = -ij$. Moreover,

$$\langle j\xi, \eta\rangle = \langle j\eta, \xi\rangle, \qquad \forall \xi, \eta \in H,$$

and $jp = (1 - q)j,\ jq = (1 - p)j,\ ja = (2 - a)j$.

Proof. 1) Let $\eta \in K$, and $i\eta = \varsigma + \varsigma^{\perp}$ be the orthogonal decomposition with respect to $H_r = K \oplus K^{\perp}$, i.e. $p(i\eta) = \varsigma$. Then $-\eta = i\varsigma + i\varsigma^{\perp}$ is the orthogonal decompositive with respect to $H_r = iK \oplus (iK)^{\perp}$, and

$$-q\eta = i\varsigma = ip(i\eta).$$

Now if $\xi, \eta \in K$, then

$$\begin{aligned} ip(\xi + i\eta) &= ip\xi + ip(i\eta) = i\xi - q\eta \\ &= q(i\xi - \eta) = qi(\xi + i\eta). \end{aligned}$$

Since $(K \dot{+} iK)$ is dense in H_r, it follows that $ip = qi$. Further, $pi = iq$.

2) From 1), a is linear on H. Clearly, a is self-adjoint on H_r, and

$$\langle a\xi, \eta \rangle = \langle a\xi, \eta \rangle_r - i\langle a(i\xi), \eta \rangle_r, \quad \forall \xi, \eta \in H.$$

Thus, a is also self-adjoint on H. Since

$$\langle a\xi, \xi \rangle = \langle a\xi, \xi \rangle_r = \langle p\xi, \xi \rangle_r + \langle q\xi, \xi \rangle_r \geq 0,$$

it follows that $0 \leq a \leq 2$. If $a\xi = 0$, then $p\xi = q\xi = 0$ from above equality, i.e., $\xi \perp (K \dot{+} iK)$ in H_r. But $(K \dot{+} iK)$ is dense in H_r, so $\xi = 0$, and 0 is not an eigenvalue of a. Let $K^\perp$ be the orthogonal complement of K in H_r. Then $K^\perp$ is also a nondegenerate closed real linear subspace of H. Considering $K^\perp$, we can see that 0 is not an eigenvalue of $(2 - a)$, i.e., 2 is not an eigenvalue of a.

3) Clearly, $(p - q)^2$ is linear on H from 1). Similar to the proof of 2), $(p - q)^2$ is positive on H. Thus, b is a positive linear operator on H. Since $(p - q)^2$ and p or q commute , it follows that b commutes with p, q and a. The equality $b = a^{\frac{1}{2}}(2 - a)^{\frac{1}{2}}$ is obvious. So 0 is not an eigenvalue of b by 2). Moreover, since $(p - q)$ is self-adjoint on H_r, $bj = jb$.

4) Since $(p - q)$ is self-adjoint on H_r and 0 is not an eigenvalue of b, it follows that j is self-adjoint and unitary on H_r. Noticing that $bi = ib, (p - q)i = -i(p - q)$, we get $ji = -ij$. For any $\xi, \eta \in H$,

$$\begin{aligned} \langle j\xi, \eta \rangle &= \langle j\xi, \eta \rangle_r + i\langle j(i\xi), \eta \rangle_r \\ &= \langle \xi, j\eta \rangle_r + i\langle i\xi, j\eta \rangle_r = \langle j\eta, \xi \rangle. \end{aligned}$$

Finally, from $bjp = (p - q)p = (1 - q)(p - q) = b(1 - q)j$ and 0 is not an eigenvalue of b, we get $ip = (1 - p)j$. Similarly, $jq = (1 - p)j$. Further, $ja = (2 - a)j$.

Q.E.D.

Lemma 8.1.3. Keep the assumptions and notations of Lemma 8.1.2, and let $\Delta = (2 - a)a^{-1} = a^{-1}(2 - a)$. Then Δ is a (unbounded) positive invertible linear operator on H,and for any everywhere finite measurable function f on $[0, +\infty), jf(\Delta)j = \overline{f}(\Delta^{-1})$.

Proof. By Lemma 8.1.2, $ja = (2 - a)j$. Thus $j\Delta j = \Delta^{-1}$. Further, by $ji = -ij$, we obtain $jf(\Delta)j = \overline{f}(\Delta^{-1})$. Q.E.D.

Lemma 8.1.4. Keep the assumptions and notations of Lemma 8.1.3, and define

$$s(\xi + i\eta) = \xi - i\eta, \quad \forall \xi, \eta \in K, \quad \mathcal{D}(s) = K \dot{+} iK;$$

$$s^+(i\xi_1 + \eta_1) = i\xi_1 - \eta_1, \quad \forall \xi_1, \eta_1 \in K^\perp, \quad \mathcal{D}(s^+) = iK^\perp \dot{+} K^\perp,$$

where $K^\perp$ is the orthogonal complement of K in H_r (it is also a nondegenerate closed real linear subspace of H). Then:

1) s and s^+ are two conjugate linear closed operators on H with a dense domain;

2) s^+ is the adjoint of s on H_r, s is the adjoint of s^+ on H_r and $jsj = s^+$;

3) $s = j\Delta^{1/2}, s^+ = j\Delta^{-1/2}$ are the polar decompositions of s, s^+ on H_r respectively. Consequently, $\mathcal{D}(\Delta^{1/2}) = K \dot{+} iK$.

Proof. 1) It is obvious.

2) Clearly, $s^+ \subset$ the adjoint of s on H_r. If ς, ς' satisfy

$$\langle \xi - i\eta, \varsigma \rangle_r = \langle \xi + i\eta, \varsigma' \rangle_r, \quad \forall \xi, \eta \in K.$$

Let $\eta = 0$. Then $(\varsigma - \varsigma') \in K^\perp$. Let $\xi = 0$. Then $i(\varsigma + \varsigma') \in K^\perp$. Thus, $\xi_1 = \frac{1}{2i}(\varsigma + \varsigma') \in K^\perp, \eta_1 = \frac{1}{2}(\varsigma - \varsigma') \in K^\perp$, and

$$\varsigma = i\xi_1 + \eta, \quad \varsigma' = i\xi_1 - \eta_1.$$

Now we can see that s^+ is the adjoint of s on H_r. Moreover, since s is closed, it follows that s is also the adjoint of s^+ on H_r.

From Lemma 8.1.2,

$$jK = jpH = (1 - q)jH = (iK)^\perp = iK^\perp.$$

Similarly, $j(iK) = K^\perp$. Thus $jsj = s^+$.

3) If $\xi_1, \eta_1 \in K^\perp$, then $p\eta_1 = 0, qi\xi_1 = ip\xi_1 = 0$ and $as^+(i\xi_1 + \eta_1) = (p - q)(i\xi_1 + \eta_1)$. Thus, $as^+ \subset p - q = jb = bj$. Since $s^+ = jsj$, it follows that $ajs \subset b$, i.e., $js \subset \Delta^{1/2}$. But js and $\Delta^{-1/2}$ are self-adjoint on H, so $s = j\Delta^{1/2}$. By Lemma 8.1.3, $s^+ = j\Delta^{-1/2}$. Now from $s^+s = \Delta, ss^+ = \Delta^{-1}$, thus $s = j\Delta^{1/2}, s^+ = j\Delta^{-1/2}$ are also the polar decompositions. Q.E.D.

Lemma 8.1.5. $\{\Delta^{it} \mid t \in \mathbb{R}\}$ is an one-parameter strongly continuous group of unitary operators on H, and satisfies the following:

$$j\Delta^{it} = \Delta^{it}j, \quad \Delta^{it}K = K, \quad \forall t \in \mathbb{R}.$$

Proof. By Lemma 8.1.3, $j\Delta^{it}j = \Delta^{it}, \forall t \in \mathbb{R}$. Moreover, from $ab = ba$, we have $\Delta^{it}b = b\Delta^{it}$. Further, Δ^{it} and $jb = p - q$ commute. Clearly, Δ^{it} and $a = p + q$ commute. Thus, $\Delta^{it}p = p\Delta^{it}$, i.e., $\Delta^{it}K = K, \forall t \in \mathbb{R}$. Q.E.D.

Definition 8.1.6. The above operators j, Δ are called the *unitary involution*, the *modular operator* (relative to the nondegenerate closed real linear subspace K of H) respectively. They will play an important role in the theory of this chapter.

Now we discuss the KMS condition.

Let K be a nondegenerate closed real linear subspace of a (complex) Hilbert space H, and keep above all notations.

Definition 8.1.7. An one-parameter strongly continuous group of unitary operators $\{u_t \mid t \in \mathbb{R}\}$ on H is called satisfying the *KMS condition* (relative to K), if for any $\xi, \eta \in K$, there is a complex function $f(z)$ which is continuous and bounded on $0 \leq \text{Im}z \leq 1$ and is analytic in $0 < \text{Im}z < 1$ such that

$$f(t) = \langle \eta, u_t\xi \rangle, \quad f(t+i) = \langle u_t\xi, \eta \rangle = \overline{f(t)}, \quad \forall t \in \mathbb{R}.$$

Clearly, this f is unique, and is called the KMS function corresponding to ξ, η.

Proposition 8.1.8. An one-parameter strongly continuous group of unitary operators $\{u_t \mid t \in \mathbb{R}\}$ on H satisfies the KMS condition (relative to K) if and only if for any $\xi, \eta \in K$, there is a complex function $f(z)$ which is continuous and bounded on $0 \leq \text{Im}z \leq 1/2$ and is analytic in $0 < \text{Im}z < 1/2$ such that

$$f(t) = \langle \eta, u_t\xi \rangle, \quad f(t+\frac{i}{2}) = \overline{f(t+\frac{i}{2})}, \quad \forall t \in \mathbb{R}.$$

Proof. The sufficiency is obvious by the Schwartz reflection principle (see [179]).

Now let f be the KMS function corresponding to ξ, η, and $g(z) = \overline{f(\overline{z-i})}$. Clearly, g is also a KMS function corresponding to ξ, η. Thus, $f = g$. In particular,

$$f(t+\frac{i}{2}) = g(t+\frac{i}{2}) = \overline{f(t+\frac{i}{2})}, \quad \forall t \in \mathbb{R}.$$

Q.E.D.

Definition 8.1.9. Let $\{u_t \mid t \in \mathbb{R}\}$ be an one-parameter strongly continuous group of unitary operators on H. $\xi(\in H)$ is said to be *analytic* (with respect to $\{u_t\}$), if there is a vector valued analytic function $\xi(z) : \mathbb{C} \to H$ such that $\xi(t) = u_t\xi, \forall t \in \mathbb{R}$.

Lemma 8.1.10. Let h be a non-negative invertible self-adjoint operator on H. For any $\delta > 0$, define

$$A(\delta) = \left\{ \xi(z) \,\middle|\, \begin{array}{l} \xi(z) \text{ is continuous and bounded from } -\delta \leq \text{Im}z \leq 0 \\ \text{to } H, \text{ and is analytic in } -\delta < \text{Im}z < 0. \end{array} \right\}$$

If $\xi \in H$, then $\xi \in D(h^\delta)$ if and only if there exists $\xi(z) \in A(\delta)$ such that $\xi(t) = h^{it}\xi, \forall t \in \mathbb{R}$. Moreover, in this case, for any z with $-\delta \leq \text{Im}z \leq 0$ we have $\xi(z)h^{iz}\xi$.

Proof. Suppose that $\xi \in D(h^\delta)$,and $z \in \mathbb{C}$ with $-\delta \leq \text{Im}z \leq 0$. Then

$$D(h^{iz}) = D(h^{-\text{Im}z}) \supset D(h^\delta)$$

and $\xi \in D(h^{iz})$. If $\{e_\lambda\}$ is the spectral family of h, then

$$\begin{aligned} &\|h^{iz}(e_n - e_{\frac{1}{n}})\xi - h^{iz}\xi\|^2 \\ = \;& \left(\int_0^{\frac{1}{n}} + \int_n^\infty\right) e^{-2\text{Im}z\cdot\ln\lambda} d\|e_\lambda\xi\|^2 \\ \leq \;& \|e_{\frac{1}{n}}\xi\|^2 + \int_n^\infty e^{2\delta\ln\lambda} d\|e_\lambda\xi\|^2 \to 0 \end{aligned}$$

uniformly for z with $-\delta \leq \text{Im}z \leq 0$. But for each $n, z \to h^{iz}(e_n - e_{\frac{1}{n}})\xi$ is an analytic function from $\mathbb{C}$ to H, thus $\xi(z) = h^{iz}\xi$ is continuous in $-\delta \leq \text{Im}z \leq 0$, and is analytic in $-\delta < \text{Im}z < 0$. Moreover,

$$\begin{aligned} \|h^{iz}\xi\|^2 = \;& \left(\int_0^1 + \int_1^\infty\right) e^{-2\text{Im}z\cdot\ln\lambda} d\|e_\lambda\xi\|^2 \\ \leq \;& \|\xi\|^2 + \|h^\delta\xi\|^2, \end{aligned}$$

$\forall z$ with $-\delta \leq \text{Im}z \leq 0$. Therefore, $\xi(z) = h^{iz}\xi \in A(\delta)$.

Now let $\xi(z) \in A(\delta)$ be such that $\xi(t) = h^{it}\xi, \forall t \in \mathbb{R}$. For any $\eta \in D(h^\delta), \eta(z) = h^{iz}\eta \in A(\Delta)$ from the preceding paragraph. Since $f(z) = \langle \xi(z), \eta \rangle$ and $g(z) = \langle \xi, h^{-i\bar{z}}\eta \rangle$ are continuous and bounded on $-\delta \leq \text{Im}z \leq 0$ and analytic in $-\delta < \text{Im}z < 0$,and $f(t) = g(t), \forall t \in \mathbb{R}$, it follows that $f = g$. Consequently,

$$\langle \xi(-i\delta), \eta \rangle = \langle \xi, h^\delta\eta \rangle, \quad \forall \eta \in D(h^\delta).$$

Therefore, $\xi \in D(h^\delta)$. Q.E.D.

Proposition 8.1.11. $\xi (\in H)$ is analytic with respect to $\{\Delta^{it}\}$ if and only if $\xi \in D$, where Δ is the modular operator (relative to K), and $D = \cap\{D(\Delta^z) \mid z \in \mathbb{C}\} = \cap\{D(\Delta^n) \mid n \in \mathbb{Z}\}$. In this case, the vector valued analytic function corresponding to ξ is $\xi(z) = \Delta^{iz}\xi$.

Proof. Using Lemma 8.1.10 to Δ and Δ^{-1}, the conclusion is clear. Q.E.D.

Proposition 8.1.12. Let $\{u_t \mid t \in \mathbb{R}\}$ be an one-parameter strongly continuous group of unitary operators on H,and $\xi \in H$. For each $r > 0$, define

$$\xi_r = \sqrt{\frac{r}{\pi}} \int_{-\infty}^{\infty} e^{-rs^2} u_s \xi ds.$$

Then ξ_r is analytic with respect to $\{u_t\}$, and $\|\xi_r - \xi\| \to 0$ as $r \to +\infty$.

Proof. Clearly $\xi_r(z) = \sqrt{\frac{r}{\pi}} \int_{-\infty}^{\infty} e^{-r(z-s)^2} u_s \xi ds$ is a vector valued analytic function: $\mathbb{C} \to H$,and

$$\xi_r(t) = \sqrt{\frac{r}{\pi}} \int_{-\infty}^{\infty} e^{-r(s-t)^2} u_t u_{s-t} \xi ds = u_t \xi_r, \quad \forall t \in \mathbb{R}.$$

Thus, ξ_r is analytic with respect to $\{u_t\}, \forall r > 0$.

For any $\varepsilon > 0$, pick $\delta > 0$ such that $\|(u_t - 1)\xi\| < \varepsilon, \forall |t| < \delta$. Then since $\sqrt{\frac{r}{\pi}} \int_{-\infty}^{\infty} e^{-rs^2} ds = 1$, it follwes that

$$\begin{aligned} \|\xi_r - \xi\| \leq & \sqrt{\frac{r}{\pi}} \int_{-\delta}^{\delta} e^{-rs^2} \|(u_s - 1)\xi\| ds + 4\|\xi\| \sqrt{\frac{r}{\pi}} \int_{\delta}^{\infty} e^{-rs^2} ds \\ < & \varepsilon + \frac{4\|\xi\|}{\sqrt{\pi}} \int_{\sqrt{r}\delta}^{\infty} e^{-s^2} ds < 2\varepsilon \end{aligned}$$

if r sufficiently large. Q.E.D.

Theorem 8.1.13. Let K be a nodegenerate closed real linear subspace of H, and Δ be the modular operator relative to K. Then $\{\Delta^{it} \mid t \in \mathbb{R}\}$ is the unique one-parameter strongly continuous group of unitary operators on H, which satisfies the KMS condition relative to K and is invariant for K.

Proof. From Lemma 8.1.5, $\{\Delta^{it}\}$ is invariant for K, i.e., $\Delta^{it} K = K, \forall t \in \mathbb{R}$. Now let $\xi, \eta \in K$. Then by Lemma 8.1.4 and Lemma 8.1.10 $K \subset \mathcal{D}(\Delta^{1/2})$ and

$$f(z) = \langle \eta, \Delta^{i\bar{z}} \xi \rangle$$

is continuous and bounded on $0 \leq \text{Im}z \leq \frac{1}{2}$ and is analytic in $0 < \text{Im}z < \frac{1}{2}$. For any $t \in \mathbb{R}$ by Lemmas 8.1.2, 8.1.4, 8.1.5,

$$f(t + \frac{i}{2}) = \langle \eta, \Delta^{it} \Delta^{\frac{1}{2}} \xi \rangle = \langle \Delta^{it} \xi, j\eta \rangle, \quad \forall t \in \mathbb{R}.$$

But $\Delta^{it}\xi \in K, j\eta \in iK^{\perp}$, so $f(t + \frac{i}{2})$ is real, $\forall t \in \mathbb{R}$. Now by Proposition 8.1.8, $\{\Delta^{it}\}$ satisfies the KMS condition relative to K.

If $\{u_t\}$ satisfies the KMS condition relative to K and is invariant for K, we need to prove $u_t = \Delta^{it}, \forall t \in \mathbb{R}$. Since $(K \dot{+} iK)$ is dense in H, it suffices to show that

$$u_t\eta = \Delta^{it}\eta, \quad \forall t \in \mathbb{R}, \quad \eta \in K.$$

From Proposition 8.1.12, we may assume that η is analytic with respect to $\{u_t\}$, and $\eta(z)$ is bounded on every horizontal strip.

Notice that $j\sqrt{\frac{r}{\pi}}\int_{-\infty}^{\infty} e^{-rs^2}\Delta^{is}\xi ds = \sqrt{\frac{r}{\pi}}\int_{-\infty}^{\infty} e^{-rs^2}\Delta^{is}j\xi ds, ji = -ij$, and $j(K \dot{+} iK)$ is dense in H. So we need only to prove

$$\langle \Delta^{it}j\xi, u_t\eta\rangle = \langle j\xi, \eta\rangle,$$

where η is as above, $\xi \in K$, and $\xi, j\xi$ are analytic with respect to $\{\Delta^{it}\}$.

Let $g(z) = \langle \Delta^{iz}j\xi, \eta(\overline{z})\rangle$. Then $g(z)$ is analytic on $\mathbb{C}$, and is bounded on every horizontal strip. If $t \in \mathbb{R}$, then $\eta(t) = u_t\eta \in K$, and $\Delta^{it}j\xi = j\Delta^{it}\xi \in jK = iK^{\perp}$. Thus, $g(t)$ is real, $\forall t \in \mathbb{R}$.

Fix $s \in \mathbb{R}$. For $\eta, \Delta^{is}\xi (\in K)$, we have a KMS function f (relative to $\{u_t\}$) such that

$$f(t) = \langle \Delta^{is}\xi, u_t\eta\rangle = \overline{f(t+i)}, \quad \forall t \in \mathbb{R}.$$

By Proposition 8.1.8, $f(t + \frac{i}{2}) = \overline{f(t + \frac{i}{2})}, \forall t \in \mathbb{R}$. Notice that

$$h(z) = \langle \Delta^{is}\xi, \eta(\overline{z})\rangle$$

is analytic on $\mathbb{C}$,and $h(t) = f(t), \forall t \in \mathbb{R}$. Using the Schwartz reflection principle to $(f - h)$, we can see that $f(z) = h(z), 0 \le \mathrm{Im}z \le 1$. In particular, $h(s + \frac{i}{2}) = f(s + \frac{i}{2})$ is real. Thus

$$g(s + \frac{i}{2}) = \langle \Delta^{is}\xi, \eta(s - \frac{i}{2})\rangle = h(s + \frac{i}{2})$$

is real, $\forall s \in \mathbb{R}$.

Now $g(z)$ is real on $\mathrm{Im}z = 0$ and $\mathrm{Im}z = \frac{i}{2}$, is continuous and bounded on $0 \le \mathrm{Im}z \le 1/2$, and is analytic in $0 < \mathrm{Im}z < 1/2$. By the Schwartz reflection principle, $g(z)$ can be extended to a bounded analytic function on $\mathbb{C}$. So $g(z)$ is a constant function. Consequently,

$$\langle \Delta^{it}j\xi, u_t\eta\rangle = g(t) = g(0) = \langle j\xi, \eta\rangle,$$

$\forall t \in \mathbb{R}$. Q.E.D.

Notes. The KMS condition was initially proposed by R. Kubo, P.C. Martin and J. Schwinger. Theorem 8.1.13 is due to M. Takesaki.

References. [93], [107], [127], [135], [174].

8.2. Tomita-Takesaki theory

In this section, let M be a VN algebra on a Hilbert space H, and $\xi_0(\in H, \|\xi_0\| = 1)$ be a cyclic-separating vector for M.

Proposition 8.2.1. Let $K = \overline{\{x\xi_0 \mid x \in M, x^* = x\}}$. Then K is a nondegenerate closed real linear subspace of H, and

$$\{x'\xi_0 \mid x' \in M', x'^* = x'\} \subset (iK)^\perp = iK^\perp,$$

where "$\perp$" is in the sense of H_r (see Section 8.1).

Proof. If $x^* = x \in M, x'^* = x' \in M'$, then $\langle x'\xi_0, x\xi_0\rangle$ is real. Thus $x'\xi_0 \in (iK)^\perp$,and

$$M'\xi_0 \subset (iK)^\perp + K^\perp = (K \cap iK)^\perp.$$

But $M'\xi_0$ is dense in H, so $K \cap iK = \{0\}$. Moreover, since $M\xi_0 \subset K \dot{+} iK$, it follows that $(K \dot{+} iK)$ is dense in H. Q.E.D.

In the following, for the above K we keep the notations of Section 8.1: $p, q, a, j, b, \Delta, s, s^+$, and etc.

Proposition 8.2.2. $q\xi_0 = 0$; $p\xi_0 = a\xi_0 = j\xi_0 = b\xi_0 = \xi_0$; $\Delta^{it}\xi_0 = \xi_0, \forall t \in \mathbb{R}$; $M\xi_0 \subset \mathcal{D}(\Delta^{1/2})$, and the operator s is the closure of the operator: $x\xi_0 \to x^*\xi_0 (\forall x \in M)$. Moreover, for each $x'^* = x' \in M$, there is $x^* = x \in M$ such that $(p - q)x'\xi_0 = x\xi_0$.

Proof. Since $\xi_0 \in K \cap (iK)^\perp$, it follows that $q\xi_0 = 0, p\xi_0 = \xi_0, a\xi_0 = \xi_0$. From $(p-q)^2\xi_0 = \xi_0$, we have also $b\xi_0 = \xi_0$. Further, $j\xi_0 = jb\xi_0 = (p-q)\xi_0 = \xi_0$. By $\xi_0 = s\xi_0 = j\Delta^{1/2}\xi_0$, we get $\Delta\xi_0 = \xi_0, \Delta^{it}\xi_0 = \xi_0, \forall t \in \mathbb{R}$. By the definition of the operator s, it is clear that $M\xi_0 \subset \mathcal{D}(\Delta^{1/2})$ and s is the closure of the operator: $x\xi_0 \to x^*\xi_0 (\forall x \in M)$.

Now let $x' \in M', 0 \le x' \le 1$, and

$$\varphi(\cdot) = \langle \cdot\xi_0, \xi_0\rangle, \quad \psi(\cdot) = \langle \cdot\xi_0, x'\xi_0\rangle.$$

Then $\varphi, \psi \in M_*$, and $0 \le \psi \le \varphi$. From Theorem 1.10.4, we have $x \in M$ with $0 \le x \le 1$ such that

$$\langle y\xi_0, x'\xi_0\rangle = \frac{1}{2}\langle (xy + yx)\xi_0, \xi_0\rangle, \quad \forall y \in M.$$

In particular, $\langle y\xi_0, x'\xi_0\rangle = \langle y\xi_0, x\xi_0\rangle_r, \forall y^* = y \in M$. Thus, $(x' - x)\xi_0 \in K^\perp$, and $x\xi_0 = px'\xi_0$. But $x'\xi_0 \in (iK)^\perp$, so $x\xi_0 = (p - q)x'\xi_0$.

Therefore, for any $x'^* = x' \in M'$ there is $x^* = x \in M$ such that $x\xi_0 = (p-q)x'\xi_0$. Q.E.D.

Lemma 8.2.3. For each $x' \in M'$ and each $\lambda \in \mathbb{C}$ with $\mathrm{Re}\lambda > 0$, there is $x \in M$ such that

$$bjx'jb = \lambda(2-a)xa + \overline{\lambda}ax(2-a).$$

Proof. We may assume that $0 \leq x' \leq 1$. Let

$$\varphi(\cdot) = \langle \cdot\xi_0, \xi_0\rangle, \quad \psi(\cdot) = \langle \cdot\xi_0, x'\xi_0\rangle.$$

Then $\varphi, \psi \in M_*$, and $0 \leq \psi \leq \varphi$. By Theorem 1.10.4, there is $x \in M_+$ such that

$$\langle y\xi_0, x'\xi_0\rangle = \langle(\lambda xy + \overline{\lambda}yx)\xi_0, \xi_0\rangle, \quad \forall y \in M.$$

Replacing y by z^*y, we have

$$\begin{aligned}\langle y\xi_0, x'z\xi_0\rangle = \ & \lambda\langle y\xi_0, zx\xi_0\rangle \\ & + \overline{\lambda}\langle yx\xi_0, z\xi_0\rangle, \quad \forall y, z \in M.\end{aligned} \tag{1}$$

For any $y'^* = y', z'^* = z' \in M'$, by Proposition 8.2.2 there are $y^* = y, z^* = z \in M$ such that

$$jby'\xi_0 = y\xi_0, \quad jbz'\xi = z\xi_0.$$

Now from (1), the property of j, and $\Delta^{1/2}b = (2-a)$, we get

$$\begin{aligned}&\langle bjx'jbz'\xi_0, y'\xi_0\rangle \\ = \ & \lambda\langle jby'\xi_0, zx\xi_0\rangle + \overline{\lambda}\langle yx\xi_0, jbz'\xi_0\rangle \\ = \ & \lambda\langle jby'\xi_0, j\Delta^{1/2}xz\xi_0\rangle + \overline{\lambda}\langle j\Delta^{1/2}xy\xi_0, jbz'\xi_0\rangle \\ = \ & \lambda\langle xz\xi_0, (2-a)y'\xi_0\rangle + \overline{\lambda}\langle(2-a)z'\xi_0, xy\xi_0\rangle \\ = \ & \lambda\langle xjbz'\xi_0, (2-a)y'\xi_0\rangle + \overline{\lambda}\langle(2-a)z'\xi_0, xjby'\xi_0\rangle,\end{aligned}$$

$\forall y'^* = y', z'^* = z' \in M'$. Since $a - jb = 2q$ and $qc'\xi_0 = 0, \forall c'^* = c' \in M'$, it follows that

$$\begin{aligned}&\langle bjx'jbz'\xi_0, y'\xi_0\rangle \\ = \ & \lambda\langle xaz'\xi_0, (2-a)y'\xi_0\rangle + \overline{\lambda}\langle(2-a)z'\xi_0, xay'\xi_0\rangle \\ = \ & \langle(\lambda(2-a)xa + \overline{\lambda}ax(2-a))z'\xi_0, y'\xi_0\rangle,\end{aligned}$$

$\forall y'^* = y', z'^* = z' \in M'$. Further, the above equality is valid for any $y', z' \in M'$. But ξ_0 is also cyclic for M', so we obtain

$$bjx'jb = \lambda(2-a)xa + \overline{\lambda}ax(2-a).$$

Q.E.D.

Lemma 8.2.4. Let $\lambda = e^{\frac{i}{2}\theta}$, with $|\theta| < \pi$,and f be an analytic function on $\mathbb{C}$ and bounded in $\{z \in \mathbb{C} \mid |\text{Re}z| \leq \frac{1}{2}\}$. Then

$$f(0) = \frac{1}{2}\int_{-\infty}^{\infty} \frac{e^{-\theta t}}{\text{ch}(\pi t)}(\lambda f(it + \frac{1}{2}) + \overline{\lambda} f(it - \frac{1}{2}))dt.$$

Proof. Consider $g(z) = \frac{\pi e^{i\theta z}}{\sin(\pi z)} f(z)$. In $\{z \in \mathbb{C} \mid |\text{Re}z| \leq 1/2\}, g(z)$ just has a pole at $z = 0$, and the residue is $f(0)$. Moreover, when $|z| \to \infty$ and $|\text{Re}z| \leq 1/2, g(z)$ converges to 0 rapidly. Thus

$$f(0) = \frac{1}{2\pi i}\int_{-\infty}^{\infty} (g(it + \frac{1}{2}) - g(it - \frac{1}{2}))idt.$$

Now through a computation, we can get the conclusion. Q.E.D.

lemma 8.2.5. Let x', λ, x be as in Lemma 8.2.3, and $\lambda = e^{\frac{i}{2}\theta}$ with $|\theta| < \pi$. Then

$$x = \frac{1}{2}\int_{-\infty}^{\infty} \frac{e^{-\theta t}}{\text{ch}(\pi t)} \Delta^{it} j x' j \Delta^{-it} dt.$$

Proof. Suppose that $\xi, \eta (\in K)$ are analytic with respect to $\{\Delta^{it}\}$. Define

$$f(z) = \langle bxb\Delta^{-z}\xi, \Delta^{\overline{z}}\eta\rangle.$$

Clearly, $f(z)$ is analytic on $\mathbb{C}$ and is bounded on every vertical strip. Since $\Delta^{\frac{1}{2}} b = 2 - a, b\Delta^{-\frac{1}{2}} = a$, it follows that

$$\begin{aligned} f(it + \tfrac{1}{2}) &= \langle bxb\Delta^{-it}\Delta^{-\frac{1}{2}}\xi, \Delta^{-it}\Delta^{\frac{1}{2}}\eta\rangle \\ &= \langle \Delta^{it}(2 - a)xa\Delta^{-it}\xi, \eta\rangle, \\ f(it - \tfrac{1}{2}) &= \langle bxb\Delta^{-it}\Delta^{\frac{1}{2}}\xi, \Delta^{-it}\Delta^{-\frac{1}{2}}\eta\rangle \\ &= \langle \Delta^{it}ax(2 - a)\Delta^{-it}\xi, \eta\rangle. \end{aligned}$$

By Lemma 8.2.3,

$$\begin{aligned} &\lambda f(it + \tfrac{1}{2}) + \overline{\lambda} f(it - \tfrac{1}{2}) \\ = &\langle \Delta^{it} b j x' j b \Delta^{-it}\xi, \eta\rangle. \end{aligned}$$

Further, from Lemma 8.2.4,

$$\begin{aligned} \langle bxb\xi, \eta\rangle &= f(0) \\ &= \frac{1}{2}\int_{-\infty}^{\infty} \frac{e^{-\theta t}}{\text{ch}(\pi t)} \langle \Delta^{it} b j x' j b \Delta^{-it}\xi, \eta\rangle dt \\ &= \langle \frac{1}{2}\int_{-\infty}^{\infty} \frac{e^{-\theta t}}{\text{ch}(\pi t)} \Delta^{it} j x' j \Delta^{-it} dt b\xi, b\eta\rangle \end{aligned}$$

since b and Δ^{it} commute. Now from Proposition 8.1.12 and since $(K\dot{+}iK)$ is dense in H and b is invertible, we can get the conclusion. Q.E.D.

Lemma 8.2.6. $\Delta^{it}jx'j\Delta^{-it} \in M, \forall x' \in M', t \in \mathbb{R}$.

Proof. Let $y' \in M', \xi, \eta \in H$, and define

$$g(t) = \langle(\Delta^{it}jx'j\Delta^{-it}y' - y'\Delta^{it}jx'j\Delta^{-it})\xi, \eta\rangle.$$

By Lemma 8.2.5, we have

$$\int_{-\infty}^{\infty} \frac{e^{-\theta t}}{\text{ch}\pi t} g(t)dt = 0, \quad \forall \theta \in \mathbb{R} \quad \text{and} \quad |\theta| < \pi.$$

Put $f(z) = \int_{-\infty}^{\infty} \frac{e^{-zt}}{\text{ch}(\pi t)} g(t)dt$. Then $f(z)$ is analytic in $|\text{Re}z| < \pi$, and $f(\theta) = 0, \forall \theta \in \mathbb{R}$ and $|\theta| < \pi$. Thus, $f = 0$. Consequently,

$$\int_{-\infty}^{\infty} \frac{e^{-ist}}{\text{ch}(\pi t)} g(t)dt = 0, \quad \forall s \in \mathbb{R}.$$

By the uniqueness of Fourier transformation, $g = 0$, i.e.,

$$\langle\Delta^{it}jx'j\Delta^{-it}y'\xi, \eta\rangle = \langle y'\Delta^{it}jx'j\Delta^{-it}\xi, \eta\rangle$$

$\forall y' \in M', \xi, \eta \in H$. Therefore, $\Delta^{it}jx'j\Delta^{-it} \in M, \forall x' \in M', t \in \mathbb{R}$. Q.E.D.

Theorem 8.2.7. $jMj = M', \Delta^{it}M\Delta^{-it} = M', \forall t \in \mathbb{R}$.

Proof. From Lemma 8.2.6 with $t = 0$, we get $jM'j \subset M$.

For any $x^* = x, y^* = y \in M$, since $x\xi_0 \in K, jy\xi_0 \in (iK)^\perp$, it follows that $\langle xjy\xi_0, \xi_0\rangle = \langle jy\xi_0, x\xi_0\rangle$ is real. Further, from the property of j, we get

$$\langle yjx\xi_0, \xi_0\rangle = \langle xjy\xi_0, \xi_0\rangle = \langle \xi_0, xjy\xi_0\rangle.$$

Generally, we have

$$\langle bja\xi_0, \xi_0\rangle = \langle \xi_0, ajb\xi_0\rangle, \quad \forall a, b \in M.$$

In particular, for $x^* = x, y^* = y \in M$, and $y' \in M'$,

$$\langle y(jy'j)jx\xi_0, \xi_0\rangle = \langle \xi_0, xjy(jy'j)\xi_0\rangle$$

since $jy'j \in M$. Noticing that $j\xi_0 = \xi_0$, we get

$$\langle xjy\xi_0, y'\xi_0\rangle - \langle jyjx\xi_0, y'\xi_0\rangle.$$

Since ξ_0 is cyclic for M', it follows that

$$xjyj\xi_0 = jyjx\xi_0, \quad \forall x^* = x, y^* = y \in M.$$

Clearly, the above equality holds also for any $x, y \in M$. Thus

$$jyjxz\xi_0 = xzjyj\xi_0 = xjyjz\xi_0, \quad \forall x, y, z \in M.$$

Since ξ_0 is cyclic for M, it follows that

$$xjyj = jyjx, \quad \forall x, y \in M,$$

i.e., $jMj \subset M'$. Therefore, $jMj = M'$.

Now from $jM'j = M$ and Lemma 8.2.6, we can see that $\Delta^{it} M \Delta^{-it} = M, \forall t \in \mathbb{R}$. Q.E.D.

Definition 8.2.8. The $*$ automorphism group $\{\sigma_t(\cdot) = \Delta^{it} \cdot \Delta^{-it} \mid t \in \mathbb{R}\}$ of M is called the *modular automorphism group* of M.

Definition 8.2.9. Let $\varphi_0(\cdot) = \langle \cdot \xi_0, \xi_0 \rangle$. Then φ_0 is a faithful normal state on M. An one-parameter strongly continuous $*$ automorphism group $\{\alpha_t(\cdot) \mid t \in \mathbb{R}\}$ of M (i.e. for each $x \in M, t \to \alpha_t(x)$ is strongly continuous) is called satisfying the *KMS condition* relative to φ_0, if for any $x, y \in M$, there is a complex function $f(z)$ which is continuous and bounded on $0 \le \text{Im}z \le 1$ and is analytic in $0 < \text{Im}z < 1$ such that

$$f(t) = \varphi_0(\alpha_t(x)y), \quad f(t+i) = \varphi_0(y\alpha_t(x)), \quad \forall t \in \mathbb{R}.$$

Clearly, this f is unique, and is called the KMS function corresponding to x, y. When $x^* = x$ and $y^* = y$, we can see that $\overline{f(t)} = \underline{f(t+i)}, \forall \in \mathbb{R}$. By Proposition 8.1.8, in this case we have also $f(t + \frac{i}{2}) = \overline{f(t + \frac{i}{2})}, \forall t \in \mathbb{R}$.

Theorem 8.2.10. φ_0 is invariant for the modular automorphism group $\{\sigma_t \mid t \in \mathbb{R}\}$ of M, i.e., $\varphi_0(\sigma_t(x)) = \varphi_0(x), \forall x \in M, t \in \mathbb{R}$. And also $\{\sigma_t \mid t \in \mathbb{R}\}$ is the unique one-parameter strongly continuous $*$ automorphism group (of M) satisfying the KMS condition relative to φ_0.

Proof. From Proposition 8.2.2, φ_0 is invariant for $\{\sigma_t \mid t \in \mathbb{R}\}$.

For any $x^* = x, y^* = y \in M$, by Theorem 8.1.3 there is a KMS function f such that

$$f(t) = \langle y\xi_0, \Delta^{it} x\xi_0 \rangle, f(t+i) = \langle \Delta^{it} x\xi_0, y\xi_0 \rangle,$$

$\forall t \in \mathbb{R}$. Since $\Delta^{-it}\xi_0 = \xi_0$, it follows that

$$f(t) = \varphi_0(\sigma_t(x)y), \quad f(t+i) = \varphi_0(y\sigma_t(x)),$$

$\forall t \in \mathbb{R}$. Further, we can see that $\{\sigma_t\}$ satisfies the KMS condition relative to φ_0.

Now let $\{\alpha_t \mid t \in \mathbb{R}\}$ be an one-parameter strongly continuous $*$ automorphism group of M, and $\{\alpha_t\}$ satisfies the KMS condition relative to φ_0.

First, we claim that φ_0 is invariant for $\{\alpha_t\}$. In fact, for $x \in M_+$ and $y = 1$, there is a KMS function f such that

$$f(t) = f(t+i) = \varphi_0(\alpha_t(x)) \geq 0, \quad \forall t \in \mathbb{R}.$$

By the Schwartz reflection principle, f can be extended to a bounded analytic function on $\mathbb{C}$. Thus f is a constant function . In particular,

$$\varphi_0(\alpha_t(x)) = f(t) = f(0) = \varphi_0(x), \quad \forall t \in \mathbb{R},$$

i.e., φ_0 is invariant for $\{\alpha_t\}$.

Now define

$$u_t x\xi_0 = \alpha_t(x)\xi_0, \qquad \forall t \in M.$$

Then u_t can be extended to a unitary operator on H, still denoted by u_t, $\forall t \in \mathbb{R}$. Clearly, $\{u_t \mid t \in \mathbb{R}\}$ is an one-parameter strongly continuous group of unitary operators on H, and $u_t K = K, \forall t \in \mathbb{R}$. We say that $\{u_t\}$ satisfies the KMS condition relative to K. In fact, for any $\xi, \eta \in K$, we can pick two sequences $\{x_n\}, \{y_n\}$ of M with $x_n^* = x_n, y_n^* = y_n, \forall n$, such that $x_n\xi_0 \to \xi, y_n\xi_0 \to \eta$. Since $\{\alpha_t\}$ satisfies the KMS condition relative to φ_0, then for each n there is a KMS function f_n such that

$$\begin{aligned} f_n(t) = &\ \varphi_0(\alpha_t(x_n)y_n) = \langle y_n\xi_0, \alpha_t(x_n)\xi_0\rangle \\ = &\ \langle y_n\xi_0, u_t x_n\xi_0\rangle, \\ f_n(t+i) = &\ \varphi_0(y_n\alpha_t(x_n)) = \langle \alpha_t(x_n)\xi_0, y_n\xi_0\rangle \\ = &\ \langle u_t x_n\xi_0, y_n\xi_0\rangle, \end{aligned}$$

$\forall t \in \mathbb{R}$. By the maximum modulus theorem (see [137]), we have

$$\begin{aligned} \sup_{0\leq \mathrm{Im} z\leq 1} |f_n(z) - f_m(z)| = &\ \sup_{t\in\mathbb{R}} |f_n(t) - f_m(t)| \\ \leq &\ \|y_n\xi_0\| \cdot \|x_n\xi - x_m\xi_0\| + \|x_m\xi_0\| \cdot \| \cdot \|y_n\xi_0 - y_m\xi_0\| \end{aligned}$$

and

$$\sup_{\substack{n \\ 0\leq \mathrm{Im} z\leq 1}} |f_n(z)| = \sup_{\substack{n \\ t\in\mathbb{R}}} |f_n(t)| \leq \sup_n \{\|y_n\xi_0\|, \|x_n\xi_0\|\}.$$

Thus, there is a KMS function f such that

$$f_n(z) \to f(z), \quad \text{uniformly for } z \text{ with } 0 \leq \mathrm{Im} z \leq 1.$$

Clearly, $f(t) = \langle \eta, u_t\xi\rangle, f(t+i) = \langle u_t\xi, \eta\rangle, \forall t \in \mathbb{R}$. Therefore, $\{u_t\}$ satisfies the KMS condition relative to φ_0.

From Theorem 8.1.13, we get $u_t = \Delta^{it}, \forall t \in \mathbb{R}$. Then for any $x \in M$ and $t \in \mathbb{R}$, we have

$$\alpha_t(x)\xi_0 = u_t x\xi_0 = \Delta^{it} x \Delta^{-it}\xi_0 = \sigma_t(x)\xi_0.$$

Since ξ_0 is separating for M, it follows that $\alpha_t(x) = \sigma_t(x), \forall x \in M, t \in \mathbb{R}$.

Q.E.D.

Notes. Theorem 8.2.7 is due to M. Tomita. The first version of Tomita's theory is given by M. Takesaki. The Proof presented here is due to M.A. Rieffel and A. Van Daele. Theorem 8.2.10 is due to M. Takesaki.

References. [127], [135], [174], [181], [182].

8.3. The modular automorphism group of a σ-finite W^*-algebra

Let M be a σ-finite W^*-algebra. Then there is a faithful normal state φ on M. Suppose that $\{\pi_\varphi, H_\varphi, \xi_\varphi\}$ is the faithful cyclic W^*-representation of M generated by φ. Clearly, ξ_φ is a cyclic-separating vector for the VN algebra $\pi_\varphi(M)$ on H_φ. Using the theory of Section 8.2 to $\{\pi_\varphi(M), H_\varphi, \xi_\varphi\}$, we have the modular operator Δ_φ (a non-negative invertible self-adjoint operator on H_φ) such that

$$\Delta_\varphi^{it}\pi_\varphi(M)\Delta_\varphi^{-it} = \pi_\varphi(M), \quad \forall t \in \mathbb{R}.$$

Since M and $\pi_\varphi(M)$ are $*$ isomorphic, we can define

$$\sigma_t^\varphi(x) = \pi_\varphi^{-1}(\Delta_\varphi^{it}\pi_\varphi(x)\Delta_\varphi^{-it}), \quad \forall x \in M, t \in \mathbb{R}.$$

Clearly, $\{\sigma_t^\varphi \mid t \in \mathbb{R}\}$ is an one-parameter $s(M, M_*)$-continuous $*$ automorphism group of M.

Definition 8.3.1. $\{\sigma_t^\varphi \mid t \in \mathbb{R}\}$ is called the *modular automorphism group* of M corresponding to φ.

By Theorem 8.2.10, φ is invariant for $\{\sigma_t^\varphi\}$, i.e., $\varphi(\sigma_t^\varphi(x)) = \varphi(x), \forall x \in M, t \in \mathbb{R}$; and $\{\sigma_t^\varphi\}$ satisfies the KMS condition relative to φ, i.e., for any $x, y \in M$, there is a KMS function f (continuous and bounded on $0 \le \text{Im}z \le 1$ and analytic in $0 < \text{Im}z < 1$) such that

$$f(t) = \varphi(\sigma_t^\varphi(x)y), f(t+i) = \varphi(y\sigma_t^\varphi(x)), \quad \forall t \in \mathbb{R}.$$

Moreover, $\{\sigma_t^\varphi \mid t \in \mathbb{R}\}$ is the unique one-parameter $s(M, M_*)$-continuous $*$ automorphism group of M satisfying the KMS condition relative to φ.

Proposition 8.3.2. Let φ be a faithful normal state on a W^*-algebra M, and

$$M^\varphi = \{x \in M \mid \sigma_t^\varphi(x) = x, \forall t \in \mathbb{R}\}.$$

Then $x \in M^{\varphi}$ if and only if $\varphi(xy - yx) = 0, \forall y \in M$.

Proof. Let $x \in M^{\varphi}$. For any $y \in M$, there is a KMS function f such that

$$f(t) = \varphi(\sigma_t^{\varphi}(x)y) = \varphi(xy),$$
$$f(t+i) = \varphi(y\sigma_t^{\varphi}(x)) = \varphi(yx), \quad \forall t \in \mathbb{R}.$$

Thus, f is a constant function. In particular,

$$\varphi(xy) = f(0) = f(i) = \varphi(yx).$$

Conversely, let $x \in M$ with $\varphi(xy - yx) = 0, \forall y \in M$. We may assume that $x^* = x$. For any $y^* = y \in M$, there is a KMS function f such that

$$f(t) = \varphi(\sigma_t^{\varphi}(x)y), \quad f(t+i) = \varphi(y\sigma_t^{\varphi}(x)), \quad \forall t \in \mathbb{R}.$$

Since $x^* = x$ and $y^* = y$, it follows that $f(t) = \overline{f(t+i)}, \forall t \in \mathbb{R}$. On the other hand, since φ is invariant for $\{\sigma_t^{\varphi}\}$, we have

$$\begin{aligned} f(t) = &\ \varphi(x\sigma_{-t}^{\varphi}(y)) = \varphi(\sigma_{-t}^{\varphi}(y)x) \\ = &\ \varphi(y\sigma_t^{\varphi}(x)) = f(t+i), \quad \forall t \in \mathbb{R}. \end{aligned}$$

Thus, by the Schwartz reflection principle f can be extended to a bounded analytic function on $\mathbb{C}$, and f is a constant function. In particular,

$$\varphi((\sigma_t^{\varphi}(x) - x)y) = f(t) - f(0) = 0, \quad \forall y^* = y \in M, t \in \mathbb{R}.$$

Since φ is faithful, it follows that $\sigma_t^{\varphi}(x) = x, \forall t \in \mathbb{R}$, i.e., $x \in M^{\varphi}$. Q.E.D.

Proposition 8.3.3. Let M be a σ-finite W^*-algebra, φ, ψ be two faithful normal states on M, and $\{\sigma_t^{\varphi}\}, \{\sigma_t^{\psi}\}$ be the modular automorphism groups of M corresponding to φ, ψ respectively. Then there exists an one-parameter $s(M, M_*)$-continuous family $\{u_t \mid t \in \mathbb{R}\}$ of unitary elements of M such that

$$\sigma_t^{\psi}(a) = u_t\sigma_t^{\varphi}(a)u_t^*, \quad u_{t+s} = u_t\sigma_t^{\varphi}(u_s),$$

$\forall a \in M, t, s \in \mathbb{R}$.

Proof. Consider the following W^*-algebra

$$M_2 = \left\{ \begin{pmatrix} a & b \\ c & d \end{pmatrix} | a, b, c, d \in M \right\},$$

and define a functional θ on M_2 as follows

$$\theta\left(\begin{pmatrix} a & b \\ c & d \end{pmatrix}\right) = \varphi(a) + \psi(d), \quad \forall a, d, c, d \in M.$$

Clearly, θ is a faithful normal state on M_2. Let $\{\sigma_t^\theta\}$ be the modular automorphism group of M_2 corresponding to θ.

Put $e_{11} = \begin{pmatrix} 1 & 0 \\ 0 & 0 \end{pmatrix}$. Then it is easy to see that $\theta(e_{11}x - xe_{11}) = 0, \forall x \in M_2$. By Proposition 8.3.2, we have $\sigma_t^\theta(e_{11}) = e_{11}, \forall t \in \mathbb{R}$. Notice that

$$\begin{aligned}\sigma_t^\theta\left(\begin{pmatrix} a & 0 \\ 0 & 0 \end{pmatrix}\right) &= \sigma_t^\theta\left(\begin{pmatrix} 1 & 0 \\ 0 & 0 \end{pmatrix}\begin{pmatrix} a & 0 \\ 0 & 0 \end{pmatrix}\begin{pmatrix} 1 & 0 \\ 0 & 0 \end{pmatrix}\right) \\ &= \begin{pmatrix} 1 & 0 \\ 0 & 0 \end{pmatrix}\sigma_t^\theta\left(\begin{pmatrix} a & 0 \\ 0 & 0 \end{pmatrix}\right)\begin{pmatrix} 1 & 0 \\ 0 & 0 \end{pmatrix},\end{aligned}$$

$\forall a \in M, t \in \mathbb{R}$. Thus for each $t \in \mathbb{R}$ we can define $\alpha_t(\cdot) : M \to M$ such that $\begin{pmatrix} \alpha_t(a) & 0 \\ 0 & 0 \end{pmatrix} = \sigma_t^\theta\left(\begin{pmatrix} a & 0 \\ 0 & 0 \end{pmatrix}\right), \quad \forall a \in M.$

It is easily verified that $\{\alpha_t \mid t \in \mathbb{R}\}$ is an one-parameter $s(M, M_*)$-continuous $*$ automorphism group of M. Since $\{\sigma_t^\theta\}$ satisfies the KMS condition relative to θ, it follows that $\{\alpha_t\}$ satisfies the KMS condition relative to φ. By Theorem 8.2.10, $\alpha_t = \sigma_t^\varphi, \forall t \in \mathbb{R}$.

By a similar discussion for $e_{22} = \begin{pmatrix} 0 & 0 \\ 0 & 1 \end{pmatrix}$, we can also get

$$\sigma_t^\theta\left(\begin{pmatrix} 0 & 0 \\ 0 & a \end{pmatrix}\right) = \begin{pmatrix} 0 & 0 \\ 0 & \sigma_t^\psi(a) \end{pmatrix}, \quad \forall a \in M, t \in \mathbb{R}.$$

Since $\begin{pmatrix} 0 & 0 \\ 1 & 0 \end{pmatrix} = \begin{pmatrix} 0 & 0 \\ 0 & 1 \end{pmatrix}\begin{pmatrix} 0 & 0 \\ 1 & 0 \end{pmatrix}\begin{pmatrix} 1 & 0 \\ 0 & 0 \end{pmatrix}$, for each t we have $u_t \in M$ such that

$$\sigma_t^\theta\left(\begin{pmatrix} 0 & 0 \\ 1 & 0 \end{pmatrix}\right) = \begin{pmatrix} 0 & 0 \\ 0 & 1 \end{pmatrix}\sigma_t^\theta\left(\begin{pmatrix} 0 & 0 \\ 1 & 0 \end{pmatrix}\right)\begin{pmatrix} 1 & 0 \\ 0 & 0 \end{pmatrix} = \begin{pmatrix} 0 & 0 \\ u_t & 0 \end{pmatrix}.$$

From

$$\begin{aligned}\begin{pmatrix} 0 & 0 \\ 0 & 1 \end{pmatrix} &= \sigma_t^\theta\left(\begin{pmatrix} 0 & 0 \\ 1 & 0 \end{pmatrix}\begin{pmatrix} 0 & 1 \\ 0 & 0 \end{pmatrix}\right) \\ &= \begin{pmatrix} 0 & 0 \\ u_t & 0 \end{pmatrix}\sigma_t^\theta\left(\begin{pmatrix} 0 & 0 \\ 1 & 0 \end{pmatrix}^*\right) = \begin{pmatrix} 0 & 0 \\ 0 & u_t u_t^* \end{pmatrix},\end{aligned}$$

we get $u_t u_t^* = 1, \forall t \in \mathbb{R}$. Similarly, $u_t u_t^* = 1, \forall t \in \mathbb{R}$. Hence, $\{u_t \mid t \in \mathbb{R}\}$ is an one-parameter $s(M, M_*)$-continuous family of unitary elements of M.

Since $\begin{pmatrix} 0 & 0 \\ 0 & a \end{pmatrix} = \begin{pmatrix} 0 & 0 \\ 1 & 0 \end{pmatrix}\begin{pmatrix} a & 0 \\ 0 & 0 \end{pmatrix}\begin{pmatrix} 0 & 1 \\ 0 & 0 \end{pmatrix}$, it follows that

$$\begin{aligned}\begin{pmatrix} 0 & 0 \\ 0 & \sigma_t^\psi(a) \end{pmatrix} &= \sigma_t^\theta\left(\begin{pmatrix} 0 & 0 \\ 0 & a \end{pmatrix}\right) \\ &= \begin{pmatrix} 0 & 0 \\ u_t & 0 \end{pmatrix}\begin{pmatrix} \sigma_t^\varphi(a) & 0 \\ 0 & 0 \end{pmatrix}\begin{pmatrix} 0 & u_t^* \\ 0 & 0 \end{pmatrix},\end{aligned}$$

i.e., $\sigma_t^\psi(a) = u_t \sigma_t^\varphi(a) u_t^*, \forall a \in M, t \in \mathbb{R}$.

Moreover, from

$$\begin{aligned}\begin{pmatrix}0 & 0\\ u_{t+s} & 0\end{pmatrix} &= \sigma_t^\theta\left(\sigma_s^\theta\left(\begin{pmatrix}0 & 0\\ 1 & 0\end{pmatrix}\right)\right) = \sigma_t^\theta\left(\begin{pmatrix}0 & 0\\ u_s & 0\end{pmatrix}\right)\\ &= \sigma_t^\theta\left(\begin{pmatrix}0 & 0\\ 1 & 0\end{pmatrix}\begin{pmatrix}u_s & 0\\ 0 & 0\end{pmatrix}\right)\\ &= \begin{pmatrix}0 & 0\\ u_t & 0\end{pmatrix}\begin{pmatrix}\sigma_t^\varphi(u_s) & 0\\ 0 & 0\end{pmatrix},\end{aligned}$$

we obtain $u_{t+s} = u_t\sigma_t^\varphi(u_s), \forall t, s \in \mathbb{R}$. Q.E.D.

Remark. Let α be a $*$ automorphism of M, α is said to be *inner*, if there is a unitary element u of M such that $\alpha(x) = u^*xu, \forall x \in M$. By Proposition 8.3.3, for each $t \in \mathbb{R}$ the innerness of the $*$ automorphism σ_t^φ of M is independent of the choice of the faithful normal state φ on M.

Lemma 8.3.4. Let φ be a faithful normal state on a W^*-algebra $M, h \in M^\varphi \cap M_+$, and the spectral family of h be $s(M, M^*)$-continuous at 0. Then $\psi(\cdot) = \varphi(h\cdot)$ is also a faithful normal positive functional on M, and

$$\sigma_t^\psi(x) = h^{it}\sigma_t^\varphi(x)h^{-it}, \quad \forall x \in M, t \in \mathbb{R},$$

where $\{\sigma_t^\psi \mid t \in \mathbb{R}\}$ is the modular automorphism group of M corresponding to ψ.

Proof. Since $h \in M^\varphi \cap M_+$, it follows from Proposition 8.3.2 that $\psi(\cdot) = \varphi(h\cdot) = \varphi(\cdot h) = \varphi(h^{1/2} \cdot h^{1/2})$. Thus ψ is also a faithful normal positive functional on M (noticing that 0 is not an eigenvalue of h).

Fix $x, y \in M$. For each positive integer n, let

$$x_n = \sqrt{\frac{n}{\pi}}\int_{-\infty}^{\infty} e^{-ns^2}\sigma_s^\varphi(x)ds.$$

Then from Proposition 8.1.12, x_n is analytic with respect to $\{\sigma_t^\varphi\}$, i.e.

$$x_n(z) = \sqrt{\frac{n}{\pi}}\int_{-\infty}^{\infty} e^{-n(s-z)^2}\sigma_s^\varphi(x)ds$$

is analytic: $\mathbb{C} \to M$ and $x_n(t) = \sigma_t^\varphi(x_n), \forall t \in \mathbb{R}$. Moreover, clearly $x_n(z)$ is bounded on every horizontal strip. Thus, the function

$$f_n(z) = \varphi(h^{iz+1}x_n(z)h^{-iz}y)$$

is continuous and bounded on $0 \le \text{Im}z \le 1$ and is analytic in $0 < \text{Im}z < 1$. Now we compute the boundary values of f_n. For $t \in \mathbb{R}$, by $h^{it} \in M^\varphi$ we have

$$f_n(t) = \varphi(hh^{it}\sigma_t^\varphi(x_n)h^{-it}y) = \psi(h^{it}\sigma_t^\varphi(x_n)h^{-it}y)$$

and

$$\begin{aligned} f_n(t+i) &= \varphi(h^{it}x_n(t+i)h^{-it}hy) \\ &= \varphi(x_n(t+i)h^{-it}hyh^{it}) \\ &= \langle \pi_\varphi(h^{-it}hyh^{it})\xi_\varphi, \pi_\varphi(x_n(t+i))^*\xi_\varphi\rangle, \end{aligned}$$

where $\{\pi_\varphi, H_\varphi, \xi_\varphi\}$ is the faithful cyclic W^*-representation of M generated by φ. Put $\eta = \pi_\varphi(x^*)\xi_\varphi(\in H_\varphi)$. Then by Proposition 8.1.12,

$$\eta_n = \sqrt{\frac{n}{\pi}}\int_{-\infty}^{\infty} e^{-ns^2}\Delta_\varphi^{is}\eta ds = \pi_\varphi(x_n^*)\xi_\varphi$$

is analytic with respect to $\{\Delta_\varphi^{it}\}$. From Proposition 8.1.11, we have

$$\sqrt{\frac{n}{\pi}}\int_{-\infty}^{\infty} e^{-n(s-z)^2}\Delta_\varphi^{is}\eta ds = \eta_n(z) = \Delta_\varphi^{iz}\eta_n, \quad \forall z \in \mathbb{C}.$$

Thus, for any $z \in \mathbb{C}$ we get

$$\begin{aligned} \pi_\varphi(x_n(z)^*)\xi_\varphi &= \sqrt{\frac{n}{\pi}}\int_{-\infty}^{\infty} \overline{e^{-n(s-z)^2}}\pi_\varphi(\sigma_s^\varphi(x^*))\xi_\varphi ds \\ &= \sqrt{\frac{n}{\pi}}\int_{-\infty}^{\infty} e^{-n(s-\bar{z})^2}\Delta_\varphi^{is}\eta ds \\ &= \eta_n(\bar{z}) = \Delta_\varphi^{i\bar{z}}\eta_n = \Delta_\varphi^{i\bar{z}}\pi_\varphi(x_n^*)\xi_\varphi. \end{aligned}$$

Further,

$$\begin{aligned} f_n(t+i) &= \langle \pi_\varphi(h^{-it}hyh^{it})\xi_\varphi, \Delta_\varphi\Delta_\varphi^{it}\pi_\varphi(x_n^*)\Delta_\varphi^{-it}\xi_\varphi\rangle \\ &= \langle \Delta_\varphi^{1/2}\pi_\varphi(h^{-it}hyh^{it})\xi_\varphi, \Delta_\varphi^{1/2}\pi_\varphi(\sigma_t^\varphi(x_n^*))\xi_\varphi\rangle \\ &= \langle \pi_\varphi(\sigma_t^\varphi(x_n))\xi_\varphi, \pi_\varphi(h^{-it}y^*hh^{it})\xi_\varphi\rangle \\ &= \varphi(h^{-it}hyh^{it}\sigma_t^\varphi(x_n)) = \varphi(hyh^{it}\sigma_t^\varphi(x_n)h^{-it}) \\ &= \psi(yh^{it}\sigma_t^\varphi(x_n)h^{-it}). \end{aligned}$$

From the maximum modulus theorem and $\|x_n - x\| \to 0$, we have

$$\sup_{0\leq \mathrm{Im} z\leq 1} |f_n(z) - f_m(z)| \leq \|\psi\| \cdot \|y\| \cdot \|x_n - x_m\| \to 0,$$

and

$$\sup_{\substack{n \\ 0\leq \mathrm{Im} z\leq 1}} |f_n(z)| \leq \|\psi\| \cdot \|y\| \cdot \sup_n \|x_n\| < \infty.$$

Hence, there is a KMS function f such that

$$f_n(z) \to f(z), \quad \text{uniformly for } z \text{ with } 0 \leq \mathrm{Im} z \leq 1,$$

and

$$f(t) = \psi(h^{it}\sigma_t^\varphi(x)h^{-it}y), \quad f(t+i) = \psi(yh^{it}\sigma_t^\varphi(x)h^{-it}),$$

$\forall t \in \mathbb{R}$. Therefore, the $*$ automorphism group $\{h^{it}\sigma_t^\varphi(\cdot)h^{-it} \mid t \in \mathbb{R}\}$ of M satisfies the KMS condition relative to ψ. Now by Theorem 8.2.10, we obtain that $\sigma_t^\psi(x) = h^{it}\sigma_t^\varphi(x)h^{-it}, \forall x \in M, t \in \mathbb{R}$. Q.E.D.

Lemma 8.3.5. Let φ be a faithful normal state on a W^*-algebra M. If $\sigma_t^\varphi(x) = x, \forall x \in M, t \in \mathbb{R}$, i.e., $M^\varphi = M$, then φ is also a trace.

Proof. For any $x \in M$, there is a KMS function f such that

$$f(t) = \varphi(\sigma_t^\varphi(x^*)x) = \varphi(x^*x) \geq 0,$$

$$f(t+i) = \varphi(x\sigma_t^\varphi(x^*)) = \varphi(xx^*) \geq 0.$$

Thus, f must be a constant. Therefore, $\varphi(x^*x) = \varphi(xx^*)$. Q.E.D.

Theorem 8.3.6. Let φ be a faithful normal state on a W^*-algebra M, and $\{\sigma_t^\varphi \mid t \in \mathbb{R}\}$ be the modular automorphism group of M corresponding to φ. Then M is semi-finite if and only if $\{\sigma_t^\varphi \mid t \in \mathbb{R}\}$ is an inner $*$ automorphism group of M, i.e., there exists an one-parameter $s(M, M_*)$-continuous group $\{u_t \mid t \in \mathbb{R}\}$ of unitary elements of M such that

$$\sigma_t^\varphi(x) = u_t x u_t^*, \quad \forall x \in M, t \in \mathbb{R}.$$

Proof. Let $\{u_t \mid t \in \mathbb{R}\}$ be an one-parameter $s(M, M_*)$-continuous group of unitary elements of M such that $\sigma_t^\varphi(x) = u_t x u_t^*, \forall x \in M, t \in \mathbb{R}$. We may assume that M is a VN algebra on a Hilbert space H. By the Stone theorem, there is a non-negative invertible self-adjoint operator h such that $u_t = h^{-it}, \forall t \in \mathbb{R}$. Since $u_t \in M^\varphi, \forall t \in \mathbb{R}$, it follows that each spectral projection of h belongs to M^φ. Let p_n be the spectral projection of h corresponding to the interval $[n^{-1}, n]$. Then $p_n, hp_n \in M^\varphi$. Define $\psi_n(\cdot) = \varphi(hp_n \cdot)$ on M_{p_n}. By Lemma 8.3.4 the modular automorphism group of $M_{p_n} = p_n M p_n$ will be

$$(hp_n)^{it}\sigma_t^\varphi(x)(hp_n)^{-it} = h^{it}u_t x u_t^* h^{-it} = x,$$

$\forall x \in p_n M p_n, t \in \mathbb{R}$. From Lemma 8.3.5, ψ_n is a faithful normal tracial state on $p_n M p_n$. Thus, p_n is a finite projection of $M, \forall n$. Clearly, $\sup_n p_n = 1$. Therefore , M is semi-finite.

Conversely, suppose that M is semi-finite. Then there is a faithful semi-finite normal trace τ on M_+.

1) Let p be a projection of M with $\tau(p) < \infty$. Then τ can be uniquely extended to a faithful normal trace on pMp. We say that there exists unique

$h \in pMp$ with $0 \le h \le 1$ such that

$$\tau((1-h)x) = \varphi(hx) = \varphi(xh), \quad \forall x \in pMp.$$

In fact, using Theorem 1.10.4 to $\tau, \varphi + \tau$ and pMp, there is $h \in pMp$ with $0 \le h \le 1$ such that

$$\begin{aligned} \tau(x) = & \frac{1}{2}(\varphi + \tau)(xh + hx) \\ = & \frac{1}{2}\varphi(xh + hx) + \tau(hx), \end{aligned}$$

i.e., $\tau((1-h)x) = \frac{1}{2}\varphi(xh + hx), \forall x \in pMp$. Then

$$\begin{aligned} \frac{1}{2}\varphi(xh^2 + hxh) = & \tau((1-h)xh) \\ = & \tau(h(1-h)x) = \tau((1-h)hx) \\ = & \frac{1}{2}\varphi(hxh + h^2x), \end{aligned}$$

i.e., $\varphi(xh^2) = \varphi(h^2x), \forall x \in pMp$. Generally, $\varphi(xh^{2n}) = \varphi(h^{2n}x), \forall n$ and $x \in pMp$. Since h can be approximated arbitrarily by the polynomials of h^2, it follows that $\varphi(xh) = \varphi(hx), \forall x \in pMp$. Therefore, we have $\tau((1-h)x) = \varphi(xh) = \varphi(hx), \forall x \in pMp$.

2) There exists unique $h \in M$ with $0 \le h \le 1$ and the spectral family of h is $s(M, M_*)$-continuous at the points 0 and 1 such that

$$\tau(1-h) < \infty, \quad \tau((1-h)x) = \varphi(hx) = \varphi(xh), \quad \forall x \in M.$$

In fact, since τ is semi-finite, we can pick an increasing net $\{p_l\}$ of projections of M such that $\sup_l p_l = 1, \tau(p_l) < \infty, \forall l$. By 1), for each l there is $h_l \in p_l M p_l$ with $0 \le h_l \le 1$ such that

$$\tau((1-h_l)x_l) = \varphi(x_l h_l) = \varphi(h_l x_l), \quad \forall x_l \in p_l M p_l.$$

By Proposition 6.5.2, we have

$$\tau((1-h_l)p_l x p_l) = \tau(p_l(1-h_l)xp_l) = \tau((1-h_l)xp_l),$$

$\forall x \in M$, and l. Thus,

$$\tau((1-h_l)xp_l) = \varphi(h_l x p_l) = \varphi(p_l x h_l), \tag{1}$$

$\forall l$ and $x \in M$. Since the closed unit ball of M is $\sigma(M, M_*)$-compact, replacing $\{h_l\}$ by a its subset in necessary case, we may assume that $h_l \to h \in M(\sigma(M, M_*))$. Clearly, $0 \le h \le 1$. If $p_l \le p_{l'}$, then by (1) we have

$$\begin{aligned} \tau((1-h_{l'})xp_l) = & \tau((1-h_{l'})xp_l p_{l'}) \\ = & \varphi(h_{l'} x p_l p_{l'}) = \varphi(h_{l'} x p_l). \end{aligned}$$

Since $\tau(p_l) < \infty$, it follows from Proposition 6.5.3 that $\varphi(\cdot p_l) \in M_*$. Now picking the limit for l', we get

$$\tau((1-h)xp_l) = \varphi(hxp_l), \quad \forall l \text{ and } x \in M. \tag{2}$$

In particular, $\varphi(hp_l) = \tau((1-h)p_l) = \tau((1-h)^{1/2}p_l(1-h)^{1/2}), \forall l$. Further, we have $\tau((1-h)) = \varphi(h) < \infty$ since τ is normal. Again by Proposition 6.5.3 and picking the limit for l in (2), we obtain

$$\tau((1-h)x) = \varphi(hx), \quad \forall x \in M. \tag{3}$$

Further from (1), we have $\tau((1-h_{l'})xp_l) = \varphi(p_{l'}xp_lh_{l'})$ when $p_{l'} \geq p_l$. By the Schwartz inequality, it is easy to see

$$\begin{aligned} & |\varphi(p_{l'}xp_lh_{l'}) - \varphi(xp_lh)| \\ \leq\ & |\varphi(xp_l(h_{l'} - h))| + |\varphi((1-p_{l'})xp_lh_{l'})| \\ \leq\ & |\varphi(xp_l(h_{l'} - h))| + \|x\|\varphi(1-p_{l'})^{1/2}. \end{aligned}$$

Then picking the limit for l', we get $\tau((1-h)xp_l) = \varphi(xp_lh), \forall l$. Now by Proposition 6.5.3 and picking the limit for l, we have

$$\tau((1-h)x) = \varphi(xh), \quad \forall x \in M. \tag{4}$$

Moreover, since φ, τ are faithful, it follows from (3) and (4) that the spectral family of h is $s(M, M_*)$-continuous at 0 and 1.

3) Let $\alpha_t(x) = h^{-it}(1-h)^{it}xh^{it}(1-h)^{-it}, \forall x \in M, t \in \mathbb{R}$. Then it suffices to show that $\sigma_t^\varphi = \alpha_t, \quad \forall t \in \mathbb{R}$.

For any $x, y \in M$, let

$$f(z) = \varphi(h^{-iz}(1-h)^{iz+1}xh^{iz+1}(1-h)^{-iz}y).$$

Clearly, $f(z)$ is continuous and bounded on $0 \leq \text{Im}z \leq 1$ and is analytic in $0 < \text{Im}z < 1$. For any $t \in \mathbb{R}$, by (3) and (4) we have

$$\left.\begin{aligned} f(t) &= \varphi(\alpha_t((1-h)xh)y), \\ f(t+i) &= \varphi(h\alpha_t(x)(1-h)y) \\ &= \tau((1-h)\alpha_t(x)(1-h)y) \\ &= \tau((1-h)y(1-h)\alpha_t(x)) \\ &= \varphi(y(1-h)\alpha_t(x)h). \end{aligned}\right\} \tag{5}$$

Pick $y = 1, x = ha(1-h)$, where $a^* = a$. Then

$$f(t) = f(t+i) = \varphi(\alpha_t(y_0ay_0))$$

is real, $\forall t \in \mathbb{R}$, where $y_0 = h(1-h)$. Thus, f is constant. In particular,

$$\varphi(\alpha_t(y_0ay_0)) = \varphi(y_0ay_0), \quad \forall a^* = a \in M, \quad t \in \mathbb{R}.$$

Let $y_0 = \int_0^1 \lambda de_\lambda$, and $p_n = e_{1-\frac{1}{n}} - e_{\frac{1}{n}}, \forall n$. Since $\{e_\lambda\}$ is $s(M, M_*)$-continuous at 0 and 1, it follows that $p_n \nearrow 1$. Put

$$a_n = \left(\int_{\frac{1}{n}}^{1-\frac{1}{n}} \frac{1}{\lambda} de_\lambda\right) a \left(\int_{\frac{1}{n}}^{1-\frac{1}{n}} \frac{1}{\lambda} de_\lambda\right), \quad \forall n.$$

Then $y_0 a_n y_0 = p_n a p_n$, and $\varphi(\alpha_t(p_n a p_n)) = \varphi(p_n a p_n), \forall n$. Picking the limit for n, we get

$$\varphi(\alpha_t(a)) = \varphi(a), \quad \forall a \in M, t \in \mathbb{R}. \tag{6}$$

For general $x, y \in M$, let

$$x_n = \left(\int_{\frac{1}{n}}^{1-\frac{1}{n}} \frac{1}{\lambda} de_\lambda\right) x \left(\int_{\frac{1}{n}}^{1-\frac{1}{n}} \frac{1}{\lambda} de_\lambda\right)$$

and

$$f_n(z) = \varphi(h^{-iz}(1-h)^{iz+1} h x_n (1-h) h^{iz+1} (1-h)^{-iz} y).$$

By (5), we have

$$f_n(t) = \varphi(\alpha_t(p_n x p_n) y),$$

$$f_n(t+i) = \varphi(y \alpha_t(p_n x p_n)), \quad \forall t \in \mathbb{R}.$$

From the maximum modulus theorem, the Schwartz inequality and (6), we get

$$\sup_{0 \le \operatorname{Im} z \le 1} |f_n(z) - f_m(z)|$$

$$\le \|y\| \cdot \varphi((p_n x p_n - p_m x p_m)^* \cdot (p_n x p_n - p_m x p_m)) \to 0$$

and

$$\sup_{0 \le \operatorname{Im} z \le \frac{1}{n}} |f_n(z)| \le \|x\| \cdot \|y\|.$$

Thus, there is a KMS function f such that

$$f_n(z) \to f(z), \quad \text{uniformly for } z \text{ with } 0 \le \operatorname{Im} z \le 1.$$

Clearly, $f(t) = \varphi(\alpha_t(x)y), f(t+i) = \varphi(y\alpha_t(x)), \forall t \in \mathbb{R}$. Therefore, $\{\alpha_t\}$ satisfies the KMS condition relative to φ, and $\alpha_t = \sigma_t^\varphi, \forall t \in \mathbb{R}$. Q.E.D.

Notes. The *Unitary cocycle theorem* (Proposition 8.3.3) is due to A. Connes. Theorem 8.3.6 is due to M. Takesaki. As noted by Takesaki, this theorem shows that every type (III) factor on a separable Hilbert space has outer automorphisms.

References. [17], [127], [174].

Chapter 9

The Connes Classification of Type (III) Factors

9.1. Preliminaries

Consider some properties of $L^1(\mathbb{R})$. Clearly, with $\|f\|_1 = \int_{\mathbb{R}} |f(s)|ds$ and the convolution

$$f * g(t) = \int_{\mathbb{R}} f(t-s)g(s)ds, \quad \forall f, g \in L^1(\mathbb{R}),$$

$L^1(\mathbb{R})$ is an abelian Banach algebra. Let

$$z_n(t) = \begin{cases} \frac{n}{2}, & -\frac{1}{n} \le t \le \frac{1}{n}, \\ 0, & \text{otherwise,} \end{cases}$$

$\forall n$. Then $z_n \ge 0, \|z_n\|_1 = 1, \forall n$, and $\{z_n\}$ is an approxianate identity for $L^1(\mathbb{R})$, i.e.,

$$\|z_n * f - f\|_1 \to 0, \quad \forall f \in L^1(\mathbb{R}).$$

For each $f \in L^1(\mathbb{R})$, define its Fourier transform $\hat{f}$:

$$\hat{f}(t) = \int_{\mathbb{R}} e^{ist} f(s)ds.$$

Then $\hat{f} \in C_0^\infty(\mathbb{R})$. If $\hat{f} \in L^1(\mathbb{R})$, then we have inversion formula:

$$f(t) = \frac{1}{2\pi} \int_{\mathbb{R}} e^{-ist} \hat{f}(s)ds \quad \text{(a.e.)}.$$

For each $f \in L^2(\mathbb{R})$, define its Fourier transform $\mathcal{F}f$:

$$(\mathcal{F}f)(t) = \frac{1}{\sqrt{2\pi}} l.i.m._{N\to\infty} \int_{-N}^{N} e^{ist} f(s)ds.$$

Then $\mathcal{F}$ is a unitary operator on $L^2(\mathbb{R})$, and

$$(\mathcal{F}^{-1}f)(t) = \frac{1}{\sqrt{2\pi}} l.i.m._{N\to\infty} \int_{-N}^{N} e^{-ist} f(s)ds$$

(the Plancherel theorem). In particular, we have the Parseval formular:

$$\langle f, g\rangle = \langle \mathcal{F}f, \mathcal{F}g\rangle, \quad \forall f, g \in L^2(\mathbb{R}).$$

Lemma 9.1.1. Let K, U be two subsets of $\mathbb{R}, K$ be compact, U be open, and $K \subset U$. Then there exists $k \in L^1(\mathbb{R})$ such that

$$0 \le \hat{k} \le 1; \quad \hat{k} = 1 \text{ on } K; \text{ and supp}\hat{k} \subset U.$$

Proof. Pick an open neighborhood V of 0 such that $K + V - V \subset U$. Let $g, h \in L^2(\mathbb{R})$ with $\mathcal{F}g = \chi_{K-V}, \mathcal{F}h = \chi_V$ respectively, and $k(\cdot) = g(\cdot)h(\cdot)/|V|$. Clearly, $k \in L^1(\mathbb{R})$ and

$$\hat{k}(t) = \frac{1}{|V|}(\chi_V * \chi_{K-V})(t) = \frac{1}{|V|}\int_V \chi_{K-V}(t-s)ds.$$

Therefore, we have

$$0 \le \hat{k} \le 1., \quad \hat{k} \equiv 1 \quad \text{on } K; \text{ and supp}\hat{k} \subset K + V - V \subset U.$$

Q.E.D.

Lemma 9.1.2. Let $t_0 \in \mathbb{R}, f \in L^1(\mathbb{R})$ with $\hat{f}(t_0) = 0, W$ be a neighborhood of t_0,and $\varepsilon > 0$. Then there exists $k \in L^1(\mathbb{R})$ such that $\|k\|_1 < 2$, supp $\hat{k} \subset W, \hat{k} \equiv 1$ on some neighborhood of t_0, and $\|f * k\|_1 < \varepsilon$.

Proof. Without loss of generality, we may assume $t_0 = 0$. Let $\delta = \varepsilon/4(1 + \|f\|_1)$, and pick a compact subset E of $\mathbb{R}$ such that

$$\int_{\mathbb{R}\setminus E} |f(s)|ds < \delta.$$

Further, pick $\eta > 0$ such that $[-3\eta, 3\eta] \subset W$, and $|1 - e^{ist}| < \delta, \forall s \in E$ and $|t| \le 3\eta$. Let $K = V = [-\eta, \eta]$, and $k = gh/2\eta$, where $g, h \in L^2(\mathbb{R})$ with $\mathcal{F}g = \chi_{K-V} = \chi_{[-2\eta,2\eta]}, \mathcal{F}h = \chi_V = \chi_{[-\eta,\eta]}$ respectively. Clearly, $K + V - V = [-3\eta, 3\eta] \subset W$. Then by Lemma 9.1.1, we have

$$0 \le \hat{k} \le 1; \quad \hat{k} \equiv 1 \quad \text{on } [-\eta, \eta]; \text{ and supp } \hat{k} \subset W.$$

Moreover,

$$\|k\|_1 \le \|g\|_2 \cdot \|h\|_2/2\eta = \|\mathcal{F}g\|_2 \cdot \|\mathcal{F}h\|_2/2\eta < 2.$$

Since $\int_{\mathbb{R}} f(s)ds = \hat{f}(0) = 0$, it follows that

$$(f * k)(t) = \int_{\mathbb{R}} f(s)[k(t-s) - k(t)]ds, \quad \forall t \in \mathbb{R}.$$

Then

$$\begin{aligned}\|f * k\|_1 \le & \int_{\mathbb{R}} |f(s)| \cdot \|k_s - k\|_1 ds \\ = & (\int_E + \int_{E'})|f(s)| \cdot \|k_s - k\|_1 ds,\end{aligned}$$

where $E' = \mathbb{R}\backslash E, k_s(\cdot) = k(\cdot - s), \forall s$. Clearly,

$$\int_{E'} |f(s)| \cdot \|k_s - k\|_1 ds \le 4\delta,$$

and

$$\int_E |f(s)| \cdot \|k_s - k\|_1 ds \le \|f\|_1 \cdot \sup_{s \in E} \|k_s - k\|_1.$$

Then by the definition of δ, it suffices to show that

$$\sup_{s\in E} \|k_s - k\|_1 \le 4\delta.$$

Since $k = gh/2\eta$, it follows that

$$2\eta(k_s - k) = g(h_s - h) + (g_s - g)h_s.$$

Notice that for any $s \in E$,

$$\|g_s - g\|_2^2 = \|\mathcal{F}g_s - \mathcal{F}g\|_2^2 = \int_{-2\eta}^{2\eta} |1 - e^{ist}|^2 dt < 4\eta\delta^2,$$

and similarly, $\|h_s - h\|_2^2 < 2\eta\delta^2$. Therefore,

$$\begin{aligned}\|k_s - k\|_1 \le & \frac{1}{2\eta}\{\|\mathcal{F}g\|_2 \cdot \|h_s - h\|_2 + \|g_s - g\|_2 \cdot \|\mathcal{F}h_s\|_2\} \\ \le & 4\delta, \qquad \forall s \in E.\end{aligned}$$

Q.E.D.

In the following, we put

$$K^1(\mathbb{R}) = \{f \in L^1(\mathbb{R}) \mid \text{supp}\hat{f} \text{ is compact}\}$$

Lemma 9.1.3. $K^1(\mathbb{R})$ is dense in $L^1(\mathbb{R})$.

Proof. Denote all continuous functions with a compat support on $\mathbb{R}$ by $K(\mathbb{R})$. Clearly, $K(\mathbb{R})$ is dense in $L^2(\mathbb{R})$. Then $\{f \in L^2(\mathbb{R}) \mid \hat{f} \in K(\mathbb{R})\}$ is dense in $L^2(\mathbb{R})$.

For any $0 \leq f \in L^1(\mathbb{R})$ and $\varepsilon > 0$, pick $g \in L^2(\mathbb{R})$ such that

$$\|g - f^{1/2}\|_2 < \varepsilon \quad \text{amd} \quad \widehat{g} \in K(\mathbb{R}).$$

Then we have

$$\begin{aligned} \|g^2 - f\|_1 \leq & \ \|g \cdot f^{1/2} - f\|_1 + \|g \cdot f^{1/2} - g^2\|_1 \\ \leq & \ \|f^{1/2}\|_2 \cdot \|g - f^{1/2}\|_2 + \|g\|_2 \cdot \|g - f^{1/2}\|_2 \\ < & \ \varepsilon(2\|f^{1/2}\|_2 + \varepsilon). \end{aligned}$$

Moreover, since $\widehat{g} * \widehat{g} \in K(\mathbb{R})$ and the inverse Fourier transform of $\widehat{g} * \widehat{g}$ is $g^2 \in L^1(\mathbb{R})$, it follows that $(g^2)^\wedge = \widehat{g} * \widehat{g} \in K(\mathbb{R})$. Q.E.D.

For a closed ideal I of $L^1(\mathbb{R})$, let

$$\begin{aligned} I^\perp = & \ \{t \in \mathbb{R} \mid \widehat{f}(t) = 0, \forall f \in I\} \\ = & \ \cap\{\mathcal{N}(\widehat{f}) \mid f \in I\}, \end{aligned}$$

where $\mathcal{N}(\widehat{f})$ is the zero point set of $\widehat{f}$. Clearly $I^\perp$ is a closed subset of $\mathbb{R}$. Conversely, for any closed subset E of $\mathbb{R}$, let

$$I(E) = \{f \in L^1(\mathbb{R}) \mid (\widehat{f}|E) \equiv 0\}.$$

Then $I(E)$ is a closed ideal of $L^1(\mathbb{R})$.

Lemma 9.1.4. 1) If E is a closed subset of $\mathbb{R}$, then $I(E)^\perp = E$.

2) If E is a compact subset of $\mathbb{R}$, then $I(E)$ is a regular closed ideal of $L^1(\mathbb{R})$, i.e., $I(E)$ is a closed ideal of $L^1(\mathbb{R})$, and $I(E)$ admits a modular unit l (i.e., $(f - f * l) \in I(E), \forall f \in L^1(\mathbb{R})$).

Proof. 1) Clearly, $E \subset I(E)^\perp$. Conversely, if $t \notin E$, then by Lemma 9.1.1 there exists $f \in L^1(\mathbb{R})$ such that $\widehat{f}(t) = 1$ and $(\widehat{f}|E) \equiv 0$. Thus $f \in I(E)$ and $t \notin I(E)^\perp$. Furhter, $I(E)^\perp \subset E$,and $I(E)^\perp = E$.

2) Since E is compact, by Lemma 9.1.1 there exists $l \in L^1(\mathbb{R})$ such that $(\widehat{l}|E) \equiv 1$. Then $(f - f * l)^\wedge|E \equiv 0$, i.e., $(f - f * l) \in I(E), \forall f \in L^1(\mathbb{R})$. Q.E.D.

Lemma 9.1.5. Let U_1, U_2 be two open subsets of $\mathbb{R}$, K be a compact subset of $\mathbb{R}$ with $K \subset U_2, f \in L^1(\mathbb{R}), I$ be a closed ideal of $L^1(\mathbb{R})$, and $f_1, f_2 \in I$ such that

$$\widehat{f_i} = \widehat{f} \quad \text{on} \quad U_i, \ i = 1, 2.$$

Then there exists $g \in I$ such that

$$\widehat{g} = \widehat{f} \quad \text{on} \quad U_1 \sqcup K.$$

Proof. Pick $e \in L^1(\mathbb{R})$ such that $\hat{e}|K \equiv 1$ and supp $\hat{e} \subset U_2$. Let $g = f_2 * e + f_1 * (1-e) = f_1 - f_1 * e + f_2 * e$. Then $g \in I$, and $\hat{g} = \hat{f}_2 = \hat{f}$ on $K(\subset U_2); \hat{g} = \hat{f}_1 = \hat{f}$ on $U_1 \backslash U_2; \hat{g} = \hat{f}_2\hat{e} + \hat{f}_1 - \hat{f}_1\hat{e} = \hat{f}\hat{e} - \hat{f} + \hat{f}\hat{e} = \hat{f}$ on $U_1 \cap U_2$. Notice that $U_1 = (U_1 \backslash U_2) \sqcup (U_1 \cap U_2)$. Therefore, we have $\hat{g} = \hat{f}$ on $U_1 \sqcup K$.

Q.E.D.

Lemma 9.1.6. Let I be a closed ideal of $L^1(\mathbb{R})$, $E = I^\perp$, and $f \in L^1(\mathbb{R})$. If $\hat{f}$ vanishes on a neighbourhood of E, then $f \in I$.

Proof. 1) Suppose that $t_0 \notin E$. We say that there is $h \in I$ such that $\hat{h} \equiv 1$ on some neighborhood of t_0.

In fact, pick a compact neighborhood K of t_0 with $E \cap K = \emptyset$. Let $J = I(K)$. Then by Lemma 9.1.4 J is a closed ideal of $L^1(\mathbb{R})$, and J admits a modular unit l with $\hat{l} \equiv 1$ on K. Clearly, $(I + J)$ is still a regular ideal of $L^1(\mathbb{R})$,and l is its modular unit. If $(I + J) \neq L^1(\mathbb{R})$, then there is a maximal regular ideal L of $L^1(\mathbb{R})$ such that $(I + J) \subset L$. Since $L = I(\{s\}) = \{g \in L^1(\mathbb{R}) \mid \hat{g}(s) = 0\}$ for some $s \in \mathbb{R}$, it follows from Lemma 9.1.4 that $s \in E \cap L$. But $E \cap K = \emptyset$, a contradiction. Thus, $I + J = L^1(\mathbb{R})$. Consequently, we can write

$$l = l_I + l_J,$$

where $l_I \in I, l_J \in J$. Now let $h = l_I \in I$. Then

$$\hat{h}|K = (\hat{l}|K) - (\hat{l}_J|K) = \hat{l}|K \equiv 1.$$

2) If $t_0 \notin E$, then there is $g \in I$ such that $\hat{g} = \hat{f}$ on some neighborhood of t_0.

In fact, pick h as in 1), and let $g = f * h$. Then g satisfies the condition.

3) Suppose that $K = \text{supp}\hat{f}$ is compact. Since $\hat{f} \equiv 0$ on some neighborhood of E, it follows that $E \cap K = \emptyset$. By 2) for each $t \in K$, there is a $g_t \in I$ and an open neighborhood U_t of t such that

$$\hat{g}_t = \hat{f} \quad \text{on} \quad U_t.$$

Further, for each $t \in K$, pick an open neighborhood V_t of t such that $t \in V_t \subset \overline{V}_t \subset U_t$ and $\overline{V}_t$ is compact. Now by the compactness of K, there are $t_1, \cdots, t_n \in K, g_i = g_{t_i} \in I, V_i = V_{t_i}, U_i = U_{t_i}, 1 \le i \le n$, such that

$$\sqcup_{i=1}^n V_i \supset K, \quad \text{and} \quad \hat{g}_i = \hat{f} \quad \text{on } U_i,\ 1 \le i \le n.$$

Further, let $0 = g_{n+1} (\in I)$. Clearly, $\hat{g}_{n+1} = \hat{f}$ on $\mathbb{R} \backslash K = U_{n+1}$.

By Lemma 9.1.5, there is $g_1' \in I$ such that

$$\hat{g}_1' = \hat{f}, \quad \text{on} \quad U_1 \sqcup \overline{V}_2 \supset V_1 \sqcup V_2.$$

Aganin by Lemma 9.1.5, there is $g_2' \in I$ such that

$$\hat{g}_2' = \hat{f}, \quad \text{on} \quad U_1 \sqcup V_2 \sqcup \overline{V}_3 \supset V_1 \sqcup V_2 \sqcup V_3.$$

$\cdots$. So we can get $g' \in I$ such that

$$\hat{g}' = \hat{f}, \quad \text{on} \quad V_1 \sqcup \cdots \sqcup V_n \supset K.$$

Now by Lemma 9.1.5, there is $g \in I$ such that

$$\hat{g} = \hat{f}, \quad \text{on} \quad U_{n+1} \sqcup K = \mathbb{R}.$$

From the uniqueness of Fourier transform, we obtain $f = g \in I$.

4) General case. Since $L^1(\mathbb{R})$ admits an approximate identity, for any $\varepsilon > 0$ we can pick $z \in L^1(\mathbb{R})$ such that $\|f - f * z\|_1 < \varepsilon/2$. From Lemma 9.1.2, there is $u \in K^1(\mathbb{R})$ such that $\|u - z\|_1 < \varepsilon/2\|f\|_1$. Then

$$\|f - f * u\|_1 \leq \|f - f * z\|_1 + \|f\|_1 \cdot \|u - z_1\| < \varepsilon.$$

Clearly, supp $(f * u)^\wedge$ ($\subset \text{supp}\hat{u}$) is compact, and $(f * u)^\wedge (= \hat{f}\hat{u})$ vanishes on a neighborhood of E. By 3), $f * u \in I$. Now since $\varepsilon (> 0)$ is arbitrary and I is closed, it follows that $f \in I$. Q.E.D.

References. [136].

9.2. The Arveson spectrum

For our purpose, we just consider a W^*-system

$$(M, \mathbb{R}, \sigma),$$

where M is a W^*-algebra; for each $t \in \mathbb{R}, \sigma_t$ is a $*$ automorphism of M; and $t \to \rho(\sigma_t(x)) = \langle \sigma_t(x), \rho \rangle$ is continuous on $\mathbb{R}, \forall x \in M, \rho \in M_*$.

Denote the collection of all bounded Radon measures on $\mathbb{R}$ by $M(\mathbb{R})$, i.e., $M(\mathbb{R}) = C_0^\infty(R)^*$. By the convolution ($(\mu * \nu)(f) = \iint f(s+t)d\mu(s)d\nu(t), \forall \mu, \nu \in M(\mathbb{R}), f \in C_0^\infty(\mathbb{R})$), $M(\mathbb{R})$ is an abelian Banach algebra with an identity $\delta_0(f) = f(0), \forall f \in C_0^\infty(\mathbb{R})$. Moreover, $L^1(\mathbb{R})$ is a closed ideal of $M(\mathbb{R})$.

Proposition 9.2.1. For each $\mu \in M(\mathbb{R})$, there exists $\sigma(\mu) \in B_\sigma(M)$ such that

$$\langle \sigma(\mu)(x), \rho \rangle = \int_{\mathbb{R}} \rho(\sigma_t(x)) d\mu(t)$$

$\forall x \in M, \rho \in M_*$, and $\|\sigma(\mu)\| \leq \|\mu\|$, where $B_\sigma(M)$ is the set of all $\sigma(M, M_*)$-$\sigma(M, M_*)$ continuous linear operators on M. In particular

$$\langle \sigma(f)(x), \rho \rangle = \int_{\mathbb{R}} \rho(\sigma_t(x)) f(t) dt, \quad \|\sigma(f)\| \leq \|f\|_1,$$

$\forall f \in L^1(\mathbb{R}), x \in M, \rho \in M_*$.

Proof. Clearly, $\|\sigma(\mu)\| \leq \|\mu\|, \quad \forall \mu \in M(\mathbb{R})$.

Now it suffices to show that for any normal positive functional ρ on M, the positive functional $\langle \sigma(\mu)(\cdot), \rho \rangle$ on M is normal, where $\mu \in M(\mathbb{R})_+$. Let $\{x_l\}$ be a bounded increasing net of M_+, and $x = \sup_l x_l$. Then for any $t \in \mathbb{R}, \langle \sigma_t(x_l), \rho \rangle \nearrow \langle \sigma_t(x), \rho \rangle$. By the Dini theorem , $\langle \sigma_t(x_l), \rho \rangle \nearrow \langle \sigma_t(x), \rho \rangle$ uniformly for $t \in K$, where K is any compact subset of $\mathbb{R}$. Now by the regularity of μ and the boundedness of $\{x_l\}$, we can see that

$$\begin{aligned} \langle \sigma(\mu)(x), \rho \rangle &= \int \rho(\sigma_t(x)) d\mu \\ &= \lim_l \int \rho(\sigma_t(x_l)) d\mu = \sup_l \int \rho(\sigma_t(x_l)) d\mu \\ &= \sup_l \langle \sigma(\mu)(x_l), \rho \rangle. \end{aligned}$$

Q.E.D.

Definition 9.2.2. Let $(M, \mathbb{R}, \sigma)$ be a W^*-system.

1) Define the *Arveson spectrum* of σ by

$$\begin{aligned} \mathrm{sp}\sigma &= \{f \in L^1(\mathbb{R}) \mid \sigma(f) = 0\}^\perp \\ &= \{t \in \mathbb{R} \mid \text{if } f \in L^1(R) \text{ with } \sigma(f) = 0, \text{ then } \hat{f}(t) = 0\} \\ &= \cap\{\mathcal{N}(\hat{f}) \mid f \in L^1(\mathbb{R}) \text{ and } \sigma(f) = 0\} \end{aligned}$$

where $\mathcal{N}(\hat{f})$ is the zero point set of $\hat{f}$, and clearly $\{f \in L^1(\mathbb{R}) \mid \sigma(f) = 0\}$ is a closed ideal of $L^1(\mathbb{R})$.

2) If $x \in M$, let

$$\begin{aligned} \mathrm{sp}_\sigma(x) &= \{f \in L^1(\mathbb{R}) \mid \sigma(f)(x) = 0\}^\perp \\ &= \{t \in \mathbb{R} \mid \text{if } f \in L^1(\mathbb{R}) \text{ with } \sigma(f)(x) = 0, \text{ then } \hat{f}(t) = 0\} \\ &= \cap\{\mathcal{N}(\hat{f}) \mid f \in L^1(\mathbb{R}) \text{ and } \sigma(f)(x) = 0\}. \end{aligned}$$

3) If E is a closed subset of $\mathbb{R}$, define the associated "*spectral subspace*" by

$$M(\sigma, E) = \{x \in M \mid \mathrm{sp}_\sigma(x) \subset E\}.$$

Clearly, by the definition $\mathrm{sp}\sigma$ is a closed subset of $\mathbb{R}$ and $0 \in \mathrm{sp}\sigma$ (since $\sigma(f)(1) = \hat{f}(0), \forall f \in L^1(\mathbb{R})$); $\mathrm{sp}_\sigma(x)$ is closed and $\mathrm{sp}_\sigma(x) \subset \mathrm{sp}\sigma, \forall x \in M; \mathrm{sp}_\sigma(0) = \emptyset$ (from Lemma 8.4.1); and $0 \in M(\sigma, E)$ for any closed $E \subset \mathbb{R}$.

Proposition 9.2.3. Let $(M, \mathbb{R}, \sigma)$ be a W^*-system. Then:
(a) $\mathrm{sp}\sigma = \cup_{x \in M} \mathrm{sp}_\sigma(x)$;
(b) $\mathrm{sp}_\sigma(x^*) = -\mathrm{sp}_\sigma(x)$, $\mathrm{sp}_\sigma(\sigma_t(x)) = \mathrm{sp}_\sigma(x), \forall x \in M, t \in \mathbb{R}$;
(c) $\mathrm{sp}_\sigma(x) = \emptyset \Longleftrightarrow x = 0$;
(d) $\mathrm{sp}_\sigma(\sigma(f)(x)) \subset \mathrm{sp}_\sigma(x) \cap \mathrm{supp}\hat{f}$, $\forall x \in M, f \in L^1(\mathbb{R})$;
(e) $\sigma_t(M(\sigma, E)) = M(\sigma, E)$, $\forall t \in \mathbb{R}$ and closed $E \subset \mathbb{R}$;
(f) for any closed $E \subset \mathbb{R}$,

$$x \in M(\sigma, E) \Longleftrightarrow \sigma(f)(x) = 0,$$

$\forall f \in L^1(\mathbb{R})$ and $\hat{f} \equiv 0$ on some neighborhood of E. Consequently, $M(\sigma, E)$ is a $\sigma(M, M_*)$-closed linear subspace of M;

(g) if $x \in M$ and $\mu \in M(\mathbb{R})$ satisfy $\hat{\mu} \equiv 0$ on some neighborhood of $\mathrm{sp}_\sigma(x)$, then $\sigma(\mu)(x) = 0$, where $\hat{\mu}(t) = \int_{\mathbb{R}} e^{ist} d\mu(s)$. Moreover, if $f \in L^1(\mathbb{R})$ satisfies either $\hat{f} \equiv 0$ or $\hat{f} \equiv 1$ on some neighborhood of $\mathrm{sp}_\sigma(x)$, then either $\sigma(f)(x) = 0$ or $\sigma(f)(x) = x$.

Proof. (b) Since $\sigma(f)(x)^* = \sigma(\overline{f})(x^*)$, it follows that

$$\begin{aligned}\mathrm{sp}_\sigma(x^*) &= \cap\{\mathcal{N}(\hat{\overline{f}}) \mid \sigma(\overline{f})(x^*) = 0\} \\ &= -\cap\{\mathcal{N}(\hat{f}) \mid \sigma(f)(x) = 0\} = -sp_\sigma(x),\end{aligned}$$

$\forall x \in M$. Moreover, by $\sigma(f)(\sigma_t(x)) = \sigma(f_t)(x), \hat{f}_t(s) = e^{ist}\hat{f}(s)$ and $\mathcal{N}(\hat{f}_t) = \mathcal{N}(\hat{f})$, where $f_t(s) = f(s-t)$, we have

$$\begin{aligned}\mathrm{sp}_\sigma(\sigma_t(x)) &= \cap\{\mathcal{N}(\hat{f}) \mid \sigma(f)(\sigma_t(x)) = 0\} \\ &= \cap\{\mathcal{N}(\hat{f}_t) \mid \sigma(f_t)(x) = 0\} = \mathrm{sp}_\sigma(x),\end{aligned}$$

$\forall t \in \mathbb{R}, x \in M$.

(e) It is immediate from (b).

(f) Let $x \in M(\sigma, E)$, and $f \in L^1(\mathbb{R})$ with $\hat{f} \equiv 0$ on some neighborhood of E. Let

$$I = \{g \in L^1(\mathbb{R}) \mid \sigma(g)(x) = 0\}.$$

Then $\mathrm{sp}_\sigma(x) = I^\perp \subset E$. Now by Lemma 9.1.6 we have $f \in I$, i.e., $\sigma(f)(x) = 0$.

Conversely, let $x \in M$ and $\sigma(f)(x) = 0$ for any $f \in L^1(\mathbb{R})$ with $\hat{f} = 0$ on some neighborhood of E. If there is $s \in (\mathrm{sp}_\sigma(x) \backslash E)$, then by Lemma 9.1.1 we can find $k \in L^1(\mathbb{R})$ with $\hat{k}(s) = 1$ and supp $\hat{k} \subset F$, where F is a closed neighborhood of s and $F \cap E = \emptyset$. Then $\sigma(k)(x) = 0$ since $\hat{k} \equiv 0$ on the open neighborhood $(\mathbb{R}\backslash F)$ of E. But $s \in \mathrm{sp}_\sigma(x)$, so $\hat{k}(s)$ must be 0, a contradiction. Therefore, $\mathrm{sp}_\sigma(x) \subset E$ and $x \in M(\sigma, E)$.

Consequently, $M(\sigma, E)$ is a linear subspace of M. Aganin by Proposition 9.2.1, $M(\sigma, E)$ is $\sigma(M, M_*)$-closed.

(c) Clearly, $\mathrm{sp}_\sigma(0) = \emptyset$. Now let $x \in M$ and $\mathrm{sp}_\sigma(x) = \emptyset$. By (f), we have $\sigma(f)(x) = 0, \forall f \in L^1(\mathbb{R})$. Thus,

$$\int_{\mathbb{R}} f(t)\rho(\sigma_t(x))dt = 0, \quad \forall f \in L^1(\mathbb{R}), \quad \rho \in M_*.$$

Since $t \to \rho(\sigma_t(x))$ is a bounded continuous function on $\mathbb{R}$, it follows that $\rho(\sigma_t(x)) = 0, \forall t \in \mathbb{R}, \rho \in M_*$. Therefore, $x = 0$.

(a) Clearly, $E = \overline{\cup_{x\in M}\mathrm{sp}_\sigma(x)} \subset \mathrm{sp}\sigma$. Now if $s \notin E$, then by Lemma 9.1.1 we can find $k \in L^1(\mathbb{R})$ such that $\hat{k}(s) = 1$, and $\hat{k} \equiv 0$ on some neighborhood of E. By (f), we have $\sigma(k)(x) = 0, \forall x \in M$, i.e., $\sigma(k) = 0$. Since $\hat{k}(s) = 1$, it follows from Definition 9.2.2 that $s \notin \mathrm{sp}\sigma$. Thus, $\mathrm{sp}\sigma \subset E$, and $E = \mathrm{sp}\sigma$.

(d) If $\sigma(g)(x) = 0$, then $\sigma(g)(\sigma(f)(x)) = \sigma(g * f)(x) = \sigma(f * g)(x) = \sigma(f)(\sigma(g)(x)) = 0$. Thus,

$$\{g \in L^1(\mathbb{R}) \mid \sigma(g)(x) = 0\} \subset \{h \in L^1(\mathbb{R}) \mid \sigma(h)(\sigma(f)(x)) = 0\},$$

and $\mathrm{sp}_\sigma(x) \supset \mathrm{sp}_\sigma(\sigma(f)(x))$.

Moreover, if $s \notin \mathrm{supp}\hat{f}$, then by Lemma 9.1.1 we can pick $k \in L^1(\mathbb{R})$ such that $\hat{k}(s) = 1$ and $\mathrm{supp}\,\hat{k} \cap \mathrm{supp}\hat{f} = \emptyset$. Thus , $\hat{k}\hat{f} = 0$ and $k * f = f * k = 0$. Further $\sigma(k)(\sigma(f)(x)) = 0$, and $\mathrm{sp}_\sigma(\sigma(f)(x)) \subset \mathcal{N}(\hat{k})$. But $\hat{k}(s) = 1$, so $s \notin \mathrm{sp}_\sigma(\sigma(f)(x))$. Therefore, $\mathrm{sp}_\sigma(\sigma(f)(x)) \subset \mathrm{supp}\hat{f}$.

(g) Since $f * \mu \in L^1(\mathbb{R})$ and $\widehat{f * \mu} = \hat{f}\hat{\mu} \equiv 0$ on some neighborhood of $\mathrm{sp}_\sigma(x)$, it follows from (f) that $\sigma(f)(\sigma(\mu)(x)) = \sigma(f * \mu)(x) = 0, \forall f \in L^1(\mathbb{R})$. Thus by the proof of (c), we have $\sigma(\mu)(x) = 0$.

Now let $f \in L^1(\mathbb{R})$ and $\hat{f} \equiv 1$ on some neighborhood of $\mathrm{sp}_\sigma(x)$. Pick $\mu = \delta_0$. Clearly, $\sigma(\mu)(x) = x$, and $\hat{\mu}(s) = 1, \forall s \in \mathbb{R}$. Thus $(f - \mu)^\wedge \equiv 0$ on some neighborhood of $\mathrm{sp}_\sigma(x)$. By the preceding paragraph, we get

$$0 = \sigma(f - \mu)(x) = \sigma(f)(x) - x.$$

Q.E.D.

Proposition 9.2.4. Let $(M, \mathbb{R}, \sigma)$ be a W^*-system, E_1 and E_2 be two closed subsets of $\mathbb{R}$, $E = \overline{E_1 + E_2}$, $x_i \in M(\sigma, E_i), i = 1, 2$, and $x = x_1 x_2$. Then $x \in M(\sigma, E)$. Consequently.

$$\mathrm{sp}_\sigma(x_1 x_2) \subset \overline{\mathrm{sp}_\sigma(x_1) + \mathrm{sp}_\sigma(x_2)}, \quad \forall x_1, x_2 \in M.$$

Proof. 1) First, we assume that $\mathrm{sp}_\sigma(x_i)$ is compact, $i = 1,2$. Replacing E_i by $\mathrm{sp}_\sigma(x_i)$, we may assume that E_i is compact, $i = 1,2$. Then $E = E_1 + E_2$ is also compact. By Proposition 9.2.3 (f), it suffices to show that

$$\sigma(f)(x) = 0$$

for any $f \in L^1(\mathbb{R})$ with $\hat{f} \equiv 0$ on some neighborhood of E.

Fix $f \in L^1(\mathbb{R})$ with $\hat{f} \equiv 0$ on $(E + V + V)$, where V is a compact neighborhood of 0. Pick $f_i \in L^1(\mathbb{R})$ such that $\hat{f}_i \equiv 1$ on some neighborhood of E_i, and $\mathrm{supp}\hat{f}_i \subset E_i + V, i = 1,2$. From Proposition 9.2.3 (g), we have

$$\sigma(f_i)(x_i) = x_i, \quad i = 1,2.$$

Then for any $\rho \in M_*$, by the Fubini theorem we get

$$\begin{aligned}
\langle \sigma(f)(x), \rho\rangle &= \int_{\mathbb{R}} \rho(\sigma_s(x)) f(s) ds \\
&= \int_{\mathbb{R}} f(s)\langle \sigma_s(\sigma(f_1)(x_1) \cdot \sigma(f_2)(x_2)),\ \rho\rangle ds \\
&= \iiint f(s) f_1(t_1) f_2(t_2) \langle \sigma_{s+t_1}(x_1) \cdot \sigma_{s+t_2}(x_2), \rho\rangle ds dt_1 dt_2 \\
&= \iiint f(s) f_1(s_1 - s) f_2(s_2 + s_1 - s) \langle \sigma_{s_1}(x_1)\sigma_{s_1+s_2}(x_2), \rho\rangle ds ds_1 ds_2.
\end{aligned}$$

Let

$$\begin{aligned}
k(s_1, s_2) &= \int f(s) f_1(s_1 - s) f_2(s_2 + s_1 - s) ds \\
&= (f * f_1 \cdot f_{2,-s_2})(s_1)
\end{aligned}$$

Fix s_2, and take Fourier taansform for s_1. Then

$$\hat{k}(t, s_2) = \hat{f}(t)(\hat{f}_1 * \hat{g})(t),$$

where $g(\cdot) = f_{2,-s}(\cdot) = f_2(\cdot + s_2)$. Noticing that

$$\begin{aligned}
\mathrm{supp}(\hat{f}_1 * \hat{g}) &\subset \mathrm{supp}\hat{f}_1 + \mathrm{supp}\hat{g} \\
&= \mathrm{supp}\hat{f}_1 + \mathrm{supp}\hat{f}_2 \subset E + V + V
\end{aligned}$$

and $\hat{f} \equiv 0$ on $E + V + V$, we have $\hat{k}(t, s_2) = 0, \forall t$. Thus for any s_2, we get $k(s_1, s_2) = 0$, a.e. for s_1. Further, by the Fubini theorem we can see that $\langle \sigma(f)(x), \rho\rangle = 0, \forall \rho \in M_*$, i.e., $\sigma(f)(x) = 0$.

2) Let $\{z_n\}$ be the approximate identity for $L^1(\mathbb{R})$ as the beginning of this section. Then it is easy to see that $\sigma(z_n)y \to y(\sigma(M, M_*)), \forall y \in M$. Thus $y \in \overline{\{\sigma(f)(y) \mid f \in L^1(\mathbb{R}), \|f\|_1 \le 1\}}^{\sigma}$, $\forall y \in M$. Now by Lemma 9.1.3 and Proposition 9.2.1, we have

$$y \in \overline{\{\sigma(f)(y) \mid f \in K^1(\mathbb{R}) \text{ and } \|f\|_1 \le 1\}}^{\sigma},$$

$\forall y \in M$. Further, from Proposition 1.2.8 and 1.2.1 we have

$$x \in \overline{\{\sigma(f)(x_1) \cdot \sigma(g)(x_2) \mid f, g \in K^1(\mathbb{R}), \|f\|_1 \text{ and } \|g\|_1 \leq 1\}}^{\sigma}.$$

Now by 1) and Proposition 9.2.3 (d),

$$\text{sp}_{\sigma}(\sigma(f)(x_1) \cdot \sigma(g)(x_2)) \subset E,$$

$\forall f, g \in K^1(\mathbb{R}), \|f\|_1$ and $\|g\|_1 \leq 1$. Finally, since $M(\sigma, E)$ is $\sigma(M, M_*)$-closed, it follows that $x \in M(\sigma, E)$. Q.E.D.

Lemma 9.2.5. Let $t \in \mathbb{R}, K$ be a compact subset of $\mathbb{R}$, and $\varepsilon > 0$. Then there exists a compact neighborhood V of t such that

$$\|\sigma_s(x) - e^{ist}x\| < \varepsilon\|x\|,$$

$\forall s \in K$, and $x \in M(\sigma, V)$.

Proof. Pick a compact neighborhood W_1 of t and $f \in K^1(\mathbb{R})$ such that $\hat{f} \equiv 1$ on W_1. For each $s \in K$, let

$$f^s(r) = f(r - s) - e^{ist}f(r).$$

Then $\hat{f^s}(t) = 0, \forall s \in K$. By Lemma 9.1.2, there is $k^s \in L^1(\mathbb{R})$ and some neighborhood W_s of t such that $\hat{k^s} \equiv 1$ on W_s and $\|f^s * k^s\|_1 < \varepsilon$. Since K is compact and $s \to f^s$ is continuous from $\mathbb{R}$ to $L^1(\mathbb{R})$, there is a compact neighborhood W_2 of t such that for each $s \in K$, we can find $k \in L^1(\mathbb{R})$ with $\hat{k} \equiv 1$ on W_2 and $\|f^s * k\|_1 < \varepsilon$.

Now let $W = W_1 \cap W_2$, and V be a compact neighborhood of t with $V \subset$ the interior of W. For each $s \in K$ and $x \in M(\sigma, V)$, pick $k \in L^1(\mathbb{R})$ such that $\hat{k} \equiv 1$ on W and $\|f^s * k\|_1 < \varepsilon$. Clearly, $\hat{f} \equiv 1$ on W, and $\widehat{f * k} \equiv 1$ on W. By Proposition 9.2.3 (g), we have $\sigma(f * k)(x) = x$. Then we obtain

$$\begin{aligned} &\|\sigma_s(x) - e^{ist}x\| \\ = &\|\sigma_s(\sigma(f * k)(x)) - e^{ist}\sigma(f * k)(x)\| \\ = &\|\sigma(f^s * k)(x)\| \leq \|f^s * k\|_1 \cdot \|x\| < \varepsilon\|x\|. \end{aligned}$$

Q.E.D.

Theroem 9.2.6. Let $(M, \mathbb{R}, \sigma)$ be a W^*-system. Then the following statements are equivalent:

1) $t \in \text{sp}\sigma$;
2) For any closed neighborhood V of $t, M(\sigma, V) \neq \{0\}$;
3) There exists a net $\{x_l\} \subset M$ with $\|x_l\| = 1, \forall l$, such that

$$\|\sigma_s(x_l) - e^{ist}x_l\| \to 0,$$

uniformaly for $s \in K$, where K is any compact subset of $\mathbb{R}$;

4) $|\hat{f}(t)| \le \|\sigma(f)\|, \forall f \in L^1(\mathbb{R})$.

Proof. 1) $\Longrightarrow$ 2). Let $t \in$ spσ. If there is a closed neighborhood V of t such that $M(\sigma, V) = \{0\}$, then pick $f \in L^1(\mathbb{R})$ with $\hat{f}(t) = 1$ and supp $\hat{f} \subset V$. By Proposition 9.2.3 (d), $\text{sp}_\sigma(\sigma(f)(x)) \subset V, \forall x \in M$, i.e., $\sigma(f)(x) \in M(\sigma, V), \forall x \in M$. Since $M(\sigma, V) = \{0\}$, it follows that $\sigma(f) = 0$. By the definition of Spσ, $\hat{f}(t)$ must be 0, a contradiction. Therefore, $M(\sigma, V) \neq \{0\}$ for each closed neighborhood V of t.

2) $\Longrightarrow$ 3). For each closed neighborhood V of t, we can pick $x_V \in M(\sigma, V)$ with $\|x_V\| = 1$ by the condition 2). From Lemma 9.2.5, the net $\{x_V \mid V\}$ satisfies the condition 3).

3)$\Longrightarrow$ 4). Let $\{x_l\}$ be as in 3), and $f \in L^1(\mathbb{R})$. Then

$$\begin{aligned}\|\sigma(f)\| \ge \; & \|\sigma(f)(x_l)\| = \|\int_{\mathbb{R}} \sigma_s(x_l) f(s) ds\| \\ \ge \; & \|\int_{\mathbb{R}} e^{ist} f(s) ds x_l\| - \int_{\mathbb{R}} \|\sigma_s(x_l) - e^{ist} x_l\| \cdot |f(s)| ds.\end{aligned}$$

Now by the condition 3) and $f \in L^1(\mathbb{R})$, we can see that $|\hat{f}(t)| \le \|\sigma(f)\|$.

4) $\Longrightarrow$ 1). It is immediate from the definition of spσ. Q.E.D.

Theorem 9.2.7. Let $(M, \mathbb{R}, \sigma)$ be a W^*-system, A be the abelian Banach subalgebra of $B(M)$ generated by $\{\sigma(f) \mid f \in L^1(\mathbb{R})\}$, and $\Omega(A)$ be the spectral space of A. Then sp$\sigma \cong \Omega(A)$.

Proof. Clearly, σ is a cotinuous homomorphism from $L^1(\mathbb{R})$ to A, and the image of σ is dense in A.

For each $\rho \in \Omega(A)$, $\langle \rho, \sigma(\cdot) \rangle$ is a non-zero multiplicative linear functional on $L^1(\mathbb{R})$. Thus, there is unique $t \in \mathbb{R}$ such that

$$\langle \rho, \sigma(f) \rangle = \hat{f}(t), \quad \forall f \in L^1(\mathbb{R}).$$

Clearly, $\hat{f}(t) = 0$ if $\sigma(f) = 0$. So $t \in$spσ. Put $t = \sigma^*(\rho)$. Then σ^* is a map from $\Omega(A)$ to spσ:

$$\sigma^*(\rho)(f) = \langle \rho, \sigma(f) \rangle = \hat{f}(t), \quad \forall f \in L^1(\mathbb{R}).$$

We say that σ^* is injective. In fact, if $\sigma^*(\rho_1) = \sigma^*(\rho_2)$ for some $\rho_1, \rho_2 \in \Omega(A)$, then

$$\langle \rho_1 - \rho_2, \sigma(f) \rangle = 0, \quad \forall f \in L^1(\mathbb{R}).$$

But $\sigma(L^1(\mathbb{R}))$ is dense in A, so $\rho_1 = \rho_2$.

Now if $t \in$ spσ, define

$$\langle \rho, \sigma(f) \rangle = \hat{f}(t), \quad \forall f \in L^1(\mathbb{R}),$$

then ρ is a non-zero multiplicative linear functional on $\sigma(L^1(\mathbb{R}))$. From Theorem 9.2.6 and $t \in \text{sp}\sigma$, we have

$$|\langle \rho, \sigma(f) \rangle| = |\hat{f}(t)| \le \|\sigma(f)\|, \quad \forall f \in L^1(\mathbb{R}).$$

Thus, ρ can be uniquely extended to a non-zero multiplicative linear functional on A, i.e., the map σ^* is also surjective.

If $\sigma^*(\rho_l) = t_l \to t = \sigma^*(\rho)$ in $\text{sp}\sigma$, then $\langle \rho_l, \sigma(f) \rangle = \hat{f}(t_l) \to \hat{f}(t) = \langle \rho, \sigma(f) \rangle, \forall f \in L^1(\mathbb{R})$. Since $\sigma(L^1(\mathbb{R}))$ is dense in A and $\|\rho\| = \|\rho_l\| = 1, \forall l$, it follows that $\rho_l \to \rho$ in $\Omega(A)$. Conversely, if $\rho_l \to \rho$ in $\Omega(A)$, then $\hat{f}(t_l) \to \hat{f}(t), \forall f \in L^1(\mathbb{R})$, where $t_l = \sigma^*(\rho_l), \forall l$, and $t = \sigma^*(\rho)$. Further, by Lemma 9.1.1 we can see that $t_l \to t$ in $\text{sp}\sigma$. Therefore, σ^* is a homeomorphism from $\Omega(A)$ onto $\text{sp}\sigma$. Q.E.D.

Theorem 9.2.8. Let $(M, \mathbb{R}, \sigma)$ be a W^*-system. Then $t \to \sigma_t$ is uniformly continuous if and only if $\text{sp}\sigma$ is compact.

Proof. Let $\text{sp}\sigma$ be compact, and pick $f \in K^1(\mathbb{R})$ such that $\hat{f} \equiv 1$ on some open neighbourhood of $\text{sp}\sigma$. By Proposition 9.2.3 (g), we have $\sigma(f)(x) = x, \forall x \in M$. Further

$$\begin{aligned}\|\sigma_t(x) - x\| &= \|\sigma(\delta_t * f)x - \sigma(f)(x)\| \\ &\le \|f_t - f\|_1 \cdot \|x\| \to 0(\text{as } t \to 0),\end{aligned}$$

uniformly for $x \in M$ with $\|x\| \le 1$, i.e., $\|\sigma_t - id\| \to 0$ as $t \to 0$.

Conversely, let $t \to \sigma_t$ be uniformly continuous, and $\{z_n\}$ be an approximate identity for $L^1(\mathbb{R})$. Then

$$\begin{aligned}\|\sigma(z_n)x - x\| &\le \int \|\sigma_t(x) - x\| \cdot z_n(t)dt \\ &\le \sup_{|t| \le \frac{1}{n}} \|\sigma_t(x) - x\| \to 0.\end{aligned}$$

i.e., $\|\sigma(z_n) - id\| \to 0$, where id is the identity operator on M. Therefore, $id \in A$, and $\Omega(A)$ is compact. Finally, by Theorem 9.2.7 $\text{sp}\sigma$ is also compact. Q.E.D.

Theorem 9.2.9. Let $(M, \mathbb{R}, \sigma)$ be a W^*-system, and $t \to \sigma_t$ be uniformly continuous (i.e. $\text{sp}\sigma$ is compact). Then there exists $h^* = h \in M$ such that $\sigma_t(x) = u_t x u_t^*$, where $u_t = e^{iht}, \forall t \in \mathbb{R}$ and $x \in M$.

Proof. 1) For any $\lambda \in \mathbb{R}$, define

$$e_\lambda = \sup \left\{ p \,\middle|\, \begin{array}{l} p \text{ is a projection of } M, \text{ and } p\sigma(f)(x) = 0, \\ \forall x \in M, f \in K^1(\mathbb{R}) \text{ and } \text{supp}\hat{f} \subset (\lambda, \infty) \end{array} \right\}.$$

Clearly, $(1-e_\lambda)$ is the minimal projection $q \in M$ such that $q\sigma(f)(x) = \sigma(f)(x), \forall x \in M, f \in K^1(\mathbb{R})$ and supp $\hat{f} \subset (\lambda, \infty)$. In other words,

$$(1-e_\lambda)M = \overline{\left[\sigma(f)(x)\cdot y \mid \begin{array}{c} x, y \in M, f \in K^1(\mathbb{R}) \text{ and} \\ \text{supp}\hat{f} \subset (\lambda, \infty) \end{array}\right]}^{\sigma}.$$

where the right side is a σ-closed right ideal of M.

2) $e_\lambda \leq e_\mu, \forall \lambda \leq \mu; e_\lambda = 1$, if $\lambda > \max\{\mu \mid \mu \in \text{sp}\sigma\}; e_\lambda = 0$, if $\lambda < \min\{\mu \mid \mu \in \text{sp}\sigma\}$. Moreover, $\lambda \to e_\lambda$ is strongly right continuous.

In fact, clearly $e_\lambda \leq e_\mu, \forall \lambda \leq \mu$. If $\lambda > \max\{\mu \mid \mu \in sp\sigma\}$, then for any $f \in K^1(\mathbb{R})$ with supp$\hat{f} \subset (\lambda.\infty), \hat{f}$ is zero on a neighborhood of spσ. From Proposition 9.2.3 (a) and (g), $\sigma(f) = 0$. Thus by 1), $e_\lambda = 1$. Similarly, $e_\lambda = 0$ if $\lambda < \min\{\mu \mid \mu \in \text{sp}\sigma\}$.

Now let $\lambda_n > \lambda, \forall n$ and $\lambda_n \to \lambda$, then $q \geq e_\lambda$, where $q = \inf_n e_{\lambda_n}$. If $f \in K^1(\mathbb{R})$ with supp$\hat{f} \subset (\lambda, \infty)$, then there is m_0 such that supp$\hat{f} \subset (\lambda_m, \infty), \forall m \geq m_0$. Thus by 1), $e_{\lambda_m}\sigma(f)(x) = 0, \forall x \in M, m \geq m_0$. Further, $q\sigma(f)(x) = 0, \forall x \in M$ and $f \in K^1(\mathbb{R})$ with supp$\hat{f} \subset (\lambda, \infty)$. Now from the definition of e_λ, we have $e_\lambda \geq q$, and $e_\lambda = q = \inf_n e_{\lambda_n}$.

3) Let $h = \int_{\mathbb{R}} \lambda de_\lambda$. Then $h^* = h \in M$. Further, let $u_t = e^{iht} = \int_{\mathbb{R}} e^{it\lambda} de_\lambda$, $\forall t \in \mathbb{R}$. Then clearly we have:

$$\begin{aligned} \int_{\mathbb{R}} f(t)u_t dt &= \int_{\mathbb{R}} \hat{f}(s) de_s, \quad \forall f \in L^1(\mathbb{R}); \\ e_\lambda \int_{\mathbb{R}} f_1(t)u_t dt &= \int_{\mathbb{R}} f_1(t)u_t dt, \quad \forall f_1 \in K^1(\mathbb{R}) \\ &\quad \text{and supp}\hat{f}_1 \subset (-\infty, \lambda], \quad \lambda \in \mathbb{R}; \\ (1-e_\lambda)\int_{\mathbb{R}} g_1(t)u_t dt &= \int_{\mathbb{R}} g_1(t)u_t dt, \quad \forall g_1 \in K^1(\mathbb{R}) \text{ and} \\ &\quad \text{supp}\hat{g}_1 \subset (\lambda, \infty), \quad \lambda \in \mathbb{R}. \end{aligned}$$

4) Let $\tau_0, \tau \in \mathbb{R}, f \in K^1(\mathbb{R})$ with supp$\hat{f} \subset (\tau, \infty)$ and $x \in M$ with sp$_\sigma(x) \subset (-\infty, \tau_0]$. Then by Proposition 9.2.3, sp$_\sigma(x^*) \subset [-\tau_0, \infty)$. Further from Proposition 9.2.4,

$$\begin{aligned} \text{sp}_\sigma(x^*\sigma(f)(y)) &\subset \overline{\text{sp}_\sigma(x^*) + \text{sp}_\sigma(\sigma(f)(y))} \\ &\subset \overline{[-\tau_0, \infty) + [\tau + \varepsilon, \infty)} \subset (\tau - \tau_0, \infty), \end{aligned}$$

where $\varepsilon > 0$ is such that supp$\hat{f} \subset [\tau + \varepsilon, \infty), \forall y \in M$. Clearly, sp$_\sigma(z)$ is compact, where $z = x^*\sigma(f)(y)$. Now pick $g \in K^1(\mathbb{R})$ with supp$\hat{g} \subset (\tau - \tau_0, \infty)$ and $\hat{g} \equiv 1$ on a neighborhood of sp$_\sigma(z)$. Then $\sigma(g)(z) = z$. By 1), we have $e_{\tau-\tau_0}\sigma(g)(z) = 0$, i.e

$$e_{\tau-\tau_0} x^*\sigma(f)(y) = 0,$$

$\forall y \in M, f \in K^1(\mathbb{R})$ with $\text{supp}\widehat{f} \subset (\tau, \infty)$. Further, from 1) we get $e_{\tau-\tau_0}x^*(1-e_\tau) = 0$, i.e.,

$$(1-e_\tau)xe_{\tau-\tau_0} = 0,$$

$\forall \tau, \tau_0 \in \mathbb{R}$ and $x \in M$ with $\text{sp}_\sigma(x) \subset (-\infty, \tau_0]$.

5) Pick $f, g \in K^1(\mathbb{R})$ with $\text{supp}\widehat{f} \subset (-\infty, 0)$, $\text{supp}\widehat{g} \subset (0, \infty)$ respectively. Let $\tau, \tau_0 \in \mathbb{R}$, and $f_1(\cdot) = f(\cdot)e^{-i\cdot(\tau-\tau_0)}, g_1(\cdot) = g(\cdot)e^{-i\cdot\tau}$. Then $f_1, , g_1 \in K^1(\mathbb{R})$ and $\text{supp}\widehat{f_1} \subset (-\infty, \tau - \tau_0)$, $\text{supp}\widehat{g}_1 \subset (\tau, \infty)$. By 3), we have $e_{\tau-\tau_0} \int f_1(s)u_s ds = \int f_1(s)u_s ds$ and $(1-e_\tau)\int g_1(t)u_t dt = \int g_1(t)u_t dt$.

Now let $M \subset B(H)$,and $x \in M$ with $\text{sp}_\sigma(x) \subset (-\infty, \tau_0]$. Then by 4) we get

$$\langle x \int f_1(s)u_s ds\xi, \int g_1(t)u_t dt\eta\rangle \equiv 0,$$

$\forall \xi, \eta \in H$. Notice that

$$\begin{aligned}
&\langle x \int f_1(s)u_s ds\xi, \int g_1(t)u_t dt\eta\rangle \\
= &\iint \langle u_t^* x u_t u_{s-t}\xi, \eta\rangle f_1(s)\overline{g_1(t)} ds ds \\
= &\iint \langle \beta_t(x)u_s\xi, \eta\rangle f_1(s-t)\overline{g_1(-t)} ds dt \\
= &\int h(s)e^{-is(\tau-\tau_0)} ds = \widehat{h}(\tau_0 - \tau),
\end{aligned}$$

where $\beta_t(x) = u_t x u_t^*$, and

$$h(s) = \int \langle \beta_t(x)u_s\xi, \eta\rangle f(s-t)\overline{g(-t)}e^{-it\tau_0} dt.$$

Since $\widehat{h}(\tau_0 - \tau) = 0, \forall \tau \in \mathbb{R}$, it follows that $h(s) = 0(a.e.)$. But $h(\cdot)$ is continuous, so $h(s) = 0, \forall s \in \mathbb{R}$. In particular,

$$0 = h(0) = \int \langle \beta_t(x)\xi, \eta\rangle k(t) dt = \langle \beta(k)(x)\xi, \eta\rangle$$

$\forall \xi, \eta \in H$, where $k(t) = f(-t)\overline{g(-t)}e^{-it\tau_0}$. Thus, $\beta(k)(x) = 0$, and $\text{sp}_\beta(x) \subset \mathcal{N}(\widehat{k})$. Notice that

$$\begin{aligned}
\widehat{k}(s) = &\int_{\mathbb{R}} \widehat{g}(r)\widehat{f}(\tau_0 - s + r) dr \\
= &\int_\varepsilon^\infty \widehat{g}(r)\widehat{f}(\tau_0 - s + r) dr,
\end{aligned}$$

where $\varepsilon > 0$ is such that $\text{supp}\widehat{g} \subset (\varepsilon, \infty)$. Since $\text{supp}\widehat{f} \subset (-\infty, 0)$, if follows that $\widehat{k}(s) = 0$ if $s < \tau_0 + \varepsilon$. Further, since $\text{sp}_\beta(x)$ is closed and ε, f, g are arbitrary, we can see that $\text{sp}_\beta(x) \subset (-\infty, \tau_0], \forall x \in M$ with $\text{sp}_\sigma(x) \subset (-\infty, \tau_0]$, i.e.,

$$M(\sigma, (\infty, \tau_0]) \subset M(\beta, (-\infty, \tau_0]), \quad \forall \tau_0 \in \mathbb{R}.$$

Now by Proposition 9.2.3 (b), we have also

$$M(\sigma,[\tau_0,\infty)) \subset M(\beta,[\tau_0,\infty)), \quad \forall \tau_0 \in \mathbb{R}.$$

6) Let $\tau \in \mathbb{R}$,and $f,g \in K^1(\mathbb{R})$ with supp $\hat{f} \subset (-\infty,0)$, supp$\hat{g} \subset (0,\infty)$ respectively. Let $f_1(\cdot) = f(\cdot)e^{-i\tau\cdot}, g_1(\cdot) = g(\cdot)e^{-i\tau\cdot}$. Then $f_1, g_1 \in K^1(\mathbb{R})$ and supp$\hat{f_1} \subset (-\infty,\tau)$, supp$\hat{g}_1 \subset (\tau,\infty)$. For any $x \in M$, since $\sigma(f_1)(x) \in M(\sigma,(-\infty,\tau]) \subset M(\beta,(-\infty,\tau])$, it follows that $\hat{g}_1 \equiv 0$ on a neighborhood of $sp_\beta(\sigma(f_1)(x))$. Thus, $\beta(g_1)(\sigma(f_1)(x)) = 0$. Notice that

$$\begin{aligned} & \beta(g_1)(\sigma(f_1)(x)) \\ = & \iint \beta_t(\sigma_s(x))g_1(t)f_1(s)dsdt \\ = & \iint \Phi_t(\sigma_{s+t}(x))g_1(t)f_1(s)dsdt \\ = & \iint \Phi_t(\sigma_s(x))f(s-t)g(t)e^{-is\tau}dsdt \\ = & \int h(s)e^{-is\tau}ds = \hat{h}(-\tau), \end{aligned}$$

where $\Phi_t = \beta_t \circ \sigma_{-t}, \forall t \in \mathbb{R}$, and

$$h(s) = \int \Phi_t(\sigma_s(x))f(s-t)g(t)dt.$$

Now since $\hat{h}(-\tau) = 0, \forall \tau \in \mathbb{R}$, and $h(\cdot)$ is continuous , it follows that $h(s) = 0, \forall s \in \mathbb{R}$. In particular,

$$0 = h(0) = \int \Phi_t(x)k(t)dt = \Phi(k)(x),$$

where $k(t) = f(-t)g(t)$. Thus, $sp_\Phi(x) \subset \mathcal{N}(\hat{k})$. Notice that

$$\hat{k}(s) = \int \hat{g}(r)\hat{f}(r-s)dr = \int_\varepsilon^\infty \hat{g}(r)\hat{f}(r-s)dr,$$

where $\varepsilon > 0$ with supp$\hat{g} = (\varepsilon,\infty)$. Since supp$\hat{f} \subset (-\infty,0)$, it follows that $\hat{k}(s) = 0$ if $s < \varepsilon$. Furhter, since $sp_\Phi(x)$ is closed, and ε, f, g are arbitrary we can see that $sp_\Phi(x) \subset (-\infty,0], \forall x \in M$. Similarly, from $\beta(f_1)(\sigma(g_1)(x)) = 0$, we have $sp_\Phi(x) \subset [0,\infty), \forall x \in M$.

Thus, $sp_\Phi(x) \subset \{0\}, \forall x \in M$. Finally, by Lemma 9.2.5 we get $\Phi_t = id$, i.e., $\sigma_t = \beta_t$, or

$$\sigma_t(x) = u_t x u_t^*, \quad \forall t \in \mathbb{R}, \quad \text{and} \quad x \in M.$$

Q.E.D.

Notes. Spectral subspaces were introduced by R. Godement. It may be viewed as an attempt to extend the Stone theorem. A systematic study of spectral subspaces and their applications to dynamical systems was presented by W.B. Arveson. Theorem 9.2.6 is due to A. Connes. And Theorem 9.2.8 and 9.2.9 are due to D. Oleson.

References. [9], [17], [57], [122], [123].

9.3. The Connes spectrum

Let $(M, \mathbb{R}, \sigma)$ be a W^*-system, and denote by M^σ the fixed point algebra:

$$M^\sigma = \{x \in M \mid \sigma_t(x) = x, \forall t \in \mathbb{R}\}.$$

Clearly, M^σ is a W^*-subalgebra of M. For a projection $e \in M^\sigma, \sigma$ induces an action σ^e on M_e such that $\sigma_t^e(exe) = e\sigma_t(x)e, \forall t \in \mathbb{R}, x \in M$. Then we obtain a W^*-system $(M_e = eMe, \mathbb{R}, \sigma^e = \sigma|M_e)$, and denote its Arveson spectrum by $\text{sp}\sigma^e$.

Definition 9.3.1. The *Connes spectrum* of W^*-system $(M, \mathbb{R}, \sigma)$ is defined by

$$\Gamma(\sigma) = \cap\{\text{sp}\sigma^e \mid 0 \neq e \in \text{Proj }(M^\sigma)\},$$

where $\text{Proj}(M^\sigma)$ is the collection of all projections of the fixed point algebra M^σ.

Clearly, $\Gamma(\sigma)$ is a closed subset of $\mathbb{R}$ and $0 \in \Gamma(\sigma)$.

Lemma 9.3.2. For any $e \in \text{Proj}(M^\sigma)$ with $e \neq 0$, and a closed subset E of $\mathbb{R}$, we have

$$M_e(\sigma^e, E) = M(\sigma, E) \cap M_e.$$

where $M_e(\sigma^e, E) = \{x \in M_e \mid \text{sp}_{\sigma^e}(x) \subset E\}$.

Proof. If $x \in M_e$, then $\sigma_t^e(x) = \sigma_t(x), \forall t \in \mathbb{R}$ and $\sigma^e(f)(x) = \sigma(f)(x), \forall f \in L^1(\mathbb{R})$. Thus by Definition 9.2.2, we have

$$\text{sp}_\sigma(x) = \text{sp}_{\sigma^e}(x), \quad \forall x \in M_e.$$

That comes to the conclusion. Q.E.D.

Proposition 9.3.3. Let $(M, \mathbb{R}, \sigma)$ be a W^*-system. Then $\Gamma(\sigma)+\text{sp}\sigma = \text{sp}\sigma$.

Proof. First, since $0 \in \Gamma(\sigma)$, it follows that

$$\text{sp}\sigma \subset \text{sp}\sigma + \Gamma(\sigma).$$

Now let $\lambda_1 \in \Gamma(\sigma), \lambda_2 \in \text{sp}\sigma$. We need to prove $\lambda = \lambda_1 + \lambda_2 \in \text{sp}\sigma$. From Theorem 9.2.6, it suffices to show that $M(\sigma, V) \neq \{0\}$ for every compact neighborhood V of λ.

Fix a compact neighborhood V of λ, and pick compact neighborhood V_i of $\lambda_i, i = 1, 2$, such that $V_1 + V_2 \subset V$.

Since $\lambda_2 \in V_2$, it follows from Theorem 9.2.6 that $M(\sigma, V_2) \neq \{0\}$. Let $x_2 \in M(\sigma, V_2)$ with $x_2 \neq 0$. Then $\sigma_t(x_2^*) \neq 0, \forall t \in I\!R$. Let $\sigma_t(x_2^*) = v_t h_t$ be the polar decomposition of $\sigma_t(x_2^*), e_t = v_t v_t^*, \forall t \in I\!R$, and $e = \sup\{e_t \mid t \in I\!R\}$. Clearly, $e_t \neq 0, \forall t \in I\!R$, and $e \neq 0$. We say that $e \in$Proj (M^σ). In fact, if $M \subset B(H)$, then $eH = \overline{[\sigma_t(x_2^*)H \mid t \in I\!R]}$. Thus $\sigma_s(e)H = eH$, i.e., $\sigma_s(e) = e, \forall s \in I\!R$.

Now $\lambda_1 \in \Gamma(\sigma) = \cap\{\text{sp}\sigma^p \mid 0 \neq p \in \text{Proj } (M^\sigma)\}$. In particular, $\lambda_1 \in \text{sp}\sigma^e$. From Theorem 9.2.6 and Lemma 9.3.2, we have

$$M(\sigma, V_1) \cap M_e = M_e(\sigma^e, V_1) \neq \{0\}.$$

Then there is $x_1 \in M(\sigma, V_1) \cap M_e$ with $x_1 \neq 0$. By the definition of e and $ex_1 = x_1 \neq 0$, there exists $t \in I\!R$ such that $e_t x_1 \neq 0$. So we can find $\xi, \eta \in H$ such that $0 \neq \langle e_t x_1 \xi, \eta\rangle = \langle x_1 \xi, e_t \eta\rangle$, where $M \subset B(H)$. But $e_t \eta \in \overline{\sigma_t(x_2^*)H}$, then there is $\varsigma \in H$ such that $\langle x_1 \xi, \sigma_t(x_2^*)\varsigma\rangle \neq 0$. Thus $\sigma_t(x_2) \cdot x_1 \neq 0$.

Put $x = \sigma_t(x_2) \cdot x_1 (\neq 0)$. By Propositions 9.2.3 (b) and 9.2.4, we get

$$\text{sp}_\sigma(x \subset \overline{\text{sp}_\sigma(x_2) + \text{sp}_\sigma(x_1)} \subset \overline{V_2 + V_1} \subset V.$$

Therefore, $x \in M(\sigma, V)$, and $M(\sigma, V) \neq \{0\}$. Q.E.D.

Proposition 9.3.4. Let $(M, I\!R, \sigma)$ be a W^*-system. Then $\Gamma(\sigma)$ is a closed subgroup of $I\!R$.

Proof. For any $e \in$Proj(M^σ) with $e \neq 0$, from Proposition 9.3.3 we have $\Gamma(\sigma^e) + \text{sp}\sigma^e = \text{sp}\sigma^e$. Clearly, $\Gamma(\sigma) \subset \Gamma(\sigma^e)$, and $\Gamma(\sigma) \subset \text{sp}\sigma^e$. Then

$$\Gamma(\sigma) + \Gamma(\sigma) \subset \Gamma(\sigma^e) + \text{sp}\sigma^e = \text{sp}\sigma^e,$$

$\forall 0 \neq e \in$Proj (M^σ). Further from definition 9.3.1, we obtain that $\Gamma(\sigma) + \Gamma(\sigma) \subset \Gamma(\sigma)$. Moreover, from Proposition 9.2.3 (a) and (b) we have

$$\overline{\bigsqcup_{x \in M_e} \text{sp}_{\sigma^e}(x)} = \text{sp}\sigma^e = \overline{\bigsqcup_{x \in M_e} \text{sp}_{\sigma^e}(x^*)} = -\text{sp}\sigma^e,$$

$\forall 0 \neq e \in$ProjM^σ. Thus, $\Gamma(\sigma) = \cap\{\text{sp}\sigma^e \mid 0 \neq e \in \text{Proj } (M^\sigma)\} = -\Gamma(\sigma)$. Finally, since $\Gamma(\sigma)$ is closed and $0 \in \Gamma(\sigma)$, $\Gamma(\sigma)$ is a closed subgroup of $I\!R$.

Q.E.D.

Lemma 9.3.5. If H is a proper closed subgroup of $I\!R$, then there is $\lambda \geq 0$ such that

$$H = Z\!\!Z\lambda.$$

Proof. We may assume that $H \neq \{0\}$. Then we claim that there is $\lambda > 0$ such that

$$H \cap (0, \lambda) = \emptyset, \quad \text{and} \quad \lambda \in H.$$

In fact, if such λ does not exist, then there is a sequence $\{\lambda_n\} \subset H$ with $\lambda_n > 0, \forall n$, and $\lambda_n \to 0$. For any $\mu \in \mathbb{R}$ and n, we can find $N(n)$ such that $N(n)\lambda_n \leq \mu \leq (N(n)+1)\lambda_n$. Thus

$$\text{dist}(\mu, H) \leq \lambda_n \to 0.$$

Since H is closed, it follows that $\mu \in H$,and $H = \mathbb{R}$, a contradiction. Hence such λ exists.

Clearly, $H \supset \mathbb{Z}\lambda$. If $\mu \in (H \backslash \mathbb{Z}\lambda)$ with $\mu > 0$, then there is n such that

$$n\lambda < \mu < (n+1)\lambda.$$

Now $0 < (\mu - n\lambda) < \lambda$ and $(\mu - n\lambda) \in H$. This contradicts $H \cap (0, \lambda) = \emptyset$. Therefore, $H = \mathbb{Z}\lambda$. Q.E.D.

Remark. From Proposition 9.3.4 and Lemma 9.3.5, the Connes spectrun $\Gamma(\sigma)$ of a W^*-system $(M, \mathbb{R}, \sigma)$ is one of the following forms:

$$\mathbb{R}, \quad \{0\}, \quad \text{and} \quad \mathbb{Z}\mu \quad (\text{some } \mu > 0).$$

So thc subgroup $e^{\Gamma(\sigma)}$ of the multiplicative group $(0, \infty)$ is one of the following forms:

$$(0, \infty), \{1\}, \quad \text{and} \quad \{\lambda^n \mid n \in \mathbb{Z}\} \quad (\text{some } \lambda \in (0,1)).$$

Lemma 9.3.6. Let $\{V_j \mid j \in \Lambda\}$ be an open cover of $\mathbb{R}$, and $x \in M$ with $x \neq 0$. Then there exists a $f \in L^1(\mathbb{R})$ and some $j \in \Lambda$ such that

$$\text{supp } \hat{f} \subset V_j, \quad \text{and} \quad \sigma(f)(x) \neq 0.$$

Proof. Let $I_0 = \{f \in K^1(\mathbb{R}) \mid \text{supp } \hat{f} \subset V_j, \text{ for some } j \in \Lambda\}$, and $I = \overline{I_0}$. Clearly, I_0 is an ideal of $L^1(\mathbb{R})$. For any $t \in \mathbb{R}$, there is $j \in \Lambda$ such that $t \in V_j$. By Lemma 9.1.1, we can find $f \in K^1(\mathbb{R})$ such that $\hat{f}(t) = 1$ and $\text{supp}\hat{f} \subset V_j$. Then $f \in I_0 \subset I$, and $t \notin I^\perp$. Thus $I^\perp = \emptyset$, and $I = L^1(\mathbb{R})$ by Lemma 9.1.6.

Moreover, by the proof of Lemma 9.2.5 $\sigma(z_n)(x) \to x \neq 0$, so there is n with $\sigma(z_n)(x) \neq 0$. Now since I_0 is dense in $L^1(\mathbb{R})$, thus we can find $f \in I_0$ such that $\sigma(f)(x) \neq 0$. Q.E.D.

Lemma 9.3.7. Let $e_1, e_2 \in \text{Proj } (M^\sigma)$,and $e_1 \neq 0, e_2 \neq 0$. If $e_1 \sim e_2$ (relative to M), then

$$\Gamma(\sigma^{e_1}) = \Gamma(\sigma^{e_2}).$$

Proof. Let $\lambda \in \Gamma(\sigma^{e_1})$. We must prove $\lambda \in \Gamma(\sigma^{e_2})$, i.e., $\lambda \in \text{sp}\sigma^{f_2}, \forall 0 \neq f_2 \in$ Proj (M^σ) and $f_2 \leq e_2$.

By Theorem 9.2.6, for $0 \neq f_2 \in$ Proj (M^σ) with $f_2 \leq e_2$ and a compact neighborhood V of λ, we need to show that

$$M(\sigma, V) \cap M_{f_2} = M_{f_2}(\sigma^{f_2}, V) \neq \{0\}.$$

Let U, W be compact neighborhoods of $\lambda, 0$ respectively such that $U + W \subset V$. Pick an open cover $\{V_j \mid j \in \Lambda\}$ such that $V_j - V_j \subset W, \forall j$. Since $e_1 \sim e_2$, there is $u \in M$ such that $u^*u = e_1, uu^* = e_2$. By $f_2uu^* = f_2e_2 = f_2 \neq 0$, we have that $f_2u \neq 0$. From Lemma 9.3.6, there exists $g \in L^1(\mathbb{R})$ such that $\text{supp}\hat{g} \subset V_j$ for some j, and $\sigma(g)(f_2u) \neq 0$. Since $f_2, e_1 \in M^\sigma$ and $f_2u = f_2 \cdot f_2u \cdot e_1$, it follows that $x = f_2xe_1$, where $x = \sigma(g)(f_2u)$. Clearly, $\text{sp}_\sigma(x) \subset \text{supp}\,\hat{g}$, and $(\text{sp}_\sigma(x) - \text{sp}_\sigma(x)) \subset (\text{supp}\hat{g} - \text{supp}\hat{g}) \subset (V_j - V_j) \subset W$. Let $M \subset B(H)$,and f_1 be the projective from H onto $\overline{[\sigma_t(x^*)H | t \in \mathbb{R}]}$. Then $f_1 \in$ Proj (M^σ) and $f_1 \neq 0$. Moreover, by $e_1\sigma_t(x^*) = \sigma_t(e_1x^*) = \sigma_t(x^*), \forall t \in \mathbb{R}$, we have $f_1 \leq e_1$. Now $\lambda \in U$ and $\lambda \in \Gamma(\sigma^{e_1}) \subset \text{sp}\sigma^{f_1}$, then from Theorem 9.2.6 and Lemma 9.3.2 we have

$$M(\sigma, U) \cap M_{f_1} = M_{f_1}(\sigma^{f_1}, U) \neq \{0\}.$$

Pick $0 \neq y \in M(\sigma, U) \cap M_{f_1}$. By the proof of Prosition 9.3.3, we can also see that

$$\sigma_{t_1}(x)yf_1 = \sigma_{t_1}(x)y \neq 0, \quad \text{for some} \quad t_1 \in \mathbb{R}.$$

Again by the definition of f_1, there is $t_2 \in \mathbb{R}$ such that $\sigma_{t_1}(x)y\sigma_{t_2}(x^*) \neq 0$. Put $z = \sigma_{t_1}(x)y\sigma_{t_2}(x^*)$. Since $f_2x = x, x^*f_2 = x^*$,and $\sigma_t(f_2) = f_2, \forall t$, it follows that $z = f_2zf_2 \in M_{f_2}$. Now by Propositions 9.2.4 and 9.2.3 we have

$$\begin{aligned}\text{sp}_\sigma(z) &\subset \overline{\text{sp}_\sigma(x) + \text{sp}_\sigma(y) - \text{sp}_\sigma(x)} \\ &\subset \overline{U + W} \subset V.\end{aligned}$$

Therefore, $M(\sigma, V) \cap M_{f_2} \neq \{0\}$. Q.E.D.

Definition 9.3.8. Two actions σ, τ of $\mathbb{R}$ on M are said to be *outer equivalent*, if there is an one-parameter strongly continuous group $\{u_t \mid t \in \mathbb{R}\}$ of unitary elements of M such that

$$u_{t+s} = u_t\sigma(u_s), \quad \tau_t(x) = u_t\sigma_t(x)u_t^*,$$

$\forall s, t \in \mathbb{R}, x \in M$.

Lemma 9.3.9. Suppose that σ and τ are outer equivalent. Then there exists an action γ of $\mathbb{R}$ on $\widetilde{M}$ such that

$$\begin{cases} \gamma_t(x \otimes e_{11}) = \sigma_t(x) \otimes e_{11}, \\ \gamma_t(x \otimes e_{22}) = \tau_t(x) \otimes e_{22}, \end{cases}$$

$\forall t \in \mathbb{R}, x \in M$, where $\widetilde{M} = M \otimes M_2(\mathbb{C}) = \left\{ \begin{pmatrix} a & b \\ c & d \end{pmatrix} | a, b, c, d \in M \right\}$, and $\{e_{ij} \mid 1 \leq i, j \leq 2\}$ is the matrix unit of $M_2(\mathbb{C})$.

Proof. Let $\{u_t \mid t \in \mathbb{R}\}$ be as in Definition 9.3.8, and define

$$\gamma_t\left(\begin{pmatrix} x_{11} & x_{12} \\ x_{21} & x_{22} \end{pmatrix}\right) = \begin{pmatrix} \sigma_t(x_{11}) & \sigma_t(x_{12})u_t^* \\ u_t\sigma_t(x_{21}) & \tau_t(x_{22}) \end{pmatrix},$$

$\forall t \in \mathbb{R}, x_{ij} \in M, 1 \leq i, j \leq 2$. It is easily verified that γ does the job.

Q.E.D.

Proposition 9.3.10. Let σ, τ be two actions of $\mathbb{R}$ on M, and let σ and τ be outer equivalent. Then $\Gamma(\sigma) = \Gamma(\tau)$.

Proof. Let $(\widetilde{M}, \mathbb{R}, \gamma)$ be as in Lemma 9.3.9. Clearly, $1 \otimes e_{11}$ and $1 \otimes e_{22} \in \widetilde{M}^\gamma$. If write $\tilde{u} = 1 \otimes e_{21}$, then

$$\tilde{u}^*\tilde{u} = 1 \otimes e_{11}, \quad \tilde{u}\tilde{u}^* = 1 \otimes e_{22}.$$

So by lemma 9.3.7, we have $\Gamma(\gamma^{1\otimes e_{11}}) = \Gamma(\gamma^{1\otimes e_{22}})$. On the other hand, it is obvious that

$$(M, \mathbb{R}, \sigma) \cong (\widetilde{M}_{1\otimes e_{11}}, \mathbb{R}, \gamma^{1\otimes e_{11}})$$

and

$$(M, \mathbb{R}, \sigma) \cong (\widetilde{M}_{1\otimes e_{22}}, \mathbb{R}, \gamma^{1\otimes e_{22}}).$$

Therefore, $\Gamma(\sigma) = \Gamma(\tau)$. Q.E.D.

Notes. The Connes spectrum was defined by A. Connes. Propositions 9.3.4 and 9.3.10 were also established by A. Connes.

References. [17].

9.4. The Connes classification of type (III) factors (σ-finite case)

Let M be a σ-finite W^*-algebra. Then there exists a faithful normal state φ on M. Let $\{\pi_\varphi, H_\varphi, \xi_\varphi\}$ be the faithful cyclic W^*-representation of M generated by φ. Clearly, ξ_φ is a cyclic-separating vector for the VN algebra $\pi_\varphi(M)$ on H_φ. From Tomita-Takesaki theory (see Section 8.2), there is the modular operator Δ_φ (a non-negative invertible self-adjoint operator on H_φ),and a W^*-system $(M, \mathbb{R}, \sigma^\varphi)$, where $\{\sigma_t^\varphi \mid t \in \mathbb{R}\}$ is the modular automorphism group of M corresponding to φ.

Proposition 9.4.1. $s \in \mathrm{sp}\,\sigma^\varphi \Longleftrightarrow e^s \in \mathrm{sp}\Delta_\varphi$. Consequently,

$$e^{\mathrm{sp}\sigma^\varphi} = \mathrm{sp}\Delta_\varphi \backslash \{0\} = \mathrm{sp}\Delta_\varphi \cap (0, \infty).$$

Proof. Let $f \in L^1(\mathbb{R})$, and $x \in M$. Then

$$\begin{aligned} \hat{f}(\ln \Delta_\varphi)\pi_\varphi(x)\xi_\varphi &= \int \Delta_\varphi^{it} f(t) dt \pi_\varphi(x)\xi_\varphi \\ &= \pi_\varphi(\sigma^\varphi(f)(x))\xi_\varphi. \end{aligned}$$

Hence,

$$\begin{aligned} \sigma^\varphi(f) = 0 \Longleftrightarrow & \ \hat{f}(\ln \Delta_\varphi)\pi_\varphi(x)\sigma_\varphi = 0, \quad \forall x \in M \\ \Longleftrightarrow & \ \hat{f}(\ln \Delta_\varphi) = 0 \\ \Longleftrightarrow & \ \hat{f}(\ln \lambda) = 0, \quad \forall 0 < \lambda \in \mathrm{sp}\Delta_\varphi. \end{aligned}$$

Since $\{\ln \lambda \mid 0 < \lambda \in \mathrm{sp}\ \Delta_\varphi\}$ is closed, it follows that $\mathrm{sp}\sigma^\varphi = \cap\{\mathcal{N}(\hat{f}) \mid \sigma^\varphi(f) = 0\} = \{\ln \lambda \mid 0 < \lambda \in \mathrm{sp}\ \Delta_\varphi\}$, i.e.,

$$e^{\mathrm{sp}\sigma^\varphi} = \mathrm{sp}\ \Delta_\varphi \backslash \{0\}.$$

Q.E.D.

Definition 9.4.2. Let M be σ-finite W^*-algebra, and define

$$\Gamma(M) = \Gamma(\sigma^\varphi),$$

where φ is a faithful normal state on M. From the Connes unitary cocycle theorem (Proposition 8.3.3) and Proposition 9.3.10, $\Gamma(M)$ is well-defined, i.e. $\Gamma(M)$ is independent of the choice of φ.

If φ is a normal state on M, let $p = \mathrm{supp}\varphi$, then φ is a faithful normal state on M_p. So there is the modular operator Δ_φ for M_p. Define

$$S(M) = \cap\{\mathrm{sp}\ \Delta_\varphi \mid \varphi \text{ is a normal state on } M\}.$$

Proposition 9.4.3. Let M be a σ-finite W^*-algebra. Then $s \in \Gamma(M) \Longleftrightarrow e^s \in S(M)$, i.e.

$$\begin{aligned} e^{\Gamma(M)} &= S(M) \cap (0, \infty) \\ &= \cap\{e^{\mathrm{sp}\sigma^\varphi} \mid \varphi \text{ is a normal state on } M\}, \end{aligned}$$

where $\{\sigma_t^\varphi \mid t \in \mathbb{R}\}$ is the modular automorphism group of M_p corresponding to φ, and $p = \mathrm{supp}\varphi$.

Proof. Let φ be a normal state on M, and $p = \text{supp}\varphi$. Pick a normal state ψ on M with $\text{supp}\psi = 1-p$. Then $\rho = \frac{1}{2}(\varphi+\psi)$ is a faithful normal state on M. Let $\{\sigma_t \mid t \in I\!R\}, \{\sigma_t^\varphi \mid t \in I\!R\}$ be the modular automorphism groups of M, M_p corresponding to ρ, φ respectively. By the KMS condition and the uniqueness of the modular automorphism group, we can see that

$$\sigma_t | M_p = \sigma_t^\varphi, \quad \forall t \in I\!R.$$

Moreover, $p \in M^\sigma$. In fact, from

$$\begin{aligned} \varphi(xp) = &\ \varphi(pxp) = \varphi(px), \\ \psi(xp) = &\ \psi(xp(1-p)) = 0 = \psi((1-p)px) = \psi(px), \end{aligned}$$

we have $\rho(xp - px) = 0, \forall x \in M$. Now by Proposition 8.3.2, we get $p \in M^\sigma$.

Noticing that

$$\Gamma(\sigma) = \cap\{\text{sp}\sigma^q \mid 0 \neq q \in \text{Proj}(M^\sigma)\}$$

and

$$\Gamma(\sigma^\varphi) = \Gamma(\sigma|M_p) = \cap\{\text{sp}\sigma^q \mid 0 \neq q \in \text{Proj}(M^\sigma), q \leq p\},$$

we have

$$\begin{aligned} \Gamma(M) = \Gamma(\sigma) \subset &\ \Gamma(\sigma^\varphi) \\ \subset &\ \text{sp}\sigma^\varphi = \{\ln\lambda \mid 0 < \lambda \in \text{sp}\Delta_\varphi\} \end{aligned}$$

by Proposition 9.4.1. Since φ is arbitrary, it follows that

$$e^{\Gamma(M)} \subset S(M) \cap (0, \infty).$$

Conversely, let $s \in I\!R$ and $e^s \in S(M)$. Pick a faithful normal state φ on M, and let $\{\sigma_t = \sigma_t^\varphi \mid t \in I\!R\}$ be the modular automorphism group of M corresponding to φ, and $0 \neq p \in \text{Proj}\ (M^\sigma)$. Clearly, $\{(\sigma_t|M_p) \mid t \in I\!R\}$ is the modular automorphism group $\{\sigma_t^\psi \mid t \in I\!R\}$ of M_p corresponding to $\psi = (\varphi|M_p)/\varphi(p)$. Then from $e^s \in S(M)$ and Proposition 8.4.26, we have $e^s \in \text{sp}\Delta_\psi$ and $s \in \text{sp}\sigma^\psi = \text{sp}\sigma^p$. Since $p \in \text{Proj}\ (M^\sigma)$ is arbitrary, we get

$$s \in \cap\{\text{sp}\sigma^p \mid 0 \neq p \in \text{Proj}(M^\sigma)\} = \Gamma(\sigma) = \Gamma(M).$$

Therefore

$$e^{\Gamma(M)} = s(M) \cap (0, \infty).$$

Q.E.D.

Remark. Let φ be a faithful normal state on M, and $\{\sigma_t \mid t \in I\!R\}$ be the modular automorphism group of M corresponding to φ. Then

$$\Gamma(M) = \Gamma(\sigma) = \cap\{\text{sp}\sigma^e \mid 0 \neq e \in \text{Proj}\ (M^\sigma)\}.$$

Since $(\varphi|M_e)$ is also faithful, there is the modular operator Δ_e for M_e, and the modular automorphism group of M_e corresponding to $(\varphi|M_e)$ is $\{\sigma_t^e = (\sigma_t|M_e) \mid t \in \mathbb{R}\}$ exactly. Thus $e^{sp\sigma^e} = \mathrm{sp}\Delta_e \cap (0,\infty)$. Further, we have

$$e^{\Gamma(M)} = \cap\{\mathrm{sp}\Delta_e \mid 0 \neq e \in \mathrm{Proj}(M^\sigma)\} \cap (0,\infty).$$

Moreover, from the remark under Lemma 9.3.5 $\Gamma(M)$ is one of following forms:

$$\mathbb{R}, \quad \{0\}, \quad \text{and} \quad \mathbb{Z}\mu \quad (\text{some} \quad \mu > 0),$$

and $e^{\Gamma(M)}$ is one of following forms:

$$(0,\infty), \quad \{1\}, \quad \text{and } \{\lambda^n \mid n \in \mathbb{Z}\} \quad (\text{some} \quad \lambda \in (0,1)).$$

Theorem 9.4.4. Let M be a σ-finite factor. Then the following statements are equivalent:

1) M is semi-finite;
2) $S(M) = \{1\}$;
3) $0 \notin S(M)$.

Proof. Let M be semi-fiinite. Clearly, $1 \in S(M)$. Now pick a non-zero finite projection p of M. From Proposition 6.3.15, there is a normal state φ on M with supp$\varphi = p$,and φ is a tracial state on M_p. Furhter, by Propositions 8.2.2 and 8.1.4, we can see that $\Delta_\varphi = 1$. Thus, $S(M) = \{1\}$.

If $S(M) = \{1\}$, then $0 \notin S(M)$ is obvious.

Now let $0 \notin S(M)$. By the definiteion of $S(M)$, there is a normal state φ on M such that $0 \notin \mathrm{sp}\Delta_\varphi$. From Lemma 8.1.3, then Δ_φ and Δ_φ^{-1} are bounded. Thus the modular automorphism group $\{\sigma_t^\varphi \mid t \in \mathbb{R}\}$ of M_p is uniformly continuous, where p =suppφ. By Theorem 9.2.9, $\{\sigma_t^\varphi \mid t \in \mathbb{R}\}$ is inner. Furhter, by Theorem 8.3.6, M_p is semi-finite. Since M is a factor, M is also semi-finite. Q.E.D.

Definition 9.4.5 Let M be a σ-finite type (III) factor.

M is said to be *type* (III_0), if $\Gamma(M) = \{0\}$, or $S(M) \cap (0,\infty) = \{1\}$.

Let $\lambda \in (0,1)$.M is said to be *type* (III_λ), if $\Gamma(M) = \mathbb{Z}\ln\lambda$, or $S(M) \cap (0,\infty) = \{\lambda^n \mid n \in \mathbb{Z}\}$.

M is said to be *type* (III_1), if $\Gamma(M) = \mathbb{R}$, or $S(M) \supset (0,\infty)$.

Now from Theorem 9.4.4, we have immediately the following.

Proposition 9.4.6. Let M be a σ-finite factor. Then:

M is type $(\mathrm{III}_0) \Longleftrightarrow S(M) = \{0,1\}$;
M is type $(\mathrm{III}_\lambda) \Longleftrightarrow S(M) = \{0, \lambda^n \mid n \in \mathbb{Z}\}, \forall \lambda \in (0,1)$;
M is type $(\mathrm{III}_1) \Longleftrightarrow S(M) = [0,\infty)$.

Notes. The theory of factors in Chapter 7 was presented by F.J. Murray and J.Von Neumann in 1930's. After near 40 year, A. Connes gave a new essential development for the classification of factors.

Reference. [17].

9.5. Examples of type (III_λ) factors

Proposition 9.5.1. Let $(M, \mathbb{R}, \sigma)$ be a W^*-system. Then we have

$$\Gamma(\sigma) = \cap\{\mathrm{sp}\sigma^e |\ 0 \neq e \in \mathrm{Proj}\ (Z(M^\sigma))\},$$

where $Z(M^\sigma)$ is the center of the fixed point algebra M^σ.

Proof. By Definition 9.3.1, it is obvious that the left sided is contained in the right side. Now it suffices to show that for any $0 \neq e \in \mathrm{Proj}(M^\sigma)$ there is $0 \neq \bar{e} \in \mathrm{Proj}(Z(M^\sigma))$ such that $\mathrm{sp}\sigma^e = \mathrm{sp}\sigma^{\bar{e}}$.

For $0 \neq e \in \mathrm{Proj}(M^\sigma)$, let

$$\bar{e} = \sup\{ueu^* \mid u \text{ is a unitary element of } M^\sigma\}.$$

Clearly, $0 \neq \bar{e} \in \mathrm{Proj}(Z(M^\sigma))$. Now by Theorem 9.2.6 and Lemma 9.3.2 we need to prove that

$$M(\sigma, E) \cap M_e \neq \{0\} \Longleftrightarrow M(\sigma, E) \cap M_{\bar{e}} \neq \{0\}$$

for any closed subset E of $\mathbb{R}$. Since $e \leq \bar{e}$, the "$\Longrightarrow$" is obvious. Suppose that $x = \bar{e}x\bar{e}$ is a non-zero element of $M(\sigma, E)$. By the definition of $\bar{e}$, there are unitary elements u, v of M^σ such that

$$(ueu^*)x(vev^*) \neq 0.$$

Let $y = eu^*xve$. Then $0 \neq y \in M_e$. Since for any $z \in M^\sigma, \sigma(f)(z) = \hat{f}(0)z, \forall f \in L^1(R)$, it follows from Definition 9.2.2 that $\mathrm{sp}_\sigma z = \{0\}, \forall z \in M^\sigma \backslash \{0\}$. Now from Proposition 9.2.4 we have that $\mathrm{sp}_\sigma y = \mathrm{sp}_\sigma x \subset E$. Thus, $M(\sigma, E) \cap M_e \neq \{0\}$. Q.E.D.

Corollary 9.5.2. Let M be a σ-finite factor, φ be a faithful normal state on M, and $\{\sigma_t \mid t \in \mathbb{R}\}$ be the modular automorphism group of M corresponding to φ. Then we have $\Gamma(M) = \cap\{\mathrm{sp}\sigma^e \mid 0 \neq e \in \mathrm{Proj}(Z(M^\sigma))\}$.

Proposition 9.5.3. Let $M, \varphi, \{\sigma_t\}$ be as in Corollary 5.2, and Δ_e be the modular operator for M_e corresponding to $(\varphi|M_e), \forall 0 \neq e \in \text{Proj}(Z(M^\sigma))$. Then we have

$$S(M) = \cap\{\text{sp } \Delta_e \mid 0 \neq e \in \text{Proj}(Z(M^\sigma))\}.$$

Proof. By Corollary 5.2, Proposition 9.4.1, Proposition 9.4.3 and its Remark, we can see that

$$\begin{aligned} e^{\Gamma(M)} = &\ S(M) \cap (0, \infty) \\ = &\ \cap\{\text{sp}\Delta_e \mid 0 \neq e \in \text{Proj}(Z(M^\sigma))\} \cap (0, \infty). \end{aligned}$$

If M is type (III), then for each $0 \neq e \in \text{Proj } (Z(M^\sigma)), M_e$ is also type (III). By Theorem 9.4.4, it follows that $0 \in S(M_e)$ and $0 \in \text{sp}\Delta_e$. Hence

$$S(M) = \cap\{\text{sp}\Delta_e \mid 0 \neq e \in \text{Proj } (Z(M^\sigma))\}.$$

If M is semi-finite, then by Theorem 9.4.4 we have $0 \notin S(M)$. We must exhibit a non-zero $e \in \text{Proj } (Z(M^\sigma))$ such that $0 \notin \text{sp}\Delta_e$, or, equivalently, such that Δ_e is bounded. From the proof of 8.3.6, we can write $\sigma_t(x) = h^{-it}xh^{it}, \forall t \in \mathbb{R}, x \in M$, where h is a non-negative invertible (maybe unbounded) operator on H (here, assume that $M \subset B(H)$),and each spectral projection of h belongs to M^σ. Further, since $xh^{it} = h^{it}, \forall t \in \mathbb{R}, x \in M^\sigma$, each spectral projection of h belongs to $\text{Proj}(Z(M^\sigma))$ indeed. Pick $n(> 1)$ such that $e = \int_{\frac{1}{n}}^{n} de_\lambda \neq 0$, where $h = \int_0^\infty \lambda de_\lambda$ is the spectral decomposition of h. Consider the functional $\psi(\cdot) = \varphi(he\cdot)$ on M_e. By Lemma 8.3.4, we have

$$\sigma_t^\psi(x) = h^{it}\sigma_t(x)h^{-it} = x, \quad \forall t \in \mathbb{R}, \quad x \in M_e.$$

Hence, ψ is a trace on M_e from Lemma 8.3.5. Consequently,

$$\begin{aligned} \varphi(x^*x) \leq &\ n\psi(x^*x) \\ = &\ n\psi(xx^*) \leq n^2\varphi(xx^*), \quad \forall x \in M_e. \end{aligned}$$

If $\{\pi_e, H_e, \xi_e\}$ is the cyclic $*$ representation of M_e generated by $(\varphi|M_e)$, then by Proposition 8.2.2 we have

$$\begin{aligned} \|\Delta_e^{1/2}\pi_e(x)\xi_e\|^2 = &\ \|j_e\Delta_e^{1/2}\pi_e(x)\xi_e\|^2 = \|\pi_e(x^*)\xi_e\|^2 \\ = &\ \varphi(xx^*) \leq n^2\varphi(x^*x) = n^2\|\pi_e(x)\xi_e\|^2, \end{aligned}$$

$\forall x \in M_e$. Therefore, Δ_e is bounded. Q.E.D.

Now let (G, Ω, μ) be a group measure space. By Definition 7.3.9 we have

$$d\mu_t(\cdot) = d\mu(t^{-1}\cdot) = r_t(\cdot)d\mu(\cdot),$$

$$0 < r_t(\cdot) < \infty,$$

$$r_{st}(\cdot) = r_t(s^{-1}\cdot)r_s(\cdot), \quad \forall s, t \in G, a.e.\mu.$$

Let $H = L^2(\Omega, \mu), M = \{m_f \mid f \in L^\infty(\Omega, \mu)\}$ be the multiplicative algebra on H, and

$$(u_t g)(\cdot) = r_t(\cdot)^{1/2} g(t^{-1}\cdot), \quad \forall g \in H,$$

$$\alpha_t(m_f) \stackrel{\bullet}{=} u_t m_f u_t^* = m_{f_{t^{-1}}},$$

where $f_{t^{-1}}(\cdot) = f(t^{-1}\cdot), \forall f \in L^\infty(\Omega, \mu), t \in G$. Then, $t \to u_t$ is a unitary representation of G on H, and (M, G, α) is a dynamical system. Let $\widetilde{H} = H \otimes l^2(G)$ and define

$$\begin{cases} (\pi(a)\xi)(t) = \alpha_{t^{-1}}(a)\xi(t), \\ (\lambda(s)\xi)(t) = \xi(s^{-1}t) \end{cases}$$

$\forall a \in M, s, t \in G, \xi(\cdot) \in \widetilde{H}$. Then we have the crossed product $M \times_\alpha G = \{\pi(M), \lambda(G)\}''$. Put $\widetilde{M} = M \times_\alpha G$ simply.

Now suppose that (G, Ω, μ) is free and ergodic (see Definition 7.3.10). Then by Lemma 7.3.11, $\pi(M)$ is maximal commutative in $\widetilde{M}$,and $\widetilde{M}$ is a factor on $\widetilde{H}$. By Lemma 7.3.4, for each $x \in \widetilde{M}$ there is unique function $b.(: G \to M)$ such that $p_s x p_t^* = u_s^* b_{st^{-1}} u_s, \forall s, t \in G$. Let $\Phi(x) = b_e$. Then Φ is a unital σ-σ continuous positive linear map from $\widetilde{M}$ to M.

Assume that φ is a faithful normal state on M. Then $\tilde{\varphi} = \varphi \circ \Phi$ is a faithful normal state on $\widetilde{M}$. Let $\{\tilde{\sigma}. \mid \cdot \in I\!R\}$ be the modular automorphism group of $\widetilde{M}$ corresponding to $\tilde{\varphi}$.

Lemma 9.5.4. $Z(\widetilde{M}^{\tilde{\sigma}}) \subset \pi(M) \subset \widetilde{M}^{\tilde{\sigma}}$.

Proof. Let $a \in M, \xi \in H, s, t \in G$. By Lemma 7.3.4 we have

$$\begin{aligned} p_s\pi(a)p_t^*\xi = & \ (\pi(a)\tilde{\xi})(s) = \delta_{s,t}\alpha_{s^{-1}}(a)\xi \\ = & \ (u_s^* a \delta_{s,t} u_s)\xi, \end{aligned}$$

where $\tilde{\xi} \in \widetilde{H}$ and $\tilde{\xi}(\cdot) = \delta_{\cdot,t}\xi$. Hence, the function $b.(: G \to M)$ corresponding to $\pi(a)$ is $b. = \delta_{\cdot,e}a$, where e is the unit of G. Let $y \in \widetilde{M}$, and $c.$ be the function $(: G \to M)$ corresponding to y, i.e.,

$$p_s y p_t^* = u_s^* c_{st^{-1}} u_s, \quad \forall s, t \in G.$$

Then

$$\begin{aligned}(\pi(a)y)_{s,t} &= \sum_{k\in G}\pi(a)_{s,k}y_{k,t}\\ &= u_s^*au_s\cdot u_s^*c_{st^{-1}}u_s = u_s^*ac_{st^{-1}}u_s,\end{aligned}$$

for $s,t\in G$. So the function $(: G\to M)$ corresponding to $\pi(a)y$ is $ac.$, and $\tilde{\varphi}(\pi(a)y) = \varphi\circ\Phi(\pi(a)y) = \varphi(ac_e)$. Similarly, $\tilde{\varphi}(y\pi(a)) = \varphi(c_e\alpha_e(a)) = \varphi(c_ea)$. Since M is abelian, it follows that $\tilde{\varphi}(\pi(a)y) = \tilde{\varphi}(y\pi(a)), \forall y\in\widetilde{M}, a\in M$. Now by Proposition 8.3.2 we obtain that $\pi(M)\subset\widetilde{M}^{\tilde{\sigma}}$. Further, we have

$$Z(\widetilde{M}^{\tilde{\sigma}})\subset\pi(M)'\cap\widetilde{M} = \pi(M)$$

since $\pi(M)$ is maximal commutative in $\widetilde{M}$. Q.E.D.

Now we assume further that: Ω is compact; μ is a probability measure on Ω; and for each $t\in G$ there are two positive constants ε_t and η_t such that $0<\varepsilon_t\le r_t(\cdot)\le\eta_t<\infty, \forall\cdot\in G$.

Define $\varphi(m_f) = \int f d\mu, \forall f\in L^\infty(\Omega,\mu)$. Clearly, φ is a faithful normal state on M. Let $\tilde{\Delta}$ be the modular operator corresponding to the faithful normal state $\tilde{\varphi} = \varphi\circ\Phi$ on $\widetilde{M}$.

Lemma 9.5.5. $\tilde{\Delta}$ is spatially isomorphic to the operator $h = \oplus_{t\in G}h_t$ on $\widetilde{H} = \sum_{t\in G} H$, where h_t is a bounded invertible positive operator on $H = L^2(\Omega,\mu)$ corresponding to multiplication by the function $r_t^{-1}(\cdot) = r_{t^{-1}}(\cdot)$, and $\oplus_{t\in G}h_t$ is a self-adjoint operator on $\widetilde{H}$ with the domain $D = \{(\xi_t)\in\widetilde{H}\mid \sum_{t\in G}\|h_t\xi_t\|^2<\infty\}$.

Proof. It is easy to check that $h = \oplus_{t\in G}h_t$ is a self-adjoint operator on $\widetilde{H}$ with the domain D.

Let $x,y\in\widetilde{M}$, and $b.,c.$ be the function $(: G\to M)$ corresponding to x,y

respectively. Then

$$
\begin{aligned}
(xy)_{s,t} &= \sum_{k\in G} x_{s,k} y_{k,t} \\
&= \sum_{k\in G} u_s^* b_{sk^{-1}} u_s \cdot u_k^* c_{kt^{-1}} u_k \\
&= u_s^* (\sum_{k\in G} b_{sk^{-1}} \cdot u_{sk^{-1}} c_{kt^{-1}} u_{sk^{-1}}^*) u_s \\
&= u_s^* (\sum_{k\in G} b_{sk^{-1}} \cdot \alpha_{sk^{-1}} (c_{kt^{-1}})) u_s \\
&= u_s^* (\sum_{k\in G} b_{st^{-1}k^{-1}} \cdot \alpha_{st^{-1}k^{-1}} (c_k)) u_s, \\
(y^*)_{s,t} &= p_s y^* p_t^* = (p_t y p_s^*)^* \\
&= (u_t^* c_{ts^{-1}} u_t)^* = u_t^* c_{ts^{-1}}^* u_t \\
&= u_s^* (u_{st^{-1}} c_{ts^{-1}}^* u_{st^{-1}}^*) u_s \\
&= u_s^* \alpha_{st^{-1}} (c_{ts^{-1}}^*) u_s,
\end{aligned}
$$

$\forall s,t \in G$. Hence, the functions $(: G \to M)$ corresponding to xy, y^* are $\sum_{t\in G} b_{\cdot k} \cdot \alpha_{\cdot k}(c_{k^{-1}}), d_{\cdot} = \alpha_{\cdot}(c_{\cdot^{-1}}^*)$ respectively. Further, the function $(: G \to M)$ corresponding to $y^* x$ is

$$\sum_{k\in G} d_{\cdot k} \cdot \alpha_{\cdot k}(b_{k^{-1}}) = \sum_{k\in G} \alpha_{\cdot k}(c_{k^{-1}\cdot^{-1}}^* b_{k^{-1}}).$$

Let $\{\tilde{\pi}, \widetilde{K}, \tilde{\xi}\}$ be the cyclic $*$ representation of $\widetilde{M}$ generated by $\tilde{\varphi}$. Then we have that

$$
\begin{aligned}
\langle \tilde{\pi}(x)\tilde{\xi}, \tilde{\pi}(y)\tilde{\xi}\rangle &= \tilde{\varphi}(y^* x) \\
&= \varphi(\sum_{k\in G} \alpha_k (c_{k^{-1}}^* b_{k^{-1}})) = \sum_{k\in G} \varphi(\alpha_{k^{-1}}(c_k^* b_k)) \\
&= \sum_{k\in G} \int_\Omega f_k(kt)\overline{g}_k(kt)d\mu(t) = \sum_{k\in G} \int_\Omega f_k(t)\overline{g}_k(t) r_k(t) d\mu(t),
\end{aligned}
$$

where $f_k, g_k \in L^\infty(\Omega,\mu)$ with $b_k = m_{f_k}, c_k = m_{g_k}, \forall k \in G$. Define

$$U\tilde{\pi}(x)\tilde{\xi} = (f_k r_k^{1/2})_{k\in G}, \quad \forall x \in \widetilde{M},$$

where $k \to b_k = m_{f_k}$ is the function $(: G \to M)$ corresponding to x. Then U can be uniquely extended to an isometry from $\widetilde{K}$ to $\widetilde{H}$. Let $\widetilde{M}_0$ be the $*$ subalgebra of $\widetilde{M}$ generated by $\{\pi(M), \lambda(G)\}$. Clearly, the number of non-zero components of $U\tilde{\pi}(x)\tilde{\xi}$ is finite, $\forall x \in \widetilde{M}_0$. Moreover, $L^\infty(\Omega,\mu)$ is dense in $H = L^2(\Omega,\mu)$. Therefore, U is a unitary operator from $\widetilde{K}$ onto $\widetilde{H}$.

Let $\widetilde{H}_0 = \{(\xi_k)_{k\in G} \in \widetilde{H} \mid \xi_k = 0$ except fiinite $k's\}$, and $h_0 = h|\widetilde{H}_0$. It is easy to see that h is the operator closure of h_0. Hence, h is the unique self-adjoint extension of h_0. Furhter, we can see that h is the operator closure and the unique self-adjoint extension of $(h|U\tilde{\pi}(\widetilde{M}_0)\tilde{\xi})$.

For any $x \in \widetilde{M}_0$ and $y \in \widetilde{M}$, let $b. = m_{f.}(f_k = 0$ except finte $k's)$ and $c. = m_{g.}$ be the functions $(: G \to M)$ corresponding to x and y respectively. Since the function $(: G \to M)$ corresponding to xy^* is $\sum_{k\in G} b_{\cdot k}\alpha_{\cdot}(c_k^*)$, it follows that

$$\begin{aligned}\langle\tilde{\Delta}^{1/2}\tilde{\pi}(x)\tilde{\xi}, \tilde{\Delta}^{1/2}\tilde{\pi}(y)\tilde{\xi}\rangle &= \langle\tilde{j}\tilde{\Delta}^{1/2}\tilde{\pi}(y)\tilde{\xi}, \tilde{j}\tilde{\Delta}^{1/2}\tilde{\pi}(x)\tilde{\xi}\rangle \\ &= \tilde{\varphi}(xy^*) = \sum_k \int f_k \bar{g}_k d\mu \\ &= \langle h(f_k r_k^{1/2}), (g_k r_k^{1/2})\rangle = \langle U^{-1}hU\tilde{\pi}(x)\tilde{\xi}, \tilde{\pi}(y)\tilde{\xi}\rangle.\end{aligned}$$

By Proposition 8.2.2, $\tilde{\Delta}^{1/2}$ is the operator closure of $(\tilde{\Delta}^{1/2}|\ \tilde{\pi}(\widetilde{M})\tilde{\xi})$. Hence,

$$\langle\tilde{\Delta}^{1/2}\tilde{\pi}(x)\tilde{\xi}, \tilde{\Delta}^{1/2}\tilde{\eta}\rangle = \langle U^{-1}hU\tilde{\pi}(x)\tilde{\xi}, \tilde{\eta}\rangle,$$

$\forall x \in \widetilde{M}_0, \tilde{\eta} \in \mathcal{D}(\tilde{\Delta}^{1/2})$. Therefore, we can see that $\tilde{\Delta}^{1/2}\tilde{\pi}(x)\tilde{\xi} \in \mathcal{D}(\tilde{\Delta}^{1/2})$ and

$$\tilde{\Delta}\tilde{\pi}(x)\tilde{\xi} = U^{-1}hU\tilde{\pi}(x)\tilde{\xi}, \quad \forall x \in \widetilde{M}_0.$$

Now $U\tilde{\Delta}U^{-1}$ is a self-adjoint extension of $(h|U\tilde{\pi}(\widetilde{M}_0)\tilde{\xi})$. By the preceding paragraph, we have $U\tilde{\Delta}U^{-1} = h$. Q.E.D.

Now let E be a Borel subset of Ω, and $\mu(E) > 0$. Then $p = m_{\chi_E}$ is a non-zero projection of M. By Lemma 9.5.4, $\tilde{p} = \pi(p)$ is a non-zero projection of $\widetilde{M}^{\sigma}$. Clearly, $\tilde{\varphi}_{\tilde{p}} = \tilde{\varphi}|\widetilde{M}_{\tilde{p}}$ is a faithful normal state on $\widetilde{M}_{\tilde{p}}$, and the modular automorphism group of $\widetilde{M}_{\tilde{p}}$ corresponding to $\tilde{\varphi}_{\tilde{p}}$ is $\tilde{\sigma}^{\tilde{p}} = \tilde{\sigma}|\widetilde{M}_{\tilde{p}}$. Suppose that $\tilde{\Delta}_{\tilde{p}}$ is the modular operator corresponding to $\tilde{\varphi}_{\tilde{p}}$. Let $x \in \widetilde{M}$ and $b. = m_{f.}$ be the function $(: G \to M)$ corresponding to x. Then it is easy to see that

$$x \in \widetilde{M}_{\tilde{p}} \Longleftrightarrow \mathrm{supp} f_k \subset E \cap kE, \quad \forall k \in G.$$

Thus, similarly we have the following.

Lemma 9.5.6. $\tilde{\Delta}_{\tilde{p}}$ is spatially isomorphic to the operator $h_E = \oplus_{t\in G} h_{t,E}$ on $\oplus_{t\in G} L^2(\Omega_t, \nu_t)$, where $\Omega_t = E \cap tE, \nu_t = \mu|\Omega_t$, and $h_{t,E}$ is the operator on $L^2(\Omega_t, \nu_t)$ corresponding to multiplication by $(r_t^{-1}(\cdot) \mid \Omega_t), \forall t \in G$.

Lemma 9.5.7. $\lambda \in S(\widetilde{M})$ if and only if for each Borel subset E of Ω with $\mu(E) > 0$ and $\varepsilon > 0$, there exists a non-zero projection q of M and $t \in G$ such that $\sup\{q, \alpha_t(q)\} \le p$ and

$$\mathrm{sp}(h_{t^{-1}}q|qH) \subset (\lambda - \varepsilon, \lambda + \varepsilon),$$

where $p = m_{\chi_E}$, and $h_{t^{-1}}$ is as in Lemma 9.5.5.

Proof. By Proposition 9.5.3 and Lemma 9.5.4, we have

$$S(\widetilde{M}) = \cap\{\mathrm{sp}\tilde{\Delta}_{\tilde{p}} \mid \tilde{p} = \pi(p),\ 0 \neq p \in \mathrm{Proj}(M)\}.$$

Hence, from Lemma 9.5.6 we obtain that

$$\begin{aligned}\lambda \in S(\widetilde{M}) &\iff \lambda \in \mathrm{sp}\tilde{\Delta}_{\tilde{p}},\ \forall \tilde{p} = \pi(p), p = m_{\chi_E}, \text{ and } \mu(E) > 0 \\ &\iff \lambda \in \mathrm{sp} h_E, \quad \forall \mu(E) > 0 \\ &\iff \lambda \in \overline{\bigcup_{t \in G} \mathrm{sp} h_{t,E}}, \quad \forall \mu(E) > 0 \\ &\iff \forall \mu(E) > 0 \text{ and } \varepsilon > 0, \text{ there exists } t \in G \\ &\qquad \text{such that } (\lambda - \varepsilon, \lambda + \varepsilon) \cap \mathrm{sp}\ h_{t^{-1},E} \neq \emptyset.\end{aligned}$$

If $\lambda \in S(\widetilde{M})$, then for each Borel subset E of Ω with $\mu(E) > 0$ and $\varepsilon > 0$, there exists $t \in G$ such that

$$(\lambda - \varepsilon, \lambda + \varepsilon) \cap \mathrm{sp}\ h_{t^{-1},E} \neq \emptyset.$$

Let $F = \{s \in E \cap t^{-1}E \mid r_t(s) \in (\lambda - \varepsilon, \lambda + \varepsilon)\}$. Then $q = m_{\chi_F}$ is a non-zero projection of M such that

$$\sup\{q, \alpha_t(q)\} \leq p, \quad \text{and} \quad \mathrm{sp}\ (h_{t^{-1}}q|qH) \subset (\lambda - \varepsilon, \lambda + \varepsilon),$$

where $p = m_{\chi_E}$. Conversely suppose that λ has the following property: for any Borel subset E of Ω with $\mu(E) > 0$ and $\varepsilon > 0$, there exists $0 \neq q \in \mathrm{Proj}(M)$ and $t \in G$ such that

$$\sup\{q, \alpha_t(q)\} \leq p, \quad \text{and} \quad \mathrm{sp}\ (h_{t^{-1}}q|qH) \subset (\lambda - \varepsilon, \lambda + \varepsilon).$$

We may assume that $q = m_{\chi_F}$ and $F \subset E$. Since $\sup\{q, \alpha_t(q)\} \leq p$, it follows that $F \subset E \cap t^{-1}E$. Moreover, $\mathrm{sp}(h_{t^{-1}}q|qH) = (\lambda - \varepsilon, \lambda + \varepsilon) \cap \mathrm{sp}\ h_{t^{-1},E} \neq \emptyset$. Therefore, $\lambda \in S(\widetilde{M})$. Q.E.D.

Definition 9.5.8. Let (G, Ω, μ) be a group measure space. Define the *ratio set* $r(G)(\subset [0, \infty))$ as follows. $\lambda(\geq 0) \in r(G)$ if and only if for any Borel subset E of Ω with $\mu(E) > 0$ and $\varepsilon > 0$, there exists a Borel subset F of Ω and $t \in G$ such that $\mu(F) > 0, F \sqcup tF \subset E$, and

$$\left|\frac{d\mu \circ t}{d\mu}(s) - \lambda\right| < \varepsilon, \quad \forall s \in F.$$

Noticing that $r_t(s) = \frac{d\mu \circ t^{-1}}{d\mu}(s)$, from Lemma 9.5.7 we have the following.

Proposition 9.5.9. Let (G, Ω, μ) be a free and ergodic group measure space. Further, suppose that Ω is compact, μ is a probability measure on Ω, and for each $t \in G$ there are two positive constants ε_t and η_t such that $0 < \varepsilon_t \leq \frac{d\mu\circ t}{d\mu}(s) \leq \eta_t < \infty, \forall s \in G$. Then there is a factor $\widetilde{M}$ such that $S(\widetilde{M}) = r(G)$.

Let $\Omega_n (n = 1, 2, \cdots)$ be the additive groups of integers, reduced mod 2, i.e., Ω_n is a compact (discrete) group composed of two elements $\{0, 1\}$ as follows: $0 + 0 = 0, 0 + 1 = 1 + 0 = 1$, and $1 + 1 = 0$. Let $\Omega = \times_{n=1}^{\infty} \Omega_n$ be the direct product of $\{\Omega_n \mid n = 1, 2, \cdots\}$. Thus Ω is a compact Hausdorff space satisfying the second countability axiom, and Ω is a compact group. Let G be the set of those $a = (a_n) \in \Omega$ for which $a_n \neq 0$ occurs for a finite number of n only. Then G is a countable group. For $b \in G$, define a homeomorphism of $\Omega : a \to b(a) = a + b (\forall a \in \Omega)$. Let μ_n be a probability measure on Ω_n with

$$\mu_n(\{0\}) = p_n, \quad \mu_n(\{1\}) = q_n,$$

where $p_n \in (0, 1)$, and $p_n + q_n = 1, \forall n$. Let $\mu = \times_{n=1}^{\infty} \mu_n$ be the infinite product measure of $\{\mu_n\}$ on Ω. Then, μ is a probability measure on Ω.

If σ is a permutation of $\{0, 1\}$, i.e., $\sigma(0) = 1, \sigma(1) = 0$, then it is easy to see that

$$\frac{d\mu_n \circ \sigma}{d\mu}(s) = \left(\frac{p_n}{q_n}\right)^{2s-1}, \quad s = 0, 1,$$

$\forall n$. Let c_n be the element of G such that the n-th component of c_n is 1 and other components of c_n are $0, \forall n$. Then we have

$$\mu \circ c_n = \times_{k \neq n} \mu_k \times (\mu_n \circ \sigma)$$

and

$$\frac{d\mu \circ c_n}{d\mu}(a) = \left(\frac{p_n}{q_n}\right)^{2a_n - 1}, \quad \forall a = (a_k)_k \in \Omega,$$

$\forall n$. For any $b \in G$ and $b \neq 0$, there is unique finite sequence $\{i_1 < \cdots < i_k\}$ of positive integers such that $b = c_{i_1} + \cdots + c_{i_k}$. Hence, we have

$$\frac{d\mu \circ b}{d\mu}(a) = \prod_{n=1}^{\infty} \left(\frac{p_n}{q_n}\right)^{(2a_n - 1)b_n},$$

$\forall a = (a_n) \in \Omega, b = (b_n) \in G$; and (G, Ω, μ) is a group measure space.

Let $b \in G$ and $b \neq 0$. Clearly, $\{a \in \Omega \mid b(a) = a\} = \emptyset$. Thus, (G, Ω, μ) is free.

For any Borel subset E of Ω, let $F = \bigcup_{b \in G} b(E)$. Clearly, F is also a Borel subset of Ω, and $b(F) = F, \forall b \in G$. Then for any $a \in F$ we have $\tilde{a} \in F$

whenever $\tilde{a}$ is obtained by changing any finitely many components of a. Thus, for any positive integer n, F has the following form:

$$F = \times_{k=1}^{n}\Omega_k \times F_n,$$

where F_n is a Borel subset of $\times_{k=n+1}^{\infty}\Omega_k$. Now if $C = C_n \times \times_{k=n+1}^{\infty}\Omega_k$ is a cylinder subset of Ω, where C_n is a (Borel) subset of $\times_{k=1}^{n}\Omega_k$, then it is obvious that

$$\mu(F \cap C) = \mu(C_n \times F_n) = \mu(C)\mu(F).$$

If K is a compact subset of Ω, then we can see that $K = \bigcap_n (\pi_n(K) \times \times_{k=n+1}^{\infty}\Omega_k)$, where π_n is the projection from Ω onto $\times_{k=1}^{n}\Omega_k$. Thus,

$$\begin{aligned}\mu(K \cap F) = &\ \lim_n \mu(\pi_n(K) \times \times_{k=n+1}^{\infty}\Omega_k)\mu(F) \\ = &\ \mu(K)\mu(F)\end{aligned}$$

Further, by the regularity of μ we have

$$\mu(C \cap F) = \mu(C)\mu(F)$$

for any Borel subset C of Ω. Now if E satisfies the following:

$$\mu(bE\Delta E) = 0, \quad \forall b \in G,$$

then we have

$$\mu(E) = \mu(F) = \mu(F \cap F) = \mu(F)^2 = \mu(E)^2,$$

and either $\mu(E) = 1$ or $\mu(E) = 0$. Therefore, (G, Ω, μ) is ergodic.

Now let $\lambda \in (0,1)$, and $p_n = \lambda(1+\lambda)^{-1}, \forall n$. Then from Definition 9.5.8 we have

$$r(G) = \{0, \lambda^n \mid n \in \mathbb{Z}\}.$$

Thus by Propositions 9.4.6 and 9.5.9 $\widetilde{M} = M \times_\alpha G$ is a type (III$_\lambda$) factor.

Proposition 9.5.10. Type (III$_\lambda$) factors $(0 < \lambda < 1)$ do exist.

Remark. By Proposition 9.5.9 and above construction, we can also obtain the examples of type (III$_0$) and (III$_1$) factors (see [165]).

Now keep the above notations: $\Omega_n = \{0,1\}, \forall n; \Omega = \times_{n=1}^{\infty}\Omega_n; G; \mu_n(\{0\}) = p_n \in (0,1), \mu_n(\{1\}) = q_n, p_n + q_n = 1, \forall n; \mu = \times_{n=1}^{\infty}\mu_n$; and the element c_n of $G, \forall n$.

Let

$$H = L^2(\Omega,\mu), \quad \widetilde{H} = H \otimes l^2(G), \quad M = \{m_f \mid f \in L^\infty(\Omega,\mu)\},$$

and

$$\begin{cases} (\pi(x)\tilde{\xi})(b) = \alpha_{b^{-1}}(x)\tilde{\xi}(b) = \alpha_b(x)\tilde{\xi}(b), \\ \alpha_b(m_f) = u_b m_f u_b^* = m_{f_b}, f_b(\cdot) = f(\cdot + b), \\ (\lambda(c)\tilde{\xi})(b) = \tilde{\xi}(b+c), \end{cases}$$

$\forall b, c \in G, x \in M, f \in L^\infty(\Omega, \mu)$, and $\tilde{\xi} \in \widetilde{H}$. Clearly, the crossed product $\widetilde{M} = M \times_\alpha G = \{\pi(M), \lambda(G)\}''$ admits a cyclic vector $\tilde{\xi}_0$:

$$\tilde{\xi}_0(b) = \begin{cases} 1, & \text{if } b = 0, \\ 0, & \text{otherwise,} \end{cases}$$

where "1" is the constant function 1 on $\Omega(\in L^2(\Omega, \mu) = H)$.

For any $a_i \in \{0, 1\}, 1 \le i \le n$, let $p(a_1, \cdots, a_n)$ be the operator on $H = L^2(\Omega, \mu)$ corresponding to multiplication by $\chi_{E(a_1, \cdots, a_n)}$, where $E(a_1, \cdots, a_n) = (a_1, \cdots, a_n) \times \times_{k=n+1}^\infty \Omega_k (\subset \Omega)$, and put

$$\tilde{p}(a_1, \cdots, a_n) = \pi(p(a_1, \cdots, a_n)), \quad \lambda_n = \lambda(e_n),$$

$\forall n$. Then it is easy to check that

$$\lambda_k \tilde{p}(a_1, \cdots, a_n) \lambda_k = \tilde{p}(a_1, \cdots, a_k + 1, \cdots, a_n),$$

$\forall a_i \in \{0, 1\}, 1 \le i \le n, 1 \le k \le n$. Thus

$$\{\tilde{p}(a_1, \cdots, a_n), \lambda_k \mid a_i \in \{0, 1\}, 1 \le i \le n, 1 \le k \le n\}$$

generates a type (I_{2^n}) subfactor of $\widetilde{M}, \forall n$.

Now define

$$\Psi\left(\begin{pmatrix} 0 & 1 \\ 1 & 0 \end{pmatrix} \otimes \bigotimes_{\substack{n=1 \\ n \neq k}}^\infty 1\right) = \lambda_k, \quad \forall k,$$

$$\Psi(e_{a_1} \otimes \cdots \otimes e_{a_n} \otimes \bigotimes_{n+1}^\infty 1) = \tilde{p}(a_1, \cdots, a_n),$$

where $e_0 = \begin{pmatrix} 1 & 0 \\ 0 & 0 \end{pmatrix}, e_1 = \begin{pmatrix} 0 & 0 \\ 0 & 1 \end{pmatrix}, \forall a_i \in \{0, 1\}, 1 \le i \le n, \forall n$. Then Ψ can be uniquely extended to a $*$ isomorphism from min-$\bigotimes_{n=1}^\infty M_2(\mathbb{C})$ into $\widetilde{M} = M \times_\alpha G$.

It is easily verified that:

$$\langle \tilde{p}(a_1, \cdots, a_n)\tilde{\xi}_0, \tilde{\xi}_0 \rangle = \mu(E(a_1, \cdots, a_n)),$$

$$\langle \lambda_k \tilde{p}(a_1, \cdots, a_n)\tilde{\xi}_0, \tilde{\xi}_0 \rangle = 0, \quad \langle \lambda_k \tilde{\xi}_0, \tilde{\xi}_0 \rangle = 0,$$

$\forall a_i \in \{0,1\}, 1 \le i \le n, 1 \le k \le n, \forall n$. Thus, we have

$$\begin{aligned}
&\langle \Psi \left(\begin{pmatrix} \alpha & \beta \\ \gamma & \delta \end{pmatrix} \otimes \bigotimes_{\substack{j=1 \\ j\neq n}}^{\infty} 1 \right) \tilde{\xi}_0, \tilde{\xi}_0 \rangle \\
= \ & \alpha \sum_{\substack{a_i=0,1 \\ 1\le i\le n-1}} \langle \tilde{p}(a_1, \cdots, a_{n-1}, 0) \tilde{\xi}_0, \tilde{\xi}_0 \rangle \\
&+\delta \sum_{\substack{a_i=0,1 \\ 1\le i\le n-1}} \langle \tilde{p}(a_1, \cdots, a_{n-1}, 1) \tilde{\xi}_0, \tilde{\xi}_0 \rangle \\
&+\beta \langle \lambda_n \tilde{\xi}_0, \tilde{\xi}_0 \rangle + (\gamma - \beta) \sum_{\substack{a_i=0,1 \\ 1\le i\le n-1}} \langle \lambda_n \tilde{p}(a_1, \cdots, a_{n-1}, 0) \tilde{\xi}_0, \tilde{\xi}_0 \rangle \\
= \ & \alpha p_n + \delta q_n,
\end{aligned}$$

$\forall \begin{pmatrix} \alpha & \beta \\ \gamma & \delta \end{pmatrix} \in M_2(\mathbb{C}), \forall n$. So $\langle \Psi(\cdot) \tilde{\xi}_0, \tilde{\xi}_0 \rangle$ is the product state $\bigotimes_{n=1}^{\infty} \varphi_n$ on $\min\text{-}\bigotimes_{n=1}^{\infty} M_2(\mathbb{C})$, where

$$\varphi_n \left(\begin{pmatrix} \alpha & \beta \\ \gamma & \delta \end{pmatrix} \right) = \alpha p_n + \delta q_n,$$

$\forall \begin{pmatrix} \alpha & \beta \\ \gamma & \delta \end{pmatrix} \in M_2(\mathbb{C}), \forall n$.

Theorem 9.5.11. Let $A = \min\text{-}\bigotimes_{n=1}^{\infty} M_2(\mathbb{C})$ be the (UHF) algebra of type $\{2^n \mid n = 1, 2 \cdots\}$, where $M_2(\mathbb{C})$ is the 2×2 matrix algebra. For any $\lambda \in (0, \frac{1}{2})$, define a product state $\psi_\lambda = \bigotimes_{n=1}^{\infty} \varphi_n$ on A as follows:

$$\varphi_n \left(\begin{pmatrix} \alpha & \beta \\ \gamma & \delta \end{pmatrix} \right) = \lambda \alpha + (1 - \lambda) \delta,$$

$\forall \begin{pmatrix} \alpha & \beta \\ \gamma & \delta \end{pmatrix} \in M_2(\mathbb{C}), \forall n$. Let R_λ be the weak closure of $\pi_\lambda(A)$, where $\{\pi_\lambda, H_\lambda\}$ is the $*$ representation of A generated by ψ_λ. Then, for any $\lambda, \mu \in (0, \frac{1}{2})$ and $\lambda \neq \mu$ the factors R_λ and R_μ are not $*$ isomorphic.

Proof. From above discussion, we have $\{\pi_\lambda, H_\lambda\} \cong \{\Psi, \widetilde{H}\}$, where $p_n = \lambda, q_n = 1 - \lambda, \forall n, \forall \lambda \in (0, \frac{1}{2})$. Thus, R_λ is indeed a type $(\mathrm{III}_{\lambda'})$ factor, where $\lambda' = \frac{\lambda}{1-\lambda}, \forall \lambda \in (0, \frac{1}{2})$. That comes to the conclusion. Q.E.D.

Notes. The ratio set $r(G)$ was introduced by W. Krieger. Theorem 9.5.11 is due to R.T. Powers. The problem of finding non-isomorphic factors on a separable Hilbert space is one of central problems in the theory of factors. P.T. Powers proved the existence fo uncountably many type (III) factors. The existence of uncountably many type (II_1) and type (II_∞) factors were obtained by D. McDuff and S. Sakai.

References. [92], [108], [130], [150], [165].

Chapter 10
Borel Structure

10.1. Polish spaces

Definition 10.1.1. A topological space is said to be *Polish*, if it is homeomorphic to a separable complete metric space.

Let E be a Polish space. A metric d on E is said to be *proper*, if (E, d) is a separable complete metric space, and the topology generated by d is equivalent to the original topology in E.

Let E be a set, and d be a metric on E. We shall denote the topology generated by d in E by *d–topology* (or d–top. simply).

Let (E, d) be a metric space, and $F \subset E$. We shall denote the *diameter* of F with respect to d by $D_d(F)$, i.e.,

$$D_d(F) = \sup\{d(x, y) | x, y \in F\}.$$

Example. Let H be a separable Hilbert space, and S be the closed unit ball of $B(H)$. Then S is a Polish space with respect to the weak (operator) toplogy, the strong (operator) topology, or the strong $*$ topology.

In fact, let $\{\xi_n\}$ be a countable dense subset of the closed unit ball of H. For any $a, b \in S$, define

$$d(a, b) = \sum_{n,m} \frac{1}{2^{n+m}} \cdot |\langle (a-b)\xi_n, \xi_m \rangle|,$$

$$\rho(a, b) = \sum_n \frac{1}{2^n} \|(a-b)\xi_n\|,$$

$$\rho^*(a, b) = \sum_n \frac{1}{2^n} (\|(a-b)\xi_n\| + \|(a-b)^*\xi_n\|).$$

Clearly, d–top, ρ–top., ρ^*–top. are equivalent to weak, strong, strong $*$ top. in S respectively. Further, since $B(H)$ is countably generated, then by the density Theorem 1.6.1 we can get the conclusions.

Proposition 10.1.2. The countable direct product and the countable direct sum of Polish spaces are still Polish.

Proof. Let $\{E_n\}$ be a sequence of Polish spaces, and d_n be a proper mectric on $E_n, \forall n$.

On $E = \cup_n E_n$ (disjoint union) , define

$$d(x,y) = \begin{cases} \min\{1, d_n(x,y)\}, & \text{if } x,y \in E_n \text{ for some } n, \\ 0, & \text{otherwise.} \end{cases}$$

Clearly, E is a Polish space with d–topology, each E_n is an open and closed subset of E, and in each E_n the d–top. is equivalent to d_n–top.

On $\times_n E_n$, define

$$d((x_n),(y_n)) = \sum_n \frac{1}{2^n} \cdot \frac{d_n(x_n,y_n)}{1+d_n(x_n,y_n)},$$

$\forall (x_n),(y_n) \in \times_n E_n$. Then we can see that $\times_n E_n$ is a Polish space with respect to the product topology. Q.E.D.

Proposition 10.1.3. Let E be a Polish space, and $F \subset E$. Then F is Polish with respect to the relative topology if and only if F is a G_δ–subset of E (a countable intersection of open subsets).

Proof. Let d be a proper metric on E. We may assume that $D_d(E) \leq 1$.

Suppose that U is an open subset of E. Let

$$\delta(x,y) = d(x,y) + |d(x,U')^{-1} - d(y,U')^{-1}|,$$

$\forall x,y \in U$, where $U' = E \backslash U$. Clearly, δ is a metric on U, and the δ–top. is equivalent to the relative top. in U. If $\{x_n\}(\subset U)$ is Cauchy with respect to δ, then it is also cauchy with respect to d. Thus , there is $x \in E$ such that $d(x_n,x) \longrightarrow 0$. Since $\{d(x_n,U')^{-1}\}$ is Cauchy and $D_d(E) \leq 1$, it follows that $\lim_n d(x_n,U') = \lambda > 0$. Thus , $d(x,U') = \lambda > 0$, i.e., $x \in U$ and $\delta(x_n,x) \longrightarrow 0$. So U as a subspace of E is also a Polish space.

Now let $F = \cap_n U_n$, where $\{U_n\}$ is a sequence of open subsets of E. Suppose that δ_n is a metric on U_n as in the preceding paragraph, $\forall n$, and define

$$d_F(x,y) = \sum_n \frac{1}{2^n} \cdot \frac{\delta_n(x,y)}{1+\delta_n(x,y)}, \forall x,y \in F.$$

Since δ_n–top. and d–top. are equivalent in F, $\forall n$, it follows that d_F–top. is equivalent to the relative top. in F. If $\{x_n\}(\subset F)$ is a Cauchy sequence with respect to d_F, then for each k there is $y_k \in U_k$ such that $\delta_k(x_n,y_k) \longrightarrow 0$.

Clearly, we have also $d(x_n, y_k) \to 0, \forall k$. Hence, there is $x \in F$ such that $y_k = x, \forall k$, and $d_F(x_n, x) \longrightarrow 0$. Therefore, F as a subspace of E is Polish.

Conversely, let $F(\subset E)$ be Polish as a subspace of E, and d_F be a proper metric on F. For each n , put

$$F_n = \left\{ x \in \overline{F} \,\middle|\, \begin{array}{c} \text{there is an open neighborhood } U \text{of } x, \\ \text{such that } D_{d_F}(U \cap F) < 1/n \end{array} \right\}.$$

Clearly , $F \subset \cap_n F_n$. Conversely, if $x \in U_n F_n$, then for each n there is an open neighborhood U_n of x such that $D_{d_F}(U_n \cap F) < 1/n$. We may assume that $U_1 \supset U_2 \supset \cdots$, and $D_d(U_n) \longrightarrow 0$. Pick $x_n \in U_n \cap F, \forall n$. Then $\{x_n\}$ is Cauchy in (F, d_F). Thus , there is $y \in F$ such that $d_F(x_n, y) \longrightarrow 0$. Clearly, $d(x_n, y) \longrightarrow 0, \cap_n U_n = \{x\}$, and $d(x_n, x) \longrightarrow 0$. Hence , $x = y \in F$, i.e., $F = \cap_n F_n$.

If $x \in F_n$, then there is an open neighborhood U of x such that $D_{d_F}(U \cap F) < 1/n$. By the definition of F_n, it is obvious that $U \cap \overline{F} \subset F_n$. Thus , F_n is an open subset of $\overline{F}$, i.e., there is an open subset G_n of E such that $F_n = \overline{F} \cap G_n, \forall n$. Put

$$U_m = \{x \in E | d(x, \overline{F}) < 1/m\}, \forall m.$$

Clearly, U_m is open , $\forall m$, and $\overline{F} = \cap_m U_m$. Therefore,

$$F = \cap_n F_n = \cap_n (G_n \cap \overline{F}) = \cap_{m,n} (G_n \cap U_m)$$

is a G_δ–subset of E. Q.E.D.

Proposition 10.1.4. Any Polish space must be homeomorphic to a G_δ–subset of $[0,1]^\infty$ (the countale infinite product of $[0,1]$).

Proof. Let E be a Polish space , d be a proper metric on E, and $\{a_n\}$ be a countable dense subset of E. Then

$$x \longrightarrow \left(\frac{d(a_n, x)}{1 + d(a_n, x)} \right)_n \qquad (\forall x \in E)$$

is a homeomorphism from E into $[0,1]^\infty$, and also by Proposition 10.1.3, its image must be a G_δ–subset of $[0,1]^\infty$. Q.E.D.

Proposition 10.1.5. Let Ω be a locally compact Hausdorff space. Then Ω is a Polish space if and only if Ω satisfies the second countability axiom.

Proof. The necessity is clear. Now let Ω satisfy the second countability axiom, and $\Omega_\infty = \Omega \cup \{\infty\}$ be the compactification of Ω. Clearly, Ω_∞ is a Polish space. Now Ω is an open subset of Ω_∞, so Ω is also a Polish space. Q.E.D.

Definition 10.1.6. The Polish space $\mathbb{N}^\infty$ is the set

$$\{n = (n_k) | n_k \text{ non-negative integer, } k = 1, 2, \cdots\}$$

with the topology generated by the metric

$$d(n,m) = \sum_k \frac{1}{2^k} \cdot \frac{|n_k - m_k|}{1 + |n_k - m_k|},$$

$\forall n = (n_k), m = (m_k) \in \mathbb{N}^\infty$.

Clearly , $\{n = (n_k)|$ the number of non-zero components of n is finite $\}$ is a countable dense subset of $\mathbb{N}^\infty$. Moreover, for any $n = (n_k) \in \mathbb{N}^\infty$,

$$\mathbb{N}^\infty_{n_1,\cdots,n_k} = \{m = (m_s) \in \mathbb{N}^\infty | m_i = n_i, 1 \le i \le k\}, k = 1, 2, \cdots$$

is a neighborhood basis of n.

Proposition 10.1.7. Let E be a Polish space. Then there exists a continuous map from $\mathbb{N}^\infty$ onto E.

Proof. Let d be a proper metric on E, and $D_d(E) \le 1$.

For $n_1 = 0, 1, \cdots$, let $F(n_1) = E$. For each n_1, pick a countable closed cover $\{F(n_1, n_2) | n_2 = 0, 1, \cdots\}$ of $F(n_1)$ such that $D_d(F(n_1, n_2)) \le 1/2, \forall n_2$. Further, for each (n_1, n_2), pick a countable closed cover $\{F(n_1, n_2, n_3) | n_3 = 0, 1, \cdots\}$ of $F(n_1, n_2)$ such that $D_d(F(n_1, n_2, n_3)) \le 1/2^2, \forall n_3, \cdots$, Generally, we have a family $\{F(n_1, \cdots, n_p) | n_i = 0, 1, \cdots, 1 \le i \le p, p = 1, 2, \cdots\}$ of closed subsets of E such that $F(n_1) = E, F(n_1, \cdots, n_p) = \cup_{k=0}^\infty F(n_1, \cdots, n_p, k)$, and $D_d(F(n_1, \cdots, n_p)) \le 2^{-(p-1)}, \forall n_1, \cdots, n_p, p = 1, 2, \cdots$.

Since (E, d) is complete, it follows that

$$\#\{\cap_{k=1}^\infty F(n_1, \cdots, n_k)\} = 1, \quad \forall n = (n_k) \in \mathbb{N}^\infty.$$

Let $\{f(n)\} = \cap_{k=1}^\infty F(n_1, \cdots, n_k), \forall n = (n_k) \in \mathbb{N}^\infty$. Clearly, f is a map from $\mathbb{N}^\infty$ onto E. Suppose that $n^{(k)} \longrightarrow n$ in $\mathbb{N}^\infty$. For any $\varepsilon > 0$, pick p such that $2^{-(p-1)} < \varepsilon$. Then we have $n_i^{(k)} = n_i, 1 \le i \le p$, if k sufficiently large. Hence , $f(n)$ and $f(n^{(k)}) \in F(n_1, \cdots, n_p)$ and $d(f(n^{(k)}), f(n)) \le 2^{-(p-1)} < \varepsilon$ if k sufficiently large. Therefore, f is also continuous. Q.E.D.

Lemma 10.1.8. Let E be a Polish space with no isolated point, and d be a proper metric on E. Then for any $\varepsilon > 0$ there is an infinite sequence $\{E_n\}$ of non-empty G_δ-subsets with no isolated point of E such that $D_d(E_n) \le \varepsilon, \forall n; E_n \cap E_m = \emptyset, \forall n \ne m$; and $\cup_n E_n = E$.

Proof. We may assume that $\varepsilon < D_d(E)$. Pick a countable open cover $\{V_n\}$ of E such that $V_n \ne \emptyset, D_d(V_n) \le \varepsilon, \forall n$. Let $E_1 = \overline{V_1}$. Clearly , E_1 is a G_δ-subset with no isolated point. By induction, define $E_n = \overline{V}_n \backslash F_n$, where $F_n =$

$\cup_{k=1}^{n-1}E_k, \forall n > 1$. Since $F_n = \cup_{k=1}^{n-1}\overline{V}_k$ is closed, it follows that E_n is a G_δ-subset. Moreover, from $(V_n\backslash F_n) \subset E_n \subset \overline{V_n\backslash F_n}, E_n$ has no isolated point. If $^{\#}\{n|E_n \neq \emptyset\} = \infty$, then $\{E_n\}$ satisfies our conditions. Otherwise, notice that $\emptyset \neq E_1 \neq E$ (since $D_d(E_1) \leq \varepsilon < D_d(E)$), then the same process can be carried to E_1 (for some ε' with $0 < \varepsilon' < D_d(E_1)$). In this way, we can complete the proof. Q.E.D.

Lemma 10.1.9. Let E be a Polish space with no isolated point. Then for any non-negative integers $n_1, \cdots, n_k$, there is a non-empty G_δ-subset $E^{(k)}_{n_1,\cdots,n_k}$ with no isolated point such that

1) if $(n_1, \cdots, n_k) \neq (m_1, \cdots, m_k)$, then

$$E^{(k)}_{n_1,\cdots,n_k} \cap E^{(k)}_{m_1,\cdots,m_k} = \emptyset;$$

2) $E^{(k)}_{n_1,\cdots,n_k} = \cup_{p=0}^{\infty} E^{(k+1)}_{n_1,\cdots,n_k,p}, \forall n_1, \cdots, n_k;$

3) if $d^{(k)}_{n_1,\cdots,n_k}$ is a proper metric on $E^{(k)}_{n_1,\cdots,n_k}$, then the diameter of $E^{(k+1)}_{n_1,\cdots,n_k,n_{k+1}}$ with respect to $(d + d^{(1)}_{n_1} + \cdots + d^{(k)}_{n_1,\cdots,n_k})$ is less than $(k+1)^{-1}, \forall n_1, \cdots, n_{k+1}$, where d is a proper metric on E.

Proof. Using Lemma 10.1.8 to (E, d) and $\varepsilon = 1$, we get $\{E^{(1)}_{n_1}|n_1 = 0, 1, \cdots\}$. Again using Lemma 10.1.8. to $(E^{(1)}_{n_1}, d + d^{(1)}_{n_1})$ and $\varepsilon = 1/2$, we get $\{E^{(2)}_{n_1,n_2}|n_2 = 0, 1, \cdots\}, \forall n_1$. Continuing this process, we can get the conclusion. Q.E.D.

Proposition 10.1.10. Let E be a non-empty Polish space. Then there exists an injective continuous map from $\mathbb{N}^\infty$ onto E, if and only if , E has no isolated point.

Proof. Since $\mathbb{N}^\infty$ has no isolated point, the necessity is obvious. Now suppose that E has no isolated point. Pick $\{E^{(k)}_{n_1,\cdots,n_k}\}$ as in Lemma 10.1.9. Since $\mathbb{N}^\infty_{n_1,\cdots,n_k}$ is homeomorphic to $\mathbb{N}^\infty$, it follows from Proposition 10.1.7 that there is a continuous map $f^{(k)}_{n_1,\cdots,n_k}$ from $\mathbb{N}^\infty_{n_1,\cdots,n_k}$ onto $E^{(k)}_{n_1,\cdots,n_k}, \forall n_1, \cdots, n_k$.

Fix k. Since $\{\mathbb{N}^\infty_{n_1,\cdots,n_k}|n_1, \cdots, n_k\}$ is a closed and open cover of $\mathbb{N}^\infty$, we can define a continuous map $f^{(k)}$ from $\mathbb{N}^\infty$ onto E, such that $f^{(k)}|E^{(k)}_{n_1,\cdots,n_k} = f^{(k)}_{n_1,\cdots,n_k}$.

For any $n = (n_k) \in \mathbb{N}^\infty$ and integers p, q with $p \leq q$, noticing that

$$f^{(q)}(n) \in E^{(q)}_{n_1,\cdots,n_q} \subset E^{(p)}_{n_1,\cdots,n_p},$$

so by Lemma 10.1.9 we have $d(f^{(p)}(n), f^{(q)}(n)) \leq p^{-1}$. Thus , $\{f^{(k)}(n)\}_k$ is a Cauchy sequence of (E, d), and there is $f(n) \in E$ such that $d(f^{(k)}(n), f(n)) \longrightarrow 0$, uniformaly for $n \in \mathbb{N}^\infty$. Further, f is continuous.

By Lemma 10.1.9, $\{f^{(p)}(n)\}_{p\geq k}$ is a Cauchy sequence of $(E^{(k)}_{n_1,\cdots,n_k}, d^{(k)}_{n_1,\cdots,n_k})$, $\forall k$. Hence , $f(n) \in \cap_{k=1}^{\infty} E^{(k)}_{n_1,\cdots,n_k}, \forall n = (n_k) \in \mathbb{N}^\infty$. But $D_d(E^{(k)}_{n_1,\cdots,n_k}) \leq$

$k^{-1}, \forall k$, so $\{f(n)\} = \cap_{k=1}^{\infty} E^{(k)}_{n_1,\cdots,n_k}, \forall n = (n_k) \in I\!N^\infty$. Now if $n = (n_k) \neq m = (m_k)$, then there is r such that $(n_1, \cdots, n_r) \neq (m_1, \cdots, m_r)$. Since$E^{(r)}_{n_1,\cdots,n_r} \cap E^{(r)}_{m_1,\cdots,m_r} = \emptyset$, it follows that $f(n) \neq f(m)$, i.e., f is injective.

Finally, for any $x \in E$, by Lemma 10.1.9 there is $n = (n_k) \in I\!N^\infty$ such that $x \in \cap_{k=1}^{\infty} E^{(k)}_{n_1,\cdots,n_k}$, i.e., $f(n) = x$. Therefore, $f(I\!N^\infty) = E$. Q.E.D.

Proposition 10.1.11. Let E be a Polish space. Then we can write $E = F \cup G$, where $F \cap G = \emptyset, G$ is a countable open subset of E, and either $F = \emptyset$ or there is an injective continuous map f from $I\!N^\infty$ to E with $f(I\!N^\infty) = F$.

Proof. Let $\{V_n\}$ be a countable basis for the topology of $E, G = \cup\{V_n | V_n$ is countable $\}$, and $F = E \backslash G$. Suppose that $F \neq \emptyset$. By Proposition 10.1.10, it sufficies to show that F has no isolated point. Let $x \in F$, and V be any neighobrhood of x. Then there is n such that $x \in V_n \subset V$. Since $x \notin G$, it follows that V_n is not countable. Now we can pick $y \in V_n \backslash G \subset V \cap F$ and $y \neq x$. Therefore, F has no isolated point. Q.E.D.

References. [10], [13], [190].

10.2. Borel subsets and Sousline subsets

Definition 10.2.1. Let E be a Polish space. A subset of E is said to be *Borel* , if it belongs to the σ-Bool algebra generated by all open subsets of E.

A subset A of E is said to be *Sousline* (or *analytic*) , if there is a continuous map f from $I\!N^\infty$ to E with $f(I\!N^\infty) = A$.

Lemma 10.2.2. Let E be a Polish space, and $\mathcal{F}$ be a family of subsets of E such that :

1) $\mathcal{F}$ contains any open subset and closed subset of E;
2) if $\{E_n\} \subset \mathcal{F}$, then $\cap_n E_n \in \mathcal{F}$;
3) if $\{E_n\} \subset \mathcal{F}$ and $E_n \cap E_m = \emptyset, \forall n \neq m$, then $\cup_n E_n \in \mathcal{F}$.

Then $\mathcal{F}$ contains any Borel subset of E.

Proof. Let $\mathcal{B} = \{V \subset E | V$ and $(E \backslash V) \in \mathcal{F}\}$. Clearly, $\mathcal{B}$ contains any open subset and closed subset of E. If $V_1, V_2 \in \mathcal{B}$, then $V_1 \backslash V_2 = V_1 \cap (E \backslash V_2) \in \mathcal{F}$, and $E \backslash (V_1 \backslash V_2) = (E \backslash V_1) \sqcup (V_1 \cap V_2) \in \mathcal{F}$. Thus, $(V_1 \backslash V_2) \in \mathcal{B}$. If $\{V_n\} \subset \mathcal{B}$ and $V_n \cap V_m = \emptyset, \forall n \neq m$, then $\cup_n V_n \in \mathcal{F}$, and $E \backslash \cup_n V_n = \cap_n (E \backslash V_n) \in \mathcal{F}$. Thus $\cup_n V_n \in \mathcal{B}$. So $\mathcal{B}$ is a σ-Bool algebra containing any open subset of E. Therefore, $\mathcal{B}$, then $\mathcal{F}$ (since $\mathcal{B} \subset \mathcal{F}$), contains any Borel subset of E. Q.E.D.

Propositon 10.2.3. Let E be a Polish space.

1) If P is a Polish space, and f is a continuous map from P to E, then $f(P)$ is a Sousline subset of E.

2) If B is a Borel subset of E, then there is a Polish space P, and an injective continuous map from P to E, such that $f(P) = B$.

3) Any Borel subset of E is Sousline.

Proof. 1) It is obvious from Proposition 10.1.7 and Definition 10.2.1.

2) Let $\mathcal{F} = \{F \subset E|$ there is a Polish space P, and an injective continuous map from P to E, such that $f(P) = F\}$. Now it suffices to check that $\mathcal{F}$ satisfies the conditions of Lemma 10.2.2.

Any open subset or closed subset of E is a Polish space itself. Thus, $\mathcal{F}$ contains any open subset and closed subset of E.

Now let $\{E_n\} \subset \mathcal{F}$. Then for each n, there is a Polish space P_n and an injective continuous map f_n from P_n to E such that $f_n(P_n) = E_n$. Define $f : \times_n P_n \longrightarrow \times_n E$ as follows:

$$f(p_1, \cdots, p_n, \cdots) = (f_1(p_1), \cdots, f_n(p_n), \cdots), \forall p_n \in P_n, n.$$

Clearly, f is continuous and injective. Put $\triangle = \{(x, \cdots, x, \cdots)|x \in E\}$. Then $\triangle$ is a closed subset of $\times_n E$. Further, $Q = f^{-1}(\triangle)$ is also a closed subset of $\times_n P_n$. Let π be the projection from $\times_n E$ onto its first component. Then $\pi \circ f$ maps injectively Q to $\cap_n E_n$. Hence , $\cap_n E_n \in \mathcal{F}$.

Finally, let $\{E_n\} \subset \mathcal{F}$, and $E_n \cap E_m = \emptyset, \forall n \neq m$. Suppose that P_n is a Polish space and f_n is an injective continuous map from P_n to E such that $f_n(P_n) = E_n, \forall n$. Define $f : P = \cup_n P_n$ (disjoint union) $\longrightarrow E$ such that $f|P_n = f_n, \forall n$. Clearly, $f(P) = \cup_n E_n$. Thus, $\cup_n E_n \in \mathcal{F}$.

Therefore, $\mathcal{F}$ satisfies the conditions of Lemma 10.2.2.

3) It is clear from the conclusions 2) and 1). Q.E.D.

Proposition 10.2.4. Let E be a Polish space, and B be a Borel subset of E. Then either B is countable, or there is an injective continuous map from $\mathbb{N}^\infty$ to E such that $f(\mathbb{N}^\infty) \subset B$ and $(B \backslash f(\mathbb{N}^\infty))$ is countable.

Proof. It is an immediate result of Propositions 10.2.3 and 10.1.11.

Q.E.D.

Proposition 10.2.5. 1) The continuous image of a Sousline subset is Sousline, i.e., if E and F are Polish spaces, f is continuous from E to F, and A is a Sousline subset of E, then $f(A)$ is a Sousline subset of F.

2) The countable intersection and the countable union of Sousline subsets are Sousline.

Proof. 1) It is obvious from Definition 10.2.1.

2) Let $\{A_n\}$ be a sequence of Sousline subsets of a Polish space E. Then for each n, there is a continuous map f_n from $\mathbb{N}^\infty$ to E such that $f_n(\mathbb{N}^\infty) = A_n$. Define $f : P = \cup_n P_n$ (disjoint union) $\longrightarrow E$ such that $f|P_n = f_n$, where $P_n = \mathbb{N}^\infty, \forall n$. Then $f(P) = \cup_n A_n$, and $\cup_n A_n$ is Sousline.

Let $\Omega = \times_n P_n$, where $P_n = \mathbb{N}^\infty, \forall n$, and

$$M = \{x = (x_n) \in \Omega | f_n(x_n) = f_m(x_m), \forall n, m\}.$$

Clearly , M is a closed subset of Ω. Then define $g(x) = f_1(x_1), \forall x \in M$. Hence, $\cap_n A_n = g(M)$ is Sousline. Q.E.D.

Definition 10.2.6. Let E be a Polish space, and A, B be two subsets of E. A and B are said to be *Borel-separated,* if there is a Borel subset F of E such that $A \subset F$ and $B \subset (E \backslash F)$.

Lemma 10.2.7. Let $\{A_n\}, \{B_m\}$ be two sequences of subsets of a Polish space E, and let A_n and B_m be Borel-separated, $\forall n, m$. Then $A = \cup_n A_n$ and $B = \cup_m B_m$ are also Borel-separated.

Proof. Let F_{nm} be a Borel subset of E such that $A_n \subset F_{nm}$ and $B_m \subset (E \backslash F_{nm}), \forall n, m$. Then

$$A_n \subset \cap_m F_{nm}, \quad B_k \subset E \backslash F_{nk} \subset E \backslash \cap_m F_{nm}, \quad \forall n, k.$$

Let $F = \cup_n \cap_m F_{nm}$. Then $A \subset F$, and $B \subset \cap_n (E \backslash \cap_m F_{nm}) = (E \backslash F)$. Therefore, A and B are Borel-separtated. Q.E.D.

Proposition 10.2.8. Let E be a Polish space, A and B be two Sousline subsets of E, and $A \cap B = \emptyset$. Then A and B are Borel-separated.

Proof. Let f, g be two continuous maps from $\mathbb{N}^\infty$ to E such that $f(\mathbb{N}^\infty) = A$ and $g(\mathbb{N}^\infty) = B$. If A and B are not Borel-separated, then by $\mathbb{N}^\infty = \cup_{k=0}^\infty \mathbb{N}_k^\infty$ and Lemma 10.2.7, there are n_1 and m_1 such that $f(\mathbb{N}_{n_1}^\infty)$ and $g(\mathbb{N}_{m_1}^\infty)$ are not Borel-separated. Continuing this process, generally we have $n = (n_k)$ and $m = (m_k) \in \mathbb{N}^\infty$ such that $f(\mathbb{N}_{n_1,\cdots,n_k}^\infty)$ and $g(\mathbb{N}_{m_1,\cdots,m_k}^\infty)$ are not Borel-separated, $\forall k$. Since $A \cap B = \emptyset$, it follows that $f(n) \neq g(m)$. Pick two open subsets U and V of E such that

$$f(n) \in U, g(m) \in V, \text{and} \quad U \cap V = \emptyset.$$

Then $f(\mathbb{N}_{n_1,\cdots,n_k}^\infty) \subset U$, $g(\mathbb{N}_{m_1,\cdots,m_k}^\infty) \subset V$, if k sufficiently large. This contradicts that $f(\mathbb{N}_{n_1,\cdots,n_k}^\infty)$ and $g(\mathbb{N}_{m_1,\cdots,m_k}^\infty)$ are not Borel-separated, $\forall k$. Therefore, A and B are Borel-separated. Q.E.D.

Proposition 10.2.9. Let $\{A_n\}$ be a disjoint sequence of Sousline subsets of a Polish space E. Then $\{A_n\}$ is Borel-separated, i.e., there is a disjoint sequence $\{B_n\}$ of Borel subsets of E such that $A_n \subset B_n, \forall n$.

Proof. By Proposition 10.2.5 and 10.2.8, for any n there is a Borel subset F_n of E such that $A_n \subset F_n, \cup_{k>n} A_k \subset (E \backslash F_n)$. Now let $B_1 = F_1$, and $B_n = F_n \backslash \cup_{i=1}^{n-1} B_i, \forall n > 1$. Then $A_n \subset B_n, \forall n$. Q.E.D.

Theorem 10.2.10. (Sousline criterion) Let E be a Polish space, and $B \subset E$. Then B is Borel if and only if B and $(E \backslash B)$ are Sousline.

Proof. The necessity is obvious. Now if B and $(E \backslash B)$ are Sousline, then by Proposition 10.2.8, B and $(E \backslash B)$ are Borel-separated. Therefore, B must be Borel. Q.E.D.

Theorem 10.2.11. Let E be a Polish space, $\mathcal{B}$ be the collection of all Borel subsets of E, and A be a Sousline subset of E. Then for any σ-finite measure ν on $\mathcal{B}$, there are B and $F \in \mathcal{B}$ such that $A \subset B, (B \backslash A) \subset F$, and $\nu(F) = 0$.

Proof. Replacing ν by an equivalent finite measure on $\mathcal{B}$, we may assume that ν is finite.

For any $S \subset E$, we say that there exists a minimal Borel cover T of S relative to ν , i.e., $S \subset T \in \mathcal{B}$, and for any $S \subset F \in \mathcal{B}$ we have $\nu(T \backslash F) = 0$. In fact, since ν is finite, we can find $\{E_n\} \subset \mathcal{B}$ such that $E_1 \supset E_2 \supset \cdots \supset S$ and $\lim_n \nu(E_n) = \inf\{\nu(F) | S \subset F \in \mathcal{B}\}$. Let

$$T = \cap_n E_n, \text{and} \quad \lambda = \inf\{\nu(F) | S \subset F \in \mathcal{B}\}.$$

Then $S \subset T \in \mathcal{B}$, and $\nu(T) = \lambda$. Now if $S \subset F \in \mathcal{B}$, we have $\nu(T) = \nu(T \cap F) = \lambda$, and $\nu(T \backslash F) = 0$.

Now we prove the theorem. By Definition 10.2.1, there is a continuous map $f : \mathbb{N}^\infty \longrightarrow E$ such that $f(\mathbb{N}^\infty) = A$. For any non-negative integers $n_1, \cdots, n_k$, let $E_{n_1,\cdots,n_k}$ be a minimal Borel cover of $f(\mathbb{N}^\infty_{n_1,\cdots,n_k})$ relative to ν, and $E_{n_1,\cdots,n_k} \subset \overline{f(\mathbb{N}^\infty_{n_1,\cdots,n_k})}$. Define

$$B = \cup_{n_1=0}^\infty E_{n_1},$$

and

$$F = \cup_{k=1}^\infty \cup_{n_1,\cdots,n_k} (E_{n_1,\cdots,n_k} \backslash \cup_{p=0}^\infty E_{n_1,\cdots,n_k,p}).$$

Then

$$A = f(\mathbb{N}^\infty) = \cup_{n_1=0}^\infty f(\mathbb{N}^\infty_{n_1}) \subset \cup_{n_1=0}^\infty E_{n_1} = B.$$

By the Definition of $E_{n_1,\cdots,n_k}$, we have

$$\nu(E_{n_1,\cdots,n_k} \backslash \cup_{p=0}^{\infty} E_{n_1,\cdots,n_k,p}) = 0, \forall n_1, \cdots, n_k.$$

Hence , $\nu(F) = 0$.

Now it suffices to show $(B\backslash A) \subset F$. Let $x \in (B\backslash A)$. Then there is n_1 such that $x \in E_{n_1}\backslash A$. Suppose that $x \notin F$. Then $x \notin (E_{n_1} \backslash \cup_{n_2=0}^{\infty} E_{n_1,n_2})$, and $x \in E_{n_1} \cap (\cup_{n_2=0}^{\infty} E_{n_1,n_2})$. Thus , there is n_2 such that $x \in E_{n_1,n_2} \cdot \cdots$, generally, we can find $n = (n_k) \in \mathbb{N}^{\infty}$ such that $x \in E_{n_1,\cdots,n_k}, \forall k$. We claim that

$$\{f(n)\} = \cap_{k=1}^{\infty} \overline{f(\mathbb{N}_{n_1,\cdots,n_k}^{\infty})}.$$

Indeed, let $y \neq f(n)$. Pick a closed neighborhood V of $f(n)$ such that $y \notin V$. Since f is continuous, it follows that $f(\mathbb{N}_{n_1,\cdots,n_k}^{\infty}) \subset V$ if k sufficiently large. Then from $\overline{f(\mathbb{N}_{n_1,\cdots,n_k}^{\infty})} \subset \overline{V} = V$ and $y \notin V$, we have $y \notin \cap_{k=1}^{\infty} \overline{f(\mathbb{N}_{n_1,\cdots,n_k}^{\infty})}$. Thus , $\{f(n)\} = \cap_{k=1}^{\infty} \overline{f(\mathbb{N}_{n_1,\cdots,n_k}^{\infty})}$. Now $x \in E_{n_1,\cdots,n_k} \subset \overline{f(\mathbb{N}_{n_1,\cdots,n_k}^{\infty})}, \forall k$, so $x = f(n) \in A$. This contradicts $x \in (B\backslash A)$. Therefore, $x \in F$, and $(B\backslash A) \subset F$.

Q.E.D.

Corollary 10.2.12. Let $E, \mathcal{B}, A, \nu$ be as in Theorem 10.2.11. Then there are C and $G \in \mathcal{B}$ such that $C \subset A, (A\backslash C) \subset G$, and $\nu(G) = 0$.

Proof. Pick B, F as in Theorem 10.2.11. Let $C = B\backslash F$, and $G = F$. Then C and G satisfy our conditions.

Q.E.D.

References. [10], [94], [190].

10.3. Borel maps and standard Borel spaces

Definition 10.3.1. $(E, \mathcal{B})$ is called a *Borel space,* if E is a set , and $\mathcal{B}$ is a σ–Bool algebra of some subsets of E. A subset B of E is called $\mathcal{B}$-*Borel* (or Borel simply if no confusion arises), if $B \in \mathcal{B}$. $\mathcal{B}$ is also called the *Borel structure* of the Borel space $(E, \mathcal{B})$.

For example, let E be a Polish space, and $\mathcal{B}$ be the collection of all Borel subsets of E (see Definition 10.2.1). Then $(E, \mathcal{B})$ is a Borel space, In the following, we understand a Polish space as a Borel space always in this meaning.

A map f from a Borel space $(E, \mathcal{B}_E)$ to another Borel space $(F, \mathcal{B}_F)$ is said to be *Borel* , if $f^{-1}(B_F) \in \mathcal{B}_E, \forall B_F \in \mathcal{B}_F$.

If f is a bijective Borel map from $(E, \mathcal{B}_E)$ onto $(F, \mathcal{B}_F)$, and f^{-1} is also Borel, then we say that Borel spaces $(E, \mathcal{B}_E)$ and $(F, \mathcal{B}_F)$ are *Borel isomorphic,* and f is called a *Borel isomorphism* from $(E, \mathcal{B}_E)$ onto $(F, \mathcal{B}_F)$.

Let $(E, \mathcal{B})$ be a Borel space, and $\mathcal{P} \subset \mathcal{B}$. $\mathcal{P}$ is called a *generated set* for $\mathcal{B}$, if $\mathcal{B}$ is the minimal σ–Bool algebra containing $\mathcal{P}$.

Proposition 10.3.2. 1) Let $(E, \mathcal{B}_E)$ and $(F, \mathcal{B}_F)$ be two Borel spaces, and $\mathcal{P}$ be a generated set for $\mathcal{B}_F$. Then a map $f : E \longrightarrow F$ is Borel if and only if $f^{-1}(B_F) \in \mathcal{B}_E, \forall B_F \in \mathcal{P}$.

2) Let f be a continuous map from a Polish space E to another Polish space F. Then f is Borel.

3) The composition of two Borel maps is still Borel.

Proof. 1) Let $\mathcal{B}'_E = \{f^{-1}(B_F) | B_F \in \mathcal{B}_F\}$. Clearly, $\mathcal{B}'_E$ is a σ–Bool algebra. If $\mathcal{B}''_E$ is a σ –Bool algebra generated by $\{f^{-1}(B_F) | B_F \in \mathcal{P}\}$, then $\mathcal{B}''_E \subset \mathcal{B}'_E$. Thus, $\mathcal{P} \subset \{B_F \in \mathcal{B}_F | f^{-1}(B_F) \in \mathcal{B}''_E\} \equiv \mathcal{B}'_F \subset \mathcal{B}_F$. Since $\mathcal{B}'_F$ is a σ–Bool algebra and $\mathcal{P}$ is a generated set for $\mathcal{B}_F$, it follows that $\mathcal{B}'_F = \mathcal{B}_F$. Further, $\mathcal{B}''_E = \mathcal{B}'_E$.

If f is Borel, then for any $B_F \in \mathcal{P}$, it is obvious that $f^{-1}(B_F) \in \mathcal{B}_E$. Conversely, let $f^{-1}(B_F) \in \mathcal{B}_E, \forall B_F \in \mathcal{P}$. Then $\mathcal{B}''_E \subset \mathcal{B}_E$. From the preceding paragraph, for any $B_F \in \mathcal{B}_F$ we get $f^{-1}(B_F) \in \mathcal{B}'_E = \mathcal{B}''_E \subset \mathcal{B}_E$. Therefore, f is Borel.

2) Since $f^{-1}(U)$ is an open subset of E for any open subset U of F, it follows from 1) that f is Borel.

3) It is obvious. Q.E.D.

Definition 10.3.3. A Borel space $(E, \mathcal{B})$ is said to be *standard* , if we can introduce a topolog τ in E such that (E, τ) is a Polish space and the collection of all Borel subsets of the Polish space (E, τ) is equal to $\mathcal{B}$.

Clearly, a Polish space regarded as a Borel space is standard. Conversely, for a standard Borel space, maybe, we can introduce several topologies such that they become different Polish spaces with original Borel structure (see Proposition 10.3.14).

Proposition 10.3.4. 1) Let $(E, \mathcal{B})$ be a standare Borel space, and f be a Borel map on E. Then $\{x \in E | x = f(x)\} \in \mathcal{B}$.

2) Let $(E, \mathcal{B}_E)$ and $(F, \mathcal{B}_F)$ be standard Borel spaces, and f be a Borel map from E to F. Then the graph $\{(x, f(x)) | x \in E\}$ of f is a Borel subset of $E \times F$, where the Borel structure of $E \times F$ is generated by $\{B_E \times B_F | B_E \in \mathcal{B}_E, B_F \in \mathcal{B}_F\}$(and $E \times F$ with this Borel structure is also a standard Borel space).

Proof. 1) We may assume that E is a Polish space. Then $\triangle = \{(x, x) | x \in E\}$ is a closed subset of $E \times E$. Define a map $(f \times id)$ from E to $E \times E : x \longrightarrow (f(x), x) (\forall x \in E)$. Clearly, $(f \times id)$ is Borel. Thus, $\{x \in E | x = f(x)\} = (f \times id)^{-1}(\triangle)$ is a Borel subset of E.

2) Define a map φ on $E \times F : (x, y) \longrightarrow (x, f(x))$. It is easily verified that φ is Borel. Now by 1), $\{(x, f(x)) | x \in E\} = \{(x, y) | (x, y) = \varphi(x, y)\}$ is a Borel

subset of $E \times F$. Q.E.D.

Proposition 10.3.5. Let E, F be Polish spaces, f be a Borel map from E to F, and A be a Sousline subset of E. Then $f(A)$ is a Sousline subset of F.

Proof. Let g be a continuous map from $\mathbb{N}^\infty$ to E such that $g(\mathbb{N}^\infty) = A$. Then $f \circ g$ is a Borel map from $\mathbb{N}^\infty$ to F, and by Proposition 10.3.4 $\{(n, f\circ g(n)) | n \in \mathbb{N}^\infty\}$ is a Borel subset of $\mathbb{N}^\infty \times F$. Define a map $g \times id$ from $\mathbb{N}^\infty \times F$ to $E \times F$: $(n, y) \longrightarrow (g(n), y), \forall n \in \mathbb{N}^\infty, y \in F$. Clearly, $g \times id$ is continuous. Then by Proposition 10.2.3, $(g \times id)(\{(n, f \circ g(n)) | n \in \mathbb{N}^\infty\}) = \{(x, f(x)) | x \in A\}$ is a Sousline subset of $E \times F$. Let π be the projection from $E \times F$ onto F, then by Proposition 10.2.5 $\pi(\{(x, f(x)) | x \in A\}) = f(A)$ is a Sousline subset of F.

Q.E.D.

Definition 10.3.6. A family $\mathcal{F}$ of some subsets of a set E is said to be *separated* (for E) , if for any $x, y \in E$ with $x \neq y$, there is $F \in \mathcal{F}$ such that either $x \in F$ and $y \notin F$ or $y \in F$ and $x \notin F$.

Lemma 10.3.7. Let $(E, \mathcal{B})$ be a Borel space, and $\mathcal{P} \subset \mathcal{B}$. Then the following statements are equivalent:

1) $\mathcal{B}$ is separated (for E) , and $\mathcal{P}$ is a generated set for $\mathcal{B}$;
2) $\mathcal{P}$ is separated (for E) , and $\mathcal{P}$ is a generated set for $\mathcal{B}$.

Proof. It suffices to show that 1) implies 2). Suppose that 1) holds. If $\mathcal{P}$ is not separated, then there are $x, y \in E$ and $x \neq y$, such that for any $F \in \mathcal{P}$ we have either $x, y \in F$ or $x, y \notin F$. Let $\mathcal{L} = \{B \in \mathcal{B} |$ either $x, y \in B$ or $x, y \notin B\}$. Clearly, $\mathcal{P} \subset \mathcal{L} \subset \mathcal{B}$. If $\{B_n\} \subset \mathcal{L}$, obviously we have $\cup_n B_n \in \mathcal{L}$. For any $B_1, B_2 \in \mathcal{L}$, one of the following relations holds: 1) $x, y \notin B_1$, 2) $x, y \in B_1 \backslash B_2$, 3) $x, y \in B_1 \cap B_2$. Thus $(B_1 \backslash B_2) \in \mathcal{L}$. Further, $\mathcal{L}$ is also a σ-Bool algebra, and $\mathcal{L} = \mathcal{B}$, and $\mathcal{B}$ is not separated, a contradiction. Therefore, $\mathcal{P}$ is separated.

Q.E.D.

Definition 10.3.8. A Borel spaec $(E, \mathcal{B})$ is said to be $\frac{1}{2}$-*standard* , if $\mathcal{B}$ is separated (for E) , and $\mathcal{B}$ contains a countable generated set.

From Lemma 10.3.7, $(E, \mathcal{B})$ is $\frac{1}{2}$ -standard if and only if $\mathcal{B}$ contains a countable generated set $\mathcal{P}$ such that $\mathcal{P}$ is separated (for E).

Clearly, A standard Borel space is $\frac{1}{2}$-standard.

Define $M = \times_n\{0, 1\} = \{a = (a_1, \cdots, a_n, \cdots) | a_n = 0$ or $1, \forall n\}$. M is the countable infinite product of the discrete compact space $\{0, 1\}$, and M is a compact Polish space.

Theorem 10.3.9. A Borel space $(E,\mathcal{B})$ is $\frac{1}{2}$ -standard if and only if it is Borel isomorphic to a subspace of M.

Proof. Let $F_n = \{a \in M|$ the n–th component of a is 1 $\}$. Then F_n is an open and closed subset of M, and $\{F_n|n\}$ is a generated set for the Borel structure of M.

If $(E,\mathcal{B})$ is $\frac{1}{2}$-standard, then there is a sequence $\{B_n\} \subset \mathcal{B}$ such that $\{B_n\}$ is separated (for E) and is a generated set for $\mathcal{B}$. Define $f : E \longrightarrow M$ as follows:

$$\text{the n–th component of } f(x) = \begin{cases} 1, & \text{if } x \in B_n, \\ 0, & \text{if } x \notin B_n, \end{cases}$$

$n = 1,2,\cdots,\forall x \in E$. Since $\{B_n\}$ is separated, it follows that f is injective. Notice that $f(B_n) = F_n \cap f(E), \forall n$. Thus by Proposition 10.3.2, f is a Borel isomorphism from E onto $f(E)$, where the Borel structure of $f(E)$ is $\{F \cap f(E)|F$ is any Borel subset of $M\}$, i.e., is generated by $\{F_n \cap f(E)|n\}$.

Conversely, suppose that $(E,\mathcal{B})$ is Borel isomorphic to a subspace of M. Since any subspace of M is $\frac{1}{2}$-standard, it follows that $(E,\mathcal{B})$ is $\frac{1}{2}$-standard.

Q.E.D.

Lemma 10.3.10. Let E be a Polish space, f be an injective continuous map from $I\!N^\infty$ to E. Then $f(I\!N^\infty)$ is a Borel subset of E.

Proof. For any k, $\{f(I\!N^\infty_{n_1,\cdots,n_k})|n_1,\cdots,n_k\}$ is a disjoint sequence of Sousline subsets of E. From Proposition 10.2.9, there is a disjoint sequence $\{F_{n_1,\cdots,n_k}|n_1,$ $\cdots,n_k\}$ of Borel subsets of E such that $f(N^\infty_{n_1,\cdots,n_k}) \subset F_{n_1,\cdots,n_k}, \forall n_1,\cdots,n_k$. By induction, define $A_{n_1} = F_{n_1}, A_{n_1,\cdots,n_k} = F_{n_1,\cdots,n_k} \cap \overline{f(I\!N^\infty_{n_1,\cdots,n_k})} \cap A_{n_1,\cdots,n_{k-1}}, \forall n_1,$ $\cdots$, n_k and $k > 1$. Then the family $\{A_{n_1,\cdots,n_k}|n_1,\cdots,n_k, k \geq 1\}$ of Borel subsets of E has the following properties:

1) $A_{n_1,\cdots,n_k} \cap A_{m_1,\cdots,m_k} = \emptyset, \forall (n_1,\cdots,n_k) \neq (m_1,\cdots,m_k)$;

2) $A_{n_1,\cdots,n_{k+1}} \subset A_{n_1,\cdots,n_k}, \forall n_1,\cdots,n_{k+1}$;

3) $f(I\!N^\infty_{n_1,\cdots,n_k}) \subset A_{n_1,\cdots,n_k} \subset \overline{f(I\!N^\infty_{n_1,\cdots,n_k})}, \forall n_1,\cdots,n_k$.

In fact , we can prove $f(I\!N^\infty_{n_1,\cdots,n_k}) \subset A_{n_1,\cdots,n_k}$ by induction, and the rest facts are obvious.

Since f is injective, it follows that $f(n) = f(\cap_{k=1}^\infty I\!N^\infty_{n_1,\cdots,n_k}) = \cap_{k=1}^\infty f(I\!N^\infty_{n_1,\cdots,n_k}), \forall n = (n_k) \in I\!N^\infty$. By the proof of Theorem 10.2.11, we have also $\{f(n)\} = \cap_{k=1}^\infty \overline{f(I\!N^\infty_{n_1,\cdots,n_k})}$. Further, from above property 3) we get $\{f(n)\} = \cap_{k=1}^\infty A_{n_1,\cdots,n_k}, \forall n = (n_k) \in I\!N^\infty$.

Now we prove that $f(I\!N^\infty) = \cap_{k=1}^\infty \cup_{n_1,\cdots,n_k} A_{n_1,\cdots,n_k}$. In fact, since $\{f(n)\} = \cap_{k=1}^\infty A_{n_1,\cdots,n_k}, \forall n = (n_k) \in I\!N^\infty$, it follows that $f(I\!N^\infty) \subset \cap_{k=1}^\infty \cup_{n_1,\cdots,n_k} A_{n_1,\cdots,n_k}$. Coversely, let $x \in \cap_{k=1}^\infty \cup_{n_1,\cdots,n_k} A_{n_1,\cdots,n_k}$. For $k = 1$, there is m_1 such that

$x \in A_{m_1}$. From above properties 1) and 2) , we have

$$x \in \cap_{k=1}^{\infty} \sqcup_{n_1,\cdots,n_k} (A_{n_1,\cdots,n_k} \cap A_{m_1})$$
$$= A_{m_1} \cap (\cap_{k=2}^{\infty} \sqcup_{n_2,\cdots,n_k} A_{m_1,n_2,\cdots,n_k}).$$

Repeating this process, there is $m = (m_k) \in \mathbb{N}^\infty$ such that $x \in \cap_{k=1}^{\infty} A_{m_1,\cdots,m_k} = \{f(m)\}$. Hence, $x = f(m) \in f(\mathbb{N}^\infty)$, and $f(\mathbb{N}^\infty) = \cap_{k=1}^{\infty} \sqcup_{n_1,\cdots,n_k} A_{n_1,\cdots,n_k}$.
Therefore, $f(\mathbb{N}^\infty)$ is Borel. Q.E.D.

Lemma 10.3.11. Let E, F be two Polish spaces, and f be an injective Borel map from E to F. Then $f(E)$ is a Borel subset of F, and f is a Borel isomorphism from E onto $f(E)$.

Proof. It suffices to show that for any Borel subset B of E, $f(B)$ is a Borel subset of F.

Fix a Borel subset B of E. Let $G = \{(x, f(x)) | x \in B\}$, d be a proper metric on F, and $\{a_n\}$ be a countable dense subset of F. Put

$$U_k^n = \{y \in F | d(y, a_k) \leq \frac{1}{2n}\}, \quad V_k^n = f^{-1}(U_k^n \cap f(B)),$$

$\forall n, k$. We claim that $G = \cap_n \cup_k (V_k^n \times U_k^n)$. Indeed, since $\sqcup_k U_k^n = F, \forall n$, it follows that $G \subset \cap_n \cup_k (V_k^n \times U_k^n)$. Conversely, let $(x, y) \in \cap_n \cup_k (V_k^n \times U_k^n)$, i.e., for each n, there is $k = k(n)$ such that $(x, y) \in V_k^n \times U_k^n$. Then $f(x) \in U_k^n \cap f(B)$. Since f is injective, we have $x \in B$ and $f(x) \in U_k^n$. By $y \in U_k^n$, we get $d(f(x), y) \leq 1/n$. But n is arbitrary, so $f(x) = y$ i.e., $(x, y) \in G$. Thus , $G = \cap_n \cup_k (V_k^n \times U_k^n)$. Further, since f is injecive, it follows that $V_k^n = f^{-1}(U_k^n) \cap B, \forall n, k$. Hence, G is a Borel subset of $(E \times F)$.

If G is countable, then $f(B)$ is Borel obviously. Now suppose that G is not countable. By Proposition 10.2.4, there is an injective continuous map g from $\mathbb{N}^\infty$ to $E \times F$ such that $g(\mathbb{N}^\infty) \subset G$, and $(G \backslash (g(\mathbb{N}^\infty))$ is countable. Denote the projection from $E \times F$ onto F by π. Since $g(\mathbb{N}^\infty) \subset G$ and f is injective, it follows that $\pi \circ g$ is injective and continuous from $\mathbb{N}^\infty$ to F. From Lemma 10.3.10, $\pi \circ g(\mathbb{N}^\infty)$ is a Borel subset of F. Therefore, $f(B) = \pi G = \pi \circ g(\mathbb{N}^\infty) \sqcup \pi(G \backslash g(\mathbb{N}^\infty))$ is a Borel subset of F. Q.E.D.

Theorem 10.3.12. Let E be a standard Borel space, F be a $\frac{1}{2}$ -standard Borel space, and f be an injective Borel map from E to F. Then $f(E)$ is a Borel subset of F, and f is a Borel isomorphism from E onto $f(E)$.

Proof. By Theorem 10.3.9, we may assume that $F \subset M$. Then by Lemma 10.3.11, we can get the conclusion. Q.E.D.

Theorem 10.3.13. Let $(E, \mathcal{B})$ be a standard Borel space. If a sequence $\{B_n\}$ of $\mathcal{B}$ is separated (for E) , then $\{B_n\}$ is a generated set for $\mathcal{B}$.

Proof. Let $\mathcal{B}_0$ be the σ -Bool algebra generated by $\{B_n\}$. Clearly, $\mathcal{B}_0 \subset \mathcal{B}$, and $(E, \mathcal{B}_0)$ is $\frac{1}{2}$-standard. Now the identity map id is an injective Borel map from $(E, \mathcal{B})$ onto $(E, \mathcal{B}_0)$. Thus by Theorem 10.3.12, we have $\mathcal{B}_0 = \mathcal{B}$. Q.E.D.

Proposition 10.3.14. Let H be a separable Hilbert space. Then the weak (operator) topology, strong (operator) topology, strong $*$ (operator) topology, $\sigma(B(H), T(H)), s(B(H), T(H)), s^*(B(H), T(H))$ and $\tau(B(H), T(H))$ in $B(H)$ will generate the same strandard Borel structure, where the Borel structure genreated by a topology means that the σ-Bool algebra is generated by all open subsets with respect to that topology. In particular, the Polish spaces of $S = \{a \in B(H) | \|a\| \le 1\}$ with respect to weak (operator) topology, strong (operator) topology, and strong $*$ (operator) topology (see the example in Section 9.1) are the same as the standard Borel spaces.

Proof. Denote one of above topologies by $\sigma, S_n = \{a \in B(H) | \|a\| \le n\}, V_1 = S_1, V_{n+1} = S_{n+1} \backslash S_n, \forall n \ge 1$. By the example in Section 9.1, (S_n, σ) is Polish , $\forall n$. Since V_n is an open subset of (S_n, σ), it follows that (V_n, σ) is also Polish. Denote the topological union of $\{(V_n, \sigma) | n \ge 1\}$ by $(B(H), \sigma')$. Clearly , $(B(H), \sigma')$ is Polish, and a subset U of $B(H)$ is σ' -open if and only if $U \cap V_n$ is an open subset of $(V_n, \sigma), \forall n$. Let $\mathcal{B}_{\sigma'}, \mathcal{B}_\sigma$ be the Borel structures of $B(H)$ generated by σ', σ respectively. Since $\sigma' \supset \sigma$, it follows that $\mathcal{B}_\sigma \subset \mathcal{B}_{\sigma'}$. On the other hand, if U is a σ'-open subset of $B(H)$, then it is obvious that $U \in \mathcal{B}_\sigma$. Thus , $\mathcal{B}_{\sigma'} = \mathcal{B}_\sigma$, and $(B(H), \mathcal{B}_\sigma) = (B(H), \mathcal{B}_{\sigma'})$ is standard.

Moreover, obviously we have $\mathcal{B}_\tau \supset \mathcal{B}_\sigma$, where $\tau = \tau(B(H), T(H))$. Then by Theorem 10.3.13, we get $\mathcal{B}_\sigma = \mathcal{B}_\tau, \forall \sigma$ (one of above topologies in $B(H)$).

Q.E.D.

Proposition 10.3.15. Let E be a standard Borel space, and $B \subset E$. Then B as a Borel subspace of E is standard if and only if B is a Borel subset of E.

Proof. Let B be standard, and id be the embedding of B into E. Then by Theorem 10.3.12, B is a Borel subset of E. Conversely, let B be a Borel subset of E. We may assume that E is a Polish space. Then by Proposition 10.2.3, there is a Polish space P and an injective continuous map f from P to E such that $f(P) = B$. Now by Theorem 10.3.12, f is a Borel isomorphism from P onto B. Therefore, B is standard. Q.E.D.

Theorem 10.3.16. The cardinal number of a standard Borel space is either countable or continuum, and the standard Borel spaces with the same cardinal number are Borel isomorphic.

Proof. By Proposition 10.1.11, it suffices to show that E and $\mathbb{R}$ are Borel isomorphic, where E is a standard Borel space and its cardinam number is continuum.

By Proposition 10.1.10, there is an injective continuous map from $\mathbb{N}^\infty$ onto $\mathbb{R}$. Then by Proposition 10.1.11 and Theorem 10.3.12, we have a Borel isomorphism f from $\mathbb{R}$ to E such that $(E\backslash f(\mathbb{R}))$ is countable. Pick a closed subset T of $\mathbb{R}$ such that $^\#T = {}^\#(E\backslash f(\mathbb{R}))$. Clearly, there is a Borel isomorphism φ from T onto $(E\backslash f(\mathbb{R}))$. From Propositon 10.1.10, we have also a Borel isomoprphism ψ from $(\mathbb{R}\backslash T)$ onto $\mathbb{R}$. Now let

$$g(t) = \begin{cases} f \circ \psi(t), & \text{if } t \in (\mathbb{R}\backslash T), \\ \varphi(t), & \text{if } t \in T. \end{cases}$$

Then g is a Borel isomorphism from $\mathbb{R}$ onto E. Q.E.D.

References. [10], [106], [190].

10.4. Borel cross sections

Lemma 10.4.1. Define a total order in $\mathbb{N}^\infty$ as follows : $n \le m$ if either $n = m$ or there exists j such that $n_k = m_k, 1 \le k < j$ and $n_j < m_j$, where $n = (n_k), m = (m_k) \in \mathbb{N}^\infty$. Then there exists a minimal element in any non-empty closed subset of $\mathbb{N}^\infty$.

Proof. Let F be a non-empty closed subset of $\mathbb{N}^\infty$. And put $\alpha_1 = \min\{n_1 | n = (n_k) \in F\}, F_1 = \{n = (n_k) \in F | n_1 = \alpha_1\}; \alpha_2 = \min\{n_2 | n = (n_k) \in F_1\}, F_2 = \{n = (n_k) \in F_1 | n_2 = \alpha_2\}; \cdots$. Then we get $F \supset F_1 \supset F_2 \supset \cdots$. Clearly, $D_d(F_j) \le 2^{-j}$ (see Definition 10.1.6) . Thus , $\cap_j F_j = \{n\}$, and this n is the minimal element of F. Q.E.D.

Theorem 10.4.2. Let E be a Polish space, and $\sim$ be an equivalent relation on E such that :

1) for any $x \in E, \{y \in E | y \sim x\}$ is closed;

2) if F is a closed subset of E , then $\tilde{F} = \{y \in E|$ there is $x \in F$ with $y \sim x\}$ is a Borel subset of E,

or replacing 2) by the following

2') if V is an open subset of E , then $\tilde{V} = \{y \in E|$ there exists $x \in V$ with $y \sim x\}$ is a Borel subset of E.

Then there exists a Borel subset B of E such that $^\#(B \cap \tilde{x}) = 1, \forall x \in E$, where $\tilde{x}$ is the equivalent class of x, i.e., $\tilde{x} = \{y \in E | y \sim x\}, \forall x \in E$.

Proof. Let d be a proper metric on E. If 1) and 2) hold, then we pick a family $\{B(n_1,\cdots,n_k)\}$ of non-empty closed subsets of E such that : (1) $E = \sqcup_{n_1=0}^{\infty} B(n_1)$, (2) $B(n_1,\cdots,n_k) = \sqcup_{p=0}^{\infty} B(n_1,\cdots,n_k,p), \forall k$, (3) $D_d(B(n_1,\cdots,n_k)) < 2^{-k}, \forall k$. If 1) and 2') hold, then we pick a family $\{B(n_1,\cdots,n_k)\}$ of non-empty open subsets of E such that : (1) , (2) , (3) are as above, and (4) $\overline{B(n_1,\cdots,n_{k+1})} \subset B(n_1,\cdots,n_k)$. Since E is Polish , the family $\{B(n_1,\cdots,n_k)\}$ can be found.

In each case (either 1) , 2) or 1) , 2') hold), define $f : \mathbb{N}^\infty \longrightarrow E$ such that $\{f(n)\} = \cap_{k=1}^{\infty} B(n_1,\cdots,n_k), \forall n = (n_k) \in \mathbb{N}^\infty$. Clearly, $f(\mathbb{N}^\infty) = E$, and f is continuous.

Let $\tilde{B}(n_1,\cdots,n_k) = \{y \in E|$ there is $x \in B(n_1,\cdots,n_k)$ with $y \sim x\}$. By the assumption, $\tilde{B}(n_1,\cdots,n_k)$ is Borel , $\forall n_1,\cdots,n_k$. Now by induction, define a family $\{A(n_1,\cdots,n_k)\}$ of Borel subsets of E as follows:

$$\begin{aligned} A(n_1) &= B(n_1) \cap [E \backslash \sqcup_{m_1<n_1} \tilde{B}(m_1)], \\ A(n_1,\cdots,n_{k+1}) &= B(n_1,\cdots,n_{k+1}) \cap A(n_1,\cdots,n_k) \\ &\quad \cap [E \backslash \sqcup_{m_{k+1}<n_{k+1}} \tilde{B}(n_1,\cdots,n_k,m_{k+1})], \end{aligned}$$

$\forall n_1,\cdots,n_{k+1}$, and $k \geq 1$.

For each equivalent class X of E, by 1) $f^{-1}(X)$ is a non-empty closed subset of $\mathbb{N}^\infty$. From Lemma 10.4.1, $f^{-1}(X)$ has a minimal element $(p_k) = p = p(X)$. Then we claim that:

a) $A(p_1,\cdots,p_k) \cap X = B(p_1,\cdots,p_k) \cap X \neq \emptyset, \forall k$;

b) if $(n_1,\cdots,n_k) \neq (p_1,\cdots,p_k)$, then $A(n_1,\cdots,n_k) \cap X = \emptyset, \forall k$.

In fact , since $f(p) \in X$ and $\{f(p)\} = \cap_{k=1}^{\infty} B(p_1,\cdots,p_k)$, it follows that $B(p_1,\cdots,p_k) \cap X \neq \emptyset, \forall k$. By the definition, $A(p_1,\cdots,p_k) \cap X \subset B(p_1,\cdots,p_k) \cap X, \forall k$. Now let $x \in B(p_1) \cap X$. Suppose that there is $m_1 < p_1$ such that $x \in \tilde{B}(m_1)$. Pick $y \in B(m_1)$ with $y \sim x$. Clearly, there is $m = (m_k) \in \mathbb{N}^\infty$ such that $y = f(m)$. Then $m \in f^{-1}(X)$. But p is the minimal element of $f^{-1}(X)$, so $p_1 \leq m_1$. This contradicts $m_1 < p_1$. Thus , $x \in B(p_1) \backslash \cup_{m_1<p_1} \tilde{B}(m_1) = A(p_1)$, i.e., $A(p_1) \cap X = B(p_1) \cap X \neq \emptyset$. Further, by induction and the similar process, we can get the conclusion a). Now let $(n_1,\cdots,n_k) \neq (p_1,\cdots,p_k)$. Suppose that there is $x \in A(n_1,\cdots,n_k) \cap X$. Then there is $n = (n_1,\cdots,n_k,\cdots) \in \mathbb{N}^\infty$ such that $f(n) = x$, i.e., $n \in f^{-1}(X)$. Thus $n \geq p$. By $(n_1,\cdots,n_k) \neq (p_1,\cdots,p_k)$, there exists $j(\leq k)$ such that $n_i = p_i, 1 \leq i < j$ and $p_j < n_j$. Then

$$\begin{aligned} &A(p_1,\cdots,p_{j-1},n_j) \cap \tilde{B}(p_1,\cdots,p_j) \\ \subset\ &[E \backslash \sqcup_{p<n_j} \tilde{B}(p_1,\cdots,p_{j-1},p)] \cap \tilde{B}(p_1,\cdots,p_j) = \emptyset. \end{aligned}$$

On the other hand, since $B(p_1,\cdots,p_j) \cap X \neq \emptyset$, it follows that $X \subset \tilde{B}(p_1,\cdots,$

p_j). Then from $j \le k$, we have

$$\begin{aligned} x &\in A(n_1, \cdots, n_j) \cap X \\ &= A(p_1, \cdots, p_{j-1}, n_j) \cap X \\ &\subset A(p_1, \cdots, p_{j-1}, n_j) \cap \tilde{B}(p_1, \cdots, p_j) = \emptyset, \end{aligned}$$

a contradiction. Thus $A(n_1, \cdots, n_k) \cap X = \emptyset, \forall (n_1, \cdots, n_k) \neq (p_1, \cdots, p_k)$, and any k.

Let $B = \cap_{k=1}^{\infty} \cup_{n_1, \cdots, n_k} A(n_1, \cdots, n_k)$. Clearly, B is a Borel subset of E. For each equivalent class X of E, by above a) and b) we have

$$\begin{aligned} B \cap X &= \cap_{k=1}^{\infty} \cup_{n_1, \cdots, n_k} (A(n_1, \cdots, n_k) \cap X) \\ &= \cap_{k=1}^{\infty} (A(p_1, \cdots, p_k) \cap X) = \cap_{k=1}^{\infty} (B(p_1, \cdots, p_k) \cap X) \\ &= (\cap_{k=1}^{\infty} (B(p_1, \cdots, p_k)) \cap X = \{f(p)\}, \end{aligned}$$

where $p = (p_k)$ is the minimal element of $f^{-1}(X)$. Therefore, we have $^{\#}(B \cap \tilde{x}) = 1, \forall x \in E$. Q.E.D.

Theorem 10.4.3. Let f be a Borel map from a Polish space E onto a Borel space F satisfying:

1) $f^{-1}(\{y\})$ is closed, $\forall y \in F$;

2) f maps any closed subset of E to a Borel subset of F,

or replacing 2) by the following

2') f maps any open subset of E to a Borel subset of F.

Then f admits a Borel cross section, i.e., there exists a Borel map g from F to E such that $f \circ g(y) = y, \forall y \in F$.

Proof. Introduce an equivalent relation $\sim$ in $E : x_1 \sim x_2$, if $f(x_1) = f(x_2)$. Then by Theorem 10.4.2, there is a Borel subset B of E such that $^{\#}(f^{-1}(\{y\}) \cap B) = 1, \forall y \in F$. Suppose that g is a map from F to E such that $\{g(y)\} = f^{-1}(\{y\}) \cap B, \forall y \in F$. Then $f \circ g(y) = y, \forall y \in F$. Let G be either any closed subset (in the case of 1) and 2)) or any open subset (in case of 1) and 2')) of E. Then $f(G)$ is a Borel subset of F. Since $g^{-1}(G) = f(G)$, it follows from Proposition 10.3.2 that g is a Borel map. Q.E.D.

Lemma 10.4.4. Let Ω be a locally compact Hausdorff space satisfying the second countability axiom, ν be a regular Borel measure on Ω, and f be a continuous map from $\mathbb{N}^{\infty}$ to Ω. Then there exists a map g from $\Omega' = f(\mathbb{N}^{\infty})$ to $\mathbb{N}^{\infty}$ and a sequence $\{K_n\}$ of compact subsets of Ω' such that

$$f \circ g(x) = x, \forall x \in \Omega', \text{ and } (\Omega' \backslash \cup_n K_n) \subset \text{some } \nu \text{ - zero subset,}$$

and for each n, g is continuous on K_n.

Proof. For any $x \in \Omega', f^{-1}(\{x\})$ is a non-empty closed subset of $\mathbb{N}^\infty$. Denote the minimal element of $f^{-1}(\{x\})$ by $g(x)$. Then $f \circ g(x) = x, \forall x \in \Omega'$.

Clearly, Ω' is a Sousline subset of Ω. By Corollary 10.2.12, the σ-finiteness of ν and the inner regularity of ν , it suffices to show that for any compact subset K of Ω' , there is a sequence $\{K_n\}$ of compact subsets such that : $K_n \subset K, g$ is continuous on $K_n, \forall n$, and $\nu(K \backslash \sqcup_n K_n) = 0$.

1) We say that a subset E of Ω has a Borel kernel relative to ν, if there exist Borel subsets F and G of Ω such that $F \subset E, (E \backslash F) \subset G$, and $\nu(G) = 0$. By Corollary 10.2.12, any Sousline subset of Ω has a Borel kernel relative to ν. Clearly, if E_n has a Borel kernel relative to $\nu, \forall n$, then $\sqcup_n E_n$ has also a Borel kernel relative to ν. Moreover, if E_i has a Borel kernel relative to $\nu, i = 1,2$, then $(E_1 \backslash E_2)$ has also a Borel kernel relative to ν. Indeed, let F_i, G_i be Borel subsets of Ω such that $F_i \subset E_i, (E_i \backslash F_i) \subset G_i$, and $\nu(G_i) = 0, i = 1,2$. Define $F = F_1 \backslash (F_2 \cup G_2)$, and $G = G_1 \cup G_2$. Then F and G are Borel, $\nu(G) = 0, F \subset (E_1 \backslash E_2)$, and $(E_1 \backslash E_2) \backslash F \subset G$.

2) Let $Q = \{n = (n_k) \in \mathbb{N}^\infty | \# \{k | n_k \neq 0\} < \infty\}$. Then Q is a countable dense subset of $\mathbb{N}^\infty$. Suppose that V is an open subset of $\mathbb{N}^\infty$. If $n = (n_k) \in V$, then $\mathbb{N}^\infty_{n_1,\cdots,n_k} \subset V$ if k sufficiently large. Let $z_1 = (n_1, \cdots, n_k, 0, \cdots)$, and $z_2 = (n_1, \cdots, n_k + 1, 0, \cdots)$. Then $n \in [z_1, z_2) = \{z \in \mathbb{N}^\infty | z_1 \leq z < z_2\} = \mathbb{N}^\infty_{n_1,\cdots,n_k}$. Since Q is countable, it follows that V is a countable union of subsets with a form $[z_1, z_2) = [0, z_2) \backslash [0, z_1)$, where $z_1, z_2 \in Q$.

3) For any $z \in Q$, we have $g^{-1}([0,z)) = f([0,z))$. In fact, if $x \in g^{-1}([0,z))$, then $x = f \circ g(x) \in f([0,z))$. Conversely, if $n \in [0,z)$, then by the definition of g, $g \circ f(n)$ is the minimal element of $f^{-1}(\{f(n)\})$. Thus, $g \circ f(n) \leq n$, and $g \circ f(n) \in [0,z)$, i.e., $f(n) \in g^{-1}([0,z))$.

4) For any $z \in Q$, clearly $[0,z)$ is a Borel subset of $\mathbb{N}^\infty$. Since f is continuous, it follows from 3) that $g^{-1}([0,z))(= f([0,z)))$ is a Sousline subset of Ω. Further by Corollary 10.2.12, $g^{-1}([0,z))$ has a Borel kernel relative to ν.

5) For any open subset V of $\mathbb{N}^\infty$ and any closed subset F of $\mathbb{N}^\infty, g^{-1}(V)$ and $g^{-1}(F)$ have Borel kernels relative to ν. In fact, from 1) , 2) , and 4) , $g^{-1}(V)$ has a Borel kernel relative to ν. Moreover, from $g^{-1}(F) = \Omega' \backslash g^{-1}(F')$ (where $F' = \mathbb{N}^\infty \backslash F$) and 1), $g^{-1}(F)$ has also a Borel kernel relative to ν.

Now let K be a compact subset of Ω with $K \subset \Omega', \{a_k\}$ be a countable dense subset of $\mathbb{N}^\infty$, and d be a proper metric on $\mathbb{N}^\infty$ as in Definition 10.1.6. Put

$$A_{k,p} = \{x \in K | d(g(x), a_k))\} \leq p^{-1}\},$$

and

$$B_{1,p} = A_{1,p}, \quad B_{k+1,p} = A_{k+1,p} \backslash \sqcup_{i=1}^k A_{i,p},$$

$\forall k, p$. Then for any p, we have

$$B_{i,p} \cap B_{j,p} = \emptyset, \quad \forall i \neq j,$$

$$\cup_{k=1}^{\infty} B_{k,p} = \cup_{k=1}^{\infty} A_{k,p} = K.$$

Further, define $g_p : K \longrightarrow \mathbb{N}^\infty$ as follows:

$$g_p(x) = a_k, \text{ if } \quad x \in B_{kp}, \forall k.$$

Clearly, $g_p(x) \longrightarrow g(x)$ in $\mathbb{N}^\infty, \forall x \in K$.

For any k, p, by the discussion 5) $A_{k,p}$ has a Borel kernel relative to ν. Thus, $B_{k,p}$ has also a Borel kernel relative to ν. Then for any m, by the regularity of ν and the definition of g_p we can find a compact subset $K_p^{(m)} \subset K$ such that g_p is continuous on $K_p^{(m)}$, and $\nu(K \backslash K_p^{(m)}) < (m \cdot 2^p)^{-1}$. Let $K^{(m)} = \cap_p K_p^{(m)}$. Then g_p is continuous on $K^{(m)}, \forall p$, and $\nu(K \backslash K^{(m)}) < m^{-1}, \forall m$. Denote the projection from $\mathbb{N}^\infty$ onto its j-th component by π_j. Clearly, $(\pi_j g_p)(x) \longrightarrow (\pi_j g)(x), \forall x \in K$. By the Egorov theorem, for any $\varepsilon > 0$ there is a compact subset $K_{mj}^{(\varepsilon)} \subset K^{(m)}$ such that

$$(\pi_j g_p)(x) \longrightarrow (\pi_j g)(x), \text{uniformly for } x \in K_{mj}^{(\varepsilon)},$$

and $\nu(K^{(m)} \backslash K_{mj}^{(\varepsilon)}) < 2^{-j}\varepsilon$. Let $K_m^{(\varepsilon)} = \cap_j K_{mj}^{(\varepsilon)}$. Then

$$g_p(x) \longrightarrow g(x) (in \mathbb{N}^\infty), \text{uniformly for } x \in K_m^{(\varepsilon)},$$

and $\nu(K^{(m)} \backslash K_m^{(\varepsilon)}) < \varepsilon$. Thus, g is continuous on $K_m^{(\varepsilon)}$. Further, pick $\{K_n\} = \{K_m^{(p^{-1})} | m, p\}$. Then $K_n \subset K, g$ is continous on $K_n, \forall n$, and $\nu(K \backslash \sqcup_n K_n) = 0$.

Q.E.D.

Theorem 10.4.5. Let E, F be two Polish spaces, ν be a σ-finite measure on all Borel subsets of F, and G be a Sousline subset of $E \times F$. If π_F is the projection from $E \times F$ onto F, then there exists a map g from $R = \pi_F(G)$ to E and a Borel subset B of F with $B \subset R$ such that

$$(g(y), y) \in G, \quad \forall y \in R,$$

and g is a Borel map from B to E, and $(R \backslash B) \subset$ some ν-zero subset.

Proof. Since ν is σ-finite, by Proposition 10.3.15 and 9.3.16 we may assume the F is a locally compact Hausdorff space satisfying the second countability axiom, and ν is a regular Borel measure on F. Now G is a Sousline subset of $E \times F$. Then there is a continuous map h from $\mathbb{N}^\infty$ to $E \times F$ such that $h(\mathbb{N}^\infty) = G$. Let $f = \pi_F \circ h$. Then $R = f(\mathbb{N}^\infty)$. By Lemma 10.4.4, we have a map η from R to $\mathbb{N}^\infty$ and a Borel subset $B \subset R$ such that $f \circ \eta(y) = y, \forall y \in R$, and η is Borel on B, and $(R \backslash B) \subset$ some ν-zero subset. Let π_E be the projection from $E \times F$ onto E, and $g = \pi_E \circ h \circ \eta$. Then $g : R \longrightarrow E$, and g is Borel on B. For any $y \in R$, since $h \circ \eta(y) \in G, \pi_E \circ h \circ \eta(y) = g(y), \pi_F \circ h \circ \eta(y) = f \circ \eta(y) = y$, it follows that

$$(g(y), y) = h \circ \eta(y) \in G.$$

Q.E.D.

Proposition 10.4.6. Let E, F be two standard Borel spaces, f be a Borel map from E onto F, and ν be a σ -finite measure on all Borel subsets of F. Then f admits a Borel cross section relative to ν, i.e., there is a Borel subset F_0 of F and a Borel map g from $(F \backslash F_0)$ to E such that $\nu(F_0) = 0$, and $f \circ g(y) = y, \forall y \in (F \backslash F_0)$.

Proof. By Proposition 10.3.4, the Graph $G = \{(x, f(x)) | x \in E\}$ of f is a Borel subset of $E \times F$. Clearly, $\pi_F(G) = F$. Now from Theorem 10.4.5, we can get the conclusion. Q.E.D.

Proposition 10.4.7. Let E, F be two standard Borel spaces, f be a Borel map from E onto F, and μ be a finite measure on all Borel subsets of E. Introduce an equivalent relation $\sim$ in $E : x_1 \sim x_2$, if $f(x_1) = f(x_2)$. Then there is a saturated Borel subset E_0 with $\mu(E_0) = 0$ such that f has a Borel cross section on $(E \backslash E_0)$.

Proof. Let $\nu = \mu \circ f^{-1}$. Clearly , ν is a finite measure on all Borel subsets of F. By Proposition 10.4.6, there is a Borel subset F_0 of F and a Borel map g from $(F \backslash F_0)$ to E such that $\nu(F_0) = 0$ and $f \circ g(y) = y, \forall y \in (F \backslash F_0)$. Now let $E_0 = f^{-1}(F_0)$. Clearly, E_0 is a saturated Borel subset of E, and $\mu(E_0) = 0$. Moreover, since $f(E \backslash E_0) = F \backslash F_0$ and $g(F \backslash F_0) \subset (E \backslash E_0)$, it follows that f has a Borel cross section on $(E \backslash E_0)$. Q.E.D.

Refereces. [10], [24], [28], [120].

Chapter 11

The Borel Spaces of Von Neumann Algebras

11.1. The standard Borel structure of $W(X^*)$

Let E be a topological spce. Denote the collection of all non–empty closed subsets of E by $\mathcal{C}(E)$.

Lemma 11.1.1. Let (E,d) be a compact metric space. For any $F_1, F_2 \in \mathcal{C}(E)$, define

$$\rho(F_1,F_2) = \max\{\sup_{x\in F_1} d(x,F_2), \sup_{y\in F_2} d(y,F_1)\}.$$

Then $(\mathcal{C}(E),\rho)$ is also a compact metric space.

Proof. It is easily verified that ρ is a metric on $\mathcal{C}(E)$. Let $\{x_n\}$ be a countable dense subset of E. For any $\varepsilon > 0$, since (E,d) is compact, there is k such that $\cup_{i=1}^k S_d(x_i,\varepsilon) = E$, where $S_d(y,\varepsilon) = \{z \in E | d(y,z) < \varepsilon\}, \forall y \in E$. We claim that

$$\cup_{I\subset\{1,\cdots,k\}} S_\rho(\{x_i\}_{i\in I},\varepsilon) = \mathcal{C}(E).$$

In fact, for any $F \in \mathcal{C}(E)$, there is $I \subset \{1,\cdots,k\}$ such that $S_d(x_i,\varepsilon) \cap F \neq \emptyset, \forall i \in I$, and $\cup_{i\in I} S_d(x_i,\varepsilon) \supset F$. Then $d(x_i,F) < \varepsilon, \forall i \in I$. On the other hand, for any $y \in F$, there is $j \in I$ such that $y \in S_d(x_j,\varepsilon)$, i.e., $d(x_j,y) < \varepsilon$, and $d(y,\{x_i\}_{i\in I}) < \varepsilon$. Therefore, $\rho(\{x_i\}_{i\in I},F) < \varepsilon$.

Now it suffices to show that $(\mathcal{C}(E),\rho)$ is complete. Let $\{F_n\} \subset \mathcal{C}(E)$ and $\rho(F_n,F_m) \longrightarrow 0$. Put

$$F = \left\{x \in E \,\middle|\, \begin{array}{l}\text{there is a subsequence } \{n_k\}, \text{and} \\ x_{n_k} \in F_{n_k}, \forall k, \text{such that } x_{n_k} \longrightarrow x\end{array}\right\}.$$

Since E is compact, it follows that $F \neq \emptyset$. Further, $F \in \mathcal{C}(E)$. For any $\varepsilon > 0$, there is n_0 such that $\rho(F_n,F_m) < \varepsilon, \forall n,m \geq n_0$. Fix $n(\geq n_0)$. If

$y \in F_n$, then from $d(y, F_m) < \varepsilon, \forall m \geq n_0$, we can find $x_m \in F_m$ such that $d(y, x_m) < \varepsilon, \forall m \geq n_0$. Since E is compact, there is a convergent subsequence of $\{x_m | m \geq n_0\}$. Suppose that its limit point is x. Then $x \in F$ and $d(y,x) \leq \varepsilon$. Thus, $d(y, F) \leq \varepsilon, \forall y \in F_n$. Conversely, let $x \in F$. Then there is a subsequence $\{n_k\}$, and $x_{n_k} \in F_{n_k}, \forall k$, such that $x_{n_k} \longrightarrow x$. Pick k sufficiently large such that $d(x_{n_k}, x) < \varepsilon$, and $n_k \geq n_0$. Since $d(x_{n_k}, F_n) < \varepsilon$, there is $y \in F_n$ with $d(x_{n_k}, y) < \varepsilon$. Further, $d(x,y) < 2\varepsilon$. Hence, $d(x, F_n) < 2\varepsilon, \forall x \in F$. Therefore, $\rho(F_n, F) < 2\varepsilon, \forall n \geq n_0$, and $\rho(F_n, F) \longrightarrow 0$. Q.E.D.

Lemma 11.1.2. Let (P, d) be a compact metric space, and E be a Polish subspace of P. Then $(C(E), \rho)$ is also a Polish space.

Proof. Denote the closure of E in P by $\overline{E}$. Then $(\overline{E}, d)$ is a compact metric space. By Lemma 11.1.1, $(C(\overline{E}), \rho)$ is also a compact metric space. Define a map $f : C(E) \longrightarrow C(\overline{E})$ as follows:

$$f(F) = \overline{F}, \quad \forall F \in C(E).$$

Clearly, $\rho(f(F_1), f(F_2)) = \rho(F_1, F_2), \forall F_1, F_2 \in C(E)$, and $f(C(E)) = \{K \in C(\overline{E}) | (K \cap E) \text{ is dense in } K\}$. Since $(C(E), \rho)$ and $(f(C(E)), \rho)$ are isometrically isomorphic, so it suffices to show that $f(C(E))$ is a Polish subspace of $(C(\overline{E}), \rho)$. By Proposition 10.1.3, we need to prove that $f(C(E))$ is a G_δ–subset of $(C(\overline{E}), \rho)$.

Write $E = \cap_n V_n$, where V_n is an open subset of $P, \forall n$. If $K \in C(\overline{E})$, and $(K \cap V_n)$ is dense in $K, \forall n$, then we have $K \in f(C(E))$. Indeed, since K is a compact subset of P, K is a Baire space. Now $(K \cap V_n)$ is an open dense subset of $K, \forall n$, it must be that $\cap_n (K \cap V_n) = K \cap E$ is dense in K, i.e., $K \in f(C(E))$. Therefore, we get

$$f(C(E)) = \{K \in C(\overline{E}) | (K \cap V_n) \text{ is dense in } K, \forall n\}.$$

Let $\mathcal{D}_n = \{K \in C(\overline{E}) | (K \cap V_n) \text{ is not dense in } K\}, \forall n$. Then $f(C(E)) = \cap_n (C(\overline{E}) \backslash \mathcal{D}_n)$. Now it suffices to prove that each $\mathcal{D}_n$ is a F_σ–subset (a countable union of closed subsets) of $C(\overline{E})$.

Fix n. Let $K \in C(\overline{E})$. If $K = \{x\}$ (some $x \in \overline{E}$), then $K \in \mathcal{D}_n \iff K \cap V_n = \emptyset \iff x \in \overline{E} \backslash V_n$. If $^\# K \geq 2$, then $K \in \mathcal{D}_n \iff$ there exists $L \in C(\overline{E})$ such that $K \cap V_n \subset L \subset K$ and $L \neq K$. Thus, we have

$$\mathcal{D}_n = \mathcal{F} \cup \pi_1(S_n \backslash \triangle),$$

where $\mathcal{F} = \{\{x\} | x \in \overline{E} \backslash V_n\}, \pi_1$ is the projection from $C(\overline{E}) \times C(\overline{E})$ onto its first component, $\triangle = \{(K, K) | K \in C(\overline{E})\}$, and

$$S_n = \{(K, L) | K, L \in C(\overline{E}), \text{and} \quad K \cap V_n \subset L \subset K\}.$$

It is easily verified that $\mathcal{F}$ is a closed subset of $(C(\overline{E}), \rho)$. Moreover, $\triangle$ is a closed subset of $C(\overline{E}) \times C(\overline{E})$ obviously.

We claim that S_n is also a closed subset of $C(\overline{E}) \times C(\overline{E})$. In fact, suppose that $\{(K_m, L_m)\} \subset S_n$, and $(K_m, L_m) \longrightarrow (K, L)$ in $C(\overline{E}) \times C(\overline{E})$. Since $L_m \subset K_m, \forall m$, it follows from the proof of Lemma 11.1.1 that $L \subset K$. Now if $x \in K \cap V_n$, then there is a subsequence $\{m_k\}$, and $x_{m_k} \in K_{m_k}, \forall k$, such that $d(x_{m_k}, x) \longrightarrow 0$. But V_n is open and $x \in V_n$, so we may assume that $x_{m_k} \in V_n, \forall k$. Thus , $x_{m_k} \in K_{m_k} \cap V_n \subset L_{m_k}, \forall k$. Further, by the proof of Lemma 11.1.1, $x \in L$, i.e. $K \cap V_n \subset L \subset K$. Therefore, $(K, L) \in S_n$.

Now $(S_n \backslash \triangle)$ is a F_σ-subset of $C(\overline{E}) \times C(\overline{E})$. So we can write $S_n \backslash \triangle = \sqcup_m G_m$, where G_m is compact in $C(\overline{E}) \times C(\overline{E}), \forall m$. Therefore,

$$\mathcal{D}_n = \mathcal{F} \sqcup \sqcup_m \pi_1(G_m)$$

is a F_σ -subset of $C(\overline{E})$. Q.E.D.

Definition 11.1.3. Let E be a Polish space, and $C(E)$ be the collection of all non-empty closed subsets of E. For any open subset U of E, put

$$u(U) = \{F \in C(E) | F \cap U \neq \emptyset\}.$$

Further, we shall denote by $\mathcal{P}$ the Borel structure of $C(E)$ generated by $\{u(U) | U \text{ is any open subset of } E\}$.

Theorem 11.1.4. Let E be a Polish space. Then $(C(E), \mathcal{P})$ is a standard Borel space.

Proof. By Proposition 10.1.4, we may assume that E is a G_δ-subset of $P = [0,1]^\infty$. Clearly, there exists a metric d on P such that (P, d) is a compact metric space. From Lemma 11.1.2, $(C(E), \rho)$ is a Polish space. Now it suffies to show that the Borel structure of $C(E)$ generated by ρ-top. is equal to $\mathcal{P}$.

First, for an open subset U of E, we say that $u(U)$ is an open subset of $(C(E), \rho)$. Indeed, let $F \in u(U)$. Then there is $x \in F \cap U$. Now if $G \in C(E)$ and $\rho(F, G)$ is very small, then $d(x, G) (\leq \rho(F, G))$ is also very small, further, $G \cap U \neq \emptyset$, i.e., $G \in u(U)$. Thus, $u(U)$ is open in $(C(E), \rho)$.

From the preceding paragraph, the Borel structure of $C(E)$ generated by ρ-top. Contains $\mathcal{P}$. By Theorem 10.3.13, It suffices to prove that $\mathcal{P}$ contains a countable separated family.

Let $\{U_n\}$ be a countable basis for the topology of E. We need only to prove that $\{u(U_n)\}_n$ is separated (for $C(E)$). If $F, G \in C(E)$ and $F \neq G$, then we may assume that there is $x \in F \backslash G$. Clearly, we can find k such that $x \in U_k$ and $U_k \cap G = \emptyset$. Thus, $F \cap U_k \neq \emptyset$ and $G \cap U_k = \emptyset$, i.e., $F \in u(U_k)$ and $G \notin u(U_k)$. Therefore, $\{u(U_n)\}_n$ is separated (for $C(E)$). Q.E.D.

Proposition 11.1.5. Let (E, d) be a separable complete metric space. Then the standard Borel structure $\mathcal{P}$ of $C(E)$ is the minimal Borel structure such that $F \longrightarrow d(x, F)$ is measurable on $\mathcal{C}(E), \forall x \in E$. In other words, $\mathcal{P}$ is generated by $\{F \in \mathcal{C}(E) | d(x, F) < \lambda\}, \forall x \in E$ and $\lambda > 0$.

Proof. First, for any $x \in E$ and $\lambda > 0$, let $U = \{y \in E | d(x, y) < \lambda\}$. Then it is easy to see that $u(U) = \{F \in \mathcal{C}(E) | d(x, F) < \lambda\}$. Thus, $\{F \in \mathcal{C}(E) | d(x, F) < \lambda\} \in \mathcal{P}$.

Now by Theorem 10.3.13, it suffices to show that the collection of $\{F \in \mathcal{C}(E) | d(x, F) < \lambda\}(\forall x \in E, \lambda > 0)$ contains a countable separated family. Let $\{x_n\}$ be a countable dense subset of $E, U_{m,n} = \{x \in E | d(x, x_n) < m^{-1}\}$, and $\theta_{m,n} = u(U_{m,n}) = \{F \in \mathcal{C}(E) | d(x_n, F) < m^{-1}\}, \forall m, n$. If $F, G \in \mathcal{C}(E)$ and $F \neq G$, then we may assume that there is $x \in F \backslash G$. Thus $d(x, G) > 2m_0^{-1}$ if m_0 sufficiently large. Pick n_0 such that $d(x, x_{n_0}) < m_0^{-1}$. Then $d(x_{n_0}, F) < m_0^{-1}$, i.e., $F \in \theta_{m_0, n_0}$. On the other hand, since $d(x_{n_0}, G) \geq d(x, G) - d(x_{n_0}, x) > m_0^{-1}$, it follows that $G \notin \theta_{m_0, n_0}$. Therefore, $\{\theta_{m,n}\}_{m,n}(\subset \mathcal{P})$ is separated (for $\mathcal{C}(E)$).

Q.E.D.

Proposition 11.1.6. Let X be a (real or complex) separable Banach space, and $C(X)$ be the collection of all closed linear subspaces of X. Then $C(X)$ is a Borel subset of $(\mathcal{C}(X), \mathcal{P})$.

Proof. Let $\{V_n\}$ be a countable basis for the topology of X. It suffices to show that

$$C(X) = \cap_{m,n}[u(V_m)' \sqcup u(V_n)' \sqcup u(V_n + V_m)] \cap \cap_{i,k}[u(V_i)' \sqcup u(\lambda_k V_i)],$$

where $\{\lambda_k\}$ is the set of all (real or complex) rational numbers, and $u(V_m)' = \mathcal{C}(X) \backslash u(V_m), \forall m$. In fact, if E belongs to the right side of above equality, then for any m, n, i, we have : i) if $E \cap V_m \neq \emptyset$ and $E \cap V_n \neq \emptyset$, then $E \cap (V_n + V_m) \neq \emptyset$; ii) if $E \cap V_i \neq \emptyset$, then $E \cap (\lambda_k V_i) \neq \emptyset, \forall k$. Thus, for any $x, y \in E$ we get $E \cap (V_n + V_m) \neq \emptyset, E \cap (\lambda_k V_n) \neq \emptyset, \forall k, m, n$ and $x \in V_m, y \in V_n$. By the closedness of E, we can see that $(x + y) \in E$ and $\lambda x \in E, \forall \lambda \in \mathbb{R}$ (or $\mathbb{C}$), i.e., $E \in C(X)$. Conversely, if $E \in C(X)$, then for any m, n, i the above properties i) and ii) hold obviously, i.e., E belongs to the right side of above equality.

Q.E.D.

Theorem 11.1.7. Let X be a (real or complex) separable Banach space, $C(X)$ be the collection of all closed linear subspaces of X, and $W(X^*)$ be the collection of all w^*-closed linear subspaces of X^*, where X^* is the conjugate space of X. Then:

1) The standard Borel structure of $C(X)$ is generated by

$$\{E \in C(X) | \|x + E\| < \lambda\}, \quad \forall x \in X, \lambda > 0,$$

2) The subsets of $W(X^*)$ with the following form

$$\{E^* \in W(X^*) | \|x + E^*_\perp\| < \lambda\},$$

where $E^*_\perp = \{y \in X | f(y) = 0, \forall f \in E^*\}, \forall x \in X, \lambda > 0$, generate a standard Borel structure of $W(X^*)$.

Proof. 1) It is obvious from Propositions 10.3.15, 11.1.6 and 11.1.5.

2) Notice that $E^* \longrightarrow E^*_\perp (\forall E^* \in W(X^*))$ is a bijection from $W(X^*)$ onto $C(X)$. Then by 1) we can get the conclusion. Q.E.D.

Proposition 11.1.8. Let H be a separable Hilbert space, and $W(H)$ be the collection of all closed linear subspaces of H. Then the subsets of $W(H)$ with the following form

$$\{E \in W(H) | \|\xi + E\| < \lambda\}, \forall \xi \in H, \lambda > 0$$

generate a standard Borel structure of $W(H)$, and $E \longrightarrow E^\perp (\forall E \in W(H))$ is a Borel isomorphism on $W(H)$.

Proof. From Theorem 11.1.7 and Proposition 10.3.2, it suffices to show that for any $\xi \in H, \lambda > 0, \{E \in W(H) | \|\xi + E^\perp\| < \lambda\}$ is a Borel subset of $W(H)$.

If $\lambda > \|\xi\|$, then we have $\{E \in W(H) | \|\xi + E^\perp\| < \lambda\} = W(H)$ obviously. Thus, we may assume $\lambda \leq \|\xi\|$. For $E \in W(H)$, let p be the projection from H onto E. Then we have

$$\|\xi + E^\perp\| = \|p\xi\|, \quad \|\xi + E\| = \|(1-p)\xi\|.$$

Let $\mu = (\|\xi\|^2 - \lambda^2)^{1/2}$. Then

$$\begin{aligned} & \{E \in W(H) \mid \|\xi + E^\perp\| < \lambda\} \\ = \; & \{E \in W(H) \mid \|\xi + E\| > \mu\} \\ = \; & W(H) \backslash \cap_n \{E \in W(H) \mid \|\xi + E\| < \frac{1}{n} + \mu\}. \end{aligned}$$

Therefore, it is a Borel subset of $W(H)$. Q.E.D.

References. [34], [177].

11.2. Sequences of Borel choice functions

First, we study the process of the Hahn-Banach theorem. Let X be real Banach space, E be a linear subspace of X, f be a linear functional on E with norm ≤ 1, and $x \in X \backslash E$. We want to extend f from E onto $(E\dot{+}[x])$ still with norm ≤ 1, i.e.,

$$|f(x+w)| \leq \|x+w\|, \quad \forall w \in E.$$

So we need to pick the value of $f(x)$ satisfying

$$-\|x+u\| - f(u) \leq f(x) \leq \|x+v\| - f(v), \quad \forall u, v \in E.$$

Then the value of $f(x)$ must satisfy the following inequality:

$$\begin{aligned} &\sup\{(-\|x+u\| - f(u))|u \in E\} \\ \leq\ & f(x) \leq \inf\{(\|x+v\| - f(v))|v \in E\}. \end{aligned}$$

Conversely, if the value of $f(x)$ satisfy the above inequality, then f is a linear functional on $E\dot{+}[x]$ still with norm ≤ 1.

Definition 11.2.1. Let X be a real Banach space, E be a linear subspace of X, and $x \in X$ (maybe $x \in E$) . For any linear functional f on E with norm ≤ 1 , define

$$L_E^{(x)}(f) = \sup\{(-\|x+u\| - f(u))|u \in E\}.$$

and

$$M_E^{(x)}(f) = \inf\{(\|x+v\| - f(v))|v \in E\}.$$

Since $\|f\| \leq 1$, it follows that $L_E^{(x)}(f) \leq M_E^{(x)}(f)$.

Lemma 11.2.2. Let X, E, x and f be as in Definiton 11.2.1.

1) If $x \in E$, then $L_E^{(x)}(f) = f(x) = M_E^{(x)}(f)$.

2) f can be extended to a linear functional on $E + [x]$ still with norm ≤ 1 if and only if the value of $f(x)$ must satisfy the inequality: $L_E^{(x)}(f) \leq f(x) \leq M_E^{(x)}(f)$.

Proof. 1) Suppose that $x \in E$. Then

$$|f(x+w)| \leq \|x+w\|, \quad \forall w \in E.$$

Further, we have

$$-\|x+u\| - f(u) \leq f(x) \leq \|x+v\| + f(v), \quad \forall u, v \in E.$$

Thus , $L_E^{(x)}(f) \leq f(x) \leq M_E^{(x)}(f)$. On the other hand, since $x \in E$, it follows that

$$L_E^{(x)}(f) \geq -f(-x) = f(x), \quad \text{and} \quad M_E^{(x)}(f) \leq -f(-x) = f(x).$$

Therefore $L_E^{(x)}(f) = f(x) = M_E^{(x)}(f)$.

2) It is obvious from 1) and the discussion of Hahn–Banch theorem. Q.E.D.

Lemma 11.2.3. Let X be a real Banach space, E be a linear subspace of X, and $x \in X$, and

$$S = \{f | f \text{ is a linear functional on } E, \text{and} \|f\| \leq 1\}.$$

Write $L_E^{(x)}(\cdot) = L(\cdot)$ and $M_E^{(x)}(\cdot) = M(\cdot)$ simply. Then $L(\cdot)$ is a convex function on S, and $L(\cdot) = -M(-\cdot)$ is continuous in the interior of S.

Proof. Let $\lambda \in [0,1]$, and $f, g \in S$. For any $u \in E$, we have

$$\begin{aligned} &-\|x+u\| - (\lambda f + (1-\lambda)g)(u) \\ &= \lambda(-\|x+u\| - f(u)) + (1-\lambda)(-\|x+u\| - g(u)) \\ &\leq \lambda L(f) + (1-\lambda)L(g). \end{aligned}$$

Thus, $L(\lambda f + (1-\lambda)g) \leq \lambda L(f) + (1-\lambda)L(g)$, i.e., $L(\cdot)$ is convex on S.

Now let $f_0 \in S$ and $\|f_0\| \leq 1-\eta$ for some $\eta \in (0,1)$. On $V = \{f \in S | \|f\| < \eta\}$, define

$$F(f) = L(f + f_0) - L(f_0), \quad \forall f \in V.$$

We need to show that $F(f)$ is continuous at $f = 0$. Clearly, $F(0) = 0, F(\cdot)$ is convex on V, and

$$F(f) \leq M(f + f_0) - L(f_0) \leq \|x\| - L(f_0), \quad \forall f \in V.$$

Put $\alpha = \|x\| - L(f_0)$. For any $\varepsilon \in (0,1)$ and $f \in S$ with $\|f\| < \eta\varepsilon$, since $f, \pm\varepsilon^{-1}f \in V$, it follows from the convexity of $F(\cdot)$ that

$$\begin{aligned} F(f) &= F((1-\varepsilon)\cdot 0 + \varepsilon \cdot \varepsilon^{-1} f) \\ &\leq \varepsilon F(\varepsilon^{-1} f) \leq \varepsilon\alpha \end{aligned}$$

and

$$\begin{aligned} 0 &= F((1+\varepsilon)^{-1} f + \varepsilon(1+\varepsilon)^{-1} \cdot (-\varepsilon^{-1} f)) \\ &\leq (1+\varepsilon)^{-1} F(f) + \varepsilon(1+\varepsilon)^{-1} F(-\varepsilon^{-1} f). \end{aligned}$$

From the second inequality, we get $F(f) \geq -\varepsilon F(-\varepsilon^{-1} f) \geq -\varepsilon\alpha$. Thus, $|F(f)| \leq \varepsilon\alpha, \forall f \in S$ with $\|f\| < \eta\varepsilon$, i.e., $F(\cdot)$ is continuous at 0. Q.E.D.

Theorem 11.2.4. Let X be a separable Banach space, and $W(X^*)$ be as in Theorem 11.1.7 (a standard Borel space). Then there is a sequence $\{f_n\}$ of Borel maps from $W(X^*)$ to $(X^*, \sigma(X^*, X))$ such that : for any $E^* \in W(X^*)$ and n, $f_n(E^*) \in (E^*)_1$ (i.e., $f_n(E^*) \in E^*$ and $\|f_n(E^*)\| \leq 1$); and $\{f_n(E^*)|n\}$ is w^*-dense in $(E^*)_1, \forall E^* \in W(X^*)$.

Proof. First, let X be real.

Suppose that $\{x_n|n = 1,2,\cdots\}$ is a dense subset of X, and fix $E^* \in W(X^*)$. Then $\{\widetilde{x_n} = x_n + E^*_\perp|n = 1,2,\cdots\}$ is dense in $X/E^*_\perp$, where $E^*_\perp = \{x \in X|f(x) = 0, \forall f \in E^*\}$. Put $B_0 = \{0\}, B_n = [\widetilde{x_1},\cdots,\widetilde{x_n}], \forall n$. These are finite dimensional linear subspace of $X/E^*_\perp$. Moreover, since $(X/E^*_\perp)^* \cong E^*$, we shall identify them in the following.

For each $t = (t_1,\cdots,t_n,\cdots)$, where $t_n \in [0,1], \forall n$, we say that there is a linear functional $f_t^{E^*}$ on $X/E^*_\perp$ such that $\|f_t^{E^*}\| \leq 1$ (i.e., $f_t^{E^*} \in (E^*)_1$) and

$$f_t^{E^*}(\widetilde{x_{n+1}}) = t_{n+1}L_n(f_t^{E^*}) + (1 - t_{n+1})M_n(f_t^{E^*}), \tag{1}$$

where $L_n(\cdot) = L_{B_n}^{(\widetilde{x_{n+1}})}(\cdot)$ and $M_n(\cdot) = M_{B_n}^{(\widetilde{x}_{n+1})}(\cdot), \forall n \geq 0$.

We prove this by induction. Assume that such $f_t^{E^*}$ exists on B_n. Put $\lambda = t_{n+1}L_n(f_t^{E^*}) + (1 - t_{n+1})M_n(f_t^{E^*})$. Since $L_n(f_t^{E^*}) \leq \lambda \leq M_n(f_t^{E^*})$, it follows from Lemma 11.2.2 that $f_t^{E^*}$ can be extended to a linear functional on B_{n+1} with norm ≤ 1 still and $f_t^{E^*}(\widetilde{x}_{n+1}) = \lambda$. Therefore, there exists a linear functional $f_t^{E^*}$ on $X/E^*_\perp$ with norm ≤ 1 and satisfying (1).

Define $Q = \{r = (r_1,\cdots,r_n,\cdots)|r_n$ is rational and $\in [0,1], \forall n$; and $^\#\{n|r_n \neq 0\} < \infty\}$, and fix $f \in E^*$ with $\|f\| < 1$.

For $n = 1$ and any $\varepsilon > 0$, since $f_r^{E^*}(\widetilde{x}_1) = (1-2r_1)\|\widetilde{x}_1\|$ and $|f(\widetilde{x}_1)| < \|\widetilde{x}_1\|$, there is a rational number $r_1^{(0)} \in [0,1]$ such that $|(f_r^{E^*} - f)(\widetilde{x}_1)| < \varepsilon, \forall r = (r_n) \in Q$ with $r_1 = r_1^{(0)}$.

For $n = 2$ and any $\varepsilon > 0$, by Lemma 11.2.3 there is $\eta > 0$ with following property : for any $g \in B_1^*$ with $|(g - f)(\widetilde{x}_1)| < \eta$ (thus $\|g - (f|B_1)\|$ is very small, and $\|g\| \leq 1$), we have

$$|L_1(f) - L_1(g)| < \varepsilon, \quad |M_1(f) - M_1(g)| < \varepsilon. \tag{2}$$

From the preceding paragraph, there is a rational number $r_1^{(0)} \in [0,1]$ such that

$$|(f_r^{E^*} - f)(\widetilde{x}_1)| < \eta, \quad \forall r = (r_n) \in Q \text{ and } r_1 = r_1^{(0)}. \tag{3}$$

By (2), we have

$$|L_1(f_r^{E^*}) - L_1(f)| < \varepsilon, \quad |M_1(f_r^{E^*}) - M_1(f)| < \varepsilon,$$

$\forall r = (r_n) \in Q$ with $r_1 = r_1^{(0)}$. Clearly , $\|(f|B_2)\| < 1$. By Lemma 11.2.2, we have $t_2 \in [0,1]$ such that

$$f(\widetilde{x}_2) = t_2L_1(f) + (1 - t_2)M_1(f). \tag{4}$$

Pick a rational number $r_2^{(0)} \in [0,1]$ satisfying

$$|[r_2^{(0)}L_1(f_r^{E^*}) + (1 - r_2^{(0)})M_1(f_r^{E^*})] - [t_2L_1(f) + (1 - t_2)M_1(f)]| < \varepsilon, \tag{5}$$

$\forall r = (r_n) \in Q$ with $r_1 = r_1^{(0)}$. We may assume $\eta \leq \varepsilon$. Now by (3), (1), (4), (5) , we get

$$|(f_r^{E^*} - f)(\tilde{x}_i)| < \varepsilon, \quad i = 1, 2,$$

$\forall r = (r_n) \in Q$ with $r_1 = r_1^{(0)}$ and $r_2 = r_2^{(0)}$.

Repeating this process, for any n and $\varepsilon > 0$ there exist rational numbers $r_1^{(0)}, \cdots, r_n^{(0)} \in [0,1]$ such that $|(f_r^{E^*} - f)(\tilde{x}_i)| < \varepsilon, 1 \leq i \leq n, \forall r = (r_k) \in Q$ with $r_k = r_k^{(0)}, 1 \leq k \leq n$.

Since above $f(\in E^*$ and $\|f\| < 1)$ is arbitrary, the set $\{f_r^{E^*} | r \in Q\}$ is w^*-dense in $(E^*)_1$.

Now for any $t = (t_1, \cdots, t_n, \cdots)$ with $t_n \in [0,1], \forall n$, we say $E^* \longrightarrow f_t^{E^*}$ is a Borel map from $W(X^*)$ to $(X^*, \sigma(X^*, X))$. It suffices to show that $E^* \longrightarrow f_t^{E^*}(\tilde{x}_n)$ is a Borel measurable function on $W(X^*), \forall n$.

For $n = 1, f_t^{E^*}(\tilde{x}_1) = (1 - 2t_1)\|\tilde{x}_1\|$ is measurable on $W(X^*)$ obviously. Now assume that $E^* \longrightarrow f_t^{E^*}(\tilde{x}_k)$ is measurable on $W(X^*), 1 \leq k \leq n$. Then $E^* \longrightarrow f_t^{E^*}(\tilde{u})$ is measurable on $W(X^*), \forall \tilde{u} \in B_n$. Moreover, by Theorem 11.1.7 $E^* \longrightarrow \|\tilde{x}_{n+1} + \tilde{u}\| = \|x_{n+1} + u + E_\perp^*\|$ is also measurable on $W(X^*), \forall \tilde{u} \in B_n$, where $u \in [x_1, \cdots, x_n]$ and $u + E_\perp^* = \tilde{u}$. Thus, $E^* \longrightarrow L_n(f_t^{E^*}) = \sup\{(-\|\tilde{x}_{n+1} + \tilde{u}\| - f_t^{E^*}(\tilde{u})) | \tilde{u} \in B_n\} = \sup\{(-\|\tilde{x}_{n+1} + \tilde{u}\| - f_t^{E^*}(\tilde{u})) | \tilde{u} = \sum_{i=1}^n r_i \tilde{x}_i,$ and r_i is rational , $1 \leq i \leq n\}$ is measurable on $W(X^*)$. Further, by (1) $E^* \longrightarrow f_t^{E^*}(\tilde{x}_{n+1})$ is measurable on $W(X^*)$.

Therefore, the theorem is proved for real case.

In the following, let X be complex. Clearly, X can be regarded as a real space, denoted by X_r. Then there is a sequence $\{f_n\}$ of Borel maps from $W(X_r^*)$ to $(X_r^*, \sigma(X_r^*, X_r))$ such that for any $E_r^* \in W(X_r^*)$ and $n, f_n(E_r^*) \in (E_r^*)_1$; and $\{f_n(E_r^*) | n\}$ is w^*-dense in $(E_r^*)_1, \forall E_r^* \in W(X_r^*)$. Now for any $E^* \in W(X^*)$ and n, define

$$g_n(E^*)(x) = f_n(\text{Re}E^*)(x) - i f_n(\text{Re}E^*)(ix), \quad \forall x \in X,$$

where $\text{Re}E^* = \{\text{Re } f | f \in E^*\} (\in W(X_r^*))$. Clearly, $g_n(\cdot)$ is a Borel map from $W(X^*)$ to $(X^*, \sigma(X^*, X))$, and $\|g_n(E^*)\| \leq 1, \forall E^* \in W(X^*)$. Let $x \in E_\perp^*$. Since $ix \in E_\perp^*$ and $E_\perp^* = (\text{Re}E^*)_\perp$, it follows that $g_n(E^*)(x) = 0$. Thus, $g_n(E^*) \in (E^*)_1, \forall E^* \in W(X^*)$. Moreover, fix $E^* \in W(X^*)$. For any $g \in (E^*)_1, y_1, \cdots, y_m \in X$, and $\varepsilon > 0$, since $\text{Re}g \in (\text{Re}E^*)_1$, we can find n such that

$$|(f_n(\text{Re } E^*) - \text{Re } g)(y_j)| < \varepsilon$$

and

$$|(f_n(\text{Re } E^*) - \text{Re } g)(iy_j)| < \varepsilon, 1 \leq j \leq m.$$

Then $|(g_n(E^*) - g)(y_j)| < \varepsilon, 1 \leq j \leq m$. Therefore, $\{g_n(E^*) | n\}$ is w^*-dense in $(E^*)_1, \forall E^* \in W(X^*)$. Q.E.D.

Theorem 11.2.5. Let X be a separable Banach space, and $(E, \mathcal{B})$ be a Borel space. Then a map $\psi : (E, \mathcal{B}) \longrightarrow W(X^*)$ is Borel if and only if there is a sequence $\{g_n\}$ of Borel maps from $(E, \mathcal{B})$ to $(X^*, \sigma(X^*, X))$ such that for each $t \in E, g_n(t) \in (\psi(t))_1, \forall n$, and $\{g_n(t)|n\}$ is a w^*-dense subset of $(\psi(t))_1$, where $(\psi(t))_1$ is the closed unit ball of $\psi(t)(\in W(X^*))$.

Proof. Suppose that $\{f_n\}$ is as in Theorem 11.2.4. If ψ is Borel, then $\{g_n = f_n \circ \psi\}$ satisfies our conditions . Conversely, if $\{g_n\}$ satisfies the conditions, then for any $x \in X, t \in E$, we have

$$\|x + \psi(t)_\perp\| = \sup_n |g_n(t)(x)|.$$

Thus , $t \longrightarrow \|x + \psi(t)_\perp\|$ is measurable on $(E, \mathcal{B}), \forall x \in X$. Now by Theorem 11.1.7, ψ is Borel . Q.E.D.

References. [35], [177].

11.3. The Borel spaces of Von Neumann algebras

Let H be a (complex) separable Hilbert space. Then $X = T(H)$ is a separable Banach space, and $X^* = B(H)$. For any $E \in W(X^*)$, let

$$E^* = \{a^* | a \in E\}, \quad E' = \{b \in B(H) | ab = ba, \forall a \in E\}.$$

Proposition 11.3.1. $E \longrightarrow E^*$ and $E \longrightarrow E'$ are Borel maps on $W(X^*)$, where the standard Borel structure of $W(X^*)$ is as in Theorem 10.1.7.

Proof. Let $\Phi(E) = E^*, \forall E \in W(X^*)$. Since $(E^*)_\perp = (E_\perp)^*$, it follows from Theorem 11.1.7 that

$$\begin{aligned} &\Phi^{-1}\{E \in W(X^*) | \|t + E_\perp\|_1 < \lambda\} \\ = \; &\{E \in W(X^*) | \|t^* + E_\perp\|_1 < \lambda\} \end{aligned}$$

is a Borel subset of $W(X^*), \forall t \in X, \lambda > 0$, where $\|\cdot\|_1$ is the trace norm of $X = T(H)$. Therefore, $E \longrightarrow E^*$ is a Borel map on $W(X^*)$.

By Theorem 11.2.4, there is a sequence $\{a_n(\cdot)\}$ of Borel maps from $W(X^*)$ to $(X^*, \sigma(X^*, X))$ such that for any $E \in W(X^*), \{a_n(E)|n\}$ is a w^*-dense subset of $(E)_1$. Then

$$E' = \{b \in X^* | \; ba_n(E) = a_n(E)b, \; \forall n\}, \forall E \in W(X^*).$$

Define

$$M = \{(x_n) | x_n \in B(H), \forall n, \text{ and } \sup_n \|x_n\| < \infty\}$$

and

$$M_* = \{(t_n) | t_n \in T(H), \forall n, \text{and } \sum_n \|t_n\|_1 < \infty\}.$$

Clearly, $M = \sum_n \oplus B(H)$ is a W^*-algebra, and M_* is the predual of M.

For any $E \in W(X^*)$, define a map $T^E : B(H) \longrightarrow M$ as follows:

$$T^E(b) = (ba_n(E) - a_n(E)b), \quad \forall b \in B(H).$$

Then $E' = \text{Ker } T^E = \{b \in B(H) | T^E(b) = 0\}$, and T^E is σ-σ continuous. Further, define a map $T_*^E : M_* \longrightarrow T(H)$ as follows

$$\begin{aligned} T_*^E((t_n))(b) &= T^E(b)((t_n)) \\ &= \sum_n tr((ba_n(E) - a_n(E)b)t_n) \end{aligned}$$

$\forall b \in B(H), (t_n) \in M_*$. Since $(T_*^E)^* = T^E$, it follows that $(E')_\perp = (\text{ Ker } T^E)_\perp = \overline{T_*^E M_*}$.

Let S be the unit ball of $B(H)$, $\{b_j\}$ be a countable dense subset of (S, σ), and $\{(t_n^{(j)})\}$ be a countable dense subset of M_*. Then for any $t \in X, E \in W(X^*)$, we have

$$\|t + (E')_\perp\|_1 = \inf_j \|t + T_*^E((t_n^{(j)}))\|_1.$$

But $\|t + T_*^E((t_n^{(j)}))\|_1 = \sup_i |tr(tb_i) + \sum_n tr((b_i a_n(E) - a_n(E)b_i)t_n^{(j)})|$, and $a_n(\cdot) : W(X^*) \longrightarrow (B(H), \sigma)$ is Borel, so $E \longrightarrow \|t + (E')_\perp\|$ is a Borel measurable function on $W(X^*)$. Therefore, $E \longrightarrow E'$ is a Borel map on $W(X^*)$. Q.E.D.

Theorem 11.3.2. Let H be a separable Hilbert space, $X = T(H)$, and $\mathcal{A}$ be the collection of all VN algebras on H. Then $\mathcal{A}$ is a Borel subset of $W(X^*)$. Consequently, the family of following subsets

$$\{M \in \mathcal{A} | \|t + M_\perp\|_1 < \lambda\}, \forall t \in X, \lambda > 0$$

will generate a standard Borel structure of $\mathcal{A}$.

Proof. By Proposition 11.3.1 and 10.3.4, $\{E \in W(X^*) | E = E^*\}$ and $\{E \in W(X^*) | E = E''\}$ are Borel subsets of $W(X^*)$. Then $\mathcal{A} = \{E \in W(X^*) | E = E^*\} \cap \{E \in W(X^*) | E = E''\}$ is also a Borel subset of $W(X^*)$. Q.E.D.

Proposition 11.3.3. Let H be a separable Hilbert space, S be the unit ball of $B(H)$, and $\mathcal{A}$ be the collection of all VN algebras on H. Then there is a sequence $\{a_n(\cdot)\}$ of Borel maps from $\mathcal{A}$ to (S,σ) such that for any $M \in \mathcal{A}, \{a_n(M)|n\}$ is a $\tau(M,M_*)$-dense subset of $(M)_1$.

Proof. By Theorem 11.2.4 and 11.3.2, there is a sequence $\{b_n(\cdot)\}$ of Borel maps from $\mathcal{A}$ to (S,σ) such that for each $M \in \mathcal{A}, \{b_n(M)|n\}$ is a weakly dense subset of $(M)_1$. Let

$$\{a_n(\cdot)|n\} = \left\{ \sum_k \lambda_k b_k(\cdot) \,\middle|\, \begin{array}{c} \lambda_k \text{ is non-negative and rational,} \\ \forall k, \text{and} \sum_k \lambda_k = 1 \end{array} \right\}$$

Now by Proposition 1.2.8, $\{a_n(\cdot)\}$ satisfies our conditions. Q.E.D.

Theorem 11.3.4. Let $(E,\mathcal{B})$ be a Borel space, and $\mathcal{A}$ be the collection of all VN algebras on a separable Hilbert space H. Then a map $\psi : E \longrightarrow \mathcal{A}$ is Borel if and only if there is a sequence $\{a_n(\cdot)\}$ of Borel maps from E to $(B(H),\sigma(B(H),T(H)))$ such that for each $t \in E$, the VN algebra $\psi(t)$ is generated by $\{a_n(t)|n\}$.

Proof. The necessity is obvious from Theorem 11.2.3. Now let such $\{a_n(\cdot)\}$ exist. By Theorem 1.6.1 and a proper treatment, we have a sequence $\{b_n(\cdot)\}$ of Borel maps from E to $(B(H),\sigma)$ such that for each $t \in E, \{b_n(t)|n\}$ is a σ-dense subset of $(\psi(t))_1$. Then by Theorem 11.2.3, $\psi : E \longrightarrow \mathcal{A}$ is Borel .
Q.E.D.

Proposition 11.3.5. $(M,N) \longrightarrow M \cap N$ and $(M,N) \longrightarrow (M \cup N)''$ are Borel maps from $\mathcal{A} \times \mathcal{A}$ to $\mathcal{A}$.

Proof. Let $\{a_n(\cdot)\}_{n\geq 1}$ be a sequence of Borel maps from $\mathcal{A}$ to $(B(H),\sigma)$ as in Proposition 11.3.3. For any $M,N \in \mathcal{A}$, let

$$b_{2n}(M,N) = a_n(M), \quad b_{2n-1}(M,N) = a_n(N), \forall n = 1,2,\cdots.$$

Then $b_n(\cdot,\cdot)$ is a Borel map from $\mathcal{A} \times \mathcal{A}$ to $(B(H),\sigma), \forall n$, and $\{b_n(M,N)|n\} = \{a_m(M), a_n(N)|m,n\}, \forall M,N \in \mathcal{A}$. Thus , $(M \cup N)''$ is generated by $\{b_n(M,N)|n\}$, $\forall M,N \in \mathcal{A}$. Now by Proposition 11.3.4, $(M,N) \longrightarrow (M \cup N)''$ is Borel from $\mathcal{A} \times \mathcal{A}$ to $\mathcal{A}$.

In the process of $(M,N) \longrightarrow (M',N') \longrightarrow (M' \cup N')'' \longrightarrow (M' \cup N')''' = M \cap N$, each map is Borel . Therefore, $(M,N) \longrightarrow (M \cap N)$ is also Borel from $\mathcal{A} \times \mathcal{A}$ to $\mathcal{A}$. Q.E.D.

Theorem 11.3.6. Let H be a separable Hilbert space, $\mathcal{A}$ be the collection of all VN algebras on H, and $\mathcal{F}$ be the collection of all factors on H. Then $\mathcal{F}$ is a Borel subset of $\mathcal{A}$. Consequently , the family of following subsets

$$\{M \in \mathcal{F} | \|t + M_\perp\|_1 < \lambda\}, \quad \forall t \in T(H), \lambda > 0$$

will generate a standard Borel structure of $\mathcal{F}$.

Proof. Since $M \longrightarrow (M, M') \longrightarrow M \cap M'$ is a Borel map from $\mathcal{A}$ to $\mathcal{A}$, $\mathcal{F} = \{M \in \mathcal{A} | M \cap M' = \mathbb{C}1_H\}$ is a Borel subset of $\mathcal{A}$. Q.E.D.

Notes. The Borel spaces of Von Neumann algebras was introduced by E.G. Effros.

References. [34], [35], [177].

11.4. Borel subsets of factorial Borel space

Let H be a separable Hilbert space, $\mathcal{A}$ be the standard Borel space of all VN algebras on H , and $\mathcal{F}$ be the standard Borel space of all factors on H.

Lemma 11.4.1. Let G be the set of all unitary operators on H. Then G is a Polish topological group with respect to strong (operator) topology.

Proof. Clearly, G is a topological group with respect to strong topology. Let S be the unit ball of $B(H)$. Then S is a Polish space with respect to strong topology. If $\{\xi_k\}$ is a dense subset of $\{\xi \in H | \|\xi\| = 1\}$, then $u \in G$ if and only if $\|u\xi_k\| = \|u^*\xi_k\| = 1, \forall k$. Thus

$$G = \cap_k \{u \in S | \|u\xi_k\| = 1\} \cap$$

$$\cap_{k,n} \sqcup_m \{u \in S \mid |\langle \xi_k, u\xi_m \rangle| > 1 - \frac{1}{n}\}$$

is a G_δ -subset of $(S, s\text{-top})$. That comes to the conclusion. Q.E.D.

Proposition 11.4.2. For any $M \in \mathcal{A}$, let $s(M) = \{uMu^* | u \in G\}$, where G is as in Lemma 10.4.1. Then $s(M)$ is a Borel subset of $\mathcal{A}, \forall M \in \mathcal{A}$.

Proof. Fix $M \in \mathcal{A}$, and put $G_0 = \{u \in G | uMu^* = M\}$. Define an equivalent relation $\sim$ in G : $u \sim v$ if $v \in uG_0$. By Lemma 11.4.1 and Theorem 10.4.2, there is a Borel subset E of G such that $^\#(E \cap uG_0) = 1, \forall u \in G$. Then , $s(M) = \{uMu^* | u \in E\}$.

We say that $u \longrightarrow uMu^*$ is a Borel map from G to $\mathcal{A}$. In fact, if $\{a_n\}$ is a countable dense subset of $((M)_1, \sigma(M, M_*))$, then $\{a_n(u) | n\}$ generates $uMu^*, \forall u \in G$, where $a_n(u) = ua_nu^* (\forall u \in G)$ is a continuous map from G to $(B(H), \sigma), \forall n$. Now by Proposition 11.3.4, $u \longrightarrow uMu^*$ is a Borel map from G to $\mathcal{A}$.

In particular, $u \longrightarrow uMu^*$ is an injective Borel map from E to $\mathcal{A}$. Therefore, by Theorem 10.3.12 $s(M)$ is a Borel subset of $\mathcal{A}$. Q.E.D.

Proposition 11.4.3. Let $M \in \mathcal{A}$. Then $a(M) = \{N \in \mathcal{A} | N$ is $*$ isomorphic to M $\}$ is a Borel subset of $\mathcal{A}$.

Proof. $\mathcal{A} = \mathcal{A}(H)$ is the collection of all VN algebras on H. We shall denote the collection of all VN algebras on $H \otimes H$ by $\mathcal{A}(H \otimes H)$. Define a map $\Phi : \mathcal{A}(H) \longrightarrow \mathcal{A}(H \otimes H)$ as follows

$$\Phi(M) = M \overline{\otimes} \mathbb{C} 1_H, \quad \forall M \in \mathcal{A}(H).$$

Let $M, N \in \mathcal{A}(H)$. Then M and N are $*$ isomorphic if and only if $\Phi(M)$ and $\Phi(N)$ are $*$ isomrorphic. Notice that $\Phi(M)' = M' \overline{\otimes} B(H)$ and $\Phi(N)' = N' \overline{\otimes} B(H)$ are properly infinite (here dim $H = \infty$; if dim $H < \infty$, then $a(M) = s(M)$ is a Borel subset of $\mathcal{A}$) . By Proposition 6.6.7, $\Phi(M)$ and $\Phi(N)$ are spatially $*$ isomorphic if $\Phi(M)$ and $\Phi(N)$ are $*$ isomorphic. Thus, $a(M) = \Phi^{-1}(s(\Phi(M)))$. From Proposition 11.3.4, it suffices to show that Φ is a Borel map.

By Proposition 11.3.3. there is a sequence $\{a_n(\cdot)\}$ of Borel maps from $\mathcal{A}$ to (S, σ), where S is the unit ball of $B(H)$, such that $\{a_n(N) | n\}$ generates $N, \forall N \in \mathcal{A}$. Then $\{a_n(\cdot) \otimes 1_H | n\}$ is a sequence of Borel maps from $\mathcal{A}$ to $(B(H \otimes H), \sigma)$, and $\{a_n(N) \otimes 1_H | n\}$ generates $\Phi(N), \forall N \in \mathcal{A}$. Now by Proposition 11.3.4, Φ is a Borel map. Q.E.D.

Proposition 11.4.4. Denote the collection of all type (I_n) factors on H by $\mathcal{F}_{\mathrm{I}_n}, \forall n$. Then $\mathcal{F}_{\mathrm{I}_n}$ is a Borel subset of $\mathcal{F}, n = \infty, 1, 2, \cdots$.

Proof. Noticing that all type (I_n) factors are $*$ isomorhpic, the conclusion is obvious from Proposition 11.4.3. Q.E.D.

Lemma 11.4.5. Denote the projection from H onto $\overline{[M'pH]}$ by $e_M(p)$. Then $(M, p) \longrightarrow e_M(p)$ is a Borel map from $\mathcal{A} \times P$ to P, where P is the collection of all projections on H, and with strong (operator) topology P is a Polish space.

Proof. Clearly, from Theorem 11.1.7 the standard Borel spaces P and $W(H)$ are Borel isomorphic. So we need to show that $(M, p) \longrightarrow \overline{[M'pH]}$ is a Borel map from $\mathcal{A} \times P$ to $W(H)$. By Theorem 11.2.5, it suffices to find a sequence $\{\eta_n(\cdot, \cdot)\}$ of Borel maps from $\mathcal{A} \times P$ to (H, w), where "w" means the weak topology in H, such that $\{\eta_n(M, p) | n\}$ is a dense subset of $\overline{[M'pH]}, \forall M \in \mathcal{A}, p \in P$.

By Proposition 11.3.3, there is a sequence $\{a_n(\cdot)\}$ of Borel maps from $\mathcal{A}$ to (S, σ) such that $\{a_n(M) | n\}$ is a $\tau(M, M_*)$ -dense subset of $(M)_1, \forall M \in \mathcal{A}$. Let

$\{\xi_k\}$ be a countable dense subset of H, and define $\varsigma_{n,k}(M,p) = a_n(M')p\xi_k, \forall$ n, k. Clearly, $\varsigma_{n,k}(\cdot,\cdot)$ is Borel from $\mathcal{A} \times P$ to $(H,w), \forall n, k$, and

$$\overline{[\varsigma_{n,k}(M,p)|n,k]} = \overline{[M'pH]}, \forall M \in \mathcal{A}, p \in P.$$

Therefore, we can find Borel maps $\eta_n : \mathcal{A} \times P \longrightarrow (H,w), \forall n$, such that $\{\eta_n(M,p)|n\}$ is dense in $\overline{[M'pH]}, \forall M \in \mathcal{A}, p \in P$. Q.E.D.

Lemma 11.4.6. Denote the collection of all infinite factors on H by $\mathcal{F}_{if}$. Then $\mathcal{F}_{if}$ is a Sousline subset of $\mathcal{F}$.

Proof. A factor M is infinite if and only if there is $v \in M$ such that $v^*v = 1$ and $vv^* \neq 1$. By Proposition 11.3.3, there are Borel maps $a_n(\cdot) : \mathcal{A} \longrightarrow (S,\sigma)(n = 1,2,\cdots)$ such that $\{a_n(N)|n\}$ is τ-dense in $(N)_1, \forall N \in \mathcal{A}$. Notice that

$$\begin{aligned} E = & \left\{(M,v)\left| \begin{array}{l} M \in \mathcal{F}, v \in S; v^*v = 1, vv^* \neq 1; \\ \text{and } a_n(M')v = va_n(M'), \forall n \end{array}\right.\right\} \\ = & (\mathcal{F} \times \{v \in S | v^*v = 1, vv^* \neq 1\}) \cap \\ & \cap_n \{(M,v)|a_n(M')v = va_n(M')\} \\ = & (\mathcal{F} \times \{v \in S | v^*v = 1, vv^* \neq 1\}) \cap \\ & \cap_{n,i,j} \{(M,v)|\langle (a_n(M')v - va_n(M'))\xi_i, \xi_j\rangle = 0\} \end{aligned}$$

is a Borel subset of $\mathcal{F} \times (S,\tau)$, where $\{\xi_j\}$ is a countable dense subset of H. If π is the projection from $\mathcal{F} \times S$ onto $\mathcal{F}$, then $\mathcal{F}_{if} = \pi E$. Therefore, $\mathcal{F}_{if}$ is a Sousline subset of $\mathcal{F}$. Q.E.D.

Proposition 11.4.7. Denote the collection of all type (II_1) factors on H by $\mathcal{F}_{\mathrm{II}_1}$. Then $\mathcal{F}_{\mathrm{II}_1}$ is a Borel subset of $\mathcal{F}$.

Proof. Denote the collection of all finite factors on H by $\mathcal{F}_f$. From Proposition 11.4.4 and Lemma 11.4.6, it suffices to prove that $\mathcal{F}_f$ is a Sousline subset of $\mathcal{F}$.

Since H is separable, a factor M on H is finite if and only if there exists a faithful normal tracial state on M, i.e. there is a sequence $\{\xi_k\}$ of H such that $\sum_k \|\xi_k\|^2 < \infty, \sum_k \langle (ab - ba)\xi_k, \xi_k\rangle = 0, \forall a, b \in M$, and $[a'\xi_k|a' \in M', k]$ is dense in H. Then by Lemma 11.4.5,

$$E = \left\{ (M,(\xi_k)) \middle| \begin{array}{l} M \in \mathcal{F}, (\xi_k) \in H_\infty; \overline{[a'\xi_k | a' \in M', k]} = H; \\ \text{and} \sum_k \langle (ab-ba)\xi_k, \xi_k \rangle = 0, \forall a,b \in M \end{array} \right\}$$

$$= \left\{ (M,(\xi_k)) \middle| \begin{array}{c} M \in \mathcal{F}, (\xi_k) \in H_\infty, \text{and } e_M(p) = 1, \\ \text{where p is the projection from H onto } \overline{[\xi_k | k]} \end{array} \right\} \cap$$

$$\cap_{n,m} \left\{ (M,(\xi_k)) \middle| \begin{array}{c} M \in \mathcal{F}, (\xi_k) \in H_\infty, \text{and} \\ \sum_k \langle (a_n(M)a_m(M) - a_m(M)a_n(M))\xi_k, \xi_k \rangle = 0 \end{array} \right\}$$

is a Borel subset of $\mathcal{F} \times H_\infty$, where $H_\infty = \sum_{n=1}^{\infty} \oplus H; a(\cdot) : \mathcal{A} \longrightarrow (S,\sigma)$ is Borel , $\forall n$, and $\{a_n(M)|n\}$ is τ–dense in $(M)_1, \forall M \in \mathcal{A}$(see Proposition 11.3.3) . Let π be the projection from $\mathcal{F} \times H_\infty$ onto $\mathcal{F}$. Then $\mathcal{F}_f = \pi E$ is a Sousline subset of $\mathcal{F}$. Q.E.D.

Lemma 11.4.8. Denote the collection of all semi-finite factors on H by $\mathcal{F}_{sf}$. Then $\mathcal{F}_{sf}$ is a Sousline subset of $\mathcal{F}$.

Proof. A factor M on H is semi–finite if and only if there exists a finite projection p of M with $c(p) = 1$, i.e., there is a projection p of M with $\overline{[MpH]} = H$, and there exists a sequence $\{\xi_k\}$ of pH with $\sum_k \|\xi_k\|^2 < \infty$ such that $\sum_k \langle \cdot \xi_k, \xi_k \rangle$ is a faithful trace on (pMp).

Consider a subset E of $\mathcal{F} \times P \times H_\infty$. $(M,p,(\xi_k)) \in E$, if $p\xi_k = \xi_k, \forall k; pa_n(M') = a_n(M')p, \forall n; \overline{[a'\xi_k | k, a' \in M']} = pH; \overline{[MpH]} = H$, and

$$\sum_k \langle (pa_n(M)pa_m(M)p - pa_m(M)pa_n(M)p)\xi_k, \xi_k \rangle = 0,$$

$\forall n,m$ where $a_n(\cdot) : \mathcal{A} \longrightarrow (S,\sigma)$ is Borel , $\forall n$, and $\{a_n(M)|n\}$ is τ–dense in $(M)_1, \forall M \in \mathcal{A}$ (see Proposition 11.3.3). By Lemma 11.4.5, the map $(M,p) \longrightarrow \overline{[MpH]}$ is Borel. Thus , E is Borel . Let π be the projection from $\mathcal{F} \times P \times H_\infty$ onto $\mathcal{F}$. Then $\mathcal{F}_{sf} = \pi E$ is a Sousline subset of $\mathcal{F}$. Q.E.D.

Lemma 11.4.9. Let M be a factor on H, and Φ be a $*$ automorphism of M. If there is a non–zero element a of M such that $\Phi(b)a = ab, \forall b \in M$, then Φ is inner, i.e., there exists a unitary element u of M such that $\Phi(b) = ubu^*, \forall b \in M$.

Proof. From $\Phi(b)a = ab$, we have $a^*\Phi(b^*) = b^*a^*, \forall b \in M$. In particular, if

b is unitary , then we get

$$b^*(a^*a)b = a^*\Phi(b^*)\cdot\Phi(b)a = a^*a,$$

$$\Phi(b)aa^*\Phi(b^*) = ab\cdot b^*a^* = aa^*.$$

Thus, a^*a and $aa^* \in M\cap M' = \mathbb{C}1_H$. Now let $u = \|a\|^{-1}a$. Then u is a unitary element of M, and $\Phi(b) = ubu^*, \forall b \in M$. Q.E.D.

Lemma 11.4.10. Let G be the group of all unitary operators on H. Then

$$E = \left\{(M,u)\,\middle|\, \begin{array}{c} M\in\mathcal{F}, u\in G; uMu^* = M, \\ \text{but } \cdot \longrightarrow u\cdot u^* \text{is not inner for M} \end{array}\right\}$$

is a Borel subset of $\mathcal{F}\times G$.

Proof. By Proposition 11.3.3, we have Borel maps $a_n(\cdot) : \mathcal{A} \longrightarrow (S,\sigma), n = 1,2,\cdots$, such that $\{a_n(M)|n\}$ is τ-dense in $(M)_1, \forall M \in \mathcal{A}$. Since $(M,u) \longrightarrow ua_n(M)u^*$ is a Borel map from $\mathcal{F}\times G$ to $(S,\sigma), \forall n$, it follows that

$$\begin{aligned} E &= \{(M,u)|M\in\mathcal{F}, u\in G, \text{and } uMu^* = M\} \\ &= \cap_{n,m}\left\{(M,u)\,\middle|\, \begin{array}{c} M\in\mathcal{F}, u\in G, \text{and} \\ ua_n(M)u^*\cdot a_m(M') = a_m(M')\cdot ua_n(M)u^* \end{array}\right\} \end{aligned}$$

is a Borel subset of $\mathcal{F}\times G$. Let d be a proper metric on (S,σ) (see Definition 10.1.1) , and consider a subset $E(j,k,m,n)$ of $\mathcal{F}\times G$. $(M,u) \in E(j,k,m,n)$, if $uMu^* = M$ and satisfies one of following conditions:

1) $d(a_j(M),0) < n^{-1}$;
2) $d(ua_k(M'),0) < n^{-1}$;
3) $d(a_j(M),0) \geq n^{-1}, d(ua_k(M'),0) \geq n^{-1}$, and $d(a_j(M), ua_k(M')) \geq m^{-1}$.

Noticing that $(M,u) \longrightarrow (a_j(M), ua_k(M'))$ is a Borel map from $\mathcal{F}\times G$ to $(S,\sigma)\times(S,\sigma)$, so $E(j,k,m,n)$ is a Borel subset of $\mathcal{F}\times G$. Now it suffices to prove that $E = \cap_n \cup_m \cap_{j,k} E(j,k,m,n)$.

If $\cdot \longrightarrow u\cdot u^*$ is an inner $*$ automorphism of M, i.e., there exists a unitary element v of M such that $uau^* = vav^*, \forall a \in M$, then $v \in uM'$. If $d(v,0) \geq 2n^{-1}$, for any m we can choose j,k such that

$$d(a_j(M),v) < (2mn)^{-1}, d(ua_k(M'),v) < (2mn)^{-1}.$$

Thus $(M,u) \notin E(j,k,m,n)$. Further, $(M,u) \notin \cap_n \cup_m \cap_{j,k} E(j,k,m,n)$.

Conversely, let $(M,u) \in \mathcal{F}\times G, uMu^* = M$, and $(M,u) \notin \cap_n \sqcup_m \cap_{j,k} E(j,k, m,n)$. Then there is n such that for any m, we have $j(m), k(m)$ and $(M,u) \notin E(j(m),k(m),m,n)$. Thus, $d(a_{j(m)}(M),0) \geq n^{-1}, d(ua_{k(m)}(M'),0) \geq n^{-1}$, and $d(a_{j(m)}(M), ua_{k(m)}(M')) < m^{-1}, \forall m$. Since the unit balls of M and M' are σ-compact, there exist σ-cluster points a, a' of $\{a_{j(m)}(M)|m\}, \{ua_{k(m)}(M')|m\}$ respectively. Hence, we get

$$a = ua', \quad d(a,0) \geq n^{-1}, d(ua',0) \geq n^{-1}.$$

Now for any $b \in M$, we have $ubu^*a = uba' = ua'b = ab$. Thus , by Lemma 11.4.9 $\cdot \longrightarrow u \cdot u^*$ is an inner $*$ automorphism of M, i.e., $(M, u) \notin E$.

Therefore, $E = \cap_n \cup_m \cap_{j,k} E(j,k,m,n)$. Q.E.D.

Lemma 11.4.11. Let G be the group of all unitary operators on H (G is a Polish topological group with respect to strong operator topology, see Lemma 11.4.1) , and $G_0 = \{u \in G | 1$ is not an eigenvalue of $u\}$. Then

$$G_0 = \cap_{n,k} \sqcup_m \cap_{1 \le j \le n} \{u \in G | \|f_m(u)\xi_j\| < k^{-1}\}$$

is a Borel subset of G, where $\{\xi_j\}$ is a countable dense subset of the unit ball of H, $\{f_m\}$ is a sequence of continuous functions on $\{z \in \mathbb{C} | |z| = 1\}$ such that : 1) $0 \le f_m \le 1$; 2) if $|1 - z| \le 2^{-m}$, then $f_m(z) = 1$; 3) if $|1 - z| \ge 2^{-m+1}$, then $f_m(z) = 0, \forall m$.

Proof. Let $u \in G$, and $\xi \in H$ with $\|\xi\| = 1$, and $u\xi = \xi$. Then $f_m(u)\xi = \xi, \forall m$. Thus , we have

$$\begin{aligned} |1 - \|f_m(u)\xi_j\|| &= |\|f_m(u)\xi\| - \|f_m(u)\xi_j\|| \\ &\le \|\xi - \xi_j\|, \forall j. \end{aligned}$$

Pick j_0 such that $\|\xi_{j_0} - \xi\| < 1/4$. Then $\|f_m(u)\xi_{j_0}\| \ge 3/4, \forall m$. Hence, if $n \ge j_0$ and $k \ge 2$, then

$$u \notin \{v \in G | \|f_m(v)\xi_{j_0}\| < k^{-1}\}, \forall m.$$

Conversely, let $u \in G_0$, and $e(\cdot)$ be the spectral measure of u on $\{z \in \mathbb{C} \mid |z| = 1\}$. Put

$$p_m = e(\{z \mid |z| = 1, |1 - z| \le 2^{-m}\}), \quad \forall m.$$

Then $p_{m+1} \le p_m, 0 \le f_{m+1}(u) \le p_m \le f_m(u), \forall m$, and $p_m \longrightarrow 0$ (strongly). For any n, k, choose m sufficiently large such that $\sup\{\|p_m\xi_j\| | 1 \le j \le n\} < k^{-1}$. Then we have

$$\|f_{m+1}(u)\xi_j\| = \|f_{m+1}(u)p_m\xi_j\| \le \|p_m\xi_j\| < k^{-1}, 1 \le j \le n.$$

Q.E.D.

Lemma 11.4.12. Let X, Z be two Polish spaces, Y be a Borel space, and f be a map from $X \times Y$ to Z such that: 1) for each $y \in Y, f(\cdot, y)$ is continuous from X to Z ; 2) for each $x \in X, f(x, \cdot)$ is Borel from Y to Z. Then f is Borel.

Proof. Let d, δ be proper metrics on X, Z respectively, and $\{x_k\}$ be a countable dense subset of X. If F is a closed subset of Z, then

$$\begin{aligned} f^{-1}(F) &= \{(x,y) \in X \times Y | f(x,y) \in F\} \\ &= \cap_n \cup_k \left\{ (x,y) \,\middle|\, \begin{matrix} d(x,x_k) < n^{-1}, \text{and} \\ \delta(f(x_k,y),F) < n^{-1} \end{matrix} \right\} \\ &= \cap_n \cup_k ((X_{k,n} \times Y) \cap (X \times Y_{k,n})) \end{aligned}$$

is a Borel subset of $X \times Y$, where $X_{k,n} = \{x \in X | d(x,x_k) < n^{-1}\}$, and $Y_{k,n} = f(x_k,\cdot)^{-1}(\{z \in Z | \delta(z,F) < n^{-1}\})$. Now by Proposition 10.3.2, f is Borel.

Q.E.D.

Lemma 11.4.13. Let G be the group of all unitary operators on H, $G_0 = \{u \in G | 1 \text{ is not an eigenvalue of } u\}$, and $f(t,z) = \exp(-t(z+1)(z-1)^{-1})), \forall t \in \mathbb{R}, z \in \mathbb{C}$ with $|z| = 1$ and $z \neq 1$. Then f is a Borel map from $\mathbb{R} \times G_0$ to G , where the Borel structures of G and G_0 are generated by strong operator topology.

Proof. Fix $u \in G_0$. Clearly , $t \longrightarrow f(t,u)$ is a continuous map from $\mathbb{R}$ to G.

Now fix $t \in \mathbb{R}$, and let $\Gamma = \{z \in \mathbb{C} || z| = 1\}$. Then $g(z) = it(z+1)(z-1)^{-1}$ is a real valued continuous function on $(\Gamma \backslash \{1\})$. Pick a sequence $\{g_n\}$ of real valued continuous functions on Γ such that $g_n(z) \longrightarrow g(z), \forall z \in (\Gamma \backslash \{1\})$. Then , $\exp(ig_n(u)) \longrightarrow \exp(ig(u)) = f(t,u)$ (strongly) , $\forall u \in G_0$. Let δ be a proper metric on the Polish space G, and F be a closed subset of G. Since $\delta(\exp(ig_n(u)), f(t,u)) \longrightarrow 0, \forall u \in G_0$, it follows that

$$\begin{aligned} f(t,\cdot)^{-1}(F) &= \{u \in G_0 | f(t,u) \in F\} \\ &= \cap_k \cup_{n \geq k} \{u \in G_0 | \delta(\exp(ig_n(u)), F) < k^{-1}\}. \end{aligned}$$

But $\exp(ig_n(\cdot))$ is continuous on $G, \forall n$, and G_0 is a Borel subset of G (Lemma 11.4.11) , thus $f(t,\cdot)^{-1}(F)$ is a Borel subset of G_0, and $f(t,\cdot)$ is a Borel map from G_0 to G.

Now by Lemma 11.4.12, f is a Borel map from $\mathbb{R} \times G_0$ to G. Q.E.D.

Lemma 11.4.14. Let E be a Sousline subset of $\mathcal{F}$. Then $s(E) = \{uMu^* | M \in E, u \in G\}$ is also a Sousline subset of $\mathcal{F}$, where G is the goup of all unitary operators on H.

Proof. By Proposition 11.3.3, we have Borel maps $a_n(\cdot) : \mathcal{F} \longrightarrow (S,\sigma), n = 1,2,\cdots$, such that $\{a_n(M)|n\}$ is τ -dense in $(M)_1, \forall M \in \mathcal{F}$. Let $b_n(M,u) = ua_n(M)u^*, \forall n$. Then $\{b_n(\cdot,\cdot)\}$ is a sequence of Borel maps from $\mathcal{F} \times G$ to

(S,σ) , and $\{b_n(M,u)|n\}$ generates $uMu^*, \forall (M,u) \in \mathcal{F} \times G$. By Proposition 11.3.4, $\psi : (M,u) \longrightarrow uMu^*$ is a Borel map from $\mathcal{F} \times G$ to $\mathcal{F}$. Therefore, by Proposition 10.3.5, $s(E) = \psi(E \times G)$ is a Sousline subset of $\mathcal{F}$. Q.E.D.

Now let $D = \{z \in \mathbb{C} | 0 \leq \text{Im} z \leq 1\}$,

$$A(D) = \left\{ f \,\middle|\, \begin{array}{c} f \text{ is complex valued, bounded and continuous on } D, \\ \text{and } f \text{ is analytic in the interior of } D \end{array} \right\},$$

$C_0^\infty(D)$ be the collection of all complex valued continuous functions on D vanishing at ∞. By maximal modulus, $C^\infty(D)$ is a Banach space.

Clearly, $f(z) \longrightarrow \exp(-|Rez|)f(z)$ is an injective map from $A(D)$ into $C_0^\infty(D)$; and this map transforms $\{f \in A(D) | |f(z)| \leq r, \forall z \in D\}$ to a closed subset of $C_0^\infty(D)$; further, $A(D)$ can be regarded as a Borel subset of $C_0^\infty(D)$. Thus, $A(D)$ admits a standard Borel structure induced by the Borel structure of $C_0^\infty(D)$.

Proposition 11.4.15. Denote the collection of all type (III) factors on H by $\mathcal{F}_{III}$. Then $\mathcal{F}_{III}$ is a Borel subset of $\mathcal{F}$.

Proof. From Lemma 11.4.8, it suffices to show that $\mathcal{F}_{III}$ is a Sousline subset of $\mathcal{F}$. Pick $\xi_0 \in H$ with $\|\xi_0\| = 1$, and consider a subset E of $\mathcal{F} \times G_0 \times \mathbb{R}$. $(M,u,s) \in E$, if : 1)$\overline{M\xi_0} = \overline{M'\xi_0} = H$; 2) for any $t \in \mathbb{R}, f(t,u)\xi_0 = \xi_0$; 3) $f(t,u)Mf(-t,u) = M, \forall t \in \mathbb{R}$; 4) $x \longrightarrow f(s,u)xf(-s,u)$ is not inner for M, where G_0 and f are as in Lemma 11.4.3.

By Proposition 11.3.3 and 11.2.3, the condition 1) determines a Borel subset of $\mathcal{F}$. For any rational number r, from the proof of Lemma 11.4.13 $f(r,\cdot)$ is a Borel map from G_0 to G. Thus , the condition 2) determines a Borel subset of G_0. By the proof of Lemma 11.4.10, $\{(M,v)|vMv^* = M\}$ is a Borel subset of $\mathcal{F} \times G$. Since it is enough that the condition 3) holds for all rational numbers, hence the condition 3) determines a Borel subset of $\mathcal{F} \times G_0$. Moreover, the condition 4) determines a Borel subset of $\mathcal{F} \times G_0 \times \mathbb{R}$ from Lemmas 11.4.10 and 11.4.13. Therefore, E is a Borel subset of $\mathcal{F} \times G_0 \times \mathbb{R}$.

From Proposition 11.3.3, we have Borel maps $a_n(\cdot) : \mathcal{F} \longrightarrow (S,\sigma), n = 1,2,\cdots$, such that $\{a_n(M)|n\}$ is τ-dense in $(M)_1, \forall M \in \mathcal{F}$. For any positive integers j,k, consider a subset $E(j,k)$ of $\mathcal{F} \times G_0 \times \mathbb{R} \times A(D)$. $(M,u,s,g) \in E(j,k)$, if : 5) $(M,u,s) \in E$; 6) $g(t) = \varphi(f(t,u)a_j(M)f(-t,u)a_k(M)), \forall t \in \mathbb{R}$; 7) $g(t+i) = \varphi(a_k(M)f(t,u)a_j(M)f(-t,u)), \forall t \in \mathbb{R}$, where $\varphi(\cdot) = \langle \cdot \xi_0, \xi_0 \rangle$.

Since it is enough that the condition 6) holds for all rational numbers , hence the condition 6) determines a Borel subset of $\mathcal{F} \times G_0 \times A(D)$. The case of the condition 7) is similar. Thus, $E(j,k)$ is a Borel subset of $\mathcal{F} \times G_0 \times \mathbb{R} \times A(D)$.

Let π_1 be the projection from $\mathcal{F} \times G_0 \times \mathbb{R} \times A(D)$ onto $\mathcal{F} \times G_0 \times \mathbb{R}$, and π_2 be the projection from $\mathcal{F} \times G_0 \times \mathbb{R}$ onto $\mathcal{F}$. Then

$$E_0 = \pi_2(\cap_{j,k} \pi_1 E(j,k))$$

is a Sousline subset of $\mathcal{F}$. By Lemma 11.4.14, $s(E_0)$ is also a Sousline subset of $\mathcal{F}$.

If $M \in \mathcal{F}_{III}$, then from Proposition 6.6.6 there is a cyclic–separating vector $\xi(\in H, \|\xi\| = 1)$ for M. Then we have $u \in G$ such that ξ_0 is cyclic –separating for uMu^*. Now it suffices to show that

$$E_0 = \{M \in \mathcal{F}_{III} | \xi_0 \text{ is cyclic–separating for } M\}.$$

Let $M \in \mathcal{F}_{III}$, and ξ_0 be cyclic –separating for M. Then $\varphi(\cdot) = \langle \cdot \xi_0, \xi_0 \rangle$ is a faithful normal state on M. Suppose that $\{\sigma_t | t \in \mathbb{R}\}$ is the modular automorphism group of M corresponding to φ. Since M is type (III), from Theorem 8.3.6 there is $s \in \mathbb{R}$ such that σ_s is not inner. By the invariance of φ for $\{\sigma_t\}$, for each $t \in \mathbb{R}$ we can define a unitary operator u_t on H such that

$$u_t a \xi_0 = \sigma_t(a)\xi_0, \quad \forall a \in M.$$

Clearly,

$$u_t \xi_0 = \xi_0, \quad u_t a u_{-t} = \sigma_t(a), \forall t \in \mathbb{R}, a \in M,$$

and $t \longrightarrow u_t$ is strongly continuous. If u is the Caley transformation of the generator of $\{u_t\}$, then $u \in G_0$ and $u_t = f(t,u), \forall t \in \mathbb{R}$. Let g_{jk} be the KMS function corresponding to $(\varphi, \{\sigma_t\}, a_j(M), a_k(M))$. Then $(M, u, s, g_{jk}) \in E(j,k), \forall j, k$. Thus , $M \in E_0$

Conversely, let $M \in E_0$, then there is a $u \in G_0$ and a $s \in \mathbb{R}$ such that

$$(M, u, s) \in \cap_{j,k} \pi_1 E(j,k).$$

Hence , $M \in \mathcal{F}, \xi_0$ is cyclic–separating for $M, \{\sigma_t(\cdot) = f(t,u) \cdot f(-t,u)\}$ is an one–parameter strongly continuous $*$ automorphism group of M, and $\sigma_s(\cdot)$ is not inner. For any $a, b \in (M)_1$, we can find

$$a_{j(n)}(M) \xrightarrow{\tau} a, \quad a_{k(n)}(M) \xrightarrow{\tau} b.$$

For each n, since $(M, u, s) \in \pi_1 E(j(n), k(n))$, there is $g_n \in A(D)$ such that

$$g_n(t) = \varphi(\sigma_t(a_{j(n)}(M))a_{k(n)}(M)),$$
$$g_n(t+i) = \varphi(a_{k(n)}(M)\sigma_t(a_{j(n)}(M))),$$

$\forall t \in \mathbb{R}$. Noticing that $f(t,u)\xi_0 = \xi_0$ and $f(t,u) \in G, \forall t \in \mathbb{R}$, from the maximum modulus theorem we have

$$|g_n(z) - g_m(z)| \longrightarrow 0, \quad \text{uniformly for} \quad z \in D.$$

Thus, there is $g \in A(D)$ such that $g_n(z) \longrightarrow g(z), \forall z \in D$. Consequently,

$$g(t) = \varphi(\sigma_t(a)b), \quad g(t+i) = \varphi(b\sigma_t(a)),$$

$\forall t \in \mathbb{R}$. By Theorem 8.2.10, $\{\sigma_t(\cdot) = f(t,u) \cdot f(-t,u) | t \in \mathbb{R}\}$ is the modular automorphism group of M corresponding to $\varphi(\cdot) = \langle \cdot \xi_0, \xi_0 \rangle$. Since σ_s is not inner, it follows from Theorem 8.3.6 that $M \in \mathcal{F}_{\text{III}}$. Q.E.D.

From above discussions, we obtain the following.

Theorem 11.4.16. $\mathcal{F}_{\text{I}_n} (n = 1, 2, \cdots), \mathcal{F}_{\text{I}_\infty}, \mathcal{F}_{\text{II}_1}, \mathcal{F}_{\text{II}_\infty}$ and $\mathcal{F}_{\text{III}}$ are Borel subsets of $\mathcal{F}$.

Notes. Proposition 11.4.15 is due to O. Nielsen. It is a key result for Theorem 11.4.16.

References. [120], [154].

Chapter 12

Reduction Theory

12.1. Measurable fields of Hilbert spaces

Let $(E, \mathcal{B})$ be a Borel space. A complex valued function f on E is said to be measurable, if it is $\mathcal{B}$–measurable.

$H(\cdot)$ is called a field of Hilbert spaces over E , if for each $t \in E, H(t)$ is a Hilbert space. $\xi(\cdot)$ is called a field of vectors (relative to $H(\cdot)$) over E, if $\xi(t) \in H(t), \forall t \in E$.

Definition 12.1.1. A field $H(\cdot)$ of Hilbert spaces over a Borel space $(E, \mathcal{B})$ is said to be *measurable,* if there is a sequence $\{\xi_n(\cdot)\}$ of fields of vectors over E such that : 1) the function $\langle \xi_n(t), \xi_m(t)\rangle_t$ is measurable on $E, \forall n, m$, where $\langle , \rangle_t$ is the inner product in $H(t), \forall t \in E$; 2) for any $t \in E, \{\xi_n(t)|n\}$ is a total subset of $H(t)$ (in particular, each $H(t)$ is separable).

In this case, a field $\xi(\cdot)$ of vectors over E is said to be *measurable* (with respect to the measurable field $H(\cdot)$ of Hilbert spaces) , if the function $\langle \xi(t), \xi_n(t)\rangle_t$ is measurable on $E, \forall n$. Denote the collection of all measurable fields of vectors by Θ.

Proposition 12.1.2. Let $H(\cdot)$ be a measurable field of Hilbert spaces over a Borel space $(E, \mathcal{B})$. Then:

1) from any $n = \infty, 0, 1, \cdots$,

$$E_n = \{t \in E|\text{dim}H(t) = n\}$$

is a Borel subset of E;

2) there exists a sequence $\{\eta_n(\cdot)\} \subset \Theta$ with following properties: (a) if $t \in E$ with dim $H(t) = \infty$, then $\{\eta_n(t)|n\}$ is an orthogonal normalized basis of $H(t)$; (b) if $t \in E$ with dim $H(t) = n < \infty$, then $\{\eta, (t), \cdots, \eta_n(t)\}$ is an

orthogonal normalized basis of $H(t)$, and $\eta_k(t) = 0, \forall k > n$; (c) $\xi(\cdot) \in \Theta$ if and only if the function $\langle \xi(t), \eta_n(t) \rangle_t$ is measurable on $E, \forall n$.

Proof. Suppose that we have $\{\eta_1(\cdot), \cdots, \eta_n(\cdot)\} \subset \Theta$ with following properties: 1) if $t \in E$ with dim $H(t) > n$, then $\langle \eta_i(t), \eta_j(t) \rangle_t = \delta_{ij}, \forall 1 \le i, j \le n$; 2) if $t \in E$ with dim $H(t) = k \le n$, then $\{\eta_1(t), \cdots, \eta_k(t)\}$ is an orthogonal normalized basis of $H(t)$, and $\eta_i(t) = 0, \forall k < i \le n$; 3) $[\eta_i(t) | 1 \le i \le n] = [\xi_i(t) | 1 \le i \le n(t)], \forall t \in E$, where $\{\xi_i(\cdot)\}$ is as in Definiton 12.1.1, and $n(t)$ is some integer with $n(t) \ge n$; 4) if $\xi(\cdot) \in \Theta$, then the function $\langle \xi(t), \eta_i(t) \rangle_t$ is measurable on E, $\forall 1 \le i \le n$.

For each $t \in E$, let $p_n(t)$ be the projection of $H(t)$ onto $[\eta_i(t) | 1 \le i \le n]$. Clearly, if $\xi(\cdot) \in \Theta$, then

$$t \longrightarrow p_n(t)\xi(t) = \sum_{i=1}^{n} \langle \xi(t), \eta_i(t) \rangle_t \eta_i(t)$$

is still a measurable field of vectors over E.

For $j \ge 1$, let

$$F_j = \left\{ t \in E \,\middle|\, \begin{array}{l} (1 - p_n(t))\xi_{n+j}(t) \ne 0, \text{and} \\ (1 - p_n(t))\xi_i(t) = 0, i < n + j \end{array} \right\}$$

and

$$\begin{aligned} F_\infty &= \{t \in E | (1 - p_n(t))\xi_i(t) = 0, \forall i\} \\ &= \{t \in E | \dim H(t) \le n\}. \end{aligned}$$

Since $\|(1 - p_n(t))\xi_i(t)\|_t$ is a measurable function on E, it follows that $\{F_\infty, F_1, F_2, \cdots\}$ is a Borel partition of E. Let

$$\eta_{n+1}(t) = \begin{cases} 0, & \text{if } t \in F_\infty, \\ \dfrac{(1 - p_n(t))\xi_{n+j}(t)}{\|(1 - p_n(t))\xi_{n+j}(t)\|}, & \text{if } t \in F_j, \quad j = 1, 2, \cdots \end{cases}$$

Clearly, $\{\eta_1(\cdot), \cdots, \eta_{n+1}(\cdot)\}$ also satisfies the conditions (1)–(4). Repeating this process, we get a sequence $\{\eta_n(\cdot) | n = 1, 2, \cdots\}$ of Θ . If $\xi(\cdot)$ is a field of vectors over E such taht $\langle \xi(t), \eta_i(t) \rangle_t$ is measurable on $E, \forall i$, then

$$\langle \xi(t), \xi_n(t) \rangle_t = \sum_t \langle \xi(t), \eta_i(t) \rangle_t \cdot \langle \eta_i(t), \xi_n(t) \rangle_t$$

is also measurable on $E, \forall n$, and further, $\xi(\cdot) \in \Theta$. Thus , the conclusion 2) is proved.

Notice that $\|\eta_i(t)\|_t$ is a measurable function on $E, \forall i$. Therefore,

$$E_n = \{t \in E | \eta_i(t) \ne 0, i \le n; \eta_j(t) = 0, j > n\}$$

is a Borel subset of $E, \forall n$. Q.E.D.

Definition 12.1.3. The $\{\eta_n(\cdot)\}$ is Propostion 12.1.2 is called an *orthogonal normalized basis* of the measurable field $H(\cdot)$. Moreover, a sequence $\{\varsigma_n(\cdot)\}$ of measurable fields of vectors is said to be *fundamental* , if $\{\varsigma_n(t)|n\}$ is a total subset of $H(t), \forall t \in E$.

Proposition 12.1.4. Let $H(\cdot)$ be a measurable field of Hilbert spaces over a Borel space $(E, \mathcal{B})$.

1) a field $\xi(\cdot)$ of vectors over E is measurable if and only if $\langle\xi(t), \varsigma_n(t)\rangle_t$ is measurable on $E, \forall n$, where $\{\varsigma_n\}$ is a fundamental sequence of measurable fields of vectors over E.

2) if $\xi(\cdot)$ is a measurable field of vectors over E, then $\|\xi(t)\|_t$ is measurable on E.

3) if $\xi(\cdot), \eta(\cdot)$ are two measurable fields of vectors over E, then $\langle\xi(t), \eta(t)\rangle_t$ is measurable on E.

4) let $\{\varsigma_m(\cdot)\} \subset \Theta$, and suppose that for each $t \in E$, there is $\varsigma(t) \in H(t)$ such that $\langle\varsigma_m(t) - \varsigma(t), \xi\rangle_t \longrightarrow 0, \forall \xi \in H(t)$. Then $\varsigma(\cdot)$ is also a measurable field of vectors over E.

Proof. 3) Let $\{\eta_n(\cdot)\}$ be an orthogonal normalized basis of $H(\cdot)$. If $\xi(\cdot), \eta(\cdot) \in \Theta$, then

$$\langle\xi(t), \eta(t)\rangle_t = \sum_t \langle\xi(t), \eta_n(t)\rangle_t \cdot \langle\eta_n(t), \eta(t)\rangle_t$$

is measurable on E.

2) It is obvious from the conclusion 3).

4) Let $\{\xi_n(\cdot)\}$ be as in Definition 12.1.1. Then

$$\langle\varsigma(t), \xi_n(t)\rangle_t = \lim_m \langle\varsigma_m(t), \xi_n(t)\rangle_t$$

is measurable on $E, \forall n$. Thus $\varsigma(\cdot) \in \Theta$.

1) The necessity is obvious from the conclusion 3). Now let $\langle\xi(t), \varsigma_n(t)\rangle_t$ be measurable on $E, \forall n$, and $\{\xi_n(\cdot)\}$ and Θ be as is in Definition 12.1.1. Let

$$\Theta' = \{\eta(\cdot)|\langle\eta(t), \varsigma_n(t)\rangle_t \text{is measurable on } E, \forall n\}.$$

Then $\xi(\cdot) \in \Theta'$, and $\{\xi_n(\cdot)\} \subset \Theta'$. Applying the conclusion 3) to Θ', $\langle\xi(t), \xi_n(t)\rangle_t$ is measurable on $E, \forall n$. Therefore, $\xi(\cdot) \in \Theta$. Q.E.D.

Example 1. The constant measurable field of Hilbert spaces.

Let $(E, \mathcal{B})$ be a Borel space, H_0 be a separable Hilbert space, and $\{\xi_n\}$ be a total subset of H_0 . Define

$$H(t) = H_0, \quad \xi_n(t) = \xi_n, \quad \forall t \in E,$$

and $\Theta = \{\xi(\cdot)|\langle\xi(t),\xi_n(t)\rangle_t = \langle\xi(t),\xi_n\rangle_0$ is measurable on $E,\forall n\}$. This measurable field of Hilbert spaces is called the constant field corresponding to H_0. Clearly, $\xi(\cdot)\in\Theta$ if and only if $\langle\xi(t),\eta\rangle_0$ is measurable on $E,\forall\eta\in H_0$. Consequently , Θ is independent of the choice of the total subset $\{\xi_n\}$.

Example 2. Let A be a separable C^*-algebra, and $S(A)$ be its state space. Consider $(S(A),\sigma(A^*,A))$ as a Borel space. For each $\rho\in S(A)$, through the GNS construction we get a Hilbert space H_ρ. If $\{a_n\}$ is a countalbe dense subset of A, then $\{(a_n)_\rho\}_n$ is dense in H_ρ, and $\langle(a_n)_\rho,(a_m)_\rho\rangle = \rho(a_m^*a_n)$ is a continuous function of ρ on $S(A),\forall n,m$. Let $H(\rho) = H_\rho,\forall\rho\in S(A)$, and $\Theta = \{\xi(\cdot)|\langle\xi(\rho),(a_m)_\rho\rangle_\rho$ is measurable on $S(A),\forall n\}$. Then we get a measurable field $H(\cdot)$ of Hilbert spaces. Clearly, $\xi(\cdot)\in\Theta$ if and only if $\langle\xi(\rho),a_\rho\rangle_\rho$ is measurable on $S(A),\forall a\in A$. Consequently, Θ is independent of the choice of the countable dense subset $\{a_n\}$.

Proposition 12.1.5. Let $H(\cdot)$ be a measurable field of Hilbert spaces over a Borel space $(E,\mathcal{B})$. For $n=\infty,0,1,\cdots$, define $E_n = \{t\in E|\dim H(t) = n\}$. Suppose that H_n is a fixed n-dimensional Hilbert space, $n=\infty,0,1,\cdots$. Then there exists $u(\cdot)$ satisfying: 1) for any $t\in E_n, u(t)$ is a unitary operator from $H(t)$ onto $H_n,\forall n$; 2) $\xi(\cdot)\in\Theta$ if and only if for any n and $\eta\in H_n,\langle u(t)\xi(t),\eta\rangle_n$ is measurable on E_n , where $\langle,\rangle_n$ is the inner prodeuct in H_n.

Proof. Let $\{\eta_k(\cdot)\}$ be an orhogonal normalized basis of the field $H(\cdot)$, and $\{\eta_k^{(n)}|1\leq k\leq n\}$ be an orthogonal normalized basis of $H_n,\forall n$. Define : $u(t)\eta_k(t) = \eta_k^{(n)},\forall t\in E_n, 1\leq k\leq n,\forall n$. Then $u(\cdot)$ satisfied the condition 1). Moreover, notice that $\xi(\cdot)\in\Theta$ if and only if $\langle\xi(t),\eta_k(t)\rangle_t$ is measurable on $E,\forall k$. Thus , $u(\cdot)$ satifies the condition 2) also . Q.E.D.

Proposition 12.1.6. Let H_0 be a countably infinite dimensional Hilbert space, and $H(\cdot)$ be a measurable field of Hilbert spaces over a Borel space $(E,\mathcal{B})$. Then there exists $u(\cdot)$ such that for each $t\in E, u(t)$ is an isometry $H(t)$ into H_0; and $t\longrightarrow u(t)H(t)$ is a Borel map from $(E,\mathcal{B})$ to $W(H_0)$ (see Proposition 10.1.8) . Moreover, $\xi(\cdot)\in\Theta$ if and only if $\langle u(t)\xi(t),\eta\rangle_0$ is measurable on $E,\forall\eta\in H_0$. Conversely , if $H(\cdot)$ is a field of Hilbert spaces on E, and for each $t\in E$ there is an isometry $u(t)$ from $H(t)$ into H_0 such that $t\longrightarrow u(t)H(t)$ is a Borel map from $(E,\mathcal{B})$ to $W(H_0)$, then $(H(\cdot),\Theta)$ is measurable, where $\Theta = \{\xi(\cdot)|\langle u(t)\xi(t),\eta\rangle_0$ is measurable on $E,\forall\eta\in H_0\}$.

Proof. Let $H(\cdot)$ be a measurable field over $(E,\mathcal{B})$, $\{\eta_n(\cdot)\}$ be an orthogonal normalized basis of $H(\cdot)$, and $\{\eta_n\}$ be an orhtogonal normalized basis of H_0.

For any $t \in E$, define

$$u(t)\eta_n(t) = \eta_n \quad \text{if} \quad n \leq \dim H(t),$$
$$u(t)\eta_n(t) = 0, \quad \text{if} \quad n > \dim H(t).$$

Then $u(t)$ is an isometry from $H(t)$ into $H_0, \forall t \in E$. If p_n is the projection from H_0 onto $[\eta_1, \cdots, \eta_n]$, then for any $\eta \in H_0$ we have

$$\|\eta + u(t)H(t)\|_0 = \|(1 - p_n)\eta\|_0, \quad \forall t \in E_n,$$

where $E_n = \{t \in E | \dim H(t) = n\}$. Thus , $\|\eta + u(t)H(t)\|_0$ is measurable on E. By Proposition 11.1.8, $t \longrightarrow u(t)H(t)$ is a Borel map from E to $W(H_0)$. Moreover, from

$$\langle u(t)\xi(t), \eta\rangle_0 = \sum_n \langle \xi(t), \eta_n(t)\rangle_t \cdot \langle u(t)\eta_n(t), \eta\rangle_0$$

and

$$\langle \xi(t), \eta_n(t)\rangle_t = \begin{cases} 0, & n > \dim H(t), \\ \langle u(t)\xi(t), \eta_n\rangle_0, & \text{if } n \leq \dim H(t), \end{cases}$$

we can see that $\xi(\cdot) \in \Theta$ if and only if $\langle u(t)\xi(t), \eta\rangle_0$ is measurable on E for any η of H_0.

Conversely, let $u(t)$ be an isometry from $H(t)$ into $H_0, \forall t \in E$, such that $t \longrightarrow u(t)H(t)$ is Borel from E to $W(H_0)$. Denote the projection from H_0 onto $u(t)H(t)$ by $p(t), \forall t \in E$. Then for each $\xi \in H_0$, the function $\|\xi + u(t)H(t)\|_0 = \|(1 - p(t))\xi\|_0$ is measurable on E. Further, $\langle p(t)\xi, \eta\rangle_0$ is measurable on E, $\forall \xi, \eta \in H_0$. Suppose that $\{\xi_n\}$ is a countable dense subset of H_0. Let $\xi_n(t) = u(t)^* p(t)\xi_n, \forall t \in E$ and n. Since $\{\xi_n(t)|n\}$ is dense in $H(t), \forall t \in E$, and $\langle \xi_n(t), \xi_m(t)\rangle_t = \langle p(t)\xi_n, \xi_m\rangle_0$ is measurable on $E, \forall n, m$, we can get a measurable field $H(\cdot)$ over E with $\{\xi_n(\cdot)\}$. Notice that $\langle \xi(t), \xi_n(t)\rangle_t = \langle u(t)\xi(t), \xi_n\rangle_0$ for any field $\xi(\cdot)$ of vectors over E. Now by the density of $\{\xi_n\}$ in H_0, we can see that $\Theta = \{\xi(\cdot) | \langle u(t)\xi(t), \eta\rangle_0$ is measurable on $E, \forall \eta \in H_0\}$. Q.E.D.

Definition 12.1.7. Let $(E, \mathcal{B})$ be a Borel space, ν be a measure on $\mathcal{B}$, and $H(\cdot)$ be a measurable field of Hilbert space over E. Let

$$\begin{aligned} H &= \int_E^{\oplus} H(t)d\nu(t) \\ &= \{\xi(\cdot) \in \Theta | \textstyle\int_E \|\xi(t)\|_t^2 d\nu(t) < \infty\} \end{aligned}$$

Proposition 12.1.8. Define an inner product in $H = \int_E^{\oplus} H(t)d\nu(t)$ as follows:

$$\langle \xi(\cdot), \eta(\cdot)\rangle = \int_E \langle \xi(t), \eta(t)\rangle_t d\nu(t).$$

Then H is a Hilbert space. Moreover, if $\xi_n(\cdot) \longrightarrow \xi(\cdot)$ in H, then there is a subsequence $\{n_k\}$ such that $\|\xi_{n_k}(t) - \xi(t)\|_t \longrightarrow 0$, a.e.$\nu$.

Proof. Let $\{\xi_n(\cdot)\}$ be a Cauchy sequence of H. Pick a subsequence $\{\xi_{n_k}(\cdot)\}$ such that $\sum_k \|\xi_{n_{k+1}}(\cdot) - \xi_{n_k}(\cdot)\| < \infty$. Put

$$\alpha_N(t) = \sum_{k=1}^{N} \|\xi_{n_{k+1}}(t) - \xi_{n_k}(t)\|_t, \quad N = 1, 2, \cdots.$$

Clearly, $\alpha_N(\cdot)$ is a non–negative measurable function on E, and

$$\left(\int \alpha_N(t)^2 d\nu(t)\right)^{1/2} \leq \sum_{k=1}^{N} \|\xi_{n_{k+1}}(\cdot) - \xi_{n_k}(\cdot)\|, \forall N.$$

Thus , $\alpha(t) = \sum_k \|\xi_{n_{k+1}}(t) - \xi_{n_k}(t)\|_t \in L^2(E, \mathcal{B}, \nu)$. Consequently, there is $F \in \mathcal{B}$ with $\nu(F) = 0$ such that $\alpha(t) < \infty, \forall t \notin F$. Let

$$\xi(t) = \begin{cases} \xi_{n_1}(t) + \sum_{k=1}^{\infty}(\xi_{n_{k+1}}(t) - \xi_{n_k}(t)), & \text{if } t \notin F, \\ 0, & \text{if } t \in F. \end{cases}$$

Then , $\xi_{n_k}(t) \longrightarrow \xi(t)$, a.e.$\nu$, and $\xi(\cdot) \in \Theta$. Moreover, since

$$\begin{aligned} &\left(\int \|\xi(t)\|_t^2 d\nu(t)\right)^{1/2} \\ \leq \; &\|\xi_{n_1}(\cdot)\| + \sum_{k\geq 1} \|\xi_{n_{k+1}}(\cdot) - \xi_{n_k}(\cdot)\| < \infty, \end{aligned}$$

it follows that $\xi(\cdot) \in H$. Noticing that

$$\|\xi_{n_k}(\cdot) - \xi(\cdot)\| \leq \sum_{j\geq k} \|\xi_{n_{j+1}}(\cdot) - \xi_{n_j}(\cdot)\| \longrightarrow 0,$$

we have $\|\xi_n(\cdot) - \xi(\cdot)\| \longrightarrow 0$ in H. Q.E.D.

Proposition 12.1.9. Let $(E, \mathcal{B}), \nu, H(\cdot)$ and H be as above, and $\{\eta_n(\cdot)\}$ be an orthogonal normalized basis of $H(\cdot)$. Then :

1) $\xi(\cdot) \in H$ if and only if

$$\sum_n \int_E |\langle \xi(t), \eta_n(t) \rangle_t|^2 d\nu(t) < \infty;$$

2) for and $\xi(\cdot), \eta(\cdot) \in H$, we have

$$\langle \xi(\cdot), \eta(\cdot) \rangle = \sum_n \int_E \langle \xi(t), \eta_n(t) \rangle_t \cdot \langle \eta_n(t), \eta(t) \rangle_t d\nu(t);$$

3) for any $\xi(\cdot) \in H$, we have $\xi(\cdot) = \sum_n \xi_n(\cdot)$, where $\xi_n(t) = \langle \xi(t), \eta_n(t) \rangle_t \eta_n(t)$, $\forall t \in E$, and $\xi_n(\cdot) \in H, \forall n$;

4) if X is a total subset of $L^2(E, \mathcal{B}, \nu)$, then $\{(f\eta_n)(\cdot) | f \in X, n\}$ is also a total subset of H.

Proof. All conclusions are obvious. Q.E.D.

Proposition 12.1.10. Let $H(\cdot), K(\cdot)$ be two measurable fields of Hilbert spaces over a Borel space $(E, \mathcal{B})$. Then there is a unique manner such that $(H \otimes K)(\cdot)$ becomes a measurable field of Hilbert spaces over E, and $(\xi \otimes \eta)(\cdot) \in \Theta((H \otimes K)(\cdot)), \forall \xi(\cdot) \in \Theta(H(\cdot))$ and $\eta(\cdot) \in \Theta(K(\cdot))$, where $(H \otimes K)(t) = H(t) \otimes K(t), (\xi \otimes \eta)(t) = \xi(t) \otimes \eta(t), \forall t \in E$.

Proof. Let $\{\xi_n(\cdot)\}, \{\eta_m(\cdot)\}$ be fundamental sequences of measurable fields of vectors of the fields $H(\cdot), K(\cdot)$ respectively . Taking $\{\xi_n \otimes \eta_m(\cdot) | n, m\}$ as fundamental fields of vectors, $(H \otimes K)(\cdot)$ becomes a measurable field of Hilbert spaces over E, and $\xi \otimes \eta(\cdot) \in \Theta((H \otimes K)(\cdot)), \forall \xi(\cdot) \in \Theta(H(\cdot))$, and $\eta(\cdot) \in \Theta(K(\cdot))$. Conversely, if $(H \otimes K)(\cdot)$ is a measurable field such that $(\xi \otimes \eta)(\cdot)$ is measurable , $\forall \xi(\cdot) \in \Theta(H(\cdot))$ and $\eta(\cdot) \in \Theta(K(\cdot))$, then $\{(\xi_n \otimes \eta_m)(\cdot) | n, m\}$ is a funcdamental measurable field of vectors. Therefore, the manner is unique. Q.E.D.

Proposition 12.1.11. Let $H(\cdot)$ be a measurable field of Hilbert spaces over a Borel space $(E, \mathcal{B})$, H_0 be a separable Hilbert space, and ν be a measure on $\mathcal{B}$. Then there is a unique isomorphism $\Phi : \int_E^{\oplus} H(t) d\nu(t) \otimes H_0 \longrightarrow \int_E^{\oplus} (H(t) \otimes H_0) d\nu(t)$ such that $(\Phi(\xi(\cdot) \otimes \eta))(t) = \xi(t) \otimes \eta, \forall t \in E, \xi(\cdot) \in \int_E^{\oplus} H(t) d\nu(t)$, and $\eta \in H_0$, where $H(\cdot) \otimes H_0$ is the tensor product of $H(\cdot)$ and the constant field corresponding to H_0.

Proof. Naturally define the map Φ. If suffices to show that the image of Φ is $\int_E^{\oplus} (H(t) \otimes H_0) d\nu(t)$. Let $\{\xi_n(\cdot)\}$ be an orthogonal normalized basis of the field $H(\cdot)$, and $\{\eta_m\}$ be an orthogonal normalized basis of H_0. Then $\{\xi_n(\cdot) \otimes \eta_m | n, m\}$ is an orhtogonal normalized basis of the field $H(\cdot) \otimes H_0$. By Proposition 12.1.9,

$$\{(f\xi_n)(\cdot) \otimes \eta_m | f \in L^2(E, \mathcal{B}, \nu), n, m\}$$

will be a total subset of $\int_E^{\oplus} (H(t) \otimes H_0) d\nu(t)$. Therefore, the image of Φ is $\int_E^{\oplus} (H(t) \otimes H_0) d\nu(t)$. Q.E.D.

Example. $L^2(E,\mathcal{B},\nu)\otimes H_0 = \int_E^{\oplus} H_0 d\nu(t) = L^2(E,\mathcal{B},\nu,H_0)$. In fact, noticing that $L^2(E,\mathcal{B},\nu) = \int_E^{\oplus} \mathbb{C} d\nu(t)$, then from Proposition 12.1.11 we can get the conclusion.

References. [28], [36], [119].

12.2. Measurable fields of operators

Definition 12.2.1. Let $(E,\mathcal{B})$ be a Borel space, and $H(\cdot), K(\cdot)$ be two measurable fields of Hilbert spaces over E. A field $a(\cdot)$ of operators from $H(\cdot)$ to $K(\cdot)$ is said to be *measurable,* if for each $t\in E, a(t)$ is a bounded linear operator form $H(t)$ to $K(t)$, and for each $\xi(\cdot)\in\Theta(H(\cdot))$ we have $(a\xi)(\cdot) = a(\cdot)\xi(\cdot)\in\Theta(K(\cdot))$.

Proposition 12.2.2. A field $a(\cdot)$ of operators from $H(\cdot)$ to $K(\cdot)$ is measurable if and only if the function

$$\langle a(t)\xi_n(t), \eta_m(t)\rangle_t$$

is measurable on $E, \forall n,m$, where $\{\xi_n(\cdot)\}, \{\eta_m(\cdot)\}$ are fundamental sequences of measurable fields of vectors of the fields $H(\cdot), K(\cdot)$ respectively. Moreover, in this case , $t\longrightarrow \|a(t)\|_t$ is a measurable function on E.

Proof. The necessity is obvious. Conversely, let $\langle a(t)\xi_n(t), \eta_m(t)\rangle_t$ is measurable on $E, \forall n,m$. Then , $a^*(\cdot)\eta_m(\cdot)\in\Theta(H(\cdot)), \forall m$, and for each $\xi(\cdot)\in\Theta(H(\cdot))$

$$\langle a(t)\xi(t), \eta_m(t)\rangle = \langle \xi(t), a(t)^*\eta_m(t)\rangle_t$$

is measurable on $E, \forall m$. Thus $a(\cdot)\eta(\cdot)\in\Theta(K(\cdot))$, and $a(\cdot)$ is measurable over E.

Moreover, if $\{\xi_n(\cdot)\}, \{\eta_m(\cdot)\}$ are the orthogonal normalized bases of $H(\cdot)$, $K(\cdot)$ respectively, then we have

$$\|a(t)\|_t = \sup\left\{\left.\frac{|\langle a(t)\sum_n \alpha_n\xi_n(t), \sum_m \beta_m\eta_m(t)\rangle_t|}{(\sum_n|\alpha_n|^2)^{1/2}\cdot(\sum_m|\beta_m|^2)^{1/2}}\right| \alpha_n,\beta_m \text{ are complex rational numbers}, \forall n,m\right\}.$$

Therefore , $t\longrightarrow\|a(t)\|_t$ is measurable on E. Q.E.D.

Proposition 12.2.3. Let $a(\cdot)$ be a measurable field of operators from $H(\cdot)$ to $K(\cdot)$, and $t\longrightarrow\|a(t)\|_t$ be eseentially bounded with respect to ν, where ν

is a measure on $(E,\mathcal{B})$. Then $(a\xi)(\cdot) = a(\cdot)\xi(\cdot)$ is a bounded linear operator from $H = \int_E^{\oplus} H(t)d\nu(t)$ to $K = \int_E^{\oplus} K(t)d\nu(t)$ (and we shall denote this operator a by $\int_E^{\oplus} a(t)d\nu(t)$). Moreover, if ν is semi-finite (i.e., for any $F \in \mathcal{B}$ with $\nu(F) > 0$, there is $G \in \mathcal{B}$ with $G \subset F$ and $0 < \nu(G) < \infty$), then $\|a\| = ess \sup \|a(t)\|_t$.

Proof. Let $\lambda = ess \sup \|a(t)\|_t$. Since $\|a(t)\xi(t)\|_t \leq \lambda \|\xi(t)\|_t$, a.e.$\nu, \forall \xi(\cdot) \in H(\cdot)$, it follows that $\|a\| \leq \lambda$.

Now suppose that ν is semi-finite. For any $\varepsilon > 0, F = \{t \in E \mid \|a(t)\|_t \geq \lambda - \varepsilon\} (\in \mathcal{B})$ is not ν-zero. Then there is $G \in \mathcal{B}$ with $G \subset F$ and $0 < \nu(G) < \infty$. If $\{\xi_n(\cdot)\}$ is an orthogonal normalized basis of the field $H(\cdot), \{\alpha_n\}$ is a sequence of complex rational numbers with $\sum_n |\alpha_n|^2 < \infty$, and $f \in L^\infty(E,\mathcal{B},\nu)$, then $\sum_n (\alpha_n f \chi_G \xi_n)(\cdot) \in H$, and

$$\|a \sum_n (\alpha_n f \chi_G \xi_n)(\cdot)\|^2 \leq \|a\|^2 \cdot \|\sum_n (a_n f \chi_G \xi_n)(\cdot)\|^2.$$

Since f is arbitrary, it follows that

$$\|a(t) \sum_n \alpha_n \chi_G(t) \xi_n(t)\|_t \leq \|a\| \cdot \|\sum_n \alpha_n \chi_G(t)\xi_n(t)\|_t, \text{a.e.}\nu.$$

Thus, $\|a(t)\|_t \leq \|a\|, \forall t \in G$, a.e.$\nu$, and $\|a\| \geq \lambda - \varepsilon$. Since ε is arbitrary, $\|a\| \geq \lambda$ and $\|a\| = \lambda = ess \sup \|a(t)\|_t$. Q.E.D.

Example. Let $H(\cdot)$ be a measurable field of Hilbert spaces over $(E,\mathcal{B})$, and H_0 be a countably infinite dimensional Hilbert space. By Proposition 12.1.6, we have a measurable field $u(\cdot)$ of isometries from $H(\cdot)$ to the constant field corresporcding to H_0. If ν is a semi-finite measure on $\mathcal{B}$, then $u = \int_E^{\oplus} u(t)d\nu(t)$ is an isometry from $H = \int_E^{\oplus} H(t)d\nu(t)$ to $\int_E^{\oplus} H_0 d\nu(t) = H_0 \otimes L^2(E,\mathcal{B},\nu)$.

Definition 12.2.4. A bounded linear operator a from $H = \int_E^{\oplus} H(t)d\nu(t)$ to $K = \int_E^{\oplus} K(t)d\nu(t)$ is said to be *decomposable,* if there is a measurable field $a(\cdot)$ of operators from $H(\cdot)$ to $K(\cdot)$ such that $a = \int_E^{\oplus} a(t)d\nu(t)$.

Proposition 12.2.5. Let ν be semi-finite, and $a_n = \int_E^{\oplus} a_n(t)d\nu(t), n = 0,1,2,\cdots$.

1) Suppose that $a_n \longrightarrow a_0$ (strongly). Then for any $F \in \mathcal{B}$ with $\nu(F) < \infty$, there is a subsequence $\{a_{n_k}\}$ such that $a_{n_k}(t) \longrightarrow a_0(t)$ (strongly), $\forall t \in F$, a.e.ν.

2) If $a_n(t) \longrightarrow a_0(t)$ (strongly) , $\forall t$, a.e.ν , and $\sup\limits_n \|a_n\| < \infty$, then $a_n \longrightarrow a_0$ (strongly).

Proof. 1) Let $\{\xi_m(\cdot)\}$ be an orthogonal normalized basis of $H(\cdot)$. Then $(\chi_F \xi_m)(\cdot) \in H$ and

$$\|a_n(\chi_F \xi_m)(\cdot) - a_0(\chi_F \xi_m)(\cdot)\| \longrightarrow 0, \quad \forall m.$$

By Proposition 12.1.8, there is a subsequence $\{a_{n_k}\}$ such that $\|a_{n_k}(t)\xi_m(t) - a_0(t)\xi_m(t)\|_t \longrightarrow 0, \forall t \in F$, a.e.$\nu$, and m. We may assume $\|a_n(t)\|_t \leq \sup\limits_m \|a_m\|, \forall n, t \in E$. Thus , $a_{n_k}(t) \longrightarrow a_0(t)$ (strongly) , $\forall t \in F$, a.e.ν.

2) Let $K = \sup\{\|a_n(t)\|_t, \|a_n\| | t \in E, n\}$. For any $\xi(\cdot) \in H$, since $f_n(t) = \|a_n(t)\xi(t) - a_0(t)\xi(t)\|_t^2 \longrightarrow 0$, a.e.$\nu$ and $|f_n(t)| \leq 4K^2\|\xi(t)\|_t^2 \in L^1(E, \mathcal{B}, \nu)$, it follows from the bounded convergence theorem that

$$\|a_n\xi(\cdot) - a_0\xi(\cdot)\|^2 = \int |f_n(t)| d\nu(t) \longrightarrow 0,$$

i.e. $a_n \longrightarrow a_0$ (strongly) . Q.E.D.

Proposotion 12.2.6. Let ν be semi-finite. Then there is a sequence $\{a_n = \int_E^\oplus a_n(t) d\nu(t)\}$ of decomposable operators on $H = \int_E^\oplus H(t) d\nu(t)$ such that $B(H(t))$ is generated by $\{a_n(t)|n\}, \forall t \in E$.

Proof. Let $E_k = \{t \in E | \dim H(t) = k\}, k = \infty, 0, 1, \cdots$. By Proposition 12.1.5, there is $u(\cdot)$ such that for each $t \in E_k, u(t)$ is a unitary operator from $H(t)$ onto H_k , where H_k is a k-dimensional Hilbert space, $\forall k$; and $\xi(\cdot) \in \Theta$ if and only if $\langle u(t)\xi(t), \eta\rangle_k$ is measurable on $E_k, \forall \eta \in H_k$ and k.

Pick $\{b_n^{(k)}|n\} \subset B(H_k)$ with $\|b_n^{(k)}\| \leq 1, \forall n$, such that $B(H_k)$ is generated by $\{b_n^{(k)}|n\}, \forall k$. Define

$$a_n(t) = u(t)^* b_n^{(k)} u(t), \forall t \in E_k, k, n.$$

Clearly, $\|a_n(t)\| \leq 1, \forall n$, and $B(H(t))$ is generated by $\{a_n(t)|n\}, \forall t \in E$. Now it suffices to show that $a_n(\cdot)$ is measurable for each n.

Let $\{\xi_n(\cdot)\}$ be an orthogonal normalized basis of $H(\cdot)$, and $\{u(t)\xi_n(t) = \xi_n^{(k)} | 1 \leq n \leq k\}$ be an orthogonal normalized basis of $H_k, \forall t \in E_k, k$ (see Proposition 12.1.5) . Thus , $\langle a_n(t)\xi_i(t), \xi_j(t)\rangle_t$ is a constant on each E_k (notice that $u(t)\xi_n(t) = 0, \forall n > k, t \in E_k), \forall i, j$, and the field $a_n(\cdot)$ is measurable, $\forall n$. Q.E.D.

Definition 12.2.7. Let $H(\cdot)$ be a measurable field of Hilbert spaces on a Borel space $(E, \mathcal{B})$ and ν be a measure on $\mathcal{B}$. For any $f \in L^\infty(E, \mathcal{B}, \nu)$, define a decomposable operator

$$m_f = \int_E^{\oplus} f(t)d\nu(t)$$

on $H = \int_E^{\oplus} H(t)d\nu(t)$, i.e., $m_f\xi(\cdot) = f(\cdot)\xi(\cdot), \forall \xi(\cdot) \in H$. The operator m_f is said to be *diagonal,* and the collection $Z = \{m_f | f \in L^\infty(E, \mathcal{B}, \nu)\}$ of all diagonal operators is called the *diagonal algebra* on $H = \int_E^{\oplus} H(t)d\nu(t)$.

Proposition 12.2.8. Let ν be semi-finite. Then $f \longrightarrow m_f$ is a bijection from $L^\infty(E, \mathcal{B}, \nu)$ onto $Z = \{m_f | f \in L^\infty(E, \mathcal{B}, \nu)\}$ if and only if $\nu(E_0) = 0$, where $E_0 = \{t \in E | \dim H(t) = 0\}$, i.e., $H(t) \neq \{0\}$, a.e.ν. In this case, we have also $\|m_f\| = \|f\|, \forall f \in L^\infty(E, \mathcal{B}, \nu)$.

Proof. It is obvious from Proposition 12.2.3. Q.E.D.

Proposition 12.2.9. Let $(E, \mathcal{B})$ be a Borel space, ν be a σ-finite measure on $\mathcal{B}$, $H(\cdot)$ be a measurable field of Hilbert spaces over E, and $H(t) \neq \{0\}$, a.e.ν . Then the diagonal algebra Z is a commutative VN algebra on $H = \int_E^{\oplus} H(t)d\nu(t)$, Z' is σ-finite, and $f \longrightarrow m_f$ is a faithful W^*-representation of $L^\infty(E, \mathcal{B}, \nu)$ on H.

Proof. Let $\{f_l\}$ be a net of $L^\infty(E, \mathcal{B}, \nu)$, and $f_l \longrightarrow 0$ with respect to the w^*-top. in $L^\infty(E, \mathcal{B}, \nu)$. If $\xi_n(\cdot), \eta_n(\cdot) \in H$ with $\sum_n(\|\xi_n(\cdot)\|^2 + \|\eta_n(\cdot)\|^2) < \infty$, then we have

$$|\sum_n \langle \xi_n(t), \eta_n(t)\rangle_t| \leq \sum_n \|\xi_n(t)\|_t \cdot \|\eta_n(t)\|_t \in L^1(E, \mathcal{B}, \nu)$$

and

$$\begin{aligned} &|\sum_n \langle m_{f_l}\xi_n(\cdot), \eta_n(\cdot)\rangle| \\ = \; &|\int_E f_l(t) \cdot \sum_n \langle \xi_n(t), \eta_n(t)\rangle_t d\nu(t)| \longrightarrow 0 \end{aligned}$$

i.e., $m_{f_l} \longrightarrow 0$ with respect to $\sigma(B(H), T(H))$. Thus , from Proposition 12.2.8 $f \longrightarrow m_f$ is a faithful W^*-representation of $L^\infty(E, \mathcal{B}, \nu)$. Consequently, Z is a commutative VN algebra on H.

Now let $E = \cup_n E_n$, where $E_n \in \mathcal{B}$ and $\nu(E_n) < \infty, \forall n$, and $\{\xi_n(\cdot)\}$ be an orhtogonal normalized basis of $H(\cdot)$. From Proposition 12.1.9,

$$\{m_f \chi_{E_n}(\cdot)\xi_m(\cdot) | n, m, f \in L^\infty(E, \mathcal{B}, \nu)\}$$

will be a total subset of H. Thus Z admits a cyclic sequence of vectors, i.e., Z' is σ-finite. Q.E.D.

Theorem 12.2.10. Let $(E, \mathcal{B})$ be a Borel space, ν be a σ-finite measure on $\mathcal{B}$, and $H_i(\cdot)$ be a measurable field of Hilbert spaces over $E, i = 1,2$. Then a bounded linear operator a from $H_1 = \int_E^{\oplus} H_1(t)d\nu(t)$ to $H_2 = \int_E^{\oplus} H_2(t)d\nu(t)$ is decomposable if and only if $am_f^{(1)} = m_f^{(2)}a, \forall f \in L^\infty(E, \mathcal{B}, \nu)$, where $m_f^{(i)}$ is the diagonal operator on H_i corresponding to $f, i = 1,2$.

Proof. The necessity is obvious . Now let $am_f^{(1)} = m_f^{(2)}a, \forall f \in L^\infty(E, \mathcal{B}, \nu)$. From the proof of Proposition 12.2.9, there is a fundamental sequence $\{\xi_n(\cdot)\}$ of measurable fields of vectors of $H_1(\cdot)$ such that $\xi_n(\cdot) \in H_1, \forall n$. Define $\eta_n(\cdot) = a\xi_n(\cdot), \forall n$. Then for any complex rational numbers $\alpha_1, \alpha_2, \cdots$ with $^\#\{n|\alpha_n \neq 0\} < \infty$ and $f \in L^\infty(E, \mathcal{B}, \nu)$, from $am_f^{(1)} = m_f^{(2)}a$ we can see that

$$\begin{aligned} &\int_E |f(t)|^2 \cdot \|\sum_n \alpha_n \eta_n(t)\|_t^2 d\nu(t) \\ \leq \ &\|a\|^2 \int_E |f(t)|^2 \cdot \|\sum_n \alpha_n \xi_n(t)\|_t^2 d\nu(t). \end{aligned}$$

Since f is arbitrary, it follows that

$$\|\sum_n \alpha_n \eta_n(t)\|_t \leq \|a\| \cdot \|\sum_n \alpha_n \xi_n(t)\|_t$$

$\forall \{\alpha_n\}$ as above and a.e.ν . Thus , there is $F \in \mathcal{B}$ with $\nu(F) = 0$ such that for each $t \notin F$, we can define a bounded linear operator $a(t)$ from $H_1(t)$ to $H_2(t)$ satisfying $a(t)\xi_n(t) = \eta_n(t), \forall n$. Further, let $a(t) = 0, \forall t \in F$. Then $a(\cdot)$ is a measurable field of operators from $H_1(\cdot)$ to $H_2(\cdot)$, and $\|a(t)\|_t \leq \|a\|, \forall t \in E$. Put $b = \int_E^{\oplus} a(t)d\nu(t)$. Then

$$bm_f^{(1)}\xi_n(\cdot) = m_f^{(2)}b\xi_n(\cdot) = m_f^{(2)}a\xi_n(\cdot) = am_f^{(1)}\xi_n(\cdot),$$

$\forall n$ and $f \in L^\infty(E, \mathcal{B}, \nu)$. But $\{m_f^{(1)}\xi_n(\cdot)|n, f \in L^\infty(E, \mathcal{B}, \nu)\}$ is a total subset of H_1, so $a = b = \int_E^{\oplus} a(t)d\nu(t)$. Q.E.D.

References. [28], [119], [158].

12.3. Measurable fields of Von Neumann algebras

Definition 12.3.1. Let $H(\cdot)$ be a measurable field of Hilbert spaces over a Borel space $(E, \mathcal{B})$. A field $M(\cdot)$ of VN algebras on $H(\cdot)$ (i.e., for each $t \in E, M(t)$ is a VN algebra on $H(t)$) is said to be *measurable,* if there is a sequence $\{a_n(\cdot)|n\}$ of measurable fields of operators on $H(\cdot)$ such that $M(t)$ is generatecd by $\{a_n(t)|n\}, \forall t \in E$.

Proposition 12.3.2. Let H_0 be the constant field over $(E, \mathcal{B})$. A field $M(\cdot)$ of VN algebras on H_0 is measurable if and only if $t \longrightarrow M(t)$ is a Borel map from E to $\mathcal{A}$, where $\mathcal{A}$ is the collection of all VN algebras on H_0.

Proof. It is obvious from Definition 12.3.1 and Proposition 11.3.4. Q.E.D.

Proposition 12.3.3. Let $M(\cdot), N(\cdot)$ be two measurable fields of VN algebras on $H(\cdot)$. Then $M(\cdot)', M(\cdot) \cap N(\cdot)$ and $(M(\cdot) \cup N(\cdot))''$ are also measurable.

Proof. Let $E_k = \{t \in E | \dim H(t) = k\}, k = \infty, 0, 1, \cdots$. Then $H(\cdot)$ can be regarded as a constant field over each E_k. Further, by Propososotion 12.3.2, 11.3.1 and 11.3.5, we can get the conclusions. Q.E.D.

Proposition 12.3.4. Let $H(\cdot)$ be a measurable field of Hilbert spaces over a Borel space $(E, \mathcal{B}), \nu$ be a σ-finitie measure on $\mathcal{B}, H = \int_E^{\oplus} H(t)d\nu(t), Z$ be the diagonal algebra on $H, \{a_n = \int_E^{\oplus} a_n(t)d\nu(t)|n\}$ be a sequence of decomposable operators on H, M be the VN algebra generated by Z and $\{a_n\}$, and $a \in B(H)$. Then $a \in M$ if and only if $a = \int_E^{\oplus} a(t)d\nu(t)$ is decomposable, and $a(t) \in M(t)$, a.e.ν, where $M(t)$ is the VN algebra on $H(t)$ generated by $\{a_n(t)|n\}, \forall t \in E$.

Proof. Let $a = \int_E^{\oplus} a(t)d\nu(t), a(t) \in M(t)$, a.e.$\nu$, and $a' \in M'$. Since $M' \subset Z'$,it follows from Theorem 12.2.10 that $a' = \int_E^{\oplus} a'(t)d\nu(t)$ is decomposable. Noticing that a' commutes with a_n and $a_n^*, \forall n$, we have $a'(t) \in M(t)', a.e.\nu$. Thus $a'(t)a(t) = a(t)a'(t), a.e.\nu$. Further, $a'a = aa', \forall a' \in M'$, and $a \in M$.

Conversely, let $a \in M$. Since $M \subset Z'$, it follows from Theorem 12.2.10 that $a = \int_E^{\oplus} a(t)d\nu(t)$ is decomposable. By Proposition 12.2.9 and $M \subset Z', M$ is σ-finite. Let M_0 be the $*$ algebra generated by Z and $\{a_n\}$. Then M_0 is strongly dense in M. From Proposition 1.14.4, there is a sequence $\{b_n = \int_E^{\oplus} b_n(t)d\nu(t)\} \subset M_0$ such that $b_n \longrightarrow a$ (strongly). Further, by Proposition 12.2.5 and the σ-finiteness of ν , we have a subsequence $\{b_{n_k}\}$ such that

$$b_{n_k}(t) \longrightarrow a(t) \quad (\text{strongly}) \text{ , a.e.}\nu.$$

Therefore , $a(t) \in M(t)$, a.e.ν . Q.E.D.

Proposition 12.3.5. Let $H(\cdot)$ be a measurable field of Hilbert spaces over a Borel space $(E,\mathcal{B}),\nu$ be a σ-finite measure on $\mathcal{B}$, and $M(\cdot)$ be a measurable field of VN algebras on $H(\cdot)$. Then

$$M = \{a = \int_E^{\oplus} a(t)d\nu(t) \in B(H) | a(t) \in M(t), a.e.\nu\}$$

is a σ-finite VN algebra on $H = \int_E^{\oplus} H(t)d\nu(t)$.

Proof. Let $\{a_n(\cdot)\}$ be a sequence of measurabel fields of operators such that $M(t)$ is generated by $\{a_n(t)|n\}, \forall t \in E$. We may assume that $\|a_n(t)\|_t \leq 1, \forall t \in E$ and n. From Proposition 12.3.4, M is the VN algebra generated by Z and $\{a_n\}$ exactly, where Z is the diagonal algebra on $H = \int_E^{\oplus} H(t)d\nu(t)$, and $a_n = \int_E^{\oplus} a_n(t)d\nu(t), \forall n$. Moreover, from $M \subset Z'$ and Proposition 12.2.9, M is also σ-finite. Q.E.D.

Definition 12.3.6. Let $H(\cdot)$ be a measurable field of Hilbert spaces over a Borel space $(E,\mathcal{B})$, and ν be a σ-finite measure on $\mathcal{B}$. A VN algebra M on $H = \int_E^{\oplus} H(t)d\nu(t)$ is said to be *decomposable,* if there is a measurable field $M(\cdot)$ of VN algebras on $H(\cdot)$ such that

$$M = \{a = \int_E^{\oplus} a(t)d\nu(t) \in B(H) | a(t) \in M(t), a.e.\nu\}.$$

In this case, we shall denote M by $\int_E^{\oplus} M(t)d\nu(t)$.

Proposition 12.3.7. A VN algebra M on $H = \int_E^{\oplus} H(t)d\nu(t)$ is decomposable if and only if M is generated by the diagonal algebra Z and a sequence $\{a_n = \int_E^{\oplus} a_n(t)d\nu(t)\}$ of decomposable operators. In this case, $M(t)$ is generated by $\{a_n(t)|n\}, a.e.\nu$, $M(\cdot)$ is unique (a.e.ν), and $Z \subset M \subset Z'$.

Proof. It is immediate from Proposition 12.3.4 and 12.3.5. Q.E.D.

Remark. From Proposition 12.2.6, $Z = \int_E^{\oplus} \mathbb{C}1_t d\nu(t)$ and $Z' = \int_E^{\oplus} B(H(t))d\nu(t)$ are decomposable VN algebras on $H = \int_E^{\oplus} H(t)d\nu(t)$.

Proposition 12.3.8. Let $M = \int_E^{\oplus} M(t)d\nu(t), M_n = \int_E^{\oplus} M_n(t)\, d\nu(t)$, $(n = 1,2,\cdots)$ be decomposable VN algebras. Then we have $M' = \int_E^{\oplus} M(t)'d\nu(t)$,

$$(\sqcup_n M_n)'' = \int_E^{\oplus} (\sqcup_n M_n(t))'' d\nu(t) \text{ and } \sqcap_n M_n = \int_E^{\oplus} (\sqcap_n M_n(t)) d\nu(t).$$

Proof. From Proposition 12.3.3, $M(\cdot)'$ is also measurable, Let $N = \int_E^{\oplus} M(t)' d\nu(t)$. Clearly, $N \subset M'$. Now let $a' \in M'$. Since $Z \subset M \subset Z', a' = \int_E^{\oplus} a'(t) d\nu(t)$ is decomposable. Suppose that M is generated by Z and $\{a_n = \int_E^{\oplus} a_n(t) d\nu(t)\}$. Then $a'(t)$ commutes with $\{a_n(t), a_n(t)^*\}_n$, $a.e.\nu$. Thus $a'(t) \in M(t)'$, a.e.ν, $a' \in N$, and $M' = N = \int_E^{\oplus} M(t)' d\nu(t)$.

For each n, suppose that M_n is generated by Z and $\{a_k^{(n)} = \int_E^{\oplus} a_k^{(n)}(t) d\nu(t) | \ k\}$, and $M_n(t)$ is generated by $\{a_k^{(n)}(t)|k\}$, $\forall t \in E$. Then $(\sqcup_n M_n)''$ is generated by Z and $\{a_k^{(n)}|n,k\}$, and $(\sqcup_n M_n(t))''$ is generated by $\{a_k^{(n)}(t)|n,k\}$. Thus $(\sqcup_n M_n)'' = \int_E^{\oplus} (\sqcup_n M_n(t))'' d\nu(t)$. Further, $\sqcap_n M_n = (\sqcup_n M_n')' = \int_E^{\oplus} (\sqcap_n M_n(t)) d\nu(t)$. Q.E.D.

Proposition 12.3.9. Let $M = \int_E^{\oplus} M(t) d\nu(t)$. Then $M \cap M' = Z$ if and only if $M(t)$ is factorial, a.e.ν.

Proof. It is immediate from $M \cap M' = \int_E^{\oplus} (M(t) \cap M(t)') d\nu(t)$ and $Z = \int_E^{\oplus} \mathbb{C} 1_t d\nu(t)$. Q.E.D.

Proposition 12.3.10. If $H = \int_E^{\oplus} H(t) d\nu(t)$ is separable, then a VN algebra M on H is decomposable if and only if $Z \subset M \subset Z'$, where Z is the diagonal algebra on H.

Proof. The necessity is obvious. Now let $Z \subset M \subset Z'$. By Theorem 12.2.10, every operator of M is decomposable. Moreover, since H is separable, M is countably generated. Thus, M is generated by Z and a sequence of decomposable operators, and M is decomposable. Q.E.D.

Remark. If $(E, \mathcal{B})$ is a standard Borel space, and ν is a σ-finite measure on $\mathcal{B}$, then $L^2(E, \mathcal{B}, \nu)$ is separable. Now by Proposition 12.1.9, $H = \int_E^{\oplus} H(t) d\nu(t)$ is also separable.

Definition 12.3.11. Let $H(\cdot), K(\cdot)$ be two measurable fields of Hilbert spaces over a Borel space $(E, \mathcal{B})$, and $M(\cdot), N(\cdot)$ be measurable fields of VN algebras on $H(\cdot), K(\cdot)$ respectively. For each $t \in E$, let $\Phi(t)$ be a $*$ homomorphism from $M(t)$ to $N(t)$. The field $\Phi(\cdot)$ of $*$ homomorphisms is said to be *measurable*, if for any measurable field $a(\cdot)$ of operators on $H(\cdot)$ with $a(t) \in M(t), \forall t \in E, \Phi(\cdot)(a(\cdot))$ is a measurable field of operators on $K(\cdot)$.

In this case, if ν is a σ-finite measure on $\mathcal{B}$, then we can define a $*$ homomorphism $\Phi = \int_E^{\oplus} \Phi(t)d\nu(t)$ from $M = \int_E^{\oplus} M(t)d\nu(t)$ to $N = \int_E^{\oplus} N(t)d\nu(t)$, i.e., $\Phi(a) = \int_E^{\oplus} \Phi(t)(a(t))d\nu(t)$, where $a = \int_E^{\oplus} a(t)d\nu(t) \in M$.

Proposition 12.3.12. Keep all notations in Definition 12.3.11.

1) If $\Phi(t)$ is normal , $\forall t \in E$, then Φ is normal .

2) If $\Phi(t)$ is a $*$ isomorphism from $M(t)$ onto $N(t), \forall t \in E$, then Φ is also a $*$ isomorphism from M onto N.

Proof. 1) From Proposition 1.12.1, it suffices to show that Φ is completely additive. By Proposition 12.3.5, M is σ-finite. Thus, we need to prove that

$$\Phi(p) = \sum_n \Phi(p_n),$$

where $\{p_n\}$ is an orthogonal sequence of projections of M, and $p = \sum_n p_n$. Let

$$p = \int_E^{\oplus} p(t)d\nu(t), \quad p_n = \int_E^{\oplus} p_n(t)d\nu(t),$$

where $p(t), p_n(t) \in M(t), \forall t \in E$ and n. From $p_n p_m = 0, \forall n \neq m$, we may assume that $p_n(t)p_m(t) = 0, \forall t \in E$ and $n \neq m$. Since ν is σ-finite, and $\{\sum_{i=1}^n p_i(t)\}_n$ is increasing $(\forall t \in E)$, it follows from Proposition 12.2.5 that $\sum_n p_n(t) = p(t)$, a.e. ν . Now by the normality of $\Phi(t) (\forall t \in E)$ and Proposition 12.2.5, we get

$$\begin{aligned} \Phi(p) &= \int_E^{\oplus} \sum_n \Phi(t)(p_n(t))d\nu(t) \\ &= \sum_n \int_E^{\oplus} \Phi(t)(p_n(t))d\nu(t) = \sum_n \Phi(p_n). \end{aligned}$$

2) Suppose that M is generated by the diagonal algebra Z_H (on H) and a sequence $\{a_n = \int_E^{\oplus} a_n(t)d\nu(t)\}_n$ of decomposable operators (on H) , and $M(t)$ is generated by $\{a_n(t)\}_n(\forall t \in E)$. Since $\Phi(Z_H) = Z_K$, where Z_K is the diagonal algebra on $K, \Phi(M)$ is generated by Z_K and $\{\Phi(a_n) = \int_E^{\oplus} \Phi(t)(a_n(t))d\nu(t)\}_n$. But $N(t)$ is generated by $\{\Phi(t)(a_n(t))\}_n, \forall t$, thus we have $\Phi(M) = N$. Moreover, Φ is injective obviously. Q.E.D.

References. [28], [35], [36], [119], [121], [158].

12.4. Decomposition of a Hilbert space into a direct integral

From Proposition 5.3.14, if Z is an abelian VN algebra on a Hilbert space H and Z admits a cyclic vector, then there is a regular Borel measure ν on the spectral space Ω (a compact Hausdorff space) of Z and a unitary operator u from H onto $L^2(\Omega,\nu)$ such that

$$\text{supp}\nu = \Omega, \quad C(\Omega) = L^\infty(\Omega,\nu), \quad u m_f u^* = \widehat{m}_f,$$

$\forall f \in L^\infty(\Omega,\nu)$, where $f \longrightarrow m_f$ is the Gelfand transformation from $C(\Omega) = L^\infty(\Omega,\nu)$ onto Z, and $\widehat{m}_f$ is the multiplicative operator on $L^2(\Omega,\nu)$ corresponding to f. In version of Hilbert integral, $L^2(\Omega,\nu) = \int_\Omega^\oplus \mathbb{C} d\nu(t)$, and $\{\widehat{m}_f | f \in L^\infty(\Omega,\nu)\}$ is the diagonal algebra on $\int_\Omega^\oplus \mathbb{C} d\nu(t)$. Thus Z is unitarily equivalent to the diagonal algebra on $\int_\Omega^\oplus \mathbb{C} d\nu(t)$.

The above case is not very interested (since $Z' = Z$ by Proposition 5.3.15) . Now let Z be an abelian VN algebra on a Hilbert space H and Z admits a cyclic sequence $\{\xi_n\}$ of vectors (i.e. Z' is σ-finite). Put

$$\eta_1 = \xi_1, \quad H_1 = \overline{Z\eta_1}, \quad p_1 : H \longrightarrow H_1,$$

$$\cdots\cdots$$

$$\eta_n = \xi_n - \sum_{k=1}^{n-1} p_k \xi_n, \quad H_n = \overline{Z\eta_n}, \quad p_n : H \longrightarrow H_n,$$

$$\cdots\cdots$$

Clearly, $p_i p_j = 0, \forall i \neq j; \sum_i p_i = 1$; and $p_i \in Z', \forall i$.

We may assume that $\|\xi_n\| \leq 1, \forall n$. Then $\|\eta_n\| \leq 1, \forall n$. Let $\eta_0 = \sum_n 2^{-\frac{n}{2}} \eta_n$. Then η_0 is a cyclic vector for Z'.

Let Ω be the spectral space of Z, and ν_n be a regular Borel measure on Ω such that

$$\langle m_f \eta_n, \eta_n \rangle = \int_\Omega f(t) d\nu_n(t), \quad \forall f \in C(\Omega),$$

$n = 0, 1, \cdots$, where $f \longrightarrow m_f$ is the Gelfemd transformation from $C(\Omega)$ onto Z. Define $\nu = \nu_0$, clearly, $\nu = \sum_{n\geq 1} 2^{-n} \nu_n$. Then for each $n = 1, 2, \cdots$, there is $h_n \in L^1(\Omega,\nu)$ with $h_n \geq 0$ such that $\nu_n = h_n \cdot \nu$.

For each $t \in \Omega$, we construct a Hilbert space $H(t)$ and a sequence $\{\varsigma_n(t)|n = 1,2,\cdots\}$ of vectors of $H(t)$ such that $\varsigma_n(t) = 0$ if $h_n(t) = 0$, and $\{\varsigma_n(t)|n$ with $h_n(t) > 0\}$ is an orthogonal normalized basis of $H(t)$. Since $\langle\varsigma_n(t), \varsigma_m(t)\rangle_t = \delta_{nm}\chi_{\text{supp}h_n}(t)$ is measurable on $\Omega, \forall n, m$, $H(\cdot)$ becomes a measurable field of Hilbert spaces over Ω with a fundamental sequence $\{\varsigma_n(\cdot)|n = 1,2,\cdots\}$ of vector fields. We say that $H(t) \neq \{0\}$, a.e.ν. In fact, if there is a Borel subset E of Ω with $\nu(E) > 0$ such that $H(t) = \{0\}, \forall t \in E$, then $h_n(t) = 0, \forall n \geq 1$ and $t \in E$. Futher, $\nu_n(E) = (h_n \cdot \nu)(E) = 0, \forall n \geq 1$ and $\nu(E) = \sum_{n\geq 1} 2^{-n}\nu_n(E) = 0$, a contradiction. Thus, $H(t) \neq \{0\}$, a.e.ν.

Let $\widehat{H} = \int_\Omega^\oplus H(t)d\nu(t)$, and define $u : H \longrightarrow \widehat{H}$ as follows

$$u m_f \eta_n = f(\cdot)h_n(\cdot)^{1/2}\varsigma_n(\cdot), \quad \forall n \geq 1 \text{ and } f \in C(\Omega).$$

Clearly, u is an isometry. If $\varsigma(\cdot) \in \widehat{H}$ satisfies

$$\langle f(\cdot)h_n(\cdot)^{1/2}\varsigma_n(\cdot), \varsigma(\cdot)\rangle = 0, \quad \forall n \geq 1 \text{ and } f \in C(\Omega),$$

then for each $n \geq 1$, we have $h_n(t)^{1/2}\langle\varsigma_n(t), \varsigma(t)\rangle_t = 0$, a.e.$\nu$. By the definition of $\{\varsigma_n(\cdot)\}$, we get $\varsigma(t) = 0$, a.e.ν . Thus , u is unitary. Moreover, it is easy to see that $u m_f u^* = \widehat{m}_f$, where $\widehat{m}_f$ is the diagonal operator on $\widehat{H}$ corresponding to $f, \forall f \in C(\Omega)$. By the proof of Theorem 5.3.1, we have also supp $\nu = \Omega$ and $C(\Omega) = L^\infty(\Omega, \nu)$. Therefore, we obtain the following.

Theorem 12.4.1. Let Z be an abelian VN algebra on a Hilbert space H, Z' be σ–finite, and Ω be the spectral space of Z. Then there is a regular Borel measure ν on Ω with supp $\nu = \Omega$, a measurable field $H(\cdot)$ of Hilbert spaces over Ω, and a unitary operator u from H onto $\widehat{H} = \int_\Omega^\oplus H(t)d\nu(t)$, such that

$$H(t) \neq \{0\}, \text{a.e.}\nu; \quad C(\Omega) = L^\infty(\Omega, \nu);$$

$$u m_f u^* = \widehat{m}_f, \quad \forall f \in L^\infty(\Omega, \nu),$$

where $f \longrightarrow m_f$ is the Gelfand trandformation from $C(\Omega)$ onto $Z, \widehat{m}_f$ is the diagonal operator on $\widehat{H}$ corresponding to f. Consequently, $uZu^* = \widehat{Z}$, where $\widehat{Z}$ is the diagonal algebra on $\widehat{H}$.

Theorem 12.4.2. Let H be a separable Hilbert space, M, Z be VN algebras on H, and $Z \subset M \cap M'$. Then there is a finite Borel measure on $\mathbb{R}$, a measurable field $H(\cdot)$ of Hilbert spaces over $\mathbb{R}$, a measurable field $M(\cdot)$ of VN algebra on $H(\cdot)$,and a unitary operator u from H onto $\widehat{H} = \int_{\mathbb{R}}^\oplus H(t)d\nu(t)$, such that

$$uZu^* = \widehat{Z}, \quad uMu^* = \int_{\mathbb{R}}^\oplus M(t)d\nu(t),$$

where $\widehat{Z}$ is the diagonal algebra on $\widehat{H}$. Moreover, if $Z = M \cap M'$, then $M(t)$ is factorial, a.e.ν .

Proof. From Theorem 5.3.7, Z can be generated by a self-adjoint operator a. Let A be the C^*-algebra generated by $\{1, a\}$. Then A is weakly dense in Z, and the spectral space of A is $\sigma(a)$. By the proof of Theorem 12.4.1, we have also $\nu, H(\cdot)$ and $u : H \longrightarrow \widehat{H} = \int_\Omega^\oplus H(t)d\nu(t)$, such that $um_f u^* = \widehat{m}_f, \forall f \in C(\Omega)$, here $\Omega = \sigma(a), f \longrightarrow m_f$ is the $*$ isomorphism from $C(\Omega)$ onto A.

From Lemma 5.4.4 and the proof of Theorem 12.4.1, we can see that $\nu \succ \nu_\xi, \forall \xi \in H$, where ν_ξ is defined by $\langle m_f \xi, \xi \rangle = \int_\Omega f(t)d\nu_\xi(t), \forall f \in C(\Omega)$. Since supp $\nu = \Omega$, $C(\Omega)$ can be embedded into $L^\infty(\Omega, \nu)$. Also by $\nu \succ \nu_\xi, \forall \xi \in H, f \rightarrow m_f(: C(\Omega) \rightarrow A)$ is $\sigma(L^\infty, L^1)$-weakly continuous. Thus, the map $f \rightarrow m_f$ can be extended to a $*$ isomorphism from $L^\infty(\Omega, \nu)$ onto Z. Further, we have $uZu^* = \widehat{Z}$.

From $Z \subset M \cap M', Z \subset M \subset Z'$, we have $\widehat{Z} \subset uMu^* \subset \widehat{Z}'$. Now since $\widehat{H} = uH$ is separable, by Proposition 12.3.10 $uMu^* = \int_\Omega^\oplus M(t)d\nu(t)$ is decomposable. Moreover, $\nu, H(\cdot)$ and $M(\cdot)$ can be trivially extended from $\Omega = \sigma(a)$ onto $\mathbb{R}$. That comes to the conclusion.

Finally, if $Z = M \cap M'$, then $\widehat{Z} = (uMu^*) \cap (uMu^*)'$. By Proposition 12.3.9, $M(t)$ is factorial, a.e.ν . Q.E.D.

Theorem 12.4.3. Suppose that Ω is a locally compact Hausdorff space, and Ω is a countable union of its compact subsets. Let ν_1, ν_2 be two regular Borel measures on Ω, and $H_1(\cdot), H_2(\cdot)$ be two non-zero measurable fields of Hilbert spaces over Ω . If there is a unitary operator u from $H_1 = \int_\Omega^\oplus H_1(t)d\nu_1(t)$ onto $H_2 = \int_\Omega^\oplus H_2(t)d\nu_2(t)$ such that $um_f^{(1)}u^* = m_f^{(2)}, \forall f \in C_0^\infty(\Omega)$, where $m_f^{(i)}$ is the diagonal operator on H_i corresponding to $f, i = 1, 2$, then $\nu_1 \sim \nu_2$, and there exists a measurable field $v(\cdot)$ of operators from $H_1(\cdot)$ to $H_2(\cdot)$ such that $v(t)$ is unitary from $H_1(t)$ onto $H_2(t)$, a.e.ν_1 , and $u = wv$, where $v = \int_\Omega^\oplus v(t)d\nu_1(t)$, and w is the canonical isomorphism from $\int_\Omega^\oplus H_2(t)d\nu_1(t)$ onto H_2, i.e., if $\nu_2 = \rho \cdot \nu_1$, then

$$w\xi(\cdot) = (\rho^{-\frac{1}{2}}\xi)(\cdot), \quad \forall \xi(\cdot) \in \int_\Omega^\oplus H_2(t)d\nu_1(t).$$

Proof. Let K be a compact subset with $\nu_1(K) = 0$. Since $H_2(\cdot)$ is non-zero , we can pick a measurable field $\eta(\cdot)$ of vectors of $H_2(\cdot)$ with $\|\eta(t)\|_t = 1, \forall t \in \Omega$.

Suppose that U is an open neighborhood of K and the closure $\overline{U}$ of U is compact. Then $(\chi_U\eta)(\cdot) \in H_2$. Put $\xi(\cdot) = u^*(\chi_U\eta(\cdot))(\in H_1)$. For any $\varepsilon > 0$, since $\nu_1(K) = 0$, we can pick an open subset V such that

$$K \subset V \subset U, \quad \text{and} \quad \int_V \|\xi(t)\|_t^2 d\nu_1(t) < \varepsilon.$$

Now let $f \in C_0^\infty(\Omega)$ with $0 \le f \le 1; f(t) = 1, \forall t \in K; f(t) = 0, \forall t \notin V$. Then we have

$$\begin{aligned}\nu_2(K) &\le \int f^2(t)d\nu_2(t) = \int \|(f\eta)(t)\|_t^2 d\nu_2(t)\\ &= \|u^* m_f^{(2)}(\chi_U\eta)(\cdot)\|_{H_1}^2 = \|m_f^{(1)}\xi(\cdot)\|_{H_1}^2 < \varepsilon.\end{aligned}$$

Since ε is arbitrary, it follows that $\nu_2(K) = 0$. Thus, $\nu_2 \prec \nu_1$. Similarly, $\nu_1 \prec \nu_2$. Hence $\nu_1 \sim \nu_2$.

Now let $\widetilde{H}_2 = \int_\Omega^\oplus H_2(t)d\nu_1(t)$. Then $\widetilde{m}_f = vm_f^{(1)}v^*$, where $v = w^*u$, and $\widetilde{m}_f$ is the diagonal operator on $\widetilde{H}_2$ corresponding to $f, \forall f \in C_0^\infty(\Omega)$. Since $C_0^\infty(\Omega)$ is w^*-dense in $L^\infty(\Omega, \nu_1)$, by Theorem 12.2.10 $v = \int_\Omega^\oplus v(t)d\nu_1(t)$ is decomposable. Moreover, v is unitary from H_1 onto $\widetilde{H}_2$. Thus, $v(t)$ is unitary from $H_1(t)$ onto $H_2(t)$, a.e.ν_1. Q.E.D.

Lemma 12.4.4. Let $(E_i, \mathcal{B}_i)$ be a standard Borel space, and ν_i be a σ-finite measure on $\mathcal{B}_i, i = 1,2$. If there exists a $*$ isomorphism π from $L^\infty(E_1, \mathcal{B}_1, \nu_1)$ onto $L^\infty(E_2, \mathcal{B}_2, \nu_2)$, then there is a Borel subset F_i of $E_i, i = 1,2$, and a Borel isomorphism Φ from $(E_2 \backslash F_2)$ onto $(E_1 \backslash F_1)$ such that

$$\nu_1(F_1) = \nu_2(F_2) = 0, \quad \nu_1 \sim \nu_2 \circ \Phi^{-1},$$

and for any $g \in L^\infty(E_1, \mathcal{B}_1, \mu_1), \pi(g)(t) = g(\Phi(t)), \forall t \in (E_2 \backslash F_2)$, a.e.$\nu_2$.

Proof. From Theorem 10.3.16, we may assume that $E_1 = E_2 = [0,1], \mathcal{B}_1$ and $\mathcal{B}_2$ are the collectionof all Borel subsets of $[0,1]$,and ν_1 and ν_2 are two probability measures on $[0,1]$. Let $f_1(t) = t(\in L^\infty(E_1, \nu_1))$ and $f_2 = \pi(f_1)(\in L^\infty(E_2, \nu_2))$. Clearly, we may assume that $0 \le f_2(t) \le 1, \forall t \in E_2$. Then $\Phi(t) = f_2(t)$ is a Borel map from E_2 to E_1.

The function 1 ($\in L^1(E_2, \nu_2)$) determines a faithful normal state ω_2 on $L^\infty(E_2, \nu_2)$, i.e.,

$$\omega_2(h) = \int_0^1 h(t)d\nu_2(t), \quad \forall h \in L^\infty(E_2, \nu_2).$$

Then $\omega_1 = \omega_2 \circ \pi$ is also a faithful normal state on $L^\infty(E_1, \nu_1)$. Thus there is unique $f \in L^1(E_1, \nu_1)$ such that $\omega_1(g) = \int_0^1 f(t)g(t)d\nu_1(t), \forall g \in L^\infty(E_1, \nu_1)$. Also we may assume that $f(t) > 0, \forall t \in E_1$.

If $p(\cdot)$ is a polynomial , then we have

$$\pi(p(f_1))(t) = p(f_2)(t) = p(\Phi(t)), \quad \forall t \in E_2.$$

For any $g \in C[0,1]$,pick a sequence $\{p_n\}$ of polynomials such that $p_n(t) \to g(t)$, uniformly for $t \in [0,1]$. Then in $L^\infty(E_2, \nu_2)$, we have $\pi(p_n(f_1)) \to \pi(g)$. Thus $\pi(g)(t) = g(\Phi(t))$, a.e.ν_2. Further,

$$\begin{aligned}\int_0^1 g(t)f(t)d\nu_1(t) = \omega_1(g) &= \omega_2(\pi(g)) \\ &= \int_0^1 g(\Phi(t))d\nu_2(t),\end{aligned}$$

$\forall g \in C[0,1]$. So $\Phi(\nu_2) = f \cdot \nu_1$, where $\Phi(\nu_2) = \nu_2 \circ \Phi^{-1}$.

For $g \in L^\infty(E_1, \nu_1)$, pick $g_n \in C[0,1]$ such that $g_n \xrightarrow{\tau} g$. Then

$$\int_0^1 |\pi(g_n)(t) - \pi(g)(t)|^2 d\nu_2(t) = \omega_2(\pi((g-g_n)^*(g-g_n)))$$

$$= \quad \omega_1((g_n - g)^*(g_n - g)) = \int_0^1 f(t)|g_n(t) - g(t)|^2 d\nu_1(t) \to 0.$$

Since $f(t) > 0, \forall t \in E$, we can find a subsequence $\{g_{n_k}\}$ such that

$$g_{n_k}(t) \to g(t), \quad a.e.\nu_1, \quad \text{and} \quad \pi(g_{n_k})(t) \to \pi(g)(t), \ a.e.\nu_2.$$

From $\pi(g_{n_k})(t) = g_{n_k}(\Phi(t)), a.e.\nu_2, \forall k$, it follows that $g_{n_k}(\Phi(t)) \to \pi(g)(t), a.e.$ ν_2. On the other hand, $g_{n_k}(t) \to g(t), a.e.\nu_1, \Phi(\nu_2) = f \cdot \nu_1$ and $g_{n_k}(\Phi(t)) \to g(\Phi(t)), a.e.\nu_2$, so we get

$$\pi(g)(t) = g(\Phi(t)), a.e.\nu_2, \quad \forall g \in L^\infty(E_1, \nu_1).$$

Replacing π by $\pi^{-1}(: L^\infty(E_2, \nu_2) \to L^\infty(E_1, \nu_1))$, there is a Borel map Ψ from E_1 to E_2 such that $\Psi(\nu_1) = \nu_1 \circ \Psi^{-1} \prec \nu_2$, and

$$\Phi^{-1}(h)(t) = h(\Psi(t)), a.e.\nu_1, \quad \forall h \in L^\infty(E_2, \nu_2).$$

Thus we have

$$\begin{cases} (g \circ \Phi)(\Psi(t)) &= \pi^{-1}(\pi(g))(t) \\ &= g(t), \quad a.e.\nu_1, \forall g \in L^\infty(E_1, \nu_1); \\ (h \circ \Psi)(\Phi(t)) &= \pi(\pi^{-1}(h))(t) \\ &= h(t), \quad a.e.\nu_2, \forall h \in L^\infty(E_2, \nu_2). \end{cases}$$

In particular, from $t \in L^\infty(E_1, \nu_1) \cap L^\infty(E_2, \nu_2)$ we get

$$\Phi \circ \Psi(t) = t, \quad a.e.\nu_1; \quad \Psi \circ \Phi(t) = t, \quad a.e.\nu_2.$$

So there is $F_2' \in \mathcal{B}_2$ with $\nu_2(F_2') = 0$ such that $\Psi \circ \Phi(t) = t, \forall t \in (E_2 \backslash F_2')$. Let $F_1' = \Psi^{-1}(F_2')$. Then $\nu_1(F_1') = 0$ since $\nu_1 \circ \Psi^{-1} \prec \nu_2$. It is easy to see that

$$\Psi(E_1 \backslash F_1') = E_2 \backslash F_2', \quad \Phi(E_2 \backslash F_2') \subset E_1 \backslash F_1', \tag{1}$$

and Φ is injective on $(E_2\backslash F_2')$. Further, pick $F_1'' \in \mathcal{B}_1$ with $F_1'' \subset (E_1\backslash F_1')$ and $\nu_1(F_1'') = 0$, such that

$$\Phi \circ \Psi(t) = t, \quad \forall t \in E_1\backslash F_1, \tag{2}$$

where $F_1 = F_1' \cup F_1''$. Clearly , $\nu_1(F_1) = 0$. Let $F_2'' = \Phi^{-1}(F_1'') \cap (E_2\backslash F_2')$. Then $\nu_2(F_2'') = 0$ since $\nu_2 \circ \Phi^{-1} \prec \nu_1$. Further, $\nu_2(F_2) = 0$, where $F_2 = F_2' \cup F_2'' = F_2' \cup \Phi^{-1}(F_1'')$. Then , Φ maps $(E_2\backslash F_2)$ into $(E_1\backslash F_1)$ injectively.

Now let $t \in (E_1\backslash F_1)$. By (1) , we have $\Psi(t) \in (E_2\backslash F_2')$. If $\Psi(t) \in F_2'' \subset \Phi^{-1}(F_1'')$, then $\Phi \circ \Psi(t) \in F_1''$. But by (2) , $\Phi \circ \Psi(t) = t \notin F_1$, a contradiction. Thus $\Psi(t) \notin F_2''$, and $\Psi(t) \in (E_2\backslash F_2)$. Further, by (2) we get $\Phi(E_2\backslash F_2) = (E_1\backslash F_1)$.

From Theorem 10.3.12 and Proposition 10.3.15, Φ is a Borel isomorphism from $(E_2\backslash F_2)$ onto $(E_1\backslash F_1)$, and its inverse is Ψ. Now by $\nu_2 \circ \Phi^{-1} \prec \nu_1, \nu_1 \circ \Psi^{-1} \prec \nu_2$, we can see that $\nu_1 \sim \nu_2 \circ \Phi^{-1}$ on $(E_1\backslash F_1)$. Q.E.D.

Theorem 12.4.5. Let $(E_i, \mathcal{B}_i)$ be a standard Borel space, ν_i be a σ-finite measure on $\mathcal{B}_i, H_i(\cdot)$ be a non-zero measurable field of Hilbert spaces over $E_i, M_i(\cdot)$ be a measurable field of VN algebras on $H_i(\cdot)$, and Z_i be the diagonal algebra on $H_i = \int_{E_i}^{\oplus} H_i(t)d\nu_i(t), i = 1, 2$. If there is a unitary operator u from H_i onto H_2 such that

$$uM_1u^* = M_2, \quad uZ_1u^* = Z_2,$$

where $M_i = \int_{E_i}^{\oplus} M_i(t)d\nu_i(t), i = 1, 2$, then there is a Borel subset F_i of E_i with $\nu_i(F_i) = 0, i = 1, 2$, a Borel isomorphism Φ from $(E_2\backslash F_2)$ onto $(E_1\backslash F_1)$, and a measurable field $u(\cdot)$ of unitary operators from $H_1(\cdot)$ to $H_2(\Phi^{-1}(\cdot))$, such that :

1) $u(t)M_1(t)u(t)^* = M_2(\Phi^{-1}(t)), \forall t \in (E_1\backslash F_1)$;
2) $\Phi(\nu_2) = \nu_2 \circ \Phi^{-1} \sim \nu_1$;
3) $u = \int_{E_1\backslash F_1}^{\oplus} \left(\frac{d\Phi(\nu_2)}{d\nu_1}(t)\right)^{1/2} u(t)d\nu_1(t)$.

Proof. By Proposition 12.2.8 and $uZ_1u^* = Z_2$, there is a $*$ isomorphism π from $L^\infty(E_1, \mathcal{B}_1, \nu_1)$ onto $L^\infty(E_2, \mathcal{B}_2, \nu_2)$. From Lemma 12.4.4, we have $F_i \in \mathcal{B}_i$ with $\nu_i(F_i) = 0, i = 1, 2$, and a Borel isomorphism Φ from $(E_2\backslash F_2)$ onto $(E_1\backslash F_1)$, such that $\Phi(\nu_2) \sim \nu_1$, and $\pi(g)(t) = g(\Phi(t)), \forall t \in (E_2\backslash F_2)$, a.e. ν_2, and $g \in L^\infty(E_1, \mathcal{B}_1, \nu_1)$. Replacing E_1, E_2, ν_1 by $(E_1\backslash F_1), (E_2\backslash F_2), \Phi(\nu_2)$ respectively, with the Borel isomorphism we can identify $(E_1, \mathcal{B}_1, \nu_1)$ with $(E_2, \mathcal{B}_2, \nu_2)$. Now our case becomes the following. Let $(E, \mathcal{B})$ be a standard Borel space, ν be a σ-finite measure on $\mathcal{B}, H_i(\cdot)$ be a measurable field of Hilbert spaces over $E, i = 1, 2$, and u be a unitary operator from $H_1 = \int_E^{\oplus} H_1(t)d\nu(t)$

onto $H_2 = \int_E^{\oplus} H_2(t)d\nu(t)$, such that

$$uM_1u^* = M_2, \quad um_f^{(1)}u^* = m_f^{(2)}, \quad \forall f \in L^\infty(E, \mathcal{B}, \nu),$$

where $M_i = \int_E^{\oplus} M_i(t)d\nu(t), m_f^{(i)}$ is the diagonal operator on H_i corresponding to $f, i = 1, 2$. By Theorem 12.2.10, $u = \int_E^{\oplus} u(t)d\nu(t)$, a.e.$\nu$. From Proposition 12.3.7, there is a sequence $\{a_n = \int_E^{\oplus} a_n(t)d\nu(t)|n\}$ of decomposable operators on H_1 such that M_1 is generated by Z_1 and $\{a_n|n\}$, and $M_1(t)$ is generated by $\{a_n(t)|n\}$, a.e.ν. Then M_2 is generated by Z_2 and $\{ua_nu^*|n\}$, and $M_2(t)$ is generated by $\{u(t)a_n(t)u(t)^*|n\}, a.e.\nu$. Therefore, $M_2(t) = u(t)M_1(t)u(t)^*, a.e.\nu$.

Q.E.D.

References. [28], [120], [158].

12.5. The relations between a decomposable Von Neumann algebra and its components

Let $(E, \mathcal{B})$ be a standard Borel space, ν be a σ-finite measure on $\mathcal{B}, H(\cdot)$ be a measruable field of Hilbert spaces over $E, H = \int_E^{\oplus} H(t)d\nu(t)$, and $M = \int_E^{\oplus} M(t)d\nu(t)$ be a decomposable VN algebra on H. In this section, we shall discuss the relations between M and $M(t)'s$.

Proposition 12.5.1. Let $p = \int_E^{\oplus} p(t)d\nu(t), p' = \int_E^{\oplus} p'(t)d\nu(t)$ be projections of M, M' respectively. Then

$$M_p = \int_E^{\oplus} M(t)_{p(t)}d\nu(t), \quad M_{p'} = \int_E^{\oplus} M(t)_{p'(t)}d\nu(t),$$

and $c(p) = \int_E^{\oplus} c(p(t))d\nu(t)$.

Proof. From Proposition 12.3.7, we can get the expressions of M_p and $M_{p'}$.

Now suppose that M is generated by the diagonal algebra Z and a sequence $\{a_n = \int_E^{\oplus} a_n(t)d\nu(t)|n\}$ of decomposable operators, and $M(t)$ is generated by $\{a_n(t)|n\}, \forall t \in E$. Through a suitable treatment, we may assume that $\{a_n(t)|n\}$ is strongly dense in $M(t), \forall t \in E$. Let $\{\xi_m(\cdot)|m\}$ be a fundamental sequence of measurable fields of vectors. Then $\{a_n(t)p(t)\xi_m(t)|n, m\}$

is a total subset of $c(p(t))H(t), \forall t \in E$. By the method in Proposition 12.1.2, we can construct an orhtogonal normalized basis $\{\eta_k(\cdot)\}$ of $c(p(\cdot))H(\cdot)$ from $\{a_n(\cdot)p(\cdot)\xi_m(\cdot)|n,m\}$. Clearly, $\eta_k(\cdot)$ is measurable, $\forall k$. Then

$$\langle c(p(t))\xi_n(t), \xi_m(t)\rangle_t = \sum_k \langle \xi_n(t), \eta_k(t)\rangle_t \cdot \langle \eta_k(t), \xi_m(t)\rangle_t$$

is measurable on $E, \forall n, m$. Thus, the field $c(p(\cdot))$ of operators is measurable. Put $z = \int_E^{\oplus} c(p(t))d\nu(t)$. Clearly,

$$z a_n p\xi(\cdot) = a_n p\xi(\cdot), \quad \forall n \text{ and } \quad \xi(\cdot) \in H.$$

So $z \geq c(p)$. Now write $c(p) = \int_E^{\oplus} q(t)d\nu(t)$, where $q(t)$ is a central projection of $M(t), \forall t \in E$. Since $c(p)a_n p = a_n p, \forall n$, it follows that $q(t)a_n(t)p(t) = a_n(t)p(t), a.e.\nu, \forall n$, i.e., $q(t) \geq c(p(t)), a.e.\nu$. Therefore, $c(p) \geq z$, and $c(p) = \int_E^{\oplus} c(p(t))d\nu(t)$. Q.E.D.

Proposition 12.5.2. If M is discrete , then $M(t)$ is also discrete , a.e.ν.

Proof. By Theorem 6.7.1, there is an abelian projection p of M with $c(p) = 1_H$. Further from Proposition 12.5.1, $p(t)$ is also an abelian projection of $M(t)$ with $c(p(t)) = 1_{H(t)}$, a.e.ν. Therefore, $M(t)$ is discrete, a.e.ν. Q.E.D.

Proposition 12.5.3. If M is properly infinite, then $M(t)$ is also properly infinite, a.e.ν.

Proof. It is immediate from Theorem 6.4.4. Q.E.D.

Proposition 12.5.4. If $M(t)$ is finite, a.e.ν, then M is also finite.

Proof. It is immediate from the definition of finite VN algebras. Q.E.D.

Proposition 12.5.5. If M is continuous, then $M(t)$ is also continuous, a.e. ν.

Proof. Put $E_k = \{t \in E | \dim H(t) = k\}$, and let z_k be the diagonal operator corresponding to $\chi_{E_k}, \forall k$. Clearly, z_k is a central projection of M, Mz_k is continuous, and $Mz_k = \int_{E_k}^{\oplus} M(t)d\nu(t), \forall k$. Thus , we may assume that $H(\cdot) = H_0$ is a constant field over E.

Suppose that M, M' are generated by the diagonal algebra Z and $\{a_n = \int_E^{\oplus} a_n(t)d\nu(t)|n\}, \{a'_n = \int_E^{\oplus} a'_n(t)d\nu(t)|n\}$ respectively, and $M(t), M(t)'$ are generated by $\{a_n(t)|n\}, \{a'_n(t)|n\}$ respectively, $\forall t \in E$. We may assume that $\|a_n\|, \|a'_n\|, \|a_n(t)\|$ and $\|a'_n(t)\| \le 1, \forall n$ and $t \in E$, and $\{a_n(t)|n\}^* = \{a_n(t)|n\}$, $\{a'_n(t)|n\}^* = \{a'_n(t)|n\}, \forall t \in E$.

Let S be the unit ball of $B(H_0)$. Clearly, S is a Polish space with respect to the strong operator topology. Consider a subset G of $S \times E$. $(a,t) \in G$, if : 1) $aa'_n(t) = a'_n(t)a, \forall n$; 2) a is a non-zero projection; 3) $aa_n(t)aa_m(t)a = aa_m(t)aa_n(t)a, \forall n, m$. Noticing Proposition 10.3.14, G is a Borel subset of $S \times E$. Let π be the projection from $S \times E$ onto E. Then from Theorem 10.4.5 there is a Borel subset $F \subset \pi G$, and a Borel map $p(\cdot)$ from F to S such that $(p(t), t) \in G, \forall t \in F$, and $(\pi G \backslash F) \subset$ some ν-zero subset.

Let $p(t) = 0, \forall t \in E \backslash F$, and $p = \int_E^{\oplus} p(t) d\nu(t)$. Then p is an abelian projection of M. But M is continuous, so $p = 0$ and $\nu(F) = 0$. Thus , $\pi(G) = \{t \in E | M(t)$ is not continouos $\}$ is contained in some ν -zero subset, i.e., $M(t)$ is continuous, a.e.ν. Q.E.D.

Proposition 12.5.6. If M is purely infinite, then $M(t)$ is also purely infinite, a.e.ν.

Proof. With the same reason as in Proposition 12.5.5, we may assume that $H(\cdot) = H_0$ is a constant field. Keep the notations of $\{a_n, a_n(t), a'_n, a'_n(t)|n\}, S$ and etc. in Proposition 12.5.5. Further, let $(H_0)_\infty = \sum_n \oplus H_0$. Consider a subset G of $S \times (H_0)_\infty \times E$. $(a, (\eta_k), t) \in G$, if : 1) $aa'_n(t) = a'_n(t)a, \forall n$; 2) a is a non-zero projection; 3) $a\eta_k = \eta_k, \forall k$; 4) $\sum_k \|\eta_k\|^2 = 1$; 5) for any finite sets $\wedge_1, \wedge_2$ of positive integers,

$$\sum_k \langle (a\Pi_{n\in\wedge_1} a_n(t) a\Pi_{m\in\wedge_2} a_m(t) a - a\Pi_{m\in\wedge_2} a_m(t) a\Pi_{n\in\wedge_1} a_n(t) a)\, \eta_k, \eta_k \rangle = 0.$$

Clearly, G is Borel subset. Let π be the projection from $S \times (H_0)_\infty \times E$ onto E. From Theorem 10.4.5, there is a Borel subset $F \subset \pi G$, and Borel maps $p(\cdot), (\eta_k(\cdot))$ from F to $S, (H_0)_\infty$, such that $(\pi G \backslash F) \subset$ some ν-zero subset , and for any $t \in F, (p(t), (\eta_k(t)), t) \in G$.

Define $p(t) = 0, \eta_k(t) = 0, \forall k$ and $t \notin F$. Then $p = \int_E^{\oplus} p(t) d\nu(t)$ is a projection of M. We may assume that $\nu(E) < \infty$. Then $\xi_k = \eta_k(\cdot) \in H, p\xi_k = \xi_k, \forall k$, and $\sum_k \|\xi_k\|^2 = \nu(F)$. By the construction, $\sum_k \langle \cdot, \xi_k, \xi_k \rangle$ is a normal trace on M_p. Since M is purely infinite, it follows that $\nu(F) = 0$, i.e., $\pi(G) \subset$ some ν-zero subset. Moreover, it is easy to see that $\pi G = \{t \in E | M(t)$ is not purely infinite $\}$. Therefore, $M(t)$ is purely infinte, a.e.ν. Q.E.D.

Proposition 12.5.7. If M is finite, then $M(t)$ is also finite, a.e.ν.

Proof. Keep the notations: $H(\cdot) = H_0, a_n, a_n(t), a'_n, a'_n(t), S$ and etc. as in Proposition 12.5.5. Consider a subset G of $S \times E$. $(v,t) \in G$, if :1) $va'_n(t) = a'_t(t)v, \forall n$; 2) $v^*v = 1, vv^* \neq 1$. Clearly, G is a Borel subset of $S \times E$. From Theorem 10.4.5, there is a Borel subset $F \subset \pi G$, where π is the projection of $S \times E$ onto E , and a Borel map $v(\cdot) : F \to S$, such that $(\pi G \backslash F) \subset$ some ν-zero subset , and $(v(t), t) \in G, \forall t \in F$.

Define $v(t) = 0, \forall t \notin F$, and $v = \int_E^{\oplus} v(t) d\nu(t)$. Then $v^*v = p$ is the diagonal operator corresponding to χ_F. Since M_p is finite, and $v(t)v(t)^* \neq 1, \forall t \in F$, it follows that $\nu(F) = 0$. Therefore , $M(t)$ is finite. a.e.ν. Q.E.D.

Proposition 12.5.8. If M is semi-finite, then $M(t)$ is also semi-finite, a.e.ν.

Proof. M contains a finite projection p with $c(p) = 1$. Write $p = \int_E^{\oplus} p(t) d\nu$ (t). From Propositon 12.5.1 and 12.5.7, $p(t)$ is a finite projection of $M(t)$, and $c(p(t)) = 1_t$, a.e.ν. Therefore, $M(t)$ is semi-finite , a.e.ν. Q.E.D.

Theorem 12.5.9. Let $(E, \mathcal{B})$ be a standard Borel space, ν be a σ-finite measure on $\mathcal{B}, H(\cdot)$ be a measurable field of Hilbert spaces over $E, H = \int_E^{\oplus} H(t) d\nu(t)$, and $M = \int_E^{\oplus} M(t) d\nu(t)$ be a decomposable VN algebra on H. If $z = \int_E^{\oplus} z(t) d\nu(t)$ is the maximal central projection of M such that Mz is finite, or semi-finite, or discrete, then $z(t)$ is also the maximal central projection of $M(t)$ such that $M(t)z(t)$ is finite, or semi-finite, or discrete, a.e.ν.

Proof. It is immediate from $Mz = \int_E^{\oplus} M(t)z(t) d\nu(t), M(1-z) = \int_E^{\oplus} M(t)$ $(1_t - z(t)) d\nu(t)$, and Propositions 12.5.3, 12.5.7, 12.5.8, 12.5.6, 12.5.2, 12.5.5 . Q.E.D.

Theorem 12.5.10 . Let $(E, \mathcal{B})$ be a standard Borel space, ν be a σ-finite measure on $\mathcal{B}, H(\cdot)$ be a measurable field of Hilbert spaces over $E, H = \int_E^{\oplus} H(t) d\nu(t)$, and $M = \int_E^{\oplus} M(t) d\nu(t)$ be a decomposable VN algebra on H. Then M is finite, semi-finite, properly infinite, purely infinite, discrete, or continuous, if and only if , $M(t)$ is finite, semi-finite, properly infinite, purely infinite , discrete, or continuous, a.e.ν.

Proof. It is immediate from above discussions. Q.E.D.

References. [21], [28], [119], [154].

12.6. The constant fields of operators and Von Neumann algebras

Lemma 12.6.1. Let A be a separable C^*-algebra with an identity, and H_0 be a separable Hilbert space. Define

$$\operatorname{Rep}(A,H_0)=\left\{\pi \,\middle|\, \begin{array}{c}\pi \text{ is a nondegenerate } * \text{ representation}\\ \text{of } A \text{ on } H_0\end{array}\right\}$$

and give $\operatorname{Rep}(A,H_0)$ a topology as follows : $\pi_l \to \pi$, if $\|\pi_l(a)\xi-\pi(a)\xi\|\to 0, \forall a\in A, \xi\in H_0$. Then $\operatorname{Rep}(A,H_0)$ is a Polish space.

Proof. Let $\{a_n\},\{\xi_m\}$ be dense subsets of the unit balls of A, H_0 respectively. For any $\pi_1,\pi_2\in\operatorname{Rep}(A,H_0)$, define

$$d(\pi_1,\pi_2)=\sum_{n,m}2^{-(n+m)}\|(\pi_1(a_n)-\pi_2(a_n))\xi_m\|.$$

Now it suffices to show that ($\operatorname{Rep}(A,H_0),d)$ is separable. Let

$$\begin{aligned} J&=\{(n,m)|n,m=1,2,\cdots\},\\ E&=\{f:J\to H_0|\sum_{n,m}2^{-(n+m)}\|f(n,m)\|<\infty\}.\end{aligned}$$

For any $f\in E$, define $\|f\|=\sum_{n,m}2^{-(n+m)}\|f(n,m)\|$. Clearly, $(E,\|\cdot\|)$ is a separable Banach space. Moreover, for any $\pi\in\operatorname{Rep}(A,H_0)$, let

$$f_\pi(n,m)=\pi(a_n)\xi_m, \forall n,m.$$

Then $\pi\to f_\pi$ is an isometric map from $\operatorname{Rep}(A,H_0)$ to $(E,\|\cdot\|)$. Therefore, ($\operatorname{Rep}(A,H_0),d)$ is separable. Q.E.D.

Lemma 12.6.2. Let $(E,\mathcal{B})$ be a Borel space . Then a map $\Phi:(E,\mathcal{B})\to\operatorname{Rep}(A,H_0)$ is Borel if and only if $t\to\langle\Phi(t)(a)\xi,\eta\rangle$ is measurable on $E, \forall a\in A,\xi,\eta\in H_0$. Here $\operatorname{Rep}(A,H_0)$ is as in Lemma 12.6.1.

Proof. It suffices to prove the sufficiency. By Lemma 12.6.1, $\operatorname{Rep}(A,H_0)$ admits a countable dense subset $\{\pi_n\}$. Now we need only to prove that for any $n,m,\Phi^{-1}(\{\pi|d(\pi,\pi_n)<m^{-1}\})\in\mathcal{B}$. But it is immediate from the sufficiency

and

$$\begin{aligned}
&\Phi^{-1}(\{\pi|d(\pi,\pi_n) < m^{-1}\}) \\
= \quad & \{t \in E|\sum_{i,j} 2^{-(i+j)}\|(\pi_n(a_i) - \Phi(t)(a_i))\xi_j\| < m^{-1}\} \\
= \quad & \{t \in E|\sum_{i,j} 2^{-(i+j)} \sup_k |\langle(\pi_n(a_i) - \Phi(t)(a_i))\xi_j, \xi_k\rangle| < m^{-1}\}.
\end{aligned}$$

Q.E.D.

Theorem 12.6.3. Let $(E, \mathcal{B})$ be a Borel space, $H(\cdot)$ be a measurable field of Hilbert spaces over $E, \wedge$ be any index set , and $a_l(\cdot)$ be a measurable field of operators on $H(\cdot), \forall l \in \wedge$. Suppose that H_0 is a separable Hilbert space, and for each $t \in E$ there is a unitary operator $u(t)$ from $H(t)$ onto H_0 such that $u(t)a_l(t)u(t)^* = b_l$, where $b_l \in B(H_0), \forall l \in \wedge$. Then there is a measurable field $v(\cdot)$ of unitary operators from $H(\cdot)$ to the constant field H_0 such that $v(t)a_l(t)v(t)^* = b_l, \forall t \in E, l \in \wedge$.

Proof. We may assume that $H(\cdot)$ is the constant field H_0 . Then for any $\xi_0, \eta \in H_0, l \in \wedge, t \to \langle u(t)^* b_l u(t)\xi, \eta\rangle$ is measurable on E. Let M be the VN algebra generated by $\{b_l|l \in \wedge\}$, and M_0 be the $*$ subalgebra of $B(H_0)$ generated by $\{b_l|l \in \wedge\}$. For any $b \in M$, since H_0 is separable, there is a sequence $\{b_n\}$ of M_0 such that $b_n \to b$ (strongly) . Thus, we can see that $t \to \langle u(t)^* b u(t), \xi, \eta\rangle$ is measurable on $E, \forall \xi, \eta \in H_0, b \in M$.

Clearly, M is countably generated. Hence, there is a separable C^* -algebra A on H_0 such that $1 \in A \subset M$, and A is strongly dense in M. Let G be the group of all unitary operators on H_0. With the strong operator topology, G is Polish (see Lemma 11.4.1) . Put

$$G_0 = \{u \in G|u^* a u = a, \forall a \in A\}.$$

Then G_0 is a closed subgroup of G. By Theorem 10.4.2, there is a Borel subset F of G such that $^{\#}(F \cap G_0 u) = 1, \forall u \in G$. For any $u \in G$, define a nondegenerate $*$ representation π_u of A on H_0:

$$\pi_u(a) = u^* a u, \quad \forall a \in A.$$

Clearly , $u \to \pi_u$ is a continuous map from G to $\text{Rep}(A, H_0)$. Denote this map by Ψ. Obviously, Ψ is injective on F. From Lemma 12.6.1, Ψ is a Borel isomorphism from F onto $\Psi(F) = \Psi(G)$. Define $(\Psi|F)^{-1} = \Phi$. For any $t \in E$, let $v(t) = \Phi \circ \Psi(u(t))$. Then we get

$$E \xrightarrow{u(\cdot)} F \xrightarrow{\Psi(\cdot)} \Psi(F) = \Psi(G) \xrightarrow{\Phi(\cdot)} F.$$

From the preceding paragraph, the function

$$\begin{aligned} t \to \langle \Psi \circ v(t)(a)\xi, \eta\rangle &= \langle v(t)^* a v(t)\xi, \eta\rangle \\ &= \langle u(t)^* a u(t)\xi, \eta\rangle \end{aligned}$$

is measurable on $E, \forall a \in A(\subset M), \xi, \eta \in H_0$. By Lemma 12.6.2, the map $\Psi \circ v(\cdot)$ is Borel from E to $\text{Rep}(A, H_0)$. Thus, $v(\cdot) = \Phi \circ \Psi \circ v(\cdot)$ is Borel from E to G. In particular, $v(\cdot)$ is a measurable field of unitary operators, and

$$v(t)^* a v(t) = u(t)^* a u(t), \quad \forall a \in A, t \in E.$$

Since A is strongly dense in M, we obtain $v(t)^* b_l v(t) = a_l(t), \forall t \in E, l \in \Lambda$.

Q.E.D.

Corollary 12.6.4. Keep the assumptions of Theorem 12.6.3, and let ν be a measure on $\mathcal{B}$. Then there is a unitary operator v from $H = \int_E^\oplus H(t) d\nu(t)$ onto $L^2(E, \mathcal{B}, \nu) \otimes H_0$ such that $v a_l v^* = 1 \otimes b_l$, where $a_l = \int_E^\oplus a_l(t) d\nu(t), \forall l$, and "1" is the identity operator on $L^2(E, \mathcal{B}, \nu)$.

Theorem 12.6.5. Let $(E, \mathcal{B})$ be a Borel space, $H(\cdot)$ be a measurable field of Hilbert spaces over E, and $M(\cdot)$ be a measurable field of VN algebras on $H(\cdot)$. Suppose that H_0 be a separable Hilbert space, and for any $t \in E$ there is a unitary operator $u(t)$ from $H(t)$ onto H_0 such that $u(t)M(t)u(t)^* = M_0$, where M_0 is a fixed VN algebra on H_0. Then there is a measurable field $v(\cdot)$ of unitary operators from $H(\cdot)$ to the constant field H_0 such that $v(t)M(t)v(t)^* = M_0, \forall t \in E$.

Proof. We may assume that $H(\cdot) = H_0$, and let G be the Polish group of all unitary operators on H_0. Let $G_0 = \{u \in G | u^* M_0 u = M_0\}$. Then there is a Borel subset F of G such that $^\#(F \cap G_0 u) = 1, \forall u \in G$. Define a map $\Psi : G \to \mathcal{A}, \Psi(u) = u^* M_0 u$, where $\mathcal{A}$ is the Borel space of all VN algebras on H_0. From the proof of Proposition 11.4.2, Ψ is Borel. Clearly, Ψ is injective on F. Thus, Ψ is a Borel isomorphism from F onto $\Psi(F) = \Psi(G)$. Define $(\Psi|F)^{-1} = \Phi$. For any $t \in E$, let $v(t) = \Phi \circ \Psi(u(t))$. Then we get

$$E \xrightarrow{v(\cdot)} F \xrightarrow{\Psi(\cdot)} \Psi(F) = \Psi(G) \xrightarrow{\Phi(\cdot)} F.$$

Notice that $\Psi \circ v(\cdot) = \Psi(u(\cdot)) = u(\cdot)^* M_0 u(\cdot) = M(\cdot)$ is a Borel map from E to $\mathcal{A}$ (see Proposition 12.3.2.) Thus , $v(\cdot) = \Phi \circ \Psi \circ v(\cdot)$ is a measurable field of unitary operators, and $v(t)M(t)v(t)^* = M_0, \forall t \in E$.

Q.E.D.

Corollary 12.6.6. Keep the assumptions of Theorem 12.6.5, and let ν be a σ-finite measure on $\mathcal{B}$. Then there is a unitary operator from $H = \int_E^{\oplus} H(t)d\nu(t)$ onto $L^2(E,\mathcal{B},\nu) \otimes H_0$ such that $vMv^* = \hat{Z}\overline{\otimes}M_0$, where $M = \int_E^{\oplus} M(t)d\nu(t)$, and $\hat{Z}$ is the multiplicative algebra on $L^2(E,\mathcal{B},\nu)$.

Proof. Pick $v(\cdot)$ as in Theorem 12.6.5, and let $v = \int_E^{\oplus} v(t)d\nu(t)$. Then $vMv^* = \hat{Z}\overline{\otimes}M_0$. Q.E.D.

References. [28], [35], [120], [173].

12.7. Borel subsets of the Borel space of Von Neumann algebras

Let H be a separable Hilbert space, and $\mathcal{A}, \mathcal{F}$ be the collections of all VN algebras, factors on H respectively. By Theorem 11.3.2, 11.3.6, $\mathcal{A}$ and $\mathcal{F}$ are two standard Borel spaces.

Proposition 12.7.1. Let $\mathcal{A}_{I_n}$ be the collection of all type (I_n) VN algebras on H. Then $\mathcal{A}_{I_n}$ is a Borel subset of $\mathcal{A}, n = \infty, 1, 2, \cdots$.

Proof. By Proposition 6.7.8, any type (I_n) VN algebra is spatially $*$ isomorphic to $Z\overline{\otimes}B(H_n)$, where Z is an abelian VN algebra, and dim $H_n = n$. Now from corollary 5.3.9 and Proposition 11.4.3, $\mathcal{A}_{I_n}$ is a Borel subset of $\mathcal{A}$. Q.E.D.

Proposition 12.7.2. Let $\mathcal{A}_I$ be the collection of all type (I) VN algebras on H. Then $\mathcal{A}_I$ is a Borel subset of $\mathcal{A}$.

Proof. As the proof of Proposition 11.4.3, define a Borel map Φ from $\mathcal{A} = \mathcal{A}(H)$ to $\mathcal{A}(H\otimes H)$: $\Phi(M) = M\overline{\otimes}\mathbb{C}1_H, \forall M \in \mathcal{A}(H)$. Then $M \in \mathcal{A}_I$ if and only if $\Phi(M)' = M'\overline{\otimes}B(H)$ is type (I_∞) (here we assume that dim $H = \infty$. If dim $H < \infty$, then $\mathcal{A}_I = \mathcal{A}$, and the conclusion is trivial). Let E be the collection of all type (I_∞) VN algebras on $H \otimes H$. Then E is a Borel subset of $\mathcal{A}(H\otimes H)$ by Proposition 12.7.1. From Proposition 11.3.1, $E' = \{N' | N \in E\}$ is also a Borel subset of $\mathcal{A}(H\otimes H)$. Then $\mathcal{A}_I = \Phi^{-1}(E')$ is a Borel subset of $\mathcal{A} = \mathcal{A}(H)$. Q.E.D.

Proposition 12.7.3. Let $\mathcal{A}_f$ be the collection of all finite VN algebras on H. Then $\mathcal{A}_f$ is a Borel subset of $\mathcal{A}$.

Proof. First, by a similar proof as Lemma 11.4.6, $(\mathcal{A}\backslash\mathcal{A}_f)$ is Sousline. Then from a similar proof as Proposition 11.4.7, $\mathcal{A}_f$ is also Sousline. Therefore, $\mathcal{A}_f$ is a Borel subset of $\mathcal{A}$. Q.E.D.

Lemma 12.7.4. Let $\mathcal{A}_{sf}$ be the collection of all semi-finite VN algebras on H. Then $\mathcal{A}_{sf}$ is a Sousline subset of $\mathcal{A}$.

Proof. It is immediate from a same proof as Lemma 11.4.8. Q.E.D.

Lemma 12.7.5. Let $\mathcal{A}_{III}$ be the collection of all type (III) VN algebras on H. Then $(\mathcal{A}\backslash\mathcal{A}_{III})$ is a Sousline subset of $\mathcal{A}$.

Proof. A VN algebra M on H is not type (III) if and only if M contains a non-zero finite projection. Consider a subset E of $\mathcal{A}\times P\times H_\infty$, where P is the collection of all projections on H, and $H_\infty=\sum_n\oplus H$. $(M,p,(\xi_k))\in E$, if : p is a non-zero projection of $M; p\xi_k=\xi_k,\forall k;\sum_k\langle\cdot\xi_k,\xi_k\rangle$ is a trace on $pMp; \{a'\xi_k|a'\in M',k\}$ is a total subset of pH. By a similar proof as Lemma 11.4.8, E is a Borel subset. Therefore, $(\mathcal{A}\backslash\mathcal{A}_{III})=\pi E$ is a Sousline subset of $\mathcal{A}$, where π is the projection from $\mathcal{A}\times P\times H_\infty$ onto $\mathcal{A}$. Q.E.D.

Lemma 12.7.6. Let $(E,\mathcal{B})$ be a standard Borel space, and $\mathcal{M}(E)$ be the collection of all non-zero finite measures on $\mathcal{B}$. Give $\mathcal{M}(E)$ the minimal Borel structure such that for any bounded measurable function f on E, $\nu\rightarrow\nu(f)=\int_E fd\nu$ is a measurable function on $\mathcal{M}(E)$. Then $\mathcal{M}(E)$ is also a standard Borel space, and its Borel structure is generated by following subsets:

$$\{\nu\in\mathcal{M}(E)|\nu(F)<\lambda\},\quad\forall F\in\mathcal{B},\lambda>0.$$

Proof. By Theorem 10.3.16, we may assume that $E=[0,1]$. Then $\mathcal{M}(E)=C(E)^*_+\backslash\{0\}$, where $C(E)^*_+$ is the collection of all continuous positive linear functionals on $C(E)$.

Notice that for any bounded measurable function f on E, there is a sequence $\{f_n\}$ of simple functions on E such that $f_n(t)\rightarrow f(t)$, uniformly for $t\in E$. Thus, the Borel sstructure of $\mathcal{M}(E)$ is generated by $\{\nu\in\mathcal{M}(E)|\nu(F)<\lambda\},\forall F\in\mathcal{B}$, and $\lambda>0$ exactly.

Since $C(E)$ is a separable Banach space, $(C(E)^*,w^*\text{-top.})$ is a standard Borel space. Hence, $\mathcal{M}(E)=C(E)^*_+\backslash\{0\}$ is also a standard Borel space with respect to the w^*-top. in $C(E)^*$. For any $f\in C(E),\nu\rightarrow\nu(f)$ is w^*-continuous on $\mathcal{M}(E)$. Further, $\nu\rightarrow\nu(F)$ is measurable on the Borel space

$(\mathcal{M}(E), w^*\text{-top.})$, $\forall F \in \mathcal{B}$. Thus the Borel structure $\mathcal{P}$ of $\mathcal{M}(E)$ generated by $\{\nu \in \mathcal{M}(E) | \nu(F) < \lambda\} (\forall F \in \mathcal{B}$ and $\lambda > 0)$ contains the standard Borel structure generated by w^*–topology.

Now by Theorem 10.3.13, it suffices to show that $\mathcal{P}$ contains a separated countable family. Let $\{r_n\}$ be all rational numbers in $[0,1]$, and $\{t_k\}$ be all rational numbers in $(0, +\infty)$. Put

$$Q_{ijk} = \{\nu \in \mathcal{M}(E) | \nu([r_i, r_j]) < t_k\}, \quad \forall i, j, k.$$

Then it is easy to see that $\{Q_{ijk} | i, j, k\}$ will be a separated family for $\mathcal{M}(E)$.

Q.E.D.

Lemma 12.7.7. Let $(E, \mathcal{B})$ be a standard Borel space, $\mathcal{M}(E)$ be as in Lemma 12.7.6, $H(\cdot)$ be a measurable field of Hilbert spaces over $E, \xi(\cdot)$ be a bounded measurable field of vectors of $H(\cdot), a(\cdot)$ be a uniformly bounded measurable field of operators on $H(\cdot)$, and $M(\cdot)$ be a measurable field of VN algebras on $H(\cdot)$. Then:

1) By the a fundament family $\{\nu \to \int_E^\oplus \eta(t) d\nu(t) | \eta(\cdot)$ is any bounded measurable field of vectors of $H(\cdot)\}$ of vector fields (where $\int_E^\oplus \eta(t) d\nu(t)$ is the element $\eta(\cdot)$ of $\int_E^\oplus H(t) d\nu(t)$),

$$\nu \to K(\nu) = \int_E^\oplus H(t) d\nu(t)$$

is a measurable field of Hilbert spaces over $\mathcal{M}(E)$;

2) $\nu \to \int_E^\oplus \xi(t) d\nu(t)$ is a measurable field of vectors of $K(\cdot)$, where $\int_E^\oplus \xi(t) d\nu(t)$ is the element $\xi(\cdot)$ of $K(\nu), \forall \nu$;

3) $\nu \to \int_E^\oplus a(t) d\nu(t)$ is a uniformly bounded measurable field of operators on $K(\cdot)$;

4) $\nu \to \int_E^\oplus M(t) d\nu(t)$ is a measurable field of VN algebras on $K(\cdot)$.

Proof. 1) We may assume that $E = [0,1]$. Let $\{\xi_n(\cdot)\}$ be an orthogonal normalized basis of $H(\cdot)$, and $\{f_m\}$ be a countable dense subset of $C(E)$. By Proposition 12.1.9, for each $\nu \in \mathcal{M}(E) \{\int_E^\oplus (f_m \xi_n)(t) d\nu(t) | n, m\}$ is a total subset of $\int_E^\oplus H(t) d\nu(t)$. Then from Lemma 12.7.6, $\nu \to K(\nu) = \int_E^\oplus H(t) d\nu(t)$ is a measurabel field of Hilbert spaces over $\mathcal{M}(E)$. Now if $\eta(\cdot)$ is a bounded measurable field of vectors of $H(\cdot)$, then $t \to \langle \eta(t), f_m(t) \xi_n(t) \rangle_t$ is bounded measurable on E. Further , from Lemma 12.7.6

$$\begin{aligned} &\nu \to \nu(\langle \eta(\cdot), f_m(\cdot)\xi_n(\cdot)\rangle.) \\ =& \langle \int_E^{\oplus} \eta(t)d\nu(t), \int_E^{\oplus} f_m(t)\xi_n(t)d\nu(t)\rangle \\ =& \int_E^{\oplus} \langle \eta(t), f_m(t)\xi_n(t)\rangle_t d\nu(t) \end{aligned}$$

is measurable on $\mathcal{M}(E), \forall m, n$. Thus , $\nu \to \int_E^{\oplus} \eta(t)d\nu(t)$ is a measurable field of vectors of $K(\cdot)$.

2) It is immediate from 1).

3) Since $a(\cdot)f_m(\cdot)\xi_n(\cdot)$ is still a bounded measurable field of vectors of $H(\cdot)$, $\forall m, n$, from Propositon 12.2.2 $\nu \to \int_E^{\oplus} a(t)d\nu(t)$ is a measurable field of operators on $K(\cdot)$. Moreover, $\| \int_E^{\oplus} a(t)d\nu(t)\| \leq \sup_t \|a(t)\|_t, \forall \nu \in \mathcal{M}(E)$.

4) Let $\{a_n(\cdot)\}$ be a sequence of measurable fields of operators on $H(\cdot)$ such that $M(t)$ is generated by $\{a_n(t)|n\}, \forall t \in E$. We may assume that $\|a_n(t)\|_t \leq 1, \forall t \in E$ and n. Then for each $\nu \in \mathcal{M}(E), \int_E^{\oplus} M(t)d\nu(t)$ is generated by $\{\int_E^{\oplus} f_m(t)1_t d\nu(t), \int_E^{\oplus} a_n(t)d\nu(t)|m, n\}$. Now from 3) and Definition 12.3.1, $\nu \to \int_E^{\oplus} M(t)d\nu(t)$ is a measurable field of VN algebras on $K(\cdot)$.

Q.E.D.

Lemma 12.7.8. Let $\mathcal{M} = \mathcal{M}(\mathcal{F}_{III} \times N) = \{\nu|\nu$ is a non–zero finite measure on $\mathcal{F}_{III} \times \mathbb{N}\}$, where $\mathcal{F}_{III}$ is the collection of all type (III) factors on H, and is a standard Borel space from Theorem 11.4.16. Suppose that $\{Z_n\}$ is a sequence of pairwise non $*$ isomorphic abelian VN algebras on H such that each abelian VN algebra on H must be $*$ isomorphic to some Z_n (see Corollary 5.3.9). Then there exists a Borel map $B(\cdot)$ from $\mathcal{M}$ to $\mathcal{A}$ such that for each $\nu \in \mathcal{M}, B(\nu)$ is spatially $*$ isomorphic to $\int_{\mathcal{F}_{III}\times\mathbb{N}}^{\oplus} (A\overline{\otimes}Z_n)d\nu(A, n)$.

Proof. From the proof of Proposition 11.4.3, $A \to A\overline{\otimes}\mathbb{C}1_H$ is a Borel map from $\mathcal{A} = \mathcal{A}(H)$ to $\mathcal{A}(H \otimes H)$. Clearly, $n \to \mathbb{C}1_H\overline{\otimes}Z_n$ is a Borel map from $\mathbb{N}$ to $\mathcal{A}(H\otimes H)$. Then by Proposition 11.3.5, $(A, n) \to A\overline{\otimes}Z_n$ is a Borel map from $\mathcal{F}_{III} \times \mathbb{N}$ to $\mathcal{A}(H \otimes H)$. Now from Lemma 12.7.7, $\nu \to \int_{\mathcal{F}_{III}\times\mathbb{N}}^{\oplus} (A\overline{\otimes}Z_n)d\nu(A, n)$ is a measurable field of VN algebra on $K(\cdot)$, where

$$K(\nu) = \int_{\mathcal{F}_{III}\times\mathbb{N}}^{\oplus} (H \otimes H)d\nu(A, n) = H \otimes H \otimes L^2(\mathcal{F}_{III} \times \mathbb{N}, \nu),$$

$\forall \nu \in \mathcal{M}$. Since $H \otimes H \otimes L^2(\mathcal{F}_{\rm III} \times \mathbb{N}, \nu)$ and H are countably infinite dimensional, from Proposition 12.3.2 there is a Borel map $B(\cdot)$ from $\mathcal{M}$ to $\mathcal{A}$ such that $B(\nu)$ is spatially $*$ isomorphic to $\int_{\mathcal{F}_{\rm III} \times \mathbb{N}}^{\oplus} (A \overline{\otimes} Z_n) d\nu(A,n), \forall \nu \in \mathcal{M}$.

Q.E.D.

Now let M be a fixed type (III) VN algebra on H. By Theorem 12.4.2, there is a finite Borel measure ν on $\mathbb{R}$, a measurable field $H(\cdot)$ of Hilbert spaces over $\mathbb{R}$ and a measurable field $M(\cdot)$ of factors on $H(\cdot)$, such that : M is spatially $*$ isomorphic to $\int_{\mathbb{R}}^{\oplus} M(t) d\nu(t)$; and this $*$ isomorphism maps $M \cap M'$ to the diagonal algebra on $\int_{\mathbb{R}}^{\oplus} H(t) d\nu(t)$. From Theorem 12.5.10, we may assume that $M(t)$ is type (III) and dim $H(t) = \infty, \forall t \in \mathbb{R}$. Further, we may assume that $H(\cdot)$ is the constant field H over $\mathbb{R}$. Then $M(\cdot)$ is a Borel map from $\mathbb{R}$ to $\mathcal{F}$ (see Proposition 12.3.2).

Lemma 12.7.9. We may assume that $M(\mathbb{R})$ is a Borel subset of $\mathcal{F}$.

Proof. Clearly, $M(\mathbb{R})$ is a Sousline subset of $\mathcal{F}$. Let $\mu = \nu \circ M^{-1}$. Clearly, μ is a finite measure on $\mathcal{F}$. By Corollary 10.2.11, there are Borel subsets E, F of $\mathcal{A}$ such that $E \subset M(\mathbb{R}), (M(\mathbb{R}) \backslash E) \subset F$, and $\mu(F) = 0$. Put $\mathbb{R}_0 = M^{-1}(\mathcal{F} \backslash E)$. Then $\nu(\mathbb{R}_0) = 0$ and $M(\mathbb{R} \backslash \mathbb{R}_0) = E$. Now change the definition of $M(\cdot)$ on $\mathbb{R}_0$ such that the new $M(\cdot)$ on $\mathbb{R}_0$ are a fixed type (III) factor on H. Then $M(\mathbb{R})$ is a Borel subset of $\mathcal{F}$.

Q.E.D.

Lemma 12.7.10. Let $M(\mathbb{R})$ be a Borel subset of $\mathcal{F}$, and introduce an equivalent relation $\sim$ in $\mathbb{R}$: $t_1 \sim t_2$ if $M(t_1) = M(t_2)$, and give $\mathbb{R}/\sim$ the quotient Borel structure. Then $\mathbb{R}/\sim$ is also a standard Borel space, and Φ is a Borel isomorphism from $\mathbb{R}/\sim$ onto $M(\mathbb{R})$, where $\Phi(\tilde{t}) = M(t)$, and $t \to \tilde{t} = \pi(t)$ is the canonical map from $\mathbb{R}$ onto $\mathbb{R}/\sim$. Moreover, we can give $M(\cdot)$ a new definition on a saturated (in the sense of $\sim$) ν -zero subset such that π admits a Borel cross section, i.e., we may assume that there is a Borel map $\sigma : \mathbb{R}/\sim \longrightarrow \mathbb{R}$ such that $\pi \circ \sigma(\tilde{t}) = \tilde{t}, \forall \tilde{t} \in \mathbb{R}/\sim$.

Proof. If F is a Borel subst of $M(\mathbb{R})$, then $M^{-1}(F)$ is a saturated Borel subset of $\mathbb{R}$. Further by the quotient Borel structure of $\mathbb{R}/\sim, \Phi^{-1}(F) = \pi \circ M^{-1}(F)$ is a Borel subset of $\mathbb{R}/\sim$. Thus , Φ is a Borel map. Clearly, Φ is injective. Now prove that Φ^{-1} is also Borel. Let $\tilde{E}$ be a Borel subset of $\mathbb{R}/\sim$. Then $E = \pi^{-1}(\tilde{E})$ is a saturated Borel subset of $\mathbb{R}$. Further, $\Phi(\tilde{E}) = M(E)$ and $\Phi((\mathbb{R}/\sim) \backslash \tilde{E}) = M(\mathbb{R} \backslash E) = M(\mathbb{R}) \backslash M(E)$ are Sousline subsets of $M(\mathbb{R})$. Thus, $\Phi(\tilde{E}) = M(E)$ is a Borel subset of $M(\mathbb{R})$, and Φ^{-1} is Borel. Now Φ is a Borel isomorphism from $\mathbb{R}/\sim$ onto $M(\mathbb{R})$. Consequently,

$\mathbb{R}/\sim$ is a standard Borel space. The rest conclusion can be obtained from Proposition 10.4.7. Q.E.D.

Lemma 12.7.11. Let $M(\mathbb{R})$ be a Borel subset of $\mathcal{F}$. Then there is a Borel map $\tilde{t} \to \nu_{\tilde{t}}$ from $\mathbb{R}/\sim$(see Lemma 12.7.10) to $\mathcal{M}(\mathbb{R})_1$ such that for any bounded measurable function f on $\mathbb{R}$ and any $\tilde{\nu}$-integrable function h on $\mathbb{R}/\sim$, where $\tilde{\nu} = \nu \circ \pi^{-1}$,

$$\begin{aligned}&\int_{\mathbb{R}} (h \circ \pi)(t) f(t) d\nu(t)\\ = &\int_{\mathbb{R}/\sim} h(\tilde{t}) d\tilde{\nu}(\tilde{t}) \int_{\mathbb{R}} f(s) d\nu_{\tilde{t}}(s),\end{aligned}$$

and supp $\nu_{\tilde{t}} \subset \pi^{-1}(\{\tilde{t}\})$, a.e.$\tilde{\nu}$.

Proof. A rational semiclosed interval means a subset of $\mathbb{R}$ with form $[a, b)$, where a and b are rational or $\pm\infty$ (for the conveninece, $[a, b) = \emptyset$ if $b \le a$, and $[a, b) = (-\infty, b)$ if $a = -\infty$). Let Σ be the collection of all disjoint finite unions of rational semiclosed interval. Then Σ is a countable Bool algebra, and any disjoint union of elements of Σ must be finite. Thus, any finitely additive finite measure on Σ must be countably additive, and can be uniquely extended to a finite Borel measure on $\mathbb{R}$.

For any $E \in \Sigma, \nu(E \cap \pi^{-1}(\cdot))$ is a finite measure on $\mathbb{R}/\sim$, and is absolutely continuous with respect to $\tilde{\nu} = \nu \circ \pi^{-1}$. Thus, there is $0 \le g_E \in L^1(\mathbb{R}/\sim, \tilde{\nu})$ such that $\nu(E \cap \pi^{-1}(\cdot)) = g_E \cdot \tilde{\nu}(\cdot)$. Since Σ is countable, then there is a $\tilde{\nu}$-zero subset $\tilde{E}_0$ of $\mathbb{R}/\sim$ such that for each $\tilde{t} \in (\mathbb{R}/\sim) \backslash \tilde{E}_0, E \to g_E(\tilde{t})$ is a finitely additive probability measure on Σ. Further, $g_\cdot(\tilde{t})$ can be uniquely extended to a probability measure on $\mathbb{R}$. Thus , we get a map $\tilde{t} \to \nu_{\tilde{t}}$ from $\mathbb{R}/\sim$ to $\mathcal{M}(\mathbb{R})_1$ such that $\nu_{\tilde{t}}(E) = g_E(\tilde{t}), \forall \tilde{t} \in (\mathbb{R}/\sim) \backslash \tilde{E}_0$ and $E \in \Sigma$, and $\nu_{\tilde{t}}$ is a fixed probability measure on $\mathbb{R}, \forall \tilde{t} \in \tilde{E}_0$. Since the Borel structure of $\mathbb{R}$ is generated by Σ, it follows from Lemma 12.7.6 that $\tilde{t} \to \nu_{\tilde{t}}$ is a Borel map from $\mathbb{R}/\sim$ to $\mathcal{M}(\mathbb{R})_1$. And for any Borel subset E of $\mathbb{R}$ and any Borel subset $\tilde{F}$ of $\mathbb{R}/\sim$, we have

$$\nu(E \cap \pi^{-1}(\tilde{F})) = \int_{\tilde{F}} \nu_{\tilde{t}}(E) d\tilde{\nu}(\tilde{t}). \tag{1}$$

Further, for any bounded measurable function f on $\mathbb{R}$ and any $\tilde{\nu}$-integrable function h on $\mathbb{R}/\sim$ we get

$$\begin{aligned}&\int_{\mathbb{R}} (h \circ \pi)(t) f(t) d\nu(t)\\ = &\int_{\mathbb{R}/\sim} h(\tilde{t}) d\tilde{\nu}(\tilde{t}) \int_{\mathbb{R}} f(s) d\nu_{\tilde{t}}(s).\end{aligned}$$

For any Borel subsets $\tilde{E}, \tilde{F}$ of $\mathbb{R}/\sim$, by (1) we have

$$\begin{aligned}\int_{\tilde{F}} \chi_{\tilde{E}}(\tilde{t}) d\tilde{\nu}(\tilde{t}) &= \tilde{\nu}(\tilde{E} \cap \tilde{F}) \\ &= \nu(\pi^{-1}(\tilde{E}) \cap \pi^{-1}(\tilde{F})) \\ &= \int_{\tilde{F}} \nu_{\tilde{t}}(\pi^{-1}(\tilde{E})) d\tilde{\nu}(\tilde{t}).\end{aligned}$$

Thus for fixed $\tilde{E}$, we get

$$\chi_{\tilde{E}}(\tilde{t}) = \nu_{\tilde{t}}(\pi^{-1}(\tilde{E})), \quad a.e.\tilde{\nu}. \tag{2}$$

Since $\mathbb{R}/\sim$ is a standard Borel space, we can find a countable family $\tilde{\Sigma}$ of Borel subsets of $\mathbb{R}/\sim$ such that for any $\tilde{t} \in \mathbb{R}/\sim$, there is $\{\tilde{E}_n\} \subset \tilde{\Sigma}$ with $\tilde{E}_1 \supset \cdots \supset \tilde{E}_n \supset \cdots$ and $\cap_n \tilde{E}_n = \{\tilde{t}\}$. Now from (2) , we can see that $1 = \nu_{\tilde{t}}(\pi^{-1}(\{\tilde{t}\}))$, a.e.$\tilde{\nu}$, i.e., supp $\nu_{\tilde{t}} \subset \pi^{-1}(\{\tilde{t}\})$, a.e.$\tilde{\nu}$. Q.E.D.

Lemma 12.7.12. Let $\tilde{t} \to \nu_{\tilde{t}}$ be as in Lemma 12.7.11. Then $\tilde{t} \to \int_{\mathbb{R}}^{\oplus} M(s) d\nu_{\tilde{t}}(s)$ is a measurable field of VN algebras on $L(\cdot)$, where $\tilde{t} \to L(\tilde{t}) = \int_{\mathbb{R}}^{\oplus} H d\nu_{\tilde{t}}(s) = H \otimes L^2(\mathbb{R}, \nu_{\tilde{t}})$ is a measurable field of Hilbert spaces over $\mathbb{R}/\sim$, and M is spatially $*$ isomophic to

$$\int_{\mathbb{R}/\sim}^{\oplus} d\tilde{\nu}(\tilde{t}) \int_{\mathbb{R}}^{\oplus} M(s) d\nu_{\tilde{t}}(s).$$

Proof. Pick a sequence $\{f_n\}$ of bounded measurable functions on $\mathbb{R}$ such that $\{f_n\}$ is dense in $L^2(\mathbb{R}, \nu)$ and is w^*-dense in $L^\infty(\mathbb{R}, \mu), \forall \mu \in \mathcal{M}(\mathbb{R})$. Let $\{\xi_m\}$ be an orthogonal normalized basis of H, and $\{a_n(\cdot)\}$ be a sequence of measurable fields of operators on the constant field H such that $M(t)$ is generated by $\{a_n(t)|n\}, \forall t \in \mathbb{R}$. Since $\tilde{t} \to \nu_{\tilde{t}}$ is Borel , by Lemma 12.7.7, 12.7.6, $\{\tilde{t} \to \int_{\mathbb{R}}^{\oplus} (f_n \xi_m)(t) d\nu_{\tilde{t}}(t) | n, m\}$ is a fundamental family of measurable fields of vectors of $L(\cdot)$, and

$$\tilde{t} \to \int_{\mathbb{R}}^{\oplus} a_n(t) d\nu_{\tilde{t}}(t), \quad \tilde{t} \to \int_{\mathbb{R}}^{\oplus} f_m(t) 1_H d\nu_{\tilde{t}}(t)$$

are measurable fields of operators on $L(\cdot), \forall n, m$ (here we may assume that $\|a_n(t)\| \le 1, \forall t \in \mathbb{R}, n)$, and for each $\tilde{t} \in \mathbb{R}/\sim$, $\int_{\mathbb{R}}^{\oplus} M(s) d\nu_{\tilde{t}}(s)$ is generated by $\{\int_{\mathbb{R}}^{\oplus} a_n(s) d\nu_{\tilde{t}}(s), \int_{\mathbb{R}}^{\oplus} f_m(s) 1_H d\nu_{\tilde{t}}(s) | n, m\}$. Thus , $\tilde{t} \to \int_{\mathbb{R}}^{\oplus} M(s) d\nu_{\tilde{t}}(s)$ is a measurable field of VN algebras on $L(\cdot)$. By Proposition 12.1.9,

$$\left\{ \tilde{t} \to h(\tilde{t}) \int_{\mathbb{R}}^{\oplus} (f_n \xi_m)(s) d\nu_{\tilde{t}}(s) \,\middle|\, \begin{array}{c} \forall n, m, \text{ and } \ h \text{ is} \\ \text{bounded measurable on } \mathbb{R}/\sim \end{array} \right\}$$

is a total subset of $\int_{\mathbb{R}/\sim}^{\oplus} d\tilde{\nu}(\tilde{t}) \int_{\mathbb{R}}^{\oplus} H d\nu_{\tilde{t}} = \int_{\mathbb{R}/\sim}^{\oplus} (H \otimes L^2(\mathbb{R}, \nu_{\tilde{t}})) d\tilde{\nu}(\tilde{t})$. Define

$$u(\tilde{t} \to h(\tilde{t}) \int_{\mathbb{R}}^{\oplus} (f_n \xi_m)(s) d\nu_{\tilde{t}}(s)) = (t \to h(\pi(t)) f_n(t) \xi_m(t)),$$

$\forall n, m, h$. From Lemma 12.7.11, u can be uniquely extended to a unitary operator from $\int_{\mathbb{R}/\sim}^{\oplus} (H \otimes L^2(\mathbb{R}, \nu_{\tilde{t}})) d\tilde{\nu}(\tilde{t})$ onto $\int_{\mathbb{R}}^{\oplus} H d\nu = H \otimes L^2(\mathbb{R}, \nu)$. Clearly, u maps the operators $\int_{\mathbb{R}/\sim}^{\oplus} h(\tilde{t}) 1_{\tilde{t}} d\tilde{\nu}(\tilde{t}), \int_{\mathbb{R}/\sim}^{\oplus} d\tilde{\nu}(\tilde{t}) \int_{\mathbb{R}}^{\oplus} a_n(s) d\nu_{\tilde{t}}(s), \int_{\mathbb{R}/\sim}^{\oplus} d\tilde{\nu}(\tilde{t})$ $\int_{\mathbb{R}}^{\oplus} f_m(s)\ 1_H d\nu_{\tilde{t}}(s)$ to $\int_{\mathbb{R}}^{\oplus} h(\pi(t)) 1_H d\nu(t), \int_{\mathbb{R}}^{\oplus} a_n(t) d\nu(t), \int_{\mathbb{R}}^{\oplus} f_m(t) 1_H d\nu(t)$ respectively, $\forall n, m, h$. Therefore,

$$u \int_{\mathbb{R}/\sim}^{\oplus} d\tilde{\nu}(\tilde{t}) \int_{\mathbb{R}}^{\oplus} M(s) d\nu_{\tilde{t}}(s) u^* = \int_{\mathbb{R}}^{\oplus} M(t) d\nu(t).$$

Q.E.D.

Lemma 12.7.13. There exists $\mu \in \mathcal{M}$ such that M is $*$ isomorphic to $B(\mu)$, where $\mathcal{M}, B(\cdot)$ are defined as in Lemma 12.7.8.

Proof. Let $\tilde{t} \to \nu_{\tilde{t}}$ be as in Lemma 12.7.11, and $\{f_n\}$ be as in the proof of Lemma 12.7.12. Then with $\{(\tilde{t} \to \int_{\mathbb{R}}^{\oplus} f_n(s) d\nu_{\tilde{t}}(s)) | n\}$ as a fundamental sequence of measurable fields of vectors, $(\tilde{t} \to L^2(\mathbb{R}, \nu_{\tilde{t}}))$ is a measurable field of Hilbert spaces over $\mathbb{R}/\sim$, and $(\tilde{t} \to Z_{\tilde{t}})$ is a measurable field of VN algebras on $(\tilde{t} \to L^2(\mathbb{R}, \nu_{\tilde{t}}))$, where $Z_{\tilde{t}}$ is the multiplicative algebra on $L^2(\mathbb{R}, \nu_{\tilde{t}})), \forall \tilde{t} \in \mathbb{R}/\sim$.

Since $L^2(\mathbb{R}, \nu_{\tilde{t}})$ is separable, there exists unique positive integer $n(\tilde{t})$ such that $Z_{\tilde{t}}$ is $*$ isomorphic to $Z_{n(\tilde{t})}, \forall \tilde{t} \in \mathbb{R}/\sim$, where $\{Z_n\}$ is defined as in Lemma 12.7.8. We say that $(\tilde{t} \to n(\tilde{t}))$ is a Borel map from $\mathbb{R}/\sim$ to $\mathbb{N}$. By Proposition 12.1.2, $\{\tilde{t} | \dim L^2(\mathbb{R}, \nu_{\tilde{t}}) = k\}$ is a Borel subset of $\mathbb{R}/\sim, \forall k$. Thus , we may assume that $\dim L^2(\mathbb{R}, \nu_{\tilde{t}})$ is a constant, $\forall \tilde{t}$. Then $(\tilde{t} \to L^2(\mathbb{R}, \nu_{\tilde{t}}))$ is unitarily isomorphic to a constant field K over $\mathbb{R}/\sim$, and $(\tilde{t} \to Z_{\tilde{t}})$ is spatially $*$ isomorphic to $(\tilde{t} \to N(\tilde{t}))$, where $N(\cdot)$ is a Borel map from $\mathbb{R}/\sim$ to $\mathcal{A}(K)$. For fixed $N_0 \in \mathcal{A}(K), a(N_0)$ is a Borel subset of $\mathcal{A}(K)$ (see Proposition 11.4.3). Then $N^{-1}(a(N_0))$ is a Borel subset of $\mathbb{R}/\sim$. Further $(\tilde{t} \to n(\tilde{t}))$ is Borel.

We can also construct a measurable field $\Psi(\cdot)$ of $*$ isomorphisms such that $\Psi(\tilde{t})$ is a $*$ isomorphism from $Z_{\tilde{t}}$ onto $Z_{n(\tilde{t})}, \forall \tilde{t} \in \mathbb{R}/\sim$. In fact, from the preceding paragraph, $(\tilde{t} \to Z_{\tilde{t}})$ is spatially $*$ isomorphic to $(\tilde{t} \to N(\tilde{t}))$, where $N(\cdot)$ is a Borel map from $\mathbb{R}/\sim$ to $\mathcal{A}(K)$. Since the function $1 (\in L^2(\mathbb{R}, \nu_{\tilde{t}}))$ is cyclic and separating for $Z_{\tilde{t}}$, $N(\tilde{t})$ admits also a cyclic–separating vector in

$K, \forall \tilde{t} \in I\!R/\sim$. By Corollary 5.3.9, we may assume that all $N(\tilde{t})(\tilde{t} \in I\!R/\sim)$ are $*$ isomorphic. Fix a $\tilde{t}_0 \in I\!R/\sim$. Then by Theorem 1.13.5, $N(\tilde{t})$ is spatially $*$ isomorphic to $N(\tilde{t}_0), \forall \tilde{t} \in I\!R/\sim$. Further, from Theorem 12.6.5 we can get the field $\Psi(\cdot)$.

By Lemma 12.7.10, Φ is a Borel isomorphism from $I\!R/\sim$ onto $M(I\!R)$. Let $k(A) = n \circ \Phi^{-1}(A), \forall A \in M(I\!R)$, and $k(A) = 1, \forall A \in (\mathcal{A} \backslash M(I\!R))$. Then $k(\cdot)$ is a Borel map from $\mathcal{A}$ to $I\!N$.

Let $\hat{\nu} = \tilde{\nu} \circ \Phi^{-1} = \nu \circ M^{-1}$. Then $\hat{\nu}$ is a finite measure on $\mathcal{A}$, and supp $\hat{\nu} \subset M(I\!R)$. In the following , the symbol " $\simeq$" means the $*$ isomorphism between VN algebras. Since Φ is a Borel isomorphism, by Proposition 12.3.12 and Lemma 12.7.10 we have

$$\begin{aligned} \int_{\mathcal{A}}^{\oplus} (A \overline{\otimes} Z_{k(A)}) d\hat{\nu}(A) &= \int_{M(I\!R)}^{\oplus} (A \overline{\otimes} Z_{k(A)}) d\hat{\nu}(A) \\ &= \int_{M(I\!R)}^{\oplus} \Phi \circ \Phi^{-1}(A) \overline{\otimes} Z_{k \circ \Phi \circ \Phi^{-1}(A)} d\tilde{\nu} \circ \Phi^{-1}(A) \\ &\simeq \int_{I\!R/\sim}^{\oplus} \Phi(\tilde{t}) \overline{\otimes} Z_{n(\tilde{t})} d\tilde{\nu}(\tilde{t}) \simeq \int_{I\!R/\sim}^{\oplus} (M(\sigma(\tilde{t})) \overline{\otimes} Z_{\tilde{t}}) d\tilde{\nu}(\tilde{t}). \end{aligned}$$

From Lemma 12.7.11, supp $\nu_{\tilde{t}} \subset \pi^{-1}(\{\tilde{t}\})$, a.e.$\tilde{\nu}$. Clearly, $M(t) = M(\sigma(\tilde{t})), \forall t \in \pi^{-1}(\{\tilde{t}\})$. Thus ,

$$\begin{aligned} \int_{I\!R}^{\oplus} M(s) d\nu_{\tilde{t}}(s) &= \int_{\pi^{-1}(\{\tilde{t}\})}^{\oplus} M(\sigma(\tilde{t})) d\nu_{\tilde{t}}(s) \\ &= M(\sigma(\tilde{t})) \overline{\otimes} Z_{\tilde{t}}, \quad \text{a.e.} \tilde{\nu}. \end{aligned}$$

Now by Lemma 12.7.12, M is $*$ isomorphic to $\int_{\mathcal{A}}^{\oplus} (A \overline{\otimes} Z_{k(A)}) d\hat{\nu}(A)$.

Noticing that supp $\hat{\nu} \subset M(I\!R) \subset \mathcal{F}_{III}$, we get

$$\begin{aligned} &\int_{\mathcal{A}}^{\oplus} (A \overline{\otimes} Z_{k(A)}) d\hat{\nu}(A) \\ &= \sum_j \oplus \int_{\mathcal{A}_j}^{\oplus} (A \overline{\otimes} Z_j) d\hat{\nu}(A) = \sum_j \oplus \int_{\mathcal{A}_j \cap \mathcal{F}_{III}}^{\oplus} (A \overline{\otimes} Z_j) d\hat{\nu}(A), \end{aligned}$$

where $\mathcal{A}_j = \{A \in \mathcal{A} | k(A) = j\}, \forall j$. Further, pick $\mu \in \mathcal{M}$ such that $\mu = \hat{\nu}$ on $(\mathcal{A}_j \cap \mathcal{F}_{III}) \times \{j\}, \forall j$. Then , M is $*$ isomorphic to $\int_{\mathcal{F}_{III} \times I\!N}^{\oplus} (A \overline{\otimes} Z_n) d\mu(A, n)$. Again by Lemma 12.7.8, M is $*$ isomorphic to $B(\mu)$. Q.E.D.

Lemma 12.7.14. Let E be a Sousline subset of $\mathcal{A}$. Then $a(E) = \{N \in \mathcal{A} | N$ is $*$ isomorphic to some element of $E\}$ is also a Sousline subset of $\mathcal{A}$.

Proof. Define $\Phi(N) = N \overline{\otimes} \mathbb{C} 1_H$. Then Φ is a Borel isomorphism from $\mathcal{A} = \mathcal{A}(H)$ into $\mathcal{A}(H \otimes H)$. From the proof of Proposition 11.4.3, we have

$$a(E) = \Phi^{-1}(s(\Phi(E))) = \Phi^{-1}(s(\Phi(E)) \cap \Phi(\mathcal{A})).$$

From Lemma 11.4.14, $s(\Phi(E))$ is a Sousline subset of $\mathcal{A}(H \otimes H)$. Then there is a Polish space P and a Borel map f from P to $\mathcal{A}(H \otimes H)$ such that $f(P) = s(\Phi(E)) \cap \Phi(\mathcal{A})$. Now $\Phi^{-1} \circ f$ is a Borel map from P to $\mathcal{A}$. Thus , by Proposition 10.3.5 $a(E) = \Phi^{-1} \circ f(P)$ is a Sousline subset of $\mathcal{A}$. Q.E.D.

Proposition 12.7.15. $\mathcal{A}_{III}$ is a Borel subset of $\mathcal{A}$.

Proof. By Lemma 12.7.5, it suffices to show that $\mathcal{A}_{III}$ is Sousline . From Lemma 12.7.8, $B(\mathcal{M})$ is a Sousline subset of $\mathcal{A}$. By Theorem 12.5.10, we have $B(\mathcal{M}) \subset \mathcal{A}_{III}$. Again by Theorem 12.7.13, $\mathcal{A}_{III} = a(B(\mathcal{M}))$. Now from Lemma 12.7.14, $\mathcal{A}_{III}$ is a Sousline subset of $\mathcal{A}$. Q.E.D.

Theorem 12.7.16. $\mathcal{A}_f$ (the collection of all finite VN algebras on H) ,$\mathcal{A}_{sf}$(the collection of all semi-finite VN algebras on H) , $\mathcal{A}_{pi}$(the collection of all properly infinite VN algebras on H) ,$\mathcal{A}_{I_k}$ (the collection of all type (I_k) VN algebras on H), $k = \infty, 1, 2, \cdots$, $\mathcal{A}_I$ (the collction of all type (I) VN algebras on H) , $\mathcal{A}_c$ (the collection of all continuous VN algebras on H), $\mathcal{A}_{II_1}$ (the collection of all type (II_1) VN algebras on H), $\mathcal{A}_{II_\infty}$ (the collection of all type (II_∞) VN algebras on H) , $\mathcal{A}_{II}$ (the collection of all type (II) VN algebras on H) , and $\mathcal{A}_{III}$ (the collection of all type (III) VN algebras on H) are Borel subsets of $\mathcal{A}$.

Proof. From Propositions 12.7.1, 12.7.2, 12.7.3 and 12.7.15, $\mathcal{A}_{I_k}, \mathcal{A}_I, \mathcal{A}_f$ and $\mathcal{A}_{III}$ are Borel subsets of $\mathcal{A}$.

For the case $\mathcal{A}_{sf}$, by Lemma 12.7.4 it suffices to prove that $(\mathcal{A} \backslash \mathcal{A}_{sf})$ is a Sousline subset of $\mathcal{A}$. Notice that $M \in (\mathcal{A} \backslash \mathcal{A}_{sf})$ if and only if $M = M_1 \oplus M_2$ and M_2 is type (III), i.e., M is $*$ isomorphic to a VN algebra $\begin{pmatrix} M_1 & \\ & M_2 \end{pmatrix}$ on $H \oplus H$, where $M_1, M_2 \in \mathcal{A}$ and M_2 is type (III) . From Proposition 11.3.3 and 11.3.4, $(M_1, M_2) \to \begin{pmatrix} M_1 & \\ & M_2 \end{pmatrix}$ is a Borel map from $\mathcal{A} \times \mathcal{A}$ to $\mathcal{A}(H \oplus H)$, and this map is injective obviously. Then by Proposition 12.7.15,

$$E = \left\{ \begin{pmatrix} M_1 & \\ & M_2 \end{pmatrix} \middle| M_1, M_2 \in \mathcal{A}, \text{ and } M_2 \text{ is type (III)} \right\}$$

is a Borel subset of $\mathcal{A}(H \oplus H)$. Let v be a unitary operator form $H \oplus H$ onto H. Similarly, from Proposition 11.3.3 and 11.3.4 $v \cdot v^*$ is a Borel isomorphism from $\mathcal{A}(H \oplus H)$ to $\mathcal{A}$. Then by Lemma 12.7.14, $(\mathcal{A} \backslash \mathcal{A}_{sf} = a(vEv^*)$ is a Sousline subset of $\mathcal{A}$.

Notice that $M \not\in \mathcal{A}_{pi}$ if and only if $M = M_1 \oplus M_2$ and M_2 is finite. Then by the same method as in the preceding paragraph, we can see that $(\mathcal{A} \backslash \mathcal{A}_{pi})$ is Sousline. Moreover, $M \in \mathcal{A}_{pi}$ if and only if there is a projcetion p of M such that $1 \sim p \sim (1-p)$. Let S be th unit ball of $B(H)$, and consider a subset E of $\mathcal{A} \times S \times S$. $(M, v_1, v_2) \in E$ if : 1) $a_n(M')v_j = v_j a_n(M'), \forall n$, and $j = 1,2$; 2) $v_j v_j^* = 1, j = 1,2$; 3) $v_1^* v_1 \cdot v_2^* v_2 = 0$; 4) $v_1^* v_1 + v_2^* v_2 = 1$, where $\{a_n(\cdot) : \mathcal{A} \to S|n\}$ is as in Proposition 11.3.3. Clearly, E is a Borel subset of $\mathcal{A} \times S \times S$. Then $\mathcal{A}_{pi}$ is Sousline. Further, $\mathcal{A}_{pi}$ is a Borel subset of $\mathcal{A}$.

By Theorem 6.8.4 and the same method, we can see that $\mathcal{A}_{II}$ is Sousline. Moreover, $M \notin \mathcal{A}_{II}$ if and only if $M = M_1 \oplus M_2$ and M_2 is either type (I) or type (III) . Then we can prove that $(\mathcal{A} \backslash \mathcal{A}_{II})$ is Sousline. Thus , $\mathcal{A}_{II}$ is a Borel subset of $\mathcal{A}$.

Further, $\mathcal{A}_{II_1} = \mathcal{A}_{II} \cap \mathcal{A}_f$ and $\mathcal{A}_{II_\infty} = \mathcal{A}_{II} \cap \mathcal{A}_{pi}$ are Borel subsets of $\mathcal{A}$.

Finally, since $\mathcal{A}_c = \mathcal{A}_{II} \cup \mathcal{A}_{III} \cup \{M \in \mathcal{A} | M = M_2 \oplus M_3$, and M_2, M_3 are type (II) , (III) respectively $\}$, $\mathcal{A}_c$ is Sousline. Moreover, $M \notin \mathcal{A}_c$ if and only if $M = M_1 \oplus M_2$, and M_1 is discrete. Thus, $(\mathcal{A} \backslash \mathcal{A}_c)$ is Sousline , and $\mathcal{A}_c$ is a Borel subset of $\mathcal{A}$. Q.E.D.

Notes. Lemma 12.7.5 was proved by J.T.Schwartz, and the hard result, Propositon 12.7.15, is due to O.Nielsen.

References. [36], [120], [121], [154], [177].

12.8. Borel subsets of the state space of a separable C^* –algebra

Let A be a separable C^*–algebra with an identity, and $S(A)$ be its state space. By $\sigma(A^*, A), S(A)$ is a compact Polish space. In fact, let $\{a_n\}$ be a countable dense subset of the unit ball of A, define $d(\varphi, \psi) = \sum_n 2^{-n} |(\varphi - \psi)(a_n)|, \forall \varphi, \psi \in S(A)$. Then d is a proper metric on $(S(A), \sigma(A^*, A))$.

For any $n = \infty, 1, 2, \cdots$, let H_n be n–dimensional Hilbert space, and R_n be the collection of all nondegenerate $*$ representations of A on H_n. In R_n , the topology is defined as follows: $\pi_l \to \pi$ if $\|(\pi_l(a) - \pi(a))\xi\| \to 0, \forall a \in A, \xi \in H_n$. By Lemma 12.6.1, R_n is a Polish space, $\forall n$.

Lemma 12.8.1. $\Phi : \pi \longrightarrow \pi(A)''$ is a Borel map from R_n to $\mathcal{A}_n$, where $\mathcal{A}_n$ is the collection of all VN algebras on $H_n, \forall n$.

Proof. Let $\{a_k\}$ be a countable dense subset of the unit ball of A, and define $a_k(\cdot) : R_n \to (B(H_n), \sigma), a_k(\pi) = \pi(a_k), \forall \pi \in R_n$ and k . For any $\xi, \eta \in H_n, \langle a_k(\cdot)\xi, \eta\rangle$ is continuous on R_n obviously. Thus, $\{a_k(\cdot)\}$ is a sequence of Borel maps from R_n to $(B(H_n), \sigma)$. Moreover, for each $\pi \in R_n, \pi(A)''$ is generated by $\{a_k(\pi)|k\}$. Now by Proposition 11.3.4, Φ is a Borel map from R_n to $\mathcal{A}_n$. Q.E.D.

Proposition 12.8.2. $R_n^{(t)} = \{\pi \in R_n | \pi(A)''$ is a type (t) VN algebra on $H_n\}$ is a Borel subset of $R_n, \forall n$, where (t) is factorial, finite, semi-finite, properly infinite, $(\mathrm{I}_k)(k = \infty, 1, 2, \cdots)$, (I) , (II $_1$), (II$_\infty$) , (II) , (III) , (c) (continuous) , or (Irr) (irreducible).

Proof. Except the case of $(t) = (Irr)$, the conclusions are obvious from Theorem 11.3.6, Theorem 12.7.16 and Lemma 12.8.1.

Now let $(t) = (Irr)$. Notice that $\pi \in R_n^{(t)}$ if and only if $\pi(A)'' = B(H_n)$, i.e., the unit ball of $\pi(A)$ is strongly dense in S, where S is the unit ball of $B(H_n)$. Let δ be a proper metric on the Polish space (S, strong top.) , $\{b_k\}$ be a countable dense subset of (S, δ), and $\{a_k\}$ be a countable dense subset of the unit ball of A. Then

$$R_n^{(t)} = \cap_{k,p} \cup_m \{\pi \in R_n | \delta(\pi(a_m), b_k) < p^{-1}\}$$

is a G_δ-subset of R_n. Q.E.D.

Now for $n = \infty, 1, 2, \cdots$, put

$$E_n = \{(\pi, \xi) | \pi \in R_n, \xi \in \Gamma_n, \text{and } \xi \text{ is cyclic for } \pi(A)\},$$

where $\Gamma_n = \{\xi \in H_n | \|\xi\| = 1\}$.

Lemma 12.8.3. E_n is a G_δ-subset of $R_n \times \Gamma_n$. Thus E_n is a Polish space with respect to the relative topology, $\forall n$.

Proof. Let $\{a_k\}$ be a countable dense subset of A, and $\{\xi_k\}$ be a countable dense subset of H_n. Then

$$E_n = \cap_{i,j} \cup_k \{(\pi, \xi) \in R_n \times \Gamma_n | \|\pi(a_k)\xi - \xi_j\| < i^{-1}\}$$

is a G_δ-subset of $R_n \times \Gamma_n$. Q.E.D.

Put $E = \sqcup\{E_n | n = \infty, 1, 2, \cdots\}$. Then E is also a Polish space, and each E_n is an open and closed subset of E. Further, define a map $\Psi : E \to S(A), \Psi(\pi, \xi)(a) = \langle \pi(a)\xi, \xi \rangle, \forall (\pi, \xi) \in E, a \in A$.

Lemma 12.8.4. Ψ is continuous from E onto $S(A)$.

Proof. The continuity is obvious. Now let $\varphi \in S(A)$. Then we have the cyclic $*$ representation $\{\pi_\varphi, H_\varphi, 1_\varphi\}$ of A. If dim $H_\varphi = n$, and u is a unitary operator from H_φ onto H_n, and let $\pi = u\pi_\varphi u^* \in R_n$, then

$$\begin{aligned} \varphi(a) = \langle \pi_\varphi(a) 1_\varphi, 1_\varphi \rangle &= \langle \pi(a)\xi, \xi \rangle \\ &= \Psi(\pi, \xi)(a), \quad \forall a \in A \end{aligned}$$

where $\xi = u1_\varphi \in \Gamma_n$. Thus , $\Psi(E) = S(A)$. Q.E.D.

Lemma 12.8.5. Let $\wedge \subset S(A)$. If $\Psi^{-1}(\wedge)$ is a Sousline or Borel subset of E, then $\wedge$ is also a Sousline or Borel subset of $S(A)$.

Proof. By Lemma 12.8.4, Ψ maps a Sousline subset of E to a Sousline subset of $S(A)$. So , if $\Psi^{-1}(\wedge)$ is a sousline subset of E, then $\wedge = \Psi(\Psi^{-1}(\wedge))$ is also a Sousline subset of $S(A)$.

Now let $\Psi^{-1}(\wedge)$ is a Borel subset of E. Then $(E\backslash\Psi^{-1}(\wedge))$ is also a Borel subset of E. From the preceding paragraph, $\wedge$ and $\Psi(E\backslash\Psi^{-1}(\wedge)) = S(A)\backslash\wedge$ are two Sousline subsets of $S(A)$. Therefore, $\wedge$ is a Borel subset of $S(A)$. Q.E.D.

Theorem 12.8.6. $S(A)(t) = \{\varphi \in S(A) | \pi_\varphi(A)''$ is a type (t) VN algebra on $H_\varphi\}$ is a Borel subset of $S(A)$, where $\{\pi_\varphi, H_\varphi\}$ is the $*$ representation of A generated by $\varphi, \forall\varphi \in S(A)$, and (t) is factorial, finite, semi-finite, properly infinite, $(\mathrm{I}_k)(k = \infty, 1, 2, \cdots)$, (I) , (II_1) , (II_∞) , (II), (III) , (c) or (Irr).

Proof. For $n = \infty, 1, 2, \cdots$, by Proposition 12.8.2 $(R_n^{(t)} \times \Gamma_n)$ is a Borel subset of $(R_n \times \Gamma_n)$. Thus , $\sqcup_n((R_n^{(t)} \times \Gamma_n) \cap E_n)$ is a Borel subset of $E, \forall(t)$.

If $(\pi, \xi) \in (R_n^{(t)} \times \Gamma_n) \cap E_n$, then π is a type (t) $*$ representation of A on H_n, and ξ is cyclic for $\pi(A)$. Clearly, the cyclic $*$ representation $\{\pi, H_n, \xi\}$ of A is unitarily equivalent to the $*$ representation of A generated by $\varphi = \Psi(\pi, \xi)$. Thus $\Psi(\pi, \xi) \in S(A)(t)$. Conversely, if $\varphi \in S(A)(t)$, then by Lemma 12.8.4 there is $(\pi, \xi) \in E_n$ (for some n) such that $\Psi(\pi, \xi) = \varphi$. Since $\langle \pi(a)\xi, \xi\rangle = \langle \pi_\varphi(a)1_\varphi, 1_\varphi\rangle, \forall a \in A$, it follows that $\{\pi, H_n, \xi\} \cong \{\pi_\varphi, H_\varphi, 1_\varphi\}$. Hence , π is a type (t) $*$ representation of A on H_n, and ξ is cyclic for $\pi(A)$, i.e, $(\pi, \xi) \in (R_n^{(t)} \times \Gamma_n) \cap E_n$. Therefore,

$$\Psi^{-1}(S(A)(t)) = \sqcup_n((R_n^{(t)} \times \Gamma_n) \cap E_n)$$

is a Borel subset of E. Now by Lemma 12.8.5, $S(A)(t)$ is a Borel subset of $S(A), \forall(t)$. Q.E.D.

Remark. Since $S(A)$ is a metrizable compact convex subset of $(A^*, \sigma(A^*, A))$, it follows from the Choquet theory (see Section 14.1) that $S(A)(Irr) = ExS(A) = P(A)$ (the pure state space of A) is a G_δ-subset of $S(A)$ indeed.

Notes. Theorem 12.8.6 is due to J. Feldman, J.T. Schwartz, and etc.

References. [45], [120], [148], [154].

Chapter 13

Type I C^*–Algebras

13.1. The spectrum of a C^*–algebra

Let A be a C^* -algebra, and Prim(A) be the collection of all primitive ideals of A (see Definiition 2.8.1.). For any $T \subset \text{Prim}(A)$, define $I(T) = \cap\{J | J \in T\}$, and $\overline{T} = \{J \in \text{Prim}(A) | J \supset I(T)\}$.

Lemma 13.1.1. For the bar "—" operation on the subsets of Prim(A), we have that

$$\overline{\emptyset} = \emptyset, \quad T \subset \overline{T}, \quad \overline{\overline{T}} = \overline{T}, \quad \text{and} \quad \overline{T_1 \cup T_2} = \overline{T_1} \cup \overline{T_2},$$

$\forall T, T_1, T_2 \subset \text{Prim}(A)$.

Proof. For any $T \subset \text{Prim}(A)$, since

$$I(\overline{T}) \subset I(T) \subset \cap\{J \in \text{Prim}(A) | J \supset I(T)\} = I(\overline{T}),$$

it follows that $\overline{\overline{T}} = \overline{T}$.

For any $T_1, T_2 \subset \text{Prim}(A)$, put $I_i = I(T_i), i = 1, 2$. Clearly, $I(T_1 \cup T_2) = I_1 \cap I_2$. Then by Lemma 2.8.7 we have that

$$\begin{aligned} J \in \overline{T_1 \cup T_2} &\Longleftrightarrow J \supset I_1 \cap I_2 \\ &\Longleftrightarrow \text{either } J \supset I_1 \quad \text{or} \quad J \supset I_2 \\ &\Longleftrightarrow \text{either } J \in \overline{T_1} \quad \text{or} \quad J \in \overline{T_2}. \end{aligned}$$

Thus, $\overline{T_1 \cup I_2} = \overline{T_1} \cup \overline{T_2}$. Q.E.D.

It follows from above lemma that there is a unique topology in Prim(A) such that for each $T \subset \text{Prim}(A), \overline{T}$ is the closure of T with respect to this topology.

Definition 13.1.2. This topology is called the *Jacobson topology* in $\text{Prim}(A)$.

Proposition 13.1.3. Let A be a C^*-algebra.

1) A subset T of $\text{Prim}(A)$ is closed $\Longleftrightarrow$ there is a subset E of A such that

$$T = \{J \in \text{Prim}(A) | J \supset E\}$$

$\Longleftrightarrow$ there is a closed two-sided ideal I of A such that $T = \{J \in \text{Prim}(A) | J \supset I\}$. Moreover, the ideal I is indeed generated by E, and is uniquely determinded by the closed subset T.

2) A subset W of $\text{Prim}(A)$ is open $\Longleftrightarrow$ there is a subset E of A such that

$$W = \{J \in \text{Prim}(A) | J \not\supset E\}$$

$\Longleftrightarrow$ there is a closed two-sided ideal I of A such that $W = \{J \in \text{Prim}(A) | J \not\supset I\}$.

3) Let $J \in \text{Prim}(A)$. Then the subset $\{J\}$ of $\text{Prim}(A)$ is closed if and only if J is a maximal closed two-sided ideal of A.

4) The space $\text{Prim}(A)$ is a To-space.

Proof. 1) Picking $E = I = I(T) = \cap\{J | J \in T\}$, the necessity is obvious. Now let $E \subset A$ and $T = \{J \in \text{Prim}(A) | J \supset E\}$. Clearly, $E \subset I(T)$. Then

$$\begin{aligned}\overline{T} &= \{J \in \text{Prim}(A) | J \supset I(T)\} \\ &\subset \{J \in \text{Prim}(A) | J \supset E\} = T \subset \overline{T}.\end{aligned}$$

So $T = \overline{T}$ is closed.

Moreover, by Proposition 2.8.2, I and the closed two-sided ideal generated by E are indeed equal to $I(T)$.

2) It is obvious from the conclusion 1).

3) Let $J \in \text{Prim}(A)$ and J be maximal. Since $I(\{J\}) = J$, so $\overline{\{J\}} = \{J\}$ is closed. Conversely, let $\overline{\{J\}} = \{J\}$, and I be a closed two-sided ideal of A with $I \supset J$. By Proposition 2.8.2, there is $J' \in \text{Prim}(A)$ with $J' \supset I$. Then $J' \in \overline{\{J\}} = \{J\}$, and $J' = I = J$. So J is maximal.

4) Let $J_1, J_2 \in \text{Prim}(A)$ and $J_1 \neq J_2$. We may assume that $J_1 \not\subset J_2$. By 1) , $T = \{J \in \text{Prim}(A) | J \supset J_1\}$ is a closed subset of $\text{Prim}(A)$. Then we have that $J_1 \in T$ and $J_2 \notin T_2$. So $\text{Prim}(A)$ is a T_0-space. Q.E.D.

Let A be a C^*-algebra, $\hat{A}$ be the collection of all unitary equivalent classes of non-zero irreducible $*$ representations of A, and ker be the nature map from $\hat{A}$ onto $\text{Prim}(A)$. Then there is a topology in $\hat{A}$ induced by the Jacobson topology in $\text{Prim}(A)$, i.e., a subset $\hat{U}$ of $\hat{A}$ is said to be open, if there is an open subset U of $\text{Prim}(A)$ such that

$$\hat{U} = \ker^{-1}(U) = \{\pi \in \hat{A} | \ker\pi \in U\}.$$

Definition 13.1.4. The *spectrum* of a C^*–algebra A is the set $\hat{A}$ endowed with the inverse image of the Jacobson topology under the natural map $\ker : \hat{A} \to \text{Prim}(A)$.

Clearly, now the map $\ker : \hat{A} \to \text{Prim}(A)$ is continuous and open.

Proposition 13.1.5. The map $I \longrightarrow \hat{A}^I = \{\pi \in \hat{A} | \ker\pi \not\supset I\}$ is a bijection from the collection of all closed two–sided ideals of A onto the collection of all open subsets of $\hat{A}$.

Moreover, for any closed two–sided ideals I, I_1 and I_2 of A, we have that $\hat{A}^I \cong \hat{I}$(homeomorphism),

$$\hat{A}^{(I_1+I_2)} = \hat{A}^{I_1} \cup \hat{A}^{I_2}, \quad \text{and} \quad \hat{A}^{(I_1 \cap I_2)} = \hat{A}^{I_1} \cap \hat{A}^{I_2}.$$

Consequently, $I_1 \cap I_2 = \{0\} \Longleftrightarrow \hat{A}^{I_1} \cap \hat{A}^{I_2} = \emptyset$.

Proof. Let I be a closed two–sided ideal of A. It is clear that

$$\hat{A}^I = \ker^{-1}(\{J \in \text{Prim}(A) | J \not\supset I\}).$$

So $\hat{A}^I$ is an open subset of $\hat{A}$.

Now let $\hat{U}$ be an open subset of $\hat{A}$. By Definition 13.1.4, we have $\hat{U} = \ker^{-1}(U)$, where U is an open subset of $\text{Prim}(A)$. Further, by Proposition 13.1.3 there is unique closed two–sided ideal I of A such that

$$\text{Prim}(A) \backslash U = \{J \in \text{Prim}(A) | J \supset I\}.$$

Hence, there is unique closed two–sided ideal I of A such that

$$\hat{U} = \{\pi \in \hat{A} | \ker\pi \not\supset I\} = \hat{A}^I.$$

The rest conclusions are easy. Q.E.D.

Proposition 13.1.6. Let A be a C^*–algebra. Then the following statements are equivalent:

1) $\hat{A}$ is a T_0–space;

2) Two irreducibe $*$ representations of A with the same kernel are unitarily equivalent, i.e., the map $\ker : \hat{A} \to \text{Prim}(A)$ is injective;

3) $\pi \to \ker\pi$ is a homeomorphism from $\hat{A}$ onto $\text{Prim}(A)$.

Proof. 2) $\Longrightarrow$ 3) . It is obvious since $\ker : \hat{A} \longrightarrow \text{Prim}(A)$ is continuous and open.

3) $\Longrightarrow$ 1). It is immediate from Proposition 13.1.3.

1) $\Longrightarrow$ 2) .Let $\hat{A}$ be a T_0 –space, and π_1, π_2 be two irreducible $*$ representations of A with $\ker\pi_1 = \ker\pi_2$. If π_1 and π_2 are not unitarily equivalent, then we may assume that there is an open subset $\hat{U}$ of $\hat{A}$ such that $\pi_1 \in \hat{U}$ and $\pi_2 \notin \hat{U}$. Suppose that $\hat{U} = \ker^{-1}(U)$, where U is an open subset of $\text{Prim}(A)$, and put $J = \ker\pi_1 = \ker\pi_2$. Then we get $J = \ker\pi_1 \in U$ and $J = \ker\pi_2 \notin U$, a contradiction. Therefore, π_1 and π_2 are unitarily equivalent.

Q.E.D.

Definition 13.1.7. A topological space X is said to be *Baire* , if $\{X_n\}$ is a sequence of open dense subsets of X, then $\cap_n X_n$ is still dense in X.

For example, a completely metric space is a Baire space, consequently, it must be second category.

Lemma 13.1.8. Let E be a Hausdorff locally convex linear space, K be a compact convex subset of E, and P be the set of extreme points of K. Then P is a Baire space with respect to the relative topology.

Proof. We may assume that E is real. For each $f \in E^*$ and $\alpha \in I\!R$, let

$$U_{f\alpha} = \{x \in K | f(x) < \alpha\}, \quad F_{f\alpha} = \{x \in K | f(x) \leq \alpha\}.$$

And for $x \in P$, put

$$L_x = \{(f, \alpha) | f \in E^*, \alpha \in I\!R, f(x) < \alpha\}.$$

We claim that

$$\{x\} = \cap\{F_{f\alpha} | (f, \alpha) \in L_x\}, \quad \forall x \in P.$$

In fact, if $(f, \alpha) \in L_x$, then $x \in U_{f\alpha}$. Thus we have

$$\{x\} \subset \cap\{F_{f\alpha} | (f, \alpha) \in L_x\}.$$

On the other hand, if $y \in E$ and $y \neq x$, then there is some $f_0 \in E^*$ and some $\alpha_0 \in I\!R$ such that $f_0(y) > \alpha_0 > f_0(x)$. So $x \in U_{f_0\alpha_0}$ and $y \notin F_{f_0\alpha_0}$. Therefore,

$$\{x\} = \cap\{F_{f\alpha} | (f, \alpha) \in L_x\}, \quad \forall x \in P.$$

Now let $x \in P$, and we prove that $\{F_{f\alpha} | (f, \alpha) \in L_x\}$ is a basis for the neighborhood system of x in K. Indeed, let U be an open subset of K, and $x \in U$. Since $\{x\} = \cap\{F_{f\alpha} | (f, \alpha) \in L_x\} \subset U$, it follows that $(K \backslash U) \subset \cup\{(K \backslash F_{f\alpha}) | (f, \alpha) \in L_x\}$. By the compactness of $(K \backslash U)$, there exist $(f_1, \alpha_1), \cdots, (f_n, \alpha_n) \in L_x$ such that $(K \backslash U) \subset \cup_{i=1}^n (K \backslash F_i)$, where $F_i = F_{f_i, \alpha_i}, 1 \leq i \leq n$, i.e., $U \supset \cap_{i=1}^n F_i$. Let $U_i = U_{f_i, \alpha_i}, K_i = K \backslash U_i, 1 \leq i \leq n$. Then $F \equiv \text{Co} \cup_{i=1}^n K_i$ is a compact subset of K. Clearly, $x \in U_i$ and $x \notin K_i, \forall 1 \leq i \leq n$. So by $x \in P$, we have $x \notin F$. Then we can find $f \in E^*$ and $\alpha \in \mathbb{R}$ such that

$$f(x) < \alpha < \inf\{f(y)|y \in F\}.$$

Hence ,$(f, \alpha) \in L_x$ and $F_{f\alpha} \cap F = \emptyset$. Further, $(U_{f\alpha} \cap K_i) \subset (F_{f\alpha} \cap F) = \emptyset$, and $U_{f\alpha} \subset U_i, 1 \leq i \leq n$. Then we have that

$$x \in U_{f\alpha} \subset F_{f\alpha} \subset \overline{\cap_i U_i} = \cap_i F_i \subset U.$$

Hence, $\{F_{f\alpha}|(f, \alpha) \in L_x\}$ is a basis for the neighborhood system of x in K.

Now let $\{V_n\}$ be a sequence of open dense subsets of P, and V be any non-empty open subset of P. We need to prove that $\cap_n V_n \cap V \neq \emptyset$.

In fact, we have open subsets $U, U_1, \cdots, U_n, \cdots$ of K such that $V = U \cap P, V_n = U_n \cap P, \forall n$. We may assume that each U_n is dense in K, and

$$U_1 \supset U_2 \supset \cdots, \quad V_1 \supset V_2 \supset \cdots.$$

Since $V \subset P$, and $\{F_{f\alpha}|(f, \alpha) \in L_x\}$ is a basis for the neighborhood system of x in $K, \forall x \in P$, we can suppose that $U = U_{f_1, \alpha_1}$ for some $f_1 \in E^*$ and $\alpha_1 \in \mathbb{R}$.

Pick $x_2 \in V \cap V_2$. Since $x_2 \in U_{f_1,\alpha_1} \cap U_2$, it follows from the preceding paragraph that there is some $(f_2, \alpha_2) \in L_{x_2}$ such that $x_2 \in U_{f_2,\alpha_2} \subset F_{f_2,\alpha_2} \subset U_{f_1,\alpha_1} \cap U_2$. Again Pick $x_3 \in V_3 \cap (U_{f_2,\alpha_2} \cap P)$, similarly we can find $(f_3, \alpha_3) \in L_{x_3}$ such that $x_3 \in U_{f_3,\alpha_3} \subset F_{f_3,\alpha_3} \subset U_{f_2,\alpha_2} \cap U_3, \cdots$. Generally, we have $\{f_n\} \subset E^*$ and $\{\alpha_n\} \subset \mathbb{R}$ such that

$$F_{f_{n+1},\alpha_{n+1}} \subset U_{f_n,\alpha_n} \cap U_{n+1}, \quad U_{f_n,\alpha_n} \cap P \neq \emptyset, \forall n.$$

Now $\{F_{f_n,\alpha_n}\}$ is a decreasing sequence of non-empty compact subsets of K. By the compactness of $K, F \equiv \cap_n F_{f_n,\alpha_n} \neq \emptyset$. Clearly, F is compact and convex, and $F \subset \cap_n U_n$. Noticing that $(K \backslash F_{f\alpha}) = \{x \in K|f(x) > \alpha\}$ is convex $(\forall f \in E^*, \alpha \in \mathbb{R})$, and the sequence $\{(K \backslash F_{f_n,\alpha_n})|n\}$ is increasing, we can see that $(K \backslash F)$ is also convex.

Since $F \cap P \subset U \cap \cap_n U_n \cap P = V \cap \cap_n V_n$, it suffices to show that $F \cap P \neq \emptyset$. Indeed, let x be an extreme point of F. If x is extremal in K, we are done. If not , let δ be a line passing through x and such that x is an interior point of the segment $K \cap \delta$, we then can show that one of the end-points of $F \cap \delta$ is an extreme point of K since F and $(K \backslash F)$ are convex. Q.E.D.

Lemma 13.1.9. Let X be a Baire space, Y be a topological space, and T be a continuous open map from X onto Y. Then Y is also a Baire space.

Proof. Let $\{V_n\}$ be a sequence of open dense subsets of Y, and V be any open subset of Y. We need to show that $\cap_n V_n \cap V \neq \emptyset$.

Let $U = T^{-1}(V), U_n = T^{-1}(V_n), \forall n$. If W is any open subset of X. Then TW is also open. Hence, $TW \cap V_n \neq \emptyset, \forall n$. Let $x \in W$ and $Tx \in V_n$. Then $x \in$

$W\cap T^{-1}(V_n) = W\cap U_n$. So $W\cap U_n \neq \emptyset$, and U_n is an open dense subset of $X, \forall n$. Now $U\cap\cap_n U_n \neq \emptyset$ since X is Baire. Therefore, $V\cap\cap_n V_n = T(U\cap\cap_n U_n) \neq \emptyset$.

Q.E.D.

Lemma 13.1.10. Let A be a C^*-algebra, $\{\{\pi_l, H_l\}|l\in\wedge\}$ be a family of nondegenerate $*$ representations of $A, I = \cap\{\ker\pi_l|l\in\wedge\}$, and ρ be a state on A with $\rho|I = 0$. Then ρ belongs to the $\sigma(A^*, A)$ –closure of following subset of A^* :

$$\mathrm{Co}\{\langle\pi_l(\cdot)\xi_l, \xi_l\rangle|\xi_l\in H_l \text{ and } \quad \|\xi_l\| = 1, l\in\wedge\}$$

Proof. Let $\pi = \oplus_{l\in\wedge}\pi_l, H = \oplus_{l\in\wedge}H_l$. Then $\{\pi, H\}$ is a faithful $*$ representation of A/I. By $\rho|I = 0, \rho$ can be regarded as a state on A/I. Now from Lemma 16.3.6 (it is easy and elementary), ρ is a $\sigma(A^*, A)$–limit of states which belong to the following subset:

$$\mathrm{Co}\{\langle\pi(\cdot)\xi, \xi\rangle|\xi\in H, \|\xi\| = 1\}.$$

That comes to the conclusion. Q.E.D.

Let A be a C^*-algebra, $P(A)$ be the pure state space of A, and $\rho_1, \rho_2 \in P(A)$.ρ_1 and ρ_2 are said to be *unitarily equivalent*, and denoted by $\rho_1\sim\rho_2$, if there exists a unitary element $u\in(A\dot{+}\mathbb{C})$ such that $\rho_1(a) = \rho_2(u^*au), \forall a\in A$.

Proposition 13.1.11. Let A be a C^*-algebra, $\rho_1, \rho_2\in P(A)$, and $\{\pi_1, H_1, \xi_1\}$, $\{\pi_2, H_2, \xi_2\}$ be the irreducible cyclic $*$ representations of A generated by ρ_1, ρ_2 respectively. Then we have that

$$\begin{aligned}\rho_1\sim\rho_2 &\Longleftrightarrow \{\pi_1, H_1\}\cong\{\pi_2, H_2\}\\ &\Longleftrightarrow \text{there is } \eta\in H_1 \text{ with } \|\eta\| = 1 \text{ such that}\\ &\qquad \rho_2(a) = \langle\pi_1(a)\eta, \eta\rangle, \quad \forall a\in A.\end{aligned}$$

Proof. Let $\rho_1(a) = \rho_2(u^*au), \forall a\in A$, where u is a unitary element of $(A\dot{+}\mathbb{C})$. Define

$$U\pi_1(a)\xi_1 = \pi_2(au)\xi_2, \quad \forall a\in A.$$

Then U can be uniquely extended to a unitary operator from H_1 onto H_2 , and

$$U\pi_1(a)U^* = \pi_2(a), \quad \forall a\in A.$$

Therefore, $\{\pi_1, H_1\}\cong\{\pi_2, H_2\}$.

Conversely, let U be a unitary operator from H_1 onto H_2 such that $U\pi_1(a)U^* = \pi_2(a), \forall a\in A$. Then we have that $\rho_2(a) = \langle\pi_1(a)\eta, \eta\rangle, \forall a\in A$, where $\eta = U^*\xi_2\in H_1$, and $\|\eta\| = \|\xi_2\| = 1$. Further, we can find a unitary operator

V on H_1 such that $V\xi_1 = \eta$. By Theorem 2.6.5, there exists a unitary element u of $(A\dot{+}\mathbb{C})$ such that $\pi_1(u)\xi_1 = \eta$. Therefore, we get

$$\rho_2(a) = \langle \pi_1(u^* a u)\xi_1, \xi_1 \rangle = \rho_1(u^* a u), \quad \forall a \in A.$$

Q.E.D.

Theorem 13.1.12. Let A be a C^*–algebra. Then $(P(A), \sigma(A^*, A)), \hat{A}$ and $\mathrm{Prim}(A)$ are Baire spaces.

Proof. Let $K = \{f \in A^* | f \geq 0, \|f\| \leq 1\}$. Clearly, K is a compact convex subset of $(A^*, \sigma(A^*, A))$. By Lemma 13.18, $(ExK, \sigma(A^*, A))$ is a Baire space. From Proposition 2.5.5, we have that $ExK = P(A) \sqcup \{0\}$. Now it is easy to see that $(P(A), \sigma(A^*, A))$ is a Baire space.

By the GNS construction, $\rho \to \pi_\rho$ is a surjective map from $P(A)$ to $\hat{A}$.

We claim that the map $\rho \to \pi_\rho$ is continuous. Indeed, by Proposition 13.1.5 any open subset of $\hat{A}$ has the form of $\hat{A}^I$, where I is some closed two–sided ideal of A. Then inverse image of $\hat{A}^I$ under that map is as follows:

$$U = \{\rho \in P(A) | \pi_\rho \in \hat{A}^I \Longleftrightarrow \pi_\rho|I \neq 0\}$$

Let $\rho \in U$. If for any $\sigma(A^*, A)$–neighborhood $U(\rho, F, 1) = \{\varphi \in P(A) | |\rho(x) - \varphi(x)| < 1, \forall x \in F\}$ of ρ in $P(A)$, where F is a finite subset of A, there is $\rho_F \in U(\rho, F, 1)\backslash U$. Then we have that $\rho_F \longrightarrow \rho$ in $\sigma(A^*, A)$ and $\pi_{\rho_F}|I = 0, \forall F$. Consequently,

$$\rho_F(axb) = 0, \quad \forall x \in I, a, b \in A, \text{and } F.$$

Thus we obtain that $\rho(axb) = 0, \forall x \in I, a, b \in A$, and $\pi_\rho|I = 0$, a contradiction. Therefore, U is open, and the map $\rho \longrightarrow \pi_\rho$ is continuous.

Now let U be an open subset of $(P(A), \sigma(A^*, A))$, and $V = \{\pi \in \hat{A}|$ there is $\rho \in U$ such that $\pi \cong \pi_\rho\}$. For each subset E of $P(A)$, put

$$\tilde{E} = \{\varphi \in P(A) | \text{there is } \rho \in E \text{ such that } \varphi \sim \rho\}.$$

Clearly , $V = \{\pi \in \hat{A}|$ there is $\rho \in \tilde{U}$ such that $\pi \cong \pi_\rho\}$. For each unitary element u of $(A\dot{+}\mathbb{C})$, let

$$u(U) = \{\rho(u^* \cdot u) | \rho \in U\}.$$

Clearly, $u(U)$ is open. By Proposition 13.1.11,

$$\tilde{U} = \sqcup\{u(U) | u \text{ is a unitary element of } (A\dot{+}\mathbb{C})\}$$

is also open. Let $F = P(A)\backslash\tilde{U}$. Then F is closed and $F = \tilde{F}$. Let $I = \sqcap\{\ker\pi_\rho | \rho \in F\}$. If $\varphi \in P(A)$ and $\ker \pi_\varphi \supset I (\Longleftrightarrow \varphi|I = 0)$, then by Lemma 13.1.10 and $F = \tilde{F}$ we have that $\varphi \in \overline{CoF}^\sigma$. Since $F = \overline{F}^\sigma$ and φ is also an extreme point of $\overline{CoF}^{\sigma}$, it follows from the Krein–Milmann theorem (see [89, Theorem 15.2]) that $\varphi \in F$. This means that

$$\{\pi \in \hat{A} | \text{ there is } \rho \in F \text{ such that } \pi \cong \pi_\rho\}$$

is a closed subset of $\hat{A}$. Therefore, V is an open subset of $\hat{A}$, and $\rho \longrightarrow \pi_\rho$ is an open map.

Now by Lemma 13.1.9, $\hat{A}$ is a Baire space.

It is well–known that ker : $\hat{A} \longrightarrow \text{Prim}(A)$ is a continuous and open surjection. So Prim (A) is also a Baire space from Lemma 13.1.9. Q.E.D.

Proposition 13.1.13. Let A be a C^*–algebra, and $x \in A$. Then $\pi \longrightarrow \|\pi(x)\|$ is a lower semicontinuous function on $\hat{A}$.

Proof. For any $k \geq 0$, we need to show that

$$\begin{aligned} E &= \{\pi \in \hat{A} | \|\pi(x)\| \leq k\} = \{\pi \in \hat{A} | \|\pi(x^*x)\| \leq k^2\} \\ &= \{\pi \in \hat{A} | \sigma(\pi(x^*x)) \subset [-k^2, k^2]\} \end{aligned}$$

is a closed subset of $\hat{A}$.

Let $a = x^*x, L = [-k^2, k^2]$, and $I = \cap \pi' | \pi' \in E\}$. Suppose that $\pi \in \overline{E}$, and there is some $\lambda \in \mathbb{R}$ with $\lambda \in \sigma(\pi(a)) \backslash L$. Pick a continuous function f on $\mathbb{R}$ such that $f|L = 0$ and $f(\lambda) \neq 0$. Then we have that $\pi'(f(a)) = f(\pi'(a)) = 0, \forall \pi' \in E$, i.e., $f(a) \in I$; and $\pi(f(a)) = f(\pi(a)) \neq 0$.

Let $T = \{J \in \text{Prim}(A) | J \supset I\}$. Then T is a closed subset of Prim(A), and $\ker^{-1}(T)$ is a closed subset of $\hat{A}$. Clearly, $E \subset \ker^{-1}(T)$. Since $\pi \in \overline{E}$, it follows that $\pi \in \ker^{-1}(T)$ and $\ker\pi \supset I$. Hence , we get $f(a) \in I \subset \ker\pi$ and $\pi(f(a)) = 0$, a contradiction. Therefore, E must be closed. Q.E.D.

Lemma 13.1.14. Let H be a Hilbert space, $\xi_1, \cdots, \xi_n \in H$, and $\varepsilon > 0$. Then there exists $\delta > 0$ with the following property: if $\eta_1, \cdots, \eta_n \in H$ such that

$$|\langle \eta_i, \eta_j \rangle - \langle \eta_i, \eta_j \rangle| \leq \delta, \quad 1 \leq i, j \leq n,$$

then we can find a unitary operator U on H with $\|U\eta_i - \xi_i\| \leq \varepsilon, 1 \leq i \leq n$.

Proof. For $n = 1$, it is easy to see that the conclusion holds , i.e., for any $\xi \in H$ and $\varepsilon > 0$ there is $\delta = \delta(\xi, \varepsilon) \in (0, \varepsilon)$ such that if $\eta \in H$ and $|\|\eta\|^2 - \|\xi\|^2| < \delta$, then we can find a unitary operator U on H with $\|U\eta - \xi\| \leq \varepsilon$.

Now we assume that for any $\xi_1, \cdots, \xi_{n-1} \in H$ and $\varepsilon > 0$ there exists $\delta = \delta(\xi_1, \cdots, \xi_{n-1}, \varepsilon) \in (0, \varepsilon)$ with the following property: if $\eta_1, \cdots, \eta_{n-1} \in H$ and $|\langle \eta_i, \eta_j \rangle - \langle \xi_i, \xi_j \rangle| \leq \delta, 1 \leq i, j \leq n-1$, then there is a unitary operator U on H such that $\|U\eta_i - \xi_i\| \leq \varepsilon, 1 \leq i \leq n-1$.

Let $\xi_1, \cdots, \xi_n \in H$ and $\varepsilon \in (0,1)$. Write $H = H' \oplus H''$, where $H' = [\xi_1, \cdots, \xi_{n-1}]$ and $H'' = (H')^\perp$, and ler P, Q be the projections from H onto H', H'' respectively, and $K = \max\{\|\xi_i\| | 1 \le i \le n\}$. Clearly , for any $\varepsilon > 0$ there exists $\delta' = \delta'(\varepsilon) \in (0, \varepsilon)$ with following property: if $\eta \in H'$ and $|\langle \eta, \xi_i \rangle| \le \delta', 1 \le i \le n-1$, then $\|\eta\| \le \varepsilon$.

Pick $\varepsilon', \varepsilon'' \in (0, \varepsilon)$ such that

$$\varepsilon''(1 + K + (1+K^2)^{1/2}) \le \delta(Q\xi_n, \xi/2), \quad \varepsilon'(1 + (1+K^2)^{1/2}) \le \delta'(\varepsilon'').$$

Let $\delta = \delta(\xi_1, \cdots, \xi_{n-1}, \varepsilon')$, and $\eta_1, \cdots, \eta_n \in H$ with

$$|\langle \eta_i, \eta_j \rangle - \langle \xi_i, \xi_j \rangle| \le \delta, \quad 1 \le i, j \le n.$$

Clearly , $\|\eta_i\| = (1+K^2)^{1/2}, 1 \le i \le n$. And by the preceding paragraph, we can find a unitary operator U on H such that

$$\| \cup \eta_i - \xi_j \| \le \varepsilon' < \varepsilon, \quad 1 \le i \le n-1.$$

Let

$$\xi_n' = P\xi_n, \quad \xi_n'' = Q\xi_n, \quad \eta_n' = PU\eta_n, \quad \eta_n'' = QU\eta_n.$$

Noticing that

$$\begin{aligned} & |\langle \xi_n' - \eta_n', \xi_i \rangle| = |\langle P(\xi_n - U\eta_n), \xi_i \rangle| \\ \le \; & |\langle \xi_n, \xi_i \rangle - \langle U\eta_n, U\eta_i \rangle| + |\langle U\eta_n, \quad U\eta_i - \xi_i \rangle| \\ \le \; & \delta + \varepsilon'(1+K^2)^{1/2} \le \varepsilon'(1 + (1+K^2)^{1/2}) \\ \le \; & \delta'(\varepsilon''), \quad 1 \le i \le n-1, \end{aligned}$$

we have that $\|\xi_n' - \eta_n'\| \le \varepsilon''$. Further,

$$\begin{aligned} & |\|\xi_n'\|^2 - \|\eta_n'\|^2| \\ \le \; & (\|\xi_n\| + \|\eta_n\|) \cdot \|\xi_n' - \eta_n'\| \le \varepsilon''(K + (1+K^2)^{1/2}). \end{aligned}$$

Thus , we have

$$\begin{aligned} & |\|\xi_n''\|^2 - \|\eta_n''\|^2| \\ \le \; & |\|U\eta_n\|^2 - \|\xi_n\|^2| + |\|\xi_n'\|^2 - \|\eta_n'\|^2| \\ \le \; & \delta + \varepsilon''(K + (1+K^2)^{1/2}) \le \varepsilon''(1 + K + (1+K^2)^{1/2}) \\ \le \; & \delta(\xi_n'', \varepsilon/2). \end{aligned}$$

Now from the preceding paragraph, there is a unitary operator V on H'' such that $\|V\eta_n'' - \xi_n''\| \le \varepsilon/2$.

Let $V = I$ on H'. Then V is unitary on H, and

$$\|VU\eta_i - \xi_i\| = \|U\eta_i - \xi_i\| \le \varepsilon, \quad 1 \le i \le n-1.$$

Moreover,

$$\|VU\eta_n - \xi_n\|^2 = \|\eta'_n - \xi'_n\|^2 + \|V\eta''_n - \xi''_n\|^2$$
$$\leq \varepsilon''^2 + \varepsilon^2/4 < \varepsilon^2,$$

i.e., $\|VU\eta_n - \xi_n\| < \varepsilon$. Q.E.D.

Lemma 13.1.15. Let A be a C^*-algebra, H be a Hilbert space such that every irreducible $*$ representation of A can be realized in a closed subspace of H, and $\{\pi_0, H_0\}$ be an irreducible $*$ representation of A, where H_0 is a closed subspace of H. Then for any $a_1, \cdots, a_p \in A, \xi_1, \cdots, \xi_n \in H_0$ and $\varepsilon > 0$, there is a neighborhood V of π_0 in $\hat{A}$ such that for any $\tau \in V$ we can find an irreducible $*$ representation $\{\pi, H_\pi\}$ of A with $\pi \cong \tau, H_\pi \subset H$ and

$$\|\pi(a_i)P_\pi\xi_j - \pi_0(a_i)\xi_j\| \leq \varepsilon,$$

$\forall 1 \leq i \leq p, 1 \leq j \leq n$, where P_π is the projection from H onto H_π.

Proof. We may assume that A has an identity 1, and $a_1 = 1$, $\|a_i\| \leq 1, 1 \leq i \leq p$.

1) Suppose that $n = 1$, and $\|\xi\| = 1$, where $\xi = \xi_1$. Consider the open neighborhood

$$W = U(\langle\pi_0(\cdot)\xi, \xi\rangle,\ a_i^* a_j,\ 1 \leq i, j \leq p,\ \varepsilon_1)$$

of $\langle\pi_0(\cdot)\xi, \xi\rangle$ in $P(A)$, where

$$\varepsilon_1 = \delta(\pi_0(a_1)\xi, \cdots, \pi_0(a_p)\xi, \varepsilon/2)$$

and $\delta(\cdots)$ is as in Lemma 12.1.14. From the proof of Theorem 13.1.12,

$$V = \{\tau \in \hat{A} | \text{there is } \rho \in W \text{ such that } \pi_\rho \cong \tau\}$$

is an open subset of $\hat{A}$. For fixed $\tau \in V$, let $\rho \in W$ be such that $\pi_\rho \cong \tau$. Suppose that $\{\pi_\rho, H_\rho, \xi_\rho\}$ is the irreducible cyclic $*$ representation of A generated by ρ and $H_\rho \subset H$. Then we have

$$|\langle\pi_\rho(a_i^* a_j)\xi_\rho, \xi_\rho\rangle - \langle\pi_0(a_i^* a_j)\xi, \xi\rangle| < \varepsilon_1,$$

i.e.,

$$|\langle\pi_\rho(a_j)\xi_\rho, \pi_\rho(a_i)\xi_\rho\rangle - \langle\pi_0(a_j)\xi, \pi_0(a_i)\xi\rangle| < \varepsilon_1,$$

$\forall 1 \leq i, j \leq p$. By Lemma 13.1.14, there is a unitary operator U on H such that

$$\|U\pi_\rho(a_i)\xi_\rho - \pi_0(a_i)\xi\| < \varepsilon/2, \quad 1 \leq i \leq p.$$

In particular,

$$\|U\xi_\rho - \xi\| = \|U\pi_\rho(a_1)\xi_\rho - \pi_0(a_1)\xi\| < \varepsilon/2.$$

Thus , we have

$$\begin{aligned}
&\|U\pi_\rho(a_i)P_\rho U^{-1}\xi - \pi_0(a_i)\xi\| \\
\le\ &\|U\pi_\rho(a_i)P_\rho U^{-1}\xi - U\pi_\rho(a_i)\xi_\rho\| + \|U\pi_\rho(a_i)\xi_\rho - \pi_0(a_i)\xi\| \\
\le\ &\|a_i\|\cdot\|P_\rho U^{-1}\xi - \xi_\rho\| + \varepsilon/2 \le \varepsilon, \quad 1\le i\le p,
\end{aligned}$$

where P_ρ is the projection from H onto H_ρ.

Now let $\pi(\cdot) = U\pi_\rho(\cdot)U^{-1}$, and $H_\pi = UH_\rho$. Then $\pi \cong \pi_\rho \cong \tau$. Since $P_\pi = UP_\rho U^{-1}$, where P_π is the projection from H onto H_π, it follows that

$$\begin{aligned}
\pi(a_i)P_\pi\xi &= U\pi_\rho(a_i)U^{-1}\cdot UP_\rho U^{-1}\xi \\
&= U\pi_\rho(a_i)P_\rho U^{-1}\xi, \quad 1\le i\le p.
\end{aligned}$$

Thus , we obtain that

$$\|\pi(a_i)P_\pi\xi - \pi_0(a_i)\xi\| \le \varepsilon, \quad 1\le i\le p.$$

2) For any n , by Theorem 2.6.5 we can find $b_1,\cdots,b_n \in A$ such that

$$\pi_0(b_i)\xi = \xi_i, \quad 1\le i\le n,$$

where $\xi = \xi_1$. From the preceding paragraph 1) , there is a neighborhood V of π_0 in $\hat{A}$ with the following property: for any $\tau\in V$ we can find an irreducible $*$ representation $\{\pi, H_\pi\}$ such that $\pi\cong\tau$, $H_\pi\subset H$, and

$$\begin{cases}
\|\pi(a_ib_j)P_\pi\xi - \pi_0(a_ib_j)\xi\| \le \varepsilon/2, \\
\|\pi(b_j)P_\pi\xi - \pi_0(b_j)\xi\| \le \varepsilon/2,
\end{cases}$$

$\forall 1\le i\le p, 1\le j\le n$. Then , we have

$$\begin{aligned}
&\|\pi(a_i)P_\pi\xi_j - \pi_0(a_i)\xi_j\| \\
\le\ &\|\pi(a_i)P_\pi(\xi_j - \pi_0(b_j)\xi)\| \\
&+\|\pi(a_i)P_\pi\pi_0(b_j)\xi - \pi(a_i)P_\pi\pi(b_j)P_\pi\xi\| \\
&+\|\pi(a_ib_j)P_\pi\xi - \pi_0(a_ib_j)\xi\| + \|\pi_0(a_ib_j)\xi - \pi_0(a_i)\xi_j\| \\
\le\ &\|a_i\|\cdot\varepsilon/2 + \varepsilon/2 \le \varepsilon,
\end{aligned}$$

$\forall 1\le i\le p, 1\le i\le n$. Q.E.D.

Proposition 13.1.16. Let A be a C^*-algebra, and $x\in A_+$. Then $\pi\longrightarrow tr\pi(x)$ is a lower semicontinuous function on $\hat{A}$.

Proof. Pick a Hilbert space H such that every irreducible $*$ representation of A can be realized in a closed subspace of H. Let $\{\pi_0, H_0\}$ be an irreducible

$*$ representation of A and $H_0 \subset H$. If $tr\pi_0(x) > \alpha$, then there is an orthogonal normalized family $\{\xi_1, \cdots, \xi_n\} \subset H_0$ such that

$$\sum_{j=1}^{n} \langle \pi_0(x)\xi_j, \xi_j \rangle - \alpha = \beta > 0.$$

By Lemma 13.1.15, there is a neighborhood V of π_0 in $\hat{A}$ such that for each $\tau \in V$ we can find an irreducible $*$ representation $\{\pi, H_\pi\}$ with $\pi \cong \tau, H_\pi \subset H$, and

$$\|\pi(x)P_\pi\xi_j - \pi_0(x)\xi_j\| \leq \beta/(n+1), \quad 1 \leq j \leq n.$$

Then we have

$$\begin{aligned} tr\tau(x) &= tr\pi(x) \geq \textstyle\sum_{j=1}^{n} \langle \pi(x)P_\pi\xi_j, P_\pi\xi_j \rangle \\ &= \textstyle\sum_{j=1}^{n} \langle \pi(x)P_\pi\xi_j, \xi_j \rangle \\ &\geq \sum_{j=1}^{n} \langle \pi_0(x)\xi_j, \xi_j \rangle - n \cdot \frac{\beta}{n+1} > \alpha. \end{aligned}$$

Therefore , for each $\alpha \geq 0$, $\{\pi \in \hat{A} | tr\pi(x) > \alpha\}$ is an open subset of $\hat{A}$, i.e., $\pi \longrightarrow tr\pi(x)$ is lower semicontinouos on $\hat{A}$. Q.E.D.

References. [27], [46], [73].

13.2. Elementary C^*–algebras and CCR (liminary) algebras

Definition 13.2.1. A C^*–algebra A is said to be *elementary* , if A is $*$ isomorphic to $C(H)$, where H is some Hilbert space.

Proposition 13.2.2. Let $A = C(H)$ be an elementary C^*–algebra.

1) Each positive linear functional on A has the following form : $\sum_i \lambda_i \langle \cdot \xi_i, \xi_i \rangle$, where $\{\xi_i\}$ is an orthogonal normalized sequence of H; $\lambda_i \geq 0, \forall i$; and $\sum_i \lambda_i < \infty$.

2) Each pure state on A must be the following form : $\langle \cdot \xi, \xi \rangle$, where $\xi \in H$ and $\|\xi\| = 1$.

3) $^\#\hat{A} = 1$.

4) A is simple , i.e., if I is a closed two–sided ideal of A , then I must be either $\{0\}$ or A.

5) If B is a C^*–subalgebra of A , and B is irreducible on H, then $B = A$.

6) If $B = C(H')$ is $*$ isomorphic to A, then there exists a unitary operator U from H onto H' such that $U^*BU = A$.

Proof. 1) It is immediate from $C(H)^* = T(H)$.

2) The conclusion is obvious by 1).

3) Let $\rho(\cdot) = \langle \cdot\xi, \xi\rangle, \sigma(\cdot) = \langle \cdot\eta, \eta\rangle$ be two pure states on A, where $\xi, \eta \in H$ and $\|\xi\| = \|\eta\| = 1$. Since A is irreducible on H, from Theorem 2.6.5 there is a unitary operator $u \in (A\dot{+}\mathbb{C})$ such that $u\xi = \eta$. Then we have $\sigma(\cdot) = \rho(u^* \cdot u)$, i.e., $\rho \sim \sigma$. By Proposition 13.1.11, $\{\pi_\rho, H_\rho\}$ and $\{\pi_\sigma, H_\sigma\}$ are unitarily equivalent. Therefore , $\#\widehat{A} = 1$.

4) If I is a proper closed two-sided ideal of A, then we have $I = \sqcap\{J \in \text{Prim}(A) | J \supset I\}$. By 3) , $\text{Prim}(A) = \{0\}$. Therefore, $I = \{0\}$.

6) Let Φ be a $*$ isomorphism from $A = C(H)$ onto $B = C(H')$. Then $\{id, H\}$ and $\{\Phi, H'\}$ are two irreducible $*$ representations of A. By 3) , Φ must be spatial.

5) Let B be a C^*-subalgebra of $A = C(H)$, and B be irreducible on H. Then B is weakly dense in $B(H)$. If $t \in T(H) = C(H)^*$ is such that $tr(tb) = 0, \forall b \in B$, then we also have $tr(tb) = 0, \forall b \in B(H)$. Since $T(H)^* = B(H)$, it follows that $t = 0$. Therefore, $B = A$. Q.E.D.

Now we consider C^*-algebras of compact operators, and the C^*-subalgebras of an elementary C^*-algebra.

Proposition 13.2.3. Let A be a C^*-subalgebra of $C(H)$, where H is some Hilbert space.

1) If p is a non-zero projection of A, then the rank of p is finite , i.e., dim $pH < \infty$; and p is an orthogonal sum of minimal projections of A. Moreover, p is minimal in A if and only if $pAp = \mathbb{C}p$.

2) $A = \overline{[\text{Proj}(A)]}$, where $\text{Proj}(A)$ is the set of all projections of A.

3) Let p be a minimal projection of A, $\xi \in pH$ with $\|\xi\| = 1$, and $K = \overline{[A\xi]}$. Then A is irreducible on K, and $A|K = C(K)$.

Proof. Since each slef-adjoint element of A has the form of $\sum_i \lambda_i p_i$, where $\{p_i\}$ is an orhtogonal sequence of projections on H; $\lambda_i \neq \lambda_j, \forall i \neq j$; and $\lambda_i \longrightarrow 0$, it follows that $A = \overline{[\text{Proj}(A)]}$.

Now if $p \in \text{Proj}(A)$, then we also have $pAp = \overline{[\text{Proj}(pAp)]}$. Hence , p is minimal in A if and only if $pAp = \mathbb{C}p$.

Now let p be a minimal projection of A, $\xi \in pH$ with $\|\xi\| = 1$, and $K = \overline{[A\xi]}$. For any $b \in (A|K)'$ and $s, t \in A$, we have

$$\begin{aligned}\langle bs\xi, t\xi\rangle &= \langle pt^*bsp\xi, \xi\rangle = \langle pt^*spb\xi, \xi\rangle \\ &= \lambda\langle b\xi, \xi\rangle = \langle s\xi, t\xi\rangle \cdot \langle b\xi, \xi\rangle,\end{aligned}$$

where $pt^*sp = \lambda p$ and $\lambda \in \mathbb{C}$ (since p is minimal) . Thus $b = \langle b\xi, \xi\rangle 1_K$, i.e., A is irreducible on K. Further, from Proposition 13.2.2 we have $A|K = C(K)$.

Q.E.D.

Let $\{A_l | l \in \mathbb{I}\}$ be a family of C^*-algebras. $\lim_l a_l = 0$, where $a_l \in A_l$, $\forall l \in \mathbb{I}$, means that for any $\varepsilon > 0$ there is a finite subset F_ε of $\mathbb{I}$ such that $\|a_l\| < \varepsilon, \forall l \notin F_\varepsilon$. The following set

$$\{(a_l)_{l\in\mathbb{I}} | a_l \in A_l, \forall l \in \mathbb{I}; \text{and } \lim_l a_l = 0\},$$

denoted by $\sum_{l\in\mathbb{I}} A_l$, is called the *direct sum* of $\{A_l | l \in \mathbb{I}\}$. Clearly , it can become a C^*-algebra with respect to the norm $\|(a_l)\| = \sup_l \|a_l\|$ and natural operations.

Proposition 13.2.4. Any C^*-algebra of compact operators is $*$ isomorphic to the direct sum of a family of elementary C^*-algebras.

Proof. Let A be a nondegenerate C^*-subalgebra of $C(H)$ on a Hilbert space H. By Proposition 13.2.3 and the Zorn lemma, we can write $H = \sum_{l\in\wedge} \oplus H_l$, where $AH_l \subset H_l$, and $A|H_l = C(H_l), \forall l \in \wedge$.

For each $l \in \wedge$, let $\pi_l(a) = a|H_l, \forall a \in A$. Then $\{\pi_l, H_l\}$ is an irreducible $*$ representation of A. Pick a subset $\mathbb{I}$ of $\wedge$ such that : for any different $l, l' \in \mathbb{I}$, the representations $\{\pi_l, H_l\}$ and $\{\pi_{l'}, H_{l'}\}$ are not unitarily equivalent; and for any $l \in \wedge$ there is $l' \in \mathbb{I}$ such that $\{\pi_l, H_l\} \cong \{\pi_{l'}, H_{l'}\}$.

Since $a \in A$ is compact, it follows that $\lim_{l\in\mathbb{I}} \pi_l(a) = 0$. Then by the property of $\mathbb{I}$ we can see that $a \longrightarrow (\pi_l(a))_{l\in\mathbb{I}}$ is a $*$ isomorphism from A into $\sum_{l\in\mathbb{I}} C(H_l)$. Now it suffices to show that

$$\{(\pi_l(a))_{l\in\mathbb{I}} | a \in A\} = \sum_{l\in\mathbb{I}} C(H_l).$$

For each $l \in \mathbb{I}$, let p_l be the projection from H onto $\overline{[A'H_l]}$. Clearly , $p_l \in A' \cap A''$. Pick a net $\{t_\alpha\} \subset A$ such that $\|t_\alpha\| \le \|p_l\|, \forall \alpha$, and $t_\alpha \to p_l($ strongly) . Then for each projection p of A, we have $\|(t_\alpha - p_l)p\| \to 0$ since $\dim pH < \infty$. Hence , $p_l p \in A, \forall p \in \text{Proj}(A)$. By Proposition 13.2.3, we get $p_l A \subset A$ and $Ap_l \subset A, \forall l \in \mathbb{I}$.

Let q_l be the projection from H onto $H_l, \forall l$. Then $q_l \in A', \forall l \in \mathbb{I}$. By Propositions 2.10.7, 2.10.4 and 1.5.9, we have $q_l A' q_{l'} = \{0\}, \forall l, l' \in \mathbb{I}$ and $l \ne l'$. Thus, $A'H_l \perp A'H_{l'}$, or $p_l p_{l'} = 0, \forall l, l' \in \mathbb{I}$ and $l \ne l'$.

Now let $(b_l)_{l\in\mathbb{I}} \in \sum_{l\in\mathbb{I}} C(H_l)$. For each $l \in \mathbb{I}$, pick $a_l \in A$ such that $\pi_l(a_l) = b_l$ and $\|a_l\| \le 2\|b_l\|$. Then , $\lim_{l\in\mathbb{I}} \|a_l\| = 0$. By the properties of $\{p_l | p \in \mathbb{I}\}$, we can define an element $a = \sum_{l \in \mathbb{I}} a_l p_l$ of A. Clearly , $\pi_l(a) = a|p_l = a_l|p_l = \pi_l(a_l) = b_l, \forall l \in \mathbb{I}$, i.e., $(\pi_l(a))_{l\in\mathbb{I}} = (b_l)_{l\in\mathbb{I}}$. Q.E.D.

Definition 13.2.5. A C^*-algebra A is said to be *CCR* (or *liminary*) , if for each irreducible $*$ representation $\{\pi, H\}$ of A and $a \in A, \pi(a) \in C(H)$.

From Propositions 13.2.2, 13.2.4 and 2.3.10, we can see that any elementary C^*-algebra, any C^*-algebra of compact operators, and any abelian C^*-algebra are CCR.

Lemma 13.2.6. Let A be a C^*-algebra, and $\{\pi, H\}, \{\pi', H'\}$ be two non-zero irreducible $*$ representations of A.

1) If $\pi(A) \cap C(H) \neq \{0\}$, then $\pi(A) \supset C(H)$.

2) If $\pi(A) \supset C(H), \pi'(A) \supset C(H')$, and $\ker\pi = \ker\pi'$, then $\{\pi, H\} \cong \{\pi', H'\}$.

3) If $\pi(A) \subset C(H)$, then $\pi(A) = C(H)$, and $\ker\pi$ is a maximal closed two-sided ideal of A.

Proof. 1) Let $I = \{x \in A | \pi(x) \in C(H)\}$. Clearly, I is a closed two-sided ideal of A, and $\pi|I \neq \{0\}$. Thus , $\pi(I)$ is also irreducible on H. By Proposition 13.2.2. 5) , we have that $\pi(A) \supset \pi(I) = C(H)$.

2) Let I be as in 1). Then $(I + \ker\pi)$ is also a closed two-sided ideal of A, and $(I + \ker\pi) \neq \ker\pi$. Thus , $\pi'(I + \ker\pi) = \pi'(I) \neq \{0\}$, and $\pi'(I)$ is irreducible on H' .

Define a $*$ representation $\{\rho, H'\}$ of $C(H)$ as follows:

$$\rho(t) = \pi'(\pi^{-1}(t)), \quad \forall t \in C(H).$$

Since $\rho(C(H)) = \pi'(I)$, $\{\rho, H'\}$ is irreducible. By Proposition 13.2.2, there is a unitary operator U from H onto H' such that

$$UtU^* = \rho(t) = \pi'(\pi^{-1}(t)), \quad \forall t \in C(H).$$

Then we have that $U\pi(x)U^* = \pi'(x), \forall x \in I$.

Let $\{u_\alpha\}$ be an approximate identity for I. Then $\pi(u_\alpha) \longrightarrow 1_H$(strongly) and $\pi'(u_\alpha) \to 1_{H'}$ (strongly) . Since $U\pi(a)U^{-1} \cdot U\pi(u_\alpha)U^{-1} = U\pi(au_\alpha)U^{-1} = \pi'(au_\alpha)) = \pi'(a)\pi'(u_\alpha), \forall \alpha$, it follows that $U\pi(a)U^{-1} = \pi'(a), \forall a \in A$, i.e, $\{\pi, H\} \cong \{\pi', H'\}$.

3) From 1) we have $\pi(A) = C(H)$. Now let $\{\rho, K\}$ be an irreduable $*$ representation of A such that $\ker\rho \supset \ker\pi$. Then $\{\rho, K\}$ can be regarded as a $*$ representation of $A/\ker\pi \cong C(H)$. Since $C(H)$ is simple , ρ is a faithful $*$ representation of $A/\ker\pi$, i.e., $\ker\pi = \ker\rho$. By Theorem 2.7.6, $\ker\pi$ must be maximal. Q.E.D.

Proposition 13.2.7. Let A be a CCR algebra. Then:

1) $\text{Prim}(A) = \{J | J$ is a maximal closed two–sided ideal of $A\}$;
2) for each $J \in \text{Prim}(A), A/J$ is a elementary C^*–algebra;
3) $\hat{A}$ is a T_1–space;
4) $\ker : \hat{A} \longrightarrow \text{Prim}(A)$ is a homeomorphism.

Proof. By Theorem 2.7.6, each maximal closed two–sided ideal of A must be primitive. Conversely, if $\{\pi, H\}$ is an irreducible $*$ representation of A, then by Lemma 13.2.6 we have $\pi(A) = C(H)$, and $\ker\pi$ is a maximal closed two–sided ideal of A. Thus, the conclusions 1) and 2) are obvious.

If π and π' are two irreducible $*$ representations of A and $\ker\pi = \ker\pi'$, then by Lemma 13.2.6 we have $\pi \cong \pi'$. From Proposition 13.1.6, $\ker : \hat{A} \longrightarrow \text{Prim}(A)$ is a homeomorphism. Moreover, from 1) Prim (A) is a T_1-space obviously. Q.E.D.

Proposition 13.2.8. Let A be a CCR algebra, B be a C^*–subalgebra of A, and I be a closed two–sided ideal of A. Then B and A/I are also CCR.

Proof. Let $\{\tilde{\pi}, H\}$ be an irreducible $*$ representation of A/I, and define $\pi(a) = \tilde{\pi}(\tilde{a})$, where $\tilde{a} = a + I, \forall a \in A$. Then $\{\pi, H\}$ is an irreducible $*$ representation of A, and $\tilde{\pi}(\tilde{a}) = \pi(a) \in C(H), \forall \tilde{a} \in A/I$ and $a \in \tilde{a}$. So A/I is CCR.

Let ρ be a pure state on B, and $\{\pi_\rho, H_\rho\}$ be the irreducible $*$ representation of B generated by ρ. ρ can be extended to a pure state on A, which is still denoted by ρ. Let $\{\overline{\pi}_\rho, \overline{H}_\rho\}$ be the irreducible $*$ representation of A generated by ρ. Then we have that $H_\rho \subset \overline{H}_\rho, \overline{\pi}_\rho(b)H_\rho \subset H_\rho$ and $\overline{\pi}_\rho(b)|H_\rho = \pi_\rho(b), \forall b \in B$. Since $\overline{\pi}_\rho(b) \in C(\overline{H}_\rho)$, it follows that $\pi_\rho(b) \in C(H_\rho), \forall b \in B$. Therefore, B is also CCR. Q.E.D.

Proposition 13.2.9. Let A be a CCR algebra, $\{\pi_1, H_1\}$ and $\{\pi_2, H_2\}$ be two irreducible $*$ representations of A which are not unitarily equivalent, and $t_i \in C(H_i), i = 1, 2$. Then there exists $a \in A$ such that $\pi_i(a) = t_i, i = 1, 2$.

Proof. Let $I_i = \ker\pi_i, i = 1, 2$. By Proposition 13.2.7, we have $I_1 \neq I_2$. Thus, $I_1 + I_2 \neq I_1$. But I_1 is maximal, it follows that $I_1 + I_2 = A$. Since $\pi_i(A) = C(H_i)$, we can find $a_i \in A$ such that $\pi_i(a_i) = t_i, i = 1, 2$. Now write

$$a_1 = a_{11} + a_{12}, \quad a_2 = a_{21} + a_{22},$$

where $a_{ij} \in I_j, 1 \leq i, j \leq 2$, and let $a = a_{12} + a_{21}$. Then we obtain that

$$\pi_1(a) = \pi_1(a_{12}) = \pi_1(a_{12} + a_{11}) = \pi_1(a_1) = t_1,$$

and

$$\pi_2(a) = \pi_2(a_{21}) = \pi_2(a_{21} + a_{22}) = \pi_2(a_2) = t_2.$$

Q.E.D.

Proposition 13.2.10. Let A be a C^*-algebra. Then

$$I = \{x \in A | \text{ for any } \pi \in \hat{A}, \quad \pi(x) \in C(H_\pi)\}$$

is the largest CCR closed two-sided ideal of A.

Proof. Clearly, I is a closed two-sided ideal of A. Let $\{\rho, H\}$ be any irreducible $*$ representation of I. Then $\{\rho, H\}$ can be extended to an irreducible $*$ repsentation $\{\pi, H\}$ of A. Hence, $\rho(x) = \pi(x) \in C(H), \forall x \in I$, and I is CCR.

Now let J be a CCR closed two-sided ideal of A, and $\{\pi, H\}$ be any irreducible $*$ representation of A. If $\pi|J = 0$, then $\pi(J) = \{0\} \subset C(H)$ obviously. If $\pi|J \neq 0$, then $\{\pi|J, H\}$ is also an irreduible $*$ representation of J. Since J is CCR, it follows that $\pi(x) \in C(H), \forall x \in J$. By the definition of I, we have that $J \subset I$. Q.E.D.

Notes. CCR (*completely continouous representations*) algebras were introduced by 1. Kaplansky.

References . [10], [27], [84].

13.3. GCR (postliminary) algebras and NGCR (antiliminary) algebras

Definition 13.3.1. A C^* -algebra A is said to be *GCR* (or *postliminary*) , if for any closed two-sided ideal I of A, A/I contains a non-zero CCR closed two-sided ideal.

Clearly, a CCR algebra must be GCR; a GCR algebra contains a non-zero CCR closed two-sided ideal.

A C^*-algebra A is said to be *NGCR* (or *antiliminary*) , if A contains no non-zero CCR closed two-sided ideal.

Clearly, a NGCR algebra also contains no non-zero GCR closed two-sided ideal.

Definition 13.3.2. Let A be a C^*-algebra. A strictly increasing family $\{I_\alpha\}$ of closed two-sided ideals of A indexed by a segment $\{0 \leq \alpha \leq \beta\}$ of the ordinals is called a *composition series* for A, if $I_0 = \{0\}, I_\beta = A$; and for each limit ordinal $\gamma(\leq \beta)$ we have

$$I_\gamma = (\cup_{\alpha<\gamma} I_\alpha)^- \qquad \text{(norm closure).}$$

Proposition 13.3.3. Let A be a C^*-algebra. Then there exists unique strictly increasing family $\{I_\alpha | 0 \leq \alpha \leq \beta\}$ of closed two-sided ideals of A satisfying:

1) $I_0 = \{0\}$; and A/I_β is NGCR ;

2) for each limit ordinal $\gamma(\leq \beta)$ we have

$$I_\gamma = (\cup_{\alpha<\gamma} I_\alpha)^- ;$$

3) if $\alpha < \beta$, then $I_{\alpha+1}/I_\alpha$ is the largest (non-zero) CCR closed two-sided ideal of A/I_α.

Proof. The family $\{I_\alpha\}$ can be constructed by induction: if α is a limit ordinal, the condition 2) defines I_α; otherwise, we have $\alpha = \alpha' + 1$, and pick I_α such that $I_\alpha/I_{\alpha'}$ is the maximal (non-zero) CCR closed two-sided ideal of $A/I_{\alpha'}$ (if $A/I_{\alpha'}$ is NGCR, then let $\beta = \alpha'$). Q.E.D.

Proposition 13.3.4. Let A be a C^*-algebra. Then the following statements are equivalent:

1) A is GCR;

2) There is a composition series $\{I_\alpha | 0 \leq \alpha \leq \beta\}$ for A such that $I_{\alpha+1}/I_\alpha$ is GCR, $\forall \alpha < \beta$;

3) There is a composition series $\{I_\alpha | 0 \leq \alpha \leq \beta\}$ for A such that $I_{\alpha+1}/I_\alpha$ is CCR , $\forall \alpha < \beta$.

Proof. 1) $\Longrightarrow$ 3) . Pick $\{I_\alpha | 0 \leq \alpha \leq \beta\}$ as in Proposition 13.3.3. Then A/I_β is NGCR. But A is GCR, so I_β must be equal to A, i.e., $\{I_\alpha | 0 \leq \alpha \leq \beta\}$ is a composition series for A. Moreover, by Proposition 12.3.3, $I_{\alpha+1}/I_\alpha$ is CCR, $\forall \alpha < \beta$.

3) $\Longrightarrow$ 2) . It is obvious.

2) $\Longrightarrow$ 1) . Let J be a proper closed two-sided ideal of A. It suffices to show that A/J contains a non-zero CCR closed two-sided ideal. Suppose that α is the minimal ordinal such that $I_\alpha \not\subset J$. Then $I_{\alpha'} \subset J, \forall \alpha' < \alpha$. If α is a limit ordinal, then we have $I_\alpha = (\cup_{\alpha'<\alpha} I_{\alpha'})^- \subset J$, a contradiction. Thus , $\alpha = \alpha' + 1$. Let $I = I_\alpha \cap J$. Then I_α/J is $*$ isomorphic to $I_\alpha/(J \cap I_\alpha) = I_\alpha/I$, and I_α/I is $*$ isomrophic to the quotient algebra of $(I_\alpha/I_{\alpha'})$ with respect to $(I/I_{\alpha'})$. Since $I_\alpha/I_{\alpha'}$ is GCR, I_α/J contains a non-zero CCR closed two-sided ideal. But I_α/J is a closed two-sided ideal of A/J, so A/J contains a non-zero CCR closed two-sided ideal . Q.E.D.

Proposition 13.3.5. Let A be a GCR algebra, B be a C^*–subalgebra of A, and J be a closed two–sided ideal of A. Then B and A/J are also GCR.

Proof. By Definition 13.3.1, A/J is GCR obviously. Now let $\{I_\alpha | 0 \leq \alpha \leq \beta\}$ be a composition series for A such that $I_{\alpha+1}/I_\alpha$ is CCR, $\forall \alpha < \beta$.

For any $\alpha(< \beta), [(I_{\alpha+1} \cap B) + I_\alpha]/I_\alpha$ is a C^*–subalgebra of $I_{\alpha+1}/I_\alpha$. Since $I_{\alpha+1}/I_\alpha$ is CCR, it follows from Proposition 13.2.8 that $[(I_{\alpha+1} \cap B) + I_\alpha]/I_\alpha$ is also CCR. But $(I_{\alpha+1} \cap B)/(I_\alpha \cap B)$ is $*$ isomorphic to $[(I_{\alpha+1} \cap B) + I_\alpha]/I_\alpha$, so $(I_{\alpha+1} \cap B)/(I_\alpha \cap B)$ is CCR , $\forall \alpha < \beta$.

Clearly , $I_0 \cap B = \{0\}$, and $I_\beta \cap B = A \cap B = B$. From Proposition 13.3.4, it suffices to show that $\{I_\alpha \cap B | 0 \leq \alpha \leq \beta\}$ is a composition series for B. By Definition 13.3.2 we need to prove that

$$I_\gamma \cap B = (\sqcup_{\alpha<\gamma}(I_\alpha \cap B))^- ,$$

where γ is a limit ordinal and $\gamma \leq \beta$. Clearly, the left side contains the right side. Conversely, let $x \in I_\gamma \cap B$, and $\varepsilon > 0$. Since $I_\gamma = (\sqcup_{\alpha<\gamma} I_\alpha))^-$, there is $\alpha < \gamma$ such that dist $(I_\alpha, x) < \varepsilon$. But $B/(I_\alpha \cap B)$ is $*$ isomorphic to $(B+I_\alpha)/I_\alpha$, and $\|\tilde{x}\| < \varepsilon$, where $\tilde{x} = x + I_\alpha$ is an element of $(B + I_\alpha)/I_\alpha$, hence we have dist $(I_\alpha \cap B, x) < \varepsilon$. Therefore, $I_\gamma \cap B = (\sqcup_{\alpha<\gamma}(I_\alpha \cap B))^-$. Q.E.D.

Proposition 13.3.6. Let A be a C^*–algebra, and $\{I_\alpha | 0 \leq \alpha \leq \beta\}$ be as in Proposition 13.3.3. Then I_β is the largest GCR closed two–sided ideal of A, and I_β is also the smallest closed two–sided ideal of A such that A/I_β is NGCR.

Proof. Clearly, $\{I_\alpha | 0 \leq \alpha \leq \beta\}$ is a composition series for I_β. From Proposition 13.3.3 and 13.3.4, I_β is GCR. Let J be any GCR closed two–sided ideal of A. Then $J/(J \cap I_\beta)$ is also GCR. Since $J/(J \cap I_\beta)$ is $*$ isomorphic to $(J + I_\beta)/I_\beta$, $(J + I_\beta)/I_\beta$ is a GCR closed two–sided ideal of A/I_β. But A/I_β is NGCR, it must be $J + I_\beta = I_\beta$, i.e., $J \subset I_\beta$.

Now let I be a closed two–sided ideal of A such that A/I is NGCR. Since $I_\beta/(I_\beta \cap I)$ is GCR , and is $*$ isomorphic to $(I_\beta + I)/I$, it follows that $(I_\beta + I)/I = \{0\}$, i.e. $I_\beta \subset I$. Q.E.D.

Proposition 13.3.7. Let A be a GCR algebra. Then:

1) for any irreducible $*$ representation $\{\pi, H\}$ of A, we have $\pi(A) \supset C(H)$;

2) ker: $\hat{A} \longrightarrow$ Prim(A) is a homeomophism;

3) $\hat{A}$ is a T_0–space.

Proof. 1) Let J be a closed two–sided ideal of A, and $\ker \pi \subset J$ such that $J/\ker \pi$ is a non–zero CCR closed two–sided ideal of $A/\ker \pi$. Then $\{\pi, H\}$

can be regarded as an irreducible $*$ representation of $J/\ker\pi$. Hence, we have

$$\pi(A) \supset \pi(J) = \pi(J/\ker\pi) = C(H).$$

2) and 3) are obvious from 1), Lemma 13.2.6 and Proposition 13.1.6.

Q.E.D.

Lemma 13.3.8. Let X be a Baire space, and f be a non-negative lower semicontinuous function on X. Then there exists $x_0 \in X$ such that f is continous at x_0.

Proof. Replacing f by $f(1+f)^{-1}$, we may assume that $f(x) \leq 1, \forall x \in X$. For each $x \in X$, let

$$\omega(x) = \inf\left\{\sup_{y\in V} f(y) - \inf_{y\in V} f(y) \mid V \text{ is a neighborhood of } x\right\}.$$

Clearly, f is continuous at $x \Longleftrightarrow \omega(x) = 0$. For each n, put

$$F_n = \{x \in X | \omega(x) \geq \frac{1}{n}\}.$$

We claim that F_n is closed, $\forall n$. In fact, if $x \in \overline{F}_n$, then for any open neighborhood V of x there exists $z \in F_n \cap V$. Since V is also a neighborhood of z, it follows that

$$\begin{aligned}\left[\sup_{y\in V} f(y) - \inf_{y\in V} f(y)\right] &\geq \inf\left\{\left[\sup_{y\in W} f(y) - \inf_{y\in W} f(y)\right] | W \text{ is open, and } z \in W\right\} \\ &= \omega(z) \geq \tfrac{1}{n}.\end{aligned}$$

Thus, $\omega(x) \geq 1/n$, and $x \in F_n$.

Suppose that the interior U of F_n is non-empty for some n, and $\alpha = \sup\{f(x)|x \in U\}$. Pick $x_0 \in U$ such that $f(x_0) > \alpha - \frac{1}{2n}$. Since f is lower semicontinuous, there is a neighborhood $V(\subset U)$ of x_0 such that $f(y) > \alpha - \frac{1}{2n}, \forall y \in V$. Of course, $f(y) \leq \alpha, \forall y \in V$. Thus, we have that $\omega(x_0) < \frac{1}{2n}$. This contradicts the fact of $x_0 \in U \subset F_n$. Therefore, $(X \backslash F_n)$ is open and dense in $X, \forall n$.

Since X is Baire, we can pick $x_0 \in \cap_n (X\backslash F_n)$. Clearly, $\omega(x_0) = 0$, i.e., f is continuous at x_0.

Q.E.D.

Lemma 13.3.9. Let A be a CCR algebra. Then

$$J = \{x \in A | \text{the rank of } \pi(x) \text{ is finite}, \forall \pi \in \hat{A}\}$$

is a $*$ dense two-sided ideal of A.

Proof. Clearly, J is a $*$ two-sided ideal of A. Now it suffices to show that $h \in \overline{J}$ for any $h^* = h \in A$ with $\|h\| \leq 1$. Pick a sequence $\{f_n\}$ of continuous functions on $\mathbb{R}$ such that $f_n(t) = 0$ if $|t| \leq 1/n$; and $f_n(t) \longrightarrow t$ uniformly for $t \in [-1,1]$. For any $\pi \in \widehat{A}$, since $\pi(h)$ is a completely continouos symmetric operator, the rank of $\pi(f_n(h)) = f_n(\pi(h))$ is finite, $\forall n$. Hence, $f_n(h) \in J, \forall n$. Clearly, $f_n(h) \longrightarrow h$. Therefore, we have $h \in \overline{J}$. Q.E.D.

Lemma 13.3.10. Let A be a C^*-algebra, $\{\pi_0, H_0\}$ be an irreducible $*$ representation of A, and $x \in A_+$. Suppose that the function $\pi \longrightarrow tr\pi(x) (\forall \pi \in \widehat{A})$ is finite and continuous at π_0.

1) If $y \in A_+$ and $y \leq x$, then $tr\pi(y)$ is also continuous at π_0.

2) Let $\pi_0(x) \neq 0$. Then there is some $z \in A_+$ and a neighborhood V of π_0 in $\widehat{A}$ such that the rank of $\pi(z)$ is one, $\forall \pi \in V$.

Proof. 1) Let $z = x - y (\in A_+)$. By Proposition 13.1.16, $tr\pi(x), tr\pi(y)$ and $tr\pi(z)$ are non-negative lower semicontinuous functions on $\widehat{A}$. Put $\alpha = tr\pi_0(x) (\in [0,+\infty)), \alpha_1 = tr\pi_0(y), \alpha_2 = tr\pi_0(z)$. Then $\alpha_1, \alpha_2 \geq 0$ and $\alpha = \alpha_1 + \alpha_2$. For any $\varepsilon > 0$, since $tr\pi(x)$ is continuous at π_0, there is a neighborhood V_1 of π_0 in $\widehat{A}$ such that

$$\alpha + \varepsilon/3 > tr\pi(x) > \alpha - \varepsilon, \quad \forall \pi \in V_1.$$

Clearly

$$V_2 = \{\pi \in \widehat{A} | \ tr\pi(y) > \alpha_1 - \frac{2}{3}\varepsilon\}$$

and

$$V_3 = \{\pi \in \widehat{A} | \ tr\pi(z) > \alpha_2 - \frac{\varepsilon}{3}\}$$

are open subsets containing π_0. Then $V = V_1 \cap V_2 \cap V_3$ is a neighborhood of π_0 in $\widehat{A}$. If $\pi \in V$ is such that $tr\pi(y) \geq \alpha_1 + \frac{2}{3}\varepsilon$, then

$$tr\pi(x) = tr\pi(y) + tr\pi(z) > \alpha_1 + \alpha_2 + \varepsilon/3 = \alpha + \varepsilon/3,$$

a contradiction. Thus, we have that

$$\alpha_1 + \frac{2}{3}\varepsilon > tr\pi(y) > \alpha_1 - \frac{2}{3}\varepsilon, \quad \forall \pi \in V.$$

Since ε is arbitrary, $tr\pi(y)$ is continuous at π_0.

2) We may assume that $\|\pi_0(x)\| = 1$. Since the operator $\pi_0(x)$ is trace class, 1 is an eigenvalue of $\pi_0(x)$. Further, there is an one rank projection p on H_0 such that $\pi_0(x)p = p$. From Lemma 13.2.6, we have $C(H_0) \subset \pi_0(A)$. So we can find $z_1 \in A_+$ such that $\pi_0(z_1) = p$. Let

$$f(t) = \begin{cases} t, & \text{if } t \in [0,1], \\ 0, & \text{if } t < 0, \\ 1, & \text{if } t > 1, \end{cases}$$

and $z_2 = f(z_1)$. Then $z_2 \in A_+, \|z_2\| \le 1$, and $\pi_0(z_2) = f(p) = p$. Further, let $z_3 = x^{1/2} z_2 x^{1/2}$. Then $0 \le z_3 \le x$, and $\pi_0(z_3) = \pi_0(x)^{1/2} p \pi_0(x)^{1/2} = p$.

From the preceding paragraph, $tr\pi(z_3)$ is continuous at π_0. So there is a neighborhood V_1 of π_0 in $\hat{A}$ such that $tr\pi(z_3) \le 5/4, \forall \pi \in V_1$. By Proposition 13.1.13, we also have a neighborhood V_2 of π_0 in $\hat{A}$ such that $\|\pi(z_3)\| > 3/4, \forall \pi \in V_2$. Let $V = V_1 \cap V_2$. Then

$$\|\pi(z_3)\| > 4/3, \quad \text{and} \quad tr\pi(z_3) \le 5/4, \quad \forall \pi \in V.$$

Since $\pi(z_3)$ is non-negative and trace class, $\pi(z_3)$ has only one eigenvalue λ_π with multiplicity 1 and $\lambda_\pi > 3/4$, and other eigenvalues of $\pi(z_3)$ belong to $[0, 1/2), \forall \pi \in V$. Now let $z = g(z_3)$, where

$$g(t) = \begin{cases} 0, & \text{if } t \le 1/2, \\ 1, & \text{if } t \ge 3/4, \end{cases}$$

and g is continuous on $I\!R$. Then $0 \le z \le z_3$, and $\pi(z) = g(\pi(z_3))$ is the spectral projection of one rank corresponding to the eigenvalue λ_π of $\pi(z_3), \forall \pi \in V$.

Q.E.D.

Proposition 13.3.11. Let A be a C^*-algebra which is not NGCR. Then there exists $0 \ne x \in A_+$ such that the rank of $\pi(x)$ is either 0 or 1 , $\forall \pi \in \hat{A}$.

Proof. Since A contains a non-zero CCR closed two-sided ideal, by Lemma 13.3.9 we can find $0 \ne y \in A_+$ such that the rank of $\pi(y)$ is finite, $\forall \pi \in \hat{A}$. Now $tr\pi(y)$ is non-negative finite lower semicontinuous function on $\hat{A}$. By $y \ne 0, \{\pi \in \hat{A} | tr\pi(y) > 0\} = U$ is a non-empty open subset of $\hat{A}$. From Proposition 13.1.5 and Theorem 13.1.12, U is also a Baire space. From Lemma 13.3.8, $tr\pi(y)$ will be continuous at some $\pi_0 \in U$. Now by Lemma 13.3.10, there is some $z \in A_+$ and some neighborhood $\hat{A}^I$ of π_0, where I is a closed two-sided ideal of A, such that the rank of $\pi(z)$ is one, $\forall \pi \in \hat{A}^I$.

We claim that $Iz \ne \{0\}$. Otherwise, let $J = \overline{[AzA]}$. Then $J \cap I = [IJ] = \{0\}$. From Proposition 13.1.5, it follows that $\hat{A}^J \cap \hat{A}^I = \emptyset$. Thus, we get $\pi(z) = 0, \forall \pi \in \hat{A}^I$, a contradiction.

Now let $wz \ne 0$ for some $w \in I$, and $x = z^* w^* w z$. Then $0 \ne x \in A_+ \cap I$, and clearly the rank of $\pi(x)$ is either 0 or 1 , $\forall \pi \in \hat{A}^I$. Further, for any $\pi \in \hat{A}$ the rank of $\pi(x)$ is also either 0 or 1 .

Q.E.D.

Proposition 13.3.12. Let A be a C^*-algebra. Then A is NGCR if and only if A satiesfies the *Glimm condition*, i.e., for any $0 \ne x \in A_+$, there is an irreducible $*$ representation $\{\pi, H\}$ of A such that $\dim \pi(x)H \ge 2$.

Proof. The sufficiency is obvious from Proposition 13.3.11. Now let A be NGCR. If there exists $0 \ne x \in A_+$ such that the rank of $\pi(x)$ is either 0 or 1

, $\forall \pi \in \hat{A}$, let $J = \overline{[AxA]}$, then J is a non-zero CCR closed two-sided ideal of A. This is impossible since A is NGCR. Therefore, A must satisfy the Glimm condition. Q.E.D.

Reference. [23], [47], [55], [84].

13.4. The existence of type (III) factorial * representations of a NGCR algebra

Proposition 13.4.1. Let A be a NGCR algebra.

1) For any $0 \neq h = h^* \in A$, there is an irreducible * representation $\{\pi, H\}$ of A such that dim $\pi(h)H \geq 2$.

2) For any $0 \neq a \in A$ and $a^*a = aa^*$, there is an irreducible * representation $\{\pi, H\}$ of A such that dim $\pi(a)H \geq 2$.

3) If A has no identity, $(A\dot{+}\mathbb{C})$ is also NGCR.

Proof. 1) Let $h = h_+ - h_-$, where h_+ and $h_- \in A_+$, and $h_+ \cdot h_- = 0$. We may assume that $h^+ \neq 0$. Then by Proposition 13.3.12, there is an irreducible * representation $\{\pi, H\}$ such that dim $\pi(h_+)H \geq 2$. Pick $\eta_1, \eta_2, \xi_1, \xi_2 \in H$ such that

$$\langle \xi_i, \xi_j \rangle = \delta_{ij}, \quad \text{and} \quad \pi(h_+)\eta_i = \xi_i, \quad 1 \leq i, j \leq 2.$$

If $\lambda, \mu \in \mathbb{C}$ are such that $\pi(h)\eta = 0$, where $\eta = \lambda\eta_1 + \mu\eta_2$, then $0 = \pi(h_+)\pi(h)\eta = \pi(h_+)^2\eta, 0 = \langle \pi(h_+)^2\eta, \eta \rangle = \|\pi(h_+)\eta\|^2$, i.e., $\lambda\xi_1 + \mu\xi_2 = 0$, so $\lambda = \mu = 0$. Thus , dim $\pi(h)H \geq \dim[\pi(h)\eta_i | i = 1, 2] = 2$.

2) Write $a = h_1 + ih_2$, where $h_i^* = h_i \in A, i = 1, 2$, and $h_1 h_2 = h_2 h_1$. We may assume that $h_1 \neq 0$. By 1), there is an irreducible * representation $\{\pi, H\}$ of A such that dim $\pi(h_1)H \geq 2$. Pick $\eta_1, \eta_2, \xi_1, \xi_2 \in H$ such that

$$\langle \xi_i, \xi_j \rangle = \delta_{ij}, \quad \text{and} \quad \pi(h_1)\eta_i = \xi_i, \quad 1 \leq i, j \leq 2.$$

If $\lambda, \mu \in \mathbb{C}$ are such that $\pi(a)\eta = 0$, where $\eta = \lambda\eta_1 + \mu\eta_2$, then

$$\begin{aligned} 0 &= \langle \pi(a)\eta, \pi(h_1)\eta \rangle \\ &= \|\pi(h_1)\eta\|^2 + i\langle \pi(h_2)\eta, \pi(h_1)\eta \rangle. \end{aligned}$$

Clearly, $\langle \pi(h_1)\eta, \pi(h_2)\eta \rangle \in \mathbb{R}$. So we have $0 = \pi(h_1)\eta = \lambda\xi_1 + \mu\xi_2$, and $\lambda = \mu = 0$. Thus , dim $\pi(a)H \geq \dim[\pi(a)\eta_i | i = 1, 2] = 2$.

3) Let I be a non-zero CCR closed two-sided ideal of $(A\dot{+}\mathbb{C})$. Clearly, it must be that $A \cap I = \{0\}$. Thus , we have $I = [e - 1]$ for some $e \in A$. Since $AI = IA = \{0\}$, e is an identity of A, a contradiction. Therefore, $(A\dot{+}\mathbb{C})$ is also NGCR. Q.E.D.

Lemma 13.4.2. Let A be a NGCR algebra with an identity 1, $d \in A_+$ with $\|d\| = 1$, and $t \in (0,1]$. Then we can find $w, w', d' \in A$ such that

1) $\|w\| = \|w'\| = \|d'\| = 1, w \geq 0, d' \geq 0,$ and $w'^* w = 0;$
2) $f_t(d)w = w, \quad f_t(d)w' = w';$
3) $w^2 d' = d', \quad w'^* w' d' = d',$

where

$$f_t(r) = \begin{cases} 0, & \text{if } r \leq 1-t, \\ \text{affine}, & \text{if } 1-t \leq r \leq 1-\frac{t}{2}, \\ 1, & \text{if } r \geq 1-\frac{t}{2} \end{cases} \quad (0 < t \leq 1)$$

Proof. Let $s = t/8$, pick $u, c \in A$ with $\|u\| \leq 1, 0 \leq c \leq 1$, and put $d_0 = f_{2s}(d) c f_{2s}(d), d_1 = f_{4s}(d) - d_0$. Clearly, $0 \leq d_0 \leq 1, -1 \leq d_1 \leq 1$. Since $f_{4s} f_{2s} = f_{2s}$, if follows that $f_{4s}(d) d_0 = d_0 = f_{4s}(d)$. Then $\{1, d_0, f_{4s}(d)\}$ can generate an abelian C^*-subalgebra B of A. Let $B \cong C(\Omega)$. By $f_{4s}(d) d_0 = d_0$, we have

$$\rho(f_{4s}(d)) = 1, \quad \rho(d_1) = 1 - \rho(d_0),$$

$\forall \rho \in \Omega$ and $\rho(d_0) \neq 0$. Hence, if $g : I\!R \longrightarrow I\!R$ is continuous and g vanishes on $[0, \frac{1}{2}]$, then by $0 \leq \rho(d_0) \leq 1$ we have $\rho(g(d_0) g(d_1)) = g(\rho(d_0)) g(\rho(d_1)) = 0, \forall \rho \in \Omega$. It follows that $g(d_1) g(d_0) = 0$, in particular, $f_{2s}(d_1) f_{2s}(d_0) = 0$.

Let $v = f_s(d_1) u f_s(d_0)$. Then we have

$$\|v\| = 1, \quad v^* v = f_s(d_0) u^* f_s(d_1)^2 u f_s(d_0).$$

and

$$f_{2s}(d_0) v^* v = v^* v = v^* v f_{2s}(d_0), \quad v^* f_{2s}(d_1) = v^*,$$

$$v^*(v^* v) = v^* f_{2s}(d_1) f_{2s}(d_0) v^* v = 0.$$

Furthermore, $f_{8s}(d) d_0 = d_0$ and $f_{8s}(d) d_1 = d_1$, hence

$$f_{8s}(d) p(d_0) = p(d_0), \quad f_{8s}(d) p(d_1) = p(d_1)$$

if p is a polynomial with no constant term, hence also if p is a continuous function vanishing at 0 . In particular $f_{8s}(d) v = v$, and $f_{8s}(d) v^* = v^*$. Finally, put

$$d' = f_{\frac{1}{4}}(v^* v), \quad w = f_{\frac{1}{2}}(v^* v)^{1/2}, \quad w' = v k(v^* v),$$

where $k : I\!R \longrightarrow I\!R$ is the function which is equal to $(f_{\frac{1}{2}}(t) t^{-1})^{1/2}$ if $t \neq 0$, and to 0 if $t = 0$. Clearly, $0 \leq d' \leq 1, 0 \leq w \leq 1$. Since $v^*(v^* v) = 0$ and $f_{1/2}(0) = 0$, it follows that $w'^* w = 0$. Since $f_{8s}(d) v = v, f_{8s}(d) v^* = v^*$, we have

$$f_{8s}(d) v^* v = v^* v, \quad f_{8s}(d) w = w, \quad f_{8s}(d) w' = w'.$$

That comes to 2) . Further,

$$w^2 = f_{1/2}(v^* v), \quad w'^* w' = k^2(v^* v) v^* v = f_{1/2}(v^* v)$$

and

$$w^2 d' = f_{1/2}(v^* v) f_{1/4}(v^* v) = d'$$

and similarly $w'^* w' = d'$. That comes to 3) . If we can choose u, c such that $\|d'\| \geq 1$, then it must be $\|d'\| = 1$. Since $w^2 \geq d'$ and $1 \geq w'^* w' \geq d'$, it follows that $\|w\| = \|w'\| = 1$. That comes to 1) , and the proof will be completed.

Now it suffices to find u, c such that $\|d'\| \geq 1$. By the Glimm condition, there is an irreducible $*$ representation $\{\pi, H\}$ of A such that dim $\pi(f_s(d))H \geq 2$. Pick $\xi, \eta \in \pi(f_s(d))H$ with $\|\xi\| = \|\eta\| = 1$ and $\langle \xi, \eta \rangle = 0$. By Theorem 2.6.5, there exists $h^* = h \in A$ such that

$$\pi(h)\xi = \xi, \quad \pi(h)\eta = 0.$$

Let $g : \mathbb{R} \longrightarrow \mathbb{R}$ be the function which is equal to 0 if $r \leq 0$, to 1 if $r \geq 1$, and to affine on $[0,1]$. Pick $c = g(h)$. Then $0 \leq c \leq 1$ and $\pi(c)\xi = \xi, \pi(c)\eta = 0$. By Theorem 2.6.5, there is also a unitary element u of A such that

$$\pi(u)\xi = \eta.$$

Since $f_{2s} f_s = f_s$ and $\xi, \eta \in \pi(f_s(d))H$, it follows that

$$\pi(f_{2s}(d))\xi = \xi, \quad \pi(f_{2s}(d))\eta = \eta.$$

Similarly, $\pi(f_{4s}(d))\eta = \eta$. Hence, we have

$$\pi(d_0)\xi = \pi(f_{2s}(d))\pi(c)\pi(f_{2s}(d))\xi = \xi, \quad \pi(d_0)\eta = 0,$$

$$\pi(d_1)\eta = \pi(f_{4s}(d))\eta - \pi(d_0)\eta = \eta,$$

$$\pi(v^* v)\xi = \pi(f_s(d_0))\pi(u)^* \pi(f_s(d_1))^2 \pi(u)\pi(f_s(d_0))\xi = \xi,$$

$$\pi(d')\xi = \pi(f_{1/4}(v^* v))\xi = f_{1/4}(1)\xi = \xi.$$

Therefore, $\|d'\| \geq \|\pi(d')\| \geq 1$. Q.E.D.

Proposition 13.4.3. Let A be a NGCR algebra with an identity 1. Then there exist non-zero elements $v(\alpha_1, \cdots, \alpha_n)$ and $b(n)$ in the unit ball of A, where $\alpha_1, \cdots, \alpha_n \in \{0,1\}, n = 0, 1, 2, \cdots$, with the following properties:

1) if $j \leq k$ and $(\alpha_1, \cdots, \alpha_j) \neq (\beta_1, \cdots, \beta_j)$, then

$$v(\alpha_1, \cdots, \alpha_j)^* v(\beta_1, \cdots, \beta_k) = 0;$$

2) if $k \geq 1$, then $v(\alpha_1, \cdots, \alpha_k) = v(\alpha_1, \cdots, \alpha_{k-1}) v(0_{k-1}, \alpha_k)$;

3) if $j < k$, then

$$v(\alpha_1, \cdots, \alpha_j)^* v(\alpha_1, \cdots, \alpha_j) v(0_{k-1}, \alpha_k) = v(0_{k-1}, \alpha_k);$$

4) $v(\emptyset) = 1, v(0_k) \geq 0$;

5) $v(\alpha_1,\cdots,\alpha_n)^*v(\alpha_1,\cdots,\alpha_n)b(n) = b(n), b(n) \geq 0$, and $\|b(n)\| = 1, n = 0,1,2,\cdots$.

Proof. For $n = 0$, put $v(\emptyset) = b(0) = 1$. Now suppose that non-zero elements $v(\alpha_1,\cdots,\alpha_j)$ in the unit ball of A and $b(j)$ of norm 1 in A_+ have been constructed for $j \leq n$ and they satisfy these properties. Using Lemma 13.4.2 to $d = b(n)$, we get w, w' and d', then let

$$v(0_{n+1}) = w, \quad v(0_n, 1) = w', \quad \text{and} \quad b(n+1) = d'.$$

By Lemma 13.4.2, we have

$$\begin{aligned} & v(\alpha_1,\cdots,\alpha_n)^*v(\alpha_1,\cdots,\alpha_n)v(0_n,\alpha_{n+1}) \\ = \; & v(\alpha_1,\cdots,\alpha_n)^*v(\alpha_1,\cdots,\alpha_n)f_\varepsilon(b(n))v(0_n,\alpha_{n+1}) \\ = \; & f_\varepsilon(b(n))v(0_n,\alpha_{n+1}) = v(0_n,\alpha_{n+1}). \end{aligned}$$

Since $b(n) = v(0_n)^*v(0_n)b(n) = v(0_n)^2b(n)$, it follows that $v(0_n)^2b(n) = b(n)v(0_n)^2, b(n)v(0_n) = v(0_n)b(n)$, and $b(n)^2 = (v(0_n)b(n))^2$. Hence, $b(n) = v(0_n)b(n)$ and

$$v(0_n)v(0_n,\alpha_{n+1}) = v(0_n)f_\varepsilon(b(n))v(0_n,\alpha_{n+1}) = v(0_n,\alpha_{n+1}).$$

Then for $j < n$ we get

$$\begin{aligned} & v(\alpha_1,\cdots,\alpha_j)^*v(\alpha_1,\cdots,\alpha_j)v(0_n,\alpha_{n+1}) \\ = \; & v(\alpha_1,\cdots,\alpha_j)^*v(\alpha_1,\cdots,\alpha_j)v(0_n)v(0_n,\alpha_{n+1}) \\ = \; & v(0_n)v(0_n,\alpha_{n+1}) = v(0_n,\alpha_{n+1}). \end{aligned}$$

Thus, $v(0_n,\alpha_{n+1})$ satisfies the conditions 2), 3), 4). Again by Lemma 13.4.2, $v(0_n,\alpha_{n+1})$ and $b(n+1)$ satisfy the condition 5), and $v(0_n,0)^*v(0_n,1) = 0$. If $j \leq n$ and $(\alpha_1,\cdots,\alpha_j) \neq (0_j)$, then

$$v(\alpha_1,\cdots,\alpha_j)^*v(0_n,\alpha_{n+1}) = v(\alpha_1,\cdots,\alpha_j)^*v(0_n)v(0_n,\alpha_{n+1}) = 0.$$

So $v(0_n,\alpha_{n+1})$ and $b(n+1)$ satisfy all conditions 1) – 5).

Now let $v(\alpha_1,\cdots,\alpha_{n+1}) = v(\alpha_1,\cdots,\alpha_n)v(0_n,\alpha_{n+1})$. Clearly, the conditions 2), 3), 4) hold for $(n+1)$. If $(\alpha_1,\cdots,\alpha_{n+1}) \neq (\beta_1,\cdots,\beta_{n+1})$, then

$$\begin{aligned} & v(\alpha_1,\cdots,\alpha_{n+1})^*v(\beta_1,\cdots,\beta_{n+1}) \\ = \; & v(0_n,\alpha_{n+1})^*v(\alpha_1,\cdots,\alpha_n)^*v(\beta_1,\cdots,\beta_n)v(0_n,\beta_{n+1}) \\ = \; & \begin{cases} 0, & \text{if } (\alpha_1,\cdots,\alpha_n) \\ & \neq (\beta_1,\cdots,\beta_n) \\ v(0_n,\alpha_{n+1})^*v(\alpha_1,\cdots,\alpha_n)^*v(\alpha_1,\cdots,\alpha_n)v(0_n,\beta_{n+1}), & \text{otherwise,} \end{cases} \\ = \; & \begin{cases} 0, & \text{if } (\alpha_1,\cdots,\alpha_n) \neq (\beta_1,\cdots,\beta_n) \\ v(0_n,\alpha_{n+1})^*v(0_n,\beta_{n+1}), & \text{otherwise} \end{cases} \\ = \; & 0. \end{aligned}$$

When $j \le n$ and $(\alpha_1, \cdots, \alpha_j) \ne (\beta_1, \cdots, \beta_j)$, it is obvious that

$$v(\alpha_1, \cdots, \alpha_j)^* v(\beta_1, \cdots, \beta_{n+1}) = v(\alpha_1, \cdots, \alpha_j)^* v(\beta_1, \cdots, \beta_n) v(0_n, b_{n+1}) = 0.$$

Thus , the condition 1) holds for $(n+1)$. Finally , if $(\alpha_1, \cdots, \alpha_n) \ne (0_n)$, then

$$\begin{aligned} & v(\alpha_1, \cdots, \alpha_{n+1})^* v(\alpha_1, \cdots, \alpha_{n+1}) b(n+1) \\ = \ & v(0_n, \alpha_{n+1})^* v(\alpha_1, \cdots, \alpha_n)^* v(\alpha_1, \cdots, \alpha_n) v(0_n, \alpha_{n+1}) b(n+1) \\ = \ & v(0_n, \alpha_{n+1})^* v(0_n, \alpha_{n+1}) b(n+1) = b(n+1). \end{aligned}$$

So the condition 5) holds also for $(n+1)$.

By induction, we can complete the proof. Q.E.D.

Denote the $2^n \times 2^n$ matrix algebra by B_n. Then B_n contains an orthogonal family $\{p(\alpha_1, \cdots, \alpha_n) | \alpha_i = 0 \text{ or } 1, 1 \le i \le n\}$ of minimal projections. Also there are partial isometries $w(\alpha_1, \cdots, \alpha_n)$ such that

$$\begin{cases} w(\alpha_1, \cdots, \alpha_n)^* w(\alpha_1, \cdots, \alpha_n) = p(0_n), \\ w(\alpha_1, \cdots, \alpha_n) w(\alpha_1, \cdots, \alpha_n)^* = p(\alpha_1, \cdots, \alpha_n), \\ w(0_n) = p(0_n), \quad \forall \alpha_i \in \{0, 1\}, 1 \le i \le n. \end{cases}$$

Clearly, if $(\alpha_1, \cdots, \alpha_n) \ne (\beta_1, \cdots, \beta_n)$, then $w^*(\alpha_1, \cdots, \alpha_n).w(\beta_1, \cdots, \beta_n) = 0$. Hence

$$w(\alpha_1, \cdots, \alpha_n)^* w(\beta_1, \cdots, \beta_n) = \prod_{i=1}^n \delta_{\alpha_i, \beta_i} p(0_n),$$

and

$$\begin{aligned} & w(\alpha_1, \cdots, \alpha_n) w(\beta_1, \cdots, \beta_n)^* w(\alpha'_1, \cdots, \alpha'_n) w(\beta'_1, \cdots, \beta'_n)^* \\ = \ & \left(\prod_{i=1}^n \delta_{\beta_i, \alpha'_i} \right) w(\alpha_1, \cdots, \alpha_n) p(0_n) w(\beta'_1, \cdots, \beta'_n)^* \\ = \ & \left(\prod_{i=1}^n \delta_{\beta_i, \alpha'_i} \right) w(\alpha_1, \cdots, \alpha_n) w(\beta'_1, \cdots, \beta'_n)^*, \end{aligned}$$

$\forall \alpha_i, \beta_i, \alpha'_i, \beta'_i \in \{0, 1\}, 1 \le i \le n$.

Now let A be a NGCR algebra on some Hilbert space H, and $1 = 1_H \in A$. Pick $v(\alpha_1, \cdots, \alpha_n), b(n)$ as in Proposition 13.4.3, and put

$$e_n = \sum_{\substack{\alpha_i \in \{0,1\} \\ 1 \le i \le n}} v(\alpha_1, \cdots, \alpha_n) v(\alpha_1, \cdots, \alpha_n)^*,$$

and $H(n) = \overline{[e(n)H]}, \forall n$.

We say that $H(n) \supset H(n+1), \forall n$. In fact, if $e(n)\xi = 0$ for some $\xi \in H$, then $v(\alpha_1, \cdots, \alpha_n)^* \xi = 0, \forall \alpha_1, \cdots, \alpha_n$. Hence, $v(\alpha_1, \cdots, \alpha_{n+1})^* \xi = v(0_n, \alpha_{n+1})^* v(\alpha_1, \cdots \alpha_n)^* \xi = 0, \forall \alpha_1, \cdots, \alpha_{n+1}$, i.e., $e(n+1)\xi = 0$.

Let $p(n)$ be the projection from H onto $H(n+1)$. Then we have $p(n) \geq p(n+1), \forall n$. For fixed n, since $v(\alpha_1, \cdots, \alpha_n)H \perp v(\beta_1, \cdots, \beta_n)H$ for $(\alpha_1, \cdots, \alpha_n) \neq (\beta_1, \cdots, \beta_n)$, it follows that

$$H(n) = p(n)H = \sum_{\substack{\alpha_i \in \{0,1\}, \\ 1 \leq i \leq n}} \oplus \overline{[v(\alpha_1, \cdots, \alpha_n)H]}, \forall n.$$

Noticing that

$$\begin{aligned} & v(\alpha_1, \cdots, \alpha_n) v(\beta_1, \cdots, \beta_n)^* v(\gamma_1, \cdots, \gamma_{n+1}) \\ = \ & v(\alpha_1, \cdots, \alpha_n)[v(\beta_1, \cdots, \beta_n)^* v(\gamma_1, \cdots, \gamma_n)] v(0_n, \gamma_{n+1}) \\ = \ & \left(\prod_{i=1}^n \delta_{\beta_i, \gamma_i} \right) v(\alpha_1, \cdots, \alpha_n, \gamma_{n+1}), \ \forall \alpha_i, \beta_i, \gamma_i, \end{aligned}$$

we have $v(\alpha_1, \cdots, \alpha_n) v(\beta_1, \cdots, \beta_n)^* H(n+1) \subset H(n+1), \forall \alpha_i, \beta_i$, i.e.,

$$v(\alpha_1, \cdots, \alpha_n) v(\beta_1, \cdots, \beta_n)^* p(n+1) = p(n+1) v(\alpha_1, \cdots, \alpha_n) v(\beta_1, \cdots, \beta_n)^*,$$

$\forall \alpha_i, \beta_i \in \{0, 1\}, 1 \leq i \leq n$. Since

$$\begin{aligned} & v(\alpha_1, \cdots, \alpha_n) v(\beta_1, \cdots, \beta_n)^* v(\alpha'_1, \cdots, \alpha'_n) v(\beta'_1, \cdots, \beta'_n)^* v(\gamma_1, \cdots, \gamma_{n+1}) \\ = \ & v(\alpha_1, \cdots, \alpha_n) v(\beta_1, \cdots, \beta_n)^* v(\alpha'_1, \cdots, \alpha'_n) v(\beta'_1, \cdots, \beta'_n)^* v(\gamma_1, \cdots, \gamma_n) \\ & \cdot v(0_n, \gamma_{n+1}) \\ = \ & \left(\prod_{i=1}^n \delta_{\beta'_i, \gamma_i} \cdot \delta_{\alpha'_i, \beta_i} \right) v(\alpha_1, \cdots, \alpha_n) v(0_n, \gamma_{n+1}) \\ = \ & \left(\prod_{i=1}^n \delta_{\alpha'_i, \beta_i} \right) v(\alpha_1, \cdots, \alpha_n)) v(\beta'_1, \cdots, \beta'_n)^* v(\gamma_1, \cdots, \gamma_{n+1}), \end{aligned}$$

it follows that

$$\begin{aligned} & v(\alpha_1, \cdots, \alpha_n) v(\beta_1, \cdots, \beta_n)^* p(n+1) \cdot v(\alpha'_1, \cdots, \alpha'_n) v(\beta'_1, \cdots, \beta'_n)^* \\ & \cdot p(n+1) \\ = \ & \left(\prod_{i=1}^n \delta_{\alpha'_i, \beta_i} \right) v(\alpha_1, \cdots, \alpha_n) v(\beta'_1, \cdots, \beta'_n)^* p(n+1), \end{aligned}$$

$\forall \alpha_i, \alpha'_i, \beta_i, \beta'_i \in \{0, 1\}, 1 \leq i \leq n$. Thus,

$$D_n = [v(\alpha_1, \cdots, \alpha_n) v(\beta_1, \cdots, \beta_n)^* p(n+1) | \alpha_i, \beta_j \in \{0, 1\}, 1 \leq i \leq n\}$$

is a C^* -algebra on H, and

$$\Phi_n(w(\alpha_1, \cdots, \alpha_n) w(\beta_1, \cdots, \beta_n)^*) = v(\alpha_1, \cdots, \alpha_n) v(\beta_1, \cdots, \beta_n)^* p(n+1)$$

$(\forall \alpha_i, \beta_i \in \{0,1\}, 1 \le i \le n)$ is a $*$ homorphism from B_n onto D_n. Since

$$\begin{aligned} & v(\alpha_1,\cdots,\alpha_{n+1})^* \cdot [v(\alpha_1,\cdots,\alpha_n)v(\alpha_1,\cdots,\alpha_n)^* p(n+1)] \cdot v(\alpha_1,\cdots,\alpha_{n+1}) \\ & \cdot b(n+1) \\ = \ & v(\alpha_1,\cdots,\alpha_{n+1})^* v(\alpha_1,\cdots,\alpha_n) \cdot [v(\alpha_1,\cdots,\alpha_n)^* v(\alpha_1,\cdots,\alpha_n) v(0_n,\alpha_{n+1})] \\ & \cdot b(n+1) \\ = \ & v(\alpha_1,\cdots,\alpha_{n+1}))^* v(\alpha_1,\cdots,\alpha_{n+1}) b(n+1) = b(n+1) \ne 0, \end{aligned}$$

it follows that $v(\alpha_1,\cdots,\alpha_n)v(\alpha_1,\cdots,\alpha_n)^* p(n+1) \ne 0$ and $\Phi_n(B_n) \ne \{0\}$. Now by the simplicity of B_n, Φ_n is a $*$ isomorphism.

Moreover,

$$\begin{aligned} & [v(\alpha_1,\cdots,\alpha_{n-1},0)v(\beta_1,\cdots,\beta_{n-1},0)^* + v(\alpha_1,\cdots,\alpha_{n-1},1)v(\beta_1,\cdots, \\ & \beta_{n-1},1)^*] v(\gamma_1,\cdots,\gamma_{n+1}) \\ = \ & \prod_{i=1}^{n-1} \delta_{\beta_k,\gamma_i} [\delta_{0,\gamma_n} v(\alpha_1,\cdots,\alpha_{n-1},0,\gamma_{n+1}) + \delta_{1,\gamma_n} v(\alpha_1,\cdots,\alpha_{n-1},1,\gamma_{n+1})] \\ = \ & \prod_{i=1}^{n-1} \delta_{\beta_i,\gamma_i} v(\alpha_1,\cdots,\alpha_{n-1},\gamma_n,\gamma_{n+1}) \\ = \ & v(\alpha_1,\cdots,\alpha_{n-1}) \cdot v(\beta_1,\cdots,\beta_{n-1})^* v(\gamma_1,\cdots,\gamma_{n-1}) v(0_{n-1},\gamma_n) \cdot v(0_n,\gamma_{n+1}) \\ = \ & v(\alpha_1,\cdots,\alpha_{n-1}) v(\beta_1,\cdots,\beta_{n-1})^* v(\gamma_1,\cdots,\gamma_{n+1}), \end{aligned}$$

hence

$$\begin{aligned} & v(\alpha_1,\cdots,\alpha_{n-1})v(\beta_1,\cdots,\beta_{n-1})^* p(n+1) \\ = \ & \sum_{i=0}^{1} v(\alpha_1,\cdots,\alpha_{n-1},i) v(\beta_1,\cdots,\beta_{n-1},i)^* p(n+1). \end{aligned}$$

Generally , from $p(t) \ge p(t+1), \forall t$, we can see that

$$\begin{aligned} & v(\alpha_1,\cdots,\alpha_n)v(\beta_1,\cdots,\beta_n)^* p(n+s) \\ = \ & \sum_{\substack{i_j \in \{0,1\} \\ 1 \le j \le s-1}} v(\alpha_1,\cdots,\alpha_n,i_1,\cdots,i_{s-1}) \\ & \cdot v(\beta_1,\cdots,\beta_n,i_1,\cdots,i_{s-1})^* p(n+s) \end{aligned} \tag{1}$$

Therefore, for any $n < r, \alpha_i, \beta_i \in \{0,1\}, 1 \le i \le n, H(r)$ is invariant under $v(\alpha_1,\cdots,\alpha_n)v(\beta_1,\cdots,\beta_n)^*$, i.e., $p(r)$ and $v(\alpha_1,\cdots,\alpha_n)v(\beta_1,\cdots,\beta_n)^*$ commute.

Now let $B(n)$ be the C^*–subalgebra of A generated by $\{1, v(\alpha_1,\cdots,\alpha_n)v(\beta_1,\cdots,\beta_n)^* | \alpha_i,\beta_i \in \{0,1\}, 1 \le i \le n\}$. Clearly, $B(n)$ is separable, and $H(r)$ is invariant under $B(n), \forall r > n$, and $D_n = B(n)p(n+1)$ is $*$ isomorphic to B_n.

Further, let $A(n)$ be the C^* -subalgebra of A generated by $\{B(i)|i \le n\}$. From (1) , we have

$$A(n)p(n+1) = B(n)p(n+1) = D_n.$$

Let $I(n) = \{x \in A(n)|xp(n+1) = 0\}$. Then $I(n)$ is a closed two-sided ideal of $A(n)$, and $A(n)/I(n)$ is $*$ isomorphic to B_n (the $2^n \times 2^n$ matrix algebra) .

Clearly, $A(n) \subset A(n+1)$, and $I(n) \subset I(n+1)$. But $A(n)/I(n)$ is simple , so that $A(n) \cap I(n+1) = I(n)$. Now let B be the C^*-sebalgebra of A generated by $\sqcup_n A(n)$, and I be the closure of $\sqcup_n I(n)$. Then I is a closed two-sided ideal of B. Moreover, since $A(n)/I(n)$ is simple, $A(n) \cap I = I(n), \forall n$.

Consider the quotient algebra B/I. By $A(n)/I(n) = A(n)/(A(n) \cap I) \cong (A(n)+I)/I, B/I = \overline{\sqcup_n (A(n)+I)/I}$ is an (UHF) algebra of type $(2^0, 2^1, \cdots, 2^n, \cdots)$. Therefore, we have the following.

Proposition 13.4.4. Let A be a NGCR algeba with an identity 1. Then there is a separable C^*-subalgebra B of A with $1 \in B$ and a closed two-sided ideal I of B such that B/I is an (UHF) algebra of type $\{2^n\}$.

Lemma 13.4.5. Let X, Y_* be two Banach spaces, $Y = (Y_*)^*$, and $B(X,Y)$ be the Banach space of all bounded linear operators from X to Y. Then through the following way

$$g(x \otimes f) = \langle Tx, f\rangle$$

$\forall x \in X, f \in Y_*, T \in B(X,Y), g \in (\gamma\text{ - }(X \otimes Y_*))^*$, where $\gamma(\cdot)$ is the largest cross norm on $X \otimes Y_*$ (see Proposition 3.1.2) , and γ- $(X \otimes Y_*)$ is the tensor product of X and Y_* with respect to $\gamma(\cdot), B(X,Y)$ is isometrically isomorphic to $(\gamma\text{ - }(X \otimes Y_*))^*$.

Proof. For any $T \in B(X,Y)$, define

$$g(u) = \sum_i \langle Tx_i, f_i\rangle,$$

where $u = \sum_i x_i \otimes f_i \in X \otimes Y_*$. Clearly, $|g(u)| \le \|T\| \sum_i \|x_i\| \cdot \|f_i\|$. Further, we have $|g(u)| \le \gamma(u)\|T\|, \forall u \in X \otimes Y_*$. So g can be uniquely extended to a linear functional on γ - $(X \otimes Y_*)$, and $\|g\| \le \|T\|$. For any $\varepsilon > 0$, pick $x \in X, f \in Y_*$ with $\|x\| = \|f\| = 1$ such that $|\langle Tx, f\rangle - \|T\|| \le \varepsilon$. Since $\gamma(x \otimes f) = 1$, it follows that $\|g\| \ge |g(x \otimes f)| = |\langle Tx, f\rangle| \ge \|T\| - \varepsilon$. Thus , we have $\|g\| = \|T\|$.

Conversely, let $g \in (\gamma\text{ - }(X \otimes Y_*))^*$. Since $|g(x \otimes f)| \le \|g\| \cdot \|x\| \cdot \|f\|, \forall x \in X, f \in Y_*$, there is $T \in B(X,Y)$ such that $g(x \otimes f) = \langle Tx, f\rangle$. Q.E.D.

Lemma 13.4.6. Let M be a hyperfinite VN algebra on a Hilbert space H, i.e., $M = (\sqcup_p M_p)''$, where $1 \in M_1 \subset \cdots \subset M_p \subset \cdots \subset M$, and for each p, M_p

is a matrix algebra. Then there exists a projection of norm one from $B(H)$ onto M'.

Proof. For any $x \in B(H)$, denote by $C(x)$ the weak closure of $Co\{u^*xu | u \in M$ and is unitary $\}$. We say that $C(x) \cap M' \neq \emptyset$. In fact, for any $p, U_p = \{u \in M_p | u$ is unitary $\}$ is a compact group. So there is an invariant Haar measure μ on U_p with $\mu(U_p) = 1$. Let $x_p = \int_{U_p} u^*xud\mu(u)$. Since $v^*x_pv = x_p, \forall v \in U_p$, it follows that $x_p \in C(x) \cap M'_p$. Thus , $C(x) \cap M'_p \neq \emptyset, \forall p$. Now $\{C(x) \cap M'_p | p\}$ is a decreasing sequence of non-empty weakly compact subsets of $B(H)$. Therefore

$$C(x) \cap M' = C(x) \cap (\cap_p M'_p) = \cap_p (C(x) \cap M'_p) \neq \emptyset,$$

$\forall x \in B(H)$.

Denote by $B(B(H))$ the Banach space of all bounded linear operators on $B(H)$. By Lemma 13.4.5, $B(B(H)) = (\gamma - (B(H) \otimes T(H))^*$. Thus , any bounded ball of $B(B(H))$ is w^*-compact. Let U be the set of all unitary elements of M, and for any $u \in U$ define $T^u \in B(B(H))$ as follows:

$$T^u x = u^*xu, \quad \forall x \in B(H).$$

Clearly, $J = \overline{Co\{T^u | u \in U\}}^{w^*}$ is a w^*-compact convex subset of $B(B(H))$. Introduce a partial order "$\leq$" into $J : T_1 \leq T_2$ if

$$C(T_1(x)) \supset C(T_2(x)), \quad \forall x \in B(H).$$

Let $\{T_l | l \in \wedge\}$ be a total ordered non-empty subset of J, and $J_l = \overline{\{T_{l'} | l' \geq l\}}^{w^*}$, $\forall l \in \wedge$. Then $\{J_l | l \in \wedge\}$ is a family of w^*-compact subsets of J, and $\{J_l | l \in \wedge\}$ has the property of finite intersection. Hence, there is $T \in \cap_l J_l$. Clearly , for any $x \subset B(H), Tx \in \overline{\{T_{l'}x | l' \geq l\}}^{w}$, $\forall l \in \wedge$. Since $T_{l'}x \in C(T_{l'}x) \subset C(T_l x)$, $\forall l' \geq l$, it follows that $Tx \in C(T_l x)$, and $C(Tx) \subset C(T_l x), \forall l \in \wedge, x \in B(H)$. Thus , $T \geq T_l, \forall l \in \wedge$. Now by the Zorn Lemma , J admits a maximal element T_0 at least.

We claim that $C(T_0 x) = \{T_0 x\}, \forall x \in B(H)$. In fact, fix $x \in B(H)$, and pick $a' \in C(T_0 x) \cap M'$. Then there is a net $\{f_\alpha(\cdot)\}$ of functions on U, where $f_\alpha(\cdot) \geq 0$, and $\sum\limits_{u \in U} f_\alpha(u) = 1, \forall \alpha$, such that

$$a' = w\text{-}\lim_\alpha \sum_{u \in U} f_\alpha(u) u^*(T_0 x) u.$$

Clearly $\{T_\alpha = \sum\limits_{u \in U} f_\alpha(u) T^u T_0\} \subset J$. Since J is w^*-compact subset of $B(B(H))$, $\{T_\alpha | \alpha\}$ admits a w^*-cluster point $T_1 (\in J)$. In particular, $T_1 \in \overline{Co\{T^u T_0 | u \in U\}}^{w^*}$. Thus, we have

$$T_1 y \in \overline{Co\{T^u T_0 y | u \in U\}}^{w} = C(T_0 y),$$

and $C(T_1y) \subset C(T_0y), \forall y \in B(H)$. That means $T_1 \geq T_0$. But T_0 is maximal, so we have $T_1 = T_0$. On the other hand, T_1 is a w^*-cluster point of $\{T_\alpha|\alpha\}$. It follows that $a' = T_1x \in M'$. Therefore, $C(T_0x) = C(T_1x) = \{a'\} = \{T_0x\} \subset C(x), \forall x \in B(H)$.

Now we have a linear map $E(x) = T_0x(\forall x \in B(H))$ from $B(H)$ to M'. Clearly, $\|E\| \leq 1$, and $E(a') = a', \forall a' \in M'$. Therefore, E is a projection of norm one from $B(H)$ onto M'. Q.E.D.

Remark. A VN algebra M on H has the *property (P)* , if $C(x) \cap M' \neq \emptyset, \forall x \in B(H)$. From the proof of Lemma 13.4.6, we can see that: there is a projection of norm one from $B(H)$ onto M' if M has the property (P); and any hyperfinite VN algebra has the property (P).

Lemma 13.4.7. Let A be a C^*-algebra with an identity 1, B be a C^*-subalgebra of A with $1 \in B$, and M be a type (III) factor on a separable Hilbert space H. If there is a linear map P from A to M satisfying :

i) $P(a) \geq 0, \quad \forall a \in A_+$;

ii) $P(b_1ab_2) = P(b_1)P(a)P(b_2), \quad \forall b_1, b_2 \in B, a \in A$;

iii) $P(B)$ is weakly dense in M,

then A admits a type (III) factorial $*$ representation.

Proof. 1) Denote by Ω the set of all linear maps Q from A to M satisfying : $Q(a) \geq 0, \forall a \in A_+; Q(b_1ab_2) = Q(b_1)Q(a)Q(b_2), \forall b_1, b_2 \in B, a \in A$; and $Q(b) = P(b), \forall b \in B$. We claim that Ω is a compact convex subset of $(B(A,M), \sigma(B(A,M), \gamma\text{ - }(A \otimes M_*)))$ (see Lemma 13.4.5) , and $Q(x^*x) \geq Q(x)^*Q(x), \forall x \in A$.

In fact, if $Q \in \Omega$, then we have $-\|h\|1_H \leq Q(h) \leq \|h\|1_H, \forall h^* = h \in A$. Hence, Ω is a bounded subset of $B(A,M)$. By $Q = P$ on $B, \forall Q \in \Omega, \Omega$ is also convex. Morevoer, it is easily verified that Ω is w^*-closed. Thus, Ω is a w^*-compact convex subset of $B(A,M)$.

For any $Q \in \Omega$, and $x \in A$, by the Kaplansky density theorem there is a net $\{c_l\} \subset P(B)$ such that $c_l \longrightarrow Q(x)$ ($*$ strongly) , and $\|c_l\| \leq \|Q(x)\|, \forall l$. For any $\varphi \in (M_*)_+$, since $Q^*\varphi \geq 0$, it follows from the Schwartz inequality that

$$\begin{aligned}
\langle Q(x)^*Q(x), \varphi\rangle &= \lim_l \langle Q(x)^*Q(b_l), \varphi\rangle \\
&= \lim_l \langle x^*b_l, Q^*\varphi\rangle \\
&\leq Q^*\varphi(x^*x)^{1/2}\overline{\lim_l}\; Q^*\varphi(b_l^*b_l)^{1/2} \\
&= \langle Q(x^*x), \varphi\rangle^{1/2}\overline{\lim_l}\langle c_l^*c_l, \varphi\rangle^{1/2} \\
&= \langle Q(x^*x), \varphi\rangle^{1/2} \cdot \langle Q(x)^*Q(x), \varphi\rangle^{1/2},
\end{aligned}$$

where $b_l \in B$ and $P(b_l) = c_l, \forall l$. Hence , $Q(x^*x) \geq Q(x)^*Q(x), \forall Q \in \Omega, x \in A$.

2) Since H is separable, there is a faithful normal state φ on M. For any $Q \in \Omega$, let

$$\varphi_Q(a) = \varphi(Q(a)), \quad \forall a \in A.$$

Then φ_Q is a state on A since $Q(1) = P(1) = 1_H$. Define

$$\mathcal{E} = \{\varphi_Q | Q \in \Omega\}.$$

Then by 1) $\mathcal{E}$ is a w^*–compact convex subset of the state space on A. Now fix $Q_0 \in \Omega$ such that $\varphi_0 = \varphi_{Q_0}$ is an extreme point of $\mathcal{E}$. Let $\{\pi_0, H_0, \xi_0\}$ be the cyclic $*$ representation of A generated by φ_0, and $N = \pi_0(A)''$(a VN algebra on H_0).

If $\pi_0(x) = 0$ for some $x \in A$, then we have $\varphi(Q_0(x^*x)) = \varphi_0(x^*x) = 0$. Since $Q_0(x^*x) \geq 0$ and φ is faithful, it follows that $Q_0(x^*x) = 0$. By 1) , $Q_0(x)^*Q_0(x) \leq Q_0(x^*x)$, hence $Q_0(x) = 0$. Now for any fixed $f \in M_*$, we can define a linear functional F on $\pi_0(A)$ as follows:

$$F(\pi_0(x)) = f(Q_0(x)), \quad \forall x \in A.$$

We say that F is strongly continuous on the unit ball of $\pi_0(A)$. In face, let $\{a_l\}$ be a net of A such that $\|\pi_0(a_l)\| \leq 1, \forall l$, and $\pi_0(a_l) \longrightarrow 0$ (strongly). Then $\varphi_0(a_l^*a_l) = \varphi(Q_0(a_l^*a_l)) \geq \varphi(Q_0(a_l)^*Q_0(a_l)) \longrightarrow 0$. We may assume that $\|a_l\| \leq 2, \forall l$. Then $\{\|Q_0(a_l)\| | l\}$ is bounded. Further, since φ is faithful and normal, so we have $Q_0(a_l) \longrightarrow 0$ with respect to s–top. of M. Hence $F(\pi_0(a_l)) = f(Q_0(a_l)) \longrightarrow 0$.

For any $a \in N = \pi_0(A)''$, we can find a net $\{a_l\} \subset \pi_0(A)$ such that $a_l \longrightarrow a$(strongly) . Then $(a_l - a_{l'}) \longrightarrow 0$ (strongly) , and from the preceding paragraph $\{F(a_l)\}$ is a Cauchy net of numbers. Hence, we can define

$$F(a) = \lim_l F(a_l),$$

and this definition is independent of the choice of $\{a_l\}$ obviously. In such way, F is extended to a linear functional on N, still denoted by F.

We claim that this extension F is strongly continuous on the unit ball of N. In fact, for any $\varepsilon > 0$ by the strong continuity of F on the unit ball of $\pi_0(A)$ we can find

$$V = V(0, \xi_1, \cdots, \xi_k, \delta) = \{a \in \pi_0(A) | \|a\| \leq 1, \|a\xi_i\| < \delta, \forall i\}$$

such that $|F(a)| < \varepsilon, \forall a \in V$, where $\xi_1, \cdots, \xi_k \in H_0$. Let

$$U = U(0, \xi_1, \cdots, \xi_k, \delta) = \{a \in N | \|a\| \leq 1, \|a\xi_i\| < \delta, \forall i\}.$$

Clearly, U is a strong neighborhood of 0 in N. For any $a \in U$, there is a net $\{a_l\} \subset \pi_0(A)$ such that $a_l \longrightarrow a$ (strongly) , and $\|a_l\| \leq 1, \forall l$. We may assume that $\|a_l\xi_i\| < \delta, \forall l, i$, i.e., $\{a_l\} \subset V$. By the definition of F on N, we get

$$|F(a)| = \lim_l |F(a_l)| \leq \varepsilon, \quad \forall a \in U.$$

Therefore, $F \in N_*$. Moreover,

$$\begin{aligned}\|F\| &= \sup\{|F(\pi_0(x))|| \ x \in A \quad \text{and} \quad \|\pi_0(x)\| \leq 1\} \\ &\leq \sup\{|f(Q_0(x))|| \ x \in A \quad \text{and} \quad \|x\| \leq 2\} \\ &\leq 2\|f\|\|Q_0\|.\end{aligned}$$

If define $\Phi(f) = F$, then Φ is a bounded linear map from M_* to N_*. Further, Φ^* is bounded and $\sigma(N, N_*)$-$\sigma(M, M_*)$ continuous from N to M.

3) The map $\Phi^* : N \longrightarrow M$ has the following properties :

I) $\Phi^*(\pi_0(x)) = Q_0(x), \quad \forall x \in A$;

II) $\Phi^*(a) \geq 0, \quad \forall a \in N_+$;

III) $\Phi^*(b_1 a b_2) = \Phi^*(b_1)\Phi^*(a)\Phi^*(b_2), \quad \forall a \in N, b_1, b_2 \in \pi_0(B)''$;

IV) $\Phi^*(a^* a) \geq \Phi^*(a^*)\Phi^*(a), \quad \forall a \in N$.

In fact, for any $f \in M_*$ and $x \in A$, we have

$$\langle \Phi^*(\pi_0(x)), f\rangle = \langle \pi_0(x), \Phi(f)\rangle = f(Q_0(x)) = \langle Q_0(x), f\rangle.$$

Hence, $\Phi^*(\pi_0(x)) = Q_0(x), \forall x \in A$.

For any $a \in N_+$, we can pick a net $\{x_l\}$ of A such that $\pi_0(x_l) \longrightarrow a^{1/2}$($*$ strongly) , and $\|\pi_0(x_l)\| \leq \|a^{1/2}\|, \forall l$. Then $\pi_0(x_l^* x_l) \longrightarrow a$(strongly) . Hence,

$$\Phi^*(a) = \sigma(M, M_*) \text{ - } \lim \Phi^*(\pi_0(x_l^* x_l)) = \sigma\text{- } \lim Q_0(x_l^* x_l).$$

By the positivity of Q_0, we have $\Phi^*(a) \geq 0, \forall a \in N_+$.

For any $y_1, y_2 \in B, x \in A$, we have

$$\begin{aligned}\Phi^*(\pi_0(y_1)\pi_0(x)\pi_0(y_2)) &= Q_0(y_1 x y_2) = Q_0(y_1)Q_0(x)Q(y_2) \\ &= \Phi^*(\pi_0(y_1))\Phi^*(\pi_0(x))\Phi^*(\pi_0(y_2))\end{aligned}$$

obviously. Then by the σ–continuity of Φ^*, we get $\Phi^*(b_1 a b_2) = \Phi^*(b_1)\Phi^*(a)\Phi^*(b_2), \forall a \in N, b_1, b_2 \in \pi_0(B)''$.

For any $a \in N$, since $P(B) = Q_0(B)$ is a weakly dense $*$ subalgebra of M, there is a net $\{x_l\} \subset B$ such that $\Phi^*(\pi_0(x_l)) = Q_0(x_l) \longrightarrow \Phi^*(a)$ (strongly) , and $\|Q_0(x_l)\| \leq \|\Phi^*(a)\|, \forall l$. Clearly, $\Phi((M_*)_+) \subset (N_*)_+$. Then for any $\psi \in$

$(M_*)_+$, we have

$$\begin{aligned}0 &\le \langle \Phi^*(a^*)\Phi^*(a), \psi\rangle = \lim\langle \Phi^*(\pi_0(1))\Phi^*(a^*)\Phi^*(\pi_0(x_l)), \psi\rangle \\ &= \lim_l \langle \Phi^*(a^*\pi_0(x_l)), \psi\rangle = \lim\langle a^*\pi_0(x_l), \Phi(\psi)\rangle \\ &\le \langle \Phi^*(a^*a), \psi\rangle^{1/2} \overline{\lim}\langle \Phi^*(\pi_0(x_l^* x_l)), \psi\rangle^{1/2} \\ &= \langle \Phi^*(a^*a), \psi\rangle^{1/2} \overline{\lim}\langle \Phi^*(\pi_0(x_l^*))\Phi^*(\pi_0(x_l)), \psi\rangle^{1/2} \\ &= \langle \Phi^*(a^*a), \psi\rangle^{1/2} \overline{\lim}\langle Q_0(x_l)^* Q_0(x_l), \psi\rangle^{1/2} \\ &= \langle \Phi^*(a^*a), \psi\rangle^{1/2} \cdot \langle \Phi^*(a^*)\Phi^*(a), \psi\rangle^{1/2}.\end{aligned}$$

Hence $\langle \Phi^*(a^*)\Phi^*(a), \psi\rangle = \langle \Phi^*(a^*a), \psi\rangle, \forall \psi \in (M_*)_+$, and $\Phi^*(a^*)\Phi^*(a) \le \Phi^*(a^*a), \forall a \in N$.

4) Now we prove that N is a factor.

In fact, let z be a central projection of N. Since

$$\Phi^*(z)Q_0(b) = \Phi^*(z\pi_0(b)) = \Phi^*(\pi_0(b)z) = Q_0(b)\Phi^*(z), \forall b \in B,$$

and $Q_0(B) = P(B)$ is weakly dense in M, $\Phi^*(z)$ is also a central element of M. But M is a factor, thus $\Phi^*(z) = \lambda(z)1_H$, where $\lambda(z) \in [0,1]$.

If $\lambda(z) \in (0,1)$, we can define

$$\begin{cases} Q_1(x) = \lambda(z)^{-1}\Phi^*(\pi_0(x)z), \\ Q_2(x) = (1-\lambda(z))^{-1}\Phi^*(\pi_0(x)(1-z)), \end{cases}$$

$\forall x \in A$. By 3) , $Q_1, Q_2 \in \Omega$, and $\lambda(z)Q_1 + (1-\lambda(z))Q_2 = Q_0$. Then $\varphi_0 = \lambda(z)\varphi_{Q_1} + (1-\lambda(z))\varphi_{Q_2}$. But φ_0 is an extreme point of $\mathcal{E}$ (see 2)) , so we have $\varphi_0 = \varphi_{Q_1} = \varphi_{Q_2}$, i.e.,

$$\lambda(z)^{-1}\langle \pi_0(x)z, \Phi(\varphi)\rangle = (1-\lambda(z))^{-1}\langle \pi_0(x)(1-z), \Phi(\varphi)\rangle,$$

$\forall x \in A$. Pick a net $\{x_l\} \subset A$ such that $\pi_0(x_l) \longrightarrow z$ (strongly) , and $\|\pi_0(x_l)\| \le 1, \forall l$. Then we get

$$\begin{aligned}1 = \lambda(z)^{-1}\langle \Phi^*(z), \varphi\rangle &= \lambda(z)^{-1}\langle z, \Phi(\varphi)\rangle \\ &= (1-\lambda(z))^{-1}\langle z(1-z), \Phi(\varphi)\rangle = 0,\end{aligned}$$

a contradiction. Hence , $\lambda(z)$ is either 0 or 1 .

If $\lambda(z) = 0$, pick $\{x_l\} \subset A$ as above, then we have

$$\begin{aligned}\langle z\xi_0, \xi_0\rangle &= \lim\langle \pi_0(x_l)\xi_0, \xi_0\rangle = \lim \varphi_0(x_l) = \lim \varphi(Q_0(x_l)) \\ &= \lim\langle \Phi^*(\pi_0(x_l)), \varphi\rangle = \langle z, \Phi(\varphi)\rangle = \lambda(z) = 0.\end{aligned}$$

Hence , $z\pi_0(A)\xi_0 = \pi_0(A)z\xi_0 = \{0\}$, i.e., $z = 0$.

If $\lambda(z) = 1$, similary we have $z = 1$. Therefore, N is a factor.

5) It suffices to show that N is not semi-finite. Then by 4), $N = \pi_0(A)''$ is a factor of type (III), and $\{\pi_0, H_0\}$ is a type (III) factorial * representation of A.

Now suppose that N is semi-finite. Let E' be the projection from H_0 onto $\overline{[\pi_0(B)\xi_0]}$.

If $\pi_0(b)E' = 0$ for some $b \in B$, then $\pi_0(b)E'\xi_0 = \pi_0(b)\xi_0 = 0$, and $0 = \|\pi_0(b)\xi_0\|^2 = \varphi_0(b^*b))$. Since φ is faithful and $Q_0(b^*b) \geq Q_0(b)^*Q_0(b)$, it follows that $Q_0(b) = 0$. Conversely, if $Q_0(b) = 0$ for some $b \in B$, then for any $c \in B$ we have

$$\begin{aligned}\|\pi_0(b)\pi_0(c)\xi_0\|^2 &= \varphi(Q_0(c^*b^*bc)) \\ &= \varphi(Q_0(c^*b^*)Q_0(b)Q_0(c)) = 0.\end{aligned}$$

Hence, $\pi_0(b)E' = 0$. Therefore, $\pi_0(b)E' \longrightarrow Q_0(b) = P(b)$ is a * isomorphic from $\pi_0(B)E'$ into M.

We say that the above * isomorphism is s-s continuous on the unit ball of $\pi_0(B)E'$. In fact, let $\{b_l\}$ be a net of B such that $\pi_0(b_l)E' \longrightarrow 0$ (strongly), and $\|\pi_0(b_l)E'\| \leq 1, \forall l$. Since $b \longrightarrow \pi_0(b)E'$ is a * homomorphism from B to $\pi_0(B)E'$, we may assume that $\|b_l\| \leq 2, \forall l$. Now if a is a w-cluster point of $\{Q_0(b_l^*b_l) = Q_0(b_l)^*Q_0(b_l)|l\}$, then

$$\varphi(a) = \lim \varphi(Q_0(b_l)^*Q_0(b_l)) = \lim \|\pi_0(b_l)\xi_0\|^2 = 0.$$

Since φ is faithful, it follows that $a = 0$. Hence, $Q_0(b_l) \longrightarrow 0$(strongly).

Conversely, if $\{\|b_l\||l\}$ is bounded, and $Q_0(b_l) \longrightarrow 0$ (strongly), then $\|\pi_0(b_l)\pi_0(c)\xi_0\|^2 = \varphi(Q_0(c^*)Q_0(b_l)^*Q_0(b_l)Q_0(c)) \longrightarrow 0$. Hence $\pi_0(b_l)E' \longrightarrow 0$ (strongly).

Therefore, the * isomorphism $\pi_0(b)E' \longrightarrow Q_0(b)(\forall b \in B)$ can be extended to a * isomorphism Γ from $\pi_0(B)''E'$ onto M, and $\pi_0(B)''E'$ is a type (III) factor.

Let F' be the central cover of E' in $\pi_0(B)'$. Then $\Psi : xF' \longrightarrow xE'$ is a * isomorphism from $\pi_0(B)''F'$ onto $\pi_0(B)''E'$, and $\Gamma \circ \Psi$ is a * isomorphism from $\pi_0(B)''F'$ onto $\pi_0(B)''E'$, and $\Gamma \circ \Psi$ is a * isomrophism from $\pi_0(B)''F'$ onto M.

Since $F' \geq E', E'\xi_0 = \xi_0$ and

$$\begin{aligned}\langle \Phi^*(F'), \varphi\rangle &= \lim\langle \pi_0(b_l), \Phi(\varphi)\rangle = \lim \varphi(Q_0(b_l)) \\ &= \lim\langle \pi_0(b_l)\xi_0, \xi_0\rangle = \langle F'\xi_0, \xi_0\rangle = 1,\end{aligned}$$

where $\{b_l\} \subset B$ such that $\pi_0(b_l) \longrightarrow F'$ (strongly), so $\Phi^*(F') \neq 0$. By the semi-finiteness of N and $F' \in \pi_0(B)'' \subset N$, we have $F' = \sup\{e \in N|e$ is a finite projection, and $e \leq F'\}$. Moreover, Φ^* is σ-σ continuous, so there is a finite projection e of N with $e \leq F'$ such that $\Phi^*(e) \neq 0$. Further, we can find a number $\lambda > 0$ and a non-zero projection p of M such that $\Phi^*(e) \geq \lambda p$.

For any $a \in M$, pick a net $\{b_l\}$ of B such that $Q_0(b_l) \longrightarrow a$ (strongly) . Then $\Psi^{-1} \circ \Gamma^{-1}(Q_0(b_l)) \longrightarrow \Psi^{-1} \circ \Gamma^{-1}(a)$ (strongly) , and $\Phi^*(\Psi^{-1} \circ \Gamma^{-1}(Q_0(b_l))) \longrightarrow \Phi^*(\Psi^{-1} \circ \Gamma^{-1}(a))$ (weakly) . On the other hand, by the definition of Γ, Ψ, and the properties of Φ^*, we have

$$\begin{aligned}\Phi^*(\Psi^{-1} \circ \Gamma^{-1}(Q_0(b_l))) &= \Phi^*(\pi_0(b_l)F') = \Phi^*(\pi_0(b_l))\Phi^*(F') \\ &= Q_0(b_l)\Phi^*(F') \longrightarrow a\Phi^*(F') \quad \text{(strongly)}.\end{aligned}$$

Thus, we obtain that

$$\Phi^*(\Psi^{-1} \circ \Gamma^{-1}(a)) = a\Phi^*(F'), \quad \Phi^*(\Psi^{-1} \circ \Gamma^{-1}(a^*)) = \Phi^*(F')a^*, \forall a \in M.$$

Now let $\{a_l\}$ be a net of Mp with $\|a_l\| \leq 1, \forall l$, and $a_l \longrightarrow 0$(strongly) . Clearly, $\Psi^{-1} \circ \Gamma^{-1}(a_l)e \longrightarrow 0$(strongly) . Since e is finite, it follows from Propsition 6.5.16 that $e\Psi^{-1} \circ \Gamma^{-1}(a_l^*) \longrightarrow 0$(strongly) . By $\Phi^*(a^*a) \geq \Phi^*(a^*)\Phi^*(a), \forall a \in N$, and the σ - σ continuity of Φ^*, we also have $\Phi^*(e\Psi^{-1} \circ \Gamma^{-1}(a_l^*)) \longrightarrow 0$(strongly) . Since $\Psi^{-1} \circ \Gamma^{-1}(M) \subset \pi_0(B)''F' \subset \pi_0(B)'', e \leq F'$, and $\{a_l\} \subset Mp$, it follows that

$$\begin{aligned}a_l^* &= \{p\Phi^*(e)p + (1-p)\}^{-1}\{p\Phi^*(e)p + (1-p)\}a_l^* \\ &= \{p\Phi^*(e)p + (1-p)\}p\Phi^*(e)a_l^* \\ &= \{p\Phi^*(e)p + (1-p)\}p\Phi^*(eF')a_l^* \\ &= \{p\Phi^*(e)p + (1-p)\}p\Phi^*(e)\Phi^*(F')a_l^* \\ &= \{p\Phi^*(e)p + (1-p)\}p\Phi^*(e)\Phi^*(\Psi^{-1} \circ \Gamma^{-1}(a_l^*)) \\ &= \{p\Phi^*(e)p + (1-p)\}p\Phi^*(e\Psi^{-1} \circ \Gamma^{-1}(a_l^*)) \longrightarrow 0(\text{strongly}).\end{aligned}$$

This means that the $*$ operation is strongly continuous on the unit ball of Mp. That contradicts the facts: $p \neq 0$, and M is type (III) (see Proposition 6.6.3).

Therefore, N is type (III) . Q.E.D.

Proposition 13.4.8. Let A be a NGCR algebra. Then A admits a type (III) factorial $*$ representation.

Proof. If A has no identity, and $\{\pi, H\}$ is a type (III) factoral $*$ representation of $(A\dot{+}\mathbb{C})$, then by Theorem 1.3.9 there is a projection $p_0 \in \pi(A)' \cap \pi(A)''$ such that $\overline{\pi(A)}^w = \pi(A)''p_0$. Hence , $\{\pi(\cdot)p_0, p_0H\}$ is a type (III) factorical $*$ representation of A. Moreover, $(A\dot{+}\mathbb{C})$ is also NGCR. Thus, we may assume that A has an identity 1.

From Proposition 13.4.4, there is a separable C^*-subalgebra B of A with $1 \in B$ and a closed two-sided ideal I of B such that B/I is a (UHF) algebra of type $\{2^n\}$.

Write $B/I = \alpha_0$ - $\otimes_n M_2^{(n)}$, where $M_2^{(n)} = M_2$ is the 2×2 matrix algebra, $\forall n$. Fix $\lambda \in (0, \frac{1}{2})$, pick a state φ on M_2 as follows:

$$\varphi\left(\begin{pmatrix} \alpha & \beta \\ \gamma & \delta \end{pmatrix}\right) = \lambda\alpha + (1-\lambda)\delta, \quad \forall \begin{pmatrix} \alpha & \beta \\ \gamma & \delta \end{pmatrix} \in M_2,$$

and let $\psi = \otimes_n \varphi_n$, where $\varphi_n = \varphi, \forall n$. By Theorem 9.5.11, The cyclic $*$ representation $\{\pi, H, \xi\}$ of B/I generated by ψ is a type (III) factorial $*$ representation. From Proposition 3.8.7, we have

$$\pi = \otimes_n \pi_n, \quad H = \otimes_n^{\xi} H_n, \quad \xi = \otimes_n \xi_n,$$

where $\{\pi_n, H_n, \xi_n\}$ is the cyclic $*$ representation of $M_2^{(n)}$ generated by $\varphi_n, \forall n$. Clearly, $\pi_n(M_2^{(n)}) \cong M_2, B(H_n) \cong M_4$, and $\pi_n(M_2^{(n)})' \cong M_2, \forall n$. Further, by Proposition 3.8.6 $\pi(B/I)'$ is generated by

$$\{(\pi_n(M_2^{(n)})' \otimes \otimes_{m \neq n} 1_m) | n\}.$$

Hence , $\pi(B/I)'$ is hyperfinite. Now from Lemma 13.4.6, there is a projection E of norm one from $B(H)$ onto $\pi(B/I)''$.

Summing the above discussion, we may assume the following: there is a state φ on B, and a projection E of norm one from $B(H)$ onto the type (III) factor $M = \pi(B)''$, where $\{\pi, H\}$ is the $*$ representation of B generated φ, and H is separable.

Pick a state $\overline{\varphi}$ on A such that $\overline{\varphi}|B = \varphi$, and let $\{\overline{\pi}, \overline{H}\}$ be the $*$ representation of A generated by $\overline{\varphi}$. Then we have $H \subset \overline{H}, \overline{\pi}(b)H \subset H$, and $\overline{\pi}(b)|H = \pi(b), \forall b \in B$. Let p be the projection from $\overline{H}$ onto H, and define $P : A \longrightarrow M$ as follows:

$$P(a) = E(p\overline{\pi}(a)p), \quad \forall a \in A.$$

By Theorem 4.1.5, $\{A, B, P, M = \pi(B)'', H\}$ satisfies the conditions of Lemma 13.4.7. Therefore, A admits a type (III) factorial $*$ representation. Q.E.D.

Remark. Let H be a separable infinite dimensional Hilbert space. Clearly, $B(H)$ is not CCR and $C(H)$ is a CCR closed tow-sided ideal of $B(H)$. By Proposition 1.1.2 and 13.3.6, the *Calkin algebra* $A = B(H)/C(H)$ must be NGCR. Thus by Proposition 13.4.8, the Calkin algebra A admits a type (III) factorial $*$ representation.

Notes. Quasi-matrix systems (Proposition 13.4.3) were introduced by J. Glimm. Proposition 13.4.8 is due to S.Sakai. The property (P) (see the Remark under Lemma 13.4.6) was introduced by J.T.Schwartz.

References. [55], [150], [153].

13.5. Type I C^*–algebras

Definition 13.5.1. A C^*–algebra A is said to be *type I* , if for any nondegenerate $*$ representation $\{\pi, H\}$ of $A, \pi(A)''$ is a type (I) VN algebra on H.

Proposition 13.5.2. If A is a GCR algebra, then A is type I.

Proof. Let $\{\pi, H\}$ be a nondegenerate $*$ representation of A, and $\{p_l\}$ be a maximal orthogonal family of non–zero central projections of $\pi(A)'$ such that $\pi(A)''p_l$ is a type (I) VN algebra on $p_l H, \forall l$.

By Proposition 6.7.2, it suffices to show that $\sum_l p_l = 1$. Suppose that $p = 1 - \sum_l p_l$ is not zero. Then $\{\pi_p, H_p\} = \{\pi|pH, pH\}$ is a non–zero $*$ representation of A. Since $\pi_p(A)$ is $*$ isomorphic to $A/\ker\pi_p$ and A is GCR, $\pi_p(A)$ is GCR. By Proposition 13.3.11, there is $0 \neq a \in \pi_p(A)_+$ such that $\dim \pi'(a)H' \leq 1$ for any irreducible $*$ representation $\{\pi', H'\}$ of $\pi_p(A)$. Thus , $a\pi_p(A)a$ is commutative. Further, aMa is also commutative, where $M = \pi_p(A)''$ (on $H_p = pH$). Let $a = \int_0^\infty \lambda de_\lambda$ be the spectral decomposition of a, and $y_\varepsilon = \int_\varepsilon^\infty \lambda^{-1} de_\lambda$. If $\varepsilon(> 0)$ is small enough, then $f = ay_\varepsilon = 1 - e_\varepsilon$ is a non–zero projection of M.

Clearly, $fa_1 f \cdot fa_2 f = a(y_\varepsilon a_1 y_\varepsilon)a \cdot a(y_\varepsilon a_2 y_\varepsilon)a = a(y_\varepsilon a_2 y_\varepsilon)a \cdot a(y_\varepsilon a_2 y_\varepsilon)a = fa_2 f \cdot fa_1 f, \forall a_1, a_2 \in M$. So fMf is commutative. Let z be the central cover of f in M. Then M'_z and M'_f are $*$ isomorphic. But $(M'_f)' = fMf$ is commutative, from Theorem 6.7.1 M'_f is type (I) . Then M'_z and $(M'_z)' = Mz$ are also type (I). Clearly, $Mz = \pi_p(A)''z = \pi(A)''pz = \pi(A)''z, z \neq 0$, and $zp_l = 0, \forall l$. This is impossible since the family $\{p_l\}$ is maximal . Therefore, $\sum_l p_l = 1$.

Q.E.D.

Definition 13.5.3. A C^*–algebra A is said to be *smooth* , if for any non–zero irreducible $*$ representation $\{\pi, H\}$ of A, we have $\pi(A) \cap C(H) \neq \{0\}$. By Lemma 13.2.6, this condition is equivalent to $C(H) \subset \pi(A)$.

Proposition 13.5.4. If A is a smooth C^*–algebra, then A is GCR.

Proof. Clearly, $(A \dot{+} \mathbb{C})$ is also smooth. Moreover, by Proposition 13.3.15, A is GCR if $(A \dot{+} \mathbb{C})$ is GCR. Thus , we may assume that A has identity 1 .

If A is not GCR, then by Proposition 13.3.6 there is a closed two-sided ideal I of A such that A/I is NGCR. Further, from Proposition 13.4.4 there is a C^*-subalgebra B of A with $1 \in B$ and a closed two-sided ideal J of B such that B/J is a (UHF) algebra of type $\{2^n\}$. By the Remark under this Proposition, B/J is simple (i.e., B/J contains no non-zero proper closed two-sided ideal) , so J is the largest proper closed two-sided ideal of B.

Fix a pure state ω on B and $\omega|J = \{0\}$. Let $\{\pi_\omega, H_\omega\}$ be the irreducible $*$ representation of B generated by ω. Then $\ker\pi_\omega = J$, and $\{\pi_\omega, H_\omega\}$ can be regarded as a faithful $*$ representation of B/J. Since B/J is infinite dimensional, $\dim H_\omega = \infty$.

Let $\mathcal{E} = \{\rho \in P(A) | \rho|B = \omega\}$. For any $\rho \in \mathcal{E}$, let $\{\pi_\rho, H_\rho\}$ be the irreducible $*$ representation of A. Clearly , $H_\omega \subset H_\rho$, so $\dim H_\rho = \infty$, and the identity operator $\pi_\rho(1) = 1_\rho$ on H_ρ is not contained in $C(H_\rho)$. Since A is smooth, it follows that $\pi_\rho(A) \supset C(H_\rho)$. Let $I(\rho) = \pi_\rho^{-1}(C(H_\rho))$. Then $I(\rho)$ is a closed two-sided ideal of A, and $1 \notin I(\rho)$. Thus $I(\rho) \not\supset B$, and $B \cap I(\rho) \subset J$.

Introduce a partial order " $\leq$" in $\mathcal{E}$: $\rho_1 \leq \rho_2$ if $I(\rho_1) \subset I(\rho_2)$. We say that $\mathcal{E}$ has no maximal element with respect to this partial order. In fact , for any $\rho \in \mathcal{E}$, since $\omega|J = \{0\}$ and $I(\rho) \cap B \subset J$, so ω can be regarded as a pure state on $B/(B \cap I(\rho))$. But $B/(B \cap I(\rho)) \cong (B + I(\rho))/I(\rho) \subset A/I(\rho)$, hence there is a pure state $\sigma \in \mathcal{E}$ such that $\sigma|I(\rho) = 0$. Since $\pi_\sigma(I(\rho)) = \{0\}$ and $I(\sigma) = \pi_\sigma^{-1}(C(H_\sigma))$, it follows that $I(\rho) \subset I(\sigma)$ and $I(\rho) \neq I(\sigma)$, i.e., $\rho \leq \sigma$ and $\rho \neq \sigma$.

On the other hand, let $\{\rho_l\}$ be a non-empty totally ordered subset of $\mathcal{E}$, and $I = \overline{\cup_l I(\rho_l)}$. Clearly, $1 \notin I$, and I is a proper closed two-sided ideal of A, and $B \cap I \subset J$. Then ω can be regarded as a pure state on $B/B \cap I$. Again by $B/(B \cap I) \cong (B + I)/I \subset A/I$, there is $\rho \in \mathcal{E}$ such that $\rho|I = \{0\}$. Then $I(\rho) \supset I \supset I(\rho_l), \forall l$, i.e., $\rho \geq \rho_l, \forall l$. By the Zorn lemma, $\mathcal{E}$ contains a maximal element at least. This contradicts the conclusion of the preceding paragraph.

Therefore, A is GCR. Q.E.D.

Remark. Let $A = \overline{\cup_n A_n}$ be a (UHF) algebra, where $1 \in A_1 \subset \cdots \subset A_n \subset \cdots$, and each A_n is a matrix algebra. If I is a proper closed two-sided ideal of A, then there is n_0 such that $I \cap A_n = \{0\}, \forall n \geq n_0$. Consider the quotient $*$ homomorphism $\Phi : A \longrightarrow A/I$. Then $\Phi|A_n$ is isometrical, $\forall n \geq n_0$. Hence, Φ is isometrical on whole A, and I must be zero. Therefore, any (UHF) algebra is simple. There is a more general result, see Lemma 15.4.1.

Theorem 13.5.5. Let A be a C^*-algebra. Then the following statements are equivalent:

1) A is type I;
2) A is GCR;
3) A is smooth;

4) Each factorial $*$ representation of A is type (I) ;
5) A has no type (III) factorial $*$ representation.

Proof. From Propositions 13.5.4 and 13.5.2, we have 3) $\Longrightarrow$ 2) $\Longrightarrow$ 1) . Moreover, 1) $\Longrightarrow$ 4) $\Longrightarrow$ 5) are obvious.

5) $\Longrightarrow$ 2) . If A is not GCR, then by Propositions 13.3.6 and 13.4.8 A admits a type (III) factorial $*$ representation. This contradicts the condition 5) .

Finally, 2) $\Longrightarrow$ 3) is obvious from Proposition 13.3.7. Q.E.D.

Lemma 13.5.6.

(i) Let A be a C^*-algebra of type I . Then $\tilde{A} = A \dot{+} \mathbb{C}$ is also a C^*-algebra of type I.

(ii) Let A_i be a C^*-algebra with an identity $1_i, i = 1,2$, and A_1 be type I. If $\alpha(\cdot)$ is a C^*-norm on $A_1 \otimes A_2$, and π is an irreducible $*$ representation of α-$(A_1 \otimes A_2)$, then we have $\pi \cong \pi_1 \otimes \pi_2$, where π_i is an irreducible $*$ representation of $A_i, i = 1,2$.

(iii) Let A be a C^*-algebra of type I, B be a C^*-algebra, and π be an irreducible $*$ representation of α_0-$(A \otimes B)$. Then we have $\pi = \pi_A \otimes \pi_B$, where π_A, π_B are irreducilbe $*$ representations of A, B respectively.

(iv) Let A_1, A_2 be two CCR algebras. Then α_0-$(A_1 \otimes A_2)$ is also CCR.

(v) Let A be a C^*-algebra of type I, B be a C^*-algebra, and J be a closed two-sided ideal of B. Then we have

$$\alpha_0\text{-}(A \otimes B)/\alpha_0\text{-}(A \otimes J) \cong \alpha_0\text{-}(A \otimes B/J).$$

Proof. (i) By Proposition 13.3.4, there is a composition series $\{I_\alpha \mid 0 \le \alpha \le \beta\}$ for A such that $I_{\alpha+1}/I_\alpha$ is CCR $\forall \alpha < \beta$. Let $I_{\beta+1} = \tilde{A}$. Then $\{I_\alpha \mid 0 \le \alpha \le \beta + 1\}$ is a composition series for $\tilde{A}$, and $I_{\alpha+1}/I_\alpha$ is CCR, $\forall \alpha < \beta + 1$. Hence $\tilde{A}$ is GCR by Proposition 13.3.4.

(ii) Clearly, $\pi(A_1 \otimes 1_2) \subset \pi(1_1 \otimes A_2)', \pi(1_1 \otimes A_2) \subset \pi(A_1 \otimes 1_2)'$, and $\pi(A_1 \otimes A_2)'' = B(H)$. Then

$$\begin{aligned} \mathbb{C} = \;& \pi(A_1 \otimes A_2)'' \cap \pi(A_1 \otimes A_2)' \\ = \;& (\pi(A_1 \otimes 1_2) \cup \pi(1_1 \otimes A_2))'' \cap \pi(A_1 \otimes 1_2)' \cap \pi(1_1 \otimes A_2)' \\ \supset \;& \pi(A_1 \otimes 1_2)'' \cap \pi(A_1 \otimes 1_2)'. \end{aligned}$$

Hence $\{\pi \mid A_1 \otimes 1_2, H\}$ is a factorial $*$ representation of A_1, and $M = \pi(A_1 \otimes 1_2)''$ is a type (I) factor on H. Let $\{p_l \mid l \in \Lambda\}$ be an orthogonal family of minimal projections of M such that $\sum_{l \in \Lambda} p_l = 1_H$. Clearly, $p_l \sim p_{l'}$ in $M, \forall l, l' \in \Lambda$. By Theorem 1.5.6, we can write that $\{M, H\} = \{B(H_1) \overline{\otimes} pMp, H_1 \otimes H_2\} =$

$\{B(H_1)\overline{\otimes}\mathbb{C}1_{H_2}, H_1 \otimes H_2\}$, where p is a minimal projection of M, $H_2 = pH$, and $\dim H_1 = {}'\Lambda$. Hence, we have

$$\pi(a_1 \otimes 1_2) = \pi_1(a_1) \otimes 1_{H_2}, \quad \forall a_1 \in A_1$$

and $\{\pi_1, H_1\}$ is an irreducible $*$ representation of A_1. On the other hand, $\pi(1_1 \otimes A_2) \subset \pi(A_1 \otimes 1_2)' = M' = \mathbb{C}1_{H_1} \otimes B(H_2)$, so

$$\pi(1_1 \otimes a_2) = 1_{H_1} \otimes \pi_2(a_2), \quad \forall a_2 \in A_2.$$

Further, by $B(H_1)\overline{\otimes}B(H_2) = B(H) = \pi(A_1 \otimes A_2)'' = \pi_1(A_1)''\overline{\otimes}\pi_2(A_2)''$, we can see that $\pi_2(A_2) = B(H_2)$, and $\{\pi_2, H_2\}$ is an irreducible $*$ representation of A_2, and $\pi = \pi_1 \otimes \pi_2$.

(iii) Clearly, α_0-$(A \otimes B)$ is a closed two-sided ideal of α_0-$(\widetilde{A} \otimes \widetilde{B})$, where $\widetilde{A} = A\dot{+}\mathbb{C}$, and $\widetilde{B} = B\dot{+}\mathbb{C}$. Then π can be extended to an irreducible $*$ representation $\widetilde{\pi}$ of α_0-$(\widetilde{A} \otimes \widetilde{B})$. By (i) and (ii), we have $\widetilde{\pi} = \widetilde{\pi}_A \otimes \widetilde{\pi}_B$, where $\widetilde{\pi}_A, \widetilde{\pi}_B$ are irreducible $*$ representations of $\widetilde{A}, \widetilde{B}$ respectively. If let $\pi_A = \widetilde{\pi}_A|A$ and $\pi_B = \widetilde{\pi}_B|B$, then π_A, π_B are irreducible $*$ representations of A, B respectively, and $\pi = \pi_A \otimes \pi_B$.

(iv) Let π be any irreducible $*$ representation of α_0-$(A_1 \otimes A_2)$. By (iii), $\pi = \pi_1 \otimes \pi_2$, where π_i is an irreducible $*$ representation of $A_i, i = 1, 2$. Since A_i is CCR, it follows that $\pi_i(A_i) \subset C(H_i), i = 1, 2$. Then $\pi(A_1 \otimes A_2) \subset C(H_1) \otimes C(H_2) \subset C(H)$. Therefore, α_0-$(A_1 \otimes A_2)$ is also CCR.

(v) Let

$$\pi_A = \oplus\{\pi_\rho \mid \rho \text{ is any pure state on } A\}$$

and

$$\pi_B = \oplus\{\pi_\sigma \mid \sigma \text{ is any pure state on } B, \text{ and } \sigma|J = 0\}.$$

Clearly, we have

$$\alpha_0\text{-}(A \otimes B/J) \cong (\pi_A \otimes \pi_B)(\alpha_0\text{-}(A \otimes B)).$$

On the other hand, if ν is a pure state on α_0-$(A \otimes B)$ and $\nu|\alpha_0$-$(A \otimes J) = 0$, then by (iii) we can write $\pi_\nu = \pi_\rho \otimes \pi_\sigma$, where ρ is a pure state on A, and σ is a pure state on B with $\sigma|J = 0$. Therefore,

$$\begin{aligned} &\alpha_0\text{-}(A \otimes B/J) \cong (\pi_A \otimes \pi_B)(\alpha_0\text{-}(A \otimes B)) \\ \cong\ & \oplus\left\{\pi_\nu \;\middle|\; \begin{array}{c}\nu \text{ is a pure state on } \alpha_0\text{-}(A \otimes B), \\ \text{and } \nu|\alpha_0(A \otimes J) = 0\end{array}\right\} \\ \cong\ & \alpha_0\text{-}(A \otimes B)/\alpha_0\text{-}(A \otimes J). \end{aligned}$$

Q.E.D.

Proposition 13.5.7.

(i) Let A be a C^*-algebra of type I, B be a C^*-subalgebra of A, and J be a closed two-sided ideal of A. Then B and A/J are type I C^*-algebras.

(ii) Let A be a C^*-algebra of type I. Then A is a nuclear C^*-algebra.

(iii) Let $A_1, \cdots, A_n$ be C^*-algebras of type I. Then $\alpha_0\text{-}\bigotimes_{i=1}^n A_i$ is also a C^*-algebra of type I.

Proof. (i) By Proposition 13.3.5 and Theorem 13.5.5 it is obvious.

(ii) Let B be any C^*-algebra, and $\alpha(\cdot)$ be a C^*-norm on $A \otimes B$. We want to prove that $\alpha(\cdot) = \alpha_0(\cdot)$ on $A \otimes B$. By Lemma 3.9.4 and Proposition 3.9.5, we may assume that A and B are unital. Then by the (ii) of Lemma 13.5.6, we have

$$\begin{aligned}
\alpha(u) &= \sup\left\{\|\pi(u)\| \,\middle|\, \begin{array}{c}\pi \text{ is any irreducible } * \text{ representation}\\ \text{of } \alpha\text{-}(A \otimes B)\end{array}\right\} \\
&\leq \sup\left\{\|\pi_A(\otimes)\pi_B(u)\| \,\middle|\, \begin{array}{c}\pi_A, \pi_B \text{ are irreducible } *\\ \text{representations of } A, B \text{ respectively}\end{array}\right\} \\
&= \alpha_0(u), \quad \forall u \in A \otimes B.
\end{aligned}$$

Therefore, A is nuclear.

(iii) First step. Let A be a CCR algebra, and B be a GCR algebra. We claim that $\alpha_0\text{-}(A \otimes B)$ is GCR.

In fact, let $\{I_\rho \mid 0 \leq \rho \leq \beta\}$ be a composition series for B such that $I_{\rho+1}/I_\rho$ is CCR, $\forall \rho < \beta$. Then by the (iv), (v) of Lemma 13.5.6,

$$\alpha_0\text{-}(A \otimes I_{\rho+1}/I_\rho) \cong \alpha_0\text{-}(A \otimes I_{\rho+1})/\alpha_0\text{-}(A \otimes I_\rho)$$

is CCR, $\forall \rho < \beta$. Clearly, $\{\alpha_0\text{-}(A \otimes I_\rho) \mid 0 \leq \rho \leq \beta\}$ is a composition series for $\alpha_0\text{-}(A \otimes B)$. Hence, $\alpha_0\text{-}(A \otimes B)$ is GCR.

Second step. Let A, B be GCR. We show that $\alpha_0\text{-}(A \otimes B)$ is GCR.

In fact, let $\{I_\rho \mid 0 \leq \rho \leq \beta\}$ be a composition serise for A such that $I_{\rho+1}/I_\rho$ is CCR, $\forall \rho < \beta$. Then by the First step, $\alpha_0\text{-}(I_{\rho+1}/I_\rho \otimes B)$ is GCR, $\forall \rho < \beta$. Clearly, $\{\alpha_0\text{-}(I_\rho \otimes B) \mid 0 \leq \rho \leq \beta\}$ is a composition series for $\alpha_0\text{-}(A \otimes B)$, and $\alpha_0\text{-}(I_{\rho+1} \otimes B)/\alpha_0\text{-}(I_\rho \otimes B) \cong \alpha_0(I_{\rho+1}/I_\rho \otimes B)$ is GCR, $\forall \rho < \beta$. Now by Proposition 13.3.4, $\alpha_0\text{-}(A \otimes B)$ is GCR.

Now let $A_1, \cdots, A_n$ be GCR (type I) algebras. We prove that $\alpha_0\text{-}\bigotimes_{i=1}^n A_i$ is GCR (type I).

The second step is the case of $n = 2$ exactly. We assume that the conclusion is true for $(n-1)$. Since $\min\text{-}\bigotimes_{i=1}^n A_i = \min\text{-}(A_1 \otimes \min - \bigotimes_{i=2}^n A_i)$, it follows from the second step and the induction that $\alpha_0\text{-}\bigotimes_{i=1}^n A_i$ is type I. Q.E.D.

Notes. Proposition 13.5.2 is due to I. Kaplansky. Theorem 13.5.5 was proved by J. Glimm for separable C^*-algebras. Then S.Sakai pointed out that this theorem holds in the absence of separability. Proposition 13.5.7 is due to W. Wulfoshn.

References. [27], [55], [84], [85], [150], [198].

13.6. Separable type I C^*-algebras

Definition 13.6.1. Let $H(\cdot)$ be a measurable field of Hibert spaces over a Borel space $(E, \mathcal{B})$, and A be a separable C^*-algebra. Suppose that $\{\pi(t), H(t)\}$ is a $*$ representation of $A, \forall t \in E$. The field $\{\pi(\cdot), H(\cdot)\}$ of $*$ representations of A over E is said to be *measurable* , if for any $a \in A, t \longrightarrow \pi(t)(a)$ is a measurable field of operators on $H(\cdot)$.

Clearly, $\|\pi(t)(a)\| \leq \|a\|, \forall t \in E, a \in A$. Thus , if the field $\{\pi(\cdot), H(\cdot)\}$ of $*$ representations of A is measurable, and ν be a σ-finite measure on $\mathcal{B}$, then $\int_E^{\oplus} \pi(t)(a)d\nu(t)$ is a bounded linear operator on $\int_E^{\oplus} H(t)d\nu(t)$ (see Proposition 12.2.3) , $\forall a \in A$. Hence, we get a $*$ representation $\{\pi, H\}$ of A, where

$$\pi(a) = \int_E^{\oplus} \pi(t)(a)d\nu(t), \quad \forall a \in A; \quad H = \int_E^{\oplus} H(t)d\nu(t).$$

Definition 13.6.2. The above $*$ representation π is called the *direct integral* of the measurable field $\pi(\cdot)$ with respect to the measure ν and denoted by $\pi = \int_E^{\oplus} \pi(t)d\nu(t)$.

Proposition 13.6.3. Let $H(\cdot)$ be a measurable field of Hilbert spaces over a Borel space $(E, \mathcal{B}), \nu$ be a σ-finite measure on $\mathcal{B}$, and A be a separable C^*-algebra.

1) Suppose that $\{\pi(\cdot), H(\cdot)\}$ is a measurable field of $*$ representations of A, and $\pi = \int_E^{\oplus} \pi(t)d\nu(t)$. Then π is nondegenerate if and only if $\pi(t)$ is nondegenerate, a.e.ν.

2) Suppose that $\{\pi_1(\cdot), H(\cdot)\}, \{\pi_2(\cdot), H(\cdot)\}$ are two measurable fields of $*$ representations of A, and $\pi_i = \int_E^{\oplus} \pi_i(t)d\nu(t), i = 1, 2$. Then $\pi_1 = \pi_2$ if and only if $\pi_1(t) = \pi_2(t)$, a.e.ν.

Proof. 1) Let Z be the diagonal algebra on $H = \int_E^{\oplus} H(t)d\nu(t)$ (Definition

12.2.7) . Since $\pi(a)$ is decomposable, it follows from Theorem 12.2.10 that $\pi(a) \in Z', \forall a \in A$. So $Z \subset \pi(A)'$. Let p be the projection from H onto the essential subspace of π (i.e., $pH = \overline{[\pi(A)H]}$). Clearly, $p \in \pi(A)'' \subset Z'$. Thus , p is decomposable, and we can write $p = \int_E^{\oplus} p(t)d\nu(t)$, where $p(\cdot)$ is a measurable field of projections on $H(\cdot)$. Since A is separable, there exists an approximate identity $\{d_n | n = 1, 2, \cdots\}$ for A. By Proposition 12.2.5, and the semi-finitness of ν, and $\pi(d_n) \longrightarrow p$ (strongly) , we have $\pi(t)(d_n) \longrightarrow p(t)$, a.e.$\nu$ (replacing $\{d_n\}$ by a subsequence if necessary) . Thus, $p(t)$ is the projection from $H(t)$ onto the essential subspace of $\pi(t)$, a.e.ν. Therefore, π is nondegenerate (i.e., $p = 1_H$) if and only if $\pi(t)$ is nondegenerate, a.e.ν.

2) The sufficiency is obvious. Now let $\pi_1 = \pi_2$, and $\{a_n\}$ be a countable dense subset of A. Then there is a Borel subset N of E such that

$$\nu(N) = 0, \text{ and } \pi_1(t)(a_n) = \pi_2(t)(a_n), \forall t \notin N, n.$$

Since $\{a_n\}$ is dense in A, it follows that $\pi_1(t)(a) = \pi_2(t)(a), \forall t \notin N, a \in A$, i.e., $\pi_1(t) = \pi_2(t)$, a.e.ν. Q.E.D.

Proposition 13.6.4. Let $H(\cdot)$ be a measurable field of Hilbert spaces over a Borel space $(E, \mathcal{B}), \nu$ be a σ-finite meausre on $\mathcal{B}, A$ be a separable C^*-algebra, and $\{\pi(\cdot), H(\cdot)\}$ be a measurable field of $*$ representations of A.

Suppose that there is a $*$ representation $\{\pi_0, H_0\}$ of A such that $\{\pi(t), H(t)\} \cong \{\pi_0, H_0\}, \forall t \in E$. Then there is a unitary operator u from $H = \int_E^{\oplus} H(t)d\nu(t)$ onto $L^2(E, \mathcal{B}, \nu) \otimes H_0$ such that

$$u\pi(a)u^* = 1 \otimes \pi_0(a), \quad \forall a \in A,$$

where $\pi = \int_E^{\oplus} \pi(t)d\nu(t)$, "1" is the identity operator on $L^2(E, \mathcal{B}, \nu)$.

Proof. Since each $H(t)$ is separable, H_0 is also separable. Now by Theorem 12.6.3 and Corollary 12.6.4, the conclusion is obvious. Q.E.D.

Proposition 13.6.5. Let $H(\cdot)$ be a measurable field of Hilbert spaces over a Borel space $(E, \mathcal{B}), \nu$ be a σ-finite measure on $\mathcal{B}$, and A be a separable C^*-algebra. If π is a $*$ representation of A on $H = \int_E^{\oplus} H(t)d\nu(t)$ such that $\pi(A) \subset Z'$, where Z is the diagonal algebra on H, then there is a measurable field $\{\pi(\cdot), H(\cdot)\}$ of $*$ representations of A such that $\pi = \int_E^{\oplus} \pi(t)d\nu(t)$.

Proof. Since A is separable, then we can construct a countable dense $*$ subalgebra B of A over the field of complex rational numbers. By $\pi(A) \subset Z', \pi(a)$ is decomposable, $\forall a \in A$. Since B is countable, we can find a Borel subset N of E such that

$$\nu(N) = 0, \quad T_a(t)T_b(t) = T_{ab}(t),$$
$$T_a(t)^* = T_{a^*}(t), \quad T_c(t) = \lambda T_a(t) + \mu T_b(t), \forall t \notin N,$$

where $a, b \in B, c = \lambda a + \mu b, \lambda, \mu$ are complex rational numbers, and $\pi(x) = \int_E^\oplus T_x(t) d\nu(T), \|T_x(t)\| \leq \|x\|, \forall t \in E, x \in B$. Since B is dense in A, for each $t \notin N$ we can get a $*$ representation $\{\pi(t), H(t)\}$ of A such that $\pi(t)(b) = T_b(t), \forall b \in B$. Further, let $\pi(t) = 0, \forall t \in N$. Now it is easy to see that $t \longrightarrow \pi(t)$ is measurable on E, and $\pi = \int_E^\oplus \pi(t) d\nu(t)$. Q.E.D.

Proposition 13.6.6. Let H be a separable Hilbert space, A be a separable C^*-algebra, π be a $*$ representation of A on H, and Z be an abelian VN algebra on H with $Z \subset \pi(A)'$. If Ω is the spectral space of Z (a compact Hausdorff space) , then there is a measurable field $H(\cdot)$ of Hilbert spaces over Ω, a regular Borel measure ν on Ω, a unitary operator u from H onto $\int_\Omega^\oplus H(t) d\nu(t)$, and a measurable field $\{\pi(\cdot), H(\cdot)\}$ of $*$ representations of A, such that uZu^* is the diagonal algebra on $\int_\Omega^\oplus H(t) d\nu(t)$, and $u\pi u^* = \int_E^\oplus \pi(t) d\nu(t)$.

Proof. It is immediate from Theorem 12.4.1 and Proposition 13.6.5. Q.E.D.

Proposition 13.6.7. Let $H(\cdot)$ be a measurable field of Hilbert spaces over a Borel space $(E, \mathcal{B}), \nu$ be a σ-finite measure on $\mathcal{B}, A$ be a separable C^*-algebra, and $\{\pi(\cdot), H(\cdot)\}$ be a measurable field of $*$ representations of A. Suppose that $Z \subset \pi(A)'$, where Z is the diagonal algebra on $H = \int_E^\oplus H(t) d\nu(t)$, and $\pi = \int_E^\oplus \pi(t) d\nu(t)$. Then Z is a maximal abelian VN subalgebra of $\pi(A)'$ if and only if $\{\pi(t), H(t)\}$ is irreducible for A, a.e.ν.

Proof. Let $M = \pi(A)''$. Clearly, Z is a maximal abelian VN subalgebra of $M' \Longleftrightarrow Z' \cap M' = Z \Longleftrightarrow Z'$ is generated by Z and M.

Now let $N = (Z \cup \pi(A))''$. Then Z is a maximal abelian VN subalgebra of M' if and only if $N = Z'$.

Let $\{a_n\}$ be a countalbe dense subset of A. Then N is generated by Z and $\{\pi(a_n) = \int_E^\oplus \pi(t)(a_n) d\nu(t) | n\}$. By Proposition 12.3.7, N is decomposable, i.e.,

$$N = \int_E^\oplus N(t) d\nu(t),$$

where $N(t)$ is a VN algebra on $H(t)$ generated by $\{\pi(t)(a_n) | n\}, \forall t \in E$.

It is well-known that $Z' = \int_E^{\oplus} B(H(t))d\nu(t)$. Therefore, Z is a maximal abelian VN subalgebra of $\pi(A)'$ if and only if $N(t) = \{\pi(t)(a_n)|n\}'' = \pi(t)(A)'' = B(H(t))$, a.e.$\nu$, i.e., $\{\pi(t), H(t)\}$ is irreducible for A, a.e.ν.

Q.E.D.

Proposition 13.6.8. Let H be a separable Hiblert space, A be a separable C^*-algebra, π be a $*$ representation of A on H, and Z be a maximal abelian VN subslagebra of $\pi(A)'$. Then there is a compact Hausdorff space Ω, a regular Borel measure ν on Ω, a measurable field $H(\cdot)$ of Hilbert spaces over Ω, a unitary operator u from H onto $\int_\Omega^{\oplus} H(t)d\nu(t)$, and a measurable filed $\{\pi(\cdot), H(\cdot)\}$ of $*$ representations of A, such that uZu^* is the diagonal algebra on $\int_\Omega^{\oplus} H(t)d\nu(t), u\pi u^* = \int_\Omega^{\oplus} \pi(t)d\nu(t)$, and $\{\pi(t), H(t)\}$ is irreducible for A, a.e.ν.

Proof. It is immediate from Propositions 13.6.6 and 13.6.7. Q.E.D.

Theorem 13.6.9. Let A be a separable C^*-algebra. Then the following statements are equivalent:

1) A is type I ;
2) $\ker : \widehat{A} \longrightarrow \text{Prim}(A)$ is a homeomorphism ;
3) $\widehat{A}$ is a T_0-space.

Proof. By Propositions 13.1.6 and 13.3.7, it suffices to show that the statement 2) implies the statement 1).

Now let $\ker : \widehat{A} \longrightarrow \text{Prim}(A)$ be a homeomorphism, and $\{\pi, H\}$ be a nondegenerate factorial $*$ representation of A. We need to prove $\pi(A)''$ is type (I) .

Pick a non-zero vector ξ of H, and let p' be the projection from H onto $\overline{[\pi(A)\xi]}$. Clearly, $p' \in \pi(A)'$. Since $p' \neq 0$ and $\pi(A)'$ is a factor, the central cover of p' in $\pi(A)'$ is 1_H . Hence, $a \longrightarrow ap'$ is a $*$ isomorphism from $\pi(A)''$ onto $\pi(A)''p'$. Clearly $p'H = \overline{[\pi(A)\xi]}$ is separable. Thus , we may assume that H is separable.

Fix a maximal abelian VN subalgebra Z of $\pi(A)'$. By Proposition 13.6.8, we may assume that :

$$H = \int_\Omega^{\oplus} H(t)d\nu(t), \quad \pi = \int_\Omega^{\oplus} \pi(t)d\nu(t),$$

Z is the diagonal algebra on $\int_\Omega^{\oplus} H(t)d\nu(t)$, and $\{\pi(t), H(t)\}$ is irreducible for $A, \forall t \in \Omega$, where Ω is compact, and ν is a regular Borel measure on Ω.

For any Borel subset B of Ω, denote by p'_B the diagonal operator corresponding to χ_B. Clearly, $p'_B \in Z \subset \pi(A)'$, and $p'_B \neq 0$ if $\nu(B) > 0$. Since $\pi(A)'$ is a factor, the central cover of p'_B in $\pi(A)'$ is 1_H. Thus , $a \longrightarrow ap'_B$ is a $*$ isomorphism from $\pi(A)''$ onto $\pi(A)''p'_B$. In particular, we have

$$\|\pi(x)\| = \|p'_B\pi(x)\|, \quad \forall x \in A.$$

Fix $0 \neq x \in A$. By Proposition 12.2.2, $t \longrightarrow \|\pi(t)(x)\|$ is a measurable function on Ω. Thus for any n

$$B_{xn} = \{t \in \Omega | \|\pi(t)(x)\| \leq \|\pi(x)\| - \frac{1}{n}\}$$

is a Borel subset of Ω. Denote by p'_{xn} the diagonal operator corresponding to $\chi_{B_{xn}}$. Then

$$\|p'_{xn}\pi(x)\| = \|\int_{B_{xn}}^{\oplus} \pi(t)(x)d\nu(t)\| \leq \|\pi(x)\| - \frac{1}{n}.$$

From the preceding paragraph, it must be that $\nu(B_{xn}) = 0, \forall x \in A, n$. Thus , we obtain that $\|\pi(t)(x)\| = \|\pi(x)\|, \forall t \notin B_x$, where $B_x = \cup_n B_{xn}$, and $\nu(B_x) = 0, x \in A$.

Suppose that D is a countable dense subset of A. Then we have

$$\|\pi(t)(x)\| = \|\pi(x)\|, \quad \forall x \in A, t \notin B$$

where $B = \cup_{x\in D} B_x$, and $\nu(B) = 0$. Consequently, ker $\pi(t) =$ ker $\pi, \forall t \notin B$.

Since $\{\pi(t), H(t)\}$ is irreducible, $\forall t \in \Omega$, by the condition 2) we may assume that $\{\pi(t), H(t)\} \cong \{\pi_0, H_0\}, \forall t \in \Omega$, where $\{\pi_0, H_0\}$ is some irreducible $*$ representation of A. Now by Proposition 13.6.4, the $*$ representation $\{\pi, H\}$ is unitarily equivalent to $\{1 \otimes \pi_0, L^2(\Omega, \nu) \otimes H_0\}$. Clearly, $\pi_0(A)'' = B(H_0)$ is type (I) factor. Therefore, $\pi(A)''$ is type (I) . Q.E.D.

Notes. Theorem 13.6.9 is due to J. Glimm. But we still don't know whether this theorem holds in the absence of separability.

References. [55].

Chapter 14

Decomposition Theory

14.1. Choguet theory of boundary integrals on compact convex sets

Let E be a (real) locally convex (Hausdorff topological linear) space, X be a nonempty compact convex subset of E, and ExX be the set of all extreme points of X. A *probability measure* on X means a non-negative regular Borel measure and its total mass is 1.

Definition 14.1.1. Let μ be a probability measure on X. A point x of X is said to be *represented* by μ, if $f(x) = \int_X f d\mu, \forall f \in E^*$.

Clearly, such x is one at most. We also say that μ is a *representing measure* of x , and x is the *barycenter* or *resultant* of μ , denoted by $r(\mu) = x$.

Proposition 14.1.2. The barycenter of a probability measure μ on X is existential.

Proof. For any $f \in E^*$, let $H_f = \{y \in E | f(y) = \int_X f d\mu\}$. Clearly, H_f is closed. We need to show that

$$\cap\{H_f \cap X | f \in E^*\} \neq \emptyset.$$

Since X is compact, it suffices to prove that for any finite set $f_1, \cdots, f_n$ of $E^*, \cap\{H_{f_i} \cap X | 1 \leq i \leq\} \neq \emptyset$. Define $T : E \longrightarrow \mathbb{R}^n$ by $Ty = (f_1(y), \cdots, f_n(y))$. Then TX is compact and convex in $\mathbb{R}^n$. Let $\alpha = (\int_X f_1 d\mu, \cdots, \int_X f_n d\mu)$, it suffices to show that $\alpha \in TX$. If $\alpha \notin TX$, then there is $\beta = (\beta_1, \cdots, \beta_n) \in \mathbb{R}^n$ such that

$$\sum_{i=1}^n \beta_i \int_X f_i d\mu > \sup \left\{ \sum_{i=1}^n \beta_i f_i(x) | x \in X \right\}.$$

Let $f = \sum_i \beta_i f_i$. Then $\int_X f d\mu > \sup\{f(x)|x \in X\}$. This is a contradiction since $\mu(X) = 1$. Therefore, $\alpha \in TX$. Q.E.D.

Proposition 14.1.3. Let $x \in X$. Then $x \in ExX \iff \delta_x$ is the unique representing measure of x.

Proof. If $x = \lambda x_1 + (1-\lambda)x_2$, where $x_1, x_2 \in X$ with $x_1 \neq x_2$, and $\lambda \in (0,1)$, then $\nu = \lambda\delta x_1 + (1-\lambda)\delta_{x_2}$ is also a representing measure of x, and $\nu \neq \delta_x$.

Conversely, let $x \in ExX$, and μ be a representing measure of x. Suppose that F is a closed subset of X with $\mu(F) > 0$ and $x \notin F$. For each $y \in F$, there is a closed convex neighborhood V_y of y such that $x \notin V_y$. Since F is compact, there is a finite set $y_1, \cdots, y_n$ of F such that $\sqcup_{i=1}^n V_i \supset F$, where $V_i = V_{y_i}, 1 \leq i \leq n$. In other words, we can find a closed convex subset K of X with $\mu(K) > 0$ and $x \notin K$. By Proposition 14.1.1 and the uniqueness of the barycenter of μ, it must be $\mu(K) < 1$. Now for each Borel subset B of X, define

$$\mu_1(B) = r^{-1}\mu(B \cap K), \quad \mu_2(B) = (1-r)^{-1}\mu(B\backslash K),$$

where $r = \mu(K)$. Then μ_1 and μ_2 are two probability measure on X. Let x_1, x_2 be the barycenters of μ_1, μ_2 respectively. Then $x = rx_1 + (1-r)x_2$, and $x \neq x_1$. This is impossible since $x \in ExX$. Therefore, supp $\mu = \{x\}$ and $\mu = \delta_x$.

Q.E.D.

Proposition 14.1.4. If X is metrizable, then ExX is a G_δ subset of X.

Proof. Let d be a proper metric on X, and let

$$F_n = \{x \in X | x = \frac{1}{2}(y+z), \text{ where } y, z \in X \text{ and } d(y,z) \geq n^{-1}\}$$

for each n. It is easily checked that each F_n is closed, and a point x of X is not extreme if and only if it is in some F_n. Thus , the complement of ExX is F_δ, and ExX is G_δ. Q.E.D.

Remark. If X is not metrizable, then ExX need not be a Borel subset of X.

A real function f on X is said to be *convex* if $f(\lambda x + (1-\lambda)y) \leq \lambda f(x) + (1-\lambda)f(y), \forall x, y \in X, 0 \leq \lambda \leq 1$; f is said to be *affine* , if $f(\lambda x + (1-\lambda)y) = \lambda f(x) + (1-\lambda)f(y), \forall x, y \subset X, 0 \leq \lambda \leq 1$. Denote by $C_r(X)$ the space of all real continuous functions on X, and $A(X) = \{f \in C_r(X) | f \text{ is affine } \}, M(X) = \{(f|_X) + r | r \in \mathbb{R}, f \in E^*\}, P(X) = \{f \in C_r(X) | f \text{ is convex }\}$. Clearly, $P(X) \cap (-P(X)) = A(X) \supset M(X)$.

Proposition 14.1.5. $A(X)$ is closed in $C_r(X)$; $M(X)$ is dense in $A(X)$; and $(P(X) - P(X))$ is dense in $C_r(X)$.

Proof. For any $g \in A(X)$ and $\varepsilon > 0$, consider the following subsets of $E \times \mathbb{R}$:

$$J_1 = \{(x,r)|x \in X, r = g(x)\}, \quad J_2 = \{(x,r)|x \in X, r = g(x) + \varepsilon\}.$$

Clearly, J_1 and J_2 are two disjoint nonempty compact convex subsets of $E \times \mathbb{R}$. By the separation theorem, there is some $L \in (E \times \mathbb{R})^*$ and some $\lambda \in \mathbb{R}$ such that

$$\sup\{L(\cdot)|\cdot \in J_1\} < \lambda < \sup\{L(\cdot)|\cdot \in J_2\}.$$

If $(x,r) \in J_1$, then $(x, r + \varepsilon) \in J_2$. Hence

$$L(x,r) < L(x, r + \varepsilon) = L(x,r) + \varepsilon L(0,1),$$

and $L(0,1) > 0$. Define $f(x) = L(x,0), \forall x \in E$. Then $f \in E^*$, and $m(\cdot) = L(0,1)^{-1}(\lambda - f(\cdot)) \in M(X)$. For any $x \in X$, we have

$$\begin{aligned} L(x, g(x)) &= L(x,0) + g(x)L(0,1) \\ &< \lambda = L(x, m(x)) = L(x,0) + m(x)L(0,1) \\ &< L(x, g(x) + \varepsilon) = L(x,0) + (g(x) + \varepsilon)L(0,1), \end{aligned}$$

i.e., $g(x) < m(x) < g(x) + \varepsilon, \forall x \in X$. Hence, $M(X)$ is dense in $A(X)$.

For any $f_1, f_2, g_1, g_2 \in P(X), \sup(f_1 - g_1, f_2 - g_2) = [\sup(f_1 + g_2, f_2 + g_1) - (g_1 + g_2)] \in (P(X) - P(X))$. So $(P(X) - P(X))$ is a vector lattice. Now by the Stone–Weierstrass theorem, $(P(X) - P(X))$ is dense in $C_r(X)$. Q.E.D.

Definition 14.1.6. Let μ, ν be two probability measures on X. We say that μ is bigger than ν in the sense of choquet–Meyer and denoted by $\nu \prec \mu$ (C. M.) , if

$$\int_X f d\mu \geq \int_X f d\nu, \quad \forall f \in P(X).$$

A probability measure μ on X is *maximal* (C. M.) , if $\nu \succ \mu$ (C. M.) for some prabability measure ν on X, then it must be $\nu = \mu$.

Proposition 14.1.7. (i) If $\nu \prec \mu$ (C. M.) and $\mu \prec \nu$ (C.M.), then it must be $\mu = \nu$;

(ii) If $\nu \prec \mu$ (C.M.) , then the barycenters of μ and ν are the same;

(iii) For any probability measure ν on X, there exists a maximal (C.M.) probability measure μ on X such that $\nu \prec \mu$ (C.M.).

Proof. (i) By the assumption, we have $\int f d\mu = \int f d\nu, \forall f \in P(X)$. Since $(P(X) - P(X))$ is dense in $C_r(X)$, it follows that $\mu = \nu$.

(ii) If $\nu \prec \mu$ (C. M.) , then for any $f \in E^*$ we have $\int f d\mu = \int f d\nu$ since $\pm f|_X \in P(X)$. Therefore, the barycenters of μ and ν are the same.

(iii) Let ν be a probability measure on X, and let

$$\mathcal{L} = \left\{ \mu \middle| \begin{array}{c} \mu \text{ is a probability measure on } X, \\ \text{and } \mu \succ \nu(\text{C.M.}) \end{array} \right\}.$$

With " $\prec$ (C.M.) ", $\mathcal{L}$ is a partially ordered nonempty set obviously.

Now let $\{\mu_l | l \in \wedge\}$ be any totally ordered nonempty subset of $\mathcal{L}$, where $\wedge$ is a totally ordered index set such that $\mu_l \prec \mu_{l'}$ (C.M.) if $l, l' \in \wedge$ and $l \leq l'$. By the weak $*$ compactness, there is a subnet $\{\lambda_\alpha | \alpha \in \mathit{I\!I}\}$ of $\{\mu_l\}$ and a probability measure μ on X such that

$$\int f d\lambda_\alpha \longrightarrow \int f d\mu, \quad \forall f \in C_r(X),$$

where $\mathit{I\!I}$ is a directed index set. Since $\{\lambda_\alpha | \alpha \in \mathit{I\!I}\}$ is a subnet of $\{\mu_l | l \in \wedge\}$, there is a map $d : \mathit{I\!I} \longrightarrow \wedge$ with following properties:

(i) for each $\alpha \in \mathit{I\!I}, \lambda_\alpha = \mu_{d(\alpha)}$;

(ii) for each $l \in \wedge$, we can find $\alpha_l \in \mathit{I\!I}$ such that $d(\alpha) \geq l, \forall \alpha \geq \alpha_l$.

Now for any $l \in \wedge$, pick α_l as in (ii) . Then $\lambda_\alpha = \mu_{d(\alpha)} \succ \mu_l$ (C.M.) , $\forall \alpha \geq \alpha_l$. Hence, we have $\mu \succ \mu_l$(C.M.) ,$\forall l \in \wedge$, i.e., μ is a upper bound of $\{\mu_l | l \in \wedge\}$ in $\mathcal{L}$. By the Zorn lemma, $\mathcal{L}$ admits a maximal element μ at least, and this μ is what we want to find . Q.E.D.

Theorem 14.1.8. Let μ be a maximal (C.M.) probability measure on X. Then μ is *pseudoconcentrated* on ExX in the sense that $\mu(B) = 0$ for each Baire subset B of X disjoint from ExX. In particular, supp $\mu \subset \overline{ExX}$, and if ExX is a Baire subset of X (for example, X is metrizable), then μ is concentrated on ExX, i.e., $\mu(ExX) = 1$.

Moreover, for any $x \in X$, there is a probability measure μ on X such that : μ is a representing measure of x; and μ is pseudoconcentrated on ExX in the above sence.

Proof. See [128]. Q.E.D.

Remark. Generally, we can't require that $\mu(B) = 0$ for each Borel subset B of X disjoint from ExX.

Definition 14.1.9. Assume that X is contained in a closed hyperplane of E which misses the origin (There is no generality lost in making this assumption, since we may embed E as the hyperplane $E\times\{1\}$ of $E\times \mathbb{R}$; the image $X\times\{1\}$ of X is affinely homeomorphic with X), and let $P=\{\alpha x|\alpha\geq 0, x\in X\}$.

X is called a *simplex* (in the sense of Choquet), if $(P-P)$ with the positive cone P is a vector lattice (i.e., if each pair x,y in $(P-P)$ has a least upper bound $\sup\{x,y\}$ in $(P-P)$) .

Such definition of a simplex coincides with usual one in case X is finite dimensional.

Theorem 14.1.10. The following statements are equivalent:

(i) X is a simplex (in the sense of Choquet);

(ii) For each $x\in X$, there is unique maximal (C.M.) probability measure μ on X such that μ is a representing measure of x.

Proof. See [128]. Q.E.D.

Notes. Theorem 14.1.8 is due to G. Choquet, P.A. Meyer, E.Bishop and K. de Leeuw, and its metrizable case is due to G.Choquet. Theorem 14.1.10 is due to G.Choquet and P.A.Meyer. Proposition 14.1.3 is due to H.Bauer.

References. [128].

14.2. The C–measure and C–isomorphism of a state

In this section, let A be a C^*-algebra with an identity 1 , $S=S(A)$ be the state space of A,φ be a fixed state on A, and $\{\pi_\varphi, H_\varphi, 1_\varphi\}$ be the cyclic $*$ representation of A generated by φ.

Clearly, S is a compact convex subset of $(A^*,\sigma(A^*,A))$. For any $a\in A$, define $\hat{a}(\rho)=\rho(a),\forall\rho\in S$. It is obvious that $a\longrightarrow\hat{a}(\cdot)$ is a positive linear map from A into $C(S)$. By Corollary 2.3.14, $E=\{\hat{a}(\cdot)|a\in A\}$ is a closed $*$ linear subspace of $C(S)$ containing the constant function 1 . Now φ can be regarded as a state on E. Then by Propostion 2.3.11 φ can be extended to a state on $C(S)$. Hence, there is a probability measure ν on S such that

$$\varphi(a)=\int_S \hat{a}(\rho)d\nu(\rho),\quad \forall a\in A. \tag{1}$$

Definition 14.2.1. Let

$$\Omega(\varphi)=\left\{\nu\left|\begin{array}{c}\nu\text{ is a probability measure on }S\\ \text{such that (1) holds.}\end{array}\right.\right\}.$$

Clearly, for each $\nu\in\Omega(\varphi)$, the barycenter (or resultant) of ν is φ, i.e., $r(\nu)=\varphi$; and ν is a representing measure of the point $\varphi(\in S)$.

Moreover, by the above discussion, $\Omega(\varphi) \neq \emptyset$. Hence, $\Omega(\varphi)$ is a non-empty compact convex subset of $(C(S)^*, \sigma(C(S)^*, C(S)))$.

Proposition 14.2.2. 1) If C is an abelian VN algebra on H_φ and $C \subset \pi_\varphi(A)', p$ is the projection from H_φ onto $\overline{C1_\varphi}$, then :

$c \longrightarrow cp$ is a $*$ isomorphism from C onto Cp; and Cp is a maximal abelian σ-finite VN algebra on pH_φ, i.e., $Cp = (Cp)' = pC'p$;

the central cover of p in C' is $1 = 1_{H_\varphi}$, i.e., $c(p) = 1$; and p is a maximal abelian projection of C';

$$p\pi_\varphi(a)p\pi_\varphi(b)p = p\pi_\varphi(b)p\pi_\varphi(a)p, \forall a, b \in A,$$

i.e., $p\pi_\varphi(A)p$ is abelian; and $p\pi_\varphi(A)p \subset pC'p = Cp$; and

$$C = \{\pi_\varphi(A), p\}'.$$

2) Suppose that p is a projection on H_φ such that $p1_\varphi = 1_\varphi$ and $p\pi_\varphi(A)p$ is abelian. Then $pH_\varphi = \overline{C1_\varphi}, C$ is abelian, and $C \subset \pi_\varphi(A)'$, where $C = \{\pi_\varphi(A), p\}'$.

3) There is a bijection between the collection

$$\left\{ C \middle| \begin{array}{c} C \text{ is an abelian VN algebra on } H_\varphi, \\ \text{and } C \subset \pi_\varphi(A)' \end{array} \right\}$$

and the collection

$$\left\{ p \middle| \begin{array}{l} p \text{ is a projection on } H_\varphi \text{such that} \\ p_{1_\varphi} = 1_\varphi \text{ and } p\pi_\varphi(A)p \text{ is abelian} \end{array} \right\}$$

such that $pH_\varphi = \overline{C1_\varphi}$ and $C = \{\pi_\varphi(A), p\}'$.

Proof. 1) By $c(p)H_\varphi = \overline{[C'pH_\varphi]} = \overline{[C'C1_\varphi]} \supset \overline{\pi_\varphi(A)1_\varphi} = H_\varphi$, we can see that $c(p) = 1$.

Since Cp is abelian and admits a cyclic vector 1_φ on pH_φ, thus Cp is maximal abelian and σ-finite from Proposition 5.3.15.

If q is an abelian projection of C' and $q \geq p$, then by Proposition 1.5.8 we have $p = c(p)q = q$, i.e., p is maximal abelian in C'.

Since $p\pi_\varphi(A)p \subset pC'p = Cp$, it follows that $p\pi_\varphi(a)p\pi_\varphi(b)p = p\pi_\varphi(b)p\pi_\varphi(a)$ p, $\forall a, b \in A$.

Now $p\pi_\varphi(A)p$ is abelian and admits a cyclic vector 1_φ on pH_φ, so by Proposition 5.3.15 we have

$$(p\pi_\varphi(A)p)'' = (p\pi_\varphi(A)p)' = (pNp)' = N'p,$$

where $N = \{\pi_\varphi(A), p\}''$. Moreover, from $Cp = (Cp)' = pC'p \supset p\pi_\varphi(A)p$, we get

$$(p\pi_\varphi(A)p)' \supset (Cp)' = Cp \supset (p\pi_\varphi(A)p)''.$$

Hence, $Cp = N'p$. Noticing that $\overline{[NpH_\varphi]} \supset \overline{\pi_\varphi(A)1_\varphi} = H_\varphi$, the central cover of p in N is also 1 . Thus, $c \longrightarrow cp(\forall c \in C)$ and $x' \longrightarrow x'p(\forall x' \in N')$ are $*$ isomorphisms from C onto Cp and from N' onto $N'p$ respectively. Now by $C \subset N'$, we must have $C = N' = \{\pi_\varphi(A), p\}'$.

2) Let $N = \{\pi_\varphi(A), p\}''$. Then the central cover of p in N is 1 , and $x' \longrightarrow x'p$ is a $*$ isomorphism from N' onto $N'p$.

Since $pNp = (p\pi_\varphi(A)p)''$ is abelian and admits a cyclic vector 1_φ on pH_φ, it follows that $pNp = (pNp)' = N'p$. Hence, $N'p$ is abelian. Further, $C = N'$ is abelian, and $C \subset \pi_\varphi(A)'$, and $\overline{C1_\varphi} = \overline{N'p1_\varphi} = \overline{pNp1_\varphi} = pH_\varphi$.

3) It is obvious from 1) and 2). Q.E.D.

Now fix an abelian VN algebra $C \subset \pi_\varphi(A)'$, and let p be the projection from H_φ onto $\overline{C1_\varphi}$.

Lemma 14.2.3. There exists unique $*$ homomorphism $\wedge : C(S) \longrightarrow C$ such that

$$\wedge(\hat{a})p = p\pi_\varphi(a)p, \quad \forall a \in A, \tag{2}$$

i.e., the following diagram is commutative:

$$\begin{array}{ccccc} C(S) & \xrightarrow{\wedge} & C & \longrightarrow & Cp \\ \uparrow & & & & \uparrow id \\ A & & & \longrightarrow & p\pi_\varphi(A)p \end{array}$$

Moreover, $\wedge C(S)$ is strongly dense in C, and

$$\wedge(\hat{a}_1 \cdots \hat{a}_n)p = p\pi_\varphi(a_1)p \cdots p\pi_\varphi(a_n)p, \quad \forall a_1, \cdots, a_n \in A.$$

Proof. Let B be the abelian C^*–algebra generated by $p\pi_\varphi(A)p$. Then $B \subset Cp$, and $B \cong C(T)$, where T is the spectral space of B. For any $t \in T$, define $\theta(t)(a) = (p\pi_\varphi(a)p)(t), \forall a \in A$. Clearly, $\theta(\cdot)$ is a continuous map from T to S. Further, for any $a_1, \cdots, a_n \in A$ and any polynomial P of n–variables, define

$$\overline{\wedge} P(\hat{a}_1, \cdots, \hat{a}_n) = P(p\pi_\varphi(a_1)p, \cdots, p\pi_\varphi(a_n)p).$$

Since

$$\begin{aligned} & \|P(p\pi_\varphi(a_1)p, \cdots, p\pi_\varphi(a_n)p)\| \\ = & \sup_{t \in T} |P((p\pi_\varphi(a_1)p)(t), \cdots, (p\pi_\varphi(a_n)p)(t))| \\ = & \sup_{t \in T} |P(\hat{a}_1(\theta(t)), \cdots, \hat{a}_n(\theta(t)))| \\ \leq & \sup_{\rho \in S} |P(\hat{a}_1(\rho), \cdots, \hat{a}_n(\rho))| = \|P(\hat{a}_1, \cdots, \hat{a}_n)\|, \end{aligned}$$

it follows from the Stone–Weierstrass theorem that $\overline{\wedge}$ can be extended to a $*$ homomorphism from $C(S)$ to $B \subset Cp$. By $C \cong Cp$, we can get a $*$

homomorphism $\wedge : C(S) \longrightarrow C$ such that

$$\wedge(\hat{a})p = p\pi_\varphi(a)p, \forall a \in A.$$

Again by the Stone–Weierstrass theorem, such $\wedge$ must be unique. Clearly, for any $a_1, \cdots, a_n \in A$ we have

$$\begin{aligned}\wedge(\hat{a}_1, \cdots, \hat{a}_n)p &= \wedge(\hat{a}_1)p \cdots \wedge (\hat{a}_n)p \\ &= p\pi_\varphi(a_1)p \cdots p\pi_\varphi(A_n)p\end{aligned}$$

since $p \in C'$.

Finally, since $Cp = pC'p = p\{\pi_\varphi(A), p\}''p = (p\pi_\varphi(A)p)''$, $\overline{\wedge} C(S) = B$ is strongly dense in Cp. Therefore, $\wedge C(S)$ is strongly dense in C. Q.E.D.

Now $\langle \wedge(\cdot)1_\varphi, 1_\varphi\rangle$ is a state on $C(S)$. Hence, there is unique probability measure μ on S such that

$$\langle \wedge(f)1_\varphi, 1_\varphi\rangle = \int_S f(\rho)d\mu(\rho), \quad \forall f \in C(S).$$

In particular, by Lemma 14.2.3 we have

$$\int_S \hat{a}_1(\rho) \cdots \hat{a}_n(\rho)d\mu(\rho) = \langle p\pi_\varphi(a_1)p \cdots p\pi_\varphi(a_n)p1_\varphi, 1_\varphi\rangle, \tag{3}$$

$\forall a_1, \cdots, a_n \in A$, and $\mu \in \Omega(\varphi)$ (see Definition 13.2.1). Clearly, the measure μ satisfying (3) is unique from the Stone–Weierstrass theorem.

Definition 14.2.4. The unique $\mu(\in \Omega(\varphi))$ determined by (3) is called the *C-measure* of the state φ.

Now we point out that $\wedge$ can be extended to a $*$ isomorphism from $L^\infty(S, \mu)$ onto C.

First, for $f \in C(S)$ we say that

$$\wedge(f) = 0 \Longleftrightarrow f = 0 \text{ a.e.}\mu.$$

This is immediate from the following equality:

$$\begin{aligned}&\langle \wedge(f)p\pi_\varphi(a_1)p \cdots p\pi_\varphi(a_n)p1_\varphi, 1\varphi\rangle \\ =\ &\langle \wedge(f) \wedge (\hat{a}_1) \cdots \wedge (\hat{a}_n)1_\varphi, 1_\varphi\rangle \\ =\ &\int f(\rho)\hat{a}_1(\rho) \cdots \hat{a}_n(\rho)d\mu(\rho),\end{aligned}$$

$\forall a_1, \cdots, a_n \in \wedge$, and by the Stone–Weierstrass theorem.

Secondly, if $\{f_l\}$ is a net of $C(S)$ with $\|f_l\| \le 1, \forall l$, such that $f_l \longrightarrow 0(\sigma(L^\infty, L^1))$, then we have

$$\begin{aligned} & \langle \wedge(f_l)p\pi_\varphi(a)1_\varphi, \pi_\varphi(b)1_\varphi\rangle \\ = & \langle \wedge(f_l)\cdot p\pi_\varphi(a)p1_\varphi, p\pi_\varphi(b)p1_\varphi\rangle \\ = & \int f_l(\rho)\hat{a}(\rho)\hat{b}(\rho)d\mu(\rho) \longrightarrow 0, \end{aligned}$$

$\forall a, b \in A$. Hence, $\wedge(f_l)p$ and $\wedge(f_l) \longrightarrow 0$ (weakly).

Then, $\wedge$ can be extended to a normal $*$ homomorphism from $L^\infty(S,\mu)$ to C, still denoted by $\wedge$, such that

$$\langle \wedge(f)1_\varphi, 1_\varphi\rangle = \int_S f(\rho)d\mu(\rho), \ \forall f \in L^\infty(S,\mu).$$

On the other hand, by $(p\pi_\varphi(A)p)'' = Cp$ (see that proof of Proposition 14.2.2) and $\wedge(C(S))p \supset p\pi_\varphi(A)p$, we have $\wedge(L^\infty(S,\mu))p = Cp$ and $\wedge(L^\infty(S,\mu)) = C$. Further, if $f \in L^\infty(S,\mu)$ is such that $\wedge(f) = 0$, then

$$\begin{aligned} & \int f(\rho)\hat{a}_1(\rho)\cdots\hat{a}_n(\rho)d\mu(\rho) \\ = & \langle \wedge(f)\wedge(\hat{a}_1,\cdots,\hat{a}_n)1_\varphi, 1_\varphi\rangle = 0, \end{aligned}$$

$\forall a_1,\cdots,a_n \in A$. Hence, $f = 0$, a.e.μ, and $\wedge$ is a $*$ isomorphism from $L^\infty(S,\mu)$ onto C.

Denote by Γ the inverse map of $\wedge$. Then Γ is a $*$ isomorphism from C onto $L^\infty(S,\mu)$, and

$$\langle x\pi_\varphi(a)1_\varphi, 1_\varphi\rangle = \int_S (\Gamma x)(\rho)\hat{a}(\rho)d\mu(\rho), \tag{4}$$

$\forall a \in A, x \in C$.

Lemma 14.2.5. Let $\nu \in \Omega(\varphi)$, and Γ_1 be a $*$ isomorphism from C onto $L^\infty(S,\nu)$ such that

$$\langle x\pi_\varphi(a)1_\varphi, 1_\varphi\rangle = \int_S (\Gamma_1 x)(\rho)\hat{a}(\rho)d\nu(\rho),$$

$\forall a \in A, x \in C$. Then $\nu = \mu$ is the C–measure, and $\Gamma_1 = \Gamma$.

Proof. For any $a \in A$ there is $x \in C$ such that $\Gamma_1 x = \hat{a}$. Then

$$\begin{aligned} & \langle xpy1_\varphi, z1_\varphi\rangle = \langle xz^*y1_\varphi, 1_\varphi\rangle \\ = & \int (\Gamma_1 xz^*y)(\rho)d\nu(\rho) = \int (\Gamma_1 z^*y)(\rho)\hat{a}(\rho)d\nu(\rho) \\ = & \langle z^*y\pi_\varphi(a)1_\varphi, 1_\varphi\rangle = \langle p\pi_\varphi(a)py1_\varphi, z1_\varphi\rangle, \end{aligned}$$

$\forall y, z \in C$. Since $\overline{C1_\varphi} = pH_\varphi$, it follows that

$$xp = p\pi_\varphi(a)p.$$

Then by $p\pi_\varphi(a)p = \wedge(\hat{a})p$ we have $x = \wedge(\hat{a})$, and $\Gamma x = \hat{a} = \Gamma_1 x$. Further,

$$\Gamma_1(x_1 \cdots x_n) = \hat{a}_1 \cdots \hat{a}_n = \Gamma(x_1, \cdots, x_n),$$

where $a_i \in A, x_i \in C$, and $\Gamma_1 x_i = \hat{a}_i, 1 \le i \le n$. Hence,

$$\begin{aligned}
&\int_S \hat{a}_1(\rho) \cdots \hat{a}_n(\rho) d\nu(\rho) = \int_S \Gamma_1(x_1 \cdots x_n)(\rho) d\nu(\rho) \\
=\ & \langle x_1 \cdots x_n 1_\varphi, 1_\varphi \rangle = \langle (x_1 p) \cdots (x_n p) 1_\varphi, 1_\varphi \rangle \\
=\ & \langle p\pi_\rho(a_1)p \cdots p\pi_\varphi(a_n)p1_\varphi, 1_\varphi \rangle,
\end{aligned}$$

$\forall a_1, \cdots, a_n \in A$. By Definition 14.2.4, $\nu = \mu$ is the C-measure. Also by $\Gamma_1(x_1 \cdots x_n) = \hat{a}_1 \cdots \hat{a}_n = \Gamma(x_1 \cdots x_n)$ we have

$$\Gamma_1^{-1}(\hat{a}_1 \cdots \hat{a}_n) = \Gamma^{-1}(\hat{a}_1 \cdots \hat{a}_n), \quad \forall a_1, \cdots, a_n \in A.$$

Since $C(S)$ is w^*-dense in $L^\infty(S, \mu)$, it follows from the Stone–Weierstrass theorem that $\Gamma_1 = \Gamma$. Q.E.D.

Definition 14.2.6. The unique $*$ isomorphism Γ from C onto $L^\infty(S, \mu)$ determined by (4) is called the *C-isomorphism* of the state φ , where μ is the C-measure of φ.

From the above discussion, we have the following.

Theorem 14.2.7. Let A be a C^*-algebra with an identity 1, $S = S(A)$ be the state space of A (a compact convex subset of $(A^*, \sigma(A^*, A)))$, $\varphi \in S$, $\{\pi_\varphi, H_\varphi, 1_\varphi\}$ be the cyclic $*$ representation of A generated by φ, and C be an abelian VN algebra on H_φ with $C \subset \pi_\varphi(A)'$. Then there is a unique probability measure μ on S (the C-measure of φ) and a unique $*$ isomorphism Γ from C onto $L^\infty(S, \mu)$ (the C-isomorphism of φ) such that $\mu \in \Omega(\varphi)$,

$$\int_S \hat{a}_1 \cdots \hat{a}_n d\mu = \langle p\pi_\varphi(a_1)p \cdots p\pi_\varphi(a_n)p1_\varphi, 1_\varphi \rangle$$

and

$$\langle x\pi_\varphi(a)1_\varphi, 1_\varphi \rangle = \int_S (\Gamma x)(\rho)\hat{a}(\rho) d\mu(\rho),$$

$\forall a, a_1, \cdots, a_n \in A$, where $\Omega(\varphi)$ is as in Definition 14.2.1, and p is the projection from H_φ onto $\overline{C1_\varphi}$.

Lemma 14.2.8. Let Ω be a compact convex subset of a locally convex topological linear space E, and μ be a probability measure on Ω. Then

$$\mu = \sigma(C(\Omega)^*, C(\Omega))\text{-}\lim_{\{h_j\}} \mu_{\{h_j\}},$$

i.e.,

$$\int_\Omega g d\mu = \lim_{\{h_j\}} \int_\Omega g d\mu_{\{h_j\}}, \quad \forall g \in C(\Omega),$$

where $0 \le h_j \in L^\infty(\Omega, \mu), \forall j, \sum_j h_j(t) = 1$, a.e.$\mu$; $\mu_{\{h_j\}} = \sum_j \alpha_j \delta_{t_j}$, for each $j, \alpha_j = \int_\Omega h_j d\mu, t_j \in \Omega$ such that

$$\alpha_j f(t_j) = \int_\Omega f(t) h_j(t) d\mu(t), \quad \forall f \in E^*$$

(notice that if $\alpha_j > 0$, then $\alpha_j^{-1} h_j \mu$ is a probability measure on Ω. Hence, from Proposition 14.1.2, there is unique $t_j \in \Omega$ such that $f(t_j) = \alpha_j^{-1} \int_\Omega f(t) h_j(t) d\mu(t), \forall f \in E^*$); $\{h'_k\}_{k=1}^m \le \{h_j\}_{j=1}^n$ if there is a partition $\{I_1, \cdots, I_m\}$ of $\{1, \cdots, n\}$ such that $h'_k = \sum_{j \in I_k} h_j, 1 \le k \le m$ (Clearly for any $\{h'_k\}$ and $\{h_j\}, \{h'_k h_j\} \ge \{h'_k\}$ and $\{h_j\}$. So that $\{\{h_j\} | 0 \le h_j \in L^\infty(\Omega, \mu), \forall j$, and $\sum_j h_j = 1, a.e.\mu\}$ is a directed set with respect to " $\le$").

Proof. First, we claim that $\mu_{\{h'_k\}} \prec \mu_{\{h_j\}}$ (C. M.) if $\{h'_k\}_{k=1}^m \le \{h_j\}_{j=1}^n$.

In fact , let $\mu_{\{h'_k\}} = \sum_k \alpha'_k \delta_{t'_k}, \mu_{\{h_j\}} = \sum_j \alpha_j \delta_{t_j}$, and $\{I_1, \cdots, I_m\}$ be a partition of $\{1, \cdots n\}$ such that $h'_k = \sum_{j \in I_k} h_j, 1 \le k \le m$. Since $\alpha'_k f(t'_k) = \int f h'_k d\mu = \sum_{j \in I_k} \int f h_j d\mu = \sum_{j \in I_k} \alpha_j f(t_j), \forall f \in E^*$, it follows that

$$\alpha'_k t'_k = \sum_{j \in I_k} \alpha_j t_j \quad \text{in } E, \quad 1 \le k \le m.$$

Then for any convex function $g \in C_r(\Omega)$ we have

$$\begin{aligned} \int g d\mu_{\{h'_k\}} &= \sum_k \alpha'_k g(t'_k) = \sum_{\alpha'_k > 0} \alpha'_k g\left(\sum_{j \in I_k} \frac{\alpha_j}{\alpha'_k} t_j\right) \\ &\le \sum_{\alpha'_k > 0} \alpha'_k \sum_{j \in I_k} \frac{\alpha_j}{\alpha'_k} g(t_j) = \sum_j \alpha_j g(t_j) = \int g d\mu_{\{h_j\}} \end{aligned}$$

(notice that if $\alpha'_k = 0$ then it must be $\alpha_j = 0, \forall j \in I_k$), i.e., $\mu_{\{h'_k\}} \prec \mu_{\{h_j\}}$ (C.M.).

Now we prove that $\mu = \sigma\text{-}\lim_{\{h_j\}} \mu_{\{h_j\}}$.

In fact, for any $g \in C(\Omega)$ and $\varepsilon > 0$, we can find a family $\{V_1, \cdots, V_n\}$ of closed convex subsets of Ω such that $\sqcup_{j=1}^n \mathrm{Int}(V_j) = \Omega$ and

$$|g(s) - g(t)| < \varepsilon, \text{ if } \quad s, t \in V_j \text{ for some } j.$$

For each $j \in \{1, \cdots, n\}$, pick $h_j \in C(\Omega)$ such that $0 \le h_j \le 1$, supp $h_j \subset$ Int (V_j), and $\sum_j h_j \equiv 1$. Suppose that $\alpha_j = \int h_j d\mu$ and $t_j \in \Omega$ such that

$$\alpha_j f(t_j) = \int f h_j d\mu, \quad \forall f \in E^*, \quad \forall j.$$

If for some $j, \alpha_j > 0$ and $t_j \notin V_j$, then there is $f \in E^*$ such that $Ref(t_j) > \sup_{t \in V_j} Ref(t)$. But

$$\begin{aligned} Ref(t_j) &= \tfrac{1}{\alpha_j} \int Ref(t) h_j(t) d\mu(t) = \frac{1}{\alpha_j} \int_{V_j} Ref \cdot h_j d\mu \\ &\le \alpha_j^{-1} \sup_{t \in V_j} Ref(t) \int h_j d\mu = \sup_{t \in V_j} Ref(t), \end{aligned}$$

so we get a contradiction. Hence, $t_j \in V_j$ if $\alpha_j > 0$. Further,

$$\begin{aligned} & |\int g d\mu - \int g d\mu_{\{h_j\}}| \\ = & \ |\int g d\mu - \sum_j \int h_j d\mu g(t_j)| \\ \le & \ \sum_j \int |g(t) - g(t_j)| h_j(t) d\mu(t) < \varepsilon. \end{aligned}$$

Now if $\{h'_l\}_{l=1}^m \ge \{h_j\}_{j=1}^n$, then there is a partition $\{I_1, \cdots, I_n\}$ of $\{1, \cdots, m\}$ such that $h_j = \sum_{l \in I_j} h'_l, 1 \le i \le n$. Thus , for any j and $l \in I_j$ we have

$$h'_l(t) = 0, \quad \text{a.e.}\mu \text{ on } \Omega \backslash \mathrm{Int}(V_j).$$

Further, $s_l \in V_j$ if $\alpha'_l > 0, \forall l \in I_j$, where $\alpha'_l = \int h'_l d\mu$, and $\alpha'_l f(s_l) = \int f h'_l d\mu, \forall f \in E^*$. Then

$$\begin{aligned} & |\int g d\mu - \int g d\mu_{\{h'_l\}}| \\ \le & \ \sum_l \int |g(t) - g(s_l)| h'_l(t) d\mu(t) \\ = & \ \sum_j \sum_{l \in I_j} \int_{V_j} |g(t) - g(s_l)| h'_l(t) d\mu(t) < \varepsilon. \end{aligned}$$

Therefore,

$$\mu(t) = \lim \mu_{\{h_j\}}(g), \quad \forall g \in C(\Omega).$$

Q.E.D.

Proposition 14.2.9. Let $A, S = S(A), \varphi \in S, \{\pi_\varphi, H_\varphi, 1_\varphi\}, C \subset \pi_\varphi(A)'$ and $pH_\varphi = \overline{C1_\varphi}$ be the same as in Theorem 14.2.7. Suppose that μ is the C -measure of φ. Then

$$\mu = \sigma(C(S)^*, C(S))\text{-} \lim_{\{B_j\}} \mu_{\{B_j\}},$$

i.e.,

$$\int_S g d\mu = \lim_{\{B_j\}} \int_S g d\mu_{\{B_j\}}, \quad \forall g \in C(S),$$

where $B_j \in C_+, \forall j, \sum_j B_j = 1; \mu_{\{B_j\}} = \sum_j \alpha_j \delta_{\rho_j}$, for each j, $\alpha_j = \langle B_j 1_\varphi, 1_\varphi\rangle, \rho_j \in S$ such that

$$\alpha_j \rho_j(a) = \langle \pi_\varphi(a) B_j 1_\varphi, 1_\varphi\rangle, \quad \forall a \in A;$$

$\{B'_k\}_{k=1}^m \leq \{B_j\}_{j=1}^n$ if there is a partition $\{I_1, \cdots, I_m\}$ of $\{1, \cdots, n\}$ such that $B'_k = \sum_{j \in I_k} B_j, 1 \leq k \leq m$. Morevoer, if $\{B'_k\} \leq \{B_j\}$, then $\mu_{\{B'_k\}} \prec \mu_{\{B_j\}} \prec \mu$ (C.M.) .

Proof. Let Γ be the C–isomorphism of φ from C onto $L^\infty(S, \mu)$. Then

$$\langle x\pi_\varphi(a)1_\varphi, 1_\varphi\rangle = \int_\varphi \hat{a}(\rho)(\Gamma x)(\rho) d\mu(\rho),$$

$\forall a \in A, x \in C$. For any $\{B_j\}$, let $h_j = \Gamma B_j, \forall j$. Now using Lemma 14.2.8 to $E = (A^*, \sigma(A^*, A)), \Omega = S$ and the C–measure μ of φ, we can get the conclusion.

Q.E.D.

Theorem 14.2.10. Let A be a C^*–algebra with an identity 1, $S = S(A)$ be the state space of $A, \varphi \in S, \{\pi_\varphi, H_\varphi, 1_\varphi\}$ be the cyclic $*$ representation of A generated by φ. Suppose that C_i is an abelian VN algebra on H_φ with $C_i \subset \pi_\varphi(A)'$, and μ_i is the C_i -measure of φ on $S, i = 1, 2$. Then the following conditions are equivalent:

(i) $\mu_1 \prec \mu_2$ (C.M.), i.e., for any convex function $g \in C_r(S)$,

$$\int_S g d\mu_1 \leq \int_S g d\mu_2;$$

(ii)

$$\int_S \hat{a}(\rho)^2 d\mu_1(\rho) \leq \int_S \hat{a}(\rho)^2 d\mu_2(\rho), \ \forall a^* = a \in A;$$

(iii) $C_1 \subset C_2$.
Consequently, if $\mu_1 = \mu_2$, then $C_1 = C_2$.

Proof. (i) $\Longrightarrow$ (ii) . It is obvious since $\hat{a}(\cdot)^2$ is convex on $S, \forall a^* = a \in A$.

(ii) $\Longrightarrow$ (iii) . Suppose that the inequality (ii) holds. Let p_i be the projection from H_φ onto $\overline{C_i 1_\varphi}$, Γ_i be the C_i–isomorphism of φ from C_i onto $L^\infty(S, \mu_i)$, and $\wedge_i = \Gamma_i^{-1}, i = 1, 2$. Notice that $\wedge_i(\hat{a})p_i = p_i \pi_\varphi(a) p_i, \forall a \in A, i = 1, 2$. For any $a = b + ic, b^* = b, c^* = c \in A$, we have

$$\begin{aligned}\langle p_1 \pi_\varphi(a) 1_\varphi, \pi_\varphi(a) 1_\varphi \rangle &= \langle p_1 \pi_\varphi(a^*) p_1 \pi_\varphi(a) p_1 1_\varphi, 1_\varphi \rangle \\ &= \langle \wedge_1(\hat{a}^* \hat{a}) 1_\varphi, 1_\varphi \rangle = \int_S [\hat{b}(\rho)^2 + \hat{c}(\rho)^2] d\mu_1(\rho) \\ &\leq \int_S [\hat{b}(\rho)^2 + \hat{c}(\rho)^2] d\mu_2(\rho) = \langle \wedge_2(\hat{a}^* \hat{a}) 1_\varphi, 1_\varphi \rangle \\ &= \langle p_2 \pi_\varphi(a) 1_\varphi, \pi_\varphi(a) 1_\varphi \rangle .\end{aligned}$$

Hence we get $p_1 \leq p_2$ since 1_φ is cyclic for $\pi_\varphi(A)$, which means that $\overline{C_1 1_\varphi} \subset \overline{C_2 1_\varphi}$.

Let U be the isometry of $L^2(S, \mu_2)$ onto $\overline{C_2 1_\varphi}$ given by $Uf = \wedge_2(f) 1_\varphi, \forall f \in L^\infty(S, \mu_2)$. For any $x \in C_1$, by $C_1 1_\varphi \subset \overline{C_2 1_\varphi}$ there exists $f \in L^2(S, \mu_2)$ such that $Uf = x 1_\varphi$. For any $n = 1, 2, \cdots$, let $E_n = \{\rho \in S \| f(\rho)| \leq n\}$ and $e_n = \wedge_2(\chi_{E_n})$. Then we have

$$\begin{aligned} e_n x 1_\varphi &= \wedge_2(\chi_{E_n}) U f \\ &= \wedge_2(\chi_{E_n}) \lim_m U(\chi_{E_m} f) \\ &= \wedge_2(\chi_{E_n}) \lim_m \wedge_2(\chi_{E_m} f) 1_\varphi = \wedge_2(\chi_{E_n} f) 1_\varphi \end{aligned}$$

since $\chi_{E_m} f \longrightarrow f$ in $L^2(S, \mu_2)$. But 1_φ is separating for $\pi_\varphi(A)'$, so we get $e_n x = \wedge_2(\chi_{E_n} f) \in C_2, \forall n$. Moreover, since $\chi_{E_n} \longrightarrow 1$ in $\sigma(L^\infty, L^1)$ and $\wedge_2$ is normal, it follows that $e_n \longrightarrow 1$ (weakly). Hence, $x \in C_2$, and $C_1 \subset C_2$.

(iii) $\Longrightarrow$ (i). Suppose that $C_1 \subset C_2$. By Proposition 14.2.9, we have

$$\mu_1 = \lim \mu_{\{B_j\}} \quad \text{and} \quad \mu_{\{B_j\}} \prec \mu_2 (\text{C.M.}),$$

$\forall B_j \in (C_1)_+ \subset (C_2)_+$, and $\sum_j B_j = 1$. Therefore, $\mu_1 \prec \mu_2$ (C.M.). Q.E.D.

Corollary 14.2.11. If $\pi_\varphi(A)'$ is abelian, let $C = \pi_\varphi(A)'$, then the C–measure μ of φ is maximal (C.M.) on $S = S(A)$. Consequently, μ is pseudoconcentrated on the pure state space $P(A)$ in the sense that $\mu(E) = 0$ for every Baire subset E of $S(A)$ disjoint from $P(A)$. Moreover, if A is separable, then μ is concentrated on $P(A)$.

Proof. Let ν be a probability measure on S, and $\nu \succ \mu$ (C.M.). By Lemma 14.2.8, it suffices to show that $\mu \succ \nu_{\{h_j\}}$ (C.M.) , $\forall 0 \leq h_j \in L^\infty(S,\nu)$, and $\sum_j h_i = 1$, a.e.ν.

Let $\nu_{\{h_j\}} = \sum_j \alpha_j \delta_{\rho_j}$, where $\alpha_j = \int h_j d\nu, \rho_j \in S$, and $\alpha_j \rho_j(a) = \int \hat{a} h_j d\nu, \forall$ $a \in A$. Since $\nu \succ \mu$(C.M.), $\nu \succ \nu_{\{h_j\}}$(C.M.) and $\mu \in \Omega(\varphi)$, it follows that $\nu, \nu_{\{h_j\}} \in \Omega(\varphi)$. Hence, $\int \hat{a} d\mu = \varphi(a) = \sum_j \alpha_j \rho_j(a), \forall a \in A$. Consequently, $\varphi \geq \alpha_j \rho_j \geq 0, \forall j$. Hence, for each j, there is $B_j \in \pi_\varphi(A)'_+ = C_+$ such that

$$\alpha_j \rho_j(a) = \langle \pi_\varphi(a) B_j 1_\varphi, 1_\varphi \rangle, \forall a \in A.$$

Clearly, $\sum_j B_j = 1$, and $\nu_{\{h_j\}} = \mu_{\{B_j\}}$. Now by Proposition 14.2.9, we obtain that

$$\mu \succ \mu_{\{B_j\}} = \nu_{\{h_j\}} (\text{C.M.}).$$

Q.E.D.

Remark. Applying Choquet theory (Section 14.1) to $S(A)$, and by $\Omega(\varphi) \neq \emptyset$(see Definition 14.2.1), there is a maximal (C.M.) probability measure μ on $S(A)$ such that

$$\varphi(a) = \int_{S(A)} \hat{a}(\rho) d\mu(\rho), \quad \forall a \in A.$$

If $\pi_\varphi(A)'$ is abelian, then by Corollary 14.2.11 we can pick that μ is the C-measure of φ, where $C = \pi_\varphi(A)'$.

μ is pseudoconcentrated on the pure state space $P(A)$. So φ seems to be an "integral" of pure states. In particular, if A is separable, then we have

$$\varphi(a) = \int_{P(A)} \hat{a}(\rho) d\mu(\rho), \quad \forall a \in A.$$

Proposition 14.2.12. Let $\mu \in \Omega(\varphi)$ (see Definition 14.2.1) . Then μ is the C-measure of φ for some abelian VN algebra $C \subset \pi_\varphi(A)'$ if and only if for each Borel subset E of S there is a projection p_E of $\pi_\varphi(A)'$ such that $\int_S \chi_E(\rho)\hat{a}(\rho) d\mu(\rho) = \langle p_E \pi_\varphi(a) 1_\varphi, 1_\varphi \rangle, \forall a \in A$.

Proof. The necessity is obvious from Theorem 14.2.7. Conversely, for any $f \in L^\infty(S,\mu)$ with $0 \leq f \leq 1$ define

$$\varphi_f(a) = \int f \hat{a} d\mu, \quad \forall a \in A.$$

Clearly, $0 \leq \varphi_f \leq \varphi$. Thus, there exists unique $x_f \in \pi_\varphi(A)'$ such that $0 \leq x_f \leq 1$ and

$$\int f\hat{a}d\mu = \langle x_f \pi_\varphi(a) 1_\varphi, 1_\varphi\rangle, \quad \forall a \in A.$$

Then we can get a bounded positive linear map $f \longrightarrow x_f$ from $L^\infty(S,\mu)$ to $\pi_\varphi(A)'$. By the condition, $x_f = p_E$ is a projection if $f = \chi_E$ for some Borel subset E of S.

Let E_1, E_2 be two Borel subsets of S, and $E_1 \cap E_2 = \emptyset$. Since $\chi_{E_1} + \chi_{E_2} = \chi_{E_1 \cup E_2}$, it follows that $p_{E_1} + p_{E_2} = p_{E_1 \cup E_2}$. By the positity, we have $p_{E_1 \cup E_2} \geq p_{E_i}, i = 1, 2$. Hence, $p_{E_1} p_{E_2} = (p_{E_1 \cup E_2} - p_{E_2}) p_{E_2} = 0$.

Now for any Borel subsets E_1, E_2 of S, by $p_{E_i} = p_{E_1 \cap E_2} + p_{E_i \setminus E_1 \cap E_2}, i = 1, 2$, we have $p_{E_1} p_{E_2} = p_{E_1 \cap E_2}$. Hence, for any simple functions f, g on S we obtain that $x_f x_g = x_{fg}$. Further, $f \longrightarrow x_f$ is a $*$ homomorphism from $L^\infty(S,\mu)$ to $\pi_\varphi(A)'$.

If $x_f = 0$ for some real $f \in L^\infty(S,\mu)$, then $x_{f_1} = x_{f_2}$, where $0 \leq f_1, f_2 \in L^\infty(S,\mu), f = f_1 - f_2$ and $f_1 f_2 = 0$. Hence, $x_{f_1^2} = x_{f_1} x_{f_2} = x_{f_1 f_2} = 0$. Consequently,

$$\int f_1^2 \hat{a} d\mu = \langle x_{f_1^2} \pi_\varphi(a) 1_\varphi, 1_\varphi\rangle = 0, \ \forall a \in A.$$

So that $f_1 = 0$. Similarly, $f_2 = 0$. Therefore, $f \longrightarrow x_f$ is a $*$ isomorphism from $L^\infty(S,\mu)$ to $\pi_\varphi(A)'$.

Now let $C = \{x_f | f \in L^\infty(S,\mu)\}$, and $\Gamma x_f = f, \forall f \in L^\infty(S,\mu)$. Since for any $x \in C, a \in A$,

$$\int_S (\Gamma x)(\rho)\hat{a}(\rho) d\mu(\rho) = \langle x\pi_\varphi(a) 1_\varphi, 1_\varphi\rangle,$$

it follows from Lemma 14.2.5 that μ is the C–measure of φ. Q.E.D.

Notes. The construction of C–measure and C–isomorphism was presented by D.Ruelle and S.Sakai. Theorem 14.2.10 is due to D.Ruelle.

References. [138], [140], [148].

14.3. Extremal decomposition and central decomposition

Theorem 14.3.1. (*Extremal decomposition*) Let A be a C^*–algebra with an identity $1, S = S(A)$ be its state space, $\varphi \in S, \{\pi_\varphi, H_\varphi, 1_\varphi\}$ be the cyclic $*$ representation of A generated by φ. If C is a maximal abelian VN subalgebra of $\pi_\varphi(A)'$, then the C–measure μ of φ is pseudoconcentrated on the pure state space $P(A)$ in the sense that $\mu(E) = 0$ for each Baire subset E of S disjoint from $P(A)$. In particular, if A is separable, then μ is concentrated on $P(A)$.

Proof. Let B be the C^*-algebra on H_φ generated by $\pi_\varphi(A)$ and C. Clearly, $B' = C$.

Let $\tilde{\varphi}(\cdot) = \langle \cdot 1_\varphi, 1_\varphi \rangle, \forall \cdot \in B$, and $\{\pi_{\tilde{\varphi}}, H_{\tilde{\varphi}}, 1_{\tilde{\varphi}}\}$ be the cyclic $*$ representation of B generated by $\tilde{\varphi}$. Clearly, there is a unitary operator U from H_φ onto $H_{\tilde{\varphi}}$ such that $U1_\varphi = 1_{\tilde{\varphi}}, U^*\pi_{\tilde{\varphi}}(x)U = x, \forall x \in B$. Hence, $\pi_{\tilde{\varphi}}(B)' = (UBU^*)' = UCU^*$ is abelian. By Corollary 14.2.10, the $\pi_{\tilde{\varphi}}(B)'$-measure $\tilde{\mu}$ of $\tilde{\varphi}$ is a maximal (C.M.) probability measure on $\tilde{S}$, where $\tilde{S}$ is the state space of B.

Let S_φ be the state space of $\pi_\varphi(A)$. We have a natural continuous map r from $\tilde{S}$ onto S_φ, i.e., $r(\tilde{\rho}) = \tilde{\rho}|\pi_\varphi(A), \forall \tilde{\rho} \in \tilde{S}$. Then $\mu_\varphi = \tilde{\mu} \circ r^{-1}$ is a probability measure on S_φ.

For each $\rho_\varphi \in S_\varphi, \rho_\varphi \circ \pi_\varphi$ is a state on A. Hence, S_φ can be regarded as a compact convex subset of S. For each Borel subset E of S, define

$$\mu(E) = \mu_\varphi(E \cap S_\varphi).$$

Then μ is a probability measure on S.

1) μ is the C-measure of φ.

In fact, for any $a_1, \cdots, a_n \in A$, by the definitions of μ and μ_φ we have

$$\begin{aligned} & \int_S \hat{a}_1(\rho) \cdots \hat{a}_n(\rho) d\mu(\rho) \\ = & \int_{S_\varphi} \hat{x}_1(\rho_\varphi) \cdots \hat{x}_n(\rho_\varphi) d\mu_\varphi(\rho_\varphi) \\ = & \int_{\tilde{S}} \hat{x}_1(r(\tilde{\rho})) \cdots \hat{x}_n(r(\tilde{\rho})) d\tilde{\mu}(\tilde{\rho}), \end{aligned}$$

where $x_i = \pi_\varphi(a_i), 1 \le i \le n$. Since $r(\tilde{\rho}) = \tilde{\rho}|\pi_\varphi(A)$ and $x_i \in \pi_\varphi(A)$, it follows that $\hat{x}_i(r(\tilde{\rho})) = r(\tilde{\rho})(x_i) = \tilde{\rho}(x_i) = \hat{x}_i(\tilde{\rho}), \forall \tilde{\rho} \in \tilde{S}, 1 \le i \le n$. Hence,

$$\int_S \hat{a}_1(\rho) \cdots \hat{a}_n(\rho) d\mu(\rho) = \int_{\tilde{S}} \hat{x}_1(\tilde{\rho}) \cdots \hat{x}_n(\tilde{\rho}) d\tilde{\mu}(\tilde{\rho}).$$

Let $\tilde{p}$ be the projection from $H_{\tilde{\varphi}}$ onto $\overline{\pi_{\tilde{\varphi}}(B)' 1_{\tilde{\varphi}}}$. Since $\tilde{\mu}$ is the $\pi_{\tilde{\varphi}}(B)'$-measure of $\tilde{\varphi}$, and $x_i \in B, \forall i$, we have

$$\int_{\tilde{S}} \hat{x}_1(\tilde{\rho}) \cdots \hat{x}_n(\tilde{\rho}) d\tilde{\mu}(\tilde{\rho}) = \langle \tilde{p}\pi_{\tilde{\varphi}}(x_1)\tilde{p} \cdots \tilde{p}\pi_{\tilde{\varphi}}(x_n)\tilde{p}1_{\tilde{\varphi}}, 1_{\tilde{\varphi}} \rangle.$$

Noticing that $U1_\varphi = 1_{\tilde{\varphi}}$; $U^*\pi_{\tilde{\varphi}}(x_i)U = x_i, \forall i$; and $U^*\tilde{p}U = p$, where p is the projection from H_φ onto $\overline{C1_\varphi}$, we obtain that

$$\begin{aligned} \int_S \hat{a}_1(\rho) \cdots \hat{a}_n(\rho) d\mu(\rho) &= \langle px_1p \cdots px_np1_\varphi, 1_\varphi \rangle \\ &= \langle p\pi_\varphi(a_1)p \cdots p\pi_\varphi(a_n)p1_\varphi, 1_\varphi \rangle. \end{aligned}$$

Now by Definition 14.2.4, μ is the C-measure of φ.

2) If $\tilde{\rho}$ is a pure state on B, then $r(\tilde{\rho})$ is a pure state on $\pi_\varphi(A)$.

In fact, let $\{\pi, H, \xi\}$ be the irreducible cyclic $*$ representation of B generated by $\tilde{\rho}$. Since C is abelian and $C \subset \pi_\varphi(A)'$, it follows that either $\pi(q) = 1_H$ or $\pi(q) = 0$ for each projection q of C. So there is a non-zero multiplicative linear functional λ on C such that $\pi(c) = \lambda(c)1_H, \forall c \in C$. Then $\{\pi, H\}$ is irreducible for $\pi_\varphi(A)$. Since ξ is cyclic for $\pi_\varphi(A)$, and $r(\tilde{\rho}))(x) = \tilde{\rho}(x) = \langle \pi(x)\xi, \xi\rangle, \forall x \in \pi_\varphi(A)$, hence the $*$ representation $\{\pi, H\}$ of $\pi_\varphi(A)$ is unitarily equivalent to the $*$ representation of $\pi_\varphi(A)$ generated by $r(\tilde{\rho})$. Therefore, $r(\tilde{\rho})$ is pure on $\pi_\varphi(A)$.

3) μ_φ is pseudoconcentrated on ExS_φ.

In face, let E_φ be a Baire subset of S_φ disjoint from ExS_φ. By 2) , $r^{-1}(E_\varphi)$ is a Baire subset of $\tilde{S}$ disjoint from $Ex\tilde{S}$. Since $\tilde{\mu}$ is maximal (C.M.), it follows that $\mu_\varphi(E_\varphi) = \tilde{\mu}(r^{-1}(E_\varphi)) = 0$.

Finally, we prove that μ is pseudoconcentrated on $P(A) = ExS$.

Let E be a Baire subset of S disjoint from ExS. Then $E \cap S_\varphi$ is a Baire subset of S_φ. By Proposition 2.4.11, $ExS_\varphi \subset ExS$. Hence, $(E \cap S_\varphi) \cap ExS_\varphi = \emptyset$. Now by 3) we obtain that

$$\mu(E) = \mu_\varphi(E \cap S_\varphi) = 0.$$

Q.E.D.

Theorem 14.3.2. (*Central decomposition*) Let A be a C^*-algebra with an identity 1, $S = S(A)$ be its state space, $\varphi \in S, \{\pi_\varphi, H_\varphi, 1_\varphi\}$ be the cyclic $*$ representation of A generated by φ, and $Z = \pi_\varphi(A)'' \cap \pi_\varphi(A)'$. Then the Z-measure μ of φ is pseudoconcentrated on the factorial state space $\mathcal{F}$ in the sense that $\mu(E) = 0$ for any Baire subset E of S disjoint from $\mathcal{F}$. In particular, if A is separable, then μ is concentrated on $\mathcal{F}$.

Proof. Let B be the C^*-algebra on H_φ generated by $\pi_\varphi(A)$ and $\pi_\varphi(A)'$. Then $B' = Z$.

Consider the state $\tilde{\varphi}(\cdot) = \langle \cdot 1_\varphi, 1_\varphi\rangle$ on B. Similar to the proof of Theorem 14.3.1, the $\pi_{\tilde{\varphi}}(B)'$-measure $\tilde{\mu}$ of $\tilde{\varphi}$ is maximal (C.M.) on $\tilde{S}$, where $\tilde{S}$ is the state space of B.

Let S_φ be the state space of $\pi_\varphi(A)$, and $r(\tilde{\rho}) = \tilde{\rho}|\pi_\varphi(A), \forall \tilde{\rho} \in \tilde{S}$. Then $\mu_\varphi = \tilde{\mu} \circ r^{-1}$ is a probability measure on S_φ. S_φ can be regarded as a compact convex subset of S, and $\mu(E) = \mu_\varphi(E \cap S_\varphi)(\forall$ Borel $E \subset S)$ is a probability measure on S. Similar to the proof of Theorem 14.3.1, μ is the Z-measure of φ.

Let $\tilde{\rho}$ be a pure state on $B, \{\pi, H, \xi\}$ be the irreduable cyclic $*$ representation of B generated by $\tilde{\rho}$, and $\alpha = \pi_\varphi(A)$. Then

$$\begin{aligned}\pi(\alpha)'' \cap \pi(\alpha)' &= \{\pi(\alpha) \cup \pi(\alpha)'\}' \\ &\subset \{\pi(\alpha) \cup \pi(\alpha')\}' = \mathbb{C}1_H.\end{aligned}$$

So $\pi(\alpha)''$ is a factor on H. Let q be the projection from H onto $\overline{\pi(\alpha)\xi}$. Clearly, $q \in \pi(\alpha)', \pi(\alpha)''q$ is a factor on qH, and $\{q\pi, qH, \xi\}$ is a factorial cyclic $*$ representation of α. Since $\langle q\pi(x)\xi, \xi\rangle = \langle \pi(x)\xi, \xi\rangle = \tilde{\rho}(x) = r(\tilde{\rho})(x), \forall x \in \alpha$, the $*$ representation $\{q\pi, qH\}$ of α is unitarily equivalent to the $*$ representation of α generated by $r(\tilde{\rho})$. Therefore, $r(\tilde{\rho})$ is a factorial state on $\pi_\varphi(A)$.

Let $\mathcal{F}_\varphi$ be the factorial state space on $\pi_\varphi(A)$, and E_φ be a Baire subset of S_φ disjoint from $\mathcal{F}_\varphi$. From the preceding paragraph, $r^{-1}(E_\varphi) \cap Ex\tilde{S} = \emptyset$. Since $\tilde{\mu}$ is maximal (C.M.) , it follows that $\mu_\varphi(E_\varphi) = \tilde{\mu}(r^{-1}(E_\varphi)) = 0$.

Finally, let E be a Baire subset of S disjoint from $\mathcal{F}$. By $\mathcal{F} \cap S_\varphi = \mathcal{F}_\varphi, E \cap S_\varphi$ is a Baire subset of S_φ disjoint from $\mathcal{F}_\varphi$. Therefore, $\mu(E) = \mu_\varphi(E \cap S_\varphi) = 0$, i.e., μ is pseudoconcentrated on $\mathcal{F}$. Q.E.D.

The $Z(= \pi_\varphi(A)'' \cap \pi_\varphi(A)')$–measure is also called the *central measure* of φ. Now we give an equivalent definition of central measure.

Definition 14.3.3. $\mu \in \Omega(\varphi)$ (see Definition 14.2.1)) is called the *central measure* of φ, if there is a normal $*$ homomorphism Ψ from the center Z^{**} of A^{**} onto $L^\infty(S, \mu)$ such that

$$\varphi(za) = \int_S \Psi(z)(\rho)\hat{a}(\rho)d\mu(\rho),$$

$\forall z \in Z^{**}, a \in A$ (notice that φ can be uniquely extended to a normal state on A^{**}).

The following proposition will show that Ψ is unique, and this Ψ is called the *central homomorphism* of φ.

Proposition 14.3.4. The μ and Ψ in Definition 14.3.3 are existential and unique, and μ is the Z–measure of φ exactly, where $Z = \pi_\varphi(A)'' \cap \pi_\varphi(A)'$.

Proof. Let μ be the Z–measure of φ, and Γ be the Z–isomorphism from Z onto $L^\infty(S, \mu)$, where $Z = \pi_\varphi(A)'' \cap \pi_\varphi(A)'$. Then we have

$$\langle x\pi_\varphi(a)1_\varphi, 1_\varphi\rangle = \int_S (\Gamma x)(\rho)\hat{a}(\rho)d\mu(\rho),$$

$\forall a \in A, x \in Z$ (see Theorem 14.2.7). By Theorem 4.2.7, the $*$ representation $\{\pi_\varphi, H_\varphi\}$ of A can be uniquely extended to a W^*–representation $\{\pi_\varphi^w, H_\varphi\}$ of A^{**}, and $\pi_\varphi^w(A^{**}) = \pi_\varphi(A)''$. For each $z \in Z^{**}$, define

$$\Psi(z) = \Gamma(\pi_\varphi^w(z)).$$

Clearly, Ψ is a normal $*$ homomorphism from Z^{**} to $L^\infty(S,\mu)$. Let z be the central projection of A^{**} such that $A^{**}z = \ker \pi_\varphi^w$. Then π_φ^w is a $*$ isomorphism from $A^{**}(1-z)$ onto $\pi_\varphi^w(A^{**}) = \pi_\varphi(A)''$. Then center of $A^{**}(1-z)$ is $Z^{**}(1-z)$(see Proposition 1.3.8.). Hence, $Z = \pi_\varphi^w(Z^{**}(1-z)) = \pi_\varphi^w(Z^{**})$, and Ψ is surjective. Moreover, for $z \in Z^{**}, a \in A$, let $x = \pi_\varphi^w(z)$. Then $x \in Z$ and

$$\begin{aligned}\varphi(za) &= \langle \pi_\varphi^w(za)1_\varphi, 1_\varphi\rangle = \langle x\pi_\varphi(a)1_\varphi, 1_\varphi\rangle \\ &= \int_S (\Gamma x)(\rho)\hat{a}(\rho)d\mu(\rho) = \int_S \Psi(z)(\rho)\hat{a}(\rho)d\mu(\rho).\end{aligned}$$

Now suppose that μ' and Ψ' satisfy the conditions in Definition 14.3.3. We need to show that μ' is equal to the Z–measure μ of φ, and Ψ' is equal to the above Ψ.

Let $z \in (Z^{**} \cap \ker \pi_\varphi^w)_+$. Then by

$$0 = \langle \pi_\varphi^w(z)1_\varphi, 1_\varphi\rangle = \varphi(z) = \int_S \Psi'(z)(\rho)d\mu'(\rho)$$

and $\Psi'(z) \geq 0$, we have $\Psi'(z) = 0$. So we can define a $*$ homomorphism Γ' from Z onto $L^\infty(S,\mu')$ as follows:

$$\Gamma'(\pi_\varphi^w(z)) = \Psi'(z), \quad \forall z \in Z^{**}.$$

If $\Psi'(z) = 0$ for some $z \in Z^{**}$, then by Definition 14.3.3 we have

$$\langle \pi_\varphi^w(z)\pi_\varphi(a)1_\varphi, 1_\varphi\rangle = \varphi(za) = \int \Psi'(z)(\rho)\hat{a}(\rho)d\mu'(\rho) = 0,$$

$\forall a \in A$. Since 1_φ is cyclic for $\pi_\varphi(A)$, it follows that $\pi_\varphi^w(z) = 0$. Hence Γ' is a $*$ isomorphism. Moreover, by $\langle x\pi_\varphi(a)1_\varphi, 1_\varphi\rangle = \langle \pi_\varphi^w(z)\pi_\varphi(a)1_\varphi, 1_\varphi\rangle = \varphi(za) = \int \Psi'(z)(\rho)\hat{a}(\rho)d\mu'(\rho) = \int (\Gamma'x)(\rho)\hat{a}(\rho)d\mu'(\rho), \forall x \in Z, a \in A$, where $z \in Z^{**}$ and $\pi_\varphi^w(z) = x$, and by Lemma 14.2.5, μ' is the Z– measure μ of φ and Γ' is the Z–isomorphism Γ of φ. Further, $\Psi' = \Psi$. Q.E.D.

Now we consdier a geometrical characterization of the central measure in $\Omega(\varphi)$.

Definition 14.3.5. Let A be a C^*–algebra with an identity 1 , and $S = S(A)$ be the state space of A.

Let π be a W^* -representation of A^{**}. The *support* of π, denoted by $s(\pi)$, is the central projection of A^{**} such that $\ker \pi = A^{**}(1 - s(\pi))$.

Let φ_1, φ_2 be two positive linear functionals on A. φ_1 and φ_2 are said to be *disjoint*, if $s(\pi_1^w) \cdot s(\pi_2^w) = 0$, where π_i^w is the W^*–representation of A^{**} generated by $\varphi_i, i = 1, 2$.

Now let ν be a probability measure on S. For each Borel subset E of S, we can define a positive functional ν_E on A as follows:

$$\nu_E(a) = \int_E \hat{a}(\rho) d\nu(\rho), \quad \forall a \in A.$$

ν is said to be *semi-central* , if for any Borel subset E of S, ν_E and $\nu_{S\setminus E}$ are disjoint.

Fix $\varphi \in S$. Let

$$\Omega_c(\rho) = \{\nu \in \Omega(\varphi) | \nu \text{ is semi-central } \},$$

where $\Omega(\varphi)$ is the same as in Definition 14.2.1.

Theorem 14.3.6. Let $Z = \pi_\varphi(A)'' \cap \pi_\varphi(A)'$. Then the Z-measure (central measure) of φ is semi-central, and there is a bijection between $\Omega_c(\varphi)$ and the collection of all abelian VN subalgebras of Z, i.e., for each $\nu \in \Omega_c(\varphi)$ there is (unique) abelian VN subalgebra C of Z such that ν is the C-measure of φ; conversely, if C is an abelian VN subalgebra of Z, then C-measure of φ is semi-central.

Proof. Let C be a VN subalgebra of Z, ν and Γ be the C-measure and C-isomorphism of φ respectively, and E be a Borel subset of S.

Let $p_1 = \Gamma^{-1}\chi_E, p_2 = \Gamma^{-1}\chi_{S\setminus E}$. Then p_1, p_2 are projections of $C, p_1p_2 = 0, p_1 + p_2 = 1$, and by Theorem 14.2.6,

$$\begin{aligned}\nu_E(a) &= \int \chi_E(\rho)\hat{a}(\rho)d\nu(\rho) = \int (\Gamma p_1)(\rho)\hat{a}(\rho)d\nu(\rho) \\ &= \langle p_1\pi_\varphi(a)1_\varphi, 1_\varphi\rangle, \\ \nu_{S\setminus E}(a) &= \langle p_2\pi_\varphi(a)1_\varphi, 1_\varphi\rangle, \quad \forall a \in A.\end{aligned}$$

Since $\pi_\varphi^w(Z^{**}) = Z \supset C$, so we can find projections z_1, z_2 of Z^{**} such that $z_1z_2 = 0, z_1 + z_2 = 1$, and $\pi_\varphi^w(z_i) = p_i, i = 1, 2$. Then

$$\nu_E(a) = \varphi(z_1a), \quad \nu_{S\setminus E}(a) = \varphi(z_2a), \forall a \in A.$$

Hence, $\nu_E(z_2) = \nu_{S\setminus E}(z_1) = 0$. Let $\pi_E, \pi_{S\setminus E}$ be the W^*-representations of A^{**} generated by $\nu_E, \nu_{S\setminus E}$ respectively. Since $z_1, z_2 \in Z^{**}$, it follows that $z_1 \in \ker\ \pi_E, z_2 \in \ker\ \pi_{S\setminus E}$. Further, $z_1 \leq (1 - s(\pi_E)), z_2 \leq (1 - s(\pi_{s\setminus E}))$, and $s(\pi_E) \cdot s(\pi_{S\setminus E}) = 0$. Therefore, ν is semi-central.

Now let $\nu \in \Omega_c(\varphi)$. By the proof of Proposition 14.2.12, there is a bounded positive linear map $f \longrightarrow x_f$ from $L^\infty(S, \nu)$ to $\pi_\varphi(A)'$ such that

$$\int f\hat{a}d\nu = \langle x_f\pi_\varphi(a)1_\varphi, 1_\varphi\rangle, \quad \forall f \in L^\infty(S, \nu), a \in A.$$

Let E be a Borel subset of S, and $f = \chi_E$. By Proposition 14.2.12, it suffices to show that x_f is a projection of Z. Let $g = \chi_{S\backslash E}$. Then

$$\int_E \hat{a} d\nu = \nu_f(a) = \langle x_f \pi_\varphi(a) 1_\varphi, 1_\varphi \rangle,$$
$$\int_{S\backslash E} \hat{a} d\nu = \nu_g(a) = \langle x_g \pi_\varphi(a) 1_\varphi, 1_\varphi \rangle, \quad \forall a \in A.$$

Further, let $\{\pi_f, H_f, 1_f\}$ and $\{\pi_g, H_g, 1_g\}$ be the cyclic W^*–representations of A^{**} generated by ν_f and ν_g respectively. Since $\nu \in \Omega_c(\varphi)$, it follows that $s(\pi_f) \cdot s(\pi_g) = 0$. Write $z = s(\pi_f), z' = s(\pi_g)$. Then z, z' is the central projections of A^{**}, and $z + z' \leq 1$.

Since $\pi_f(z) = I_f$ (the identity operator on H_f) and $0 \leq x_f \leq 1$, for any $a \in A_+$ we have

$$\begin{aligned} \langle \pi_\varphi^w(za) 1_\varphi, 1_\varphi \rangle &\geq \langle \pi_\varphi^w(za) x_f 1_\varphi, 1_\varphi \rangle \\ &= \nu_f(za) = \langle \pi_f(za) 1_f, 1_f \rangle \\ &= \langle \pi_f(a) 1_f, 1_f \rangle = \nu_f(a) = \langle x_f \pi_\varphi(a) 1_\varphi, 1_\varphi \rangle. \end{aligned}$$

Hence , $\langle (\pi_\varphi^w(z) - x_f) \pi_\varphi(a) 1_\varphi, 1_\varphi \rangle \geq 0, \forall a \in A_+$. Further, $\pi_\varphi^w(z) \geq x_f$. Similarly, $\pi_\varphi^w(z') \geq x_g$. Since $z + z' \leq 1$, and $x_f + x_g = x_{f+g} = 1$, it follows that

$$1 - \pi_\varphi^w(z) \geq \pi_\varphi^w(z') \geq x_g = 1 - x_f, \quad \text{and} \quad x_f \geq \pi_\varphi^w(z).$$

Hence, $x_f = \pi_\varphi^w(z)$ is a projection of Z since $\pi_\varphi^w(Z^{**}) = Z$. Q.E.D.

Corollary 14.3.7. Let $\mu \in \Omega(\varphi)$. Then μ is the central measure of φ if and only if $\mu \in \Omega_c(\varphi)$ and μ is the largest (C.M.) measure of $\Omega_c(\varphi)$.

Proof. It is immediate from Theorems 14.2.10 and 14.3.6. Q.E.D.

Notes. The formulation of the central decomposition of a state was first given by S.Sakai. Theorem 14.3.2 in non separable cases is due to W.Wils.

References. [148], [197].

14.4 Ergodic decomposition and tracial decomposition

Let (A, G, α) be a dynamical system, where A is a C^*–algebra with an identity 1 , G is a group, and α is a homomorphism from G to the $*$ automorphism group Aut(A) of A. Further, let $S = S(A)$ be the state space of A, and

$$S_G = \{\varphi \in S | \varphi(\alpha_s(a)) = \varphi(a), \forall s \in G, a \in A\},$$

i.e., S_G is the set of all *G–invariant states*. Clearly, S_G is a closed (compact) convex subset of $(S, \sigma(A^*, A))$.

Fix $\varphi \in S_G$, let $\{\pi_\varphi, H_\varphi, 1_\varphi\}$ be the cyclic $*$ representation of A generated by φ , and define

$$u_\varphi(s)\pi_\varphi(a)1_\varphi = \pi_\varphi(\alpha_{s^{-1}}(a))1_\varphi,$$

$\forall a \in A, s \in G$. Then $u_\varphi(s)$ can be uniquely extended to a unitary operator on H_φ, denoted by $u_\varphi(s)$ still, $\forall s \in G$. Clearly, $s \longrightarrow u_\varphi(s)$ is a unitary representation of G on H_φ, and $\{\pi_\varphi, u_\varphi, H_\varphi\}$ is a covariant representation of (A, G, α), i.e.,

$$u_\varphi(s)\pi_\varphi(a)u_\varphi(s)^* = \pi_\varphi(\alpha_{s^{-1}}(a)),$$

$\forall s \in G, a \in A$. Moreover, $u_\varphi(s)1_\varphi = 1_\varphi, \forall s \in G$.

Let $E_\varphi = \{\xi \in H_\varphi | u_\varphi(s)\xi = \xi, \forall s \in G\}$, and p_φ be the projection from H_φ onto E_φ. Clearly, E_φ is a closed linear subspace of H_φ, and $1_\varphi \in E_\varphi$.

Proposition 14.4.1. With the above notations, we have that:

(i) $p_\varphi u_\varphi(s) = u_\varphi(s)p_\varphi = p_\varphi, \forall s \in G$, and $p_\varphi \in \overline{Co u_\varphi(G)}^s$, where " $-\,s$" means the strong closure;

(ii) $\{\pi_\varphi(A), p_\varphi\}' = \{\pi_\varphi(A), u_\varphi(G)\}'$;

(iii) Let $M_\varphi = \{\pi_\varphi(A), p_\varphi\}'' = \{\pi_\varphi(A), u_\varphi(G)\}''$. Then the central cover of p_φ in M_φ is 1 , and $x' \longrightarrow x'p_\varphi$ is a $*$ isomorphism from M'_φ onto $M'_\varphi p_\varphi$;

(iv) Let $N_\varphi = (p_\varphi\pi_\varphi(A)p_\varphi)''$ (a VN algebra on $p_\varphi H_\varphi = E_\varphi$). Then $N'_\varphi = (p_\varphi\pi_\varphi(A)p_\varphi)' = M'_\varphi p_\varphi$.

Proof. (i) We need to show that for any $\eta_1, \cdots, \eta_n \in H_\varphi$ and $\varepsilon > 0, U(p_\varphi, \eta_1, \cdots, \eta_n, \varepsilon) \cap Co u_\varphi(G) \neq \emptyset$, where $U(p_\varphi, \eta_1, \cdots, \eta_n, \varepsilon) = \{x \in B(H_\varphi) | \|(p_\varphi - x)\eta_i\| < \varepsilon, 1 \le i \le n\}$ is a strong neighborhood of p_φ in $B(H_\varphi)$.

Let $H_\varphi = E_\varphi \oplus E_\varphi^\perp$. Since $u_\varphi(s)E_\varphi \subset E_\varphi$, it follows that $u_\varphi(s)E_\varphi^\perp \subset E_\varphi^\perp, \forall s \in G$. For any $\eta \in E_\varphi^\perp$, let $\Gamma_\eta = \overline{Co\{u_\varphi(s)\eta | s \in G\}}$. Γ_η is a closed convex subset of H_φ. Hence, there is unique $\eta_0 \in \Gamma_\eta$ such that $\|\eta_0\| = \min\{\|\xi\| | \xi \in \Gamma_\eta\}$. Since $u_\varphi(s)\Gamma_\eta \subset \Gamma_\eta$ and $\|u_\varphi(s)\eta_0\| = \|\eta_0\|$, it follows from the uniqueness of η_0 that $u_\varphi(s)\eta_0 = \eta_0, \forall s \in G$. Hence, $\eta_0 \in E_\varphi \cap E_\varphi^\perp = \{0\}$. From this fact, for η_1 we can find $\lambda_j^{(1)} > 0, s_j^{(1)} \in G, \forall j$, and $\sum_j \lambda_j^{(1)} = 1$ such that

$$\|\sum_j \lambda_j^{(1)} u_\varphi(s_j^{(1)})(1 - p_\varphi)\eta_1\| < \varepsilon.$$

For η_2 , similarly there are $\lambda_k^{(2)} > 0, s_k^{(2)} \in G, \forall k$, and $\sum_k \lambda_k^{(2)} = 1$ such that

$$\|\sum_k \lambda_k^{(2)} u_\varphi(s_k^{(2)}) \sum_j \lambda_j^{(1)} u_\varphi(s_j^{(1)})(1 - p_\varphi)\eta_2\| < \varepsilon.$$

And we also have

$$\|\sum_k \lambda_k^{(2)} u_\varphi(s_k^{(2)}) \sum_j \lambda_j^{(1)} u_\varphi(s_j^{(1)})(1-p_\varphi)\eta_1\| < \varepsilon.$$

$\cdots$. Generally, we can find $x \in Cou_\varphi(G)$ such that

$$\|x(1-p_\varphi)\eta_i\| < \varepsilon, \quad 1 \le i \le n.$$

By $u_\varphi(s)p_\varphi = p_\varphi, \forall s \in G$, it must be $xp_\varphi = p_\varphi$. Therefore,

$$\|(x-p_\varphi)\eta_i\| = \|x(1-p_\varphi)\eta_i\| < \varepsilon, \quad 1 \le i \le n,$$

i.e., $U(p_\varphi, \eta_1, \cdots, \eta_n, \varepsilon) \cap Cou_\varphi(G) \neq \emptyset$.

(ii) Since $p_\varphi \in u_\varphi(G)''$, it follows that $\{\pi_\varphi(A), u_\varphi(G)\}' \subset \{\pi_\varphi(A), p_\varphi\}'$. Conversely, let $x \in \{\pi_\varphi(A), p_\varphi\}'$. Then for any $s \in G, a \in A$ we have

$$\begin{aligned} xu_\varphi(s)\pi_\varphi(a)1_\varphi &= x\pi_\varphi(\alpha_{s^{-1}}(a))1_\varphi = \pi_\varphi(\alpha_{s^{-1}}(a))x1_\varphi \\ &= u_\varphi(s)\pi_\varphi(a)u_s^* xp_\varphi 1_\varphi = u_\varphi(s)\pi_\varphi(a)u_\varphi(s)^* p_\varphi x1_\varphi \\ &= u_\varphi(s)\pi_\varphi(a)p_\varphi x1_\varphi = u_\varphi(s)x\pi_\varphi(a)1_\varphi. \end{aligned}$$

Hence, $u_\varphi(s)x = xu_\varphi(s), \forall s \in G$, and $x \in \{\pi_\varphi(A), u_\varphi(G)\}'$.

(iii) By $\overline{[M_\varphi p_\varphi H_\varphi]} \supset \pi_\varphi(A)1_\varphi = H_\varphi$, the central cover of p_φ in M_φ is 1 .

(iv) $M'_\varphi p_\varphi = (p_\varphi M_\varphi p_\varphi)' = (p_\varphi \pi_\varphi(A) p_\varphi)' = N'_\varphi$. Q.E.D.

Theorem 14.4.2. $p_\varphi\pi_\varphi(A)p_\varphi$ is commutative if and only if for any $a_1, a_2 \in A, \xi \in E_\varphi, \varepsilon > 0$, there are $\lambda_i > 0, s_i \in G$, and $\sum_i \lambda_i = 1$ such that

$$|\langle \xi, [\sum_i \lambda_i \pi_\varphi(\alpha_{s_i}(a_1)), \pi_\varphi(a_2)]\xi\rangle| < \varepsilon,$$

where $[x, y] = xy - yx, \forall x, y \in B(H_\varphi)$.

Proof. For any $a_1, a_2 \in A$ with $\|a_1\| \le 1, \|a_2\| \le 1$, and $\xi_1, \xi_2 \in p_\varphi H_\varphi$ with $\|\xi_1\| \le 1, \|\xi_2\| \le 1$, and $\delta > 0$, by Proposition 14.4.1 (i) we can pick $\lambda_i > 0, s_i \in G$, and $\sum_i \lambda_i = 1$ such that

$$\|(\sum_i \lambda_i u_\varphi(s_i) - p_\varphi)\pi_\varphi(a_2)\xi_2\| < \delta/2$$

and

$$\|(\sum_i \lambda_i u_\varphi(s_i) - p_\varphi)\pi_\varphi(a_2^*)\xi_1\| < \delta/2.$$

Then by $u_\varphi(s)\xi_i = u_\varphi(s)p_\varphi\xi_i = \xi_i, i = 1, 2, \forall s \in G$, we have

$$\begin{aligned}
&|\langle\xi_1, (\pi_\varphi(a_1)p_\varphi\pi_\varphi(a_2) - \pi_\varphi(a_2)p_\varphi\pi_\varphi(A_1))\xi_2\rangle \\
&-\langle\xi_1, [\sum_i \lambda_i\pi_\varphi(\alpha_{s_i}(a_1)), \pi_\varphi(a_2)]\xi_2\rangle| \\
\le\ & |\langle\xi_1, \pi_\varphi(a_1)p_\varphi\pi_\varphi(a_2)\xi_2\rangle - \langle\xi_1, \sum_i \lambda_i u_\varphi(s_i)^*\pi_\varphi(a_1)u_\varphi(s_i)\pi_\varphi(a_2)\xi_2\rangle| \\
&+|\langle\xi_1, \pi_\varphi(a_2)p_\varphi\pi_\varphi(a_1)\xi_2\rangle - \langle\xi_1, \sum_i \lambda_i\pi_\varphi(a_2)u_\varphi(s_i)^*\pi_\varphi(a_1)u_\varphi(s_i)\xi_2\rangle| \\
\le\ & \|(p_\varphi - \sum_i \lambda_i u_\varphi(s_i))\pi_\varphi(a_2)\xi_2\| \\
&+|\langle p_\varphi\pi_\varphi(a_2^*)\xi_1, \pi_\varphi(a_1)\xi_2\rangle - \langle\sum_i \lambda_i u_\varphi(s_i)\pi_\varphi(a_2^*)\xi_1, \pi_\varphi(a_1)\xi_2\rangle| \\
\le\ & \|(p_\varphi - \sum_i \lambda_i u_\varphi(s_i))\pi_\varphi(a_2)\xi_2\| + \|(p_\varphi - \sum_i \lambda_i u_\varphi(s_i))\pi_\varphi(a_2^*)\xi_1\| < \delta.
\end{aligned}$$

Let $p_\varphi\pi_\varphi(A)p_\varphi$ be commutative. For any $\xi \in E_\varphi, a_1, a_2 \in A$ and $\varepsilon > 0$, put $\xi_1 = \xi_2 = \xi$, and $\delta = \frac{\varepsilon}{\|a_1\|\|a_2\|\|\xi\|^2}$. Then by the above inequality there are $\lambda_i > 0, s_i \in G$ and $\sum_i \lambda_i = 1$ such that

$$|\langle\xi, [\sum_i \lambda_i\pi_\varphi(\alpha_{s_i}(a_1)), \pi_\varphi(a_2)]\xi\rangle| < \varepsilon$$

since $\langle\xi, (\pi_\varphi(a_1)p_\varphi\pi_\varphi(a_2) - \pi_\varphi(a_2)p_\varphi\pi_\varphi(a_1))\xi\rangle = 0$.

Conversely, let $a_1, a_2 \in A, \xi \in E_\varphi$ and $\varepsilon > 0$. By Proposition 14.4.1 (i) , pick $\mu_j > 0, t_j \in G$ and $\sum_j \mu_j = 1$ such that

$$\|(\sum_j \mu_j u_\varphi(t_j) - p_\varphi)\pi_\varphi(a_2)\xi\| = \|\sum \mu_j u_\varphi(t_j)(1 - p_\varphi)\pi_\varphi(a_2)\xi\| < \varepsilon/2$$

and

$$\|(\sum_j \mu_j u_\varphi(t_j) - p_\varphi)\pi_\varphi(a_2^*)\xi\| = \|\sum_j \mu_j u_\varphi(t_j)(1 - p_\varphi)\pi_\varphi(a_2^*)\xi\| < \varepsilon/2.$$

By the assumption, for $\sum_j \mu_j\alpha_{t_j}(a_1), a_2, \xi$ and $\varepsilon > 0$, there are $\lambda_i > 0, s_i \in G$ and $\sum_i \lambda_i = 1$ such that

$$|\langle\xi, [\sum_i \lambda_i\pi_\varphi(\sum_j \mu_j\alpha_{s_it_j}(a_1)), \pi_\varphi(a_2)]\xi\rangle| < \varepsilon.$$

Since

$$\begin{aligned}
&\|(\sum_{ij} \lambda_i\mu_j u_\varphi(s_it_j) - p_\varphi)x\xi\| = \|\sum_{ij} \lambda_i\mu_j u_\varphi(s_it_j)(1 - p_\varphi)x\xi\| \\
=\ & \|\sum_i \lambda_i u_\varphi(s_i)\sum_j \mu_j u_\varphi(t_j)(1 - p_\varphi)x\xi\| < \varepsilon/2,
\end{aligned}$$

where $x = \pi_\varphi(a_2)$ or $\pi_\varphi(a_2^*)$, it follows that

$$\begin{aligned}
&|\langle\xi,(\pi_\varphi(a_1)p_\varphi\pi_\varphi(a_2)-\pi_\varphi(a_2)p_\varphi\pi_\varphi(a_1))\xi\rangle \\
&-\langle\xi,[\sum_{ij}\lambda_i\mu_j\pi_\varphi(\alpha_{s_it_j}(a_1)),\pi_\varphi(a_2)]\xi\rangle| \\
\le\ &|\langle\xi,\pi_\varphi(a_1)p_\varphi\pi_\varphi(a_2)\xi\rangle-\langle\xi,\sum_{ij}\lambda_i\mu_j u_\varphi(s_it_j)^*\pi_\varphi(a_1)u_\varphi(s_it_j)\pi_\varphi(a_2)\xi\rangle| \\
&+|\langle\xi,\pi_\varphi(a_2)p_\varphi\pi_\varphi(a_1)\xi\rangle-\langle\xi,\sum_{ij}\lambda_i\mu_j\pi_\varphi(a_2)u_\varphi(s_it_j)^*\pi_\varphi(a_1)u_\varphi(s_it_j)\xi\rangle| \\
\le\ &\|\xi\|\|a_1\|\|(p_\varphi-\sum_{ij}\lambda_i\mu_j u_\varphi(s_it_j))\pi_\varphi(a_2)\xi\| \\
&+\|\xi\|\|a_1\|\|(p_\varphi-\sum_{ij}\lambda_i\mu_j u_\varphi(s_it_j))\pi_\varphi(a_2^*)\xi\|<\varepsilon\|\xi\|\|a_1\|.
\end{aligned}$$

Hence

$$\begin{aligned}
&|\langle\xi,(\pi_\varphi(a_1)p_\varphi\pi_\varphi(a_2)-\pi_\varphi(a_2)p_\varphi\pi_\varphi(a_1))\xi\rangle| \\
\le\ &\varepsilon+\varepsilon\|\xi\|\|a_1\|.
\end{aligned}$$

Since $a_1, a_2 \in A, \varepsilon > 0$ and $\xi \in E_\varphi$ are arbitrary, $p_\varphi\pi_\varphi(A)p_\varphi$ is commutative.

Q.E.D.

Corollary 14.4.3. If for any $a_1, a_2 \in A$, $\xi \in E_\varphi$, we have

$$\inf_{s\in G}|\langle\xi,\pi_\varphi([\alpha_s(a_1),a_2])\xi\rangle| = 0,$$

then $p_\varphi\pi_\varphi(A)p_\varphi$ is commutative.

Theorem 14.4.4. Let $p_\varphi\pi_\varphi(A)p_\varphi$ be commutative, and $C = \{\pi_\varphi(A), p_\varphi\}'$. Then C is abelian, $C \subset \pi_\varphi(A)'$ and $E_\varphi = \overline{C1_\varphi}$. Suppose that μ is the C-measure of φ. Then supp $\mu \subset S_G$, and $\nu \prec \mu$ (C.M.) for each probability ν on S_G with $\int_{S_G}\hat{a}(\rho)d\nu(\rho) = \varphi(a), \forall a \in A$. Consequently, μ is pseudoconcentrated on ExS_G in the sense that $\mu(E) = 0$ for each Baire subset E of S_G disjoint from ExS_G. Moreover, if A is separable, then $\mu(ExS_G) = 1$.

Proof. By the 2) of Proposition 14.2.2, C is abelian, $C \subset \pi_\varphi(A)'$ and $p_\varphi H_\varphi = E_\varphi = \overline{C1_\varphi}$.

From Proposition 14.2.8, $\mu = \sigma$ - $\lim \mu_{\{B_j\}}$, where $B_j \in C_+, \forall j, \sum_j B_j = 1; \mu_{\{B_j\}} = \sum_j \lambda_j\delta_{\rho_j} \prec \mu$ (C.M.), for each $j, \lambda_j = \langle B_j1_\varphi, 1_\varphi\rangle, \rho_j \in S$ such that $\lambda_j\rho_j(a) = \langle\pi_\varphi(a)B_j1_\varphi, 1_\varphi\rangle, \forall a \in A$. If $\lambda_j > 0$, then for each $s \in G, a \in A$ we have

$$\begin{aligned}
\rho_j(\alpha_s(a)) &= \lambda_j^{-1}\langle\pi_\varphi(a)u_\varphi(s)B_jp_\varphi1_\varphi, u_\varphi(s)p_\varphi1_\varphi\rangle \\
&= \lambda_j^{-1}\langle\pi_\varphi(a)u_\varphi(s)p_\varphi B_j1_\varphi, u_\varphi(s)p_\varphi1_\varphi\rangle = \rho_j(a).
\end{aligned}$$

Hence, $\rho_j \in S_G, \forall j$, and supp $\mu \subset S_G$.

Now let ν be any probability measure on S_G with $\int_{S_G} \hat{a}(\rho)d\nu(\rho) = \varphi(a), \forall a \in A$. By Lemma 14.2.7, $\nu = \sigma$ - $\lim \mu_{\{h_j\}}$, where $0 \le h_j \in L^\infty(S_G, \nu), \forall j, \sum_j h_j = 1(a.e.\nu); \nu_{\{h_j\}} = \sum_j \lambda_j \delta_{\rho_j}$, for each j, $\lambda_j = \int_{S_G} h_j d\nu, \rho_j \in S_G$ such that $\lambda_j \rho_j(a) = \int_{S_G} \hat{a}(\rho)h_j(\rho)d\nu(\rho), \forall a \in A$. Now it suffices to show that $\nu_{\{h_j\}} \prec \nu$ (C.M.) on S_G.

Since $\sum_j \lambda_j \rho_j(a) = \int \hat{a}(\rho)d\nu(\rho) = \varphi(a), \forall a \in A$, $\varphi \ge \lambda_j\rho_j(\ge 0)$ on $A, \forall j$. Then for each j, there is $B_j \in \pi_\varphi(A)'_+$ such that

$$\lambda_j \rho_j(a) = \langle \pi_\varphi(a)B_j 1_\varphi, 1_\varphi \rangle, \quad \forall a \in A.$$

By $\rho_j \in S_G$ and $\overline{\pi_\varphi(A)1_\varphi} = H_\varphi$ we can see that

$$B_j 1_\varphi = u_\varphi(s)B_j 1_\varphi, \quad \forall s \in G, \text{and } j.$$

On the other hand,

$$\begin{aligned} \pi_\varphi(a)u_\varphi(s)B_j u_\varphi(s)^* &= u_\varphi(s)\pi_\varphi(\alpha_s(a))B_j u_\varphi(s)^* \\ &= u_\varphi(s)B_j\pi_\varphi(\alpha_s(a))u_\varphi(s)^* \\ &= u_\varphi(s)B_j u_\varphi(s)^*\pi_\varphi(a), \quad \forall a \in A. \end{aligned}$$

Hence, $u_\varphi(s)B_j u_\varphi(s)^* \in \pi_\varphi(A)'$. Further

$$u_\varphi(s)B_j u_\varphi(s)^* 1_\varphi = u_\varphi(s)B_j 1_\varphi = B_j 1_\varphi.$$

But 1_φ is separating for $\pi_\varphi(A)'$, so we get $u_\varphi(s)B_j u_\varphi(s)^* = B_j, \forall s \in G$, i.e., $B_j \in u_\varphi(G)'$. Further, by Proposition 14.4.1, $B_j \in \pi_\varphi(A)' \cap u_\varphi(G)' = \{\pi_\varphi(A), u_\varphi(G)\}' = C, \forall j$. Clearly, $B_j \ge 0, \sum_j B_j = 1$, and $\nu_{\{h_j\}} = \mu_{\{B_j\}}$. Therefore, $\mu \succ \mu_{\{B_j\}} = \nu_{\{h_j\}}$ (C.M.) , $\forall \{h_j\}$, and $\mu \succ \nu$(C.M.) on S_G.

Q.E.D.

Definition 14.4.5. $\varphi \in S_G$ is said to be *ergodic* , if φ is an extreme point of S_G, i.e., $\varphi \in ExS_G$.

Proposition 14.4.6. Let $\varphi \in S_G$. Then φ is ergodic if and only if $\{\pi_\varphi(A), p_\varphi\}' = \mathbb{C}1_{H_\varphi}$.

Moreover, if $\dim E_\varphi = 1$, then φ is ergodic.

Proof. Let φ be ergodic, and $h \in \{\pi_\varphi(A), p_\varphi\}' = \{\pi_\varphi(A), u_\varphi(G)\}'$ with $0 \le h \le 1$. Then φ_h is G–invariant, and $0 \le \varphi_h \le \varphi$, where $\varphi_h(a) = \langle \pi_\varphi(a) h 1_\varphi, 1_\varphi \rangle, \forall a \in A$. Since $\varphi \in ExS_G$, it follows that $\varphi_h = \lambda\varphi$ for some $\lambda \in [0,1]$. Now by $\langle (h-\lambda)\pi_\varphi(a) 1_\varphi, 1_\varphi \rangle = 0, \forall a \in A$, we get $h = \lambda$. Therefore, $\{\pi_\varphi(A), p_\varphi\}' = \mathbb{C} 1_{H_\varphi}$.

Now let $\{\pi_\varphi(A), p_\varphi\}' = \{\pi_\varphi(A), u_\varphi(G)\}' = \mathbb{C} 1_{H_\varphi}$. If φ is not ergodic, then there is some $\rho \in S_G$ and some $\lambda \in (0,1)$ such that $\varphi \ge \lambda\rho$ and $\varphi \ne \rho$. Further, we can find $h \in (\pi_\varphi(A)' \backslash \mathbb{C} 1_{H_\varphi})_+$ such that

$$\rho(a) = \langle \pi_\varphi(a) h 1_\varphi, 1_\varphi \rangle, \quad \forall a \in A.$$

By $\rho \in S_G, u_\varphi(s) h 1_\varphi = h 1_\varphi, \forall s \in G$. On the other hand,

$$\begin{aligned} u_\varphi(s) h u_\varphi(s)^* \pi_\varphi(a) &= u_\varphi(s) h \pi_\varphi(\alpha_s(a)) u_\varphi(s)^* \\ &= \pi_\varphi(a) u_\varphi(s) h u_\varphi(s)^*, \quad \forall a \in A, s \in G, \end{aligned}$$

i.e., $u_\varphi(s) h u_\varphi(s)^* \in \pi_\varphi(A)'$. Further, since $u_\varphi(s) h u_\varphi(s)^* 1_\varphi = u_\varphi(s) h 1_\varphi = h_\varphi, \forall s \in G$, and 1_φ is separating for $\pi_\varphi(A)'$, it follows that $h \in u_\varphi(G)'$. Hence, we get $h \in \pi_\varphi(A)' \cap u_\varphi(G)' = \{\pi_\varphi(A), u_\varphi(G)\}' = \mathbb{C} 1_{H_\varphi}$, a contradiction. Therefore, φ is ergodic.

Finally, let dim $E_\varphi = 1$. Then by Proposition 14.4.1, $\{\pi_\varphi(A), p_\varphi\}' p_\varphi = (p_\varphi \pi_\varphi(A) p_\varphi)' = C p_\varphi$. Moreover, $x' \longrightarrow x' p_\varphi$ is a $*$ isomorphism from $\{\pi_\varphi(A), p_\varphi\}'$ onto $\{\pi_\varphi(A), p_\varphi\}' p_\varphi$ (see Proposition 14.4.1). Therefore, $\{\pi_\varphi(A), p_\varphi\}' = \mathbb{C} 1_{H_\varphi}$, and φ is ergoduic. Q.E.D.

Definition 14.4.7. The system (A, G, α) is said to be G*–abelian,* if for any $\varphi \in S_G, p_\varphi \pi_\varphi(A) p_\varphi$ is commutative.

Proposition 14.4.8. Let (A, G, α) be G–abelian, and $\varphi \in S_G$.

(i) $(p_\varphi \pi_\varphi(A) p_\varphi)'' = (p_\varphi \pi_\varphi(A) p_\varphi)' = \{\pi_\varphi(A), p_\varphi\}' p_\varphi$ is a maximal abelian VN algebra on $p_\varphi H_\varphi = E_\varphi$;

(ii) $\{\pi_\varphi(A), p_\varphi\}' = \{\pi_\varphi(A), u_\varphi(G)\}'$ is abelian;

(iii) φ is ergodic $\Longleftrightarrow \{\pi_\varphi(A), p_\varphi\}' = \mathbb{C} 1_{H_\varphi}$
$\Longleftrightarrow \dim E_\varphi = 1$.

Proof. (i) Since $(p_\varphi \pi_\varphi(A) p_\varphi)''$ is an abelian VN algebra on $p_\varphi H_\varphi$, and it admits a cyclic vector 1_φ, hence $(p_\varphi \pi_\varphi(A) p_\varphi)'' = (p_\varphi \pi_\varphi(A) p_\varphi)'$ is maximal abelian on $p_\varphi H_\varphi$.

(ii) By Proposition 14.4.1, $(p_\varphi \pi_\varphi(A) p_\varphi)' = \{\pi_\varphi(A), p_\varphi\}' p_\varphi$ is abelian, and $x' \longrightarrow x' p_\varphi$ is a $*$ isomorphism from $\{\pi_\varphi(A), p_\varphi\}'$ onto $\{\pi_\varphi(A), p_\varphi\}' p_\varphi$. Thus , $\{\pi_\varphi(A), p_\varphi\}' = \{\pi_\varphi(A), u_\varphi(G)\}'$ is abelian.

(iii) Let φ be ergodic. By Proposition 14.4.6, we have $\{\pi_\varphi(A), p_\varphi\}' = \mathbb{C} 1_{H_\varphi}$. Then from (i) ,

$$(p_\varphi \pi_\varphi(A) p_\varphi)'' = (p_\varphi \pi_\varphi(A) p_\varphi)' = \mathbb{C} p_\varphi.$$

Therefore, dim $E_\varphi = 1$. Now by Proposition 14.4.6, the conclusion is obvious.
Q.E.D.

Theorem 14.4.9. The system (A, G, α) is G-abelian if and only if for any $\varphi \in S_G, a_i^* = a_i \in A, i = 1, 2,$

$$\inf_{a_1' \in Co\{\alpha_s(a_1) | s \in G\}} |\varphi([a_1', a_2])| = 0.$$

Proof. Th necessity is obvious from Theorem 14.4.2.

Conversely, it suffices to show that

$$\langle \xi, (\pi_\varphi(a_1) p_\varphi \pi_\varphi(a_2) - \pi\delta_\varphi(a_2) p_\varphi \pi_\varphi(a_1))\xi\rangle = 0,$$

$\forall \varphi \in S_G; a_i^* = a_i \in A$ with $\|a_i\| \leq 1, i = 1, 2$; and $\xi \in p_\varphi H_\varphi = E_\varphi$ with $\|\xi\| = 1$. For any $\varepsilon > 0$, by Proposition 14.4.1 there are $\lambda_i > 0, s_i \in G$ and $\sum_i \lambda_i = 1$ such that

$$\|(\sum_i \lambda_i u_\varphi(s_i) - p_\varphi)\pi_\varphi(a_1)\xi\| < \varepsilon/2.$$

Let $a_1' = \sum_i \lambda_i \alpha_{s_i^{-1}}(a_1)$, then by $u_\varphi(t) p_\varphi = p_\varphi (\forall t \in G)$ we have

$$\|p_\varphi \pi_\varphi(a_1)\xi - u_\varphi(s)\pi_\varphi(a_1')\xi\| = \|(p_\varphi \pi_\varphi(a_1) - \pi_\varphi(a_1'))\xi\| < \varepsilon/2,$$

$\forall s \in G$. By the sufficient condition, there are $\mu_j > 0, t_j \in G$, and $\sum_j \mu_j = 1$ such that

$$|\psi([\sum_j \mu_j \alpha_{t_j}(a_1'), a_2])| < \varepsilon,$$

where $\psi(\cdot) = \langle \pi_\varphi(\cdot)\xi, \xi\rangle (\in S_G)$. Then

$$\begin{aligned}
&|\langle \xi, (\pi_\varphi(a_1) p_\varphi \pi_\varphi(a_2) - \pi_\varphi(a_2) p_\varphi \pi_\varphi(a_1))\xi\rangle| \\
\leq\ & |\langle \xi, [\sum_j \mu_j \pi_\varphi(\alpha_{t_j}(a_1')), \pi_\varphi(a_2)]\xi\rangle| \\
&+ |\langle \pi_\varphi(a_2)\xi, p_\varphi \pi_\varphi(a_1)\xi - \sum_j \mu_j \pi_\varphi(\alpha_{t_j}(a_1'))\xi\rangle| \\
&+ |\langle p_\varphi \pi_\varphi(a_1)\xi - \sum_j \mu_j \pi_\varphi(\alpha_{t_j}(a_1'))\xi, \pi_\varphi(a_2)\xi\rangle| \\
\leq\ & 2\|\pi_\varphi(a_2)\xi\| \cdot \|p_\varphi \pi_\varphi(a_1)\xi - \sum_j \mu_j \pi_\varphi(\alpha_{t_j}(a_1'))\xi\| \\
&+ |\langle \xi, [\sum_j \mu_j \pi_\varphi(\alpha_{t_j}(a_1')), \pi_\varphi(a_2)]\xi\rangle| \\
\leq\ & \varepsilon + |\psi([\sum_j \mu_j \alpha_{t_j}(a_1'), a_2])| < 2\varepsilon.
\end{aligned}$$

Since $\varepsilon(>0)$ is arbitrary, it follows that

$$\langle \xi, (\pi_\varphi(a_1)p_\varphi\pi_\varphi(a_2) - \pi_\varphi(a_2)p_\varphi\pi_\varphi(a_1))\xi\rangle = 0.$$

Q.E.D.

Remark. Clearly, if A is abelian, then (A, G, α) is G–abelian.

Moreover, if for any $a_i^* = a_i \in A, i = 1,2$ and $\varphi \in S_G$, $\inf\{|\varphi([\alpha_s(a_1), a_2])| | s \in G\} = 0$, then by Theorem 14.4.9 (A, G, α) is G–abelian.

Theorem 14.4.10. (*Ergodic decomposition*) Let (A, G, α) be a G–abelian system. Then S_G is a simplex (in the sense of Choquet). Therefore, for any $\varphi \in S_G$ there is a unique probability measure μ on S_G such that

$$\int_{S_G} \hat{a}_1(\rho)\cdots\hat{a}_n(\rho)d\mu(\rho) = \langle p_\varphi\pi_\varphi(a_1)p_\varphi\cdots p_\varphi\pi_\varphi(a_n)p_\varphi 1_\varphi, 1_\varphi\rangle,$$

$\forall a_1, \cdots, a_n \in A$, and $\nu \prec \mu$ (C.M.) for each probability measure ν on S_G with $\int_{S_G} \hat{a}(\rho)d\nu(\rho) = \varphi(a), \forall a \in A$. Consequently, μ is pseudoconcentrated on ExS_G in the sense that $\mu(E) = 0$ for each Baire subset E of S_G disjoint from ExS_G. Moreover, if A is separable, then $\mu(ExS_G) = 1$.

Proof. For any $\varphi \in S_G$, let μ be the C–measure of φ, where $C = \{\pi_\varphi(A), p_\varphi\}'$. By Theorem 14.4.4, μ is the unique largest (C.M.) probability measure on S_G such that $\mu \succ \nu$ (C.M.) for each ν as above. Now by Theorem 14.4.10, S_G is a simplex, and μ is pseudoconcentrated on ExS_G. Moreover, since μ is the C–measure of φ and supp $\mu \subset S_G$, it follows from Theorem 14.2.6 that

$$\int_{S_G} \hat{a}_1(\rho)\cdots\hat{a}_n(\rho)d\mu(\rho) = \langle p_\varphi\pi_\varphi(a_1)p_\varphi\cdots p_\varphi\pi_\varphi(a_n)p_\varphi 1_\varphi, 1_\varphi\rangle,$$

$\forall a_1, \cdots, a_n \in A$ Q.E.D.

Definition 14.4.11. Let A be a C^*–algebra with an identity 1 , G be the group of all unitary elements of A, and $S = S(A)$ be the state space of A. For each $v \in G$, define $\alpha_v(a) = vav^*, \forall a \in A$. Then (A, G, α) is a dynamical system. Clearly, S_G is the ***tracial state space*** $\mathcal{T} = \mathcal{T}(A)$ of A, i.e.,

$$\mathcal{T} = \mathcal{T}(A) = \{\varphi \in S | \varphi(ab) = \varphi(ba), \forall a, b \in A\}.$$

Proposition 14.4.12. Let (A, G, α) be the same as in Definition 14.4.11. Then it is G–abelian.

Proof. It is immediate from Theorem 14.4.9. Q.E.D.

Theorem 14.4.13. (*Tracial decomposition*) Let A be a C^*-algebra with an identity , and $\mathcal{T} = \mathcal{T}(A)$ be its tracial state space (a compact convex subset of $(A^*, \sigma(A^*, A))$). Then $\mathcal{T}$ is a simplex (in the sense of Choquet). Therefore, for any $\varphi \in \mathcal{T}$ there is unique probability measure μ on $\mathcal{T}$ such that

$$\int_{\mathcal{T}} \hat{a}_1(\rho)\cdots\hat{a}_n(\rho)d\mu(\rho) = \langle p_\varphi\pi_\varphi(a_1)p_\varphi\cdots p_\varphi\pi_\varphi(a_n)p_\varphi 1_\varphi, 1_\varphi\rangle,$$

$\forall a_1, \cdots, a_n \in A$, and $\mu \succ \nu$(C.M.) for each pprobability measure ν on $\mathcal{T}$ with $\int_{\mathcal{T}} \hat{a}(\rho)d\nu(\rho) = \varphi(a), \forall a \in A$. Moreover, μ is indeed the central measure of φ, and μ is pseudoconcentrated on $Ex\mathcal{T} = \mathcal{T} \cap \mathcal{F}$ in the sense that $\mu(E) = 0$ for any Baire subset E of $\mathcal{T}$ disjoint from $Ex\mathcal{T}$, where $\mathcal{F} = \mathcal{F}(A)$ is the factorial state space of A. In particular, if A is separable, then $\mu(\mathcal{T} \cap \mathcal{F}) = 1$.

Proof. By Proposition 14.4.12 and Theorem 14.4.10, it suffices to show that $Ex\mathcal{T} = \mathcal{T} \cap \mathcal{F}$, and $\{\pi_\varphi(A), p_\varphi\}' = \pi_\varphi(A)'' \cap \pi_\varphi(A)'$ for each $\varphi \in \mathcal{T}$.

Let $\varphi \in \mathcal{T}, Z = \pi_\varphi(A)'' \cap \pi_\varphi(A)'$. For any $x \in Z$, pick a net $\{a_l\} \subset A$ such that $\pi_\varphi(a_l) \longrightarrow x$(strongly). Then for any unitary element v of A we have

$$\begin{aligned} u_\varphi(v)x1_\varphi &= \lim u_\varphi(v)\pi_\varphi(a_l)u_\varphi(v)^*1_\varphi \\ &= \lim \pi_\varphi(\alpha_{v^{-1}}(a_l))1_\varphi = \lim \pi_\varphi(v^*a_lv)1_\varphi \\ &= \pi_\varphi(v)^*x\pi_\varphi(v)1_\varphi = x1_\varphi. \end{aligned}$$

Hence, $Z1_\varphi \subset E_\varphi$. Conversely, for any $a \in A$,

$$\begin{aligned} p_\varphi\pi_\varphi(a)1_\varphi &= \lim_l \sum_j \lambda_j^{(l)} u_\varphi(v_j^{(l)})\pi_\varphi(a)u_\varphi(v_j^{(l)})^*1_\varphi \\ &= \lim_l \sum_j \lambda_j^{(l)}\pi_\varphi(v_j^{(l)*}av_j^{(l)})1_\varphi, \end{aligned}$$

where $\lambda_j^{(l)} > 0, \sum_j \lambda_j^{(l)} = 1, \forall l, v_j^{(l)} \in G$ (the gropu of unitary elements of A) such that $p_\varphi = s\text{-}\lim_l \sum_j \lambda_j^{(l)} u_\varphi(v_j^{(l)})$ (see Proposition 14.4.1) . We may assume that

$$\sum_j \lambda_j^{(l)}\pi_\varphi(v_j^{(l)*}av_j^{(l)}) \xrightarrow{w} x \in \pi_\varphi(A)''$$

(replacing $\{\sum_j \lambda_j^{(l)}\pi_\varphi(v_j^{(l)*}av_j^{(l)})\}_l$ by a subnet if necessary). Then

$$\langle x1_\varphi, \pi_\varphi(b)1_\varphi\rangle = \langle p_\varphi\pi_\varphi(a)1_\varphi, \pi_\varphi(b)1_\varphi\rangle$$

$\forall b \in A$, and $p_\varphi \pi_\varphi(a) 1_\varphi = x 1_\varphi$. Pick $\{a_l\} \subset A$ such that $\pi_\varphi(a_l) \longrightarrow x$ (strongly). Then by $p_\varphi \pi_\varphi(a) 1_\varphi \in E_\varphi$ we have

$$\begin{aligned} x1_\varphi &= u_\varphi(v) x 1_\varphi = \lim u_\varphi(v) \pi_\varphi(a_l) u_\varphi(v)^* 1_\varphi \\ &= \lim \pi_\varphi(v^* a_l v) 1_\varphi = \pi_\varphi(v)^* x \pi_\varphi(v) 1_\varphi, \end{aligned}$$

$\forall v \in G$. Since $\varphi \in \mathcal{T}$, it is easy to see that 1_φ is also separating for $\pi_\varphi(A)''$. Hence,

$$x = \pi_\varphi(v)^* x \pi_\varphi(v),$$

$\forall v \in G$, and $x \in Z$. Therefore, $p_\varphi \pi_\varphi(a) 1_\varphi \in Z 1_\varphi, \forall a \in A$, and $E_\varphi = \overline{Z 1_\varphi}$. By Proposition 14.2.2, we have

$$\pi_\varphi(A)'' \cap \pi_\varphi(A)' = \{\pi_\varphi(A), p_\varphi\}', \quad \forall \varphi \in \mathcal{T}.$$

Now let $\varphi \in Ex\mathcal{T}$. By Proposition 14.4.8, dim $E_\varphi = 1$. From the preceding paragraph, $\mathbb{C} p_\varphi = E_\varphi = \overline{Z 1_\varphi}$, and $z 1_\varphi = \lambda_z 1_\varphi$, where $\lambda_z \in \mathbb{C}, \forall z \in Z$. Further,

$$z \pi_\varphi(a) 1_\varphi = \pi_\varphi(a) z 1_\varphi = \lambda_z \pi_\varphi(a) 1_\varphi, \forall a \in A,$$

i.e. $z = \lambda_z 1_{H_\varphi}, \forall z \in Z$. Therefore, $\pi_\varphi(A)''$ is a factor, and $\varphi \in \mathcal{T} \cap \mathcal{F}$. Conversely, let $\varphi \in \mathcal{T} \cap \mathcal{F}$. Then $E_\varphi = \overline{Z 1_\varphi} = \mathbb{C} 1_\varphi$, and dim $E_\varphi = 1$. From Proposition 14.4.8, $\varphi \in Ex\mathcal{T}$. Therefore, $Ex\mathcal{T} = \mathcal{T} \cap \mathcal{F}$. Q.E.D.

Notes. Theorem 14.4.10 is due to O. Lanford and D. Ruelle. Theorem 14.4.9 is due to D.Ruelle. Proposition 14.4.8 is due to G.G. Emch.

References. [44], [98], [139].

Chapter 15

(AF)-Algebras

15.1. The definition of (AF)-algebras

Definition 15.1.1. A C^*-algebra A is said to be *approximately finite - dimensional*, or (AF) simply, if there is an increasing sequence $\{A_n\}$ of finite dimensional $*$ subalgebras of A such that $\sqcup_n A_n$ is dense in A, i.e., $\overline{\sqcup_n A_n} = A$.

Proposition 15.1.2. Let $A = \overline{\sqcup_n A_n}$ be an (AF)-algebra. Then A has an identity 1 if and only if there exists n_0 such that $1_n = 1, \forall n \geq n_0$, where 1_n is the identity of $A_n, \forall n$.

Proof. Suppose that A has an identity 1. If there is a subsequence $\{n_k\}$ with $1_{n_k} \neq 1$, then $1_n \neq 1, \forall n$, since $\{A_n\}$ is increasing. On the other hand by $\overline{\sqcup_n A_n} = A$, we have n_0 and $x \in A_{n_0}$ such that $\|x - 1\| < 1$. We may assume that $A \subset B(H)$, and 1 is the identity operator on H. Pick $\xi \in (1 - 1_{n_0})H$ with $\|\xi\| = 1$. Then we get

$$1 > \|1 - x\| \geq \|\xi - x\xi\| = \|\xi - x1_{n_0}\xi\| = \|\xi\| = 1,$$

a contradiction. Therefore, 1_n must be equal to 1 for all enough large n.

Q.E.D.

Lemma 15.1.3. For any $\varepsilon \in (0, \frac{1}{4})$, there exists $\gamma = \gamma(\varepsilon) > 0$ with the following property: if A is a C^*-algebra on a Hilbert space H, p is a projection on H, and $a \in A$ with $\|a - p\| < \gamma$, then we have a projection q of A with $\|p - q\| < \varepsilon$.

Proof. We may assume that $a^* = a$. Let $\delta \in (0, \frac{\varepsilon}{2})$, and $m(> 0)$ be the minimal value of the function $|\lambda^2 - \lambda|$ on the following set:

$$[-2, 2] \backslash [(-\delta, \delta) \sqcup (1 - \delta, 1 + \delta)].$$

Now pick $\gamma = \gamma(\varepsilon) > 0$ such that $\gamma^2 + 3\gamma \leq \min\{\frac{3}{2}\varepsilon, \frac{m}{2}\}$. Notice that

$$\begin{aligned}
&\max\{|\lambda^2 - \lambda| \mid \lambda \in \sigma(a)\} = \|a^2 - a\| \\
\leq\ & \|a^2 - ap - pa + p\| + \|p(a-p)\| + \|(a-p)p\| + \|p - a\| \\
\leq\ & \|(a-p)^2\| + 3\|a - p\| < \gamma^2 + 3\gamma \leq \min\{\frac{3}{2}\varepsilon, \frac{m}{2}\}
\end{aligned}$$

and

$$\begin{aligned}
&|\lambda^2 - \lambda| > 1 \quad \text{if} \quad |\lambda| > 2, \\
&|\lambda^2 - \lambda| \geq m \quad \text{if} \quad \lambda \in [-2,2] \backslash [(-\delta,\delta) \sqcup (1-\delta, 1+\delta)].
\end{aligned}$$

Thus, $\sigma(a) \subset (-\delta,\delta) \sqcup (1-\delta,1+\delta)$. Pick a continuous function f on $\mathbb{R}$ such that $f(\lambda) = 0$ if $\lambda \in (-\delta,\delta)$ and $f(\lambda) = 1$ if $\lambda \in (1-\delta,1+\delta)$. Then $q = f(a)$ is a projection of A, and $\|p - q\| \leq \|p - a\| + \|a - q\| < \gamma + \delta < \varepsilon$. Q.E.D.

Lemma 15.1.4. Let $\varepsilon > 0$, and n be a positive integer. Then there exists $\delta_1 = \delta_1(\varepsilon, n) > 0$ with the following property: if A is a C^*-algebra, and $p_1, \cdots, p_n$ are projections of A satisfying $\|p_i p_j\| < \delta_1, \forall 1 \leq i \neq j \leq n$, then we have projections $q_1, \cdots, q_n$ of A satisfying $q_i q_j = 0, \|p_i - q_i\| < \varepsilon, \forall 1 \leq i \neq j \leq n$.

Proof. For $n = 1$, it is obvious. Now assume that the conclusion holds for n. For $(n+1)$ and $\varepsilon > 0$, let

$$\delta_1(\varepsilon, n+1) = \min\left\{\frac{\gamma(\varepsilon)}{6n}, \delta_1(\frac{\gamma(\varepsilon)}{6n}, n)\right\},$$

where $\gamma(\varepsilon)$ is as in Lemma 15.1.3, and we may assume that $\varepsilon \in (0, \frac{1}{4})$ and $\gamma(\varepsilon) \leq \varepsilon$. If A is a C^*-algebra, and $p_1, \cdots, p_{n+1}$ are projections of A satisfying $\|p_i p_j\| < \delta_1(\varepsilon, n+1), \forall 1 \leq i \neq j \leq n+1$, then we have $\|p_i p_j\| < \delta_1(\frac{\gamma(\varepsilon)}{6n}, n), \forall 1 \leq i \neq j \leq n$. By the inductive assumption, there are projections $q_1, \cdots, q_n$ of A with $q_i q_j = 0, \forall 1 \leq i \neq j \leq n$, and

$$\|p_i - q_i\| < \gamma(\varepsilon)/6n, \quad 1 \leq i \leq n.$$

Let $q = \sum_{i=1}^n q_i$. Then

$$\begin{aligned}
&\|p_{n+1} - (1-q)p_{n+1}(1-q)\| \\
\leq\ & 3\|p_{n+1}q\| \leq 3\sum_{i=1}^n \|p_{n+1}q_i\| \\
\leq\ & 3(\sum_{i=1}^n \|p_{n+1}p_i\| + \frac{\gamma(\varepsilon)}{6}) < 3n\delta_1(\varepsilon, n+1) + \frac{\gamma(\varepsilon)}{2} \leq \gamma(\varepsilon).
\end{aligned}$$

If B is the abelian C^*-subalgebra of A generated by $\{q_1, \cdots, q_n, (1-q)p_{n+1}(1-q)\}$, then from Lemma 15.1.3 there is a projection q_{n+1} of B such that $\|p_{n+1} - q_{n+1}\| < \varepsilon$. Clearly, $q_{n+1}q_i$ is still a projection, and

$$\|q_{n+1}q_i\| \leq \|p_{n+1}q_i\| + \varepsilon < \|p_{n+1}p_i\| + 2\varepsilon < 1,$$

$1 \leq i \leq n$. So $q_{n+1}q_i = 0, 1 \leq i \leq n$, and $\{q_1, \cdots, q_{n+1}\}$ is what we want to find.

Q.E.D.

Lemma 15.1.5. Let A be a C^*-algebra, $\{p_1, \cdots, p_n\}$ and $\{q_1, \cdots, q_n\}$ be two orthogonal families of projections of A, and $\|p_i - q_i\| < 1, 1 \leq i \leq n$. Then there exists a partial isometry w of A such that

$$(p_i w q_i)^* \cdot (p_i w q_i) = q_i, \quad (p_i w q_i) \cdot (p_i w q_i)^* = p_i, \quad 1 \leq i \leq n,$$

$$w^* w = \sum_{i=1}^n q_i, \quad \text{and} \quad w w^* = \sum_{i=1}^n p_i.$$

Proof. Pick $\delta \in (0,1)$ such that $\|p_i - q_i\| < \delta, 1 \leq i \leq n$. Let f be a continuous function on $\mathbb{R}$ with:

$$f(\lambda) = 0 \quad \text{if} \quad \lambda \leq \frac{1}{2}(1-\delta); \quad f(\lambda) = \lambda^{-1} \quad \text{if} \quad \lambda \geq 1 - \delta;$$

and f is affine on $[\frac{1}{2}(1-\delta), 1-\delta]$. For $1 \leq i \leq n$, let B_i be the abelian C^*-subalgebra of A generated by $\{p_i, p_i q_i p_i\}$, and Ω_i be its spectral space, i.e., $B_i \cong C(\Omega_i)$.

Fix $i \in \{1, \cdots, n\}$. Since $\rho(p_i) = 1$ and $\|p_i q_i p_i - p_i\| \leq \|p_i - q_i\| < \delta$, it follows that $\rho(p_i q_i p_i) \in (1-\delta, 1], \forall \rho \in \Omega_i$, and hence from the definition of the function f, we have that

$$f(\rho(p_i q_i p_i)) \cdot \rho(p_i q_i p_i) = 1 = \rho(p_i), \quad \forall \rho \in \Omega_i.$$

Thus, $p_i = f(p_i q_i p_i) \cdot p_i q_i p_i$. Put $x_i = f(p_i q_i p_i)^{1/2}$. Then $p_i = p_i q_i p_i x_i^2 = x_i p_i q_i p_i x_i = p_i x_i q_i x_i p_i$ since B_i is commutative.

Changing p_i and q_i, we can see that $q_i = y_i^2 q_i p_i q_i$, where $y_i = f(q_i p_i q_i)$. Now let $w_i = p_i x_i q_i$. Then

$$\begin{aligned} w_i w_i^* &= p_i x_i q_i x_i p_i = p_i, \\ w_i^* w_i &= q_i x_i p_i x_i q_i = y_i^2 q_i p_i q_i \cdot p_i x_i^2 q_i \\ &= y_i^2 q_i (p_i q_i p_i x_i^2) q_i = y_i^2 q_i p_i q_i = q_i, \end{aligned}$$

$1 \leq i \leq n$. Finally, let $w = \sum_{i=1}^n w_i$. Then this w is what we want to find.

Q.E.D.

Lemma 15.1.6. Let $\varepsilon \in (0,1]$, and n be a positive integer. Then there exists $\delta_2 = \delta_2(\varepsilon, n) > 0$ with the following property: if $\{p_1, \cdots, p_n\}$ and $\{q_1, \cdots, q_n\}$ are two orthogonal families of projections of a C^*-algebra A, and $\|p_i - q_i\| < \delta_2, 1 \le i \le n$, then we can find a partial isometry w of A such that

$$(p_i w q_i)^* \cdot (p_i w q_i) = q_i, \quad (p_i w q_i) \cdot (p_i w q_i)^* = p_i,$$
$$\|p_i - p_i w q_i\| < \varepsilon, \quad 1 \le i \le n,$$

and $w^* w = \sum_{i=1}^n q_i, ww^* = \sum_{i=1}^n p_i$. Moreover, if A has an identity 1 and $\sum_{i=1}^n p_i = 1$, then w can be chosen such that $\|1 - w\| < \varepsilon$.

Proof. Take $\delta_2(\varepsilon, n) = \frac{\varepsilon}{4n}$, and keep all notations of Lemma 15.1.5 ($\delta = \delta_2$). Then we have

$$\begin{aligned} \|p_i - w_i\| = & \ \|p_i(p_i - x_i q_i)\| \\ \le & \ \|x_i q_i - x_i p_i\| + \|(p_i - x_i) p_i\| \\ \le & \ \|x_i\| \cdot \|q_i - p_i\| + \|x_i - p_i\|, \quad 1 \le i \le n. \end{aligned}$$

Since $\rho(p_i q_i p_i) \in (1 - \delta_2, 1]$, it follows that $\rho(x_i) \in [1, (1-\delta_2)^{-\frac{1}{2}}), \forall \rho \in \Omega_i$. Then $\|x_i\| < (1-\delta_2)^{-1}$, and $\|p_i - x_i\| < (1-\delta_2)^{-1} - 1 = (1-\delta_2)^{-1}\delta_2$. Thus, we get

$$\|p_i - p_i w q_i\| = \|p_i - w_i\| < 2\delta_2(1-\delta_2)^{-1} < \varepsilon, \quad 1 \le i \le n.$$

If A has an identity 1 and $\sum_{i=1}^n p_i = 1$, then from the preceding paragraph we have

$$\|1 - w\| \le \sum_{i=1}^n \|p_i - w_i\| < 2n\delta_2(1-\delta_2)^{-1} < \varepsilon.$$

Q.E.D.

Lemma 15.1.7. For any $\varepsilon \in (0,1]$, there exists $\delta_3 = \delta_3(\varepsilon) > 0$ with the following property: if A is a C^*-algebra on a Hilbert space H, p_1 and p_2 are two projections of A, q_1 and q_2 are two projections on $H, a \in A$ and $v \in B(H)$ such that

$$\|p_i - q_i\| < \delta_3, \quad i = 1, 2, \quad v^* v = q_1,$$
$$vv^* = q_2, \quad \text{and} \quad \|v - a\| < \delta_3,$$

then there is $u \in A$ with $u^* u = p_1, uu^* = p_2$ and $\|u - v\| < \varepsilon$.

Proof. Let $\delta_3 = \delta_3(\varepsilon) = \frac{\varepsilon}{128}$, B be the abelian C^*-subalgebra of A generated by $\{p_2, p_2 a a^* p_2\}$, and Ω be the spectral space of B, i.e., $B \cong C(\Omega)$. Noticing that

$$\begin{aligned}\|p_2 - p_2aa^*p_2\| \le & \ \|p_2 - aa^*\| \le \|p_2 - q_2\| + \|q_2 - aa^*\| \\ \le & \ \|p_2 - q_2\| + \|vv^* - va^*\| + \|va^* - aa^*\| < 4\delta_3 < 1,\end{aligned}$$

we get

$$|\rho(p_2 - p_2aa^*p_2)| < 4\delta_3 < 1, \quad \forall \rho \in \Omega. \tag{1}$$

Pick $x \in B_+$ such that

$$\rho(x^2) = \rho(p_2aa^*p_2)^{-1}, \quad \forall \rho \in \Omega, \tag{2}$$

i.e.,

$$x^2p_2aa^*p_2 = p_2. \tag{3}$$

From (1), we have

$$\begin{aligned}\rho(x^2) = & \ (1 - \rho(p_2 - p_2aa^*p_2))^{-1} \\ = & \ \sum_{n=0}^{\infty} \rho(p_2 - p_2aa^*p_2)^n, \quad \forall \rho \in \Omega.\end{aligned}$$

Further, by (1) (2) we can see that

$$\begin{aligned}\|x^2 - p_2\| = & \ \max\{|1 - \rho(x^2)| \mid \rho \in \Omega\} \\ \le & \ \max_{\rho\in\Omega} \sum_{n=1}^{\infty} |\rho(p_2 - p_2aa^*p_2)|^n \le \frac{4\delta_3}{1 - 4\delta_3} < 8\delta_3,\end{aligned} \tag{4}$$

$$\|x\| \le (1 - 4\delta_3)^{-\frac{1}{2}} \le 2, \quad \|x^2\| \le (1 - 4\delta_3)^{-1} \le 2, \tag{5}$$

and

$$\begin{aligned}\|x - p_2\| = & \ \max_{\rho\in\Omega}|\rho(x + p_2)^{-1} \cdot |\rho(x^2 - p_2)|| \\ \le & \ \|x^2 - p_2\| < 8\delta_3.\end{aligned} \tag{6}$$

Let $w = xp_2a$. Then $ww^* = p_2$ by (3). Thus, w^*w is a projection of A. Further, by $q_1 = v^*q_2v$ and (4), (5), we have

$$\begin{aligned}\|w^*w - p_2\| \le & \ \|a^*p_2x^2p_2a - v^*p_2x^2p_2a\| + \|q_1 - p_1\| \\ & +\|v^*p_2x^2p_2a - v^*p_2x^2p_2v\| + \|v^*p_2x^2p_2v - q_1\| \\ \le & \ \|a - v\| \cdot \|a\| \cdot \|x^2\| + \|x^2\| \cdot \|a - v\| + \|p_2x^2p_2 - p_2\| \\ & +\|p_2 - q_2\| + \|p_1 - q_1\| \\ < & \ 2\delta_3(1 + \delta_3) + 2\delta_3 + 8\delta_3 + 2\delta_3 \\ < & \ 16\delta_3 = \delta_2(\tfrac{\varepsilon}{2}, 1),\end{aligned}$$

where $\delta_2(\frac{\varepsilon}{2},1)=\frac{\varepsilon}{8}$ (see the proof of Lemma 15.1.6). Now from Lemma 15.1.6 (for $w^*w, p_1, \frac{\varepsilon}{2}$ and $n=1$), there is a partial isometry w_1 of A such that

$$\begin{cases} w_1^*w_1 = p_1, w_1w_1^* = w^*w \\ \|w_1 - w^*w\| < \varepsilon/2 \end{cases} \tag{7}$$

Let $u = ww_1$. Then $u^*u = p_1, uu^* = p_2$. Finally, by $w = xp_2a, w = ww^*w, v = q_2v$ and (5), (6), (7), we get

$$\begin{aligned} \|u-v\| &\leq \|ww_1 - w\| + \|w - v\| \\ &\leq \|w_1 - w^*w\| + \|v - p_2v\| + \|p_2v - xp_2v\| + \|xp_2v - xp_2a\| \\ &\leq \|w_1 - w^*w\| + \|q_2 - p_2\| + \|p_2 - x\| + \|x\|\cdot\|v - a\| \\ &< \frac{\varepsilon}{2} + \delta_3 + 8\delta_3 + 2\delta_3 < \varepsilon. \end{aligned}$$

Q.E.D.

Lemma 15.1.8. For any $\varepsilon > 0$ and positive integer n, there exists $\delta_4 = \delta_4(\varepsilon, n) > 0$ with the following property: Let A be a C^*-algebra on a Hilbert space H, and $\{e_{ij}^{(k)} \mid 1 \leq i,j \leq n_k, 1 \leq k \leq m\}$ be a matrix unit on H, i.e.,

$$e_{ij}^{(k)} \in B(H), \quad e_{ij}^{(k)*} = e_{ji}^{(k)}, \quad e_{ij}^{(k)} e_{i'j'}^{(k')} = \delta_{kk'}\delta_{ji'}e_{ij'}^{(k)},$$

$\forall i,j,k,i',j',k'$, where $n_1 + \cdots + n_m = n$. If there is a subset $\{a_{ij}^{(k)}\}$ of A such that $\|e_{ij}^{(k)} - a_{ij}^{(k)}\| < \delta_4, \forall i,j,k$, then we have a matrix unit $\{q_{ij}^{(k)}\}$ of A such that

$$\|e_{ij}^{(k)} - q_{ij}^{(k)}\| < \varepsilon, \quad \forall 1 \leq i,j \leq n_k, 1 \leq k \leq m.$$

Moreover, if $\sum_{i,k} e_{ii}^{(k)} = 1_H$, then $1_H \in A$ and $\{q_{ij}^{(k)}\}$ can be chosen such that $\sum_{i,k} q_{ii}^{(k)} = 1_H$.

Proof. First, let $\delta_4 \leq \gamma(\varepsilon_1)$, here $\gamma(\cdot)$ is as in Lemma 15.1.3, and ε_1 will be chosen later. Then we have projection $\{p_{ii}^{(k)}\}$ of A such that $\|e_{ii}^{(k)} - p_{ii}^{(k)}\| < \varepsilon_1, \forall i,k$. So $\|p_{ii}^{(k)}p_{jj}^{(l)}\| < 2\varepsilon_1, \forall (i,k) \neq (j,l)$. Pick $\varepsilon_1 < \delta_1(\varepsilon_2, n)$, here $\delta_1(\cdot,\cdot)$ is as in Lemma 15.1.4, and ε_2 will be determined in the following. Then there is an orthogonal family $\{q_{ii}^{(k)}\}$ of projections of A such that $\|q_{ii}^{(k)} - p_{ii}^{(k)}\| < \varepsilon_2, \forall i,k$. Hence, $\|q_{ii}^{(k)} - e_{ii}^{(k)}\| < \varepsilon_1 + \varepsilon_2, \forall i,k$. Notice that

$$\begin{aligned} &e_{1i}^{(k)*}e_{1i}^{(k)} = e_{ii}^{(k)}, \quad e_{1i}^{(k)}e_{1i}^{(k)*} = e_{11}^{(k)}, \\ &\|q_{ii}^{(k)} - e_{ii}^{(k)}\| < \varepsilon_1 + \varepsilon_2, \quad \|q_{11}^{(k)} - e_{11}^{(k)}\| < \varepsilon_1 + \varepsilon_2, \\ &\|a_{1i}^{(k)} - e_{1i}^{(k)}\| < \delta_4 \leq \gamma(\varepsilon_1) < \varepsilon_1 + \varepsilon_2, \quad \forall i,k. \end{aligned}$$

Now pick $\varepsilon_1+\varepsilon_2<\delta_3(\varepsilon_3)$, here $\delta_3(\cdot)$ is as in Lemma 15.1.7, and ε_3 will be chosen later. Then there exists a subset $\{q_{1i}^{(k)}\}$ of A such that

$$q_{1i}^{(k)*}q_{1i}^{(k)}=q_{ii}^{(k)},\quad q_{1i}^{(k)}q_{1i}^{(k)*}=q_{11}^{(k)},$$
$$\|q_{1i}^{(k)}-e_{1i}^{(k)}\|<\varepsilon_3,\quad \forall i,k.$$

Let $q_{ij}^{(k)}=q_{1i}^{(k)*}q_{1j}^{(k)}$. Then $\{q_{ij}^{(k)}\}$ is a matrix unit of A. When $(\varepsilon_1+\varepsilon_2+\varepsilon_3)$ is small enough, we get

$$\|q_{ij}^{(k)}-e_{ij}^{(k)}\|<\varepsilon,\quad \forall i,j,k.$$

Moreover, if $\sum_{i,k}e_{ij}^{(k)}=1_H$, we may assume that $n\varepsilon<1$, then

$$\|\sum_{i,k}q_{ii}^{(k)}-1_H\|\le n\varepsilon<1.$$

Thus, $\sum_{i,k}q_{ii}^{(k)}=1_H$. Q.E.D.

Lemma 15.1.9. Let A be a C^*-algebra with an identity 1. Then A is (AF) if and only if the following conditions are satisfied:

1) A is separable;

2) for any finite subset $\{a_1,\cdots,a_n\}$ of A and $\varepsilon>0$, there exists a finite dimensional $*$ subalgebra B of A and a subset $\{b_1,\cdots,b_n\}$ of B such that $\|a_i-b_i\|<\varepsilon, 1\le i\le n$.

Moreover, if A is (AF) and D is a finite dimensional $*$ subalgebra of A, then we can find an increasing sequence $\{A_n\}$ of finite dimensional $*$ subalgebras of A such that $D\subset A_1; 1\in A_n,\forall n$; and $\overline{\cup_n A_n}=A$.

Proof. The necessity is obvious. Now let A satisfy the conditions 1) and 2), let $\{x_n\}$ be a countable dense subset of $\{x\in A \mid \|x\|<1/2\}$ with $x_1=0$, and D be either $\{0\}$ or the given finite dimensional $*$ subalgebra of A. Suppose that $A_1=D+\mathbb{C}$. Then $1\in A_1$, and there is $a_1^{(1)}\in A_1$ such that $\|a_1^{(1)}-x_1\|<2^{-1}$ (for example, $a_1^{(1)}=0$). Now assume that we have finite dimemsional $*$ subalgebras $A_1\subset\cdots\subset A_n$, such that for each $l\in\{1,\cdots,n\}$, there are $a_1^{(l)},\cdots,a_l^{(l)}\in A_l$ with $\|a_i^{(l)}-x_i\|<2^{-l}, 1\le i\le l$. Let $\{e_{ij}^{(k)} \mid 1\le i,j\le n_k, 1\le k\le m\}$ be a matrix unit of A_n, and $n_1^2+\cdots+n_m^2=\dim A_n$. Clearly, $\sum_{i,k}e_{ii}^{(k)}=1$. By the condition 2), there is a finite dimensional $*$ subalgebra B of A, and $\{b_{ij}^{(k)}, b_l \mid 1\le i,j\le n_k, 1\le k\le m, 1\le l\le n+1\}\subset B$ such that

$$\|b_{ij}^{(k)}-e_{ij}^{(k)}\|<\delta_4(\varepsilon,\dim A_n),\quad \forall i,j,k,$$

and $\|b_i - x_i\| < \varepsilon, 1 \le i \le n+1$, where $\varepsilon > 0$ will be determined in the following. Then by Lemma 15.1.8, we have a matrix unit $\{f_{ij}^{(k)}\}$ of B with $\sum_{i,k} f_{ii}^{(k)} = 1$ such that $\|e_{ij}^{(k)} - f_{ij}^{(k)}\| < \varepsilon, \forall i,j,k$. Let $\varepsilon < \delta_2(\eta, m)$, where $\eta > 0$ will be chosen later. Then by Lemma 15.1.6, there is a partial isometry w of A such that

$$\begin{cases} (e_{11}^{(k)} w f_{11}^{(k)})^* \cdot (e_{11}^{(k)} w f_{11}^{(k)}) = f_{11}^{(k)}, \\ (e_{11}^{(k)} w f_{11}^{(k)}) \cdot (e_{11}^{(k)} w f_{11}^{(k)})^* = e_{11}^{(k)}, \\ \|e_{11}^{(k)} - e_{11}^{(k)} w f_{11}^{(k)}\| < \eta, \quad 1 \le k \le m. \end{cases} \tag{1}$$

Put $u = \sum_{j,k} e_{j1}^{(k)} w f_{j1}^{(k)}$. Then u is a unitary element of A, and $u f_{ij}^{(k)} u^* = e_{ij}^{(k)}, \forall i,j,k$. Suppose that $A_{n+1} = uBu^*$. Then $A_n \subset A_{n+1}$. Let $a_i^{(n+1)} = u b_i u^*$. Then

$$\begin{aligned} \|a_i^{(n+1)} - x_i\| \le & \ \|b_i - x_i\| + \|u b_i u^* - b_i\| \\ < & \ \varepsilon + \Big\| \sum_{j,k,l,m} [e_{j1}^{(k)} w f_{j1}^{(k)} b_i f_{l1}^{(m)} w^* e_{1l}^{(m)} - f_{jj}^{(k)} b_i f_{ll}^{(m)}] \Big\| \\ \le & \ \varepsilon + (\dim A_n)^2 \sup_{j,k,l,m} \|f_{jj}^{(k)} b_i f_{ll}^{(m)} - e_{j1}^{(k)} w f_{j1}^{(k)} b_i f_{l1}^{(m)} w^* e_{1l}^{(m)}\|, \end{aligned}$$

$1 \le i \le n+1$. We may assume that $\varepsilon < \frac{1}{2}$. Since $\|x_i\| < \frac{1}{2}$ and $\|b_i - x_i\| < \frac{1}{2}$, it follows that $\|b_i\| < 1, 1 \le i \le n+1$. Thus,

$$\begin{aligned} & \|f_{jj}^{(k)} b_i f_{ll}^{(m)} - e_{j1}^{(k)} w f_{j1}^{(k)} b_i f_{l1}^{(m)} w^* e_{1l}^{(m)}\| \\ \le & \ \|f_{jj}^{(k)} b_i f_{ll}^{(m)} - f_{jj}^{(k)} b_i f_{l1}^{(m)} w^* e_{1l}^{(m)}\| \\ & + \|(f_{jj}^{(k)} - e_{j1}^{(k)} w f_{1j}^{(k)}) b_i f_{l1}^{(m)} w^* e_{1l}^{(m)}\| \\ \le & \ 2 \sup_{s,t} \|f_{ss}^{(t)} - e_{s1}^{(t)} w f_{1s}^{(t)}\|, \quad \forall j,k,l,m. \end{aligned}$$

By $\|f_{ss}^{(t)} - e_{ss}^{(t)}\| < \varepsilon, \|e_{1s}^{(t)} - f_{1s}^{(t)}\| < \varepsilon, \forall s,t$, and (1), we have

$$\begin{aligned} \|f_{ss}^{(t)} - e_{s1}^{(t)} w f_{1s}^{(t)}\| < & \ \varepsilon + \|e_{s1}^{(t)} e_{11}^{(t)} e_{1s}^{(t)} - e_{s1}^{(t)} e_{11}^{(t)} w f_{11}^{(t)} f_{1s}^{(t)}\| \\ < & \ 2\varepsilon + \|e_{11}^{(t)} f_{1s}^{(t)} - e_{11}^{(t)} w f_{11}^{(t)} f_{1s}^{(t)}\| < 2\varepsilon + \eta, \quad \forall s,t. \end{aligned}$$

Now suppose that ε and η are small enough. Then we get $\|a_i^{(n+1)} - x_i\| < 2^{-n-1}, 1 \le i \le n+1$. By induction, $\{A_n | n \ge 1\}$ can be found. Q.E.D.

Lemma 15.1.10. Let A be a C^*-algebra. Then A is (AF) if and only if $(A \dot{+} \mathbb{C})$ is (AF).

Proof. The necessity is obvious. Conversely, let $(A\dot{+}\mathbb{C})$ is (AF). Then by Proposition 15.1.2, there is an increasing sequence $\{B_n\}$ of finite dimensional $*$ subalgebras of $(A\dot{+}\mathbb{C})$, such that $1 \in B_n, \forall n$, and $\overline{\sqcup_n B_n} = A\dot{+}\mathbb{C}$. Let $A_n = A \cap B_n, \forall n$. Clearly, $\{A_n\}$ is an increasing sequence of finite dimensional $*$ subalgebras of A. For any $a \in A$, we can find $b_n \in B_n, \forall n$, such that $b_n \to a$. Since $B_n = A_n\dot{+}\mathbb{C}$, it follows that $b_n = a_n + \lambda_n$, where $a_n \in A_n, \lambda_n \in \mathbb{C}, \forall n$. Then $a_n \to a$ and $\lambda_n \to 0$. Therefore, $A = \overline{\sqcup_n A_n}$ is (AF). Q.E.D.

Theorem 15.1.11. Let A be a C^*-algebra. Then A is (AF) if and only if the following conditions are satisfied:

1) A is separable;

2) for any finite subset $\{a_1, \cdots, a_n\}$ of A, and $\varepsilon > 0$, there exists a finite dimensional $*$ subalgebra B of A and a subset $\{b_1, \cdots, b_n\}$ of B such that $\|a_i - b_i\| < \varepsilon, 1 \le i \le n$.

Moreover, if A is (AF) and D is a finite dimensional $*$ subalgebra of A, then we can find an increasing sequence $\{A_n\}$ of finite dimensional $*$ subalgebras of A such that $D \subset A_1$, and $\overline{\sqcup_n A_n} = A$.

Proof. It is immediate from Lemmas 15.1.9 and 15.1.10. Q.E.D.

Proposition 15.1.12. Let A be an (AF)-algebra, and p be a projection of A. Then pAp is also (AF).

Proof. For any $x_1, \cdots, x_n \in pAp$ and $\varepsilon > 0$, by Theorem 15.1.11 there is a finite dimensional $*$ subalgebra B of A and $a, y_1, \cdots, y_n \in B$ such that $\|x_i - y_i\| < \varepsilon, 1 \le i \le n$, and $\|a - p\| < \gamma(\delta_2(\varepsilon, 1))$, where $\gamma(\cdot)$ and $\delta_2(\cdot,\cdot)$ are as in Lemmas 15.1.3 and 15.1.6 respectively. Then there exists a projection q of B with $\|p - q\| < \delta_2(\varepsilon, 1)$. Further, there is a partial isometry w of A such that

$$ww^* = q, \qquad w^*w = p, \quad \text{and} \quad \|p - w\| < \varepsilon.$$

Let $C = w^*Bw$. Clearly, C is a finite dimensional $*$ subalgebra of pAp. Moreover, we have

$$\begin{aligned}
\|x_i - w^*y_iw\| &\le \|x_i - py_ip\| + \|py_ip - w^*y_iw\| \\
&\le \|x_i - y_i\| + \|(p - w^*)y_ip\| + \|w^*y_i(p - w)\| \\
&< \varepsilon + 2\varepsilon\|y_i\| \le \varepsilon + 2\varepsilon(\|x_i\| + \varepsilon),
\end{aligned}$$

$1 \le i \le n$. Therefore, by Theorem 15.1.11 pAp is an (AF)-algebra. Q.E.D.

Theorem 15.1.13. Let A be a C^*-algebra with an identity 1. Then A is a (UHF)-algebra if and only if the following conditions are satisfied:

1) A is separable,

2) for any $a_1, \cdots, a_n \in A$ and $\varepsilon > 0$, there is a finite dimensional subfactor B of A with $1 \in B$ and $b_1, \cdots, b_n \in B$ such that $\|a_i - b_i\| < \varepsilon, 1 \le i \le n$.

Moreover, if A is (UHF) and D is a finite dimensional subfactor of A with $1 \in D$, then there exists an increasing sequence $\{A_n\}$ of finite dimensional subfactors of A such that $A_1 = D, 1 \in A_n, \forall n$, and $A = \overline{\cup_n A_n}$.

Proof. It is similar to the proof of Lemma 15.1.9. Q.E.D.

Notes. The theory of (AF)-algebra was introducted by O. Bratteli. It is a generalization of the theory of Glimm (UHF) algebras.

References. [15], [26], [54].

15.2. Dimensions and isomorphic theorem

Consider a $*$ algebra A over $\mathbb{C}$. $p \in A$ is called a projection, if $p^* = p = p^2$. And we denote the set of all projections of A by $\mathrm{Proj}(A)$. Moreover, we assume that if $a \in A$ and $a^*a = 0$, then $a = 0$.

Definition 15.2.1. $p, q \in \mathrm{Proj}(A)$ are *equivalent*, denoted by $p \sim q$, if there exists $v \in A$ such that $v^*v = p$ and $vv^* = q$.

In this case, we have $vp = v$ and $v^*q = v^*$. Indeed, since $(v - vp)^*(v - vp) = v^*v - v^*vp - pv^*v + pv^*vp = 0$, it follows that $v - vp = 0$, i.e., $v = vp$. Similarly, $v^*q = v^*$.

Thus, $\sim$ is an equivalent relation on $\mathrm{Proj}(A)$. For any $p \in \mathrm{Proj}\ (A)$, denote its equivalent class by $\tilde{p}$.

Definition 15.2.2. Denote all equivalent classes of Proj (A) by $E(A)$, i.e., $E(A) = \mathrm{Proj}\ (A)/\sim$. The canonical map $d(\cdot)$ from $\mathrm{Proj}(A)$ onto $E(A)$ is called *dimensions*.

Definition 15.2.3. Elements $\tilde{p}_1, \cdots, \tilde{p}_n \in E(A)$ is said to be *additive*, if there exists a $p_i \in \tilde{p}_i, 1 \le i \le n$, such that $p_i p_j = 0, \forall i \ne j$. In this case, we define $\tilde{p}_1 + \cdots, + \tilde{p}_n = (p_1 + \cdots p_n)^\sim$.

We claim that the *partial addition* on $E(A)$ is well-defined. In fact, if we have another $q_i \in \tilde{p}_i, 1 \le i \le n$, such that $q_i q_j = 0, \forall i \ne j$. For $i \in \{1, \cdots n\}$, there is $v_i \in A$ with $v_i^* v_i = p_i, v_i v_i^* = q_i$. Let $v = v_1 + \cdots + v_n$. Since

$$v_i^* v_j = v_i^* q_i q_j v_j = 0, \quad v_j v_i^* = v_j p_j p_i v_i^* = 0,$$

$\forall i \ne j$, it follows that $v^*v = p_1 + \cdots + p_n, vv^* = q_1 + \cdots + q_n$, i.e., $(p_1 + \cdots + p_n)^\sim = (q_1 + \cdots + q_n)^\sim$.

Definition 15.2.4. Let A, B be two $*$ algebra as above. $E(A)$ and $E(B)$ are said to be *isomorphic*, if there is a bijective map Ψ from $E(A)$ onto $E(B)$ such that $\{\tilde{p}_1, \cdots, \tilde{p}_n\}(\subset E(A))$ is additive if and only if $\{\Psi(\tilde{p}_1), \cdots, \Psi(\tilde{p}_n)\}(\subset E(B))$ is additive; and $\Psi(\tilde{p}_1 + \cdots + \tilde{p}_n) = \Psi(\tilde{p}_1) \cdots + \Psi(\tilde{p}_n)$.

Clearly, if Φ is a $*$ isomorphism from A onto B, let $\Psi(\tilde{p}) = \widetilde{\Phi(p)}$, then Ψ is an isomorphism from $E(A)$ onto $E(B)$.

Definition 15.2.5. A $*$ algebra A is called a (LF) algebra, if $A = \sqcup_n A_n$ with $A_1 \subset \cdots \subset A_n \subset \cdots$, and every A_n is a finite dimensional C^*-algebra.

In particular, any (AF)-algebra contains a dense (LF) $*$ subalgebra.

Lemma 15.2.6. Let $A = \sqcup_n A_n$ be a (LF)-algebra, B and C be two finite dimensional C^*-subalgebras of A, and Φ be a $*$ isomorphism from B onto C, such that $p \sim \Phi(p), \forall p \in \text{Proj}(B)$. Then there exists a unitary element u of A (if A has an identity) or $(A\dot{+}\mathbb{C})$ (if A has no identity) such that

$$\Phi(b) = ubu^*, \quad \forall b \in B.$$

Proof. Let $\{e_{ij}^{(k)} \mid 1 \le i, j \le n_k, 1 \le k \le m\}$ be a matrix unit of B, and $n_1^2 + \cdots + n_m^2 = \dim B$. Then $\{f_{ij}^{(k)} = \Phi(e_{ij}^{(k)})\}$ is also a matrix unit and a basis of C. By the assumption, for each $k \in \{1, \cdots, m\}$ there is $v_k \in A$ such that $v_k^* v_k = e_{11}^{(k)}, v_k v_k^* = f_{11}^{(k)}$. Let

$$w = \sum_{j,k} f_{j1}^{(k)} v_k e_{1j}^{(k)}.$$

Clearly, $w^* w = e, ww^* = \Phi(e)$, and $wbw^* = \Phi(b), \forall b \in B$, where $e = \sum_{j,k} e_{jj}^{(k)}$ is the identity of $B, \Phi(e) = \sum_{j,k} f_{jj}^{(k)}$ is the identity of C.

Let n be large enough such that $w, w^* \in A_n$. Then $e \sim \Phi(e)$ relative to A_n. By Proposition 6.3.2, we have $w' \in A_n$ such that

$$w'^* w' = 1_n - e, w' w'^* = 1_n - \Phi(e),$$

where 1_n is the identity of A_n. If 1 is the identity of A (if A has an identity) or $(A\dot{+}\mathbb{C})$ (if A has no identity), then $u = w + w' + (1 - 1_n)$ is a unitary element of A or $(A\dot{+}\mathbb{C})$, and we have

$$ubu^* = wbw^* = \Phi(b), \quad \forall b \in B.$$

Q.E.D.

Lemma 15.2.7. Let A_i be a $*$ algebra, $E_i = E(A_i), i = 1,2$, and Ψ be an isomorphism from E_2 onto E_1. If B_2 is a finite dimensional C^*-subalgebra of A_2, then there exists a finite dimensional C^*-subalgebra B_1 of A_1 and a $*$ isomorphism Φ from B_1 onto B_2 such that $\Psi(\widetilde{\Phi(p)}) = \tilde{p}, \forall p \in$Proj (B_1).

Proof. Let $\{f_{ij}^{(k)} \mid 1 \le i,j \le n_k, 1 \le k \le m\}$ be a matrix unit and a basis of B_2. By the property of Ψ, there is an orthogonal family $\{e_{ii}^{(k)}\}$ of projections of A_1 such that

$$\Psi(\widetilde{f_{ii}^{(k)}}) = \widetilde{e_{ii}^{(k)}}, \quad \forall i,k.$$

Since $f_{ii}^{(k)} \sim f_{11}^{(k)}$, it follows that $e_{ii}^{(k)} \sim e_{11}^{(k)}$ and there is $e_{i1}^{(k)} \in A_1$ such that

$$e_{i1}^{(k)*} e_{i1}^{(k)} = e_{11}^{(k)}, \quad e_{i1}^{(k)} e_{i1}^{(k)*} = e_{ii}^{(k)}, \quad \forall i,k.$$

Let $e_{ij}^{(k)} = e_{i1}^{(k)} e_{j1}^{(k)*}, \Phi(e_{ij}^{(k)}) = f_{ij}^{(k)}, \forall i,j,k$, and $B_1 = [e_{ij}^{(k)} \mid i,j,k]$. Then B_1 is a finite dimensional c^*-subalgebra of A_1, and Φ is a $*$ isomorphism from B_1 onto B_2. Now if p is a projection of B_1, then there is a subset Λ of $\{(i,k) \mid 1 \le i \le n_k, 1 \le k \le m\}$ such that $p \sim \sum_{(i,k)\in\Lambda} e_{ii}^{(k)}$. By the additivity of Ψ, we get

$$\begin{aligned}\Psi(\widetilde{\Phi(p)}) &= \Psi(\widetilde{\sum_{(i,k)\in\Lambda} f_{ii}^{(k)}}) \\ &= \sum_{(i,k)\in\Lambda} \Psi(\widetilde{f_{ii}^{(k)}}) = \sum_{(i,k)\in\Lambda} \widetilde{e_{ii}^{(k)}} = \tilde{p}.\end{aligned}$$

Q.E.D.

Theorem 15.2.8. Let $A = \sqcup_n A_n$ and $A' = \sqcup_n A'_n$ be two (LF)-algebras. If $E = E(A)$ and $E' = E(A')$ are isomorphic, then A and A' are $*$ isomorphic.

Proof. Let's try to find two subsequences $\{m_k\}, \{n_k\}$ of positive integers, and a sequences $\{B_k\}$ of finite dimensional C^*-subalgebras of A, and a $*$ isomorphism Ψ_k from B_k onto A'_{n_k} for each k, such that

$$A_{m_1} \subset B_1 \subset A_{m_2} \subset B_2 \subset \cdots \subset A_{m_k} \subset B_k \subset \cdots,$$

$$\Psi_{k+1}|B_k = \Psi_k, \quad \forall k,$$

and the following diagram is commutative:

$$\begin{array}{ccccccccc} B_1 & \hookrightarrow & B_2 & \hookrightarrow & \cdots & \hookrightarrow & B_k & \hookrightarrow & \cdots \\ \downarrow\Psi_1 & & \downarrow\Psi_2 & & & & \downarrow\Psi_k & & \\ A'_{n_1} & \hookrightarrow & A'_{n_2} & \hookrightarrow & \cdots & \hookrightarrow & A'_{n_k} & \hookrightarrow & \cdots, \end{array}$$

where "$\hookrightarrow$" represents the embedding map. Then we can see that A and A' are $*$ isomorphic.

The process and principle are as follows:

$$\begin{array}{ccccccccccc}
A_1 = A_{m_1} & \hookrightarrow & B_1 & \hookrightarrow & A_{m_2} & \hookrightarrow & B_2 & \hookrightarrow & A_{m_3} & \hookrightarrow & \cdots \\
 & & u_1\downarrow & & G_2\downarrow & & u_2\downarrow & & \downarrow & & \\
\Phi_1\downarrow & & C_1 & & C_2' & & C_2 & & C_3' & & \\
 & & F_1\downarrow & & u_2'\downarrow & & F_2\downarrow & & \downarrow & & \\
B_1' & \hookrightarrow & A'_{n_1} & \hookrightarrow & B_2' & \hookrightarrow & A'_{n_2} & \hookrightarrow & B_3' & \hookrightarrow & \cdots.
\end{array}$$

Now we begin the proof.

Let Ψ be an isomorphism from E onto E' and $m_1 = 1$. By Lemma 15.2.7, there exists a finite dimensional C^*-subalgebra B_1' of A' and a $*$ isomorphism Φ_1 from A_{m_1} onto B_1' such that

$$\Psi(\tilde{p}_1) = \widetilde{\Phi_1(p_1)}, \quad \forall p_1 \in \text{Proj }(A_{m_1}). \tag{1}$$

Pick n_1 such that $B_1' \subset A'_{n_1}$. Similarly by Lemma 15.2.7, we can find a $*$ subalgebra C_1 of A and a $*$ isomorphism F_1 from C_1 onto A'_{n_1} such that

$$\Psi(\tilde{q}_1) = \widetilde{F_1(q_1)}, \quad \forall q_1 \in \text{Proj }(C_1). \tag{2}$$

Then the following diagram is commutative:

$$\begin{array}{ccc}
A_{m_1} & \overset{F_1^{-1}\circ\Phi_1}{\longrightarrow} & C_1 \\
\Phi_1\downarrow & & F_1\downarrow \\
B_1' & \hookrightarrow & A'_{n_1},
\end{array}$$

and by (1), (2), we have

$$F_1^{-1}\circ\widetilde{\Phi_1(p_1)} = \Psi^{-1}(\widetilde{\Phi_1(p_1)}) = \tilde{p}_1, \quad \forall p_1 \in \text{Proj }(A_{m_1}).$$

Using Lemma 15.2.6 to $A, A_{m_1}, F_1^{-1}\circ\Phi_1(A_{m_1})$, there is a unitary element u_1 of $(A\dotplus\mathbb{C})$ such that

$$u_1 a u_1^* = F_1^{-1}\circ\Phi_1(a), \quad \forall a \in A_{m_1}.$$

Let $B_1 = u_1^* C_1 u_1$. Then $B_1 \supset u_1^*(F_1^{-1}\circ\Phi_1(A_{m_1}))u_1 = A_{m_1}$. Thus, $\Psi_1(\cdot) = F_1(u_1 \cdot u_1^*)$ is a $*$ isomorphism from B_1 onto A'_{n_1}, and by (2) we have

$$\Psi(\tilde{r}_1) = \Psi(\widetilde{u_1 r_1 u_1^*}) = F_1(\widetilde{u_1 r_1 u_1^*}) = \widetilde{\Psi_1(r_1)}. \tag{3}$$

Pick $m_2(> m_1)$ such that $B_1 \subset A_{m_2}$. From Lemma 15.2.7, there is a $*$ subalgebra C_2' of A' and a $*$ isomorphism G_2 from A_{m_2} onto C_2' such that

$$\Psi(\tilde{p}_2) = \widetilde{G_2(p_2)}, \quad \forall p_2 \in \text{Proj }(A_{m_2}). \tag{4}$$

Then we have the following commutative diagram:

$$\begin{array}{ccc}
B_1 & \hookrightarrow & A_{m_2} \\
\Psi_1\downarrow & & G_2\downarrow \\
A'_{n_1} & \overset{G_2\circ\Psi_1^{-1}}{\longrightarrow} & C_2'
\end{array}$$

and by (3), (4),

$$G_2 \circ \widetilde{\Psi_1^{-1}}(p'_1) = \Psi(\widetilde{\Psi_1^{-1}(p'_1)}) = \widetilde{p'_1}, \quad \forall p'_1 \in \text{Proj } (A'_{n_1}).$$

Using Lemma 15.2.6 to $A', A'_{n_1}, G_2 \circ \Psi_1^{-1}(A'_{n_1})$, there is a unitary element u'_2 of $(A'\dot{+}\mathbb{C})$ such that

$$u'_2 a' u'^*_2 = G_2 \circ \Psi_1^{-1}(a'), \quad \forall a' \in A'_{n_1}.$$

Let $B'_2 = u'_2 C_2 u'^*_2$. Then $B'_2 \supset u'^*_2(G_2 \circ \Psi_1^{-1}(A'_{n_1}))u'_2 = A'_{n_1}$. Thus we have the following commutative diagram:

$$\begin{array}{ccc} B_1 & \hookrightarrow & A_{m_2} \\ \Psi_1\downarrow & & \Phi_2\downarrow \\ A'_{n_1} & \hookrightarrow & B'_2, \end{array}$$

where $\Phi_2(a) = u'^*_2 G_2(a) u'_2, \forall a \in A_{m_2}$, and by (4),

$$\Phi_2\widetilde{(p_2)} = G_2\widetilde{(p_2)} = \Psi(\tilde{p}_2), \quad \forall p_2 \in \text{Proj } (A_{m_2}). \tag{5}$$

Pick $n_2(> n_1)$ such that $B'_2 \subset A'_{n_2}$. From Lemma 15.2.7, there is a $*$ subalgebra C_2 of A and a $*$ isomorphism F_2 from C_2 onto A'_{n_2} such that

$$\Psi(\tilde{q}_2) = F_2\widetilde{(q_2)}, \quad \forall q_2 \in \text{Proj } (C_2). \tag{6}$$

Similar to the preceding paragraph, we have a unitary element u'_2 of $(A'\dot{+}\mathbb{C})$ such that

$$F_2^{-1} \circ \Phi_2(a) = u_2 a u_2^*, \quad \forall a \in A_{m_2}.$$

Let $B_2 = u_2^* C_2 u_2$. Then $B_2 \supset A_{m_2}$, and $\Psi_2(\cdot) = F_2(u_2 \cdot u_2^*)$ is a $*$ isomorphism from B_2 onto A'_{n_2}, and by (6),

$$\Psi_2\widetilde{(r_2)} = \Psi(\tilde{r}_2), \quad \forall r_2 \in \text{Proj } (B_2). \tag{7}$$

Moreover, we have the following commutative diagram:

$$\begin{array}{ccccccc} A_{m_1} & \hookrightarrow & B_1 & \hookrightarrow & A_{m_2} & \hookrightarrow & B_2 \\ \Phi_1\downarrow & & \Psi_1\downarrow & & \Phi_2\downarrow & & \Psi_2\downarrow \\ B'_1 & \hookrightarrow & A'_{n_1} & \hookrightarrow & B'_2 & \hookrightarrow & A'_{n_2}. \end{array}$$

Going on this way, we can complete the proof. Q.E.D.

Theorem 15.2.9. Two (AF)-algebras $A = \overline{\sqcup_n A_n}$ and $B = \overline{\sqcup_n B_n}$ are $*$ isomorphic if and only if the (LF)-algebras $\sqcup_n A_n$ and $\sqcup_n B_n$ are $*$ isomorphic.

Proof. The sufficiency is obvious. Now let A and B be $*$ isomorphic. By Theorem 15.2.8, it suffices to show that $E = E(A)$ and $E' = E\ (\sqcup_n A_n)$ are isomorphic.

First, E' can be embedded into E naturally. In fact, let $p, q \in \text{Proj}\,(\sqcup_n A_n)$, and $v \in A$ with $v^*v = p, vv^* = q$. Pick n and $a \in A_n$ such that $p, q \in A_n$, and $\|a - v\| < \delta_3(\frac{1}{2})$, where $\delta_3(\cdot)$ is as in Lemma 15.1.7. Then there is $u \in A_n$ such that $p = u^*u$ and $q = uu^*$. So p and q are also equivalent in $\sqcup_n A_n$.

Moreover, the above embedding is also surjective. In fact, let p be any projection of A. By Lemmas 15.1.3 and 15.1.6, we can find a projection q of $\sqcup_n A_n$, such that $p \sim q$.

Clearly, the above embedding keeps the partial addition. Conversely, let $\{p_1, \cdots, p_m\} \subset \text{Proj}\,(A)$ with $p_i p_j = 0, \forall i \neq j$. From Lemmas 15.1.3, 15.1.4 and 15.1.6, there is an orthogonal family $\{q_1, \cdots, q_m\}$ of $\text{Proj}(\sqcup_n A_n)$ such that $p_1 \sim q_i, 1 \leq i \leq n$. Therefore, E and E' are isomorphic. Q.E.D.

Theorem 15.2.10. Two (AF)-algebras $A = \overline{\sqcup_n A_n}$ and $B = \overline{\sqcup_n B_n}$ are $*$ isomorphic if and only if there exists a subsequence $\{A_{n_k}\}$ of $\{A_n\}$ and a $*$ subalgebra B'_k of A_{n_k} for each k such that:

1) $B'_1 \subset \cdots \subset B'_k \subset \cdots$, and there is a $*$ isomorphism Φ from $\sqcup_n B'_k$ onto $\sqcup_n B_k$ with $\Phi(B'_k) = B_k, \forall k$;

2) for each n, A_n is contained in some B'_k.

Proof. The sufficiency is obvious from Theorem 15.2.9. Now let A and B be $*$ isomorphic. By Theorem 15.2.9, we have a $*$ isomorphism Φ from $\sqcup_n A_n$ onto $\sqcup_n B_n$. For any k, let $B'_k = \Phi^{-1}(B_k)$. Clearly, $B'_1 \subset \cdots \subset B'_k \subset \cdots$, and for each $k, B'_k \subset A_{n_k}$ for some n_k. We may assume that $n_1 < n_2 < \cdots$. Finally, for each n, since $\sqcup_j B'_j = \Phi^{-1}(\sqcup_j B_j) = \sqcup_j A_j$, it follows that $A_n \subset B'_k$ for some k.

Q.E.D.

Definition 15.2.11. Let $\{p_n\}$ be a sequence of positive integers with $p_n | p_{n+1}, \forall n$. A sequence $\{r_n\}$ of prime numbers is said to be determined by $\{p_n\}$, if there is a sequence $\{s_1 < s_2 < \cdots\}$ of positive integers such that

$$\prod_{i=1}^{s_1} r_i = m_1, \quad \prod_{i>s_1}^{s_2} r_i = m_2, \quad \prod_{i>s_2}^{s_3} r_i = m_s, \cdots,$$

where $m_1 = p_1, m_n = p_n/p_{n-1}, \forall n \geq 2$.

Theorem 15.2.12. Let A_i be a (UHF)-algebra of type $\{p_n^{(i)}\}$, and $\{r_n^{(i)}\}$ be a sequence of prime numbers determined by $\{p_n^{(i)}\}, i = 1, 2$. Then A_1 and A_2 are $*$ isomorphic if and only if for any prime number r, the times of r appeared in $\{r_n^{(1)}\}$ and $\{r_n^{(2)}\}$ are the same.

Proof. By Proposition 3.8.3, it suffices to prove the necessity. Let A_1 and A_2 be $*$ isomorphic. From Theorem 15.2.10, we can find two sequences

$\{k_1 < k_2 < \cdots\}, \{n_{k_1} < n_{k_2} < \cdots\}$, a subfactor $B_{k_j}^{(1)}$ of A_1 for each j, and a $*$ isomrophism Φ from $\sqcup_j B_{k_j}^{(1)}$ onto $\sqcup_j A_{k_j}^{(2)}$, such that

$$B_{k_1}^{(1)} \subset A_{n_{k_1}}^{(1)} \subset B_{k_2}^{(1)} \subset A_{n_{k_2}}^{(1)} \subset \cdots \subset B_{k_j}^{(1)} \subset A_{n_{k_j}}^{(1)} \subset \cdots,$$

and

$$\Phi(B_{k_j}^{(1)}) = A_{k_j}^{(2)}, \quad \forall j,$$

where $1_i \in A_1^{(i)} \subset \cdots \subset A_n^{(i)} \subset \cdots \subset A_i, A_i = \overline{\sqcup_n A_n^{(i)}}$, and $A_n^{(i)}$ is $p_n^{(i)} \times p_n^{(i)}$ matrix algebra, $\forall n, i = 1,2$. Now by Proposition 3.8.3, we have $p_{k_j}^{(2)} | p_{n_{k_j}}^{(1)}$ and $p_{n_{k_j}}^{(1)} | p_{k_j}^{(2)}, \forall j$. That comes to the conclusion. Q.E.D.

Notes. C^*-algebraic dimension theory was first explicitly strudied by J. Diximer, who used it to classify the matroid algebras. G.A. Elliott then extended this theory to the (AF)-algebras. Theorem 15.2.10 is due to G.A. Elliott. And Theorem 15.2.12 is due to J. Glimm.

References. [15], [26], [42], [54].

15.3. The Bratteli diagrams of (AF)-algebras

Let $A = \overline{\sqcup_n A_n}$ be an (AF)-algebra. Then we can see A as the inductive limite of the increasing sequence $\{A_n\}$ of finite dimensional C^*-algebra. In fact, let Φ_n be the embedding map from A_n into A_{n+1}, and for any $m > n$, $\Phi_{mn} = \Phi_{m-1} \circ \cdots \circ \Phi_n$. By Corollary 3.7.4, A is $*$ isomorphic to $\varinjlim\{A_n, \Phi_{mn} \mid m > n\} = \varinjlim_n\{A_n, \Phi_n\}$. Conversely, let $\{A_n\}$ be a sequence of finite dimensional C^*-algebras, and for each n, suppose that Φ_n is a $*$ isomorphism from A_n into A_{n+1}. Then the inductive limit $\{A_n, \Phi_n | n\}$ is an (AF)-algebra.

So it is important to consider the $*$ isomorphism from a finite dimensional C^*-algebra into another finite dimensional C^*-algebra.

Let $A = \oplus_{j=1}^n A_j, B = \oplus_{i=1}^m B_i$ be two finte dimensional C^*-algebras, where A_j, B_i are $*$ isomrophic to matrix algebras, $\forall i, j$, and let Φ be a $*$ isomorphism from A into B. If p is a minimal (i.e. rank one) projection of A_j, then $\Phi(p)z_i$ is a projection of B_i, where z_i is a central projection of B with $B_i = Bz_i$. Suppose that the rank of $\Phi(p)z_i$ in B_i is s_{ij} (a nonnegative integer), i.e., $\Phi(p)z_i$ is a sum of an orthogonal family Λ of minimal projections of B_i, and $\#\Lambda = s_{ij}$. We claim that the non-negative integer s_{ij} is independent of the choice of minimal projection p of A_j. In fact, if q is another minimal projection of A_j, then there is $v \in A_j$ such that $v^*v = p$ and $vv^* = q$. Thus we have

$$\Phi(p)z_i = (\Phi(v)z_i)^* \cdot (\Phi(v)z_i), \quad \Phi(q)z_i = (\Phi(v)z_i) \cdot (\Phi(v)z_i)^*,$$

i.e. the ranks of $\Phi(p)z_i$ and $\Phi(q)z_i$ in B_i are the same: s_{ij}. Therefore, Φ determines a unique $m \times n$ embedding matrix $(s_{ij})_{1 \le i \le m, 1 \le j \le n}$ of non-negative integers. Clearly, we have

$$\sum_{j=1}^{n} s_{ij} (\dim A_j)^{1/2} \le (\dim B_i)^{1/2}, \quad 1 \le i \le m; \tag{1}$$

and (1) becomes an equality for any i if and only if $\Phi(1_A) = 1_B$. Moreover, since $\Phi(p) \neq 0$ for any minimal projection p of A_j, it follows that

$$\sum_{i=1}^{m} s_{ij} > 0, \quad 1 \le j \le m. \tag{2}$$

Conversely, if a $m \times n$ matrix (s_{ij}) of non-negative integers satisfies (1) and (2), then we can construct a $*$ isomorphism Φ from A into B such that the embedding matrix determined by Φ as above is (s_{ij}) exactly. In fact, let $\{e_{st}^{(j)} \mid 1 \le s,t \le (\dim A_j)^{1/2}\}, \{f_{st}^{(i)} \mid 1 \le s,t \le (\dim B_i)^{1/2}\}$ be matrix units of A_j, B_i respectively, $1 \le j \le n, 1 \le i \le m$, and define

$$\Phi(e_{st}^{(j)})z_i = \begin{cases} 0, & \text{if } k = 0, \\ f^{(i)}_{r+(s-1)k+1, r+(t-1)k+1} + \cdots + f^{(i)}_{r+sk, r+tk}, & \text{if } k > 0, \end{cases}$$

where $k = s_{ij}, r = \sum_{l=1}^{j-1} s_{il} (\dim A_l)^{1/2}, \forall 1 \le s,t \le (\dim A_j)^{1/2}, 1 \le j \le n, 1 \le i \le m$. Then this Φ satisfies our condition.

Lemma 15.3.1. Let Φ, Ψ be two $*$ isomorphisms from A into B, where A and B are two finite dimensional C^*-algebras. If the embedding matrices determined by Φ, Ψ are the same, then there exists a unitary element u of B such that

$$u^* \Phi(a) u = \Psi(a), \quad \forall a \in A.$$

Proof. Replacing Φ, Ψ by $\Phi_i(\cdot) = \Phi(\cdot)z_i, \Psi_i(\cdot) = \Psi(\cdot)z_i$ respectively $(1 \le i \le m)$, we may assume that $B = B(H)$, where H is a finite dimensional Hilbert space (dim H =(dim $B_i)^{1/2}$).

For $j \in \{1, \cdots, n\}$, let $\{e_{st}^{(j)} \mid 1 \le s,t \le (\dim A_j)^{1/2}\}$ be a matrix unit of A_j. By the assumption, we have $\dim \Phi(e_{11}^{(j)})H = \dim \Psi(e_{11}^{(j)})H$. Let $\{\xi_1^{(j)}, \cdots, \xi_k^{(j)}\}$ and $\{\eta_1^{(j)}, \cdots, \eta_k^{(j)}\}$ be orthogonal normalized bases of $\Phi(e_{11}^{(j)})H$ and $\Psi(e_{11}^{(j)})H$ respectively. Since $e_{11}^{(j)} = e_{s1}^{(j)*} e_{s1}^{(j)}$ and $e_{ss}^{(j)} = e_{s1}^{(j)} e_{s1}^{(j)*}$, $\{\Phi(e_{s1}^{(j)})\xi_l^{(j)} \mid 1 \le l \le k\}$

and $\{\Psi(e_{s1}^{(j)})\eta_l^{(j)} \mid 1 \le l \le k\}$ are orthogonal normalized bases of $\Phi(e_{ss}^{(j)})H$ and $\Psi(e_{ss}^{(j)})H$ respectively, $1 \le s \le (\dim A_j)^{1/2}$. Clearly, $\langle \Phi(e_{s1}^{(j)})\xi_l^{(j)}, \Phi(e_{t1}^{(j)})\xi_l^{(j)}\rangle = \delta_{st}, \langle \Psi(e_{s1}^{(j)})\eta_l^{(j)}, \Psi(e_{t1}^{(j)})\eta_l^{(j)}\rangle = \delta_{st}, \forall s,t,l$. Thus we can find a unitary operator u on H such that

$$u\Psi(e_{s1}^{(j)})\eta_l^{(j)} = \Phi(e_{s1}^{(j)})\xi_l^{(j)}, \quad \forall s,l,j.$$

Noticing that

$$\begin{aligned} u^*\Phi(e_{st}^{(j)})u\Psi(e_{r1}^{(j)})\eta_l^{(j)} &= \delta_{rt}\Psi(e_{s1}^{(j)})\eta_l^{(j)} \\ &= \Psi(e_{st}^{(j)})\Psi(e_{r1}^{(j)})\eta_l^{(j)}, \end{aligned}$$

$\forall s,t,r,l,j$, we have $u^*\Phi(a)u = \Psi(a), \forall a \in A$. Q.E.D.

From the above Lemma, a $*$ isomorphism Φ form A into B is determined completely by the embedding matrix (s_{ij}). So it is reasonable to write $\Phi = (s_{ij})$. More intuitively, we use the following diagram to describe Φ:

$$\begin{array}{ccc} A_1 & \xrightarrow{s_{11}} & B_1 \\ \vdots & & \vdots \\ A_j & \xrightarrow{s_{ij}} & B_i \qquad \Phi = (s_{ij})_{\substack{1\le i\le m \\ 1\le j\le n}}. \\ \vdots & & \vdots \\ A_n & \xrightarrow{s_{mn}} & B_m \end{array}$$

Lemma 15.3.2. Let A_i, B_i be two finite dimensional C^*-algebras, θ_i be a $*$ isomorphism from A_i onto $B_i, i = 1,2, \Phi, \Psi$ be $*$ isomporphisms from A_1 into A_2, B_1 into B_2 respectively, and the embedding matrices determined by Φ, Ψ be the same. Then there exists a $*$ isomorphism $\tilde{\theta}_2$ from A_2 onto B_2 such that the following diagram is commutative:

$$\begin{array}{ccc} A_1 & \xrightarrow{\Phi} & A_2 \\ \theta_1 \downarrow & & \tilde{\theta}_2 \downarrow \\ B_1 & \xrightarrow{\Psi} & B_2. \end{array}$$

Proof. First, notice the following two facts:

1) Let C_1, C_2, C_3 be finite dimensional C^*-algebras, Φ_1, Φ_2 be $*$ isomorphisms from C_1 into C_2, C_2 into C_3 respectively, and $(s_{ij}^{(1)}), (s_{kl}^{(2)})$ be the embedding matrices determined by Φ_1, Φ_2 respectively. Then the embedding matrix determined by $\Phi_2 \circ \Phi_1$ is $(s_{kl}^{(2)}) \cdot (s_{ij}^{(1)})$.

2) Let C_1, C_2 be two finite dimensional C^*-algebras, and θ be a $*$ isomorphism from C_1 onto C_2. Then the embedding matrix determined by θ is the unit matrix.

Their proofs are easy.

Now consider $*$ isomorphisms $\Phi, \theta_2^{-1} \circ \Psi \circ \theta_1$ from A_1 into A_2. From above facts, their embedding matrices are the same. So by Lemma 15.3.1, there is a unitary element u of A_2 such that

$$u^* \Phi(\cdot) u = \theta_2^{-1} \circ \Psi \circ \theta_1(\cdot), \quad \forall \cdot \in A_1.$$

Define $\tilde{\theta}_2(\cdot) = \theta_2(u^* \cdot u), \forall \cdot \in A_2$, then we get

$$\Psi \circ \theta_1 = \tilde{\theta}_2 \circ \Phi.$$

Q.E.D.

Definition 15.3.3. Let a be an (AF)-algebra, and $\{A_n\}$ be an increasing sequence of finitie dimensional C^*-subalgebras of A with $\overline{\cup_n A_n} = A$. A *Bratteli diagram* of A corresponding to $\{A_n\}$ is

$$\mathcal{D}(A, \{A_n\}) = \mathcal{D} = (D, d, \mathcal{U}),$$

where $D = \{(n,m) | m = 1, \cdots, r(n), n = 1, 2, \cdots\}$ ($r(n)$ is the number of minimal central projections of A_n); $d : D \to \mathbb{N}, d(n,m) = \dim\,(A_n z_m^{(n)})^{1/2}$ (where $\{z_1^{(n)}, \cdots, z_{r(n)}^{(n)}\}$ is the set of all minimal central projections of A_n), $\forall (n,m) \in D; \mathcal{U} = \{\Phi_1, \cdots, \Phi_n, \cdots\}$ (where Φ_n is the embedding matrix from A_n into $A_{n+1}, \forall n$).

Example 1. The Bratteli diagram of a (UHF)-algebra of type $\{p_n\}$.

In this case, $r(n) = 1, d(n,1) = p_n, \Phi_n = (m_n)$, where $m_n = p_n^{-1} p_{n+1}, \forall n$, i.e.,

$$\begin{array}{ccccccccccc} \cdot & \xrightarrow{m_1} & \cdot & \xrightarrow{m_2} & \cdot & \longrightarrow & \cdots & \cdot & \xrightarrow{m_n} & \cdot & \cdots \\ p_1 & & p_2 & & p_3 & & & p_n & & p_{n+1} & \end{array}$$

Example 2. Let H be a separable infinite dimensional Hilbert space, $K = C(H)$ be the set of all compact linear operators on H, and

$$A = K \dot{+} \mathbb{C} 1_H.$$

If $\{\xi_n\}$ is an orthogonal normalized basis of H, and p_n, q_n, r_n are the projections from H onto $[\xi_1, \cdots, \xi_n], [\xi_1, \cdots, \xi_n]^\perp, [\xi_n]$ respectively, $\forall n$, then we can write $A = \overline{\cup_n A_n}$, where

$$A_n = p_n B(H) p_n \dot{+} \mathbb{C} 1_H = p_n B(H) p_n \oplus \mathbb{C} q_n, \forall n.$$

Clearly, any minimal projection of $p_n B(H) p_n$ is still a minimal projection of $p_{n+1} B(H) p_{n+1}, q_n = q_{n+1} + r_{n+1}$, and r_{n+1} is a minimal projection of $p_{n+1} B(H) \cdot p_{n+1}, \forall n$. So $\mathcal{D}(A, \{A_n\})$ is as follows

$$r(n) = 2, d(n,1) = n, d(n,2) = 1, \Phi_n = \begin{pmatrix} 1 & 1 \\ 0 & 1 \end{pmatrix},$$

$\forall n$.

Example 8. The GICAR (*gauge invariant canonical anticommutation relation*) algebra. It has a diagram with $r(n) = n+1, d(n,m) = C_n^m$, and

$$\Phi_n = \begin{pmatrix} 1 & & & \\ 1 & 1 & & 0 \\ & 1 & \ddots & \\ & & \ddots & 1 \\ 0 & & \ddots & 1 \end{pmatrix}, \quad (n+2)\times(n+1),$$

$\forall n = 0, 1, 2 \cdots, m = 0, 1, \cdots, n$, i.e., the Pascal's triangle.

Proposition 15.3.4. Let $A = \overline{\sqcup_n A_n}, B = \overline{\sqcup_n B_n}$ be two (AF)-algebras, and $\mathcal{D}(A, \{A_n\}) = \mathcal{D}(B, \{B_n\})$. Then A and B are $*$ isomorphic.

Proof. Let Φ_n, Ψ_n be the $*$ isomorphisms from A_n into A_{n+1}, B_n into B_{n+1} respectively. By the assumption and Lemma 15.3.2, for each n we can construct a $*$ isomorphism θ_n from A_n onto B_n such that the following diagram is commutative:

$$\begin{array}{ccccccccccc} A_1 & \xrightarrow{\Phi_1} & A_2 & \xrightarrow{\Phi_2} & A_3 & \longrightarrow & \cdots & \longrightarrow & A_n & \xrightarrow{\Phi_n} & A_{n+1} & \longrightarrow \cdots \\ \theta_1\downarrow & & \theta_2\downarrow & & \theta_3\downarrow & & & & \theta_n\downarrow & & \theta_{n+1}\downarrow & \\ B_1 & \xrightarrow{\Psi_1} & B_2 & \xrightarrow{\Psi_2} & B_3 & \longrightarrow & \cdots & \longrightarrow & B_n & \xrightarrow{\Psi_n} & B_{n+1} & \longrightarrow \cdots. \end{array}$$

Thus we get a $*$ isomorphism θ from $\sqcup_n A_n$ onto $\sqcup_n B_n$ such that $\theta|A_n = \theta_n, \forall n$. Therefore, A and B are $*$ isomorphic. Q.E.D.

Clearly, any (AF)-algebra has a Bratteli diagram at least. Conversely, let $\mathcal{D} = \{D, d, u\}$ be a diagram , and each $\Phi_n(\in \mathcal{U})$ satisfy (1), (2), i.e.,

$$d(n+1, i) \geq \sum_{j=1}^{r(n)} s_{ij}^{(n)} d(n, j), \quad 1 \leq i \leq r(n+1),$$

and

$$\sum_{i=1}^{r(n+1)} s_{ij}^{(n)} > 0, \quad 1 \leq j \leq r(n),$$

where $\Phi_n = (s_{ij}^{(n)})_{1\leq i\leq r(n+1), 1\leq j\leq r(n)}, \forall n$. Then we can construct an (AF)-algebra $A = \overline{\sqcup_n A_n}$ such that $\mathcal{D}(A, \{A_n\}) = \mathcal{D}$. Moreover, from Proposition 15.3.4 such (AF)-algebra A is unique up to $*$ isomorphism.

The Bratteli diagrams of an (AF)-aglebra depends on not only the algebra itself but also the choice of the dense increasing sequence of finite dimensional C^*-subalgebras. So for $*$ isomorphic (AF)-algebras, their Bratteli diagrams may be very different.

Notes. The Bratteli diagrams of (AF)-algebras was introduced by O. Bratteli. It is an important tool for studying the constructions of (AF)-algebras.

Let A, B be two finite dimensional C^*-algebras, and Φ be a $*$ isomoprhism from A into B. Then the embedding matrix is exactly the homomorphism $K_0(\Phi) : K_0(A) \to K_0(B)$ from the point of view of K-theory.

References. [11], [15], [99].

15.4. Ideals of (AF)-algebras

Lemma 15.4.1. Let $A = \overline{\cup A_n}$ be an (AF)-algebra, and J be a closed two-sided ideal of A. Then $J = \overline{\cup_n(J \cap A_n)}$.

Proof. Let $J_n = J \cap A_n, \forall n$. Clearly, J_n is a two-sided ideal of $A_n, \forall n$, and $\overline{\cup_n J_n} \subset J$.

Let $x \in A \backslash \overline{\cup_n J_n}$, and $\varepsilon = \inf\{\|x - y\| | y \in \cup_n J_n\}$. Clearly, $\varepsilon > 0$. Pick $x_n \in A_n, \forall n$, such that $x_n \to x$. Then we have n_0 with $\|x_n - x\| < \varepsilon/2, \forall n \geq n_0$. So

$$\|x_n - y\| \geq \|x - y\| - \|x_n - x\| > \varepsilon/2,$$

$\forall y \in J_n$, and $n \geq n_0$. Let $a \to \tilde{a}$ be the canonical map from A onto A/J. Since A_n/J_n can be naturally embedded into A/J, it follows that

$$\|\tilde{x}_n\| = \inf\{\|x_n - y\| | y \in J_n\} \geq \varepsilon/2, \quad \forall n \geq n_0,$$

and $\|\tilde{x}\| \geq \varepsilon/2, x \notin J$. Therefore, we have $J = \overline{\cup_n J_n}$ Q.E.D.

Definition 15.4.2. Let $\mathcal{D} = \{D, d, \mathcal{U}\}$ be a diagram of an (AF)-algebra, $D = \cup_n D_n, D_n = \{(n, m) | 1 \leq m \leq r(n)\}, \forall n, \mathcal{U} = \{\Phi_n = (s_{ij}^{(n)}) | n\}$. A point $(n+1, i)$ is called a *descendant* of a point (n, j), if $s_{ij}^{(n)} > 0$. In general, a point $y \in D_m$ is called a *descendant* of a point $x \in D_n$, which is denoted by $x \to y$, if $m > n$, and there exist points $x_k \in D_k, n \leq k \leq m$, such that $x_n = x, x_m = y$, and x_{k+1} is a descendant of $x_k, n \leq k \leq m - 1$.

Let $x = (n, j), y = (m, i)$. Clearly, $x \to y$ if any only if the (i, j)-element of the matrix $(\Phi_{m-1} \cdots \Phi_n)$ is not zero.

Definition 15.4.3. Let $\mathcal{D} = \{D, d, \mathcal{U}\}$ be a diagram of an (AF)-algebra. A subset E of D is called an *ideal*, if : 1) any descendant of x belongs to E, $\forall x \in E$; 2) suppose that $x \in D_n$ and $\{y \in D_{n+1} | y$ is a descendant of $x\} \subset E$, then $x \in E$.

Lemma 15.4.4. Let $A = \overline{\sqcup_n A_n}$ be an (AF)-algebra, and $\mathcal{D}(A, \{A_n\}) = \mathcal{D} = \{D, d, \mathcal{U}\}$ be the corresponding diagram. If J is a two-sided ideal of $\sqcup_n A_n$, then there exists an ideal subset E of D such that

$$J = \sqcup_n \oplus \{A_{n,k} | (n,k) \in E\}, \tag{1}$$

where each $A_{n,k}$ is a matrix algebra, and $A_n = \oplus_{k=1}^{r(n)} A_{n,k}, \forall n$.

Conversely, if E is an ideal subset of D, then $\sqcup_n \oplus \{A_{n,k} | (n,k) \in E\}$ determines a two sided ideal J of $\sqcup_n A_n$, and $J \cap A_n = \oplus\{A_{n,k} | (n,k) \in E\}, \forall n$.

Proof. Let J be a two-sided ideal of $\sqcup_n A_n$. Since $J = \sqcup_n (J \cap A_n)$, there is a subset E of D such that (1) holds. Now we must prove that E is an ideal subset of D.

Let $(n,k) \in E$, and $(n,k) \to (n+1,l)$. Clearly, if p is a minimal projection of $A_{n,k}$, then $pz \neq 0$, where z is the minimal central projection of A_{n+1} with $A_{n+1}z = A_{n+1,l}$. Since $pz \in J$ and $pz \in A_n z \subset A_{n+1,l}$, it follows that $J \cap A_{n+1,l} \neq \{0\}$. But $A_{n+1,l}$ is a matrix algebra, hence $A_{n+1,l} \subset J$, i.e., $(n+1,l) \in E$.

Now let $(n,k) \in D$, and $\{(n+1,j) | (n,k) \to (n+1,j)\} \subset E$. Then

$$\begin{aligned} A_{n,k} &\subset \oplus\{A_{n+1,j} | (n,k) \to (n+1,j)\} \\ &\subset \oplus\{A_{n+1,j} | (n+1,j) \in E\} \subset J \end{aligned}$$

and $(n,k) \in E$. Therefore, E is an ideal subset.

Conversely, let E be an ideal subset of D, and $J = \sqcup_n \oplus \{A_{n,k} | (n,k) \in E\}$. Put $J_n = \oplus\{A_{n,k} | (n,k) \in E\}, \forall n$. If $(n,k) \in E$, then $A_{n,k} \subset \oplus\{A_{n+1,j} | (n,k) \to (n+1,j)\} \subset J_{n+1}$. So $J_n \subset J_{n+1}, \forall n$, and $J = \sqcup_n J_n$ is a two-sided ideal of $\sqcup_n A_n$. Moreover, if $A_{n,k} \subset J$, then there is $m (> n)$ such that $A_{n,k} \subset J_m$. Thus $\{(m,r) | (n,k) \to (m,r)\} \subset E$. Since E is an ideal subset , it follows that $(n,k) \in E$. Therefore, $J \cap A_n = J_n, \forall n$. Q.E.D.

Theorem 15.4.5. Let $A = \overline{\sqcup_n A_n}$ be an (AF)-algebra, and $\mathcal{D}(A, \{A_n\}) = \{D, d, \mathcal{U}\}$ be the corresponding diagram. Then there are bijections between the following collections:

1) the collection of all closed two-sided ideals of A;
2) the collection of all two-sided ideals of $\sqcup_n A_n$;
3) the collection of all ideal subsets of D.

Proof. By Lemma 15.4.4, there is a bijection between the collections of 2) and 3).

From Lemma 15.4.1, $J \to \overline{J}$ is a map from the collection 2) onto the collection 1). Now let J_1, J_2 be two different two-sided ideals of $\sqcup_n A_n$. We must prove that $\overline{J}_1 \neq \overline{J}_2$. By Lemma 15.4.4, we may assume that there is $A_{n,k} \subset J_1$, but $A_{n,k} \cap J_2 = \{0\}$. Thus $z \notin J_2$, where z is the minimal central projection of A_n such that $A_n z = A_{n,k}$. Further, for any $m \geq n, z + (J_2 \cap A_m)$ is a non-zero projection of $A_m/(J_2 \cap A_m)$, i.e.,

$$\inf\{\|z - y\| | y \in A_m \cap J_2\} = 1, \quad \forall m \geq n.$$

Since $J_2 = \sqcup_{m \geq n}(J_2 \cap A_m)$, it follows that

$$\inf\{\|z - y\| | y \in J_2\} = 1,$$

and $z \notin \overline{J}_2$. Therefore, $\overline{J}_1 \neq \overline{J}_2$. Q.E.D.

Remark. Let $A = \overline{\sqcup_n A_n}$ and $B = \overline{\sqcup_n B_n}$ be two (AF)-algebras, and $A_n = B_n, \forall n$. But they can pick different Bratteli diagrams such that they have different sets of two-sided ideals. Hence, the structure of an (AF)-algebra $A = \overline{\sqcup_n A_n}$ depends on not only each A_n but also each embedding way from A_n into A_{n+1}.

Proposotion 15.4.6. Let $A = \overline{\sqcup_n A_n}$ be an (AF)-algebra, and $\mathcal{D} = \{D, d, \mathcal{U}\}$ be the corresponding diagram. If J is a closed two-sided ideal of A, then J and A/J are also (AF)-algebras, and they have diagrams:

$$\{E, d|E, \mathcal{U}|E\}, \quad \{D\backslash E, d|(D\backslash E), \mathcal{U}|(D\backslash E)\}$$

respectively, where E is the ideal subset of D corresponding to J. Moreover, if $\mathcal{U} = \{U_n = (s_{ij}^{(n)})_{1 \leq i \leq r(n+1), 1 \leq j \leq r(n)} | n\}$, then

$$\mathcal{U}|E = \{V_n = (s_{ij}^{(n)})_{(n+1,i) \in E, (n,j) \in E} | n\}$$

and

$$\mathcal{U}|(D\backslash E) = \{W_n = (s_{ij}^{(n)})_{(n+1,i) \notin E, (n,j) \notin E} | n\}.$$

Proof. From Lemma 16.4.1 and Lemma 16.4.4, clearly J is an (AF)-algebra, and has a diagram $\{E, d|E, \mathcal{U}|E\}$.

Since $A/J = \overline{\sqcup_n (A_n/J)}$, A/J is also an (AF)-algebra. Notice that for each n,

$$\begin{aligned} A_n/J &= \oplus\{A_{n,k}/J | (n,k) \notin E\} \\ &\subset A_{n+1}/J = \oplus\{A_{n+1,j}/J | (n+1,j) \notin E\}, \end{aligned}$$

and $A_{n,k}/J \cong A_{n,k}, A_{n+1,j}/J \cong A_{n+1,j}, \forall (n,k)$ and $(n+1,j) \notin E$. Thus, $A/J = \overline{\sqcup_n (A_n/J)}$ has a diagram $\{D\backslash E, d|(D\backslash E), \mathcal{U}|(D\backslash E)\}$. Q.E.D.

Definition 15.4.7. Let $\mathcal{D} = \{D, d, \mathcal{U}\}$ be a diagram of an (AF)-algebra. An ideal subset E of D is said to be *prime* , if it follows from $x, y \notin E$ that there is $z \notin E$ such that $x \to z, y \to z$.

Theorem 15.4.8. Let $A = \overline{\cup_n A_n}$ be an (AF)-algebra, $\mathcal{D} = \{D, d, \mathcal{U}\}$ be its diagram, and J be a closed two-sided ideal of A. Then the following statements are equivalent:

1) J is primitive;
2) J is prime;
3) The ideal subset E of D corresponding to J is prime.

Proof. 1) $\Longrightarrow$ 2). it is obvsious from Proposition 2.8.8.

Replacing A by A/J and from Proposiiton 15.4.6, we may assume that $J = \{0\}$ and $E = \emptyset$.

2) $\Longrightarrow$ 3). For any points $x = (n,k), y = (m,l) \in D$, let

$$J_1 = \cup_{p>n} \oplus \{A_{pr} | x \to (p,r)\}, J_2 = \cup_{p>m} \oplus \{A_{pr} | y \to (p,r)\}.$$

By the condition 2), we have $\overline{J}_1 \cap \overline{J}_2 \neq \{0\}$. Further, $J_1 \cap J_2 = \overline{J}_1 \cap \overline{J}_2 \cap (\cup_n A_n) \neq \{0\}$ from Lemma 15.4.1. Thus there exists $(p,r) \in D$ such that $A_{pr} \subset J_1 \cap J_2$. By the definition of J_1, we can find $(p_1, r_1) \in D$ with $x \to (p_1, r_1)$ and $(p,r) \to (p_1, r_1)$. Thus, $A_{p_1, r_1} \subset J_1 \cap J_2$. Again by the definition of J_2, there is $(p_2, r_2) \in D$ with $y \to (p_2, r_2)$ and $(p_1, r_1) \to (p_2, r_2)$. Let $z = (p_2, r_2)$. Then $x \to z$ and $y \to z$.

3) $\Longrightarrow$ 1). Since any finite subset of D has a common descendant, we can find a subsequence $\{n_k\}$ and a function $j(\cdot)$ such that for each $k, j(k) \in \{1, \cdots, r(n_{k+1})\}$ and

$$(n_k, i) \longrightarrow (n_{k+1}, j(k)), \quad 1 \leq i \leq r(n_k).$$

Now we may assume that

$$(n,i) \longrightarrow (n+1, 1), \quad \forall 1 \leq i \leq r(n) \text{ and } n.$$

Thus , there is a minimal pojection p_n of $A_n, \forall n$, such that $p_n \geq p_{n+1}, \forall n$. If write

$$p_n a p_n = \rho_n(a) p_n, \quad \forall a \in A_n,$$

then ρ_n is a pure state on $A_n, \forall n$. Since

$$\rho_{n+1}(a) p_{n+1} = p_{n+1} a p_{n+1} = p_{n+1} p_n a p_n p_{n+1} = \rho_n(a) p_{n+1},$$

$\forall a \in A_n$, it follows that $\rho_{n+1} | A_n = \rho, \forall n$. Hence there is a state ρ on A such that $\rho | A_n = \rho_n, \forall n$. Moreover, it is easily verified that ρ is pure. Now it sufficcs to show that ker $\pi_\rho = \{0\}$, where π_ρ is the irreducible $*$ representation of A generated by ρ. From Lemma 15.4.1, it is equivalent to prove that ker $\pi_\rho \cap A_n = \{0\}, \forall n$. Let z be a minimal central projection, and $A_n z = A_{nk}$. Since $(n,k) \to (n+1, 1)$, it follows that $p_{n+1} \precsim p$ relative to A_{n+1}, where p is a minimal projection of A_{nk}. Thus , there is $v \in A_{n+1}$ such that $v^*v = p_{n+1}$, and $vv^* \leq p \leq z$. Further, we have

$$\begin{aligned} 1 = \rho_{n+1}(p_{n+1}) &= \rho(v^*v) = \rho(v^*zv) \\ &= \langle \pi_\rho(z)\pi_\rho(v)\xi_\rho, \pi_\rho(v)\xi_\rho \rangle, \end{aligned}$$

and $\pi_\rho(z) \neq 0, z \notin \ker \pi_\rho$. Therefore, $\ker \pi_\rho \cap A_n = \{0\}, \forall n$. Q.E.D.

Notes. From Proposition 15.4.6, we have the extension problem for (AF)-algebras. This problem was first studied by G. Elliott, and may be stated as follows. Given a C^*-algebra A and a closed two-sided J of A such that J and A/J are both (AF)-algebras, does it follow that A is itself an (AF)-algebra? Elliott was able to prove that this would be the case if one could show that any projection of A/J is the image of projection of A. Then L. Brown solved affirmatively this problem in terms of K-theory.

Referneces. [15], [40], [99]

15.5. Dimension groups

Definition 15.5.1. (G, P) is called an *ordered group,* if G is an abelian group, and P is a subset of G satisfying: (1) $P + P \subset P$; (2) $P \cap (-P) = \{0\}$; (3) $P - P = G$; (4) unperforated, i.e., if $a \in G$ and $na \in P$ for some $n \in \mathbb{N}$, then $a \in P$.

P is also called the *positive part* of the ordered group G, and denoted by G_+ sometimes. We shall write $a \geq 0$ if $a \in P$; and $a \geq b$ if $(a - b) \in P$.

Definition 15.5.2. An element u of an ordered group (G, P) is called an *order unit* , if any $a \in P$, there exists $n \in \mathbb{N}$ such that $a \leq nu$.

Since P is unperforated, it follows that $u \in P$.

An homomorphism ρ from an ordered group (G, P) to $\mathbb{R}$ is called a *state* relative to an order unit u, if ρ is positive (denoted by $\rho \geq 0$), i.e., $\rho(a) \geq 0, \forall a \in P$; and $\rho(u) = 1$. We shall denote by $S_u(G)$ the set of all states relative to u.

Clearly, $\mathbb{R}^G = \times_G \mathbb{R}$ with product topology is a locally convex Hausdorff topologyical linear space. With the embedding: $\rho \longrightarrow (\rho(a))_{a \in G}, S_u(G)$ is a closed convex subset of $\mathbb{R}^G$ obviously. We claim that $S_u(G)$ is also a compact subset of $\mathbb{R}^G$. In fact, for any $a \in G$ there is $n_a \in \mathbb{N}$ such that $-n_a u \leq a \leq n_a u$. Thus, we have

$$S_u(G) \subset \times_{a \in G}[-n_a, n_a].$$

Clearly, $\times_{a \in G}[-n_a, n_a]$ is a compact subset of $\mathbb{R}^G$. Therefore, $S_u(G)$ is a compact convex subset of $\mathbb{R}^G$.

Moreover, if $v \in P$ is another order unit, then $\rho \to \rho(v)^{-1}\rho$ is an affine homeomorphism from $S_u(G)$ onto $S_v(G)$ obviously.

Lemma 15.5.3. Let (G,P) be an ordered group, $u(\in P)$ be an order unit, H be a subgroup of G, and $u \in H$. Then $(H, H_+ = P \cap H)$ is also an ordered group with an order unit u. Moreover, if $\rho \in S_u(H)$, and $a \in G\backslash H$, then ρ can extended to a state on $(H + \mathbb{Z}a)$. Consequtly, ρ can be extended to a state on (G, P, u).

Proof. It suffices to define $\rho(a) = \lambda$ such that

$$\rho(h + ma) = \rho(h) + m\lambda \geq 0 \tag{1}$$

for any $h \in H$ and $m \in \mathbb{Z}$ with $(h + ma) \in P$.

If $h + na \geq 0, h' - n'a \geq 0$, where $h, h' \in H, n, n' \in \mathbb{N}$, then

$$nh' \geq nn'a \geq -n'h$$

and $-\rho(h)/n \leq \rho(h')/n'$. Since $u \in H$, we can pick

$$\lambda \in \left[\sup\left\{\frac{-\rho(h)}{n}\,\middle|\, \begin{matrix} h \in H, n \in \mathbb{N}, \\ (h+na) \geq 0 \end{matrix}\right\},\quad \inf\left\{\frac{\rho(h)'}{n'}\,\middle|\, \begin{matrix} h' \in h, n' \in \mathbb{N}, \\ (h'-n'a) \geq 0 \end{matrix}\right\}\right].$$

Clearly, such λ satisfies the condition (1). Q.E.D.

Definition 15.5.4. Let (G,P) be an ordered group. A subgroup J of G is called an *order ideal* , if $J = J_+ - J_+$, where $J_+ = J \cap P$, and for any $a, b \in P$ with $a \leq b$ and $b \in J_+$ we have also $a \in J_+$.

An order ideal J is said to be *prime* , if J_1 and J_2 are two order ideals with $J = J_1 \cap J_2$, then either $J = J_1$ or $J = J_2$.

Definition 15.5.5. A group G with the following form

$$G = \varinjlim\{\mathbb{Z}^{r(n)}, \Phi_n\}$$

is called a *dimension group,* where $\Phi_n = (s_{ij}^{(n)})$ is a $r(n+1) \times r(n)$ matrix of non-negative integers with $\sum_{l=1}^{r(n+1)} s_{ij}^{(n)} > 0, 1 \leq j \leq r(n), \forall n$. In detail, every element of G has the following form:

$$\Phi_{n\infty}(t_n) = (0, \cdots, 0, t_n, t_{n+1}, \cdots) + I,$$

where $t_s \in \mathbb{Z}^{r(s)}$ and $t_{s+1} = \Phi_s(t_s), \forall s \geq n$, and

$$I = \{(t_1, \cdots, t_m, 0, \cdots, 0, \cdots) | m \geq 1, t_i \in \mathbb{Z}^{r(i)}\}.$$

Proposition 15.5.6. Let $G = \underrightarrow{\lim}\{\mathbb{Z}^{r(n)}, \Phi_n\}$ be a dimension group, and $P = \sqcup_n \Phi_{n\infty}(\mathbb{Z}_+^{r(n)})$. Then (G, P) is a countable ordered group, and (G, P) has also the *Riesz interpolation property* , i.e., if $a, b, c, d \in G$ with $a, b \le c, d$, then there exists $e \in G$ such that $a, b \le e \le c, d$.

Proof. Notice that Φ_n keeps the order, and $(\mathbb{Z}^{r(n)}, \mathbb{Z}_+^{r(n)})$ has the Riesz interpolation property, $\forall n$. Thus the conclusions are obvious. Q.E.D.

Definition 15.5.7. Let $G = \underrightarrow{\lim}\{\mathbb{Z}^{r(n)}, \Phi_n\}$ be a dimension group, $D_n = \{(n, m)|1 \le m \le r(n)\}, D = \sqcup_n D_n$, and $\mathcal{U} = \{\Phi_n|n\}$. Then $\{D, \mathcal{U}\}$ is called a *diagram* of G.

Clearly, any (AF)-algebra admits a dimension gruop. Conversely, if G is a dimension group with a diagram $(D, \mathcal{U})$, the we can construct an (AF)-algebra A such that A admits a diagram $\{D, d, \mathcal{U}\}$. Indeed, it suffices to pick $\{d(n, i)\}$ such that

$$d(n + 1, i) \ge \sum_{j=1}^{r(n)} s_{ij}^{(n)} d(n, j), \quad 1 \le i \le r(n + 1),$$

where $\Phi_n = (s_{ij}^{(n)}), \forall n$.

Proposition 15.5.8. Let $G = \underrightarrow{\lim}\{\mathbb{Z}^{r(n)}, \Phi_n\}$ be a dimension group with a diagram $\{D, \mathcal{U}\}$. Then there is a bijection between the collection of all order ideals of G and the collection of all ideal subsets of D (see Definition1.4.3).

Moreover, if J is an order ideal of G and E is the ideal subset of D corresponding to J, then J is a dimension group with a diagram $\{E, \mathcal{U}|E\}$.

Proof. Let J be an order ideal of G, and $E = \{(n, k)|\Phi_{n\infty}(e_k^{(n)}) \in J\}$, where $\{e_k^{(n)}|1 \le k \le r(n)\}$ is the canonical basis of $\mathbb{Z}^{r(n)}$. If $(n, k) \in E$ and $(n, k) \to (m, p)$, by Definition 15.4.2 we have

$$\begin{aligned} \Phi_{n\infty}(e_k^{(n)}) &= \Phi_{m\infty}(\Phi_{m-1} \circ \cdots \circ \Phi_n(e_k^{(n)})) \\ &\ge \Phi_{m\infty}(e_p^{(m)}) \ge 0. \end{aligned}$$

Since J is an order ideal, it follows that $(m, p) \in E$. Now let $x = (n, k) \in D_n$, and $\{y \in D_{n+1}|x \to y\} \subset E$. Noticing that $J_+ + J_+ \subset J_+$ and

$$\Phi_{n\infty}(e_k^{(n)}) = \sum_{\substack{i \\ (n+1,i) \in E}} s_{ik}^{(n)} \Phi_{n+1,\infty}(e_i^{(n+1)}),$$

we have $\Phi_{n\infty}(e_k^{(n)}) \in J_+$, i.e., $(n, k) \in E$. Therefore, E is an ideal subset of D.

Further, let $J(E)$ be the subgroup of G generated by $\{\Phi_{n\infty}(e_k^{(n)})|(n,k) \in E\}$. Clearly, $J(E) \subset J$, and $J(E)$ is a dimension group with a diagram $\{E, \mathcal{U}|E\}$. Let $a \in J_+$. By $G_+ = \sqcup_n \Phi_{n\infty}(\mathbb{Z}_+^{r(n)})$ there are non-negative integers $\lambda_1, \cdots, \lambda_{r(n)}$ such that $a = \Phi_{n\infty}(\sum_k \lambda_k e_k^{(n)})$. Since J is an order ideal, it follows that $\Phi_{n\infty}(e_k^{(n)}) \in J_+$ if $\lambda_k > 0$, i.e., $(n,k) \in E$ if $\lambda_k > 0$. Therefore, $a \in J(E)$, and $J = J(E)$.

Conversely, let E be an ideal subset of D, and define $J = J(E)$ as above. We claim that J is an order ideal of G. In fact, since E is an ideal subset , it follows that

$$\Phi_{n\infty}\left(\left\{\sum_{\substack{k\\(n,k)\in E}} \lambda_k e_k^{(n)} | \lambda_k \in \mathbb{Z}\right\}\right) \subset \Phi_{n+1,\infty}\left(\left\{\sum_{\substack{i\\(n+1,i)\in E}} \lambda_i e_i^{(n+1)} | \lambda_i \in \mathbb{Z}\right\}\right),$$

$\forall n$. Thus , we have

$$J = \sqcup_n \Phi_{n\infty}\left(\left\{\sum_{\substack{k\\(n,k)\in E}} \lambda_k e_k^{(n)} | \lambda_k \in \mathbb{Z}\right\}\right),$$

$$J_+ = \sqcup_n \Phi_{n\infty}\left(\left\{\sum_{\substack{k\\(n,k)\in E}} \lambda_k e_k^{(n)} | \lambda_k \in \mathbb{Z}_+\right\}\right),$$

and $J = J_+ - J_+$. If $a, b \in G_+$ with $a \leq b$ and $b \in J_+$, from the expression of J_+ we can see that $a \in J_+$. Thus J is an order ideal of G. Moreover, if $(n,k) \in D$ with $\Phi_{n\infty}(e_k^{(n)}) \in J_+$, then we can write

$$\Phi_{n\infty}(e_k^{(n)}) = \sum_{\substack{j\\(m,j)\in E}} \lambda_j \Phi_{m\infty}(e_j^{(m)}),$$

where $m > n$ and $\lambda_j \in \mathbb{Z}_+, \forall j$. Thus, there is p with $p > m, n$ such that

$$\begin{aligned}\Phi_{np}(e_k^{(n)}) &= \Phi_{mp}\Big(\sum_{\substack{j\\(m,j)\in E}} \lambda_j e_j^{(m)}\Big)\\ &\in \left\{\sum_{\substack{i\\(p,i)\in E}} \mu_i e_i^{(p)} | \mu_i \in \mathbb{Z}\right\}.\end{aligned}$$

This means that every descendant of (n,k) in D_p belongs to E. Since E is an ideal subset of D, it follows that $(n,k) \in E$. Q.E.D.

Proposition 15.5.9. Let $G = \varinjlim\{\mathbb{Z}^{r(n)}, \Phi_n\}$ be a dimension group with a diagram $\{D, \mathcal{U}\}$, and $J = J(E)$ be an order ideal of G, where E is an ideal subset of D. Then G/J is also a dimension group with a diagram $\{D\backslash E, \mathcal{U}|(D\backslash E)\}$.

Proof. For any n, let

$$\mathbb{Z}^{r(n)} = \mathbb{Z}^{p(n)} \dot{+} \mathbb{Z}^{q(n)},$$

where $\mathbb{Z}^{p(n)} = [e_k^{(n)} | (n,k) \in E], \mathbb{Z}^{q(n)} = [e_k^{(n)} | (n,k) \notin E]$, and $\{e_k^{(n)} | 1 \le k \le r(n)\}$ is the canonical basis of $\mathbb{Z}^{r(n)}$. By this decomposition, we have projectons $P_n : Z^{r(n)} \to \mathbb{Z}^{p(n)}$ and $Q_n = (1 - P_n) : \mathbb{Z}^{r(n)} \to \mathbb{Z}^{q(n)}, \forall n$. Further, let $\Psi_n = Q_{n+1}(\Phi_n | \mathbb{Z}^{q(n)}), \forall n$. Then the dimension group $\varinjlim\{\mathbb{Z}^{q(n)}, \Psi_n\}$ admits a diagram $\{D\backslash E, \mathcal{U}|(D\backslash E)\}$. We have that:

$$\eta_n = \eta_{n+1} \circ \Psi_n$$

where $\eta_n : \mathbb{Z}^{q(n)} \to G/J$ and $\eta_n(t_n) = \Phi_{n\infty}(t_n) + J, \forall t_n \in \mathbb{Z}^{q(n)}$, and n. Indeed, since E is an ideal subset of D, it follows that

$$\begin{aligned} \eta_n(t_n) &= \Phi_{n+1,\infty} Q_{n+1} \Phi_n(t_n) + \Phi_{n+1,\infty} P_{n+1} \Phi_n(t_n) + J \\ &= \Phi_{n+1,\infty} \Psi_n(t_n) + J = \eta_{n+1}(\Psi_n(t_n)), \end{aligned}$$

$\forall t_n \in \mathbb{Z}^{q(n)}$ and n. Hence, we can define a map $\eta : \varinjlim\{\mathbb{Z}^{q(n)}, \Psi_n\} \longrightarrow G/J$ as follows:

$$\eta(\Psi_{n\infty}(t_n)) = \Phi_{n\infty}(t_n) + J, \quad \forall t_n \in \mathbb{Z}^{q(n)} \text{ and } n.$$

If $t_n \in \mathbb{Z}^{q(n)}$ with $\Phi_{n\infty}(t_n) \in J$, then $t_n = 0$ since E is an ideal subset ,i.e., η is injective. From

$$\begin{aligned} G/J &= \sqcup_n(\Phi_{n\infty}(\mathbb{Z}^{r(n)}) + J) \\ &= \sqcup_n(\Phi_{n\infty}(\mathbb{Z}^{q(n)}) + J), \end{aligned}$$

η is also surjective. Moreover, since η and η^{-1} keep the order, G/J is order isomoprphic to $\varinjlim\{\mathbb{Z}^{q(n)}, \Psi_n\}$. Q.E.D.

Proposition 15.5.10. Let G be a dimension group with a diagram $\{D, \mathcal{U}\}$, and $J = J(E)$ be an order ideal of G, where E is an ideal subset of D. Then J is prime if and only if E is prime (see Definition 15.4.7.)

Proof. By Proposition 15.5.9 and replacing G by G/J, we may assume that $J = \{0\}$ and $E = \emptyset$.

Let the order ideal $\{0\}$ be prime. For any $x_i \in D$, put $E_i = \{z \in D | x_i \to z$, i.e., z is a descendant of $x_i\}, J_i = J(E_i), i = 1, 2$. Suppose that F_i is the ideal subset of D generated by $E_i, i = 1, 2$. Then $J_i = J(F_i), i = 1, 2$. By the assumption, we have $J_1 \cap J_2 \neq \{0\}$. Thus, $F_1 \cap F_2 \neq \emptyset$. Pick $y \in F_1 \cap F_2$. By x_1

and $y \in F_1$, there is $z_1 \in D$ such that $x_1 \to z_1$ and $y \to z_1$. From $y \in F_2$, we have also $z_1 \in F_2$. Further, by z_1 and $x_2 \in F_2$, there is $z \in D$ such that $z_1 \to z$ and $x_2 \to z$. Therfore, $x_1 \to z$ and $x_2 \to z$, i.e., $\emptyset$ is a prime subset of D.

Conversely, let $\emptyset$ be prime. Suppose that $J_i = J(E_i)$ is a non-zero order ideal of G, where E_i is an ideal subset of $D, i = 1, 2$. Since $E_1 \cap E_2 \neq \emptyset$, it follows that $J_1 \cap J_2 \neq \{0\}$. Therefore, $\{0\}$ is a prime order ideal of G. Q.E.D.

Example 1. The CAR (*canonical anticommutation relation*) algebra i.e., the (UHF)-algebra of type $\{2^n\}$.

It has a diagram as follows:

$$\underset{2}{\cdot} \xrightarrow{2} \underset{2^2}{\cdot} \xrightarrow{2} \underset{2^3}{\cdot} \cdots \cdots \underset{2^n}{\cdot} \xrightarrow{2} \underset{2^{n+1}}{\cdot} \cdots\cdots$$

Thus we need to consider the dimension group:

$$G = \varinjlim\{\mathbb{Z}^{r(n)}, \Phi_n\},$$

where $r(n) = 1, \Phi_n = [2], \forall n$. Define a map:

$$(0, \cdots, 0, t_n, t_{n+1}, \cdots) + I \longrightarrow \frac{t_n}{2^n},$$

where $t_n \in \mathbb{Z}^{r(n)}, t_{n+r} = 2^r t_n, \forall r$. Then we can see that G is order isomorphic to the dyadic rationals $\{\frac{k}{2^n} | k \in \mathbb{Z}, n = 1, 2, \cdots\} = \mathbb{Z}[1/2]$ (relative ordering in $\mathbb{R}$).

Example 2. Let H be a separable infinite dimensional Hilbert space and $K = C(H)$. From Section 1.3, K has a diagram as follows:

$$\underset{1}{\cdot} \xrightarrow{1} \underset{2}{\cdot} \xrightarrow{1} \underset{3}{\cdot} \longrightarrow \cdots \cdots \underset{n}{\cdot} \xrightarrow{1} \underset{n+1}{\cdot} \cdots\cdots$$

So we have a dimension group $G = \varinjlim\{\mathbb{Z}^{r(n)}, \Phi_n\}$, with $r(n) = 1$ and $\Phi_n = [1], \forall n$. Clearly, $G \cong \mathbb{Z}$ (usual ordering).

Example 3. The dimension group of the GICAR algebra.

From Example 3 of Section 1.3, its dimension group G will be the inductive limit of the following system:

$$\mathbb{Z} \xrightarrow{\begin{bmatrix}1\\1\end{bmatrix}} \mathbb{Z} \oplus \mathbb{Z} \xrightarrow{\begin{bmatrix}1&0\\1&1\\0&1\end{bmatrix}} \mathbb{Z} \oplus \mathbb{Z} \oplus \mathbb{Z} \longrightarrow \cdots,$$

i.e.

$$G = \varinjlim\{\mathbb{Z}^{n+1}, \Phi_n | n \geq 0\} = \sqcup_{n \geq 0} \Phi_n(\mathbb{Z}^{n+1}),$$

where

$$\Phi_n = \begin{bmatrix} 1 & & 0 \\ 1 & \ddots & \\ & \ddots & 1 \\ 0 & & 1 \end{bmatrix} \qquad (n+2)\times(n+1), \forall n \geq 0.$$

Lemma 15.5.11. Let $a, b \in \mathbb{R}$ with $4b > a^2$. Then there exists a positive integer N such that all coefficients of the polynomial $(x+1)^N(x^2 - ax + b)$ are non-negative.

Proof. Clearly, $b > 0$. So we may assume that $a > 0$. Write

$$(x+1)^N(x^2 - ax + b) = \sum_{i=-2}^{N} \frac{N!}{(i+2)!(N-i)!} C_{i+2} x^{i+2}.$$

Then for $0 \leq i \leq N-2$ we have

$$\begin{aligned} C_{i+2} = & \frac{(i+2)!(N-i)!}{N!}\{C_N^i - aC_N^{i+1} + bC_N^{i+2}\} \\ = & (i+1)(i+2) - a(i+2)(N-i) + b(N-i)(N-i-1) \\ = & (1+a+b)(i - (b+\tfrac{a}{2})(1+a+b)^{-1}N)^2 \\ & +(b - \tfrac{a^2}{4})(1+a+b)^{-1}N^2 - (2a+b)(N-i) + (3i+2) \\ \geq & (b - \tfrac{a^2}{4})(1+a+b)^{-1}N^2 - (2a+b)N. \end{aligned}$$

Moreover,

$$C_0 = \tfrac{(N+2)!}{N!}b, \quad C_1 = \tfrac{(N+1)!}{N!}(Nb - a),$$
$$C_{N+1} = \tfrac{(N+1)!}{N!}(N-a), \quad C_{N+2} = \tfrac{(N+2)!}{N!}.$$

Therefore, if N is large enough, any coefficient of the polynomial $(x+1)^N(x^2 - ax + b)$ is non-negative. Q.E.D.

Theorem 15.5.12. The dimension group $G = \varinjlim\{\mathbb{Z}^{n+1}, \Phi_n | n \geq 0\}$ of the GICAR algebra is order isomorphic to

$$(P_{\mathbb{Z}}([0,1]), \quad P_{\mathbb{Z}}^+([0,1])),$$

where $P_{\mathbb{Z}}([0,1])$ is the additive group of all polynomials on $[0,1]$ with integer coefficients , and

$$P_{\mathbb{Z}}^+([0,1]) = \{f \in P_{\mathbb{Z}}([0,1]) | f(t) > 0, \forall t \in (0,1)\} \cup \{0\}.$$

Proof. Let $u \in G$. Then there is $n(\geq 0)$ and an element $(a_0, \cdots, a_n)$ of $\mathbb{Z}^{n+1}$ such that

$$u = \Phi_{n\infty}((a_0, \cdots, a_n)).$$

From $(a_0, \cdots, a_n)$, we have unique $(b_0, \cdots, b_n) \in \mathbb{Z}^{n+1}$ such that

$$a_0 x^n + \cdots + a_n = b_0(x+1)^n + \cdots + b_n, \quad \forall x \geq 0.$$

Define a homomorphism $\Phi : G \to P_{\mathbb{Z}}([0,1])$ as follows:

$$\Phi(u) = p(t) = b_0 + b_1 t - \cdots + b_n t^n,$$

where $(b_0, \cdots, b_n) \in \mathbb{Z}^{n+1}$ is determined by u as above.

First, we must show that Φ is well-defined. If $(a'_0, \cdots, a'_{n+1}) = \Phi_n((a_0, \cdots, a_n))$, then

$$a'_0 = a_0;\ a'_j = a_{j-1} + a_j, 1 \leq j \leq n;\ a'_{n+1} = a_n.$$

Thus $a'_0 x^{n+1} + \cdots + a'_{n+1} = (x+1)(a_0 x^n + \cdots + a_n), \forall x$. Let $(b'_0, \cdots, b'_{n+1})$, in $\mathbb{Z}^{n+2}$, satisfy $b'_0(x+1)^{n+1} + \cdots + b'_{n+1} = a'_0 x^{n+1} + \cdots + a'_{n+1}, \forall x$. Then we have

$$b'_0(x+1)^{n+1} + \cdots + b'_{n+1} = (x+1)(a_0 x^n + \cdots + a_n) = (x+1)(b_0(x+1)^n + \cdots + b_n),$$

$\forall x$. So , $b'_j = b_j, 0 \leq j \leq n$, and $b'_{n+1} = 0$. Thus, the definition of Φ is independent of the choice of n. Moreover, if $\Phi_{n\infty}((a_0, \cdots, a_n)) = 0$, then there is $m(> n)$ such that $\Phi_{nm}((a_0, \cdots, a_n)) = 0$, where $\Phi_{nm} = \Phi_{m-1} \circ \cdots \circ \Phi_n$. Since each Φ_k is injective, it follows that $a_0 = \cdots = a_n = 0$. Therefore, Φ is well-defined.

Clearly, Φ is an isomorphism from G onto $P_{\mathbb{Z}}([0,1])$. So it suffices to show that Φ is also an order isomorphism.

Let $u \in G_+ \backslash \{0\}$. Then there is $(a_0, \cdots, a_n) \in \mathbb{Z}_+^{n+1} \backslash \{0\}$ such that $u = \Phi_{n\infty}((a_0, \cdots, a_n))$. Thus

$$b_0(x+1)^n + \cdots + b_n = a_0 x^n + \cdots + a_n > 0, \quad \forall x > 0.$$

Therefore, we have

$$p(t) = \left. \frac{b_0(x+1)^n + \cdots + b_n}{(x+1))^n} \right|_{t=\frac{1}{x+1}} > 0, \quad \forall t \in (0.1).$$

Conversely, let $p(t) = b_0 + \cdots + b_n t^n \in P^+_{\mathbb{Z}}([0,1]) \backslash \{0\}$. Then

$$f(x) = b_0(x+1)^n + \cdots + b_n = a_0 x^n + \cdots + a_n > 0, \quad \forall x > 0.$$

We need to prove that $\Phi_{n\infty}((a_0, \cdots, a_n)) \in G_+$, or to show that there exists $m(> n)$ such that

$$\Phi_{nm}((a_0, \cdots, a_n)) \in \mathbb{Z}_+^{m+1}.$$

Clearly, it is equivalent to prove that there exists a positive integer N such that all coefficients of the polynomial $(x+1)^N f(x)$ are non-negative.

Since $f(x) > 0, \forall x > 0$, we can write

$$f(x) = C\prod_i(x+\lambda_i)\prod_j(x-\alpha_j)(x-\overline{\alpha}_j),$$

where $C > 0, \lambda_i \geq 0$, and $\alpha_j \in \mathbb{C}\backslash\mathbb{R}, \forall i,j$. Now applying Lemma 15.5.11 to each $(x-\alpha_j)(x-\overline{\alpha}_j)$, we can find a positive integer N such that all coefficients of the polynomial $(x+1)^N f(x)$ are non-negative. Q.E.D.

Notes. Proposition 15.5.6 is indeed a characterization of dimension groups. We have the following Effros-Handelman-Shen theorem: if G is a countable ordered group, and G satisfies the Riesz interpolation propperty, then G is a dimension group.

Let A be an (AF)-algebra with a diagram $\{D, d, \mathcal{U}\}$. If $D = \sqcup_n D_n$; $D_n = \{(n,m)|1 \leq m \leq r(n)\}$; $\mathcal{U} = \{\Phi_n|n\}$, then the dimension group $G = \varinjlim\{\mathbb{Z}^{r(n)}, \Phi_n\}$ is indeed the K_0-group of A.

References. [39], [40], [43], [59], [103], [133], [160].

15.6. Scaled dimension groups and stablly isomorphic theorem

Definition 15.6.1. Let G be a dimension group. A subset Γ of G_+ is called a *scale* for G, if: 1) G_+ is generatred by Γ, i.e., $G_+ = \Gamma + \Gamma + \cdots$; 2) for any $a, b \in G_+$ with $a \leq b$ and $b \in \Gamma$, we have also $a \in \Gamma$.

In this case, we say that $(G, G_+ = P, \Gamma)$ is a *scaled dimension group.*

For example, if G has an order unit u, then $\Gamma = [0,u] = \{v \in G | 0 \leq v \leq u\}$ is a scale for G.

In a scale Γ, we can define a *partial addition* , i.e., $a, b \in \Gamma$ is said to be *additive* , if $(a+b) \in \Gamma$.

Lemma 15.6.2. Let $\alpha_i, \beta_j \in \mathbb{Z}_+, 1 \leq i \leq r, 1 \leq j \leq s$, and $\alpha_1 + \cdots + \alpha_r = \beta_1 + \cdots + \beta_s$. Then there is a subset $\{\gamma_{ij}|1 \leq i \leq r, 1 \leq j \leq s\}$ of $\mathbb{Z}_+$ such that

$$\alpha_i = \sum_{k=1}^{s}\gamma_{ik}, \beta_j = \sum_{k=1}^{r}\gamma_{kj}, \quad 1 \leq i \leq r, 1 \leq j \leq s.$$

Proof. If $\beta_1 \geq \alpha_1$, let $\gamma_{11} = \alpha_1, \gamma_{1j} = 0, 2 \leq j \leq s$, then we need to find $\{\gamma_{ij}|\ 2 \leq i \leq r, 1 \leq j \leq s\} (\subset \mathbb{Z}_+)$ such that

$$\alpha_i = \sum_{k=1}^{s}\gamma_{ik}, \quad 2 \leq i \leq r, \quad \sum_{k=2}^{r}\gamma_{k1} = \beta_1 - \alpha_1,$$

and

$$\beta_j = \sum_{k=2}^{r} \gamma_{kj}, \quad 2 \le j \le s.$$

If $\alpha_1 \ge \beta_1$, let $\gamma_{11} = \beta_1, \gamma_{i1} = 0, \quad 2 \le i \le r$, then we need to find $\{\gamma_{ij} | 1 \le i \le r, 2 \le j \le s\} (\subset \mathbb{Z}_+)$ such that

$$\sum_{k=2}^{s} \gamma_{1k} = \alpha_1 - \beta_1, \quad \sum_{k=2}^{s} \gamma_{ik} = \alpha_i, \quad 2 \le i \le r$$

and

$$\sum_{k=1}^{r} \gamma_{kj} = \beta_j, \quad 2 \le j \le s.$$

Repeating this process, we can get the proof. Q.E.D.

Proposition 15.6.3. Let G_i be a dimension group, Γ_i be a scale for $G_i, i = 1,2$, and Φ be an isomorphism from Γ_1 onto Γ_2, i.e., Φ and Φ^{-1} keep the partial addition . Then Φ can be uniquely extended to an order isomorphism from G_1 onto G_2.

Proof. Let $a_i, b_j \in \Gamma_1$ and

$$a_1 + \cdots + a_r = b_1 + \cdots + b_s.$$

Then by Lemma 15.6.2 we have $\{c_{ij}\} \subset (G_1)_+$ such that

$$a_i = \sum_{k=1}^{s} c_{ik}, \quad b_j = \sum_{k=1}^{r} c_{kj}, \quad 1 \le i \le r, 1 \le j \le s.$$

Since Γ_1 is a scale, it follows that $c_{ij} \in \Gamma, \forall i, j$, and $\{c_{ik}|k\}, \{c_{kj}|k\}$ are additive in $\Gamma_1, \forall i, j$. Then $\{\Phi(c_{ik})|k\}, \{\Phi(c_{kj})|k\}$ are also additive in Γ_2, and

$$\Phi(a_i) = \sum_{k=1}^{s} \Phi(c_{ik}), \quad \Phi(b_j) = \sum_{k=1}^{r} \Phi(c_{kj}), \quad \forall i, j.$$

Hence , we have

$$\Phi(a_1) + \cdots + \Phi(a_r) = \Phi(b_1) + \cdots + \Phi(b_s)$$

in G_2. Moreover, since Γ_1 is a scale for G_1 and $G_1 = (G_1)_+ - (G_1)_+$, Φ can be uniquely extended to a homomorphism from G_1 to G_2 . We shall still denote this extension by Φ.

From $\Phi(\Gamma_1) = \Gamma_2$ and Γ_2 is a scale for G_2, it follows that $\Phi(G_1) = G_2, \Phi((G_1)_+) = (G_2)_+$. Now it suffices to show that Φ is injective.

Let $a, b \in (G_1)_+$ and $\Phi(a) = \Phi(b)$. Write

$$a = a_1 + \cdots + a_r, \quad b = b_1 + \cdots + b_s,$$

where $a_i, b_j \in \Gamma_1, \forall i, j$. From the preceding paragraph, we have

$$\Phi(a_1) + \cdots + \Phi(a_r) = \Phi(b_1) + \cdots + \Phi(b_r).$$

By Lemma 15.6.2, there is a subset $\{d_{ij} | 1 \leq i \leq r, 1 \leq j \leq s\} \subset (G_2)_+$ such that

$$\Phi(a_i) = \sum_{k=1}^{s} d_{ik}, \quad \Phi(b_j) = \sum_{k=1}^{r} d_{kj}, \quad \forall i, j.$$

Since $\Phi(a_i), \Phi(b_j) \in \Gamma_2, \forall i, j$, it follows that $d_{ij} \in \Gamma_2$, and $\{d_{ik}|k\}, \{d_{kj}|k\}$ are additive in $\Gamma_2, \forall i, j$. Pick $c_{ij} \in \Gamma_1$ with $\Phi(c_{ij}) = d_{ij}, \forall i, j$. Since Φ is an isomorphism from Γ_1 onto Γ_2, $\{c_{ik}|k\}$ and $\{c_{kj}|k\}$ are also additive in Γ_1 , and

$$\Phi(\sum_{k=1}^{s} c_{ik}) = \sum_{k=1}^{s} \Phi(c_{ik}) = \Phi(a_i), \quad 1 \leq i \leq r,$$
$$\Phi(\sum_{k=1}^{r} c_{kj}) = \sum_{k=1}^{r} \Phi(c_{kj}) = \Phi(b_j), \quad 1 \leq j \leq s.$$

Further, we have

$$\sum_{k=1}^{s} c_{ik} = a_i, \quad \sum_{k=1}^{r} c_{kj} = b_j, \quad \forall i, j.$$

Therefore, $a = a_1 + \cdots a_r = b_1 + \cdots + b_s = b$, i.e., Φ is injective. Q.E.D.

Propostion 15.6.4. Let $A = \overline{\sqcup_n A_n}$ be an (AF)-algebra with a diagram $\{D, d, \mathcal{U}\}$, and $G = \varinjlim\{\mathbb{Z}^{r(n)}, \Phi_n\}$ be its dimension group, where $D = \sqcup_n D_n, D_n = \{(n, m) | 1 \leq m \leq r(n)\}, \forall n$, and $\mathcal{U} = \{\Phi_n | n\}$. For each n, let $t_n = (d(n, 1), \cdots, d(n, r(n)))(\in \mathbb{Z}_+^{r(n)})$, and $[0, t_n] = \{t'_n \in \mathbb{Z}_+^{r(n)} | t'_n \leq t_n\}$. Then $\Gamma = \sqcup_n \Phi_{n\infty}([0, t_n])$ is a scale for G. Moreover, if E is the dimension range of A (all equivalent classes of Proj(A), see Definition 1.2.2), define

$$\Psi(\tilde{p}) = \Phi_{n\infty}((\lambda_1, \cdots, \lambda_{r(n)})), \quad \forall \tilde{p} \in E,$$

where λ_k is the rank of p_k in $A_{n,k}, 1 \leq k \leq r(n)$, and $p = p_1 + \cdots + p_{r(n)} \in \tilde{p} \cap A_n$ (see the proof of Theorem 1.2.9), then Ψ is an isomorphism from E onto Γ.

Proof. It is obvious. Q.E.D.

Theorem 15.6.5. Let A, B be two (AF)-algebras. Then A and B are $*$ isomorphic if and only if their scaled dimension groups (see Proposition 1.6.4) are scaled isomophic.

Proof. The result follows immediately from Propositions 15.6.4, 15.6.3 and Theorems 15.2.8, 15.2.9. Q.E.D.

Corollary 15.6.6. Dimension groups of $*$ isomorphic (AF)-algebras are order isomorphic. Consequently, the dimension group of an (AF)-algebra is uniquely determined up to order isomorphism.

Now we consider the (AF)-algebra $K = C(H)$, where H is a separable infinite dimensional Hilbert space. From Sections 15.3 and 15.5, $K = \overline{\cup_n K_n}$, where K_n is $n \times n$ matrix algebra, and the embedding matrix $\Phi_n : K_n \to K_{n+1}$ is $\Phi_n = [1], \forall n$. Clearly, for any subsequece $n_1 < n_2 < \cdots$, we have also $K = \overline{\cup_k K_{n_k}}$ and a diagram

$$\underset{n_1}{\cdot} \xrightarrow{1} \underset{n_2}{\cdot} \xrightarrow{1} \underset{n_3}{\cdot} \quad \cdots \quad \underset{n_k}{\cdot} \xrightarrow{1} \underset{n_{k+1}}{\cdot} \quad \cdots\cdots$$

Moreover, for any C^*-algebra A, there is only one spatial C^*-norm $\alpha_0(\cdot)$ on $A \otimes K_n$, and α_0-$(A \otimes K_n) = A \otimes K_n, \forall n$(see Lemma 3.6.1). Thus, there is only one spatial C^*-norm $\alpha_0(\cdot)$ on $A \otimes K$, i.e. K is a nuclear C^*-algebra.

Lemma 15.6.7. Let $A = \overline{\cup_n A_n}$ be an (AF)-aglebra, and $G(A)$ be its dimension group. Then there exists a subsequence $\{n_1 < n_2 < \cdots\}$ of positive integers such that the scale Γ of the dimension group corresponding to α_0-$(A \otimes K) = \overline{\cup_k (A_k \otimes K_{n_k})}$ (see Proposition 1.6.4.) is $G(A)_+$. Moreover, the dimension group of α_0-$(A \otimes K)$ is order isomoprhic to $G(A)$.

Proof. For any $n_1 < n_2 < \cdots$, it is obvious that α_0-$(A \otimes K) = \overline{\cup_k (A_k \otimes K_{n_k})}$. Since the embedding matrix from $A_k \otimes K_{n_k}$ into $A_{k+1} \otimes K_{n_{k+1}}$ is $\Phi_k \otimes [1]$, where Φ_k is the embedding matrix from A_k into $A_{k+1}, \forall k$, it follows that the dimension group $G(\alpha_0$-$(A \otimes K))$ of α_0-$(A \otimes K)$ is order isomorphic to $G(A)$.

Let $\{D, d, \mathcal{U}\}$ be the diagram of $A = \overline{\cup_k A_k}$. Then the diagram of α_0-$(A \otimes K) = \overline{\cup_k (A_k \otimes K_{n_k})}$ is $\{D, d', \mathcal{U}\}$, where $d'(k, i) = n_k d(k, i), \forall 1 \le i \le r(k)$ and k. So the scale Γ of the dimension group corresponding to $\overline{\cup_k (A_k \otimes K_{n_k})}$ is

$$\Gamma = \cup_k \Phi_{k\infty}([0, n_k t_k]),$$

where $t_k = (d(k,1), \cdots, d(k, r(k))), \forall k$. Now it suffices to show that there exists $\{n_k\}$ such that $\Gamma = G(A)_+$.

Let $\mathcal{U} = \{\Phi_k = (s_{ij}^{(k)}) | k\}, s_k = \sum_{i,j} s_{ij}^{(k)}$, and $n_1 = 1, n_k = 2^{k-1} s_1 \cdots s_{k-1}, \forall k \ge 2$. Moreover, put $1_k = (1, \cdots, 1) \in Z^{r(k)}, \forall k$. Then $t_k \ge 1_k, \forall k$.

For any $a \in G(A)_+$, there exist N and n such that $\Phi_{n\infty}(N 1_n) \ge a$. Pick $m (> n)$ with $2^{m-1} \ge N$. Then $n_m = 2^{m-1} s_1 \cdots s_{m-1} \ge N s_n \cdots s_{m-1}$. Since $\Phi_k(1_k) \le s_k 1_{k+1}, \forall k$, it follows that

$$n_m t_m \ge n_m 1_m \ge \Phi_{m-1} \circ \cdots \circ \Phi_n(N 1_n).$$

Then from $0 \le a \le \Phi_{n\infty}(N 1_n) \le \Phi_{m\infty}(n_m t_m) \in \Gamma$, we have $a \in \Gamma$. Therefore, $\Gamma = G(A)_+$. Q.E.D.

Theorem 15.6.8. Let $A = \overline{\sqcup_n A_n}, B = \overline{\sqcup_n B_n}$ be two (AF)-algebras, and $G(A), G(B)$ be their dimension groups respectively. Then $G(A)$ and $G(B)$ are order isomorphic if and only if α_0-$(A \otimes K)$ and α_0-$(B \otimes K)$ are $*$ isomorphic, where $K = C(H)$ and H is a separable infinite dimensional Hilbert space.

Proof. The sufficiency is obvious from Corollary 15.6.6 and $G(A) = G(\alpha_0\text{-}(A \otimes K)), G(B) = G(\alpha_0\text{-}(B(\otimes K))$.

Now let $G(A)$ and $G(B)$ be order isomorphic. Pick $\{n_1 < n_2 < \cdots\}$ and $\{m_1 < m_2 < \cdots\}$ such that the scales of the dimension groups corresponding to $\overline{\sqcup_k A_k \otimes K_{n_k}}, \overline{\sqcup_k (B_k \otimes K_{m_k})}$ are $G(A)_+, G(B)_+$ respectively. Thus, the scaled dimension groups of $\overline{\sqcup_k (A_k \otimes K_{n_k})}$ and $\overline{\sqcup_k (B_k \otimes K_{m_k})}$ are scaled isomorphic obviously. By Theorem 15.6.5, α_0-$(A \otimes K)$ and α_0-$(B \otimes K)$ are $*$ isomorphic.

Q.E.D.

References. [11], [38], [40], [103], [160].

15.7. The tracial state space on an (AF)-algebra

Let $A = \overline{\sqcup_n A_n}$ be an (AF)-algebra with an identity e and a diagram as follows:

$$e \in A_1 \overset{\Phi_1}{\hookrightarrow} A_2 \overset{\Phi_2}{\hookrightarrow} \cdots\cdots,$$

where $A_n = M(t^{(n)}) = \oplus_{k=1}^{r(n)} M_{t_k^{(n)}}, t^{(n)} = (t_1^{(n)}, \cdots, t_{r(n)}^{(n)}) \in \mathbb{N}^{r(n)}, \forall n$, and M_k is the $(k \times k)$ matrix algebra, $\forall k$. Then we have the scaled dimension group:

$$(G, G_+ = P, \Gamma),$$

where

$$G = \varinjlim\{\mathbb{Z}^{r(n)}, \Phi_n\},$$

$$P = \sqcup_n \Phi_{n\infty}(\mathbb{Z}_+^{r(n)}),$$

and

$$\begin{aligned} \Gamma &= [0, u] = \{v \in G | 0 \le v \le u\} \\ &= \sqcup_n \Phi_{n\infty}([0, t^{n)}]), \end{aligned}$$

where $u = \Phi_{n\infty}(t^{(n)})(\forall n)$ is an order unit of G (notice that the definition of u is independent of n since $e \in A_1$ and $t^{(n+1)} = \Phi_n(t^{(n)}), \forall n$).

Let τ be a *tracial state* on A, i.e., τ be a state on A with $\tau(ab) = \tau(ba), \forall a, b \in A$. For any $v \in \Gamma$ we can find some $(s_1, \cdots, s_{r(n)}) \in \mathbb{Z}_+^{r(n)}$

with $s_k \leq t_k^{(n)} (1 \leq k \leq r(n))$ such that $v = \Phi_{n\infty}((s_1, \cdots, s_{r(n)}))$. Let q be a projection of A_n such that $q = q_1 \oplus \cdots \oplus q_n$, where q_k is a rank s_k projection of $M_{t_k^{(n)}}, 1 \leq k \leq r(n)$. In equivalent sense, q is uniquely determined by v. Thus, we can define

$$\rho(v) = \tau(q).$$

Then ρ is an additive function from Γ to $[0,1]$. From the proof of Proposiiton 1.6.3 , ρ can be extended to a state on the ordered group (G, P, u).

Conversely, let $\rho \in S_u(G)$, and

$$\lambda_k^{(n)} = \rho(\Phi_{n\infty}((0, \cdots, 0, t_k^{(n)}, 0, \cdots, 0))),$$

$1 \leq k \leq r(n)$. Then $\lambda_k^{(n)} \geq 0, \forall k$, and

$$\lambda_1^{(n)} + \cdots + \lambda_{r(n)}^{(n)} = \rho(\Phi_{n\infty}(t^{(n)})) = \rho(u) = 1.$$

For any $a \in A_n$, we can uniquely write

$$a = a_1 + \cdots + a_{r(n)},$$

where $a_k \in M_{t_k^{(n)}}, 1 \leq k \leq r(n)$. Then define

$$\tau_n(a) = \sum_{k=1}^{r(n)} \lambda_k^{(n)} tr(a_k),$$

where $tr(\cdot)$ is the unique cononical tracial state on $M_k (\forall k)$. Clearly, τ_n is a tracial state on $A_n, \forall n$. We claim that

$$\tau_{n+1}|A_n = \tau_n, \quad \forall n.$$

Fix n and $k \in \{1, \cdots, r(n)\}$. It suffices to show that

$$\tau_{n+1}(p) = \tau_n(p) = \lambda_k^{(n)} / t_k^{(n)},$$

where p is a minimal projection of $M_{t_k^{(n)}}$.

In fact, let $\Phi_n = (s_{ij}^{(n)})_{1 \leq i \leq r(n+1), 1 \leq j \leq r(n)}$. Then $\Phi_n((0, \cdots, 0, 1, 0, \cdots, 0)) = (s_{1k}^{(n)}, \cdots, s_{r(n+1)k}^{(n)})$. Hence $\lambda_k^{(n)} = t_k^{(n)} \sum_{j=1}^{r(n+1)} \lambda_j^{(n+1)} s_{jk}^{(n)} / t_j^{(n+1)}$, and

$$\tau_n(p) = \frac{\lambda_k^{(n)}}{t_k^{(n)}} = \sum_{j=1}^{r(n+1)} \frac{\lambda_j^{(n+1)} s_{jk}^{(n)}}{t_j^{(n+1)}} = \tau_{n+1}(p).$$

Therefore, we have a tracial state τ on $\cup_n A_n$ such that $\tau|A_n - \tau_n, \forall n$. Clearly,

$$|\tau(a)| \leq \sum_{k=1}^{r(n)} \lambda_k^{(n)} |tr(a_k)| \leq \|a\|,$$

$\forall a = a_1 + \cdots + a_{r(n)} \in A_n, \forall n$. It follows that τ can be uniquely extended to a tracial state on A.

From the above discussion, we have the following.

Theorem 15.7.1. With the above assumptions and notations, the tracial state space $\mathcal{T}(A)$ on A is affinely homeomorphic to the state space $S_u(G)$ on G.

Example. The tracial state space of the GICAR algebra.

Let A be the GICAR algebra (see Section 15.3, Example 3). By the above discussion, any tracial state on A is determined by $\{\lambda_i^{(n)} | 0 \le i \le n, n \ge 0\}$. Let $\mu_i^{(n)} = \lambda_i^{(n)}/C_n^i$. Then we have an *inverse Pascal's triangle:*

$$\mu_i^{(n)} = \mu_i^{(n+1)} + \mu_{i+1}^{(n+1)} \ge 0,$$

$\forall 0 \le i \le n, n \ge 0$. If put $r_n = \mu_0^{(n)}$, then we need to find all sequences $\{r_n | n \ge 0\}$ of non-negative numbers with $r_0 = 1$ such that

$$\mu_{i+1}^{(n+1)} = \sum_{j=0}^{i+1} (-1)^j C_{i+1}^j r_{n-i+j} \ge 0,$$

$\forall 0 \le i \le n$, and $n \ge 0$.

By Theorem 15.5.12, the dimension group of A is

$$(G, P) = (P_{\mathbb{Z}}([0,1]), P_{\mathbb{Z}}^+([0,1])),$$

and function $1 = u$ is an order unit. Since $G = \sqcup_{n \ge 0} H_n$, where $H_n = \mathbb{Z} \dot{+} \mathbb{Z}t \dot{+} \cdots \dot{+} \mathbb{Z}t^n, \forall n \ge 0$, each state $\rho \in S_u(G)$ is determined by

$$r_n = \rho((1-t))^n), \quad \forall n \ge 0.$$

By Lemma 15.5.3, $\{r_n\}$ satisfies the following:

$$\sup \left\{ -\frac{\sum_{i=0}^n a_i r_i}{m} \;\middle|\; \begin{array}{c} a_i, m \in \mathbb{Z}, m > 0, \text{and} \\ \sum_{i=0}^n a_i t^i + m t^{n+1} > 0, \forall t \in (0,1) \end{array} \right\}$$

$$\le r_{n+1}$$

$$\le \inf \left\{ \frac{\sum_{i=0}^n a_i r_i}{m} \;\middle|\; \begin{array}{c} a_i, m \in \mathbb{Z}, m > 0, \text{and} \\ \sum_{i=0}^n a_i t^i - m t^{n+1} > 0, \forall t \in (0,1) \end{array} \right\},$$

$\forall n \geq 0$, and $r_0 = 1$.

Lemma 15.7.2. For any $r \in [0,1]$ and $n \geq 2$, we have

$$\sup \left\{ -\frac{\sum_{i=0}^{n-1} a_i r^i}{m} \,\middle|\, \begin{array}{l} a_i, m \in \mathbb{Z}, m > 0, \text{and} \\ \sum_{i=0}^{n-1} a_i t^i + m t^n > 0, \forall t \in (0,1) \end{array} \right\} = r^n.$$

Proof. First, let $n = 2$.

If $(a + bt + mt^2) > 0, \forall t \in (0,1)$, then $(a + br + mr^2) \geq 0$, and $r^2 \geq -m^{-1}(a + br)$. Hence, we have

$$r^2 \geq \sup \left\{ -\frac{a + br}{m} \,\middle|\, \begin{array}{l} a, b, m \in \mathbb{Z}, m > 0, \text{and} \\ (a + bt + mt^2) > 0, \forall t \in (0,1) \end{array} \right\}.$$

Now it suffices to show that for any $\varepsilon > 0$,

$$\sup\{\cdots\} > r^2 - \varepsilon,$$

or to find $a, b, m \in \mathbb{Z}$ and $m > 0$ such that

$$-m^{-1}(a + br) > r^2 - \varepsilon, \text{ and } (a + bt + mt^2) > 0, \quad \forall t \in (0,1).$$

Clearly, it must be $a \geq 0$ and $b < 0$. So we want to find $a, b, m \in \mathbb{Z}_+$ and $b > 0, m > 0$ such that

$$\frac{a}{m} - \frac{b}{m} r + r^2 < \varepsilon$$

and $(a - bt + mt^2) > 0, \forall t \in (0,1)$. Suppose that $4ma > b^2$. Then automatically,

$$a > 0, \text{ and } (a - bt + mt^2) > 0, \forall t \in (0,1).$$

Now the problem is to find $a, b, m \in \mathbb{N}$ such that

$$4ma > b^2, \text{ and } \left(\frac{a}{m} - \frac{b}{m} r + r^2\right) < \varepsilon.$$

Pick $p, q \in \mathbb{N}$ with $q \geq p$ such that

$$2\left|\frac{p}{q} - r\right| + \left|\frac{p^2}{q^2} - r^2\right| < \varepsilon/2.$$

Then the problem becomes to find $a, b, m \in \mathbb{N}$ such that

$$4ma > b^2, \; b \leq 2m, \quad \left|\frac{a}{m} - \frac{b}{m} \cdot \frac{p}{q} + \frac{p^2}{q^2}\right| < \varepsilon/2.$$

Now take $a, k \in \mathbb{N}$ such that

$$\frac{a}{k} > \frac{p^2}{q}, \text{ and } (\frac{a}{k} - \frac{p^2}{q}) \cdot \frac{1}{q} < \varepsilon/2,$$

and let $m = kq, b = 2kp$. Then we have

$$4ma > b^2, \quad b \leq 2m,$$

and

$$|\frac{a}{m} - \frac{b}{m} \cdot \frac{p}{q} + \frac{p^2}{q^2}| = (\frac{a}{k} - \frac{p^2}{q}) \cdot \frac{1}{q} < \varepsilon/2.$$

Hence, the conclusion holds for $n = 2$.

For general $n \geq 2$, it is obvious that

$$\begin{aligned} r^n &\geq \sup \left\{ -\frac{\sum_{i=1}^{n-1} a_i r^i}{m} \;\middle|\; \begin{array}{l} a_i, m \in \mathbb{Z}, m > 0, \text{and} \\ \sum_{i=1}^{n-1} a_i t^i + mt^n > 0, \forall t \in (0,1) \end{array} \right\} \\ &\geq \sup \left\{ -\frac{ar^{n-2} + br^{n-1}}{m} \;\middle|\; \begin{array}{c} a, b, m \in \mathbb{Z}, m > 0, \text{and} \\ (at^{n-2} + bt^{n-1} + mt^n) > 0, \forall t \in (0,1) \end{array} \right\} \\ &= r^{n-2} \cdot \sup \left\{ -\frac{a + br}{m} \;\middle|\; \begin{array}{c} a, b, m \in \mathbb{Z}, m > 0, \text{and} \\ (a + bt + mt^2) > 0, \forall t \in (0,1) \end{array} \right\} \\ &= r^{n-2} \cdot r^2 = r^n \end{aligned}$$

from the preceding paragraph. That comes to the conclusion. Q.E.D.

Lemma 15.7.3. For each $r \in [0, 1], \rho_r$ is an extreme point of $S_u(G)$, where $\rho_r((1-t)^n) = r^n, \forall n \geq 0$.

Proof. Let $\rho, \sigma \in S_u(G)$ and $\lambda \in (0, 1)$ be such that

$$\rho_r = \lambda\rho + (1 - \lambda)\sigma$$

Then we have

$$r^n = \lambda s_n + (1 - \lambda)t_n,$$

where $s_n = \rho((1-t)^n), t_n = \sigma((1-t)^n), \forall n \geq 0$. Since

$$s_2 \geq \sup \left\{ -\frac{a + bs_1}{m} \;\middle|\; \begin{array}{c} a, b, m \in \mathbb{Z}, m > 0, \text{and} \\ (a + bt + mt^2) > 0, \forall t \in (0,1) \end{array} \right\},$$

it follows from Lemma 15.7.2 that $s_2 \geq s_1^2$. Similarly, $t_2 \geq t_1^2$. Now from $r = \lambda s_1 + (1-\lambda)\sigma_1$ and $r^2 = \lambda s_2 + (1-\lambda)t_2$, it must be $s_1 = t_1 = r, s_2 = t_2 = r^2$. We assume that $s_k = t_k = r^k, 1 \leq k \leq n-1 (n \geq 3)$. Then

$$s_n \geq \sup \left\{ -\frac{\sum_{i=0}^{n-1} a_i s_i}{m} \,\middle|\, \begin{array}{l} a_i, m \in \mathbb{Z}, m > 0, \text{and} \\ (\sum_{i=0}^{n-1} a_i t^i + m t^n) > 0, \forall t \in (0,1) \end{array} \right\}$$

$$= \sup \left\{ -\frac{\sum_{i=0}^{n-1} a_i r^i}{m} \,\middle|\, \cdots \right\} = r^n,$$

and $t_n \geq r^n$. But $r^n = \lambda s_n + (1-\lambda)t_n$, it follows that $s_n = t_n = r^n$. Therefore, we have $\rho = \sigma = \rho_r$, and ρ_r is an extreme point of $S_u(G)$. Q.E.D.

Proposition 15.7.4. Let A be the GICAR algebra, and G be its dimension group. Then we have

$$ExS_u(G) = \{\rho_r | 0 \leq r \leq 1\},$$

where $\rho_r((1-t)^n) = r^n, \forall n \geq 0, \forall r \in [0,1]$.

Proof. First, we show that

$$S_u(G) = \overline{Co\{\rho_r | 0 \leq r \leq 1\}}^{\sigma}$$

in $(\mathbb{R}^G)^*$. In fact, if there exists a $\rho \in S_u(G) \backslash \overline{Co\{\rho_r | 0 \leq r \leq 1\}}^{\sigma}$, then by the separation theorem we can find a finite subset F of $G = P_{\mathbb{Z}}([0,1])$ and $\lambda_p \in \mathbb{R}, \forall p \in F$ such that

$$\sum_{p \in F} \lambda_p \rho(p) > \sup_{0 \leq r \leq 1} \sum_{p \in F} \lambda_p \rho_r(p)$$
$$= \sup_{0 \leq r \leq 1} \sum_{p \in F} \lambda_p p(r).$$

Using rational numbers to approximate each λ_p ,we can obtain $q \in P_{\mathbb{Z}}([0,1])$ such that

$$\rho(q) > \sup_{0 \leq r \leq 1} q(r).$$

Pick $m \in \mathbb{N}, n \in \mathbb{Z}$ such that

$$\rho(q) > \frac{n}{m} > \sup_{0 \leq r \leq 1} q(r).$$

Then $(n - mq) \in G_+$. By $\rho \in S_u(G)$ we get

$$\rho(n - mq) \geq 0, \text{ and } \rho(q) \leq n/m,$$

a contradiction. Therefore, $S_u(G) = \overline{Co\{\rho_r | 0 \leq r \leq 1\}}^{\sigma}$.

Moreover, $r \to \rho_r$ is continuous from $[0, 1]$ to $S_u(G)$ obviously . Hence, $\{\rho_r | 0 \leq r \leq 1\}$ is a closed subset of $S_u(G)$. Now by the Krein-Milmann theorem we have

$$ExS_u(G) \subset \{\rho_r | 0 \leq r \leq 1\}.$$

Further, from Lemma 15.7.3 we obtain that $ExS_u(G) = \{\rho_r | 0 \leq r \leq 1\}$.

Q.E.D.

Remark. The tracial state space $\mathcal{T}(A)$ of the GICAR algebra A is affinely homeomorphic to $S_u(G)$. So $\mathcal{T}(A)$ is a "*Bauer simplex* " (i.e., its extreme points subset is closed, see [128]) in $(A^*, \sigma(A^*, A))$, and the set of all extreme points of $\mathcal{T}(A)$ is homeomorphic to the connected space [0,1].

References. [103], [128].

Chapter 16

Crossed Products

16.1. W^*-crossed products

In Sections 7.3, 9.5, a discrete crossed product $M \times_\alpha G$ is defined, where G is a discrete group. Also in Section 8.2, a W^*-system $(M, \mathbb{R}, \sigma)$ appears. Now we consider the general case.

Definition 16.1.1. (M, G, α) is called a VN-*dynamical system* , if M is a VN-algebra on a Hibert space H, G is a locally compact group, α is a homomorphism from G into Aut(M), where Aut(M) is the group of all $*$ automorphisms of M, and $t \longrightarrow \langle \alpha_t(x)\xi, \eta\rangle$ is continuous on $G, \forall x \in M$, and $\xi, \eta \in H$.

In this case, we claim that for each $x \in M$ $t \longrightarrow \alpha_t(x)$ is also continuous from G to $(M, \tau(M, M_*))$. In fact, since $\|\alpha_t(x)\| = \|x\|, \forall t$, and $\alpha_t(x)^* = \alpha_t(x^*)$, it suffices to show that $t \longrightarrow \alpha_t(x)$ is continuous from G to $(M$, strong top.) . We may assume that $x = u$ is unitary. Let a net $t_l \rightarrow t$ in G. Then

$$\begin{aligned}\|(\alpha_{t_l}(u) - \alpha_t(u))\xi\|^2 = \; & 2\|\xi\|^2 - \langle \alpha_{t_l}(u)\xi, \alpha_t(u)\xi\rangle \\ & -\langle \alpha_t(u)\xi, \alpha_{t_l}(u)\xi\rangle \longrightarrow 0,\end{aligned}$$

$\forall \xi \in H$, i.e., $\alpha_{t_l}(u) \longrightarrow \alpha_t(u)$ strongly.

Definition 16.1.2. Let (M, G, α) be a VN-system . For any $x \in M$, define

$$(\pi(x)f)(s) = \alpha_{s^{-1}}(x)f(s), \quad \forall f \in L^2(G, H),$$

where H is the action space of M, i.e., $M \subset B(H)$; $L^2(G, H) = H \otimes L^2(G), L^2(G) = L^2(G, \mu)$, and μ is a left Haar measure on G. We shall write $d\mu(s) = ds$ simply.

It is easily verified that $\|\pi(x)\| \leq \|x\|$, and $\{\pi, L^2(G, H)\}$ is a $*$ representation of M.

Proposition 16.1.3. Let (M, G, α) be a VN-system with $M \subset B(H)$. Then $\{\pi, L^2(G, H)\}$ is a faithful W^*-representation of M. Moreover, $\pi(M) \subset M\overline{\otimes}Z$, where $Z = \{m_f | f \in L^\infty(G, \mu)\}$ is the multiplicative algebra on $L^2(G)$ (see Definition 5.3.11).

Proof. Let $\pi(x) = 0$ for some $x \in M$. Then for any $\xi \in H$ and compact subset K of G, we have

$$0 = \langle \pi(x)\chi_K \otimes \xi, \chi_K \otimes \xi\rangle = \int_K \langle \alpha_{s^{-1}}(x)\xi, \xi\rangle ds$$

Since $s \longrightarrow \langle \alpha_{s^{-1}}(x)\xi, \xi\rangle$ is continuous on G, and K is arbitrary, it follows that $\langle \alpha_s(x)\xi, \xi\rangle = 0, \forall s \in G, \xi \in H$. Thus, $x = 0$, i.e., π is faithful.

Now let $\{x_l\}$ be a bounded increasing net of M_+ and $x = \sup_l x_l$. Clearly $\pi(x_l) \nearrow y = \sup_l \pi(x_l)$ strongly. We need to prove that $y = \pi(x)$. By the Dini theorem, for any $\xi \in H$ and compact subset K of G we have $\langle \alpha_{s^{-1}}(x_l)\xi, \xi\rangle \nearrow \langle \alpha_{s^{-1}}(x)\xi, \xi\rangle$ uniformly for $s \in K$. Then we can see that

$$\langle \pi(x_l)f, f\rangle \longrightarrow \langle \pi(x)f, f\rangle, \quad \forall f \in L^2(G, H).$$

Therefore, $y = \pi(x)$, and $\{\pi, L^2(G, H)\}$ is a faithful W^*-representation of M.

Fix $x \in M$. Suppose that $\xi^{(k)}(\cdot); G \longrightarrow H$ is continuous and there is a compact subset K of G such that

$$\text{supp}\xi^{(k)}(\cdot) \subset K, \quad 1 \leq k \leq n.$$

For any $\varepsilon > 0$, noticing that

$$\begin{aligned} & \|\alpha_s^{-1}(x)\xi^{(k)}(t) - \alpha_t^{-1}(x)\xi^{(k)}(t)\| \\ \leq\ & \|(\alpha_s^{-1}(x) - \alpha_t^{-1}(x))\xi^{(k)}(s)\| + \|(\alpha_s^{-1}(x) - \alpha_t^{-1}(x))(\xi^{(k)}(s) - \xi^{(k)}(t))\| \\ \leq\ & \|(\alpha_s^{-1}(x) - \alpha_t^{-1}(x))\xi^{(k)}(s)\| + 2\|x\| \cdot \|\xi^{(k)}(s) - \xi^{(k)}(t)\|, \end{aligned}$$

there is an opern neighborhood U_s of s such that

$$\|\alpha_s^{-1}(x)\xi^{(k)}(t) - \alpha_t^{-1}(x)\xi^{(k)}(t)\| < \eta, \quad \forall t \in U_s, 1 \leq k \leq n,$$

where $\eta(> 0)$ satisfies $\eta^2\mu(K) = \varepsilon$. By the compactness of K, there are $s_1, \cdots, s_m \in K$ and $U_i = U_{s_i}, 1 \leq i \leq m$, such that $\sqcup_{i=1}^m U_i \supset K$. Pick continuous functions $g_1, \cdots, g_m$ on G such that

$$0 \leq g_i(t) \leq 1, \quad \text{supp}g_i \subset U_i, \quad \forall 1 \leq i \leq m, \quad t \in G;$$

and

$$\sum_{i=1}^m g_i(t) \leq 1, \forall t \in G; \quad \sum_{i=1}^n g_i(t) = 1, \forall t \in K.$$

Let $a = \sum_{i=1}^{m} \alpha_{s_i}^{-1}(x) \otimes m_{g_i} \in M\overline{\otimes}Z$. Since

$$\begin{aligned}\|a\xi\|^2 &= \int \|\sum_i g_i(t)\alpha_{s_i}^{-1}(x)\xi(t)\|^2 dt \\ &\leq \|x\|^2 \int (\sum_i g_i(t))^2 \|\xi(t)\|^2 dt \leq \|x\|^2 \cdot \|\xi\|^2,\end{aligned}$$

$\forall \xi \in L^2(G, H)$, it follows that $\|a\| \leq \|x\|$. Further, we have

$$\begin{aligned}&\|(a - \pi(x))\xi^{(k)}\| \\ =& \int \|\sum_i g_i(t)\alpha_{s_i}^{-1}(x)\xi^{(k)}(t) - \alpha_t^{-1}(x)\xi^{(k)}(t)\|^2 dt \\ =& \int_K \|\sum_i g_i(t)(\alpha_{s_i}^{-1}(x) - \alpha_t^{-1}(x))\xi^{(k)}(t)\|^2 dt \\ <& \int_K (\sum_i g_i(t)\eta)^2 dt = \eta^2 \mu(K) = \varepsilon, \quad 1 \leq k \leq n.\end{aligned}$$

Therefore, $\pi(x) \in M\overline{\otimes}Z$. Q.E.D.

Definition 16.1.4. Let (M, G, α) be a VN-system with $M \subset B(H)$. For any $t \in G$, define

$$(\lambda(t)f)(s) = f(t^{-1}s), \quad \forall f \in L^2(G, H).$$

Clearly, $t \to \lambda(t)$ is the strongly continuous unitary representation of G on $L^2(G, H)$.

Proposition 16.1.5. Let (M, G, α) be a VN-system with $M \subset B(H)$. Then $\{\pi, \lambda, L^2(G, H)\}$ is a *covariant representation* of the system (M, G, α), i.e.,

$$\lambda(t)\pi(x)\lambda(t)^* = \pi(\alpha_t(x)), \quad \forall x \in M, t \in G.$$

Proof.

$$\begin{aligned}&(\lambda(t)\pi(x)\lambda(t)^* f)(s) = (\pi(x)\lambda(t^{-1})f)(t^{-1}s) \\ =& \alpha_{s^{-1}t}(x)(\lambda(t^{-1})f)(t^{-1}s) = \alpha_{s^{-1}t}(x)f(s) \\ =& \alpha_{s^{-1}}(\alpha_t(x))f(s) = (\pi(\alpha_t(x))f)(s),\end{aligned}$$

$\forall s, t \in G, x \in M, f \in L^2(G, H)$. Q.E.D.

If we consider the VN-system $(\pi(M), G, \beta)$, where $\beta_t(\pi(x)) = \pi(\alpha_t(x)), \forall t \in G, x \in M$, then the action β on $\pi(M)$ is *unitarily implemented* by $\lambda(\cdot)$, i.e., $\beta_t(\pi(x)) = \lambda(t)\pi(x)\lambda(t)^*, \forall t \in G, x \in M$.

Let (M,G,α) be a W^**-system* (i.e., M is a W^*-algebra , $\alpha : G \to \text{Aut}(M)$ is a homomorphism , and $t \to \alpha_t(x)$ is continuous from G to $(M,\sigma(M,M_*)), \forall x \in M$). From Proposition 1.1.5, we can find a faithful W^*-representation of M such that the action α will be unitarily implemented.

Definition 16.1.6. Let (M,G,α) be a VN-system with $M \subset B(H)$. The VN algebra

$$\{\pi(x), \lambda(t) | x \in M, t \in G\}''$$

on $L^2(G,H)$ is called the *crossed product* of M by the action α of G, denoted by $M \times_\alpha G$.

When G is discrete, this definition coincides with Definition 7.3.3.

Noticing that

$$\pi(x)\lambda(t)\pi(y)\lambda(s) = \pi(x\alpha_t(y))\lambda(ts),$$

$$(\pi(x)\lambda(t))^* = \pi(\alpha_t^{-1}(x^*))\lambda(t^{-1}),$$

$\forall x,y \in M, s,t \in G$, we have $M \times_\alpha G = \overline{[\pi(x)\lambda(t) | x \in M, t \in G]}^{\sigma}$.

Later, we shall show that the definition of $M \times_\alpha G$ is independent of H up to $*$ isomorphism.

Now consider the case of unitary implement.

Proposition 16.1.7. Let (M,G,α) be a VN-system with $M \subset B(H)$ and $s \to u_s$ be a strongly continuous unitary representation of G on H such that $u_s M u_s^* = M$ and $\alpha_s(x) = u_s x u_s^*, \forall s \in G, x \in M$. Define

$$(Wf)(s) = u_s f(s), \quad \forall f \in L^2(G,H).$$

Then W is a unitary operator on $L^2(G,H)$, and

$$\pi(x) = W^*(x \otimes 1)W, \quad \lambda(s) = W^*(u_s \otimes \lambda_s)W,$$

$\forall x \in M, s \in G$, where $s \to \lambda_s$ is the left regular reperesentation of G on $L^2(G)$. Consequently, $M \times_\alpha G$ is spatially $*$ isomorphic to the VN algebra

$$\{x \otimes 1, u_s \otimes \lambda_s | x \in M, s \in G\}''$$

on $L^2(G,H)$.

Proof. Clearly, W is unitary, and

$$(W^* f)(s) = u_s^* f(s), \quad \forall s \in G, f \in L^2(G,H).$$

Moreover,

$$(\pi(x)f)(s) = \alpha_{s^{-1}}(x)f(s) = u_s^* x u_s f(s) = (W^*(x \otimes 1)Wf)(s),$$

and

$$(W^*(u_s \otimes \lambda_s)Wf)(t) = u_t^*((u_s \otimes \lambda_s)Wf)(t)$$
$$= u_t^* u_s (Wf)(s^{-1}t) = f(s^{-1}t) = (\lambda(s)f)(t),$$

$\forall s, t \in G, x \in M, f \in L^2(G, H)$. Q.E.D.

Lemma 16.1.8. Let $(M_1, G, \alpha^{(1)}), (M_2, G, \alpha^{(2)})$ be two VN-systems. Then $(M_1 \overline{\otimes} M_2, G, \alpha)$ is also a VN-system, where $\alpha_t = \alpha_t^{(1)} \otimes \alpha_t^{(2)}, \forall t \in G$.

Proof. Let $M_i \subset B(H_i), i = 1, 2$. Suppose that $t \longrightarrow u_t^{(i)}$ is a strongly continuous unitary representation of G on H_i, and $\alpha_t^{(i)}(x_i) = u_t^{(i)} x_i u_t^{(i)*}, \forall t \in G, x_i \in M_i, i = 1, 2$. Clearly, $t \longrightarrow u_t^{(1)} \otimes u_t^{(2)}$ is a strongly continuous unitary representation of G on $H_1 \otimes H_2$. Thus $(M_1 \overline{\otimes} M_2, G, \alpha)$ is a VN-system, where

$$\alpha_t(\cdot) = (\alpha_t^{(1)} \otimes \alpha_t^{(2)})(\cdot) = (u_t^{(1)} \otimes u_t^{(2)}) \cdot (u_t^{(1)} \otimes u_t^{(2)})^*,$$

$\forall t \in G$.

For general case, by Proposition 16.1.5 we have VN-systems:

$$(\pi_i(M_i), G, \tilde{\alpha}^{(i)}), i = 1, 2,$$

where $\{\pi_i, L^2(G, H_i)\}$ is a faithful W^*-representation of M_i (see Proposition 16.1.3.), and

$$\tilde{\alpha}_t^{(i)}(\pi_i(x_i)) = \pi_i(\alpha_t^{(i)}(x_i)) = \lambda_i(t)\pi_i(x)\lambda_i(t)^*,$$

$\forall t \in G, x_i \in M_i, i = 1, 2$. From the preceding paragraph, $(\pi_1(M_1)\overline{\otimes}\pi_2(M_2), G, \tilde{\alpha})$ is a VN-system , where $\tilde{\alpha} = \tilde{\alpha}^{(1)} \otimes \tilde{\alpha}^{(2)}$. By Theorem 4.3.4, $\pi_1 \otimes \pi_2$ is also a faithful W^*-representation of $M_1 \overline{\otimes} M_2$, and

$$\pi_1(M_1)\overline{\otimes}\pi_2(M_2) = (\pi_1 \otimes \pi_2)(M_1 \overline{\otimes} M_2).$$

Notice that

$$\tilde{\alpha}_t((\pi_1 \otimes \pi_2)(a))$$
$$= (\lambda_1(t) \otimes \lambda_2(t)) \cdot (\pi_1 \otimes \pi_2)(a) \cdot (\lambda_1(t) \otimes \lambda_2(t))^*$$
$$= (\pi_1 \otimes \pi_2)(\alpha_t^{(1)} \otimes \alpha_t^{(2)}(a)) = (\pi_1 \otimes \pi_2)(\alpha_t(a)),$$

$\forall a \in M_1 \overline{\otimes} M_2$. Now since $(\pi_1 \otimes \pi_2)^{-1}$ is σ-σ continuous, $t \to \alpha_t(a)$ is continuous from G to $(M_1 \overline{\otimes} M_2, \sigma), \forall a \in M_1 \overline{\otimes} M_2$. Therefore, $(M_1 \overline{\otimes} M_2, G, \alpha)$ is also a VN-system. Q.E.D.

Definition 16.1.9. Let $t \longrightarrow \lambda_t, \rho_t$ be the left, right regular representation of G on $L^2(G)$, i.e.,

$$(\lambda_t \xi)(x) = \xi(t^{-1}s), \quad (\rho_t \xi)(s) = \Delta(t)^{1/2} \xi(st),$$

$\forall s,t\in G,\xi\in L^2(G)$. Clearly, $(B(L^2(G)),G,ad\rho)$ is a VN-system, where

$$ad\rho_t(a)=\rho_t a\rho_t^*,\quad \forall a\in B(L^2(G)),t\in G.$$

Now let (M,G,α) be a VN-system with $M\subset B(H)$. By Lemma 16.1.8, $(M\overline{\otimes}B(L^2(G)),G,\theta)$ is also a VN-system, where $\theta_t=\alpha_t\otimes ad\rho_t,\forall t\in G$. From Proposition 16.1.3, $\pi(M)\subset M\overline{\otimes}Z\subset M\overline{\otimes}B(L^2(G))$. Moreover, $\lambda(s)=1\otimes\lambda_s\in M\overline{\otimes}B(L^2(G)),\forall s\in G$. Thus, we get

$$M\times_\alpha G\subset M\overline{\otimes}B(L^2(G)).$$

In the following, we shall prove that $M\times_\alpha G$ is the *fixed point algebra* of the system $(M\overline{\otimes}B(L^2(G)),G,\theta)$,
i.e.,

$$M\times_\alpha G=\{a\in M\overline{\otimes}B(L^2(G))|\theta_t(a)=a,\forall t\in G\}.$$

We shall identity Z with $L^\infty(G)$. Then $1\otimes g\in M\otimes B(L^2(G))$, and

$$\theta_t(1\otimes g)=1\otimes ad\rho_t(g)=1\otimes g_t,$$

where $g_t(\cdot)=g(\cdot t),\forall g\in L^\infty(G)$, and $1=1_H$.

Lemma 16.1.10. For any $h\in L^1(G)\cap L^\infty(G)$, we have

$$\int\theta_t(1\otimes h)dt=\int h(t)dt\cdot 1_{H\otimes L^2(G)}.$$

in the σ-weak topology.

Proof. For any $\xi,\eta\in L^2(G,H)$, by the Fubini theorem we have

$$\begin{aligned}\int\langle\theta_t(1\otimes h)\xi,\eta\rangle dt &=\int dt\int h(st)\langle\xi(s),\eta(s)\rangle ds\\ &=\int h(t)dt\cdot\int\langle\xi(s),\eta(s)\rangle ds.\end{aligned}$$

That comes to the conclusion. Q.E.D.

Lemma 16.1.11. Denote all continuous functions on G with a compact support by $K(G)$. Let $f,g\in K(G)$, and $x\in M\overline{\otimes}B(L^2(G))$. Then $\int\theta_t((1\otimes f)x(1\otimes g))dt\in M\overline{\otimes}B(L^2(G))$ in the σ-weak topology, and it is σ-continuous for x.

Proof. For any $\xi,\eta\in L^2(G,H)$, notice that

$$\begin{aligned}&|\langle\theta_t((1\otimes f)x(1\otimes g))\xi,\eta\rangle|\\ =\ &\|x\|\cdot\|\theta_t(1\otimes g)\xi\|\cdot\|\theta_t(1\otimes\overline{f})\eta\|\end{aligned}$$

and

$$\|\theta_t(1\otimes g)\xi\|^2 = \langle\theta_t(1\otimes \overline{g}g)\xi,\xi\rangle, \|\theta_t(1\otimes \overline{f})\eta\|^2 = \langle\theta_t(1\otimes \overline{f}f\rangle\eta,\eta\rangle.$$

From Lemma 16.1.10, we have

$$\begin{aligned} &|\int\langle\theta_t((1\otimes f)x(1\otimes g))\xi,\eta\rangle dt| \\ \leq\ & \|x\|\cdot(\int\|\theta_t(1\otimes g)\xi\|^2dt)^{1/2}\cdot(\int\|\theta_t(1\otimes \overline{f})\eta\|^2dt)^{1/2} \\ =\ & \|x\|\cdot\|f\|_2\cdot\|g\|_2\cdot\|\xi\|\cdot\|\eta\|, \quad \forall\xi,\eta\in L^2(G,H). \end{aligned}$$

Hence, $\int\theta_t(1\otimes f)x(1\otimes g))dt\in M\overline{\otimes}B(L^2(G))$ in the σ-weak topology, and $\|\int\theta_t((1\otimes f)x(1\otimes g))dt\|\leq\|x\|\cdot\|f\|_2\cdot\|f\|_2$.

Now let $g=\overline{f}$, and let $\{x_l\}$ be a bounded increasing net of $(M\overline{\otimes}B(L^2(G)))_+, x=\sup\limits_l x_l$, and $\xi\in L^2(G,H)$ with a compact support. Since

$$0\leq\langle\theta_t((1\otimes f)x_l(1\otimes g))\xi,\xi\rangle\nearrow\langle\theta_t((1\otimes f)x(1\otimes g))\xi,\xi\rangle$$

$\forall t\in G$, by the Dini theorem the convergence is uniform for $t\in K$, where K is any compact subset of G. Since $(\theta_t(1\otimes g)\xi)(x)=g(st)\xi(s)$ and suppξ is compact, it follows that

$$\int\langle\theta_t((1\otimes f)x_l(1\otimes g))\xi,\xi\rangle dt\nearrow\int\langle\theta_t((1\otimes f)x(1\otimes g))\xi,\xi\rangle dt.$$

Further, from $\|\int\theta_t((1\otimes f)y(1\otimes g))dt\|\leq\|y\|\cdot\|f\|_2\cdot\|g\|_2(\forall y\in M\overline{\otimes}B(L^2(G)))$ and $\{\xi\in L^2(G,H)|\text{supp}\xi \text{ is compact }\}$ is dense in $L^2(G,H)$, we can see that

$$\int\theta_t((1\otimes f)x_l(1\otimes g))dt\longrightarrow\int\theta_t((1\otimes f))x(1\otimes g))dt$$

in the σ-weak topology.

Then conclusion holds also for any $f,g\in K(G)$ by polarization. Q.E.D.

Lemma 16.1.12. If $x\in M$ and $f\in K(G)$, then

$$\int\theta_t(x\otimes f)dt=\int f(t)\pi(\alpha_t(x))dt$$

in the σ-weak topology.

Proof. For any $\xi,\eta \in L^2(G,H)$, by the Fubini theorem we have

$$\int \langle \theta_t(x\otimes f)\xi,\eta\rangle dt = \int \langle \alpha_t(x)\otimes f_t\xi,\eta\rangle dt$$
$$= \int dt \int f(st)\langle \alpha_t(x)\xi(s),\eta(s)\rangle ds$$
$$= \int ds \int f(t)\langle \alpha_{s^{-1}}(\alpha_t(x))\xi(s),\eta(s)\rangle dt$$
$$= \int ds \int f(t)\langle (\pi(\alpha_t(x))\xi)(s),\eta(s)\rangle dt$$
$$= \int \langle f(t)\pi(\alpha_t(x))\xi,\eta\rangle dt.$$

That comes to the conclusion. Q.E.D.

Lemma 16.1.13. For any $f,g \in K(G)$ and $a \in M\overline{\otimes}B(L^2(G))$, we have that $\int \theta_t((1\otimes f)a(1\otimes g))dt \in M \times_\alpha G$.

Proof. First, let $a = x\otimes h\lambda_s$, where $x\in M, h\in K(G), s\in G$. Then

$$\theta_t((1\otimes f)(x\otimes h\lambda_s)(1\otimes g)) = \theta_t(x\otimes fh\lambda_s g)$$
$$= \theta_t(x\otimes k)\theta_t(1\otimes\lambda_s) = \theta_t(x\otimes k)\lambda(s),$$

where $k(\cdot) = f(\cdot)h(\cdot)g(s^{-1}\cdot)$. By Lemma 16.1.12, we have

$$\int \theta_t((1\otimes f)(x\otimes h\lambda_s)(1\otimes g))dt$$
$$= \int k(t)\pi(\alpha_t(x))\lambda(s)dt \in M\times_\alpha G.$$

Since $\{h\lambda_s | s\in G, h\in K(G)\}'' = B(L^2(G))$, it follows from Lemma 16.1.11 that $\int \theta_t((1\otimes f)a(1\otimes g))dt \in M\times_\alpha G, \forall a\in M\overline{\otimes}B(L^2(G))$. Q.E.D.

Lemma 16.1.14. Let $h\in K(G)$, and K be a compact subset of G. Let $\phi_K(s) = \int_K h(st)dt$. Then $\phi_K\in K(G), 1\otimes\phi_K = \int_K \theta_t(1\otimes h)dt$, and

$$\sigma\text{-}\lim_K(1\otimes\phi_K) = \int h(t)dt\cdot 1_{H\otimes L^2(G)}.$$

Proof. For any $\xi,\eta\in L^2(G,H)$, by the Fubini theorem we have

$$\int_K \langle \theta_t(1\otimes h)\xi,\eta\rangle dt = \int_K dt\int h(st)\langle\xi(s),\eta(s)\rangle ds$$
$$= \int \langle\xi(s),\eta(s)\rangle ds \int_K h(st)dt = \int \phi_K(s)\langle\xi(s),\eta(s)\rangle ds$$
$$= \langle (1\otimes\phi_K)\xi,\eta\rangle.$$

Thus, $\int_K \theta_t(1\otimes K)dt = 1\otimes\phi_K$. Further, from Lemma 16.1.10 we get w - $\lim_K(1\otimes\phi_K) = \int h(t)dt\cdot 1_{H\otimes L^2(G)}$. Moreover, since $|\phi_K(s)|\le\|h\|_1$ and $\|1\otimes\phi_K\|\le\|h\|_1, \forall K$, we have also

$$\sigma\text{-}\lim_K(1\otimes\phi_K) = \int h(t)dt\cdot 1_{H\otimes L^2(G)}.$$

Q.E.D.

Theorem 16.1.15. Let (M,G,α) be a VN-system with $M\subset B(H)$. Then we have

$$M\times_\alpha G = \{a\in M\overline{\otimes}B(L^2(G))|\theta_t(a) = a, \forall t\in G\},$$

where $\theta_t = \alpha_t\otimes ad\rho_t, \forall t\in G$.

Proof. Clearly $\theta_t(\lambda(s)) = (\alpha_t\otimes ad\rho_t)(1\otimes\lambda_s) = 1\otimes\lambda_s = \lambda(s), \forall s,t\in G$. For any $x\in M$ and $f\in K(G)$, since the integral in Lemma 16.1.12 exists in the σ-weak topology and $*$ automorphisms are σ-weakly continuous, it follows that $\int\theta_s(\theta_t(x\otimes f))dt = \int f(t)\theta_s(\pi(\alpha_t(x)))dt$. On the other hand, we have

$$\int\theta_s(\theta_t(x\otimes f)))dt = \int\theta_{st}(x\otimes f)dt = \int\theta_t(x\otimes f)dt$$

obviously. Thus, by Lemma 16.1.12 and $f\in K(G)$ is arbitrary, we get $\theta_s(\pi(\alpha_t(x)))) = \pi(\alpha_t(x)), \forall s,t\in G$. In particular, $\theta_s(\pi(x)) = \pi(x), \forall x\in M$. Therefore, $M\times_\alpha G\subset\{a\in M\overline{\otimes}B(L^2(G))|\theta_t(a) = a, \forall t\in G\}$.

Now let $a\in M\overline{\otimes}B(L^2(G))$ be such that $\theta_t(a) = a, \forall t\in G$. Pick $h\in K(G)$ with $h\ge 0$ and $\int h(t)dt = 1$. For any compact subset K of G, let $\phi_K(s) = \int_K h(st)dt$. Then by Lemma 16.1.13, we have

$$a_K = \int\theta_t((1\otimes\phi_K)a(1\otimes h))dt\in M\times_\alpha G.$$

We claime that

$$a_K(1\otimes g)\longrightarrow a(1\otimes g),\quad (1\otimes g)a_K\longrightarrow(1\otimes g)a$$

in the σ-weak topology, $\forall g\in K(G)$. In fact, let $L = (\text{supp}g)^{-1}(\text{supp}h)$. Clearly, L is compact, and the function $h(\cdot t)g(\cdot)$ is 0 if $t\notin L$. Thus, we have

$$\begin{aligned} a_K(1\otimes g) &= \int\theta_t(1\otimes\phi_K)\cdot a\cdot\theta_t(1\otimes h)\cdot(1\otimes g)dt \\ &= \int_L\theta_t((1\otimes\phi_K)a(1\otimes h))dt(1\otimes g)\end{aligned}$$

Similar to Lemma 16.1.11, we can see that $x \longrightarrow \int_L \theta_t(x)dt$ is a $\sigma-\sigma$ continuous map on $M\overline{\otimes}B(L^2(G))$. Therefore , from Lemma 16.1.14 an 16.1.10 we have

$$\begin{aligned} a_K(1\otimes g) \longrightarrow \quad & a\int_L \theta_t(1\otimes h)dt(1\otimes g) = a\int_L \theta_t(1\otimes h)(1\otimes g)dt \\ & = a\int \theta_t(1\otimes h)\cdot(1\otimes g)dt = a(1\otimes g) \end{aligned}$$

in the σ-weak topology.

By Lemma 16.1.14 and the Fubini theorem, we can see that

$$\begin{aligned} (1\otimes g)a_K &= \int (1\otimes g)\cdot\theta_t(1\otimes\phi_K)\cdot a\cdot\theta_t(1\otimes h)dt \\ &= \int (1\otimes g)\cdot\int_K \theta_{ts}(1\otimes h)ds\cdot a\cdot\theta_t(1\otimes h)dt \\ &= \int_K ds\int (1\otimes g)\triangle(s)^{-1}\cdot\theta_t(1\otimes h)\cdot a\cdot\theta_{ts^{-1}}(1\otimes h)dt \\ &= \int dt\int_{K^{-1}}(1\otimes g)\cdot\theta_t(1\otimes h)\cdot a\cdot\theta_{ts}(1\otimes h)ds \\ &= \int (1\otimes g)\cdot\theta_t(1\otimes h)\cdot a\cdot\theta_t(1\otimes\phi_{K^{-1}})dt. \end{aligned}$$

Now by the same discussion in the preceding paragraph, we get

$$(1\otimes g)a_K \longrightarrow \int (1\otimes g)\cdot\theta_t(1\otimes h)adt = (1\otimes g)a$$

in the σ-weak topology.

For any $a'\in(M\times_\alpha G)'$, since $a_K\in M\times_\alpha G$, it follows that

$$\begin{aligned} (1\otimes g)a'a(1\otimes g) &= \sigma\text{-}\lim_K (1\otimes g)a'a_K(1\otimes g) \\ = \quad \sigma\text{-}\lim_K &(1\otimes g)a_Ka'(1\otimes g) = (1\otimes g)aa'(1\otimes g), \end{aligned}$$

$\forall g\in K(G)$. Pick a net $\{g_l\}\subset K(G)$ such that $1\otimes g_l \longrightarrow 1_{H\otimes L^2(G)}$ in the σ-weak topology. Then $a'a = aa', \forall a'\in(M\times_\alpha G)'$, and $a\in M\times_\alpha G$. Q.E.D.

Proposition 16.1.16. Let (M,G,α) be a VN-system with $M\subset B(H)$, and α be spatial, i.e., there is a strongly continuous unitary representation $s\to u_s$ of G on H such that $\alpha_s(x)=u_sxu_s^*, \forall x\in M, s\in G$. Then we have

$$(M\times_\alpha G)' = \{x'\otimes 1, u_s\otimes\rho_s | x'\in M', s\in G\}''.$$

Proof. Since $\theta_t(a)=a \Longleftrightarrow (u_t\otimes\rho_t)a = a(u_t\otimes\rho_t)$, it follows from Theorem 16.1.15 that

$$M\times_\alpha G = (M\overline{\otimes}B(L^2(G)))\cap\{u_s\otimes\rho_s | s\in G\}'.$$

Therefore, we have

$$(M \times_\alpha G)' = \{x' \otimes 1, u_s \otimes \rho_s | x' \in M', s \in G\}''.$$

Q.E.D.

Proposition 16.1.17. Let $(M, G, \alpha), (N, G, \beta)$ be two VN-systems with $M \subset B(H), N \subset B(K)$ respectively. If τ is a $*$ isomophism from M on N with $\tau(\alpha_t(x)) = \beta_t(\tau(x)), \forall t \in G, x \in M$, then $\tilde{\tau}$ is a $*$ isomorphism from $M \times_\alpha G$ onto $N \times_\beta G$, and $\tilde{\tau}(\pi_\alpha(x)) = \pi_\beta(\tau(x)), \forall x \in M$, where $\tilde{\tau} = \tau \otimes id : M\overline{\otimes}B(L^2(G)) \longrightarrow N\overline{\otimes}B(L^2(G))$, and $\{\pi_\alpha, L^2(G, H)\}, \{\pi_\beta, L^2(G, K)\}$ are faithful W^* - representations of M, N respectively as in Propositon 16.1.3.

Proof. For any $x \in M, h \in K(G)$, by Lemma 16.1.12 we have

$$\int \alpha_t(x) \otimes \rho_t h \rho_t^* dt = \int h(t)\pi_\alpha(\alpha_t(x))dt.$$

Thus

$$\begin{aligned} &\int h(t)\tilde{\tau} \circ \pi_\alpha(\alpha_t(x))dt \\ = &\int \beta_t(\tau(x)) \otimes \rho_t h \rho_t^* dt = \int h(t)\pi_\beta(\beta_t(\tau(x)))dt. \end{aligned}$$

Since h is arbitrary, it follows that

$$\tilde{\tau}(\pi_\alpha(x)) = \pi_\beta(\tau(x)), \quad \forall x \in M.$$

Clearly, $\tilde{\tau}(\lambda(s)) = \lambda(s), \forall s \in G$. So $\tilde{\tau}(M \times_\alpha G) = N \times_\beta G$. Moreover, $\tilde{\tau}$ is a $*$ isomorphism from $M\overline{\otimes}B(L^2(G))$ onto $N\overline{\otimes}B(L^2(G))$ by Theorem 4.3.4. Therefore, $\tilde{\tau}$ is also a $*$ isomorphism from $M \times_\alpha G$ onto $N \times_\beta G$. Q.E.D.

Definition 16.1.18. (M, G, α) is called a *W^*-dynamical system*, if M is a W^*-algebra, α is a homomorphism from G to Aut(M), and $t \to \alpha_t(x)$ is continuous from G to $(M, \sigma(M, M_*)), \forall x \in M$.

Clearly, $t \longrightarrow \alpha_t(x)$ is also continuous from G to $(M, \tau(M, M_*)), \forall x \in M$.

Definition 16.1.19. Let (M, G, α) be a W^*-system. The *fixed point algebra* of the W^*-system $(M\overline{\otimes}B(L^2(G)), G, \theta), \{a \in M\overline{\otimes}B(L^2(G)) | \theta_t(a) = a, \forall t \in G\}$, is called the *crossed product* of M by the action α of G, and denoted by $M \times_\alpha G$, where $M\overline{\otimes}B(L^2(G))$ is the W^*-tensor product of M and $B(L^2(G))$ (see Section 4.3), and $\theta_t = \alpha_t \otimes ad\rho_t, \forall t \in G$.

Now let $\{\pi, H\}$ be a faithful W^*-representation of M. Then $(\pi(M), G, \alpha^{(\pi)})$ is a VN-system, where

$$\alpha_t^{(\pi)}(\pi(x)) = \pi(\alpha_t(x)), \forall x \in M, t \in G.$$

Futher, we have VN-system $(\pi(M)\overline{\otimes}B(L^2(G)), G, \theta^{(\pi)})$, where $\theta^{(\pi)} = \alpha^{(\pi)} \otimes adp$. Clearly, $(\pi \otimes id)$ is a $*$ isomorphism from $M\overline{\otimes}B(L^2(G))$ onto $\pi(M)\overline{\otimes}B(L^2(G))$, and

$$(\pi \otimes id)\theta = \theta^{(\pi)}(\pi \otimes id).$$

Therefore, $M \times_\alpha G$ is $*$ isomorphic to $\pi(M) \times_{\alpha^{(\pi)}} G$.

Notes. In full generality W^*-crossed products first appeared in M. Takesaki's paper. The version presented here is taken from A.Van Daele completely.

References. [176], [189].

16.2. Takesaki's duality theorem

In this section, we assume that the locally compact group G is abelian, and denote its dual group by $\widehat{G}$. Let $\mu, \hat{\mu}$ be the Haar measures on $G, \widehat{G}$ respectively, such that the Fourier transform

$$\hat{f}(p) = \int \langle s, p\rangle f(s)ds, \quad \forall f \in L^2(G)$$

is a unitary operator from $L^2(G)$ onto $L^2(\widehat{G})$. Moreover, write $d\mu(t) = dt, d\hat{\mu}(p)$ dp simply.

Now for each $p \in \widehat{G}$, define a unitary operator V_p on $L^2(G)$ as follows:

$$(V_p f)(s) = \overline{\langle s, p\rangle} f(s), \quad \forall f \in L^2(G).$$

Clearly, $p \longrightarrow V_p$ is a strongly continuous unitary representation of $\widehat{G}$ on $L^2(G)$.

Lemma 16.2.1. Let (M, G, α) be a VN-system with $M \subset B(H)$. Then $(1 \otimes V_p)a(1 \otimes V_p)^* \in M \times_\alpha G, \forall a \in M \times_\alpha G$, where $1 = 1_H$.

Proof. Since $(V_p \lambda_s V_p^* f)(t) = \overline{\langle t, p\rangle}(V_p^* f)(s^{-1}t) = \overline{\langle s, p\rangle} f(s^{-1}t) = \overline{\langle s, p\rangle}(\lambda_s f)(t), \forall f \in L^2(G)$, it follows that $V_p \lambda_s V_p^* = \overline{\langle s, p\rangle}\lambda_s, \forall s \in G, p \in \widehat{G}$. Moreover, by

$$\begin{aligned}((1 \otimes V_p)\pi(x)\xi)(s) &= \overline{\langle s, p\rangle}(\pi(x)\xi)(x) = \overline{\langle s, p\rangle}\alpha_{s^{-1}}(x)\xi(s) \\ &= \alpha_{s^{-1}}(x)\overline{\langle s, p\rangle}\xi(s) = (\pi(x)(1 \otimes V_p)\xi)(s),\end{aligned}$$

$\forall \xi \in L^2(G, H)$, we have $(1 \otimes V_p)\pi(x) = \pi(x)(1 \otimes V_p), \forall x \in M, p \in \widehat{G}$. Therefore, $(1 \otimes V_p)(M \times_\alpha G)(1 \otimes V_p)^* = M \times_\alpha G$. Q.E.D.

Definition 16.2.2. Let (M, G, α) be a VN-system with $M \subset B(H)$, and define

$$\hat{\alpha}_p(a) = (1 \otimes V_p)a(1 \otimes V_p)^*, \quad \forall a \in M \times_\alpha G, p \in \hat{G}.$$

Then the VN-system $(M \times_\alpha G, \hat{G}, \hat{\alpha})$ is called the *dual system* of (M, G, α), and $\hat{\alpha}$ is called the *dual action* of α.

Proposition 16.2.3. Let $(M, G, \alpha), (N, G, \beta)$ be two VN-systems, and τ be a $*$ isomorphism from M onto N such that

$$\tau(\alpha_t(x)) = \beta_t(\tau(x)), \quad \forall x \in M, t \in G.$$

Then $\tilde{\tau} = \tau \otimes id$ (see Proposition 1.1.17) is also a $*$ isomorphism from $M \times_\alpha G$ onto $M \times_\beta G$ with

$$\tilde{\tau}(\hat{\alpha}_p(a)) = \hat{\beta}_p(\tilde{\tau}(a)), \quad \forall a \in M \times_\alpha G, p \in \hat{G},$$

where $\hat{\alpha}, \hat{\beta}$ are the dual actions of α, β respectively. Consequently, $(M \times_\alpha G) \times_{\hat{\alpha}} \hat{G}$ and $(N \times_\beta G) \times_{\hat{\beta}} \hat{G}$ are $*$ isomorphic.

Proof. Notice that

$$\begin{aligned}\tilde{\tau}(\hat{\alpha}_p(a)) &= \tilde{\tau}(1 \otimes V_p) \cdot \tilde{\tau}(a) \cdot \tilde{\tau}(1 \otimes V_p)^* \\ &= (1 \otimes V_p)\tilde{\tau}(a)(1 \otimes V_p)^* = \hat{\beta}_p(\tilde{\tau}(a)),\end{aligned}$$

$\forall a \in M \times_\alpha G, p \in \hat{G}$. Now the conclusion is obvious from Proposition 16.1.17.

Q.E.D.

In the following lemmas, let (M, G, α) be a VN-system with $M \subset B(H)$, and α be spatial , i.e., there is a strongly continuous unitary representation $s \to u_s$ of G on H such that $\alpha_s(x) = u_s x u_s^*, \forall x \in M, s \in G$.

Lemma 16.2.4. Let $R_0 = (M \times_\alpha G) \times_{\hat{\alpha}} \hat{G} (\subset B(H) \otimes L^2(G) \otimes L^2(\hat{G}))$. Then R_0 is spatially $*$ isomorphic to R_1, where

$$R_1 = \{x \otimes 1 \otimes 1, u_s \otimes \lambda_s \otimes 1, 1 \otimes V_p \otimes \lambda_p | x \in M, s \in G, p \in \hat{G}\}''$$

is a VN algebra on $H \otimes L^2(G) \otimes L^2(\hat{G})$, and $s \to \lambda_s, p \longrightarrow \lambda_p$ are the left regular representations of $G, \hat{G}$ on $L^2(G), L^2(\hat{G})$ respectively.

Proof. Since $\hat{\alpha}$ is spatial, it follows from Proposition 16.1.7 that $(M \times_\alpha G) \times_{\hat{\alpha}} \hat{G}$ is spatially $*$ isomorphic to

$$\{a \otimes 1, 1 \otimes V_p \otimes \lambda_p | a \in M \times_\alpha G, p \in \hat{G}\}''.$$

Define a unitary operator W on $L^2(G, H)$ as follows:

$$(Wf)(s) = u_s f(s), \quad \forall f \in L^2(G, H).$$

From Proposition 16.1.7 we have

$$W(M \times_\alpha G)W^* = \{x \otimes 1, u_s \otimes \lambda_s | x \in M, s \in G\}''.$$

Since $W(1 \otimes V_p) = (1 \otimes V_p)W, \forall p \in \widehat{G}$, it follows that

$$\begin{aligned} & (W \otimes 1)\{a \otimes 1, 1 \otimes V_p \otimes \lambda_p | a \in M \times_\alpha G, p \in \widehat{G}\}''(W \otimes 1)^* \\ = & \{x \otimes 1 \otimes 1, u_s \otimes \lambda_s \otimes 1, 1 \otimes V_p \otimes \lambda_p | x \in M, s \in G, p \in \widehat{G}\}''. \end{aligned}$$

Therefore, $(M \times_\alpha G) \times_2 \widehat{G}$ is spatially $*$ isomorphic to R_1. Q.E.D.

Lemma 16.2.5. R_1 is spatially $*$ isomorphic to R_2, where

$$R_2 = \{x \otimes 1 \otimes 1, u_s \otimes \lambda_s \otimes 1, 1 \otimes V_p \otimes V_p | x \in M, s \in G, p \in \widehat{G}\}''$$

is a VN algebra on $H \otimes L^2(G) \otimes L^2(G)$.

Proof. Let $\mathcal{F}^*$ be the inverse Fourier transform from $L^2(\widehat{G})$ onto $L^2(G)$:

$$(\mathcal{F}^* f)(S) = \int \overline{\langle s, p \rangle} f(p) dp, \quad \forall f \in L^2(\widehat{G}).$$

Then $1 \otimes 1 \otimes \mathcal{F}^*$ is a unitary operator from $H \otimes L^2(G) \otimes L^2(\widehat{G})$ onto $H \otimes L^2(G) \otimes L^2(G)$. Now it suffices to show that

$$\mathcal{F}^* \lambda_p \mathcal{F} = V_p, \quad \forall p \in \widehat{G}.$$

For any $f \in K(\widehat{G})$, we have

$$\begin{aligned} (\mathcal{F}^* \lambda_p f)(s) &= \int \overline{\langle s, q \rangle} (\lambda_p f)(q) dq = \int \overline{\langle s, q \rangle} f(p^{-1} q) dq \\ &= \overline{\langle s, p \rangle} \int \overline{\langle s, q \rangle} f(q) dq = \overline{\langle s, p \rangle} (\mathcal{F}^* f)(s) \\ &= (V_p \mathcal{F}^* f)(s), \quad \forall s \in G, p \in \widehat{G}. \end{aligned}$$

Therefore, $\mathcal{F}^* \lambda_p \mathcal{F} = V_p, \forall p \in \widehat{G}$. Q.E.D.

Lemma 16.2.6. R_2 is spatially $*$ isomorphic to $R_3 \overline{\otimes} \mathbb{C} 1_{L^2(G)}$, where

$$R_3 = \{x \otimes 1, u_s \otimes \lambda_s, 1 \otimes V_p | x \in M, s \in G, p \in \widehat{G}\}''$$

is a VN algebra on $H \otimes L^2(G)$.

Proof. Define a unitary operator U on $L^2(G) \otimes L^2(G)$ as follows :

$$(Uf)(s,t) = f(st,t), \quad \forall f \in L^2(G) \otimes L^2(G).$$

It is easily verified that

$$U^*(V_p \otimes V_p)U = V_p \otimes 1, \quad U^*(\lambda_s \otimes 1)U = \lambda_s \otimes 1,$$

$\forall s \in G, p \in \hat{G}$. Therefore, we have

$$\begin{aligned} &(1 \otimes U^*)R_2(1 \otimes U) \\ =\ &(1 \otimes U^*)\{x \otimes 1 \otimes 1, u_s \otimes \lambda_s \otimes 1, 1 \otimes V_p \otimes V_p | \cdots\}''(1 \otimes U) \\ =\ &\{x \otimes 1 \otimes 1, u_s \otimes \lambda_s \otimes 1, 1 \otimes V_p \otimes 1 | x \in M, s \in G, p \in \hat{G}\}'' \\ =\ &R_3 \overline{\otimes} \mathbb{C} 1_{L^2(G)}. \end{aligned}$$

Q.E.D.

Lemma 16.2.7. R_3 is spatially $*$ isomophic to $M \overline{\otimes} B(L^2(G))$.

Proof. Define a unitary operator W on $H \otimes L^2(G)$ as follows:

$$(Wf)(s) = u_s f(s), \quad \forall f \in L^2(G, H).$$

From Proposition 16.1.7, we have

$$W^*(u_s \otimes \lambda_s)W = 1 \otimes \lambda_s, \quad W^*(x \otimes 1)W = \pi(x),$$

$\forall x \in M, s \in G$. Clearly, $W(1 \otimes V_p) = (1 \otimes V_p)W, \forall p \in \hat{G}$.

1) $\{V_p | p \in \hat{G}\}''$ is the multiplicative algebra Z on $L^2(G)$.

In fact, V_p is the multiplicative operator associated with the function $\overline{\langle \cdot, p \rangle}$ on $G, \forall p \in \hat{G}$. If $h \in L^1(G)$ is such that $\int \overline{\langle s, p \rangle} h(s) ds = 0, \forall p \in \hat{G}$, then by the uniqueness of Fourier transform we have $h = 0$. This means that $[\overline{\langle \cdot, p \rangle} | p \in \hat{G}]$ is w^*-dense in $L^\infty(G)$. Therefore, we have $\{V_p | p \in \hat{G}\}'' = Z$.

2) $\{V_p, \lambda_s | s \in G, p \in \hat{G}\}'' = B(L^2(G))$.

This is obvious since Z is a maximal abelian VN algebra on $L^2(G)$.

3) $W^*(M \overline{\otimes} Z)W = M \overline{\otimes} Z$.

In fact, by Propositon 16.1.3, we have $W^*(x \otimes 1)W = \pi(x) \in M \overline{\otimes} Z, \forall x \in M$. Moreover, $W^*(1 \otimes V_p)W = 1 \otimes V_p, \forall p \in \hat{G}$. Thus, $W^*(M \overline{\otimes} Z)W \subset M \overline{\otimes} Z$. On the other hand, it is easy to see that $W(x \otimes 1)W^* \cdot (x' \otimes 1) = (x' \otimes 1) \cdot W(x \otimes 1)W^*, \forall x' \in M', x \in M$. Thus

$$W(x \otimes 1)W^* \in (M' \otimes 1)' = M \overline{\otimes} B(L^2(G)), \forall x \in M.$$

Clearly, $W(x \otimes 1)W^* \in \{1 \otimes V_p | p \in \hat{G}\}' = B(H) \overline{\otimes} Z, \forall x \in M$. By Proposition 1.4.14, we have $W(M \otimes 1)W^* \subset M \overline{\otimes} Z$. Therefore, $W(M \overline{\otimes} Z)W^* \subset M \overline{\otimes} Z$, and $W^*(M \overline{\otimes} Z)W = M \overline{\otimes} Z$.

Finally, R_3 is spatially $*$ isomorphic to

$$\begin{aligned} & W^*\{x\otimes 1, 1\otimes V_p, u_s\otimes\lambda_s | x\in M, s\in G, p\in\widehat{G}\}''W \\ = & \{W^*(M\overline{\otimes}Z)W, W^*\{u_s\otimes\lambda_s|s\in G\}W\}'' \\ = & \{M\overline{\otimes}Z, 1\otimes\lambda_s|s\in G\}'' = M\overline{\otimes}B(L^2(G)). \end{aligned}$$

Q.E.D.

Let $adV_p(A) = V_paV_p^*, \forall a\in B(L^2(G)), p\in\widehat{G}$. By the proof of Lemma 16.2.1, we can see that

$$(ad\lambda_s)(adV_p) = (adV_p)(ad\lambda_s)$$

on $B(L^2(G)), \forall s\in G, p\in\widehat{G}$.

Now let (M,G,α) be a W^*-system. Then $(id\otimes adV)$ is an action of $\widehat{G}$ on $M\overline{\otimes}B(L^2(G))$, and

$$\theta(id\otimes adV) = (id\otimes adV)\theta,$$

where $\theta = \alpha\otimes ad\rho = \alpha\otimes ad\lambda$. By Definition 16.1.19, $((id\otimes adV)|M\times_\alpha G)$ is an action of $\widehat{G}$ on $M\times_\alpha G$.

Definition 16.2.8. Let (M,G,α) be a W^*-system, and G be abelian. Then the W^*-system $(\widehat{M},\widehat{G},\widehat{\alpha})$ is called the *dual system* of (M,G,α), and $\widehat{\alpha}$ is called the *dual action* of α, where $\widehat{M} = M\times_\alpha G$, and $\widehat{\alpha} = (id\otimes adV)|M\times_\alpha G$.

Two W^*-systems (M,G,α) and (N,G,β) are called isomorphic, if there is a $*$ isomorphism τ from M onto N such that $\tau(\alpha_t(x)) = \beta_t(\tau(x)), \forall t\in G, x\in M$.

Proposition 16.2.9. Suppose that two W^*-systems (M,G,α) and (N,G,β) are isomorphic. Then their dual systems $(\widehat{M},\widehat{G},\widehat{\alpha})$ and $(\widehat{N},\widehat{G},\widehat{\alpha})$ are also isomorphic.

Proof. Let τ be a $*$ isomorphism from M onto N with $\tau(\alpha_t(x)) = \beta_t(\tau(x))$, $\forall\, x\in M, t\in G$. Then $\tilde{\tau} = \tau\otimes id$ is a $*$ isomorphism from $M\overline{\otimes}B(L^2(G))$ onto $N\overline{\otimes}B(L^2(G))$, and

$$\tilde{\tau}(id\otimes adV_p) = (id\otimes adV_p)\tilde{\tau}, \quad \tilde{\tau}\theta_t^{(\alpha)} = \theta_t^{(\beta)}\tilde{\tau},$$

$\forall p\in\widehat{G}, t\in G$, where $\theta^{(\alpha)} = \alpha\otimes ad\lambda, \theta^{(\beta)} = \beta\otimes ad\lambda$. Therefore, $\tilde{\tau}$ is an isomorphism from $(M\times_\alpha G,\widehat{G},\widehat{\alpha})$ to $(N\times_\beta G,\widehat{G},\widehat{\beta})$. Q.E.D.

Theorem 16.2.10. Let (M,G,α) be a W^*-system, and G be abelian. Then the double dual system $(\widehat{\widehat{M}},\widehat{\widehat{G}} = G,\widehat{\widehat{\alpha}})$ is isomorphic to $(M\overline{\otimes}B(L^2(G)), G, \theta = \alpha\otimes ad\rho)$.

Proof. By Propositions 16.2.9 and 16.1.5 , we may assume that (M, G, α) is a VN-system with $M \subset B(H)$, and there is a strongly continuous unitary representation $s \longrightarrow u_s$ of G on H such that $\alpha_s(x) = u_s x u_s^*, \forall s \in G, x \in M$.

Let $(\widehat{M}, \widehat{G}, \hat{\alpha})$ be the dual system of (M, G, α). Then $\widehat{M} = M \times_\alpha G \subset B(H \otimes L^2(G))$, and $\hat{\alpha}_p = ad(1 \otimes V_p), \forall p \in \widehat{G}$. By Proposition 16.1.7, let

$$(\widetilde{W} f)(s,p) = ((1 \otimes V_p) f)(p)(s) = \overline{\langle s, p\rangle} f(s,p)$$

$\forall f \in H \otimes L^2(G) \otimes L^2(\widehat{G})$. Then $\widetilde{W}$ is a unitary operator, and

$$\widetilde{W} R_0 \widetilde{W}^* = \widetilde{R}_0 = \{a \otimes 1, 1 \otimes V_p \otimes \lambda_p | p \in \widehat{G}\}'',$$

where $R_0 = \widehat{\widehat{M}} = (M \times_\alpha G) \times_{\hat{\alpha}} \widehat{G}$. Since $\hat{\hat{\alpha}}_s = ad(1 \otimes 1 \otimes V_s)$, where $(V_s g)(p) = \overline{\langle s, p\rangle} g(p), \forall g \in L^2(\widehat{G}), s \in G$, and $\widetilde{W}$ commutes with $(1 \otimes 1 \otimes \hat{V}_s), \forall s \in G$, so $(R_0, G, \hat{\hat{\alpha}})$ is isomorphic to $(\widetilde{R}_0, G, \hat{\hat{\alpha}})$.

Clearly, $(W \otimes 1)$ commutes with $(1 \otimes 1 \otimes V_s), \forall s \in G$. By Lemma 16.2.4, $(\widetilde{R}_0, G, \hat{\hat{\alpha}})$ is isomorphic to $(R_1, G, \hat{\hat{\alpha}})$, where $\hat{\hat{\alpha}}_s = ad(1 \otimes 1 \otimes V_s), \forall s \in G$.

By $\mathcal{F}^* V_s \mathcal{F} = \lambda_s, \forall s \in G$, and Lemma 16.2.5, $(R_1, G, \hat{\hat{\alpha}})$ is isomorphic to $(R_2, G, ad(1 \otimes 1 \otimes \lambda))$.

By Lemma 16.2.6 and $U^*(1 \otimes \lambda_s) U = \lambda_s \otimes \lambda_s, (R_2, G, ad(1 \otimes 1 \otimes \lambda))$ is isomorphic to $(R_3 \overline{\otimes} \mathbb{C}, G, ad(1 \otimes \lambda \otimes \lambda))$.

Clearly, $(R_3 \overline{\otimes} \mathbb{C}, G, ad(1 \otimes \lambda \otimes \lambda))$ is isomorphic to $(R_3, G, ad(1 \otimes \lambda))$.

Finally, by Lemma 16.2.7 and $W^*(1 \otimes \lambda_s) W = u_s \otimes \lambda_s, \forall s \in G, (R_3, G, ad(1 \otimes \lambda))$ is isomorphic to $(M \overline{\otimes} B(L^2(G)), G, ad(u \otimes \lambda) = \theta)$. Q.E.D.

Remark. Let (M, G, α) be a W^*-system, and G be alelian. On $M \overline{\otimes} B(L^2(G)) \overline{\otimes} B(L^2(\widehat{G}))$, we have three actions : $\alpha_s^{(1)} = \alpha_s \otimes ad\lambda_s \otimes id (s \in G); \alpha_p^{(2)} = \hat{\alpha}_p \otimes ad\lambda_p = id \otimes adV_p \otimes ad\lambda_p (p \in \widehat{G})$; and $\alpha_s^{(3)} = id \otimes id \otimes adV_s (s \in G)$. Clearly, these actions commute each other; the fixed point algebra of $(M \overline{\otimes} B(L^2(G)) \overline{\otimes} B(L^2(\widehat{G})), G, \alpha^{(1)})$ is $(M \times_\alpha G) \overline{\otimes} B(L^2(\widehat{G}))$; the fixed point algebra of $((M \times_\alpha G) \overline{\otimes} B(L^2(\widehat{G})), \widehat{G}, \alpha^{(2)})$ is $(M \times_\alpha G) \times_{\hat{\alpha}} \widehat{G} = \widehat{\widehat{M}}$, and $\hat{\hat{\alpha}} = \alpha^{(3)} | \widehat{\widehat{M}}$.

Now let M be a σ-finite VN algebra on a Hilbert space H, φ and ψ be two faithful normal states on M, $\{\sigma_t^\varphi | t \in \mathbb{R}\}$ and $\{\sigma_t^\psi | t \in \mathbb{R}\}$ be the modular automorphism groups of M corresponding to φ and ψ reprectively (see Definition 8.3.1). Then we get VN-systems $(M, \mathbb{R}, \sigma^\varphi)$ and $(M, \mathbb{R}, \sigma^\psi)$.

Proposition 16.2.11. VN algebras $M \times_{\sigma^\varphi} \mathbb{R}$ and $M \times_{\sigma^\psi} \mathbb{R}$ on $L^2(\mathbb{R}, H)$ are spatially $*$ isomorphic.

Proof. By the unitary cocycle theorem (Proposition 8.3.3), there is an one-parameter $s(M, M_*)$-continuous family $\{u_t | t \in \mathbb{R}\}$ of unitary elements of M

such that

$$\sigma_t^\psi(x) = u_t\sigma_t^\varphi(x)u_t^*, \quad u_{t+s} = u_t\sigma_t^\varphi(u_s),$$

$\forall s,t \in I\!R, x \in M$. Define a unitary operator U on $L^2(I\!R, H)$ as follows

$$(Uf)(s) = u_{-s}f(s), \quad \forall f \in L^2(I\!R, H).$$

Clearly, $(U^*f)(s) = u_{-s}^*f(s), \forall f \in L^2(I\!R, H)$. Let π_φ, π_ψ be the faithful W^*-representations of M on $L^2(I\!R, H)$ corresponding to φ, ψ respectively (see Proposition 16.1.3). Then it is easily verified that

$$U\pi_\varphi(x)U^* = \pi_\psi(x), \quad U\lambda(t)U^* = \pi_\psi(u_t^*)\lambda(t),$$

$\forall x \in M, t \in I\!R$. Thus, we have $U(M \times_{\sigma^\varphi} I\!R)U^* \subset (M \times_{\sigma^\psi} I\!R)$.

Similarly, it is easy to see that

$$U^*\lambda(t)U = \pi_\varphi(u_t)\lambda(t), U^*\pi_\psi(x)U = \pi_\varphi(x),$$

$\forall x \in M, t \in I\!R$. Therefore, we obtain that $U^*(M \times_{\sigma^\psi} I\!R)U \subset (M \times_{\sigma^\psi} I\!R)$, and $U(M \times_{\sigma^\varphi} I\!R)\ U^* = M \times_{\sigma^\psi} I\!R$. Q.E.D.

Proposition 16.2.12. The dual systems $(M \times_{\sigma^\varphi} I\!R, I\!R, \hat{\sigma}^\varphi)$ and $(M \times_{\sigma^\psi} I\!R, I\!R, \hat{\sigma}^\psi)$ are isomorphic.

Proof. Let $\gamma(a) = UaU^*, \forall a \in M \times_{\sigma^\varphi} I\!R$, where U is defined as in Proposition 16.2.11. It suffices to show that

$$\hat{\sigma}_t^\psi(\gamma(a)) = \gamma(\hat{\sigma}_t^\varphi(a)), \quad \forall t \in I\!R, a \in M \times_{\sigma^\varphi} I\!R.$$

Since $\hat{\sigma}_t^\varphi$ and $\hat{\sigma}_t^\psi$ are unitarily implemented by $1 \otimes V_t$(on $L^2(I\!R, H) = H \otimes L^2(I\!R)$), and $(1 \otimes V_t)U = U(1 \otimes V_t), \forall t \in I\!R$, the conclusion is obvious.

Q.E.D.

Remark. For a σ-finite W^*-algebra M, by Propositions 16.2.11 and 16.2.12 it is reasonable to write a W^*-system $(M \times_\sigma I\!R, I\!R, \hat{\sigma})$. Moreover, we can prove that the W^* -algebra $M \times_\sigma I\!R$ is semi-finite.

References. [127], [165], [176], [189].

16.3. Group algebras and Group C^*-algebras

This section is a survey on group algebras and group C^*-algebras, and we shall not give the proofs for most of conclusions in this section. Indeed, this section is the preliminaries of next section: C^*-crossed products.

Lacally compact groups

Let G be a locally compact group , and μ_l be a left invariant *Haar measure* on G, i.e., μ_l be a regular Borel measure on G with $\mu_l(tE) = \mu_l(E), \forall t \in G$ and Borel subset E of G. It is well-known that:

1) μ_l is uniquely determined up to multiplication by a positive constant; and $\mu_l(U) > 0$ for any Borel open subset U of G;

2) G is compact if and only if $\mu_l(G) < \infty$;

3) G is discrete if and only if $\mu_l(\{e\}) > 0$, where e is the unit of G.

Let $\mu_r(E) = \mu_l(E^{-1})$ for any Borel subset E of G. Then μ_r is a right invariant Haar measure on G i.e., $\mu_r(Et) = \mu_r(E), \forall t \in G$ and Borel subset E of G. Since μ_l and μ_r are equivalent, we can write that $\mu_l = \triangle \cdot \mu_r$. The function $\triangle(\cdot)$ is called the *modular function* of G, and it is positive and continuous on G, and

$$\triangle(e) = 1, \triangle(st) = \triangle(s) \triangle (t), \quad \triangle(s^{-1}) = \triangle(s)^{-1}, \forall s, t \in G.$$

If write $d\mu_l(s) = ds$ simply, then we have

$$d(ts) = ds, \quad d(st) = \triangle(t) ds, ds^{-1} = \triangle(s)^{-1} ds, \forall t \in G,$$

or

$$\int_G f(ts)ds = \int_G f(s)ds, \quad \int_G f(s^{-1})ds = \int_G \frac{f(s)}{\triangle(s)}ds,$$
$$\triangle(t) \int_G f(s)ds = \int_G f(s)d(st) = \int_G f(st^{-1})ds,$$

$\forall t \in G, f \in K(G)$, where $K(G)$ is the set of all continuous functions on G with a compact support.

G is said to be *unimodular*, if $\triangle(\cdot) \equiv 1$. In this case, $\mu_l = \mu_r$ is an invariant measure on G. For examples, compact groups, abelian groups and discrete groups are unimodular.

Measure algebras and group algebras

Let G be a locally compact group, and denote the collection of all bounded Radon measures on G by $M(G)$. It is well-known that $M(G) = C_0^\infty(G)^*$. In $M(G)$, define the multiplication

$$\int_G f(s)d(\mu * \nu)(s) = \int_G \int_G f(st)d\mu(s)d\nu(t)$$

and $*$ operation

$$\int_G f(s)d\mu^*(s) = \overline{\int_G \overline{f(s^{-1})} d\mu(s)},$$

$\forall f \in C_0^\infty(G), \mu, \nu \in M(G)$. Then $M(G)$ is a Banach $*$ algebra with an identity δ_e, and $\|\mu^*\| = \|\mu\|, \forall \mu \in M(G)$. This Banach $*$ algebra is called the *measure algebra* of G.

$L^1(G) = L^1(G, ds)$ is called the *group algebra* of G, its multiplication and $*$ operation are as follows:

$$(f * g)(t) = \int_G f(s)g(s^{-1}t)ds = \int_G f(ts)g(s^{-1})ds,$$
$$f^*(t) = \Delta(t)^{-1}\overline{f(t^{-1})}, \forall f, g \in L^1(G).$$

Clearly with the norm $\|\cdot\|_1, L^1(G)$ is also a Banach $*$ algebra and $\|f^*\|_1 = \|f\|_1, \forall f \in L^1(G)$.

With the map : $f \longrightarrow f(s)ds, L^1(G)$ can be $*$ isometrically embedded into $M(G)$, and becomes a closed $*$ two-sided ideal of $M(G)$. We have the formulas:

$$(\nu * f)(t) = \int_G f(s^{-1}t)d\nu(s), (f * \nu)(t) = \int_G f(ts^{-1})\frac{d\nu(s)}{\Delta(s)},$$

$\forall \nu \in M(G), f \in L^1(G)$. In particular,

$$(\delta_s * f)(t) = f(s^{-1}t), \quad (f * \delta_s)(t) = \Delta(s)^{-1}f(ts^{-1}),$$

$\forall s \in g, f \in L^1(G)$. Moreover, it is well-known that:

1) G is abelian $\Longleftrightarrow L^1(G)$ is abelian $\Longleftrightarrow M(G)$ is abelian;

2) $L^1(G)$ has an identity if and only if G is discrete;

3) Let U be any neighborhood of e in G , and $z_U \in K(G)$ with supp $z_U \subset U, z_U \geq 0$, and $\int_G z_U(t)dt = 1$. Then $\{z_U|U\}$ is an approximate identity for $L^1(G)$, i.e.,

$$\|z_U * f - f\|_1 \longrightarrow 0, \|f * z_U - f\|_1 \longrightarrow 0, \forall f \in L^1(G);$$

4) $M(G)$ and $L^1(G)$ are semi-simple, but in general they are not hermitian.

Positive linear functionals, the GNS construction, and $*$ representations

A linear functional ρ on$L^1(G)$(or $M(G)$) is said to be *positive*, denoted by $\rho \geq 0$, if $\rho(a^*a) \geq 0, \forall a \in L^1(G)$(or $M(G)$).$\{\pi, H\}$ is a $*$ representation of $L^1(G)$ (or $M(G)$), if H is a Hilbert space, and π is a $*$ homomorphism from the Banach $*$ algebra $L^1(G)$ (or $M(G)$) to $B(H)$. We have the following.

1) Let ρ be a positive linear functional on $L^1(G)$ (or $M(G)$). Since $L^1(G)$ admits an approximate identity, it follows from the Cohn factorization theorem (see F.F. Bonsall and J. Duncan, Complete normed algebras, Berlin, Springer, 1973.) that ρ is bounded and hermitian. Moreover, we have

$$\|\rho\| = \lim_U \rho(z_U^* * z_U)$$

and

$$|\rho(b^*a)| \leq \rho(a^*a)^{1/2} \cdot \rho(b^*b)^{1/2}, \quad |\rho(a)| \leq \|\rho\|\nu(a^*a)^{1/2},$$

$\forall a, b \in L^1(G)$ (or $M(G)$), where $\nu(\cdot)$ is the function of spectral radius.

2) If π is $*$ representation of $L^1(G)$ (or $M(G)$), then $\|\pi\| \leq 1$.

3) If $\{\pi, H\}$ is a nondegenerate $*$ representation of $L^1(G)$, then it can be uniquely extended to a $*$ representation of $M(G)$. It suffices to define that

$$\pi(\nu)\pi(f)\xi = \pi(\nu * f)\xi$$

or

$$\pi(\nu)\xi = \lim_U \pi(\nu * z_U)\xi,$$

$\forall f \in L^1(G), \xi \in H$.

4) A positive linear functional ρ on $L^1(G)$ is called a *state* , if $\|\rho\| = 1$. For each state ρ on $L^1(G)$, by the GNS construction there is a cyclic $*$ reperentation $\{\pi_\rho, H_\rho, \xi_\rho\}$ of $L^1(G)$ such that

$$\rho(a) = \langle \pi_\rho(a)\xi_\rho, \xi_\rho\rangle, \forall a \in L^1(G).$$

Then ρ can be extended to a state on $M(G)$.

Clearly, each nondegenerate $*$ representation of $L^1(G)$ is a direct sum of a family of cyclic $*$ representations, and each cyclic $*$ representation of $L^1(G)$ is unitarily equivalent to the $*$ representation generated by a state.

5) Let ρ be a stete on $L^1(G)$. Then ρ is a *pure state* (an extreme point of the state space on $L^1(G)$) if and only if the $*$ representation $\{\pi_\rho, H_\rho\}$ generated by ρ is toplogically irreducible.

6) For each non-zero $a \in L^1(G)$, there exists a toplogically irreducible $*$ representation π of $L^1(G)$ such that $\pi(a) \neq 0$.

(7) The left regular representation $\{\lambda, L^2(G)\}$ of $L^1(G)$ is faithful, where $\lambda(f)g = f * g, \forall f \in L^1(G), g \in L^2(G)$.

Indeed, let $\lambda(f) = 0$ for some $f \in L^1(G)$. Since $z_U \in L^1(G) \cap L^2(G)$ for any compact neighborhood U of e and $\lambda(f)z_U = f * z_U = 0$, it follows that $f = \|\cdot\|_1 \text{ - } \lim_U f * z_U = 0$.

Unitary representations of G and nondegenerate $*$ representations of $L^1(G)$

Let G be a locally compact group. $\{u_\cdot, H\}$ is called a *unitary representation* of G, if u_s is a unitary oerator on H for each $s \in G, u_{st} = u_s u_t, u_e = 1$, and $s \longrightarrow \langle u_s\xi, \eta\rangle$ is a continuous function on $G, \forall \xi, \eta \in H$. In this case, $s \rightarrow u_s$ is also continuous from G to $(B(H), \tau(B(H), T(H)))$. Moreover, we have the following facts.

1) Let $\{u_\cdot, H\}$ be a unitary representation of G. If for any $\nu \in M(G)$ define

$$\pi(\nu) = \int_G u_s d\nu(s),$$

i.e., $\langle\pi(\nu)\xi,\eta\rangle = \int_G \langle u_s\xi,\eta\rangle d\nu(s), \forall \xi,\eta \in H$, then $\{\pi, H\}$ is a $*$ representation of $M(G)$, and

$$\pi(\delta_s) = u_s, \quad \forall s \in G.$$

Moreover , $f \longrightarrow \pi(f) = \int_G f(s)u_s ds$ is a nondegenerate $*$ representation of $L^1(G)$.

Conversely, let $\{\pi, H\}$ be a nondegenerate $*$ representation of $L^1(G)$. Then it can be uniquely extended to a $*$ representation of $M(G)$, still denoted by $\{\pi, H\}$. Further $\{u. = \pi(\delta.), H\}$ is a unitary representation of G.

Therefore, there is a bijection between the collection of all unitary representations of G and the collection of all nondegenerate $*$ representations of $L^1(G)$.

2) Let $\{u., H\}$ be a unitary representation of G, and $\{\pi, H\}$ be the nondegenerate $*$ representation of $M(G)$ correspoinding to $\{u., H\}$, i.e.,

$$\pi(f) = \int f(s)u_s ds, \quad \pi(\nu) = \int u_s d\nu(s),$$

$\forall f \in L^1(G), \nu \in M(G)$. Then we hvae

$$\pi(M(G))'' = \pi(L^1(G))'' = \{u_s | s \in G\}''.$$

3) For each $s \in G$ with $s \neq e$, there exists a topologically irreducible unitary representation $\{u., H\}$ such that $u_s \neq 1_H$.

4) Let $\{\lambda., L^2(G)\}$ be the left regular representation of G, i.e.,

$$(\lambda_s f)(t) = f(s^{-1}t), \forall s \in G, f \in L^2(G).$$

Then

$$\lambda(f) = \int f(s)\lambda_s ds, \quad \forall f \in L^1(G)$$

is exactly the left regular representation of $L^1(G)$.

Positive linear functionals and continuous positive-definite functions

A function φ on G is said to be *positive-definite* , if $\sum_{k,l=1}^{n} \overline{\lambda_l}\lambda_k\varphi(s_l^{-1}s_k) \geq 0, \forall s_1, \cdots, s_n \in G, \lambda_1, \cdots, \lambda_n \in \mathbb{C}$.

We have the following facts.

1) Let φ be a positive-definite function on G. Then we have

$$\varphi(e) \geq 0, \varphi(s^{-1}) = \overline{\varphi(s)}, |\varphi(s)| \leq \varphi(e), \forall s \in G.$$

2) Let $\{u., H\}$ be a unitary representation of G, and $\xi \in H$. Then $\varphi(\cdot) = \langle u.\xi, \xi\rangle$ is a continuous positive-definite function on G.

Conversely, if φ is a continuous positive-definite function on G, then there exists a cyclic unitary representation $\{u., H, \xi\}$ of G such that

$$\varphi(\cdot) = \langle u., \xi, \xi\rangle.$$

3) Let ρ be a positive linear functional on $L^1(G), \{\pi_\rho, H_\rho, \xi_\rho\}$ be the cyclic $*$ representation of $L^1(G)$ generated by ρ, and $\{u^{(\rho)}_\cdot, H_\rho\}$ be the unitary representation of G corresponding to $\{\pi_\rho, H_\rho\}$. Then $\varphi(\cdot) = \langle u^{(\rho)}_\cdot \xi_\rho, \xi_\rho\rangle$ is a continuous positive-definite function on G, and

$$\rho(f) = \int f(s)\varphi(s)ds, \forall f \in L^1(G); \|\rho\| = \|\varphi\|_\infty = \varphi(e).$$

Conversely, if φ is a continuous positive-definite function on G, then $\rho(f) = \int f(s)\varphi(s)ds(\forall f \in L^1(G))$ is a positive linear functional on $L^1(G)$.

In particular, there is a bijection between $\{\rho|\rho$ is a state on $L^1(G)\}$ and $\{\varphi|\varphi$ is continuous and positive-definite on G, and $\varphi(e) = 1\}$.

4) Let φ_1, φ_2 be two continuous positive-definite functions on G. Then $\varphi_1\varphi_2$ is still positive-definite.

5) Let $\{\varphi_l\}$ be a net of continuous positive-definite functions with $\varphi_l(e) = 1, \forall l$, and ρ_l be the state on $L^1(G)$ corresponding to $\varphi_l, \forall l$. Then $\{\rho_l\}$ converges to a state ρ in w^*-toplogy, i.e., there is a continuous positive-definite function φ on G with $\varphi(e) = 1$ such that $\int f(s)\varphi_l(s)ds \longrightarrow \int f(s)\varphi(s)ds, \forall f \in L^1(G)$, if and only if , $\varphi_l(s) \longrightarrow \varphi(s)$ uniformly for $s \in K$, where K is any compact subset of G.

6) (R.Godement's theorem) Let φ be a continuous positive-definite function on G, and $\varphi \in L^2(G)$. Then there exists $\psi \in L^2(G)$ such that $\varphi(\cdot) = \langle \lambda.\psi, \psi\rangle$.

The enveloping C^*-algebra of a Banach $*$ algebra

Let A be a Banach $*$ algebra, and suppose that A admits a bounded approximate identity $\{a_l\}$, and $\|a^*\| = \|a\|, \forall a \in A$.

A positive linear functional ρ on A is continuous automatically, and $\|\rho\| = \sup\{\rho(a^*a)|a \in A, \|a\| \leq 1\} = \lim_l \rho(a_l) = \lim_l \rho(a_l^* a_l)$.

For any $*$ representation $\{\pi, H\}$ of A , we have also $\|\pi\| \leq 1$.

Let ρ be a state on A (i.e., $\rho \geq 0$ and $\|\rho\| = 1$). By the GNS construction, there is a cyclic $*$ representation $\{\pi_\rho, H_\rho, \xi_\rho\}$ such that $\rho(a) = \langle \pi_\rho(a)\xi_\rho, \xi_\rho\rangle, \forall a \in A$. Moroever, ρ is pure if and only if π_ρ is topologically irreductible.

For any $a \in A$, we define

$$\|a\|_c = \sup\{\|\pi(a)\| | \pi \text{ is a } * \text{ representation of } A\}.$$

Then we can prove that

$$\begin{aligned}\|a\|_c &= \sup\{\|\pi(a)\| | \pi \text{ is topologically irreducible }\} \\ &= \sup\{\rho(a^*a)^{1/2} | \rho \text{ is a state on } A\} \\ &= \sup\{\rho(a^*a)^{1/2} | \rho \text{ is a pure state on } A\} \\ &= \sup\{\alpha(a^*a)^{1/2} | \alpha \text{ is a } C^*\text{-seminorm on } A\} \leq \|a\|,\end{aligned}$$

$\forall a \in A$. In other wored, $\|\cdot\|_c$ is the largest C^*-seminorm on A . Let

$$N = \{a \in A | \|a\|_c = 0\}.$$

Clearly, N is a closed two-sided ideal of A, and $\|\cdot\|_c$ can become a C^*-norm on A/N.

Then completion of $(A/N, \|\cdot\|_c)$ is called the *enveloping* C^* *-algebra* of A, and denoted by $C^*(A)$.

Now let A admit a faithful $*$ representation. Then $N = \{0\}, \|\cdot\|_c$ is the largest C^*-norm on A, and $C^*(A)$ is the completion of $(A, \|\cdot\|_c)$. Moreover, since $\|a\|_c \leq \|a\|, \forall a \in A$, $\{a_l\}$ is still an approximate identity for $C^*(A)$. If ρ is a state on A, then by $|\rho(a)| = |\langle\pi_\rho(a)\xi_\rho, \xi_\rho\rangle| \leq \|\pi_\rho(a)\| \leq \|a\|_c, \forall a \in A$, and $\rho(a_l) \to 1, \rho$ can be uniquely extended to a state on $C^*(A)$. Conversely, if ρ is a state on $C^*(A)$, by $\rho(a_l) \to 1$ then $(\rho|A)$ is a state on A. Therefore, the state spaces of A and $C^*(A)$ are the same.

Group C^*-algebras and reduced group C^*-algebras

Definition 16.3.1. Let G be a locally compact group, and $\|\cdot\|_c$ be the largest C^*-norm on $L^1(G)$ (notice that the left regular representation of $L^1(G)$ is faithful). Then the enveloping C^*-algebra of $L^1(G)$, i.e., the completion of $(L^1(G), \|\cdot\|_c)$, is called *the C^*-algebra of the group G,* and denoted by $C^*(G)$.

From the preceding paragraph, $\{z_U\}$ is still an approximate identity for $C^*(G)$; and the state space of $C^*(G)$ is equal to the state space of $L^1(G)$.

Proposition 16.3.2. Let G be abelian. Then $C^*(G)$ is $*$ isomrophic to $C_0^\infty(\hat{G})$, where $\hat{G}$ is the dual of G.

Proof. Since $C^*(G)$ is abelian, so the spetral space of $C^*(G)$ is the pure state space on $C^*(G)$. But pure state spaces of $C^*(G)$ and $L^1(G)$ are the same. Therefore, the spactral space of $C^*(G)$ is $\hat{G}$, the spectral space of $L^1(G)$. Q.E.D.

Definition 16.3.3. Let G be a locally compact group, and $\{\lambda, L^2(G)\}$ be the left regular representation of $L^1(G)$. Then $\|f\|_r = \|\lambda(f)\| (\forall f \in L^1(G))$ is a C^* -norm on $L^1(G)$. The completion of $(L^1(G), \|\cdot\|_r)$ is called *the reduced C^*-algebra of the group G,* and denoted by $C_r^*(G)$.

Clearly, $\|f\|_r \le \|f\|_c, \forall f \in L^1(G)$. So the identity map on $L^1(G)$ induces a $*$ homomorhpism from $C^*(G)$ onto $C_r^*(G)$. Therefore, $C_r^*(G)$ is $*$ isomorphic to a quotient C^*-algebra of $C^*(G)$.

Moreover, the VN algebra $R(G) = \{\lambda_s \mid s \in G\}''$ on $L^2(G)$ is called *the VN algebra of the group G.*

Amenable groups

Definition 16.3.4. Let G be a locally compact group. G is said to be *amenable,* if there exists a left invariant mean m on $L^\infty(G)$, i.e., m is a state on $L^\infty(G)$ (see $L^\infty(G)$ as a C^*-algebra) and

$$m({}_sf) = m(f),$$

where ${}_sf(\cdot) = f(s^{-1}\cdot), \forall s \in G, f \in L^\infty(G)$.

If G is amenable, then we can prove that there exists also a right invariant mean and a two-sided invariant mean on $L^\infty(G)$.

Remark. If G is discrete, then G is amenable if and only if $R(G)$ has the property (P) (see the Remark under Lemma 13.4.6 and [153]).

Example 1. If G is a compact group, then G is amenable.
Indeed, we have an invariant Haar measure μ on G with $\mu(G) = 1$. Define

$$m(f) = \int f(s)d\mu(s), \quad \forall f \in L^\infty(G).$$

Clearly, m is an invariant mean on $L^\infty(G)$.

Example 2. If G is abelian, then G is amenable.
In fact, let M be the mean (state) space on $L^\infty(G)$. Clearly, M is a compact convex subset of $(L^\infty(G)^*, w^*\text{- }top.)$. For any $s \in G$, define

$$(T_s m)(f) = m({}_sf), \quad \forall m \in M, f \in L^\infty(G).$$

Then T_s is an affine continuous map from M to $M, \forall s \in G$. Since G is abelian, it follows that $T_sT_t = T_tT_s, \forall s, t \in G$. Now by the Markov-Kakutani fixed point theorem, there exists $m_0 \in M$ such that $T_s m_0 = m_0, \forall s \in G$. Clearly, m_0 is an invariant mean on $L^\infty(G)$.

Example 3. Let F_2 be the free group of two generators u, v with discrete topology. We say that F_2 is not amenable.

In fact, if m is a left invariant mean on $l^\infty(F_2)$, let E_x be the set of elements in F_2 beginning with $x, \forall x \in \{u, v, u^{-1}, v^{-1}\}$, then

$$1 = m(G) = m(\{e\}) + m(E_u) + m(E_{u^{-1}}) + m(E_v) + m(E_{v^{-1}}).$$

On the other hand, by the left invariance of m we have

$$\begin{aligned} 1 = m(G) &= m(E_u) + m(uE_{u^{-1}}) = m(E_u) + m(E_{u^{-1}}) \\ &= m(E_v) + m(vE_{v^{-1}}) = m(E_v) + m(E_{v^{-1}}). \end{aligned}$$

This is a contradiction. Therefore, F_2 is not amenable.

For amenability, there are many classical descriptions. But for our purpose, it suffices to point out the following Godement' condition:

G is amenable if and only if there is a net $\{\psi_l\} \subset L^2(G)$ such that

$$\langle \psi_l, \lambda_t \psi_l \rangle \longrightarrow 1 \quad \text{uniformly for } t \in K,$$

where K is any compact subset of G.

Main theorem of this section

Definition 16.3.5. Let A be a C^*-algebra, and $\{\pi, H\}$ be a $*$ representation of A. A state (or positive linear functional) ρ on A is said to be *associated with* π, if there exists $\xi \in H$ such that

$$\rho(a) = \langle \pi(a)\xi, \xi \rangle, \forall a \in A.$$

Now let π_1, π_2 be two $*$ representations of A. We say that π_1 is *weakly contained* in π_2 (or π_2 weakly contains π_1), if $\ker \pi_2 \subset \ker \pi_1$.

Lemma 16.3.6. Let H be a Hilbert space, and ρ be a state on the C^*-algebra $B(H)$. Then ρ belongs to the $\sigma(B(H)^*, B(H))$ -closure of

$$Co\{\langle \cdot \xi, \xi \rangle | \xi \in H, \|\xi\| = 1\}.$$

Proof. If the conclusion is not true, then by the separation theorem there is $a \in B(H)$ such that

$$Re\rho(a) > \sup\{Re\langle a\xi, \xi \rangle | \xi \in H, \|\xi\| = 1\}.$$

Let $h = \frac{1}{2}(a + a^*)$, then we have

$$\begin{aligned} \rho(h) &> \sup\{\langle h\xi, \xi \rangle | \xi \in H, \|\xi\| = 1\} \\ &= \max\{\lambda | \lambda \in \sigma(h)\}. \end{aligned}$$

On the other hand, it is obvious that

$$\rho(h) \leq \max\{\lambda | \lambda \in \sigma(h)\},$$

a contradiction. Therefore, the conclusion holds. Q.E.D.

Proposition 16.3.7. Let A be a C^*-algebra, and $\{\pi_1, H_1\}, \{\pi_2, H\}$ be two $*$ representations of A. Then the following statementas are equivalent:

1) π_1 is weakly contained in π_2;

2) Each positive functional on A associated with π_1 is a w^*-limit of sums of positive functionals associated with π_2;

3) Each state on A associated with π_1 is a w^*-limit of states which are sums of positive functionals associated with π_2.

Proof. 1) $\Longrightarrow$ 3). Let ρ be a state on A associated with π_1. Since $\ker \pi_2 \subset \ker \pi_1$, ρ can become a state on $A/\ker \pi_2$ (Proposition 2.4.11). Clearly, we may assume that $A/\ker \pi_2 \subset B(H_2)$. Since ρ can be extended to a state on $B(H_2)$, then by Lemma 16.3.6 we have the statement 3).

3) $\Longrightarrow$ 2). It is obvious.

2) $\Longrightarrow$ 1). For any $a \in \ker \pi_2$, and $\xi \in H$, by the condition 2) we have $\langle \pi_1(a)\xi, \xi \rangle = 0$. Therefore, $a \in \ker \pi_1$, and $\ker \pi_2 \subset \ker \pi_1$. Q.E.D.

Theorem 16.3.8. Let G be a locally compact group. Then the following statements are equivalent:

1) G is amenable;

2) Any $*$ representation of $C^*(G)$ is weakly contained in its left regular representation, where the left regular representation of $C^*(G)$ is the unique extension of the left regular representation of $L^1(G)$;

3) The left regular representation of $C^*(G)$ is faithful;

4) $C^*(G) = C_r^*(G)$.

Proof. Clearly, the statements 2), 3) and 4) are equivalent.

1) $\Longrightarrow$ 2). Let G be amenable. By Godement's condition, there is a net $\{\psi_l\} \subset L^2(G)$ such that $\langle \lambda_t, \psi_l, \psi_l \rangle \longrightarrow 1$ uniformly on any compact subset of G. Since $K(G)$ is dense in $L^2(G)$, we may assume that $\psi_l \in K(G), \forall l$. Clearly, $\langle \lambda.\psi_l, \psi_l \rangle \in K(G), \forall l$. If ρ is any positive functional on $C^*(G)$, then there exists unique continuous positive-definite function φ on G such that $\rho(g) = \int f(s)\varphi(s)ds, \forall f \in L^1(G)$. Clearly, $\varphi(t)\langle \lambda_t\psi_l, \psi_l \rangle \to \varphi(t)$ uniformly on any compact subset of G, and for each $l, \varphi(\cdot)\langle \lambda.\psi_l, \psi_l \rangle \in L^2(G)$ and $\varphi(\cdot)\langle \lambda.\psi_l, \psi_l \rangle$ is continuous positive-definite. By the Godement's theorem, we can write

$$\varphi(\cdot)\langle \lambda.\psi_l, \psi_l \rangle = \langle \lambda.\varphi_l, \varphi_l \rangle,$$

where $\varphi_l \in L^2(G), \forall l$. Since $\{\|\varphi_l\|_2 | l\}$ is bounded, it follows that $\rho_l \longrightarrow \rho$ in $\sigma(A^*, A)$, where $A = C^*(G)$ and

$$\rho_l(f) = \int f(s)\langle \lambda_s \varphi_l, \varphi_l\rangle ds = \langle \lambda(f)\varphi_l, \varphi_l\rangle,$$

$\forall f \in L^1(G)$ and l. Therefore, any positive functional on $C^*(G)$ is a w^*-limit of positive functionals associated with the left regular representation. By Proposition 16.3.7 , we have the statement 2).

2) $\Longrightarrow$ 1). By Proposition 16.3.7, for any continuous positive-definite function φ on G with $\varphi(e) = 1$ there are $\varphi_1^{(l)}, \cdots, \varphi_{n_l}^{(l)} \in L^2(G)$ such that

$$\varphi_l(t) = \sum_i \langle \lambda_t \varphi_i^{(l)}, \varphi_i^{(l)}\rangle \longrightarrow \varphi(t)$$

uniformly on any compact subset of G and $\varphi_l(e) = 1, \forall l$. Since $K(G)$ is dense in $L^2(G)$, we may assume that $\varphi_i^{(l)} \in K(G), \forall l, i$. By $\langle \lambda . \varphi_i^{(l)}, \varphi_i^{(l)}\rangle \in L^2(G), \forall i$, and the Godement's theorem, we can write

$$\varphi_l(t) = \langle \lambda_t \psi_l, \psi_l\rangle, \quad \forall t \in G,$$

where $\psi_l \in L^2(G), \forall l$. Picking $\varphi \equiv 1$ and by Godement's condition, G is amenable. Q.E.D.

References. [27], [58], [61], [125], [127].

16.4. C^*-crossed products

Definition 16.4.1. (A, G, α) is called a *C^*-dynamical system* , if A is a C^*-algebra, G is a locally compact group, α is a homomorphism from G into $\mathrm{Aut}(A)$, where $\mathrm{Aut}(A)$ is the group of all $*$ isomorphisms of A, and $t \longrightarrow \alpha_t(a)$ is continuous from G to $A, \forall a \in A$.

Definition 16.4.2. Let (A, G, α) be a C^*-system. Define

$$L^1(G, A, \alpha) = \left\{ f \,\middle|\, \begin{array}{c} f \text{ is measurable from } G \text{ to } A, \\ \text{and } \int_G \|f(s)\|_A ds < \infty \end{array} \right\}.$$

By the norm

$$\|f\|_1 = \int_G \|f(s)\|_A ds,$$

the multiplication

$$(f * g)(t) = \int_G f(s)\alpha_s(g(s^{-1}t))ds,$$

and the $*$ operation

$$f^*(t) = \Delta(t)^{-1}\alpha_t(f(t^{-1}))^*,$$

$\forall f, g \in L^1(G, A, \alpha), L^1(G, A, \alpha)$ becomes a Banach $*$ algebra, and $\|f^*\|_1 = \|f\|_1, \forall f \in L^1(G, A, \alpha)$.

Clearly, $L^1(G) \otimes A$ is dense in $L^1(G, A, \alpha)$; and $L^1(G, \mathbb{C}, id) = L^1(G)$.

Proposition 16.4.3. Let (A, G, α) be a C^*-system, $\{z_U\}$ be an approximate identity for $L^1(G)$ as in section 1.3, and $\{a_l\}$ be an approximate identity for A. Then $\{z_U(t)\alpha_t(a_l)|(U, l)\}$ is an approximate identity for $L^1(G, A, \alpha)$.

Proof. It suffices to show that

$$\|z_U(\cdot)\alpha.(a_l) * ga - ga\|_1 \longrightarrow 0, \text{ and } \|ga * z_U(\cdot)\alpha.(a_l) - ga\|_1 \longrightarrow 0,$$

$\forall g \in L^1(G), a \in A$. Notice that

$$\begin{aligned}
&\|ga * z_U(\cdot)\alpha.(a_l) - ga\|_1 \\
&= \int_G dt \| \int_G g(s) a \alpha_s(z_U(s^{-1}t)\alpha_{s^{-1}t})(a_l)) ds - g(t)a\| \\
&= \int_G dt \| \int_G g(s) z_U(s^{-1}t) a \alpha_t(a_l) ds - g(t)a\| \\
&= \int dt |(g * z_U)(t) - g(t)| \cdot \|a\| \\
&\quad + \int dt |(g * z_U)(t)| \cdot \|a\alpha_t(a_l) - a\|
\end{aligned}$$

and

$$\begin{aligned}
&\int dt |(g * z_U)(t)| \cdot \|a\alpha_t(a_l) - a\| \\
&\leq \int_K |(g * z_U)(t)| \cdot \|\alpha_{t^{-1}}(a) \cdot a_l - \alpha_{t^{-1}}(a)\| dt + 2\|a\| \int_{G \setminus K} |(g * z_U)(t)| dt,
\end{aligned}$$

where K is a compact subset of G. Since $\{\alpha_{t^{-1}}(a) | t \in K\}$ is a compact subset of A, and $\|ba_l - b\| \to 0$ uniformly for $b \in B$, where B is any compact subset of A, it follows that

$$\|ga * z_U(\cdot)\alpha.(a_l) - ga\|_1 \longrightarrow 0, \forall g \in L^1(G), a \in A.$$

Moreover, by $\int z_U(s) ds = 1$ we have

$$\begin{aligned}
&\|z_U(\cdot)\alpha.(a_l) * ga - ga\|_1 \\
&= \int_G dt \| \int_G z_u(s) g(s^{-1}t) \alpha_s(a_l a) ds - \int_s z_U(s) g(t) ds\| \\
&\leq \int_G dt \int_G z_U(s) \cdot |g(s^{-1}t) - g(t)| ds \cdot \|a_l a - a\| \\
&\quad + \|g\|_1 \int_U z_U(s) \|\alpha_s(a) - a\| ds.
\end{aligned}$$

Thus, $\|z_U(\cdot)\alpha.(a_l) * ga - ga\|_1 \longrightarrow 0, \forall g \in L^1(G), a \in A.$ Q.E.D.

Remark. For any $f \in L^1(G, A, \alpha)$ and $\varepsilon > 0$, we can find $g_i \in L^1(G), a_i \in A$ such that

$$\|\sum_i g_i a_i - \tilde{f}\|_1 < \varepsilon,$$

where $\tilde{f}(t) = \alpha_{t^{-1}}(f(t)), \forall t \in G$. Thus $[g(\cdot)\alpha.(a)|g \in L^1(G), a \in A]$ is also dense in $L^1(G, A, \alpha)$. Then we can prove that $\{z_U a_l|(U, l)\}$ is also an approximate identity for $L^1(G, A, \alpha)$.

Lemma 16.4.4. $L^1(G, A, \alpha)$ admits a faithful $*$ representation.

Proof. We may assume that $A \subset B(H)$ for some Hilbert space. Define a $*$ representation $\{\pi, L^2(G, H)\}$ of $L^1(G, A, \alpha)$ as follows:

$$(\pi(f)\xi)(t) = \int_G \alpha_{t^{-1}}(f(s))\xi(s^{-1}t)ds,$$

$\forall f \in L^1(G, A, \alpha), \xi \in L^2(G, H)$.

Now let $f \in L^1(G, A, \alpha)$ be such that $\pi(f) = 0$. Then for any $g, h \in K(G)$ and $\xi, \eta \in H$, we have

$$\begin{aligned} 0 &= \langle \pi(f) g \otimes \xi, h \otimes \eta \rangle \\ &= \int\int \langle \alpha_{t^{-1}}(f(s))\xi, \eta\rangle g(s^{-1}t)\overline{h(t)}dsdt. \end{aligned}$$

Since $h \in K(G)$ is arbitrary, it follows that

$$\int \langle \alpha_{t^{-1}}(f(s))\xi, \eta\rangle g(s^{-1}t)ds = 0, a.e., \forall \xi, \eta \in H.$$

Notice that

$$\begin{aligned} &|\int \langle \alpha_{t^{-1}}(f(s))\xi, \eta\rangle g(s^{-1}t)ds - \int \langle \alpha_{r^{-1}}(f(s))\xi, \eta\rangle g(s^{-1}r)ds| \\ \le &\int |\langle(\alpha_{t^{-1}} - \alpha_{r^{-1}})(f(s))\xi, \eta\rangle| \cdot |g(s^{-1}t)|ds \\ &+ \int |\langle \alpha_{r^{-1}}(f(s))\xi, \eta\rangle| \cdot |g(s^{-1}t) - g(s^{-1}r)|ds \\ \le &\int \|f(s)\| \cdot \|\xi\| \cdot \|\eta\| \cdot |g(s^{-1}t) - g(s^{-1}r)|ds \\ &+ 2\|\xi\| \cdot \|\eta\| \cdot \|g\|_\infty \cdot \|f - \sum_i f_i \otimes a_i\|_{L^1(G,A,\alpha)} \\ &+ \sum_i \int |\langle(\alpha_{t^{-1}} - \alpha_{r^{-1}})(a_i)\xi, \eta\rangle| \cdot |f_i(s) g(s^{-1}t)|ds, \end{aligned}$$

where $f_i \in L^1(G), a_i \in A, \forall i$. Thus, $t \longrightarrow \int \langle \alpha_{t^{-1}}(f(s))\xi, \eta\rangle g(s^{-1}t)ds$ is continuous on G. Further, we have

$$\int \langle \alpha_{t^{-1}}(f(s))\xi, \eta\rangle g(s^{-1}t)ds = 0,$$

$\forall t \in G, \xi, \eta \in H, g \in K(G)$. In particular,

$$\int \langle f(s)\xi, \eta\rangle g(s^{-1})ds = 0, \forall g \in K(G), \xi, \eta \in H.$$

Therefore, we get

$$\langle f(s)\xi, \eta\rangle = 0, a.e., \forall \xi, \eta \in H.$$

Now let H be the Hilbert space of the universal $*$ representation of A. Then we can see that

$$F(f(s)) = 0, a.e., \forall F \in A^*.$$

Since $f(G)$ can be contained in a separable linear subspace of A, we may assume that A is separable. Let $\{F_n\}$ be a countable w^*-dense subset of $\{F \in A^* | \|F\| \leq 1\}$. Then there is a Borel subset E of G with $\mu_l(E) = 0$ such that

$$F_n(f(s)) = 0, \forall n \text{ and } s \notin E,$$

i.e., $f(s) = 0$, a.e. Therefore, $L^1(G, A, \alpha)$ admits a faithful $*$ representation.

Q.E.D.

From Proposition 16.4.3., Lemma 16.4.4, and the general theory in Section 16.3, we have the following facts: each positive functional on $L^1(G, A, \alpha)$ is bounded and hermitian automatically; $\|\pi\| \leq 1$ if π is a $*$ representation of $L^1(G, A, \alpha)$; there exists the GNS construction for each positive functional on $L^1(G, A, \alpha)$; and there exists the largest C^*-norm $\|\cdot\|$ on $L^1(G, A, \alpha)$ with $\|\cdot\| \leq \|\cdot\|_1$.

Definition 16.4.5. Let (A, G, α) be a C^*-system. The completion of $(L^1(G, A, \alpha), \|\cdot\|)$ is called the *crossed product* of A by the action α of G, and denoted by $A \times_\alpha G$, where $\|\cdot\|$ is the largest C^* -norm on $L^1(G, A, \alpha)$.

Clearly, an bounded approximate identity of $L^1(G, A, \alpha)$ is also an approximate identity for $A \times_\alpha G$; the state spaces of $L^1(G, A, \alpha)$ and $A \times_\alpha G$ are the same.

Example . If $\alpha = id$, then we have

$$(f * g)(t) = \int f(s)g(s^{-1}t)ds, \quad f^*(t) = \Delta(t)^{-1}f(t^{-1})^*,$$

$\forall f, g \in L^1(G, A, id)$. Of course, $A \otimes L^1(G)$ is dense in $A \times_{id} G$. Thus , $A \times_{id} G$ is the completion of $A \otimes L^1(G)$ with respect to the norm

$$\sup\{\|\pi(\cdot)\| \ |\pi \text{ is a } * \text{ representation of } A \otimes L^1(G)\}.$$

By Proposition 3.3.2, we have that

$$A \times_{id} G = \max\text{-}(A \otimes C^*(G)),$$

where max-$(A \otimes B)$ means the projective tensor product, and $\|\cdot\|_{\max}(= \alpha_1(\cdot)$ in Chapter 3) is the maximal C^*-norm on $A \otimes B$.

Definition 16.4.6. $\{\pi, u, H\}$ is called a *covariant representation* of a C^*-system (A, G, α), if $\{\pi, H\}$ is a nondegenerate $*$ representation of the C^*-algebra $A, \{u, H\}$ is a strongly continuous unitary representation of the group G, and

$$\pi(\alpha_s(x)) = u_s \pi(x) u_s^*, \quad \forall x \in A, s \in G.$$

With a covariant representation $\{\pi, u, H\}$ of (A, G, α), we can define a $*$ representation $\{\pi \times u, H\}$ of $L^1(G, A, \alpha)$ as follows:

$$(\pi \times u)(f) = \int \pi(f(t)) u_t dt, \quad \forall f \in L^1(G, A, \alpha),$$

i.e.,

$$\langle (\pi \times u)(f)\xi, \eta\rangle = \int_G \langle \pi(f(t)) u_t \xi, \eta\rangle dt, \quad \forall \xi, \eta \in H.$$

Theorem 16.4.7. Let (A, G, α) be a C^*-system. Then $\{\pi, u\} \longrightarrow \pi \times u$ is a bijection between the collection of all covariant representations of (A, G, α) to the collection of all nondegenerate $*$ representations of $L^1(G, A, \alpha)$.

Proof. First, we prove that $\{\pi \times u, H\}$ is nondegenerate. In fact, let $\xi \in H$ be such that $(\pi \times u)(f)\xi = 0, \forall f \in L^1(G, A, \alpha)$. In particular, we have

$$\int g(t) \langle \pi(a) u_t \xi, \eta\rangle dt = 0, \forall g \in L^1(G), a \in A, \eta \in H.$$

Then $\langle \pi(a) u_t \xi, \eta\rangle = 0, \forall t \in G, a \in A, \eta \in H$. Further, $\pi(a)\xi = 0, \forall a \in A$. But π is nondegenerate, so $\xi = 0$, i.e., $(\pi \times u)$ is nondegenerate.

Now let $\{\rho, H\}$ be a nondegenerate $*$ representation of $L^1(G, A, \alpha)$, and $\{g_l\}$ be an bounded approximate identity for $L^1(G, A, \alpha)$. For any $x \in A$ and $f \in L^1(G, A, \alpha)$, put $(xf)(t) = x \cdot f(t), \forall t \in G$. Clearly, $xg_l * f = x(g_l * f) \longrightarrow xf$ in $L^1(G, A, \alpha)$. Then we can define

$$\pi(x) = s\text{-}\lim_l \rho(x g_l), \quad \forall x \in A.$$

In particular, $\pi(x)\rho(f)\xi = \rho(xf)\xi, \forall x \in A, f \in L^1(G,A,\alpha)$. It is easy to see that $\{\pi, H\}$ is a $*$ representation of A. Further, define

$$u_r = s\text{-}\lim_l \rho(\alpha_r(g_l(r^{-1}\cdot)), \quad \forall r \in G.$$

Noticing that

$$\begin{aligned} &(\alpha_r(g_l(r^{-1}\cdot)) * f)(t) \\ = &\int \alpha_r(g_l(r^{-1}s))\alpha_s(f(s^{-1}t))ds \\ = &\alpha_r((g_l * f)(s^{-1}t)) \longrightarrow \alpha_r({}_rf) \quad \text{in } L^1(G,A,\alpha), \end{aligned}$$

we have

$$u_r\rho(f)\xi = \rho(\alpha_r({}_rf))\xi, \quad \forall f \in L^1(G,a,\alpha), \xi \in H.$$

Since $f^* * f = \alpha_r({}_rf)^* * \alpha_r({}_rf)$, it follows that u_r is unitary, $\forall r \in G$. Moreover, $r \to \alpha_r({}_rf)$ is continuous from G to $L^1(G,A,\alpha), \forall f \in K(G,A)$. Thus, $\{u,H\}$ is a strongly continuous unitary representation of G. Since

$$\begin{aligned} &u_r\pi(x)u_r^*\rho(f)\xi = u_r\rho(x\alpha_{r^{-1}}({}_{r^{-1}}f))\xi \\ = &\rho(\alpha_r(x)f)\xi = \pi(\alpha_r(x))\rho(f)\xi, \end{aligned}$$

$\forall x \in A, r \in G, f \in L^1(G,A,\alpha), \xi \in H$, $\{\pi, u, H\}$ is a covariant representation of (A,G,α).

Finally , for any $f,g,h \in L^1(G,A,\alpha), \xi,\eta \in H$, we have

$$\begin{aligned} &\int \langle \pi(f(t))u_t\rho(g)\xi, \rho(h)\eta\rangle dt \\ = &\int \langle \rho(f(t)\alpha_t({}_tg))\xi, \rho(h)\eta\rangle dt \\ = &\langle \rho(f * g)\xi, \rho(h)\eta\rangle = \langle \rho(f)\rho(g)\xi, \rho(h)\eta\rangle. \end{aligned}$$

Therefore, $\rho = \pi \times u$. Moreover, since ρ is nondegenerate and

$$\rho(f) = \int u_t\pi(\alpha_{t^{-1}}(f(t))dt, \quad \forall f \in L^1(G,A,\alpha),$$

π is also nondegenerate. Q.E.D.

Remark. It is easily verified that

$$(\pi \times u)(A \times_\alpha G)'' = \{\pi(x), u_s | x \in A, s \in G\}''.$$

Definition 16.4.8. Let (A,G,α) be a C^*-system. A map $\Phi : G \longrightarrow A^*$ is said to be *positive-definite*, if

$$\sum_{i,j=1}^n \Phi(s_i^{-1}s_j)(\alpha_{s_i^{-1}}(a_i^*a_j)) \geq 0$$

$\forall n, s_1, \cdots, s_n \in G$, and $a_1, \cdots, a_n \in A$.

$\Phi : G \longrightarrow A^*$ is said to be *continuous positive-definite*, if Φ is positive-definite, and $t \longrightarrow \Phi(t)(x)$ is continuous on $G, \forall x \in A$.

Proposition 16.4.9. Let (A, G, α) be a C^*-system, and $\Phi : G \longrightarrow A^*$ be positive-definite.

1) $t \longrightarrow \Phi(t)(x^*\alpha_t(x))$ is a positive-definite function on $G, \forall x \in A$. Consequently, $\Phi(e) \geq 0$ on A.

2) $\|\Phi(t)\| \leq 2\|\Phi(e)\|, \forall t \in G$.

3) If $\varphi(\cdot)$ is a positive-definite function on G, then $\varphi(\cdot)\Phi(\cdot) : G \longrightarrow A^*$ is also positive-definite.

4) For a covariant representation $\{\pi, u, H\}$ of (A, G, α), and $\xi \in H$, let

$$\Phi(t)(x) = \langle \pi(x) u_t \xi, \xi \rangle, \forall x \in A, t \in G.$$

Then $\Phi : G \longrightarrow A^*$ is continuous positive-definite.

Proof. 1) For any $s_1, \cdots, s_n \in G, \lambda_1, \cdots, \lambda_n \in \mathbb{C}$, we have

$$\begin{aligned} &\sum_{i,j} \Phi(s_i^{-1}s_j)(x^*\alpha_{s_i^{-1}s_j}(x))\overline{\lambda}_i\lambda_j \\ = &\sum_{i,j} \Phi(s_i^{-1}s_j)(\alpha_{s_i^{-1}}(\alpha_{s_i}(\lambda_i x)^* \cdot \alpha_{s_j}(\lambda_j x))) \geq 0. \end{aligned}$$

So $t \longrightarrow \Phi(t)(x^*\alpha_t(x))$ is positive-definite on $G, \forall x \in A$.

2) Define $[y, x]_t = \Phi(t)(x^*\alpha_t(y)), \forall x, y \in A$. By 1) and Section 1.3, we have

$$|[z, z]_t| \leq [z, z]_e = \Phi(e)(z^*z), \forall z \in A.$$

Then from polarization, we can see that

$$\begin{aligned} &|\Phi(t)(x^*\alpha_t(y))| = |[y, x]_t| \\ \leq \ &\Phi(e)(y^*y + x^*x) \leq \|\Phi(e)\|(\|x\|^2 + \|y\|^2). \end{aligned}$$

Let $\{a_l\}$ be an approximate identity for A. Then

$$\begin{aligned} |\Phi(t)(x^*)| &= \lim_l |\Phi(t)(x^*\alpha_t(a_l))| \\ &\leq \|\Phi(e)\|(\|x\|^2 + 1). \end{aligned}$$

Further, $\|\Phi(t)\| = \sup_{\|x\| \leq 1} |\Phi(t)(x^*)| \leq 2\|\Phi(e)\|, \forall t \in G$.

3) For $s_1, \cdots, s_n \in G, a_1, \cdots, a_n \in A$, write

$$\sum_{i,j} \varphi(s_i^{-1}s_j)\Phi(s_i^{-1}s_j)(\alpha_{s_i^{-1}}(a_i^*a_j)) = \sum_{i,j} \lambda_{ij}\mu_{ij},$$

where $\lambda_{ij} = \varphi(s_i^{-1}s_j), \mu_{ij} = \Phi(s_i^{-1}s_j)(\alpha_{s_i^{-1}}(a_i^* a_j)), \forall i,j$. Since (λ_{ij}) and (μ_{ij}) are two $n \times n$ positive matrices, it follows that

$$\begin{aligned}\sum_{i,j} \lambda_{ij}\mu_{ij} &= \sum_{i,j,k} \lambda_{ij}\overline{\gamma}_{ki}\gamma_{kj} \\ &= \sum_k (\sum_{ij} \lambda_{ij}\overline{\gamma}_{ki}\gamma_{kj}) \geq 0,\end{aligned}$$

where $(\mu_{ij}) = (\gamma_{ij})^* \cdot (\gamma_{ij})$. Thus $\varphi\Phi$ is also positive-definite.

4) It is obvious. Q.E.D.

Theorem 16.4.10. Let (A, G, α) be a C^*-system. Then there is a bijection between the collection of all positive linear functionals on $L^1(G, A, \alpha)$ and the collection of all continuous positive-definite maps from G to A^*.

In detail, let $\Phi : G \longrightarrow A^*$ be continuous positive-definite, and let

$$F(f) = \int \Phi(t)(f(t))dt, \quad \forall f \in L^1(G, A, \alpha).$$

Then F is positive on $L^1(G, A, \alpha)$. Conversely, let F be a positive liear functional on $L^1(G, A, \alpha), \{\rho, H, \xi\}$ be the cyclic $*$ representation of $L^1(G, A, \alpha)$ generated by F, and $\{\pi, u, H\}$ be the covariant representation of (A, G, α) such that $\rho = \pi \times u$. Define

$$\Phi(t)(x) = \langle \pi(x)u_t\xi, \xi\rangle, \forall t \in G, x \in A.$$

Then $\Phi : G \longrightarrow A^*$ is continuous positive-definite, and

$$F(f) = \int \Phi(t)(f(t))dt, \quad \forall f \in L^1(G, A, \alpha).$$

Moreover, we have that $\|\Phi(t)\| \leq \|\Phi(e)\| = \|F\| (\forall t \in G)$ in above correspondence.

Proof. Let F be a positive linear functional on $L^1(G, A, \alpha)$, and $\{\rho, H, \xi\}$, $\{\pi, u, H\}$ and Φ be as above. Then

$$\begin{aligned}F(f) = &\ \langle \rho(f)\xi, \xi\rangle = \langle (\pi \times u)(f)\xi, \xi\rangle \\ = &\int \langle \pi(f(t))u_t\xi, \xi\rangle dt = \int \Phi(t)(f(t))dt,\end{aligned}$$

$\forall f \in L^1(G, A, \alpha)$. If $\{f_l\}$ is a bounded approximate identity for $L^1(G, A, \alpha)$, then

$$\|F\| = \lim_l F(f_l) = \lim_l \langle \rho(f_l)\xi, \xi\rangle = \|\xi\|^2.$$

On the other hand, since $\{\pi, H\}$ is a nondegenerate $*$ representation of A, it follows that

$$\begin{aligned} \|\Phi(e)\| &= \sup\{|\Phi(e)(x)| | x \in A, \|x\| \le 1\} \\ &= \sup\{|\langle\pi(x)\xi, \xi\rangle| | x \in A, \|x\| \le 1\} \\ &= \lim_l |\langle\pi(a_l)\xi, \xi\rangle| = \|\xi\|^2 = \|F\| \ge \|\Phi(t)\|, \forall t \in G, \end{aligned}$$

where $\{a_l\}$ is an approximate identity for A.

Now let $\Phi : G \longrightarrow A^*$ be continuous positve-definite. Since $L^1(G) \otimes A$ is dense in $L^1(G, A, \alpha)$, $t \longrightarrow \Phi(t)(f(t))$ is measurable on $G, \forall f \in L^1(G, A, \alpha)$. By Proposition 16.4.9, we have $|\Phi(t)(f(t))| \le 2\|\Phi(e)\| \cdot \|f(t)\|, \forall f \in L^1(G, A, \alpha)$. Thus , we can define a linear functional

$$F(f) = \int \Phi(t)(f(t))dt, \quad \forall f \in L^1(G, A, \alpha).$$

For any $g_1, \cdots, g_k \in K(G), a_1, \cdots, a_k \in A, s_1, \cdots, s_l \in G$, notice that

$$\begin{aligned} &\sum_{ijnm} \Phi(s_n^{-1}s_m)(a_i^*\alpha_{s_n^{-1}s_m}(a_j))\overline{g_i(s_n)}g_j(s_m) \\ = &\sum_{(i,n),(j,m)} \Phi(s_{in}^{-1}s_{jm})(\alpha_{s_{in}^{-1}}(a_{in}^*a_{jm})) \ge 0 \end{aligned}$$

where $s_{in} = s_n, a_{in} = g_i(s_n)\alpha_{s_n}(a_i), \forall i, n$. It follows that

$$\sum_{i,j} \int\int \Phi(s^{-1}t)(a_i^*\alpha_{s^{-1}t}(a_j))\overline{g_i(s)}g_j(t)dsdt \ge 0$$

i.e.,

$$\sum_{i,j} \int \Phi(t)(a_i^*\alpha_t(a_j))(g_i^* * g_j)(t)dt \ge 0,$$

$\forall g_1, \cdots, g_k \in K(G), a_1, \cdots, a_k \in A$. Therefore, F is positive on $L^1(G, A, \alpha)$.

Q.E.D.

Corollary 16.4.11. There is a bijection between the state space of $A \times_\alpha G$ and

$$\left\{\Phi \,\middle|\, \begin{array}{l} \Phi : G \longrightarrow A^* \text{ is continuous positive-definite,} \\ \text{and } \|\Phi(e)\| = 1, \text{i.e., } \Phi(e) \text{ is a state on } A \end{array}\right\}.$$

Let (A, G, α) be a C^*-system, and $\{\pi, H\}$ be a $*$ reperesentation of A. Define

$$\begin{cases} (\overline{\pi}(x)\xi)(t) = \pi(\alpha_{t^{-1}}(x))\xi(t), \\ (\lambda(s)\xi)(t) = \xi(s^{-1}t), \end{cases}$$

$\forall x \in A, s \in G, \xi \in L^2(G, H)$. Then $\{\overline{\pi}, \lambda, L^2(G, H)\}$ is a covariant representation of (A, G, α), i.e.,

$$\lambda(s)\overline{\pi}(x)\lambda(s)^* = \overline{\pi}(\alpha_s(x)), \quad \forall x \in A, s \in G$$

(it is similar to Proposition 16.1.5). Further, we have a $*$ representation $\{\bar{\pi}\times\lambda, L^2(G,H)\}$ of $A\times_\alpha G$:

$$(\bar{\pi}\times\lambda)(f)=\int_G \bar{\pi}(f(t))\lambda(t)dt,\quad \forall f\in L^1(G,A,\alpha).$$

Definition 16.4.12. The $*$ representation $\{\bar{\pi}\times\lambda, L^2(G,H)\}$ is called the *regular representation* of $A\times_\alpha G$ induced by the $*$ representation $\{\pi,H\}$ of A, and denoted by $\{\mathrm{Ind}\pi, L^2(G,H)\}$, i.e.,

$$\mathrm{Ind}\pi(f)=\int_G \bar{\pi}(f(s))\lambda(s)ds$$

or

$$(\mathrm{Ind}\pi(f)\xi)(t)=\int_G \pi\circ\alpha_{t^{-1}}(f(s))\xi(s^{-1}t)ds,$$

$\forall f\in L^1(G,A,\alpha), \xi\in L^2(G,H)$.

Now we make the following discussions.

1) Let $\{\pi,H\}$ be a $*$ representation of $A, \{\xi_i|i\in\wedge\}(\subset H)$ be a cyclic set of vectors for $\pi(A)$, and $\{f_j|j\in\tilde{\wedge}\}(\subset L^2(G))$ be a cyclic set of vectors for $\lambda(L^1(G))$, where $\{\lambda, L^2(G)\}$ is the left regular representation of $L^1(G)$. Then $\{f_j\otimes\xi_i|i,j\}$ is cyclic for $\mathrm{Ind}\pi(A\times_\alpha G)$.

In fact, let $\xi\in\{\mathrm{Ind}\pi(A\times_\alpha G)(f_j\otimes\xi_i)|i,j\}^\perp$. Then for any $f\in K(G)$ and $x\in A$, we have

$$\begin{aligned}0&=\langle\mathrm{Ind}\pi(f\otimes x)(f_j\otimes\xi_i),\xi\rangle\\&=\int\int\langle\pi(\alpha_{t^{-1}}(x))f(s)f_j(s^{-1}t)\xi_i,\xi(t)\rangle dsdt\\&=\int\langle\pi\circ\alpha_{t^{-1}}(x)\xi_i,\xi(t)\rangle(f*f_j)(t)dt.\end{aligned}$$

Since $\{f*f_j=\lambda(f)f_j|f\in K(G),j\}$ is total in $L^2(G)$, it follows that

$$\langle\pi\circ\alpha_{t^{-1}}(x)\xi_i,\xi(t)\rangle=0,\text{a.e.},\ \forall x\in A,i.$$

For any compact subset E of G with $|E|>0$, and $\varepsilon>0$, by the Lusin theorem there is a compact subset F of G with $F\subset E$ and $|E\backslash F|<\varepsilon$ such that $\xi(\cdot):F\longrightarrow H$ is continuous, where $|B|=\int\chi_B(t)dt$ for any Borel subset B of G. From Proposition 5.1.2 and $0<|F|<\infty$, we can write

$$F=\sqcup_n K_n\sqcup N,$$

where $|N|=0,\{K_n\}$ is a disjoint sequence of compact subsets with the following property: for any opern subset U of G, if $U\cap K_n\neq\emptyset$ for some n, then $|U\cap K_n|>0$. Now it is easily verified that

$$\langle\pi\circ\alpha_{t^{-1}}(x)\xi_i,\xi(t)\rangle=0,\forall t\in\sqcup_n K_n, x\in A, i.$$

Since $\{\pi(x)\xi_i | x \in A, i\}$ is total in H , it follows that $\xi(t) = 0, \forall t \in \sqcup_n K_n$. But $\varepsilon(> 0)$ is arbitrary, so

$$\xi(t) = 0, \quad a.e., \text{on } E$$

for any compact subset E of G, i.e., $\xi = 0$ in $L^2(G, H)$.

2) Let φ be a positive linear functional on A, and $f_1, f_2 \in K(G)$. Define

$$\tilde{\varphi}_{f_1 f_2}(x) = \int\int f_1(s^{-1}t)\overline{f_2(t)}\varphi \circ \alpha_{t^{-1}}(x(s))dsdt, \tag{1}$$

$\forall x \in L^1(G, A, \alpha)$. If $\{\pi_\varphi, H_\varphi, \xi_\varphi\}$ is the cyclic $*$ representation of A generated by φ, then

$$\begin{aligned} &\langle \text{Ind}\pi_\varphi(x) f_1 \otimes \xi_\varphi, f_2 \otimes \xi_\varphi \rangle \\ = &\int \langle \int \pi_\varphi \circ \alpha_{t^{-1}}(x(s)) f_1(s^{-1}t)\xi_\varphi ds, f_2(t)\xi_\varphi \rangle dt \\ = &\int\int \varphi \circ \alpha_{t^{-1}}(x(s)) f_1(s^{-1}t)\overline{f_2(t)}dsdt = \tilde{\varphi}_{f_1 f_2}(x), \end{aligned} \tag{2}$$

$\forall x \in L^1(G, A, \alpha)$.

Now notice the following fact:

$$\{\text{Ind}\pi_\varphi(y) f \otimes \xi_\varphi | f \in K(G), y \in K(G, A)\}$$

is dense in $L^2(G, H_\varphi)$, where $y \subset K(G, A)$ means that $y(\cdot) : G \longrightarrow A$ is continuous and supp y is compact.

Indeed, for any $g \in K(G)$ and $a \in A$, we have

$$(\text{Ind}\pi_\varphi(g \otimes a) z_U \otimes \xi_\varphi)(t) = (g * z_U)(t)\pi_\varphi \circ \alpha_{t^{-1}}(a)\xi_\varphi,$$

where $\{z_U\}$ is an approximate identity for $L^1(G)$ as in Section 16.3. Moreover,

$$\begin{aligned} &\|(g * z_U(\cdot)\pi_\varphi \circ \alpha_{\cdot^{-1}}(a)\xi_\varphi - g(\cdot)\pi_\varphi \circ \alpha_{\cdot^{-1}}(a)\xi_\varphi\|^2_{L^2(G,H_\varphi)} \\ = &\int \|\pi_\varphi \circ \alpha_{t^{-1}}(a)\xi_\varphi\|^2 \cdot |g * z_U(t) - g(t)|^2 dt \\ \leq &\ \|a\|^2 \cdot \|\varphi\| \cdot \int |g * z_U(g) - g(t)|^2 dt \\ \leq &\ \|a\|^2 \cdot \|\varphi\| \cdot (\|g\|_\infty + \|g\|_\infty \cdot \sup_{s \in U} |\triangle(s)|^{-1})\|g * z_U - g\|_1 \longrightarrow 0. \end{aligned}$$

Thus , the closure of

$$\{\text{Ind}\pi_\varphi(\sum_i g_i \otimes a_i) z_U \otimes \xi_\varphi | a_i \in A, g_i \in K(G), U\}$$

in $L^2(G, H_\varphi)$ contains the following subset $\mathcal{L}$;

$$\mathcal{L} = \{\sum_i g_i(\cdot)\pi_\varphi \circ \alpha_{\cdot^{-1}}(a_i)\xi_\varphi | a_i \in A, g_i \in K(G)\}.$$

Similar to 1), $\mathcal{L}$ is dense in $L^2(G, H_\varphi)$. Therefore, $\{\mathrm{Ind}\pi_\varphi(y)f \otimes \xi_\varphi | f \in K(G), y \in K(G,A)\}$ is dense in $L^2(G, H_\varphi)$.

Now from (1) , (2) and above fact, we get

$$\begin{aligned} \|\mathrm{Ind}\pi_\varphi(x)\| = & \sup\left\{\frac{\|\mathrm{Ind}\pi_\varphi(xy)f \otimes \xi_\varphi\|}{\|\mathrm{Ind}\pi_\varphi(y)f \otimes \xi_\varphi\|} \middle| \begin{array}{c} f \in K(G), y \in K(G,A), \text{ and} \\ \|\mathrm{Ind}\pi_\varphi(y)f \otimes \xi_\varphi\| > 0 \end{array}\right\} \\ = & \sup\left\{\left|\frac{\tilde{\varphi}_{ff}(y^*x^*xy)^{1/2}}{\tilde{\varphi}_{ff}(y^*y)^{1/2}}\right| \begin{array}{c} y \in K(G,A), f \in K(G), \\ \text{and } \tilde{\varphi}_{ff}(y^*y) > 0 \end{array}\right\}. \end{aligned} \tag{3}$$

3) For any $y \in K(G,A)$ and $f \in K(G)$, define

$$\Omega(y,f) = \left\{\varphi \middle| \begin{array}{c} \varphi \text{ is a positive functional on } A, \\ \text{and } \tilde{\varphi}_{ff}(y^*y) > 0 \end{array}\right\}. \tag{4}$$

And for any $*$ representation π of A, let

$$\mathcal{M}(\pi) = \left\{\varphi \middle| \begin{array}{c} \varphi \geq 0 \text{ on } A, \text{ and } \pi_\varphi \text{ is weakly} \\ \text{contained in } \pi, \text{ i.e., } \ker\pi \subset \ker\pi_\varphi \end{array}\right\}. \tag{5}$$

Clearly, there is a subset $\wedge$ of $\mathcal{M}(\pi)$ such that

$$\pi = \oplus_{\varphi\in\wedge}\pi_\varphi \oplus 0.$$

Then $\mathrm{Ind}\pi = \oplus_{\varphi\in\wedge}\mathrm{Ind}\pi_\varphi \oplus 0$, and by (3), (4) we have

$$\begin{aligned} \|\mathrm{Ind}\pi(x)\| & = \sup\{\|\mathrm{Ind}\pi_\varphi(x)\| | \varphi \in \wedge\} \\ & \leq \sup\left\{\frac{\tilde{\varphi}_{ff}(y^*x^*xy)^{1/2}}{\tilde{\varphi}_{ff}(y^*y)^{1/2}} \middle| \begin{array}{l} y \in K(G,A), f \in K(G), \\ \text{and } \varphi \in \Omega(y,f) \cap \mathcal{M}(\pi) \end{array}\right\} \end{aligned}$$

$\forall x \in L^1(G,A,\alpha)$.

Conversely, for any $\varphi \in \mathcal{M}(\pi)$, we want to prove that $\|\mathrm{Ind}\pi(x)\| \geq \|\mathrm{Ind}\pi_\varphi (x)\|, \forall x \in L^1(G,A,\alpha)$. This is divided into three steps.

(i) Let $\varphi(\cdot) = \langle\pi(\cdot)\xi, \xi\rangle$ (some $\xi \in H_\pi$). Pick $\{\xi_l\} \subset H_\pi$ such that $\xi \in \{\xi_l\}$, and $H_\pi = \oplus_l H_l \oplus H_0$, where $H_l = \overline{[\pi(A)\xi_l]}, \forall l$, and $\pi|H_0 = 0$. Let $\pi_l = \pi|H_l, \forall l$. Then $\pi = \oplus_l \pi_l \oplus 0$, and $\mathrm{Ind}\pi = \oplus_l \mathrm{Ind}\pi_l \oplus 0$. Thus, $\|\mathrm{Ind}\pi(x)\| \geq \|\mathrm{Ind}\pi_l(x)\|, \forall l$, in particular, $\|\mathrm{Ind}\pi(x)\| \geq \|\mathrm{Ind}\pi_\varphi(x)\|, \forall x \in L^1(G,A,\alpha)$.

(ii) Let $\varphi = \sum_{i=1}^{n} \varphi_i$, where $\varphi_i(\cdot) = \langle\pi(\cdot)\xi_i, \xi_i\rangle$, and $\xi_i \in H_\pi, 1 \leq i \leq n$. We claim that

$$\|\mathrm{Ind}\pi_\varphi(x)\| \leq \max_i \|\mathrm{Ind}\pi_i(x)\|, \quad \forall x \in L^1(G,A,\alpha),$$

where $\pi_i = \pi_{\varphi_i}, 1 \leq i \leq n$. From this inequality and (i) , we shall get $\|\mathrm{Ind}\pi_\varphi(x)\| \leq \|\mathrm{Ind}\pi(x)\|, \forall x \in L^1(G,A,\alpha)$.

In fact, for any $x \in L^1(G,A,\alpha), y \in K(G,A), f \in K(G)$ and $\tilde{\varphi}_{ff}(y^*y) > 0$, let

$$\alpha_i = \tilde{\varphi}^{(i)}_{ff}(y^*y), \beta_i = \tilde{\varphi}^{(i)}_{ff}(y^*x^*xy),$$

where $\tilde{\varphi}^{(i)}_{ff} = \widetilde{(\varphi_i)}_{ff}, 1 \le i \le n$. Since $\tilde{\varphi}^{(i)}_{ff}$ is positive on $L^1(G,A,\alpha)$, it follows that $\beta_i = 0$ if $\alpha_i = 0$. Then by (3),

$$\begin{aligned} \frac{\tilde{\varphi}_{ff}(y^*x^*xy)}{\tilde{\varphi}_{ff}(y^*y)} &= \frac{\beta_1 + \cdots + \beta_n}{\alpha_1 + \cdots + \alpha_n} \le \max\left\{\frac{\beta_i}{\alpha_i} | i \text{ with } \alpha_i > 0\right\} \\ &\le \max_i \|\text{Ind}\pi_i(x)\|^2. \end{aligned}$$

Further, by (3) we have that $\|\text{Ind}\pi_\varphi(x)\| \le \max_i \|\text{Ind}\pi_i(x)\|, \forall x \in L^1(G,A,\alpha)$.

(iii) Let $\varphi \in \mathcal{M}(\pi)$ with $\|\varphi\| = 1$. From Proposition 16.3.7, there is a net $\{\varphi_l\}$ of states on A such that each φ_l is a sum of positive functionals associated with π, and $\varphi_l(a) \longrightarrow \varphi(a), \forall a \in A$.

Clearly, for any $a \in A, \varphi_l(\alpha_t(a)) \longrightarrow \varphi(\alpha_t(a))$ uniformly on any compact subset of G. Thus

$$\tilde{\varphi}^{(l)}_{ff}(g \otimes a) \longrightarrow \tilde{\varphi}_{ff}(g \otimes a),$$

$\forall f, g \in K(G), a \in A$, where $\tilde{\varphi}^{(l)}_{ff} = \widetilde{(\varphi_l)}_{ff}, \forall l$. By (1) , $\|\tilde{\varphi}^{(l)}_{ff}\| \le \|f\|_\infty \cdot \|f\|_1, \forall l$. It follows that

$$\tilde{\varphi}^{(l)}_{ff}(z) \longrightarrow \tilde{\varphi}_{ff}(z), \forall z \in L^1(G,A,\alpha).$$

Now from (3) and (ii) , we get

$$\|\text{Ind}\pi_\varphi(x)\| \le \sup_l \|\text{Ind}\pi_{\varphi_l}(x)\| \le \|\text{Ind}\pi(x)\|,$$

$\forall x \in L^1(G,A,\alpha)$.

Therefore, we have that

$$\begin{aligned} \|\text{Ind}\pi(x)\| &= \sup\{\|\text{Ind}\pi_\varphi(x)\| | \varphi \in \mathcal{M}(\pi)\} \\ &= \sup\left\{\frac{\tilde{\varphi}_{ff}(y^*x^*xy)^{1/2}}{\tilde{\varphi}_{ff}(y^*y)^{1/2}} \middle| \begin{array}{c} y \in K(G,A), f \in K(G), \\ \varphi \in \mathcal{M}(\pi) \cap \Omega(y,f) \end{array}\right\}, \end{aligned} \tag{6}$$

$\forall x \in L^1(G,A,\alpha)$.

From the proof of Lemma 16.4.4, if $\{\pi_u, H_u\}$ is the universal $*$ representation of A, then $\{\text{Ind}\pi_u, L^2(G,H_u)\}$ is a faithful $*$ representatio of $L^1(G,A,\alpha)$. Thus, $\|x\|_r = \sup\{\|\text{Ind}\pi(x)\| | \pi$ is a $*$ representation of $A\}$ will be a C^*-norm on $L^1(G,A,\alpha)$.

Definition 16.4.13. Let (A,G,α) be a C^*-system. The completion of $(L^1(G,A,\alpha), \|\cdot\|_r)$ is called the *reduced crossed product* of A by the action α

of G, and denoted by $A \times_{\alpha r} G$, where $\|x\|_r = \sup\{\|\text{Ind}\pi(x)\| | \pi$ is a $*$ representation of $A\}, \forall x \in L^1(G, A, \alpha)$.

By (6), we have that

$$\|x\|_r = \sup\left\{ \frac{\tilde{\varphi}_{ff}(y^* x^* x y)^{1/2}}{\tilde{\varphi}_{ff}(y^* y)^{1/2}} \middle| \begin{array}{c} y \in K(G, A), f \in K(G), \\ \varphi \in \Omega(y, f) \end{array} \right\}, \tag{7}$$

$\forall x \in L^1(G, A, \alpha)$.

Lemma 16.4.14. Let $\{\pi, H\}$ be a $*$ representation of A, and $t \in G$. Then the covariant representations $\{\overline{\pi}, \lambda, L^2(G, H)\}$ and $\{\overline{\pi \circ \alpha_t}, \lambda, L^2(G, H)\}$ of (A, G, α) are unitary equivalent, and the $*$ representations $\{\text{Ind}\pi, L^2(G, H)\}$ and $\{\text{Ind}\pi \circ \alpha_t, L^2(G, H)\}$ of $A \times_\alpha G$ are unitary equivalent. In detail, if define $(U_t\xi)(\cdot) = \Delta(t)^{1/2}\xi(\cdot t), \forall \xi \in L^2(G, H)$, then we have

$$U_t^* \overline{\pi}(a) U_t = \overline{\pi \circ \alpha_t}(a), \quad U(t)\lambda(s) = \lambda(s)U(t),$$
$$U_t^* \text{Ind}\pi(x) U_t = \text{Ind}(\pi \circ \alpha_t)(x),$$

$\forall a \in A, s, t \in G, x \in A \times_\alpha G$.

Proof. For any $\xi \in L^2(G, H), a \in A, r, s, t \in G, x \in L^1(G, A, \alpha)$, we have

$$\begin{aligned}(U_t^* \overline{\pi}(a) U_t \xi)(s) &= (\overline{\pi}(a) U_t \xi)(st^{-1}) \triangle (t)^{-1/2} \\ &= \pi \circ \alpha_{ts^{-1}}(a)(U_t\xi)(st^{-1}) \triangle (t)^{-1/2} \\ &= (\pi \circ \alpha_t) \circ \alpha_{s^{-1}}(a)\xi(s) = (\overline{\pi \circ \alpha_t}(a)\xi)(s),\end{aligned}$$

and

$$(U_t^* \text{Ind}\pi(x) U_t \xi)(r) = \Delta(t)^{-1/2}(\text{Ind}\pi(x) U_t \xi)(rt^{-1})$$
$$= \int \pi \circ \alpha_{tr^{-1}}(x(s))(U_t\xi)(s^{-1}rt^{-1}) ds \triangle (t)^{-1/2}$$
$$= \int (\pi \circ \alpha_t) \circ \alpha_{r^{-1}}(x(s))\xi(s^{-1}r) ds = (\text{Ind}(\pi \circ \alpha_t)(x)\xi)(r).$$

that comes to the conclusion. Q.E.D.

Theorem 16.4.15. Let (A, G, α) be a C^*-system, and $\{\pi, H\}$ be a $*$ representation of A. Then the following statements are equivalent:

1) $\sum\limits_{t \in G} \oplus \pi \circ \alpha_t$ is a faithful $*$ representation of A;

2) $\|\text{Ind}\pi(x)\| = \|x\|_r, \forall x \in L^1(G, A, \alpha)$.

Consequently, if $\{\pi, H\}$ is a faithful $*$ representation of A, then $\{\text{Ind}\pi, L^2(G, H)\}$ can be uniquely extended to a faithful $*$ representation of $A \times_{\alpha r} G$.

Proof. Suppose that $\|\text{Ind}\pi(x)\| = \|x\|_r, \forall x \in L^1(G,A,\alpha)$. If $a \in A$ satisfies $\pi \circ \alpha_t(a) = 0, \forall t \in G$, then

$$(\text{Ind}\pi(g \otimes a)\xi)(t) = \int g(s)\pi \circ \alpha_{t^{-1}}(a)\xi(s^{-1}t)ds = 0,$$

$\forall g \in K(G), \xi \in L^2(G,H)$. Thus we have

$$\|g \otimes a\|_r = \|\text{Ind}\pi(g \otimes a)\| = 0, \forall g \in K(G),$$

and $a = 0$. So $\sum_{t \in G} \oplus \pi \circ \alpha_t$ is a faithful $*$ representation of A.

Conversely, let $\sum_{t\in G} \oplus \pi \circ \alpha_t$ be a faithful $*$ representation of A. By Proposition 16.3.7, any state on A is a w^*-limit of states which are sums of positive functionals associated with $\{\pi \circ \alpha_t | t \in G\}$. By Lemma 16.4.14, we have

$$\|\text{Ind}\pi(x)\| = \|\text{Ind}(\pi \circ \alpha_t)(x)\|, \quad \forall t \in G, x \in L^1(G,A,\alpha).$$

Then from the above discussion 3) (ii) and (iii), and the formulas (6) , (7), we get

$$\begin{aligned}
&\|\text{Ind}\pi(x)\| \\
&= \sup\left\{ \frac{\tilde{\varphi}_{ff}(y^*x^*xy)^{1/2}}{\tilde{\varphi}_{ff}(y^*y)^{1/2}} \,\middle|\, \begin{array}{l} y \in K(G,A), f \in K(G), \text{ and} \\ \varphi \in \Omega(y,f) \cap (\sqcup_{t\in G} \mathcal{M}(\pi \circ \alpha_t)) \end{array} \right\} \\
&= \sup\left\{ \frac{\tilde{\varphi}_{ff}(y^*x^*xy)^{1/2}}{\tilde{\varphi}_{ff}(y^*y)^{1/2}} \,\middle|\, \begin{array}{l} y \in K(G,A), f \in K(G), \varphi \in \Omega(y,f), \\ \varphi = \varphi_1 + \cdots + \varphi_n, \varphi_i \in \mathcal{M}(\pi \circ \alpha_{t_i}), \forall i \end{array} \right\} \\
&= \sup\left\{ \frac{\tilde{\varphi}_{ff}(y^*x^*xy)^{1/2}}{\tilde{\varphi}_{ff}(y^*y)^{1/2}} \,\middle|\, \begin{array}{c} y \in K(G,A), f \in K(G), \text{ and} \\ \varphi \in \Omega(y,f) \end{array} \right\} \\
&= \|x\|_r, \quad \forall x \in L^1(G,A,\alpha).
\end{aligned}$$

Q.E.D.

Example. Let $\alpha = id$, and π be a faithful $*$ representation of A. Then for any $g \in L^1(G)$ and $a \in A$ we have

$$\text{Ind}\pi(g \otimes a) = \lambda(g) \otimes \pi(a)$$

on $L^2(G,H) = L^2(G) \otimes H$. Therefore, we get

$$A \times_{\alpha r} G = \min\text{-}(A \otimes C_r^*(G)),$$

where $\|\cdot\|_{\min} (= \alpha_0(\cdot)$ in Chapter 3) is the spatial C^*-norm.

Proposition 16.4.16. Let (A, G, α) be a C^*-system, and B be a C^*-subalgebra of A with $\alpha_t(B) = B, \forall t \in G$. Then we have that $B \times_{\alpha r} G \hookrightarrow A \times_{\alpha r} G$.

Proof. It suffices to show that

$$\|x\|_{B\times_{\alpha r}G} = \|x\|_{A\times_{\alpha r}G}, \forall x \in L^1(G, B, \alpha).$$

Fix a $x \in L^1(G, B, \alpha)$.

Clearly, $\|x\|_{A\times_{\alpha r}G} \leq \|x\|_{B\times_{\alpha r}G}$.

Conversely, let $\{\pi, H\}$ be a $*$ representation of B. Since each state on B can be extended to a state on A, and $\pi \cong \oplus_{\varphi\in\Lambda}\pi_\varphi$, where Λ is a subset of the state space on B, there is a $*$ representation $\{\rho, K\}$ of A such that $H \subset K, \rho(b)H \subset H$, and $\rho(b)|H = \pi(b), \forall b \in B$. Then we have

$$\begin{aligned}(\text{Ind}\rho(x)\xi)(t) &= \int_G \rho \circ \alpha_{t^{-1}}(x(s))\xi(s^{-1}t)ds \\ &= \int_G \pi \circ \alpha_{t^{-1}}(x(s))\xi(s^{-1}t)ds = (\text{Ind}\pi(x)\xi)(t),\end{aligned}$$

$\forall \xi \in L^2(G, H) \subset L^2(G, K)$. Thus,

$$\|x\|_{A\times_{\alpha r}G} \geq \|\text{Ind}\rho(x)\| \geq \|\text{Ind}\pi(x)\|.$$

Further, we get $\|x\|_{A\times_{\alpha r}G} \geq \|x\|_{B\times_{\alpha r}G}$, and $\|x\|_{B\times_{\alpha r}G} = \|x\|_{A\times_{\alpha r}G}$. Q.E.D.

Theorem 16.4.17. Let (A, G, α) be a C^*-system, and G be amenable. Then $A \times_\alpha G = A \times_{\alpha r} G$.

Proof. It suffices to show that $\|x\|_r \geq \|x\|, \forall x \in L^1(G, A, \alpha)$.

For any state φ on $L^1(G, A, \alpha)$, by the GNS construction there is a cyclic $*$ representation $\{\pi_\varphi, H_\varphi, \xi_\varphi\}$ of $A \times_\alpha G$ with $\|\xi_\varphi\| = 1$. By Theorem 16.4.7 , we have unique covariant representation $\{\rho, u, H_\varphi\}$ of (A, G, α) such that $\pi_\varphi = \rho \times u$.

Since G is amenable, by Godement's condition there is a net $\{g_l\} \subset L^2(G)$ such that

$$\langle \lambda_t g_l, g_l \rangle \longrightarrow 1$$

uniformly on any compact subset of G. We may assume that $\|g_l\|_2 = 1, \forall l$.

Let $\varphi_l(y) = \langle \text{Ind}\rho(y)\xi_l, \xi_l \rangle, \forall y \in L^1(G, A, \alpha)$, where $\xi_l(s) = g_l(s)\, u_{s^{-1}}\xi_\varphi, \forall s \in$

G, l. Clearly, $\xi_l \in L^2(G, H_\varphi)$ and $\|\xi_l\| = 1, \forall l$. Notice that

$$\begin{aligned}
\varphi_l(y) &= \int \langle \overline{\rho}(y(t))\lambda(t)\xi_l, \xi_l\rangle dt \\
&= \int\int \langle \rho \circ \alpha_{s^{-1}}(y(t))\xi_l(t^{-1}s), \xi_l(x)\rangle ds dt \\
&= \int\int g_l(t^{-1}s)\overline{g_l(s)}\langle u_s(\rho \circ \alpha_{s^{-1}}(y(t)))u_s^* u_t \xi_\varphi, \xi_\varphi\rangle ds dt \\
&= \int \langle \lambda_t g_l, g_l\rangle \cdot \langle \rho(y(t))u_t\xi_\varphi, \xi_\varphi\rangle dt \\
&\longrightarrow \int \langle \rho(y(t))u_t\xi_\varphi, \xi_\varphi\rangle dt = \varphi(y),
\end{aligned}$$

$\forall y \in L^1(G, A, \alpha)$. Further,

$$\varphi(x^*x) = \lim_l \varphi_l(x^*x) = \lim_l \langle \mathrm{Ind}\rho(x^*x)\xi_l, \xi_l\rangle \le \|x\|_r^2,$$

$\forall x \in L^1(G, A, \alpha)$. Therefore, we get

$$\begin{aligned}
\|x\|^2 &= \sup\{\varphi(x^*x) | \varphi \text{ is a state on } L^1(G, A, \alpha)\} \\
&\le \|x\|_r^2, \quad \forall x \in L^1(G, A, \alpha).
\end{aligned}$$

Q.E.D.

Corollary 16.4.18. Let (A, G, α) be a C^*-system, G be amenable, and B be a C^*-subalgebra of A with $\alpha_t(B) = B, \forall t \in G$. Then $B \times_\alpha G \hookrightarrow A \times_\alpha G$.

Now we discuss the embedding of A and G into $M(A \times_\alpha G)$ for a C^*-system (A, G, α), where $M(A \times_\alpha G)$ is the multiplier algebra of $A \times_\alpha G$ (see Section 2.1.2).

Let (A, G, α) be a C^*-system, and $\{\rho, H\}$ be a nondegenerate faithful $*$ representation of $A \times_\alpha G$. Clearly, $\{\rho, H\}$ is still nondegenerate and faithful for $L^1(G, A, \alpha)$. By Theorem 16.4.7, there is a covariant representation $\{\pi, u, H\}$ of (A, G, α) such that $\rho = \pi \times u$. Since

$$\rho(f) = \int_G \pi(f(s))u_s ds = \int_G u_s \pi(\alpha_{s^{-1}}(f(s)))ds,$$

$\forall f \in L^1(G, A, \alpha)$ and

$$\rho(g \otimes a) = \int_G \pi(g(s)a)u_s ds = \pi(a)\int_G g(s)u_s ds,$$

$\forall g \in L^1(G)$ and $a \in A$, it follows that $\{\pi, H\}$ is also faithful and nondegenerate for A.

For any $a \in A$ and $f \in L^1(G, A, \alpha)$, define

$$(L_a f)(t) = af(t), \quad (R_a f)(t) = f(t)\alpha_t(a), \quad \forall t \in G.$$

Clearly, L_a and R_a are bounded on $L^1(G, A, \alpha)$ with norm $\leq \|a\|$. Further, we have

$$\rho(L_a f) = \pi(a)\rho(f), \quad \text{and} \quad \rho(R_a f) = \rho(f)\pi(a).$$

Hence,

$$\|L_a f\|_{A\times_\alpha G} = \|\rho(L_a f)\| \leq \|a\| \cdot \|f\|_{A\times_\alpha G},$$

and $\|R_a f\|_{A\times_\alpha G} \leq \|a\| \cdot \|f\|_{A\times_\alpha G}$. Then, L_a and R_a can be extended to bounded linear operators on $A \times_\alpha G$ with $\|L_a\|, \|R_a\| \leq \|a\|$, and

$$L_a x = \rho^{-1}(\pi(a)\rho(x)), \quad R_a x = \rho^{-1}(\rho(x)\pi(a)),$$

$\forall a \in A, x \in A \times_\alpha G$. Of course , these expressions are independent of the choice of $\{\rho, H\}$. Moreover, it is easily verififed that

$$L_a(xy) = (L_a x)y, \quad R_a(xy) = x(R_a(y)), \quad x(L_a y) = (R_a x)y,$$

$\forall x, y \in A \times_\alpha G$. So that (L_a, R_a) is a double centralizer of $A \times_\alpha G, \forall a \in A$ (see Definition 2.12.5). Then we get a $*$ isomorphism $a \rightarrow (L_a, R_a)$ from A into $M(A \times_\alpha G)$. By Section 2.12, we can write that

$$a = (\sigma\text{- or } s\text{-}) \lim(L_a f_l) = (\sigma\text{- or } s\text{-}) \lim(R_a f_l),$$

$\forall a \in A$, where $\{f_l\}$ is an approximate identity for $L^1(G, A, \alpha)$ (then for $A \times_\alpha G$), and σ- or s-topology is in $(A \times_\alpha G)^{**}$.

Now for any $s \in G$ and $f \in L^1(G, A, \alpha)$, define

$$(L_s f)(t) = \alpha_s(f(s^{-1}t)), \quad (R_s f)(t) = \Delta(s)^{-1} f(ts^{-1}), \quad \forall t \in G.$$

Clearly, L_s, R_s are bounded on $L^1(G, A, \alpha)$. Further, we have

$$\rho(L_s f) = u_s \rho(f), \quad \rho(R_s f) = \rho(f) u_s.$$

Hence, L_s and R_s can be extended to bounded linear operators on $A \times_\alpha G$, and

$$L_s x = \rho^{-1}(u_s \rho(x)), \quad R_s x = \rho^{-1}(\rho(x) u_s),$$

$\forall s \in G, x \in A \times_\alpha G$. It is easily verified that (L_s, R_s) is a double centralizer of $A \times_\alpha G$, and (L_s, R_s) is a unitary element of $M(A \times_\alpha G)$. Then we get a faithful representation $s \rightarrow (L_s, R_s)$ of G into the group of all unitary elements of $M(A \times_\alpha G)$. By Section 2.12, we can write that

$$t = (\sigma\text{- or } s\text{-}) \lim(L_t f_l) = (\sigma\text{- or } s\text{-}) \lim(R_t f_l),$$

$\forall t \in G$.

We say that $s \to (L_s, R_s)$ is continuous with respect to the strict topology in $M(A \times_\alpha G)$ (see Definition 2.12.11). In fact, it suffices to show that

$$\|L_s f - f\| \to 0, \quad \text{and} \quad \|R_s f - f\| \to 0$$

in $A \times_\alpha G$ as $s \to e$ in $G, \forall f \in K(G, A)$. But it is obvious since

$$\|L_s f - f\|_1 \le \int_G \|f(s^{-1}t) - f(t)\|_A dt + \int_G \|(\alpha_s - 1)(f(t))\|_A dt,$$

$$\|R_s f - f\|_1 \le \int_G |\Delta(s^{-1})| \cdot \|f(ts^{-1}) - f(t)\|_A dt + \int_G |\Delta(s)^{-1} - 1| \cdot \|f(t)\|_A dt,$$

and $\|g\|_{A \times_\alpha G} \le \|g\|_1, \forall g \in L^1(G, A, \alpha)$.

Finally, regarding A and G as subset of $M(A \times_\alpha G)$, and regarding $\{\rho = \pi \times u, H\}$ as a faithful $*$ representation of $M(A \times_\alpha G)$ (see Proposition 2.12.9), we have that

$$\rho(a) = \pi(a), \quad \rho(s) = u_s, \quad \text{and} \quad sas^{-1} = \alpha_s(a),$$

$\forall a \in A, s \in G$. In fact, the equalities $\rho(a) = \pi(a)$ and $\rho(s) = u_s$ are obvious. Further, by

$$\rho(sas^{-1}) = u_s \pi(a) u_s^* = \pi(\alpha_s(a)) = \rho(\alpha_s(a))$$

we have $sas^{-1} = \alpha_s(a)$ in $M(A \times_\alpha G)$.

Summing up the above discussion, we have the following.

Proposition 16.4.19. Let (A, G, α) be a C^*-system, and $M(A \times_\alpha G)$ be the multiplier algebra of $A \times_\alpha G$. Then A and G can be embedded into $M(A \times_\alpha G)$ such that

$$ss^* = s^*s = 1, \quad sas^{-1} = \alpha_s(a),$$

$$(sf)(t) = (L_s f)(t) = \alpha_s(f(s^{-1}t)),$$

$$(fs)(t) = (R_s f)(t) = \Delta(s)^{-1} f(ts^{-1}),$$

$$(af)(t) = (L_a f)(t) = af(t),$$

$$(fa)(t) = (R_a f)(t) = f(t)\alpha_t(a),$$

in the sense of $M(A \times_\alpha G), \forall s, t \in G, a \in A$, and $f \in L^1(G, A, \alpha)$.

Moreover, if $\{\pi, u, H\}$ is a covariant representation of (A, G, α) such that $\{\pi \times u, H\}$ is faithful and nondegenerate for $A \times_\alpha G$, then we have

$$\rho(a) = \pi(a), \rho(s) = u_s, \quad \forall a \in A, s \in G,$$

where $\{\rho, H\}$ is the faithful $*$ extension of $\{\pi \times u, H\}$ on $M(A \times_\alpha G)$.

Notes. Crossed products of C^*-algebras with discrete group were introduced by T.Turumaru. Later G.Zeller-Meyer carried out a penetrating analysis. General crossed products were defined by S.Doplicker, D.Kastler and D.W.Robinson, and Theorem 16.4.7 is also due to them. The reduced crossed products were defined by G.Zeller-meyer for discrete groups, and generalized by H.Takai.

References. [29], [127], [166], [186], [202].

16.5. Takai's duality theorem

Definition 16.5.1. Two C^*-systems (A, G, α) and (B, G, β) are said to be *isomophic*, denoted by $(A, G, \alpha) \cong (B, G, \beta)$, if there is a $*$ isomorphism Φ from A onto B such that $\Phi \circ \alpha_t \circ \Phi^{-1} = \beta_t, \forall t \in G$.

Proposition 16.5.2. If $(A, G, \alpha) \cong (B, G, \beta)$, then $A \times_\alpha G$ and $B \times_\beta G$ are $*$ isomorphic.

Proof. Let Φ be a $*$ isomorphism from A onto B with $\Phi \circ \alpha_t = \beta_t \circ \Phi, \forall t \in G$. Define

$$\Psi(f)(t) = \Phi(f(t)), \forall f \in L^1(G, A, \alpha).$$

Clearly, Ψ is a $*$ isomorphism from $L^1(G, A\alpha)$ onto $L^1(G, B, \beta)$. Moreover, if π is a $*$ representation of $L^1(G, B, \beta)$, then $\pi \circ \Psi$ is a $*$ representation of $L^1(G, A, \alpha)$ obviously. Conversely, for any $*$ represetation ρ of $L^1(G, A, \alpha)$, there is a $*$ represntation π of $L^1(G, B, \beta)$ such that $\rho = \pi \circ \Psi$. Therefore, we have

$$\begin{aligned} \|\Psi(f)\| &= \sup\{\|\pi \circ \Psi(f)\| | \pi \text{ is a } * \text{ represntation of } L^1(G, B, \beta)\} \\ &= \sup\{\|\rho(f)\| | \rho \text{ is a } * \text{ representation of } L^1(G, A, \alpha)\} \\ &= \|f\|, \quad \forall f \in L^1(G, A, \alpha), \end{aligned}$$

and Ψ can be uniquely extended to a $*$ isomorphism from $A \times_\alpha G$ onto $B \times_\beta G$.

Q.E.D.

Proposition 16.5.3. Let (A, G, α) and (B, G, β) be two C^*-systems. Then there is a (tensor product) C^*-system $(\min\text{-}(A \otimes B), G, \alpha \otimes \beta)$ such that

$$(\alpha \otimes \beta)_t(a \otimes b) = \alpha_t(a) \otimes \beta_t(b),$$

$\forall t \in G, a \in A, b \in B$, where $\min\text{-}(A \otimes B)$ means the injective tensor product, and $\|\cdot\|_{\min} (= \alpha_0(\cdot)$ in Chapter 3) is the spatial C^*-norm on $A \otimes B$.

Proof. It suffices to show that

$$\|\sum_{i=1}^{n}\alpha_t(a_i)\otimes\beta_t(b_i)\|_{\min}=\|\sum_{i=1}^{n}a_i\otimes b_i\|,$$

$\forall t\in G, a_i\in A, b_i\in B, 1\leq i\leq n$. Bu that is obvious from Theorem 3.2.5.

Q.E.D.

Lemma 16.5.4. Let G be a locally compact group, and $t\to\rho_t$ be the right regular representation of G on $L^2(G)$, i.e., $(\rho_t\xi)(s)=\triangle(t)^{1/2}\xi(st),\forall\xi\in L^2(G)$. Then $(C(L^2(G)),G,ad\rho)$ is a C^* -system, where $ad\rho_t(a)=\rho_t a\rho_t^*,\forall t\in G,a\in C(L^2(G))$.

Proof. It suffices to show that

$$\|\rho_t a\rho_t^*-a\|=\|\rho_t a-a\rho_t\|\longrightarrow 0\quad\text{as}\quad t\to e\text{ in }G,$$

$\forall a\in C(L^2(G))$. Since the subset of all finite rank operators is dense in $C(L^2(G))$, we may assume that a is an one-rank operator $\xi\otimes\eta$, where $\xi,\eta\in L^2(G)$.

For any $\varsigma\in L^2(G)$ with $\|\varsigma\|_2\leq 1$, we have

$$\begin{aligned}&\|(\rho_t a-a\rho_t)\varsigma\|_2=\|\langle\varsigma,\eta\rangle\rho_t\xi-\langle\rho_t\varsigma,\eta\rangle\xi\|_2\\ \leq\ &\|\langle\varsigma,\eta\rangle(\rho_t\xi-\xi)\|_2+\|\langle\varsigma,\eta\rangle\xi-\langle\varsigma,\rho_{t^{-1}}\eta\rangle\xi\|_2\\ \leq\ &\|\eta\|_2\cdot\|\rho_t\xi-\xi\|_2+\|\xi\|_2\cdot\|\eta-\rho_{t^{-1}}\eta\|\longrightarrow 0\end{aligned}$$

as $t\longrightarrow e$ in G. That comes to the conclusion. Q.E.D.

Now let (A,G,α) be a C^*-system, and

$$\begin{aligned}C_0^\infty(G,A)&=\min\text{-}(C_0^\infty(G)\otimes A)\\ &=\{f:G\to A|\ f\text{ is continuous, and }\|f(\cdot)\|\in C_0^\infty(G)\}.\end{aligned}$$

By Proposition 16.5.3, $(C_0^\infty(G,A),G,\gamma)$ is a C^*-system, where $\gamma=\lambda\otimes\alpha$, and $(\lambda_t g)(s)=g(t^{-1}s),\forall g\in C_0^\infty(G)$ (notice that g is uniformly continuous on G).

Lemma 16.5.5. Let (A,G,α) be a C^*-system. Then

$$(C_0^\infty(G,A)\times_\gamma G,G,\rho)$$

is also a C^*-system, where

$$(\rho(t)f)(s)=\rho_t f(s),\forall t\in G,f\in L^1(G,C_0^\infty(G,A),\gamma),$$

and $(\rho_t g)(s) = g(st), \forall t \in G, g \in C_0^\infty(G,A)$.

Proof. Since $\rho_t\gamma_s = \gamma_s\rho_t$ on $C_0^\infty(G,A)$ and $\rho(s)\rho(t) = \rho(st), \forall s,t \in G$, ρ_t is a $*$ isomrophism of $L^1(G, C_0^\infty(G,A),\gamma), \forall t \in G$.

If $\{\pi, u\}$ is a covariant representation of $(C_0^\infty(G,A), G, \gamma)$, then $\{\pi\circ\rho_t, u\}$ is also a covariant representation of $(C_0^\infty(G,A), G, \gamma), \forall t \in G$. Thus, we have

$$\begin{aligned}
&\|\rho(t)f\| \\
&= \sup\left\{\|\int \pi\circ\rho_t(f(s))u_s ds\| \,\middle|\, \begin{array}{c}\{\pi,u\} \text{ is a covariant}\\ \text{representation of } (C_0^\infty(G,A),G,\gamma)\end{array}\right\} \\
&= \sup\{\|\int \pi(f(s))u_s ds\| \,|\, \{\pi,u\} \text{is as above}\} = \|f\|,
\end{aligned}$$

$\forall f \in L^1(G, C_0^\infty(G,A),\gamma)$, and $\rho(t)$ can be uniquely extended to a $*$ automorphism of $C_0^\infty(G,A)\times_\gamma G, \forall t \in G$.

If $f = g \otimes h$, where $g \in L^1(G), h \in K(G,A)$, then

$$\begin{aligned}
\|\rho(t)f - f\| &\le \|\rho(t)f - f\|_1 \\
&= \int \|\rho_t f(s) - f(s)\|_{C_0^\infty(G,A)} ds \\
&= \int |g(s)| ds \cdot \sup_{r\in G} \|h(rt) - h(r)\|_A \longrightarrow 0
\end{aligned}$$

as $t \to e$ in G. Therefore, $(C_0^\infty(G,A)\times_\gamma G, G, \rho)$ is a C^* -system. Q.E.D.

Theorem 16.5.6. Let (A,G,α) be a C^*-system. Then the C^*-systems (min-$(A \otimes C(L^2(G)), G, \alpha\otimes ad\rho)$ and $(C_0^\infty(G,A)\times_\gamma G, G, \rho)$ are isomorphic.

Proof. We may assume that $A \subset B(H)$. Define a faithful $*$ representation $\{\pi, L^2(G,H)\}$ of $C_0^\infty(G,A)$ as follows:

$$(\pi(f)\xi)(t) = f(t)\xi(t), \quad \forall f \in C_0^\infty(G,A), \xi \in L^2(G,H).$$

It generates a faithful $*$ representation $\{Ind\pi = \tilde{\pi}\times\lambda, L^2(G\times G, H)\}$ of $C_0^\infty(G,A)\times_\gamma G$, i.e.,

$$\begin{aligned}
&(\mathrm{Ind}\pi(z)\xi)(s,t) = (\int (\tilde{\pi}(z(r,\cdot))\lambda(r)\xi)(s,\cdot)dr)(t) \\
&= (\int \pi\circ\gamma_{s^{-1}}(z(r,\cdot))\xi(r^{-1}s,\cdot)dr)(t) \\
&= \int \alpha_{s^{-1}}(z(r,st))\xi(r^{-1}s,t)dr,
\end{aligned}$$

$\forall z \in K(G, C_0^\infty(G,A)), \xi \in L^2(G\times G, H)$. Define a unitary operator w on

$L^2(G \times G, H)$ as follows:

$$\begin{cases} (w\xi)(x,t) = \triangle(t)^{1/2}\xi(st,t), \\ (w^*\xi)(s,t) = \triangle(t)^{-1/2}\xi(st^{-1},t), \forall \xi \in L^2(G \times G, H). \end{cases}$$

Then we have

$$\begin{aligned} (w^*\mathrm{Ind}\pi(z)w\xi)(s,t) &= (\mathrm{Ind}\pi(z)w\xi)(st^{-1},t) \cdot \triangle(t)^{-1/2} \\ &= \int \alpha_{ts^{-1}}(z(r,s))(w\xi)(r^{-1}st^{-1},t)dr \triangle(t)^{-1/2} \\ &= \int \alpha_{ts^{-1}}(z(r,s))\xi(r^{-1}s,t)dr, \end{aligned}$$

$\forall z \in K(G, C_0^\infty(G,A)), \xi \in L^2(G \times G, H)$. In particular, if $z(r,s) = \alpha_s(a)f(r^{-1}s)\, g(s)$, where $a \in A, f, g \in K(G)$, then

$$\begin{aligned} (w^*\mathrm{Ind}\pi(z)w\xi)(s,t) &= g(s)\alpha_t(a)\int f(r^{-1}s)\xi(r^{-1}s,t)dr \\ &= g(s)\alpha_t(a)\int f(r)\xi(r,t)dr/\triangle(r), \end{aligned} \tag{1}$$

$\forall \xi \in L^2(G \times G, H)$. Further, define a faithful $*$ representation $\{\tilde{\pi}, L^2(G,H)\}$ of A as follows :

$$(\tilde{\pi}(a)\xi)(t) = \alpha_t(a)\xi(t), \quad \forall \xi \in L^2(G,H), a \in A.$$

Then we have

$$w^*\mathrm{Ind}\pi(z)w = \tilde{\pi}(a) \otimes v_{f'g} \in \tilde{\pi}(A) \otimes C(L^2(G)), \tag{2}$$

where $z(r,s) = \alpha_s(a)f(r^{-1}s)g(s), f, g \in K(G), f'(r) = f(r)/\triangle(r)$, and $v_{f'g}$ is an one-rank operator on $L^2(G)$:

$$v_{f'g}\eta = \langle \eta, \overline{f'} \rangle g, \quad \forall \eta \in L^2(G).$$

Now we claim that

$$[f(r^{-1}s)g(s)\alpha_s(a) | a \in A, f, g \in K(G)]$$

is dense in $C_0^\infty(G,A) \times_\gamma G$. In fact, define $\Phi : K(G \times G, A) \to K(G \times G, A)$ as follows:

$$(\Phi y)(r,s) = y(r^{-1}s,s), \quad (\Phi^{-1}y)(r,s) = y(sr^{-1},s),$$

$\forall y \in K(G \times G, A)$. Clearly, Φ is a bijection, and keeps the maximal norm. For any $y \in K(G \times G, A)$ and $\varepsilon > 0$, there is a compact subset F of G such that

$$\mathrm{supp} y \subset F \times F.$$

We may assume that $e \in F$ and $F = F^{-1}$. Pick $\delta = \varepsilon/|F^2|$, and $f_i, g_i \in K(F), a_i \in A$ such that

$$\max_{r,s\in G} \|y(r,s) - \sum_i f_i(r)g_i(s)\alpha_s(a_i)\|_A < \delta.$$

Then we have

$$\max_{r,s\in G} \|(\Phi y)(r,s) - z(r,s)\|_A < \delta,$$

where $z(r,s) = \sum_i f_i(r^{-1}s)g_i(s)\alpha_s(a_i)$. Clearly, supp Φy and supp $z \subset F^2 \times F^2$.

Thus

$$\|\Phi y - z\|_{C_0^\infty(G,A)\times_\gamma G} \leq \|\Phi y - z\|_{L^1(G,C_0^\infty(G,A),\gamma)}$$

$$= \int_G \max_{s\in G} \|(\Phi y)(r,s) - z(r,s)\|_A dr < \delta|F^2| = \varepsilon.$$

Of course, $K(G \times G, A)$ is dense in $C_0^\infty(G,A) \times_\gamma G$. Therefore, $[f(r^{-1}s)\ g(s)\ \alpha_s(a)|a \in A, f, g \in K(G)]$ is also dense in $C_0^\infty(G,A) \times_\gamma G$.

Then by (2) , we obtain that

$$w^*\text{Ind}\pi(C_0^\infty(G,A) \times_\gamma G)w = (\tilde{\pi} \times id)\ \text{min-}(A \otimes C(L^2(G))).$$

Further, for any $\sigma \in G, z(r,s) = f(r^{-1}s)g(s)\alpha_s(a)$, where $a \in A, f, g \in K(G)$, by Lemma 16.5.5,

$$(\rho(\sigma)z)(r,s) = z(r,s\sigma) = \alpha_s(\alpha_\sigma(a))g_\sigma(s)f_\sigma(r^{-1}s).$$

Then by (1) , we have

$$(w^*\text{Ind}\pi(\rho(\sigma)z)w\xi)(s,t)$$

$$= g(s\sigma)\alpha_{t\sigma}(a)\int f(r\sigma)\xi(r,t)dr/\triangle(r)$$

$$= (\tilde{\pi}(\alpha_\sigma(a)) \otimes \rho_\sigma v_{f'g}\rho_\sigma^*\xi)(s,t),$$

$\forall \xi \in L^2(G \times G, H)$, i.e,

$$w^*\text{Ind}\pi(\rho(\sigma)z)w = (\tilde{\pi} \otimes id)(\alpha_\sigma \otimes ad\rho_\sigma)(a \otimes v_{f'g}).$$

Therefore, the C^*-systems $(C_0^\infty(G,A)\times_\gamma G, G, \rho)$ and $(\text{min-}(A \otimes C(L^2(G))), G, \alpha \otimes ad\rho)$ are isomorphic. Q.E.D.

In the following, let G be abelian, and $\widehat{G}$ be its dual. Let ds, dp be the Haar measures on $G, \widehat{G}$ respectively such that the Fourier transform

$$\hat{f}(p) = \int \langle s,p\rangle f(s)ds, \quad \forall f \in L^2(G),$$

is a unitary opertor from $L^2(G)$ onto $L^2(\widehat{G})$. Let (A,G,α) be a C^*-system, and define

$$\widehat{\alpha}_p(f)(t) = \overline{\langle t,p\rangle} f(t), \quad \forall p \in \widehat{G}, t \in G, f \in L^1(G,A,\alpha).$$

Then $\widehat{\alpha}_p$ is a $*$ isomorphism of $L^1(G,A,\alpha)$, and $\widehat{\alpha}_p\widehat{\alpha}_{p'} = \widehat{\alpha}_{pp'}, \forall p,p' \in \widehat{G}$. Moreover, since

$$\begin{aligned}\|\widehat{\alpha}_p(f)\| &= \sup\{\|\pi \circ \widehat{\alpha}_p(f)\| | \pi \text{ is a } * \text{represnetation of } L^1(G,A,\alpha)\} \\ &= \sup\{\|\pi(f)\| | \pi \quad \text{is as above}\} = \|f\|, \quad \forall f \in L^1(G,A,\alpha),\end{aligned}$$

then $\widehat{\alpha}_p$ can be uniquely extended to a $*$ automorphism of $A \times_\alpha G$. For any $f \in L^1(G,A,\alpha)$, we have

$$\begin{aligned}\|\widehat{\alpha}_p(f) - f\| &\le \|\widehat{\alpha}_p(f) - f\|_1 \\ &\le \int |\langle p,t\rangle - 1| \cdot \|f(t)\| dt \longrightarrow 0\end{aligned}$$

as $p \to \hat{e}$ (the unit of $\widehat{G}$) in $\widehat{G}$ since $\langle p,t\rangle \to 1$ uniformly on any compact subset of G as $p \to \hat{e}$ in $\widehat{G}$. Therefore, we obtain a new C^*-system $(A \times_\alpha G, \widehat{G}, \widehat{\alpha})$.

Definition 16.5.7. Let (A,G,α) be a C^*-system, and G be abelian. The above C^*-system $(A \times_\alpha G, \widehat{G}, \widehat{\alpha})$ is called the *dual system* of (A,G,α), and $\widehat{\alpha}$ is called the *dual action* of α.

Proposition 16.5.8. If $(A,G,\alpha) \cong (B,G,\beta)$, and G is abelian, then we have

$$(A \times_\alpha G, \widehat{G}, \widehat{\alpha}) \cong (B \times_\beta G, \widehat{G}, \widehat{\beta}).$$

Proof. Let Φ be a $*$ isomorphism from A onto B with $\Phi \circ \alpha_t = \beta_t \circ \Phi, \forall t \in G$. Then Ψ is a $*$ isomorphism from $A \times_\alpha G$ into $B \times_\beta G$ by Proposition 16.5.2. Now it suffices to show that $\Psi \circ \widehat{\alpha}_p \circ \Psi^{-1} = \widehat{\beta}_p, \forall p \in \widehat{G}$.

For any $p \in \widehat{G}$ and $f \in L^1(G,B,\beta)$, let $g = \Psi^{-1}(f)(\in L^1(G,A,\alpha))$. Then

$$\begin{aligned}\Psi \circ \widehat{\alpha}_p \circ \Psi^{-1}(f)(t) &= \Psi \circ \widehat{\alpha}_p(g)(t) \\ &= \Phi(\widehat{\alpha}_p(g)(t)) = \overline{\langle t,p\rangle}\Phi(g(t)) \\ &= \overline{\langle p,t\rangle} f(t) = \widehat{\beta}_p(f)(t).\end{aligned}$$

That comes to the conclusion. Q.E.D.

Theorem 16.5.9. Let (A,G,α) be a C^*-system, and G be abelian. Then we have that

$$((A \times_\alpha G) \times_{\widehat{\alpha}} \widehat{G}, G, \widehat{\widehat{\alpha}}) \cong (\min\text{-}(A \otimes C(L^2(G))), G, \alpha \otimes ad\rho).$$

Proof. 1) Let $\{\pi, H\}$ be a faithful $*$ representatio of A. By Theorem 16.4.15, $\{\mathrm{Ind}\pi = \pi \times \lambda, L^2(G,H)\}$ is a faithful $*$ representation of $A \times_\alpha G$. Further, $\{\hat{\pi}, L^2(\hat{G} \times G, H)\}$ is a faithful $*$ representation of $(A \times_\alpha G) \times_{\hat{\alpha}} \hat{G}$, where $\hat{\pi} = \mathrm{Ind}(\mathrm{Ind}\pi) = \overline{\mathrm{Ind}\pi} \times \lambda$, and $\{\overline{\mathrm{Ind}\pi}, \lambda, L^2(\hat{G} \times G, H)\}$ is a covariant representation of $(A \times_\alpha G, \hat{G}, \hat{\alpha})$.

Now let $f \in K(\hat{G} \times G, A), \xi \in L^2(\hat{G} \times G, H)$, and consider $\hat{\pi}(f)\xi$. For any $p \in \hat{G}, t \in \hat{G}$, we have

$$
\begin{aligned}
&(\hat{\pi}(f)\xi)(p,t) = ((\overline{\mathrm{Ind}\pi} \times \lambda)(f)\xi)(p,t) \\
= \ & (\int \overline{\mathrm{Ind}\pi}(f(q,\cdot))\lambda_q \xi dq)(p,t) \\
= \ & (\int \mathrm{Ind}\pi \circ \hat{\alpha}_{p^{-1}}(f(q,\cdot))\xi(q^{-1}p,\cdot)dq)(t) \\
= \ & \int_{\hat{G}} dq \int_G ds \pi \circ \alpha_{t^{-1}}(\hat{\alpha}_{p^{-1}}(f(q,\cdot))(s)\xi(q^{-1}p, s^{-1}t) \\
= \ & \int_{\hat{G}} dq \int_G ds \langle s, p\rangle \pi \circ \alpha_{t^{-1}}(f(q,s))\xi(q^{-1}p, s^{-1}t)
\end{aligned}
$$

Thus , we can write that

$$
\hat{\pi}(f) = \int\int_{\hat{G} \times G} f(q,s)u(s)v(q)dsdq,
$$

$\forall f \in K(\hat{G} \times G, A)$, where

$$
\begin{cases}
(a\xi)(p,t) = \pi \circ \alpha_{t^{-1}}(a)\xi(p,t), \\
(u(s)\xi)(p,t) = \langle s,p\rangle \xi(p, s^{-1}t), \\
(v(q)\xi)(p,t) = \xi(q^{-1}p, t),
\end{cases}
$$

$\forall a \in A, s, t \in G, p, q \in \hat{G}, \xi \in L^2(\hat{G} \times G, H)$.

2) Define a unitary operator on $L^2(\hat{G} \times G, H)$ as follows:

$$
(J\xi)(p,t) = \overline{\langle t,p\rangle}\xi(p,t), \quad (J^*\xi)(p,t) = \langle t,p\rangle \xi(p,t),
$$

$\forall t \in G, p \in \hat{G}, \xi \in L^2(\hat{G} \times G, H)$. By 1), we have

$$
\begin{aligned}
&(J\hat{\pi}(f)J^*\xi)(p,t) \\
= \ & \overline{\langle t,p\rangle} \int\int_{\hat{G} \times G} dqds \langle s,p\rangle \pi \circ \alpha_{t^{-1}}(f(q,s))(J^*\xi)(q^{-1}p, s^{-1}t) \\
= \ & \int\int_{\hat{G} \times G} dqds \langle s,q\rangle \overline{\langle t,q\rangle} \pi \circ \alpha_{t^{-1}}(f(q,s))\xi(q^{-1}p, s^{-1}t).
\end{aligned}
$$

Thus, we can write that

$$
J\hat{\pi}(f)J^* = \int\int_{\hat{G} \times G} f(q,s)u'(s)v'(q)dsdq,
$$

$\forall f \in K(\widehat{G} \times G, A)$, where

$$\begin{cases} (a\xi)(p,t) = \pi \circ \alpha_{t^{-1}}(a)\xi(p,t), \\ (u'(s)\xi)(p,t) = \xi(p, s^{-1}t), \\ (v'(q)\xi)(p,t) = \overline{\langle t, q\rangle}\xi(q^{-1}p, t), \end{cases}$$

$\forall a \in A, s, t \in G, p, q \in \widehat{G}, \xi \in L^2(\widehat{G} \times G, H)$.

3) Consider the C^*-system $(A, \widehat{G}, id)$. Clearly,

$$A \times_{id} \widehat{G} \cong \min\text{-}(A \otimes C_0^\infty(G)) = C_0^\infty(G, A).$$

Define a homomorphism $\beta : G \longrightarrow Aut(L^1(\widehat{G}, A, id))$ as follows:

$$\beta_t(\hat{f})(p) = \langle t, p\rangle \alpha_t(\hat{f}(p)),$$

$\forall t \in G, p \in \widehat{G}, \hat{f} \in L^1(\widehat{G}, A, id)$. If $\{L, \nabla\}$ is a covariant representation of $(A, \widehat{G}, id)$, then $L(a)\nabla_p = \nabla_p L(a), \forall a \in A, p \in \widehat{G}$. Thus $\{L \circ \alpha_t, \langle t, \cdot\rangle \nabla_\cdot\}$ is also a covariant representation of $(A, \widehat{G}, id), \forall t \in G$, and

$$\begin{aligned} \|\beta_t(\hat{f})\| &= \sup\left\{\left\| \int L(\beta_t(\hat{f})(p)) \nabla_p \, dp\right\| \;\middle|\; \begin{array}{c}\{L, \nabla\} \text{ is a covariant} \\ \text{representation of } (A, \widehat{G}, id)\end{array}\right\} \\ &= \sup\{\| \int L \circ \alpha_t(\hat{f}(p))\langle t, p\rangle \nabla_p \, dp\| | \{L, \nabla\} \text{is as above}\} \\ &= \|\hat{f}\|, \quad \forall f \in L^1(\widehat{G}, A, id), t \in G \end{aligned}$$

Further, β_t can be uniquely extended to a $*$ automorphism of $A \times_{id} \widehat{G}, \forall t \in G$. Moreover, notice that

$$\begin{aligned} \|\beta_t(\hat{f}) - \hat{f}\| &\leq \|\beta_t(\hat{f}) - \hat{f}\|_1 \\ &= \int_{\widehat{G}} \|\langle t, p\rangle \alpha_t(\hat{f}(p)) - \hat{f}(p)\| dp \\ &\leq \int_{\widehat{G}} |\langle t, p\rangle - 1| \cdot \|\hat{f}(p)\| dp + \int_{\widehat{G}} \|\alpha_t(\hat{f}(p)) - \hat{f}(p)\| dp, \end{aligned}$$

$\forall \hat{f} \in L^1(\widehat{G}, A, id)$. As $t \to e$ in G, since $\langle t, p\rangle \to 1$ uniformly on any compact subset of $\widehat{G}$, we can see that the first term $\int_G |\langle t, p\rangle - 1| \cdot \|\hat{f}(p)\| dp \longrightarrow 0$; if pick $\hat{f}(p) = \hat{g}(p)a$, where $\hat{g} \in L^1(\widehat{G})$ and $a \in A$, then we also have

$$\int_{\widehat{G}} \|\alpha_t(\hat{f}(p)) - \hat{f}(p)\| dp \longrightarrow 0 \quad \text{as } t \to e \text{ in } G.$$

Therefore, $(A \times_{id} \widehat{G}, G, \beta)$ is a C^*-system.

4) Let $\{\pi, H\}$ be a faithful $*$ representation of A. Then $\{(\pi \otimes id) \times \lambda, L^2(\widehat{G}, H)\}$ is a faithful $*$ representation of $A \times_{id} \widehat{G}$. Further, $\{\tilde{\pi}, L^2(\widehat{G} \times$

$G,H)\}$ will be a faithful $*$ representation of $(A\times_{id}\widehat{G})\times_\beta G$, where $\tilde{\pi} = \mathrm{Ind}((\pi\otimes id)\times\lambda)$. For any $f\in K(\widehat{G}\times G,A)$ and $\xi\in L^2(\widehat{G}\times G,H)$, compute $\tilde{\pi}(f)\xi$ as follows:

$$\begin{aligned}
&(\tilde{\pi}(f)\xi)(p,t)\\
=\ &\left(\int_G \overline{(\pi\otimes id)\times\lambda}(f(\cdot,s))\lambda_s\xi ds\right)(p,t)\\
=\ &\left(\int_G ((\pi\otimes id)\times\lambda)\circ\beta_{t^{-1}}(f(\cdot,s))\xi(\cdot,s^{-1}t)ds\right)(p)\\
=\ &\int\int_{\widehat{G}\times G} dqds\langle t,q\rangle\pi\circ\alpha_{t^{-1}}(f(q,s))\xi(q^{-1}p,s^{-1}t).
\end{aligned}$$

Thus, we can write that

$$\tilde{\pi}(f) = \int\int_{\widehat{G}\times G} f(q,s)v'(q)u'(s)dqds,$$

$\forall f\in K(\widehat{G}\times G,A)$, where

$$\begin{cases}
(a\xi)(p,t) = \pi\circ\alpha_{t^{-1}}(a)\xi(p,t),\\
(u'(s)\xi)(p,t) = \xi(p,s^{-1}t),\\
(v'(q)\xi)(p,t) = \langle t,q\rangle\xi(q^{-1}p,t),
\end{cases}$$

$\forall a\in A, s,t\in G, p,q\in\widehat{G}, \xi\in L^2(\widehat{G}\times G,H)$. Since $v'(q)u'(s) = \overline{\langle s,q\rangle}u'(s)v'(q)$, it follows from 2) that

$$\tilde{\pi}(f) = \int\int_{\widehat{G}\times G} g(q,s)u'(s)v'(q)dqds = J\tilde{\pi}(g)J^*,$$

where $g(q,s) = \overline{\langle s,q\rangle}f(q,s), \forall f\in K(\widehat{G}\times G,A)$.

5) Notice that $K(\widehat{G}\times G,A)$ is dense in $(A\times_{id}\widehat{G})\times_\beta G$ and $(A\times_\alpha G)\times_{\widehat{\alpha}}\widehat{G}$. So if define

$$\Phi(f)(p,t) = \overline{\langle t,p\rangle}f(p,t),\quad \forall f\in K(\widehat{G}\times G,A),$$

then by 2), 4) Φ can be uniquely extended to a $*$ isomorphism from $(A\times_{id}\widehat{G})\times_\beta G$ onto $(A\times_\alpha G)\times_{\widehat{\alpha}}\widehat{G}$.

6) It is well-known that the Fourier transform

$$\hat{f}\in L^1(\widehat{G})\longrightarrow f(s) = \int_{\widehat{G}}\overline{\langle s,p\rangle}\hat{f}(p)dp\in C_0^\infty(G)$$

can be uniquely extended to a $*$ isomorphism from $C^*(\widehat{G})$ onto $C_0^\infty(G)$. Then

$$\hat{f}\otimes a\longrightarrow f\otimes a,\quad \forall\hat{f}\in L^1(\widehat{G}), a\in A$$

can be uniquely extended to a $*$ isomorphism Ψ from min-$(C^*(\hat{G}) \otimes A) = A \times_{id} \hat{G}$ onto min-$(C_0^\infty(G) \otimes A) = C_0^\infty(G, A)$. Morevoer, by 3) we have

$$\begin{aligned}\Psi \circ \beta_t \circ \Psi^{-1}(f \otimes a) &= \Psi \otimes \beta_t(\hat{f} \otimes a) \\ &= \Psi(\langle t, \cdot\rangle \hat{f}(\cdot)\alpha_t(a)) \\ &= \int \overline{\langle \cdot, p\rangle}\langle t, p\rangle \hat{f}(p)dp \otimes \alpha_t(a) \\ &= f(t^{-1}\cdot) \otimes \alpha_t(a),\end{aligned}$$

$\forall f \in C_0^\infty(G), a \in A$. Thus Ψ is an isomorphism from $(A \times_{id} \hat{G}, G, \beta)$ onto $(C_0^\infty(G, A), G, \gamma)$, where $\gamma = \lambda \otimes \alpha$. By Proposition 16.5.2, the Fourier transformation

$$\hat{f} \in K(\hat{G} \times G, A) \rightarrow f(s,t) = \int_{\hat{G}} \overline{\langle s, p\rangle} \hat{f}(p,t)dp$$

can be uniquely extended to a $*$ isomorphism from $(A \times_{id} \hat{G}) \times_\beta G$ onto $C_0^\infty(G, A) \times_\gamma G$.

7) By 5) and 6), the map

$$\hat{\Phi} : z(s,t) = af(s)g(t) \longrightarrow \overline{\langle s, p\rangle} af(s)\hat{g}(p),$$

$\forall a \in A, f \in K(G), \hat{g} \in L^1(\hat{G})$ and $g(t) = \int_{\hat{G}} \langle t, p\rangle \hat{g}(p)dp \in C_0^\infty(G)$, can be uniquely extended to a $*$ isomorphism from $C_0^\infty(G, A) \times_\gamma G$ onto $(A \times_\alpha G) \times_{\hat{\alpha}} \hat{G}$.

Further, for any $z(s,t) = af(s)g(t)$, and $r \in G$, since

$$\begin{aligned} & \hat{\Phi}(\rho(r)z)(s,p) = \hat{\Phi}(af(\cdot)g_r(\cdot))(s,p) \\ = \ & \overline{\langle s, p\rangle \langle r, p\rangle} af(s)\hat{g}(p) = \overline{\langle r, p\rangle} \hat{\Phi}(z)(s,p) \\ = \ & \hat{\hat{\alpha}}_r \circ \hat{\Phi}(z)(s,p),\end{aligned}$$

$\forall a \in A, f \in K(G), \hat{g} \in L^1(\hat{G})$ and $g(\cdot) = \int_{\hat{G}} \langle \cdot, p\rangle \hat{g}(p)dp$, it follows that $\hat{\Phi} \circ \rho(r) = \hat{\hat{\alpha}}_r \circ \hat{\Phi}, \forall r \in G$. Therefore, $(C_0^\infty(G, A) \times_\gamma G, G, \rho) \cong ((A \times_\alpha G) \times_{\hat{\alpha}} \hat{G}, G, \hat{\hat{\alpha}})$.

Finally, from Theorem 16.5.6, we obtain that

$$((A \times_\alpha G) \times_{\hat{\alpha}} \hat{G}, G, \hat{\hat{\alpha}}) \cong (\text{min-}(A \otimes C(L^2(G))), G, \alpha \otimes ad\rho).$$

Q.E.D.

Notes. In the study of the structure of Von Neumann algebras of type (III), M. Takesaki obtained a duality theorem for crossed products of Von Neumann algebras (Theorem 16.2.10). At the same time, he also conjectured about its C^*-algebra version. Then H. Takai showed that conjecture is affirmative.

References. [127], [166], [176].

16.6. Some examples of crossed products

1) The relation between C^*-and W^*-crossed products.

Let (M, G, α) be a W^*-system. By Definition 16.1.19, the W^*-crossed product is

$$M \times_\alpha G = \{a \in M \overline{\otimes} B(L^2(G)) | \theta_t(a) = a, \forall t \in G\},$$

where $\theta_t = \alpha_t \otimes ad\rho_t, \forall t \in G$. If pick a faithful W^*-representation $\{\pi, H\}$ of M, and let

$$(\overline{\pi}(x)\xi)(s) = \alpha_{s^{-1}}(x)\xi(s), (\lambda(t)\xi)(s) = \xi(t^{-1}s)$$

$\forall \xi(\cdot) \in L^2(G, H), s, t \in G$. Then $M \times_\alpha G$ is $*$ isomorphic to the VN algebra $\{\overline{\pi}(M), \lambda(G)\}''$ on $H \otimes L^2(G) = L^2(G, H)$. Hence, $M \times_\alpha G$ is the W^*-subalgebra of $M \overline{\otimes} B(L^2(G))$ generated by $\{(\pi \otimes id)^{-1}\overline{\pi}(x), 1 \otimes \lambda_s | x \in M, s \in G\}$.

Now let A be a C^*-subalgebra of M, A be $\sigma(M, M_*)$-dense in M, $\alpha_t(A) \subset A, \forall t \in G$, and (A, G, α) be a C^*-system. Further, we assume that G is amenable. If $\{\pi, H\}$ is a faithful W^*-representation of M, then by Theorem 16.4.15 $\{\text{Ind}\pi = \overline{\pi} \times \lambda, L^2(G, H)\}$ is a faithful $*$ representation of $A \times_\alpha G$. From the norm on $M \overline{\otimes} B(L^2(G))$, the largest C^*-norm on $L^1(G, A, \alpha)$ is as follows:

$$\|(\pi \otimes id)^{-1} \int_G \overline{\pi}(f(s))\lambda(s)ds\|, \quad \forall f \in L^1(G, A, \alpha).$$

Moreover, $(\overline{\pi} \otimes \lambda)(A \times_\alpha G)'' = \{\overline{\pi}(M), \lambda(G)\}''$. Hence, $A \times_\alpha G$ is the norm closure of $\{(\pi \otimes id)^{-1} \int_G \overline{\pi}(f(s))\lambda(s)ds | f \in L^1(G, A, \alpha)\}$ in $M \overline{\otimes} B(L^2(G))$, and the σ-closure of $A \times_\alpha G$ in $M \overline{\otimes} B(L^2(G))$ is $M \times_\alpha G$.

2) Let G be an amenable group, $M = L^\infty(G), A = C_0^\infty(G)$, and $\alpha_t(f)(s) = f(t^{-1}s), \forall f \in L^\infty(G), t, s \in G$. Then A is σ-dense in $M, (M, G, \alpha)$ is a W^*-system, and (A, G, α) is a C^*-system . By the discussion 1), $A \times_\alpha G$ is σ-dense in $M \times_\alpha G$.

Put $H = L^2(G)$, and

$$\pi(f)g = fg, \quad \forall f \in L^\infty(G), g \in L^2(G).$$

Then $M \times_\alpha G$ is $*$ isomorphic to the VN algebra $\{\overline{\pi}(M), \lambda(G)\}''$ on $L^2(G \times G)$, where

$$\begin{cases} (\overline{\pi}(f)\xi)(s,t) = f(st)\xi(s,t), \\ (\lambda(r)\xi)(s,t) = \xi(r^{-1}s, t), \end{cases}$$

$\forall \xi \in L^2(G \times G), s, t, r \in G$. Define a unitary operator W on $L^2(G \times G)$ as follows:

$$\begin{cases} (W\xi)(s,t) = \triangle(t)^{1/2}\xi(st,t), \\ (W^*\xi)(s,t) = \triangle(t)^{-1/2}\xi(st^{-1},t), \end{cases}$$

$\forall \xi \in L^2(G \times G)$. Then it is easy to check that

$$W^*\overline{\pi}(f)W = \pi(f) \otimes 1, W^*\lambda(r)W = \lambda_r \otimes 1$$

on $L^2(G) \otimes L^2(G) = L^2(G \times G), \forall f \in L^\infty(G), r \in G$. Since $\pi(M)$ is maximal commutative in $B(L^2(G))$, it follows that $M \times_\alpha G$ is $*$ isomorphic to $B(L^2(G))$.

Moreover, if $z(r,s) = f(r^{-1}s)g(s)(\in L^1(G, A, \alpha))$, where $f, g \in K(G)$, then we have

$$W^*(\overline{\pi} \times \lambda)(z)W = v_{f'g} \otimes 1,$$

where $v_{f'g}$ is one rank operator on $L^2(G)$, i.e., $v_{f'g}\eta = \langle \eta, \overline{f'} \rangle g, \forall \eta \in L^2(G)$ (see the proof of Theorem 16.5.6). Therefore, $A \times_\alpha G$ is $*$ isomorphic to $C(L^2(G))$.

3) *Every (UHF) algebra can be represented as a C^*-crossed product.*

(i) Let $m(\cdot)$ be a function from $\mathbb{N}$ to $\mathbb{N}$. For each $n \in N$, put $G_n = \mathbb{Z}_{m(n)}$. Clearly, G_n is a finite abelian group, and $\widehat{G}_n = G_n$, i.e.,

$$\langle j, k \rangle = e^{2\pi i jk/m(n)}, \quad \forall j \in \widehat{G}_n, k \in G_n,$$

$\forall n$. Now consider the discrete abelian group

$$G = \oplus_{n=1}^\infty G_n = \left\{ s = (s_1, \cdots, s_n, 0, \cdots, 0, \cdots) \middle| \begin{array}{l} n = 1, 2, \cdots, \\ s_i \in G_i, 1 \le i \le n. \end{array} \right\}$$

Then

$$\widehat{G} = \prod_{n \ge 1}^\infty G_n = \{\sigma = (\sigma_k)_{k \ge 1} | \sigma_k \in G_k, \forall k\}$$

is a compact abelian group (product topology), and

$$\langle \sigma, s \rangle = \prod_{k \ge 1} \langle \sigma_k, s_k \rangle, \quad \forall \sigma \in \widehat{G}, s \in G.$$

(ii) For each n, denote by $C_n = C(G_1 \times \cdots \times G_n)$ the set of all complex functions on $G_1 \times \cdots \times G_n$. Clearly, $C_n = C(G_1) \otimes \cdots \otimes C(G_n)$ (C^*-tensor product). If identify C_n with $C_n \otimes 1_{n+1}$, then we have $C_n \hookrightarrow C_{n+1}$, i.e.,

$$f(s_1, \cdots, s_n, s_{n+1}) = f(s_1, \cdots, s_n),$$

$\forall f \in C_n, s_i \in G_i, 1 \le i \le n+1$. Further, C_n can be embedded into $C(\widehat{G})$, i.e.,

$$f(s) = f(s_1, \cdots, s_n), \forall f \in C_n, s = (s_k) \in G.$$

We claim that $C(\widehat{G}) = \overline{\sqcup_n C_n} = \alpha_0\text{-} \otimes_{n=1}^{\infty} C(G_n)$ (infinite C^*-tensor product). In fact, for any $f \in C(\widehat{G})$ and $\varepsilon > 0$, since f is uniformly continuous on $\widehat{G}$, there is a neighborhood $W = W_1 \times \cdots \times W_n \times G_{n+1} \times G_{n+2} \times \cdots$ of 0 (here $0_i \in W_i \subset G_i, 1 \le i \le n$) such that

$$|f(s) - f(t)| < \varepsilon, \forall s, t \in \widehat{G} \text{ and } (s - t) \in W.$$

If define $f_n(s_1, \cdots, s_n) = f(s_1, \cdots, s_n, 0, \cdots), \forall n, s_i \in G_i, 1 \le i \le n$, then $f_n \in C_n$ and $\|f_n - f\|_\infty < \varepsilon$.

(iii) For each n, let $H_n = l^2(G_n)$. Then H_n is $m(n)$ dimensional Hiblert space. Put

$$\xi_0^{(n)}(s_n) = \delta_{s_n, o_n}, \forall s_n \in G_n,$$

where 0_n is the zero element of $G_n, \forall n$, and $\xi_0 = \otimes_n \xi_0^{(n)}$. Then we have $l^2(G) = \otimes_n^{\xi_0} H_n$ (see Section 3.8). Now the (UHF) algebra

$$A_\infty = \alpha_0\text{-} \otimes_n B(H_n) = \alpha_0\text{-} \otimes_n M_{m(n)}$$

is a C^*-algebra on $H = l^2(G)$. Define

$$(u_\sigma \xi)(s) = \langle \sigma, s \rangle \xi(s),$$

$\forall \xi \in l^2(G) = H, \sigma \in \widehat{G}, s \in G$. Clearly, $\sigma \to u_\sigma$ is a strongly continuous unitary representation of G on H, and $u_\sigma = \otimes_n u_{\sigma_n}, \forall \sigma = (\sigma_n) \in \widehat{G}$, where $(u_{\sigma_n} g)(s_n) = \langle \sigma_n, s_n \rangle g(s_n), \forall g \in H_n = l^2(G_n), s_n, \sigma_n \in G_n, \forall n$. Hence , $(A_\infty, \widehat{G}, \widehat{\alpha})$ is a C^*-system, where

$$\widehat{\alpha}_\sigma(t) = u_\sigma t u_\sigma^*, \quad \forall t \in A_\infty, \sigma \in \widehat{G}.$$

For each $f \in C(\widehat{G})$, define

$$(m_f \xi)(s) = f(s)\xi(s), \quad \forall \xi \in H = l^2(G), s \in G.$$

(notice that G can be regraded as a dense subset of $\widehat{G}$), and let $A = \{m_f | f \in C(\widehat{G})\}$. We claim that A is the fixed point algebra of the system $(A_\infty, \widehat{G}, \widehat{\alpha})$.

Clearly, $\widehat{\alpha}_\sigma(m_f) = u_\sigma m_f u_\sigma^* = m_f, \forall \sigma \in \widehat{G}, f \in C(\widehat{G})$. Conversely, suppose that $x \in A_\infty$ and $\widehat{\alpha}_\sigma(x) = x, \forall \sigma \in \widehat{G}$. For $\varepsilon > 0$, pick $y \in \otimes_{k=1}^n B(H_k) \otimes \otimes_{j>n} 1_j$ such that $\|y - x\| < \varepsilon$. Define

$$z = (m(1) \cdots m(n))^{-1} \sum_{\sigma \in G_1 \times \cdots \times G_n} \widehat{\alpha}_\sigma(y).$$

Then $\widehat{\alpha}_\sigma(z) = z, \forall \sigma \in \widehat{G}, z \in \otimes_{k=1}^n B(H_k)$, and $\|x - z\| < \varepsilon$. Since $z u_\sigma = u_\sigma z, \forall \sigma \in G_1 \times \cdots \times G_n$, it follows that $z = m_f$ for some $f \in C_n \subset C(\widehat{G})$. $\varepsilon (> 0)$ is arbitrary. Hence, $x \in A$.

(iv) Let $s \to \lambda_s$ be the left regular represectation of G on $l^2(G)$, and define

$$\alpha_s(x) = \lambda_s x \lambda_s^*, \forall x \in B(l^2(G)).$$

Then (A, G, α) is a C^*-system, and

$$\alpha_s(m_f) = m_{{}_s f},$$

where $({}_s f)(t) = f(t-s), \forall s \in G, t \in \hat{G}, f \in C(\hat{G})$.

We say that the (UHF) algebra A_∞ is $*$ isomorphic to the crossed product $A \times_\alpha G$.

Define

$$\begin{aligned}(\pi(m_f)\xi)(s,t) &= (\alpha_{s^{-1}}(m_f)\xi(s))(t)\\ &= f(s+t)\xi(s,t),\\ (\lambda(r)\xi)(s,t) &= \xi(s-r,t),\end{aligned}$$

$\forall f \in C(\hat{G}), \xi \in l^2(G,H) = l^2(G \times G), r,s,t \in G$. Since G is abelian, $\{\pi \times \lambda, l^2(G,H)\}$ is a faithful $*$ representation of $A \times_\alpha G$. In particular

$$\begin{aligned}((\pi \times \lambda)(z)\xi)(s,t) &= (\sum_{r\in G} \alpha_{s^{-1}}(z(r,\cdot))(\lambda(r)\xi)(s))(t)\\ &= \sum_{r\in G} z(r,s+t)\xi(s-r,t),\end{aligned}$$

$\forall z \in K(G, C(\hat{G})), \xi \in l^2(G \times G)$. Define a unitary operator W on $l^2(G \times G)$ as follows:

$$\begin{cases}(W\xi)(s,t) = \xi(s+t,t),\\ (W^*\xi)(s,t) = \xi(s-t,t),\end{cases}$$

$\forall \xi \in l^2(G \times G)$. Then we have

$$W^*(\pi \times \lambda)(z)W = (\sum_{r\in G} m_{g_r}\lambda_r) \otimes 1,$$

where $g_r(\cdot) = z(r,\cdot) \in C(\hat{G}), \forall z \in K(G, C(\hat{G}))$. Hence, $A \times_\alpha G$ is $*$ isomorphic to the C^*-algebra on $l^2(G)$ generated by $\{m_f, \lambda_r | f \in C(\hat{G}), r \in G\}$. On the other hand, $\otimes_{k=1}^n B(H_k)$ is generated by $\{\lambda_r, m_f | f \in C_n, r \in G_1 \times \cdots \times G_n\}$ obviously, $\forall n$. Since $C(\hat{G}) = \overline{\cup_n C_n}$ and $G = \oplus_n G_n$, A_∞ and $A \times_\alpha G$ are $*$ isomorphic.

4) *Semi-direct product.*

(i) Let H, K be two groups, and ρ be a homomorphism from K to $\mathrm{Aut}(H)$, where $\mathrm{Aut}(H)$ is the automorphism group of H. Define

$$(h,k)(h',k') = (h\rho(k)(h'), kk'),$$

$\forall h, h' \in H, k, k' \in K$. By this multiplication, $H \times K$ becomes a group. This group is called the *semi-direct product* of (H, K, ρ) , and denoted by $G = H \times_\rho K$.

Now if H, K are two locally compact groups, and the map $(k, h) \to \rho(k)h$ is continuous from $K \times H$ to H, then we have the following facts:

(1) $G = H \times_\rho K$ is also locally compact with respect to product topology;

(2) If dh is the left invariant Haar measure on H, then for any $k \in K$ there is a positive constant $\delta(k)$ such that $d\rho(k)(h) = \delta(k)dh$. Moreover, $\delta(\cdot)$ is continuous on $K, \delta(e_K) = 1$ and $\delta(k_1k_2) = \delta(k_1)\delta(k_2), \forall k_1, k_2 \in K$;

(3) If dk is the left invariant Haar measure on K, then $d(h,k) = \delta(k)^{-1}dhdk$ is the left invariant Haar measure on $G = H \times_\rho K$;

(4) If $\triangle_H, \triangle_K, \triangle_G$ are the modular functions on H, K, G respectively, then we have

$$\triangle_G(h,k) = \delta(k)^{-1} \triangle_H (h) \triangle_K (k), \quad \forall h \in H, k \in K.$$

Moreover , $\triangle_H$ is ρ-invariant, i.e.,

$$\triangle_H(\rho(k)(h)) = \triangle_H(h), \quad \forall h \in H, k \in K.$$

The proof of these facts can be found in [70].

(ii) Let $G = H \times_\rho K$. If define

$$(\alpha_k f)(h) = \delta(k)^{-1} f(\rho(k)^{-1}(h)),$$

$\forall k \in K, h \in H, f \in L^1(H)$, then we can obtain a C^*-system $(C^*(H), K, \alpha)$.

Moreover, let

$$\Phi(f)(k, \cdot) = \delta(k)^{-1} f(\cdot, k), \quad \forall f \in L^1(G), k \in K, \cdot \in H.$$

Then Φ is a $*$ isomorphism from $L^1(G)$ onto $L^1(K, L^1(H), \alpha)$.

The proof is easy.

(iii) Let $G = H \times_\rho K$. Then we have

$$C^*(G) \cong C^*(H) \times_\alpha K.$$

In fact, by $L^1(G) \cong L^1(K, L^1(H), \alpha) \subset L^1(K, C^*(H), \alpha) \subset C^*(H) \times_\alpha K$, it suffices to show that every $*$ representation of $L^1(K, L^1(H), \alpha)$ can be extended to a $*$ representaton of $L^1(K, C^*(H), \alpha)$.

Put $B = L^1(H), A = C^*(H)$. Let $\{g_l\}$ be a bounded approximate identity for $L^1(G)$. Then $\{f_l = \Phi(g_l)\}$ is a bounded approximate identity for $L^1(K, B, \alpha)$. Since $L^1(K, B, \alpha)$ is dense in $L^1(K, A, \alpha)$, $\{f_l\}$ is also a bounded approximate identity for $L^1(K, A, \alpha)$.

Now let $\{\rho, \overline{H}\}$ be a nondegenerate representation of $L^1(K, B, \alpha)$, and define

$$\pi(b) = s\text{-}\lim_l \rho(bf_l), u_k = s\text{-}\lim \rho(\alpha_k(f_l(k^{-1}\cdot))),$$

(notice that $f_l(\cdot) \in B, \forall \cdot \in K), \forall b \in B, k \in K$. Then $\{\pi, u, \overline{H}\}$ is a covariant representastion of (B, K, α), and $\rho = \pi \times u$. Clearly, $\{\pi, \overline{H}\}$ can be uniquely extended to a $*$ representation of A, and $\{\pi, u, \overline{H}\}$ is also a covariant representation of (A, K, α). Hence $\{\rho = \pi \times u, \overline{H}\}$ can be extended to a $*$ representation of $L^1(K, A, \alpha)$.

Example. Let G be a locally compact group, and define $\rho : G \to \mathrm{Aut}(G)$ as follows:

$$\rho(s)(t) = sts^{-1}, \quad \forall s, t \in G.$$

Then

$$(\alpha_s f)(t) = \Delta(s) f(s^{-1}ts), \quad \forall f \in L^1(G), s, t \in G,$$

and $(C^*(G), G, \alpha)$ is a C^*-system. By above discussion, we have

$$C^*(G) \times_\alpha G \cong C^*(G \times_\rho G).$$

5) *The periodic action and mapping torus.*

Let $(A, \mathbb{Z}, \alpha)$ be a C^*-system, and $\alpha^n = id$, where n is a fixed positive integer, and α is a $*$ automorphism of A.

(i) Let $\mathcal{A}$ be the closure of

$$l^1_n(\mathbb{Z}, A, \alpha) = \{f \in l^1(\mathbb{Z}, A, \alpha) | f(k) = 0, \forall k \not\equiv 0 (mod\ n)\}$$

in $A \times_\alpha \mathbb{Z}$. Then $\mathcal{A}$ is a C^*-subalgebra of $A \times_\alpha \mathbb{Z}$, and

$$f \in \mathcal{A} \cap l^1(\mathbb{Z}, A, \alpha) \longrightarrow \hat{f} \in l^1(\mathbb{Z}, A, id)$$

can be uniquely extended to a $*$ isomorphism from $\mathcal{A}$ onto $A \times_{id} \mathbb{Z} \cong C(T, A)$, where $\hat{f}(k) = f(kn), \forall k \in \mathbb{Z}$, and T is the group of unit circle, i.e., $T = \{z \in \mathbb{C} \mid |z| = 1\}$.

Proof. Since $(fg)(k) = \sum_{m \in \mathbb{Z}} f(m)\alpha^m(g(k-m)), f^*(k) = \alpha^k(f(-k)^*), \forall f, g \in l^1(\mathbb{Z}, A, \alpha)$ and $k \in \mathbb{Z}$, $\mathcal{A}$ is a C^*-subalgebra of $A \times_\alpha \mathbb{Z}$, and $g \to \check{g}$ is a $*$ homomorphism from $l^1(\mathbb{Z}, A, id)$ to $l^1_n(\mathbb{Z}, A, \alpha)$, where

$$\check{g}(k) = \begin{cases} 0, & \text{if } k \not\equiv 0 (mod\ n), \\ g(m), & \text{if } k = mn. \end{cases}$$

Then $g \to \check{g}$ can be extended to a $*$ homomorphism from $A \times_{id} \mathbb{Z}$ to $\mathcal{A}$, and we have

$$\| \check{g} \|_{A \times_\alpha \mathbb{Z}} \leq \|g\|_{A \times_{id} \mathbb{Z}}, \forall g \in l^1(\mathbb{Z}, A, id).$$

Now suppose that $A \subset B(H)$ for some Hilbert space H. By Theorem 16.4.15, there is a faithful $*$ representation $\{\rho, l^2(\mathbb{Z}, H)\}$ of $A \times_\alpha \mathbb{Z}$ (noticing that $\mathbb{Z}$ is amenable):

$$(\rho(f)\xi)(k) = \sum_{j\in\mathbb{Z}} \alpha^{-k}(f(j))\xi(k-j),$$

$\forall f \in l^1(\mathbb{Z}, A, \alpha), \xi \in l^2(\mathbb{Z}, H)$, and $k \in \mathbb{Z}$. Let

$$l^2(\mathbb{Z}, H) = l^2_n(\mathbb{Z}, H) \oplus l^2_\times(\mathbb{Z}, H)$$

be an orthogonal decomposition of $l^2(\mathbb{Z}, H)$, where

$$l^2_n(\mathbb{Z}, H) = \{\xi \in l^2(\mathbb{Z}, H) | \xi(k) = 0, \forall k \not\equiv 0 (mod\ n)\}$$

and

$$l^2_\times(\mathbb{Z}, H) = \{\xi \in l^2(\mathbb{Z}, H) | \xi(kn) = 0, \forall k \in \mathbb{Z}\}.$$

Clearly, if $g \in l^1(\mathbb{Z}, A, id)$, then $l^2_n(\mathbb{Z}, H)$ and $l^2_\times(\mathbb{Z}, H)$ are invariant for $\rho(\check{g})$. Hence , we get

$$\| \check{g} \|_{A\times_\alpha Z} = \|\rho(\check{g})\| \geq \|\rho(\check{g})|l^2_n(\mathbb{Z}, H)\|.$$

Also by Theorem 16.4.15, there is a faithful $*$ representation $\{\sigma, l^2(\mathbb{Z}, H)\}$ of $A \times_{id} \mathbb{Z}$:

$$(\sigma(g)\xi)(k) = \sum_{j\in\mathbb{Z}} g(j)\xi(k-j),$$

$\forall g \in l^1(\mathbb{Z}, A, id), \xi \in l^2(\mathbb{Z}, H)$, and $k \in \mathbb{Z}$. Define a unitary operator U from $l^2(\mathbb{Z}, H)$ onto $l^2_n(\mathbb{Z}, H)$ as follows:

$$(U\xi)(k) = \begin{cases} \xi(j), & \text{if } k = nj \text{ for some } j\ , \\ 0, & \text{otherwise,} \end{cases}$$

$\forall \xi \in l^2(\mathbb{Z}, H)$. Clearly, we have

$$U\sigma(g)U^* = \rho(\check{g})|l^2_n(\mathbb{Z}, H), \quad \forall g \in l^1(\mathbb{Z}, A, id).$$

Thus,

$$\|g\|_{A\times_{id}\mathbb{Z}} = \|\sigma(g)\| = \|\rho(\check{g})|l^2_n(\mathbb{Z}, H)\| \leq \| \check{g} \|_{A\times_\alpha\mathbb{Z}}.$$

Further, we obtain that

$$\|g\|_{A\times_{id}\mathbb{Z}} = \| \check{g} \|_{A\times_\alpha\mathbb{Z}}, \quad \forall g \in l^1(\mathbb{Z}, A, id),$$

and $\mathcal{A}$ is $*$ isomorphic to $A \times_{id} \mathbb{Z}$. Q.E.D.

(ii) If define $\alpha^k(a + \lambda) = \alpha^k(a) + \lambda, \forall k \in \mathbb{Z}, a \in A, \lambda \in \mathbb{C}$, then we have C^*-system $(A\dot{+}\mathbb{C}, \mathbb{Z}, \alpha)$. Since $l^1(\mathbb{Z}, A, \alpha)$ is a $*$ two-sided ideal of $l^1(\mathbb{Z}, A\dot{+}\mathbb{C}, \alpha)$, and $\mathbb{Z}$ is amenable, it follows from Proposition 16.4.16 that $A \times_\alpha \mathbb{Z}$ is a closed two-sided ideal of $(A\dot{+}\mathbb{C}) \times_\alpha \mathbb{Z}$. Put $\lambda \in l^1(\mathbb{Z}, A\dot{+}\mathbb{C}, \alpha)$ with

$$\lambda(k) = 0 \text{ if } k \neq 0, \text{ and } \lambda(1) = 1.$$

Then λ is an invertible element of $(A\dot{+}\mathbb{C}) \times_\alpha \mathbb{Z}$, and

$$\lambda^j(k) = \delta_{j,k}, \quad (f\lambda^j)(k) = f(k-j),$$

$\forall j, k \in \mathbb{Z}, f \in l^1(\mathbb{Z}, A, \alpha)$. Hence $\mathcal{A}\lambda^j$ is the closure of $\{f \in l^1(\mathbb{Z}, A, \alpha) | f(k) = 0, \forall k \not\equiv j (mod\ n)\}$ in $A \times_\alpha \mathbb{Z}, \forall j \in \mathbb{Z}$. We claim that

$$A \times_\alpha \mathbb{Z} = \mathcal{A}\dot{+}\mathcal{A}\lambda\dot{+}\cdots\dot{+}\mathcal{A}\lambda^{n-1}.$$

In fact, for any $f \in l^1(\mathbb{Z}, A, \alpha)$ and $j \in \{0, \cdots, n-1\}$, let

$$f_j(k) = \begin{cases} f(k), & \text{if } k \equiv j(mod\ n), \\ 0, & \text{otherwise.} \end{cases}$$

Then $f_j \in \mathcal{A}\lambda^j$, and $f = f_0 + \cdots + f_{n-1}$. Hence

$$l^1(\mathbb{Z}, A, \alpha) \subset \mathcal{A} + \mathcal{A}\lambda + \cdots + \mathcal{A}\lambda^{n-1}.$$

Notice the following fact. If $f \in l^1_n(\mathbb{Z}, A, \alpha), g \in l^1(\mathbb{Z}, A, \alpha)$ and $g(kn) = 0, \forall k \in \mathbb{Z}$, then

$$\|f+g\|_{A\times_\alpha\mathbb{Z}} \geq \|f\|_{A\times_\alpha\mathbb{Z}}.$$

Indeed, along the notations : $\rho, l^2(\mathbb{Z}, H) = l^2_n(\mathbb{Z}, H) \oplus l^2_\times(\mathbb{Z}, H)$ in (i) , we have $\rho(g) l^2_n(\mathbb{Z}, H) \subset l^2_\times(\mathbb{Z}, H)$. Hence,

$$\|\rho(f+g)\xi\|^2 = \|\rho(f)\xi\|^2 + \|\rho(g)\xi\|^2 \geq \|\rho(f)\xi\|^2,$$

$\forall \xi \in l^2_n(\mathbb{Z}, H)$. By the proof of (i) we obtain

$$\|f+g\|_{A\times_\alpha\mathbb{Z}} = \|\rho(f+g)\| \geq \|\rho(f)|l^2_n(\mathbb{Z}, H)\| = \|f\|_{A\times_\alpha\mathbb{Z}}.$$

Now let $a_i \in \mathcal{A}, 0 \leq i \leq n-1$, be such that

$$a_0 + a_1\lambda + \cdots + a_{n-1}\lambda^{n-1} = 0.$$

For each $j \in \{0, \cdots, n-1\}$, pick a sequence $\{f_k^{(j)}\}$ of $l^1_n(\mathbb{Z}, A, \alpha)$ such that $f_k^{(j)} \to a_j$ in $A \times_\alpha \mathbb{Z}$. Then

$$\sum_{j=0}^{n-1} f_k^{(j)}\lambda^j \to 0 \text{ as } k \to \infty.$$

By the fact of the preceding paragraph, we have $f_k^{(0)} \to 0$ in $A \times_\alpha \mathbb{Z}$. Hence, $a_0 = 0$. Similarly, from $(a_1 + a_2\lambda + \cdots + a_{n-1}\lambda^{n-2})\lambda = 0$ we have $a_1 = 0$. Generally, $a_j = 0, \forall 0 \leq j \leq n-1$. So $\mathcal{A} + \mathcal{A}\lambda + \cdots + \mathcal{A}\lambda^{n-1} = \mathcal{A}\dot{+}\mathcal{A}\lambda\dot{+}\cdots\dot{+}\mathcal{A}\lambda^{n-1}$.

Finally, for any $a \in A \times_\alpha \mathbb{Z}$, pick sequences $\{f_k^{(j)}|k\}(0 \le j \le n-1)$ of $l_n^1(\mathbb{Z}, A, \alpha)$ such that

$$\sum_{j=0}^{n-1} f_k^{(j)} \lambda^j \longrightarrow a \quad \text{as} \quad k \longrightarrow \infty.$$

Then $\sum_{j=0}^{n-1}(f_k^{(j)} - f_l^{(j)})\lambda^j \longrightarrow 0$ as $k, l \longrightarrow \infty$. From the preceding paragraph, it must be

$$(f_k^{(j)} - f_l^{(j)}) \longrightarrow 0 \quad \text{as} \quad k, l \to \infty, 0 \le j \le n-1.$$

So for each $j \in \{0, \cdots, n-1\}$ there is $a_j \in \mathcal{A}$ such that $f_k^{(j)} \to a_j$. Therefore, $a = a_0 + a_1\lambda + \cdots + a_{n-1}\lambda^{n-1}$, and $A \times_\alpha \mathbb{Z} = \mathcal{A} \dot{+} \mathcal{A}\lambda \dot{+} \cdots \dot{+} \mathcal{A}\lambda^{n-1}$.

(iii) By (i) and Proposition 16.3.2, the Fourier transformation:

$$f \in l_n^1(\mathbb{Z}, A, \alpha) \to F(z) = \sum_{k \in Z} f(nk) z^k \in C(T, A)$$

can be uniquely extended to a $*$ isomorphism from $\mathcal{A}$ onto $C(T, A)$. Denote this $*$ isomorphism by Φ.

Now on $C(T, A) \times \cdots \times C(T, A)$($n$ times) define multiplication, $*$ operation and norm as follows:

$$(F_j)_{0 \le j \le n-1} \cdot (G_j)_{0 \le j \le n-1}$$

$$= (\sum_{k=0}^{j} F_k \cdot \alpha^k G_{j-k} + \sum_{k=j+1}^{n-1} F_k \cdot \alpha^k G_{n+j-k} \cdot z)_{0 \le j \le n-1},$$

$$(F_j)^*_{0 \le j \le n-1} = (F_0^*, \alpha F_{n-1}^* \cdot \bar{z}, \alpha^2 F_{n-1}^* \cdot \bar{z}, \cdots, \alpha^{n-1} F_1^* \cdot \bar{z}),$$

$$\|(F_j)_{0 \le j \le n-1}\| = \|\sum_{j=0}^{n-1} \Phi^{-1}(F_j)\lambda^j\|_{A \times_\alpha \mathbb{Z}},$$

where $(F \cdot \alpha^k G)(z) = F(z)\alpha^k(G(z)), (F \cdot \alpha^k G \cdot z)(z) = zF(z)\alpha^k(G(z)), (\alpha^k F^* \cdot \bar{z})(z) = \bar{z}\alpha^k(F(z)^*), \forall F, G \in C(T, A), z \in T, k \in \mathbb{Z}$. Then $C(T, A) \times \cdots \times C(T, A)$ (n times) is a C^*-algebra, and through the following map Ψ :

$$\sum_{j=0}^{n-1} a_j \lambda^j \xrightarrow{\Psi} (\Phi(a_j))_{0 \le j \le n-1}$$

$(\forall a_j \in \mathcal{A}, 0 \le j \le n-1), A \times_\alpha \mathbb{Z}$ and $C(T, A) \times \cdots \times C(T, A)$ (n times) are $*$ isomorphic.

In fact, by (i), (ii) Ψ is a linear isomorphism from $A \times_\alpha \mathbb{Z}$ onto $C(T, A) \times \cdots \times C(T, A)$($n$ times) obviously. Further , from the definition of norm Ψ is an isometry. For any $f_0, \cdots, f_{n-1}, g_0, \cdots, g_{n-1} \in l^1_n(\mathbb{Z}, A, \alpha)$ we have

$$\begin{aligned}&(\sum_{j=0}^{n-1} f_j\lambda^j)\cdot(\sum_{k=0}^{n-1} g_k\lambda^k)\\ =& \sum_{j=0}^{n-1}\left\{\sum_{k=0}^{j} f_k\cdot\alpha^k g_{j-k} + \sum_{k=j+1}^{n-1} f_k\cdot\alpha^k g_{n+j-k}\cdot\lambda^n\right\}\lambda^j\end{aligned}$$

and

$$(\sum_{j=0}^{n-1} f_j\lambda^j)^* = f_0^* + (\alpha f_{n-1}^*\cdot\lambda^{-n})\lambda + \cdot + (\alpha^{n-1}f_1^*\cdot\lambda^{-n})\lambda^{n-1},$$

where $(\alpha^k f)(j) = \alpha^k(f(j)), \forall f \in l^1(\mathbb{Z}, A, \alpha), j, k \in \mathbb{Z}$. If write $F_j = \Phi(f_j), G_k = \Phi(g_k)$, then we obtain

$$\Psi\left(\sum_{j=0}^{n-1} f_j\lambda^j\cdot\sum_{k=0}^{n-1} g_k\lambda^k\right) = \left(\sum_{k=0}^{j} F_k\cdot\alpha^k G_{j-k} + \sum_{k=j+1}^{n-1} F_k\cdot\alpha^k G_{n+j-k}\cdot z\right)_{0\le j\le n-1}$$

and

$$\Psi((\sum_{j=0}^{n-1} f_j\lambda^j)^*) = (F_0^*, \alpha F_{n-1}^*\cdot\bar{z}, \cdots, \alpha^{n-1}F_1^*\cdot\bar{z}).$$

Therefore, Ψ is a $*$ isomorphism from $A\times_\alpha \mathbb{Z}$ onto $C(T, A)\times\cdots\times C(T, A)$ (n times) .

(iv) Since $\alpha^n = id$, we can also consider the C^*-system $(A, \mathbb{Z}_n, \alpha)$.

$l^1(\mathbb{Z}_n, A, \alpha)$ is a Banach $*$ algebra. For any $a = (a_s)_{0\le s\le n-1}, b = (b_s)_{0\le s\le n-1}$, we have the following formulas:

$$a^* = (\alpha^s(a_{-s}^*))_{0\le s\le n-1},$$

$$ab = (\sum_{t=0}^{n-1} a_t\alpha^t(b_{s-t}))_{0\le s\le n-1},$$

$$\|a\|_1 = \|a_0\| + \cdots + \|a_{n-1}\|,$$

where the foot index of any integer is understood in the sense of $(mod\ n)$. Let $\|\cdot\|$ be the largest C^*-norm on $l^1(\mathbb{Z}_n, A, \alpha)$. Then the crossed product $A\times_\alpha \mathbb{Z}_n$ is the completion of $(l^1(\mathbb{Z}_n, A, \alpha), \|\cdot\|)$.

Assume that $A\subset B(H)$ for some Hilbert space H. Then we have a faithful $*$ representation $\{\rho, H_n\}$ of $A\times_\alpha \mathbb{Z}_n$ as follows:

$$\rho(a)\xi = (\sum_{t=0}^{n-1}\alpha^{-s}(a_t)\xi_{s-t})_{0\le s\le n-1},$$

$\forall a = (a_s)_{0\le s\le n-1} \in l^1(\mathbb{Z}_n, A, \alpha)$, $\xi = (\xi_s)_{0\le s\le n-1} \in H_n$, where $H_n = H \oplus \cdots \oplus H(n \text{ times })$. Consequently,

$$\begin{aligned}\|a\|^2 &= \|\rho(a)\|^2 \\ &= \sup_{\xi\in H_n, \|\xi\|\le 1} \sum_{s=0}^{n-1} \|\sum_{t=0}^{n-1} \alpha^{-s}(a_t)\xi_{s-t}\|^2 \\ &\ge \sup_{\xi_0\in H, \|\xi_0\|\le 1} \sum_{s=0}^{n-1} \|\alpha^{-s}(a_s)\xi_0\|^2 \\ &\ge \|\alpha^{-s}(a_s)\| = \|a_s\|, \quad \forall s \in \{0, \cdots, n-1\},\end{aligned}$$

i.e.,

$$\|a\|_1 \ge \|a\| \ge \max_{0\le s\le n-1} \|a_s\|,$$

$\forall a = (a_s)_{0\le s\le n-1} \in l^1(\mathbb{Z}_n, A, \alpha)$. Hence, $l^1(\mathbb{Z}_n, A, \alpha)$ is C^*-equivalent (see Definition 2.14.21), i.e., as linear spaces we have

$$A \times_\alpha \mathbb{Z}_n = l^1(\mathbb{Z}_n, A, \alpha) = A \times \cdots \times A(n \text{ times}).$$

Moreover, $A \times_\alpha \mathbb{Z}_n$ admits a matrix representation as follows. Define

$$\theta : a = (a_s)_{0\le s\le n-1} \longrightarrow (\alpha^{-i}(a_{i-j}))_{0\le i,j\le n-1},$$

$\forall a = (a_s)_{0\le s\le n-1} \in A \times_\alpha \mathbb{Z}_n$. Then it is easy to see that θ is a $*$ isomorphism from $A \times_\alpha \mathbb{Z}_n$ into $M_n(A) = A \otimes M_n = \{(a_{ij})_{0\le i,j\le n-1} | a_{ij} \in A, \forall i, j\}$. Further, let

$$U = \begin{Bmatrix} 0 & \cdot & \cdots & 1 \\ 1 & \ddots & & 0 \\ & \ddots & \ddots & \vdots \\ 0 & & 1 & 0 \end{Bmatrix} (\in M_n).$$

Then an element $(a_{ij})_{0\le i,j\le n-1}$ of $M_n(A)$ belongs to $\theta(A \times_\alpha \mathbb{Z})$ if and only if $U(a_{ij})U^* = (\alpha(a_{ij}))$. In fact, first it is easy to check that

$$U(\alpha^{-i}(a_{i-j}))U^* = (\alpha(\alpha^{-i}(a_{i-j}))).$$

Conversely, if $U(a_{ij})U^* = (\alpha(a_{ij}))$, then

$$U^k(a_{ij})U^{*k} = (\alpha^k(a_{ij})), \quad \forall k \ge 0.$$

Let $b_{n-j} = a_{oj}, 0 \le j \le n-1$, and $b_o = b_n$. We need to show that

$$\alpha^{-i}(b_{i-j}) = a_{ij} \quad \forall i, j.$$

It is equivalent to prove that

$$\alpha^i(a_{ij}) = b_{i-j} = a_{0,j-i}, \quad \forall i, j,$$

where b_k and $a_{0,k}$ for any $k \in \mathbb{Z}$ are understood in the sense of (*mod* n). Notice that $\alpha^i(a_{ij})$ is the (i,j) element of $U^i(a_{kl})U^{*i}, \forall i,j$. Then by the form of U, we can obtain the conclusion.

Now consider the dual sysem $(A \times_\alpha \mathbb{Z}_n, \mathbb{Z}_n, \hat{\alpha})$ of $(A, \mathbb{Z}_n, \alpha)$. By Definition 16.5.7, we have

$$\hat{\alpha}((a_s)_{0\le s\le n-1}) = (e^{-2\pi is/n}a_s)_{0\le s\le n-1},$$

$\forall (a_s)_{0\le s\le n-1} \in A \times_\alpha \mathbb{Z}_n$. The *mapping torus* of $\hat{\alpha}$ on $A \times_\alpha Z_n$ is defined as follows:

$$M_{\hat{\alpha}}(A \times_\alpha \mathbb{Z}_n) = \left\{ (\tilde{F}_j(t))_{0\le j\le n-1} \,\middle|\, \begin{array}{l} t \to \tilde{F}_j(t) \text{ is continuous from } [0,1] \\ \text{to } A, \forall j, \text{ and } (\tilde{F}_j(1))_j = \hat{\alpha}(\tilde{F}_j(0))_j \end{array} \right\}.$$

(v) $A \times_\alpha \mathbb{Z}$ is $*$ isomorphic to $M_{\hat{\alpha}}(A \times_\alpha \mathbb{Z}_n)$.
In fact, define

$$\sum_{j=0}^{n-1} a_j\lambda^j \longrightarrow (e^{-2\pi itj/n}F_j(e^{-2\pi it}))_{0\le j\le n-1},$$

where $F_j = \Phi(a_j), 0 \le j \le n-1$, and Φ is the $*$ isomorphism from $\mathcal{A}$ onto $C(T,A)$ (see (iii)), $\forall a_j \in \mathcal{A}, 0 \le j \le n-1$. Then it is a $*$ isomorphism from $A \times_\alpha \mathbb{Z}$ to $M_{\hat{\alpha}}(A \times_\alpha \mathbb{Z}_n)$ obviously. Conversely, if $(\tilde{F}_j(t))_{0\le j\le n-1} \in M_{\hat{\alpha}}(A \times_\alpha \mathbb{Z}_n)$, then

$$\tilde{F}_j(1) = e^{-2\pi ij/n}\tilde{F}_j(0), \quad 0 \le j \le n-1.$$

So the function $e^{2\pi itj/n}\tilde{F}_j(t)(t \in [0,1])$ picks the same values at the points $t = 0$ and $t = 1, \forall j$. So for any $j \in \{0, \cdots, n-1\}$ there is $F_j \in C(T,A)$ such that

$$F_j(e^{-2\pi it}) = e^{2\pi itj/n}\tilde{F}_j(t), \quad \forall t \in [0,1].$$

Let $a_j \in \mathcal{A}$ be such that $\Phi(a_j) = F_j, \forall j$. Then

$$\begin{aligned} \sum_{j=0}^{n-1} a_j\lambda^j \longrightarrow \quad & (e^{-2\pi itj}F_j(e^{-2\pi it}))_{0\le j\le n-1} \\ & = (\tilde{F}_j(t))_{0\le j\le n-1}. \end{aligned}$$

Therefore, $A \times_\alpha \mathbb{Z}$ and $M_{\hat{\alpha}}(A \times_\alpha \mathbb{Z}_n)$ are $*$ isomorphic.

(vi) $A \times_\alpha \mathbb{Z}$ also admits a matrix representation.
In fact, define

$$(F_j)_{0\le j\le n-1} \stackrel{\theta}{\longrightarrow} (\alpha^{-i}F_{i-j} \cdot z^{\kappa(i-j)})_{0\le i,j\le n-1},$$

$\forall F_j \in C(T,A), 0 \le j \le n-1$, where $\kappa(\cdot)$ is a function on $\mathbb{Z}$ as follows:

$$\kappa(k) = \begin{cases} 1, & \text{if } k < 0, \\ 0, & \text{if } k \ge 0, \end{cases} \quad \forall k \in \mathbb{Z}.$$

Then by (iii) θ is a $*$ isomorphism from $A \times_\alpha \mathbb{Z}$ into $M_n(C(T,A)) = C(T,A) \otimes M_n$. Moreover, let

$$U(z) = \begin{pmatrix} 0 & \cdots & \cdots & z \\ 1 & \ddots & & \vdots \\ & \ddots & & \vdots \\ 0 & & 1 & 0 \end{pmatrix} \quad (n \times n)$$

$\forall z \in T$. By the discussion of (iv) we have

$$\begin{aligned} &\theta(A \times_\alpha \mathbb{Z}) \\ = \ &\{(F_{ij}(z)) \in M_n(C(T,A)) | U(z)(F_{ij}(z))U(z)^* = (\alpha F_{ij}(z)), \forall z \in T\}\,. \end{aligned}$$

Notes. The example 3 appeared in [175]. To understant it, M. Takesaki had to develop the duality theory.

The proof about mapping torus (the example 5) presented here follows from B.R. Li and Q. Lin.

References. [11], [70], [127], [175].

Chapter 17

Jones Index Theory

17.1. The coupling constant

Suppose that M is a VN algebra on a separable Hilbert space H. By Proposition 1.14.3, M_* is separble and M is σ-finite. Let $(M)_1$ be the closed unit ball of M. From the example under Definition 10.1.1, $((M)_1, \sigma(M, M_*)), ((M)_1, s(M, M_*))$ and $((M)_1, s^*(M, M_*)) = ((M)_1, \tau(M, M_*))$ are Polish spaces. If φ is any faithful normal state on M, then by Lemma 1.11.2 a proper metric $d(\cdot,\cdot)$ for the Polish space $((M)_1, s(M, M_*))$ is $d(a,b) = \varphi((a-b)^*(a-b))^{1/2}, \forall a, b \in (M)_1$. Moreover, if $\{x_n\}$ is a countable dense subset of the Polish space $((M)_1, \sigma(M, M_*))$, then $\{x_n\}$ generates M (i.e., $\{x_n, x_n^* | n\}'' = M$) obviously. Hence, M is countably generated.

Proposition 17.1.1. Let M be a W^*-algebra. Then the following statements are equivalent:

1) M admits a faithful W^*-representation on some separable Hilbert space;

2) M is σ-finite and countably generated;

3) $((M)_1, \sigma(M, M_*))$ is a (compact) Polish space, where $(M)_1$ is the closed unit ball of M.

Proof. From the above discussion, we have that 1) implies 3).

Now let $((M)_1, \sigma(M, M_*))$ be a Polish space. Then M is countably generated obviously. Moreover, since $((M)_1, \sigma(M, M_*))$ is metrizable, M_* is separable. Hence by Proposition 1.14.3 M is also σ-finite.

Finally, let M be σ-finite and countably generated. Suppose that $\{x_n\}$ is a countable generated subset for M, and $\{\pi, H\}$ is a faithful nondepenerate W^*-representation of M. Since M is σ-finite, there is a sequence $\{\xi_n\}$ of H such that $p_n p_m = \delta_{nm} p_n, \forall n, m$, and $\sum_n p_n = 1_H$, where p_n is the projection from H onto $\pi(M)'\xi_n, \forall n$. Clearly , $p_n \in \pi(M), \forall n$, and $\{\xi_n\}$ is a separating sequence of vectors for $\pi(M)$. Let p' be the projection from H onto $\overline{[\pi(M)\xi_n|n]}$, and A be the $*$ subalgebra over the field of complex rational numbers generated by $\{\pi(x_n), \pi(x_m)^*|n,m\}$. Then A is countable and strongly dense in $\pi(M)$, and $p'H = \overline{[A\xi_n|n]}$ is separable. If $c(p')$ is the central cover of p' in $\pi(M)'$, then we have

$$c(p')H = \overline{[\pi(M)'p'H]} \supset \overline{[\pi(M)'\xi_n|n]} = H,$$

and $c(p') = 1_H$. Hence, $a \longrightarrow ap'$ is a $*$ isomorphism from $\pi(M)$ onto $\pi(M)p'$. Consequently, $\{p'\pi, p'H\}$ is also a faithful W^*–representation of M, and $p'H$ is separable. Q.E.D.

In this chapter, every W^* -algebra is assumed to admit a faithful W^*-representation on some separable Hilbert space, and every Hilbert space is separable. For simplification, we shall not repeat these assumptions.

Now let M be a finite factor. Then there is unique faithful normal tracial state τ on M. Using this τ, we define an inner product on M:

$$\langle x, y\rangle = \tau(y^*x), \quad \forall x, y \in M.$$

Denote by $L^2(M)$ the completion of $(M, \langle,\rangle)$, and by $\|\cdot\|_2$ the norm on $L^2(M)$, i.e., $\|x\|_2 = \tau(x^*x)^{1/2}, \forall x \in M$. Since τ is a trace, the $\|\cdot\|_2$–topology is equivalent to $\tau(M, M_*)$ in the closed unit ball $(M)_1$, and $((M)_1, \tau(M, M_*))$ is a Polish space. Thus it follows that $L^2(M)$ is a separable Hilbert space.

Clearly, $\|x\|_2 = \|x^*\|_2, \forall x \in M$. Hence, $J : x \longrightarrow x^*(\forall x \in M)$ can be uniquely extended to a conjugate linear isometry on $L^2(M)$, still denoted by J. Then we have $\langle Ja, Jb\rangle = \langle b, a\rangle, \forall a, b \in L^2(M)$.

For each $x \in M$, let $\lambda(x)y = xy, \forall y \in M$. Clearly, $\|\lambda(x)y\|_2 \leq \|x\| \cdot \|y\|_2, \forall y \in M$. Thus , $\lambda(x)$ can be uniquely extended to a bounded linear operator on $L^2(M)$, still denoted by $\lambda(x)$. Then we obtain a faithful $*$ representation $\{\lambda, L^2(M)\}$ of M. We claim that this is also a W^*–representation. It suffices to show that for any $y, z \in M, \langle\lambda(\cdot)y, z\rangle$ is a continuous function on $((M)_1, \sigma(M, M_*))$. But it is obvious from the normality of the trace τ . Hence, $\lambda(M)$ is a VN algebra on $L^2(M)$, and is $*$ isomorphic to M.

Consider the opposite algebra M^{opp} of M (i.e., $x \circ y \equiv yx, \forall x, y \in M$). It is also a finite factor, and $\{\rho, L^2(M)\}$ is a faithful W^*–representation of M^{opp}, where

$$\rho(x)y = yx, \quad \forall x \in M^{opp}, y \in M.$$

Hence , $\rho(M^{opp})$ is a VN algebra on $L^2(M)$, and is $*$ isomorphis to M^{opp}.

Proposition 17.1.2. With the above notations, we have

$$\lambda(M)' = \rho(M) = J\lambda(M)J, \quad \rho(M)' = \lambda(M) = J\rho(M)J,$$

$$J\lambda(x)J = \rho(x^*), \quad J\rho(x)J = \lambda(x^*), \quad \forall x \in M.$$

Proof. Clearly, $\lambda(x)\rho(y) = \rho(y)\lambda(x), J\lambda(x)J = \rho(x^*)$, and $J\rho(x)J = \lambda(x^*)$, $\forall\ x,\ y \in M$. Thus , we have

$$\lambda(M) \subset \rho(M)', \quad \rho(M) \subset \lambda(M)'; \quad J\lambda(M)J = \rho(M), J\rho(M)J = \lambda(M).$$

Let

$$B_l = \left\{ a \in L^2(M) \,\middle|\, \begin{array}{c}\text{there exists } \lambda(a) \in B(L^2(M)) \text{ such that} \\ \lambda(a)x = \rho(x)a, \forall x \in M\end{array} \right\}$$

and

$$B_r = \left\{ b \in L^2(M) \,\middle|\, \begin{array}{c}\text{there exists } \rho(b) \in B(L^2(M)) \text{ such that} \\ \rho(b)x = \lambda(x)b, \forall x \in M\end{array} \right\}$$

It is easily verified that $\lambda(a) \in \rho(M)', \forall a \in B_l$; and $\rho(b) \in \lambda(M)', \forall b \in B_r$. Conversely, if $t \in \rho(M)'$, then $ta \in B_l$ and $\lambda(ta) = t\lambda(a), \forall a \in B_l$. Consequently, $t = \lambda(t1)$, where 1 is the identity of $M(\subset L^2(M))$. Similarly, if $s \in \lambda(M)'$, then $sb \in B_r$ and $\rho(sb) = s\rho(b), \forall b \in B_r$. Consequently, $s = \rho(s1)$. Thus , we have

$$\rho(M)' = \{\lambda(a) | a \in B_l\}, \quad \text{and} \quad \lambda(M)' = \{\rho(b) | b \in B_r\}.$$

Now it suffices to show that $\lambda(a)\rho(b) = \rho(b)\lambda(a), \forall a \in B_l, b \in B_r$. In fact, let $a \in B_l, b \in B_r$. Since $\lambda(a)^* \in \rho(M)'$ and $\rho(b)^* \in \lambda(M)'$, there is some $c \in B_l$ and some $d \in B_r$ such that $\lambda(a)^* = \lambda(c), \rho(b)^* = \rho(d)$. Noticing that

$$\begin{aligned} \langle a, x\rangle &= \langle a, \rho(1)x\rangle = \langle \rho(1)a, x\rangle \\ &= \langle \lambda(a)1, x\rangle = \langle 1, \lambda(c)x\rangle = \langle 1, \rho(x)c\rangle \\ &= \langle \rho(x^*)1, c\rangle = \langle x^*, c\rangle = \langle Jc, x\rangle, \end{aligned}$$

$\forall x \in M$, we get $a = Jc$. Similarly, we have $b = Jd$. Pick $\{x_n\}, \{y_n\} \subset M$ such that $x_n \longrightarrow a, y_n \longrightarrow b$ in $L^2(M)$. Then $x_n^* = Jx_n \longrightarrow Ja = c, y_n^* = Jy_n \longrightarrow b = d$. Further,

$$\begin{aligned} & \langle \lambda(a)\rho(b)x, y\rangle = \langle \rho(b)x, \lambda(c)y\rangle \\ = \ & \langle \lambda(x)b, \rho(y)c\rangle = \lim\langle \lambda(x)y_n, \rho(y)x_n^*\rangle \\ = \ & \lim\langle xy_n, x_n^*y\rangle = \lim\langle x_nx, yy_n^*\rangle \\ = \ & \lim\langle \rho(x)x_n, \lambda(y)y_n^*\rangle = \langle \rho(x)a, \lambda(y)d\rangle \\ = \ & \langle \lambda(a)x, \rho(d)y\rangle = \langle \rho(b)\lambda(a)x, y\rangle, \quad \forall x, y \in M. \end{aligned}$$

Therefore, we have $\lambda(a)\rho(b) = \rho(b)\lambda(a)$. Q.E.D.

Remark. We also have $B_l = B_r = M$. In fact, from the above proof it follows that $\lambda(B_l) = \rho(M)' = \lambda(M)$. Now if $\lambda(a) = \lambda(x)$, where $a \in B_l, x \in M$, then $x = \lambda(x)1 = \lambda(a)1 = \rho(1)a = a$. Hence, $B_l = M$. Similarly, $B_r = M$.

Moreover, $\lambda(M)$ and $\lambda(M)' = \rho(M)$ are finite factor on the separable Hilbert space $L^2(M)$, and they admit a cyclic–separating vector $1(\in M \subset L^2(M))$.

Proposition 17.1.3. Let M be a finite factor, and $\{\pi, H\}$ be a faithful nondegenerate W^*-representation of M. Then there is a projection p' of $(\lambda(M) \otimes 1_K)'$, and a unitary operator u from H onto $p'(L^2(M) \otimes K)$ such that

$$u\pi(x) = (\lambda(x) \otimes 1_K)u, \quad \forall x \in M,$$

i.e., the VN algebras $\pi(M)$ (on H) and $(\lambda(M) \otimes 1_K)p'$ are spatially $*$ isomorphic, where K is a countably infinite dimensional Hilbert space. Consequently, $u\pi(M)u^* = p'(\lambda(M) \otimes 1_K)$ and $u\pi(M)'u^* = p'(\lambda(M) \otimes 1_K)'p'$.

Proof. $\Phi : \lambda(x) \longrightarrow \pi(x)(\forall x \in M)$ is a $*$ isomorphism from $\lambda(M)$ onto $\pi(M)$. Now the conclusion follows immediately from Theorem 1.12.4 and the separability of H. Q.E.D.

Definition 17.1.4. Let M be a finite factor, and $\{\pi, H\}$ be a faithful nondegenerate W^*-representation of M. Generally, $\pi(M)'$ is not finite. Of course, $\pi(M)'$ is semi–finite. Then there exist faithful semi–finite normal traces on $\pi(M)'_+$ (uniquely determined up to multiplication by a positive constant, see Proposition 7.1.2) . Now we define a *natural trace* Tr'_H on $\pi(M)'_+$ as follows.

(i) If $\{\pi, H\} = \{\lambda \otimes 1_K, L^2(M) \otimes K\}$, where K is a countably infinite dimensional Hilbert space, then $(\lambda(M) \otimes 1_K)' = J\lambda(M)J\overline{\otimes}B(K)$ is infinite. Pick an orhtogonal normalized basis $\{e_i\}$ of K. Then for each $t' \in (\lambda(M)\otimes 1_K)'$ we can uniquely write $t' = (J\lambda(x_{ij})J)$, where $x_{ij} \in M, \forall i, j$. Define the natrual trace as follows:

$$Tr'_{L^2(M)\otimes K}(t') = \sum_i \tau(x_{ii}), \quad \forall t' = (J\lambda(x_{ij})J) \in (\lambda(M) \otimes 1_K)'_+,$$

where τ is the unique faithful normal tracial state on M. Clearly, $Tr'_{L^2(M)\otimes K}$ is a faithful semi–finite normal trace on $(\lambda(M)\otimes 1_K)'_+$. If $\{f_i\}$ is another orthogonal normalized basis of K, and $t' = (J\lambda(x_{ij})J)$ is the matrix representation of t' in the decomposition $L^2(M) \otimes K = \sum_i \oplus[L^2(M) \otimes e_i]$, then the matrix representation of t' in the decomposition $L^2(M) \otimes K = \sum_i \oplus[L^2(M) \otimes f_i]$ is $(t_{ij})(J\lambda(x_{ij})J)(t_{ij})^*$, where $t_{ij} = \langle e_j, f_i \rangle, \forall i, j$. Since (t_{ij}) is unitary, the definition of $Tr'_{L^2(M)\otimes K}$ is independent of the choice $\{e_i\}$.

(ii) For general faithful nondegenerate W^*-representation $\{\pi, H\}$ of M, by Proposition 17.1.3 there is a projection p' of $(\lambda(M) \otimes 1_K)'$ and a unitary operator u from H onto $p'(L^2(M) \otimes K)$ such that

$$u\pi(x)u^* = (\lambda(x) \otimes 1_K)p', \quad \forall x \in M, \quad \text{and} \quad u\pi(M)'u^* = p'(\lambda(M) \otimes 1_K)'p'.$$

Then we define the natural trace as follows:

$$Tr'_H(t') = Tr'_{L^2(M)\otimes K}(ut'u^*), \quad \forall t' \in \pi(M)'_+.$$

Here we must prove that this definition is independent of the choice of u and p'. In fact, if u_1 and u_2 are two unitary operators from H onto $p'_1(L^2(M) \otimes K)$ and $p'_2(L^2(M) \otimes K)$ respectively such that $u_i\pi(x)u_i^* = (\lambda(x) \otimes 1_K)p'_i, i = 1, 2$, then $(\lambda(x) \otimes 1_K)u_1u_2^* = u_1\pi(x)u_2^* = u_1u_2^*p'_2(\lambda(x) \otimes 1_K) = u_1u_2^*(\lambda(x) \otimes 1_K), \forall x \in M$. Thus $u_1u_2^* \in (\lambda(M) \otimes 1_K)'$. Similarly, $u_2u_1^*$ and $u_1t'u_2^* = u_1t'u_1^* \cdot u_1u_2^* \in (\lambda(M) \otimes 1_K)', \forall t' \in \pi(M)'_+$. If $t' \in \pi(M)'_+$ is such that $Tr'_{L^2(M)\otimes K}(u_1t'u_1^*) < \infty$, then $u_1t'u_1^* \in \mathcal{M}$, where $\mathcal{M}$ is the definition ideal of $Tr'_{L^2(M)\otimes K}$. By Proposition 6.5.2, we have $u_1t'u_2^* = u_1t'u_1^* \cdot u_1u_2^* \in \mathcal{M}$, and

$$\begin{aligned} Tr'_{L^2(M)\otimes K}(u_1t'u_1^*) &= Tr'_{L^2(M)\otimes K}(u_1t'u_2^* \cdot u_2u_1^*) \\ &= Tr'_{L^2(M)\otimes K}(u_2u_1^* \cdot u_1t'u_2^*) = Tr'_{L^2(M)\otimes K}(u_2t'u_2^*). \end{aligned}$$

Similarly, if $t' \in \pi(M)'_+$ is such that $Tr'_{L^2(M)\otimes K}(u_2t'u_2^*) < \infty$, then $Tr'_{L^2(M)\otimes K}(u_2t'u_2^*) = Tr'_{L^2(M)\otimes K}(u_1t'u_1^*)$. Therefore, we have

$$Tr'_{L^2(M)\otimes K}(u_1t'u_1^*) = Tr'_{L^2(M)\otimes K}(u_2t'u_2^*), \quad \forall t' \in \pi(M)'_+.$$

Moreover, Tr'_H is a faithful normal trace on $\pi(M)'_+$ obviously. Now if $t'(\neq 0) \in \pi(M)'_+$, then by the semi-finiteness of $Tr'_{L^2(M)\otimes K}$ there is $0 \neq a' \in (\lambda(M) \otimes 1_K)'_+$ with $a' \leq ut'u^*$ such that $Tr'_{L^2(M)\otimes K}(a') < \infty$. Clearly $a' \in p'(\lambda(M) \otimes 1_K)'p' = u\pi(M)'u^*$. So $s' = u^*a'u \in \pi(M)'_+$ and $s' \leq u^*(ut'u^*)u = t'$, and $Tr'_H(s') = Tr'_{L^2(M)\otimes K}(a') < \infty$. Therefore, Tr'_H is also semi-finite on $\pi(M)'_+$.

Definition 17.1.5. Let M be a finite factor, and $\{\pi, H\}$ be a faithful nondegenerate W^*-representation of M. Then

$$\dim_M(H) = Tr'_H(1)$$

is called the *coupling constant* between $\pi(M)$ and $\pi(M)'$, where $Tr'_H(\cdot)$ is the natural trace on $\pi(M)'_+$.

Since $\dim_M(H) = Tr'_{L^2(M)\otimes K}(uu^*)$ and $H \cong uu^*(L^2(M) \otimes K)$, $\dim_M(H)$ is also the " dimension " of H as a M-module (see Section 7.1).

Proposition 17.1.6. Let M be a finite factor.

(i) If $\{\pi_1, H_1\}$ and $\{\pi_2, H_2\}$ are two faithful nondegenerate W^*-representations of M, then

$$\dim_M(H_1) = \dim_M(H_2) \Longleftrightarrow \{\pi_1, H_2\} \cong \{\pi_2, H_2\};$$

(ii) If $\{\pi_i, H_i\}, i = 1, 2, \cdots$, is a sequence of faithful nondegenerate W^*-representations of M, then

$$\dim_M(\sum_i \oplus H_i) = \sum_i \dim_M(H_i);$$

(iii) The natural trace $Tr'_{L^2(M)}$ is exactly the restriction of the unique faithful normal tracial state τ' of $\lambda(M)' = \rho(M)$ on $\lambda(M)'_+$, and $\dim_M(L^2(M)) = 1$;

(iv) If $\{\pi, H\}$ is a faithful nondegenerate W^*-representation of M, then

$$\begin{aligned} \pi(M)' \text{ is finite } &\Longleftrightarrow Tr'_H \text{ is finite on } \pi(M)'_+ \\ &\Longleftrightarrow \dim_M(H) < \infty. \end{aligned}$$

(In this can, Tr'_H can be uniquely extended onto $\pi(M)'$, but it is not necessarily normalized) .

Proof. (i) First, we notice the following fact. Suppose that N is a σ-finite and semi-finite factor. If p is an infinite projection of N and p_0 is a non-zero finite projection of N, then there is an orthogonal sequence $\{p_n \mid n \geq 1\}$ of projections of N such that

$$p \sim \sum_{n=1}^{\infty} p_n, \quad \text{and} \quad p_n \sim p_0, \forall n \geq 1.$$

In particular, all infinite projections of N are equivalent (relative to N).

In fact, by the Zorn lemma and the σ-finiteness of N, there is a maximal orthogonal sequence $\{p_n \mid n \geq 1\}$ of projections of N such that

$$p_n \leq p, \quad p_n \sim p_0, \quad \forall n \geq 1.$$

Let $q = p - \sum_{n=1}^{\infty} p_n$. Since N is a factor and $\{p_n\}$ is maximal, it follows that $q \preceq p_0$. Then it is easily verified that $\sum_{n=1}^{\infty} p_n \sim \sum_{n=1}^{\infty} p_n + q = p$.

Now let u_i be the unitary operator from H_i onto $p'_i(L^2(M) \otimes K)$ such that $\dim_M(H_i) = Tr'_{L^2(M)\otimes K}(u_i u_i^*), p'_i = u_i u_i^* \in (\lambda(M) \otimes 1_K)', i = 1, 2$. Then by the above fact we can see that: $\dim_M(H_1) = \dim_M(H_2)$ if and only if $p'_1 = u_1 u_1^* \sim p'_2 = u_2 u_2^*$ (relative to $(\lambda(M) \otimes 1_K)'$). Since $\{\pi_i, H_i\} \cong \{(\lambda \otimes 1_K)p'_i, p'_i(L^2(M) \otimes K)\}, i = 1, 2$, it follows that

$$\dim_M(H_1) = \dim_M(H_2) \Longleftrightarrow \{\pi_1, H_1\} \cong \{\pi_2, H_2\}.$$

(ii) Let $\pi = \sum_i \oplus \pi_i, H = \sum_i \oplus H_i$. Clearly, $\{\pi, H\}$ is still a faithful nondegenerate W^*-representation of M. Let u be a unitary operator from H onto $p'(L^2(M) \otimes K)$ such that

$$u\pi(x) = (\lambda(x) \otimes 1_K)u, \quad \forall x \in M,$$

and $p' = uu^* \in (\lambda(M) \otimes 1_K)'$. Then $\dim_M(H) = Tr'_H(1) = Tr'_{L^2(M)\otimes K}(p')$.

Fix i, and let $u_i = uq_i, p'_i = u_i u_i^*$, where q_i is the projection from H onto H_i. Clearly, $q_i \in \pi(M)'$, and $u_i^* u_i = q_i$. We claim that

$$u_i \pi_i(x) = (\lambda(x) \otimes 1_K)u_i, \quad \forall x \in M, \quad \text{and} \quad p'_i \in (\lambda(M) \otimes 1_K)'.$$

In fact, $u_i\pi_i(x) = u\pi(x)q_i = (\lambda(x) \otimes 1_K)uq_i = (\lambda(x) \otimes 1_K)u_i, \forall x \in M$. Moreover, since $p'_i = u_i u_i^* = uq_i u^*$ and

$$\begin{aligned} p'_i(\lambda(x) \otimes 1_K)p'_i = &\ uq_i \cdot u^*(\lambda(x) \otimes 1_K)u \cdot q_i u^* \\ = &\ uq_i\pi(x)q_i u^* = u\pi(x) \cdot q_i u^* \\ = &\ (\lambda(x) \otimes 1_K) \cdot uq_i u^* = (\lambda(x) \otimes 1_K)p'_i, \end{aligned}$$

$\forall x \in M$, it follows that $p'_i \in (\lambda(M) \otimes 1_K)'$.

Now we have that $\dim_M(H_i) = Tr'_{L^2(M)\otimes K}(p'_i), \forall i$.

Clearly, $p'_i p'_j = \delta_{ij} p'_j, \forall i, j$ and $\sum_i p'_i = p'$. Then by the complete additivity of $Tr'_{L^2(M)\otimes K}$ we have

$$\begin{aligned} \sum_i \dim_M(H_i) = &\ \sum_i Tr'_{L^2(M)\otimes K}(p'_i) \\ = &\ Tr'_{L^2(M)\otimes K}(p') = \dim_M(\sum_i \oplus H_i). \end{aligned}$$

(iii) Let $u(a) = a \otimes \xi, \forall a \in L^2(M)$, where $\xi \in K$ and $\|\xi\| = 1$. Then u is unitary from $L^2(M)$ onto $L^2(M) \otimes [\xi]$. If p' is the projection from $L^2(M) \otimes K$ onto $L^2(M) \otimes [\xi]$, then $p' \in (\lambda(M) \otimes 1_K)'$ and $u\lambda(x)u^* = (\lambda(x) \otimes 1_K)p', \forall x \in M$. Let $\{e_i\}$ be an orthogonal normalized basis of K with $e_1 = \xi$. Then p' has a matrix representation $(J\lambda(x_{ij})J)$ such that $x_{11} = 1, \quad x_{ij} = 0, \forall (i,j) \neq (1,1)$
Thus, we have

$$\dim_M(L^2(M)) = Tr'_{L^2(M)}(1) = Tr'_{L^2(M)\otimes K}(p') = 1,$$

and $Tr'_{L^2(M)}$ can be uniquely extended to the faithful normal tracial state τ' on $\lambda(M)'$.

(iv) By Proposition 7.1.2, we can see that

$$\begin{aligned} \pi(M)' \quad \text{is finite} \quad &\Longleftrightarrow p' = uu^* \text{ is finite in } (\lambda(M) \otimes 1_K)' \\ &\Longleftrightarrow Tr'_{L^2(M)\otimes K}(p') < \infty \\ &\Longleftrightarrow \dim_M(H) = Tr'_H(1) < \infty. \end{aligned}$$

Q.E.D.

Proposition 17.1.7. Let M be a semi-finite factor on a Hilbert space H. Then there exist faithful normal semi-finite traces ρ, ρ' on M_+, M'_+ respectively such that

$$\rho(e_\xi) = \rho'(e'_\xi), \quad \forall 0 \neq \xi \in H,$$

where e_ξ, e'_ξ are the cyclic projections from H onto $\overline{M'\xi}, \overline{M\xi}$ respectively. In particular, if σ, σ' are any faithful normal semi-finite traces on M_+, M'_+ respectively, then there exists a positive constant c such that $\sigma(e_\xi) = c\sigma'(e'_\xi), \forall 0 \neq \xi \in H$. Consequently, if M, M' are finite, and τ, τ' are the unique faithful normal tracial states on M, M' respectively, then the number

$$c_M = \tau(e_\xi)/\tau'(e'_\xi)$$

is independent of the choice of $\xi (\neq 0)$.

Proof. Let φ be a faithful normal state on M, and $\{\pi_\varphi, H_\varphi, \xi_\varphi\}$ be the faithful cyclic W^*-representation of M generated by φ. Clearly, ξ_φ is also separating for $N = \pi_\varphi(M)$. Since $((M)_1, s(M, M_*))$ is a Polish space and a proper metric on $((M)_1, s(M, M_*))$ is $d(a, b) = \varphi((a-b)^* \cdot (a-b))^{1/2} (\forall a, b \in (M)_1)$, the Hilbert space H_φ is separable. Now by Theorem 1.12.4, M is spatially $*$ isomorphic to $(N \otimes 1_K)p'$, where K is a countably infinite dimensional Hilbert space, p' is a projection of $(N \otimes 1_K)'$. Fix a vector $\eta_0 \in K$ with $\|\eta_0\| = 1$, and let p be the projection from $H_\varphi \otimes K \otimes K$ onto $H_\varphi \otimes K \otimes [\eta_0]$. Clearly, $p \in N \otimes 1_K \overline{\otimes} B(K)$, and $N \otimes 1_K$ is spatally $*$ isomorphic to $p(N \otimes 1_K \overline{\otimes} B(K))p$. Further, M is spatially $*$ isomorphic to $p(N \otimes 1_K \overline{\otimes} B(K))p \cdot (p' \otimes 1_K)$.

Since ξ_φ is cyclic-separating for N, by Theorem 8.2.7 there is a conjugate linear isometry j on H_φ with $j^2 = 1$ such that $jNj = N'$. We may assume that $K = l^2$. Define

$$t((\lambda_n) \otimes (\mu_n)) = ((\overline{\mu}_n) \otimes (\overline{\lambda}_n)), \forall (\lambda_n), (\mu_n) \in l^2.$$

Then t can be uniquely extended to a conjugate linear isometry, still denoted by t, on $K \otimes K$ with $t^2 = 1$. Clearly, $t(1_K \otimes B(K))t = B(K) \otimes 1_K = (1_K \otimes B(K))'$. Let

$$J = j \otimes t, \quad \text{and} \quad T = N \otimes 1_K \overline{\otimes} B(K).$$

Then J is a conjugate linear isometry on $(H_\varphi \otimes K \otimes K)$ with $J^2 = 1, T$ and $T' = N' \overline{\otimes} B(K) \otimes 1_K$ are semi-finite factors on $(H_\varphi \otimes K \otimes K)$, and $JTJ = T'$.

Let Φ be a faithful normal semi-finite trace on T_+, and define

$$\Phi'(t') = \Phi(Jt'J), \quad \forall t' \in T'_+.$$

Then Φ' is a faithful normal semi-finite trace on T'_+. For any $0 \neq \varsigma \in (H_\varphi \otimes K \otimes K)$, denote by $P_\varsigma, P'_\varsigma$ the cyclic projections from $(H_\varphi \otimes K \otimes K)$ onto

$\overline{T'\varsigma}, \overline{T\varsigma}$ respectively. It is easy to see that $JP'_\varsigma J = P_{J\varsigma}$. Since T is a factor, it follows that either $P_{J\varsigma} \preceq P_\varsigma$ or $P_\varsigma \preceq P_{J\varsigma}$. Suppose that $P_{J\varsigma} \preceq P_\varsigma$. By Theorem 1.13.2 we have $P'_{J\varsigma} \preceq P'_\varsigma$. Then $P_\varsigma = JP'_{J\varsigma}J \preceq JP'_\varsigma J = P_{J\varsigma}$. So we obtain that $P_\varsigma \sim P_{J\varsigma}$ and

$$\Phi'(P'_\varsigma) = \Phi(JP'_\varsigma J) = \Phi(P_{J\varsigma}) = \Phi(P_\varsigma),$$

$\forall 0 \neq \varsigma \in (H_\varphi \otimes K \otimes K)$.

Since T is a factor, the central cover of p in T must be 1 . Hence $\theta(t') = t'p(\forall t' \in T')$ is a $*$ isomorphism from T' onto $T'p$. Let

$$\Phi_1 = \Phi|(pTp)_+, \quad \Phi'_1 = \Phi' \circ \theta^{-1}.$$

Then Φ_1, Φ'_1 are faithful normal semi-finite traces on $(pTp)_+, (T'p)_+$ respectively. For any $0 \neq \varsigma \in p(H_\varphi \otimes K \otimes K) = (H_\varphi \otimes K \otimes [\eta_0])$, denote by $Q_\varsigma, Q'_\varsigma$ the cyclic projections from $(H_\varphi \otimes K \otimes [\eta_0])$ onto $\overline{T'p\varsigma}, \overline{pTp\varsigma}$ respectively. Clearly, $Q_\varsigma = P_\varsigma, Q'_\varsigma = pP'_\varsigma$. Thus , from the preceding paragraph we have

$$\Phi'_1(Q'_\varsigma) = \Phi'(P'_\varsigma) = \Phi(P_\varsigma) = \Phi_1(Q_\varsigma),$$

$\forall 0 \neq \varsigma \in p(H_\varphi \otimes K \otimes K)$.

Similarly, there are faithful normal semi-finite traces Φ_2, Φ'_2 on $(pTp \cdot p' \otimes 1_K)_+, (p' \otimes 1_K \cdot T'p \cdot p' \otimes 1_K)_+$ respectively such that

$$\Phi'_2(R'_\varsigma) = \Phi_2(R_\varsigma), \quad \forall 0 \neq \varsigma \in (p' \otimes 1_K)p(H_\varphi \otimes K \otimes K),$$

where $R_\varsigma, R'_\varsigma$ are the cyclic projections from $(p' \otimes 1_K)p(H_\varphi \otimes K \otimes K)$ onto $\overline{(p' \otimes 1_K)T'\varsigma}, \overline{pT\varsigma}$ respectively .

Therefore, we can find faithful normal semi-finite traces ρ, ρ' on M_+, M'_+ respectively such that

$$\rho(e_\xi) = \rho'(e'_\xi), \quad \forall 0 \neq \xi \in H.$$

Finally, since the faithful normal semi-finite traces on M_+ and M'_+ are uniquely determined up to multiplication by a positive constant , the rest conclusions are obvious. Q.E.D.

Proposition 17.1.8. Let M be a finite factor on a Hilbert space H, M' be also finite, and τ, τ' be the unique faithful normal tracial states on M, M' respectively. Then we have that:

(i) $\dim_{Mp'}(p'H) = \tau'(p')\dim_M(H)$, where p' is any non-zero projection of M';

(ii) Let $0 \neq \xi \in H$, and e_ξ, e'_ξ be the cyclic projections from H onto $\overline{M'\xi}, \overline{M\xi}$ respectively. Then

$$\dim_M(H) = c_M = \tau(e_\xi)/\tau'(e'_\xi)$$

and c_M is independent of the choice of $\xi(\neq 0)$;

(iii) $\dim_M(H)\dim_{M'}(H) = 1$;

(iv) $\dim_{pMp}(pH) = \tau(p)^{-1}\dim_M(H)$, where p is any non-zero projection of M;

(v) If L is a finite dimensional Hilbert space, then $\dim_M(H \otimes L) = \dim L \cdot \dim_M(H)$.

Proof. (i) Let u be a unitary operator from H onto $uu^*(L^2(M) \otimes K)$ such that $ux = (\lambda(x) \otimes 1_K)u, \forall x \in M$. Then for any non-zero projection p' of M' we have

$$\begin{aligned}\dim_{Mp'}(p'H) &= Tr'_{L^2(M)\otimes K}(up' \cdot (up')^*) \\ &= Tr'_H(p').\end{aligned}$$

Since M' is finite, it follows that $Tr'_H(\cdot) = \tau'(\cdot)Tr'_H(1)$. Therefore, we obtain that

$$\dim_{Mp'}(p'H) = \tau'(p')Tr'_H(1) = \tau'(p')\dim_M(H).$$

(ii) By Proposition 17.1.7, there is a positive constant c such that

$$c \cdot Tr'_{L^2(M)\otimes K}(f'_\eta) = tr(f_\eta), \quad \forall 0 \neq \eta \in L^2(M) \otimes K,$$

where f_η, f'_η are the cyclic projections from $L^2(M) \otimes K$ onto $\overline{(\lambda(M) \otimes 1_K)'\eta}$, $\overline{(\lambda(M) \otimes 1_K)\eta}$ respectively, tr is the unique faithful normal tracial state on $\lambda(M) \otimes 1_K$ i.e., $tr(\lambda(x) \otimes 1_K) = \tau(x), \forall x \in M$. Now we compute the constant c. Pick $\eta = 1 \otimes k$ with $0 \neq k \in K$. Then , $f_\eta(L^2(M) \otimes k) = \overline{(\lambda(M) \otimes 1_K)'(1 \otimes k)} = \overline{(\rho(M) \otimes B(K))(1 \otimes k)} = L^2(M) \otimes K$, i.e., $f_\eta = 1$; and $f'_\eta(L^2(M) \otimes K) = \overline{(\lambda(M) \otimes 1_K)(1 \otimes k)} = L^2(M) \otimes [k]$. Thus , we have $tr(f_\eta) = 1 = Tr'_{L^2(M)\otimes K}(f'_\eta)$, and $c = 1$.

Let u be a unitary operator from H onto $p'(L^2(M) \otimes K)$, where $p' = uu^* \in (\lambda(M) \otimes 1_K)'$ such that $ux = (\lambda(x) \otimes 1_K)u, \forall x \in M$. For any $0 \neq \xi \in H$, it is easy to see that

$$ue_\xi u^* = p' f_{u\xi}, \quad \text{and} \quad ue'_\xi u^* = f'_{u\xi}.$$

Since $f_{u\xi} \in \lambda(M) \otimes 1_K$ and $ue_\xi = f_{u\xi}u$, it follows that $f_{u\xi} = \lambda(e_\xi) \otimes 1_K$. From the preceding paragraph, we have

$$Tr'_{L^2(M)\otimes K}(f'_{u\xi}) = tr(f_{u\xi}) = \tau(e_\xi).$$

Moreover,

$$\begin{aligned}Tr'_{L^2(M)\otimes K}(f'_{u\xi}) &= Tr'_{L^2(M)\otimes K}(ue'_\xi u^*) \\ &= Tr'_H(e'_\xi) = \tau'(e'_\xi)Tr'_H(1) = \tau'(e'_\xi)\dim_M(H).\end{aligned}$$

Therefore, we obtain that $\dim_M(H) = c_M$.

(iii) It is immediate from (ii) .

(iv) From (iii) and (i) we have

$$\begin{aligned}\dim_{pM_p}(pH) &= [\dim_{M'_p}(pH)]^{-1}\\ &= [\tau(p)\dim_{M'}(H)]^{-1} = \tau(p)^{-1}\dim_M(H).\end{aligned}$$

(v) Pick $0 \neq \xi \in H$ and $0 \neq l \in L$. Clearly,

$$\overline{(M \otimes 1_L)(\xi \otimes l)} = (e'_\xi \otimes p_l)(H \otimes L),$$

and

$$\overline{(M \otimes 1_L)'(\xi \otimes l)} = (e_\xi \otimes 1_L)(H \otimes L),$$

where p_l is the projection from L onto $[l]$. Since $M \otimes 1_L$ and $(M \otimes 1_L)' = M'\overline{\otimes}B(L)$ are finite , it follows from (ii) that

$$\begin{aligned}\dim_M(H \otimes L) &= \frac{tr_{M\otimes 1_L}(e_\xi \otimes 1_L)}{tr'_{M'\overline{\otimes}B(L)}(e'_\xi \otimes p_l)}\\ &= \frac{\tau(e_\xi)}{(\dim L)^{-1}\tau'(e'_\xi)} = \dim L \cdot \dim_M(H).\end{aligned}$$

Q.E.D.

Proposition 17.1.9. Let M be a finite factor on a Hilbert space H, and M' be finite too. Then

$$\dim_M(H) = c_M \geq 1 \Longleftrightarrow M \text{ admits a separating vector },$$

and

$$\dim_M(H) = c_M \leq 1 \Longleftrightarrow M \text{ admits a cyclic vector.}$$

Consequently, $c_M = 1 \Longleftrightarrow M$ admits a cyclic–separating vector.

Proof. The sufficiency is obvious.

We claim that either M or M' admits a cyclic vector. In fact, since H is separable, by the Zorn lemma we can find a maximal sequence $\{\xi_i\}$ of non–zero elements of H such that $e_ie_j = \delta_{ij}e_i$, and $e'_ie'_j = \delta_{ij}e'_i, \forall i,j$, where e_i, e'_i are the cyclic projections from H onto $\overline{M'\xi_i}, \overline{M\xi_i}$ respectively, $\forall i$. Put $f = 1 - \sum_i e_i, f' = 1 - \sum_i e'_i$. If $ff' \neq 0$, pick $0 \neq \xi \in ff'H$, then we have $e_\xi e_i = 0$ and $e'_\xi e'_i = 0, \forall i$. It is impossible since the family $\{\xi_i\}$ is maximal . Thus, $ff' = 0$. Suppose that $f' \neq 0$. Since M' is a factor, $x \longrightarrow xf'$ is a $*$ isomorphism from M onto Mf'. Hence, $f = 0$, i.e., $\sum_i e_i = 1$. We may assume that $\|\xi_i\| \leq 2^{-i}, \forall i$, and let $\xi = \sum_i \xi_i$. Then $\overline{M'\xi} \supset \overline{M'e'_i\xi} = \overline{M'\xi_i} = e_iH, \forall i$. Hence ξ is a cyclic vector for M' . Similarly, if $f \neq 0$, then ξ is cyclic for M.

Now the necessity is also obvious. Q.E.D.

Remark. Let M be a type (I_n) factor on a m-dimensional Hilbert space H, where $n, m < \infty$. Clearly, $n|m$, and $m = np$. Then we can write $H = H_n \otimes H_p$ and $M = B(H_n) \otimes 1_p$, where H_n, H_p are n-dimensional, p-dimensional Hilbert spaces respectively. Pick $0 \neq \xi \in H_n, 0 \neq \eta \in H_p$. Then $(B(H_n) \otimes 1_p)(\xi \otimes \eta) = H_n \otimes \eta, (B(H_n) \otimes 1_p)'(\xi \otimes \eta) = (1_n \otimes B(H_p))(\xi \otimes \eta)) = \xi \otimes H_p$. Thus $e_{\xi\otimes\eta} = p_\xi \otimes 1_p$ and $e'_{\xi\otimes\eta} = 1_n \otimes p_\eta$, where p_ξ, p_η are the projections from H_n, H_p onto $[\xi], [\eta]$ respectively. Therefore, we have

$$\dim_M(H) = \tau(p_\xi)/\tau'(p_\eta) = p/n,$$

where τ, τ' are the canonical tracial states on $B(H_n), B(H_p)$ respectively.

Notes. Except for the presentation, all the material of this section comes from the original papers by F.J. Murray and J.Von Neumann.

References. [28], [60], [75], [111], [112], [113].

17.2. Index for subfactors

Definition 17.2.1. Let M be a finite factor. N is called a *subfactor* of M, if N is a W^*-subalgebra of M with the common identity, and is a factor (so N must be also finite) . The *index* of N in M, denoted by $[M : N]$, is $\dim_N(L^2(M))$.

Lemma 17.2.2. Let M be a finite factor on a Hilbert space H, and M' be finite too. Then there is a finite subset $\{\xi_1, \cdots, \xi_n\}$ of H such that $H = \sum_{i=1}^{n} \oplus \overline{[M\xi_i]}$, and there exists a common positive integer m such that for any $\sigma(M, M_*)$ -continuous positive linear functional φ on M, we can find $\eta_1, \cdots, \eta_m \in H$ and $\varphi(x) = \sum_{j=1}^{m} \langle x\eta_j, \eta_j \rangle, \forall x \in M$. Consequently, the weak topology and $\sigma(M, M_*)$ -topology in M are equivalent.

Proof. Pick a l-dimensional Hilbert space $L (l < \infty)$ such that $\dim_{M'}(H \otimes L) \geq 1$. Then by Proposition 17.1.9 $M' \otimes 1_L$ adimits a separating vector $\varsigma = (\varsigma_1, \cdots, \varsigma_l) \in H \otimes L$. Thus , ς is cyclic for $(M' \otimes 1_L)' = M \otimes B(L)$. Now it is easy to see that $\{\varsigma_1, \cdots, \varsigma_l\}$ is a cyclic subset for M. Pick $\xi_1 = \varsigma_1$, and let $\varsigma'_i = (1 - p'_1)\varsigma_i, 2 \leq i \leq l$, where p'_1 is the projection from H onto $\overline{[M\xi_1]}$. Clearly, we have $(1-p_1')H = \overline{[M\varsigma_i'|2 \le i \le l]}$. In this way, we can find $\{\xi_1, \cdots, \xi_n\} \subset H$ such that $H = \sum_{i=1}^{n} \oplus \overline{[M\xi_i]}$.

Similarly, there is a m-dimensional Hilbert space K such that $\dim_M(H \otimes K) \ge 1$. Thus $M \otimes 1_K$ admits a separating vector in $H \otimes K$. By Proposition 1.13.6, for any $\varphi \in (M_*)_+$ there exists $\eta = (\eta_1, \cdots, \eta_m) \in H \otimes K$ such that

$$\varphi(x) = \langle (x \otimes 1_K)\eta, \eta \rangle = \sum_{j=1}^{m} \langle x\eta_j, \eta_j \rangle, \forall x \in M.$$

Q.E.D.

Corollary 17.2.3. Let M_i be a finite factor on a Hilbert space H_i, and M_i' be finite too, $i = 1, 2$. If Φ is a $*$ isomorphism from M_1 onto M_2 , then we can write

$$\Phi = \Phi_3 \circ \Phi_2 \circ \Phi_1,$$

where $\Phi_1(x) = x \otimes 1_K, \forall x \in M_1$, and K is a finite dimensional Hilbert space; $\Phi_2(\cdot) = \cdot p', \forall \cdot \in M_1 \otimes 1_K$, and p' is a non-zero projection of $(M_1 \otimes 1_K)'$; and Φ_3 is a spatial $*$ isomorphism from $(M_1 \otimes 1_K)p'$ onto M_2 .

Proof. It is immediate from Lemma 17.2.2 and the proof of Theorem 1.12.4.

Q.E.D.

Proposition 17.2.4. Let M be a finite factor, N be a subfactor of M, and $\{\pi, H\}$ be a faithful nondegenerate W^*-representation of M. If $\dim_M(H) < \infty$ (i.e., $\pi(M)'$ is finite) , then we have that

$$[M : N] = \dim_N(H)/\dim_M(H).$$

In particular, $[M : N] < \infty$ if and only if for some (then for any) faithful nondegenerate W^*-representation $\{\pi, H\}$ of M with $\dim_M(H) < \infty$ we have $\dim_N(H) < \infty$.

Proof. Let $\{\pi_1, H_1\}, \{\pi_2, H_2\}$ be two faithful nondegenerate W^*-representations of M, and $\dim_M(H_i) < \infty, i = 1, 2$. By Corollary 17.2.3, there is a finite dimensional Hilbert space K and a non-zero projection p' of $(\pi_2(M) \otimes 1_K)'$ such that

$$\{\pi_1, H_1\} \cong \{(\pi_2 \otimes 1_K)p', p'(H_2 \otimes K)\}.$$

Thus , we have $\dim_N(H_1) = \dim_N(p'(H_2 \otimes K))$. If $\dim_N(H_2) < \infty$, then by Proposition 17.1.8 we have $\dim_N(H_2 \otimes K) < \infty$, and $\dim_N(H_1) \le \dim_N(H_2 \otimes K) < \infty$. Hence,

$$\dim_N(H_1) < \infty \Longleftrightarrow \dim_N(H_2) < \infty.$$

Consequently, $[M:N]=\dim_N(L^2(M))<\infty \Longleftrightarrow \dim_N(H)<\infty$, where $\{\pi, H\}$ is some (then any) faithful nondegenerate W^*-representation of M and $\dim_M(H)<\infty$.

Now let $\dim_M(H)<\infty$. By Corollary 17.2.3, there is a finite dimensional Hilbert space K and a non-zero projection p' of $(\lambda(M)\otimes 1_K)'$ such that

$$\{\pi, H\}\cong\{(\lambda\otimes 1_K)p', p'(L^2(M)\otimes K)\}.$$

By Proposition 17.1.8 we have

$$\begin{aligned}\dim_N(H) &= \dim_N(p'(L^2(M)\otimes K))\\ &= \tau'(p')\dim_N(L^2(M)\otimes K)\\ &= \tau'(p')\dim K\dim_N(L^2(M)),\end{aligned}$$

where τ' is the unique faithful normal tracial state on $(\lambda(M)\otimes 1_K)'=\lambda(M)'\overline{\otimes} B(K)$; and

$$\dim_M(H)=\dim_M(p'(L^2(M)\otimes K))=\tau'(p')\dim K.$$

Therefore, $[M:N]=\dim_N(L^2(M))=\dim_N(H)/\dim_M(H)$. Q.E.D.

Remark. If $N=B(H_n)\otimes 1_p, M=B(H_n)\otimes B(H_p)=B(H_m)$, where $m=np$, then $\dim L^2(M)=m^2$. By the end of Section 17.1 we have

$$[M:N]=\dim_N(L^2(M))=\frac{m^2}{n^2}=p^2.$$

Lemma 17.2.5. Let M be a finite factor, P be a subfactor of M, and $a\in M$. If there is a sequence $\{b_n\}\subset P$ such that $b_n\longrightarrow a$ in $L^2(M)$, then $a\in P$. Consequently, $L^2(P)\cap M=P$ (regard $L^2(P)$ and M as linear subspaces of $L^2(M)$).

Proof. Since $\|b_n-a\|_2\longrightarrow 0$ and $b_n\in P$, it follows that $xa\in L^2(P), \forall x\in P$. Thus , $\rho(a)P\subset L^2(P)$, and $L^2(P)$ is invariant under $\rho(a)$. Put $t=\rho(a)|L^2(P)$. Then t is a bounded linear operator on $L^2(P)$, and $tx=xa=\lambda(x)a, \forall x\in P$. Hence

$$a\in B_r=\left\{b\in L^2(P)\,\middle|\,\begin{array}{c}\text{there exists } \rho(b)\in B(L^2(P)) \text{ such that}\\ \rho(b)x=\lambda(x)b, \forall x\in P\end{array}\right\}$$

Now by the Remark under Proposition 17.1.2, we have $a\in P$. Q.E.D.

Proposition 17.2.6. Let M be a finite factor.

(i) If N is a subfactor of M, then we have

$$[M:M]=1, \quad \text{and} \quad [M:N]\geq 1;$$

(ii) If N is a subfactor of $M, M \subset B(H)$, and N' is finite, then we have $[M:N] = [N':M'] < \infty$;

(iii) If Q is a subfactor of M, P is a subfactor of Q, then we have

$$[M:P] = [M:Q] \cdot [Q:P], \quad \text{and} \quad [M:P] \geq [M:Q];$$

(iv) If Q is a subfactor of M, P is a subfactor of $Q, [M:P] < \infty$, and $[M:P] = [M:Q]$, then we have $Q = P$.

Proof. (i) $\lambda(M)$ admits a cyclic–separating vector $1 (\in M \subset L^2(M))$. So 1 is also separating for $\lambda(N)$. Now by Proposition 17.1.9 we have

$$[M:N] = \dim_N(L^2(M)) \geq 1.$$

(ii) Since M, N' are finite, N, M' are also finite. Now by Proposition 17.1.6 we have $\dim_M(H) < \infty, \dim_{M'}(H) < \infty, \dim_N(H) < \infty$ and $\dim_{N'}(H) < \infty$. Further, from Propositions 17.1.8 and 17.2.4 we obtain that

$$\begin{aligned}[M:N] &= \dim_N(H)/\dim_M(H) \\ &= \dim_{N'}(H)^{-1}/\dim_{M'}(H)^{-1} \\ &= \dim_{M'}(h)/\dim_{N'}(H) = [N':M'] < \infty.\end{aligned}$$

(iii) If $[M:Q] < \infty$, then we can pick a faithful nondegenerate W^*–representation $\{\pi, H\}$ of M such that $\pi(M)'$ is finite and $\dim_Q(H) < \infty$ (i.e., $\pi(Q)'$ is finite). Hence

$$[M:Q] \cdot [Q:P] = \frac{\dim_Q(H)}{\dim_M(H)} \cdot \frac{\dim_P(H)}{\dim_Q(H)} = \frac{\dim_P(H)}{\dim_M(H)} = [M:P].$$

Let $[M:Q] = \infty$, and pick a faithful nondegenerate W^*–representation $\{\pi, H\}$ of M such that $\pi(M)'$ is finite. Then $\pi(Q)'$ is infinite, and $\pi(P)' (\supset \pi(Q)')$ is also infinite. Hence $\dim_P(H) = \infty$ and $[M:P] = \infty$.

(iv) By (iii) it suffices to show that $M = P$ if $[M:P] = 1$.

Since $[M:P] = 1 = \dim_P(L^2(M))$, $\lambda(P)'$ is finite. Let τ, τ' be the unique faithful normal tracial states on $\lambda(P), \lambda(P)'$ respectively, and e, e' be the cyclic projections from $L^2(M)$ onto $\overline{\lambda(P)'1}, \overline{\lambda(P)1}$ respectively, where 1 is the identity of $M (\subset L^2(M))$. Since 1 is cyclic–separating for $\lambda(M)$, and $\lambda(P)' \supset \lambda(M)'$, it follows that $e = 1_{L^2(M)}$. Now from

$$1 = [M:P] = \dim_P(L^2(M)) = \tau(e)/\tau'(e'),$$

we have $\tau'(e') = 1$ and $e' = 1_{L^2(M)}$, i.e., $\overline{\lambda(P)1} = \overline{P1} = L^2(M)$. Hence, for any $a \in M$ there is $\{b_n\} \subset P$ such that $b_n = b_n 1 \longrightarrow a$ in $L^2(M)$. Then by Lemma 17.2.5 we have $a \in P$. Therefore, $P = M$. Q.E.D.

Proposition 17.2.7. Let M_i be a finite factor, and N_i be a subfactor of $M_i, i = 1, 2$. Then we have

$$[M_1 \overline{\otimes} M_2 : N_1 \overline{\otimes} N_2] = [M_1 : N_1] \cdot [M_2 : N_2].$$

Proof. Let τ_1, τ_2 be the faithful normal tracial states on M_1, M_2 respectively. Then $\tau_1 \otimes \tau_2$ is the faithful normal tracial state on $M_1 \overline{\otimes} M_2$. Hence we have $L^2(M_1 \overline{\otimes} M_2) = L^2(M_1) \otimes L^2(M_2), \lambda(M_1 \overline{\otimes} M_2) = \lambda(M_1) \overline{\otimes} \lambda(M_2)$, and $\lambda(N_1 \overline{\otimes} N_2) = \lambda(N_1) \overline{\otimes} \lambda(N_2)$. By Theorem 6.9.12, $\lambda(N_1 \overline{\otimes} N_2)'$ is finite $\Longleftrightarrow$ both $\lambda(N_1)'$ and $\lambda(N_2)'$ are finite.

Now we may assume that $\lambda(N_1)'$ and $\lambda(N_2)'$ are finite. Let τ_1', τ_2' be the faithful normal tracial states on $\lambda(M_1)', \lambda(M_2)'$ respectively, and $\xi_i = 1_i$ be the identity of $M_i, i = 1, 2$. Then ξ_i is cyclic-separating for $\lambda(M_i)$ (in $L^2(M_i)), i = 1, 2$. Clearly, $\overline{\lambda(N_i)'\xi_i} = L^2(M_i), i = 1, 2$, and $\overline{\lambda(N_1 \overline{\otimes} N_2)'(\xi_1 \otimes \xi_2)} = L^2(M_1 \overline{\otimes} M_2)$. On the other hand, we have $\overline{\lambda(N_1 \overline{\otimes} N_2)(\xi_1 \otimes \xi_2)} = (e_1' \otimes e_2')(L^2(M_1) \otimes L^2(M_2))$, where e_i' is the cyclic projection from $L^2(M_i)$ onto $\overline{\lambda(N_i)\xi_i}, i = 1, 2$. Therefore, we have

$$\begin{aligned}
[M_1 \overline{\otimes} M_2 : N_1 \overline{\otimes} N_2] &= \dim_{N_1 \overline{\otimes} N_2}(L^2(M_1 \overline{\otimes} M_2)) \\
&= \frac{(\tau_1 \otimes \tau_2)(1)}{(\tau_1' \otimes \tau_2')(e_1' \otimes e_2')} = \frac{1}{\tau_1'(e_1')} \cdot \frac{1}{\tau_2(e_2')} \\
&= \dim_{N_1}(L^2(M_1)) \cdot \dim_{N_2}(L^2(M_2)) \\
&= [M_1 : N_1] \cdot [M_2 : N_2].
\end{aligned}$$

Q.E.D.

Example. Let G be an ICC group, and H be an ICC subgroup of G. Suppose that $s \longrightarrow \lambda_s$ is the left regular representation of G on $l^2(G)$. Then $R(G) = M$ is a type (II_1) factor on $l^2(G)$, and the faithful normal tracial state τ on $R(G)$ is as follows:

$$\tau(x) = \langle x\delta_e, \delta_e \rangle, \quad \forall x \in R(G),$$

where $\delta_e(s) = \delta_{e,s}, \forall s \in G$, and e is the unit of G. Clearly, $\{\delta_s | s \in G\}, \{\lambda_s | s \in G\}$ are orthogonal normalized bases of $l^2(G), L^2(M)$ respectively. Then U is a unitary operator from $l^2(G)$ onto $L^2(M)$, where $U\delta_s = \lambda_s, \forall s \in G$. Thus, we have

$$\dim_{R(G)}(l^2(G)) = \dim_M(L^2(M)) = 1.$$

Similarly, we have $\dim_{R(H)}(l^2(H)) = 1$. Pick a subset E of G such that $Hg' \neq Hg'', \forall g', g'' \in E$ and $g' \neq g''$; and $\{Hg | g \in G\} = \{Hg | g \in E\}$. Then we have

$$l^2(G) = \sum_{g \in E} \oplus l^2(Hg).$$

Now by Proposition 17.1.6 we obtain that

$$\begin{aligned}[R(G):R(H)] &= \dim_{R(H)}(L^2(R(G))) \\ &= \#\{Hg | g \in E\} = [G:H]\end{aligned}$$

Notes. Index for subfactors was inlroduced by V.F.R.Jones, and he showed that this definition agrees with the ring-theoretic one.

References.[60], [75], [76].

17.3. The fundamental construction

Let M be a finite factor, τ be the unique faithful normal tracial state on M, and N be a subfactor of M.

For each $h \in M$ with $0 \le h \le 1$, define

$$\varphi_h(y) = \tau(hy) = \tau(h^{1/2}yh^{1/2}), \quad \forall y \in N.$$

Clearly, $\varphi_h \in (N_*)_+$. If $y \in N_+$, then $\varphi_h(y) = \tau(hy) = \tau(y^{1/2}hy^{1/2})$. So we have $0 \le \varphi_h \le (\tau|N)$. By Theorem 1.10.3 there exists $t_0 \in N$ with $0 \le t_0 \le 1$ such that

$$\tau(hy) = \varphi_h(y) = \tau(t_0yt_0) = \tau(t_0^2y), \quad \forall y \in N.$$

Let $E(h) = t_0^2 (\in N)$. Then we obtain that $\tau(hy) = \tau(E(h)y), \forall y \in N$. Generally, for any $x \in M$ there is $E(x) \in N$ such that

$$\tau(xy) = \tau(E(x)y), \quad \forall y \in N.$$

We claim that $E(x)$ is uniquely determined by x. In fact, if $z \in N$ is such that $\tau(zy) = 0, \forall y \in N$, then we have $\tau(zz^*) = 0$ and $z = 0$ since τ is faithful. Hence, $x \longrightarrow E(x)$ is a linear map from M onto N, and clearly,

$$E(z) = z, \forall z \in N, \quad \text{and} \quad E^2(x) = E(x), \forall x \in M.$$

From the above discussion, we also have $0 \le E(h) \le \|h\|$ for $h \in M_+$. So $E(x^*) = E(x)^*, \forall x \in M$. Moreover, since

$$\tau(E(aE(b))y) = \tau(a \cdot E(b)y) = \tau(E(a)E(b)y),$$

$\forall y \in N$, it follows that $E(aE(b)) = E(a) \cdot E(b), \forall a, b \in M$. Similarly, by $\tau(E(E(a)b)y) = \tau(E(a)by) = \tau(b \cdot yE(a)) = \tau(E(b)yE(a)) = \tau(E(a)E(b)y), \forall y \in N$, we have $E(E(a)b) = E(a)E(b), \forall a, b \in M$. Hence

$$0 \le E((x - E(x))^*(x - E(x))) = E(x^*x) - E(x)^*E(x),$$

and

$$0 \le E(x)^* E(x) \le E(x^* x) \le \|x^* x\| = \|x\|^2, \quad \forall x \in M.$$

Therefore, $E(\cdot)$ is a projection of norm one from M onto N.

If $h \in M_+$ is such that $E(h) = 0$, then $\tau(hy) = 0, \forall y \in N$. In particular, $\tau(h) = 0$. But τ is faithful, so $h = 0$, i.e., $E(\cdot)$ is faithful.

Now suppose that $\{x_l\} \subset M, x_l \longrightarrow 0$ in $\sigma(M, M_*)$, and $\|x_l\| \le 1, \forall l$. Let $\{\pi, H, \xi\}$ be the faithful cyclic W^*-representation of N generated by $(\tau|N)$. Then we have

$$\begin{aligned} &\langle \pi(E(x_l))\pi(y)\xi, \pi(z)\xi\rangle = \tau(z^* E(x_l) y) \\ = \;& \tau(E(x_l) y z^*) = \tau(x_l y z^*) \longrightarrow 0, \quad \forall y, z \in N. \end{aligned}$$

Since $\|\pi(E(x_l))\| \le 1, \forall l$, it follows that $\pi(E(x_l)) \longrightarrow 0$ (weakly) . Further, $E(x_l) \longrightarrow 0$ in $\sigma(N, N_*)$. Therefore, $E(\cdot)$ is $\sigma(M, M_*)$ - $\sigma(N, N_*)$ continuous.

By Theorem 4.1.5 we have the following.

Proposition 17.3.1. Let M be a finite factor, and N be a subfactor of M. Then there is a linear map $E(\cdot) : M \longrightarrow N$ such that

(i) $E(\cdot)$ is a projection of norm one from M onto N;

(ii) $E(\cdot)$ is faithful, i.e., if $E(x) = 0$ for some $x \in M_+$, then $x = 0$;

(iii) $E(\cdot)$ is $\sigma(M, M_*)$ - $\sigma(N, N_*)$ continuous;

(iv) $E(\cdot)$ is completely positive, in particular,

$$E(M_+) = N_+, \text{and } E(x^*) = E(x)^*, \forall x \in M;$$

(v) $E(aE(b)) = E(E(a)b) = E(a)E(b), \forall a, b \in M$;

(vi) $E(x)^* E(x) \le E(x^* x), \forall x \in M$;

(vii) $\tau(E(x)y) = \tau(xy), \forall x \in M, y \in N$. In particular, $E(\cdot)$ keeps $\tau(\cdot)$.

Proposition 17.3.2. Let M be a finite factor, τ be the unique faithful normal tracial state on M, and N be a subfactor of M. Using τ, we construct the Hilbert spaces $L^2(M)$ and $L^2(N)$. Naturally, $L^2(N)$ can be regarded as a closed linear subspace of $L^2(M)$ (i.e., $L^2(N) = \overline{N}$, the closure of N in $L^2(M)$). Let P be the projection from $L^2(M)$ onto $L^2(N)$, and $E(\cdot)$ be as in Proposition 17.3.1. Then we have the following:

(i) $P(x) = E(x), \forall x \in M$;

(ii) $P\lambda(x)P = \lambda(E(x))P, \forall x \in M$;

(iii) If $x \in M$, then $x \in N \Longleftrightarrow P\lambda(x) = \lambda(x)P$;

(iv) $\lambda(N)' = \{\lambda(M)', P\}'' = \{\rho(M), P\}''$, and $\lambda(N) = \{\rho(M), P\}'$;

(v) $JP = PJ$, and $JPJ = P$.

Proof. (i) For any $x \in M$, by Proposition 17.3.1 we have

$$\begin{aligned}\langle x - E(x), y\rangle &= \tau(y^*x) - \tau(y^*E(x)) \\ &= \tau(xy^*) - \tau(E(x)y^*) = 0, \quad \forall y \in N.\end{aligned}$$

Thus, $(x - E(x)) \perp \overline{N} = L^2(N) = P(L^2(M)), \forall x \in M$. Further, by $E(x) \in N \subset L^2(N)$ we get $P(x) = E(x), \forall x \in M$.

(ii) For any $x, y \in M$, by (i) we have

$$\begin{aligned}P\lambda(x)Py &= P(xE(y)) = E(xE(y)) \\ &= E(x)E(y) = \lambda(E(x))Py.\end{aligned}$$

Hence, $P\lambda(x)P = \lambda(E(x))P, \forall x \in M$.

(iii) If $x \in N$, then for any $y \in M$,

$$P\lambda(x)y = E(xy) = xE(y) = \lambda(x)Py.$$

Hence, $P\lambda(x) = \lambda(x)P$. Conversely, let $P\lambda(x) = \lambda(x)P$. Then we have

$$x = \lambda(x)1 = \lambda(x)P1 = P\lambda(x)1 = Px = E(x) \in N,$$

where 1 is the common identity of M and N.

(iv) Clearly, $\lambda(N) \subset \rho(M)'$. Further, by (iii) we have

$$\lambda(N) \subset \{\rho(M), P\}'.$$

Conversely, if $t \in \{\rho(M), P\}'$, then $t \in \rho(M)' = \lambda(M)$. So $t = \lambda(x)$ for some $x \in M$. Now by $\lambda(x)P = P\lambda(x)$ and (iii) , we get $x \in N$. Thus , $t = \lambda(x) \in \lambda(N)$.

(v) For any $x \in M$, it is obvious that

$$JPx = JE(x) = E(x^*) = Px^* = PJx.$$

Since M is dense in $L^2(M)$, it follows that $JP = PJ$. Q.E.D.

Definition 17.3.3. Keep the assumptions and notations in Proposition 17.3.2. Then, define

$$\langle M, P\rangle = \{\lambda(M), P\}''.$$

It is a VN algebra on $L^2(M)$.

Proposition 17.3.4. Using the assumptions and notations in Proposition 17.3.2 and Definition 17.3.3, we have the following:

(i) $\langle M, P\rangle = J\lambda(N)'J$, and $\langle M, P\rangle$ is a factor;

(ii) $\{\sum_i \lambda(x_i)P\lambda(y_i) | x_i, y_i \in M\}$ is a weakly dense $*$ subalgebra of $\langle M, P\rangle$;

(iii) $x \longrightarrow \lambda(x)P$ is a $*$ isomrophism from N onto $P\langle M,P\rangle P$;
(iv)

$$\langle M,P\rangle \text{ is finite } \Longleftrightarrow \lambda(N)' \text{ is finite} \Longleftrightarrow [M:N]<\infty;$$

(v) If M and N are type (II_1), and $[M:N]<\infty$, then $\langle M,P\rangle$ is also type (II_1).

Proof. (i) By Proposition 17.3.2, we have

$$\begin{aligned} J\lambda(N)'J &= J\{\lambda(M)',P\}''J \\ &= \{J\rho(M)J, JPJ\}'' = \{\lambda(M),P\}'' = \langle M,P\rangle. \end{aligned}$$

Moreover, since $\lambda(N)'$ is a factor, $\langle M,P\rangle$ is also a factor.

(ii) Noticing that

$$\begin{aligned} \lambda(x)P\lambda(y)P\lambda(z)a &= xE(yE(za)) \\ &= xE(y)E(za) = \lambda(xE(y))P\lambda(z)a, \end{aligned}$$

$\forall a,x,y,z \in M$, we can see that

$$X = \{\lambda(x_0) + \sum_i \lambda(x_i)P\lambda(y_i) | x_0, x_i, y_i \in M\}$$

is a $*$ subalgebra of $\langle M,P\rangle$. Since $\lambda(M) \subset X$ and $P \in X$, it follows that $\overline{X}^{\sigma} = \langle M,P\rangle$. Now let

$$Y = \{\sum_i \lambda(x_i)P\lambda(y_i) | x_i, y_i \in M\}.$$

Clearly, Y is a $*$ two-sided ideal of X. Thus, $\overline{Y}^{\sigma}$ is a σ -closed two-sided ideal of $\langle M,P\rangle$. But $\langle M,P\rangle$ is a facotr, so by Proposition 1.7.1 Y is weakly dense in $\langle M,P\rangle$.

(iii) For any $x \in N, y \in M$, we have

$$P\lambda(x)Py = E(xE(y)) = xE(y) = \lambda(x)Py,$$

i.e., $P\lambda(x)P = \lambda(x)P, \forall x \in N$. So $\lambda(N)P \subset P\langle M,P\rangle P$. Since $\lambda(x)P = P\lambda(x), \forall x \in N$ (see Proposition 17.3.2), $x \longrightarrow \lambda(x)P$ is a $*$ homomorphism from N to $P\langle M,P\rangle P$. It is also injective. Indeed, if $\lambda(x)P = 0$ for some $x \in N$, then $xy = 0, \forall y \in N$. So $x = 0$.

Now by (ii) , it suffices to show that

$$P\lambda(x)P\lambda(y)P \in \lambda(N)P, \quad \forall x,y \in M.$$

It is equivalent to prove that $P\lambda(x)P \in \lambda(N)P, \forall x \in M$. But it is immediate by $P\lambda(x)P = \lambda(E(x))P \in \lambda(N)P, \forall x \in M$.

(iv) Since $\langle M,P\rangle = J\lambda(N)'J$ is $*$ isomorphic to $\lambda(N)'$, it follows that

$$\begin{aligned}\langle M,P\rangle \text{ is finite } &\Longleftrightarrow \lambda(N)' \text{ is finite}\\ &\Longleftrightarrow [M:N] = \dim_N(L^2(M)) < \infty.\end{aligned}$$

(v) By (iv), $\lambda(N)'$ is a finite factor. Since $\lambda(M)' = J\lambda(M)J$ is continuous and $\lambda(N') \supset \lambda(M)'$, it follows that $\lambda(N)'$ is type (II_1). Now since $\langle M,P\rangle$ is $*$ isomorphic to $\lambda(N)', \langle M,P\rangle$ is also type (II_1). Q.E.D.

Proposition 17.3.5. Let M be a finite factor, N be a subfactor of M with $[M:N] < \infty$, and τ be the unique faithful normal tracial state on M. Then τ has the *Markov property of modulus* $\beta = [M:N]$, i.e.,

$$tr(\lambda(x)) = \tau(x), \quad \beta tr(\lambda(x)P) = \tau(x), \quad \forall x \in M,$$

where $tr(\cdot)$ is the unique faithful normal tracial state on $\langle M,P\rangle$ (noticing that $\langle M,P\rangle$ is also a finite factor by Proposition 17.3.4). In particular, $tr(P) = [M:N]^{-1}$.

Moreover, we have $[\langle M,P\rangle : M] = [M:N]$.

Proof. Since $tr \circ \lambda(\cdot)$ is also a faithful normal tracial state on M, it follows that $tr(\lambda(x)) = \tau(x), \forall x \in M$. By Proposition 17.3.4, $x \longrightarrow \lambda(x)P$ is a $*$ isomorphism from N onto $\lambda(N)P$, and $P \in \lambda(N)'$. So we have

$$tr(\lambda(x)P) = k\tau(x), \quad \forall x \in N,$$

where k is some positive constant. Clearly, $k = tr(P)$.

From $[M:N] < \infty, \lambda(N)'$ is finite. Then

$$[M:N] = \dim_N(L^2(M)) = tr(e)/tr'(e'),$$

where tr' is the unique faithful normal tracial state on $\lambda(N)'$, and e, e' are the cyclic projections from $L^2(M)$ onto $\overline{\lambda(N)'1}, \overline{\lambda(N)1}$ respectively. Since $\lambda(N)' \supset \lambda(M)'$, it follows that $e = 1_{L^2(M)}$. Moreover, $\overline{\lambda(N)1} = \overline{N} = L^2(N)$. Hence, $[M:N] = tr'(P)^{-1}$. Further, by $\langle M,P\rangle = J\lambda(N)'J$ we have $[M:N] = tr'(P)^{-1} = tr(JPJ)^{-1} = tr(P)^{-1}$. Then,

$$\begin{aligned}tr(\lambda(x)P) &= tr(P\lambda(x)P) = tr(\lambda(E(x))P)\\ &= [M:N]^{-1}\tau(E(x)) = [M:N]^{-1}\tau(x), \quad \forall x \in M.\end{aligned}$$

Finally, by Proposition 17.2.4 we obtain that

$$\begin{aligned}[\langle M,P\rangle : M] &= \dim_M(L^2(M))/\dim_{\langle M,P\rangle}(L^2(M))\\ &= \dim_{\lambda(N)'}(L^2(M))^{-1}\\ &= \dim_N(L^2(M)) = [M:N].\end{aligned}$$

Q.E.D.

Now let M be a finite factor, N be a subfactor of M, and $[M : N] < \infty$. By Proposition 17.3.5, we obtain a new pair of finite factors $\lambda(M)$ and $\langle M, P\rangle = \{\lambda(M), P\}''$ on $L^2(M)$. Let $M_0^{(0)} = N; M_0^{(1)} = N, M_1^{(1)} = M; M_1^{(2)} = \lambda(M_1^{(1)}), P_1 = P$, and

$$M_2^{(2)} = \langle M, P\rangle = \langle M_1^{(2)}, P_1\rangle = \{M_1^{(2)}, P_1\}''.$$

Since $[M_2^{(2)} : M_1^{(2)}] = [M_1^{(1)} : M_0^{(1)}] = [M : N] < \infty$, we can continue this process.

Thus, we have the following:

$$\begin{array}{llll}
M_0^{(0)} & & & \\
\downarrow & & & \\
M_0^{(1)} \subset & M_1^{(1)} & & \\
& \downarrow \lambda & & \\
& M_1^{(2)} \subset & M_2^{(2)} = \langle M_1^{(2)}, P_1\rangle & \\
& & \downarrow \lambda & \\
& & \cdots\cdots & \\
& & \quad\quad \downarrow \lambda & \\
& & \quad\quad M_k^{(k+1)} \subset & M_{k+1}^{(k+1)} = \langle M_k^{(k+1)}, P_k\rangle \\
& & & \cdots\cdots\cdots
\end{array}$$

Hence, there is an inductive system of finite facotrs:

$$M_0^{(0)} \xrightarrow{\Phi_0} M_1^{(1)} \xrightarrow{\Phi_1} \cdots \xrightarrow{\Phi_k} M_{k+1}^{(k+1)} = \langle M_k^{(k+1)}, P_k\rangle \longrightarrow \cdots,$$

where $M_{k+1}^{(k+1)}$ is a finite factor on $L^2(M_k^{(k)})$, P_k is the projection from $L^2(M_k^{(k)})$ onto $L^2(M_{k-1}^{(k)})$, and Φ_k is a $*$ isomorphism from $M_k^{(k)}$ onto $M_k^{(k+1)}, \forall k \geq 1$.

We consider the inductive limit of $\{M_k^{(k)}, \Phi_k | k\}$. By Section 3.7, let

$$\times_k M_k^{(k)} = \{(a_k)_{k\geq 0} | a_k \in M_k^{(k)}, \forall k \geq 0\},$$

$$A = \left\{(a_k) \in \times_k M_k^{(k)} \,\middle|\, \begin{array}{c}\text{there exists } k_0 \text{ such that} \\ a_{s+1} = \Phi_s(a_s), \forall s \geq k_0\end{array}\right\},$$

and

$$I = \left\{(a_k) \in A \,\middle|\, \begin{array}{c}\text{there exists } k_0 \text{ such that} \\ a_s = 0, \forall s \geq k_0\end{array}\right\}.$$

Then $\times_k M_k^{(k)}$ is a $*$ algebra in a natural way, A is a $*$ subalgebra of $\times_k M_k^{(k)}$, I is a $*$ two-sided ideal of A , and

$$\varinjlim\{M_k^{(k)}, \Phi_k | k\} = A/I.$$

Let

$$M_k = \left\{ \tilde{a} \in A/I \,\middle|\, \begin{array}{c} \text{there exists } (0,\cdots,0,a_k,a_{k+1},\cdots) \in \tilde{a}, \\ \text{where } a_{s+1} = \Phi_s(a_s),\ \forall s \geq k \end{array} \right\},$$

$\forall k \geq 0$. Then we have the following commutative diagram:

$$\begin{array}{ccccccccccc} 1 \in & M_0^{(0)} & \xrightarrow{\Phi_0} & M_1^{(1)} & \xrightarrow{\Phi_1} & M_2^{(2)} & \xrightarrow{\Phi_2} & \cdots & \longrightarrow & M_k^{(k)} & \longrightarrow & \cdots \\ & \downarrow \Psi_0 & & \downarrow \Psi_1 & & \downarrow \Psi_2 & & & & \downarrow \Psi_k & & \\ 1 \in & M_0 & \subset & M_1 & \subset & M_2 & \subset & \cdots & \subset & M_k & \subset & \cdots, \end{array}$$

where Ψ_k is a $*$ isomorphism from $M_k^{(k)}$ onto M_k as follows: for any $a_k \in M_k^{(k)}$, let $\Psi_k(a_k) = \tilde{a}$, then

$$(0,\cdots,0,a_k,\Phi_k(a_k),\cdots) \in \tilde{a},$$

$\forall k \geq 0$. Define $e_k = \Psi_{k+1}(P_k), \forall k \geq 1$. Since $M_{k+1}^{(k+1)} = \langle \Phi_k(M_k^{(k)}), P_k \rangle$, it follows that $M_{k+1} = \langle M_k, e_k \rangle$, i.e., M_{k+1} is generated by M_k and a projection $e_k, \forall k \geq 1$. Clearly, M_k is a finite factor, and

$$[M_{k+1} : M_k] = [M : N], \quad \forall k \geq 0.$$

Definition 17.3.6. The above chain $1 \in M_0 \subset \cdots \subset M_{k+1} = \langle M_k, e_k \rangle \subset \cdots$ is called the *tower* of finite factors induced by a pair $\{N \subset M\}$ of finite factors with $[M : N] < \infty$.

Theorem 17.3.7. Let M be a finite factor, N be a subfactor of M with $[M : N] < \infty$, and $1 \in M_0 \subset M_1 \subset \cdots \subset M_{k+1} = \langle M_k, e_k \rangle \subset \cdots\cdots$ be the tower of finite factors induced by $\{N \subset M\}$. Then we have the following:

(i) the pair $M_0 \subset M_1$ is $*$ isomorphisc to $N \subset M$;

(ii) M_k is a finite factor, and $[M_{k+1} : M_k] = [M : N], \forall k \geq 0$;

(iii) If τ_k is the unique faithful normal tracial state on M_k, then τ_k has the Markov property of modulus of $\beta = [M : N]$, i.e., $\tau_{k+1}|M_k = \tau_k$, and

$$\beta \tau_{k+1}(x e_k) = \tau_k(x), \quad \forall x \in M_k,$$

$\forall k \geq 1$. In particular, $\tau(e_k) = \beta^{-1}, \forall k \geq 1$, where τ is the tracial state on $\sqcup_k M_k$ such that $\tau|M_k = \tau_k, \forall k$;

(iv) $M_{k+1} = \langle M_k, e_1, \cdots, e_k \rangle$, i.e., M_{k+1} is generated by M_k and $\{e_1, \cdots, e_k\}, \forall k \geq 1$;

(v) the sequence $\{e_k | k \geq 1\}$ of projections satisfies the following relations:

$$e_k e_{k+1} e_k = \beta^{-1} e_k, \quad e_{k+1} e_k e_{k+1} = \beta^{-1} e_{k+1}, \forall k \geq 1;$$

and $e_k e_j = e_j e_k$, if $|k - j| \geq 2$.

Proof. It suffices to prove (v) .

Let $(k-j) \geq 2$ and $j \geq 1$. Notice that

$$\begin{array}{ccccc} M_{k-1}^{(k-1)} & \xrightarrow{\Phi_{k-1}} & M_k^{(k)} & \xrightarrow{\Phi_k} & M_{k+1}^{(k+1)} \\ \downarrow \Psi_{k-1} & & \downarrow \Psi_k & & \downarrow \Psi_{k+1} \\ M_{k-1} \subset & & M_k \subset & & M_{k+1} \end{array}$$

and $e_j \in M_{k-1}, e_k \in M_{k+1}$. So the relation $e_k e_j = e_j e_k$ is equivalent to that $\Phi_k \circ \Phi_{k-1} \circ \Psi_{k-1}^{-1}(e_j)$ and $\Psi_{k+1}^{-1}(e_k)$ commute. Define $\overline{N} = \Phi_{k-1}(M_{k-1}^{(k-1)}), \overline{M} = M_k^{(k)}, \overline{P} = P_k$, and $\overline{M}_1 = M_{k+1}^{(k+1)}$. Then $\overline{P}$ is the projection from $L^2(\overline{M})$ onto $L^2(\overline{N})$, and $\overline{M}_1 = \langle \lambda(\overline{M}), \overline{P} \rangle = \{\lambda(\overline{M}), \overline{P}\}''$. Since $\Phi_k \circ \Phi_{k-1} \circ \Psi_{k-1}^{-1}(e_j) = \Phi_k \circ \Phi_{k-1}(P_j) \in \lambda(\overline{N})$, and $\Psi_{k+1}^{-1}(e_k) = \overline{P}$, it follows from Proposition 17.3.2 (iii) that $\Phi_k \circ \Phi_{k-1}(P_j)$ and $\overline{P}$ commute. Therefore, we have that $e_k e_j = e_j e_k, \forall |k-j| \geq 2$.

For $k \geq 1$, let $\overline{N} = M_{k-1}^{(k)}, \overline{M} = M_k^{(k)}, \Phi = \Phi_k(= \lambda); \overline{M}_1 = M_{k+1}^{(k+1)}, \Psi = \Phi_{k+1}(= \lambda)$; and $\overline{M}_2 = M_{k+2}^{(k+2)}$. Then we have

$$(\overline{N} \subset)\overline{M} \xrightarrow{\Phi=\lambda} \langle \Phi(\overline{M}), P_1 \rangle = \overline{M}_1 \xrightarrow{\Psi=\lambda} \langle \Psi(\overline{M}_1), P_2 \rangle = \overline{M}_2,$$

where P_1 is the projection from $L^2(\overline{M})$ onto $L^2(\overline{N}), \overline{M}_1$ is a finite factor on $L^2(\overline{M})$ generated by $\Phi(\overline{M}) = \lambda(\overline{M})$ and P_1; P_2 is the projection from $L^2(\overline{M}_1)$ onto $L^2(\Phi(\overline{M}))$, and $\overline{M}_2$ is a finite factor on $L^2(\overline{M}_1)$ generated by $\Psi(\overline{M}_1) = \lambda(\overline{M}_1)$ and P_2. Denote by E, F the conditional expectation from $\overline{M}$ onto $\overline{N}, \overline{M}_1$ onto $\Phi(\overline{M})$ respectively, i.e., $E = (P_1|\overline{M}), F = (P_2|\overline{M}_1)$. Then the relations: $e_k e_{k+1} e_k = \beta^{-1} e_k$, and $e_{k+1} e_k e_{k+1} = \beta^{-1} e_{k+1}$ are equivalent to

$$P_2 \Psi(P_1) P_2 = \beta^{-1} P_2, \quad \text{and} \quad \Psi(P_1) P_2 \Psi(P_1) = \beta^{-1} \Psi(P_1).$$

First we show that $F(P_1) = \beta^{-1}1$. Indeed, since $F(P_1) = P_2(P_1)$, it follows that

$$\begin{aligned} F(P_2) = \beta^{-1}1 &\iff (P_1 - \beta^{-1}1) \perp L^2(\Phi(\overline{M})) \quad \text{in} \quad L^2(\overline{M}_1) \\ &\iff \langle (P_1 - \beta^{-1}1), \lambda(x^*) \rangle = 0, \forall x \in \overline{M}. \end{aligned}$$

If tr is the canonical tracial state on $\overline{M}_1$, then the equality $F(P_1) = \beta^{-1}1$ is equivalent to the following:

$$\beta tr(\lambda(x) P_1) = tr(\lambda(x)) = \tau(x), \quad \forall x \in \overline{M},$$

where τ is the canonical tracial state on $\overline{M}$. This is exactly the Markov property of modulus β of τ (see Proposition 17.3.5). Hence, $F(P_1) = \beta^{-1}1$.

Now for any $x \in \overline{M}_1(\subset L^2(\overline{M}_1))$, we have

$$\begin{aligned} P_2 \Psi(P_1) P_2 x &= F(P_1 \cdot F(x)) = F(P_1) F(x) \\ &= \beta^{-1} F(x) = \beta^{-1} P_2 x. \end{aligned}$$

Hence, $P_2\Psi(P_1)P_2 = \beta^{-1}P_2$.

By Proposition 17.3.4 and Theorem 1.6.1, the set

$$\{\sum_i \Phi(x_i)P_1\Phi(y_i)|x_i, y_i \in \overline{M}\}$$

is dense in $L^2(\overline{M}_1)$. Now it suffices to show that

$$\Psi(P_1)P_2\Psi(P_1)(\Phi(x)P_1\Phi(y)) = \beta^{-1}\Psi(P_1)(\Phi(x)P_1\Phi(y)),$$

$\forall x, y \in \overline{M}$.

Fix $x, y \in \overline{M}$. Since

$$\begin{aligned} P_1\Phi(x)P_1\Phi(y)z &= E(xE(yz)) \\ &= E(x)E(yz) = \Phi(E(x))P_1\Phi(y)z, \end{aligned}$$

$\forall z \in \overline{M}(\subset L^2(\overline{M}))$, it follows that $P_1\Phi(x)P_1\Phi(y) = \Phi(E(x))P_1\Phi(y)$. Then

$$\begin{aligned} &\Psi(P_1)P_2\Psi(P_1)(\Phi(x)P_1\Phi(y)) \\ = \ &P_1F(P_1\Phi(x)P_1\Phi(y)) = P_1F(\Phi(E(x))P_1\Phi(y)) \\ = \ &P_1\Phi(E(x))F(P_1)\Phi(y) = \beta^{-1}P_1\Phi(E(x))\Phi(y). \end{aligned}$$

Noticing that

$$\begin{aligned} P_1\Phi(E(x))z &= E(E(x)z) = E(x)E(z) \\ &= E(xE(z)) = P_1\Phi(x)P_1z, \end{aligned}$$

$\forall z \in \overline{M}(\subset L^2(\overline{M}))$, we have

$$\begin{aligned} &\Psi(P_1)P_2\Psi(P_1)(\Phi(x)P_1\Phi(y)) \\ = \ &\beta^{-1}P_1\Phi(x)P_1\Phi(y) = \beta^{-1}\Psi(P_1)(\Phi(x)P_1\Phi(y)), \end{aligned}$$

$\forall x, y \in \overline{M}$. Therefore, $\Psi(P_1)P_2\Psi(P_1) = \beta^{-1}\Psi(P_1)$. Q.E.D.

Remark. Let M be a type (I_m) factor, and N be a type (I_n) subfactor of M. Then it must be $n|m$. From Chapter 15, there is a Bratteli diagram as follows:

$$\underset{n}{\cdot} \xrightarrow{p} \underset{m}{\cdot},$$

where $m = np$. Clearly, $L^2(M) = M, L^2(N) = N$, and $P = E : L^2(M) \longrightarrow L^2(N)$. By Proposition 17.3.4, we have $\langle M, P\rangle = \{\lambda(M), P\}'' = \{\sum_i \lambda(x_i)P\ \lambda(y_i)|x_i, y_i \in M\}$. Further,

$$\begin{aligned} \langle M, P\rangle &= \mathrm{End}_N^r(M) \\ &= \left\{t : M \longrightarrow M \,\middle|\, \begin{array}{c} t \text{ is linear, and} \\ t(xy) = (tx)y, \forall x \in M, y \in N. \end{array}\right\}. \end{aligned}$$

In fact, it is easy to see that $\mathrm{End}^r_N(M) = \rho(N)'$. Since $P = E$ is the conditional expectation from M onto N, it follows that $P\rho(y) = \rho(y)P, \forall y \in N$. Hence

$$\begin{aligned}\langle M, P\rangle &= \{\sum_i \lambda(x_i)P\lambda(y_i) | x_i, y_i \in M\}\\ &\subset \rho(N)' = \mathrm{End}^r_N(M).\end{aligned}$$

Conversely, if $t \in \langle M, P\rangle'$, then

$$tx = t\lambda(x)1 = \lambda(x)t1 = \rho(t1)x, \quad \forall x \in M,$$

and $t = \rho(t1)$. Moreover, since $P(t1) = tP1 = t1$, it follows that $(t1) \in N$. Hence, $t \in \rho(N)$, i.e., $\langle M, P\rangle' \subset \rho(N)$. Therefore, we obtain that $\langle M, P\rangle = \rho(N)' = \mathrm{End}^r_N(M)$.

By $[\langle M, P\rangle : M] = [M : N] = m^2/n^2, \langle M, P\rangle$ must be a type (I_l) factor, where $l = m^2/n$.

Let

$$1 \in M_0 \subset M_1 \subset M_2 = \langle M_1, e_1\rangle \subset \cdots\cdots \subset M_{k+1} = \langle M_k, e_k\rangle \subset \cdots\cdots$$

be the tower of finite factors induced by $\{N \subset M\}$. Then M_k is a type (I_{np^k}) factor, $\forall k \geq 0$. Thus, we have a Bratteli diagram as follows:

$$\underset{n}{\cdot} \xrightarrow{p} \underset{np}{\cdot} \xrightarrow{p} \underset{np^2}{\cdot} \xrightarrow{p} \cdots\cdots \xrightarrow{p} \underset{np^k}{\cdot} \xrightarrow{p} \cdots\cdots$$

Notes. The technique of towers of algebra and Theorem 17.3.7 are due to V.F.R.Jones. About Proposiiton 17.3.4(ii), M.Pimsner and S.Popa proved that $\langle M, P\rangle = \{\sum_i \lambda(x_i)P\lambda(y_i) | x_i, y_i \in M\}$ indeed.

References. [60], [75], [129].

17.4. The values of index for subfactors

Define a sequence $\{P_k(\lambda) | k = 0, 1, 2, \cdots\}$ of polynomials as follows:

$$P_0 = 1, P_1 = 1, P_{k+1}(\lambda) = P_k(\lambda) - \lambda P_{k-1}(\lambda), \forall k \geq 1.$$

In particular,

$$P_2(\lambda) = 1 - \lambda, \quad P_4(\lambda) = 1 - 3\lambda + \lambda^2, \quad P_6(\lambda) = 1 - 5\lambda + 6\lambda^2 - \lambda^3,$$

$$P_3(\lambda) = 1 - 2\lambda, \quad P_5(\lambda) = 1 - 4\lambda + 3\lambda^2, \quad P_7(\lambda) = 1 - 6\lambda + 10\lambda^2 - 4\lambda^3.$$

Lemma 17.4.1. Consider an integer $k \geq 0$, and set $m = [\frac{k}{2}]$. Then

(i) The polynomial P_k is of degree m. Its leading coefficient is $(-1)^m$ if $k = 2m$ is even and $(-1)^m(m+1)$ if $k = 2m+1$ is odd. Consequently, $\lim_{\lambda \to -\infty} P_k(\lambda) = +\infty$.

(ii) P_k has m distinct roots which are given by $\dfrac{1}{4\cos^2(\frac{j\pi}{k+1})}$ for $j = 1, 2, \cdots, m$.

(iii) Assume $k \geq 1$. Let λ be a real number with

$$\frac{1}{4\cos^2(\frac{\pi}{k+2})} < \lambda < \frac{1}{4\cos^2(\frac{\pi}{k+1})}.$$

Then $P_1(\lambda) > 0, P_2(\lambda) > 0, \cdots, P_k(\lambda) > 0$, and $P_{k+1}(\lambda) < 0$.

Proof. Claim (i) is easily checked by induction.

For (ii), we consider $\lambda > 1/4$ and let

$$\mu_1 = \frac{1}{2}(1 + i(4\lambda - 1)^{1/2}), \quad \mu_2 = \frac{1}{2}(1 - i(4\lambda - 1)^{1/2}).$$

We say that $P_k(\lambda) = (\mu_1 - \mu_2)^{-1}(\mu_1^{k+1} - \mu_2^{k+1})$ for each $k \geq 0$. In fact, it holds for $k = 0, 1$. Now assume the conclusion holds for each $j \leq k$. Then by $\lambda = \mu_1\mu_2$ we have that

$$\begin{aligned} P_{k+1}(\lambda) &= P_k(\lambda) - \lambda P_{k-1}(\lambda) \\ &= (\mu_1 - \mu_2)^{-1}[\mu_1^{k+1} - \mu_2^{k+1} - \mu_1\mu_2(\mu_1^k - \mu_2^k)] \\ &= (\mu_1 - \mu_2)^{-1}(\mu_1^{k+2} - \mu_2^{k+2}). \end{aligned}$$

Hence, we have $P_k(\lambda) = (\mu_1 - \mu_2)^{-1}(\mu_1^{k+1} - \mu_2^{k+1}), \forall \lambda > 1/4$ and $k \geq 0$. Consider now a real number θ with $\theta \in (0, \frac{\pi}{2})$ and set $\lambda = \dfrac{1}{4\cos^2\theta}(> 1/4)$, so that $\mu_1 = \dfrac{e^{i\theta}}{2\cos\theta}$ and $\mu_2 = \dfrac{e^{-i\theta}}{2\cos\theta}$. Then

$$P_k(\lambda) = \frac{\sin((k+1)\theta)}{2^k \cos^k(\theta) \sin\theta}$$

which vanishes when $\theta = \dfrac{j\pi}{k+1}$ with $j = 1, \cdots, m$.

Claim (iii) is obvious for $k = 1$, and we may assume $k \geq 2$. For $l \in \{2, \cdots, k\}$, the smallest root of P_l is $\dfrac{1}{4\cos^2(\frac{\pi}{l+1})}$, and $P_l(\lambda) > 0$ for $\lambda < \dfrac{1}{4\cos^2(\frac{\pi}{l+1})}$. As $\lambda < \dfrac{1}{4\cos^2(\frac{\pi}{k+1})} \leq \dfrac{1}{4\cos^2(\frac{\pi}{l+1})}$, we have $P_l(\lambda) > 0$. The two smallest roots of P_{k+1} are

$$\lambda_1 = \frac{1}{4\cos^2(\frac{\pi}{k+2})}, \quad \lambda_2 = \frac{1}{4\cos^2(\frac{2\pi}{k+2})},$$

and $P_{k+1} < 0$ on (λ_1, λ_2). As $\lambda_1 < \lambda < \frac{1}{4\cos^2(\frac{\pi}{k+1})} < \lambda_2$, we obtain in particular $P_{k+1}(\lambda) < 0$. Q.E.D.

Now let $\{\varepsilon_j | j \geq 1\}$ be a sequence of non-zero projections on a Hilbert space H such that

$$\beta\varepsilon_i\varepsilon_j\varepsilon_i = \varepsilon_i, \text{ if } |i-j| = 1; \quad \varepsilon_i\varepsilon_j = \varepsilon_j\varepsilon_i, \text{ if } |i-j| \geq 2,$$

where β is a constant with $\beta \geq 1$.

Assume that

$$P_j(t) \neq 0, \quad 0 \leq j \leq k-1$$

where $k \geq 1, t = \beta^{-1}$, and the polynomials $\{P_j\}$ are as above. Define inductively operators on H by

$$\delta_1 = 1$$

$$\delta_2 = 1 - \varepsilon_1$$

$$\cdots\cdots$$

$$\delta_{m+1} = \delta_m - \frac{P_{m-1}(t)}{P_m(t)}\delta_m\varepsilon_m\delta_m \quad (m \leq k-1).$$

Lemma 17.4.2. For $m \in \{1, \cdots, k\}$, we have:

(i)

$$(\varepsilon_m\delta_m)^2 = \frac{P_m(t)}{P_{m-1}(t)}\varepsilon_m\delta_m;$$

(ii)

$$(\delta_m\varepsilon_m)^2 = \frac{P_m(t)}{P_{m-1}(t)}\delta_m\varepsilon_m.$$

Furthermore, if $m \leq k-1$:

(iii) δ_{m+1} is a projection, and $(1-\delta_{m+1})$ is a linear combination of monomials in $\{\varepsilon_1, \cdots, \varepsilon_m\}$;

(iv) $\varepsilon_j\delta_{m+1} = \delta_{m+1}\varepsilon_j = 0, 1 \leq j \leq m$;

(v) $\delta_{m+1} = 1 - \sup\{\varepsilon_1, \cdots, \varepsilon_m\}$.

Proof. Recall that the integer k is fixed in this discussion; we may assume that $k \geq 2$. Since the lemma is obvious for $m = 1$, we proceed by induction on

m. Thus assume that $m \geq 2$ and conclusions (i) to (v) hold for $1, 2, \cdots, m-1$. Put

$$\alpha = (\varepsilon_m \delta_m)^2 = \varepsilon_m(\delta_{m-1} - \frac{P_{m-2}(t)}{P_{m-1}(t)}\delta_{m-1}\varepsilon_{m-1}\delta_{m-1})\varepsilon_m \delta_m.$$

Since $(1 - \delta_{m-1})$ is a linear combination of monomials in $\{\varepsilon_1, \cdots, \varepsilon_{m-2}\}$, it follows that $\delta_{m-1}\varepsilon_m = \varepsilon_m \delta_{m-1}$. Applying (v) for the values $(m-1)$ and $(m-2)$, we have $\delta_m \leq \delta_{m-1}$. Hence

$$\begin{aligned}\alpha &= \varepsilon_m \delta_m - \frac{P_{m-2}(t)}{P_{m-1}(t)}\delta_{m-1}\varepsilon_m \varepsilon_{m-1}\varepsilon_m \delta_m \\ &= \varepsilon_m \delta_m - \frac{P_{m-2}(t)}{P_{m-1}(t)} t \delta_{m-1}\varepsilon_m \delta_m = \frac{P_m(t)}{P_{m-1}(t)}\varepsilon_m \delta_m.\end{aligned}$$

The conslusion (i) follows. Using $*$ operation, we get the conclusion (ii).

From (i) we have

$$(\delta_m \varepsilon_m \delta_m)^2 = \delta_m (\varepsilon_m \delta_m)^2 = \frac{P_m(t)}{P_{m-1}(t)}\delta_m \varepsilon_m \delta_m.$$

Hence,

$$\begin{aligned}\delta_{m+1}^2 &= \delta_m^2 + \frac{P_{m-1}(t)^2}{P_m(t)^2}(\delta_m \varepsilon_m \delta_m)^2 - 2\frac{P_{m-1}(t)}{P_m(t)}\delta_m \varepsilon_m \delta_m \\ &= \delta_{m+1}.\end{aligned}$$

So (iii) follows.

For $j \in \{1, \cdots, m-1\}$, we know by induction that $\varepsilon_j \delta_m = \delta_m \varepsilon_j = 0$, so that $\varepsilon_j \delta_{m+1} = \delta_{m+1}\varepsilon_j = 0$ by definition of δ_{m+1}. Moreover

$$\varepsilon_m \delta_{m+1} = \varepsilon_m \delta_m - \frac{P_{m-1}(t)}{P_m(t)}(\varepsilon_m \delta_m)^2 = 0$$

by (i). Take $*$ operation, then $\delta_{m+1}\varepsilon_m = 0$. So (iv) holds.

Finally, if p is any projection on H with $p \geq \varepsilon_j, 1 \leq j \leq m$, then we have $p(1-\delta_{m+1}) = (1-\delta_{m+1})p = (1-\delta_{m+1})$ since $(1-\delta_{m+1})$ is a linear combination of monomials in $\{\varepsilon_1, \cdots, \varepsilon_m\}$, i.e., $(1 - \delta_{m+1}) \leq p$. By (iv) it is obvious that $(1 - \delta_{m+1}) \geq \varepsilon_j, 1 \leq j \leq m$. Therefore, we have $(1 - \delta_{m+1}) = \sup\{\varepsilon_1, \cdots, \varepsilon_m\}$.

Q.E.D.

Lemma 17.4.3. Assume that $k \geq 3$.

(i) If $\delta_{k-1} = \delta_k$, then $\varepsilon_{k-1} \leq 1 - \delta_{k-2}$;

(ii) If $k > 4$ and $\delta_{k-1} = \delta_k$, then $\delta_{k-3}\varepsilon_i \varepsilon_j = 0, \forall i, j \geq k-2$ and $|i-j| \geq 2$.

Proof. (i) Let $p = \varepsilon_{k-1}(\delta_{k-2} - \delta_{k-1})$. By Lemma 17.4.2 (iv) and $\delta_{k-1} = \delta_k$, we have $\varepsilon_{k-1}\delta_{k-1} = \varepsilon_{k-1}\delta_k = 0$. Since δ_{k-2} is a linear combination of monomials in $\{1, \varepsilon_1, \cdots, \varepsilon_{k-3}\}$, it follows that $\varepsilon_{k-1}\delta_{k-2} = \delta_{k-2}\varepsilon_{k-1}$. Hence, $p = \varepsilon_{k-1}\delta_{k-2}$ is a projection. Further,

$$\begin{aligned} p = pp^* = \; & \varepsilon_{k-1}(\delta_{k-1} - \delta_{k-2})(\delta_{k-1} - \delta_{k-2})\varepsilon_{k-1} \\ &= \varepsilon_{k-1}(\frac{P_{k-3}(t)}{P_{k-2}(t)}\delta_{k-2}\varepsilon_{k-2}\delta_{k-2})^2\varepsilon_{k-1} \\ &= \frac{P_{k-3}(t)^2}{P_{k-2}(t)^2}\varepsilon_{k-1}\delta_{k-2}(\varepsilon_{k-2}\delta_{k-2})(\varepsilon_{k-2}\delta_{k-2})\varepsilon_{k-1} \\ &= \frac{P_{k-3}(t)}{P_{k-2}(t)}\varepsilon_{k-1}\delta_{k-2}\varepsilon_{k-2}\delta_{k-2}\varepsilon_{k-1} \\ &= \frac{tP_{k-3}(t)}{P_{k-2}(t)}\delta_{k-2}\varepsilon_{k-1}\delta_{k-2} = (1 - \frac{P_{k-1}(t)}{P_{k-2}(t)})p. \end{aligned}$$

Since $P_{k-1}(t) \neq 0$, it follows that $p = 0$, i.e., $\varepsilon_{k-1}\delta_{k-2} = 0$, and $\varepsilon_{k-1} \leq 1 - \delta_{k-2}$.

(ii) Put $q = \delta_{k-3}\varepsilon_{k-2}\varepsilon_k$. Clearly, q is a projection. Since

$$\varepsilon_k\delta_{k-2} = \beta\varepsilon_k\varepsilon_{k-1}\delta_{k-2}\varepsilon_k = 0,$$

it follows that $q = \varepsilon_k\varepsilon_{k-2}(\delta_{k-3} - \delta_{k-2})$. Then

$$\begin{aligned} q = qq^* \; &= \varepsilon_k\{\varepsilon_{k-2}(\delta_{k-3} - \delta_{k-2})^2\varepsilon_{k-2}\}\varepsilon_k \\ &= \frac{P_{k-4}(t)^2}{P_{k-3}(t)^2}\varepsilon_k\{\varepsilon_{k-2}(\delta_{k-3}\varepsilon_{k-3})(\delta_{k-3}\varepsilon_{k-3})\delta_{k-3}\varepsilon_{k-2}\}\varepsilon_k \\ &= \frac{P_{k-4}(t)}{P_{k-3}(t)}\varepsilon_k\{\varepsilon_{k-2}\delta_{k-3}\varepsilon_{k-3}\delta_{k-3}\varepsilon_{k-2}\}\varepsilon_k \\ &= \frac{tP_{k-4}(t)}{P_{k-3}(t)}\varepsilon_k\delta_{k-3}\varepsilon_{k-2}\delta_{k-3}\varepsilon_k = (1 - \frac{P_{k-2}(t)}{P_{k-3}(t)})q. \end{aligned}$$

As $P_{k-2}(t) \neq 0$, we get $q = 0$, i.e., $\delta_{k-3}\varepsilon_{k-2}\varepsilon_k = 0$.

If $j \geq k$, let $v_j = \beta^{\frac{j-k}{2}}\varepsilon_k\varepsilon_{k+1}\cdots\varepsilon_j$, then

$$v_j^*\varepsilon_k v_j = \varepsilon_j, \quad \text{and} \quad v_j(\delta_{k-3}\varepsilon_{k-2}) = (\delta_{k-3}\varepsilon_{k-2})v_j.$$

Hence, $\delta_{k-3}\varepsilon_{k-2}\varepsilon_j = v_j^*\delta_{k-3}\varepsilon_{k-2}\varepsilon_k v_j = 0$.

Finally, if $k-2 \leq i \leq j-2$, let $u_i = \beta^{\frac{i-k+2}{2}}\varepsilon_{k-2}\cdots\varepsilon_i$, then $u_i^*\varepsilon_{k-2}u_i = \varepsilon_i$, and $u_i(\delta_{k-3}\varepsilon_j) = (\delta_{k-3}\varepsilon_j)u_i$. Hence , $\delta_{k-3}\varepsilon_i\varepsilon_j = u_i^*\delta_{k-3}\varepsilon_{k-2}\varepsilon_j u_i = 0$. Q.E.D.

Lemma 17.4.4. Assume that $k \geq 4$ and

$$4\cos^2\frac{\pi}{k-1} < \beta < 4\cos^2\frac{\pi}{k}.$$

Then $P_j(t) \neq 0, \forall j \leq k-1$, and $\delta_{k-1} = \delta_k$.

Proof. By Lemma 17.4.1 (iii) we can see that $P_j(t) \neq 0, \forall j \leq k-1$, and $P_{k-2}(t) > 0, P_{k-1}(t) < 0$. Then from Lemma 17.4.2 (i) we have

$$\begin{aligned} 0 \leq (\delta_{k-1}\varepsilon_{k-1}\delta_{k-1})^2 &= \delta_{k-1}(\varepsilon_{k-1}\delta_{k-1})^2\delta_{k-1} \\ &= \frac{P_{k-1}(t)}{P_{k-2}(t)}\delta_{k-1}\varepsilon_{k-1}\delta_{k-1} \leq 0. \end{aligned}$$

Hence, $0 = \delta_{k-1}\varepsilon_{k-1}\delta_{k-1} = (\varepsilon_{k-1}\delta_{k-1})^*(\varepsilon_{k-1}\delta_{k-1})$, and $\varepsilon_{k-1}\delta_{k-1} = 0 = \delta_{k-1}\varepsilon_{k-1}$. Since $\delta_{k-1} = 1 - \sup\{\varepsilon_1, \cdots, \varepsilon_{k-2}\}$, it follows that $\varepsilon_{k-1} \leq \sup\{\varepsilon_1, \cdots, \varepsilon_{k-2}\}$. Therefore, we obtain that

$$\delta_k = 1 - \sup\{\varepsilon_1, \cdots, \varepsilon_{k-1}\} = 1 - \sup\{\varepsilon_1, \cdots, \varepsilon_{k-2}\} = \delta_{k-1}.$$

Q.E.D.

Theorem 17.4.5. Let $\{\varepsilon_j | j \geq 1\}$ be an infinite sequence of non-zero projections on a Hilbert space H such that

$$\beta\varepsilon_i\varepsilon_j\varepsilon_i = \varepsilon_i, \text{if} |i-j| = 1; \quad \varepsilon_i\varepsilon_j = \varepsilon_j\varepsilon_i, \text{if} |i-j| \geq 2,$$

where β is a constant with $\beta \geq 1$. Then it must be either $\beta = 4\cos^2\frac{\pi}{q}$ for some integer $q \geq 3$, or $\beta \geq 4$.

Proof. Suppose that $\beta \in (0,1)$ but $\beta \notin \{4\cos^2\frac{\pi}{q} | q = 4, 5, \cdots\}$. Then we can find a contradiction.

In fact, pick an integer $k \geq 4$ such that

$$4\cos^2\frac{\pi}{k-1} < \beta < 4\cos^2\frac{\pi}{k}.$$

Then there are oeprators $\delta_1, \cdots, \delta_k$ on H with $\delta_{k-1} = \delta_k$ by Lemma 17.4.4. Clearly, $\delta_1 \neq \delta_2$ since $\varepsilon_1 \neq 0$. If $1 - \varepsilon_1 = \delta_2 = \delta_3 = 1 - \sup\{\varepsilon_1, \varepsilon_2\}$, then $\varepsilon_1 \geq \varepsilon_2$. Further, by $\varepsilon_2 = \varepsilon_2\varepsilon_1\varepsilon_2 = \beta^{-1}\varepsilon_2$ we get $\beta = 1$, a contradiction. Hence, $\delta_2 \neq \delta_3$. Let l be the smallest value in $\{1, \cdots, k\}$ such that $\delta_{l-1} \neq \delta_l = \delta_{l+1}$. Clearly, $3 \leq l \leq k-1$.

For $m \geq 1$, let $\gamma_m = \delta_{l-2}\varepsilon_{l-2+m}$. Since $\delta_{l-2}\varepsilon_{l-2+m} = \varepsilon_{l-2+m}\delta_{l-2}$, γ_m is a projection. Moreover,

$$\begin{aligned} \beta\gamma_m\gamma_n\gamma_m &= \beta\delta_{l-2}\varepsilon_{l-2+m}\varepsilon_{l-2+n}\varepsilon_{l-2+m} \\ &= \delta_{l-2}\varepsilon_{l-2+m} = \gamma_m, \text{ if } |m-n| = 1; \\ \gamma_m\gamma_n &= \gamma_n\gamma_m, \text{ if } |m-n| \geq 2. \end{aligned}$$

Replacing $\{\varepsilon_j\}$ by $\{\gamma_j\}$ we can obtain that

$$\gamma_{k-1} \leq \sup\{\gamma_1, \cdots, \gamma_{k-3}\}$$

from Lemmas 17.4.4 and 17.4.3 (i). On the other hand, since $(l+1) \geq 4$ and $\delta_{(l+1)-1} = \delta_{l+1}$, it follows from Lemma 17.4.3 (ii) that

$$\gamma_m \gamma_{k-1} = \delta_{(l+1)-3} \varepsilon_{l-2+m} \varepsilon_{l-2+k-1},$$

$\forall 1 \leq m \leq k-3$. Hence, $\gamma_{k-1} = 0$. Further, by $\beta \gamma_m \gamma_{m+1} \gamma_m = \gamma_m, m = k-2, \cdots, 1$, we can see that $0 = \gamma_1 = \delta_{l-2} \varepsilon_{l-1}$. Then, $\varepsilon_{l-1} \leq \sup\{\varepsilon_1, \cdots, \varepsilon_{l-3}\}$. Further

$$\sup\{\varepsilon_1, \cdots, \varepsilon_{l-2}\} = \sup\{\varepsilon_1, \cdots, \varepsilon_{l-1}\},$$

and $\delta_{l-1} = \delta_l$. This contradicts the definition of l. Q.E.D.

Theorem 17.4.6. Let M be a finite factor, and N be a subfactor of M with $[M : N] < \infty$. Then it must be either

$$[M : N] = 4 \cos^2 \frac{\pi}{q} \text{ for some integer } q \geq 3$$

or $[M : N] \geq 4$.

Proof. It is immediate from Theorems 17.3.7 and 17.4.5. Q.E.D.

Notes. Theorem 17.4.6 is due to V.F.R.Jones, and the proof presented here is taken from H.Wenzl.

For any $\beta \geq 4$ or $\beta = 4\cos^2 \frac{\pi}{q}$ (some integer $q \geq 3$), the infinite sequence $\{\varepsilon_j | j \geq 1\}$ of non-zero projections on a separable Hilbert as in Theorem 17.4.5 does exist. And V.F.R. Jones proved that: $R = \{1, \varepsilon_1, \varepsilon_2, \cdots\}''$ is a hyperfinite (II_1) factor; and $[R : R_\beta] = \beta$, where $R_\beta = \{1, \varepsilon_2, \varepsilon_3, \cdots\}''$.

Now let R be a hyperfinite (II_1) factor on a separable Hilbert space H. We may assume that $H = L^2(R)$. By Theorem 6.9.12, $R \overline{\otimes} R$ is a hyperfinite (II_1) factor on $H \otimes H$, and $(R \otimes 1)' = R' \overline{\otimes} B(H)$ is an infinite factor on $H \otimes H$

From Propositions 4.3.6 and 1.8.10, we have $L^2(R \overline{\otimes} R) = L^2(R) \otimes L^2(R)$. Therefore, by Proposition 17.1.6 we obtain that

$$[R \overline{\otimes} R : R \otimes 1] = \dim_{R \otimes 1}(L^2(R \overline{\otimes} R)) = \infty.$$

Moreover, the index theory, Markov property, the tower of algebras and etc. can be generatized to the cases: a pair of finite dimensional C^*-algebras, and a pair of finite VN algebras with a finite dimensional center.

References. [60], [75], [195].

Appendix

Weak Topology and Weak $*$ Topology

Let X be a Banach space, and X^* be its conjugate space. Denote by $\sigma(X, X^*)$ and $\sigma(X^*, X)$ the weak topology in X and weak $*$ topology in X^* respectively. Clearly, $(X, \sigma(X, X^*))$ and $(X^*, \sigma(X^*, X))$ are locally convex Hausdorff linear topological spaces. This appendix is a survey about $\sigma(X, X^*)$ and $\sigma(X^*, X)$. All proofs can be found from the references.

Theorem A.1. (Representation of σ–continuous linear functionals) We have that

$$(X, \sigma(X, X^*))^* = X^*, \quad \text{and} \quad (X^*, \sigma(X^*, X))^* = X.$$

Theorem A.2. (Asoli–Mazur, Separation theorem)

(i) Let K be a (norm–) closed convex subset of X, and $x_0 \in X \backslash K$. Then there is $f \in X^*$ such that

$$Ref(x_0) > \sup\{Ref(x) | x \in K\}.$$

Consequently, a convex subset of X is (norm –) closed if and only if it is $\sigma(X, X^*)$–closed.

(ii) Let K^* be a $\sigma(X^*, X)$–closed convex subset of X^*, and $f_0 \in X^* \backslash K^*$. Then there is $x \in X$ such that

$$Ref_0(x) > \sup\{Ref(x) | f \in K^*\}.$$

Theorem A.3. (Bipolar theorem)

(i) Let A be a subset of X. Then

$$A^0 = \{f \in X^* | |f(x)| \leq 1, \forall x \in A\}$$

is a circled convex $\sigma(X^*, X)$–closed subset of X^*, and

$$(A^0)_0 = \{x \in X \| f(x)| \leq 1, \forall f \in A^0\}$$

is the circled convex $\sigma(X, X^*)$–closure of A.

(ii) Let A^* be a subset of X^*. Then

$$A_0^* = \{x \in X \| f(x)| \leq 1, \forall f \in A^*\}$$

is a circled convex $\sigma(X, X^*)$–closed subset of X, and

$$(A_0^*)^0 = \{f \in X^* \| f(x)| \leq 1, \forall x \in A_0^*\}$$

is the circled convex $\sigma(X^*, X)$–closure of A^*.

Corollary A.4.

(i) Let E be a linear subspace of X. Then

$$E^0 = E^\perp = \{f \in X^* | f(x) = 0, \forall x \in E\}$$

is a $\sigma(X^*, X)$ –closed linear subspace of X^*, and

$$(E^0)_0 = (E^\perp)_\perp = \{x \in X | f(x) = 0, \forall f \in E^\perp\}$$

is the norm– or $\sigma(X, X^*)$–closure of E. In particular, E is nonm–or $\sigma(X, X^*)$–closed if and only if $E = (E^\perp)_\perp$.

(ii) Let F be a linear subspace of X^*. Then

$$F_0 = F_\perp = \{x \in X | f(x) = 0, \forall f \in F\}$$

is a norm – or $\sigma(X, X^*)$–closed linear subspace of X, and

$$(F_0)^0 = (F_\perp)^\perp = \{f \in X^* | f(x) = 0, \forall x \in F_\perp\}$$

is the $\sigma(X^*, X)$–closure of F. In particular, F is $\sigma(X^*, X)$ –closed if and only if $F = (F_\perp)^\perp$.

Remark. If E is a (norm–) closed linear subspace of X , it is well–known that

$$E^* = X^*/E_\perp, \quad \text{and} \quad (X/E)^* = E^\perp.$$

Now if M is a $\sigma(X^*, X)$–closed linear subspace of X^*, then we have

$$M = (M_\perp)^\perp, \quad \text{and} \quad (X/M_\perp)^* = (M_\perp)^\perp = M.$$

Hence M is the conjugate space of the Banach space $M_* = X/M_\perp$, i.e., $M = (M_*)^* = (X/M_\perp)^*$, and $M_* = X/M_\perp$ is a predual of M.

Theorem A.5. (Banach –Alaoglu) Let S^* be the closed unit ball of X^*, i.e., $S^* = \{f \in X^* | \|f\| \leq 1\}$. Then $(S^*, \sigma(X^*, X))$ is a compact Hausdorff space.

Theorem A.6. (Goldstine) Let S, S^{**} be the closed unit balls of X, X^{**} respectively. Then by the canonical embedding , S is $\sigma(X^{**}, X^*)$–dense in S^{**}.

Remark. By Theorems A.5 and A.6, we can see that the topological space $(S, \sigma(X, X^*))$ is compact if and only if X is reflexive.

Theorem A.7. (Metrizability)

(i) Let S^* be the closed unit ball of X^*. Then $(S^*, \sigma(X^*, X))$ is metrizable if and only if X is separable.

(ii) Let S be the closed unit ball of X. Then $(S, \sigma(X, X^*))$ is metrizable if and only if X^* is separable.

Remark. We also have the following fact. $(X, \sigma(X, X^*))$ or $(X^*, \sigma(X^*, X))$ is metriable if and only if $\dim X < \infty$.

Theorem A.8. (Krein–Smulian) Let K^* be a convex subset of X^*. Then K^* is $\sigma(X^*, X)$–closed if and only if $K^* \cap \lambda S^*$ is $\sigma(X^*, X)$–closed , $\forall \lambda > 0$, where S^* is the closed unit ball of X^*.

Consequently, a linear subspace F of X^* is $\sigma(X^*, X)$–closed if and only if $F \cap S^*$ is $\sigma(X^*, X)$–closed. Moreover, a linear functional θ on X^* is $\sigma(X^*, X)$ –continuous if and only if θ is $\sigma(X^*, X)$–continuous on S^*.

Theorem A.9. (Banach –Alaoglu)

(i) Let U be a $\sigma(X, X^*)$–neighborhood of 0 in X. Then

$$U^0 = \{f \in X^* | |f(x)| \leq 1, \forall x \in U\}$$

is a circled convex $\sigma(X^*, X)$–compact subset of X^*.

(ii) Let U^* be a $\sigma(X^*, X)$–neighborhood of 0 in X^*. Then

$$U_0^* = \{x \in X | |f(x)| \leq 1, \forall f \in U^*\}$$

is a circled convex $\sigma(X, X^*)$–compact subset of X.

Theorem A.10. (Eberlian –Smulian) Let A be a subset of X. Then the following statements are equivalent;

(i) The $\sigma(X, X^*)$–closure $\overline{A}^{\sigma}$ of A is $\sigma(X, X^*)$–compact;

(ii) A is weakly sequentially compact, i.e., any sequence in A admits a subsequence which converges weakly to an element of X;

(iii) Every countably infinite subset of A has a weak cluster point in X, i.e., a point such that every weak neighborhood contains an element in the infinite subset ;

(iv) A is (norm–) bounded, and for any pair of sequences $\{x_n\} \subset A$ and $\{f_m\} \subset X^*$ with $\|f_m\| \leq 1, \forall m$, we have

$$\lim_n \lim_m f_m(x_n) = \lim_m \lim_n f_m(X_n)$$

whenever both of the limits exist;

(v) If $\{C_n\}$ is a decreasing sequence of closed convex subsets of X such that $\overline{A}^\sigma \cap C_n \neq \emptyset, \forall n$, then $\overline{A}^\sigma \cap (\cap_n C_n) \neq \emptyset$.

Remark. As a corollary, we have the Kakutani theorem: If X is reflexive, then $(S, \sigma(X, X^*))$ and $(S^*, \sigma(X^*, X))$ are weakly sequentially compact, where S, S^* are the closed unit balls of X, X^* respectively.

Theorem A.11. (James) Let A be a (norm–) bounded and $\sigma(X, X^*)$–closed subset of X. Then the following statements are equivalent:

(i) A is $\sigma(X, X^*)$–compact;

(ii) For each $f \in X^*, Ref(\cdot)$ attains its supremum on A;

(iii) For each $f \in X^*, |f(\cdot)|$ attains its supremum on A.

Corollary A.12. (Krein–Smulian) Let A be a $\sigma(X, X^*)$–compact subset of X. Then its closed convex hull $\overline{CoA}$ is also $\sigma(X, X^*)$ –compact.

Remark. About norm topology, we have the Mazur theorem: Let A be a (norm –) compact subset of X. Then $\overline{CoA}$ is also (norm –) compact.

Theorem A.13. (Mackey)

(i) Let J be a locally convex Hausdorff linear topology in X. Then $(X, J)^* = X^*$ if and only if $\sigma(X, X^*) \subset J \subset$ norm topology, i.e., the Mackey toplogy in X is the norm topology.

(ii) Let J^* be a locally convex Hausdorff linear topology in X^*. Then $(X^*, J^*)^* = X$ if and only if $\sigma(X^*, X) \subset J^* \subset \tau(X^*, X)$, where $\tau(X^*, X)$ is the Mackey topology in X^*, i.e., the uniformly convergent topology on any $\sigma(X, X^*)$–compact subset of X, or $\{A^0 | A$ is any $\sigma(X, X^*)$–compact subset of $X\}$ is a neighborhood basis of 0 with respect to $\tau(X^*, X)$.

References. [31], [71], [74], [89].

References

[1] J.F. Aarnes, On the Mackey topology for a Von Neumann algebra, Math. Scand., **22** (1968), 87–107.

[2] C.A. Akemann, The dual space of an operator algebra, Trans. Amer. Math. Soc., **126** (1967), 268–302.

[3] C.A. Akemann, Sequential convergence in the dual of a W^*–algebra, Comm. Math. Phys., **7** (1968), 222–224.

[4] C.A. Akemann and G.K. Pedersen, Complications of semicontinuity in C^*–algebra theory, Duke Math. J., **40** (1973), 785–795.

[5] C. A. Akemann, G. K. Pedersen and J. Tomiyama, Multipliers of C^*–algebras, J. Functional Anal., **13** (1973), 277–301.

[6] H. Araki and G.A. Elliott, On the definition of C^*–algebras, Pub. R.I.M.S., Kyoto Univ., **9** (1973), 93–112.

[7] R. Arens, Representations of $*$ algebras, Duke Math. J., **14** (1947), 269–282.

[8] W.B. Arveson, Subalgebras of C^*–algebras, Acta Math., **123** (1969), 141–224.

[9] W.B. Arveson, On groups of automorphisms of operator algebras, J. Functional Anal., **15** (1974) 217–243.

[10] W.B. Arveson, An invitation to C^*–algebras, Springer–Verlag, 1976.

[11] B. Blackadar, K–Theory for operator algebras, Springer–Verlag, 1986.

[12] N. Bourbaki, Intégration, Act. Sc. Ind. nos. 1175, 1244 and 1281, Paris, Hermann, 1965, 1967 and 1959.

[13] N. Bourbaki, Elements of Mathematics, General topology, Part 2, Reading Mass.: Addison–Wesley, 1966.

[14] R.C. Busby, Double centralizers and extensions of C^*–algebras, Trans. Amer. Math. Soc., **132** (1968), 79–99.

[15] O. Bratteli, Inductive limits of finite dimensional C^* – algebras, Trans. Amer. Math. Soc., **171** (1972), 195–234.

[16] P. Civin and B. Yood, The second conjugate space of a Banach algebra as an algebra, Pacific J. Math., **11** (1961), 847–870.

[17] A. Connes, Une classification des facteurs de type (III), Ann. Sci. Ecole Norm. Sup., Paris (4), **6** (1973), 133–252.

[18] J. Dixmier, Les anneaux d'operateurs de classe finie, Ann. Ec. Norm. Sup., **66** (1949), 209–261.

[19] J. Dixmier, Les fonctionelles lineaires sur l'ensemble des operateurs bornes d'un espace de Hilbert, Ann. Math., **51** (1950), 387–408.

[20] J. Dixmier, Sur certains espaces consideres par M. H. Stone, Summa Brail. Math. (2), **11** (1951), 151–182.

[21] J. Dixmier, Sur la reduction des anneaux d' operateurs, Ann. Ec. Norm. Sup., **68** (1951), 185–202.

[22] J. Dixmier, Forms lineaires sur un anneau d' operateurs, Bull. Soc. Math. France, **81** (1953), 9–39.

[23] J. Dixmier, Sur les C^*–algébras, Bull. Soc. Math. France, **88** (1960), 95–112.

[24] J. Dixmier, Dual et quasi–dual d'une algebre de Banach involutive, Trans. Amer. Math. Soc., **104** (1962), 278–283.

[25] J. Dixmier, Traces sur les C^*–algébres, Ann. Inst. Fourier, **13** (1963), 219–262.

[26] J. Dixmier, On some C^*–algebras, considered by Glimm, J. Functional Anal., **1** (1967), 182–203.

[27] J. Dixmier, C^*–Algebras, North–Holland, 1977.

[28] J. Dixmier, Von Neumann algebras, North–Holland, 1981.

[29] S. Doplicher, D. Kastler and D.W. Robinson, Coveriance algebras in field theory and statistical mechanics, Comm. Math. Phys., **3** (1966), 1–28.

[30] R. S. Doran and V. A. Belfi, Characterizations of C^* – algebras: The Gelfand–Naimark Theorems, Marcel Dekker Inc., 1986.

[31] N.Dunford and J. T. Schwartz, Linear operators, Part I, New York Interscience, 1958.

[32] H.A. Dye, The Radon–Nikodym theorem for finite rings of operators, Trans. Amer. Math. Soc., **72** (1952), 243–280.

[33] E. G.Effros, Order ideals in a C^*–algebra, Duke Math. J., **30** (1963), 391–412.

[34] E. G. Effros, Convergence of closed subsets in a topological space, Proc. Amer. Math. Soc., **16** (1965), 929–931.

[35] E. G. Effros, The Borel space of Von Neumann algebras on a separable Hilbert space, Pacific J. Math., **15** , 1153–1164.

[36] E. G. Effros, Global structure in Von Neumann algebras, Trans. Amer. Math. Soc., **121** (1966), 434–454.

[37] E.G. Effros and C. Lance, Tensor products of operator algebras, Adv. Math., **25** (1977), 1–34.

[38] E.G. Effros and J.Rosenberg, C^*–algebras with approximately inner flip, Pacific J.Math., **77** (1978), 417–443.

[39] E.G. Effros, D. Handelman and C.L. Shen, Dimension groups and their affine representations, Amer. J. Math., **102** (1980), 385–407.

[40] E.G. Effros, Dimensiona and C^*–algebras, Amer. Math. Soc., 1981.

[41] G.A. Elliott, A weakening of the axioms for a C^*–algebra, Math. Ann., **189** (1970) , 257–260.

[42] G. A. Elliott, On the classification of inductive limits of sequences of semisimple finite dimensional algebras, J. Alg., **38** (1976), 29–44.

[43] G. A. Elliot, On totally ordered groups and K_0, Lecture Notes in Math., No. **734**, Springer, 1979.

[44] G. G. Emch, Algebraic methods in statistical mechanics and quantum field theory, Wiley–Interscience, New York, 1972.

[45] J. Feldman, Borel sets of states and reppresentations, Michigan Math. J., **12** (1965), 363–365.

[46] J.M.G. Fell, The dual spaces of C^*–algebras, Trans. Amer. Math. Soc., **94** (1960), 365–403.

[47] J.M.G. Fell, C^*–algebras with smooth dual, Illinois J. Math., **4** (1960), 221–230.

[48] J. W. M. Ford, A sequare root lemma for Banach $*$ algebras, J. London Math. Soc., **42** (1967), 521–522.

[49] B. Fuglede and R.V. Kadison, On a conjecture of Murray and Von Neumann, Proc. Nat. Acad. Sc. U.S.A., **37** (1951), 420–425.

[50] M. Fukamiya, On a theorem of Gelfand and Naimark and the B^*–algebra, Kumamoto J.Sci., **1** (1952), 17–22.

[51] I.M. Gelfand, On normed rings, Dokl. Akad. Nauk SSSR, **23** (1939), 430–432.

[52] I.M. Gelfand and M. A. Naimark, On the imbdeeing of normed rings into the ring of operators in Hilbert space, Mat. Sb., **12** (1943), 197–213.

[53] B. W. Glickfeld, A metric characterization of $C(X)$ and its generalization to C^*–algebras, Illinois J. Math., **10** (1966), 547–546.

[54] J. Glimm, On a certain class of operator algebras, Trans. Amer. Math. Soc., **95** (1960), 216–244.

[55] J. Glimm, Type I C^*–algebras, Ann. Math., **73** (1961), 572–612.

[56] J. Glimm and R.V. Kadison, Unitary operators in C^*–algebras, Pacific J.Math., **10** (1960), 547–556.

[57] R. Godement, Théorèmes taubériens et théorie spectrale, Ann. Sci. Ecole Norm. Sup. (3), **63** (1947), 119–138.

[58] R. Godement, Les fonctions de types positif et la théorie de groupes, Trans. Amer. Math. Soc., **63** (1948), 1–84.

[59] K. Goodearl and D.Handelman, Rank functions and K_0 of regular rings, J. Pure Appl. Alg., **7**(1976), 195–216.

[60] F.M. Goodman, P. de la Harpe and V. F. R. Jones, Coxeter graphs and towers of algebras, Springer–Verlag, 1989.

[61] F. P. Greenleaf, Invariant means on topological groups, Van Nostrand, 1969.
[62] A. Grothendieck, Sur les applications lineaires faiblement compactes d' espaces de type $C(K)$, Canad. J. Math., **5** (1953), 129–173.
[63] A. Grothendieck, Produits tensoriels topologiques et espace nucleaires, Mem. Amer. Math. Soc., **16** (1955).
[64] a. Grothendieck, Un résultat sur le dual d'une C^*–algébre, J. Math. Pures Appl., **36** (1957), 97–108.
[65] A. Guichardel, Tensor products of C^*–algebras, Dokl. Akad. Nauk SSSR, **160** (1965), 986–989.
[66] L. A. Harris, Banach algebras with involution and Möbius transofrmations, J. Functional Anal., **11** (1972), 1–16.
[67] P. R. Halmos, Measure theory, New York, Springer, 1950.
[68] P. R. Halmos and J. Von Neumann, Operator methods in classical mechanics, II, Ann. Math., **43** (1942), 332–350.
[69] R. Herman and M. Takesaki, The comparability theorem for cyclic projecitons, Bull. London Math. Soc., **9** (1977), 186–187.
[70] E. Hewitt and K.A. Ross, Abstract harmonic analysis, I, Springer–Verlag, 1963.
[71] R. B. Holmes, Geometric functional analysis and its applications, Springer–Verlag, 1975.
[72] L. Ingelstam, Real Banach algebras, Ark. Math., **5** (1964), 239–270.
[73] N. Jacobson, A topology for the set of primitive ideals in an arbitrary ring, Proc. Nat. Acad. Sci. U.S.A., **31** (1945), 333–338.
[74] R. C. James, Weakly compact sets, Trans. Amer. Math. Soc., **113** (1964), 129–140.
[75] V.F. R. Jones, Index for subfactors, Inventiones Math., **72** (1983), 1–25.
[76] V. F. R. Jones, Index for subrings of rings, Contemp. Math., **43** (Amer. Math. Soc. 1985), 181–190.
[77] R. V. Kadison, Isometries of operator algebras, Ann. Math., **54** (1951), 325–338.
[78] R.V.Kadison, On the additivity of the trace in finite factors, Proc. Nat. Acad. Sci. U.S.A., **41** (1955), 385–387.
[79] R.V. Kadison, Irreducible operator algebras, Proc. Nat. Acad. Sci. U.S.A., **43** (1957), 273–276.
[80] R.V. Kadison and J. R. Ringrose, Fundamentals of the theory of operator algebras, Academic Press, I, 1983; II, 1986.
[81] I. Kaplansky, Normed algebras, Duke Math. J., **16** (1949), 399–418.
[82] I. Kaplansky, Projections in Banach algebras, Ann. Math., **53** (1951), 235–249.
[83] I. Kaplansky, A theorem on rings of operators, Pacific J. Math., **1** (1951), 227–232.

[84] I. Kaplansky, The structure of certain operator algebras, Trans. Amer. Math. Soc., **70** (1951), 219–255.

[85] I. Kaplansky, Group algebras in the large, Tôhoku Math. J., **3** (1951), 249–256.

[86] I. Kaplansky, Algebras of type I, Ann. Math., **56** (1952), 460–472.

[87] I. Kaplansky, Math. Rev., **14** (1953), 884.

[88] I. Kaplansky, Modules over operator algebras, Amer. J. Math., **75** (1953), 839–853.

[89] J. L. Kelley and I. Namioka, Linear topological spaces, Princeton, N.J., Van Nostrand, 1963.

[90] J. L. Kelley and R. L. Vaught, The positive cone in Banach algebras, Trans. Amer. Math. Soc., **74** (1953), 44–45.

[91] A. A. Kirillov, Elements of the theory of representations, Springer, 1976.

[92] W. Krieger, On the Araki–Woods asymtotic ratio set and non-singular transformations of a measure space, Lecture Notes in Math., **160** , Springer 1970, PP. 158–177.

[93] R. Kubo, Statistical-mechanical theory of irreversible processes, I–General theory and simple applications to magnetic and conduction problems, J. Phys. Soc. Japan, **12** (1967), 570–588.

[94] K. Kuratowski, Topology, I, New York, Academic Press, 1966.

[95] C. Lance, On nuclear C^*–algebras, J. Functional Anal., **12** (1973), 157–176.

[96] C. Lance, Tensor products of non–unital C^*–algebras, J. London Math. Soc., (2) **12** (1976), 160–168.

[97] C. Lance, Tensor products and nuclear C^*–algebras, Operator algebras and applications, (I), 379–399, 1982.

[98] O. E. Lanford and D. Ruelle, Integral representations of invariant states on B^*–algebras, J. Math. Phys., **8** (1967), 1460–1463.

[99] A. J. Lazar and D.C. Taylor, Approximately finite dimensional C^*–algebras and Bratteli diagrams, Trans. Amer. Math. Soc., **259** (1980), 599–619.

[100] B. R. Li, Real C^*–algebras (in Chinese), Acta Math. Sinica, **18** (1973), 216–218.

[101] B. R. Li, Real operator algebras, Scientia Sinica, **22** (1979), 733–746.

[102] B. R. Li, Tensor products of C^*–algebras, Scientia Sinica, Special Issue (II) (1979), 232–248.

[103] B. R. Li and X. H. Jiang, Traces on (AF)–algebras, Lecture Notes in Contemp. 88–99, Inst. of Math. Academia Sinica 1989, Science Press, Beijing, China.

[104] G. W. Mackey, Induced representations of locally compact grups, II: The Frobenius reciprecity theorem, Ann. Math., **58** (1953), 193 221.

[105] G. W. Mackey, The theory of group representations, Mimeographed Notes, Univ. of Chicago, Chicago, Ill., 1955.

[106] G. W. Mackey, Borel structures in groups and their duals. Trans. Amer. Math. Soc., **85** (1957), 134–165.

[107] P. C. Martin and J. Schwinger, Theory of many particles systems, I, Phys. Rev. **115** (1959), 1342–1344.

[108] D. Mcduff, Uncountably many (II_1) factos, Ann. Math. **90** (1969), 372–377.

[109] Y. Misonou, On the direct product of W^*-algebras, Tôhoku Math. J., **6** (1954), 189–204.

[110] Y. Misonou, Generalized approximately finite operator algebras, Tôhoku Math. J., **7** (1954), 192–205.

[111] F. J. Murray and J. Von Neumann, On rings of operators, Ann. Math., **37** (1936), 116–229.

[112] F. J. Murray and J. Von Neumann, On rings of operators, II, Trans. Amer. Math. Soc., **41** (1937), 208–248.

[113] F. J. Murray and J. Von Neumann, On rings of operators, IV, Ann. Math., **44** (1943), 716–808.

[114] J. Von Neumann, Zur algebra der funktional operationen und theorie der normalen operatoren , Math. Ann., **102** (1929), 307–427.

[115] J. Von Neumann, On a certain topology for rings of operators, Ann. Math., **37** (1936), 111–115.

[116] J. Von Neumann, On infinite direct products, compisitio Math., **6** (1938), 1–77.

[117] J. Von Neumann, On rings of operators, III, Ann. Math., **41** (1940), 94–161.

[118] J. Von Neumann, On some algebraical properties of operator rings, Ann. Math., **44** (1943), 709–715.

[119] J. Von Neumann, On rings of oeprators: Reduction theory, Ann. Math., **50** (1949), 401–485.

[120] O. A. Nielsen, Borel sets of Von Neumann algebras, Amer. J. math., **95** (1973), 145–164.

[121] O. A. Nielsen, Direct integral theory, Marcel Dekker Inc., 1980.

[122] D. Olesen, On norm - continuity and compactness of spectrum, Math. Scand., **35** (1974), 223–236.

[123] D. Olesen, On spectral subspaces and their applications to automorphism groups, Symposia math., **20** (1976), 253–296.

[124] T. W. Palmer, Real C^*-algebras, Pacific J. Math., **35** (1970), 195–204.

[125] A. T. Paterson, Amensbility, Math surveys and monographs No. **29** , Amer. Math. Soc., 1988.

[126] G. K. Padersen, Some operator monotone functions, Proc. Amer. Math. Soc., **36** (1972), 309–310.

[127] G. K. Pedersen, C^*-Algebras and their automorphism groups, Academic Press, 1979.

[128] R. R. Phelps, Lectures on Choquet's theorem, Princeton, N. J., D. Van Nostrand, 1966.

[129] M. Pimsner and S. Popa, Entropy and index for subfactors, Ann. Sci. Ecole Norm. Sup. (Pairs), ser. 4, **19** (1986), 57–106.

[130] R. T. Powers, Representations of uniformly hyperfinite algebras and the associated Von Neumann rings, Ann. Math., **86** (1967), 138–171.

[131] V. Pták, On the spectral radius in Banach algebras with involution, Bull. London Math. Soc., **2** (1970), 327–334.

[132] L. Pukanszky, Some examples of factos, Pub. Math. Debrecen, **4** (1958), 135–156.

[133] J. Renault, A groupoid approach to C^*-algebras, Lecture Notes in Math. **793**, Springer–Verlag, 1980.

[134] M. A. Rieffel and A. Van Daele, The commutation theorem for tensor products of Von Neumann algebras, Bull. London Math. Soc., **7** (1975), 257–260.

[135] M. A. Rieffel and A Van Daele, A bounded operator approach to Tomita–Takesaki theory, Pacific J. Math., **69** (1977), 187–221.

[136] W. Rudin, Fourier analysis on groups, Interscience Publ., New York, 1962.

[137] W. Rudin, Real and complex analysis, McGraw–Hill, New York, 1966.

[138] D. Ruelle, States of physical system, Comm. Math. Phys., **3** (1966), 133–150.

[139] D. Ruelle, Statistical mechanics, Rigorous results, New York; Benjamin, 1969.

[140] D. Ruelle, Integral representation of states on a C^*–algebra, J. Functional Anal., **6** (1970), 116–151.

[141] B. Russo and H. A. Dye, A note on unitary operators in C^*–algebras, Duke Math. J., **33** (1966), 413–416.

[142] T. Saitô, Generators of Von Neumann algebras, Lecture Notes in Math., Springer–Verlag, **247** (1972), 435–531.

[143] S. Sakai, A characterization of W^*–algebras, Pacific J. Math., **6** (1959), 763–773.

[144] S. Sakai, On topological properties of W^*–algebras, Proc. Japan Acad., **33** (1957), 439–444.

[145] S. Sakai, On linear functionals of W^*–algebras, Proc. Japan Acad., **34** (1958), 571–574.

[146] S. Sakai, On topologies of finite W^*–algebras, Illinois J. Math., **9** (1965), 236–241.

[147] S. Sakai, A Radon – Nikodym theorem in W^* – algebras, Bull. Amer. Math. Soc., **71** (1965), 149–151.

[148] S. Skai, On the central decompositon for positive functionals on C^* – algebras, Trans. Amer. Math. Soc., **118** (1965), 406–419.

[149] S. Sakai, On the tensor product of W^*-algebras, Amer. J. Math., **90** (1968), 335–341.
[150] S. Sakai, C^*-Algebras and W^*-algebras, Springer–Verlag, New York, 1971.
[151] R. Schatten, A theory of cross-space, Ann. Math. Studies No.**26**, Princeton Univ. Press, Princeton, New jersey, 1950.
[152] R. Schatten, Norm ideals of completely continuous operators, Ergebnisse der Mathematik, No.**27**, Springer–Verlag, Berlin and New York, 1970.
[153] J. T. Schwartz, Two finite, non - hyperfinite, non–isomorphic factors, Comm. Pure Appl. Math., **16** (1963), 111–120.
[154] J. T. Schwartz, Type II factors in a central decomposition. Comm. Pure Appl. Math., **16** (1963), 247–252.
[155] I. E. Segal, Irreducible representations of operator algebras, Bull. Amer. Math. Soc., **61** (1947), 69–105.
[156] I. E. Segal, Two–sided ideals in operator algebras, Ann. Math., **50** (1949), 856–865.
[157] I. E. Segal, Equivalence of measure spaces, Amer. J. Math., **73** (1951), 275–313.
[158] I. E. Seqal, Decomposition of operator algebras, Mem. Amer. Math. Soc., **9** (1951), 1–67.
[159] I. E. Seqal, A non–commutative extension of abstract integration, Ann. Math., **57** (1953), 401–457.
[160] C. L. Shen, On the classification of the ordered groups associated with the approximately finite dimensional C^*–algebras, Duke Math. J., **46** (1979), 613–633.
[161] S. Sherman, The second adjoint of a C^*–algebra, Proc. Intern. Congr. Math., Cambridge, **1** (1950), 470.
[162] S. Shirali and J. W. M. Ford, Symmetry in complex involutory Banach algebras (II) , Duke Math. J., **37** (1970), 275–280.
[163] W. F. Stinespring, Positive functions on C^* – algebras, Proc. Amer. Math. Soc., **6** (1955), 211–216.
[164] M. h. Stone, Applications of the theory of Boolean rings to general topology, Trans. Amer. Math. Soc., **41** (1937), 375–381.
[165] V. S. Sunder, An invitation to Von Neumann algebras , Springer–Verlag, 1987.
[166] H. Takai, On a duality for crossed products of C^*–algebras, J. Functional Anal., **19** (1975), 25–39.
[167] Z. Takeda, Inductive limit and infinite direct product of operator algebras, Tôhoku Math. J., **7** (1955), 67–86.
[168] Z. Takeda, Conjugate spaces of operator algebras, Proc. Japan Acad., **28** (1954), 90–95.
[169] M. Takesaki, On the conjugate space of an operator algebra, Tohoku Math. J., **10** (1958), 194–203.

[170] M. Takesaki, On the singularity of a positive linear functional on an operator algebra, Proc. Japan. Acad., **35** (1959), 365–366.

[171] M. Takesaki, On the cross–norm of the direct product of C^*–algebras, Tohoku Math. J., **16** (1964), 111–122.

[172] M. Takesaki, Covariant representations of C^*–algebras and their locally compact automorphism groups , Acta Math., **119** (1967), 273–303.

[173] M. Takesaki, Remarks on the reduction theory of Von Neumann algebras, Proc. Amer. Math. Soc., **20** (1969), 434–438.

[174] M. Takesaki, Tomita's theory of modular Hilbert algebras and its applications, Lacture Notes in Math., No. **128**, Springer–Verlag, 1970.

[175] M. Takesaki, A liminal crossed product of a uniformly hyperfinite C^*–algebra by a compact abelian automorphism group, J. Functional Anal., **7** (1971), 140–146.

[176] M. Takesaki, Duality for crossed products and the structure of Von Neumann algebras of type (III), Acta Math., **131** (1973), 249–310.

[177] M. Takesaki, Theory of operator algebras, I, Springer–Verlag, New York, 1979.

[178] A. E. Taylor, Introduction to functional analysis, John wiley and Sons, Inc., 1958.

[179] E. C. Titchmarsh, The theory of functions, Oxford, 1952.

[180] M. Tomita, Spectral theory of operator algebras, I, Math. Okayama Univ., **9** (1959), 63–98.

[181] M. Tomita, Quasi–standard Von Neumann algebras, Mimeographed Notes, Kyushu Univ., 1967.

[182] M. Tomita, standard forms of Von Neumann algebras, The Vth functional Anal. symposium of the Math. Soc. of Japan, Sendai, 1967.

[183] J. Tomiyama, On the projections of norm one in W^*—algebras, Proc. Japan Acad., **33** (1957), 608–612.

[184] J. Tomiyama, Tensor products and projections of norm one in Von Neumann algebras, Lecture Notes. Univ. of Copenhagen, 1970.

[185] T. Turumaru, On the direct product of operator algebras, Tohoku Math. J., **4** (1952), 242–251; II, Tohoku Math. J., **5** (1953), 1–7; III, Tohoku Math. J., **6** (1954), 208–211; IV, Tohoku Math. J., **8** (1956), 281–285.

[186] T. Turumaru, Crossed products of operator algebras, Tohoku Math. J., **10** (1958), 355–365.

[187] H. Umegaki, Conditional expectations in an operator algebra, I, Tohoku Math. J., **6** (1954), 177–181.

[188] H. Umegaki, Weak compactness in an operator space, Kodai Math. Sem. Rep., **8** (1956), 145–151.

[189] A. Van Daele, Continuous crossed products and type (III) Von Neumann algebras, London Math. Soc. Lecture Notes Series **31**, Cambridge Univ. Press, 1978.

[190] F. H. Vasilescu et al., Theoria operatorilor si algebre de operatori, Bucuresti, 1973.

[191] B. J. Vowden, On the Gelfand–Naimark theorem, J. London Math. Soc., **42** (1967), 725–731.

[192] B. J. Vowden, A new proof in the spatial theory of Von Neumann algebras, J. London Math. Soc., **44** (1969), 429–432.

[193] B. J. Vowden, C^*–Norm and tensor product of C^*–algebras, J. London Math. Soc., (2), **7**(1974), 595–596.

[194] S. Wassermann, On tensor products of certain group C^*–algebra, J. Functional Anal., **23** (1976), 239–254.

[195] H. Wenzl, On sequences of projections, C. R. Math. Rep. Acad. Sci. Canada, **9** (1987), 5–9.

[196] H. Widom, Approximately finite algebras, Trans. Amer. Math. Soc., **83** (1956), 170–178.

[197] W. Wils, Dés intégration central des forms positives sur les C^*–algébras, C. R. Acad. Sci. Paris, **267** (1968), 810–812.

[198] W. Wulfsohn, Le produit tensoriel de C^*–algébres, Bull. Sci. Math., **87** (1963), 13–27.

[199] F. J. Yeadon, A new proof of the existence of a trace in a finite Von Neumann algebra, Bull. Amer. Math. Soc., **77** (1971), 257–260.

[200] F. J. Yeadon, A note on the Mackey topolog of a Von Neumann algebra, J. Math. Anal. Appl., **45** (1974), 721–722.

[201] B. Yood, Faithful $*$ representations of normed algebras, Pacific J. Math., **10** (1960), 345–363.

[202] G. Zeller–Meyer, Produits croisés d'une C^*–algébre par un groupe d' automorphismes, J. Math. Pures Appl., **47** (1968), 101–239.

Natation Index

Subject Index

www.ingramcontent.com/pod-product-compliance
Lightning Source LLC
Chambersburg PA
CBHW050632190326
41458CB00008B/2239